Engin
Performance Diagnosis and Tune-Up

Third Edition

By Chek-Chart Publications

George T. Clark, Editor
Roger L. Fennema, Contributing Editor
Richard K. DuPuy, Contributing Editor
William J. Turney, Contributing Editor

Acknowledgments

In producing this series of textbooks for automobile mechanics and technicians, Chek-Chart has drawn extensively on the technical and editorial knowledge of the nation's automakers, suppliers, and after-market equipment manufacturers. Automotive design is a technical, fast-changing field, and we gratefully acknowledge the help of the following companies in allowing us to present the most up-to-date information, illustrations and photographs possible:

Allen Testproducts
American Motors Corporation
Borg-Warner Corporation
Caldo Automotive Supply
Champion Spark Plug Company
Chrysler Corporation
Coats Diagnostic (Hennessy Industries)
Ferret Instruments, Inc.
Fluke Corporation
Ford Motor Company
Fram Corporation, A Bendix Company
General Motors Corporation
 AC-Delco Division
 Delco-Remy Division
 Rochester Products Division
 Saginaw Steering Gear Division
 Buick Motor Division
 Cadillac Motor Car Division
 Chevrolet Motor Division
 Oldsmobile Division
 Pontiac Division
Interro Systems, Inc.
Jaguar Cars, Inc.
Marquette Mfg. Co. (Bear Mfg. Co.)
Mazda Motor Corporation
Nissan Motors
The Prestolite Company, An Eltra Co.
 Robert Bosch Corporation
Snap-on Incorporated
Sun Electric Corporation
TIF Instruments, Inc.
Toyota Motor Corporation
Vetronix Corp.
Volkswagen of America

The comments, suggestions, and assistance of the following contributors were invaluable:

Jerry Mullen; DeAnza Community College, Automotive Technology Department, Cupertino, Calif.
Ralph Birnbaum, Dave Crippen, David Weiland; technical consultants, Calif.
Les Clark, Russ Suzuki, Angel Santiago, Al Bauer, and Bryan Wilson; General Motors Training Center, Burbank, Calif.
Pete Egus; AC-Delco, Los Angeles, Calif.
Robert Baier, Dan Rupp, Nick Backer, and Bill Takayama; Chrysler Training Center, Ontario, Calif.
Robert Van Antwerp, Jim Milum, and Ed Moreland; Ford Training Center, La Mirada, Calif.
Bob Kruze; Merry Oldsmobile, San Jose, Calif.

At Chek-Chart, Daniel Doornbos managed the production of this book. Copyedited by Angela Smith and Brenda Woo. New illustrations provided by Dave Douglass. Electronic graphic production by Gerald McEwan Graphics.

Library of Congress Cataloging-in-Publication Data

Engine performance diagnosis and tune-up/by Chek-Chart Publications
 George T Clark, editor—3rd ed.
  p. cm.—(Addison-Wesley/Chek-Chart automotive series)
 Includes index.
 Contents: [1] Classroom manual—[2] Shop manual
 ISBN 0-673-98102-9
 1. Automobiles—Motors—Maintenance and Repair.
I. Clark, George T. II. Chek-Chart Publications (Firm)
III. Series.
TL210.E52 1997
629.25'04'0288—dc20

96-8650
CIP

Contents

On the Cover:
Front—Using the Fluke 98 Automotive Scopemeter to test throttle position sensor voltage, courtesy of the Fluke Corporation.
Rear—Monitoring engine control functions using the inSight Personal Diagnostic Assistant, courtesy of Matco and Interro Systems, Inc.

Introduction to Engine Performance Diagnosis and Tune-Up

Engine Performance Diagnosis and Tune-Up is part of the Chek-Chart Automotive Series. The package for each course has two volumes, a *Classroom Manual* and a *Shop Manual*. Other titles in this series include:

- *Automotive Electrical and Electronic Systems*
- *Fuel Systems and Emission Controls*
- *Automatic Transmissions and Transaxles*
- *Automotive Brake Systems*
- *Automotive Engine Repair and Rebuilding*
- *Automotive Steering, Suspension, and Wheel Alignment*
- *Automotive Heating, Ventilation, and Air Conditioning Systems*

Each book is written to help the instructor teach students how to become excellent professional automobile technicians. The two-manual texts are the core of a complete learning system that leads a student from basic theories to actual hands-on experience.

The entire series is job-oriented, especially designed for students who intend to work in the automotive service profession. A student will be able to use the knowledge gained from these books and from the instructor to get and keep a job. Learning the material and techniques in these volumes is a big step toward a satisfying and rewarding career.

The books are divided into *Classroom* and *Shop Manuals* for an improved presentation of the descriptive information and study lessons, along with the practical testing, repair, and overhaul procedures. The manuals are designed to be used together: the descriptive chapters in the *Classroom Manual* correspond to the application chapters in the *Shop Manual*.

Each book is divided into several parts, and each of these parts is complete by itself. Instructors will find the chapters to be complete, readable, and well thought-out. Students will benefit from the many learning aids included, as well as from the thoroughness of the presentation.

The series was researched and written by the editorial staff of Chek-Chart. For over 65 years, Chek-Chart has provided automotive and equipment manufacturer's service specifications to the automotive service field. Chek-Chart's complete, up-to-date automotive data bank was used extensively in preparing this textbook series.

Because of the comprehensive material, the hundreds of high-quality illustrations, and the inclusion of the latest automotive technology, instructors and students alike will find that these books keep their value over the years. In fact, they will form the core of the master technician's professional library.

How To Use This Book

Why Are There Two Manuals?

This two-volume text —*Engine Performance Diagnosis and Tune-Up*— is not like most other textbooks. It is actually two books, a *Classroom Manual* and a *Shop Manual* that should be used together. The *Classroom Manual* will teach you what you need to know about ignition, fuel, and electronic control systems in a late-model vehicle. The *Shop Manual* will show you how to fix and adjust those systems.

The *Classroom Manual* will be valuable in class and at home, for study and for reference. It has text and pictures that you can use for years to refresh your memory about the basics of automotive systems affecting driveability.

In the *Shop Manual*, you will learn about test procedures, troubleshooting, and overhauling the systems and parts you are studying in the *Classroom Manual*. These procedures are explained in an easy-to-understand manner, and some use step-by-step photo-sequences as a guide. Use the two manuals together to fully understand how the parts work and how to fix them when they do not work.

What Is In These Manuals?

There are several aids in the *Classroom Manual* that will help you learn more:

1. The text is broken into short sections with subheads for easier understanding and review.
2. Each chapter is fully illustrated with line drawings and photographs.
3. Key words are printed in boldface type and are defined on the same or the facing page, and also in a glossary at the end of the manual.
4. Review questions following each chapter allow you to test your knowledge of the material covered.
5. A brief summary at the end of each chapter helps you review for exams.
6. Every few pages you will find short blocks of "nice to know" information that supplement the main text.
7. At the back of the *Shop Manual* is a sample test similar to those given for Automotive Service Excellence (ASE) certification. Use it to study and prepare yourself when you are ready to become certified as an expert in the area of engine performance.

In addition to detailed instructions on overhaul, test, and service procedures, this is what you will find in the *Shop Manual:*

1. Information on how to use and maintain shop tools and test equipment.
2. Detailed safety precautions.

3. Troubleshooting charts and tables to help you locate ignition, fuel or electronic control system problems.
4. Professional repair tips to help you perform repairs more rapidly and accurately.
5. A complete index to help you quickly find what you need.

Where Should I Begin?

If you already know something about automotive ignition, fuel, or electronic control systems and how to repair them, you may find that parts of this book are a helpful review, while others address new technologies you are not yet familiar with. If you are just starting in car repair, these manuals will assist you in becoming an automotive professional.

Your instructor will design a course to take advantage of what you already know, and what facilities and equipment are available to work with. You may be asked to take certain chapters of these manuals out of order. That is fine; the important thing is to fully understand each subject before you move on to the next. Study the vocabulary words, and use the review questions to help you understand the material.

While reading in the *Classroom Manual*, refer to your *Shop Manual* and relate the descriptive text to the service procedures. When working on the vehicle, look back at the *Classroom Manual* to keep basic information fresh in your mind. Working on a complicated late-model system is not always easy. Take advantage of the information in the *Classroom Manual*, the procedures of the *Shop Manual*, and the knowledge of your instructor to help you.

Remember that the *Shop Manual* is a good book for work, not just a good workbook. Keep it on hand while you are working on an ignition, fuel or electronic control system. For ease of use, the *Shop Manual* will fold flat on the workbench or under the car, and it can withstand quite a bit of rough handling.

To perform many engine performance repair and servicing procedures, you will also need an accurate source of specifications. Most shops have either automakers' service manuals, which lists these specifications, or an independent book such as the **Chek-Chart** ***Car Care Guide.*** This unique book, which is updated yearly, provides engine performance maintenance intervals needed to work on specific cars.

PART ONE

Engine Operation and Tune-Up Requirements

Chapter

Engine Operating Principles

The modern automobile engine has many systems and parts that must work together efficiently for the best operating performance, economy, and emission control. The following systems work to meet these requirements:

- Engine mechanical parts
- Ignition system
- Fuel system
- Engine lubrication system
- Cooling system
- Emission control systems
- Starting system
- Charging system.

Poor performance or a breakdown in any one of these areas can affect total engine operation. Automotive technicians must make sure that the systems are operating correctly and, if any problems are found, either correct the problems or warn the car owner of their findings. This process of diagnosis, preventive maintenance, and corrective maintenance is commonly called an engine tune-up.

Before you can test these areas of engine operation, you must understand how all of the engine systems work and how they relate to each other. Only then will you be able to test the systems quickly and accurately and spot faults before they become major problems.

The first five chapters of this *Classroom Manual* examine the various engine systems as they relate to proper operation and engine tune-up. Some material is a preview of subjects covered in detail in Parts Two, Three, Four, and Five of this manual. You may already have studied some of the subjects covered in the first five chapters. If so, these chapters will be a quick review to refresh your basic understanding. If this material is new to you, however, study it thoroughly. The information in the first five chapters is a necessary foundation for the conclusions reached in the chapter entitled,"What Is a Tune Up".

This Chapter introduces the concept of a *complete* tune-up and diagnosis. It examines the trends and requirements that affect today's tune-up technician. In a way, it is the heart of this *Classroom Manual*. To get the most out of it you need a thorough understanding of all the systems involved in tune-ups, beginning with engine operating principles.

ENGINE OPERATION

All engines use some kind of fuel to produce mechanical power. The oldest "engine" known to man is the simple lever. Food "fuels" the muscle pushing the lever to move objects that the muscle alone could never budge. In a similar way, the automotive engine uses fuel to perform work, figure 1-1.

Figure 1-1. Today's automotive engines produce high power with low fuel consumption and low exhaust emissions—not an easy design task.

It is common to think of the automotive engine as a gasoline engine; most of the automotive engines in the world are fueled by gasoline. But the correct name for the automotive engine is **internal combustion engine,** which can be designed to run on any fuel that **vaporizes** easily or on any flammable gas.

The automotive engine is called an internal combustion engine because the fuel it uses is burned inside the engine. **External combustion engines** burn the fuel outside the engine. A common example of an external combustion engine

Internal Combustion Engine: An engine, such as a gasoline or diesel that burns fuel inside the engine.

Vaporize: To change from a solid or liquid into a gaseous state.

External Combustion Engine: An engine, such as a steam engine that burns fuel outside of the engine.

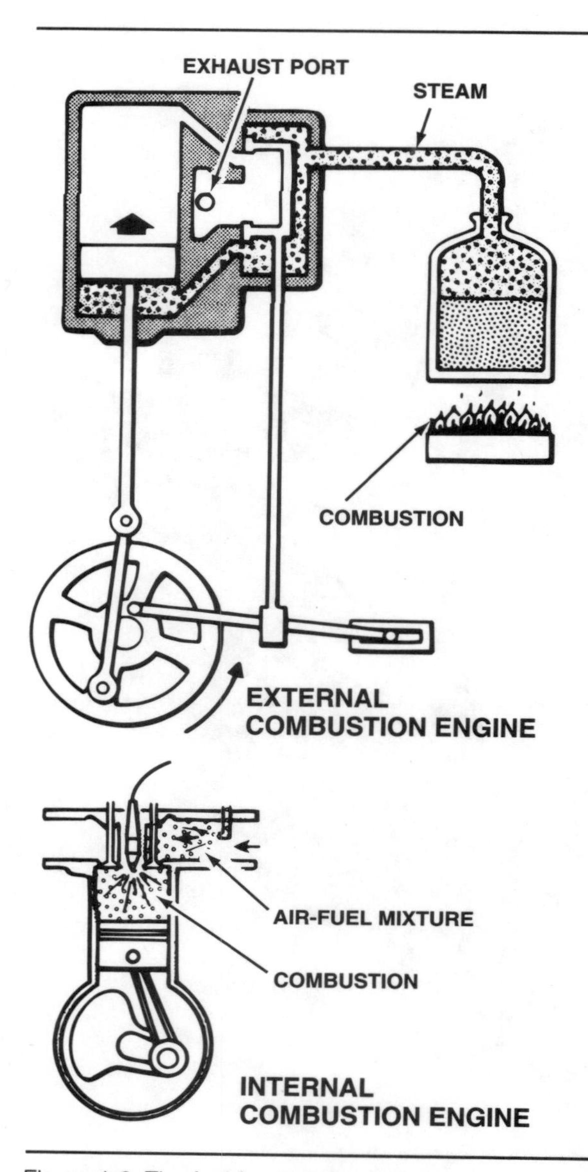

Figure 1-2. The fuel for an internal combustion engine is burned inside the engine. The fuel for an external combustion engine is burned outside the engine.

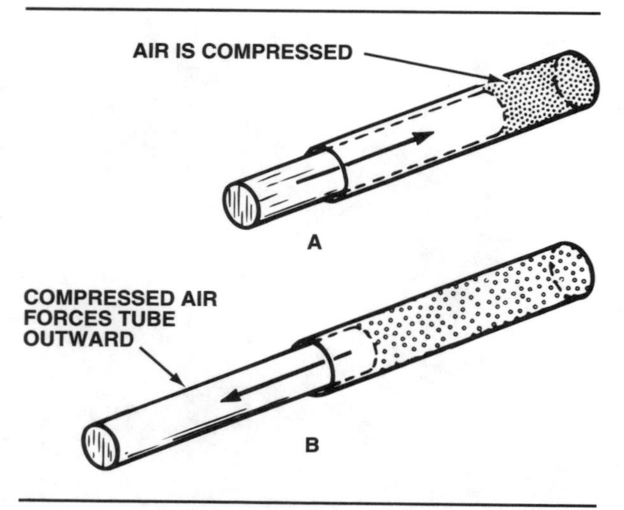

Figure 1-3. Push the inner tube in rapidly, and air is compressed (A). Release the inner tube quickly, and the compressed air forces it out (B).

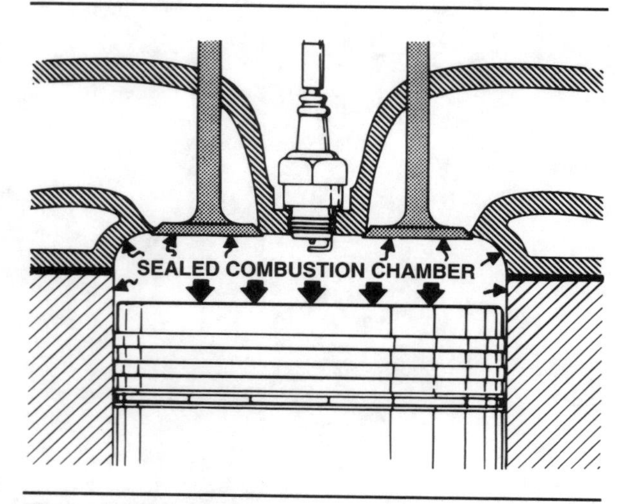

Figure 1-4. For combustion to produce power in an engine, the combustion chamber must be sealed.

is the steam engine. Fuel is burned to produce heat to make steam, but this burning takes place anywhere from a few feet to several miles away from the engine. Figure 1-2 shows the basic differences between internal and external combustion engines.

The internal combustion engine burns its fuel inside a combustion chamber. One side of this chamber is open to a piston. When the fuel burns, the hot gases expand rapidly and push the piston away from the combustion chamber. This basic action of heated gases expanding and pushing is the source of power for all internal combustion engines. This includes piston, rotary, and turbine engines.

Compression and Combustion

Gasoline by itself will not burn; it must be mixed with oxygen in the air. If fuel burns in the open air, it produces no power because it is not confined. If the same amount of fuel is enclosed and burned, it will expand with some force. To get the most force from the burning of a liquid fuel, it must be vaporized, mixed with air, and compressed to a small volume before it is burned. This compression and combustion is the most efficient way of releasing the energy stored in the air-fuel mixture.

In a piston engine, a piston moving in a cylinder provides compression. An example of this type of compression can be found in a two-section mailing tube with metal ends, figure 1-3. Push the inside tube in very quickly and it will

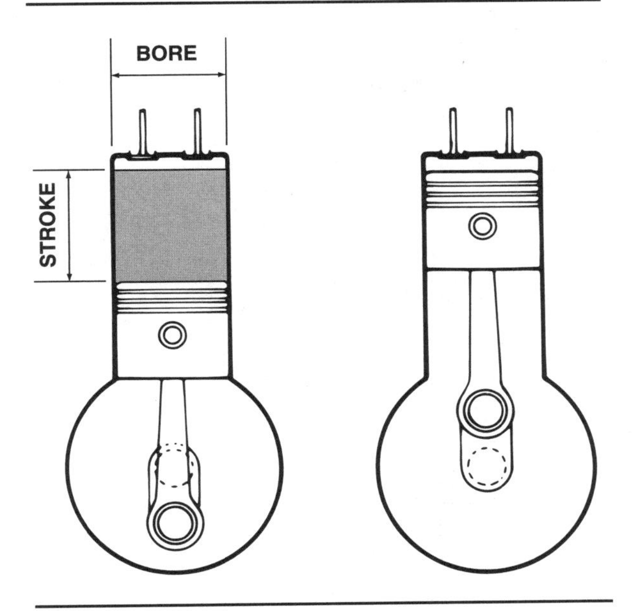

Figure 1-5. One top-to-bottom or bottom-to-top movement of the piston is called a stroke. One piston stroke performs 180 degrees of crankshaft rotation; two strokes perform 360 degrees of crankshaft rotation.

compress the air inside. Release the inside tube quickly and it will fly out. In a similar way, the piston compresses the air-fuel mixture in the cylinder.

Internal combustion engines are designed to compress and burn the vaporized air-fuel mixture in a sealed chamber, figure 1-4. Here, the combustion energy can work on the movable piston to produce mechanical energy. When heat from the burning fuel causes the fuel vapor, air, and exhaust gases inside the cylinder to expand, it produces much more power than was required to compress it. The burning, expanding gases push the piston to the other end of the cylinder.

Vacuum

The air-fuel mixture enters the combustion chamber past an intake valve. The suction or vacuum that pulls the mixture into the cylinder is created by the descending piston. You can create this same suction in the example of the two-piece mailing tube shown in figure 1-3. Draw the assembled tubes apart quickly and let go. The suction tends to pull the tubes together. If an open intake valve were in the end of the outer tube, pulling the inside tube would draw air past the valve; this suction is known as engine vacuum. Suction exists because of a difference in air pressure between the two areas. We will study vacuum and air pressure in more detail in a later chapter.

THE FOUR-STROKE CYCLE

The piston creates a vacuum by moving down through the cylinder **bore.** The distance the piston travels from one end of the cylinder to the

■ Nicolaus August Otto (1832–91)—Creator of the Four-Stroke Internal Combustion Engine

Nicolaus Otto was born in 1832, in a small hamlet in Germany near the Rhine river. Poor economic conditions forced him to drop out of secondary school to become a grocery clerk, and he ended up as a salesman of tea, sugar, and kitchenwares.

Otto was intrigued by the Lenoir internal combustion engine, the first internal combustion engine commercially available in 1860. Handicapped by his lack of education, he spent three years and all his own money (and much of that of his friends) trying to improve the engine. In 1864, Otto formed a company with Eugen Langen, a technologically minded speculator who provided much-needed capital. Their company, today called Klöckner-Humboldt-Deutz AG, is the first and oldest internal combustion engine manufacturing company in the world. The company and its refined engines were immensely successful.

The first four-stroke engine was not built until 1876. All previous internal combustion engines—including Otto's—were noncompression, in that fuel and air were drawn into the cylinder during part of a piston's downward stroke and then ignited. The expanding gases then pushed the piston down the remainder of its stroke. Many inventors used this design to make the pistons double-acting, with a power stroke each way. Otto's new engine used a downward stroke of the piston to draw in an intake charge, and a second upward stroke to compress it. Two more strokes were also required to extract power and push out exhaust gases—the four-stroke engine cycle. Otto's competitors were dubious, believing that to waste three piston strokes for a single power stroke must surely outweigh any advantages of compressing the mixture.

They were quite wrong. Comparing Otto's new engine with his earlier best-seller showed that the four-stroke weighed one-third as much, could run almost twice as fast, and needed only seven percent of the cylinder displacement to produce the same horsepower, with almost identical fuel consumptions. Within 10 years a four-stroke engine powered the first motorcycle, and soon after that the engine appeared in what would be called the horseless carriage.

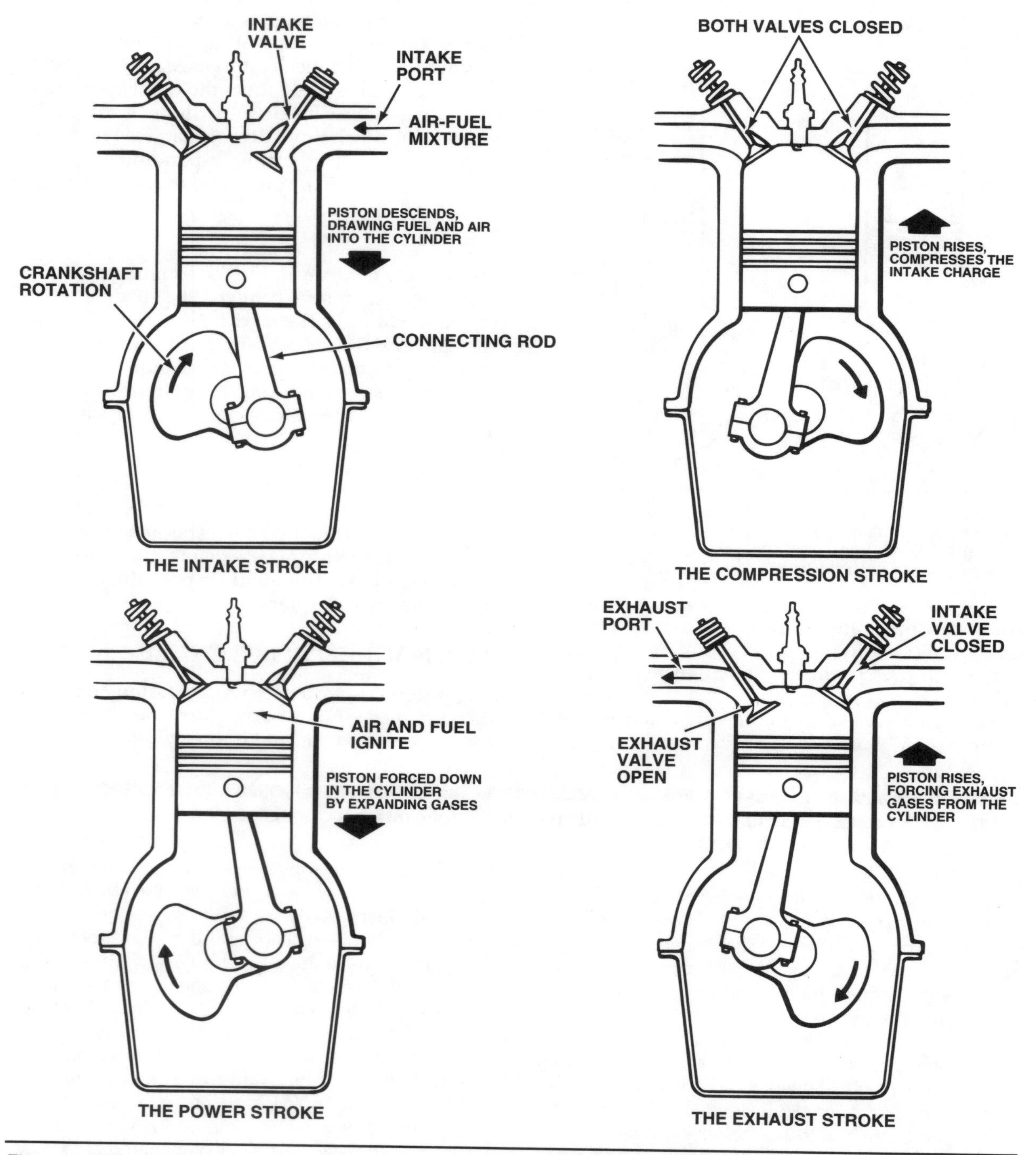

Figure 1-6. The downward movement of the piston draws the air-fuel mixture into the cylinder through the intake valve on the intake stroke. On the compression stroke, the mixture is compressed by the upward movement of the piston with both valves closed. Ignition occurs at the end of the compression stroke, and combustion drives the piston downward to produce power. On the exhaust stroke, the upward-moving piston forces the burned gases out the open exhaust valve.

other is called a **stroke,** figure 1-5. After the piston reaches the end of the cylinder, it will move back to the other end. As long as the engine is running, the piston continues to move, or stroke, back and forth in the cylinder.

An internal combustion engine must go through four separate actions to complete one operating sequence or cycle. Depending on the type of reciprocating engine, a complete operating cycle may require two or four strokes. In the

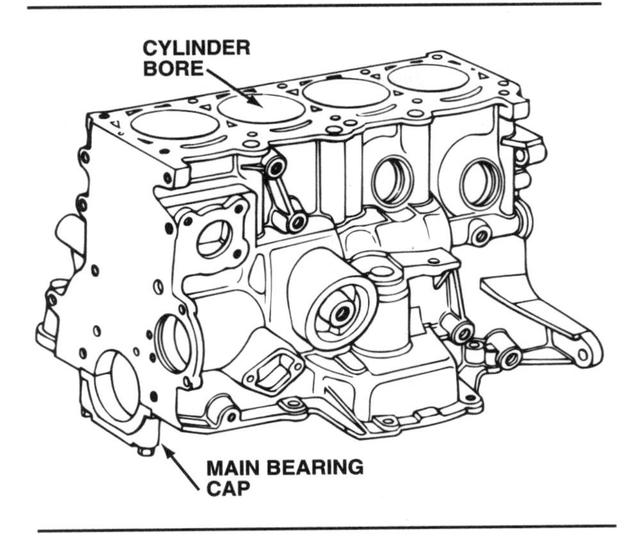

Figure 1-7. Most in-line engines position the cylinders vertically, like this Chrysler 4-cylinder.

four-stroke engine, four strokes of the piston in the cylinder are needed to complete one full operating cycle. Each stroke is named after the action it performs—intake, compression, power, and exhaust—in that order, figure 1-6.

1. Intake stroke: As the piston moves down, the mixture of vaporized fuel and air is drawn into the cylinder past the open intake valve.
2. Compression stroke: The intake valve closes, the piston returns up, and the mixture is compressed within the combustion chamber.
3. Power stroke: The mixture is ignited by a spark, and the expanding gases of combustion force the piston down in the cylinder. The exhaust valve opens near the bottom of the stroke.
4. Exhaust stroke: The piston returns up with the exhaust valve open, and the burned gases are pushed out to prepare for the next intake stroke. The intake valve usually opens just before the top of the exhaust stroke. This four-stroke cycle is continuously repeated in every cylinder as long as the engine is running.

Engines that use the four-stroke sequence are known as four-stroke engines. This four-stroke-cycle engine is also called the Otto-cycle engine after its inventor, Dr. Nikolaus Otto, who built the first successful four-stroke engine in 1876. Most automobile engines are four-stroke, spark-ignition engines. Other types of engines include two-stroke and compression-ignition (diesel).

Diesel engines do not use a spark to ignite the air-fuel mixture. Heat from its high compression ignites the fuel. (Diesel engine operation is explained later in this chapter.) Aside from the differences in ignition and fuel systems, diesel and gasoline engines are physically quite similar.

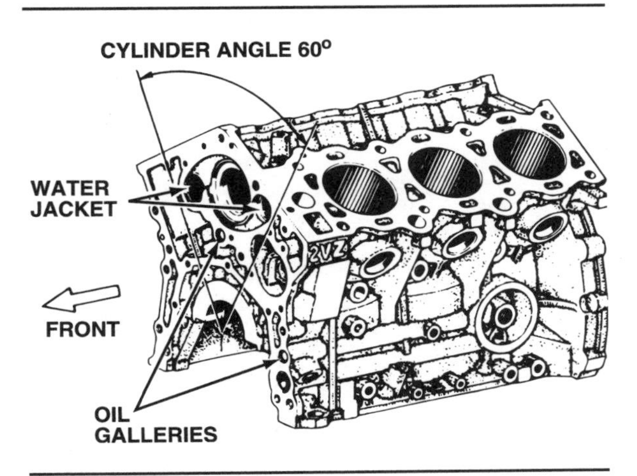

Figure 1-8. This Lexus V-type cylinder block has its cylinders inclined at 60 degrees.

Reciprocating Engine

Except for the Wankel rotary engine, all production automotive engines are **reciprocating,** or piston type. Reciprocating means "up and down" or "back and forth". It is the up-and-down action of a piston in a cylinder that gives the reciprocating engine its name. Power is produced by the in-line motion of a piston in a cylinder. However, this *linear* motion must be changed to rotating motion to turn the wheels of a car or truck.

CYLINDER ARRANGEMENT

Up to this point we have been talking about a single piston in a single cylinder. While single-cylinder engines are common in motorcycles, outboard motors, and small agricultural machines, automotive engines have more than

Bore: The diameter of an engine cylinder; to enlarge the diameter of a drilled hole.

Stroke: One complete top-to-bottom or bottom-to-top movement of an engine piston.

Four-Stroke Engine: The Otto-cycle engine. An engine in which a piston must complete four strokes to make up one operating cycle. The strokes are intake, compression, power, and exhaust.

Reciprocating Engine: An engine in which the pistons move up and down or back and forth as a result of combustion of an air-fuel mixture at one end of the piston cylinder. Also called a piston engine.

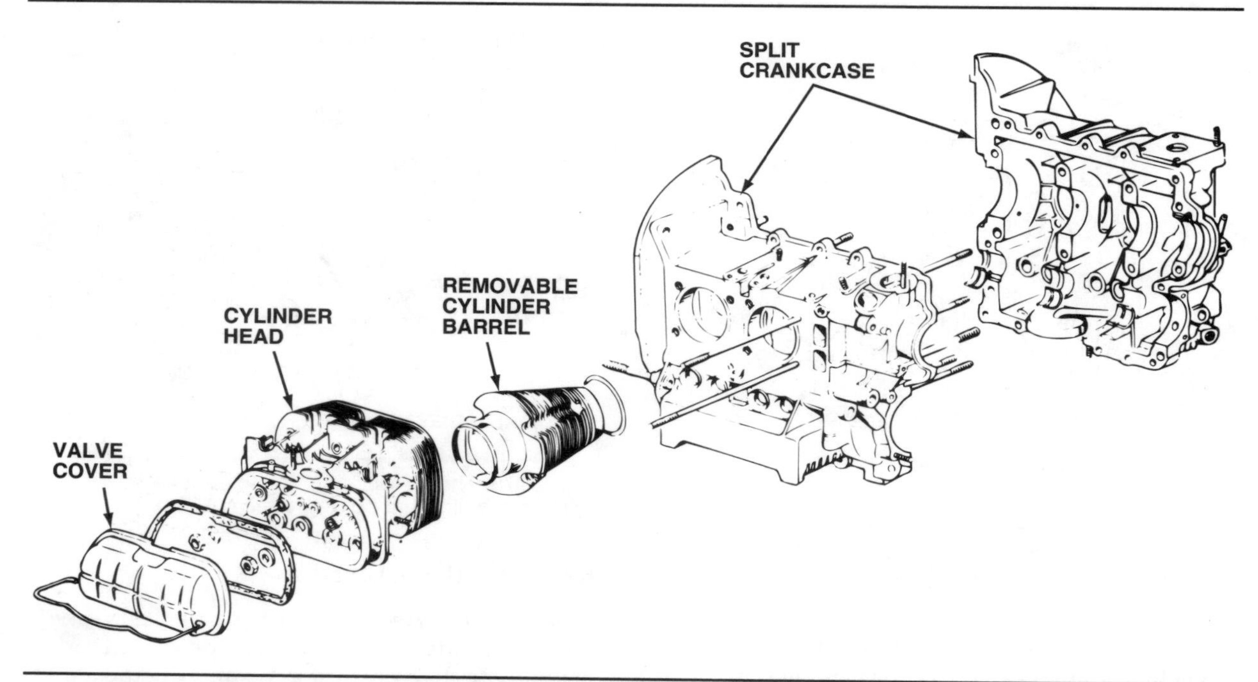

Figure 1-9. This Volkswagen horizontally opposed engine is built with a split crankcase. Individual cylinder castings bolt to the split crankcase, and cylinder heads attach to each pair of cylinders.

one cylinder. Most car engines have 4, 6, or 8 cylinders, although engines with 3, 5, 10, and 12 cylinders are also produced as we write this book. Within the engine block, the cylinders are arranged in one of three ways:

- In-line engines have a single bank of cylinders arranged in a straight line. Typically, the engine bank consists of vertical cylinders, as shown in figure 1-7. However, there are exceptions where they are inclined to either side. Most in-line engines have 4 or 6 cylinders, but many in-line engines with 3, 5, and 8 cylinders have been built.
- V-type engines, figure 1-8, have two banks of cylinders, usually inclined either 60 or 90 degrees from each other. Most V-type engines have 6 or 8 cylinders, but V2 (or V-twin), V4, V12, and V16 engines have been built.
- Horizontally opposed, "flat", or "boxer" engines have two banks of cylinders 180 degrees apart, figure 1-9. These engine designs are often air-cooled, and are found in the Chevrolet Corvair and some Ferrari, Porsche, Subaru, and Volkswagen models. Ferrari and Subaru designs are liquid-cooled. Late-model Volkswagen vans use a version of the traditional air-cooled VW horizontally opposed engine with liquid-cooled cylinder heads. Most opposed engines have 2, 4, or 6 cylinders, but flat engines with 8, 12, and 16 cylinders have been built.

The camshaft controls the opening and closing of the valves in a piston engine and is driven by the engine crankshaft. Lobes on the camshaft push each valve open as the shaft rotates, figure 1-10. A spring closes each valve when the lobe is not holding it open.

Once during each revolution of the camshaft, the lobe will push the valve open. The timing of the valve opening is critical to engine operation. Intake valves must be opened just before the beginning of the intake stroke. Exhaust valves must be opened just before the beginning of the exhaust stroke. Because the intake and exhaust valves open only once during every two revolutions of the crankshaft, the camshaft must run at half the crankshaft speed.

Turning the camshaft at half the crankshaft speed is accomplished by using a gear or sprocket on the camshaft that is twice the diameter of the crankshaft gear or sprocket, figure 1-11. If you count the teeth on each gear or sprocket, you will find exactly twice as many on the camshaft as on the crankshaft.

Engine Balance

When an engine has more than one cylinder, the crankshaft is usually made so that the firing impulses are evenly spaced. Engine speed is measured in revolutions per minute, or in degrees of crankshaft rotation. It takes two crankshaft revolutions or 720 degrees of crank-

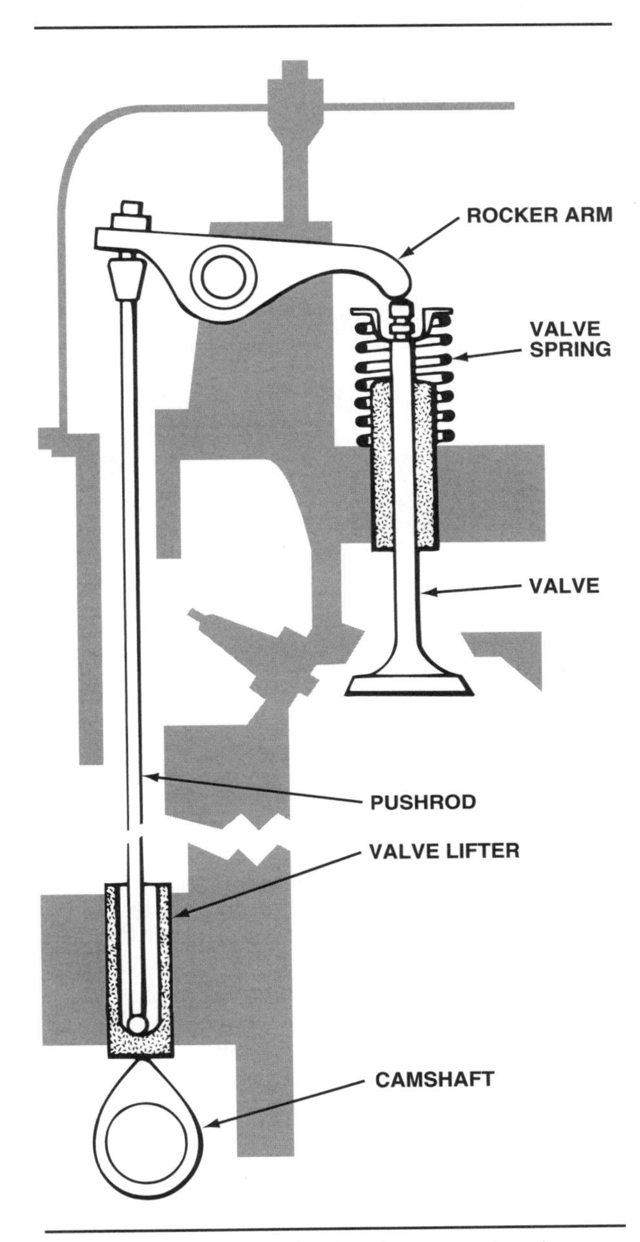

Figure 1-10. Valve mechanism for an overhead-valve engine.

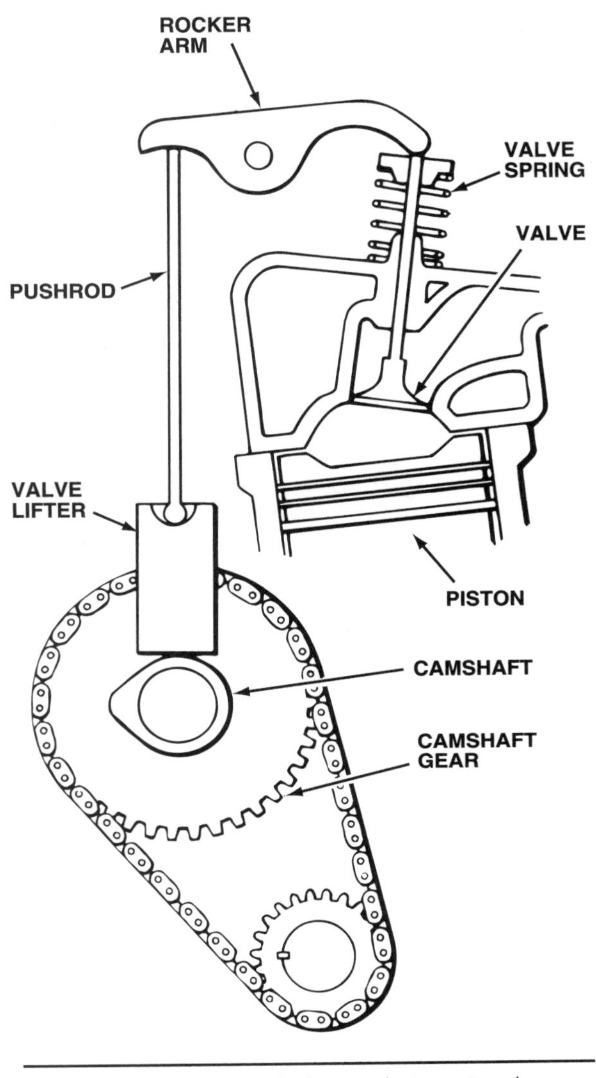

Figure 1-11. The crankshaft sprocket must make two revolutions for each revolution of the camshaft's timing gear.

shaft rotation to complete the four-stroke sequence. If a four-stroke engine has 2 cylinders, the firing impulses and crankshaft throws can be spaced so that there is a power impulse every 360 degrees. On an in-line 4-cylinder engine, the crankshaft is designed to provide firing impulses every 180 degrees. An in-line 6-cylinder crankshaft is built to fire every 120 degrees. In an 8-cylinder engine, either in-line or V-type, the firing impulses occur every 90 degrees of crankshaft rotation.

As you can see, the more cylinders an engine has, the closer the firing impulses. On 6- and 8-cylinder engines, for example, the firing impulses are close enough that power strokes overlap slightly. In other words, a new power stroke begins before the power stroke that preceded it ends. This provides a smooth transition from one firing pulse to the next. A 4-cylinder engine has no overlap of its power strokes, which makes it a relatively rough-running engine compared to engines with more cylinders. Therefore, an 8-cylinder engine runs smoother than a 4-cylinder engine.

Figure 1-12 shows common crankshaft arrangements and firing impulse frequencies. Other arrangements are possible, such as V4, opposed 4-cylinder, and several combinations of 2-cylinder layouts. The cylinder arrangement,

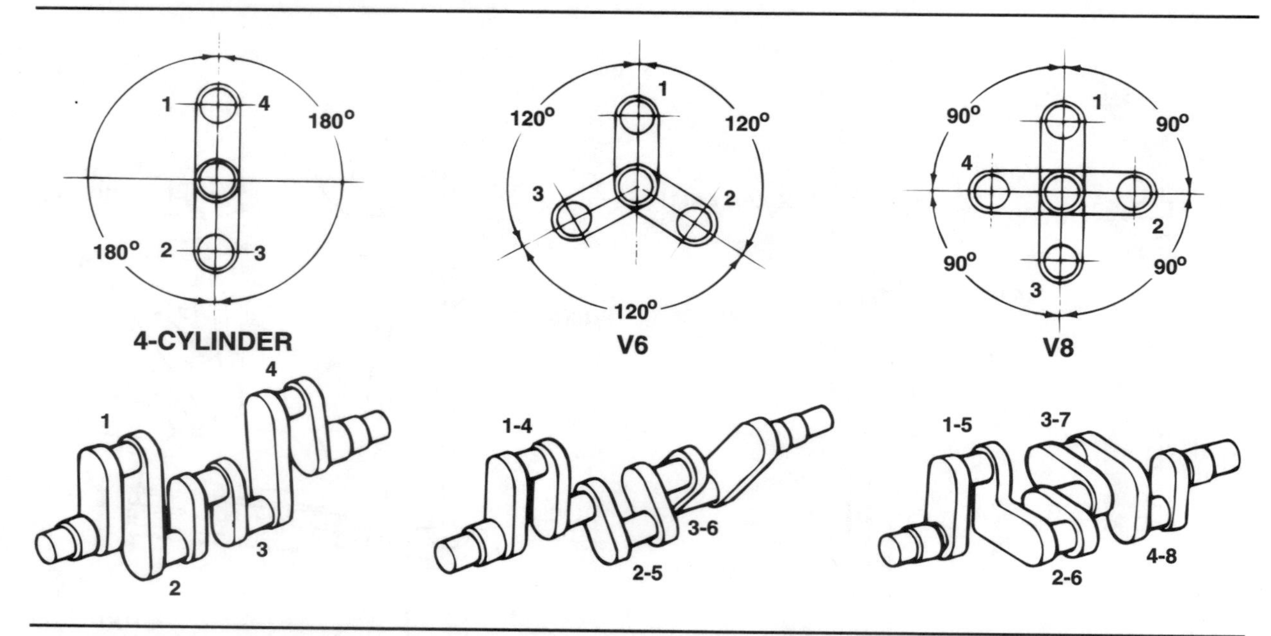

Figure 1-12. The more cylinders an engine has, the closer together the firing impulses. Here are common crankshaft designs for 4-, 6-, and 8-cylinder engines.

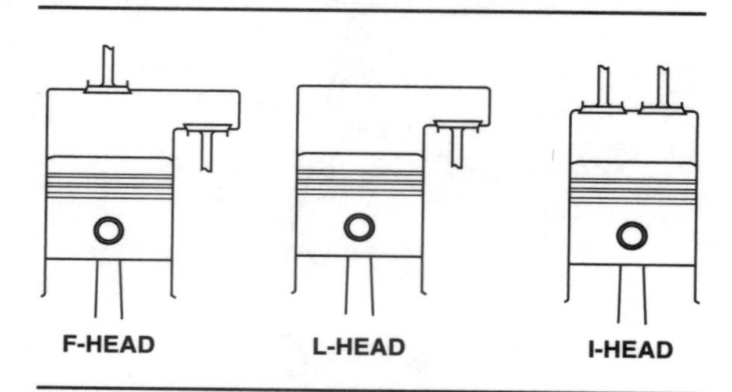

Figure 1-13. Historically, designers have used these three valve designs for four-stroke engines.

cylinder numbering order, and crankshaft design all determine the firing order of an engine, which we will study at the end of this chapter.

VALVE ARRANGEMENT

All modern automotive piston engines use **poppet valves,** figure 1-11, which work by linear motion. Most water faucet valves operate with a circular motion. The poppet valve is opened simply by pushing on it. The poppet valve must have a seat on which to rest and from which it closes off a passageway. There must also be a spring to hold the valve against the seat. In operation, a push on the end of its stem opens the valve; when the force is removed, the spring closes the valve.

Intake and exhaust valves on modern engines are located in the cylinder head. Because the valves are "in the head", this basic arrangement is called an I-head design. In the past, poppet valves have been arranged in three different ways, figure 1-13.

- The L-head design positions both valves side by side in the engine block. Because the cylinder head is rather flat and contains only the combustion chamber, water jacket, and spark plugs, L-head engines are also called "flatheads". Still very common on lawn mowers, this valve arrangement has not been used in a domestic automotive engine since the mid-1960's.
- The F-head design positions the intake valve in the cylinder head and the exhaust valve in the engine block. The F-head, a compromise between the L-head and I-head designs, was last used in the 1971 Jeep.
- The I-head design, in overhead-valve or overhead-camshaft form, positions the intake and exhaust valves in the cylinder head. All modern automotive engines use this design.

In the overhead-valve engine, the camshaft is in the engine block and the valves are opened by valve lifters, pushrods, and rocker arms, figure 1-14. In the overhead-camshaft engine, the camshaft is mounted in the head, either above or to one side of the valves, figure 1-15. This improves valve action at higher engine speeds. The valves may open directly by means of valve lifters or camshaft followers, or through rocker arms. The double overhead-camshaft engine has two camshafts, one on each side of the valves. One camshaft operates the intake valves, the other the exhaust valves.

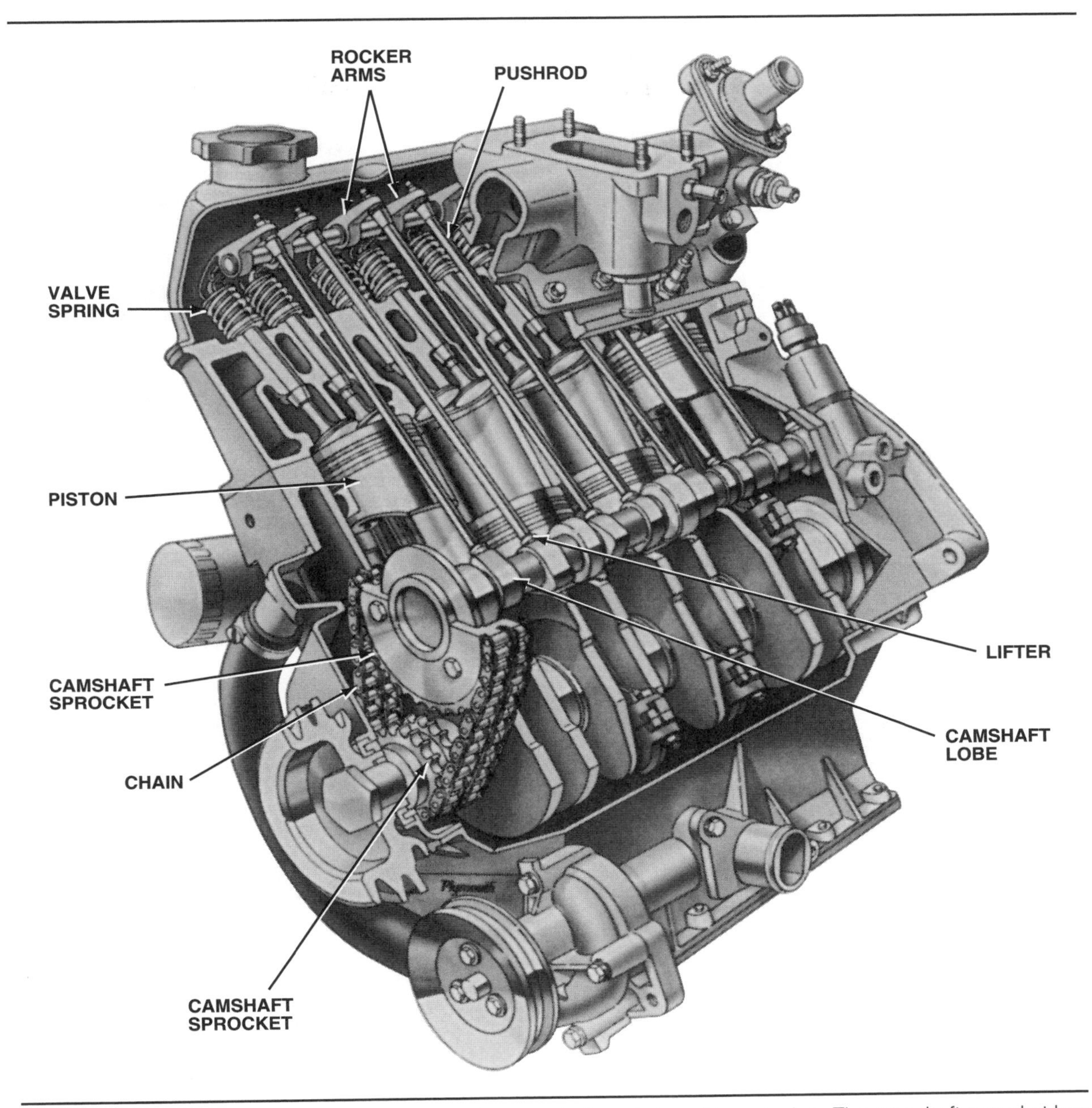

Figure 1-14. Each valve is opened by a lobe on the camshaft and closed by a spring. The camshaft sprocket has twice as many teeth as the crankshaft sprocket, causing the camshaft to rotate at one-half crankshaft speed.

Valve Lifters

Valve lifters can be mechanical or hydraulic, as shown in figure 1-16. A mechanical valve lifter is solid metal. A hydraulic lifter is a metal cylinder containing a plunger that rides on oil. A chamber below the plunger fills with engine oil through a feed hole and, as the camshaft lobe lifts the lifter, the chamber is sealed by a check valve. The trapped oil transmits the lifting motion of the camshaft lobe to the valve pushrod. Hydraulic lifters are generally quieter than mechanical lifters and do not normally need to be adjusted, because the amount of oil in the chamber varies to keep the valve adjustment correct.

Number of Valves

As you saw in figure 1-6, most automobile engines have one intake and one exhaust valve per cylinder. This means that a 4-cylinder engine

Poppet Valve: A valve that plugs and unplugs its opening by linear movement.

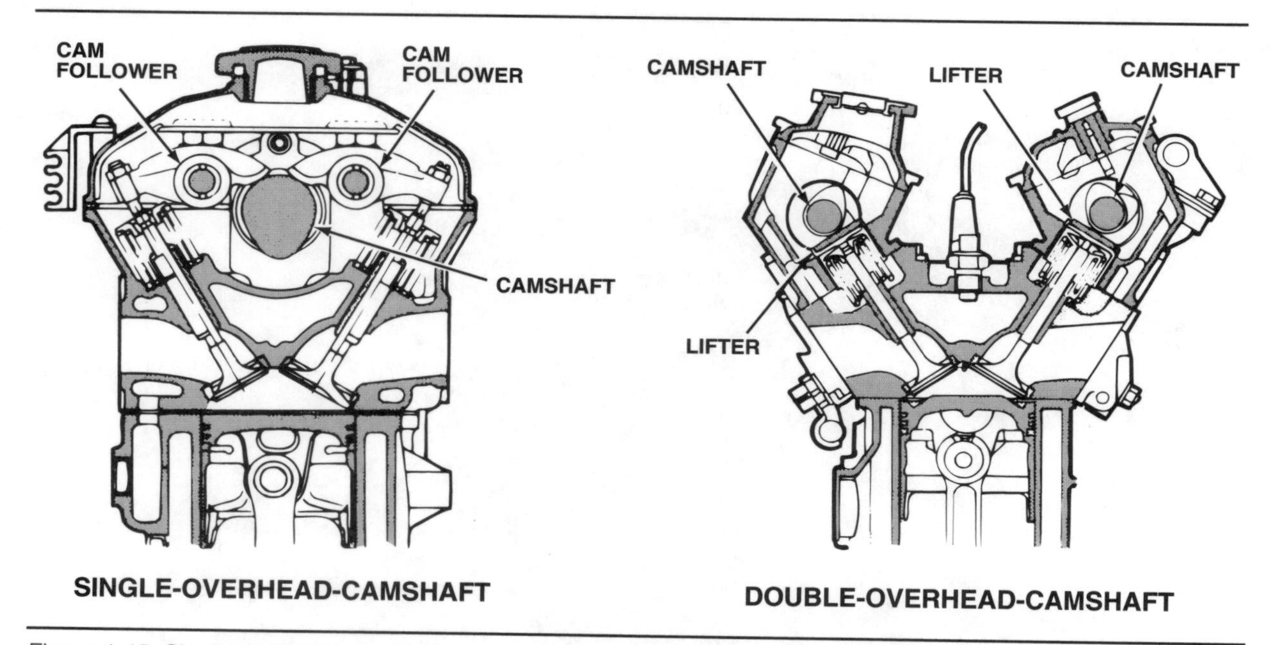

Figure 1-15. Single-overhead-camshaft engines usually require an additional component, such as a rocker arm, to operate all the valves. Double-overhead-camshaft engines actuate the valves directly.

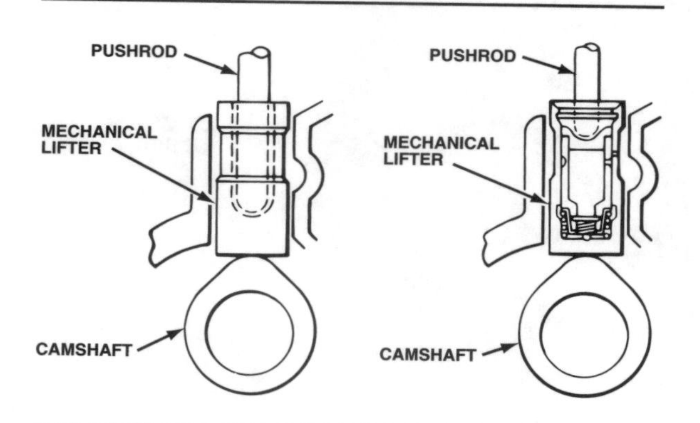

Figure 1-16. Mechanical lifters are solid metal. Hydraulic lifters use engine oil to take up clearance and transmit motion.

has 8 valves, a 6-cylinder engine has 12 valves, and a V8 has 16 valves.

Many engines have been built, however, with more than two valves per cylinder. Engines with three valves per cylinder are currently used in several Japanese-production cars. Engines with four valves per cylinder have been used in racing engines since the 1912 Peugeot Grand Prix cars. Street motorcycle engines with five valves per cylinder have been built in recent years. In spite of performance advantages, the higher costs and greater complexity of engines with more than two valves per cylinder kept such designs from being common in production engines until the mid-1980s, figure 1-17.

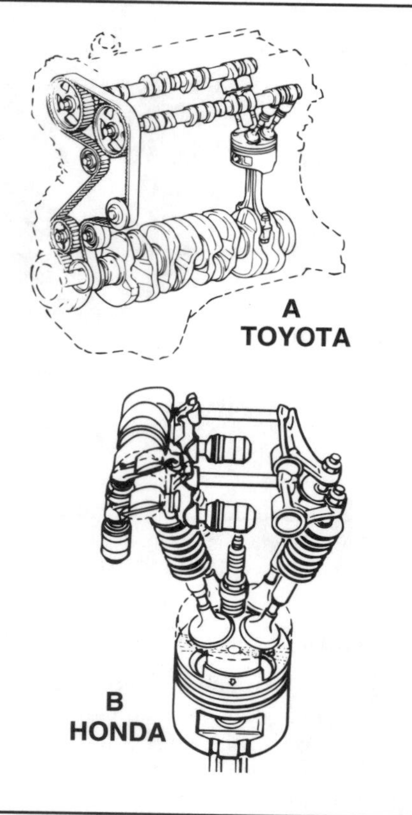

Figure 1-17. All four-valve-per-cylinder production engines use overhead camshafts. Most have separate intake and exhaust camshafts as in the Toyota (A). Some have a single camshaft, such as the Honda Acura V6 design (B) that operates the exhaust valves through short pushrods.

Modern Combustion Chamber Design

Engineers have worked with combustion chamber design since the automobile was first invented. In the years before exhaust emission controls, much of the experimentation and design work was done with racing engines in an effort to make them go ever faster. The need to reduce exhaust emissions, however, refocused attention on the combustion chamber. Efforts were made to promote rapid, uniform burning of the air-fuel charge to control emissions and improve fuel economy.

Combustion of the air-fuel charge in a cylinder is not an instantaneous explosion, but rather, a controlled burning of the charge by the spark from the spark plug. When the spark ignites the air-fuel mixture, a flame front spreads out across the combustion chamber to consume the mixture. Movement of the flame front is called burn time and requires about three milliseconds.

However, combustion chamber design, high engine temperature, pressure, or poor gasoline quality can cause an unwanted, violent explosion of the air-fuel charge. A single one of these factors or a combination of them may cause this abnormal combustion, which takes two forms: **detonation** or **preignition.** Detonation occurs if the air-fuel charge ignites before the flame front reaches it *after* the spark plug fires; this secondary explosion, figure 1-18, is popularly known as "knock-

Detonation: An unwanted explosion of an air-fuel mixture caused by high heat and compression. Occurs *after* the spark plug fires. Also called knocking or pinging.

Preignition: A premature ignition of the air-fuel mixture *before* the spark plug fires. It is caused by excessive heat or pressure in the combustion chamber.

■ Isaac de Rivaz and His Self-Propelled Carriage

The first self-propelled vehicles using an internal combustion engine were those built by Isaac de Rivaz, a Swiss engineer and government official. De Rivaz built the first version in 1805, but his improved model of 1813 was more impressive. This "great mechanical chariot" was 17 ft long by 7 ft wide (5.2 m by 2.1 m), and weighed 2,100 lb (950 kg). Its top speed was about 3 mph (5 km/h) on a level road, and it could climb a 12-percent grade.

The engine that drove this marvel used a single cylinder, open at the top, with a bore and stroke of 14.4 in. by 59 in. (36.5 cm by 150 cm). A long rod attached to the piston was connected by a chain to a drum outside the cylinder. When the piston descended, the chain would rotate the drum, turning the drive axle through a rope and pulley. When the piston rose, a ratchet let the rope and pulley freewheel.

Coal gas (a burnable mixture of hydrogen and methane) was stored in a collapsible leather bladder and pumped into a mixing chamber in a carburetor by a bellows for each stroke of the piston. To start an engine cycle, the driver pulled on a lever, which first dropped the floor of the cylinder a few inches to draw in an intake charge, and next closed an electric circuit to cause a spark to jump a gap between two wires in the combustion chamber. The burning, expanding gases shot the piston up to the top of its stroke, as the ratchet mechanism freewheeled. After combustion stopped, the gases in the cylinder cooled and contracted, and atmospheric pressure pushed the piston down. This engaged the ratchet and rolled the carriage forward about 16 to 20 ft (5 to 6 m). Exhaust was expelled when the driver raised the floor of the cylinder again in preparation for the next stroke of the engine.

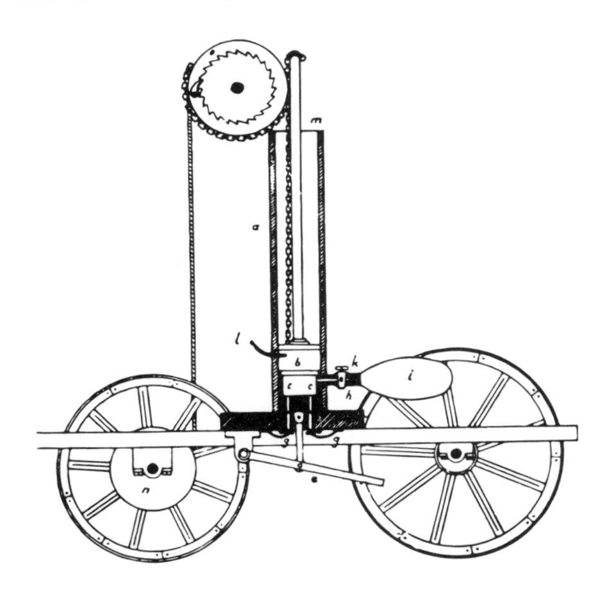

De Rivaz's carriage was not very practical, as trips were limited to the 2-mi (3-km) fuel capacity of the leather bag, and the driver had to pull and push the lever for each piston stroke. However, his use of a fuel-mixing carburetor, spark ignition, and a portable fuel tank were all engineering feats that would eventually become common on the internal combustion engines in modern day "carriages".

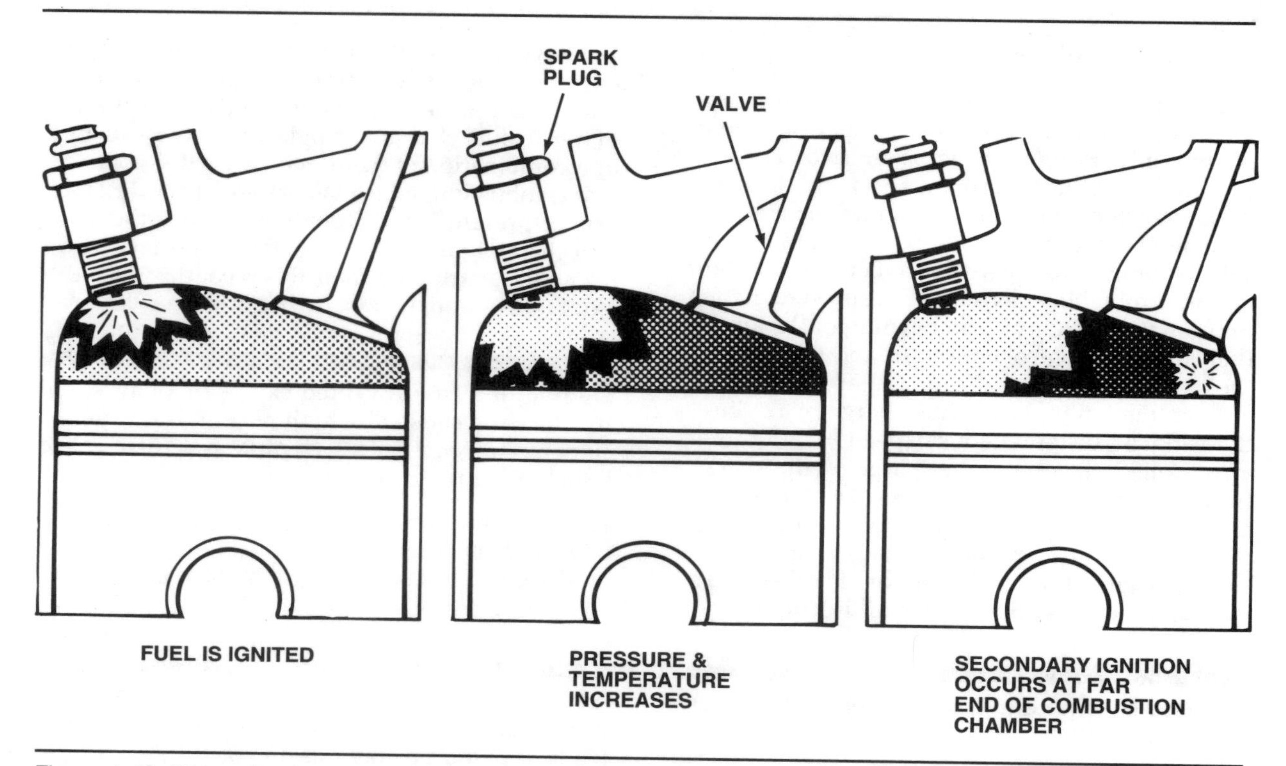

Figure 1-18. Detonation is a secondary ignition of the air-fuel mixture caused by high cylinder temperatures. It is commonly called "pinging" or "knocking".

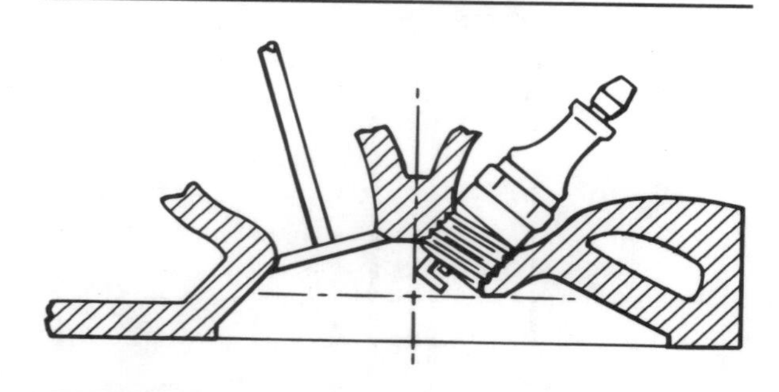

Figure 1-19. As shown in this Ford illustration, a centralized location of the spark plug in the combustion chamber reduces the distance the flame front must travel to reach the cylinder wall.

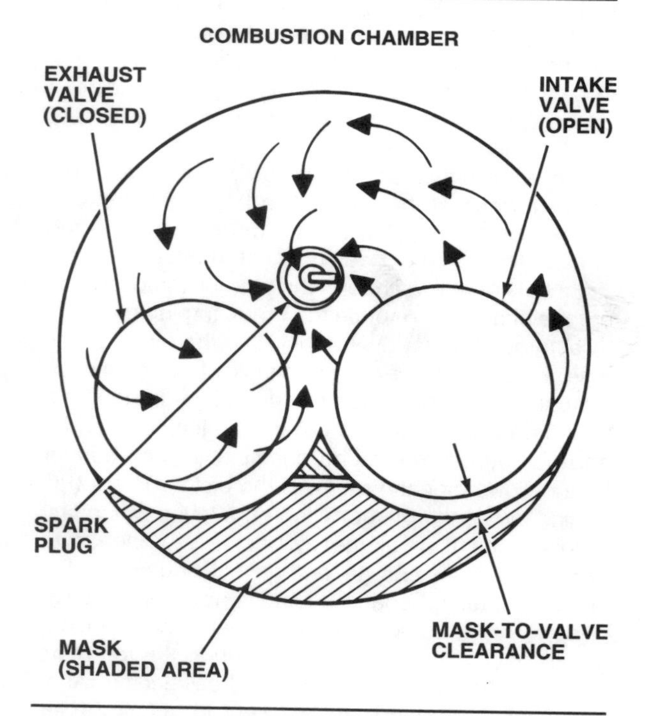

Figure 1-20. This Ford illustration shows how the air-fuel charge can be directed into the combustion chamber with a swirling motion to promote more even distribution. Masking (shrouding) the area around the intake valve with additional metal produces this effect.

ing" or "pinging". The distinctive sound is caused by the two flame fronts colliding. Detonation causes a loss of power and overheating of valves, pistons, and spark plugs. The overheating in turn causes more detonation and may eventually damage the engine. Preignition occurs when the air-fuel charge is ignited *before* the spark plug fires. During preignition, combustion usually comes from excessive combustion chamber temperatures, heat caused by extended detonation, or a single "hot-spot" in the combustion chamber.

In the ideal combustion chamber design, the entire air-fuel charge would burn completely,

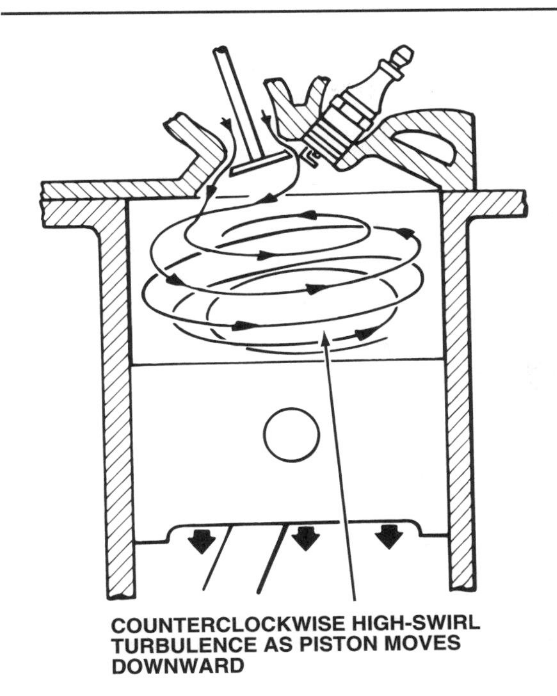

Figure 1-21. This Ford illustration shows that the downward piston movement in cylinders with shrouded valves causes a high-swirl turbulence of the air-fuel charge.

leaving no unburned areas to be exhausted and eliminating the possibility of detonation. In actual practice, however, there is always some part of the mixture that does not completely burn.

Current combustion chamber design favors the **fast-burn** or **high-swirl combustion chamber** in which the combustion process is completed in a shorter period of time. This design usually incorporates the following features:

- *Compact combustion chamber*—By providing a smaller amount of surface area for a given chamber volume, the flame front is reduced and the time required for combustion is shortened.
- *Centralized spark plug location*—Positioning the spark plug electrode closer to the center of the combustion chamber, figure 1-19, reduces the distance the flame must travel to the edges of the chamber. This also shortens the combustion period.
- *Masked or shrouded intake port*—By masking or shrouding the intake valve area in the combustion chamber, figure 1-20, the air-fuel mixture is directed in a concentrated stream as it is drawn in through the valve and subjected to turbulence or swirl as the piston moves downward, figure 1-21.
- *Higher compression ratio*—Positioning the spark plug deeper in the combustion chamber reduces the chamber volume and increases the compression ratio. In some designs, this is advantageous. In other designs, the piston crown is dished to offset plug positioning and to maintain the compression ratio at a point where no decrease in spark advance is required. The desired end result of using a higher compression ratio is to obtain a more densely compressed and fully atomized air-fuel charge for more complete combustion in less time. Fast-burn combustion chambers with four valves per cylinder have become common in late-model engine design.

ENGINE DISPLACEMENT AND COMPRESSION RATIO

In any discussion of engines, the term "engine size" comes up often. This does not refer to the outside dimension of an engine, but to its displacement. As the piston strokes in the cylinder, it moves through or displaces a specific volume. Another important engine measurement term is compression ratio. Displacement and compression ratio are related to each other, as you will learn in the following paragraphs.

Engine Displacement

Engine **displacement** is a measurement of engine volume. The number of cylinders is a factor in determining displacement, but the arrangement of cylinders is not. Engine displacement is calculated by multiplying the piston displacement of one cylinder by the number of cylinders. Total engine displacement is the volume displaced by all the pistons.

We learned earlier that the bore and stroke are important engine dimensions, and are necessary

Fast-Burn Combustion Chamber: A compact combustion chamber with a centrally located spark plug. The chamber is designed to shorten the combustion period by reducing the distance of flame front travel.

High-Swirl Combustion Chamber: A combustion chamber in which the intake valve is shrouded or masked to direct the incoming air-fuel charge and create turbulence that will circulate the mixture more evenly and rapidly.

Displacement: A measurement of engine volume. It is calculated by multiplying the piston displacement of one cylinder by the number of cylinders. The total engine displacement is the volume displaced by all the pistons.

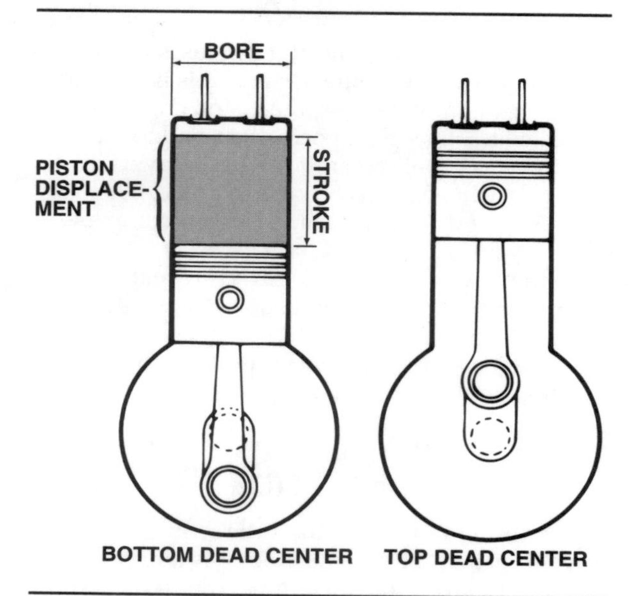

Figure 1-22. Basic engine dimensions.

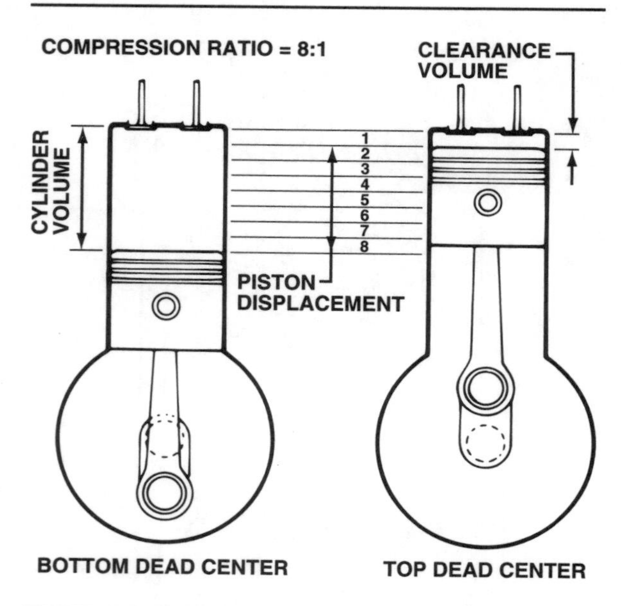

Figure 1-23. Compression ratio is the ratio of the total cylinder volume to the clearance volume.

measurements for calculating engine displacement. The displacement of one cylinder is the volume through which the piston's top surface moves as it travels from the bottom of its stroke **(bottom dead center or BDC)** to the top of its stroke **(top dead center or TDC)**, figure 1-22. Piston displacement is computed as follows:

$$\text{Piston displacement} = \left(\frac{\text{Bore}}{2}\right)^2 \times 3.1416 \times \text{Stroke}$$

1. Divide the bore (cylinder diameter) by 2. This gives you the radius of the bore.
2. Square the radius (multiply it by itself).
3. Multiply the square of the radius by 3.1416 to find the area of the cylinder cross section.
4. Multiply the area of the cylinder cross section by the length of the stroke.
5. You now know the piston displacement for one cylinder. Multiply this by the number of cylinders to determine the total engine displacement.

For example, to find the displacement of a 6-cylinder engine with a 3.80-in. bore and a 3.40-in. stroke:

1. $\frac{3.80}{2} = 1.9$
2. $1.9 \times 1.9 = 3.61$
3. $3.61 \times 3.1416 = 11.3412$
4. $11.3412 \times 3.40 = 38.56$
5. $38.56 \times 6 = 231.36$

The displacement is 231 cubic inches. Fractions of an inch are usually not included.

Metric displacement specifications

When stated in U.S. customary values, displacement is given in cubic inches (cu in.). The engine's cubic inch displacement is abbreviated "cid". When stated in metric values, displacement is given in cubic centimeters (cc) or in liters (L) (1 L equals 1000 cc). To convert engine displacement specifications from one value to another, use the following formulas:

- To change cubic centimeters to cubic inches, multiply by 0.061 (cc × 0.061 = cu in.).
- To change cubic inches to cubic centimeters, multiply by 16.39 (cu in. × 16.39 = cc).
- To change liters to cubic inches, multiply by 61.02 (L × 61.02 = cu in.).

Our 231-cu in. engine from the previous example is also a 3786-cc engine (231 × 16.39 = 3786). When expressed in liters, this figure is rounded up to 3.8 L.

Metric displacement in cubic centimeters can be calculated directly with the displacement formula, using centimeter measurements instead of inches. Here is how it works for the same engine with a bore equaling 96.52 mm (9.652 cm) and a stroke equaling 86.36 mm (8.636 cm):

1. $\frac{9.652}{2} = 4.826$
2. $4.826 \times 4.826 = 23.29$
3. $23.29 \times 3.1416 = 73.16$
4. $73.16 \times 8.636 = 631.81$
5. $631.81 \times 6 = 3791$ cc

This figure is a few cubic centimeters different than the 3786-cc displacement we got by converting 231 cu in. directly to cubic centimeters. This is due to rounding. Again, the engine displacement can be rounded up to 3.8 L.

Compression Ratio

The **compression ratio** compares the total cylinder volume when the piston is at BDC to the volume of the combustion chamber when the piston is at TDC, figure 1-23. Total cylinder volume may seem to be the same as piston displacement, but it is not. Total cylinder volume is the piston displacement plus the combustion chamber volume. The combustion chamber volume with the piston at TDC is sometimes called the **clearance volume.**

Compression ratio is the total volume of a cylinder divided by the clearance volume. If the clearance volume is 1/8 the total cylinder volume, the compression ratio is 8 to 1. The formula is as follows:

$$\frac{\text{Total volume}}{\text{Clearance volume}} = \text{Compression ratio}$$

To determine the compression ratio of an engine in which each piston displaces 510 cc and has a clearance volume of 64 cc:

1. 510 + 64 = 574 cc (total cylinder volume)
2. $\frac{574}{64} = 8.968$

The compression ratio is 8.986 to 1. This would be rounded and expressed as a compression ratio of 9 to 1, and this can also be written 9:1.

A higher compression ratio is desirable because it increases the efficiency of the engine, making the engine develop more power from a given quantity of fuel. A higher compression ratio increases cylinder pressure, which packs the fuel molecules more tightly together. The flame of combustion then travels more rapidly and across a shorter distance.

THE IGNITION SYSTEM

After the air-fuel mixture is drawn into the cylinder and compressed by the piston, it must be ignited. The ignition system creates a high

Bottom Dead Center (BDC): The point when the piston is at the bottom of its stroke in the cylinder.

Top Dead Center (TDC): The point when the piston is at the top of its stroke in the cylinder.

Compression Ratio: A ratio of the total cylinder volume when the piston is at BDC to the volume of the combustion chamber when the pis ton is at TDC.

Clearance Volume: The combustion chamber volume with the piston at TDC.

■ William Cecil's Internal Combustion Engine

Internal combustion engines are by no means new. The first internal combustion engine that would operate continuously by itself was built in 1820 by William Cecil.

Cecil's engine was constructed with three cylinders arranged in a "T", and connected at their intersection by a rotating valve. The vertical cylinder contained a piston that descended to draw in a charge of air mixed with hydrogen—the two horizontal cylinders remained empty during the intake stroke. At the bottom of the intake stroke, the central valve rotated to briefly open a passage connecting the cylinder to a small flame, igniting the hydrogen. The valve then rotated again, permitting the burning, expanding gases to fill up the two cylinders forming the crosspiece of the "T", pushing out the air they contained through flapper exhaust valves. As the burned gases cooled and contracted, they pulled the flapper valves shut. Atmospheric pressure then pushed the piston back up into the cylinder for the power stroke, and the cycle repeated.

Cecil's engine had a cylinder capacity of about 30 cu in. (500 cc), and used a flywheel that weighed 50 lb (23 kg). It would run evenly on a air-fuel mixture of 1:4, although best power was obtained with a richer mixture of 1.25:1. Because of the flame-type ignition, top speed was limited to about 60 rpm—above this speed the flame could not light the hydrogen reliably.

Cecil demonstrated a working model of his engine to the Cambridge Philosophical Society in 1820, but chose not to pursue its development further.

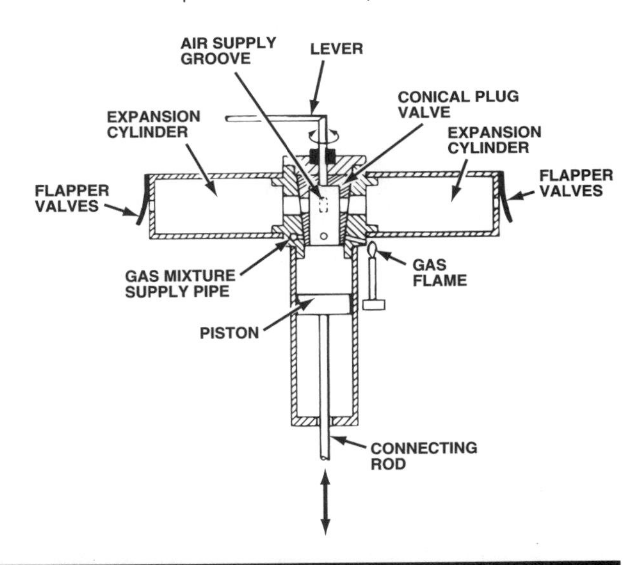

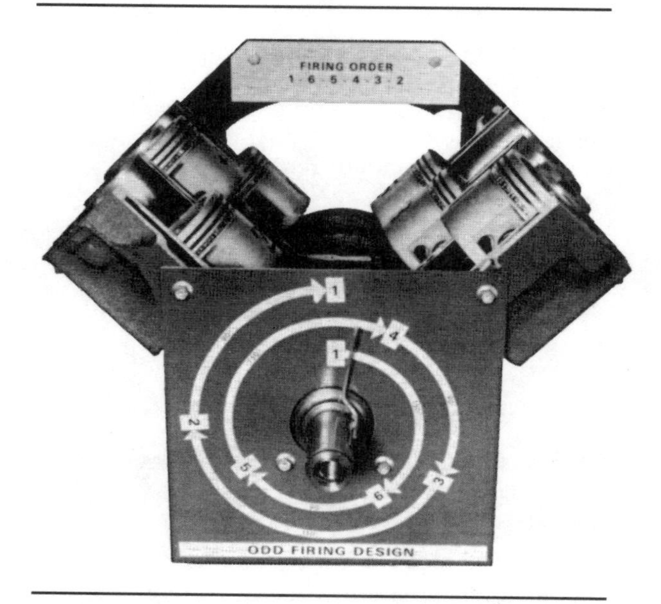

Figure 1-24. Buick's early V6 engine required uneven firing intervals to accommodate a 90-degree angle between the cylinder banks.

electrical potential or voltage. This voltage jumps a gap between two electrodes in the combustion chamber. The arc (spark) between the electrodes ignites the compressed mixture. The two electrodes are part of the spark plug, a major part of the ignition system.

Ignition Interval

The timing of the spark is critical to proper engine operation, and must occur near the start of the power stroke. If the spark occurs too early or too late, full power will not be obtained from the burning air-fuel mixture.

As we have seen, every two strokes of a piston rotate the crankshaft 360 degrees, and there are 720 degrees of rotation in a complete four-stroke cycle. During the four strokes of the cycle, the spark plug for each cylinder fires only once. In a single-cylinder engine, there would be only one spark every 720 degrees. These 720 degrees are called the **ignition interval** or **firing interval.** It is the number of degrees of crankshaft rotation that occur between ignition sparks.

Common ignition intervals

Since a 4-cylinder engine has four power strokes during 720 degrees of crankshaft rotation, one power stroke must occur every 180 degrees ($720 \div 4 = 180$). The ignition system must produce a spark for every power stroke, so it produces a spark every 180 degrees of crankshaft

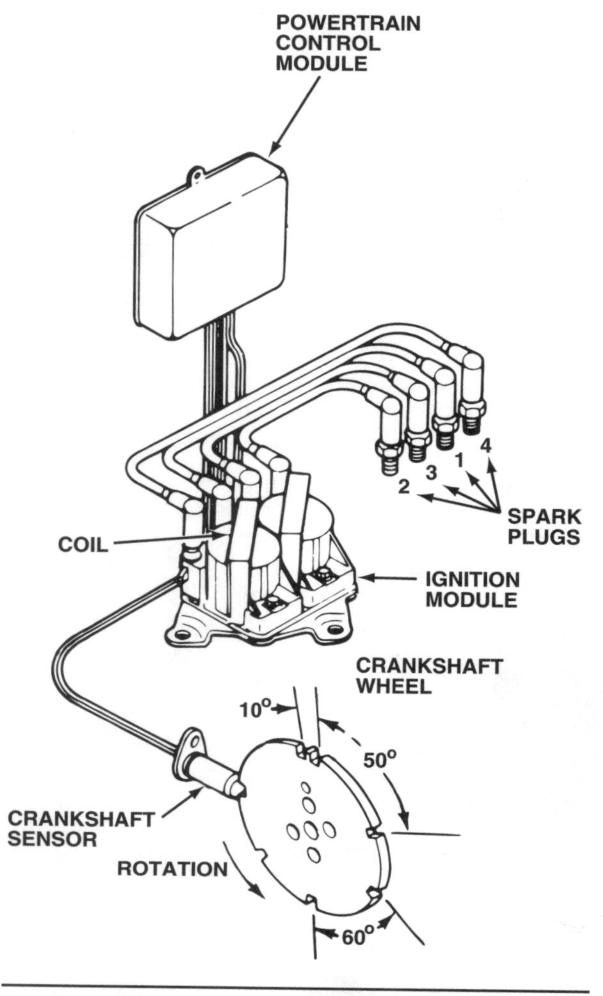

Figure 1-25. The spark plug wires must be connected in the proper sequence, or the engine will run poorly or not at all.

rotation. This means that a 4-cylinder engine has an ignition interval of 180 degrees.

An in-line 6-cylinder engine has six power strokes during every 720 degrees of crankshaft rotation, for an ignition interval of 120 degrees ($720 \div 6 = 120$). An 8-cylinder engine has an ignition interval of 90 degrees ($720 \div 8 = 90$).

Unusual ignition intervals

Some companies, such as Jaguar, Ferrari, and BMW, produce 12-cylinder engines with a 60-degree firing interval. Audi and Mercedes have 5-cylinder engines with a 144-degree firing interval. Suzuki produces a 3-cylinder engine with a 240-degree firing interval used in Chevrolet's Sprint; Subaru markets a similar engine.

Other firing intervals result from unusual engine designs. General Motors has produced two different V6 engines from V8 engine blocks.

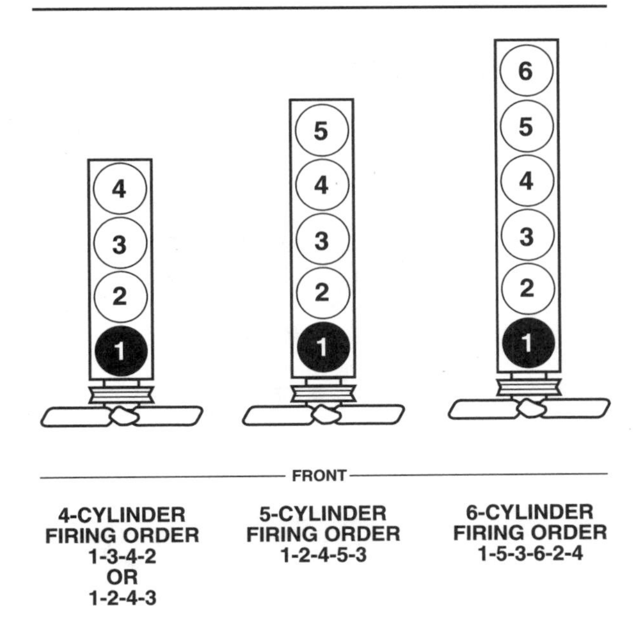

Figure 1-26. These are customary cylinder numbering and possible firing orders of in-line engines.

Figure 1-27. The domestic 90-degree V6 engines and several Japanese 60-degree engines are numbered this way. However, they use different firing orders.

The Buick version, developed in the 1960's, has alternating 90- and 150-degree firing intervals, figure 1-24. The uneven firing intervals resulted from building a V6 with a 90-degree crankshaft and block. This engine was modified in mid-1977 by redesigning the crankshaft to provide uniform 120-degree firing intervals, as in an in-line 6. In 1978, Chevrolet introduced a V6 engine that fires at alternating 108- and 132-degree intervals.

Spark frequency

In a spark-ignition engine, each power stroke is begun by a spark igniting the air-fuel mixture. Each power stroke needs an individual spark. An 8-cylinder engine, for example, requires four sparks per engine revolution (Remember that there are two 360-degree engine revolutions in each 720-degree operating cycle). When the engine is running at 1,000 rpm, the ignition system must deliver 4,000 sparks per minute. At high speed, about 4,000 rpm, the ignition system must deliver 16,000 sparks per minute. The ignition system must perform precisely to meet these demands.

Firing Order

The order in which the air-fuel mixture is ignited within the cylinders is called the **firing order,** and varies according to the crankshaft design. Firing orders reduce the vibration and imbalance created by the power strokes of the pistons.

Engine designers number the cylinders for identification. However, the cylinders seldom fire in the order in which they are numbered. Consult a service manual if you are unsure of an engine's firing order.

The ignition system must deliver a spark to the correct cylinder at the correct time. To get the correct firing order, the spark plug cables must be attached to the distributor cap, or to the coil on distributorless ignition systems, in the proper sequence, figure 1-25.

In-line engines

Straight or in-line engines are numbered from front to rear, figure 1-26. The most common firing order for domestic and imported 4-cylinder engines is 1-3-4-2. That is, the number one cylinder power stroke is followed by the number three cylinder power stroke, then the number four power stroke and, finally, the number two cylinder power stroke. Then, the next number one power stroke occurs again. Some in-line 4-cylinder engines have been built with firing orders of 1-2-4-3. One of these two firing orders is necessary due to the geometry of the engine.

The 5-cylinder, spark-ignited engine built by Audi and the 5-cylinder diesel engine built by Mercedes are also numbered front to rear. They share the same firing order: 1-2-4-5-3.

Ignition Interval (Firing Interval): The number of degrees of crankshaft rotation between ignition sparks. Sometimes called firing interval.

Firing Order: The sequence by cylinder number in which combustion occurs in the cylinders of an engine.

Figure 1-28. General Motors 60-degree V6 engines and many Japanese V6 engines are numbered and fired this way.

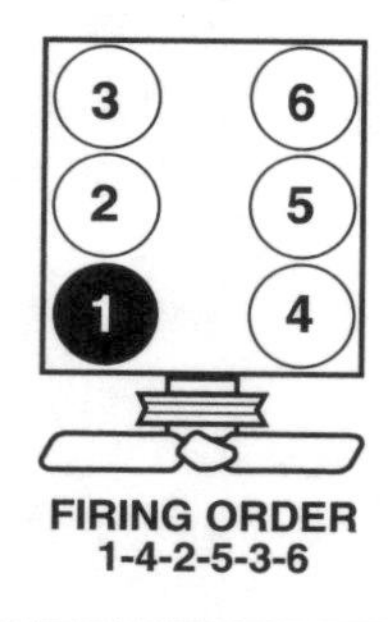

Figure 1-29. Ford and Acura V6 engines are numbered and fired this way.

The cylinders of an in-line 6-cylinder engine are numbered from front to rear. The firing order for all in-line 6-cylinder engines, domestic and imported, is 1-5-3-6-2-4.

V-type engines

The V-type engine structure allows designers greater freedom in selecting a firing order and still producing a smooth-running power plant. Consequently, there is a great variety of cylinder numbering and firing orders for V-type engines, too many to cover completely. Figures 1-27 through 1-31 show a representative sampling of common cylinder numbering styles and firing orders for these engines.

ENGINE-IGNITION SYNCHRONIZATION

During the engine operating cycle, the intake and exhaust valves open and close at specific times. The ignition system delivers a spark when the piston is near the top of the compression stroke and both valves are closed. These actions must all be coordinated, or engine damage can occur.

Figure 1-30. Most Chrysler and General Motors V8s as well as the 1990 Lexus V8s are numbered and fired this way. The 1990 Infiniti V8 is numbered this way, but fires differently.

Figure 1-31. Ford V8 engines are numbered this way but use two different firing orders, depending on the engine. Audi and Mercedes-Benz V8 engines are numbered like the Fords, but use a third firing order.

Distributor Drive

The distributor must supply one spark to each cylinder during each cylinder's operating cycle. The distributor cam has as many lobes as the engine has cylinders, or in an electronic ignition system, the trigger wheel has as many teeth as the engine has cylinders. One revolution of the distributor shaft will deliver one spark to each cylinder. Since each cylinder needs only one spark for each two crankshaft revolutions, the distributor shaft must turn at only one-half

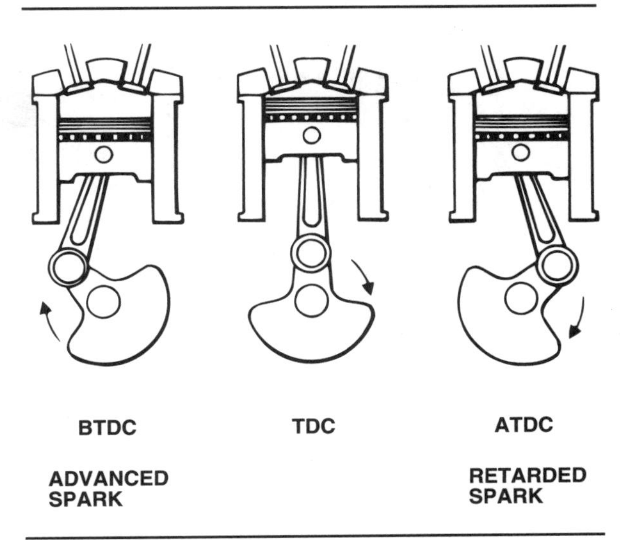

Figure 1-32. Piston position is identified in terms of crankshaft position.

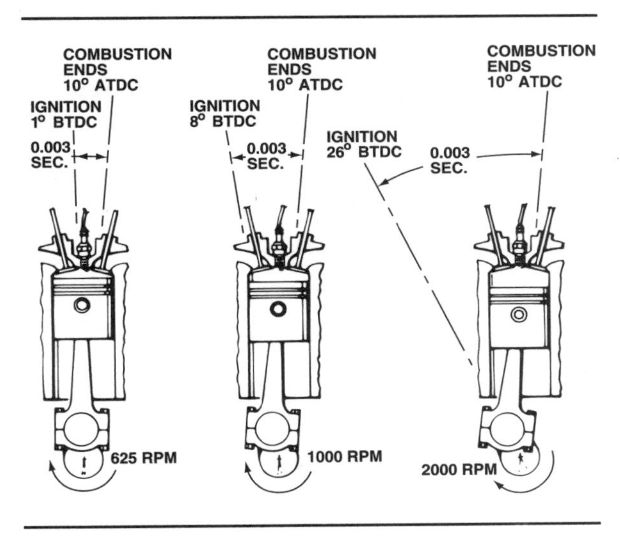

Figure 1-33. As engine speed increases, ignition timing must be advanced.

engine crankshaft speed. Therefore, the distributor is driven by the camshaft, which also turns at one-half crankshaft speed.

Crankshaft Position

As you learned earlier, the exact bottom of the piston stroke is called bottom dead center (BDC). The exact top of the piston stroke is called top dead center (TDC). The ignition spark occurs near TDC, as the compression stroke is ending. As the piston approaches the top of its stroke, it is said to be before top dead center (BTDC). A spark that occurs BTDC is called an advanced spark, figure 1-32. As the piston passes TDC and starts down, it is said to be after top dead center (ATDC). A spark that occurs ATDC is called a retarded spark.

Burn Time

The instant the air-fuel mixture ignites until its combustion is complete is called the burn time. For pump gasoline, burn time requires only a few milliseconds. The exact duration of burn time varies, depending on a number of factors, the most important of which include fuel type, compression ratio, bore size, spark plug location, and combustion chamber shape.

As its name implies, burn time is a function of time and not of piston travel or crankshaft degrees. The ignition spark must occur early enough so that the combustion pressure reaches its maximum just when the piston is beginning its downward power stroke. Combustion should be completed by about 10 degrees ATDC. If the spark occurs too soon BTDC, the rising piston will be opposed by combustion pressure. If the spark occurs too late, the force on the piston will be reduced. In either case, power will be lost. In extreme cases, the engine could be damaged. Ignition must start at the proper instant for maximum power and efficiency. Engineers perform many tests with running engines to determine the proper time for ignition to begin.

Engine Speed

As engine speed increases, piston speed increases. If the air-fuel ratio remains relatively constant, the fuel burning time will remain relatively constant. However, at greater engine speed the piston will travel farther during this burning time. The spark must occur earlier to ensure that maximum combustion pressure occurs at the proper piston position. Making the spark occur earlier is called spark advance or ignition advance.

For example, consider an engine, figure 1-33, that requires 0.003 second for the fuel charge to burn and that achieves maximum power if the burning is completed at 10 degrees ATDC.

- At an idle speed of 625 rpm, position A, the crankshaft rotates about 11 degrees in 0.003 second. Therefore, timing must be set at 1 degree BTDC to allow ample burning time.
- At 1,000 rpm, position B, the crankshaft rotates 18 degrees in 0.003 second. Ignition should begin at 8 degrees BTDC.
- At 2,000 rpm, position C, the crankshaft rotates 36 degrees in 0.003 second. Spark timing must be advanced to 26 degrees BTDC.

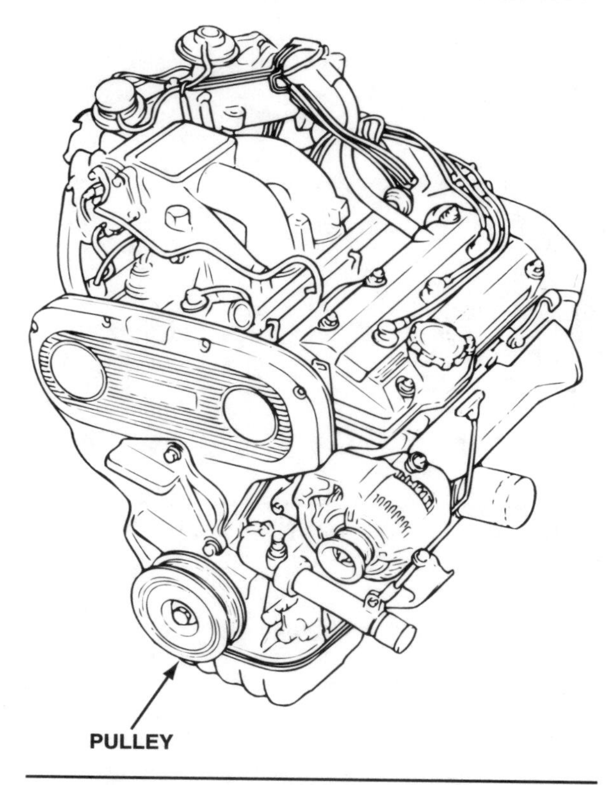

Figure 1-34. Most engines have a pulley bolted to the front end of the crankshaft.

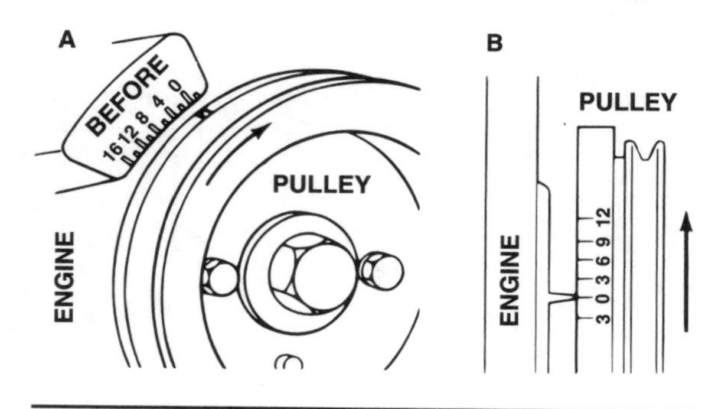

Figure 1-35. These are common types of timing marks.

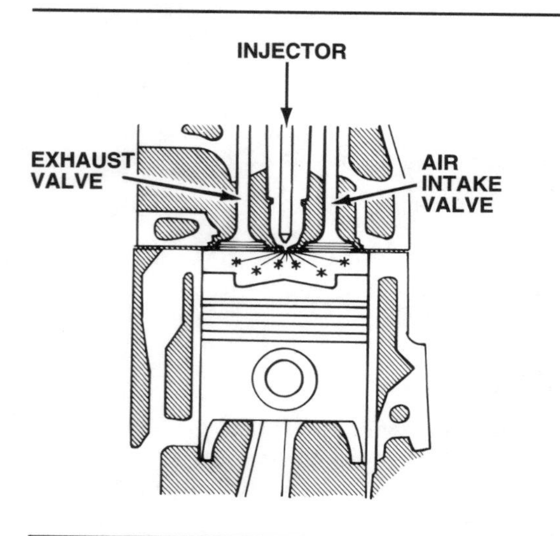

Figure 1-36. Diesel combustion occurs when fuel is injected into the hot, highly compressed air in the cylinder.

INITIAL TIMING

As we have seen, ignition timing must be set correctly for the engine to run at all. This is called the engine's initial, or base, timing. Initial timing is the correct setting at a specified engine speed. In figure 1-33, initial timing was 1 degree BTDC. Initial timing is usually within a few degrees of TDC. For many years, most engines were timed at the specified slow-idle speed for the engine. However, some engines built since 1974 require timing at speeds either above or below the slow-idle speed.

Timing Marks

We have seen that base timing is related to crankshaft position. To properly time the engine, we must be able to determine crankshaft position. The crankshaft is completely enclosed in the engine block, but most cars have a pulley and vibration damper bolted to the front of the crankshaft, figure 1-34. This pulley rotates with the crankshaft and can be considered an extension of the shaft.

Marks on the pulley show crankshaft position. For example, when a mark on the pulley is aligned with a mark on the engine block, the number one piston is at TDC.

Timing marks vary widely, even within a manufacturer's product line. There are two common types of timing marks, figure 1-35:

- A mark on the crankshaft pulley, and marks representing degrees of crankshaft position on the engine block, position A
- Marks on the pulley representing degrees of crankshaft position, and a pointer on the engine block, position B.

Some cars may also have a notch on the engine flywheel and a scale on the transmission cover or bellhousing. Many late-model Ford cars, in addition to a conventional timing mark, have a special test socket for electromagnetic engine timing. Technicians time these cars with a conventional timing light. The special test probe that fits into the socket is used on the assembly line in the manufacture of the car, not during a normal tune-up.

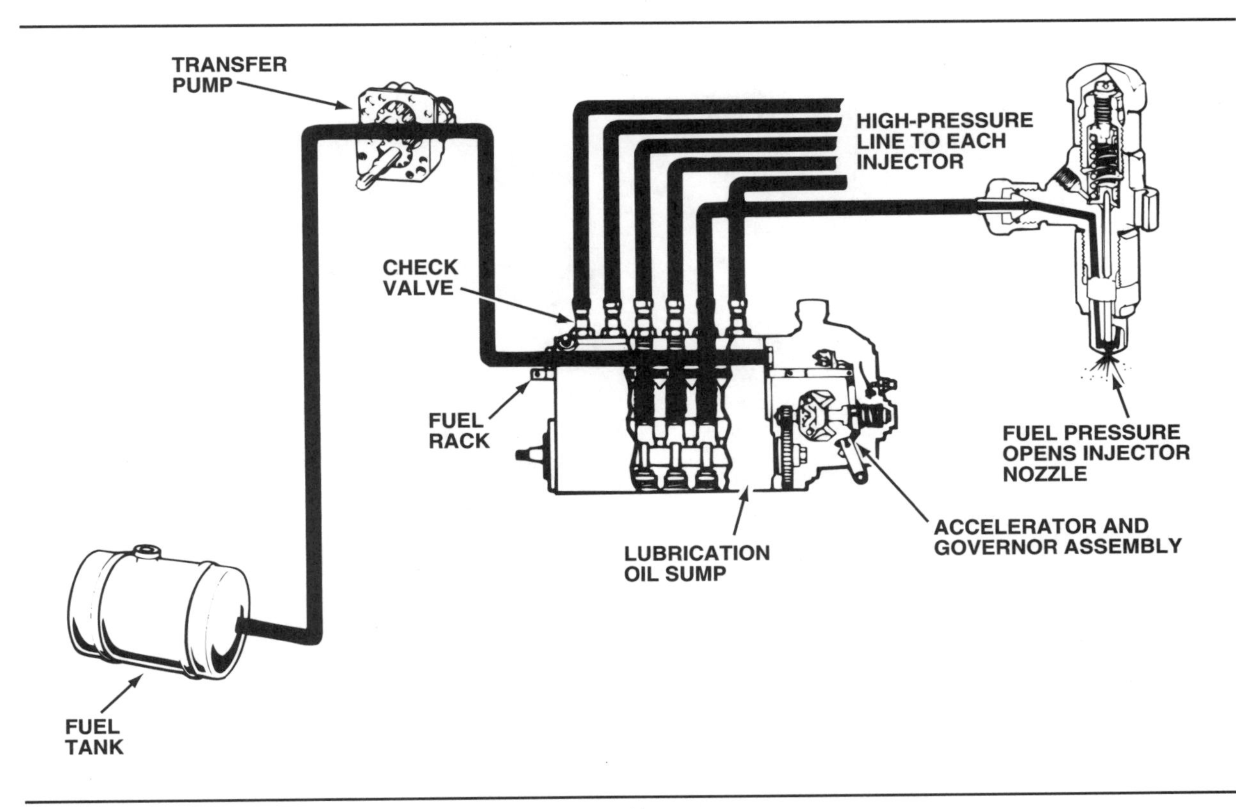

Figure 1-37. This is a typical automotive diesel fuel injection system.

OTHER ENGINE TYPES

Other engine types have been installed in automobiles over the years. Three used with success include the diesel, rotary, and stratified-charge engines.

The Diesel Engine

In 1892, a German engineer named Rudolf Diesel perfected the compression-ignition engine that bears his name. The diesel engine uses heat created by compression to ignite the fuel, so it requires no spark-ignition system.

The diesel engine requires compression ratios of 16:1 and higher. Incoming air is compressed until its temperature reaches about 1,000°F (540°C). As the piston reaches the top of its compression stroke, fuel is injected into the cylinder where it is ignited by the hot air, figure 1-36. As the fuel burns, it expands and produces power.

Diesel engines differ from gasoline-burning engines in other ways. Instead of a carburetor to

■ **Rudolf Diesel**

The theory of a diesel engine was first set down on paper in 1893 when Rudolf Christian Karl Diesel wrote a technical paper, "Theory and Construction of a Rational Heat Engine". Diesel was born in Paris in 1858. After graduating from a German technical school, he went to work for the refrigeration pioneer, Carl von Linde. Diesel was a success in the refrigeration business, but at the same time was developing his theory on the compression ignition engine. The theory is really very simple; air, when compressed, gets very hot. If you compress it enough, say on the order of 20:1, and squirt fuel into the compressed air, it will ignite.

In its first uses, the diesel engine was used to power machinery in shops and plants. By 1910, it was used in ships and locomotives, and a 4-cylinder engine was even used in a delivery van. The engine proved too heavy at that time to be practical for automobile use. It was not until 1927, when Robert Bosch invented a small fuel injection mechanism, that the use of the diesel engine in trucks and cars became practical.

Unlike many of the early inventors, Diesel did make a great deal of money from his invention, but in 1913 he disappeared from a ferry crossing the English Channel, and was presumed to have committed suicide.

mix the fuel with air, a diesel uses a precision **injection pump** and individual fuel injectors. The pump delivers fuel to the injectors at a high pressure and at timed intervals. Each injector measures the fuel exactly, spraying it into the combustion chamber at the precise moment required for efficient combustion, figure 1-37. The injection pump and injector system thus perform the tasks of the carburetor and distributor in a gasoline engine.

The air-fuel mixture of a gasoline engine remains nearly constant—changing only within a narrow range—regardless of engine load or speed. But in a diesel engine, *air* remains constant and the amount of *fuel* injected is varied to control power and speed. The air-fuel mixture of a diesel can vary from as lean as 85:1 at idle to as rich as 20:1 at full load. This higher air-fuel ratio and the increased compression pressures make the diesel more fuel efficient than a gasoline engine.

Like gasoline engines, diesel engines are built in both two-stroke and four-stroke versions. The most common two-stroke diesels are the truck and industrial engines made by the Detroit Diesel Allison Division of General Motors. In these engines, air intake is through ports in the cylinder wall. Exhaust is through poppet valves in the head. A blower box blows air through the intake port to supply air for combustion and to blow the exhaust gases out of the exhaust valves. Crankcase fuel induction cannot be used in a two-stroke diesel.

For many years, diesel engines were used primarily in trucks and heavy equipment. Mercedes-Benz, however, has built diesel cars since 1936, and the energy crises of 1973 and 1979 focused attention on the diesel as a substitute for gasoline engines in automobiles.

In the late 1970s, General Motors and Volkswagen developed diesel engines for cars. They were followed quickly by most major automakers in offering optional diesel engines for their vehicles. By the early 1980s many automakers were predicting diesel power for more than 30 percent of the domestic auto population. These predictions did not come true.

In the mid-1980s, increasing gasoline supplies and lower prices, in combination with the disadvantages of noise and higher purchasing costs, reduced the incentives for customers to buy diesel automobiles. Stringent diesel emission regulations added even more to manufacturing costs and made the engines harder to certify for sale. A few automobile diesel engines, such as the Oldsmobile V8, were derived from gasoline power plants and suffered reliability problems. In the late 1980s, diesel engine use in the United States and Canada was mostly limited to truck and industrial applications, although diesel automobiles continue to sell well in Europe, where gasoline costs remain high.

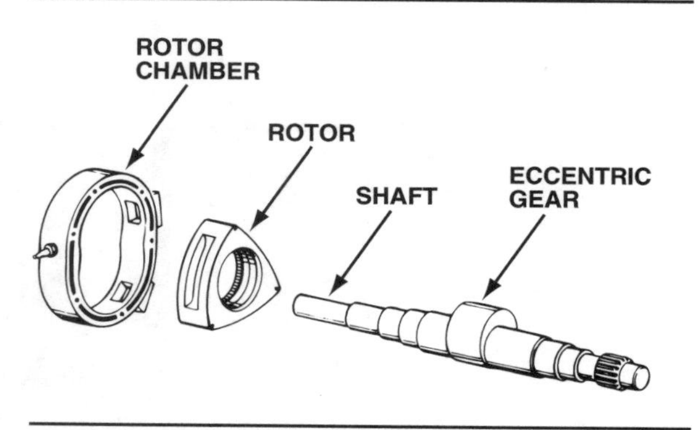

Figure 1-38. The main parts of a Wankel rotary engine are the rotor chamber, the three-sided rotor, and the shaft with an eccentric gear.

The Rotary (Wankel) Engine

The reciprocating motion of a piston engine is complicated and inefficient. For these reasons, engine designers have spent decades attempting to devise engines in which the working parts would all rotate on an axis. The major problem with this rotary concept has been the sealing of the combustion chamber. Of the various solutions proposed, only the rotary design of Felix Wankel—as later adapted by NSU, Curtiss-Wright, and Toyo Kogyo (Mazda)—has proven practical.

Although the same sequence of events occur in a rotary and a reciprocating engine, the rotary is quite different in design and operation. A curved triangular rotor moves on an **eccentric,** or off-center, geared portion of a shaft within an elliptical chamber, figure 1-38. As it turns, seals on the rotor's corners follow the housing shape. The rotor thus forms three separate chambers whose size and shape change constantly during rotation. The intake, compression, power, and exhaust functions occur within these chambers, figure 1-39. Wankel engines can be built with more than one rotor. Mazda-production engines, for example, are 2-rotor engines.

One revolution of the rotor produces three power strokes or pulses, one for each face of the rotor. In fact, each rotor face can be considered the same as one piston. Each pulse lasts for about three-quarters of a rotor revolution. The combination of rotary motion and longer overlapping power pulses results in a smooth-running engine.

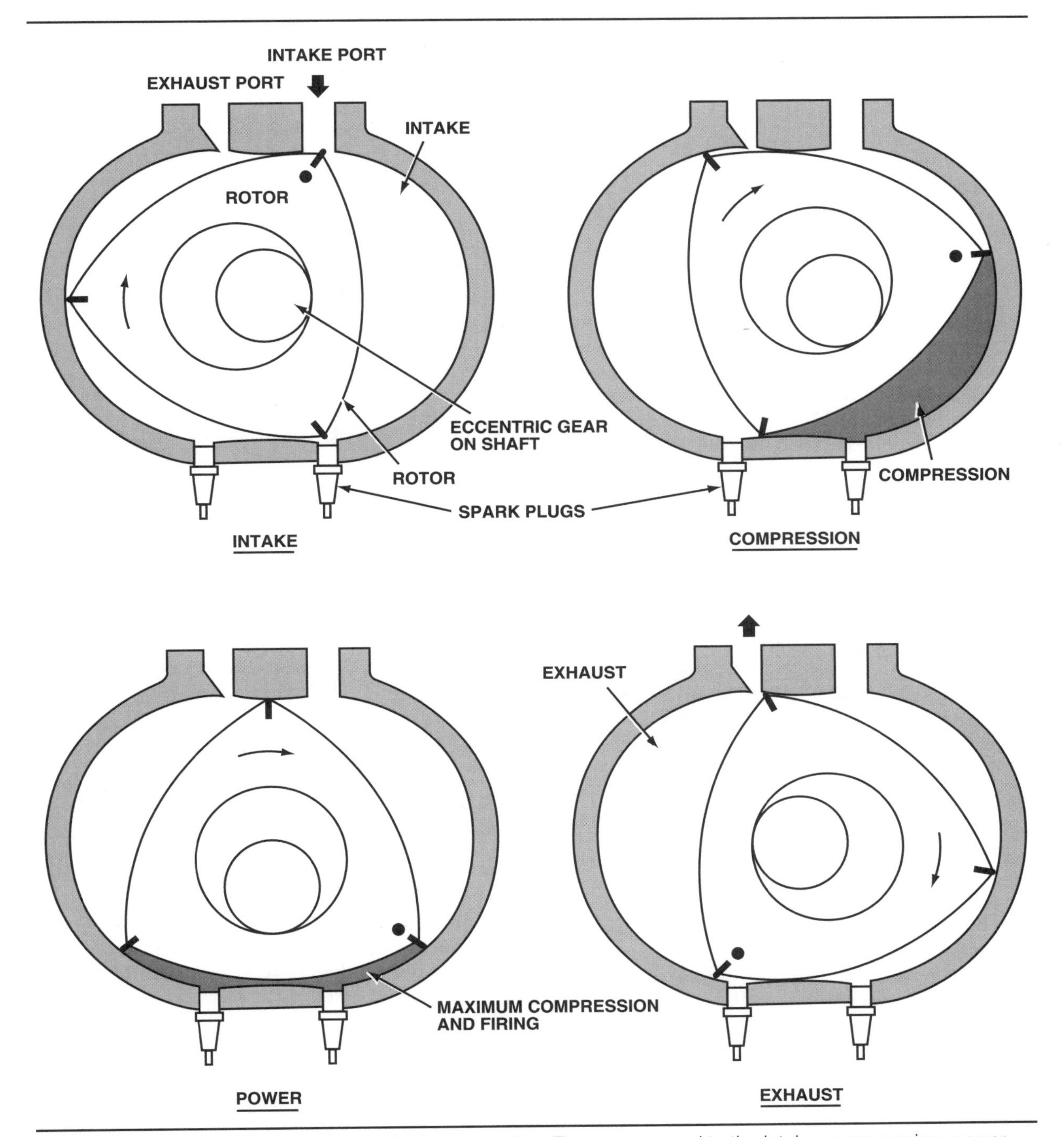

Figure 1-39. These are the four stages of rotary operation. They correspond to the intake, compression, power, and exhaust strokes of a four-stroke reciprocating engine. The sequence is shown for only one rotor face, but each face of the rotor goes through all four stages during each rotor revolution.

About equivalent in power output to that of a 6-cylinder piston engine, a 2-rotor engine is only one-third to one-half the size and weight. With no pistons, connecting rods, valves, lifters, and other reciprocating parts, the rotary engine has 40 percent fewer parts than a piston engine.

Injection Pump: A pump used on diesel engines to deliver fuel under high pressure at precisely timed intervals to the fuel injectors.

Eccentric: Off center. A shaft lobe that has a center different from that of the shaft.

While the rotary overcomes many of the disadvantages of the piston engine, it has its own disadvantages. It is basically a very "dirty" engine; in other words, it gives off a high level of emissions, and requires additional external devices to clean up the exhaust.

Stratified-Charge Combustion Chambers

The concept of a **stratified-charge** engine has been around in many forms for many years. However, Honda's Compound Vortex Controlled Combustion (CVCC) design was the first stratified-charge gasoline engine used in a mass-produced car. Introduced in 1975, it continued through the 1987 model year.

The CVCC engine has a separate small precombustion chamber located above the main combustion chamber, and contains a tiny additional valve, figure 1-40. Except for this feature, the CVCC is a conventional four-stroke piston engine. However, it uses a two-stage combustion process.

CVCC engines use 3-barrel carburetors. The third barrel provides a rich air-fuel mixture to the small precombustion chamber. This rich mixture is easy to ignite. The other two barrels in the carburetor supply a lean mixture to the main combustion chamber. This lean mixture is difficult to ignite with a spark but easily lit by the flame begun in the small chamber. The lean mixture in the main chamber allows good fuel economy and low exhaust emissions.

Figure 1-41 shows the stages in the operating cycle. The first stage is precombustion, in which the air-fuel mixture is ignited in the precombustion chamber. In the second stage, the flame front created moves down into the main combustion chamber to ignite a mixture containing less fuel. The stratified-charge engine takes its name from this layering or stratification of the air-fuel mixture just before combustion. At that time, there is a rich mixture (with lots of fuel) near the spark plug, a moderate mixture in the auxiliary combustion chamber, and a lean mixture (with little fuel) in the main chamber. The result is a more complete combustion of the air-fuel mixture, which keeps unburned fuel and emissions to a minimum.

Charge stratification is applied to reciprocating diesel and rotary gasoline engines. Most diesel engines used in cars have a precombustion chamber into which the fuel is injected. This allows the combustion to occur in two stages: in the precombustion chamber and in the main chamber. This improves cold starting and combustion efficiency, and reduces engine noise and vibration.

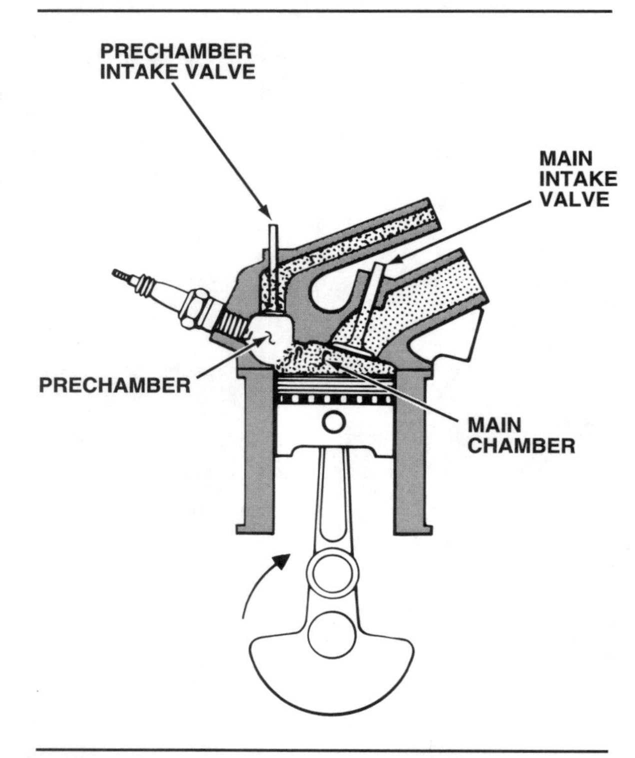

Figure 1-40. Honda CVCC cylinder head, showing the main combustion chamber and precombustion chamber (prechamber) with extra intake valve.

SUMMARY

Most automobile engines are internal combustion, reciprocating four-stroke engines. An air-fuel mixture is drawn into sealed combustion chambers by a vacuum created by the downward stroke of a piston. The mixture is ignited by a spark.

Valves at the top of the cylinder open and close to admit the air-fuel mixture and release the exhaust. These valves are driven by a camshaft and are synchronized to the crankshaft. The sequence in which the cylinders fire is the firing order. Several valve designs have been used, including I-head, F-head, and L-head. Most engine cylinders have two valves, but some engines have three, four, or five per cylinder.

Displacement and compression ratio are two frequently used engine specifications. Displacement indicates engine size, and compression ratio compares total cylinder volume to compression chamber volume.

The ignition interval is the number of degrees between ignition sparks. A 4-cylinder engine commonly has an ignition interval of 180 degrees; a V8, 90 degrees; an in-line 6-cylinder,

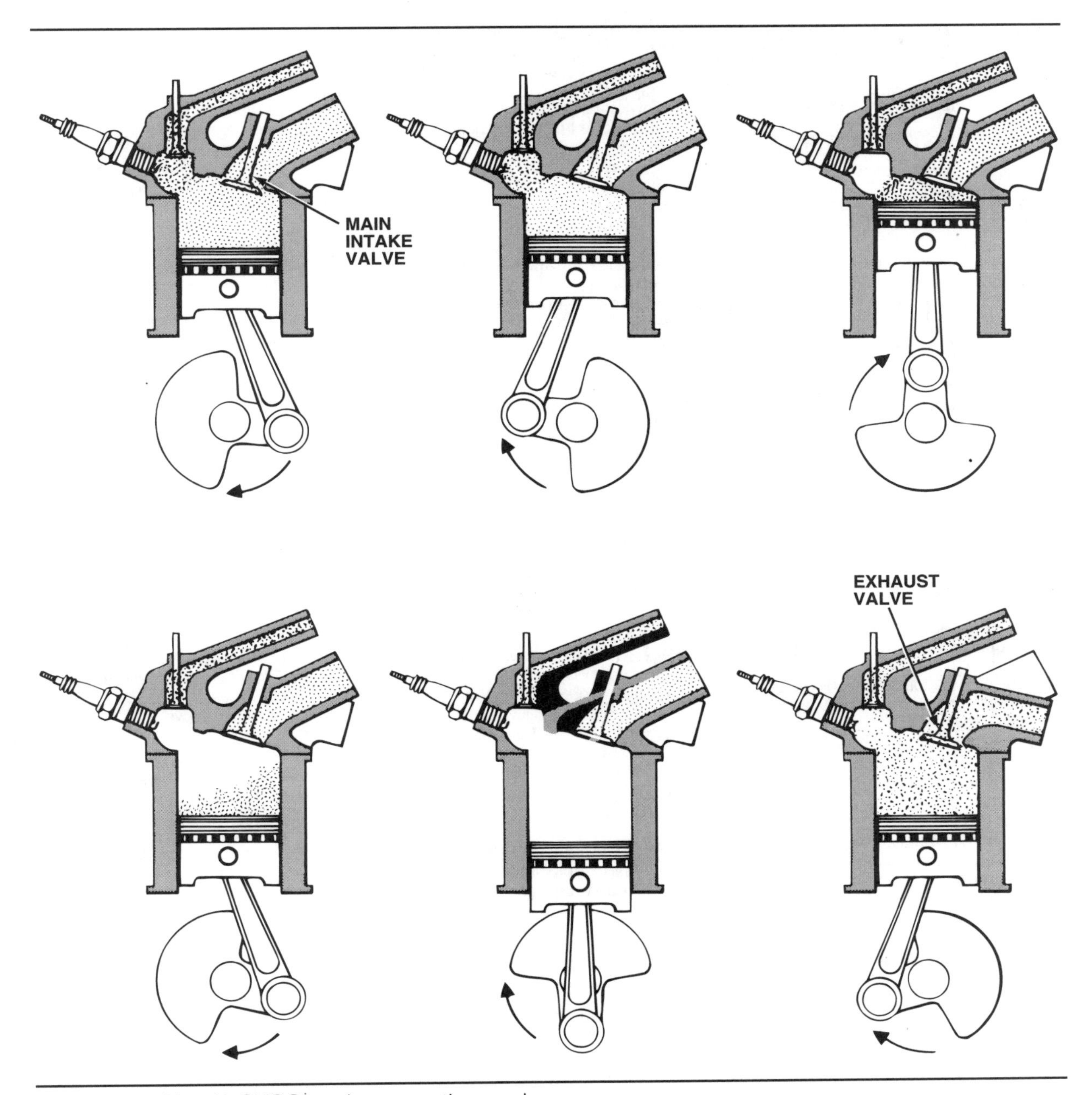

Figure 1-41. Honda CVCC engine operating cycle.

120 degrees; although many intervals have been used over the years. Still, each cylinder fires once every 720 degrees of crankshaft rotation. The ignition spark is provided by the distributor or electronic ignition, and is synchronized with the crankshaft rotation.

Correct ignition timing is essential for engine operation. Timing marks on the front of the engine block or the pulley indicate crankshaft position and can be used to alter the timing from the engine's base or initial timing.

Some of the most successful automobile engine concepts used besides four-stroke gasoline engines include the diesel, rotary (Wankel), and stratified-charge.

Stratified-Charge Engine: An engine that combusts the air-fuel mixture in two stages. The first occurs when a rich intake charge burns in the precombustion chamber; the second, when a leaner intake charge burns in the main combustion chamber.

Review Questions

Choose the single most correct answer.
Compare your answers to the correct answers on page 507.

1. An automotive internal combustion engine:
   a. Uses energy released when a compressed air-fuel mixture is ignited
   b. Has pistons driven downward by explosions in the combustion chambers
   c. Is better than an external combustion engine
   d. All of the above

2. A spark plug fires near the end of the:
   a. Intake stroke
   b. Compression stroke
   c. Power stroke
   d. Exhaust stroke

3. The camshaft rotates at:
   a. The same speed as the crankshaft
   b. Double the speed of the crankshaft
   c. Half the speed of the crankshaft
   d. It depends on the engine design

4. Which of the following is never in direct contact with the engine valves?
   a. Pushrods
   b. Valve springs
   c. Rocker arms
   d. Valve seats

5. The bore is the diameter of the:
   a. Connecting rod
   b. Cylinder
   c. Crankshaft
   d. Combustion chamber

6. The four-stroke cycle operates in which order?
   a. Intake, exhaust, power, compression
   b. Intake, power, exhaust, compression
   c. Compression, power, intake, exhaust
   d. Intake, compression, power, exhaust

7. Diesel engines:
   a. Have no valves
   b. Produce ignition by heat of compression
   c. Have low compression
   d. Use special carburetors

8. How many strokes of a piston are required to turn the crankshaft through 360 degrees?
   a. One
   b. Two
   c. Three
   d. Four

9. A "retarded spark" is one that occurs:
   a. At top dead center
   b. Before top dead center
   c. After top dead center
   d. At bottom dead center

10. An internal combustion engine is an efficient use of available energy because:
    a. The combustion of the fuel produces more energy than is required to compress and fire it
    b. Gasoline is so easy to burn
    c. Internal combustion engines turn with so little friction
    d. Compression ratios are high enough to vaporize gasoline

11. Which of the following shows the typical ignition interval of an in-line 6-cylinder engine?
    a. 600 degrees 4 6 = 100 degrees
    b. 360 degrees 4 3 = 120 degrees
    c. 90 degrees 4 6 = 540 degrees
    d. 720 degrees 4 6 = 120 degrees

12. How many sparks (spark plug firings) per crankshaft revolution are required by a typical V8 engine?
    a. 8
    b. 6
    c. 4
    d. 2

13. The firing order of an engine is:
    a. The same on all V8s
    b. Is stated in reference books and on many cylinder blocks
    c. Can be deduced by common sense
    d. None of the above

14. Ignition timing:
    a. Requires a basic setting specified in degrees BTDC or ATDC, depending on the vehicle
    b. Varies while engine is running, synchronizing combustion with proper piston position
    c. Is indicated by the alignment of markings on the crankshaft pulley or flywheel, with other markings or a pointer fixed to the cylinder block
    d. All of the above

15. In a four-stroke engine, the piston is driven down in the cylinder by expanding gases during the ___ stroke.
    a. Compression
    b. Exhaust
    c. Power
    d. Intake

16. The ignition interval of an engine is the number of degrees of crankshaft rotation that:
    a. Occur between ignition sparks
    b. Are required to complete one full stroke
    c. Take place in a four-stroke engine
    d. All of the above

17. The rotary engine:
    a. Operates without pistons, connecting rods, or poppet valves
    b. Is not a reciprocating engine
    c. Produces three power strokes per revolution of its rotor
    d. All of the above

18. Which of the following is *not* used in calculating engine displacement?
    a. Stroke
    b. Bore
    c. Number of cylinders
    d. Valve arrangement

19. To change cubic centimeters to cubic inches, multiply by:
    a. 0.061
    b. 16.39
    c. 61.02
    d. 1000

20. Compression ratio is:
    a. Piston displacement plus clearance volume
    b. Total volume times number of cylinders
    c. Total volume divided by clearance volume
    d. Stroke divided by bore

Chapter

2

Engine Air-Fuel Requirements

Automobile engines run on a mixture of gasoline and air. Gasoline has several advantages as a fuel, including that it:

- Vaporizes (evaporates) easily
- Burns quickly but under control when mixed with air and ignited
- Has a high heat value and produces a large amount of heat energy
- Is easy to store, handle, and transport. Gasoline also has certain disadvantages as a fuel. In particular, burning gasoline creates harmful pollutants, which enter the atmosphere with the engine exhaust. For the time being, however, no better alternative than gasoline is available to fuel automotive engines.

To understand how the fuel system works in an engine, we must understand the engine's air-fuel requirements. This chapter discusses those requirements, and describes how the fuel gets from the fuel tank to the combustion chamber.

AIR PRESSURE—HIGH AND LOW

You can think of an internal combustion engine as a big air pump. As the pistons move up and down in the cylinders, they pump in air and fuel for combustion and pump out exhaust gases. They do this by creating difference in air pressure. The air outside an engine has weight and exerts pressure, as does the air inside an engine.

As a piston moves down on an intake stroke with the intake valve open, it creates a larger area inside the cylinder for the air to fill. This lowers the air pressure within the engine. Because the pressure inside the engine is lower than the pressure outside, air flows into the engine to fill the low-pressure area and equalize the pressure.

The low pressure within the engine is called **vacuum**. You can think of the vacuum as sucking air into the engine, but it is really the higher pressure on the outside that forces air into the low-pressure area inside. The difference in pressure between the two areas is called a **pressure differential.** The pressure differential principle has many applications in automotive fuel and emission systems.

An engine pumps exhaust out of its cylinders by creating pressure as a piston moves upward on the exhaust stroke. This creates high pressure in the cylinder, which forces the exhaust toward the lower-pressure area outside the engine.

Pressure differential can be applied to liquids as well as to air. Fuel pumps work on this principle. The pump creates a low-pressure area in the fuel system that allows the higher pressure of the air and fuel in the tank to force the fuel through the lines to the carburetor or the injection system.

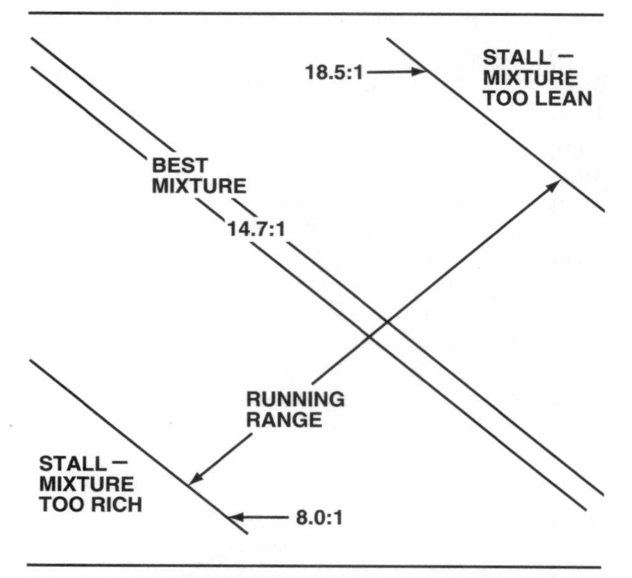

Figure 2-1. An engine can run without stalling on an air-fuel mixture with a ratio between 8:1 and 18.5:1.

AIRFLOW REQUIREMENTS

All gasoline automobile engines share certain air-fuel requirements. For example, a four-stroke engine can take in only so much air at any time, and how much fuel it consumes depends on how much air it takes in. Engineers calculate engine airflow requirement using these three factors:

- Engine displacement
- Engine revolutions per minute (rpm)
- Volumetric efficiency.

The airflow number represents cubic feet per minute (cfm) or cubic meters per minute (cmm). The designer sizes the carburetor or fuel injection intake airflow capacity to match the engine's maximum requirement. Volumetric efficiency and how it relates to engine airflow is described in the following paragraphs.

Volumetric Efficiency

Volumetric efficiency is a comparison of the actual volume of air-fuel mixture drawn into an engine to the theoretical maximum volume that could be drawn in. Volumetric efficiency is expressed as a percentage, and changes with engine speed. For example, an engine might have 75 percent volumetric efficiency at 1,000 rpm. The same engine might be rated at 85 percent at 2,000 rpm and 60 percent at 3,000 rpm.

If the engine takes in the airflow volume slowly, a cylinder might fill to capacity. It takes a definite amount of time for the airflow to pass through all the curves of the intake manifold and valve port. Therefore, manifold and port design directly relate to engine "breathing", or volumetric efficiency. Camshaft timing and exhaust tuning also are important.

If the engine is running fast, the intake valve does not stay open long enough for full volume to enter the cylinder. At 1,000 rpm, the intake valve might be open for one-tenth of a second. As engine speed increases, this time decreases to a point where only a small airflow volume can enter the cylinder. Therefore, volumetric efficiency decreases as engine speed increases. At high speed, it may drop to as low as 50 percent.

The average street engine never reaches 100 percent volumetric efficiency. With a street engine, you can expect a volumetric efficiency of about 75 percent at maximum speed, or 80 percent at the torque peak. A high-performance street engine will be about 85 percent, or a bit more efficient at peak torque. A race engine will usually have 95 percent or better volumetric efficiency. These figures apply only to naturally aspirated engines, however. Turbocharged and supercharged engines can easily achieve more than 100 percent volumetric efficiency.

AIR-FUEL RATIOS

Fuel burns best when the intake system turns it into a fine spray and mixes it with air before sending it into the cylinders. In carbureted engines, the fuel becomes a spray and mixes with air in the carburetor. In fuel injected engines, the fuel becomes a fine spray as it leaves the tip of the injectors and mixes with air in the intake manifold. In both cases, there is a direct relationship between engine airflow and fuel requirements. This relationship is called the **air-fuel ratio**.

The air-fuel ratio is the proportion by weight of air and gasoline that the carburetor or injection system mixes as needed for engine combustion. This ratio is important, since there are limits to how rich (more fuel) or lean (less fuel) the mixture can be and remain combustible. The mixtures with which an engine can operate without stalling range from 8 to 1 to 18.5 to 1, figure 2-1. These ratios are usually stated this way: 8 parts of air by weight combined with 1 part of gasoline by weight (8:1) is the richest mixture that an engine can tolerate and still fire regularly; 18.5 parts of air mixed with 1 part of gasoline (18.5:1) is the leanest. Richer or leaner air-fuel ratios cause the engine to misfire badly or not run at all.

Air-fuel ratios are calculated by weight rather than volume. If it requires 14.7 pounds or kilograms of air to burn one pound or kilogram of gasoline, the air-fuel ratio is 14.7:1.

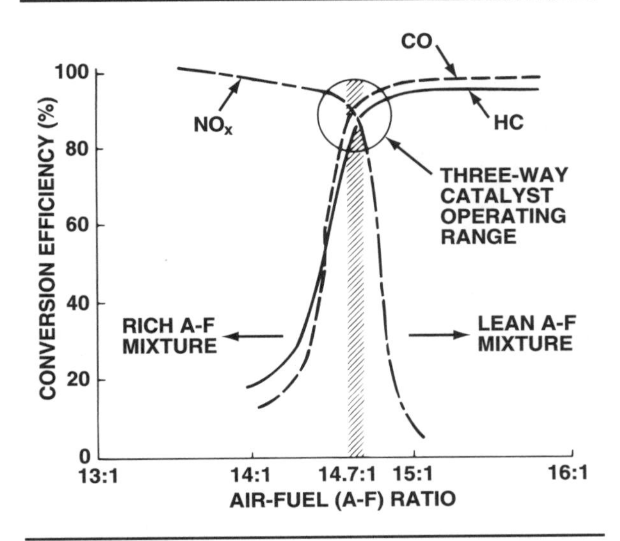

Figure 2-2. With a three-way catalytic converter, emission control is optimum with an air-fuel ratio between 14.65:1 and 14.75:1.

Stoichiometric Air-Fuel Ratio

The ideal mixture or ratio at which all the fuel combines with all of the oxygen in the air and *burns completely* is called the **stoichiometric ratio**—a chemically perfect combination. In theory, this ratio is an air-fuel mixture of 14.7:1. In reality, the exact ratio at which perfect mixture and combustion occurs depends on the molecular structure of gasoline, which varies somewhat. The stoichiometric ratio is a compromise between maximum power and maximum economy. Late-model computerized vehicles are designed to regulate the air-fuel ratio at stoichiometric.

Emission control is also optimum at this ratio, if the exhaust system uses a three-way oxidation-reduction catalytic converter. As the mixture gets richer, hydrocarbon (HC) and carbon monoxide (CO) conversion efficiency falls off. With leaner mixtures, oxides of nitrogen (NO_x) conversion efficiency also falls off. As figure 2-2 shows, the conversion efficiency range is very narrow—14.65:1 to 14.75:1. A fuel system without feedback control cannot maintain this narrow range.

Engine Air-Fuel Requirements

An automobile engine will work with the air-fuel mixture ranging from 8:1 to 18.5:1. The ideal ratio provides the most power and the most economy, while producing the least emissions. Such a ratio does not exist because engine fuel requirements vary widely, depending on temperature, load, and speed conditions.

Research proves that a 15:1 to 16:1 ratio provides the best fuel economy, while a 12.5:1 to 13.5:1 ratio gives maximum power output. An engine needs a rich mixture for idle, heavy load, and high-speed conditions and a leaner mixture for normal cruising and light-load conditions. No single air-fuel ratio provides the best fuel economy and the maximum power output at the same time.

Just as outside conditions such as speed, load, temperature, and atmospheric pressure change the engine's fuel requirements, other forces at work inside the engine cause additional variations. Here are two examples:

- The mixture may be imperfect because the fuel may not vaporize completely
- The mixture from a carburetor or a throttle body does not distribute equally through the intake manifold to each cylinder, so some cylinders get a richer or leaner mixture than others.

For an engine to run well under a variety of outside and inside conditions, the carburetor or injection system must vary the air-fuel mixture quickly to give the best mixture possible for engine requirements at any given moment.

Power Versus Economy

If the goal is to get the most power from an engine, all the oxygen in the mixture must burn because the power output of any engine is limited by the amount of air it can pull in. For the oxygen to combine completely with the available fuel, there should be extra fuel available. This makes the air-fuel ratio richer, with the result that some fuel does not burn.

Vacuum: Low pressure within an engine created by a downward piston intake stroke with the intake valve open.

Pressure Differential: The difference between the atmospheric pressure outside of the engine and the area of low pressure inside the engine created by the downward piston intake stroke.

Volumetric Efficiency: The actual volume of air-fuel mixture an engine draws in compared to the theoretical maximum it could draw in, written as a percentage.

Air-Fuel Ratio: The ratio of air to gasoline in the air-fuel mixture that enters an engine.

Stoichiometric Ratio: An ideal air-fuel mixture for combustion, in which all oxygen and all fuel will burn completely.

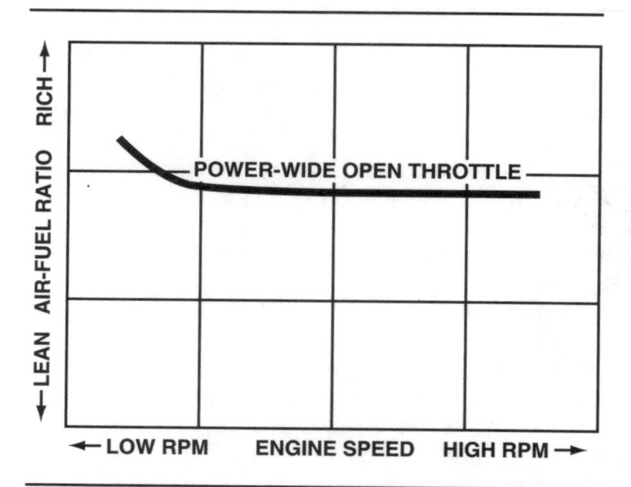

Figure 2-3. The air-fuel ratio that provides the most power is rich overall, and slightly richer at low engine speeds.

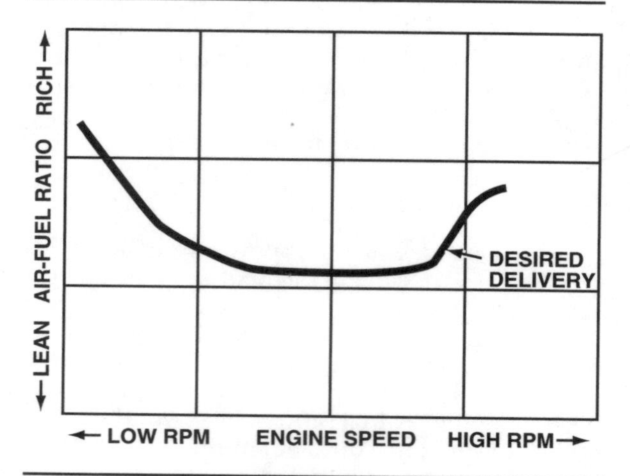

Figure 2-4. The air-fuel ratio that provides the best economy is lean overall, but slightly richer at both higher and lower engine speeds.

To get the best fuel economy and the lowest emissions, the gasoline must burn as completely as possible in the combustion chamber. This means combustion with the least amount of leftover waste material, or emissions, provides the greatest economy. The intake system must provide more air to make sure that enough oxygen is available to combine with the gasoline. This results in a leaner air-fuel mixture than the ideal.

The air-fuel ratio required to provide maximum power changes very little, except at low speeds, figure 2-3. Reducing speed reduces the airflow into the engine, resulting in poorer mixing of the air and fuel and less efficiency in distributing it to the cylinders. The mixture must be slightly richer at low speeds to make up for this.

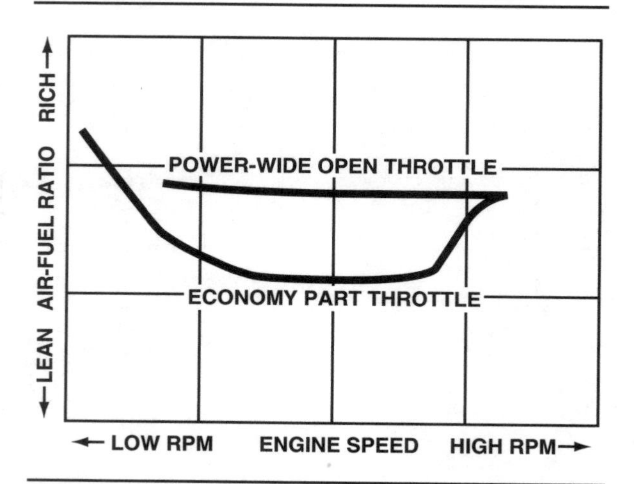

Figure 2-5. The carburetor or fuel injection system must determine the best air-fuel ratio to balance power and economy.

The same is true for maximum fuel economy—the leaner air-fuel ratio remains virtually the same throughout most of the operating range, figure 2-4. However, the mixture must be richer during idle and low speeds, and during higher speeds and under load—two conditions that require more power.

Sometimes the air-fuel mixture gets richer when it is not required or wanted, as with high-altitude driving. As altitude increases, **atmospheric pressure** drops and the air becomes thinner. The same volume of air weighs less and contains less oxygen at higher altitudes, so an engine takes in fewer pounds or kilograms of air and less oxygen. The result is a richer air-fuel ratio, which must be corrected to achieve efficient high-altitude engine operation. Altitude-compensating carburetors and fuel injection airflow and pressure sensors solve this problem. The corrected air-fuel mixture lets the engine burn fuel efficiently, but total horsepower is less, corrected or not. With the intake system providing less fuel to the cylinder to mix with the lower oxygen content of the high-altitude air, fewer calories of energy are available to do work.

For these reasons, the carburetor or injection system must deliver fuel so that the best mileage is provided during normal cruising, with maximum power available when the engine is under load, accelerating, or at high speed, figure 2-5.

INTRODUCTION TO ELECTRONIC ENGINE CONTROLS

Electronic engine controls first appeared on vehicles manufactured for the 1977 model year. Early control systems regulated only a single

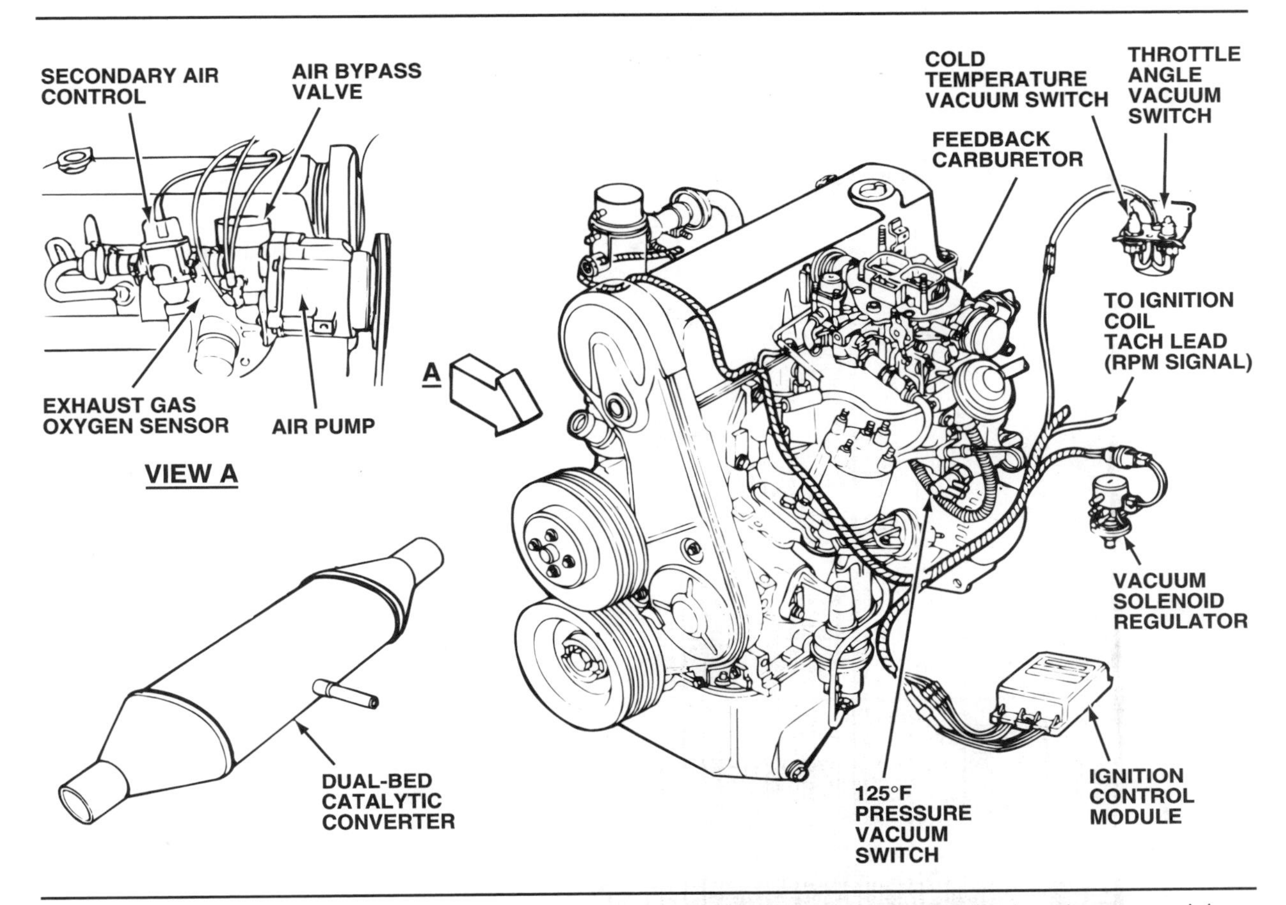

Figure 2-6. The first electronically controlled fuel management systems included a feedback carburetor, an injection control module, an O2S, and a catalytic converter.

function, either ignition timing or fuel metering. However, automakers rapidly expanded them to include control over both systems, and other engine functions.

As illustrated in figure 2-6, the basic parts of the first electronically controlled fuel management systems included:

- A feedback carburetor
- An ignition control module (ICM), powertrain control module (PCM), or both
- An exhaust gas oxygen sensor (O2S) mounted in the exhaust manifold
- A catalytic converter.

Carbureted engines used either direct or indirect actuators to control the fuel-metering rods and air bleeds. With the direct method, a solenoid or stepper motor mounted on or in the carburetor operated the fuel-metering rods or the air bleeds, or both. With indirect control, a remote-mounted, solenoid-actuated vacuum valve regulated the carburetor vacuum diaphragms that operated the fuel-metering rods and air bleeds.

The PCM constantly monitored the oxygen content of the exhaust gas through signals received from the O2S. The PCM sent a pulsed voltage signal to the control device, varying the ratio of on-time to off-time according to the signals received from the O2S. As the percentage of on-time increased or decreased, the mixture became leaner or richer.

With fuel injection systems, the PCM controls the air-fuel mixture by switching one or more injectors on and off, at a rate based on engine speed. The PCM also varies the length of time the injectors remain open to establish the air-fuel ratio. As the microprocessor receives data from system sensors, it lengthens or shortens the pulse width (on-time) according to engine operating and load conditions. We will study more about fuel injection systems and how they work in Part Four.

Atmospheric Pressure: The pressure that the earth's atmosphere exerts on objects. At sea level, this pressure is 14.7 psi (101 kPa) at 32°F (0°C).

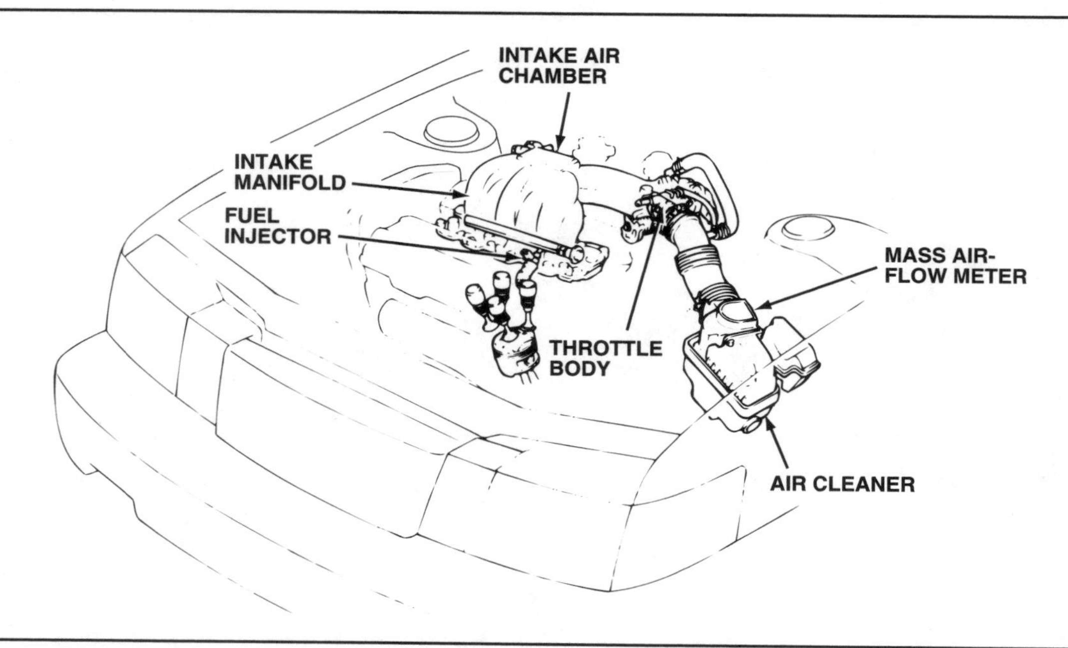

Figure 2-7. The two basic parts of an engine intake system are the intake manifold and the fuel delivery system.

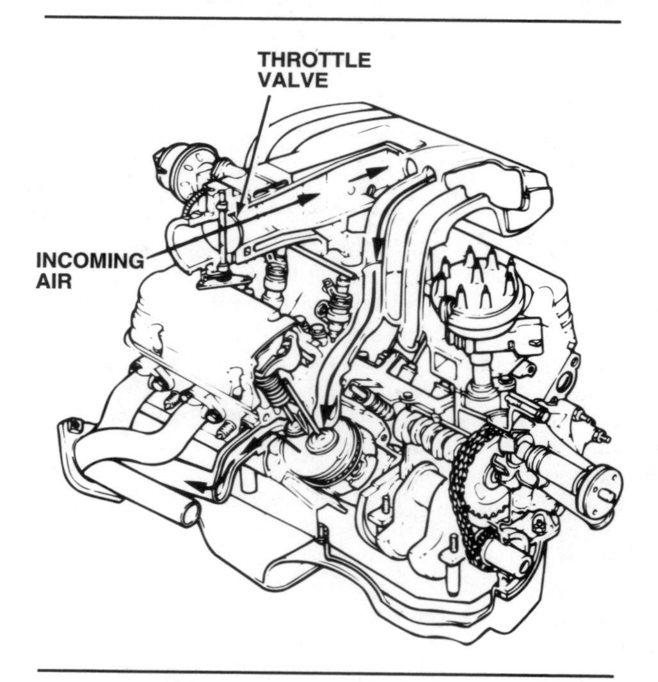

Figure 2-8. Atmospheric pressure pushes air into the intake system.

THE INTAKE SYSTEM

Primary parts of the intake system are a manifold and a fuel delivery system, figure 2-7. The manifold is a casting with a series of enclosed passages that route the air-fuel mixture from the carburetor, or throttle body, to the intake ports in the engine's cylinder head. With port fuel injection systems, the intake manifold routes only air, and the fuel injectors deliver the fuel directly to the intake port.

Airflow

We have pointed out that you can think of an automobile engine as a large air pump. As the pistons move up and down in the cylinders, they draw in air and fuel for combustion and pump out burned exhaust. They do this by creating a difference in air pressure.

You recall that the downward movement of the piston intake stroke creates a partial vacuum, lowering the air pressure in the combustion chamber and intake manifold. Opening the carburetor or fuel injection throttle valve, figure 2-8, lets air move from the higher-pressure area outside the engine, through the intake, to the lower-pressure area of the manifold. The throttle valve opening determines how much and how fast the air travels. When the intake valve in the cylinder head opens, the vacuum draws the air-fuel mixture from the intake manifold into the combustion chamber, figure 2-9.

Air-fuel mixture control

Before gasoline can do its job as a fuel, it must be metered, atomized, vaporized, and distributed to each cylinder in the form of a burnable mixture. To do this, a metering device mixes the gasoline with air in the correct ratio and distributes the mixture as required by engine load, speed, throt-

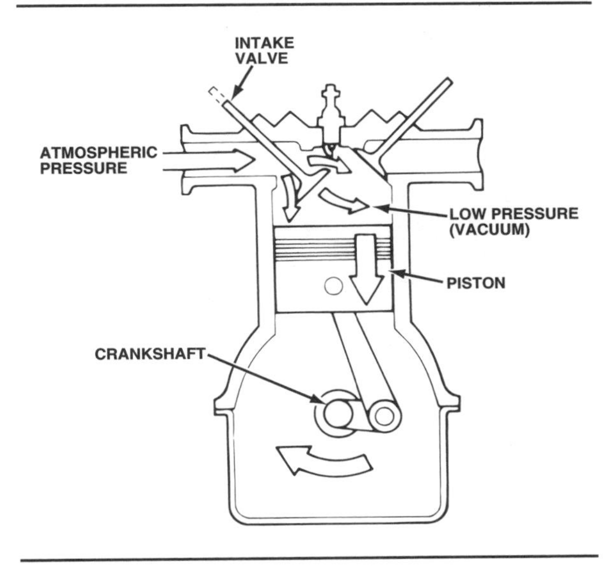

Figure 2-9. Atmospheric pressure pushes the air-fuel mixture into the combustion chamber.

Figure 2-10. The carburetor is made of light-weight metal and draws fuel from a float bowl through the barrels, atomizing and metering the gasoline.

tle valve position, and operating temperature. On most late-model engines, a fuel injection system does the job of air-fuel mixing done by a carburetor on older engines.

As of 1991 no cars have carburetors. A limited number of import trucks still have carburetors, but the manufacturers are phasing them out. However, we will study carburetors because there are still vehicles using them. Understanding how carburetors work improves your understanding of fuel injection and its advantages.

Carburetors

Carburetors consist of lightweight metal bodies that have one, two, three, or four bores or barrels, figure 2-10. Airflow through the barrels draws fuel out of a float bowl. As air rushes

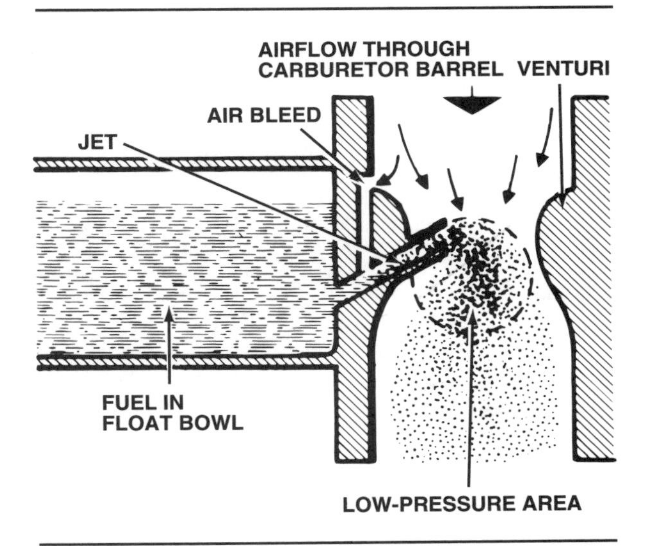

Figure 2-11. Low pressure in the carburetor barrel draws fuel from the float bowl.

through the barrel, pressure inside the barrel drops. A passage or jet connects the barrel to the fuel-filled float bowl. The higher atmospheric pressure in the float bowl forces the fuel through the jet and into the low-pressure stream of air rushing through the barrel, figure 2-11. The airflow speed through the carburetor affects how much fuel comes through the jet. The higher the airflow velocity, the lower the pressure in the carburetor barrel and the more fuel flows through the jet from the float bowl.

Carburetor barrels do not have straight sides because they work better with a constriction, or venturi. When air flows through the venturi constriction, it speeds up. The increase in speed lowers the air pressure inside the venturi and draws more fuel into the airflow.

Besides mixing liquid fuel with air, the carburetor must also atomize the liquid as much as possible. A small opening, called an "air bleed" in the fuel inlet passage helps break up the liquid fuel for better atomization. The air bleed mixes air with the liquid fuel as it is drawn into the main airflow. The air bleed opening is *above* the venturi, where high pressure causes the air to flow in.

The carburetor must also change the air-fuel mixture automatically. It has circuits that deliver a richer mixture for starting, idle, and acceleration, and a leaner mixture for part-throttle operation. Even with all its various circuits, a carburetor is a mechanical device that is neither totally accurate nor particularly fast in responding to changing engine requirements. Adding electronic feedback mixture control improves a carburetor's fuel-metering capabilities under some circumstances, but many mechanical jets,

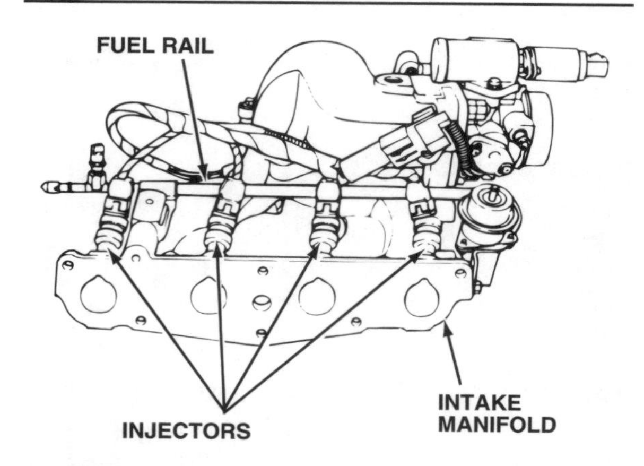

Figure 2-12. Port fuel injection places an injector near each intake port in the engine.

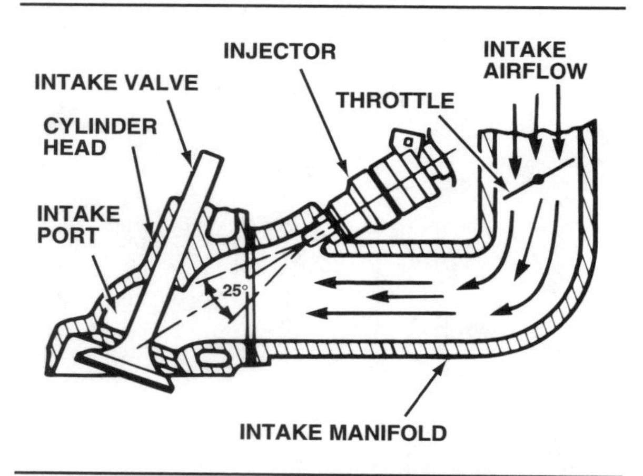

Figure 2-13. The injector sprays atomized fuel into the intake airflow.

passages, and air bleeds still do most of the work. Adding feedback controls and other emission-related devices result in very complex carburetors that are expensive to repair or replace.

The intake manifold also is a device that, when teamed with a carburetor, results in less than ideal air-fuel control. If there is only one carburetor for several cylinders, it is difficult to position the carburetor an equal distance from all cylinders and remain within the space limitations under the hood. In the longer intake manifold passages, the fuel tends to fall out of the airstream. Bends in the passageways slow the flow or cause puddles of fuel to collect. The manifold runners have to be kept as short as possible to minimize fuel delivery lag, and there cannot be any low points where fuel might puddle. These restrictions severely limit the amount of manifold tuning possible, and even the best designs still have problems with fuel condensing on cold manifold walls.

Multiple carburetors helped eliminate some compromises in carburetor and manifold design by improving mixture distribution. Few multiple-carburetor systems are used today because fuel injection is a better solution. The first fuel injection systems were mechanical, and they relied on an injection pump, mechanically driven by the crankshaft and timed to engine rpm, to allow the injectors to fire at the proper time.

Electronic fuel injection

The development of reliable solid-state components during the 1970s made the best type of fuel injection system, **electronic fuel injection (EFI)**, practical. EFI provides precise mixture control over all speed ranges and under all operating conditions. Its fuel delivery components are simpler and often less expensive than a feedback carburetor. Most designs allow a wider range of manifold designs. Equally important is that EFI offers the potential of highly reliable and very precise electronic control. With few exceptions, modern engines use one of two types of electronically controlled fuel injection:

- Port injection
- Throttle body injection.

With **port fuel injection**, (also called multipoint or multiport injection) individual injection nozzles for each cylinder are located in the ends of the intake manifold passages, near the intake valves, figure 2-12. Only air flows through the manifold up to the location of the injectors. At that point, the system injects a precise amount of atomized gasoline into the airflow. The fuel vaporizes immediately before it passes around the valve and into the combustion chamber, figure 2-13.

With **throttle body fuel injection (TBI)**, a throttle body on top of the intake manifold houses one or two injection nozzles, figure 2-14. The throttle body looks like the lower part of a carburetor, but does not have a fuel bowl or any of the fuel-metering circuits of a carburetor. The system injects fuel under pressure into the airflow as it passes through the throttle body.

Fuel atomization and vaporization

Proper fuel distribution depends on six factors:

- Correct fuel **volatility**
- Proper fuel **atomization**
- Complete fuel **vaporization**
- Intake manifold passage design
- Intake throttle valve angle
- Carburetor, throttle body, or fuel injector location on the intake manifold.

Volatility is a measure of gasoline's ability to change from a liquid to a vapor and is affected

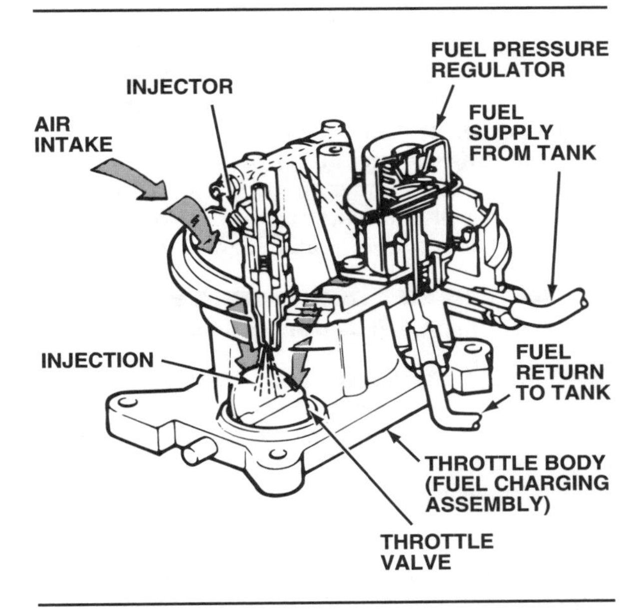

Figure 2-14. Throttle body injection meters fuel better than a carburetor, but provides less-precise delivery than port injection.

by temperature and altitude. The more volatile it is, the more efficiently the gasoline vaporizes. Efficient vaporization promotes even fuel distribution to all engine cylinders for complete combustion.

Refiners control volatility by blending different hydrocarbons that have different boiling points. In this way, it is possible to produce a fuel with a high boiling point for use in warm weather, and one with a lower boiling point for cold-weather driving. Such blending involves some guesswork about weather conditions in various regions, so severe and unexpected temperature changes can cause a number of temperature-related problems ranging from hard starting to vapor lock.

Several factors contribute to changing gasoline from a liquid into a combustible vapor. Gasoline must atomize, or break up into a fine mist, before it can properly vaporize. In a carbureted engine, the liquid fuel first enters the carburetor where it is sprayed into the incoming air and atomized, figure 2-15. The resulting atomized air-fuel mixture then moves into the intake

Electronic Fuel Injection (EFI): A computer-controlled fuel injection system that precisely controls mixture for all operating conditions and at all speed ranges.

Port Fuel Injection: A fuel injection system in which individual injectors are installed in the intake manifold at a point close to the intake valve. Air passing through the manifold mixes with the injector spray just as the intake valve opens.

Throttle Body Fuel Injection (TBI): A fuel injection system in which one or two injectors are installed in a carburetor-like throttle body mounted on a conventional intake manifold. Fuel is sprayed at a constant pressure above the throttle plate to mix with the incoming air charge.

Volatility: The ease with which a liquid changes from a liquid to a gas or vapor.

Atomization: Breaking down into small particles or a fine mist.

Vaporization: Changing from a liquid into a gas (vapor).

■ Birth of the Internal Combustion Engine

Like most inventions, the internal combustion engine did not spring full-grown from one person's mind. It developed gradually, as various inventors built on each other's work from 1690 to 1876. Historians generally credit Denis Papin's atmospheric engine concept as the first step toward today's modern gasoline burner. (Papin also invented the pressure cooker.) The principles Papin put forth for an external combustion engine led directly to James Watt and Richard Trevithick's basic steam engine.

Meanwhile, Sadi Carnot, a French physicist, postulated theories in 1824 that advanced the science of heat-exchange thermodynamics. During the 1850s, companies refined volatile fuels from petroleum, and by 1860 a French engineer named J. J. E. Lenoir modified a steam engine that used illuminating gas as fuel. Alphonse Beau De Rochas published a treatise on four-stroke-engine theory in 1862, but never built an engine.

Things really started moving when the firm of Otto and Langen produced an engine in 1867. Their design used a rack-and-gear device to transmit power from a free-moving piston to a shaft and flywheel. A freewheeling clutch in the gear allowed it to rotate freely in one direction and transmit power in the other. The same firm built the first modern internal combustion engine, the Otto Silent Engine, in 1876. That four-stroke design was manufactured in the United States after 1878, and is said to have inspired the early research of Henry Ford and other automobile pioneers.

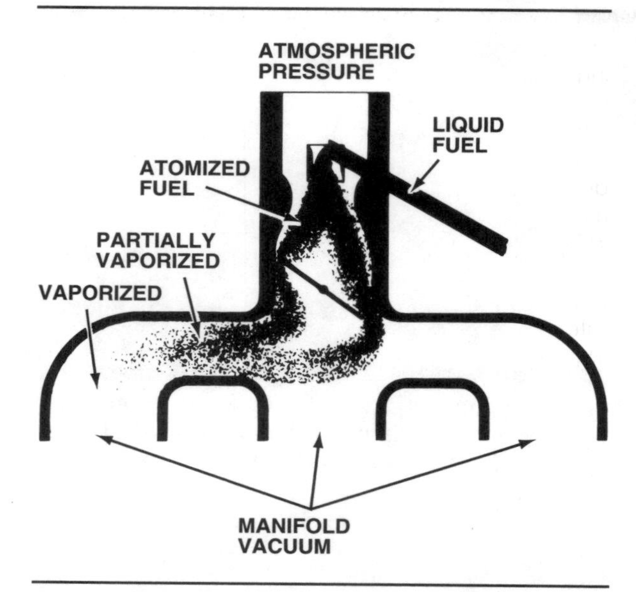

Figure 2-15. In a carburetor, the fuel atomizes as it enters the barrel.

manifold where manifold heat vaporizes the many fine droplets of the atomized fuel.

Rapid vaporization occurs only when the fuel is hot enough to boil. The boiling point is related to pressure: the higher the pressure, the higher the boiling point; the lower the pressure, the lower the boiling point. Because intake manifold pressure is usually quite a bit less than atmospheric pressure, the boiling point of gasoline drops when it enters the manifold, and the fuel quickly begins to vaporize. Fuel-injected engines spray the fuel into the incoming air under much greater pressure than do carburetors, so the fuel atomizes more thoroughly and vaporizes more quickly.

Heat from the intake manifold floor combines with heat absorbed from air particles surrounding the fuel particles to speed vaporization. The higher the temperature, the more complete the vaporization, so raising the intake manifold temperature helps vaporization. Several things can cause poor vaporization:

- Low mixture velocity or low fuel injection pressure
- Insufficient fuel volatility
- A cold manifold
- Poor manifold design
- Cold incoming air
- Low manifold vacuum.

When poor vaporization occurs, too much liquid fuel reaches the cylinders. Some of this additional fuel escapes through the exhaust as unburned hydrocarbons, and some washes oil from the cylinder walls, causing engine wear. Blowby gases carry the rest past the piston rings.

Carburetors can be precise and consistent, but fuel injection can always provide the *correct* amount of fuel for all conditions—cold start, wide open throttle, etc. Carburetion is a compromise. Also, injecting the fuel under pressure provides good vaporization under all loads and speeds. Of the six causes of poor vaporization listed above all apply to carburetors, but only the first four have

Sonic Tuning of Intake and Exhaust Systems

What exactly is a "tuned exhaust"? Why do racers spend so much time and money cutting and welding different header systems for their cars? And what about "tuned intakes"? Why are the velocity stacks on a racing fuel injection induction system sometimes short, sometimes long, and sometimes even different lengths for different venturis in the same system? There is obviously horsepower to be found in fiddling with the intake and exhaust system, but where does it come from, and how does it work?

Answers to these questions lie in understanding resonance, or sonic tuning. Sonic tuning in the exhaust is a way to scavenge the exhaust gases from a cylinder more efficiently than by piston action alone. It also can help the induction system, like a sort of natural supercharging that gets more mixture into the cylinder than does engine vacuum. Most street engines get around 80 to 85 percent volumetric efficiency at peak torque. With a tuned intake and exhaust system, on the other hand, a race engine can get better than 100 percent volumetric efficiency over a narrow portion of its power band.

Here is how it works in the exhaust system. As the exhaust valve opens, the pressurized gases burst out of the cylinder into the exhaust port, forming a high-pressure pulse in the exhaust gases already in the header primary pipe. This pressure pulse runs through the header pipe at the speed of sound—about 1,700 feet per second at exhaust temperature. The speed of the pulse does not depend on the speed of the gases. In fact, the pulse reaches the end of the header pipe much sooner than the gases do. When it reaches the end, it inverts—becoming a negative pressure pulse, or vacuum—and rushes back up the pipe towards the engine.

During this time the rising piston has been pushing exhaust gases through the exhaust port. When the negative pulse reaches the exhaust port, the extra vacuum helps scavenge the gases left in the cylinder, actually pulling them out past the valve. Because the intake valve is also open (the valves are at overlap), the vacuum also helps pull fresh mixture into the cylinder before the piston even starts its downstroke.

The end result is more horsepower, because the cylinder charge is both denser and less diluted with residual exhaust gases. Obviously, the way to optimize the system is to adjust the length of the header pipes so that the negative pressure pulse arrives at the exhaust valve at just the right time. And equally obviously, a tuned exhaust is only going to work at its best over a narrow range of engine rpm, perhaps several hundred rpm. Making the pipes shorter lets the pressure pulse reach the exhaust port sooner, tuning the pipes for some higher rpm. Making the pipes longer does the opposite, tuning the pipes for some lower rpm. Racers can find noticeable differences in horsepower when they change the length of their pipes by as little as three inches.

Pipes with megaphones at the ends make the pressure pulse longer and less distinct, spreading the effect over a wider rpm than a straight, cut-off pipe, which decreases the effect. A reverse-cone megaphone (one that increases and then decreases in diameter) can produce an additional power boost by sending a positive pressure pulse up the exhaust pipe, chasing the vacuum pulse. At some engine speeds, this pressure pulse catches the fresh mixture escaping from the exhaust port and "stuffs" it back into the combustion chamber.

Pressure pulses resonate back and forth in the induction tract, too. They are caused by the repeated opening and closing of the intake valve in the path of the moving column of air-fuel mixture. Racers sometimes use sonic tuning in the induction system, but this is much trickier than in the exhaust. Ideally, a positive pulse should arrive late in the intake stroke to reduce reverse pumping, and a negative pulse should arrive just as the intake valve closes to ease the disruption the closing valve has on the moving

column of mixture. A third positive pulse can pressurize the port just ahead of the valve, letting the intake charge burst into the cylinder the instant the valve opens.

In practice, however, sonic tuning in the intake is difficult to control. The temperature of the intake charge is hard to predict, because it is simultaneously heated by the hot manifold and cooled by the evaporating gasoline. In addition, opening and closing the throttle valves raises and lowers the pressure in the intake manifold and port, which has a strong effect on pulse speed. And finally, a standard log-type or common plenum chamber manifold gives a designer little room for tuning, because the pulses mix together and cancel or reinforce each other wherever the passages meet.

Port fuel injection systems do allow some intake tuning, though, because the long runners are separate between the intake plenum and the intake valve. The manufacturers can adjust the length of the runners to move the power boost up and down the rpm range of the engine, perhaps to fill a "hole" in the power band, or to add a little more low- or high-rpm torque, depending on the expected use of the engine.

Aftermarket induction systems with a separate carburetor or injection nozzle for each cylinder can also be tuned by making the velocity stacks longer or shorter. Some tuners use different-length stacks on the same engine to widen the power band by making different cylinders resonate at different rpm. In general, long runners seem to help low-rpm horsepower, and short runners seem to help high-rpm horsepower, but in practice tuners usually find the best length for a given engine and application by trial and error.

Sonic tuning has the most effect on piston-port two-stroke engines, which require it for cylinder scavenging. Four-strokes with poppet valves gain much less from sonic tuning, because they have a camshaft to force the intake and exhaust gases to move at certain times. Nonetheless, resonance still has a profound effect on racing four-stroke engines. With an intake and exhaust system working in harmony, the designer can specify a camshaft with much more duration than the engine could otherwise tolerate. This means that higher valve lifts are possible while still keeping valve acceleration low for reliability. The result is greater volumetric efficiency from the camshaft, made possible by a tuned intake and exhaust system.

If you are contemplating modifying a street engine, the benefits of sonic tuning are less than you might hope. Big power gains are only possible over a narrow range of engine speeds—useful at the track, but not in traffic. A better approach is to plan a system concentrating on low restriction and high velocity in both the intake and exhaust. Using small-diameter intake runners and header pipes helps horsepower with small-displacement engines and at low rpm, keeping gas velocities high enough so that induction and scavenging are more efficient. For high-rpm use or larger engines, larger intakes and header diameters keep the passages from restricting gas flow.

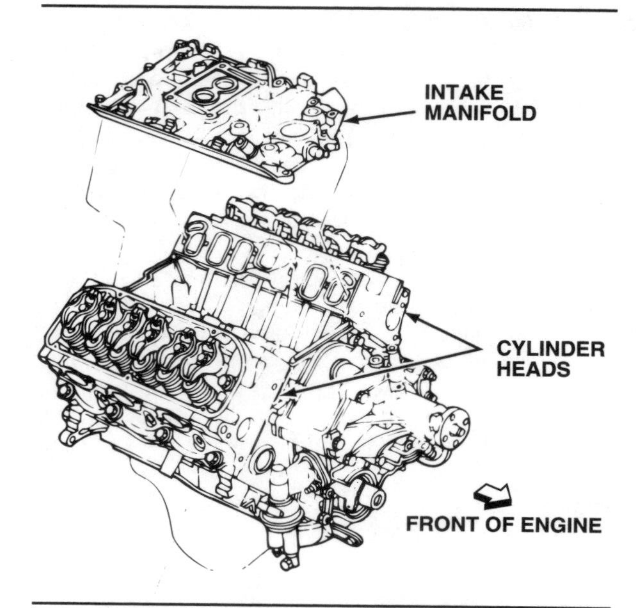

Figure 2-16. The intake manifold is located between the cylinder banks of a V-type engine.

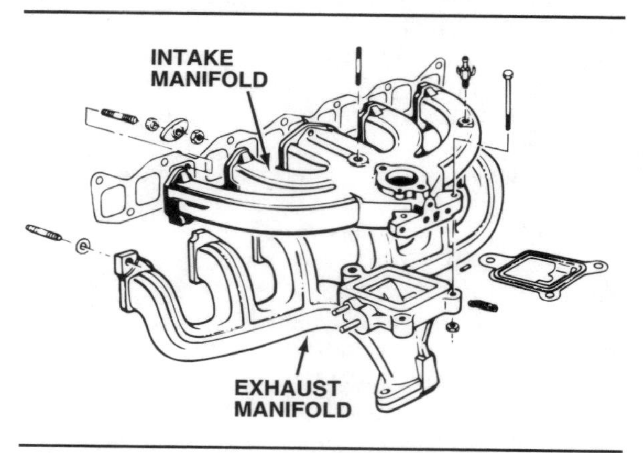

Figure 2-17. Most in-line engines have the intake and exhaust manifolds on the same side.

any effect on TBI systems, and only the first two on port fuel injection systems. Fuel injection, accurate metering, precise fuel control, and improved vaporization provide more power across the speed range of the engine, better fuel economy, and reduced exhaust emissions.

With TBI, manifold temperature still affects vaporization. As the mixture travels the length of the manifold, some of the air-fuel mixture separates and fuel condenses on the walls of the manifold. Since TBI more consistently atomizes the fuel under all conditions, separation is less likely than with a carburetor. However, with port fuel injection, the manifold carries only air, so the air-fuel mixture does not separate as it travels through the manifold.

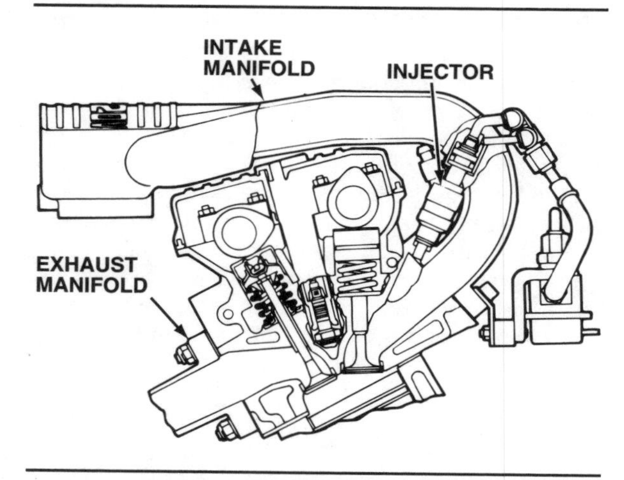

Figure 2-18. High-performance in-line engines use a crossflow design.

The Intake Manifold

The air-fuel mixture flowing from the carburetor or throttle body must be evenly distributed to each cylinder. The intake manifold does this with a series of carefully designed passages that connect the carburetor or throttle body with the engine's intake valve ports. To do its job, the intake manifold must provide efficient vaporization and air-fuel delivery. The manifolds for port fuel-injected engines carry only air, and designers size, or tune, them to do this efficiently.

Intake manifolds on older engines are usually cast iron, but modern engines often have aluminum manifolds because aluminum conducts heat better and is lighter. Better heat conductivity helps transfer heat to the air-fuel mixture faster and more uniformly.

The intake manifold on a V-type engine is in the valley between the cylinder banks, figure 2-16, and it is on the side of an in-line engine. The average in-line engine has the intake and exhaust manifolds on the same side, figure 2-17. High-performance in-line engines have two manifolds mounted on opposite sides of the cylinder head. This is called a crossflow design, figure 2-18.

Most intake manifolds are separate pieces that unbolt from the cylinder head, figure 2-19. However, Ford 3.3- and 4.1-liter (200- and 250-cu in.) car engines had the intake manifold cast into the head to reduce cost and simplify overall engine manufacturing. The last of the in-line Chevrolet 6-cylinder engines in the late 1970s had intake manifolds that were integral with the heads. That design was an attempt to improve mixture temperature and distribution, which are hard to control uniformly for an in-line six.

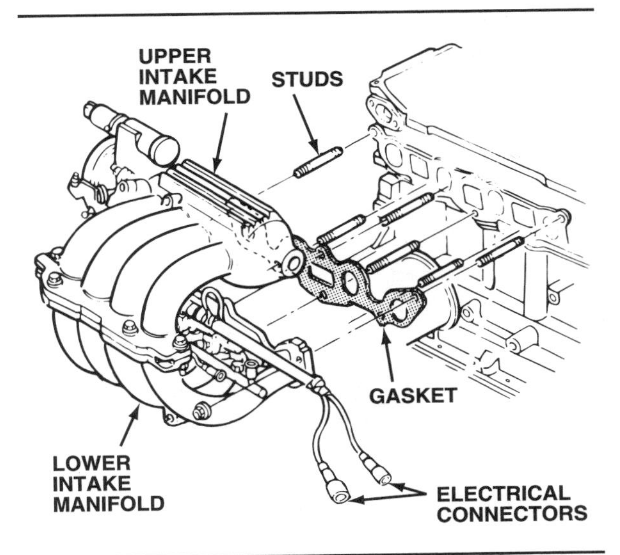

Figure 2-19. Most intake manifolds can be unbolted from the engine.

FUEL COMPOSITION

Gasoline is a clear, colorless liquid—a complex blend of various basic hydrocarbons (hydrogen and carbon). As a fuel, it has good vaporization qualities, and is capable of producing tremendous power when combined with oxygen and ignited. Yet, it is impossible to predict accurately how a certain blend of gasoline will perform in a particular engine, since no two engines are identical. Even mass-produced engines have variations that can affect fuel efficiency.

In laboratory tests, oil refiners calculate and measure the most important gasoline characteristics for a specific job. They blend fuels for particular temperature and altitude conditions. The gasoline you use during the summer is not the same blend available in the winter, nor is the gasoline sold in Denver the same as that sold in Death Valley. Besides temperature and altitude, refiners must consider several other things during the blending process:

- Volatility
- Chemical impurities
- Octane rating
- Additives.

Volatility

Volatility is a measure of gasoline's ability to change from a liquid to a vapor and is related to temperature and altitude. The more volatile it is, the more efficiently the gasoline vaporizes. As we have seen, efficient vaporization promotes even fuel distribution to the engine cylinders, and helps provide complete combustion.

Chemical Impurities

Gasoline is refined from crude oil and contains a number of impurities that can harm engines and fuel systems. For example, if gasoline has a high sulfur content, some sulfur may reach the engine crankcase, where it combines with water and forms sulfuric acid. Sulfuric acid corrodes engine parts, although proper crankcase ventilation helps avoid damage. Another impurity is gum, which forms sticky deposits that eventually clog carburetor and injector passages and cause sticking piston rings and valves.

To a large extent, the amount of chemical impurities in gasoline depends on the type of crude oil used, the refining process, and the oil refiner's desire to keep production costs low. The more-expensive process of **catalytic cracking** usually produces lower-sulfur gasoline than

■ Vapor Lock

Vapor lock occurs when gasoline "gas" bubbles form in the fuel lines, reducing fuel flow to the engine. Partial vapor lock makes the air-fuel mixture leaner and reduces engine power. Complete vapor lock causes an engine to stall, and makes restarting impossible until the bubbles in the fuel system disperse—usually when the fuel system cools. Any one or combination of these four factors can cause vapor lock:

- High gasoline temperature due to engine overheating or hot weather
- Low fuel system pressure
- Gasoline with too high a volatility
- Low air pressure due to driving at high altitude.

Vapor lock is an uncommon problem today for several reasons. Oil companies are good at reducing the vapor-locking tendencies of gasoline by adjusting its volatility to weather and geographic requirements, generally blending lower-volatility fuels for summer or high-altitude use and higher-volatility fuels in winter. In fact, the trend is toward year-round use of lower-volatility fuel because it evaporates more slowly, putting fewer hydrocarbons into the air.

Vapor lock is theoretically possible in fuel injected engines, but the high fuel-line pressures normal with fuel injection make vapor lock unlikely. The higher the pressure, the higher the boiling point of the fuel, making vapor difficult to form. On today's cars an electric fuel pump is usually mounted at the fuel tank, pressurizing 98 percent of the fuel system. Older cars with mechanical fuel pumps mounted on the engine actually created a low-pressure condition in the entire fuel line from the tank to the pump, increasing the likelihood of vapor lock.

the less-expensive and more-common **thermal cracking** method.

Octane Rating

When engine compression pressure reaches a certain level, the pressurized air-fuel mixture generates a great deal of heat. This can cause a secondary air-fuel explosion known as detonation, which was discussed in the "Engine Operating Principles" chapter.

To prevent detonation, gasoline must have a certain **antiknock value**. This characteristic derives from the type of crude oil and the refining processes used to extract the gasoline. An **octane rating** indicates the antiknock value. Gasoline with a high octane rating resists detonation, while one with a low octane value does not.

Additives

Certain chemicals not normally present in gasoline are added during refining to improve its performance:

- Anti-icers are specially treated alcohols that prevent moisture in the air from freezing in the carburetor or throttle body at low temperatures.
- Antioxidant inhibitors prevent gum formation.
- Phosphorus compounds prevent spark plug misfiring and preignition.
- Metal deactivators prevent gasoline from reacting chemically with metal storage containers in which it is stored and transported.
- Cleaners and detergents prevent the formation or accumulation of compounds that could clog the small passages or orifices in a carburetor or fuel injector.

Major gasoline refiners use different proprietary chemicals as detergents or cleaners. Aftermarket cleaners and detergents may be a good idea when fuel quality is below standard.

Tetraethyl lead (TEL) was once a gasoline additive that prevented detonation and provided lubrication for valve seats and many other moving parts. However, TEL is a highly toxic substance, and it remains as a particulate in the exhaust. Lead particulate emissions create a health hazard if a large amount collects in a small geographic area. Lead also destroys catalytic converter effectiveness. For these reasons, the U.S. EPA phased out lead use in gasoline. Refiners ceased producing leaded gasoline on January 1, 1988.

Octane boosters and lead substitutes sold in the automotive aftermarket may provide some octane increase. Refiners must select appropriate additives with care since many are only alcohol solutions, a common but undesirable octane booster. Octane boosters and lead substitutes are expensive and impractical for everyday use. Because of the U.S. EPA and health regulations, no aftermarket fuel additive contains actual TEL.

Alcohol Additives and Fuel Quality

Gasoline blended with alcohol is widely available, although many states do not legally require labelling it as such. A mixture of 10 percent ethanol (ethyl alcohol) and 90 percent unleaded gasoline is called "gasohol" or oxygenated fuel. **Gasohol** is a generic term, however, and there are no set standards for the type and amount of alcohol it contains. Several companies sell premium fuels that use ethanol as the octane booster.

Alcohol improperly blended with gasoline can cause numerous and serious problems with an automotive fuel system, including:

- Corrosion inside fuel tanks, steel fuel lines, fuel pumps, carburetors, and fuel injectors
- Deterioration of the plastic liner used in some fuel tanks, eventually plugging the in-tank filter
- Deterioration and premature failure of fuel line hoses and synthetic rubber or plastic materials such as o-ring seals, diaphragms, inlet needle tips, accelerator pump cups, and gaskets
- Hard starting, poor fuel economy, lean surge, vapor lock, and other driveability problems.

Fuels with an alcohol content tend to absorb moisture from the air. Once the moisture content of the fuel reaches approximately one percent, it combines with the alcohol and separates from the fuel. This water-alcohol mixture then settles at the bottom of the fuel tank. The fuel pickup carries it into the fuel line, to the carburetor or fuel injectors, causing lean surge.

All alcohols are solvents. While **ethanol** is relatively mild, **methanol** (methyl alcohol) is highly corrosive. It attacks fuel system components unless properly mixed with corrosion inhibitors and appropriate suspension agents or cosolvents to keep the water-alcohol combination from separating from the gasoline.

Gasohol has a cleaning effect on service station storage tanks, as well as the vehicle's fuel tank. Because of this cleaning action, a combination of rust, sludge, and metallic particles pass into the automotive fuel system. These substances reduce fuel flow through the filter and eventually plug the carburetor or injector passageways.

Fuel economy and driveability are other areas of concern with alcohol-gasoline blends. Alcohols contain fewer British thermal units of energy per gallon or liter than gasoline, which can result in reduced fuel mileage. Besides their lower energy content, alcohols are less volatile than gasoline; they require higher temperatures before they ignite and burn. Since the stoichiometric ratio for alcohol is 6.5:1 rather than the 14.7:1 of gasoline, an alcohol-gasoline blend creates a lean mixture. In turn, this creates or worsens lean surge in some driving conditions and increases the probability of vapor lock.

The problem of alcohol improperly blended with gasoline was so common in the United States during the mid-1980s that automotive tool manufacturers offered alcohol detection kits to determine fuel quality. The detection procedure required water as a reacting agent, but if the alcohol-blending process had used cosolvents as a suspension agent, the test would not show the presence of alcohol unless you used ethylene glycol (antifreeze) instead of water. Consequently, technicians often tested gasoline samples twice, first with water and then with ethylene glycol. The procedure could not differentiate between types of alcohol (ethanol or methanol), nor was it absolutely accurate. However, it was accurate enough to detect whether there was enough alcohol in the fuel for the user to take precautions. Since the mid-1980s, refiners have solved most of the alcohol-blending problems that created a demand for these kits. Today, it is not generally necessary to test for harmful amounts of alcohol in gasoline, although the kits are still on the market.

Methanol as an Alternative Fuel

In spite of some disadvantages, federal law encourages automakers to develop methanol as an alternative fuel. Ford and General Motors have manufactured variable-fuel vehicles that can operate on varying percentages of methanol and gasoline, as well as 100 percent gasoline. These vehicles have been in fleet service for some years as a means of testing their practicality. In addition, federal regulations require the use of oxygenated and reformulated fuels in certain areas of the country during periods of severe pollution. Among other things, these fuels contain a higher-than-normal amount of alcohol.

Methanol contains approximately one-half the energy of regular unleaded gasoline. Therefore, a much larger amount of methanol fuel is needed for the engine to work well. Variable-fuel injection systems operate on the same basic principles as gasoline engine injection systems. However, methanol requires the use of several different and redesigned components with considerable changes in materials. See a later chapter for a complete discussion on alternative-fueled systems.

SUMMARY

We measure engine size by displacement, or the total volume of all the cylinders. Displacement is a determining factor in the engine's airflow requirement. Stock engines operate below volumetric efficiency, or the maximum volume of air they could take in. Engines operate efficiently on an air-fuel ratio of 8 :1 to 18.5:1, but the exact ratio at which perfect mixture and complete combustion occur, or the stoichiometric ratio, is closer to 14.7:1. The stoichiometric ratio is also the best compromise between power and economy, as well as producing the least emissions. Electronic fuel management systems have been designed to provide the engine with the proper air-fuel ratio according to engine speed and load demands.

Gasoline needs to be metered, atomized, vaporized, and distributed to each cylinder by a metering device. The metering device on most later-model engines is a fuel injector. Older cars used the fuel-metering circuits in a carburetor. Proper fuel distribution depends on fuel volatility, atomization, and vaporization; intake manifold passage design, intake throttle valve angle;

Catalytic Cracking: An oil refining process that uses a catalyst to break down (crack) the larger components of the crude oil. The gasoline produced by this method usually has a lower sulfur content than gasoline produced by thermal cracking.

Thermal Cracking: A common oil refining process that uses heat to break down (crack) the larger components of the crude oil. The gasoline produced by this method usually has a higher sulfur content than gasoline produced by catalytic cracking.

Antiknock Value: The characteristic of gasoline that helps prevent detonation.

Octane Rating: The measurement of the antiknock value of a gasoline.

Tetraethyl Lead (TEL): A gasoline additive that helps prevent detonation.

Gasohol: A blend of ethanol and unleaded gasoline, usually at a 1:9 ratio. It is also referred to as an oxygenated fuel.

Ethanol: Ethyl alcohol distilled from grain or sugar cane.

Methanol: Methyl alcohol distilled from wood or made from natural gas.

and on carburetor, throttle body, or fuel injector location on the intake manifold.

Gasoline needs proper refining to remove chemical impurities and blending with additives to prevent preignition, detonation, carburetor icing, varnish formation, and misfiring. Tetraethyl lead was once the major octane booster, but the U.S. EPA severely limited its use.

A fuel blend made up of alcohol and gasoline is called oxygenated fuel. If this fuel is misblended, as sometimes happened in the mid-1980s, it can permanently damage a fuel system.

However, blending problems are less common than when these fuels first appeared. Ford and GM have developed flexible-fuel vehicles for fleet use, which can operate on varying percentages of methanol and gasoline. Cars must run on oxygenated fuel, containing a higher proportion of methanol, in certain areas that have severe pollution to reduce CO emissions. Methanol has less energy than gasoline, so the engine needs more of it to run. Using methanol requires different fuel system components and materials.

Review Questions

Choose the single most correct answer.
Compare your answers to the correct answers on page 507.

1. A disadvantage of gasoline is that it:
   a. Vaporizes easily
   b. Burns quickly
   c. Produces pollutants upon combustion
   d. Has a high heat value

2. Which is not a factor in determining airflow requirement?
   a. Engine displacement
   b. Maximum rpm
   c. Carburetor size
   d. Volumetric efficiency

3. Volumetric efficiency:
   a. Is the theoretical maximum air volume that can be drawn into an engine
   b. Increases as engine speed increases to maximum efficiency at maximum rpm
   c. Is expressed in cubic feet per minute or cubic meters per minute
   d. Is the ratio of air entering the engine to engine displacement

4. At maximum speed, the volumetric efficiency of a stock engine is approximately:
   a. 75 percent
   b. 50 percent
   c. 100 percent
   d. Better than 100 percent

5. The richest air-fuel ratio that an internal combustion engine can tolerate is about:
   a. 4:1
   b. 2.5:1
   c. 8:1
   d. 18.5:1

6. To burn one pound of gasoline with maximum efficiency requires about:
   a. 8 pounds of air
   b. 15 pounds of air
   c. 18 pounds of air
   d. 17 pounds of air

7. A rich air-fuel mixture is needed for:
   a. Cruising speed
   b. Light load
   c. Better driveability
   d. Acceleration

8. An internal engine condition affecting fuel requirements is:
   a. Engine load
   b. Mixture distribution
   c. Atmospheric pressure
   d. Engine speed

9. Obtaining maximum power results in:
   a. No change in air-fuel ratios
   b. Leaner mixtures
   c. Unburned oxygen
   d. Excess unburned fuel

10. Maximum fuel economy requires:
    a. Less air
    b. Leaner air-fuel mixtures
    c. Richer air-fuel mixtures
    d. Higher temperatures

11. For maximum power, the air-fuel ratio becomes:
    a. Leaner at low speeds
    b. Richer at low speeds
    c. Richer at high speeds
    d. Leaner during acceleration

12. For maximum fuel economy, the air-fuel ratio becomes:
    a. Leaner during acceleration
    b. Leaner at high speeds
    c. Leaner at low speeds
    d. Leaner for middle speeds

13. Different fuel-blending techniques are used by refiners to:
    a. Increase fuel volatility
    b. Decrease the octane rating
    c. Replace impurities with additives
    d. Add sulfuric acid

14. Technician A says carburetors use two, three, or four barrels.
    Technician B says carburetors are not found on later-model vehicles.
    Who is right?
    a. A only
    b. B only
    c. Both A and B
    d. Neither A nor B

15. Technician A says a TBI system uses one injector at each cylinder port.
    Technician B says the TBI unit is installed on the intake manifold where a carburetor would be.
    Who is right?
    a. A only
    b. B only
    c. Both A and B
    d. Neither A nor B

16. Technician A says port fuel injection is also called multipoint injection.
    Technician B says a port fuel injection unit looks like the lower part of a carburetor.
    Who is right?
    a. A only
    b. B only
    c. Both A and B
    d. Neither A nor B

17. Which of the following is not a quality of tetraethyl (TEL) lead?
    a. Prevents detonation
    b. Lubricates valve seats
    c. Can destroy a catalytic converter
    d. Is a nontoxic substance

18. Gasohol is generally regarded as a blend of:
    a. 10 percent ethanol and 90 percent gasoline
    b. 90 percent ethanol and 10 percent gasoline
    c. 25 percent ethanol and 75 percent gasoline
    d. 75 percent ethanol and 25 percent gasoline

Chapter

3

Engine Lubrication

An automotive technician must understand the lubrication system and the motor oil that runs through it. The quality of lubrication an engine receives affects the life and performance of that engine. Even if you perform the finest service, an engine will provide inferior service, or may even fail, if you neglect the lubrication system or use the improper lubricants in the engine.

In this chapter we first cover the purpose of motor oil, how it is rated, and what additives do to help motor oil. Then, we detail how the lubrication system works and the relationship between lubrication performance, economy, and emission control.

PURPOSES OF MOTOR OIL

Motor oil in a car engine does five major jobs:

- It reduces friction between moving parts, which lessens both wear and heat.
- It acts as a coolant, removing heat from the metal of the engine.
- It carries dirt and wear particles away from moving surfaces, cleaning the engine.
- It helps seal the combustion chamber by forming a film around the valve guides, and between the piston rings and the cylinder wall.
- It acts as a shock absorber, cushioning engine parts to protect them from the force of combustion.

These jobs help keep the engine running smoothly and efficiently. If the motor oil fails at any one of them, performance suffers or engine damage results.

MOTOR OIL COMPOSITION AND ADDITIVES

Petroleum-based motor oils consist of hydrogen and carbon molecules. They are complex hydrocarbon compounds, as is gasoline. Petroleum-based motor oil is a product of crude oil, which oil companies extract from underground oil-bearing rock. Oil companies transport the crude oil to refineries where it is separated into fractions, the various components of the crude. The lighter fractions become products like gasoline or kerosene, and heavier fractions usually become the base stock for petroleum-based motor oil. Synthetic oils may use only a small amount of petroleum products or none at all in their makeup, so we cover synthetic motor oil in a later section of this chapter.

Motor Oil Additives

Fifty years ago, motor oil consisted of only the base stock. Change intervals every 1,000 miles were common—as were stuck piston rings and rapid engine wear. To improve the performance

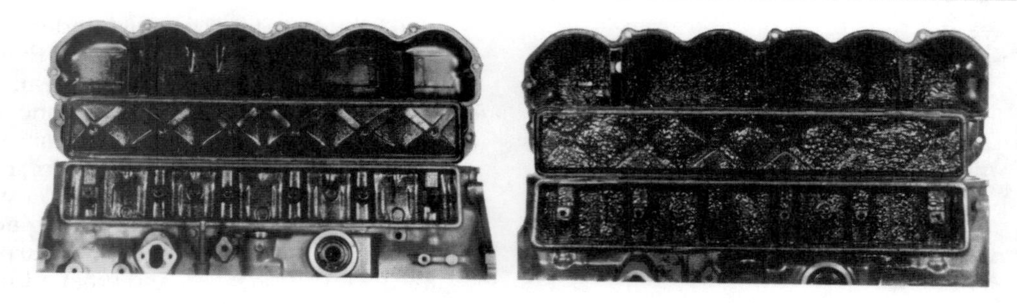

Figure 3-1. Sludge deposits usually occur because of lubrication system neglect.

of a motor oil, manufacturers blend in chemical additive packages. These packages are secret, propriety formulations, but a common additive is zinc diorganodithiophosphate (ZDDP). Additives make up about 25 percent of motor oil. A typical SG/CD 10W-30 oil may be 75 percent oil base stock, 11 percent viscosity improver, and 14 percent other additives.

The purpose of a motor oil additive can be to:

- Replace a property of the oil lost during refining
- Strengthen a natural quality already in the oil
- Add a property that the oil did not naturally have.

A few of the common motor oil additives and their jobs are described in the following paragraphs.

Oxidation occurs when oil reacts chemically with oxygen, just as rust occurs when iron and oxygen react. Heat, and contaminates such as copper and glycol, accelerate the oxidation reaction in motor oil. This process of oxidation can leave hard carbon and **varnish** deposits in the engine. **Antioxidants** reduce this problem by preventing the buildup of acids, destroying chemicals that produce undesirable oxidation by-products, and interrupting the oxidation chain reaction.

Varnish deposits that do develop are fought with **detergent** additives. For example, detergents clean piston ring grooves and keep the rings free to seal with maximum effectiveness, which maintains peak engine performance.

In any engine, some combustion chamber gases get past the piston rings and enter the crankcase. This is called **blowby**. Blowby gases contain water vapor, carbon dioxide, and unburned fuel that form acids that rust or corrode engine parts. Rust and corrosion preventives are added to motor oil to neutralize acids.

Both the water vapor and the fuel in blowby gases can mix with cold oil and form **sludge,** figure 3-1. This thick black deposit clogs oil passages and reduces engine lubrication. **Dispersants** reduce sludge formation by keeping sludge particles suspended in the oil, to be removed when the oil and filter are changed.

The oil in an engine is constantly being churned by moving parts. This can mix air and other gases with the oil, causing the oil to foam. Oil foam does not form a good oil barrier between moving parts, and may allow metal-to-metal contact. In extreme cases, foam can cause oil pumps to gas lock, or lose their prime, and stop pumping oil. Foam inhibitors reduce the oil's surface tension, the property that allows bubbles to form and stay formed. Oil with low surface tension allows the bubbles to burst, releasing the gas, and reducing the formation of foam.

Viscosity is the tendency of a liquid to resist flowing. Some additives help an oil to flow under wide temperature ranges. These additives are called **viscosity index (VI) improvers** and **pour-point depressants,** and are among the primary ingredients in multigrade oils. VI improvers help the oil resist thinning at high temperatures. Since VI improvers actually increase viscosity at low and high temperatures, a 10W-30 multigrade motor oil starts out as slightly less than a 10-weight base stock oil. Oil companies then add sufficient VI improvers so that the oil is able to perform like a 30-weight oil at high temperatures.

Pour-point depressants help the oil flow in colder temperatures. These additives lower the temperature at which the oil solidifies by blocking wax crystal growth in the oil base stock. Surprisingly, some VI improvers also function as pour-point depressants, so the oil company need not add supplemental pour-point depressants.

Additive precautions

Once it was not considered good practice to mix oil brands in the same engine. Experts thought that because different manufacturers use different additives, the chemicals could oppose each other and decrease the cleaning and lubricating abilities of the oil. Today, it is not as great a concern because most oils are compatible. There are only about half a dozen additive packages available that allow an oil to meet current automakers' recommendations. To be on the safe side, it is still a good practice to always add the same oil, but it is not critical.

Many oil manufacturers advise against the use of other oil additives sold separately. A quality motor oil already contains all the additives it needs. The extra additives are simply not necessary, and in some cases can cause problems. Adding a high-detergent additive to an engine with a great deal of sludge may break loose large chunks of that sludge. The oil filter will be quickly filled and as the oil bypasses the filter, large amounts of sludge and dirt will flow through the engine unchecked.

Synthetic Motor Oils

Synthetic-based lubricants have been used in avia-aviation and special-purpose engines for decades. Since the late 1970s, **synthetic motor oils** have attracted much attention and have slowly gained in popularity because they offer several advantages over petroleum-based engine oils.

Synthetic oils are created by reacting various complex molecules to form a new molecular structure. Petroleum, several types of acids, and alcohols may be used in varying proportions to make synthetic-based oils. The process involved is complex; consequently, synthetic oils may sell for three to five times the price of their petroleum-based counterparts.

Synthetics contain less wax than petroleum oils so they tend to remain liquid when cold. This permits faster lubrication in very cold temperatures, which is perhaps the greatest advantage of synthetic engine oils.

Some manufacturers of synthetic oils claim that the increased slipperiness of their oil will decrease engine friction, thus contributing to small increases in both fuel economy and power. Although these claims have been substantiated in some cases, you must weigh the savings in fuel economy and the increase in performance against the higher price of the oil. For example, the higher cost of the synthetic oil may more

Varnish: An undesirable deposit, usually on the engine pistons, formed by oxidation of fuel and motor oil.

Antioxidants: Chemicals or compounds added to motor oil to reduce oil oxidation, which leaves carbon and varnish in the engine.

Detergent: A chemical compound added to motor oil that removes dirt or soot particles from surfaces, especially piston rings and grooves.

Blowby: Combustion gases that leak past the piston rings into the crankcase; these include water vapor, acids, and unburned fuel.

Sludge: A thick black deposit caused by the mixing of blowby gases and oil.

Dispersant: A chemical added to motor oil that keeps sludge and other undesirable particles picked up by the oil from gathering and forming deposits in the engine.

Viscosity: The tendency of a liquid, such as oil, to resist flowing.

Viscosity Index (VI) Improvers: Chemical compounds added to motor oil to help the oil resist thinning at high temperatures.

Pour-Point Depressants: Chemical compounds added to motor oil to help the oil flow at colder temperatures.

Synthetic Motor Oils: Lubricants formed by artificially combining molecules of petroleum and other materials.

■ PCV System Service

When a PCV system becomes restricted or clogged, the cause is usually an engine problem or lack of proper maintenance. For example, scored cylinder walls or badly worn rings and pistons will allow too much blowby. Start-and-stop driving requires more-frequent maintenance and causes PCV problems more quickly than highway driving, as will any condition allowing raw fuel to reach the crankcase. Using the wrong grade of oil, or not changing the crankcase oil at periodic intervals, will also cause the ventilation system to clog.

When a PCV system begins to clog, the engine may stall, idle roughly, or overheat. As ventilation becomes more restricted, burned plugs or valves, bearing failure, or scuffed pistons can result. Also, look for an oil-soaked distributor or points, or leaks around valve covers or other gaskets. Do not overlook the PCV system while troubleshooting. A clogged PCV valve, or one of incorrect capacity, may be the cause of poor engine performance.

than equal any money saved due to improved fuel economy.

Beyond these benefits, many synthetic oils have been shown to deteriorate at a slower rate than comparable petroleum-based oils. Therefore, they can remain in service longer between oil changes. However, synthetic oils hold contaminants in suspension no better than petroleum-based oils, so you should not assume that the use of a synthetic oil alone justifies postponing scheduled oil and filter changes.

While synthetic oil base stocks and additives may differ chemically from petroleum oils, they undergo the same tests and receive the same SAE viscosities and API Service Classifications. Thus, you select them in the same manner, according to the engine or car manufacturers' specifications.

MOTOR OIL DESIGNATIONS

Because engines and operating conditions vary greatly, oil refiners blend and sell different types of motor oil. The oil used in a heavy-duty diesel truck is different from that used in a high-performance car engine. For example, diesel engine oils must resist the acidity of sulfur in the diesel fuel, which is not a factor in gasoline engines. Both engines need lubrication, and as we shall see, an oil may or may not meet all the requirements of both engines.

Engine oil is commonly identified in two ways: by American Petroleum Institute (API) Service Classification and by Society of Automotive Engineers (SAE) Viscosity Number.

Another method for identifying oils is by military specification numbers or "MIL-specs". While they are rarely used to specify engine oils for cars and trucks, they often appear on oil containers.

API Service Classification

The **API Service Classification** rates engine oils on their ability to lubricate, resist oxidation, prevent high- and low-temperature engine deposits, and protect the engine from rust and corrosion. API has organized a system of letter classifications with two categories, the S-series and the C-series. The S-series service classification emphasizes oil properties critical to gasoline engines, while the C-series emphasizes oil properties for diesel engines. In order for an oil formulation to be given a particular classification, the oil is run through a series of tests in specific engines. Some oil companies conduct their own tests; others pay independent laboratories to perform the tests for them. If the oil's performance meets the minimum standard, the oil can be sold bearing that API Service Classification.

The S-Series Oils

The S-series oils come under the following classifications: SA, SB, SC, SD, SE, SF, SG, and SH. No performance tests are required to meet the SA classification.

- The SB classification requires that the oil provide some antiscuff capability and resistance to oil oxidation. This type of protection dates back to the 1930s.
- The SC classification requires the oil to provide control of high- and low-temperature deposits, wear, rust, and corrosion in gasoline engines. This level of protection was required by new car warranties from 1964 through 1967.
- The SD classification requires increased control of high- and low-temperature deposits, wear, rust, and corrosion in gasoline engines over SC oils. This classification met new car warranties from 1968 through 1970, and some in 1971.
- The SE classification requires still higher levels of control of high- and low-temperature deposits, wear, rust, and corrosion in gasoline engines over SC and SD oils, and some anti-wear performance. This level of protection was required for some cars and trucks beginning in 1971, and continued through 1979.
- The SF classification requires increased oxidational stability and anti-wear performance relative to SE oils. This classification met new-car warranties from 1980 through 1988.
- The SG classification requires increased control of engine deposits, oil oxidation, and engine wear relative to the other oils in this series. SG oils exceed the performance of SF and CC oils. This classification meets new-car warranty requirements beginning in 1989.
- The SH classification requires increased control of engine deposits, oil oxidation, engine wear, rust, and corrosion relative to the SG classification. This classification meets new-car warranty requirements beginning in 1994.

As these designations progress alphabetically they increase in levels of protection. Each classification replaces the one before it, with SH offering the most protection. Just as an SB oil would not provide enough protection for a 1990 engine, an SG oil would be better for a 1930s engine than the SB oils available for it in its day.

The C-Series Oils

The C-series oils are CA, CB, CC, CD, CD-II, CE, CF-4, and CG-4.

- The CA classification requires that the oil protect against bearing corrosion and ring-belt deposits in naturally aspirated (nonsuper-

charged) diesel engines running on fuel with a minimum sulfur content of 0.35 percent. The requirements of this classification date back to the late 1940s. These oils meet military specification MIL-L-2104A.
- The CB designation requires the same protection as CA, but the test engine is run on fuel with a minimum of 0.95 percent sulfur. These oils meet the obsolete "Supplement 1" specification added to MIL-L-2104A, and were used in gasoline engines as well.
- The CC designation requires that the oil provide protection from high temperature deposits in lightly supercharged diesel engines running on low-sulfur-content fuel, and also protection from rust, corrosion, and low-temperature deposits in gasoline engines. This designation was introduced in 1961, and meets the MIL-L-2104B specification.
- The CD classification requires oils to provide protection from bearing corrosion and high-temperature deposits in supercharged diesel engines running on fuels of various qualities. Oils meeting the CD classification meet the Caterpillar Series-3 specification introduced in 1955, and MIL-L-45199.
- The CD-II classification requires the same performance as CD oils, and is designated for two-stroke diesel engines where deposit and wear control are critical. Two-stroke diesels are most commonly used in intercity and highway buses. This classification came into being in 1988.
- The CE classification also requires the same performance as CD oils, plus passing the engine oil performance tests for the Mack EO-K/2 specification and the Cummins NTC-400 test for piston deposits and oil consumption. This classification was introduced in 1987, although oils meeting its individual requirements were already available for some time. CE oils are intended for heavy-duty diesel engines, such as those used in large on- and off-highway trucks and construction equipment.
- The CF-4 classification requires performance equal to CE oils in many respects, with significantly improved control of oil consumption and piston deposits. This classification is designed for high-speed, four-stroke diesel engines, particularly those used in on-highway, heavy-duty trucks. Introduced in late 1990, CF-4 oils are part of the overall engine design required to meet the 1991 U.S. EPA emission regulations affecting heavy trucks. CF-4 oils may be used where CC, CD, and CE oils are recommended.
- The CG-4 classification requires performance exceeding CF-4 oils, with improved control of high-temperature piston deposits, corrosion, foaming, oxidation, and soot accumulation. This oil has been especially designed to meet 1994 U.S. EPA exhaust emission standards, and may also be used in engines requiring API service classifications CD, CE, and CF-4. Introduced in 1995, this classification is designed for high-speed, four-stroke diesel engines, particularly those used in highway and off-road applications.

While each S-series classification exceeds the one before it, this is not necessarily the case in the C-series. For instance, the CB tests are run on higher-sulfur fuel than the CC tests. As a result, a CC oil may not offer equivalent protection from the effects of sulfur on the engine as a CB oil. On the other hand, CD and CE oils must pass tests running on high-sulfur-content fuels, and thus would offer similar protection in this area.

Some oils are identified with dual service classifications separated by a slash, for example, SF/CD or SG/CE. These oils meet the requirements of both service classifications shown, and can be used in any engine that calls for one or the other. Many manufacturers specify oils that meet both "S" and "C" service classifications; this is particularly true for heavy-duty and turbocharged engines. Where an oil with a dual service classification is specified, do not use oils that meet only a single classification or the engine warranty may be voided.

SAE Viscosity Grades

The **SAE Viscosity Grade,** expressed as a number, refers to an oil's resistance to flow. Typical oil viscosity grade numbers are 5W, 10W, 15W, 20W, 20, 30, 40, and 50. Lower viscosity grade numbers indicate thinner oils that flow more easily. Higher numbers refer to thicker oils with greater resistance to flow.

To make sure oils are capable of providing adequate lubrication, they are tested for viscosity at both high and low temperatures. Viscosity numbers without a "W" suffix indicate that the oil meets certain viscosity requirements at 212°F (100°C) only. Grades with a "W" suffix meet the requirements at 212°F (100°C). They also meet

API Service Classification: A system of letters signifying an oil's performance. The classification is assigned by the American Petroleum Institute.

SAE Viscosity Grade: A system of numbers signifying an oil's viscosity at a specific temperature. The viscosity grade is assigned by the Society of Automotive Engineers.

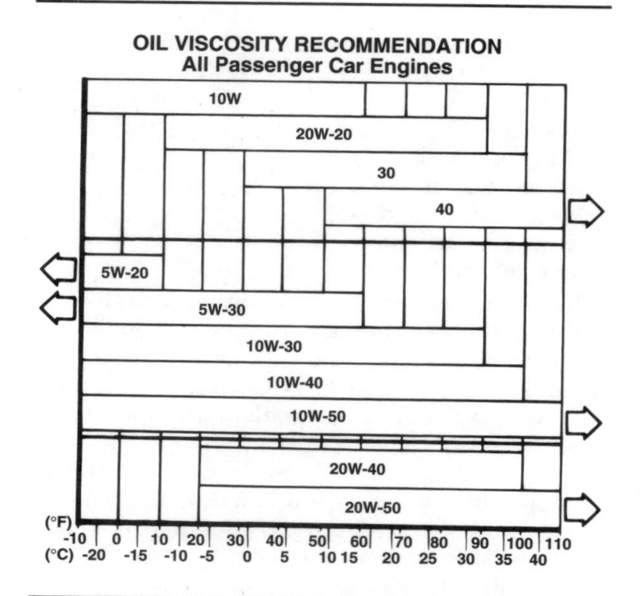

Figure 3-2. Engine oil viscosity recommendations based on ambient temperature. There is much overlapping of the temperatures at which different multigrade oils work.

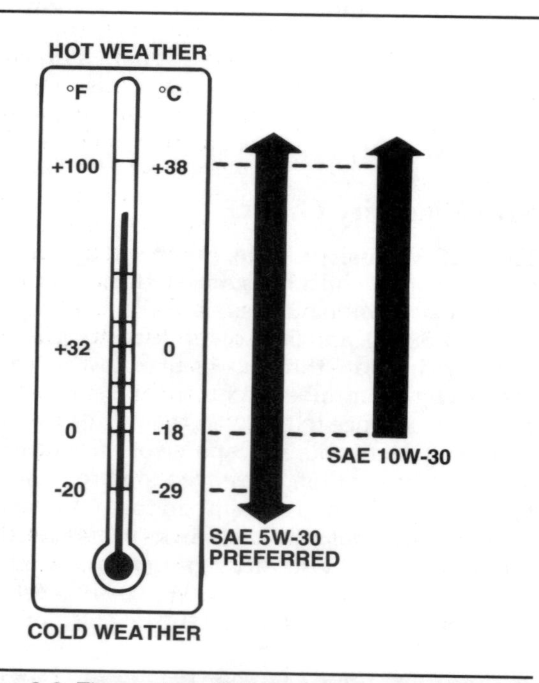

Figure 3-3. The engine oil viscosity recommendations for 1990 Chevrolet car engines shown here are typical of recommendations for many late-model cars.

certain minimum flow rates at temperatures ranging from 23°F (-5°C) down to -22°F (-30°C).

Oils with only one SAE viscosity number (**single-grade** oils) were once widely recommended. However, you must change the oil with the seasons to ensure that the oil in the

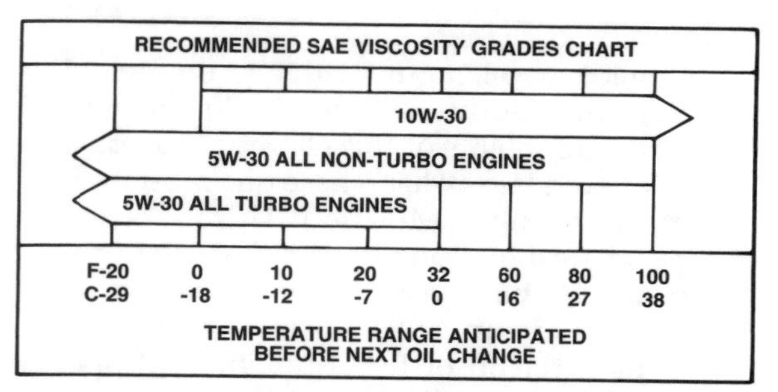

Figure 3-4. Recommended viscosity grades for 1990 Chrysler front-wheel-drive passenger cars. These recommendations reflect the current popularity of lighter oil grades, which promote good fuel economy.

crankcase strikes the proper balance between high- and low-temperature operation. Engines require low-viscosity oils in winter, while high-viscosity oils are needed in summer. Today, some manufacturers still recommend single-grade oils in heavy-duty and diesel engine applications.

Most modern oils have dual Viscosity Grade numbers separated by a hyphen (dash) such as 5W-30, 10W-30, 15W-40, or 20W-50. These **multigrade** oils meet the low- and high-temperature specifications for both grades of oil indicated. They contain VI improver additives that allow them to flow well at low temperatures, yet still resist thinning at high temperatures. These properties allow you to use multigrade oils throughout the year with less concern about ambient temperature fluctuations. Figure 3-2 is an example of one manufacturer's gasoline engine oil viscosity recommendations based on ambient temperature. The multigrade oils allow use in the widest range of temperatures.

Vehicle manufacturers today recommend multigrade oils for most of their engines; however, not all multigrade oils are recommended for all cars. For example, General Motors and Chrysler do not recommend 10W-40 oils. Also, most automakers warn against the use of low-viscosity multigrade (and single-grade) oils, such as 5W-20 and 10W, for sustained high-speed driving, trailer towing, or any other circumstance in which the engine is placed under constant heavy load. Automakers specify the exact kind of oil to use under certain operating conditions. Figures 3-3 and 3-4 show engine oil charts prepared by two different manufacturers.

Energy-Conserving Oils

Besides their service classification and viscosity grade, the API designates certain oils as "Energy Conserving". Oil makers specially formulate these oils to reduce internal engine friction, and

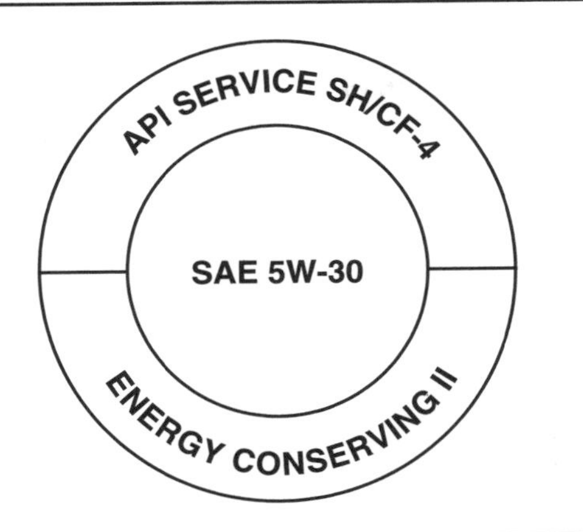

Figure 3-5. The API Engine Service Classification symbol, or "donut", appears on every container of motor oil. Learn to read this symbol because it gives you essential information about the oil.

thus improve fuel economy. Laboratories test candidate oils to determine if they really promote fuel economy. Oils achieving a 1.5 percent fuel economy increase may be labelled Energy Conserving. Those achieving a 2.7 percent increase may be labelled Energy Conserving II. These oils may contain friction-reducing additives and usually have lower viscosity than non-Energy Conserving oils.

Engine Oil Identification

To allow easy identification of engine oils, the API established the Engine Service Classification Symbol, or "donut", figure 3-5. The symbol appears as a label printed on oil containers. The upper half of the symbol displays the API Service Classifications of the oil, the center displays the SAE Viscosity Grade of the oil, and the lower half contains the words "Energy Conserving" if the oil is formulated to meet those requirements. When selecting an oil, always make sure the oil quality information on the API donut conforms with the vehicle manufacturer's specifications.

ENGINE OILING SYSTEM AND PRESSURE REQUIREMENTS

All modern car engines have a pressurized lubrication system. The parts of this system, as shown in figure 3-6, are:

- Oil reservoir and its ventilation
- Oil pump and pickup
- Pressure relief valve
- Filter
- Galleries and lines
- Indicators.

Oil Reservoir

There must be enough oil in the engine to circulate throughout the system, plus some reserve so that the oil can cool before being recirculated. All of this oil is kept in the engine oil pan, or sump. Because the oil pan is at the bottom of the engine, the oil drains into the pan after passing through the engine.

The flow of air past the pan when the car is moving helps cool the oil. If this cooling effect is not sufficient, the manufacturer may add an external oil cooler, which is common on diesel, turbocharged, or air-cooled engines, figure 3-7. Heat is removed from oil passing through the cooler by airflow or coolant circulation, depending on cooler design.

Engine oil pan capacities vary greatly; 4-cylinder engines usually hold three to five quarts of oil; most V-type engines hold four to five quarts. Some high-performance gasoline engines and diesel engines require six, seven, or more quarts. Oil is put in the crankcase through a capped oil filler hole at the top of the engine.

Ventilation

We have seen that blowby gases enter the crankcase from the combustion chambers. The gases can mix with motor oil and cause engine damage. Because the combustion chamber gases are under very high pressure, they increase the pressure within the crankcase. If the crankcase is not vented, the pressure will force oil out of the engine at loosely sealed points such as the oil filler cap and the junction of the oil pan and block. This is messy, wasteful, and can be dangerous as a fire hazard if the oil gets on the brake or clutch materials. To prevent this, and to reduce hydrocarbon emissions, the crankcase is ventilated.

On older cars, the oil pan was vented directly to the atmosphere through a **road draft tube,** figure 3-8. Until the 1960s, this was the most common type of crankcase ventilation.

Single Grade: An oil that has been tested at only one temperature, and designated by one SAE viscosity number.

Multigrade: An oil that meets viscosity requirements at more than one test temperature, designated by dual SAE viscosity numbers.

Road Draft Tube: The earliest type of crankcase ventilation; it vents blowby gases to the atmosphere.

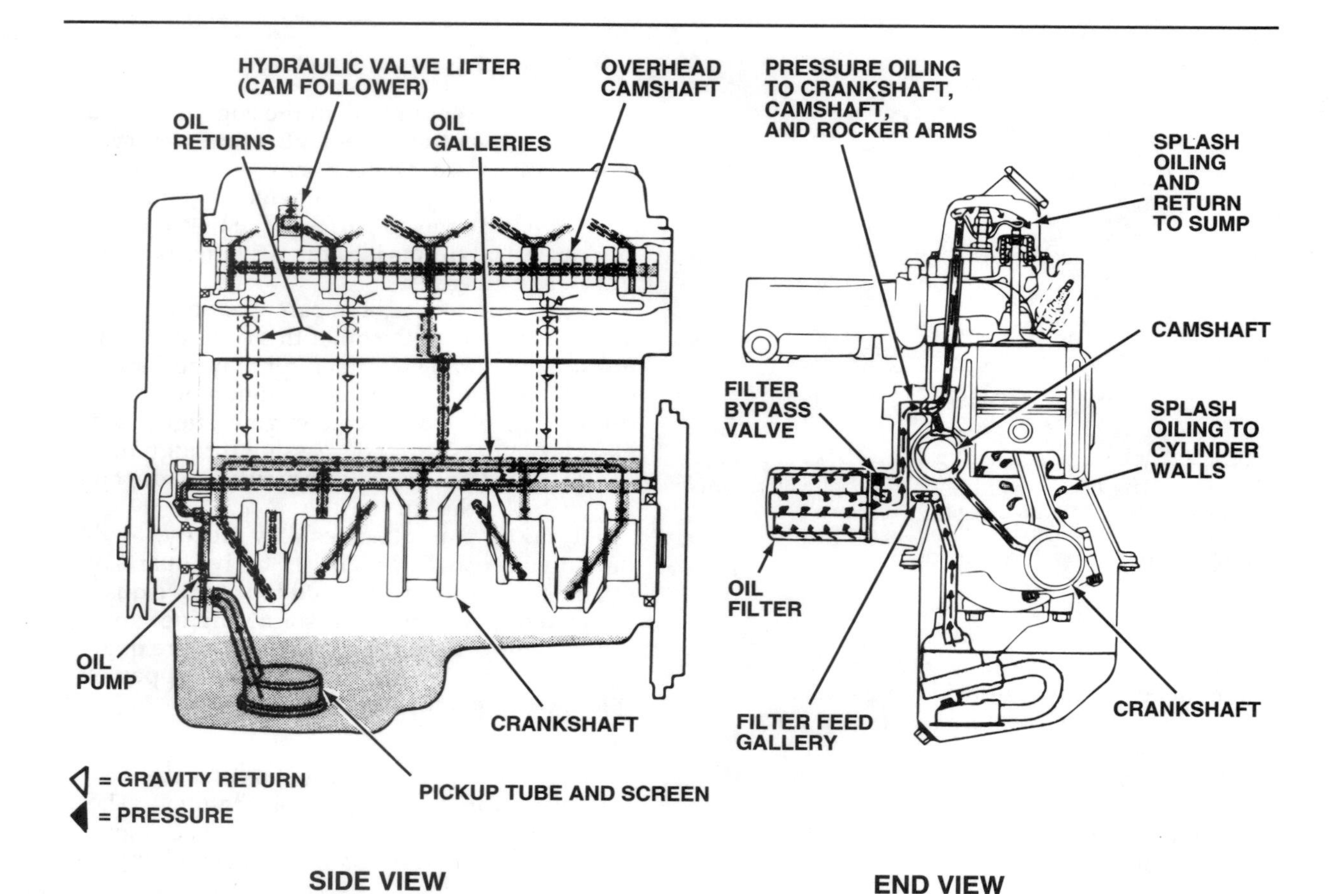

Figure 3-6. Modern engine design uses both pressure and splash methods of oiling. Oil travels under pressure through galleries to reach the top end of the engine. Gravity flow or splash oiling lubricates many parts. A bypass valve is used to prevent oil starvation if the filter clogs.

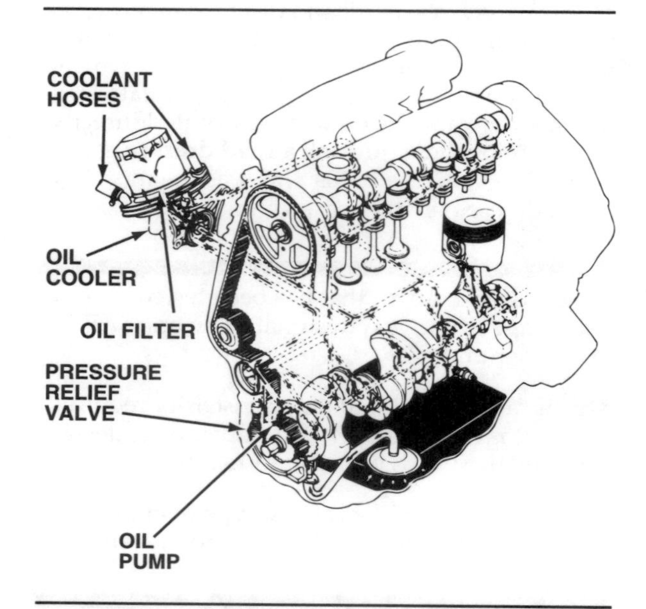

Figure 3-7. A separate coolant jacket allows engine coolant to circulate around the oil cooler of this diesel engine to help reduce oil temperature.

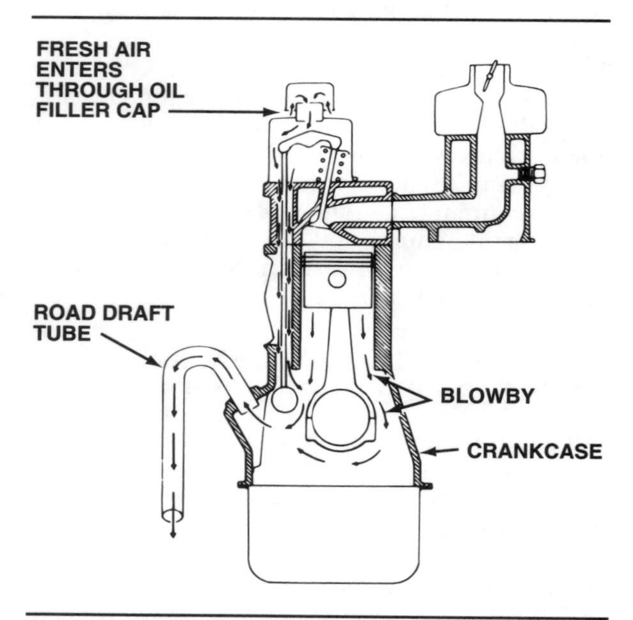

Figure 3-8. Open crankcase ventilation systems used road draft tubes. These systems were efficient at ventilating the crankcase, but added to air pollution.

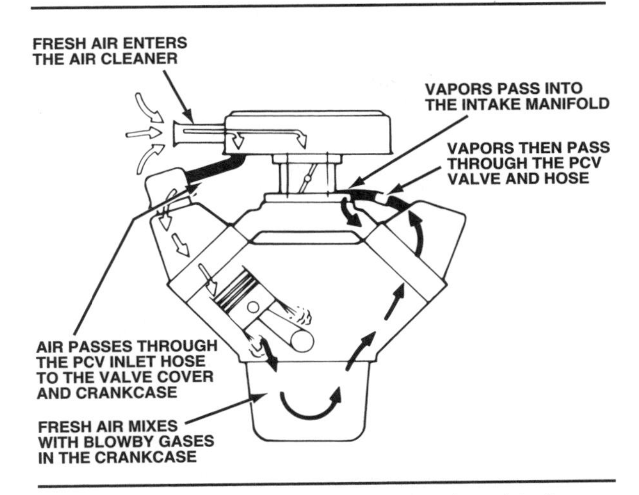

Figure 3-9. Modern PCV systems are closed to the atmosphere.

All late-model cars have **positive crankcase ventilation** (PCV) systems, figure 3-9, the first pollution control device. They were required on all California cars in 1964, and were standard nationwide by 1968. These systems are closed to the atmosphere, which keeps blowby from polluting the air. In a PCV system, clean filtered air is drawn into the crankcase, and crankcase vapors are recycled to the intake manifold. PCV systems also provide better crankcase ventilation than the road draft tube. Since the crankcase is vented to the intake manifold, the engine vacuum works to evacuate the crankcase of vapors. A later chapter will cover PCV systems in greater detail.

Oil Pump and Pickup

The oil pump is a mechanical device that forces motor oil to circulate through the engine. On most overhead-valve engines and some overhead-camshaft engines, the pump is driven by the camshaft through an extension of the distributor shaft, figure 3-10. On many late-model overhead-camshaft engines, the front of the crankshaft drives the oil pump, figure 3-11. Some overhead-camshaft engines use a separate or intermediate shaft to drive the oil pump, figure 3-12.

Traditionally, the most common pump is the gear type, figure 3-13. One oil pump gear is driven by gears and a shaft from the camshaft and is

Positive Crankcase Ventilation (PCV): Later-model crankcase ventilation systems that return blowby gases to the combustion chambers.

Dry-Sump Lubrication

The vast majority of automobile engines carry their oil supply in the oil pan or sump at the bottom of the engine. These "wet-sump" lubrication systems work very well on most street engines. However, some motorcycles, exotic street cars, and racing cars carry their oil in a separate tank, away from the engine. When the engine sump is freed of oil reservoir duties, it is comparatively dry, hence the term "dry sump".

The key part of dry-sump lubrication systems is an oil pump with two sections. A scavenging section draws oil from the engine and sends it to a remote tank reservoir; a pressure section draws oil from the tank and pumps it to the blocked oil passages. The scavenging section works in direct relation to its shaft speed; the faster it turns, the more oil volume it pumps. The pressure section has a pressure relief valve to limit the maximum delivery pressure.

With a dry-sump system, designers can mount the engine much lower in the chassis because there are not four or more quarts of oil carried beneath it. The oil reservoir is located away from engine heat, reducing oil temperature. Oil control is improved, and this provides the two largest benefits of dry-sump lubrication. First, the crankshaft is no longer splashing around in a sea of oil, which reduces oil drag on the crank and can significantly increase horsepower. Second, on wet-sump systems the centrifugal force imposed by high cornering speeds stacks the oil against one side of the oil pan. If the oil is out of reach of the wet-sump oil pickup tube, the system

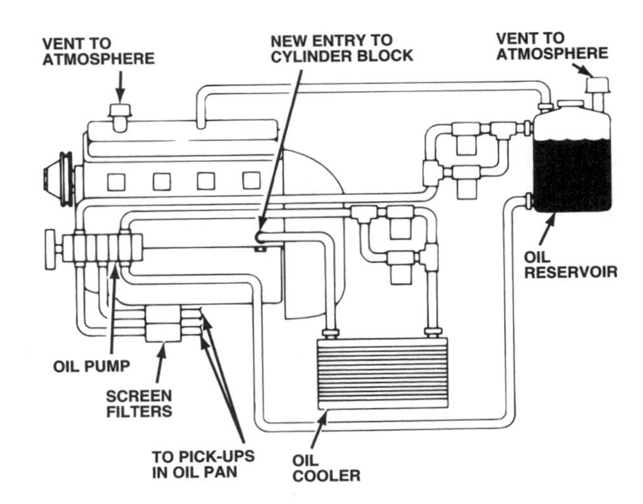

will draw air. The engine will starve for oil while the air bubble works its way through the system. Dry-sump systems avoid this problem and deliver oil to the engine despite the cornering loads, improving engine life in racing conditions.

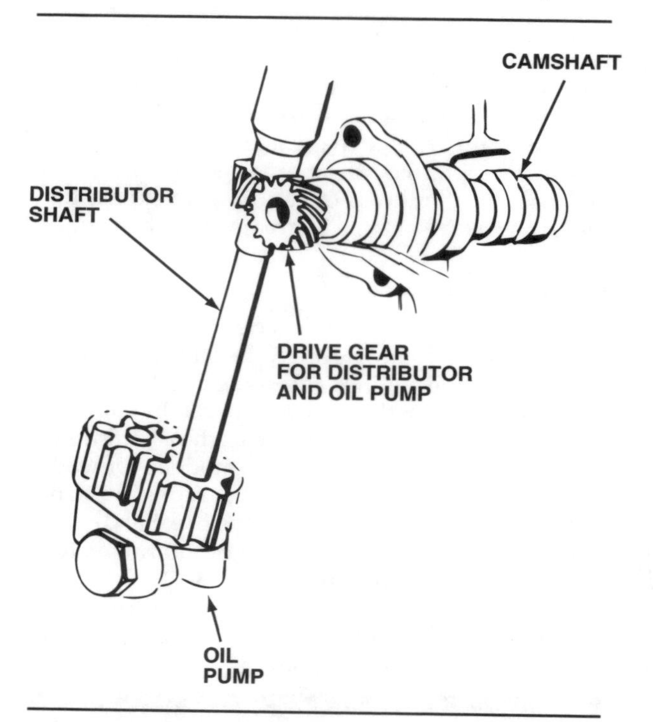

Figure 3-10. This oil pump is driven by the camshaft.

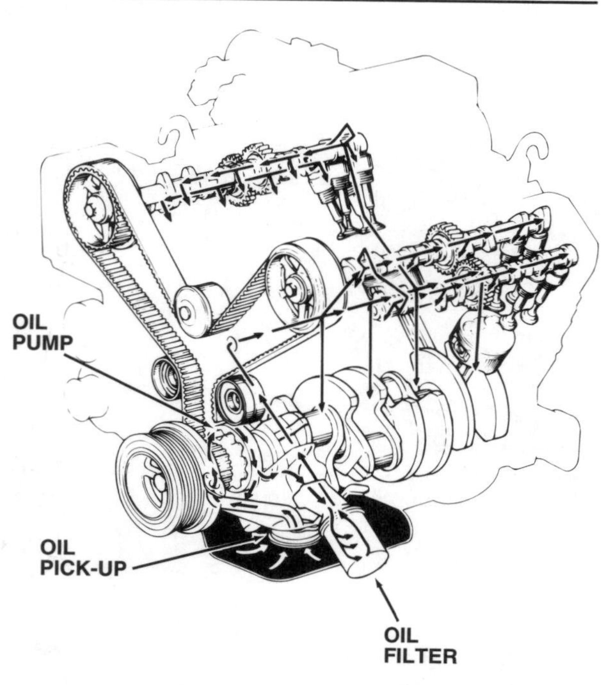

Figure 3-11. The front of the crankshaft drives a rotor-type oil pump on this Lexus LS 400 V8 engine.

called the drive gear. When the drive gear turns it forces the second oil pump gear, called the idler gear, to turn. As the two gears turn, the oil between the gear teeth is carried along. At the point where the two gears mesh, there is very little room for oil, so the oil is forced out of the area under pressure.

The rotor-type pump is a newer design. It works on the same principle of carrying oil from a large area into a smaller area to create pressure. An inner rotor can be mounted off-center within an outer rotor, figure 3-14, or can be driven through gears by the camshaft or from the nose of the crankshaft as in figure 3-11. The inner rotor drives the outer rotor. As the rotors turn, they carry oil from the areas of large clearance to the areas of small clearance, figure 3-15, forcing the oil to flow from the pump under pressure.

The oil entering the pump comes from the oil pan. An oil pickup tube extends from the pump to the bottom of the pan. A screen at the bottom end of the pickup tube keeps sludge and large particles from entering the oil pump and damaging it, or clogging the oil lines and galleries.

Pressure Relief Valve

The oil leaving the pump passes through a pressure relief valve, or pressure-regulating valve. The valve limits the maximum engine oil pressure. It consists of a spring-loaded ball or piston set into an opening in the valve body, figure 3-16. Oil pressure forces the ball or piston to move against spring tension and open the hole in the valve body. Oil escapes through this hole to decrease the overall system pressure. The strength of the spring determines the maximum oil pressure the valve allows.

The pressure relief valve is usually built into the oil pump housing. The oil that escapes through the relief hole is sent back to the inlet side of the pump. This arrangement of oil flow from the relief valve reduces oil foaming and agitation so that the pump puts out a steady stream of oil.

Filter

Oil leaving the pressure relief valve flows through the oil filter before reaching the rest of the system. The filter consists of paper or cloth fibers that will pass liquid oil but trap dirt. If dirt plugs the filter, oil flow is restricted.

To prevent a clogged filter from completely stopping oil flow and damaging the engine, the filter or the filter housing contains a bypass valve. It allows oil to flow around the filter element instead of through it when the oil is too thick to go through the filter because of its low temperature, or when the filter outlet pressure drops. You must replace oil filters at specific

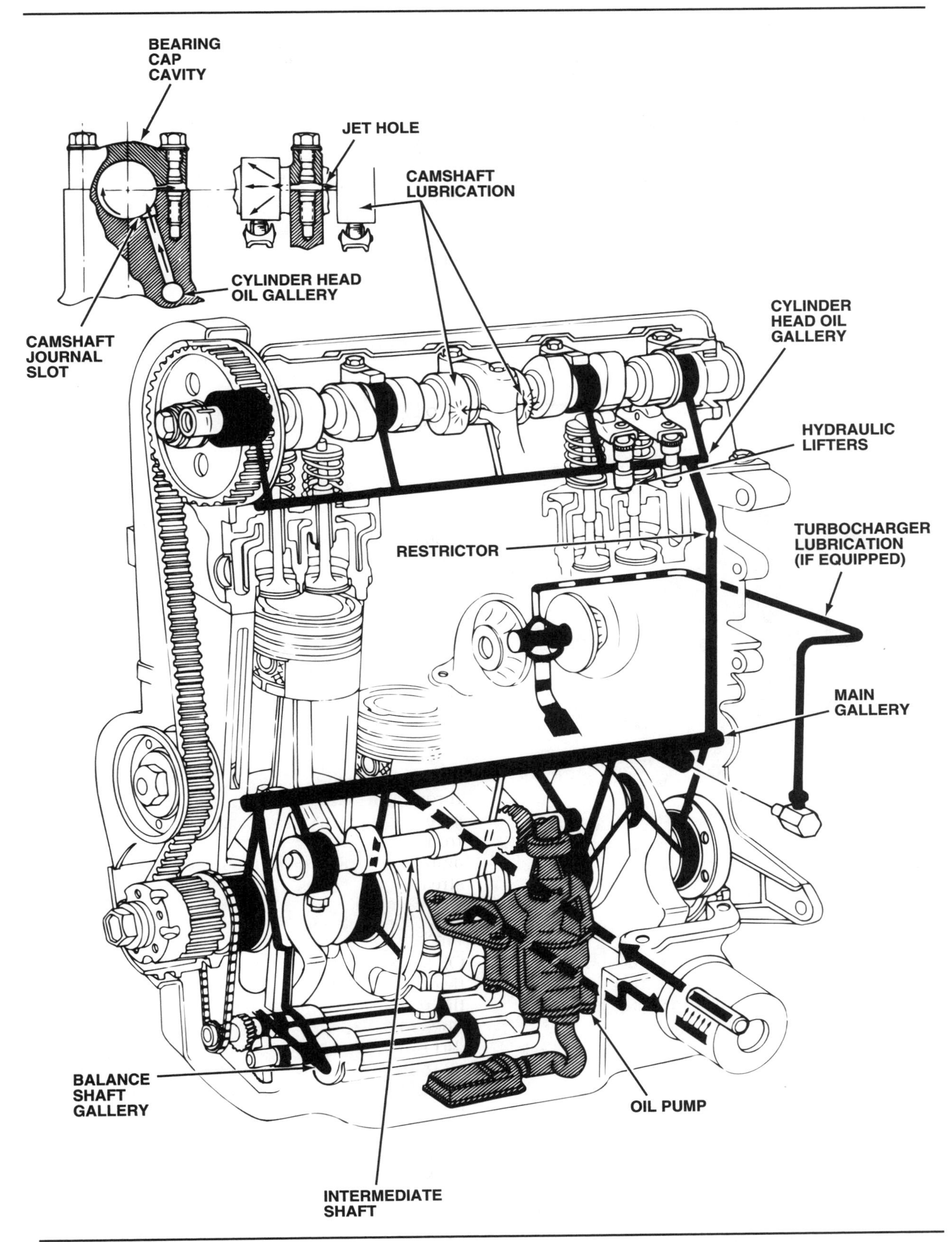

Figure 3-12. On this overhead-camshaft Chrysler engine, an intermediate shaft drives the oil pump.

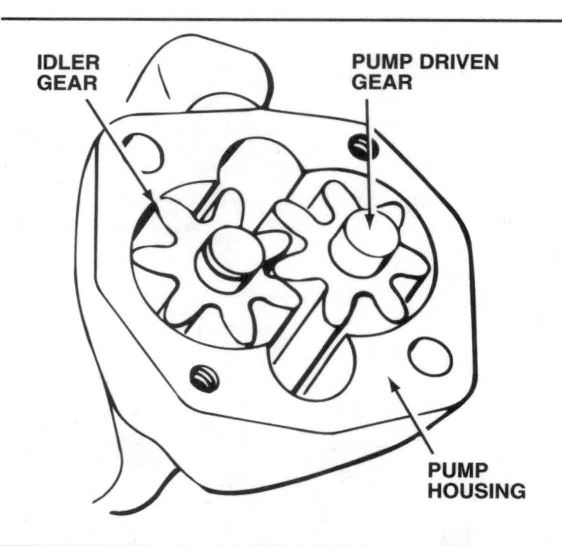

Figure 3-13. A gear-type oil pump uses two spur-type gears.

Figure 3-14. A rotor-type oil pump usually contains a single rotor.

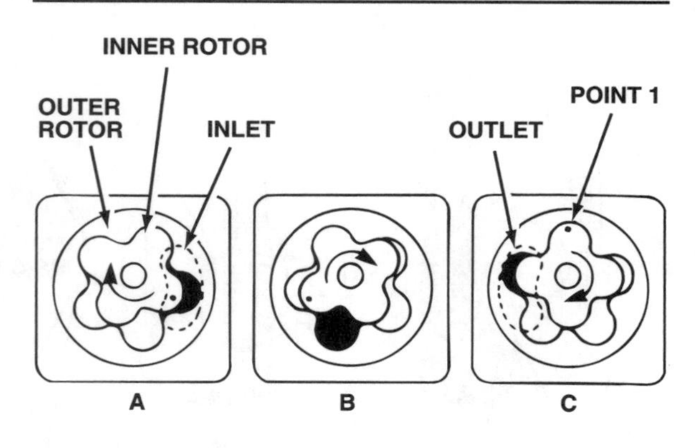

Figure 3-15. The operating principle of the rotor-type oil pump.

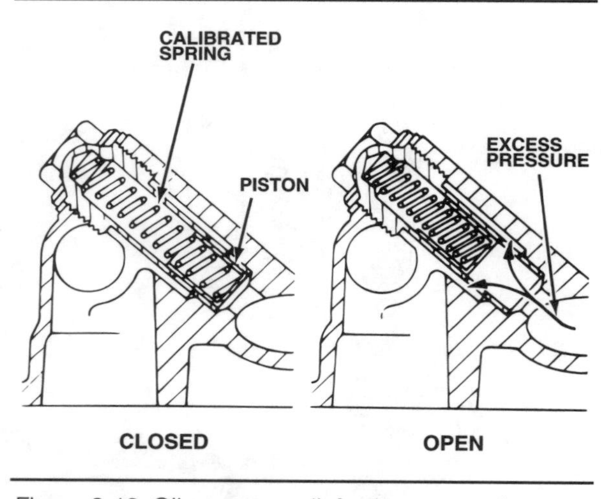

Figure 3-16. Oil pressure relief valves are spring loaded.

intervals to prevent clogging. Older cars have replaceable oil filter elements inside permanent housings. Most modern engines use spin-on filters that are completely disposable, figure 3-17.

Galleries, Lines, and Drillways

A modern engine has many areas that must have a constant supply of oil under pressure. To ensure that they do, the lubrication system includes a network of passages that direct oil to these parts of the engine, figure 3-6, 3-11, and 3-12. The main oil passages are called oil galleries, and are cast into the engine block during manufacture. They can also be separate tubes, called oil lines, connected to the engine. Not all engines have separate oil lines, but all have oil galleries.

From the filter, oil flows to a large main gallery. In-line engines and some V-type engines have one main gallery. Other V-type engines have two main galleries, one for each bank of cylinders. From the main gallery, smaller passages drilled in the block, called drillways, direct oil to the camshaft bearings and to the crankshaft main bearings. Oil must also reach the crankshaft connecting rod bearings. The manufacturer drills holes through the crankshaft, figure 3-18, so that oil at the main bearings can also flow to the rod bearings.

There are two methods for oiling the cylinder

Figure 3-17. Spin-on oil filters are completely disposable and are usually discarded at each oil change.

walls. Some engines rely on oil splash created by the crankshaft and connecting rod to throw oil on the cylinder walls, figure 3-6. On other engines the rod bearing caps have small holes that line up with similar holes in the crankshaft and rod bearings. When the holes in the cap and bearing align with the hole in the crankshaft, figure 3-19, a small stream of oil squirts through and hits the lower cylinder wall. This helps lubricate the cylinder wall and piston.

Some turbocharged gasoline engines and many diesel engines have an oil jet that directs a shot of oil directly to the underside of the piston crown, figure 3-20. This spray of oil not only lubricates, but helps reduce piston crown temperatures. Combustion chamber temperatures are always high in turbocharged and diesel engines. The oil spray helps reduce the chance of detonation and preignition in turbocharged gasoline engines and piston damage due to high temperatures in both types of engines.

Different engine designs have different ways of getting oil from the main gallery to the head and valve assemblies. There can be galleries drilled through the block and the head, or the oil can travel through hollow valve train pushrods. Also, oil may flow through an enlarged head-bolt hole. From the valve assembly, the oil drains down through the engine into the oil pan. Engine designers may place the drain holes so that the dripping oil helps lubricate the camshaft.

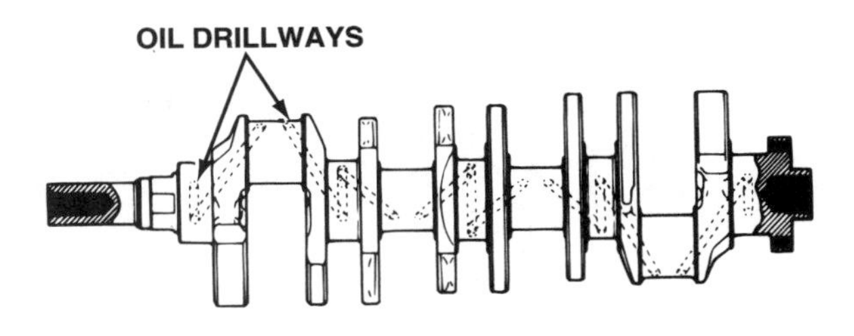

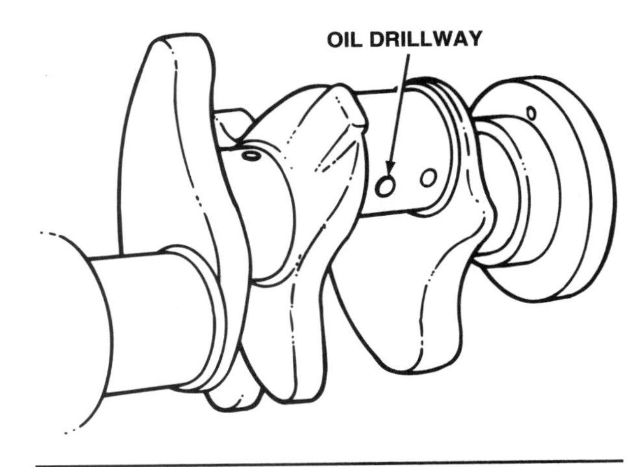

Figure 3-18. So that oil can reach the main and rod bearings, crankshafts have a network of drilled passages called drillways.

Indicators

Manufacturers install a dipstick in the oil pan as the primary indicator of engine oil level. Some cars have an electronic oil level sensor that indicates when the oil level is low.

To keep the driver informed of engine oil pressure while the engine is running, automakers equip cars with a low-oil-pressure warning lamp or a gauge that shows the pressure at all times.

Bypass Oil Systems

Before the full-flow oil filtering system was invented, engines that used oil filters had a bypass system. The filter was mounted on a bracket attached to the engine or anywhere in the engine compartment. Oil lines were connected to a tapped hole in the side of the engine block, and to a drain on the pan or block. The filter received pressurized oil, and drained it back into the oil pan after filtration.

Since the oil did not have to go through the engine filter before getting to the bearings, a piece of dirt could, theoretically, circulate through the oil system indefinitely until it happened to get into the filter and be trapped. This catch-as-catch-can system was discontinued in favor of the full-flow system.

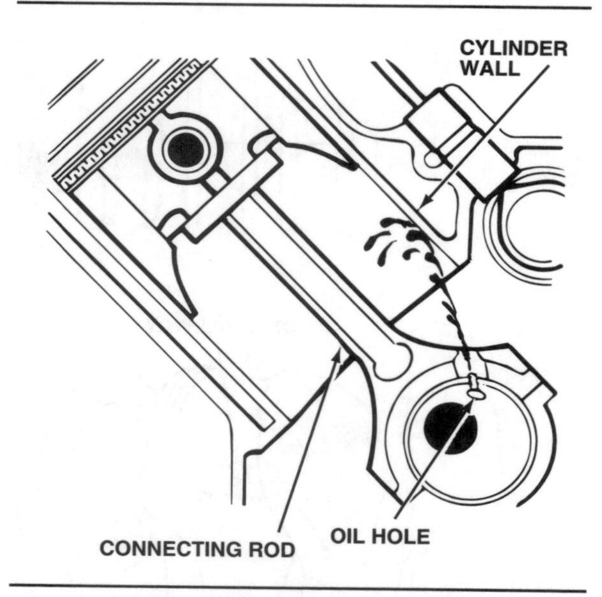

Figure 3-19. Some oil spurts through the connecting rod cap onto the cylinder walls.

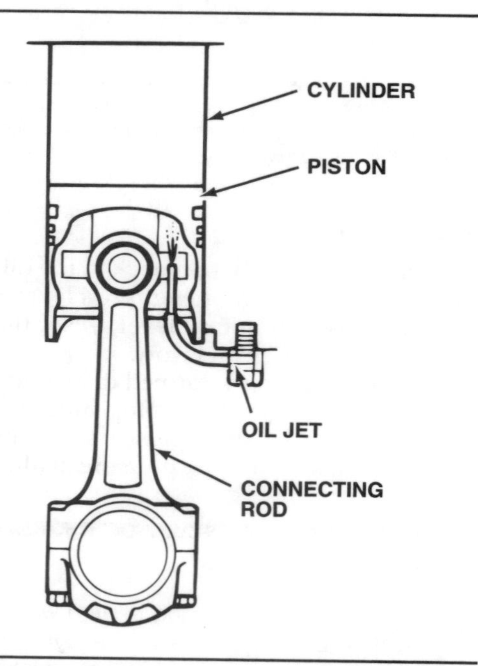

Figure 3-20. Oil jets in diesel and turbocharged engines cool piston crowns.

Dipstick

When the engine is at rest, almost all of the oil drains into the oil pan. All engines have a measuring rod, called a dipstick, that extends from outside the engine into the pan, figure 3-21. The dipstick has markings on it that indicate the maximum and minimum oil levels for that engine. When you pull the dipstick out of the pan, you can see a film of oil on the stick. The level of the film relative to the markings on the stick indicates how much oil is in the pan, figure 3-22.

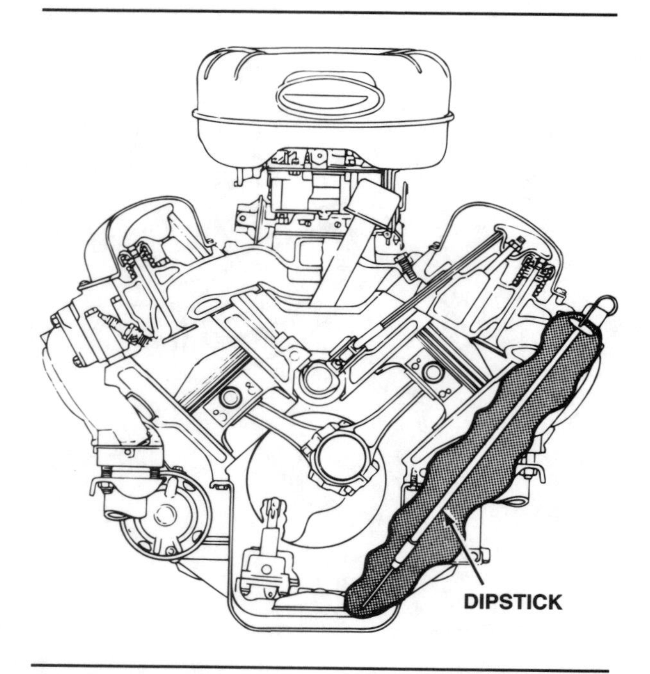

Figure 3-21. The engine dipstick is the main oil level indicator.

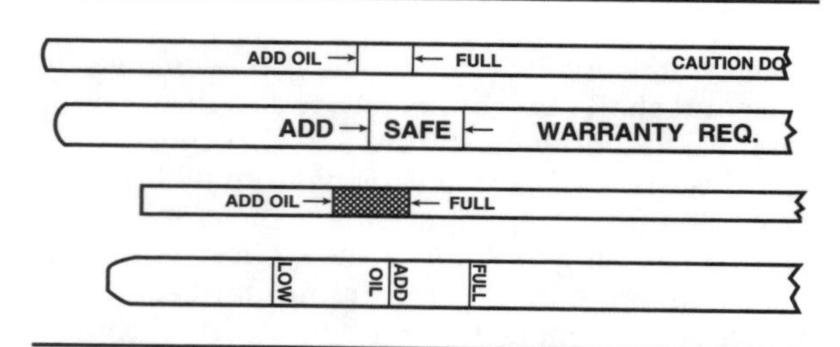

Figure 3-22. Dipstick markings vary from automaker to automaker. Here are several different types.

Oil level sensor

As a convenience item, some cars have an oil level sender that warns the driver when the level is low. A small float and electronic sending unit are installed in the oil pan, figure 3-23. When the oil is below a certain level, the float is low enough to complete the electric circuit, illuminating a warning lamp on the instrument panel.

Oil pressure warning lamp

The oil pressure warning lamp lights when the oil pressure is less than a set level. This happens during cranking, and when there is a problem in the lubricating system.

Warning lamps light when a set of electrical contacts close, figure 3-24. The contacts are con-

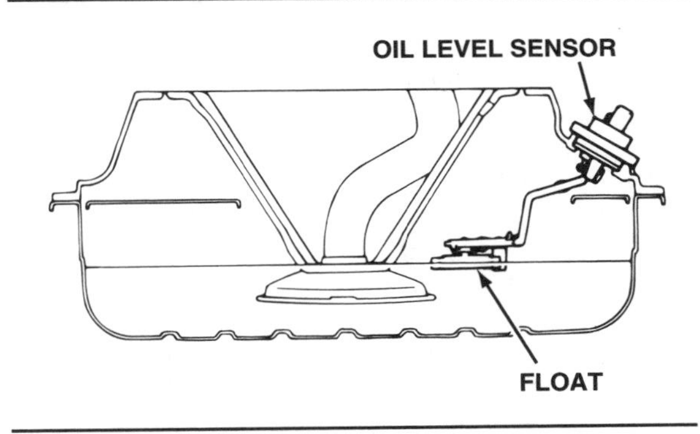

Figure 3-23. A small float in the oil pan indicates the oil level through an electrical sending unit.

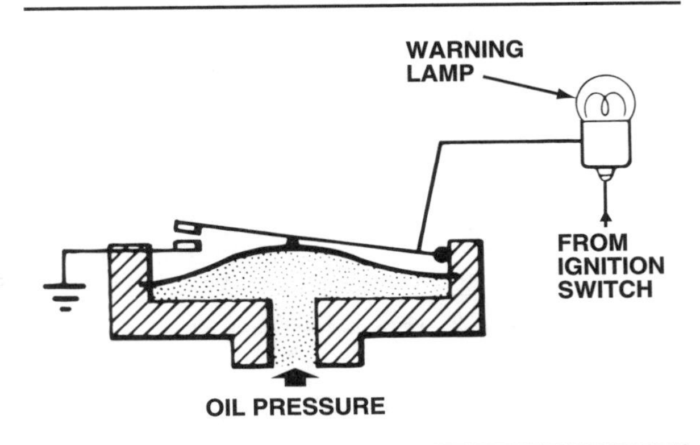

Figure 3-24. The oil pressure switch is connected to a warning lamp that alerts the driver of low oil pressure.

trolled by a movable diaphragm exposed on one side to engine oil pressure. The contacts close when oil pressure is low, and the lamp lights. When pressure increases, the diaphragm moves and the contacts open, turning off the lamp.

The contacts and diaphragm are contained within a sending unit. The sending unit is usually threaded into the side of the engine block so that it reaches an oil passage. The sending unit may be threaded into the oil filter adapter where the filter attaches to the engine.

Oil pressure gauge

Oil pressure gauges operate electrically or mechanically. An electric gauge has a sending unit similar to the warning lamp sending unit. A movable diaphragm varies the current flow through the gauge in proportion to oil pressure. The current flow determines the position of the gauge needle.

Mechanical gauges, figure 3-25, have an oil line running from the engine to a flexible, curved tube inside the gauge. When there is engine oil pressure, oil is forced through the oil line to the tube in the gauge. Oil pressure causes the tube

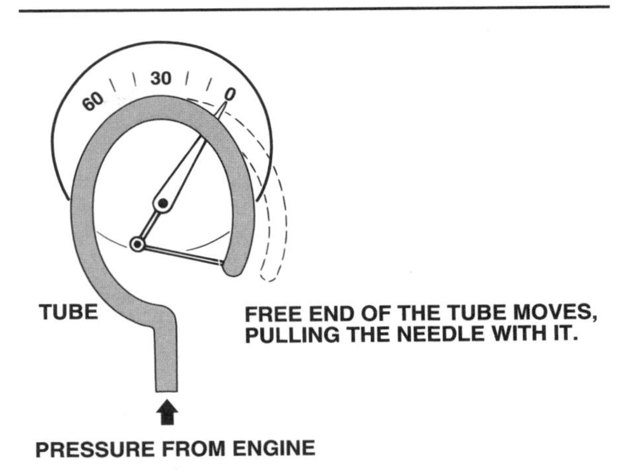

Figure 3-25. A typical mechanical oil pressure gauge uses a flexible tube that reacts to oil pressure.

to straighten out, and the higher the pressure the more the tube moves. Since the tube is connected to the gauge needle, engine oil pressure determines the position of the gauge needle.

ENGINE LUBRICATION EFFECTS ON MECHANICAL WEAR

When an engine is new, the moving parts fit together very closely. The distance between two moving parts is called clearance. As the engine wears, the clearance increases. After a small amount of wear an engine is considered "broken in". At this point internal friction diminishes, the piston rings seal well against the cylinder walls, and performance improves. When an engine wears far beyond this point, or gets "loose", it does not operate as efficiently as it once did. When the piston rings no longer seal the combustion chamber well, power is wasted as blowby. Camshaft and valve assembly wear changes valve timing, and performance suffers.

The lubrication system helps keep engine wear to a minimum. Using the proper API-rated oil will reduce engine wear because the oil is formulated to match the engine's requirements. Changing the oil and filter at or before recommended intervals will minimize engine wear by replacing used additives, and by removing harmful dirt and wear particles from circulation.

The SAE oil viscosity rating will affect engine wear and performance as well. If oil with too low a viscosity is used during high-speed or high-temperature operation, the oil film between moving parts may become so thin that it breaks down, allowing metal-to-metal contact and rapid wear.

If oil with too high a viscosity is used during low-temperature operation, the oil may be so thick that it is difficult to pump through the oil passages. The result may be inferior lubrication and high wear because not enough oil reaches key critical areas. Also, power will be wasted in forcing engine parts to overcome the resistance of the thick oil. In extreme cases, a combination of cold temperature and high-viscosity oil may prevent the starting system from cranking the engine at all.

SUMMARY

Motor oil has five major jobs in an engine: reducing friction, cooling, cleaning, sealing, and absorbing shock. Additives mixed with the oil help it to do these jobs. Two oil rating systems are generally used: the API Service Classification and the SAE Viscosity Grade. The API number rates an oil's performance in a laboratory engine, and the SAE viscosity number rates the oil's thickness. An oil may have one or more API ratings. An oil's service classification and viscosity rating must be matched to an engine's requirements for best engine life, performance, and economy.

An engine lubrication system includes the pan, filter, pump, oil galleries or oil lines (or both), dipstick, and pressure warning lamp or pressure gauge. The oil is stored in the oil pan. The pump takes it from the pan, pressurizes it, and sends it through the oil galleries or oil lines to lubricate the moving parts of the engine. The filter traps dirt and other particles that the oil holds in suspension, so the oil and filter must be changed periodically. A dipstick is used to measure the oil level in the pan when the engine is not running, and a pressure warning lamp (for low oil pressure) or a pressure gauge (for continuous pressure reading) is used when the engine is running.

The main purpose of engine lubrication is control of mechanical wear. Excessive wear will hurt performance, economy, and emission control. The use of the proper oil, and regular oil and filter changes, will keep engine wear to a minimum.

Review Questions

Choose the single most correct answer.
Compare your answers to the correct answers on page 507.

1. Which of the following is not a primary job of motor oil:
   a. Cooling the engine
   b. Reducing friction
   c. Reducing exhaust emissions
   d. Cleaning the engine

2. Additives in motor oil can:
   a. Replace qualities lost during refining
   b. Strengthen natural qualities
   c. Add qualities not naturally present
   d. All of the above

3. Blowby gases come from the:
   a. Crankcase
   b. Combustion chamber
   c. Oil filter
   d. Carburetor

4. The oil used in a 1991 passenger-vehicle gasoline engine should have an API service classification of at least:
   a. SD
   b. CD
   c. CC
   d. SG

5. The "W" in SAE viscosity grade indicates that the oil:
   a. Meets viscosity requirements at 0°F (-18° C)
   b. Is for winter use
   c. Has no additives
   d. None of the above

6. Typically, the highest viscosity number given to automotive motor oils is:
   a. SAE 40
   b. SAE 50
   c. SAE 100
   d. SAE 70

7. The term "multigrade" means that an oil:
   a. Has been given two API service classifications
   b. Has many additives
   c. Meets viscosity requirements when both hot and cold
   d. Can be used in gasoline or diesel engines

8. Synthetic motor oils:
   a. May deteriorate at a slower rate than conventional petroleum oils
   b. Cost less than conventional oils
   c. Contain more wax than conventional petroleum oils
   d. Require more frequent drain intervals

9. If the crankcase is not ventilated:
   a. Exhaust emissions will increase
   b. The oil will foam and not protect the engine
   c. Too much oil will enter the combustion chambers
   d. Blowby gases will pressurize the crankcase and force oil out

10. The oil pump is often driven by the:
    a. Pistons
    b. Camshaft
    c. Fuel pump
    d. Valves

11. The oil pressure relief valve:
    a. Is usually mounted in the oil pump housing
    b. Limits the maximum pressure in the oiling system
    c. Contains a spring-loaded ball or piston
    d. All of the above

12. When an oil filter becomes clogged:
    a. A bypass valve allows dirty oil to lubricate the engine
    b. The oil pump must slow down
    c. Engine operating temperature increases
    d. All of the above

13. To direct oil to critical points, the oiling system includes:
    a. Internal galleries
    b. Connecting rod squirt holes
    c. Passages drilled in the crankshaft
    d. All of the above

14. Most dipsticks will show you:
    a. What grade of oil to use
    b. If the oil is oxidized
    c. The maximum and minimum oil levels
    d. None of the above

15. Oil pressure warning lamps have a ____ installed on the engine:
    a. Sending unit
    b. Lamp bulb
    c. Oil tube
    d. Pressure gauge

16. Lubrication's greatest effect on an engine is:
    a. Making the engine run cooler
    b. Keeping the compression ratio steady
    c. Controlling engine mechanical wear
    d. Increasing engine horsepower

17. Technician A says that oil pumps use a rotor to force the oil through the oil system.
    Technician B says that oil pumps use gears for that purpose.
    Who is right?
    a. A only
    b. B only
    c. Both A and B
    d. Neither A nor B

18. Technician A tells his customers to use the oil recommended the year their engine was manufactured. If SE oil was recommended in 1978 for the customer's Chevrolet Caprice, you should use a SE oil after the engine is rebuilt in 1991 by a machine shop.
    Technician B tells his customers to use the oil with the latest specifications regardless of the year the engine was manufactured. His customer's 1978 Caprice should use SG oil in 1991.
    Who is right?
    a. A only
    b. B only
    c. Both A and B
    d. Neither A nor B

Chapter

4

Cooling and Exhaust Systems

When the air-fuel mixture in a combustion chamber burns, the cylinder temperature can soar to 6,000°F (3300°C). During the complete four-stroke cycle, the average temperature is about 1,500°F (800°C). Only about one-forth to one-third of this heat energy is turned into mechanical energy by the engine. The remaining heat must be removed by other means to maintain engine efficiency and prevent overheating.

About half of this waste heat remains in the combustion gases, and is removed through the exhaust system. The other half is absorbed by the metal of the engine block, which is then cooled by the cooling system. The engine cooling system, in turn, can recycle much of this waste heat through the passenger compartment heating system.

Proper operation of the cooling system is essential. If the excess heat is allowed to build up in the engine:

- Engine oil temperature will rise, reducing viscosity and engine lubrication, and decomposing the oil.
- The incoming air-fuel mixture will become too hot and reduce engine efficiency.
- Excessive cylinder head temperatures can cause damaging preignition or detonation.
- Metal engine parts can expand to the point of damage, seizure, or total engine failure.

COOLING SYSTEM FUNCTION

We have said that much of the total heat energy from combustion is absorbed by the engine's metal. This is both good and bad. It is good because an engine that is too cool will have poor fuel vaporization, poor lubrication, excessive acids in the blowby gases, and high hydrocarbon (HC) emissions. It is bad because an engine that is too hot will have poor volumetric efficiency, poor lubrication, high oxides of nitrogen (NO_x) emissions, and in extreme cases, the fuel may detonate or preignite.

Obviously, there must be a "just-right" engine operating temperature that will minimize these problems. This temperature is slightly different from engine to engine, depending on the design. Before extensive use of emission controls, engine coolant temperatures averaged about 180°F (82°C). Later-model, emission-controlled engines average about 10° to 15°F (6° to 8°C) hotter to reduce exhaust emissions and improve fuel economy. These average temperatures can vary depending on driving conditions. Automotive engines use either an air or liquid cooling system to maintain the proper operating temperature.

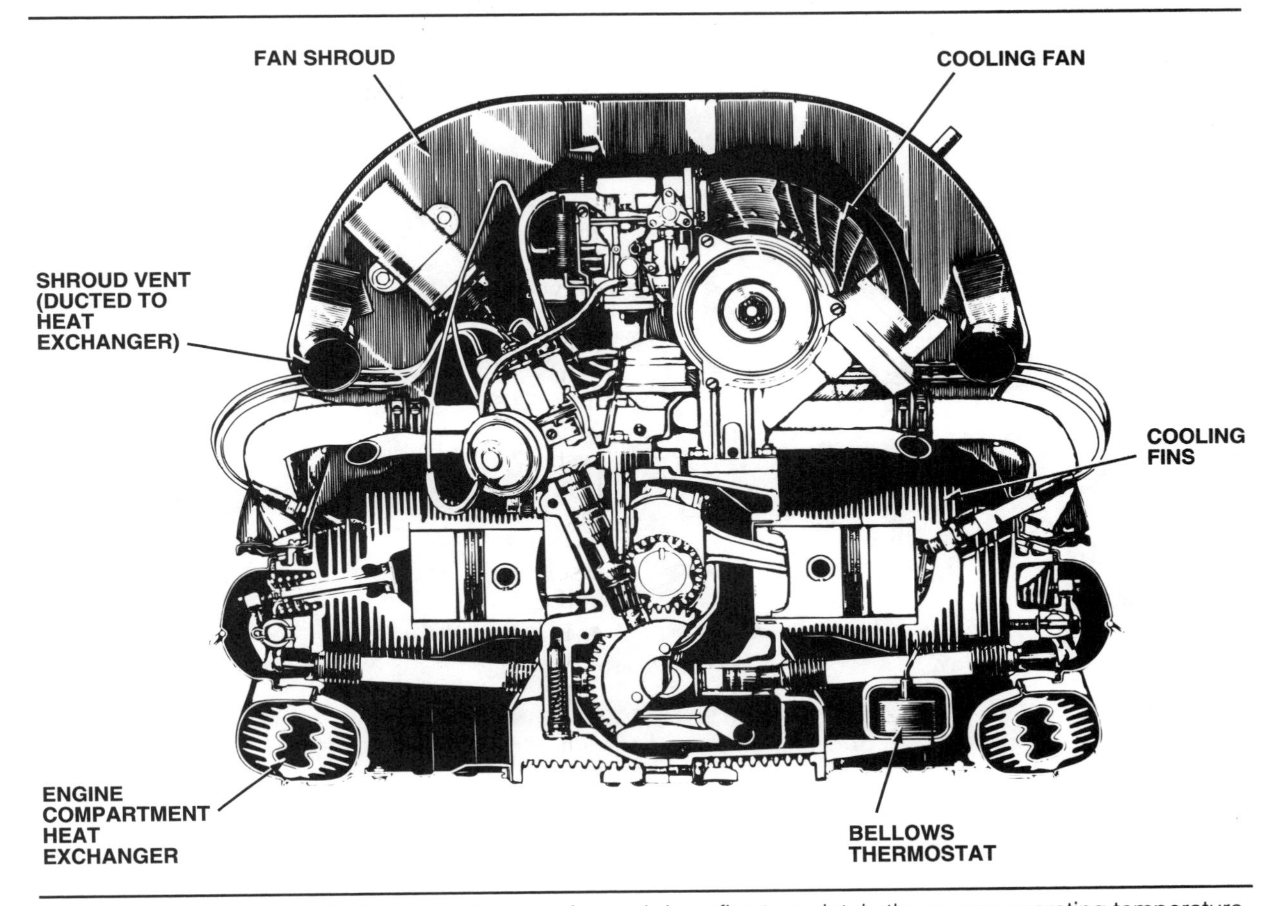

Figure 4-1. A typical air-cooled engine relies on a fan and deep fins to maintain the proper operating temperature.

Air cooling systems

Some automotive engines rely on a flow of air over their outer surfaces to reduce and stabilize engine temperature, figure 4-1. These air-cooled engines have deeply finned heads and cylinders. The fins provide more metal mass to absorb heat and draw it away from the cylinders and combustion chambers. The fins also expose more surface area to the air to help dissipate the heat. The oil crankcases of air-cooled engines have internal, as well as external, fins that reduce oil temperature to help keep overall engine temperature under control. The internal fins speed heat absorption from the oil to the crankcase and the external fins dissipate the heat to the air. However, when an engine is installed in a car and surrounded by the body, the fins alone are not sufficient to keep engine temperature in the correct zone.

Automotive air-cooled engines require an auxiliary fan to furnish an adequate supply of air for cooling the engine. A shroud directs the airflow so that it circulates around the heads and cylinders efficiently. A thermostat opens and closes either the fan's air intake or outlet (depending on design) so that the engine warms up quickly and the operating temperature does not vary over a wide range.

Most automotive air-cooled engines require an oil cooler for additional engine heat reduction. These oil coolers are small radiators through which the oil circulates after being pumped through the engine, before returning to the crankcase. Air-cooled engines depend on their oil supply even more heavily than their liquid-cooled counterparts to maintain the correct engine temperature. It is doubly important that the crankcases of air-cooled engines are kept full and the oil changed regularly.

Air-cooled engines tend to be a little lighter and have better power-to-weight ratios than liquid-cooled engines of the same power. Air cooling does not require a water pump, so the engine does not expend energy circulating coolant through its cylinder block. Air-cooled engines also tend to warm up quickly, and can be very fuel efficient since they spend less time running on choke.

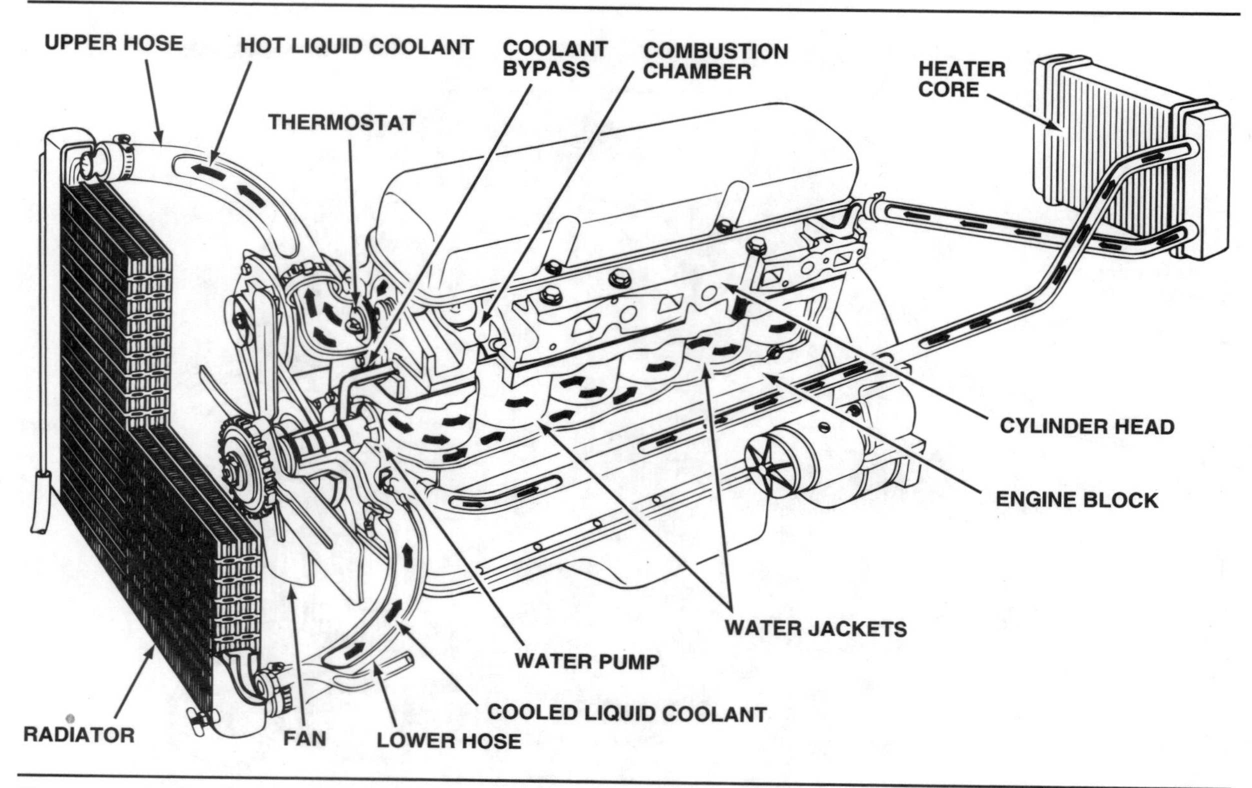

Figure 4-2. A liquid cooling system uses the circulation of liquid coolant to maintain the proper engine operating temperature.

Liquid cooling systems

Most automotive engines use a liquid cooling system, figure 4-2. Liquid coolant constantly circulates through the engine, absorbing heat from the engine block and cylinder head. The coolant then circulates outside of the engine and is exposed indirectly to the air by a radiator. The air absorbs heat from the coolant, so the coolant can go back into the engine and absorb more heat. The greater the difference in temperature between the coolant and the air, the more heat will be absorbed by the air.

Liquid cooling systems give more uniform engine temperatures than air-cooled systems, so thermal stresses are reduced and piston ring and valve seal is theoretically better. These are the main reasons most race engines are liquid cooled. The consistent temperatures also allow very precise fuel metering, which helps keep exhaust emissions low. The water jackets around liquid-cooled engines tend to silence some of the mechanical operating noise, making them quieter running.

Circulation patterns

Within the engine, coolant circulates in passages called **water jackets**, figure 4-3. Outside the engine, the coolant circulates through hoses and the radiator. To keep the engine running at its ideal temperature, there are two patterns of coolant circulation. The two patterns are controlled by the thermostat, which is a temperature-sensitive valve that opens and closes a passage between the engine and the radiator.

When the engine is cold, it must be warmed quickly to its ideal temperature. With the cooling system in full operation, it would take a long time for this to happen. To speed the warmup, the thermostat is closed when the coolant is cold, figure 4-4A. This keeps the coolant in the engine from circulating through the radiator. The coolant does, however, circulate inside the engine, so the coolant warms uniformly. Because there is no heat transfer between the coolant in the engine and the air passing the radiator, the heat of combustion quickly warms the engine.

As the coolant within the engine gets warm, the thermostat opens, figure 4-4B. This allows full coolant circulation between the engine and the radiator. The operating temperature of the engine is determined by the thermostat, which opens between 180° and 210°F (80° and 100°C), depending on its design. The amount of coolant in the system and the size of the radiator determine the cooling capacity of the system.

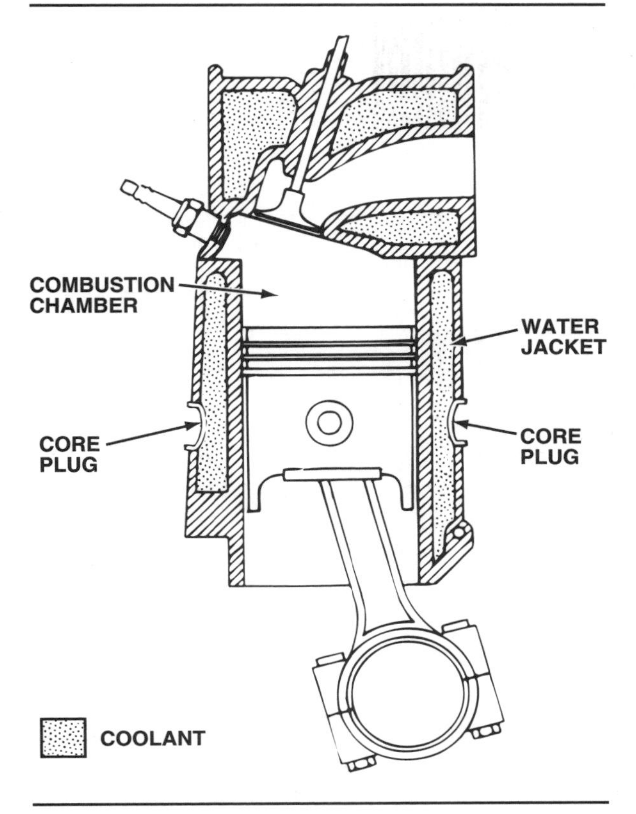

Figure 4-3. Coolant circulates through the water jackets in the engine block and head.

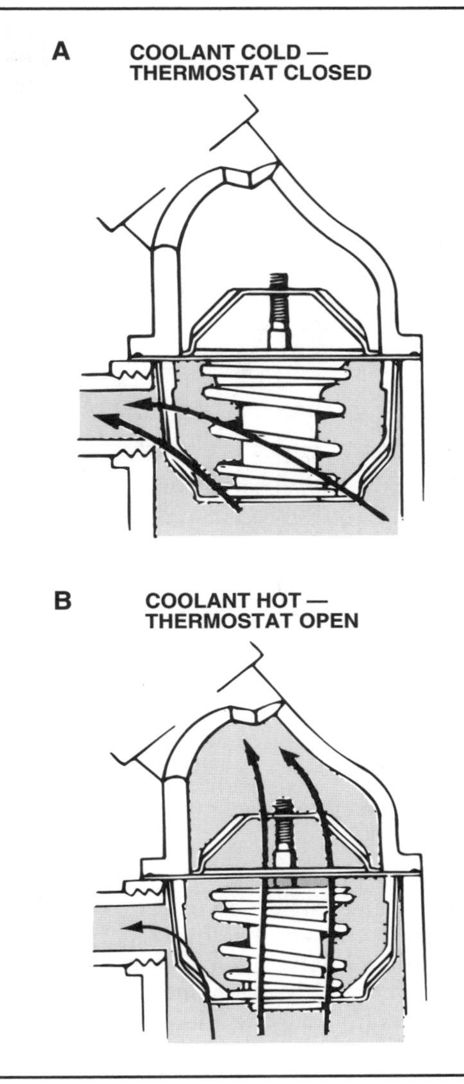

Figure 4-4. With a cold engine, the thermostat stays closed (A). As the engine warms, the thermostat opens (B).

ENGINE COOLANT

Early cooling systems used plain water as engine coolant. With its high specific heat and thermal conductivity, plain water can transfer heat quite well. However, it has disadvantages when used in a cooling system:

- Iron, steel, and aluminum engine parts react with water to form rust and other types of corrosion, which weaken parts, clog water jackets, and slow heat transfer from the coolant to the engine castings.
- Using hard water introduces minerals to the cooling system, which clogs water jackets with thick scale deposits, and also reduces heat transfer.
- When the car is not running and the system is not pressurized, water will freeze at its normal freezing point of 32°F (0°C). The expansion of water as it freezes can crack radiators, fittings, and even engine blocks.

To prevent freezing, early motorists added many substances to the water in their cooling systems. Many of these did work, but had various drawbacks. Salt, calcium chloride, and soda were used as antifreezes, but formed acids in the coolant and corroded the engine and radiator. Adding sugar or honey to the cooling system also prevented freezing, but only at concentrations so high that the syrup was difficult to circulate through the engine. Kerosene and engine oil were sometimes used because of their low freezing points, but were flammable, and caused the rubber hoses to deteriorate.

Methanol (wood alcohol) or ethanol (grain alcohol) were much more successful. Alcohols work well as antifreezes; a 50/50 mixture of either methanol or ethanol and water will pro-

Water Jackets: Passages in the head and block that allow coolant to circulate throughout the engine.

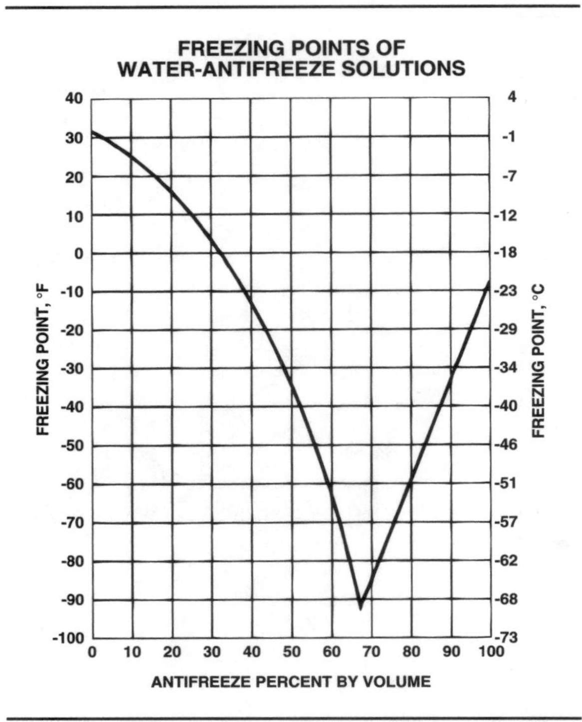

Figure 4-5. Ethylene glycol antifreeze and water should be mixed 17:8, respectively. At this ratio, where the ethylene glycol concentration is 68 percent, the coolant provides the maximum level of lower temperature protection.

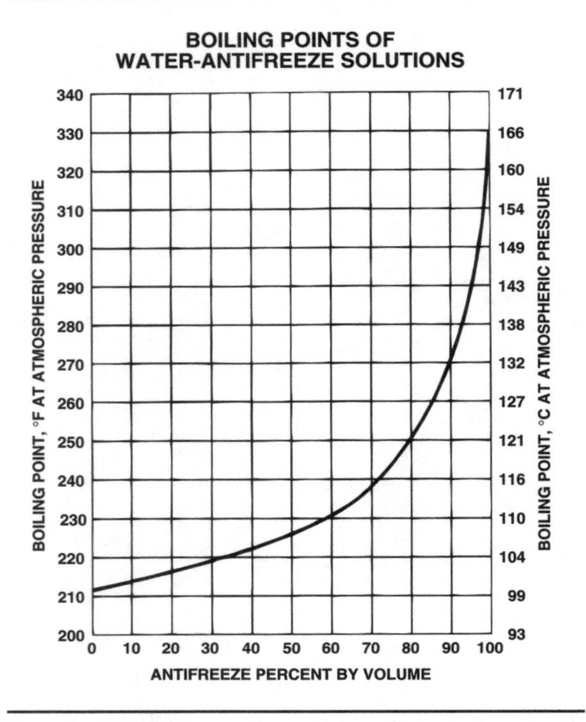

Figure 4-6. Adding antifreeze also extends the upper limit of engine coolant by protecting against boiling, even without a radiator pressure cap.

tect against freezing down to -20°F (-30°C). However, alcohols boil at lower temperatures than water and will therefore evaporate from the engine fairly quickly. To maintain freeze protection, the alcohol concentration must be tested periodically and enough fresh alcohol added to bring the protection up to the necessary level.

At one time, glycerine was also popular for use in cooling systems, as the first "permanent antifreeze" (permanent in that it did not evaporate like the alcohols). Concentrated glycerine boils at 227°F (110°C), and a 50/50 mixture protected against freezing as well as alcohol. Glycerine antifreezes were marketed partially diluted with water and with added corrosion inhibitors.

Modern coolant

Most commercially available coolants for cars and light-duty trucks are made up of 95-percent ethylene glycols. Typically, this coolant consists of 87 percent monoethylene glycol and eight percent diethylene glycol. Water accounts for approximately three percent of the total makeup, and trace amounts of dye give coolant its distinctive color. Other chemicals account for the final two percent.

These other chemicals are generically referred to as inhibitors. They may include as many as 12 different chemicals (in a single coolant) to combat various types of corrosion and foaming. The coolant may also contain small particles designed to seal minor leaks in the cooling system. Some late-model ethylene glycol solutions with revised inhibitor packages are chemically incompatible with earlier ethylene glycol solutions and cooling systems. The revised inhibitor packages provide several advantages: protection against galvanic corrosion of the engine's internal aluminum surfaces, longer water pump seal life, improved radiator protection, and longer service intervals. Carefully read the factory manual for coolant service to prevent health risks, and to maintain the automaker's original service interval.

Propylene glycol, another type of coolant, has traditionally been used only in heavy-duty equipment. In 1995, General Motors (GM) approved this coolant's use in passenger cars and trucks built prior to and including the 1994 model year. Do not mix propylene glycol with the more-common ethylene glycol; the mixture's freeze/boiling-point protection level would be impossible to determine. Propylene requires specially calibrated test instruments, such as a hydrometer, discussed in the *Shop Manual*.

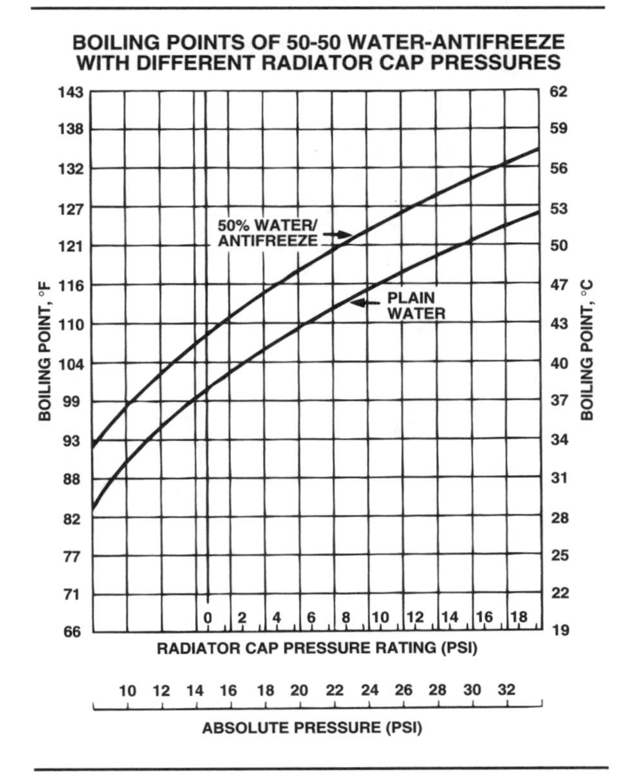

Figure 4-7. Pressurizing the cooling system provides protection against boilover for both water- and antifreeze-based engine coolant.

Mixing antifreeze and water

Modern cooling systems use a mixture of water and antifreeze, producing a much better coolant than either water or antifreeze alone. Ethylene glycol does not transfer heat as well as water, but the greater temperature range within which an ethylene glycol-water mixture remains a liquid more than makes up for this disadvantage.

Concentrated ethylene glycol freezes at about -8°F (-20°C), figure 4-5, but a 50/50 mix of ethylene glycol and water resists freezing down to a temperature of -34°F (-40°C). Maximum protection against freezing is given by a solution of 68 percent antifreeze. Concentrating the antifreeze any more than this causes the coolant freezing point to rise, instead of fall. The freezing protection of concentrated, 100-percent antifreeze is no better than that of a 35-percent solution in water.

Because of the high temperature at which antifreeze vaporizes, mixing it with water in the coolant also protects against coolant boilover. Boiling is harmful because the gases form steam pockets in the water jacket. Steam in the cooling system absorbs much less heat than liquid, so localized hot spots form that can warp or crack castings. A cooling system filled with water and ethylene glycol resists boiling to significantly higher temperatures than one filled with plain water. Concentrated antifreeze boils at 330°F (165°C), and a 50/50 mix resists boiling up to 226°F (110°C), figure 4-6. Pressurizing the system with a sealed radiator cap further improves the system. In fact, many later-model temperature warning lamps are calibrated for high temperatures obtainable only with antifreeze-based coolant. Plain water used with these systems can boil without activating the warning light.

Although 50-percent antifreeze is the industry standard, some manufacturers recommend only a 33-percent solution or less. Less than a 50-percent solution protects from freezing but does not give the corrosion protection available with a 50-percent solution.

Backwards by Tradition

Virtually all modern automobile engines have cooling systems with similar coolant circuits. Hot coolant from the top of the engine travels through an upper hose to the top of the radiator, cools, and returns to the engine through a lower hose at the radiator bottom.

This pattern originated early this century, when coolant circulated through the engine powered by no more than changes in its own density. These *thermo-siphon* systems worked because cool water sinks and hot water rises. As the water entered the radiator, it cooled, sank to the bottom, and entered the engine block through the lower hose. Inside the engine, it warmed and circulated up through the water jacket, finally escaping at the top through the upper radiator hose. Thermo-siphon systems were barely adequate even in the days of the Model T, and water pumps became universal during the 1920s. However, the basic cooling pattern of block-first, cylinder head-second remained for over 70 years.

Aftermarket kits are now available to racers to reverse the direction of coolant flow in common high-performance engines such as the small-block Chevrolet V8. The kits include a modified water pump and external tubing that delivers coolant from the radiator straight to the cylinder heads, and only then back through the engine block and radiator. On a racing engine, reverse cooling would permit more spark advance, leaner jetting, and higher compression. On a street engine, reverse cooling would permit not only more horsepower, but lower emissions, better fuel economy, and better lubrication.

It remains to be seen whether reverse cooling will take over in original manufacturers' designs. In late 1990, both Dow Chemical and General Motors (GM) were exploring reverse cooling systems as original equipment. GM later introduced reverse-flow cooling on the 5.7-liter LT1 engine.

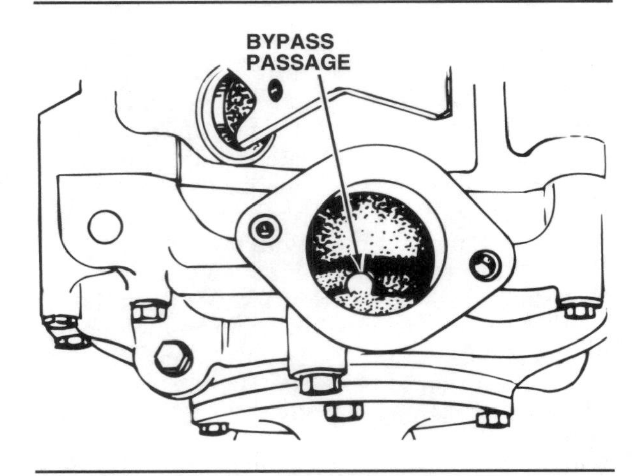

Figure 4-8. This internal passage in the thermostat housing directs cold coolant to the water pump.

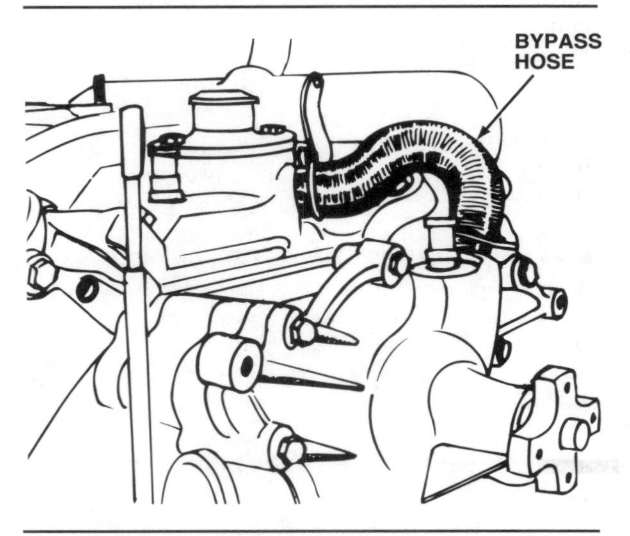

Figure 4-9. This external bypass hose connects the thermostat housing to the water pump.

COOLING SYSTEM PRESSURE

There are two reasons for pressurizing a cooling system: to increase water pump efficiency and to raise the coolant's boiling point.

Pump efficiency is affected by pressure. Without a pressure cap installed in the system, a water pump is only about 85-percent efficient. With a 14-psi (97-kPa) pressure cap, the pump becomes almost 100-percent efficient, because the pressure decreases **cavitation**. Cavitation is the forming of low-pressure (vacuum) bubbles by the water pump blades. A pressurized system makes it difficult for these bubbles to form. Cavitation is undesirable because the bubbles implode (collapse) with enough power to blast small cavities in any metal surface nearby.

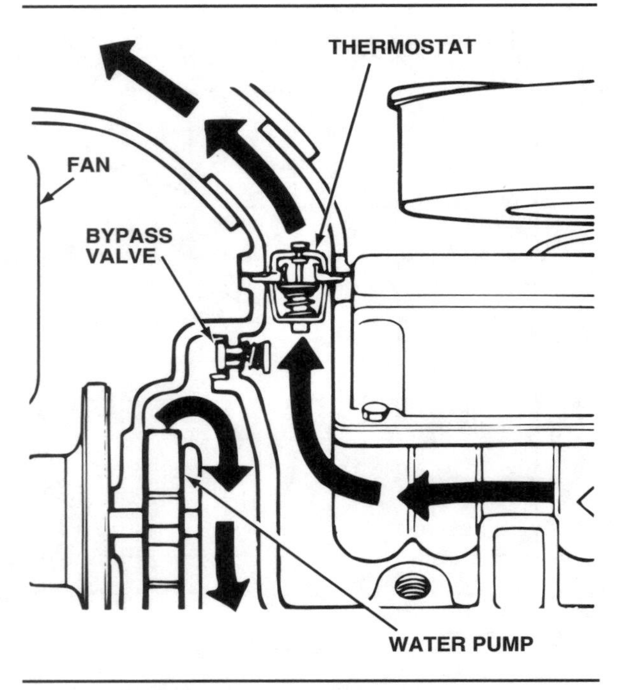

Figure 4-10. Some engines use a spring-loaded bypass valve that opens under coolant pressure when the thermostat is closed.

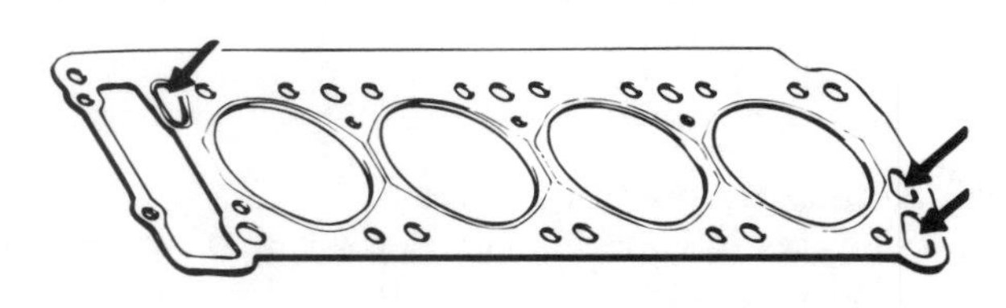

Figure 4-11. These head gasket holes show the placement of coolant passages that carry engine coolant the full length of the cylinder head.

Unchecked, cavitation will erode the water pump blades and housing.

Coolant boiling-point control is equally important. Water boils at 212°F (100°C) under atmospheric pressure (14.7 psi or 100 kPa) at sea level. A 50-percent solution of ethylene glycol and water boils at around 223°F (110°C) under the same conditions. Even this higher figure may be dangerously close to, or even below, the operating temperature of later-model engines. Pressurizing the system raises the coolant's boiling point, figure 4-7. Each 1-psi (7-kPa) increase in pressure raises the boiling point about 3°F (1.7°C). Thus, coolant under 15-psi (103-kPa) pressure boils at approximately 268°F (130°C), 45°F (25°C) higher than without the cap. Most cooling systems are pressurized at 12 to 17 psi (83 to 117 kPa). Remember that when removing the radiator pressure cap on a hot engine, as the pressure is released, the coolant will boil instantly.

COOLING SYSTEM COMPONENTS

A typical liquid cooling system, figure 4-2, includes the:

- Water jackets
- Thermostat-controlled bypass
- Core plugs
- Radiator
- Water pump
- Radiator fan
- Thermostat
- Radiator pressure cap
- Coolant hoses
- Drive belt
- Coolant recovery system.

Water Jackets, Bypass, and Core Plugs

Water jackets and coolant passages are cast into the cylinder block and head. They are designed so that the coolant will circulate freely, and not remain stationary in pockets.

The thermostat bypass circuit permits coolant to flow through the water pump and back into the engine before it has circulated through the radiator. The bypass can be an internal passage in the block, figure 4-8, or an external hose or tube, figure 4-9. In some designs, the thermostat closes the bypass circuit as it opens the passage to the radiator, forcing all the coolant through the radiator when the engine is warm. Other systems leave the bypass open, and split the flow between the radiator and bypass when the engine is running and warmed up.

Some engines have a spring-loaded bypass circuit, figure 4-10. When the cold engine is first started, the thermostat remains closed. Coolant pressure builds up enough to open a spring-loaded bypass valve installed just below the thermostat. This allows the coolant to circulate through the bypass channel and back into the engine block. When engine operating temperature is reached, the thermostat opens, and coolant circulates through the radiator. When the thermostat opens, it lowers coolant pressure, allowing the spring-loaded bypass valve to close.

The size of coolant passages aids in coolant distribution. The front coolant passage inlet holes in the head are frequently smaller than the rear holes. This forces the coolant to flow to the rear of the block, figure 4-11. Because more coolant goes to the back of the block, it means more coolant will flow the full length of the cylinder head. This is necessary for better circulation because the coolant outlet is at, or near, the front of the cylinder head.

Core plugs are pressed into the sides of the cylinder block to close the holes left by the foundry where the block and heads are cast, figure 4-12. They are called core plugs because

Cavitation: An undesirable condition in the cooling system where the water pump blades are allowed to form air bubbles capable of forming small cavities in metal surfaces. Maintaining proper cooling system pressure prevents cavitation.

■ Nucleate Boiling

When most technicians think about an automotive cooling system, they think of it as a means of preventing engine coolant from boiling. After all, the purpose of radiator pressure caps and antifreeze is to raise the temperature a cooling system can withstand before the coolant boils.

In fact, things are not quite so simple. Some of the coolant in a normally operating engine is actually boiling all of the time. Were the process to stop, the engine would rapidly overheat, and either seize or detonate, becoming scrap metal.

Nucleate boiling is the key to low engine temperature. Ordinary ethylene glycol and water coolant boils at around 263°F (128°C). In the cylinder head, however, the temperature of the combustion chamber or exhaust port walls may approach 500°F (250°C). As individual droplets of coolant flow through the water jackets against the other sides of these hot surfaces, they boil into a gas and are instantly washed away, to be replaced by fresh liquid coolant. The superheated bubbles are swept into the lower-temperature coolant flowing through the middle of the passage, where they cool off and condense back to liquid.

Boiling the coolant into a gas removes heat from the metal surface much more efficiently than would merely warming the coolant. With nucleate boiling, the *overall* temperature of the coolant still stays below its boiling point—only the small nuclei of coolant actually touching the metal flash into a gas. The end result is much better heat transfer between the metal and the coolant, and lower operating temperatures.

Nucleate boiling requires a strong, turbulent coolant flow to work. If the coolant were to stop moving, steam pockets and steam-blanketed areas would form very quickly. This would lead to local overheating, detonation, boilover, and warped or cracked castings. Many engine blocks and cylinder heads have special passages called *steam holes* that permit bubbles from boiling coolant to escape when the engine is operated at low speeds or turned off.

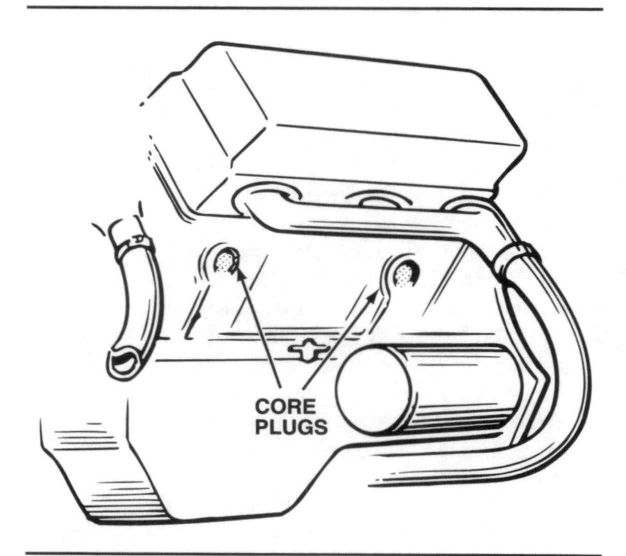

Figure 4-12. Core plugs are pressed into the cylinder block.

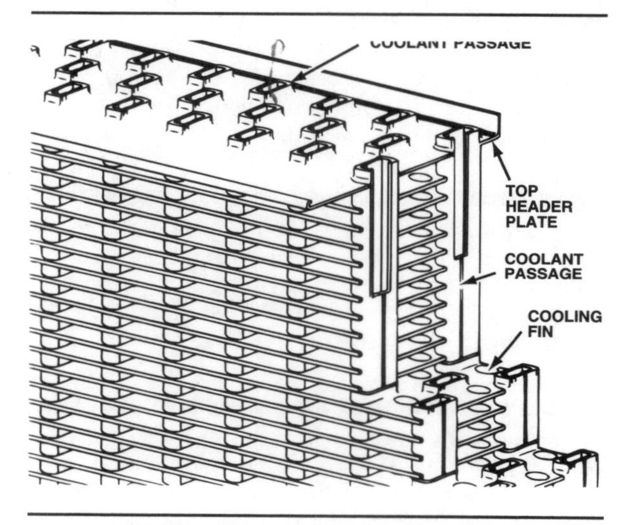

Figure 4-13. The tubes and cooling fins of the radiator core.

the holes are used by the foundry to remove core sand from the inside of the block. Core plugs are also called freeze plugs. Some people mistakenly believe that the plugs are there to relieve pressure in the block if the water in the engine freezes. Core plugs will frequently pop out if the water freezes, but that is not their purpose. If one does pop out, the chances are that the same freeze also cracked the block. Core plugs are definitely not insurance against freeze damage.

Radiators

The purpose of the radiator is to expose a large amount of surface area to the surrounding air. Radiators are made of thin metal tubes, usually

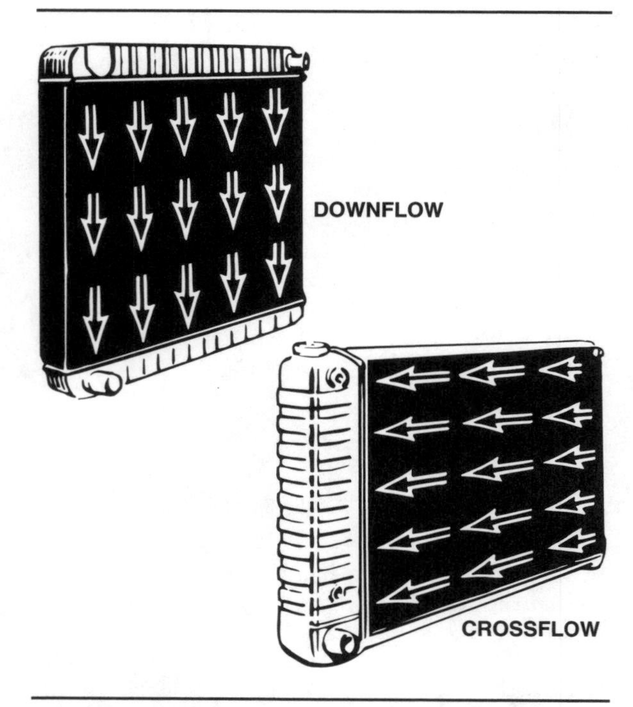

Figure 4-14. A radiator may be either a downflow or a crossflow type.

brass or aluminum, with cooling fins attached, figure 4-13. Coolant flows through the tubes and transfers its heat to the radiator metal. The air flowflowing past the radiator tubes and fins absorbs this heat. The number of tubes and fins in a radiator determines the unit's heat-transferring capacity.

The assembly of tubes and fins is called the radiator core. To keep a steady stream of coolant flowing into and out of the core, two larger metal or plastic tanks are connected to opposite ends of the tubes. There are two types of radiator construction for cars and light trucks, figure 4-14:

- Downflow radiators have vertical tubes and a tank at the top and bottom of the core.
- Crossflow radiators have horizontal tubes and a tank on both sides of the core. Cross-flow radiators usually allow a lower hood line on the car. Other than the flow direction, there is no difference in operation between the two types.

One of the radiator tanks, either the top or a side tank, has an open neck sealed with a pressure cap. On cars with automatic transmissions, one of the tanks contains a cooler for the transmission fluid, figure 4-15. The fluid and the coolant do not mix, but the extra heat of the transmission fluid adds to the work of the cooling system. For this reason systems on automatic-transmission cars are usually built with a greater capacity than those on manual-transmission cars.

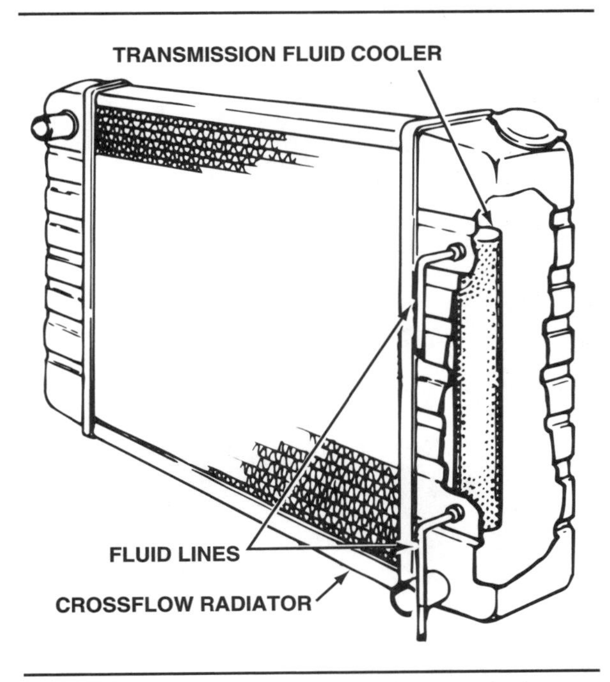

Figure 4-15. This Harrison automatic transmission fluid cooler is installed in one of the radiator tanks.

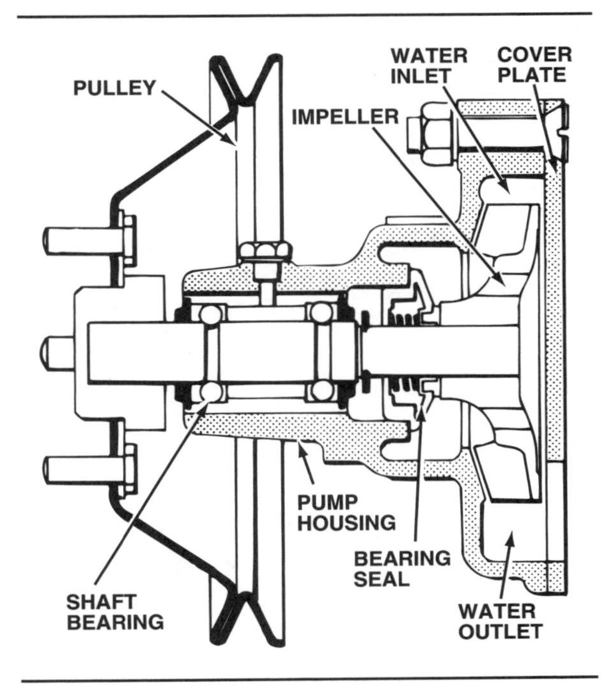

Figure 4-16. Cutaway of a typical water pump, showing the impeller, seals, and bearings.

Water Pumps

The water pump uses **centrifugal force** to circulate the coolant. It consists of a fan-shaped impeller, figure 4-16, set in a round chamber with curved inlet and outlet passages. The chamber is called a scroll because of these curved areas, figure 4-17. The impeller is driven by a belt from the crankshaft pulley, and spins within the scroll. Coolant from the radiator or the bypass enters the inlet and is picked up by the impeller blades. Centrifugal force flings the coolant outward, and the scroll walls direct it to the outlet passage and the engine. Because the water pump is driven by the engine, it keeps coolant circulating whenever the engine is running. Centrifugal pumps must turn rapidly to be efficient. Worn or loose belts slip, causing the pump to turn slower and lose efficiency, and possibly result in engine overheating. Early water pumps had to be externally lubricated, but modern coolant mixtures contain a water pump lubricant. The pump bearings must be well-sealed to prevent lubricant or coolant leaks.

Radiator Fan

When the car is traveling at speeds above 25 mph (40 km/h), airflow through the radiator core is great enough to absorb all excess heat. However, at low speed or idle there is not enough natural airflow. To increase the amount

Centrifugal Force: A force exerted by a rotating object that moves it away from the center of rotation.

Engines Without Water Pumps

All modern liquid-cooled engines have a pump to circulate the coolant. But there was a time when the water pump was something that only the more expensive cars had. The 1909 Model T Ford had a water pump, but Ford deleted it in the interest of simplicity and low cost. In addition to the Model T, many automotive engines had cooling systems operated by the thermo-syphon principle. This means, simply, that hot water rises and cold water sinks. As the water in the block was heated, it rose to the top of the radiator. Once in the radiator, it started to cool, and slowly sank. This heating and cooling created enough circulation to keep the engine from overheating, as long as the weather was cool. In hot weather, or when pulling hard up a hill, the engines would often overheat. Because overheating was common, motorists accepted it as just one more hazard of motoring, like flat tires or getting stuck in the mud. By the late 1920s car buyers were beginning to expect more sophistication and trouble-free operation, even in low-price automobiles. In 1928, Ford brought out its first all-new model in almost 20 years, the Model A. It had—and kept—a water pump.

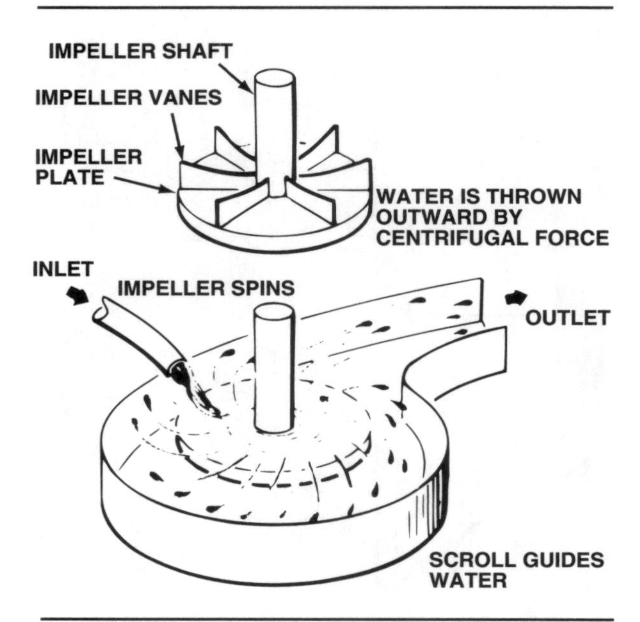

Figure 4-17. The water pump impeller is driven by the engine. Its spinning blades apply centrifugal force to the engine coolant.

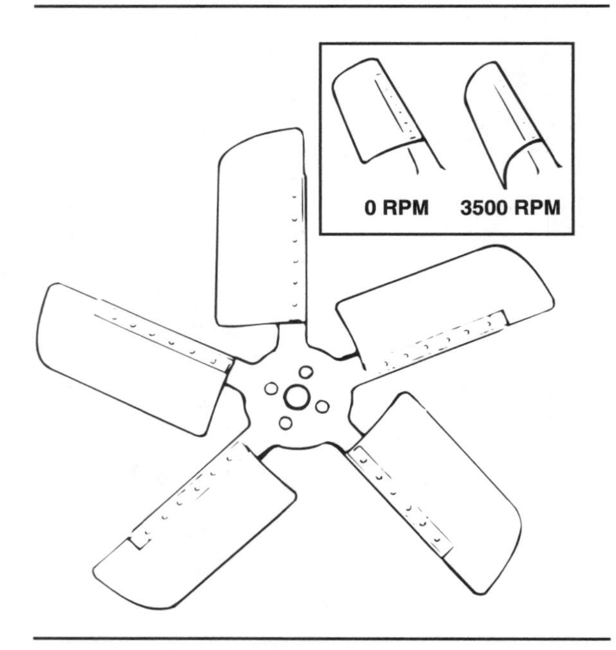

Figure 4-19. The flexible blades of this fan change shape as engine speed changes.

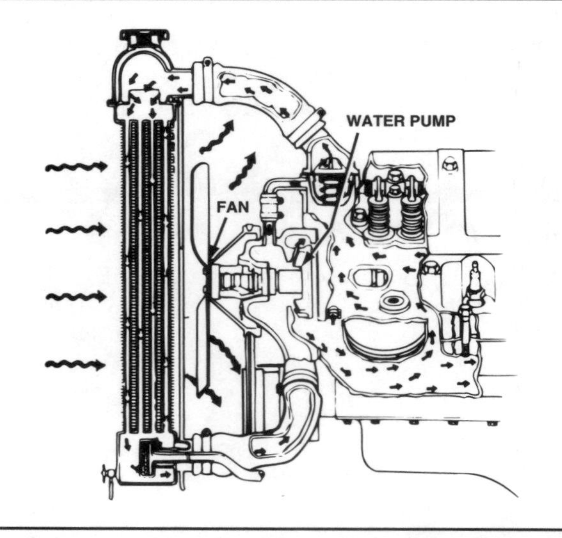

Figure 4-18. The engine-driven fan blows air into the engine compartment through the radiator core.

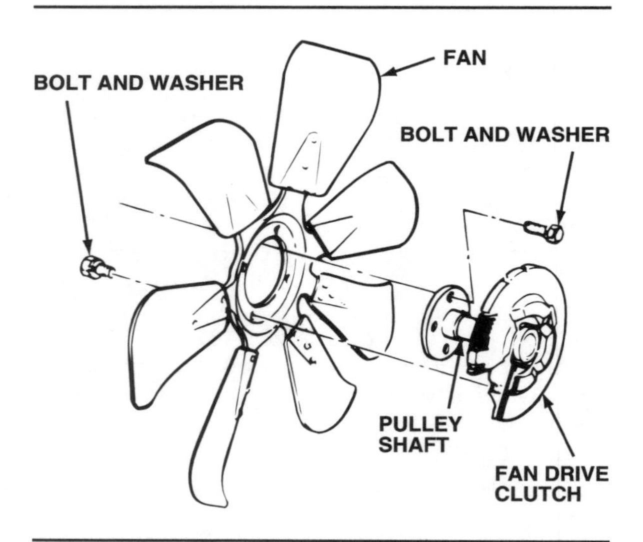

Figure 4-20. This Ford fan has a fluid clutch controlling the speed of rotation.

of air passing through the radiator, a fan is mounted in the engine compartment, figure 4-18. The fan is often mounted on the same shaft as the water pump impeller. The fan blades pull extra air through the radiator. Many late-model cars have shrouds around the fan to increase the fan's cooling efficiency by channeling all airflow through the radiator.

If engine power drives the fan, this power is wasted at high speeds when natural airflow will do the job. There are three ways to avoid this waste:

- Flexible fan blades
- Clutch fans
- Electric fans.

Flexible fan blades change their angle, or pitch, with changing engine speeds, figure 4-19. At low speed, the blades are sharply angled; this uses a lot of engine power. At high speeds, centrifugal force flattens the fan blades. They pull less air

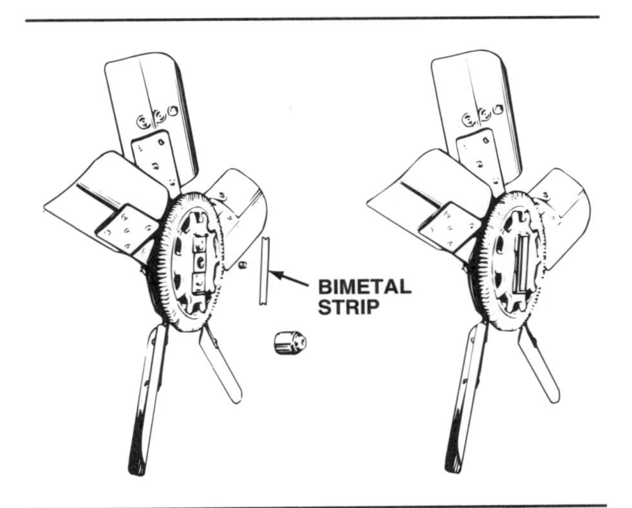

Figure 4-21. The bimetal temperature sensor spring controls the amount of silicone that is allowed into the drive and that, in turn, controls the speed of the fan.

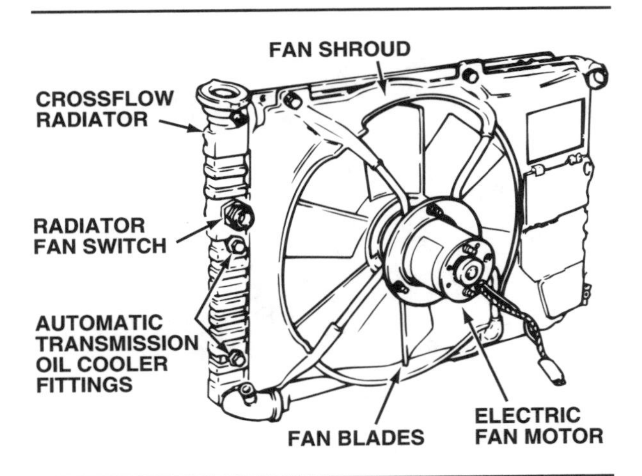

Figure 4-22. In this Chrysler car, an electric motor drives the radiator fan.

and require less power. This also reduces fan noise at high speeds.

Clutch fans are heavier than conventional fans, but the power saved makes their extra weight worthwhile. The drive, figure 4-20, is a clutch assembly that depends on a silicone fluid to transmit motion from the drive belt pulley to the fan itself. These are also called fluid drive fans. When there is a lot of fluid present, the drive pulley and fan rotate at the same speed. When there is little fluid present, the drive slips and the fan turns at less than drive pulley speed. When the drive is slipping, less engine power is used to drive the fan.

The amount of silicone fluid present is controlled by a **bimetal temperature sensor** spring exposed to the airflow from the radiator, figure 4-21. When the airflow is cool, the spring lets

Figure 4-23. If the water pump is located in a place where the fan cannot be installed, as is sometimes the case with transverse-mounted engines, then the fan is powered with electricity.

very little silicone into the drive, and the fan turns slowly or idles. When the airflow is hot, the spring allows a lot of silicone into the drive, and the fan turns faster.

Electric fans are not driven by the engine but by an electric motor, figure 4-22. The motor is switched on and off by a temperature-sensitive switch that senses engine coolant temperature. Many later-model automobiles and light trucks use electric fans; in fact, all later-model cars with transverse-mounted engines have electric fans, figure 4-23. When you work on a car with an electric fan, disconnect the battery ground cable or the fan power lead before getting near the fan, figure 4-24. The fan can switch on, whether or not the engine is running, on many cars.

Thermostats

The thermostat regulates coolant flow between the engine and the radiator. When the coolant is cold, the thermostat is completely closed. As the coolant warms, the thermostat begins to open. The thermostat's position varies during normal operation, but it rarely opens completely, unless the engine is working very hard or the weather is hot.

Bimetal Temperature Sensor: A sensor or switch that reacts to changes in temperature. It is made of two strips of metal welded together that expand differently when heated or cooled, causing the strip to bend.

Figure 4-24. Always disconnect the fan power lead before working near an electric fan.

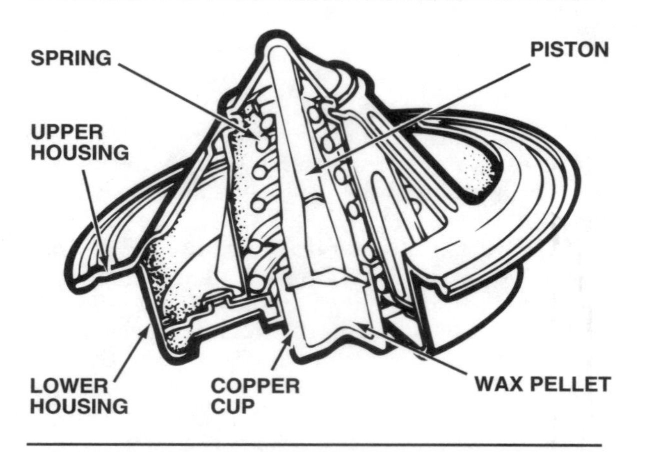

Figure 4-25. A cross section of a typical, wax-actuated thermostat made by Stant, showing the position of the wax pellet and the spring.

All late-model thermostats are wax-actuated, figure 4-25. A chamber in the thermostat is filled with wax that is solid when cold, but melts and expands when heated. When the wax is cold, a spring holds the thermostat closed. As the coolant temperature increases, the wax expands and forces the thermostat to open against the spring tension. This type of thermostat will work reliably under normal system pressure. Some early thermostats were an **aneroid bellows** design, figure 4-26. These thermostats only work reliably in nonpressurized cooling systems.

The thermostat's opening temperature is stamped on the outside. Late-model thermostats usually open between 180° and 190°F (82° and 88°C), higher than thermostats used on older vehicles, to meet stricter emissions requirements. Replacement thermostats must have the correct opening temperature for the system to

Figure 4-26. These bellows-actuated thermostats are no longer used in vehicles because coolant pressure affects bellows operation.

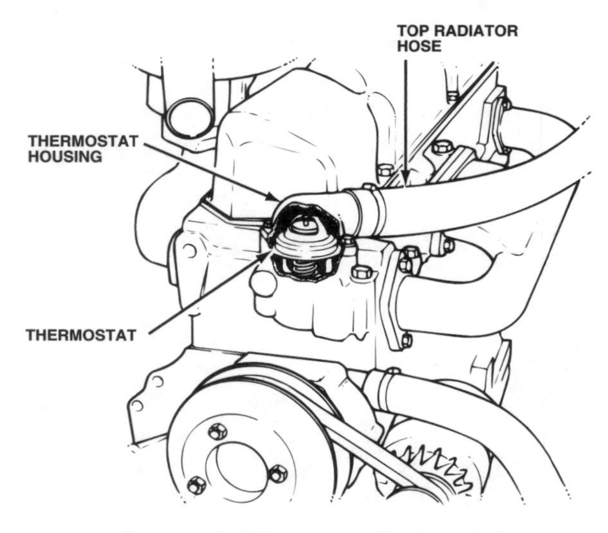

Figure 4-27. A typical thermostat location at the front of the cylinder head.

work properly. The thermostat is mountedin a metal housing at the top front of the engine, figure 4-27.

Radiator Pressure Cap

To improve the performance of the entire cooling system, it is sealed and allowed to build pressure. A very small part of the pressure comes from the water pump. The rest of the pressure comes from coolant expanding as it warms.

The radiator cap must keep the system pressurized, vent any excess pressure, and allow atmospheric pressure to reenter the system as the coolant cools and contracts after the engine is shut off.

To do this, radiator caps have two valves, figure 4-28. A pressure valve relieves the excess pressure inside the system during engine operation. A vacuum valve allows outside air to enter

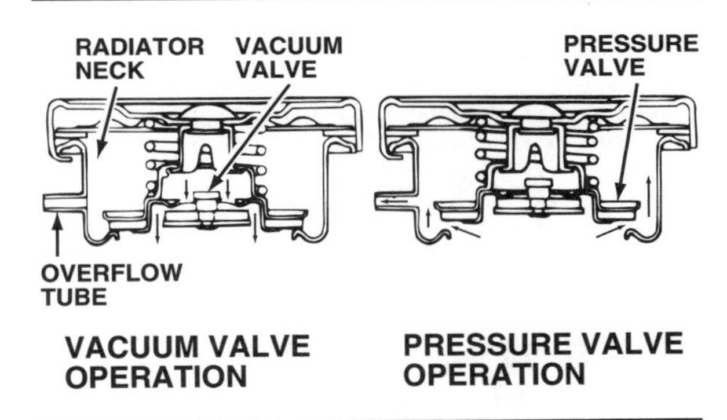

Figure 4-28. A crosssection of a radiator pressure cap used in a recovery system.

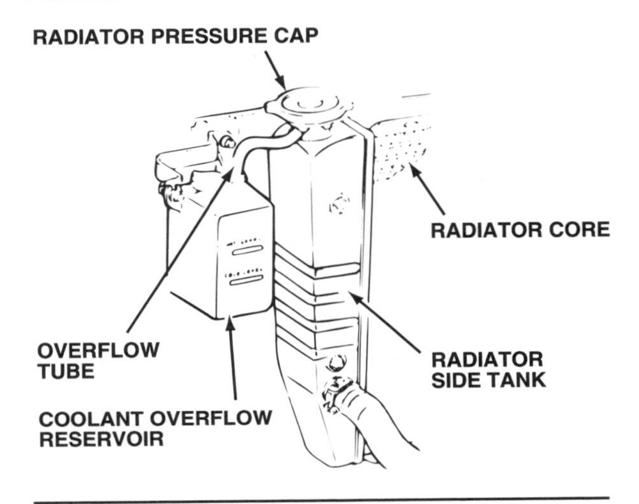

Figure 4-29. Most radiator caps are mounted on one of the radiator tanks, similar to this Ford system.

the system when the engine is off. These pressure caps usually are on the top or side tank, figure 4-29, but occasionally can be found in unusual locations, such as on a special hose section, figure 4-30.

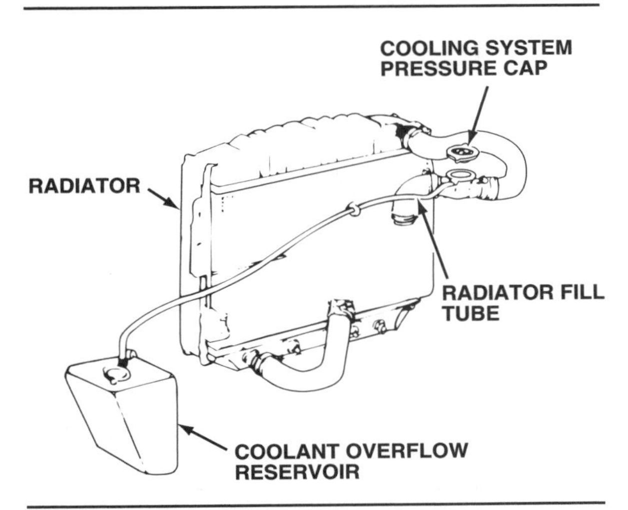

Figure 4-30. Some radiator caps are mounted away from the radiator to shorten the radiator profile.

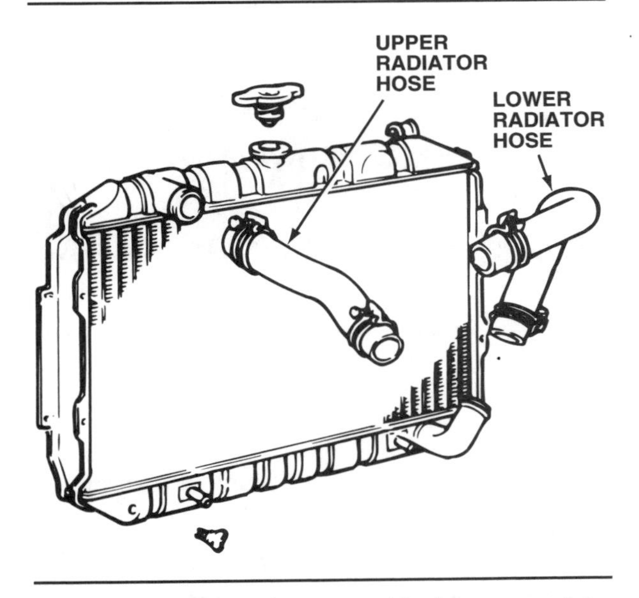

Figure 4-31. Radiator hoses must be large enough to carry the coolant when the thermostat is fully open.

Radiator Hoses and Drive Belts

Radiator hoses connect the cooling system passages to the radiator tanks, figure 4-31. The hoses are flexible and will absorb the motion between the vibrating engine and the stationary radiator. The hoses are made of synthetic rubber, and are often reinforced with a wire coil. The hoses can be molded into a specific shape or they can be ribbed, flexible lengths that adapt to different installations.

The hose ends fit over necks on the engine and radiator, and are held in place with hose clamps, figure 4-32. Hose clamps can be held in place by spring tension or by a screw.

For many years, the drive belt that operated the radiator fan and water pump was a reinforced rubber V-belt, figure 4-33. It fit tightly over the crankshaft pulley, the fan pulley, and usually, the alternator pulley. Belt tension was usually adjusted by moving the alternator in its mounting.

Aneroid Bellows: An accordion-shaped bellows that responds to changes in coolant pressure or atmospheric pressure by expanding or contracting.

Figure 4-32. Hose clamps come in different sizes and designs for different applications.

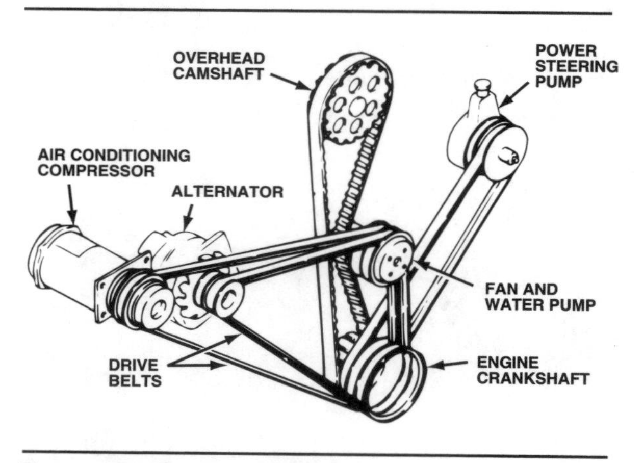

Figure 4-33. On V-belt drive systems, most accessories have separate drive belts.

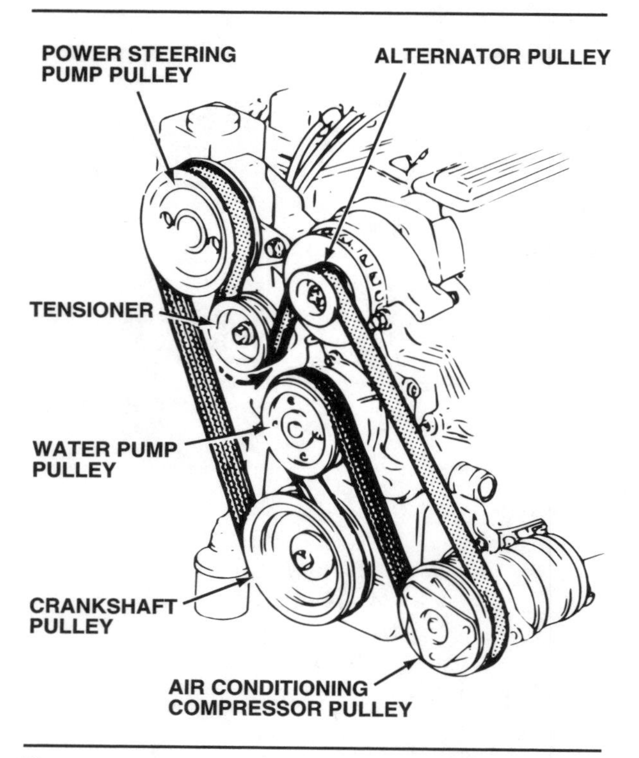

Figure 4-34. A serpentine drive belt system uses a single belt to drive all accessories.

Today, most new domestic and many new imported automobiles use a single serpentine belt to drive all accessories, figure 4-34. The belt is made of reinforced rubber, and has several small V-shaped grooves that fit corresponding grooves in the accessory drive pulleys. On some serpentine belt systems, belt tension is adjustable. Other serpentine systems have a spring-loaded automatic tensioner assembly that maintains spring tension at all times.

Coolant Recovery System

All late-model cars have a closed cooling system. Instead of a pressure cap that vents to the atmosphere, the pressure cap is connected to a coolant tank or overflow reservoir, figure 4-35. When system temperature increases and the coolant expands, the extra coolant flows out of the radiator neck, through an overflow hose, and into a coolant overflow tank. When the system cools and a vacuum develops in the system, coolant is drawn back into the radiator through the same hose. On earlier systems, expanding coolant went out the overflow hose and onto the ground. This coolant loss made frequent coolant top-ups necessary, and the presence of air in the cooling system increased rust and corrosion.

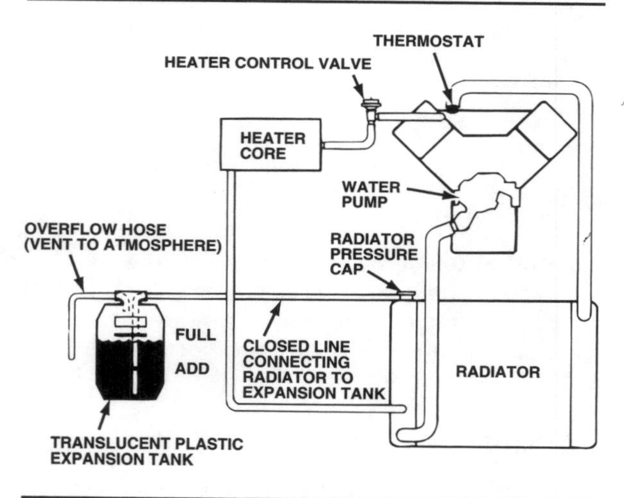

Figure 4-35. The level in the coolant recovery system rises and lowers with engine temperature.

A coolant recovery system is even more necessary in late-model cars, which tend to run hotter because of emission control equipment, automatic transmissions, air conditioning, and

Figure 4-36. When adding coolant to the recovery system reservoir, be sure to take note of the "hot" and "cold" fill marks and fill accordingly.

other power-consuming accessories. The coolant tank ensures that there will always be enough coolant to fill the system because none is lost if the radiator overflows. Coolant recovery systems are said to increase cooling system efficiency about 10 percent.

When more coolant must be added to a recovery system, it is added to the overflow tank, not directly to the radiator. The tank is marked with "hot" and "cold" fill levels, figure 4-36.

ENGINE TEMPERATURE EFFECTS ON PERFORMANCE, ECONOMY, AND EMISSIONS

When the engine is too cold, it suffers poor fuel vaporization. The liquid fuel entering the cylinders tends to wash oil off the cylinder walls, reducing lubrication. The liquid fuel runs past the piston rings and into the crankcase, where it dilutes the oil and further hinders lubrication. A cold engine also does not vaporize moisture that condenses inside the crankcase. This moisture mixes with the oil to form sludge, which increases engine friction, robs it of power, and increases mechanical wear.

Emission control systems monitor coolant temperature to determine when to rely on the exhaust oxygen sensor to determine the air-fuel mixture. If the coolant does not reach the proper temperature, the emission control system will supply the engine with a richer-than-normal air-fuel mixture, causing high levels of HC and carbon monoxide emissions. Later chapters, discussing open- and closed-loop operation, explain this relationship in greater detail.

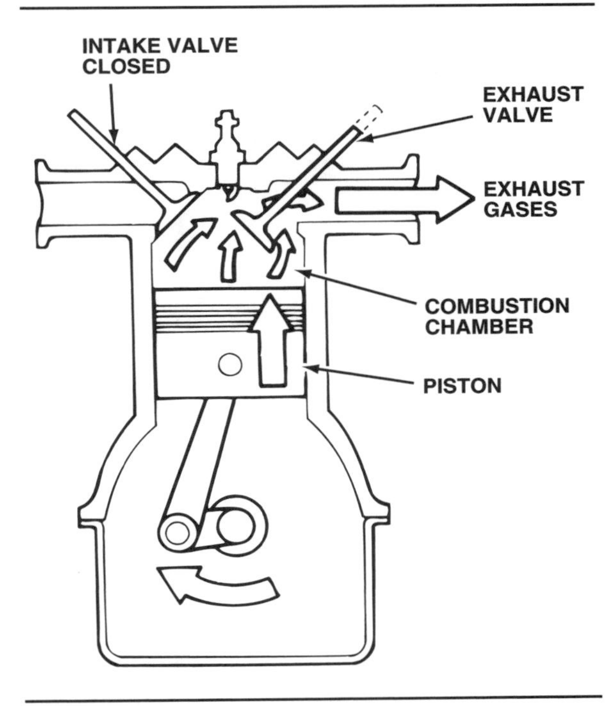

Figure 4-37. Burned gases are pushed out of the combustion chamber by the piston's upward exhaust stroke.

An engine that is too hot forms carbon and varnish deposits that affect engine operation. Excessive heat also thins the engine oil to the point that it no longer lubricates well, increasing mechanical wear. The thin oil will also be drawn into the combustion chamber in greater amounts, causing carbon buildup, raising the compression ratio, and increasing the chance of detonation.

THE EXHAUST SYSTEM

After the air-fuel mixture is burned and escapes through the exhaust valve, it must continue its exit away from the car. The exhaust system routes engine exhaust gases to the rear of the car, quiets the exhaust noise, and, since the mid-1970s, also helps reduce pollutants in the exhaust.

Backpressure

The exhaust system's design has a marked effect on engine performance. Proper extraction of exhaust gases through the exhaust manifold design is as important as air-fuel flow through the intake manifold. The flow of exhaust gases from the engine should be as smooth as possible. As the piston moves upward during the exhaust stroke, it forces combustion gas through the open exhaust valve and out of the cylinder, figure 4-37. The exhaust gases are under pressure, a force that helps expel the gases from the com-

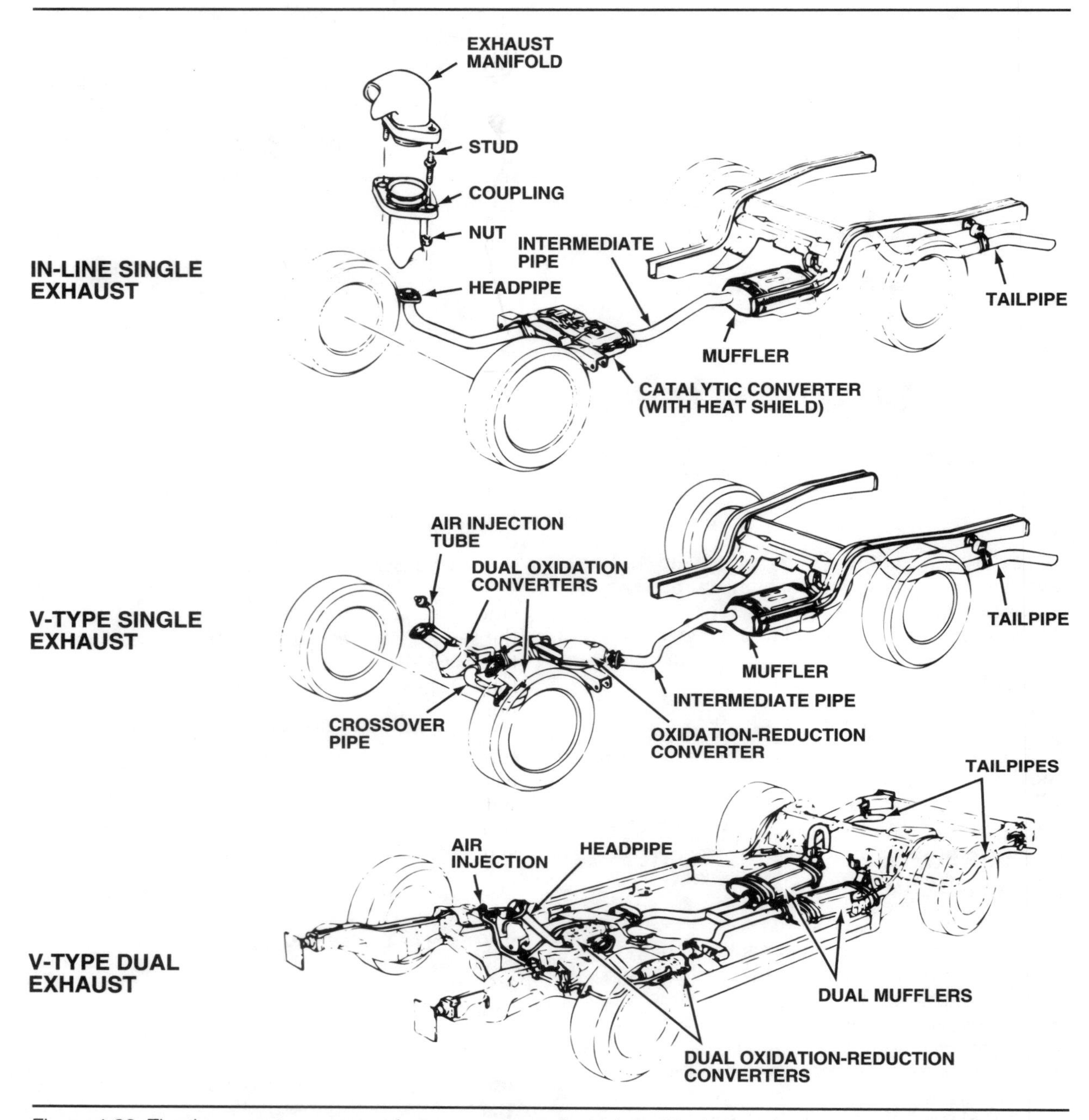

Figure 4-38. The three most common exhaust system configurations. The in-line single exhaust system is used with an in-line engine. The V-type single exhaust system uses a Y-pipe to connect the two cylinder banks to a single exhaust pipe. The V-type dual exhaust is essentially two separate exhaust systems for a V-type engine.

bustion chamber past the exhaust valve, and into the exhaust manifold.

Whenever gas is pushed through a passageway, turbulence and friction along the sides of the passage cause a resistance called **backpressure**. A piston encounters backpressure each time it comes up on the exhaust stroke. Too much backpressure will not allow all the exhaust to leave the combustion chamber before the intake stroke starts. This preheats and leans the incoming air-fuel mixture, reducing efficiency and power. It can also cause engine mechanical failures, such as burned valves.

Backpressure in the exhaust manifold and exhaust system can also contaminate the intake manifold's fresh air-fuel mixture. A brief **camshaft overlap** period, combined with backpressure in

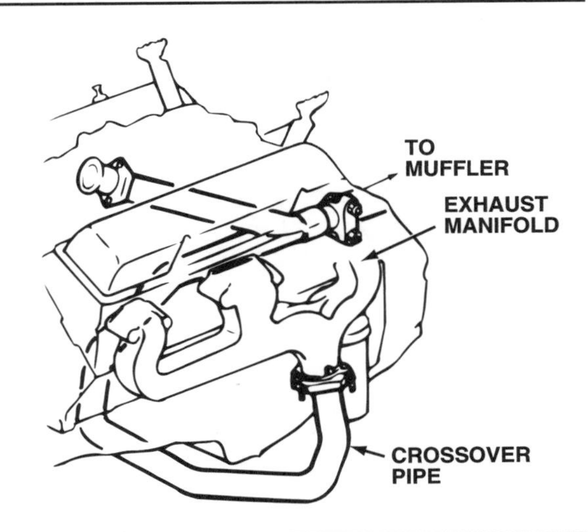

Figure 4-39. This V-type engine has a crossover pipe connecting the two banks to a single exhaust.

an exhaust manifold, can cause one cylinder's intake stroke to draw exhaust gases from a nearby cylinder. A direct and unrestricted flow of exhaust gas causes less backpressure and prevents this unnecessary power loss or engine damage.

Exhaust System Arrangement

The exhaust system is usually arranged to suit the engine design. There are three basic exhaust system configurations, figure 4-38. Since an in-line engine usually has a single exhaust manifold, a single exhaust system generally is used. The exhaust pipe connects directly to the exhaust manifold flange. Each cylinder bank on a V-type engine, however, has its own exhaust manifold. When a single exhaust system is used, the two manifolds will connect with a Y-pipe or a crossover pipe, figure 4-39. With dual exhaust systems, each manifold connects to its own exhaust pipe, muffler, and tailpipe.

Dual exhaust systems reduce exhaust backpressure by splitting the exhaust gas flow into two outlet lines. Since the exhaust manifold is at the beginning of the flow, manifold design is the most important factor in reducing backpressure. However, good muffler design also can help to minimize restrictions.

Exhaust System Components

Major parts of the exhaust system include:

- Exhaust manifolds
- Pipes
- Mufflers
- Resonators
- Catalytic converters.

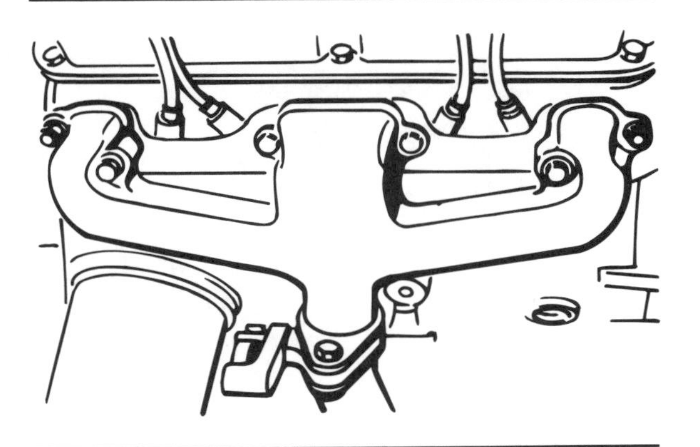

Figure 4-40. Cast-iron exhaust manifolds are quiet and durable, but not the most efficient design for producing power.

Exhaust manifolds

Exhaust manifolds attach directly to the side of the engine cylinder head, matching their ports to the exhaust ports on the cylinder head, figure 4-40. Exhaust manifolds are the first pieces outside of the engine that absorb the intense heat of the exhaust, so manifolds are usually made of thick and sturdy cast iron. The thick cast iron is long lasting and helps to damp out combustion noise.

Some exhaust manifolds are fabricated from tubular steel, which is lighter than iron, and easier to assemble into an efficient manifold with low backpressure. For greater corrosion resistance, other exhaust manifolds are made of stainless steel.

High-performance exhaust manifolds

Although the specialized science of exhaust tuning was once the province of the hot rodder and high-performance vehicles, the problem of exhaust backpressure is an important factor in controlling the emissions of late-model engines. Automotive engineers design backpressure-

Backpressure: The resistance, caused by turbulence and friction, that is created when a gas or liquid is forced through a restrictive passage.

Camshaft Overlap: The period, measured in degrees of crankshaft rotation, during which both the intake and exhaust valves are open. It occurs at the end of the exhaust stroke and the beginning of the intake stroke.

Bifurcated: Separated into two parts. A bifurcated exhaust manifold has four primary runners that converge into two secondary runners; these converge into a single outlet into the exhaust system.

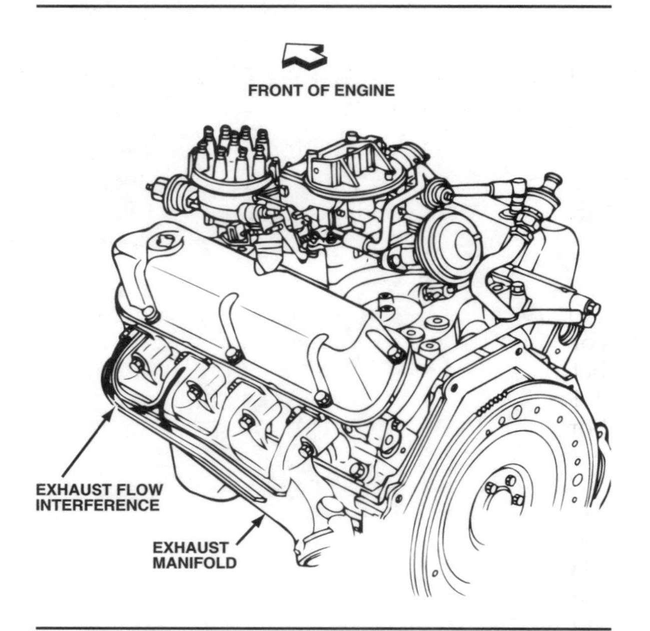

Figure 4-41. Exhaust gas interference from adjacent cylinders increases backpressure.

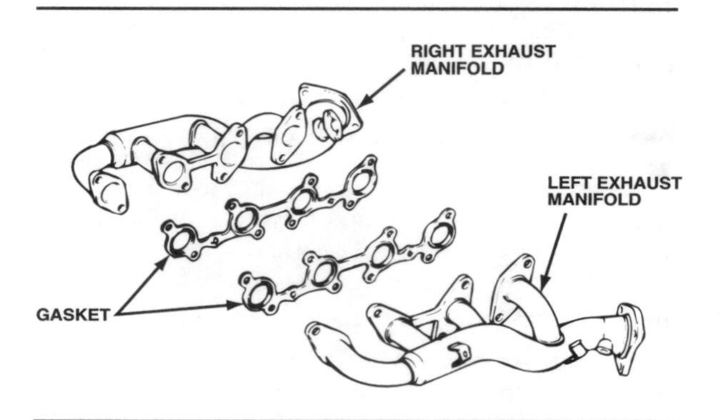

Figure 4-42. Streamlining the exhaust manifold is the first step in producing high-performance exhausts that reduce backpressure.

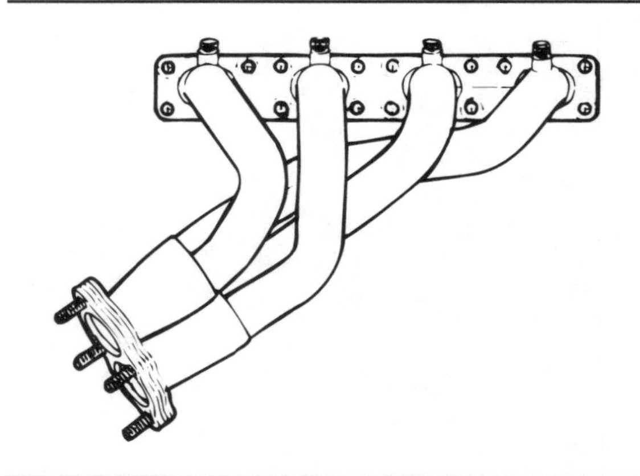

Figure 4-43. The 1991 BMW 318 engine uses an example of the most efficient type of exhaust manifold. The system is streamlined, smoothing exhaust flow. Each cylinder has its own pipe, eliminating exhaust gas interference from other cylinders.

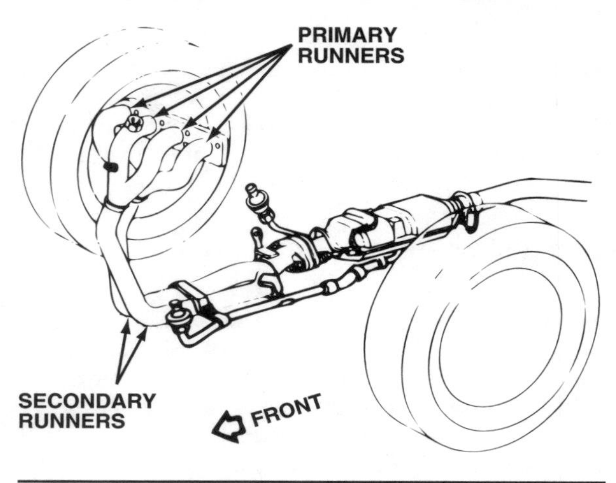

Figure 4-44. Ford's bifurcated exhaust manifold is essentially a set of stainless-steel headers.

reducing exhaust systems that can affect the volumetric efficiency of an engine, and yet still leave enough backpressure to allow backpressure exhaust gas recirculation (EGR) valves, on applicable engines, to properly function. The EGR system reduces NO_x and prevents detonation.

Sharp turns and narrow passages in an exhaust manifold will slow the flow of gases from the exhaust ports, and increase the amount of backpressure in the system. Cast-iron exhaust manifolds also increase backpressure because they release exhaust gases from each cylinder into a common chamber. The pressure of gases from one cylinder interferes with the flow from other cylinders, figure 4-41. Many original equipment manifold designs have duplicated high-performance manifolding by using the swept-back manifold design, figures 4-42 and 4-43.

Some Ford 4-cylinder engines use a **bifurcated** stainless-steel exhaust manifold, figure 4-44. The low-restriction, tuned tubular header design consists of four primary runners. These form into two secondary runners and converge into a single outlet that connects to the exhaust system. By routing pressure pulses from adjacent firing cylinders through different pipes, backpressure is reduced by approximately 30 percent.

On in-line 6-cylinder engines, Ford uses two separate cast iron exhaust manifolds, one for the three front cylinders and the other for the three rear cylinders. This design solves two former problems: The single manifold previously used did not seal well against the head and also tended to fail prematurely due to cracking. Like the bifurcated manifold design, the two manifolds connect to a single outlet that leads into the catalytic converter and muffler.

Exhaust Pipes

In the interest of reducing backpressure, exhaust pipes should be as straight as possible, without sharp turns and restrictions. Pipes must also be able to withstand the constant presence of hot, corrosive exhaust gases, salted roads during winter, and undercar hazards such as rocks.

To improve their strength, the exhaust pipes on most late-model cars are formed with an inner and outer skin. Occasionally, the inner skin will collapse and form a restriction in the system. From the outside, the exhaust pipe will look normal even when the inside is partially or almost completely blocked. Close inspection is necessary to locate such a defect.

To further improve the exhaust pipe's resistance to corrosion, most exhaust pipes are now made of stainless steel. However, stainless-steel pipes do not last forever. They are more brittle and prone to fatigue breakage than ordinary mild steel. Exhaust pipe inspection and replacement will continue to be important jobs.

Mufflers and Resonators

The muffler is an enclosed chamber that contains baffles, small chambers, and pipes to direct exhaust gas flow. The gas route through the muffler is full of twists and turns, figure 4-45. This quiets the exhaust flow but also creates backpressure. For this reason mufflers must be carefully matched to the engine and exhaust system.

Some mufflers consist of a straight-through perforated pipe surrounded by sound-deadening material, usually fiberglass. These "glass-pack" mufflers, as they are often called, reduce backpressure but are not nearly as quiet as conventional mufflers.

Some engines also add resonators to the system. These are small mufflers specially designed to "fine tune" the exhaust and give it a pleasant, quiet, "resonant" tone.

Catalytic Converters

In order to meet tightening exhaust emission standards, automakers turned to the catalytic converter in 1975. These were oxidation converters that changed HC and CO to harmless CO_2 and H_2O. Reduction converters first appeared on some 1978 cars, and also helped reduce NO_X emissions.

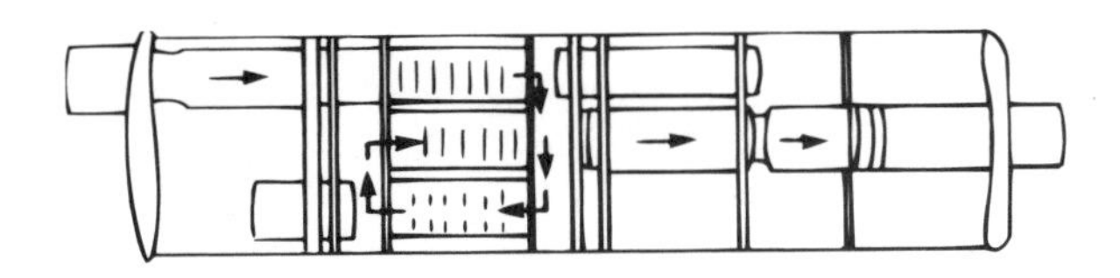

Figure 4-45. Exhaust gases must twist and turn to travel through this muffler.

In the simplest arrangement, one catalytic converter is installed in the exhaust system between the manifold and the muffler, figure 4-46. Many cars, however, use two converters. Some cars have converters bolted directly to the exhaust manifold.

Catalytic converters are simple. The catalyst inside the converter combines with the exhaust gases, and causes a chemical reaction to take place. This amounts to a "realignment" of molecules that results in less emissions to the atmosphere. Catalytic converters have no moving parts and never need adjusting. Most converters have a guaranteed life span of at least 50,000 miles (80,467 km), if not 70,000 miles (112,654 km), and may outlast the car. The pellets in early GM converters can be removed and replaced by

■ Electronic Mufflers?

Just when you thought electronics had penetrated every aspect of automotive technology that it possibly could, along came electronic mufflers. Some manufacturers actually have prototype electronic mufflers in the testing stages.

Although it sounds like a joke, the concept is sound and actually pretty simple. Electronic mufflers use sensors and microphones to pick up the pressure waves of sound emitted by the exhaust pipe. A computer analyzes the sound and produces a mirror-image pattern of sound pulses, instantly sending them through a set of speakers mounted near the exhaust outlet. These computer-produced sounds are 180 degrees out of phase with the engine-produced sounds. The sounds from the speakers could be termed "antinoise waves", and they help cancel out the total amount of noise. Since the system is computer controlled, it can be tailored to work over any rpm range or load condition.

With the better sound control that this system affords, the mufflers themselves can be less restrictive, resulting in greater power and fuel economy. With a little tuning of this system, muffler engineers may even be able to give a car a more pleasant exhaust note.

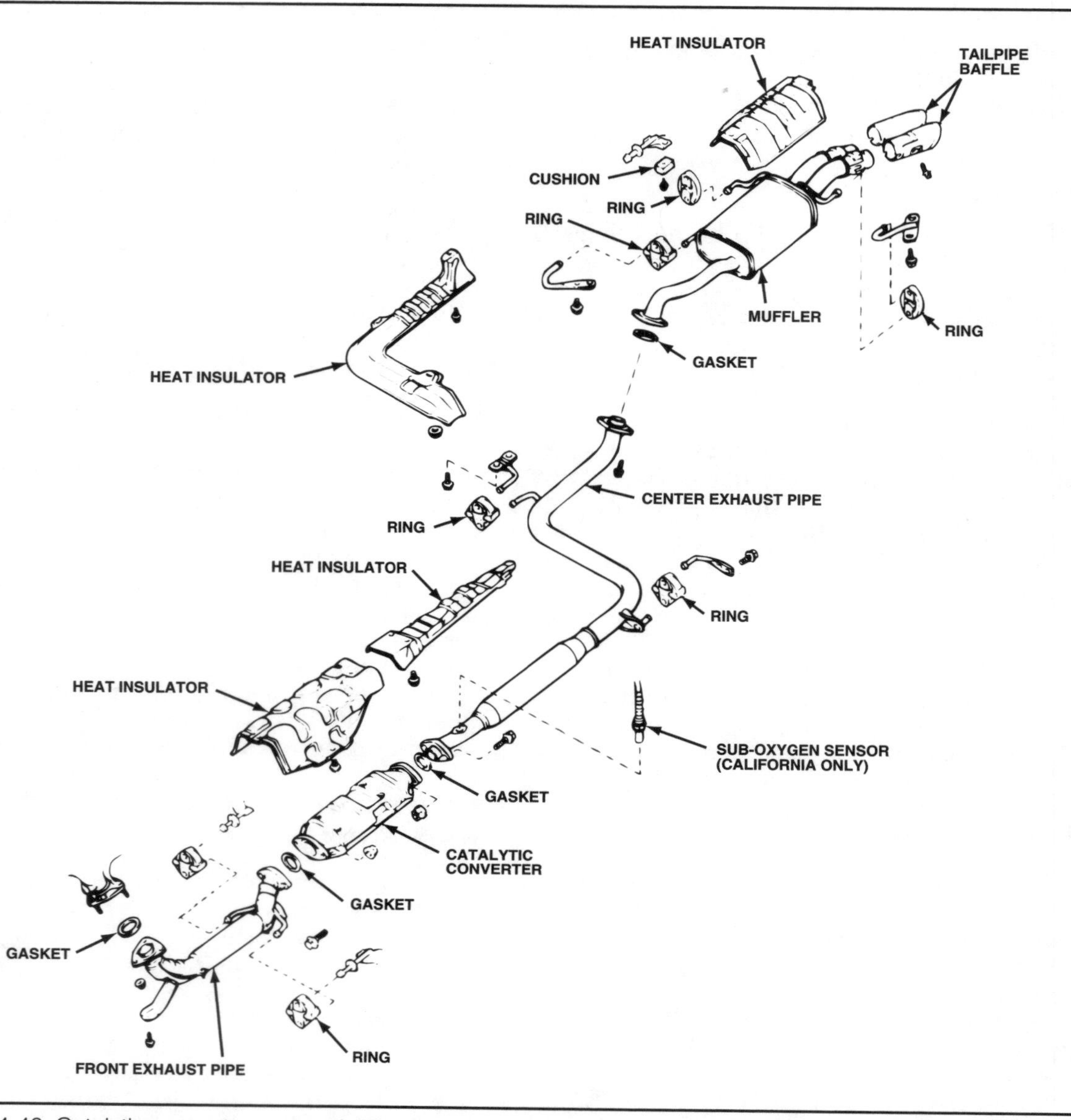

Figure 4-46. Catalytic converters are typically placed in the exhaust system between the exhaust manifold and muffler.

removing a plug in the bottom of the unit. Tetraethyl lead, the gasoline octane booster described in an earlier chapter, will coat or "poison" the catalyst and reduce its efficiency. This may make converter replacement necessary.

The link between a converter and a well-tuned engine is important. An engine that is misfiring or improperly tuned can also destroy a catalytic converter because the converter cannot accept exhaust temperatures above 1,500°F (815°C), nor will it work correctly if the air-fuel mixture is too rich. Two spark plugs misfiring in succession for a prolonged time will raise the temperature in the converter and shorten its life.

SUMMARY

The cooling and exhaust systems must remove about one-third of the heat produced by a car engine. Some car engines use a flow of air over their outer surfaces to reduce engine temperature. Most cooling systems use a liquid coolant to remove this heat. It circulates through the engine block where it picks up the heat, then goes through the radiator where the outside air flowing past removes heat. Most coolants consist of a solution of water and ethylene glycol, which permits a greater range of engine operating temperatures.

The coolant circulates through the system by a water pump. As the coolant heats, it expands, often overflowing the radiator. Newer systems recover this excess coolant in a separate reservoir, where it can be reused as the engine cools. An electric- or engine-driven fan cools the radiator by drawing air through it at low speeds and idle. A pressure cap keeps the system pressurized, and a thermostat regulates the flow of coolant by sensing engine temperature. Engine operating temperatures that are either too hot or too cold will produce greater mechanical engine wear.

Exhaust systems carry the exhaust away from the combustion chambers through a series of pipes, mufflers, resonators, and catalytic converters. The system must be as straight as possible to reduce backpressure in the cylinders, and generally must dampen engine exhaust noise.

The exhaust system is arranged to suit the engine design. The three major types are in-line, single V-type, and dual V-type. When an engine's cylinders are in-line all of the exhaust valves are on the same side of the engine. An exhaust system pipe connects to this side of the engine at the exhaust manifold. A V-type engine has exhaust valves on both cylinder banks, so it requires two separate exhaust systems, one for each bank. Regardless of exhaust pipe arrangement, the pipes connect to one or more units that may include a muffler and a catalytic converter.

Review Questions

Choose the single most correct answer.
Compare your answers to the correct answers on page 507.

1. Which of the following terms describes a type of radiator?
   a. Downflow
   b. Backflow
   c. Thruflow
   d. Acrossflow

2. The automobile engine converts about ____ of its total heat energy into mechanical energy.
   a. one-quarter
   b. one-third
   c. one-half
   d. two-thirds

3. All of the following conserve power at high engine speeds except:
   a. Flexible blade fans
   b. Multiblade fans
   c. Clutch fans
   d. Electric fans

4. Which mixture of ethylene glycol coolant and water gives protection down to the lowest temperatures?
   a. 100 percent ethylene glycol
   b. 73 percent ethylene glycol and 27 percent water
   c. 68 percent ethylene glycol and 32 percent water
   d. 50 percent ethylene glycol and 50 percent water

5. Core plugs are installed in the engine block to:
   a. Relieve pressure if the coolant should freeze
   b. Allow the block to be flushed during engine overhaul
   c. Provide inspection points for internal block condition
   d. None of the above

6. An engine that is too hot will have:
   a. Poor lubrication
   b. Poor cooling system circulation
   c. High oxides of carbon
   d. High thermal inefficiency

7. A 50 percent water and a 50 percent ethylene glycol antifreeze mixture in a cooling system with a 14-psi (97-kPa) pressure cap will boil at:
   a. 248°F (120°C)
   b. 255°F (124°C)
   c. 263°F (128°C)
   d. 270°F (132°C)

8. The water pump uses ____ to circulate the coolant.
   a. Positive pressure
   b. Gravitational force
   c. Centripetal force
   d. Centrifugal force

9. The greater the difference in temperature between the coolant and the air, the:
   a. Hotter the engine will run
   b. Greater the cooling system pressure
   c. More heat will be absorbed by the air
   d. Faster the coolant will circulate through the system

10. A pressurized cooling system will:
    a. Reduce engine temperature
    b. Increase coolant warmup rate
    c. Reduce the coolant boiling point
    d. Increase the cooling system boiling point

11. If the cooling system makes the engine run too cool, the engine will:
    a. Make more torque
    b. Have less efficiency
    c. Preignite
    d. Produce higher HC emissions

12. A disadvantage of ethylene glycol when used as a coolant is that:
    a. Its freezing point is much lower than water
    b. It does not transfer heat as well as water
    c. Its boiling point is much higher than water
    d. None of the above

13. The thermostat regulates coolant flow between the engine and the:
    a. Cylinder head
    b. Heater
    c. Radiator
    d. Transmission cooler

14. Which of the following is not a part of the water pump?
    a. Pulley
    b. Impeller
    c. Stator
    d. Scroll

15. For every one psi increase in pressure, the coolant boiling point is raised about:
    a. 3°F (1.7°C)
    b. 2°F (1.1°C)
    c. 1°F (0.6°C)
    d. 4°F (2.2°C)

16. Technician A says that thermostats have their opening temperature stamped on the outside.
    Technician B says that modern thermostats are wax actuated.
    Who is right?
    a. A only
    b. B only
    c. Both A and B
    d. Neither A nor B

17. Technician A says an engine that runs too cool will have poor lubrication.
    Technician B says an engine that runs too warm will have poor lubrication.
    Who is right?
    a. A only
    b. B only
    c. Both A and B
    d. Neither A nor B

18. Technician A says that water jackets are cast into the block and head of liquid cooled engines.
    Technician B says the internal size of water jackets makes no difference to coolant distribution.
    Who is right?
    a. A only
    b. B only
    c. Both A and B
    d. Neither A nor B

19. Technician A say that during the complete four-stroke cycle, the average combustion chamber temperature is 750°F (400°C).
    Technician B says that the average temperature is about 1,500°F (800°C).
    Who is right?
    a. A only
    b. B only
    c. Both A and B
    d. Neither A nor B

20. Excessive exhaust backpressure:
    a. Causes power loss
    b. Can contaminate air-fuel mixture in the intake manifold
    c. Can be minimized by exhaust manifold streamlining
    d. All of the above

21. High-performance exhaust manifolds:
    a. Increase horsepower by 30 percent
    b. Cause engines to operate at higher temperatures
    c. Reduce backpressure
    d. All of the above

22. A major type of exhaust system is the:
    a. In-line
    b. Single V
    c. Dual V
    d. All of the above

23. Oxidation catalytic converters help change:
    a. HC and CO to H_2O and CO
    b. HC and CO to CO_2 and H_2O
    c. HC and CO to HC_2O
    d. HC and CO to H_2O and C

Chapter

5

Introduction to Emission Controls

The combustion process in automotive engines produces harmful by-products that are discharged from the engine and become air pollutants. Emission control systems are necessary to minimize the formation and discharge of these pollutants.

When emission control requirements were first introduced, automakers and car owners were able to comply by installing add-on or "hang-on" devices that were not an integral part of engine and vehicle design. As regulations became more strict, automakers had to include emission controls in basic engine design.

The first emission control regulation was adopted in California in 1961. Today, almost four decades later, emission control regulations are still being tightened and new control systems developed. Sophisticated computer-controlled systems appear on most cars, and emission control requirements are important considerations in the design and operation of all parts of the fuel system. The ignition system, which provides the spark for combustion, plays an equally important role in emission control.

How did these great changes in automotive emission controls come about? What exactly *is* air pollution, and how does the automobile contribute to it? This chapter examines air pollution and automotive emissions, including the legislation controlling emissions and the ways in which automakers have met the regulations.

AIR POLLUTION—A PERSPECTIVE

We can define air pollution as the introduction of contamination into the atmosphere in an amount large enough to injure human, animal, or plant life. There are many types and causes of air pollution, but they all fall into two general groups: natural and man-made. Natural pollution is caused by such things as the organic plant life cycle, forest fires, volcanic eruptions, and dust storms. While pollution from such sources is often beyond our control, we *can* control man-made pollution from industrial plants and automobiles.

Most urban and large industrial areas around the world suffer periodic air pollution. During the late 1940s, a unique form of air pollution was identified in the Los Angeles area. When certain pollutants are exposed to sunlight, irritating chemical compounds called **photochemical smog** form. As this phenomenon increased both in intensity and frequency, it posed more of a problem. California took the lead in combating it by becoming the first state to place controls on motor vehicle emissions, figure 5-1. As smog gradually began to appear in other parts of the country, the federal government moved into the

Figure 5-1. During the 1960s, high levels of airborne pollutants in the Los Angeles basin prompted the state of California to enact emission control regulations.

area of regulation. To understand why, we must look at the automobile-produced elements that form air pollution and smog.

MAJOR POLLUTANTS

An internal combustion engine emits three major gaseous pollutants into the air: **hydrocarbons (HC)**, **carbon monoxide (CO)**, and **oxides of nitrogen (NO_x)**, figure 5-2. In addition, an automobile engine gives off many small liquid or solid particles, such as lead, carbon, sulfur, and other **particulate matter (PM_{10})**, which con-contribute to pollution. By themselves, all these emissions are not smog, but simply air pollutants.

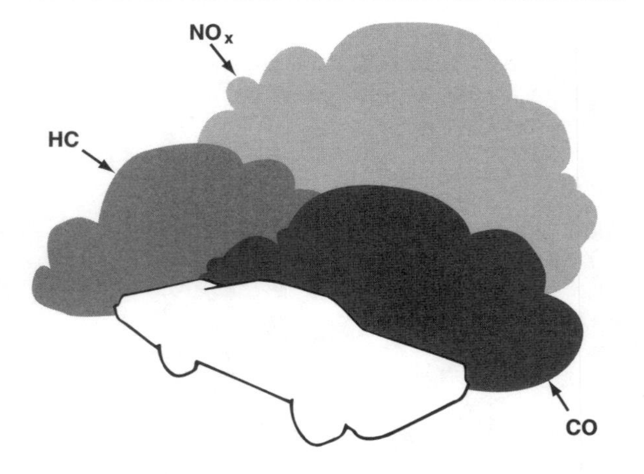

Figure 5-2. Hydrocarbons (HC), carbon monoxide (CO), and oxides of nitrogen (NO_x) are the three major automotive pollutants.

Hydrocarbons

Gasoline is a HC compound. Unburned HC's given off by an automobile are mostly unburned fuel. Over 200 different varieties of HC pollutants come from automotive sources. While most come from the fuel system and the engine exhaust, others are oil and gasoline fumes from the crankcase. Even a car's tires, paint, and upholstery emit tiny amounts of HC's. Figure 5-3 shows the three major sources of HC emissions from an automobile:

- Fuel system evaporation—20 percent
- Crankcase vapors—20 percent
- Engine exhaust—60 percent.

HC's are the only major automotive air pollutant

that come from sources *other* than engine exhaust. HC's molecules of all types are changed into other compounds by combustion. If an automobile engine burned gasoline completely, there would be no HC's in the exhaust, only water and carbon dioxide (CO_2). But when the vaporized and compressed air-fuel mixture is ignited, combustion occurs so rapidly that gasoline near the sides of the combustion chamber may not get burned. This unburned fuel then passes out with the exhaust gases. The problem is worse with engines that misfire or are not properly tuned.

Carbon Monoxide

Although not part of photochemical smog, CO is also found in automobile exhaust in large amounts. A deadly poison, CO is both odorless and colorless. CO is absorbed by the red corpuscles in the body, displacing the oxygen (O_2). In a small quantity, it causes headaches, vision difficulties, and delayed reaction times. In larger quantities, it causes vomiting, coma, and death.

Because it is a product of incomplete combustion, the amount of CO produced depends on the way in which HC's burn. When the air-fuel mixture burns, its HC's combine with O_2. If the air-fuel mixture contains too much fuel, there is not enough O_2 to complete this process, so CO forms. Using an air-fuel mixture with less fuel makes combustion more complete. The leaner mixture increases the ratio of O_2, which reduces the formation of CO by producing CO_2 instead.

CO_2, although not considered a pollutant affecting public health, does contribute to global warming. Currently, scientists are studying the relationship between CO_2 levels and global warming.

Oxides of Nitrogen

Air is made up of about 78 percent nitrogen, 21 percent O_2, and one percent other gases. When the combustion chamber temperature reaches 2,500°F (1370°C) or greater, the nitrogen and O_2 in the air-fuel mixture combine to form large quantities of NO_X. NO_X also is formed at lower temperatures, but in far smaller amounts. One component of NO_X, nitrogen dioxide, is extremely toxic to humans. In addition, NO_X combines with other elements in the air to form nitric acid, a contributor to acid rain. In the presence of sunlight, NO_X and HC's combine to form **ozone**, a powerful oxidizer and primary component of smog. Ozone makes breathing difficult, causes headaches, lowers your immunity to diseases, and injures many types of vegetation.

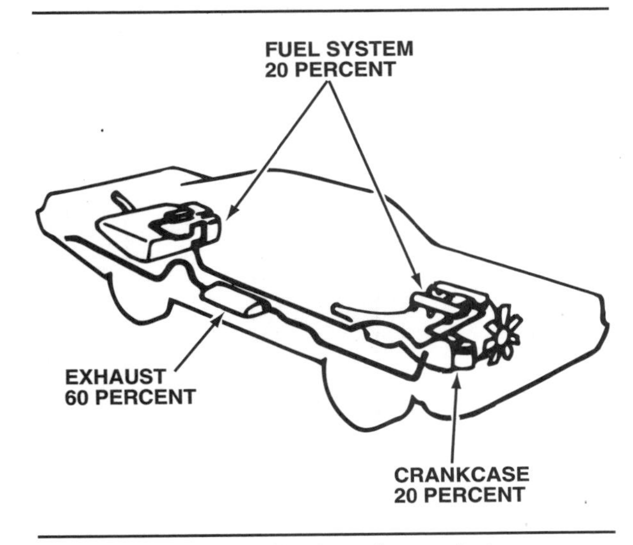

Figure 5-3. Sources of hydrocarbon emissions.

Lowering the engine's combustion temperature reduces NO_X formation. However, it also makes the air-fuel mixture burn less efficiently, creating large amounts of HC and CO. To combat this problem, automakers use various emission control systems, which we will study in later chapters.

Photochemical Smog: A combination of pollutants that form harmful chemical compounds when exposed to by sunlight.

Hydrocarbon (HC): A major pollutant containing hydrogen and carbon produced by internal combustion engines. Gasoline is a hydrocarbon compound.

Carbon Monoxide (CO): An odorless, colorless, tasteless poisonous gas. A major pollutant given off by an internal combustion engine.

Oxides of Nitrogen (NO_X): Chemical compounds of nitrogen given off by an internal combustion engine. NO_X combines with hydrocarbons to produce ozone, a primary component of smog.

Ozone: A gas with a penetrating odor, and a primary component of smog. Ground-level ozone forms when HC's and NO_X, in certain proportions, react in the presence of sunlight. Ozone irritates the eyes, damages the lungs, and aggravates respiratory problems.

Particulate Matter (PM_{10}): Microscopic particles of materials such as lead and carbon, given off by an internal combustion engine as pollution.

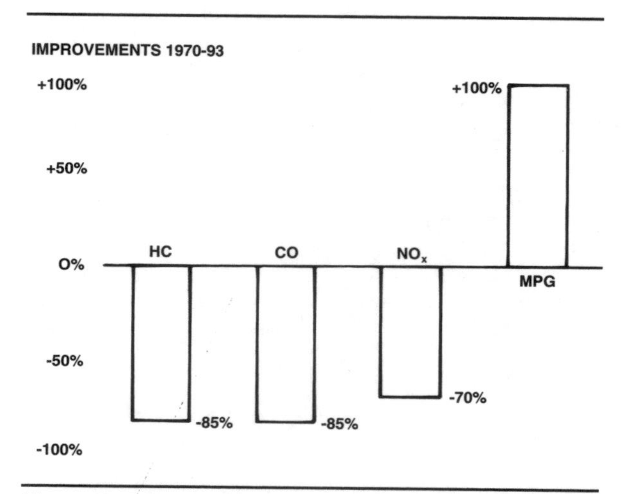

Figure 5-4. Since 1970, the average car's pollution has fallen, and its fuel efficiency has increased greatly. (Information courtesy of the U.S. EPA)

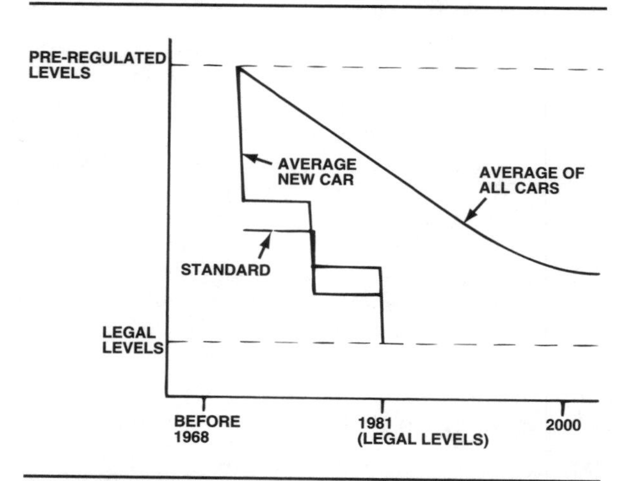

Figure 5-5. Clean air legislation has lead to dramatic decreases in the amount of pollution cars create. (Information courtesy of the U.S. EPA)

Particulate Matter

PM_{10} consists of microscopic—ten microns (0.000039 in.) or smaller—solid particles, such as dust, soot, and smoke that remain in the atmosphere for a long time. PM_{10} is a prime cause of secondary pollution. For example, particulates such as lead and carbon tend to collect in the atmosphere. Large amounts of these substances can penetrate deep into our lungs, causing lung disease.

Particulates produced by automobiles are a small percentage of the total particulates in the atmosphere. Most come from fixed sources, such as factories. While automobiles *do* produce particulates, the amount has decreased considerably in the past few decades because the fuel industry has eliminated additives such as lead from gasoline and changed other characteristics of the fuel. As a result, the amount and types of additives used in gasoline are carefully controlled.

Sulfur Oxides

Sulfur in gasoline and other fossil fuels (coal and oil) enters the atmosphere in the form of **sulfur oxides (SO_x)**. As SO_x breaks down, they combine with water in the air to form corrosive sulfuric acid, which is a secondary pollutant. Starting in the 1980s, there has been a lot of publicity about this type of pollution—commonly referred to as "acid rain"—especially in the northeastern United States and in Canada.

POLLUTION AND THE AUTOMOBILE

A single car actually gives off only microscopic amounts of pollutants, but there are millions of automobiles in use in the United States. Multiply each car's contribution toward air pollution by the number of cars, and you have the potential for a staggering amount of pollution. Without emission controls, automobiles would create almost as much air pollution as all other sources combined.

Great progress has been made in reducing automobile air pollution since 1966. HC emissions from the engine crankcase and fuel system have been almost totally eliminated. The other 60 percent of HC emissions—from the exhaust—has been lowered considerably. From 1970–93, the HC and CO emissions of a typical car each decreased about 85 percent, figure 5-4. Figure 5-5 shows how dramatically the emissions of a typical car have fallen in the decades since regulations began.

However, some of these gains are offset by an equally dramatic increase in the number of cars on the road and the number of miles people drive their cars. The United States Environmental Protection Agency (U.S. EPA) estimates that vehicle travel has doubled since 1970, due in great part to urban sprawl. So, although each car pollutes less, there are many more cars contributing to pollution.

SMOG—CLIMATIC REACTION WITH AIR POLLUTANTS

Smog and air pollution are not the same thing: smog is a form of air pollution, but air pollution is not necessarily smog. Although all three major pollutants are by-products of combustion, each is created in a different way. HC's come mostly from unburned fuel, CO from air-fuel mixtures that contain too much fuel, and NO_x from high com-

Figure 5-6. Smog engulfs the Los Angeles Civic Center in the 1960s. When the base of the temperature inversion is only 1,500 feet (457 meters) above the ground, the inversion layer—a layer of warm air above a layer of cool air—prevents the natural dispersion of air contaminants into the upper atmosphere.

bustion chamber temperatures. HC and NO_x are the two principal materials that combine in the atmosphere in sunlight to form smog, or ozone.

The Photochemical Reaction

Although scientists can create smog in laboratory experiments, they do not completely understand what it is. They do know, however, that three things must be present for smog to form in the atmosphere:

- Sunlight
- Relatively still air
- A high concentration of HC and NO_x.

When these three elements coexist, the sunlight causes a chemical reaction between the HC and NO_x and creates smog.

Temperature Inversion

Normally, air temperature decreases at higher altitudes. Warm air near the ground rises and becomes cooler by contact with the cooler air above it. When nature follows this normal pattern, smog and other pollutants are carried away. But some areas experience a natural weather pattern called a **temperature inversion**. When this occurs, a layer of warm air prevents the upward movement of cooler air near the ground. This inversion acts as a "lid" over the stagnant air. Since the air cannot rise, smog and pollution collect.

When the inversion layer is several thousand feet high, the smog can rise enough for reasonable visibility. But when the inversion layer is within 1,000 feet (300 meters) of the ground, it traps the smog. This reduces visibility, making the distant landscape impossible to see, figure 5-6, and caus-

Sulfur Oxides (SO_x): Sulfur given off by processing and burning gasoline and other fossil fuels. As it decomposes, sulfur dioxide combines with water to form sulfuric acid, or "acid rain".

Temperature Inversion: A weather pattern in which a layer, or "lid", of warm air keeps a layer of cooler air trapped beneath it.

CALIFORNIA PASSENGER CAR NEW VEHICLE STANDARDS

YEAR	HC	CO	NO_x
1966-69	275 ppm	1.5%	-
1970	2.2 g/m	23 g/m	-
1971	2.2 g/m	23 g/m	4.0 g/m
1972	3.2 g/m	39 g/m	3.2 g/m
1974	3.2 g/m	39 g/m	3.0 g/m
1975-76	0.9 g/m	9 g/m	2.0 g/m
1977-79	0.41 g/m	9 g/m	1.5 g/m
1980	0.41 g/m	9 g/m	1.0 g/m
1981-92	0.41 g/m	7 g/m	0.7 g/m

Figure 5-7. This chart of California exhaust emission limits for new cars shows how standards became more stringent in the wake of the Clean Air Act. California standards are stricter than federal. (Information courtesy of the California Air Resources Board)

ing people to experience eye irritation, headaches, and difficulty in breathing. Temperature inversion was first noted in the Los Angeles area, which provides a classic example of this phenomenon. Surrounding mountains in that area form a natural basin in which temperature inversion is present to some extent for more than 300 days a year.

TIMETABLE FOR IMPLEMENTATION OF CERTAIN PROVISIONS OF THE 1990 CLEAN AIR ACT

1992	Limits on maximum gasoline vapor pressure go into effect nationwide.
	Regulations for minimum oxygen content of gasoline go into effect in 39 areas.
1993	Production of vehicles requiring leaded gasoline becomes illegal.
1994	Phase-in of tighter tailpipe standards and cold-temperature CO standards for light-duty vehicles begins.
	Expansion of I/M programs begins in certain cities.
	Requirement for new cars to be equipped with on-board diagnostics systems takes effect.
1995	Reformulated gasoline must be sold in the nine smoggiest cities in the U.S.
	New warranty provisions on emission control systems take effect.
1996	Phase-in of California Pilot Program begins.
	Lead banned from use in motor vehicle fuel.
	All new vehicles must meet tighter tailpipe standards and cold-temperature CO standards.
1997	Federal "Clean Fleet" program begins in 19 states in areas with excessive ozone and CO.
2001	Second phase of California Pilot Program and Federal "Clean Fleet" program begins.

Figure 5-8. Various parts of the Clean Air Act Amendments of 1990 take effect through the 1990s and in the early 21st century.

AIR POLLUTION LEGISLATION AND REGULATORY AGENCIES

Once the problem of air pollution was recognized, it became the subject of intense research and investigation. By the early 1950s, scientists believed that smog in Los Angeles was caused by the photochemical process. California, as we noted earlier, became the first state to enact legislation designed to limit automotive emissions. Standards established by California one year often became U.S. federal standards the next year.

Regulatory Agencies

The U.S. EPA is the federal agency responsible for enforcing the Clean Air and Air Quality Acts passed by Congress. It was formed as part of the Department of Health, Education, and Welfare. The California Air Resources Board (CARB) has authority roughly parallel to that of the U.S. EPA, but extends only to vehicles sold in or brought into California, figure 5-7. Canadian vehicle emission standards are established by the national Ministry of Transport.

Emission Control Legislation

Beginning with the 1961 new cars, California required control over crankcase emissions. This became standard for the rest of the United States with new 1963 cars. That year, domestic automakers voluntarily equipped their new models with a blowby device that virtually eliminated crankcase emissions on all cars.

California followed by requiring that 1966 and later new cars sold within its boundaries have exhaust emission controls. The use of exhaust emission control systems was extended nationwide during the 1967– 68 model years.

The first federal air pollution research program began in 1955. In 1963, Congress passed the Clean Air Act, providing the states with money to develop air pollution control programs. This law was amended in 1965 to give the federal government authority to set emission standards for new cars, and was amended again in 1977–78. Under this law, emission standards were first applied nationwide to 1968 models.

In addition to the Clean Air Act, the federal government took a new approach to air pollution in 1967 with the Air Quality Act. This act and its major amendments of 1970, 1974, and 1977, instituted changes designed to turn piecemeal programs into a unified attack on pollution of all kinds. Canada attacked its own smog problem with vehicle emission requirements established by the Ministry of Transport that took effect with 1971 models.

The 1990 Clean Air Act

In 1990, the U.S. Congress amended and updated the Clean Air Act for the first time in 13 years. Besides making tailpipe standards more stringent and expanding vehicle Inspection and Maintenance (I/M) Programs, the 1990 law focused attention on fuel itself, in addition to vehicle technology. Some key aspects of the 1990 Clean Air Act include:

- Tighter tailpipe standards
- CO control
- Ozone control
- Reformulated gasoline
- Other controls.

Figure 5-8 shows a timetable for implementing certain provisions of the 1990 Clean Air Act.

Tighter tailpipe standards

Tailpipe standards for 1990–93 were 0.41 gram per mile (g/m) HC, 3.4 g/m CO, and 1.0 g/m NO_X. The 1990 Clean Air Act requires phasing in lower-limited standards for HC and NO_X between 1994–96—0.25 g/m of nonmethane HC and 0.4 g/m NO_X. The law also requires the U.S. EPA to study whether even stricter standards are necessary, feasible, and economical. If the U.S. EPA determines by 1999 to lower the standards, they will be halved beginning in the 2004 model year.

Carbon monoxide control

The U.S. EPA places primary blame for CO pollution on mobile sources (including cars and trucks, as well as vehicles such as bulldozers and construction equipment) in 39 U.S. cities.

CO emissions from cars are particularly high during cold weather, when vehicles operate less efficiently. While previous CO standards applied only at 75°F (24°C), the 1990 Clean Air Act establishes an additional CO standard of 10 g/m at 20°F (-7°C). If CO levels are still excessive in six or more cities by 1997, the standard will be tightened to 3.4 g/m, to be met beginning with the 2002 model year.

The 1990 law also requires a higher O_2 content in the gasoline sold during the winter in the 39 CO-saturated cities, a requirement that took effect in 1992. O_2 leans the air-fuel mixture, reducing CO emissions, lowering fuel economy, and increasing CO_2 emissions.

Ozone control

Ozone—the combination of HC and NO_X, the primary components of smog—is the most wide-widespread air quality problem in the United States. The U.S. EPA rates the following areas as having a "severe" ozone problem:

- New York City and Long Island in New York State, and north New Jersey
- Baltimore, Maryland
- Muskegon, Michigan
- Chicago, Illinois; Gary, Indiana; and Lake County, Wisconsin
- Milwaukee, Wisconsin
- Houston, Galvaston, and Brazoria, Texas
- San Diego, California
- Southeast desert, California
- Ventura County, California.

Los Angeles, California, receives its very own ozone rating: "extreme". Approximately 15 other cities and urban areas across the United States are rated as having "serious" ozone levels, and about 30 more have "moderate" ozone problems.

How New Cars Are Emission-Certified

Whether you can buy a particular new car each year—and where you can buy it—depend on the automaker's success in the emission certification process. The U.S. Environmental Protection Agency (U.S. EPA) performs a constant-volume sampling test for each car, called the Federal Test Procedure (FTP), or the Federal Vehicle Certification Standard Procedure. Some vehicles are also tested by the California Air Resources Board.

As each new model year approaches, automakers build prototype, or emission-data, cars for U.S. EPA use. The automakers are responsible for conducting a 50,000-mile (80,000-km) durability test of their emission control systems. Before the U.S. EPA test, the automakers drive the test cars for 4,000 miles (6500 km) to stabilize the emission systems.

The automaker preconditions the car before the U.S. EPA test, then it stands for 12 hours at an air temperature of 73°F (22°C) to simulate a cold start. The actual test is done on a chassis dynamometer, using a driving cycle that represents urban driving conditions. The car's exhaust is mixed with air to a constant volume and analyzed for harmful pollutants.

The entire test requires about 41 minutes. The first 23 minutes are a cold-start driving test. The next 10 minutes are a waiting, or hot-soak period. The final eight minutes are a hot-start test, representing a short trip in which the car is stopped and started several times while hot. If the emission test results for all data cars are equal to or lower than the HC, CO, and NO_X standards, the U.S. EPA grants certification for the engine "family", and the manufacturer can sell the car to the public.

This certification process explains why some engines disappeared in the wake of clean-air legislation—they were too "dirty" and could not be "cleaned up". It also explains why some powertrain combinations may not be available in California—which has different requirements—when they can be purchased in other states.

PASSENGER CARS		LIGHT-DUTY TRUCKS		
MODEL YEAR	MPG	4 x 2 MPG	4 x 4 MPG	COMBINED MPG
1978	18	-	-	-
1979	19	17.2	15.8	-
1980	20	16	14	-
1981	22	16.7	15	-
1982	24	18	16	17.5
1983	26	19.5	17.5	19
1984	27	20.3	18.5	20
1985	27.5	19.7	18.9	19.5
1986	26	20.5	19.5	20
1987	26	21	19.5	20.5
1988	26	21	19.5	20.5
1989	26.5	21.5	19	20.5
1990	27.5	20.5	19	20
1991	27.5	20.7	19.1	20.2
1992	27.5	-	-	20.2
1993	27.5	-	-	20.4
1994	27.5	-	-	20.5
1995	27.5	-	-	20.6

Figure 5-9. This chart shows how CAFE standards have progressed from the late 1970s into the 1990s. (Information courtesy of the U.S. EPA)

Gasoline vapors are a major source of HC in ozone, so the 1990 Clean Air Act calls for reducing evaporative emissions by improved engine and fuel system vapor traps and wider use of systems to capture vapors during refueling in smoggy cities. In addition, the Clean Air Act places a cap on fuel volatility—reducing its tendency to evaporate.

Reformulated gasoline (RFG)

The 1990 Clean Air Act established standards for the content of gasoline sold in the nine worst ozone areas. This "cleaner" gasoline must meet or exceed a certain minimum O_2 content, and it must not exceed a certain maximum level of benzene. Also, the gasoline must reduce toxic and smog-forming emissions 15 percent by 1995, and 20 to 25 percent by 2000, without increasing NO_x emissions. Other cities may choose to use this clean fuel too, and the U.S. EPA expects many to do so.

Other controls

The 1990 law calls for the U.S. EPA to monitor toxic emissions such as benzene and formaldehyde, regulating them as needed. Both California and the federal government are mandating clean car programs. The California Pilot Program requires that, starting in 1996, car manufacturers supply at least 150,000 "clean" cars for sale in California. The qualifications of a California Pilot Program clean car are that it emits no more than 0.125 g/m HC, 3.4 g/m CO, and 0.4 g/m NO_x. In 1999, the number of clean cars required increases to 300,000.

Meanwhile, starting in 1998, the federal government requires fleets in certain very polluted cities to have 30 percent of their new cars be clean cars. For these areas, a clean car is one that can use clean fuel and can meet extra-low emission standards. By 2000, the proportion rises to 70 percent of the fleet. This federal program affects 22 metropolitan areas in 19 states.

Corporate Average Fuel Economy (CAFE) Standards

The 1973 energy crisis focused national attention on fuel economy and resulted in the establishment of Corporate Average Fuel Economy (CAFE) standards, figure 5-9, as a part of the Federal Energy Act of 1975. The U.S. EPA and the Department of Transportation (DOT) are responsible for administering the CAFE standards.

The combination of CAFE and clean air laws gives automotive engineers conflicting goals—reducing emissions while improving fuel economy. If the CAFE standards are not reached each year, automakers pay a penalty on each vehicle sold—the so-called "gas guzzler" tax. However, automakers who *exceed* the yearly average on a corporate basis gain a credit that can be applied to later years.

By downsizing vehicles, using smaller engines, and paying particular attention to reducing weight and improving aerodynamic efficiency, the domestic automotive industry transformed itself in the early 1980s. During this period, there was a concerted national effort to conserve fuel because of high prices at the pump and a desire to reduce the nation's dependence on foreign oil.

This national effort worked so well that oil prices, which had stabilized by 1983, collapsed in 1985–86. At the same time, car owners indicated a desire to return to larger cars by ignoring the fuel-efficient smaller cars and purchasing the less efficient cars with bigger engines and more room. This led General Motors (GM) and Ford to petition the U.S. EPA for a "rollback" of CAFE standards to avoid massive fines resulting from meeting customer demand. Over the objections of Chrysler and other small automakers who met the standards on time, a temporary return to the 1983 level was enacted for 1986, with the 1987 standard returning to that of 1985. However, in the early 1990s, as gas prices climbed again, CAFE standards returned to the original levels established by Congress.

On-Board Diagnostics (OBD) Systems Regulation

In the late 1980s, CARB mandated that all cars sold in the state of California must have an **on-board diagnostics (OBD) system.** These initial regulations, known as OBD-I, were minimal, and most of the major vehicle manufacturers

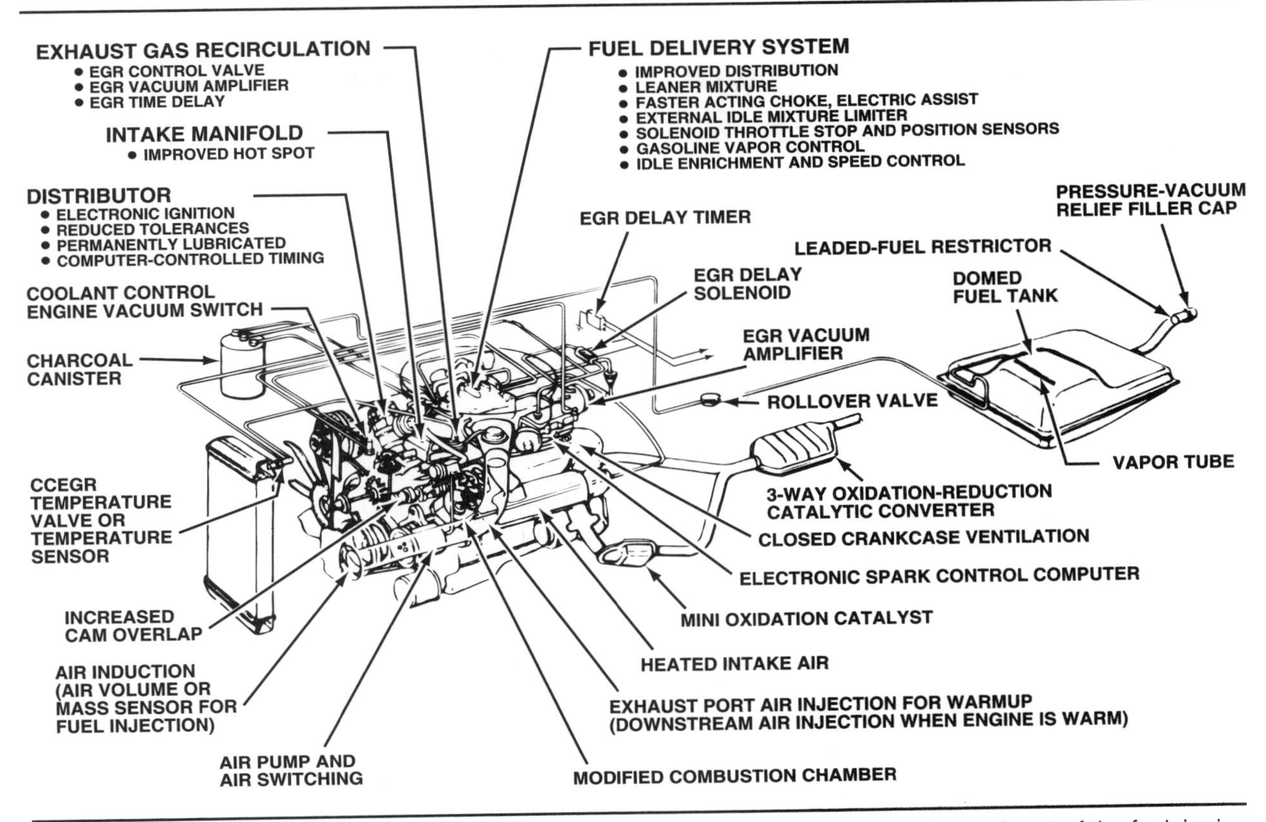

Figure 5-10. The complex emission controls on a modern automobile engine are an integral part of the fuel, ignition, and exhaust systems. (Chrysler)

had systems already in production that complied to the standards. In general, three elements were required by OBD-I:

1. An instrument panel warning lamp to alert the driver of certain control system failures
2. The ability of the system to record and transmit diagnostic trouble codes (DTC) for emission-related failures
3. Electronic system monitoring of the exhaust gas recirculation valve, fuel system, and some other emission-related components.

The purpose of the regulations was twofold: provide the driver an early warning on an emissions failure, and make the technician's task of locating the source of the failure easier. The ultimate goal was a reduction of tailpipe pollutants, which would result in improved air quality.

OBD-I may have been founded with good intent, but implementation left much to be desired. The manufacturers were free to interpret the rules as they saw fit, and the result was a vast array of systems. Rather than simplify the job of locating and repairing a failure, the technician was now faced with a tangled network of procedures that often required the use of expensive special test equipment and information available only to the dealers.

The overall reduction of tailpipe emissions was marginal. It soon became apparent that more-stringent measures were needed if the ultimate goal, breathable air in Southern California, was to be achieved. This led CARB to develop OBD-II, a second generation of more-stringent legal requirements.

OBD-II development

For their second set of regulations, CARB desired more-precise control and monitoring of emission-related components, as well as standard service procedures without the use of dedicated special tools. Because the technology required to develop this new program was outside its field of expertise, CARB enlisted the aid

On-Board Diagnostics (OBD) System: A type of automotive diagnostic system mandated by the California Air Resources Board (CARB) or the U.S. EPA. Although the system's abilities vary depending on the specific version, generally, OBD seeks to have a polluting vehicle serviced sooner, and to improve the technician's ability to repair emission-related problems.

of the Society of Automotive Engineers (SAE). CARB would establish the guidelines, and SAE would develop the technology.

The new regulations, OBD-II, were gradually phased into production beginning with the 1994 model year. By 1996, most vehicles sold in California will be required to have an OBD-II-compliant system.

On a national level, the U.S. EPA has adapted the system implemented by California. As it stands, the federal government requires all vehicles to comply with OBD-II standards established by CARB through the 1997 model year. In 1998, new federal standards established by the U.S. EPA will take effect. Later chapters detail OBD-II-compliant engine control systems.

AUTOMOTIVE EMISSION CONTROLS

Early researchers dealing with automotive pollution and smog began work with the idea that *all* pollutants were carried into the atmosphere by the car's exhaust pipe. But auto manufacturers doing their own research discovered that the fuel tank and engine crankcase also give off pollutants. The total automotive emission system, figure 5-10, contains three different types of controls. The emission controls on a modern automobile are not a separate system, but an integral part of an engine's fuel, ignition, and exhaust systems.

In order to service a car's fuel system and emission controls, you must have a basic understanding of the internal combustion engine and how it works. The chapters of this book will cover engine operating principles and air-fuel requirements, emission controls as they relate to different parts of the fuel system, and major emission controls that can be studied separately from the fuel system.

Automotive emission controls can be grouped into major families:

- Crankcase emission controls
- Evaporative emission controls
- Exhaust emission controls.

Positive crankcase ventilation (PCV) systems control HC emissions from the engine crankcase. Evaporative emission control (EEC or EVAP) systems control the evaporation of HC vapors from the fuel tank, pump, and fuel injection system. Various systems and devices control HC, CO, and NO_x emissions from engine exhaust. We can divide exhaust emission controls into the following categories:

- Air injection systems
- Catalytic converters
- Engine modifications
- Spark timing controls
- Exhaust gas recirculation.

Air injection systems add air to the exhaust to help burn HC and CO and to aid catalytic conversion. The first catalytic converters installed in the exhaust systems of 1975–76 cars helped the chemical oxidation of the exhaust—that is, burning HC and CO. Later catalytic converters also promote the chemical reduction of NO_x emissions.

Automakers have made a variety of changes in engine design and fuel and ignition system operation to help eliminate all three major pollutants. Various systems to delay or retard ignition spark timing help control HC and NO_x emissions. Early spark-timing controls modified the distributor vacuum advance, but later-model cars use electronic engine control systems that eliminate the need for mechanical or vacuum timing devices.

Recirculating a small amount of exhaust gas back to the intake manifold to dilute the incoming air-fuel mixture is an effective way to control NO_x emissions. This is called an exhaust gas recirculation (EGR) system.

SUMMARY

The automobile is a major source of air pollution, resulting from gasoline burned in the engine and vapors escaping from the crankcase, fuel tank, and the rest of the fuel system. The major automobile-produced pollutants are unburned HC's, CO, and NO_x. The use of emission controls in recent decades has reduced automotive pollutants by 65 to 98 percent.

Emission controls began as separate "add-on" components and systems but are now integrated into engine and vehicle design. The major emission control systems are PCV systems, evaporative control systems, air injection, spark timing controls, EGR, catalytic converters, and electronic engine control systems.

Increasingly, stringent emission standards, combined with the CAFE regulations imposed beginning in the late 1970s, and throughout the 1980s and 1990s, have reshaped the domestic automotive industry. The end result has been smaller, more fuel-efficient vehicles that produce less pollution. Automakers are still working to improve emission control and gas mileage.

Review Questions

Choose the single most correct answer.
Compare your answers to the correct answers on page 507.

1. Smog:
   a. Is a natural pollutant
   b. Cannot be controlled
   c. Is created by a photochemical reaction
   d. Was first identified in New York City

2. The three major pollutants in automobile exhaust are:
   a. Sulfates, particulates, carbon dioxide
   b. Sulfates, carbon monoxide, nitrous oxide
   c. Carbon monoxide, oxides of nitrogen, hydrocarbons
   d. Hydrocarbons, carbon dioxide, nitrous oxide

3. Fuel evaporation accounts for what percentage of total HC emissions?
   a. 10 percent
   b. 60 percent
   c. 33 percent
   d. 20 percent

4. Carbon monoxide is a result of:
   a. Incomplete combustion
   b. A lean mixture
   c. Excess oxygen
   d. Impurities in the fuel

5. Which of the following is *not* true?
   a. High engine temperatures increase NO_X emissions
   b. High engine temperatures reduce HC and CO emissions
   c. Low engine temperatures reduce HC and CO emissions
   d. Low engine temperatures reduce NO_X emissions

6. Particulate matter is/are:
   a. A by-product of photochemical smog
   b. Created only by diesel engines
   c. Caused by the chemical reaction of CO and NO_X
   d. Microscopic particles suspended in the atmosphere

7. Sulfur oxides are harmful because they:
   a. Damage three-way catalytic converters
   b. Combine with water to form sulfuric acid
   c. Are primary automotive pollutants
   d. Are visible in bright sunlight

8. Ozone is created by a combination of sunlight, still air, and:
   a. High levels of CO and HC
   b. High levels of CO and NO_X
   c. High levels of HC and NO_X
   d. High levels of HC and sulfur oxides

9. A temperature inversion increases air pollution by:
   a. Pushing cool air up
   b. Forming a "lid" over stagnant air
   c. Pushing warm air down
   d. Decreasing wind force

10. U.S. federal emission limits are established by the:
    a. California Air Resources Board
    b. Ministry of Transport
    c. Environmental Protection Agency
    d. Department of Transportation

11. Corporate Average Fuel Economy (CAFE) standards were rolled back in the:
    a. 1960s
    b. 1970s
    c. 1980s
    d. 1990s

12. Sulfur by-products of combustion can combine with water to form:
    a. Sulfates
    b. Particulates
    c. Oxides of sulfur
    d. Sulfuric acid

13. The only major type of pollutant that comes from a vehicle source other than the exhaust is:
    a. HC
    b. CO
    c. CO_2
    d. NO_X

14. OBD-I seeks to decrease levels of automotive emissions by:
    a. Alerting the driver to a potential problem
    b. Monitoring the efficiency with which certain systems are operating
    c. Measuring the levels of HC and CO coming from the tailpipe
    d. Both a and b

6 What Is a Tune-Up?

The previous chapters explained why engine maintenance is necessary for the best performance, economy, and emission control. Now we can look more closely at the combination of engine maintenance services commonly called a tune-up. This chapter describes:

- The difference between traditional and modern ideas of tune-up
- Automakers' and equipment suppliers' tune-up recommendations
- The effects of emission control regulations on tune-up services.

TUNE-UP'S CHANGING MEANING

The automobile has undergone many changes since the first emission control regulations appeared in 1961. Engine equipment has changed, and the services necessary to maintain that equipment have changed. Today, compared to 20 years ago, this means that:

- Fewer items need to be replaced.
- Service intervals for replacement items have lengthened.
- More testing procedures are required than replacement and adjustment procedures.

However, the goal of today's tune-up is still the same as yesterday's—to restore a vehicle's performance, fuel economy, and emissions level by returning the engine as nearly as possible to manufacturer's specifications.

Periodic Service—The Traditional Tune-Up

Before and during the infancy of emission control regulations, an engine tune-up was done at regular intervals based on the items that needed to be replaced, inspected, or adjusted, figure 6-1. This was usually called preventive maintenance or periodic maintenance. The tune-up included:

- Testing the engine's general mechanical condition to find any major problems
- Replacing certain parts, such as filters and ignition system parts
- Adjusting tolerances, especially in the ignition and fuel systems.

Diagnosis and Correction—The Modern Tune-Up

As the number of serviceable items decreased and service intervals lengthened, tune-up technicians could no longer simply replace parts and make a few adjustments. They had to be aware of problems that could appear in systems where maintenance is not normally needed. The emphasis of a tune-up changed from routine services to a

TUNE-UP RELATED SERVICES REQUIRED DURING FOUR YEARS OR 50,000 MILES OF OPERATION 1968 DODGE	
Change crankcase oil	12 times
Replace crankcase oil filter	6 times
Lubricate manifold heat valve	12 times
Check PCV system	12 times
Replace PCV valve and clean system	4 times
Clean air cleaner filter	12 times
Replace air cleaner filter	2 times
Clean carburetor choke shaft, fast-idle cam, and pivot pin	8 times
Oil distributor shaft	8 times
Clean and oil the oil filler cap	8 times
Change engine coolant	4 times
Lubricate distributor cam	4 times
Engine tune-up (replace or adjust as needed):	
• Spark plugs	4 times
• Ignition points	4 times
• Distributor cap and rotor	4 times
• Ignition coil	4 times
• Ignition cables	4 times
• Carburetor choke	4 times
• Idle speed and fuel mixture	4 times
Replace fuel filter	2 times

Figure 6-1. These tune-up services focus on replacing or adjusting items at regular intervals.

concentration on testing. Extensive checks of overall performance and individual system performance are now needed to pinpoint problems that could have been missed during a traditional replace-and-adjust tune-up. Introduced in the previous chapter, On-Board Diagnostics Generation II (OBD-II) illustrates the increased level of trouble shooting skill necessary to diagnose and repair vehicles. A vehicle equipped with an OBD-II system can alert the driver of an emission-related problem sooner and more accurately than previously possible. Prior to OBD-II, these problems would have been addressed when a driveability problem manifested itself, or during a service visit for regular maintenance. Testing, diagnosis, and correction, combined with those periodic services that remain, make up current tune-up procedures.

AUTOMAKERS' MAINTENANCE RECOMMENDATIONS

Automakers have always published servicing guidelines to inform the owner and technician how the car should be maintained. Recommended service schedules allow the automaker to place under warranty various vehicle items based on a specific program of maintenance. Today's maintenance schedules represent the minimum amount of service the car will require when used in "normal" driving. The following paragraphs explain how and why automakers' maintenance recommendations have changed.

Services

Most tune-up-related services have been eliminated from later-model cars. Compare the typical

TUNE-UP RELATED SERVICES REQUIRED DURING 4 YEARS OR 50,000 MILES OF OPERATION 1995 DODGE	
Change crankcase oil	8 times
Replace crankcase oil filter	4 times
Replace air filter	2 times
Replace PCV valve	2 times
Replace fuel filter	As needed
Change engine coolant	1 time
Replace spark plugs	2 times
Inspect and adjust drive belts	1 time
Check cooling system	4 times

Figure 6-2. These tune-up requirements illustrate the reduced servicing needs of late-model cars.

CHEVROLET	1968 Service Intervals		1995 Service Intervals	
	Months	Miles	Months	Miles
REPLACE				
Crankcase oil	4	6,000	12	7,500
Crankcase oil filter	8	12,000	12	7,500
PCV valve	12	12,000	—	30,000
Ignition points		12,000	None	
Spark plugs		12,000	—	30,000
Fuel filter	12	12,000	30	30,000
Coolant	24		30	30,000
Air cleaner filter				
Rotate		12,000	Not required	
Replace		24,000	—	30,000
ADJUST				
Valves		12,000	Not required	
Idle speed		12,000	Not required	
Rotate distributor cam lubricator		12,000	Not required	
Ignition timing		12,000	—	60,000
CHECK				
Ignition condenser		12,000	None	
PCV system	4	6,000	—	50,000
Ignition cables		12,000	—	60,000
Carburetor choke		Not required	None	
Vacuum advance		12,000	None	

Figure 6-3. A comparison of the tune-up services and intervals for a 1968 and a 1995 Chevrolet.

later-model tune-up requirements shown in figure 6-2 with those listed in figure 6-1. For another example, between 1972 and 1982 10 routine maintenance items were dropped from Ford's maintenance schedule. Not all of these were tune-up related, but one item dropped was distributor cap and rotor replacement:

- The 1972 model has a 24,000-mile (38,400-km) replacement interval.
- The 1982 model does not have a specified replacement interval.

Figure 6-3 shows tune-up services recommended by Chevrolet for its 1968 and 1995 models. You can see that eight services required in 1968 are not called for in 1995; some of these service parts are not on 1995 models, while other service requirements were simply eliminated:

- Ignition points replacement
- Air cleaner filter rotation
- Valve adjustment (on some engines)
- Distributor cam lubricator rotation
- Ignition condenser testing
- Carburetor choke adjustment
- Idle speed
- Vacuum advance adjustment.

Several emission-related checks that were nonexistent in 1968 are now incorporated into state Inspection and Maintenance (I/M) Programs. These checks, conducted at regular intervals, include, but are not limited to:

- Evaporative fuel control
- Exhaust gas recirculation (EGR)
- Thermostatically controlled air cleaner
- Malfunction indicator lamp (MIL).

These system checks are a good example of the changing nature of the tune-up, from replacement and adjustment to testing and diagnosis.

Intervals

The other major change in manufacturers' maintenance recommendations has been the lengthening of the time and mileage intervals between required services. For instance, the service intervals on 1995 Ford products with V8 engines have been extended for more than 25 items, including:

- Spark plug replacement—from every 12,000 miles (19,200 km) in 1972 to every 60,000 miles (96,600 km) in 1995
- Engine oil replacement—from every 4,000 miles (6,400 km) in 1972 to every 5,000 miles (8,000 km) in 1995
- Engine coolant replacement—from every 24 months in 1972 to every 36 months in 1995.

The Chevrolet servicing table in figure 6-3 shows that almost every 1995 tune-up service interval has been lengthened compared with 1968 intervals.

Legal Requirements

There are two main reasons behind these changes in automakers' tune-up services and intervals:

1. The 1990 Amendments to the federal Clean Air Act require that automakers provide a five-year, 50,000-mile (80,000-km) warranty on their products' emission control devices.
2. During the U.S. EPA's new-car certification procedures, recommended maintenance schedules *must* be followed. These schedules are then given to the new-car buyer. This makes the car owner responsible for maintaining the emission control devices so they will operate properly for five years or 50,000 miles (80,000 km), whichever comes first.

The rule that maintenance schedules be followed during U.S. EPA certification tests has an interesting result. The complete maintenance schedules that must be published for consumers has become a sales issue. Manufacturers compare their schedules with those of their competition, and also with the schedules recommended in earlier years. For example, Ford's 1982 sales literature pointed out that the changes in maintenance requirements meant a savings in maintenance costs. When comparisons are made between competing car models, it is in the automaker's interest to have a cheaper maintenance schedule. This s one of the reasons that manufacturers have worked to eliminate service items and to extend intervals.

PARTS AND EQUIPMENT SUPPLIERS' RECOMMENDATIONS

Automakers supply specific maintenance schedules for their products. More general recommendations are made by:

- Manufacturers of replacement parts
- Manufacturers of tune-up testing equipment.

Manufacturers recommendations differ from automakers' schedules for several reasons. Automakers tend to base their schedules on the least-demanding vehicle operating conditions in order to reduce services and lengthen intervals. Parts and equipment suppliers base their recommendations on more-severe operating conditions, which are closer to what the average car encounters. This tends to shorten their recommended service intervals. Of course, they are also interested in selling as many parts and as much equipment as possible, so they recommend more-frequent replacement and testing.

SEVERE SERVICE EFFECTS ON TUNE-UP REQUIREMENTS

If the term "severe service" or "short trip/city" makes you think of police cars, taxicabs, and other special-duty vehicles, you may be surprised at today's severe service definition. A car is in severe service if:

- Most trips are less than 5 to 10 miles (8 to 16 km).
- It is used in stop-and-go driving.
- Its engine is idled for long periods.
- It is driven in dusty or sandy conditions.
- It is driven in very hot or very cold weather.
- It is used to pull a trailer, or uses a top-mounted carrier.

Other severe service factors include individual driving habits such as the use of air conditioning, extended high-speed driving, and rapid acceleration.

Today's common definition of normal driving is actually an *ideal* condition: long-distance driving on relatively dust-free roads. Almost any other driving condition qualifies as severe service.

Automakers include severe service recommendations in their maintenance schedules. These usually call for cutting certain service intervals in half. A few tune-up items are affected by severe service recommendations, including:

- Air cleaner filters (affected by dusty or sandy conditions)
- Crankcase oil and filter
- Spark plugs.

Transmission, transaxle, differential, and wheel bearing services are also affected by severe service operation.

EXHAUST EMISSION INSPECTION STANDARDS

All cars manufactured for sale in the United States have had to meet U.S. federal exhaust emission standards since 1968. These standards apply to new cars as they come off the assembly line. Because of ongoing air quality problems, the California Air Resources Board (CARB) requires that all new vehicles sold in that state meet an even more-stringent set of emission standards.

As discussed earlier, vehicle manufacturers must warrant the emission performance of their cars for five years or 50,000 miles (80,000 km), whichever comes first. Beyond this point it is the responsibility of the vehicle owner to maintain the emission control system in good condition. Because owners tend to ignore vehicle maintenance requirements, and in some cases actively disconnect or remove emission control devices with resulting increase in tailpipe pollutants, the U.S. EPA has been given the power to withhold federal highway funds from cities and states that fail to meet certain standards of ambient air quality. Once a car is in use and mechanical wear has begun, its emission control systems start to lose efficiency.

The Clean Air Act Amendments of 1990 created two types of I/M Programs: Enhanced and Basic. Areas with higher carbon monoxide and ozone pollution receive an Enhanced I/M Program. At the time of this publication's printing, details regarding the new Program are undecided. In general, the new I/M Program includes an emission test designed to approximate the emission test conducted during the Federal Test Procedure (FTP), discussed in the previous chapter. Cost to the public and the automotive service industry, along with the political climate, will determine the test's form and how the Program is implemented.

Depending on the city or state in which you live and work, servicing cars to meet I/M Program standards will be an important part of the tune-up process.

TUNE-UP MERCHANDISING

Few car owners consistently have their automobiles serviced according to factory recommendations. Many people bring their cars in for service when they "think" it is time, or they get an annual tune-up and lube job. Some individuals wait until there is something wrong with the car to bring it in for service. You can do your customers a favor by pointing out the factory-recommended services their cars require. For the customer, it is better service—for you, greater profit.

Many companies supply promotional items that can help a tune-up shop sell its services. These companies include automakers, replacement-parts suppliers, test-equipment makers, and magazines aimed at the automotive service trade. The promotional products range from eye-catching signs and banners to ideas and suggestions for bringing tune-ups to the car owner's attention. Some ideas include:

- Displaying new and used parts side-by-side to emphasize the wear caused by everyday use
- Asking the car owners what fuel mileage their cars get, then showing them the appropriate U.S. EPA mileage estimate, to emphasize what the car might be getting if properly tuned
- Setting up specific appointments for the work to be done
- After the tune-up, showing the customer the parts that were replaced and explaining why they were replaced
- Showing the customer before-and-after test results in writing.

Our definitions of an engine tune-up do not tell you what specific services should be included. The services contained in your tune-up package should be clearly explained to the car owner. First, of course, you must decide what items will be serviced. Many shops offer a one-price service that covers an overall engine test and the replacement of a few common parts. If any unusual problems are found, they will be fixed only at additional cost. A similar tune-up package charges a fixed labor rate for engine testing, then adds the cost of any needed replacement items. These methods are often used in high-volume tune-up specialty shops. Other shops usually charge for the actual amount of time spent working on the car plus the cost of the parts replaced.

THE TOTAL TUNE-UP

According to the Ignition Manufacturers Institute (IMI), a trade association of major parts manufacturers, only a technician with the following elements can provide a competent tune-up:

- Up-to-date training
- Dependable test equipment
- Accurate specifications
- Ability to perform the appropriate test procedure
- Quality replacement parts.

This definition of an engine tune-up agrees with our earlier explanation. When tuning a late-model car, the technician must be able to understand the information obtained from the car's condition and from test equipment. The equipment must be dependable and accurate, or the test results are meaningless. A vital part of diagnosis is having accurate specifications so that you can compare how the car *should* be operating with how it is *actually* operating. A specific test procedure ensures that all engine performance areas are thoroughly checked in the proper order. Using quality replacement parts reduces the risk of installing new parts that are already defective or might wear out in less than the recommended service interval.

The remaining chapters in this *Classroom Manual* cover the operation of the various engine systems involved in tune-up testing and service. The relationship of each of these systems to engine performance, economy, and emission control is also explained.

SUMMARY

Changes in the automobile caused by emission control regulations have in turn caused changes in engine tune-up ideas and procedures. The traditional tune-up's replace/adjust/test structure has shifted to the modern emphasis on comprehensive testing and fault diagnosis. In general, today's service recommendations contain fewer replacement items, longer service intervals, and more testing of engine systems.

Automakers' recommendations are based on ideal driving conditions, which very few cars actually meet. Shorter servicing intervals are often recommended by replacement-parts manufacturers and test-equipment suppliers. Automakers recognize this difference and include severe-service recommendations in their schedules.

Almost all areas of the country now require periodic I/M Program emission control system and tailpipe emission inspections to ensure that cars meet federal emission control standards. Like any other service, however, tune-up service still has to be *sold* to consumers. Large companies supply advertising aids and tune-up procedure checklists to help the tune-up technician attract and keep business. The total tune-up is an important service if late-model vehicles are to be serviced properly.

Review Questions

Compare your answers to the correct answers on page 507.

1. Which of the following is generally *not* true for a modern tune-up?
   a. Fewer items need to be replaced
   b. The service intervals have gradually lengthened
   c. Tune-ups are no longer as important as they used to be
   d. More testing procedures are required now than before

2. Which of the following is true of the automakers' service recommendations?
   a. They represent the minimum service under normal driving
   b. They represent the maximum service under normal driving
   c. They represent the maximum service under severe driving
   d. They represent the minimum service under severe driving

3. Which of the following is *not* a part of severe service driving?
   a. Most of the trips are 10 miles (16 km) or less
   b. The car frequently is used for trailer towing
   c. The car is driven at moderate speeds for extended periods of time on highways
   d. The car is driven in dusty or sandy conditions

4. Which of the following are included as normal service driving?
   a. Long distance driving
   b. Driving at cruising speeds
   c. Driving on dust-free highways
   d. All of the above

5. What two types of Inspection and Maintenance programs does the 1990 Clean Air Act Amendments mandate?
   a. Basic and Severe
   b. Basic and Moderate
   c. Severe and Enhanced
   d. Enhanced and Basic

6. A properly tuned-up vehicle:
   a. Can pass an I/M test procedure
   b. Has had all the components called for by the automaker checked, adjusted, or replaced
   c. Has no driveability problems
   d. All of the above

PART TWO

Electrical Systems

7 Battery, Charging, and Cranking Systems

At a glance, you may think that the ignition system is the only automotive electrical system that needs to be considered during a tune-up. If you ignore other systems, however, you may miss the causes of engine performance problems. This chapter contains a brief explanation of the battery, the charging, and the cranking systems, and shows how they relate to engine performance.

To examine the car's electrical systems we should be familiar with some electrical terms and ideas. Electricity can be defined as the flow of electrical current through a **conductor,** which is a material that allows easy current flow. An **insulator** is a material that does not allow current flow. The amount of current flow through a conductor is called the **amperage**, and it is measured in **amperes,** commonly called amps. The force that causes current flow is called **voltage,** and is measured in **volts.** For current to flow, there must be a complete path of conductors to carry the current. This path is called a **circuit.** Although there can be no current flow if the circuit is not complete, there *can* be voltage present in the incomplete circuit. All circuits contain some amount of **resistance** to current flow, which is measured in **ohms.**

The voltage for a car's electrical circuits comes from the battery. The circuits are made up of:

- Wiring
- The engine, frame, and body of the car.

The engine, frame, and body are called the **ground.** The cable from one battery post, or terminal, is bolted to the car engine or frame, figure 7-1. This is called the **ground cable.** The cable from the other battery terminal provides current for all the car's electrical loads. This is called the **insulated,** or **hot cable.** The insulated side of every circuit in the vehicle is the wiring running from the battery to the devices in the circuit. The ground side of every circuit is the vehicle chassis, figure 7-1. This is called a single-wire, or a ground-return, system.

We will be showing you pictures of circuits called circuit diagrams, which show the wires and devices in the circuit. They do not show the car engine, frame, or body. Instead, the ground symbol, figure 7-1, is used to show that the wire or device is bolted to the car chassis. You can think of the ground symbol as showing a direct connection back to the battery ground cable.

BATTERY OPERATION AND CONSTRUCTION

The automotive battery does not actually store electricity, as is often believed. It converts electrical energy into chemical energy, which is stored until the battery is connected to an external cir-

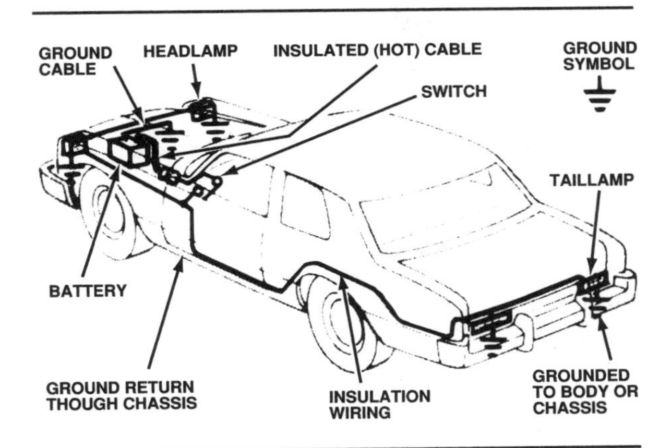

Figure 7-1. Half of the automotive electrical system is the ground path through the vehicle chassis.

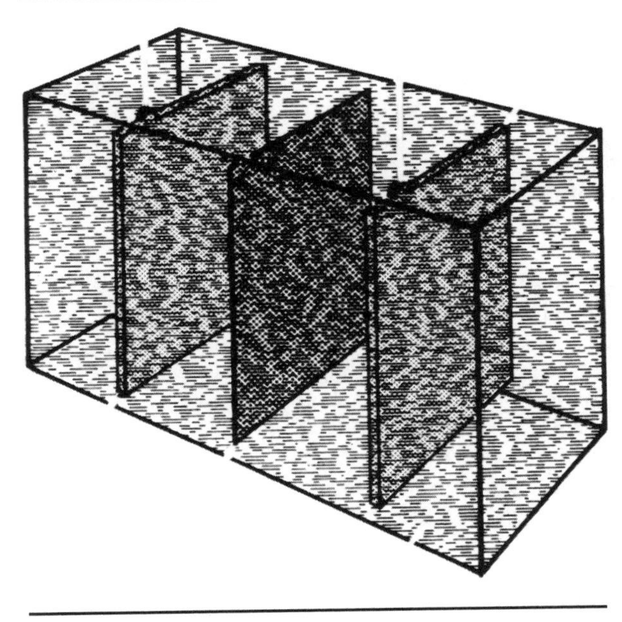

Figure 7-2. The potential difference between the two plates of a battery can cause current to flow in an outside circuit. (Chevrolet)

cuit. The stored chemical energy is then converted back to electrical energy, which flows from one battery terminal, through the circuit, and back to the other battery terminal.

We will begin our study of batteries by listing their functions, and looking at the chemical action and construction of a battery.

An automotive battery:

- Operates the starter motor
- Provides current for the ignition system during cranking
- Supplies power for the lighting systems and electrical accessories when the engine is not operating
- Acts as a voltage stabilizer for the complete electrical system
- Provides current when the electrical demand of the vehicle exceeds the output of the charging system.

Electrochemical Action

All automotive wet-cell batteries operate because of the chemical action of two dissimilar metals in the presence of a conductive and reactive solution called an **electrolyte**. Because this chemical action produces electricity, it is called electrochemical action. The chemical action of the electrolyte causes electrons to be removed from one metal and added to the other. This loss and gain of electrons causes the metals to be oppositely charged, and a potential difference, or voltage, exists between them.

Conductor: A material that allows for the easy flow of electrons.

Insulator: A material that will not conduct electricity.

Amperage: The amount of current flow through a conductor.

Ampere: The unit for measuring the rate of electric current flow.

Voltage: The electromotive force that causes current flow. The potential difference in electrical force between two points when one is negatively charged and the other is positively charged.

Volt: The unit for measuring the amount of electrical force.

Circuit: A circle or unbroken path of conductors through which an electric current can flow.

Resistance: Opposition to electrical current flow.

Ohm: The unit for measuring electrical resistance.

Ground: The connection of an electrical circuit or unit to the engine or chassis to return electrical current to the battery.

Ground Cable: The battery cable that provides a ground connection from the vehicle chassis to the battery.

Insulated (Hot) Cable: The battery cable that conducts battery current to the automotive electrical system.

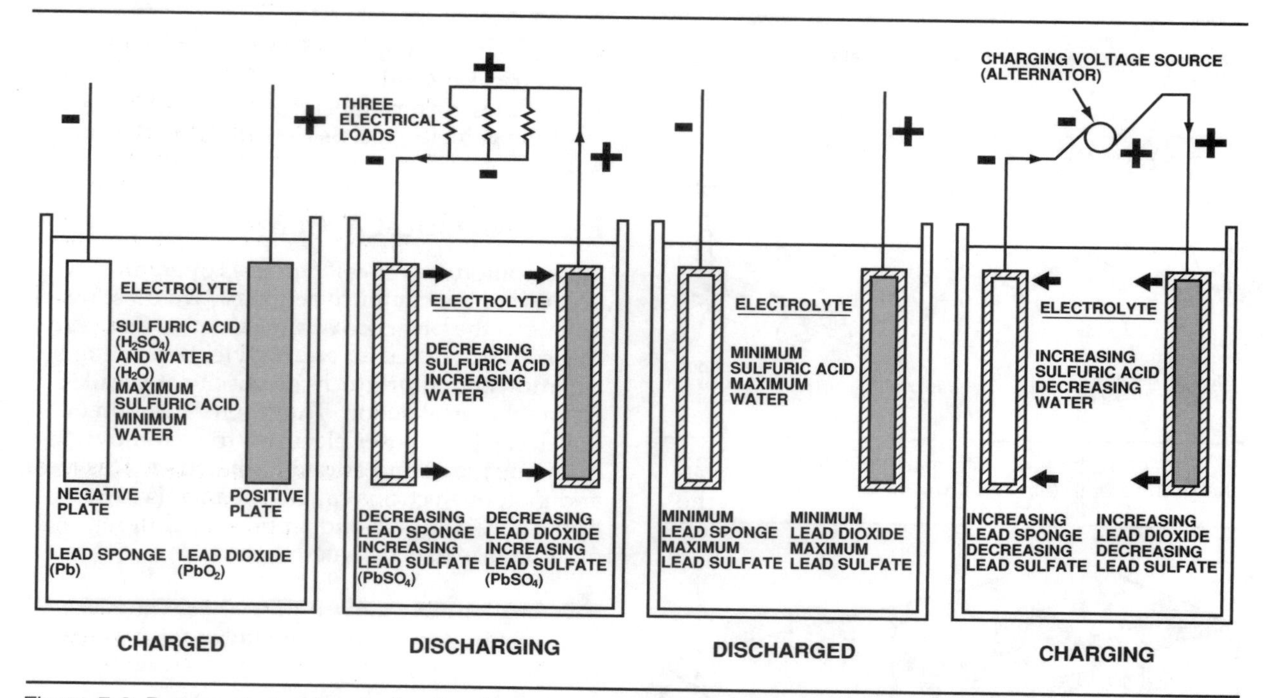

Figure 7-3. Battery electrochemical action from charged to discharged and back to charged.

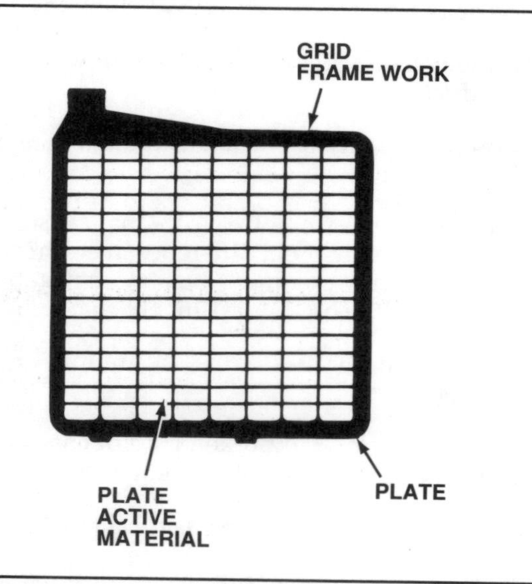

Figure 7-4. The grid provides a support for the plate active material.

The metal piece that has lost electrons is positively charged, and is called the positive plate. The piece that has gained electrons is negatively charged, and is called the negative plate. If a conductor and a load are connected between the two plates, current will flow through the conductor, figure 7-2. For simplicity, battery current flow is assumed to be conventional current flow (+ to -) through the external circuit connected to the battery.

Primary and secondary batteries

There are two general types of batteries: primary and secondary.

The action within a **primary battery** causes one of the metals to be totally destroyed after a period of time. When the battery has delivered all of its voltage to an outside circuit, it is useless and must be replaced. Many small dry-cell batteries, such as those for flashlights and radios, are primary batteries.

In **secondary batteries**, both the electrolyte and the metals change their atomic structure as the battery supplies voltage to an outside circuit. This is called discharging. The action can be reversed, however, by applying an outside current to the battery terminals and forcing current to flow through the battery in the opposite direction. This current flow causes chemical action that restores the battery materials to their original condition, and the battery can again supply voltage. This is called charging the battery. The condition of the battery materials is called the battery's state of charge.

Electrochemical action in automotive batteries

A fully charged automotive battery contains a series of negative plates of chemically active sponge lead (Pb), positive plates of lead dioxide (PbO_2), and an electrolyte of sulfuric acid (H_2SO_4) and water (H_2O), figure 7-3.

As the battery discharges, figure 7-3, the chemical action taking place reduces the acid content in the electrolyte and increases the water content. At the same time, both the negative and positive plates gradually change to lead sulfate ($PbSO_4$).

A discharged battery, figure 7-3, has a very weak acid solution because most of the electrolyte has changed to water. Both series of plates are mostly lead sulfate. The battery now stops functioning because the plates are basically two similar metals in the presence of water, rather than two dissimilar metals in the presence of an electrolyte.

During charging, figure 7-3, the chemical action is reversed. The lead sulfate on the plates gradually decomposes, changing the negative plates back to sponge lead and the positive plates to lead dioxide. The sulfate is redeposited in the water, which increases the sulfuric acid content and returns the electrolyte to full strength. The battery is now again able to supply voltage.

This electrochemical action and battery operation from fully charged to discharged and back to fully charged is called **cycling**.

A safety note is important here. Hydrogen and oxygen gases are formed during battery charging. Hydrogen gas is explosive. Never strike a spark or bring a flame near a battery, particularly during or after charging. *This could cause the battery to explode.*

Battery Construction

There are four types of automotive batteries currently in use:

- Vent-cap (requires maintenance)
- Low-maintenance (requires limited maintenance)
- Maintenance-free (requires no maintenance)
- Recombinant (requires no maintenance).

The basic physical construction of all types of automotive batteries is similar, but not the materials. We will look at traditional vent-cap construction first, and then explain how the other battery types differ.

Vent-cap batteries

Battery construction begins with the positive and negative plates. The plates are built on grids of conductive materials, figure 7-4, which act as a framework for the dissimilar metals. These dissimilar metals are called the active materials of the battery. The active materials, sponge lead and lead dioxide, are pasted onto the grids. When dry, the active materials are very porous so that the electrolyte can easily penetrate and react with them.

A number of similar plates, all positive or all negative, are connected into a plate group, figure 7-5. The plates are joined by welding them to a plate strap through a process called lead burning. The plate strap has a connector or a terminal post for attaching plate groups to each other.

A positive and a negative plate group are interlaced so that their plates alternate, figure 7-6. The negative plate group normally has one more plate than the positive group. To reduce the possibility of a short between plates of the two groups, they are separated by chemically inert separators, figure 7-6. Separators are usually made of plastic or fiberglass. The separators have ribs on one side next to the positive plates. These ribs hold electrolyte near the positive plates for efficient chemical action.

Electrolyte: The chemical solution in a battery that conducts electricity and reacts with the plate materials.

Primary Battery: A battery in which chemical processes destroy one of the metals necessary to create electrical energy. Primary batteries cannot be recharged.

Secondary Battery: A battery in which chemical processes can be reversed. A secondary battery can be recharged so that it will continue to supply voltage.

Cycling: Battery electrochemical action and operation where one complete cycle is from fully charged to discharged and back to fully charged.

■ Other Secondary Cells

The Edison (nickel-iron alkali) cell and the silver cell are two other types of secondary cells. The positive plate of the Edison cell is made of pencil shaped perforated steel tubes that contain nickel hydroxide. These tubes are held in a steel grid. The negative plate has pockets that hold iron oxide. The electrolyte used in this cell is a solution of potassium hydroxide and a small amount of lithium hydroxide.

An Edison cell weighs about ½ as much as a lead-acid cell of the same amp-hour capacity. This cell has a long life and is not damaged by short circuits or overloads. However, it is more costly than a lead-acid cell.

The silver cell has a positive plate of silver oxide and a negative plate of zinc. The electrolyte is a solution of potassium hydroxide or sodium. For its weight, this cell has a high amp-hour capacity. It can withstand large overloads and short circuits. It, too, is more expensive than a lead-acid cell.

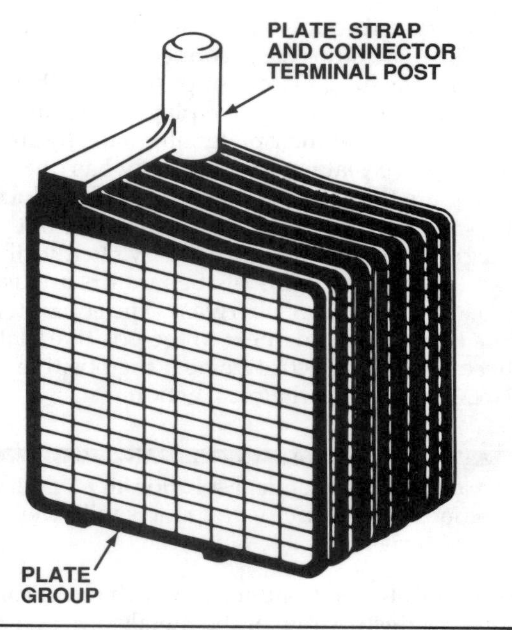

Figure 7-5. A number of plates are connected into a group.

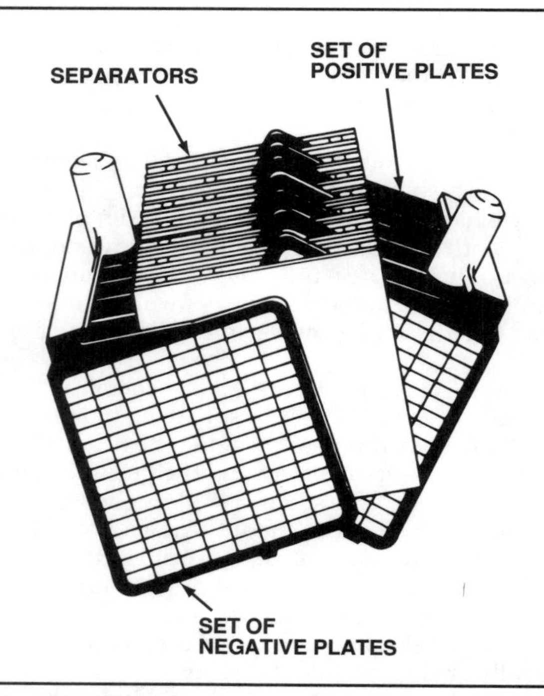

Figure 7-6. Two groups are interlaced to form a battery element.

A complete assembly of positive plates, negative plates, and separators is called an **element**. It is placed in a cell of a battery case. Because each **cell** provides approximately 2.1 volts, a 12-volt battery has six cells and actually produces approximately 12.6 volts when fully charged.

The elements are separated from each other by cell partitions and rest on bridges at the bottom of the case, which form chambers where sediment can collect. These bridges prevent accumulated sediment from shorting across the bottoms of the plates. Once installed in the case, the elements (cells) are connected to each other by connecting straps that pass over or through the cell partitions, figure 7-7. The cells are connected alternately in series (positive to negative to positive to negative, etc.), and the battery top is bonded onto the case to form a watertight container.

Vent caps in the battery top provide an opening for adding electrolyte and for the escape of gases that form during charging and discharging. The battery is connected to the car's electrical system by two external terminals. These terminals are either tapered posts on top of the case or internally threaded connectors on the side. The terminals, which are connected to the ends of the series of elements inside the case, are marked positive (+) or negative (-) according to the end of the series each terminal represents.

Low-maintenance and maintenance-free batteries

Most new batteries today are either semisealed low-maintenance or sealed maintenance-free batteries. Low-maintenance batteries provide some method of adding water to the cells, such as:

- Individual slotted vent caps installed flush with the top of the case
- Two vent panel covers, each of which exposes three cells when removed
- A flush-mounted strip cover that is peeled off to reveal the cell openings.

Maintenance-free batteries have only small gas vents that prevent pressure buildup in the case. A low-maintenance battery requires that water be added much less often than with a traditional vent-cap battery, while a maintenance-free battery will never need to have water added during its lifetime.

These batteries differ from vent-cap batteries primarily in the materials used for the plate grids. For decades, automotive batteries used antimony as the strengthening ingredient of the grid alloy. In low-maintenance batteries, the amount of antimony is reduced to about three percent. In maintenance-free batteries, the antimony is eliminated and replaced by calcium or strontium.

Reducing the amount of antimony or replacing it with calcium or strontium alloy results in lowering the battery's internal heat, and reduces the amount of gassing that occurs during charg-

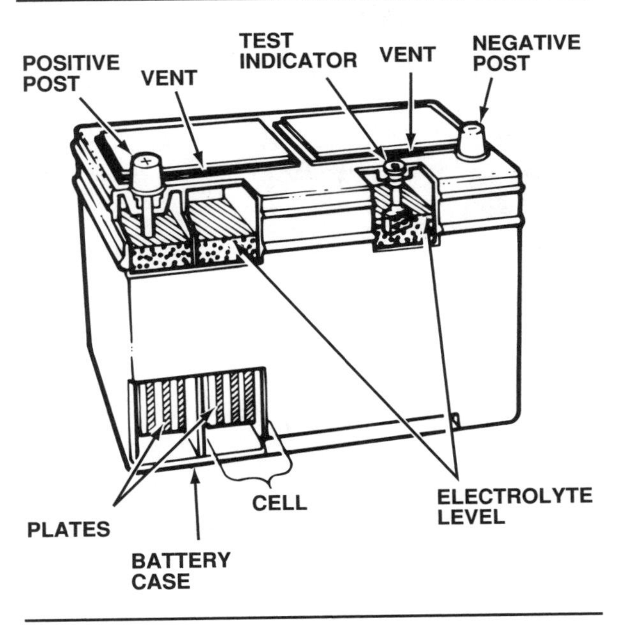

Figure 7-7. A cutaway view of an assembled battery. (Chrysler)

ing. Since these are the principal reasons for battery water loss, these changes reduce or eliminate the need to periodically add water. Reduced water loss also minimizes terminal corrosion, since the major cause of this corrosion is condensation from normal battery gassing.

In addition, nonantimony lead alloys have better conductivity, so a maintenance-free battery has about a 20-percent higher cranking performance rating than a traditional vent-cap battery of comparable size.

Sealed maintenance-free batteries

More recently, completely sealed maintenance-free batteries were introduced. These new batteries do not require—and do not have—the small gas vent used on previous maintenance-free batteries. Although these batteries are basically the same kind of lead-acid voltage cells used in automobiles for decades, a slight change in plate and electrolyte chemistry reduces hydrogen generation to almost nothing. During charging, a vent-cap or maintenance-free battery releases hydrogen at the negative plates and oxygen at the positive plates. Most of the hydrogen is released through electrolysis of the water in the electrolyte near the negative plates as the battery reaches full charge. In the sealed maintenance-free design, the negative plates never reach a fully charged condition and therefore cause little or no release of hydrogen. Oxygen *is* released at the positive plates, but it passes through the separators and recombines with the negative plates. The overall effect is virtually no gassing from the battery. Because the oxygen released by the electrolyte recombines with the negative plates, some manufacturers call these batteries "recombination" or **recombinant** electrolyte batteries.

Recombinant batteries

Recombination electrolyte technology and improved grid materials allow some sealed maintenance-free batteries to develop fully charged, open-circuit voltage of approximately 2.2 volts per cell, or a total of 13.2 volts for a six-cell battery. Microporous fiberglass separators reduce internal resistance and contribute to higher voltage and current ratings.

In addition, the electrolyte in these new batteries is contained within plastic envelope-type separators around the plates, figure 7-8. The entire case is not flooded with electrolyte. This eliminates the possibility of damage due to sloshing or acid leaks from a cracked battery. This design feature reduces battery damage during handling and installation, and allows a more-compact case design. Because the battery is not vented, terminal corrosion from battery gassing and electrolyte spills or spray is also eliminated.

The envelope design also catches active material as it flakes off the positive plates during discharge. By holding the material closer to the plates, envelope construction ensures that it will be more completely redeposited during charging.

Although recombinant batteries are examples of advanced technology, test and service requirements are basically the same as for other maintenance-free lead-acid batteries. Some manufacturers caution, however, that fast charging at high current rates may overheat the battery and can cause damage. Always check the manufacturer's instructions for test specifications and charging rates before servicing one of these batteries.

Element: A complete assembly of positive plates, negative plates, and separators making up one cell of a battery.

Cell: A case enclosing one element in an electrolyte. Each cell produces approximately 2.1 to 2.2 volts. Cells are connected in series.

Recombinant: A nongassing battery design in which the oxygen released by the electrolyte recombines with the negative plates.

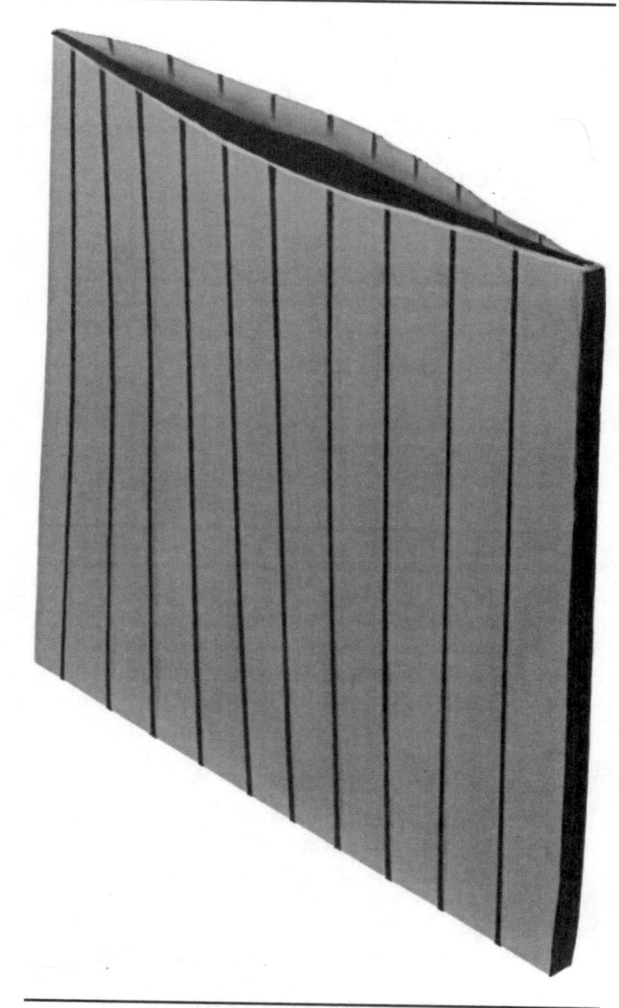

Figure 7-8. Many maintenance-free batteries have envelope-type separators that hold active material near the plates.

Battery Electrolyte

For the battery to become chemically active, it must be filled with an electrolyte solution. The electrolyte in an automotive battery is a solution of sulfuric acid and water. In a fully charged battery, the solution is approximately 35 to 39 percent acid by weight (25 percent by volume) and 61 to 65 percent water by weight. The state of charge of a battery can be measured by checking the **specific gravity** of the electrolyte.

Specific gravity is the weight of a given volume of liquid divided by the weight of an equal volume of water. Since acid is heavier than water, and water has a specific gravity of 1.000, the specific gravity of a fully charged battery is approximately 1.260 when weighed in a hydrometer. As the battery discharges, the specific gravity of the electrolyte decreases because

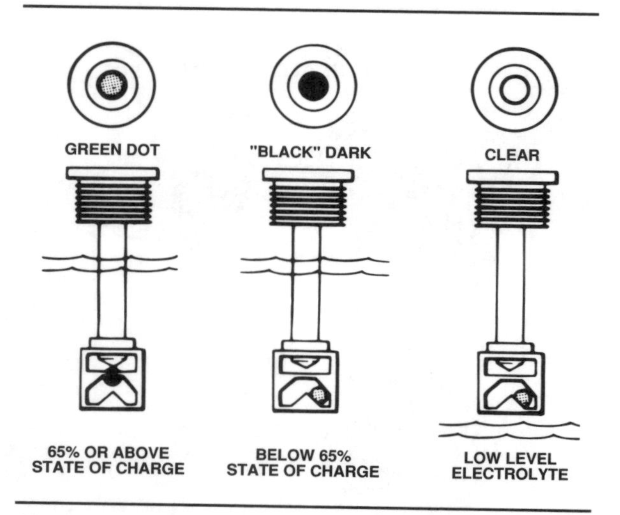

Figure 7-9. Delco "Freedom" batteries have this integral hydrometer built into their tops.

the acid is changed into water. The specific gravity of the electrolyte can tell you approximately how discharged the battery has become:

1.265 specific gravity	100% charged
1.225 specific gravity	75% charged
1.190 specific gravity	50% charged
1.155 specific gravity	25% charged
1.120 specific gravity or lower	discharged

These values may vary slightly according to the design factors of a particular battery.

Specific gravity measurements are based on a standard temperature of 80°F (27°C). At higher temperatures, specific gravity is lower. At lower temperatures, specific gravity is higher. For every change of 10°F (6°C), specific gravity changes by four points (0.004). That is:

- For every 10°F (6°C) above 80°F (27°C), add 0.004 to the specific gravity reading.
- For every 10°F (6°C) below 80°F (27°C), subtract 0.004 from the specific gravity reading.

When you study battery service in the *Shop Manual*, you will learn to measure specific gravity of a vent-cap battery with a hydrometer.

State-of-Charge Indicators

Many low-maintenance and maintenance-free batteries have a visual state-of-charge indicator installed in the battery top. The indicator shows whether the electrolyte has fallen below a minimum level, and also functions as a go/no-go hydrometer.

The indicator, figure 7-9, is a plastic rod inserted in the top of the battery and extending into the electrolyte. In the design used by Delco, a green plastic ball is suspended in a cage from the bottom of the rod. Depending on the specific gravity of the electrolyte, the ball will float or sink in the cage, changing the appearance of the indicator "eye" from green to dark. When the eye is dark, the battery should be recharged.

Other manufacturers either use the "Delco Eye" under license, or one of several variations of the design. One variation contains a red and blue ball side by side in the cage. When the specific gravity is high, only the blue ball can be seen in the "eye". As the specific gravity falls, the blue ball sinks in the cage, allowing the red ball to take its place. When the battery is recharged, the increasing specific gravity causes the blue ball to move upward, forcing the red ball back into the side of the cage.

Another variation is the use of a small red ball on top of a larger blue ball. When the specific gravity is high, the small ball is seen as a red spot surrounded by blue. As the specific gravity falls, the blue ball sinks, leaving the small ball to be seen as a red spot surrounded by a clear area. The battery then should be recharged.

If the electrolyte drops below the level of the cage in batteries using a state-of-charge indicator, the "eye" will appear clear or light yellow. This means that the battery must be replaced because it has lost too much electrolyte.

Wet and Dry-Charged Batteries

Batteries may be manufactured and sold as either wet-charged or dry-charged batteries. Before maintenance-free batteries became widely used, dry-charged batteries were very common. A wet-charged battery is completely filled with an electrolyte when it is built. A dry-charged battery is shipped from the factory without electrolyte. During manufacture, the positive and negative plates are charged and then completely washed and dried. The battery is then assembled and sealed to keep out moisture. It will remain charged as long as it is sealed, and it can be stored for a long time in any reasonable environment. A dry-charged battery is put into service by adding electrolyte, checking the battery state of charge, and charging if needed.

Even when a wet-charged battery is not in use, a slow reaction occurs between the plates and the electrolyte. This is a self-discharging reaction, and will eventually discharge the battery almost completely. Because this reaction occurs faster at higher temperatures, wet-charged batteries should be stored in as cool a place as possible when not in use. A fully charged battery stored at a room temperature of 100°F (38°C) will almost completely discharge after 90 days. If the battery is stored at a temperature of 60°F (16°C), very little discharge will take place.

Battery Charging Voltage

A battery is charged by forcing current to flow through it in the direction opposite to its discharge current. In an automobile, this charging current is supplied by the generator or alternator. The battery offers some resistance to this charging current, because of the battery's chemical voltage and the resistance of the battery's internal parts. The battery's chemical voltage is a form of **counterelectromotive force (CEMF)**, which *opposes* and *slows down* the *increase* in current.

When a battery is fully charged, its CEMF is very high. Very little charging current can flow through it. When the battery is discharged, its CEMF is very low, and charging current flows freely. For charging current to enter the battery, the charging voltage must be higher than the battery's CEMF *plus* the voltage drop caused by the battery's internal resistance.

Understanding this relationship of CEMF to the battery state of charge is helpful. When the battery is nearly discharged it needs, and will accept, a lot of charging current. When the battery is fully charged, the high CEMF will resist charging current. Any additional charging current could overheat and damage the battery materials.

The temperature of the battery affects the charging voltage because temperature affects the resistance of the electrolyte. Cold electrolyte has higher resistance than warm electrolyte, so a colder battery is harder to charge. The effects of temperature must be considered when servicing automotive charging systems and batteries, as we will see later in this chapter.

Battery Selection and Rating Methods

Automotive batteries are 12-volt wet-cell lead-acid batteries and are available in a variety of sizes, shapes, and current ratings. They are

■ Do Not Pull the Plugs

Do you make a practice of removing the vent plugs from a battery before charging it? Prestolite says you should not, at least with many late-model batteries. "A great number of batteries manufactured today will have safety vents", says Prestolite. "If these are removed, the batteries are open to external sources of explosion ignition". Prestolite recommends that the vent plugs should be left in place when charging on batteries with safety vents.

Figure 7-10. The most common type of top terminal battery clamp.

called "starting batteries", and are designed to deliver a large current output for a brief time to start an engine. After starting, the charging system takes over to supply most of the current required to operate the car. The battery acts as a system stabilizer and provides current whenever the electrical loads exceed the charging current output. An automotive battery must provide good cranking power for the car's engine and adequate reserve power for the electrical system in which it is used.

Manufacturers also make 12-volt automotive-type batteries that are not designed for automotive use. These are called "cycling batteries", and are designed to provide a power source for a vehicle or accessory without continual recharging. Cycling batteries provide a constant low current for a long period of time. They are designed for industrial, marine, and recreational vehicle (RV) use. Most of their current capacity is exhausted in each cycle before recharging.

The brief high-current flow required of a starting battery is produced by using relatively thin plates, compared to those used in a cycling battery. The thicker plates of the cycling battery will provide a constant current drain for several hours. Using a starting battery in an application calling for a cycling battery will shorten its life considerably, as we will see later in the chapter. The use of a cycling battery to start and operate a car will cause excessive internal heat from the brief but high current draw, resulting in a shorter service life.

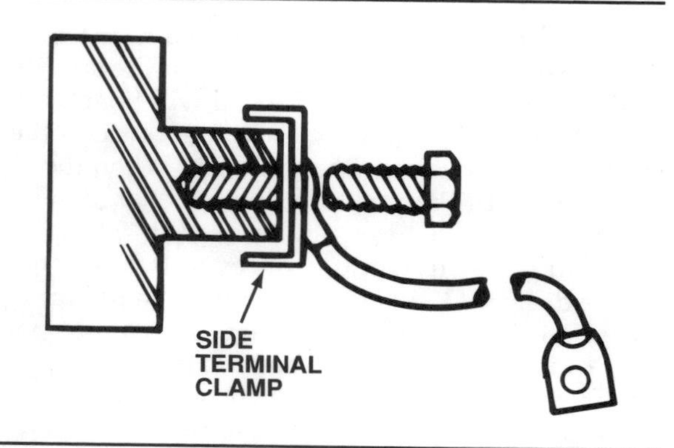

Figure 7-11. This Chrysler side terminal clamp is attached with a bolt.

Test standards and rating methods devised by the Battery Council International (BCI) and the Society of Automotive Engineers (SAE) are designed to measure a battery's ability to meet the requirements for which it is to be used.

The BCI publishes application charts that list the correct battery for any car. Optional heavy-duty batteries are normally used in cars with air conditioning, several major electrical accessories, or in cars operated in cold climates. To ensure adequate cranking power and to meet all other electrical needs, a replacement battery may have a higher rating, but never a lower rating, than the original unit. The battery must also be the correct size for the car, and have the correct type of terminals. BCI standards include a coding system called the group number. BCI battery rating methods are explained in the following paragraphs.

Ampere-hour rating

The oldest battery rating method, no longer used to rate batteries, was the ampere-hour rating. This rating method was the industry standard for decades. It was replaced, however, years ago by the cranking performance and reserve capacity ratings, which provide better indications of a battery's performance.

The ampere-hour method was also called the 20-hour discharge rating method. This rating represented the steady current flow that a battery delivered at a temperature of 80°F (27°C) without cell voltage falling below 1.75 volts (a total of 10.5 volts for a 12-volt battery). For example, a battery that continuously delivered 3 amps for 20 hours was rated as a 60 ampere-hour battery (3 amps x 20 hours = 60 amp hours).

Cranking performance rating

The **cranking performance rating** indicates the power a battery can supply for an engine cranking at 0°F (-18°C). The rating figures for car and light truck batteries range from 165 to 1,050 cold-cranking amps. These figures represent the current flow a battery can deliver for 30 seconds at 0°F (-18°C) while maintaining at least 1.2 volts per cell, for a minimum terminal voltage of 7.2 volts for a 12-volt battery.

Reserve capacity rating
The **reserve capacity rating** indicates the long-term power available from a battery for ignition, lighting, and accessories required in emergencies. Reserve capacity (listed in minutes) is the time a fully charged battery at 80°F (27°C) can deliver 25 amps and maintain at least 1.75 volts at every cell, or 10.5 volts total for a 12-volt battery. Battery reserve capacity ratings range from 30 to 175 minutes, and correspond approximately to the length of time a vehicle can be driven after the charging system has failed.

Group number
Carmakers provide a designated amount of space in the engine compartment to accommodate the battery. Since battery companies build batteries of various current-capacity ratings in a variety of sizes and shapes, it is useful to have a guide when replacing a battery, because it must fit into the space provided. The BCI size **group number** identifies a battery in terms of its length, width, height, terminal design, and other physical features.

Battery Installation Components

Selecting and maintaining properly designed battery installation components are a prerequisite for good battery operation and service life.

Connectors, carriers, and holddowns
Battery cables are made up of multistrand wires, usually number 6 to 0 gauge. Gasoline engine vehicles generally use the number 6, while diesel engine vehicles use the larger number 0. A new battery cable should always be the same gauge as the one being replaced.

Battery terminals may be tapered posts on the top or internally threaded terminals on the side of the battery. To prevent accidental reversal of

Specific Gravity: The weight of a volume of liquid divided by the weight of the same volume of water at a given temperature and pressure. Water has a specific gravity of 1.000.

Counterelectromotive Force (CEMF): An induced voltage that opposes the source voltage and any increase or decrease in source current flow.

Cranking Performance Rating: A battery rating based on the amperes of current that a battery can supply for 30 seconds at 0°F (-18°C) with no battery cell falling below 1.2 volts.

Reserve Capacity Rating: A battery rating based on the number of minutes a battery can supply 25 amps for 30 seconds at 0°F (-18°C) with no battery cell falling below 1.75 volts.

Group Number: A battery identification number that indicates battery dimensions, terminal design, holddown location, and other physical features.

■ Parasitic Losses

Parasitic losses are small current drains required to operate automotive accessories, such as the clock. These systems continue to work when the car is parked and the ignition is off. Usually the current demands are small and not likely to cause a problem.

With the advent of computer controls, problems arising from excessive parasitic losses have become more frequent. Many late-model cars have computers to control such diverse items as engine operation, radio tuning, suspension leveling, and climate control. Each microprocessor contains random access memory (RAM) that stores information relevant to its job. To "remember", RAM requires a constant supply of power, and therefore puts a continuous drain on the car's electrical system.

The combined drain of several computer memories can discharge a battery to the point where there is insufficient cranking power after only a few weeks. These vehicles may require more-frequent battery charging when driven infrequently compared to older cars with lower parasitic voltage losses.

Because of the higher parasitic current drains on late-model cars, the old test of removing a battery cable connection and tapping it against the terminal while looking for a spark is both dangerous and no longer a valid check for excessive current drain. Furthermore, every time the power source to the computer is interrupted, the information programmed into memory is lost and will have to be reprogrammed when the battery is reconnected.

On engine control systems with learning capability, like GM's Computer Command Control, driveability may also be affected until the computer relearns the engine calibration modifications that were erased from its memory when the battery was disconnected.

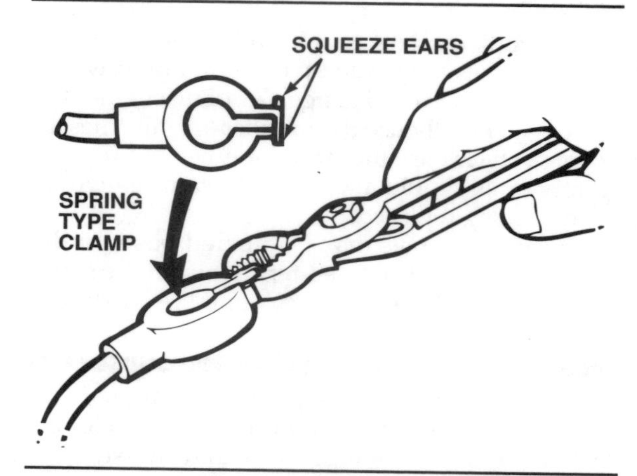

Figure 7-12. The spring-type clamp generally is found on nondomestic cars.

battery polarity (incorrectly connecting the cables), the positive terminal is slightly larger than the negative terminal. Three basic styles of connectors are used to attach the battery cables to the battery terminals:

- A bolt-type clamp is used on top-terminal batteries, figure 7-10. The bolt passes through the two halves of the cable end into a nut. When tightened, it squeezes the cable end against the battery post.
- A bolt-through clamp is used on side-terminal batteries. The bolt threads through the cable end and directly into the battery terminal, figure 7-11.
- A spring-type clamp is used on some top-terminal batteries. A built-in spring holds the cable end on the battery post, figure 7-12.

Batteries are usually mounted on a shelf or tray in the engine compartment, although some manufacturers place the battery in the trunk, under the seat, or elsewhere in the vehicle. The shelf or tray that holds the battery is called the carrier, figure 7-13. The battery is mounted on the carrier with brackets called holddowns, figures 7-13 and 7-14. These keep the battery from tipping over and spilling acid. A battery must be held securely in its carrier to protect it from vibration that can damage the plates and internal plate connectors.

Battery heat shields

Many late-model cars use battery heat shields, figure 7-15, to protect batteries from high underhood temperatures. Most heat shields are made of plastic, and some are integral with the battery holddown. Integral shields are usually large plastic plates that sit alongside the battery.

Heat shields do not require removal for routine battery inspection and testing, but must be removed for battery replacement.

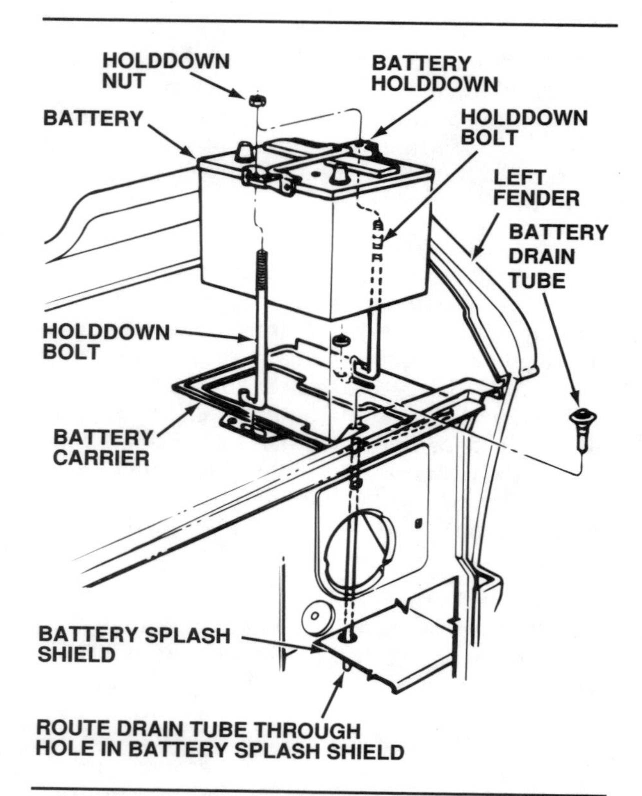

Figure 7-13. A common type of battery carrier and holddown.

Battery Life and Performance Factors

All batteries have a limited life, but certain conditions can shorten that life. The important factors that affect battery life are discussed in the following paragraphs.

Electrolyte level

As we have seen, the design of maintenance-free batteries has minimized the loss of water from electrolyte so that battery cases can be sealed. Given normal use, the addition of water to such batteries is not required during their service life. However, even maintenance-free batteries will lose some of their water to high temperature, overcharging, deep cycling, and recharging—all factors in battery gassing and resulting water loss.

With vent-cap batteries, and to some extent low-maintenance batteries, water is lost from the electrolyte during charging in the form of hydrogen and oxygen gases. This causes the electrolyte level to drop. If the level drops below the top of the plates, active material will be exposed to the air. The material will harden and resist electrochemical reaction. Also, the remaining electrolyte will have a high concentration of acid, which can cause the plates to deteriorate quickly. Even the addition of water will not restore such hardened plates to a fully active condition.

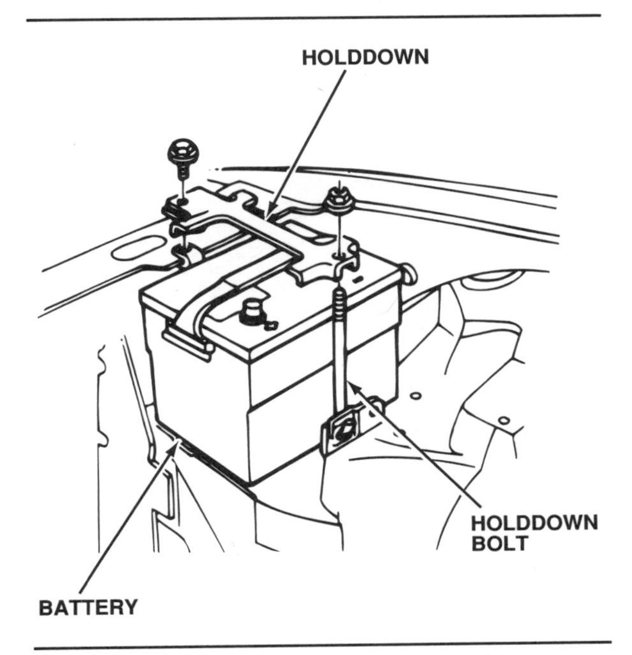

Figure 7-14. Another common battery holddown.

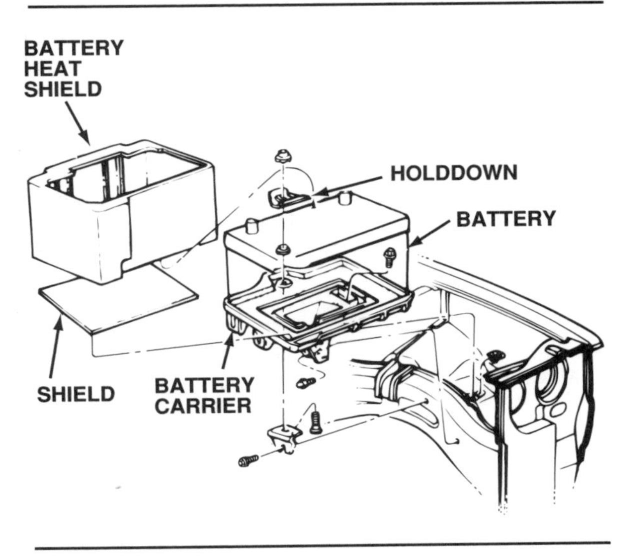

Figure 7-15. A molded heat shield that fits over the battery is used by Chrysler and some other carmakers.

Corrosion
Battery corrosion is caused by spilled electrolyte and electrolyte condensation from gassing. The sulfuric acid attacks and can destroy not only connectors and terminals, but metal holddowns and carriers as well. Corroded connectors increase resistance at the battery connections. This reduces the applied voltage for the car's electrical system. Corrosion also can cause mechanical failure of the holddowns and carrier, which can damage the battery. Spilled electrolyte and corrosion on the battery top also can create a current leakage path, which can allow the battery to discharge.

Overcharging
Batteries can be overcharged by the automotive charging system or by a separate battery charger. In either case, there is a violent chemical reaction in the battery. The water in the electrolyte is rapidly broken down into hydrogen and oxygen gases. These gas bubbles can wash active material off the plates, as well as lower the level of the electrolyte. Overcharging can also cause excessive heat, which can oxidize the positive grid material and even buckle the plates.

Undercharging and sulfation
If an automobile is not charging its battery, either because of stop-and-start driving or a fault in the charging system, the battery will be constantly discharged. As we saw in the explanation of electrochemical action, a discharged plate is covered with lead sulfate. The amount of lead sulfate on the plate will vary according to the state of charge. As the lead sulfate builds up in a constantly undercharged battery, it can crystallize and not recombine with the electrolyte. This is called battery **sulfation**. The crystals are difficult to break down by normal recharging and the battery becomes useless. Despite the chemical additives sold as "miracle cures" for sulfation, a completely sulfated battery cannot be effectively recharged.

Cycling
As we learned at the beginning of this chapter, the operation of a battery from charged to discharged and back to charged is called cycling. Automotive batteries are not designed for continuous deep-cycle use (although special marine and RV batteries are). If an automotive battery is repeatedly cycled from a fully charged condition to an almost discharged condition, the active material on the positive plates may shed and fall into the bottom of the case. If this happens, the material cannot be restored to the plates. Cycling thus reduces the capacity of the battery and shortens its useful service life.

Temperature
Temperature extremes affect battery service life and performance in a number of ways. High temperature, caused by overcharging or excessive engine heat, increases electrolyte loss and shortens battery life.

Sulfation: The crystallization of lead sulfate on the plates of a constantly discharged battery.

COMPARISON OF CRANKING POWER AVAILABLE FROM FULLY CHARGED BATTERY AT VARIOUS TEMPERATURES

Temperature	Cranking power
80°F (26.7°C)	100%
32°F (0°C)	65%
0°F (-17.8°C)	40%

Figure 7-16. This graph represents the increased power required to crank an engine at low temperatures.

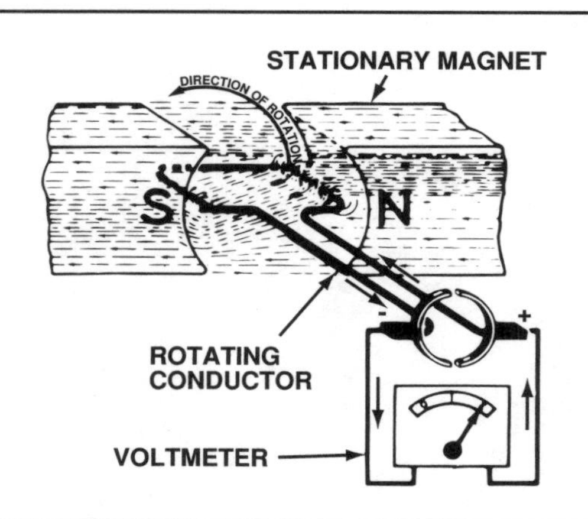

Figure 7-17. A simplified generator. (Prestolite)

Low temperatures in winter can also harm a battery. If the electrolyte freezes, it can expand and break the case, ruining the battery. The freezing point of electrolyte depends on its specific gravity and thus, on the battery's state of charge. A fully charged battery with a specific gravity of 1.265 to 1.280 will not freeze until its temperature drops below -60°F (-51°C). A discharged battery with an electrolyte that is mostly water can freeze at -18°F (-8°C).

As we saw earlier, cold temperatures make it harder to keep the battery fully charged, yet this is when a full charge is most important. Figure 7-16 compares the energy levels available from a fully charged battery at various temperatures. As you can see, the colder a battery, the less energy it can supply. Yet the colder an engine gets, the more energy it requires for cranking. This is why battery care is especially important in cold weather.

Vibration

As mentioned earlier, a battery must be securely mounted in its carrier to protect it from vibration. Vibration can shake the active materials off the plates and severely shorten a battery's life. Vibration can also loosen the plate connections to the plate strap and damage other internal connections. Some manufacturers now build batteries with plate straps and connectors in the center of the plates to reduce the effects of vibration. Severe vibration can even crack a battery case and loosen cable connections.

CHARGING SYSTEM OPERATION

The charging system converts the engine's mechanical energy into electrical energy. This electrical energy is used to maintain the battery's state of charge and to operate the loads of the automotive electrical system. During our study of the charging system, we will use the conventional theory of current flow (+ to -).

During cranking, all electrical energy for the car is supplied by the battery. Once the engine is running, the charging system must produce enough electrical energy to recharge the battery and to supply the demands of other loads in the electrical system.

If the cranking system is in poor condition and draws too much current, or if the charging system cannot recharge the battery and supply the additional loads, more energy must be drawn from the battery for short periods of time.

Voltage Sources

The two devices that convert mechanical energy to electrical energy are the generator or the alternator. Both units use the principle of **induction** to create an electrical voltage. When the lines of force of a magnetic field cut through a conductor (wire), a voltage will appear in the wire. The voltage will cause current flow in a complete circuit such as the automobile electrical system.

The generator is constructed with a stationary magnet, figure 7-17, and loops of wire that rotate within the magnet's field. The loops of wire are turned by a drive belt from the engine. When the loops turn, the lines of force from the magnet cut through them, and a voltage is induced within the loops. This voltage is used to recharge the battery and operate the rest of the car's electrical system.

The current that recharges the battery must be **direct current** (dc), that is, current that always flows in the same direction. Induction as used in generators and alternators will cause **alternating current** (ac), a current that flows first in one direction and then in the other. This alternating current must be **rectified,** or changed to direct current, before it can be used in the car's electrical system.

The generator is constructed so that the current is rectified mechanically. The alternator can-

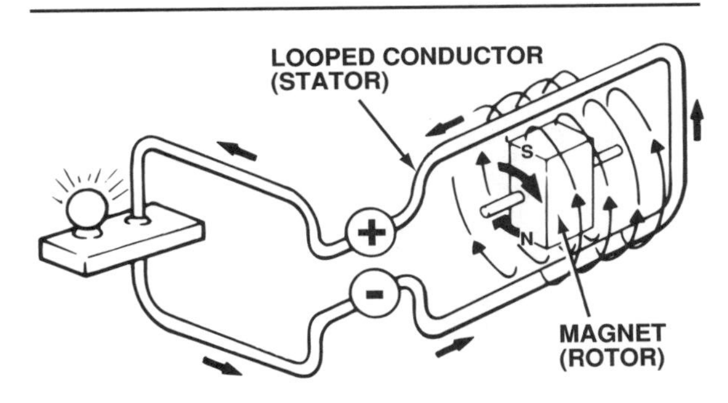

Figure 7-18. A simplified alternator.

not use a mechanical rectifier. Until a practical way was developed to rectify alternator output, generators were used in cars. Today, alternators use **solid-state** electronic parts as rectifiers. These are called **diodes,** and they act as one-way electrical check valves. Current can flow through a diode in one direction but not in the other.

The alternator, figure 7-18, works on the same principle of induction as the generator. As the lines of force from the magnet cut through the conductor, voltage is induced.

Charging Voltage

Although the automotive electrical system is called a 12-volt system, the alternator must produce more than 12 volts. We learned earlier in this chapter that each battery cell produces about 2.1 volts when fully charged. This means that the open-circuit voltage of a fully charged 12-volt battery (six cells) is approximately 12.6 volts. If the alternator cannot produce more than 12 volts, it cannot charge the battery until system voltage drops under 12 volts. This would leave nothing extra to serve the other electrical demands put on the system by lights, air conditioning, and power accessories.

Alternating-current charging systems are generally regulated to produce a maximum output of 14.5 volts. Output of more than 16 volts will overheat the battery electrolyte and shorten its life. High voltage also can damage components that rely heavily on solid-state electronics, such as fuel injection and engine control systems. On the other hand, low voltage output will cause the battery to become sulfated. As you can see, the charging system must be maintained within the voltage limits specified by the carmaker if the vehicle is to perform properly.

Alternating current charging system components

Late-model automotive charging systems, figure 7-19, contain:

- A battery, which provides the initial field current required to operate the alternator and, in turn, is charged and maintained by the alternator.
- An alternator, which is belt-driven by the engine and converts mechanical motion into charging voltage and current. A simple alternator, figure 7-20, consists of a magnet rotating inside a fixed-loop stator, or conductor. The alternating current produced in the conductor is rectified by diodes for use by the electrical system.
- A regulator, which limits the field current flow and thus the alternator's voltage output according to the electrical system demand. A regulator can be either an electromechanical or a solid-state device. Some late-model solid-state regulators are part of the vehicle's on-board computer.

Induction: The production of an electrical voltage in a conductor, or coil, by moving the conductor or coil through a magnetic field, or by moving the magnetic field past the conductor or coil.

Direct Current (dc): A flow of electricity in one direction through a conductor.

Alternating Current (ac): A flow of electricity through a conductor, first in one direction, then in the opposite direction.

Rectified: Electrical current changed from ac to dc.

Solid-State: A method of controlling electrical current flow, using parts primarily made of semiconductor materials.

Diodes: Electronic devices made of P-material and N-material bonded at a junction. Diodes allow current flow in one direction and block it in the other.

Deep-Cycle Service

Some batteries, like those in golf carts and electric vehicles, are used for deep-cycle service. This means that as they provide electrical power they go from a fully charged state to an almost fully discharged state, and are then recharged and used again.

Maintenance-free batteries should never be used in deep-cycle service. Deep-cycle service promotes shedding of the active materials from the battery plates. This action drastically reduces the service life of a maintenance-free battery.

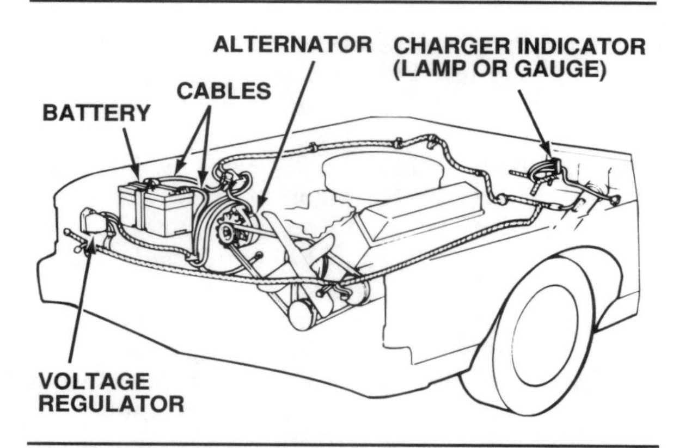

Figure 7-19. The major components of an automotive charging system. (Chrysler)

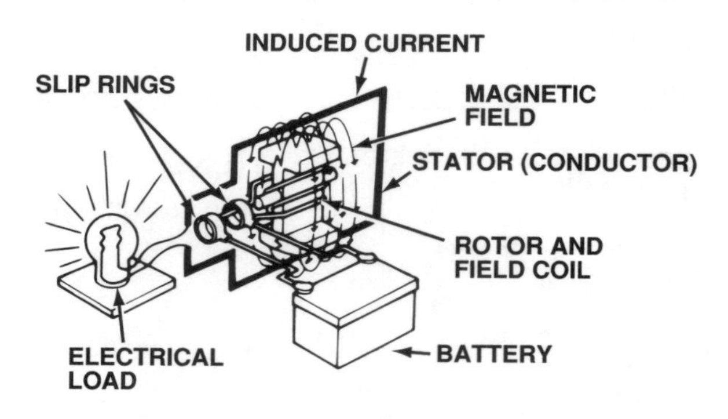

Figure 7-20. An alternator is based on the rotation of a magnet inside a fixed-loop conductor. (Chrysler)

- An ammeter, a voltmeter, or an indicator warning lamp mounted on the instrument panel to give a visual indication of charging system operation.

Charging system circuits
The charging system consists of two major circuits, figure 7-21:

- The **field circuit**, which delivers current to the alternator field
- The **output circuit**, which sends voltage and current to the battery and other electrical components.

Single-Phase Current

Alternators induce voltage by rotating a magnetic field inside a fixed conductor. The greatest current output is produced when the rotor is parallel to the stator with its magnetic field at right angles to the stator, figure 7-20. When the rotor makes one-quarter of a revolution and is at right angles to the stator with its magnetic field parallel to the stator, figure 7-22, there is no current output. Figure 7-23 shows the voltage levels

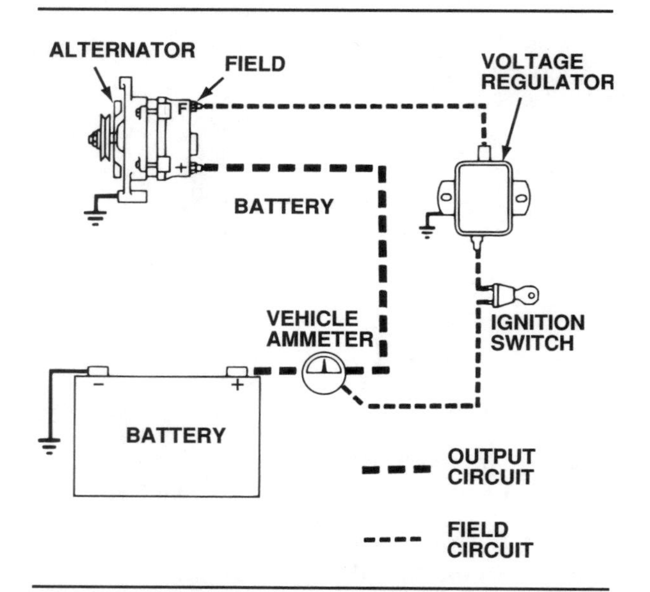

Figure 7-21. The output circuit and the field circuit make up the automotive charging system. (Prestolite)

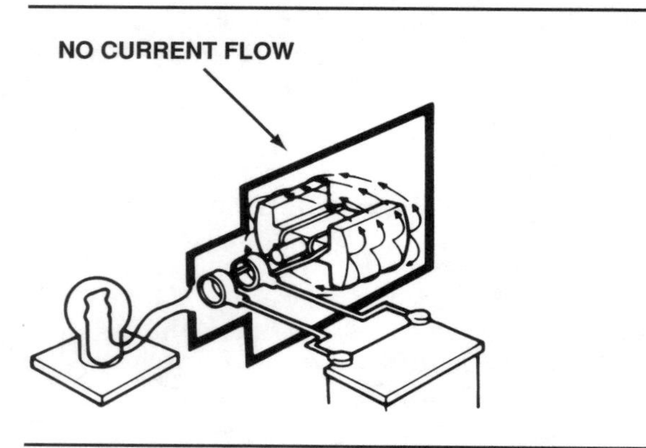

Figure 7-22. No current flows when the rotor's magnetic field is parallel to the stator. (Chrysler)

induced across the upper half of the looped conductor during one revolution of the rotor.

The constant change of voltage, first to a positive peak and then to a negative peak, produces a **sine wave voltage**. This name comes from the trigonometric sine function. The wave shape is controlled by the angle between the magnet and the conductor. The sine wave voltage induced across one conductor by one rotor revolution is called a **single-phase voltage**. Positions 1 through 5 of figure 7-23 show complete sine wave single-phase voltage.

This single-phase voltage causes alternating current to flow in a complete circuit, because the voltage switches from positive to negative as the rotor turns. The alternating current caused by a single-phase voltage is called **single-phase current**.

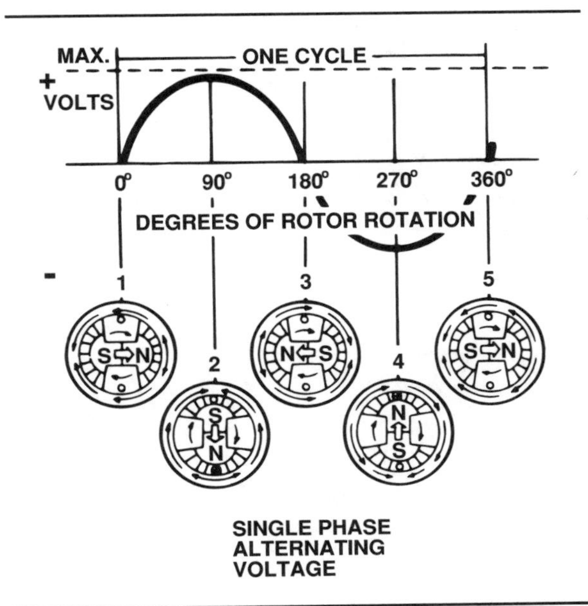

Figure 7-23. These are the voltage levels induced across the upper half of the conductor during one rotor revolution. (Chrysler)

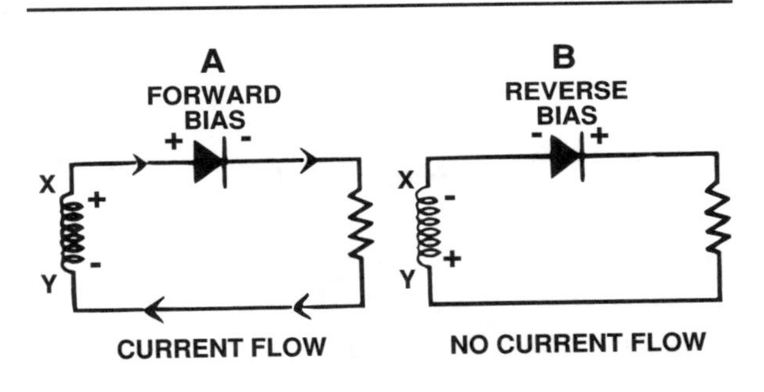

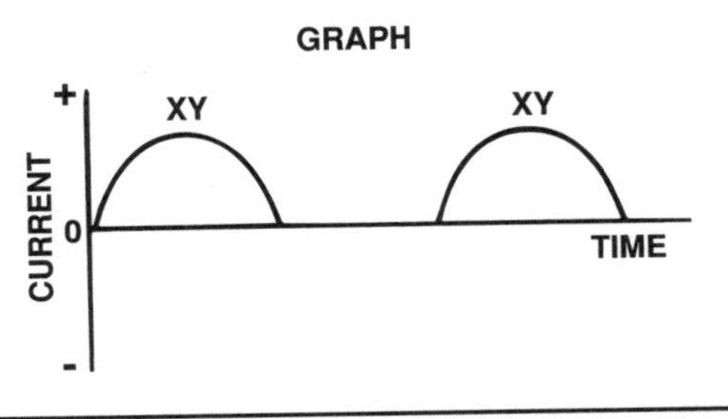

Figure 7-24. A single diode in the circuit results in half-wave rectification. (Delco-Remy)

Diode Rectification

If the single-phase voltage shown in figure 7-23 made current flow through a simple circuit, the current would flow first in one direction and then in the opposite direction. As long as the rotor turned, the current would reverse its flow with every half revolution. We know that the battery cannot be recharged with alternating current. Alternating current must be rectified to direct current; this is done with diodes.

A diode acts as a one-way electrical valve. If we insert a diode in a simple circuit, figure 7-24, one-half of the ac voltage will be blocked. That is, the diode will allow current to flow from X to Y, as shown in position A. In position B, the current cannot flow from Y to X because it is blocked by the diode. The graph in figure 7-24 shows the total current flow.

The first half of the current flow, from X to Y, was allowed to pass through the diode. It is shown on the graph as curve XY. The second half of the flow, from Y to X, was not allowed to pass through the diode. It does not appear on the graph because it never flowed through the circuit. When the voltage reverses at the start of the next rotor revolution, the current is again allowed through the diode from X to Y.

Field Circuit: The charging system circuit that delivers current to the alternator field.

Output Circuit: The charging system circuit that sends voltage and current to the battery and other electrical systems and devices.

Sine Wave Voltage: The constant change, first to a positive peak and then to a negative peak, of an induced alternating voltage in a conductor.

Single-Phase Voltage: The sine wave voltage induced within one conductor by one revolution of an alternator rotor.

Single-Phase Current: Alternating current caused by a single-phase voltage.

Heat Sinks

The term "heat sink" is commonly used to describe the block of aluminum or other material in which the alternator diodes are mounted. The job of the heat sink is to absorb and carry away heat from the diodes generated by the electrical current flowing through them. This action keeps the diodes cool and prevents damage.

An internal combustion engine is also a heat sink. The engine absorbs and then radiates the heat caused by combustion to the surrounding area.

Although they are not thought of as heat sinks, many individual parts of an automobile—such as brake drums—are designed so that they will also do this important job.

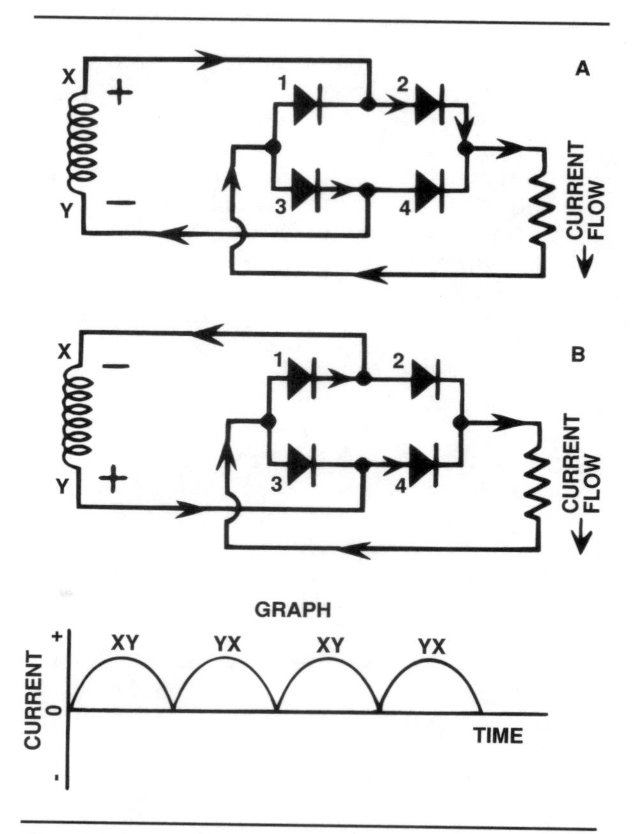

Figure 7-25. More diodes are needed for full-wave rectification. (Delco-Remy)

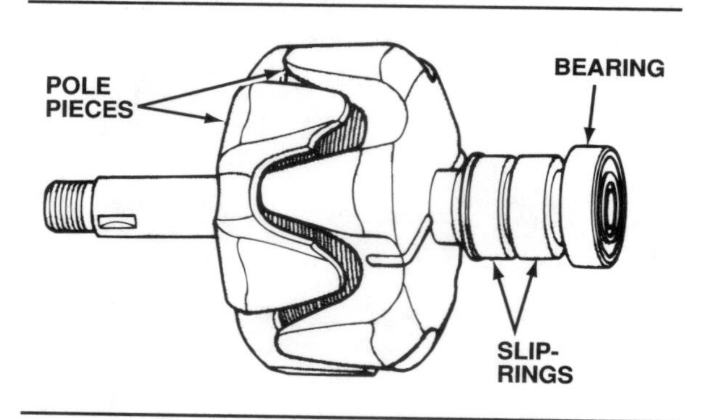

Figure 7-26. A typical 12-pole rotor. (Chrysler)

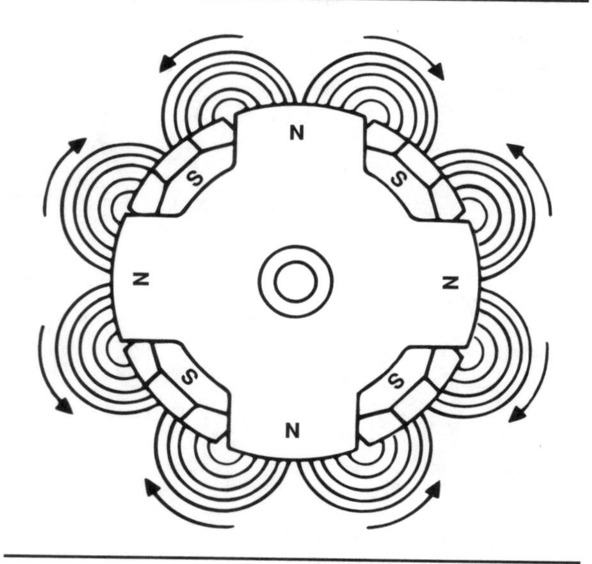

Figure 7-27. The flux lines surrounding an eight-pole rotor. (Prestolite)

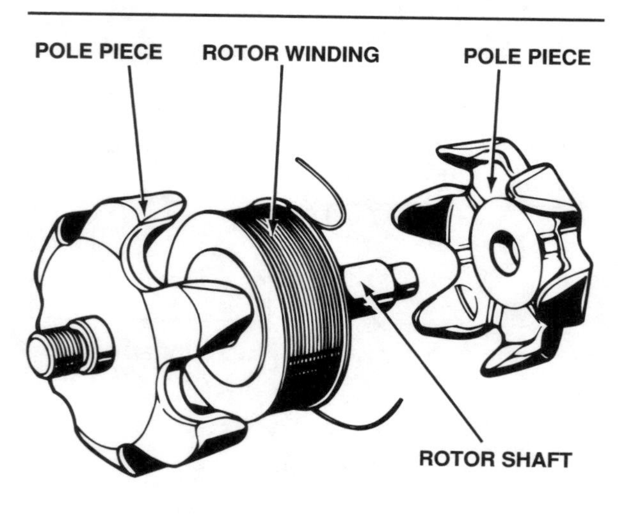

Figure 7-28. The magnetic field of the rotor is caused by current flow through the rotor winding. (Bosch)

An alternator with only one conductor and one diode would show this current output pattern. However, this output would not be very useful, because half of the time there is no current available. This is called **half-wave rectification**, since only half of the ac sine wave voltage produced by the alternator is allowed to flow as dc voltage.

If we add more diodes to the circuit, figure 7-25, more of the ac voltage can be rectified to dc. In position A, current is flowing from X to Y. It flows from X, through diode 2, through the load, through diode 3, and back to Y. In position B, current is flowing from Y to X. It flows from Y, through diode 4, through the load, through diode 1, and back to X.

Notice that in both cases, current flowed through the load in the same direction. The alternating current has been rectified to direct current. The graph in figure 7-25 shows the current output of an alternator with one conductor and four diodes. There is more current available because all of the voltage has been rectified. This is called **full-wave rectification**. There are still

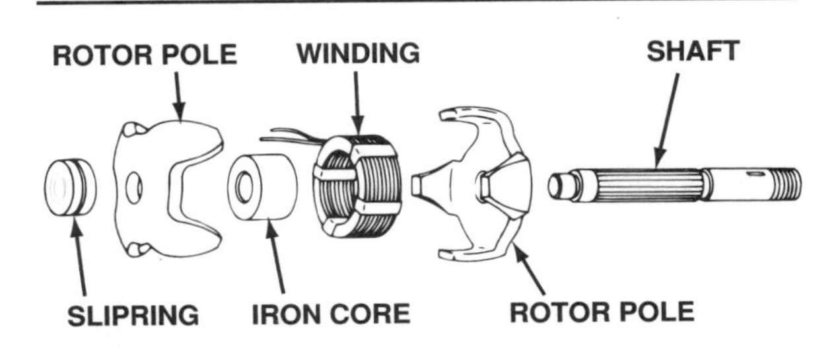

Figure 7-29. An exploded view of the parts of the complete rotor assembly. (Prestolite)

moments, however, when current flow is at zero. Most automotive alternators use three conductors and six diodes to produce overlapping current waves so that current output is *never* at zero.

Alternator Construction

Our simple illustrations have shown the principles of alternator operation. To provide enough direct current for an automobile, alternators must have a more complex design. But no matter how the design varies, the principles of operation remain the same.

The design of an alternator limits the maximum current output of the alternator. To change this maximum value for different applications, manufacturers change the design of the stator, rotor, and other components. The following paragraphs describe the major parts of an automotive alternator.

Rotor

The rotor carries the alternator's magnetic field. Unlike a generator, which usually has only two magnetic poles, the alternator rotor has several North (N) and South (S) poles. This increases the number of flux lines within the alternator and increases the voltage output. A typical automotive rotor, figure 7-26, has 12 poles: 6 N and 6 S. The rotor consists of two steel rotor halves, or pole pieces, with fingers that interlace. These fingers are the poles. Each pole piece has either *all* N or *all* S poles. The magnetic flux lines travel between adjacent N and S poles, as you can see from the eight-pole rotor shown in figure 7-27.

As you look along the outside of the rotor, note that the flux lines point first in one direction and then in the other. This means that as the rotor spins inside the alternator, the fixed conductors are being cut by flux lines that point in alternating directions. The induced voltage will alternate, just as in our example of a simple alternator with only two poles. Automotive alternators may have any number of poles, as long as they are placed N-S-N-S. Common alternator designs use 8 to 14 poles.

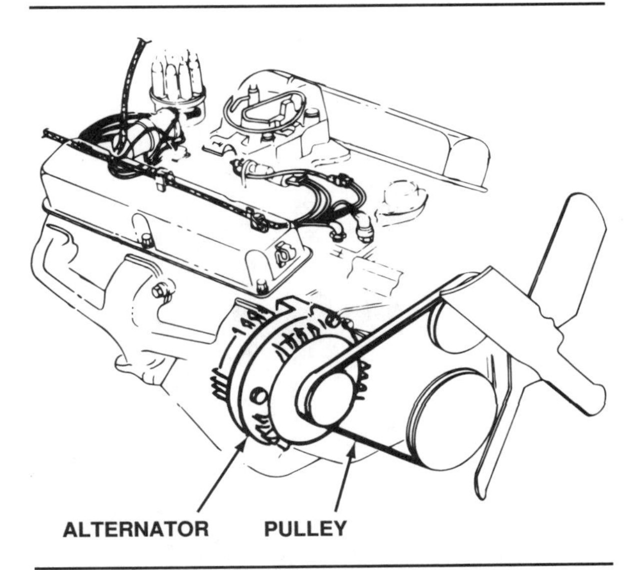

Figure 7-30. The alternator and drive pulley. (Chrysler)

The rotor poles may retain some magnetism when the alternator is not in operation, but this residual magnetism is not strong enough to induce any voltage across the conductors. The magnetic field of the rotor is produced by current flow through the rotor winding, a coil of wire between the two pole pieces, figure 7-28. This is also called the excitation, or field, winding. Varying the amount of field current flow through the rotor winding will vary the strength of the magnetic field. This affects the voltage output of the alternator.

A soft iron core is mounted inside the rotor winding, figure 7-29. One pole piece is attached to either end of the core. When field current flows through the winding, the iron core is magnetized. The pole pieces take on the magnetic polarity of the end of the core to which they are attached. Current is supplied to the winding through sliprings and brushes.

The combination of a soft iron rotor core and steel rotor halves provides better localization and permeability of the magnetic field. The rotor pole pieces, winding, core, and sliprings are pressed onto a shaft. The ends of this shaft are

Half-Wave Rectification: A process by which only one-half of an ac sine wave voltage is rectified and allowed to flow as dc.

Full-Wave Rectification: A process by which all of an ac sine wave voltage is rectified and allowed to flow as dc.

Figure 7-31. An alternator stator.

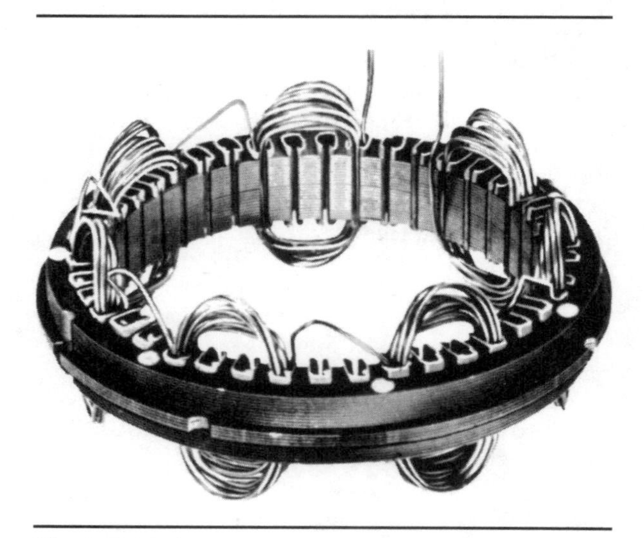

Figure 7-32. A stator with only one conductor installed. (Delco-Remy)

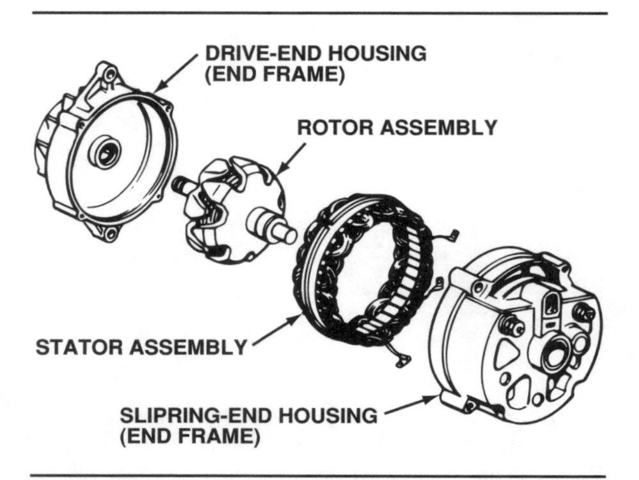

Figure 7-33. The alternator housing encloses the rotor and stator.

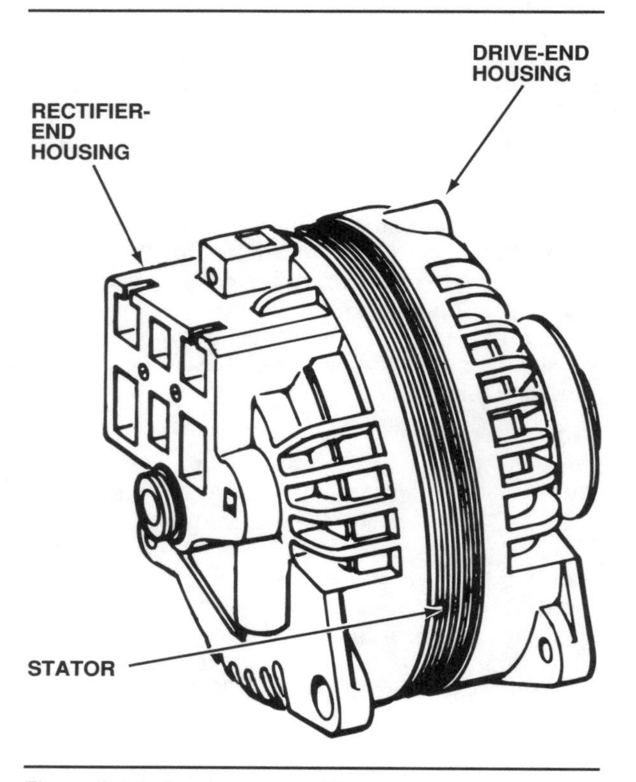

Figure 7-34. An alternator with an exposed stator core.

held by bearings in the alternator housing. Outside the housing, a drive pulley is attached to the shaft, figure 7-30. A belt from the automobile engine passes around this pulley to turn the alternator shaft and rotor assembly.

Stator

The three alternator conductors are wound onto a cylindrical laminated core. The lamination prevents unwanted eddy currents from forming in the core. The assembled piece is called a stator, figure 7-31. Each conductor, called a stator winding, is formed into a number of coils spaced evenly around the core. There are as many coils in each conductor as there are pairs of N-S rotor poles. Figure 7-32 shows an incomplete stator with only one of its conductors installed. In this example, there are seven coils in the conductor, so the matching rotor would have seven pairs of N-S poles (a total of 14 poles).

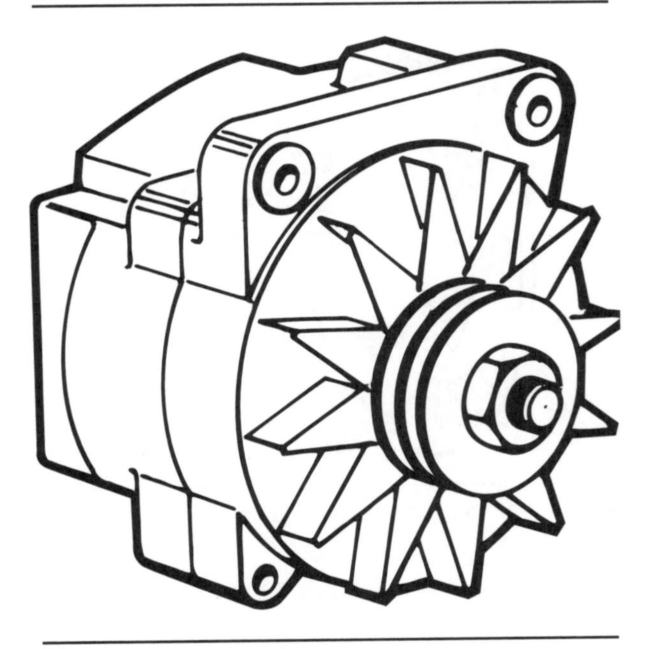

Figure 7-35. An alternator with the stator core enclosed.

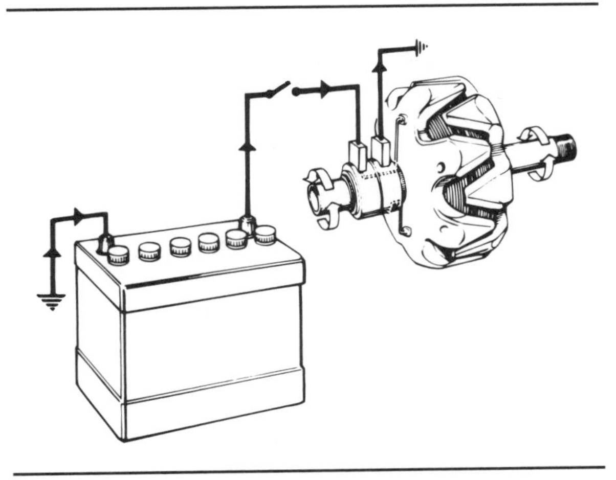

Figure 7-36. The sliprings and brushes carry current to the rotor windings.

Housing

The alternator housing, or frame, is made of two pieces of cast aluminum, figure 7-33. Aluminum is lightweight, nonmagnetic, and conducts heat well. One housing piece holds a bearing for the end of the rotor shaft where the drive pulley is mounted. This is often called the drive-end housing, or front housing, of the alternator. The other end holds the diodes, the brushes, and the electrical terminal connections. It also holds a bearing for the slipring end of the rotor shaft. This is often called the slipring-end housing, or rear housing. Together, the two pieces totally enclose the rotor and the stator windings.

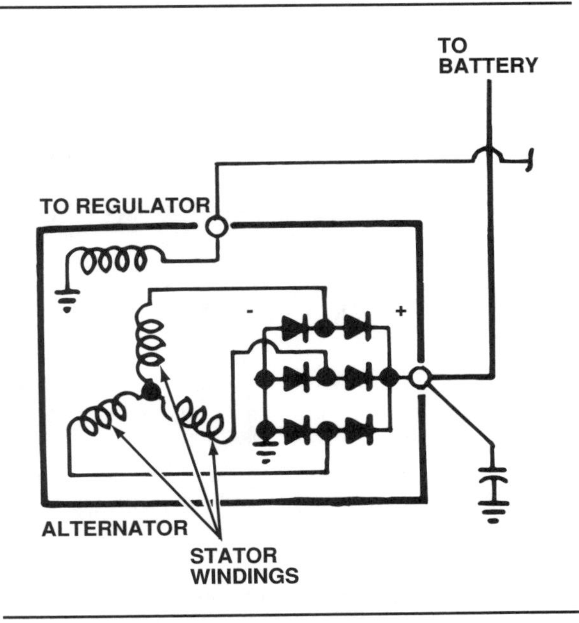

Figure 7-37. Each conductor is attached to one positive and one negative diode.

The end housings are bolted together. Some stator cores have an extended rim that is held between the two housings, figure 7-34. Other stator cores provide holes for the housing bolts, but do not extend to the outside of the housings, figure 7-35. In both designs the stator is bolted rigidly in place inside the alternator housing.

Because the alternator housing is bolted directly to the engine, it is part of the electrical ground path. Anything connected to the housing, which is not insulated from the housing, is grounded.

Sliprings and brushes

The sliprings and brushes conduct current to the rotor winding. Most automotive alternators have two sliprings mounted on the rotor shaft. The sliprings are insulated from the shaft and from each other. One end of the rotor winding is connected to each slipring, figure 7-36. One brush rides on each ring, under spring tension from its brush holder, to carry current to and from the winding. The brushes are connected in parallel with the alternator output circuit. They draw some of the alternator current output and route it through the rotor winding. Current flow through the winding must be direct current. We will see later how this is supplied.

Field current in an alternator is usually about 1.5 to 3.0 amps. Because the brushes carry so little current, they do not require as much maintenance as generator brushes, which must conduct all of the generator's current output.

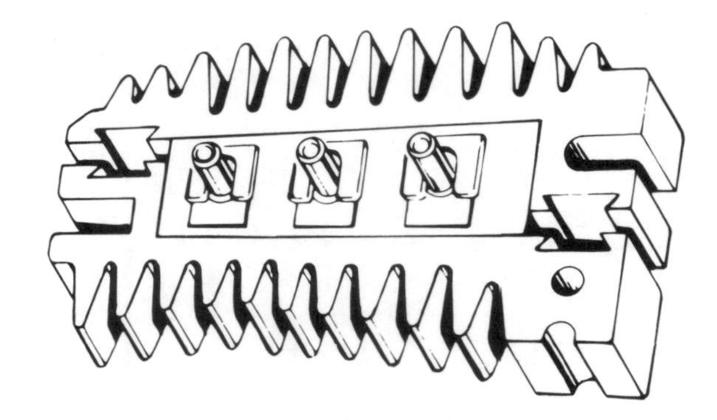

Figure 7-38. Positive and negative diodes can be mounted in a heat sink for protection.

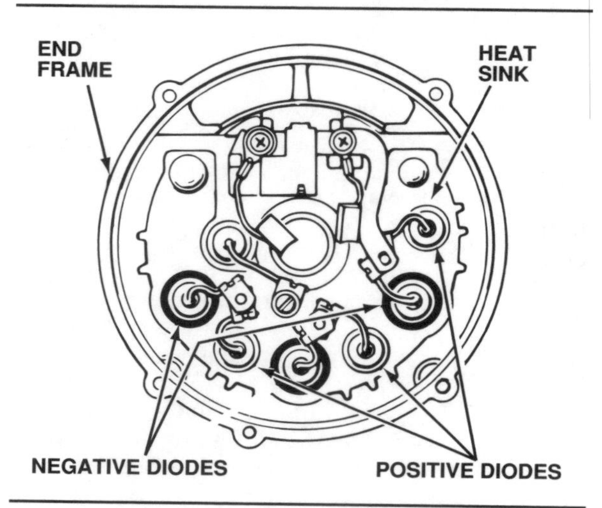

Figure 7-39. Negative diodes may be pressed into the rear housing.

Diode installation

Automotive alternators that have three stator windings generally use six diodes to rectify the current output. The connections between the conductors and the diodes vary slightly, but each conductor is connected to one positive and one negative diode, figure 7-37.

The three positive diodes are always insulated from the alternator rear housing. They are connected to the insulated terminal of the battery and to the rest of the automotive electrical system. The battery cannot discharge through this connection because the bias of the diodes blocks any current flow from the battery. The positive diodes will conduct only that current flowing from the conductors toward the battery. The positive diodes are mounted together on a conductor called a heat sink, figure 7-38. The heat sink carries heat away from the diodes, just as the radiator carries heat away from the engine. Too much heat from current flow could damage the diodes.

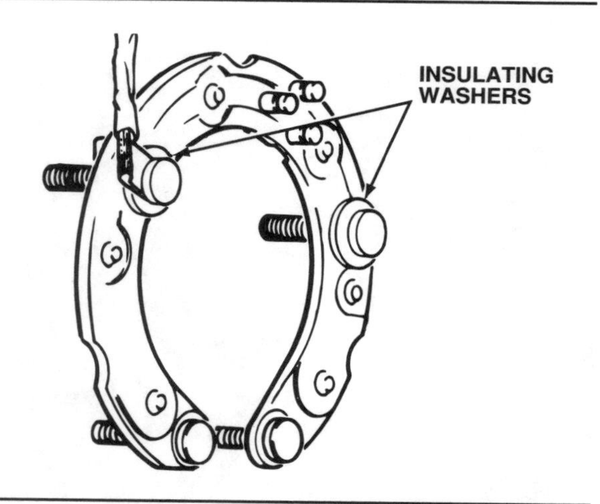

Figure 7-40. As shown on this Ford application, all diodes and connections may be included in a single printed circuit board.

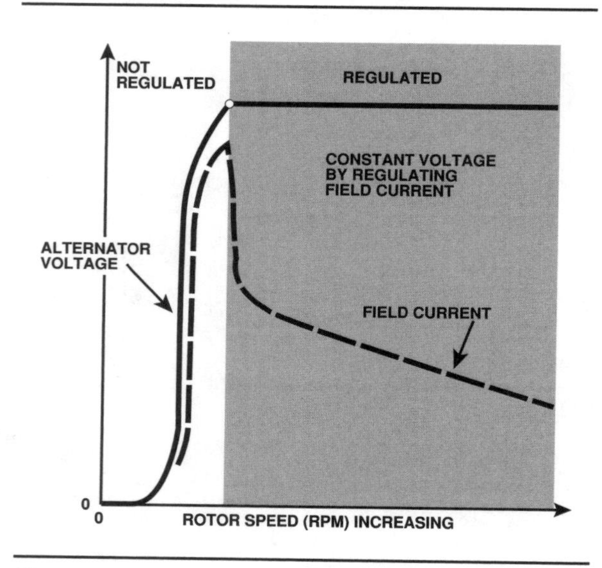

Figure 7-41. Field current is decreased as rotor speed increases to keep alternator output voltage at a constant level.

The three negative diodes may be pressed or threaded into the alternator rear housing, figure 7-39. On high-output alternators, they may be mounted in a heat sink for added protection. In either case, the connection to the alternator housing is a ground path. The negative diodes will conduct only that current flowing from ground into the conductors.

Each group of three or more negative or positive diodes can be called a diode bridge, a diode trio, or a diode plate.

Some manufacturers use complete rectifier assemblies containing all the diodes and connections on a printed circuit board, figure 7-40. This

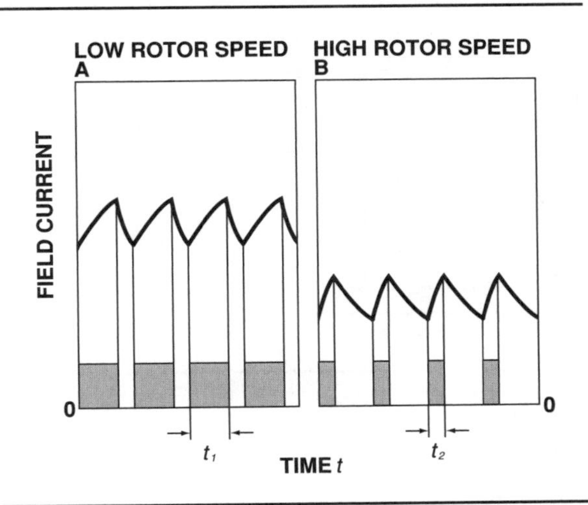

Figure 7-42. The field current flows for longer periods of time at low speeds (t_1) than at high speeds (t_2). (Bosch)

assembly is replaced as a unit if any of the individual components fail.

Each stator winding connects to its proper negative diode through a circuit in the rectifier. A capacitor generally is installed between the output terminal at the positive diode's heat sink to ground at the negative diode's heat sink. This capacitor is used to eliminate voltage switching transients at the stator, to smooth out the ac voltage fluctuations, and to reduce electromagnetic interference.

Voltage Regulation

The regulator used with an alternator has only one job: to limit the voltage strength of the alternator's output. The car's electrical system is designed to work on about 14 volts. If a greater voltage is applied to the battery and electrical circuits, they could be damaged. As a general rule, the regulator should prevent the alternator from producing more than 15 volts during all operating conditions.

The alternator uses induction to limit its current output. When an induced voltage causes current flow in a conductor, that conductor creates its own magnetic field. As current flow increases, the magnetic field expands. This expanding magnetic field induces another voltage within the same conductor. This countervoltage will tend to oppose any change in the original current flow.

When induced voltage causes current flow in the conductors of the stator, a countervoltage also appears. The countervoltage opposes any increased current flow in the stator. The more current the alternator puts out, the greater this

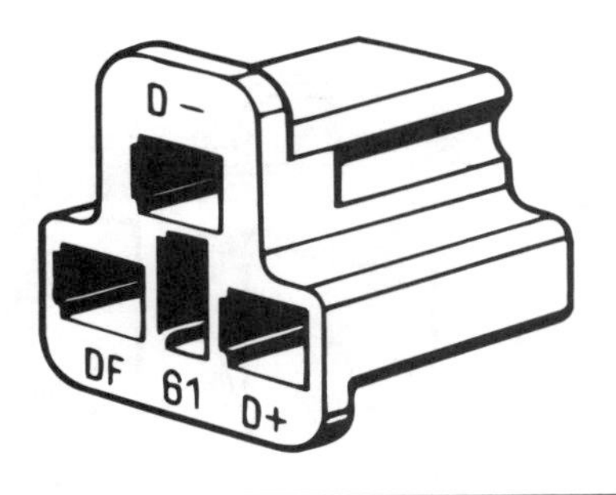

Figure 7-43. Most regulators use a multiple connector plug to ensure that connections are properly made. (Bosch)

countervoltage becomes. The alternator reaches its maximum current output when the countervoltage is great enough to totally stop any further increase in the *current*. However, because the two *voltages* will continue to increase as alternator speed increases, a method of regulating alternator voltage is required.

Alternator output voltage is directly related to field strength and rotor speed. An increase in either factor will increase voltage output. Similarly, a decrease in either factor will decrease voltage output. Rotor speed is controlled by engine speed and cannot be changed simply to control the alternator. Field strength can be changed by controlling the field current in the rotor windings. This is how both alternator and generator voltage regulators work. Figure 7-41 shows how the field current (dashed line) is lowered to keep alternator voltage (solid line) at a constant maximum, even when the rotor speed increases.

At low rotor speeds, the field current is allowed to flow at full strength for relatively long periods of time, and is reduced only for short periods, figure 7-42, position A. This causes a high average field current. At high rotor speeds, the field current is reduced for long periods of time and flows at full strength only for short periods, position B. This causes a low average field current.

The field circuit can be controlled by an electromagnetic regulator. In the last decade, however, semiconductor technology has made solid-state voltage regulators possible. Because they are smaller and have no moving parts, solid-state regulators have replaced the older electromagnetic type in ac charging systems.

Some solid-state regulators are mounted on the inside or outside of the alternator. This eliminates exposed wiring and connections that could be damaged. Remotely mounted voltage regulators often use a multiple-plug connector, figure 7-43, to ensure that all connections are properly made. In the mid-1980s, many carmakers moved

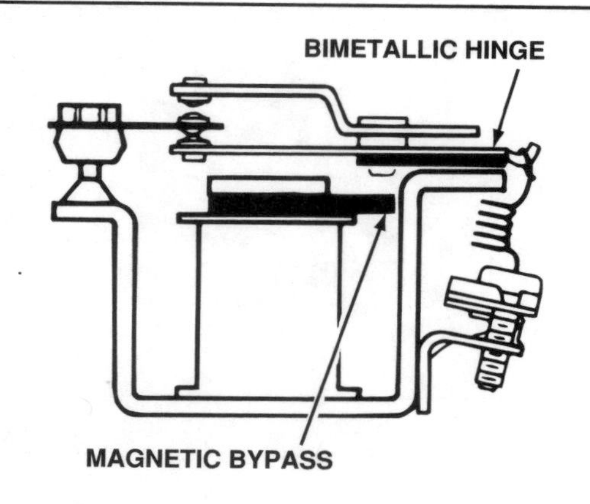

Figure 7-44. A Delco-Remy electromagnetic voltage regulator.

Figure 7-45. Late-model vehicles use integrated circuits to regulate field current. On this Ford design, the integral alternator regulator (IAR) and brush holder are combined.

the regulator function into the engine control computer of fully integrated electronic engine control systems.

Electromagnetic regulators
An electromagnetic regulator is a mechanical switch that is opened and closed by magnetism. It has a coil, figure 7-44, and a hinged armature. Current flow through the coil creates a magnetic field that moves the armature. The current flow through the coil is caused by the alternator's voltage. As the alternator's voltage varies, the current flow through the regulator coil varies. Field current for the alternator must flow through contact points on the regulator armature. The position of the armature determines how much field current can reach the alternator. As the alternator's voltage rises, the coil's magnetic field increases, changing the position of the armature, and thus decreasing field current flow.

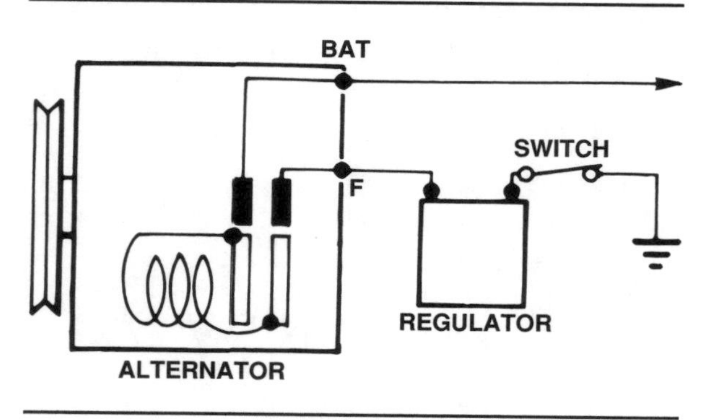

Figure 7-46. An A-circuit.

Solid-state regulators
Solid-state regulators have completely replaced the older electromagnetic design on late-model cars. They are compact, have no moving parts, and are not seriously affected by temperature changes. The early solid-state designs combined transistors with the electromagnetic field relay. A later and more-compact design, the integrated-circuit (IC) regulator, combines all control circuitry and components on a single silicon chip, figure 7-45. Attaching terminals are added, and the chip is sealed in a small plastic module that mounts inside, or on the back of, the alternator. Because of their construction, however, all solid-state regulators are nonserviceable and must be replaced if defective. No adjustments are possible.

These are the major components of a solid-state regulator:

- Diodes
- Zener diodes
- Transistors
- Thermistors
- Capacitors

Diodes are one-way electrical check valves. A zener diode is specially doped to act as a one-way electrical check valve until a specific reverse voltage level is reached. At that point, the zener diode reverse current will be conducted. Transistors act as relays.

The electrical resistance of a **negative temperature coefficient resistor** decreases as its temperature increases. Such resistors used in automotive applications are called a negative temperature coefficient (NTC). The thermistor in a solid-state regulator reacts to temperature to ensure proper battery charging voltage.

A capacitor is used by some manufacturers to smooth out any abrupt voltage surges and protect the regulator from damage. Diodes can also be used as circuit protection.

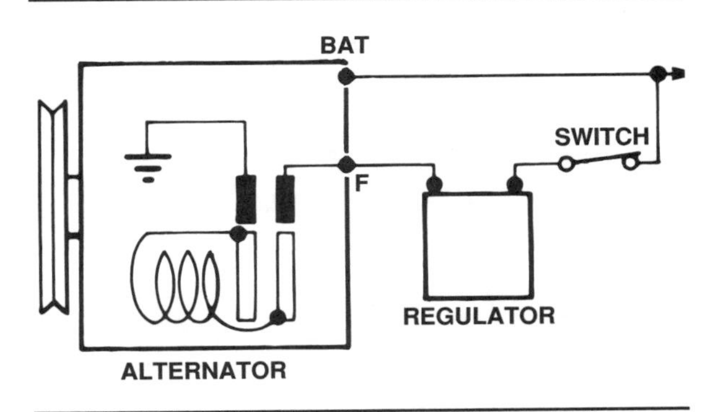

Figure 7-47. A B-circuit.

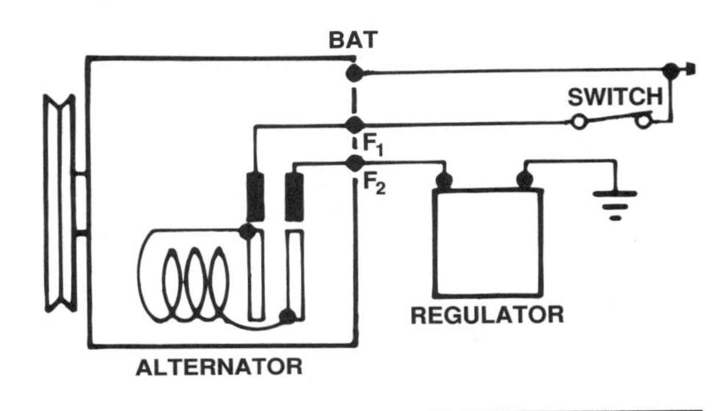

Figure 7-48. An isolated-field circuit.

Field Circuit Control

There are different types of field circuits, depending on where the voltage regulator and power source are connected. The three most common types are:

- A-circuit (externally grounded field)
- B-circuit (internally grounded field)
- Isolated-field circuit.

A-circuit

The A-circuit alternator, figure 7-46, also can be called an externally grounded field alternator. Both brushes are insulated from the alternator housing. One brush is connected to the voltage regulator, where it is grounded. The second brush is connected to the alternator output circuit within the alternator, where it draws current for the rotor winding. The regulator lies between the rotor field winding and ground. This type of circuit is often used with solid-state regulators, which can be small enough to be mounted on the alternator housing.

B-circuit

The B-circuit alternator, figure 7-47, also can be called an internally grounded field alternator. One brush is grounded within the alternator housing. The other brush is insulated from the housing and connected through the insulated

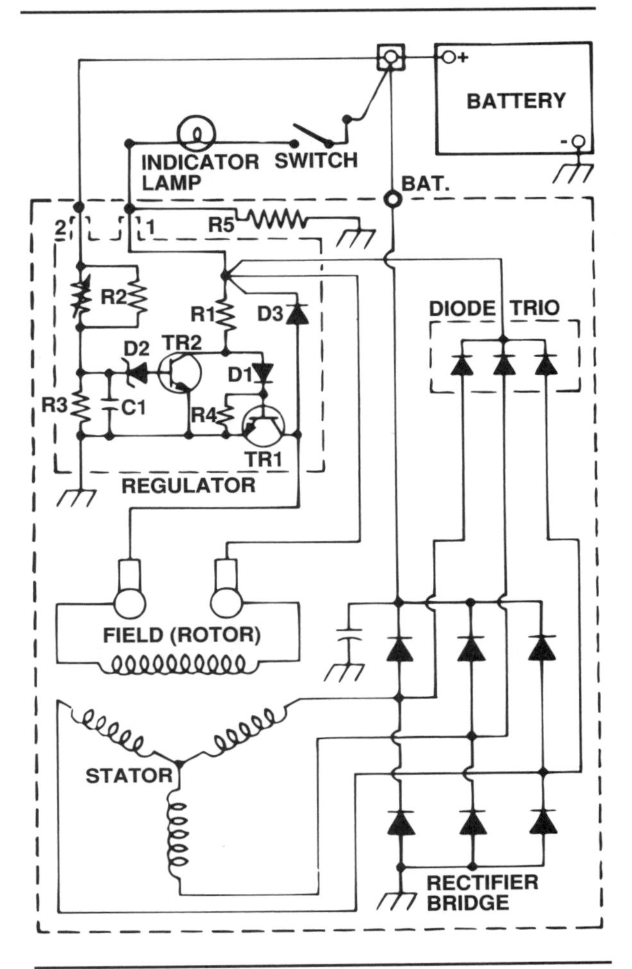

Figure 7-49. Circuit diagram of a Delco-Remy solid-state voltage regulator.

voltage regulator to the alternator output circuit. The rotor field winding is between the regulator and ground. This type of circuit is most often used with electromagnetic voltage regulators, which are mounted away from the alternator housing.

Isolated-field circuit

The **isolated-field circuit**, figure 7-48, is a variation of the A-circuit. The rotor winding is again

Negative Temperature Coefficient Resistor: A resistor specially constructed so that its resistance decreases as its temperature increases.

Isolated Field Circuit: A variation of the A-circuit. Field current is drawn from the alternator output outside of the alternator and sent to an insulated brush. The other brush is grounded through the voltage regulator.

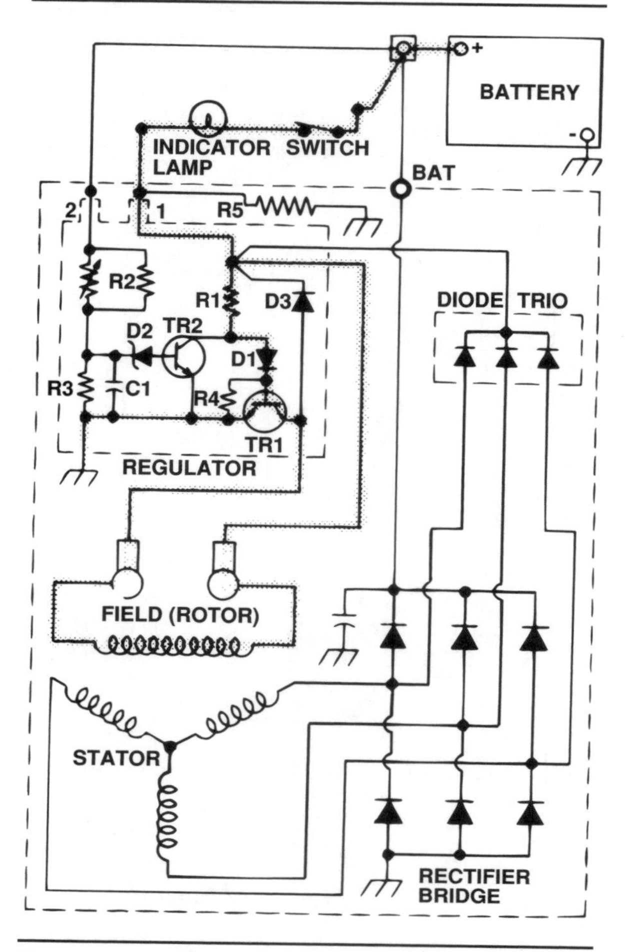

Figure 7-50. Field current in a Delco-Remy solid-state regulator during starting.

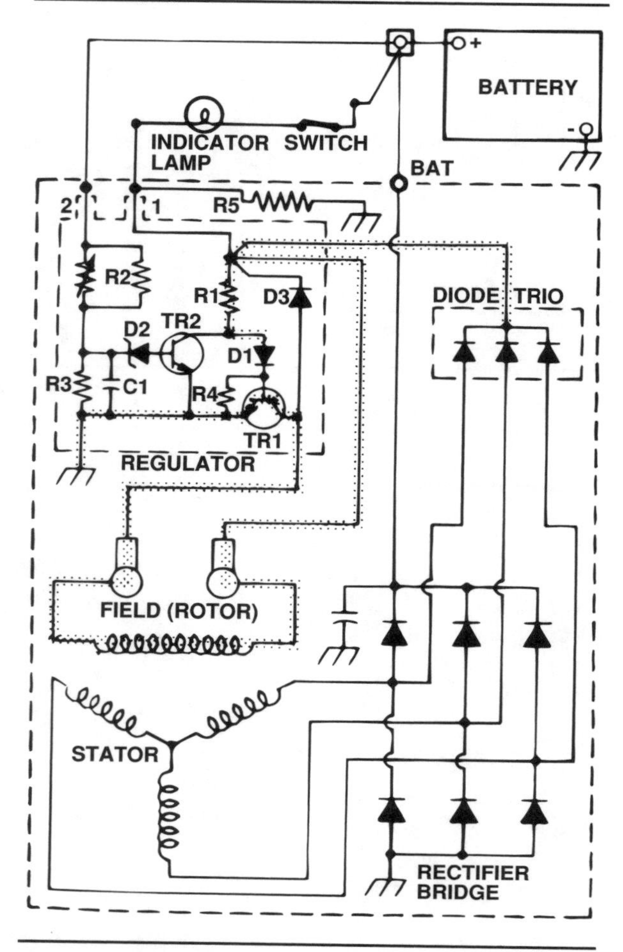

Figure 7-51. Field current being drawn from alternator output. (Delco-Remy)

grounded through the voltage regulator. Current for the rotor winding is drawn from the alternator output circuit outside the alternator housing, and comes through a third terminal on the housing. Current flow through this third terminal can be controlled by an external switch or a field relay. This type of circuit is used mainly by Chrysler Corporation.

General field circuit operation

Figure 7-49 is a simplified circuit diagram of a solid-state regulator. This A-circuit regulator is contained within the alternator housing. The number 2 terminal on the alternator is always connected to the battery, but battery discharge is limited by the high resistance of R2 and R3. The circuit allows the regulator to sense battery voltage.

When the ignition switch is closed, figure 7-50, current flows from the battery to ground through the base of TR1. This makes TR1 conduct current through its emitter-collector circuit from the battery to the low-resistance rotor winding. The alternator field is energized, and the warning lamp is turned on.

When the alternator begins to produce current, figure 7-51, field current is drawn from unrectified alternator output and rectified by the diode trio. No current flows through the warning lamp.

When the alternator has charged the battery to a maximum safe voltage level, figure 7-52, the battery voltage between R2 and R3 causes zener diode D2 to conduct current. TR2 is turned on, and field current flows directly to ground. TR1 is turned off, so no current reaches the rotor winding.

With TR1 off, the field current decreases and system voltage drops. D2 then blocks current flow in the base of TR2, and TR1 turns back on. The field current and system voltage increase. This cycle repeats many times per second to limit the alternator voltage to a predetermined value.

The other components within the regulator perform various functions. Capacitor C1 pro-

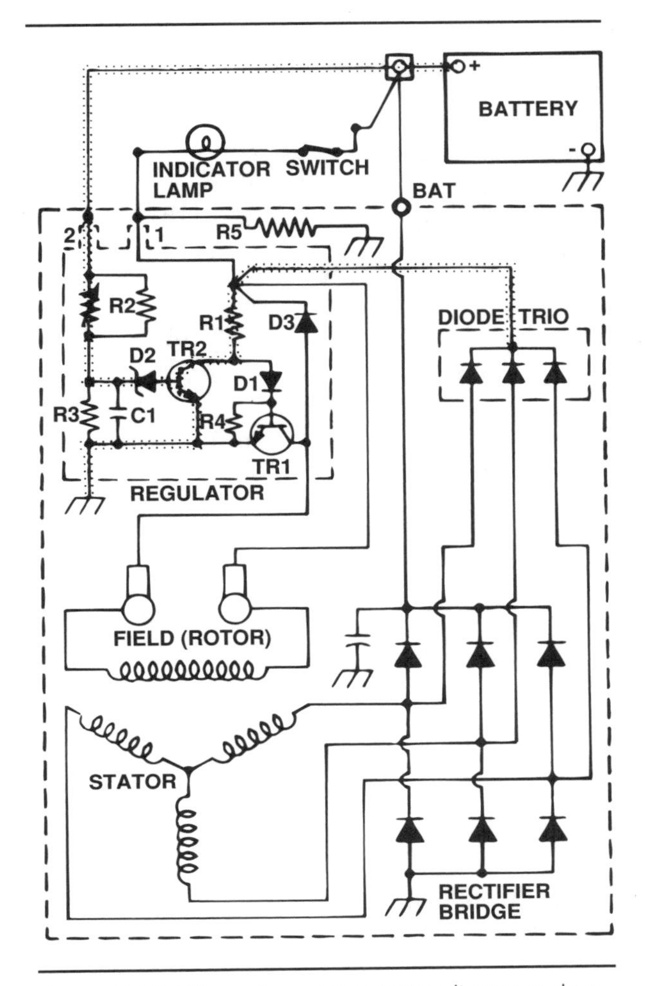

Figure 7-52. When alternator output voltage reaches a maximum safe level, no current is allowed in the rotor winding. (Delco-Remy)

vides smooth voltage across R3, and R4 prevents excessive current through TR1 at high temperatures. To prevent circuit damage, D3 bypasses high voltages induced in the field windings when TR1 turns off. Resistor R2 is a thermistor that causes the regulated voltage to vary with temperature.

Computer-Controlled Regulation

Chrysler Corporation eliminated the separate regulator by moving its function to the engine control computer in 1985. When the ignition is turned on, the engine controller logic module or logic circuit checks battery temperature to determine the control voltage, figure 7-53. A predriver transistor in the logic module or logic circuit then signals the power module or power circuit driver transistor to turn the alternator field current on, figure 7-54. The logic module or logic circuit continually reads battery temperature

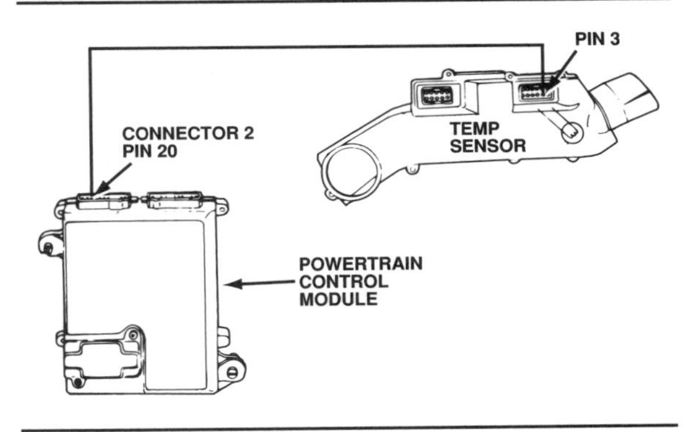

Figure 7-53. Battery temperature circuit of the Chrysler computer-regulated charging system. (Chrysler)

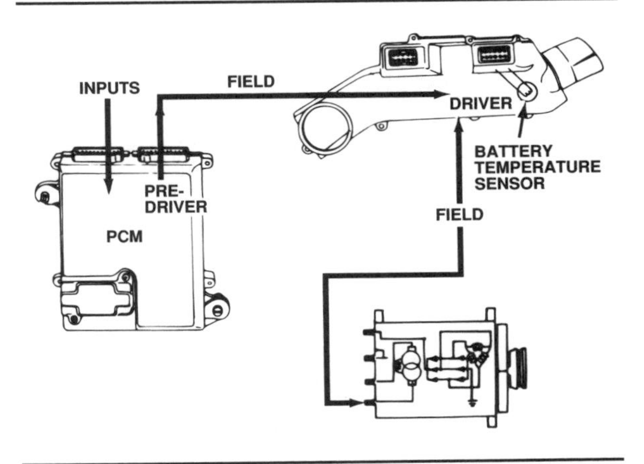

Figure 7-54. Chrysler's method of regulating voltage through the field windings using computer control.

and system voltage. At the same time, it instructs the power module or power circuit driver to adjust the field current as required to maintain output voltage between 13.6 and 14.8 volts, +/- 0.3 volt. Figure 7-55 shows the complete circuitry involved.

General Motors has taken a different approach to regulating voltage in the Delco-Remy CS-series alternators. Turning the ignition switch to the run position supplies voltage to alternator terminals L and F (or F/I). This voltage activates a solid-state digital regulator that uses pulse width modulation (PWM) to supply rotor current and thus control output voltage. The rotor current is proportional to the PWM pulses from the digital regulator. With the ignition on, narrow width pulses are sent to the rotor, creating a weak magnetic field. As the engine starts, the regulator senses alternator rotation through ac voltage detected on an internal wire. Once the engine is running, the regulator switches the

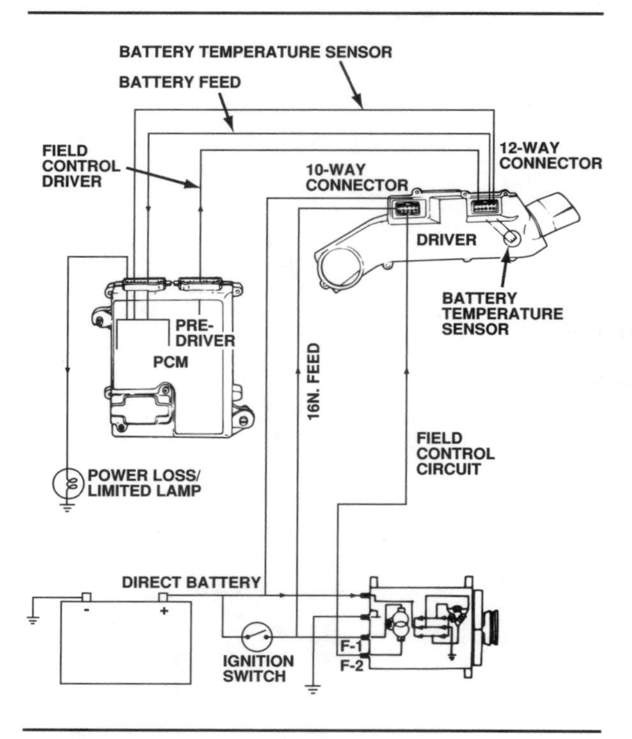

Figure 7-55. Chrysler computer-regulated charging system voltage control.

field current on and off at a fixed frequency of about 40 0 cycles per second (Hertz). By changing the pulse width, or on-off time of each cycle, the regulator provides a correct average field current for proper system voltage control.

A lamp driver in the digital regulator controls the indicator warning lamp, turning on the bulb when it detects an under- or overvoltage condition, or a nonrotating alternator.

The powertrain control module (PCM) does not directly control charging system voltage, as in the Chrysler application. However, it does monitor battery and system voltage through an ignition switch circuit. If the PCM reads a voltage above 17 volts, or less than 9 volts for longer than 10 seconds, it sets a code 16 in memory and turns on the service engine soon or malfunction indicator lamp (MIL).

Fault codes

On late-model Chrysler vehicles, the on-board diagnostic system (OBS) capability of the engine control system detects charging system problems and can record up to five fault codes in the system memory. Some codes will light a power loss, power limited, or MIL on the instrument panel; others will not. Chrysler fault codes and their use are described in the test procedures in Chapter Six of the *Shop Manual*. Problems in the General Motors CS charging system cause the PCM to turn on the MIL and set a single code in memory.

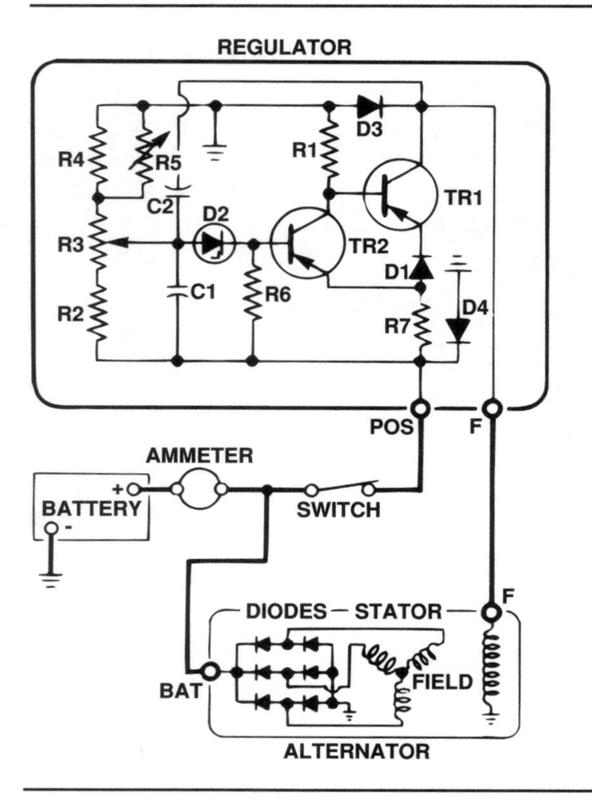

Figure 7-56. A Delco-Remy ammeter installation.

Indicators

A charging system failure cripples an automobile. Therefore, most manufacturers provide some way for the driver to monitor the system operation. The indicator may be an ammeter, a voltmeter, or an indicator (warning) lamp.

Ammeter

An instrument panel ammeter measures charging system current into and out of the battery and the rest of the electrical system, figure 7-56. When current is flowing from the alternator into the battery, the ammeter needle moves in the CHARGE direction. When the battery takes over the electrical system's load, current flows in the opposite direction and the needle moves into the DISCHARGE zone. The ammeter simply indicates if the battery or the alternator is doing the most work in the electrical system. Some ammeters are graduated to indicate the approximate current in amps, such as 5, 10, or 20. Others simply show an approximate rate of charge or discharge, such as high, medium, or low.

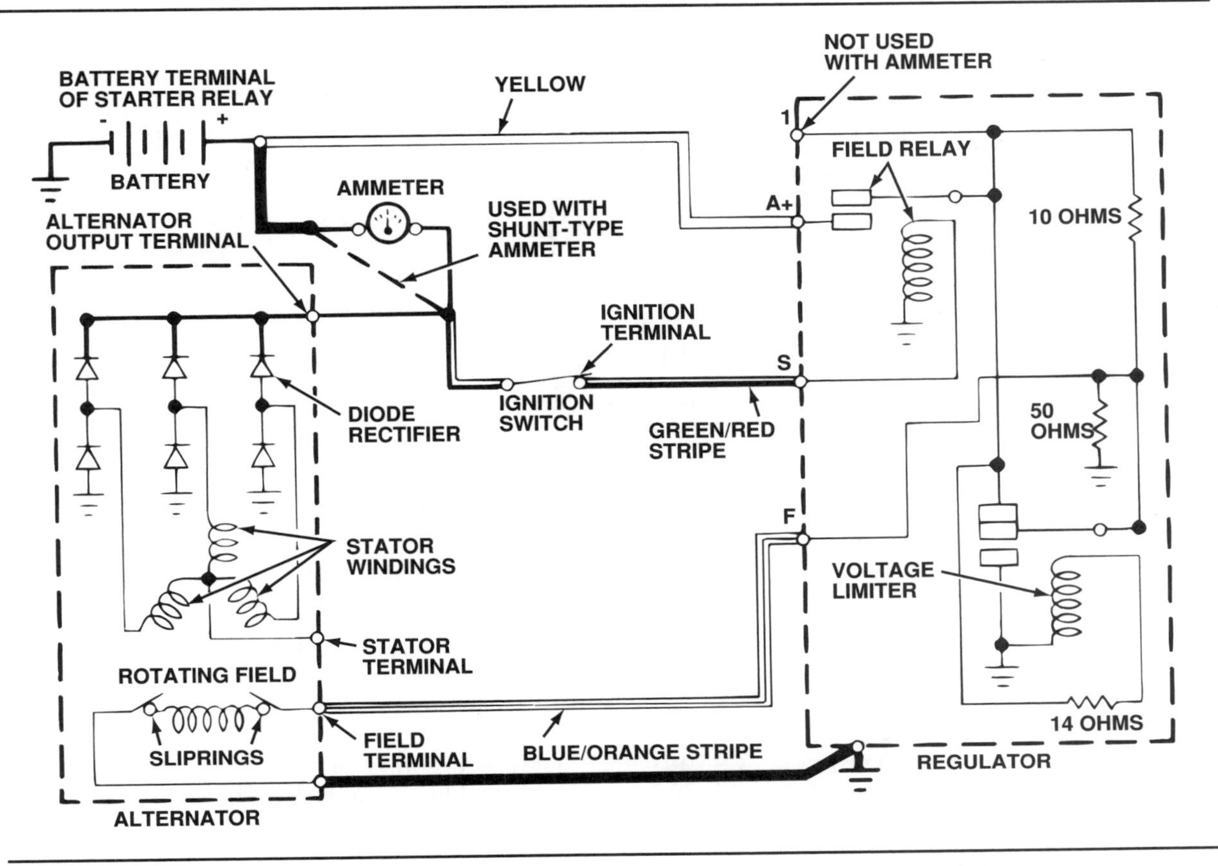

Figure 7-57. A Ford ammeter installation

Some ammeters have a resistor in parallel so that the meter does not carry all of the current. These are called shunt ammeters.

While the ammeter tells the driver whether the charging system is functioning normally, it does not give a good picture of the battery condition. Even when the ammeter indicates a charge, the current output may not be high enough to fully charge the battery while supplying other electrical loads. Figure 7-57 shows a typical ammeter circuit, including a shunt resistor.

Voltmeter

The instrument panels of many later-model cars contain a voltmeter instead of an ammeter, figure 7-58. A voltmeter measures electrical pressure, and indicates regulated alternator voltage output or battery voltage, whichever is greater. System voltage is applied to the meter through the ignition switch contacts. Figure 7-59 shows a typical voltmeter circuit.

The voltmeter tells a driver more about the condition of the car's electrical system than an ammeter. When a voltmeter begins to indicate lower-than-normal voltage, it is time to check the battery and the voltage regulator.

Figure 7-58. An automotive voltmeter for the instrument panel.

Indicator lamps

Most charging systems use an instrument panel indicator or warning lamp to show general charging system operation. Although the lamp usually does not warn the driver of an overcharged battery or high charging voltage, it will light to show an undercharged battery or low voltage from the alternator.

The lamp also lights when the battery supplies field current before the engine starts. The lamp is often connected in parallel with a resistor, so that field current will flow even if the

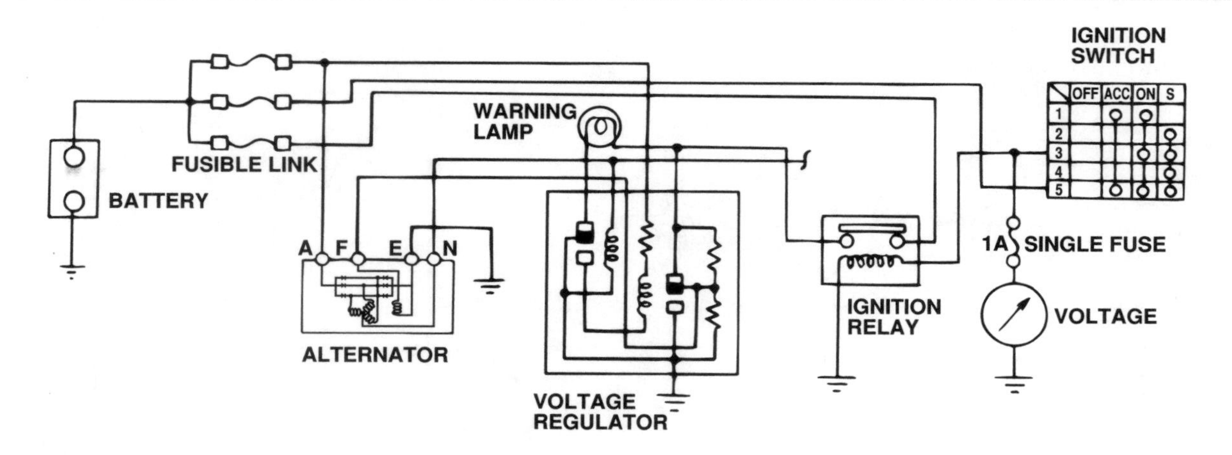

Figure 7-59. The circuit diagram of a typical voltmeter installation.

bulb fails. The lamp is wired so that it lights when battery current flows through it to the alternator field. When the alternator begins to produce voltage, this voltage is applied to the side of the lamp away from the battery. When the two voltages are equal, there will be no voltage drop across the lamp and it will go out.

The indicator lamp for a Delco-Remy CS system works differently from most others. It lights if charging voltage is either too low or too high. Any problem in the charging system causes the lamp to light at full brilliance.

CRANKING SYSTEM OPERATION

The automotive starting system is also called the cranking system. Its only job is to crank the engine fast enough for the engine to fire and run. The ignition and fuel systems must supply the spark and fuel for the engine to start and run, but the starting system cranks the engine to get it going.

The starting system draws a large amount of current from the battery to power the starter motor. The starter circuit on a large gasoline-powered V8 engine must carry as much as 300 amps under some conditions. Large V8 diesel engines require even more current. To handle this current safely and with a minimum voltage loss from resistance, the cables must be the correct size, and all connections must be clean and tight.

The starting system is controlled by the driver through the ignition switch. If the heavy cables that carry current to the starter were routed to the instrument panel and the switch, they would be so long that the starter would not get enough current to operate properly. To avoid such a voltage drop, the starting system has two circuits, figure 7-60:

- The starter circuit
- The control circuit.

Starter Motor Circuit

The starter circuit, or motor circuit, shown as the solid lines of figure 7-60, consists of:

- The battery
- A magnetic switch
- The starter motor
- Heavy-gauge cables.

The circuit between the battery and the starter motor is controlled by a magnetic switch (relay or solenoid). Switch design and function vary from system to system.

A gear on the starter motor armature engages with gear teeth on the engine flywheel. When current reaches the starter motor, it begins to turn. This turns the car's engine, which can quickly fire and run by itself. If the starter motor remained engaged to the engine flywheel, the starter motor would be spun by the engine at a very high speed. This would damage the starter motor. To avoid this, there must be a mechanism to disengage the starter motor from the engine. There are several designs that will do this, as we will see later in this chapter.

Control Circuit

The control circuit is shown by the dashed lines in figure 7-60. It allows the driver to use a small amount of battery current, about three to five amps, to control the flow of a large amount of battery current to the starter motor. Control circuits usually consist of an ignition switch connected through normal-gauge wiring to the battery and the magnetic switch. When the ignition switch is in the start position, a small amount of current flows through the coil of the magnetic switch. This closes a set of large contact points within the magnetic switch, and allows battery current to flow directly to the starter motor.

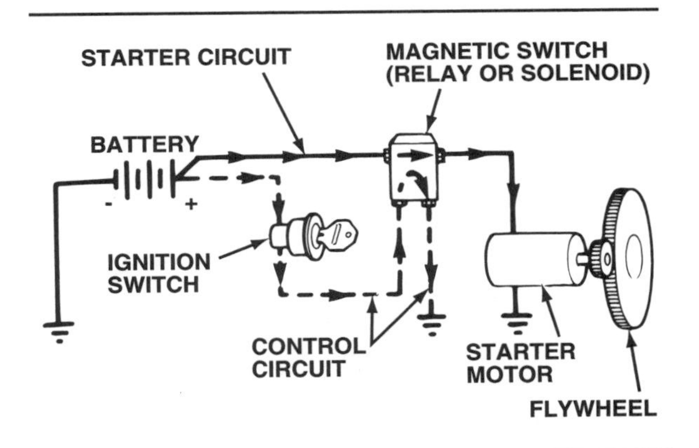

Figure 7-60. In this diagram of the starting system, the starter circuit is shown as a solid line and the control circuit as a dashed line. (Delco-Remy)

Basic System Parts

We have already studied the battery, which is an important part of the starting system. The other circuit parts are the:

- Ignition switch
- Starting safety switch (on some systems)
- Relays or solenoids (magnetic switches)
- Starter motor
- Wiring.

Ignition switch

The ignition switch has jobs other than controlling the starting system. The ignition switch normally has at least four positions:

- Accessories
- Off
- On (Run)
- Start.

Switches on late-model cars also have a lock position to lock the steering wheel. All positions except start are **detented**. That is, the switch will remain in that position until moved. When the ignition key is turned to start and released, it will return to the on (run) position. The start position is the actual starter switch part of the ignition switch. It applies battery voltage to the magnetic switch.

There are two types of ignition switches in use. On older cars, the switch is mounted on the instrument panel and contains the contact points, figure 7-61. The newer type, used on cars with locking steering columns, is usually mounted on the column. Many column-mounted switches operate remotely mounted contact points through a rod, figure 7-62. Other column-mounted switches operate directly on contact points, figure 7-63. Older domestic and imported cars sometimes used separate push-button switches or cable-operated switches that controlled the starting system separately from the ignition switch.

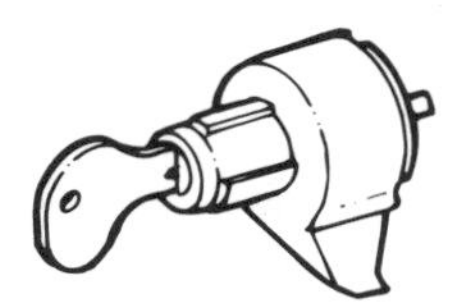

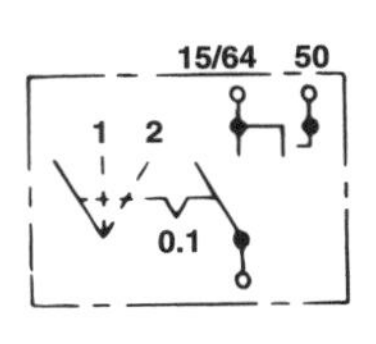

Figure 7-61. This ignition switch acts directly on the contact points. (Bosch)

Starting safety switch

The **starting safety switch** is also called a neutral-start switch. It is a normally open switch that prevents the starting system from operating when the automobile's transmission is in gear. If the car has no starting safety switch, it is possible to spin the engine with the transmission in gear. This will make the car lurch forward or backward, which could be dangerous. Safety switches or interlock devices are now required by law with all automatic and manual transmissions.

Starting safety switches can be connected in two places within the starting system control circuit. The safety switch can be placed between the ignition switch and the magnetic switch, figure 7-64, so that the safety switch must be closed before current can flow to the magnetic switch. The safety switch also can be connected between

Detented: Positions in a switch that allow the switch to stay in that position. In an ignition switch, the ON, OFF, LOCK, and ACCESSORY positions are detented.

Starting Safety Switch: A neutral start switch. It keeps the starting system from operating when a car's transmission is in gear.

■ The Beginning

The automobile became common in America in the early 1900s. The first automobiles had no electric starting systems. The driver had to insert a crank into the front of the engine and turn it by hand until the engine fired and ran. This required both skill and physical strength. Without a self-starting system, engines were limited in displacement and compression ratios. Operating an automobile was very inconvenient for those who could not crank the engine themselves.

In 1910, Charles F. Kettering began work on a practical automotive self-starter. The system first appeared on the 1912 Cadillac, and was quickly adopted by other manufacturers.

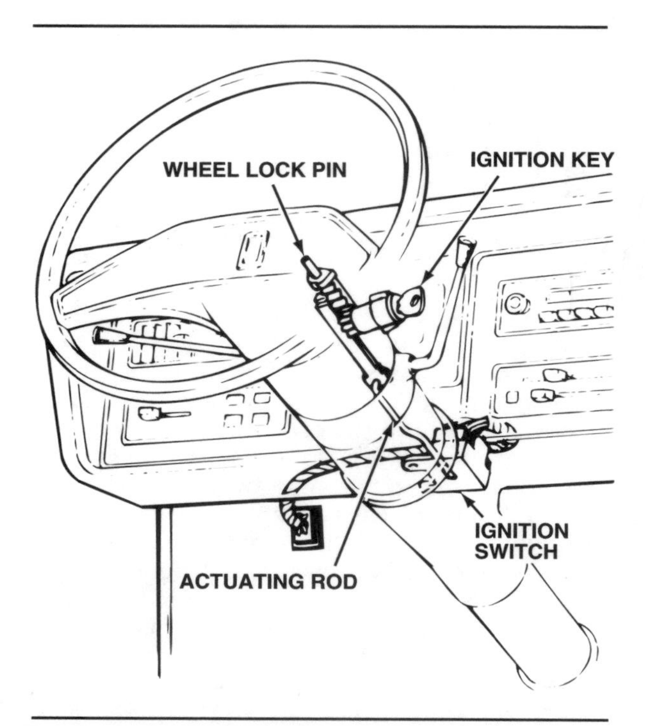

Figure 7-62. Many column-mounted ignition switches act on the contact points through the movement of a rod. (Ford)

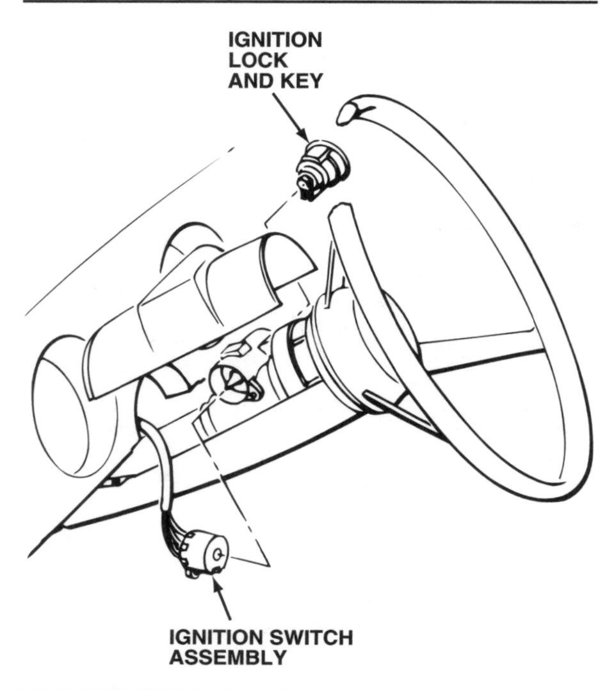

Figure 7-63. This column-mounted ignition switch acts directly on the contact points.

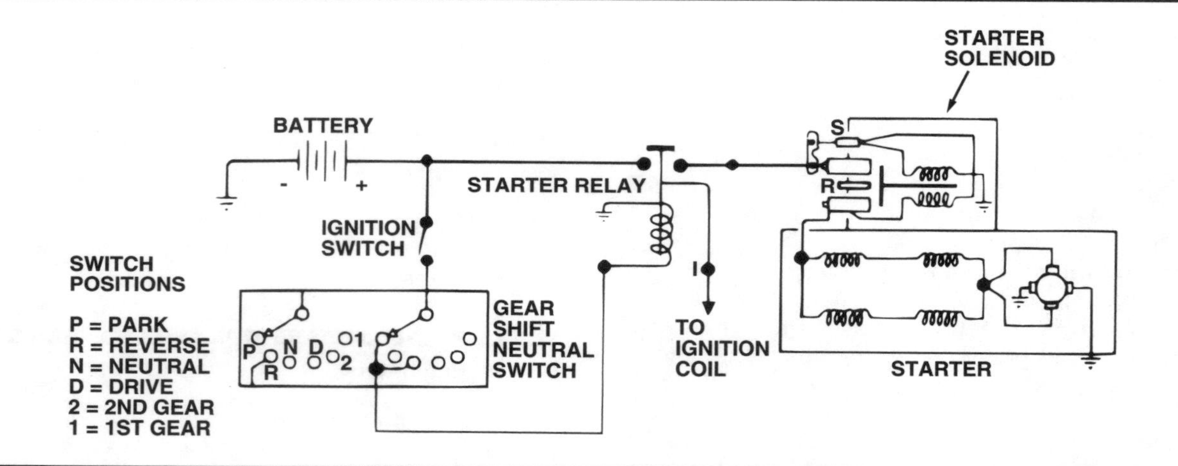

Figure 7-64. This starting safety switch must be closed before battery current can reach the magnetic switch. (Ford)

the magnetic switch and ground, figure 7-65, so that the switch must be closed before current can flow from the magnetic switch to ground.

Where the starting safety switch is installed depends on the type of transmission used and whether the gear shift lever is column- or floor-mounted.

Automatic transmissions

The safety switch used with an automatic transmission can be an electrical switch or a mechanical device. Electrical switches have contact points that are closed only when the gear lever is in Park or Neutral, as shown in figure 7-64. The contacts are in series with the control circuit, so that no current can flow through the magnetic switch unless the transmission is out of gear.

Many manufacturers use an additional circuit in the neutral-start switch to light the backup lamps when the transmission is placed in Reverse, figure 7-65. Ford vehicles equipped with an electronic automatic transmission or transaxle use an additional circuit in the neutral safety switch to inform the microprocessor of the

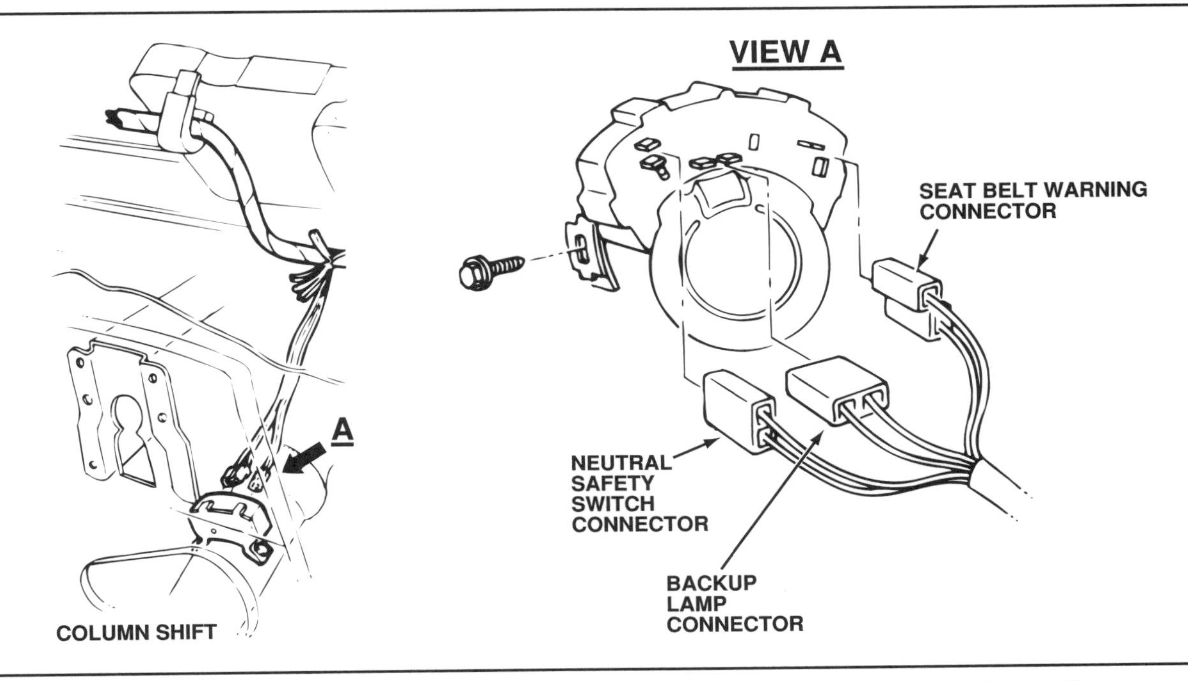

Figure 7-65. An electrical safety switch installed near the column-mounted gear shift lever. (Chevrolet)

Figure 7-66. The Ford starter relay or magnetic switch.

position of the manual lever shaft. This signal is used to determine the desired gear and electronic pressure control. The switch is now called a transmission range (TR) sensor.

General Motors has done essentially the same as Ford, renaming the Park/Neutral switch used on its THM 4T60-E transaxle. It now is called either a PRNDL switch or a TR sensor, and provides input to the PCM regarding torque converter clutch slip. This input allows the PCM to make the necessary calculations to control clutch apply and release feel.

Manual transmissions

The starting safety switch used with a manual transmission on older vehicles is usually a neutral safety switch similar to those shown in figure 7-65. A clutch pedal position (CPP) switch, or an interlock switch, is commonly used with manual transmissions and transaxles on late-model vehicles. This is an electric switch mounted on the floor or firewall near the clutch pedal. Its contacts are normally open, and close only when the clutch pedal is fully depressed.

Relays and solenoids

A magnetic switch in the cranking system allows the control circuit to open and close the starter circuit. The switch can be a:

- Relay, which uses the electromagnetic field of a coil to attract an armature and close the contact points
- Solenoid, which uses the electromagnetic field of a coil to pull a plunger into the coil and close the contact points.

■ Crank Lost?

In 1935, if a car's electric starter would not work, and if the starting crank had been misplaced, motorists were advised that they could jack up a rear wheel, put the clutch in, place the transmission gears in "high", let the clutch out, and start the engine by turning the rear wheel.

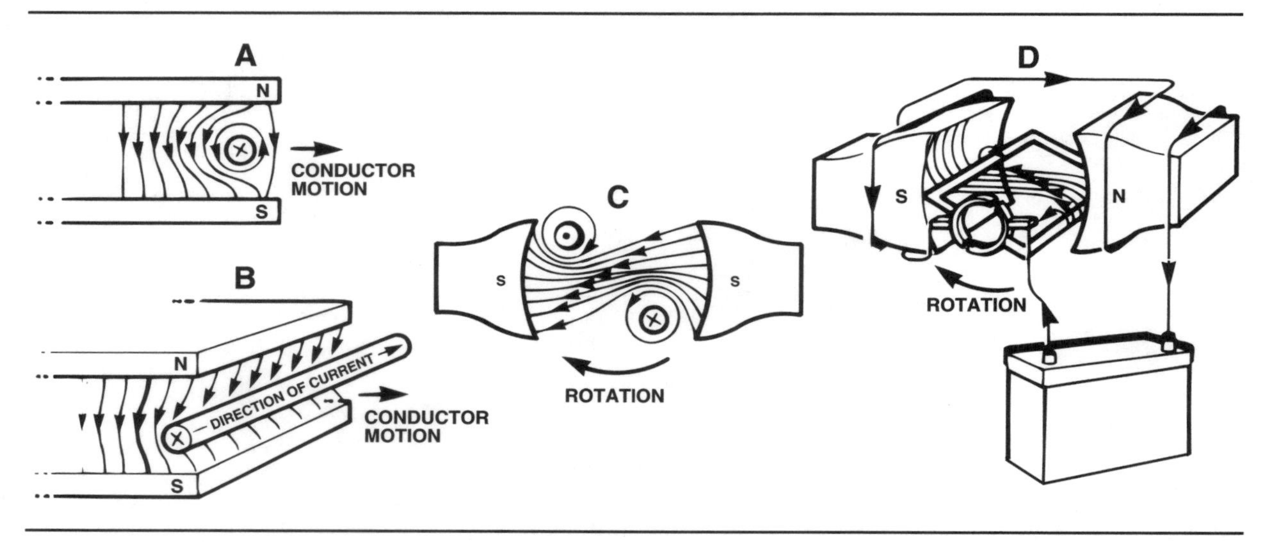

Figure 7-67. The motor principle as illustrated by Prestolite.

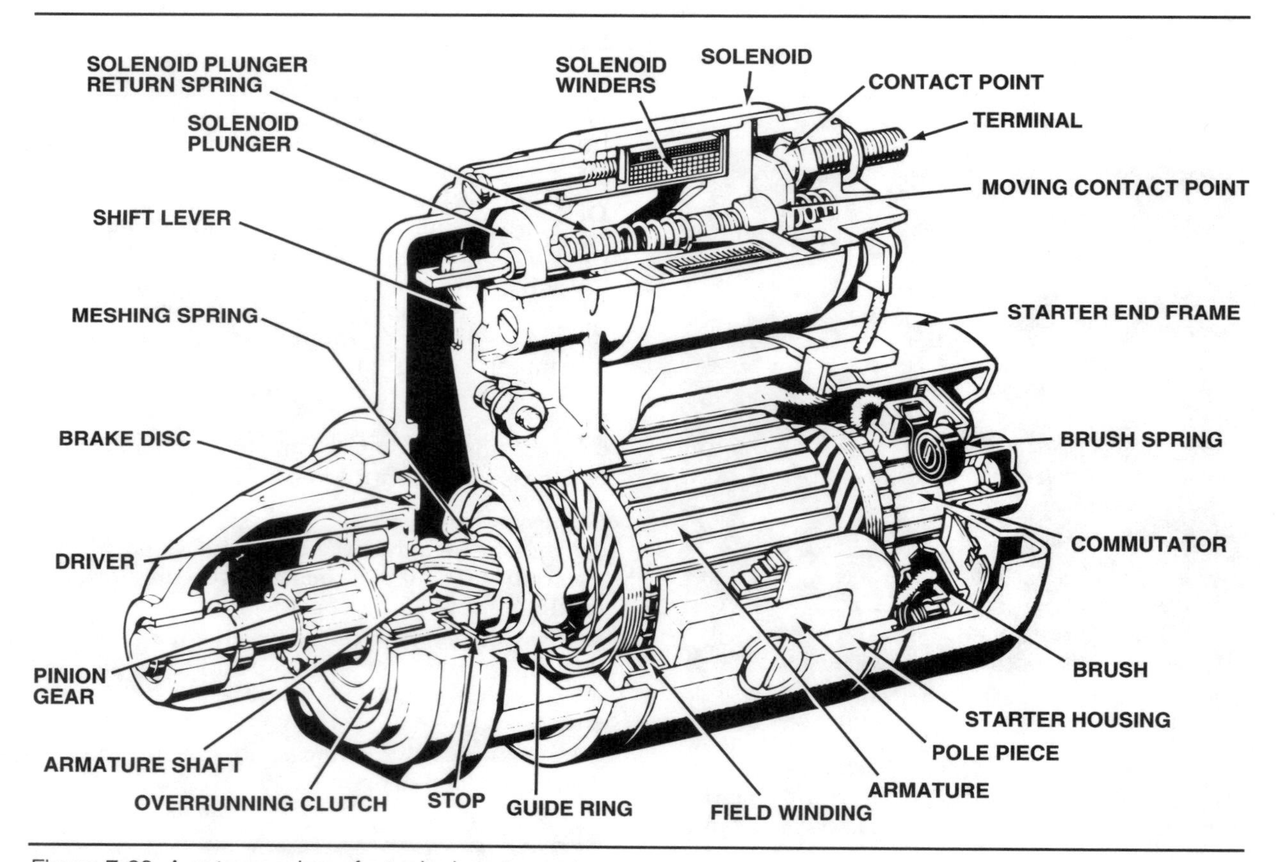

Figure 7-68. A cutaway view of a typical starter motor.

In addition to closing the contact points, solenoid-equipped circuits often use the movement of the solenoid to engage the starter motor with the engine flywheel. We will explain this in the next chapter.

The terminology used with relays and solenoids is often confusing. Technically, a relay operates with a hinged armature and does only an electrical job. A solenoid operates with a movable plunger and usually does a mechanical job. Sometimes, a solenoid is used only to open and close an electric circuit. The movement of the plunger is not used for any mechanical work. Manufacturers sometimes call these sole-

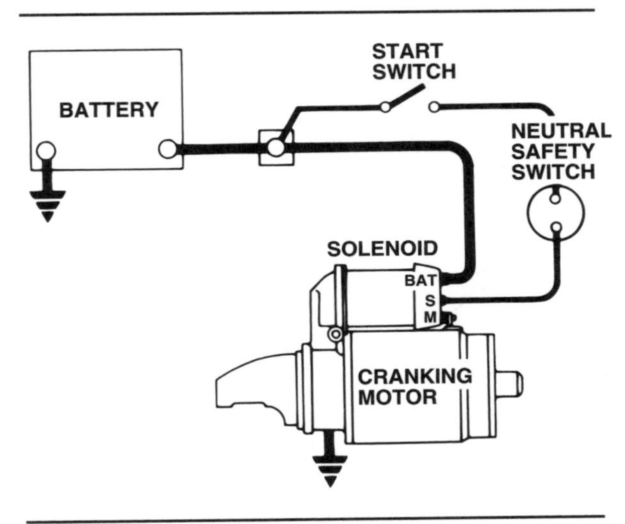

Figure 7-69. A typical GM starting circuit. (Delco-Remy)

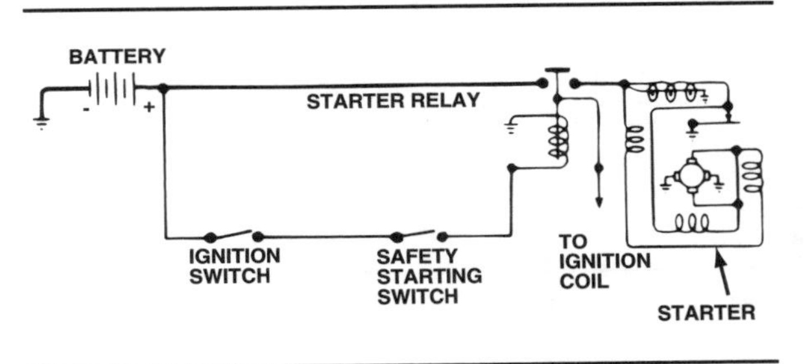

Figure 7-70. The Ford starting system with the positive engagement starter.

noids "starter relays". Figure 7-66 shows a commonly used Ford starter relay. We will continue to use the general term "magnetic switch", and will tell you if the manufacturer uses a different name for the device.

Starter motor

The starter motor converts electrical energy from the battery into mechanical energy to turn the engine. It does this through the interaction of magnetic fields, figure 7-67. When current flows through a conductor, a magnetic field is formed around the conductor. If the conductor is placed in another magnetic field, the two fields will be weakened at one side and strengthened at the other side, position A. The conductor will tend to move from the strong field into the weak field, position B. Positions C and D show how a simple motor can use this movement to make the conductors rotate. An automotive starter motor has many conductors, and uses a lot of current to create enough rotational force to crank the engine. Figure 7-68 shows a cutaway view of a starter motor.

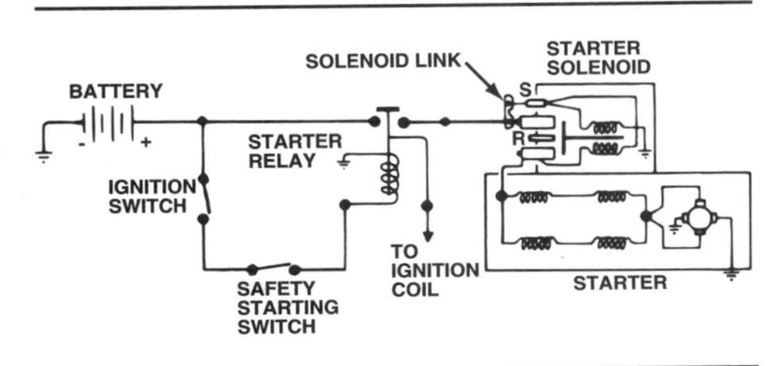

Figure 7-71. The Ford solenoid-actuated system uses the starter-mounted solenoid only for a mechanical job, not for an electrical job.

Wiring

The starter motor circuit uses heavy-gauge wiring to carry current to the starter motor. The control circuit carries less current and thus uses lighter-gauge wires.

Specific Cranking Systems

Various manufacturers use different starting system components. The following paragraphs briefly describe the circuits used by major manufacturers.

Delco-Remy and Bosch

Delco-Remy and Bosch starter motors are used by General Motors. The most commonly used Delco-Remy and Bosch automotive starter motor depends on the movement of a solenoid both to control current flow in the starter circuit and to engage the starter motor with the engine

■ Vacuum Control Switches

Magnetic switches are not the only control devices that have been used in the starting system's control circuit. Before the 1950s, some GM cars had a vacuum control switch mounted in the carburetor. To start the car, the driver turned on the ignition switch and depressed the accelerator pedal. The pedal movement was transmitted through a linkage to the vacuum control switch, closing its contacts. Current flowed from the battery, through the vacuum switch contacts, to the starter motor.

When the engine started and ran, carburetor vacuum opened the vacuum switch contacts to stop cranking. If the carburetor vacuum was not great enough to do this, a secondary system would stop the cranking. Generator voltage was applied to the ground side of the motor. When generator voltage equalled battery voltage, there would be no voltage drop across the motor and the motor would not crank.

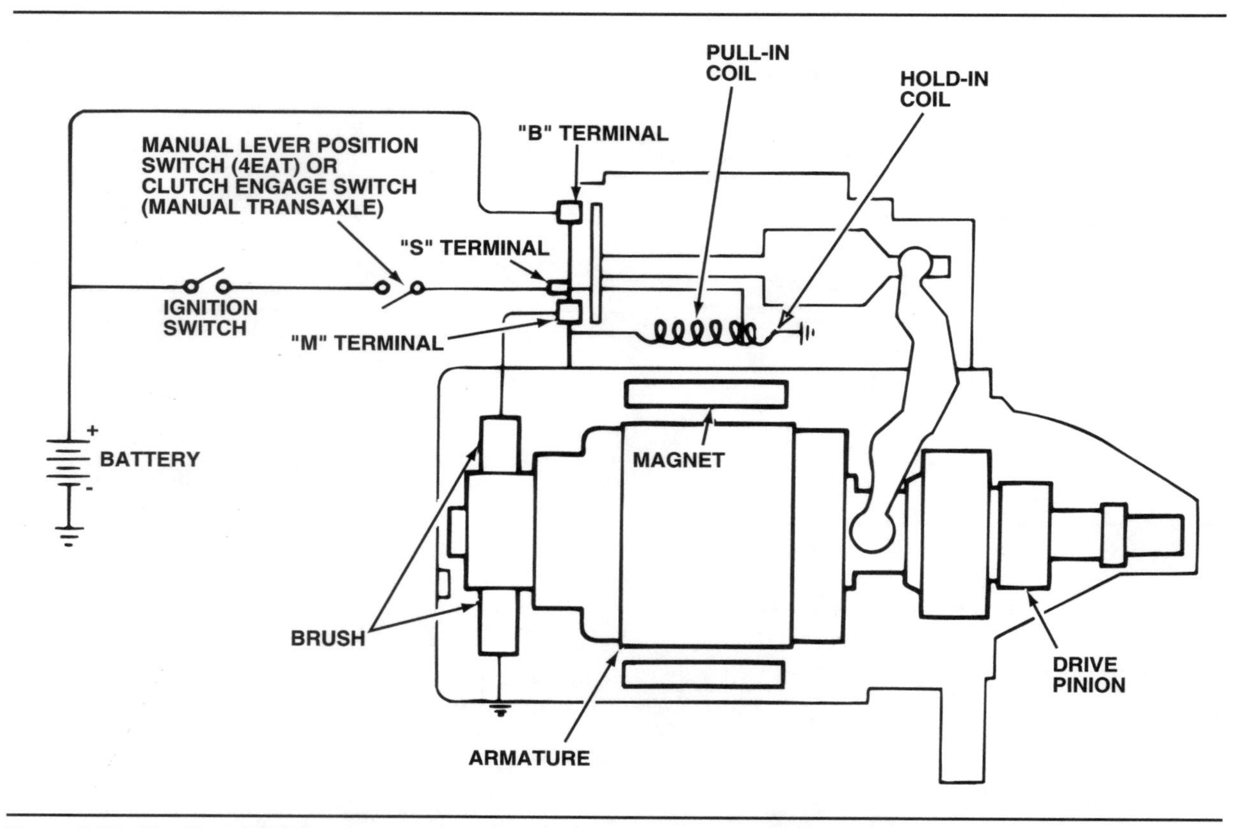

Figure 7-72. The Ford PMGR starter circuit may use only a solenoid instead of a solenoid and a relay.

flywheel. This is called a **solenoid-actuated starter**. The solenoid is mounted on, or enclosed with, the motor housing, figure 7-69.

Motorcraft

Ford has used three types of starter motors and therefore has several different starting system circuits.

The Motorcraft positive engagement starter has a movable pole shoe that uses electromagnetism to engage the starter motor with the engine. This motor does not use a solenoid to *move* anything, but uses a solenoid to open and close the starter circuit as a magnetic switch, figure 7-70. Ford calls this solenoid a starter relay, figure 7-66.

The Motorcraft solenoid-actuated starter is very similar to the Delco-Remy and depends on the movement of a solenoid to engage the starter motor with the engine, figure 7-71. The solenoid is mounted within the motor housing and receives battery current through the same type of starter relay used in the positive engagement system. Although the motor-mounted solenoid could do the job of this additional starter relay, the second relay is installed on many Ford automobiles to make the cars easier to build.

Motorcraft solenoid-actuated starters were used on Ford cars and trucks with large V8 engines.

The Motorcraft permanent magnet gear-reduction (PMGR) starter is a solenoid-actuated design that operates much like the Motorcraft solenoid-actuated starter described above. However, the starter circuit may or may not use a starter relay, depending on the car model, figure 7-72.

Rear-wheel-drive (RWD) Ford automobiles with manual transmissions have no starting safety switch. Front-wheel-drive (FWD) models with manual transaxles have a clutch interlock switch. If a Ford car with an automatic transmission has a column-mounted shift lever, a blocking interlock device prevents the ignition key from turning when the transmission is in gear. If the automatic transmission shift lever is mounted on the floor, an electrical switch prevents current from flowing to the starter relay when the transmission is in gear. The switch may be mounted on the transmission case or near the gear shift lever.

Chrysler

Chrysler uses a solenoid-actuated starter motor. The solenoid is mounted inside the motor housing and receives battery current through a starter relay.

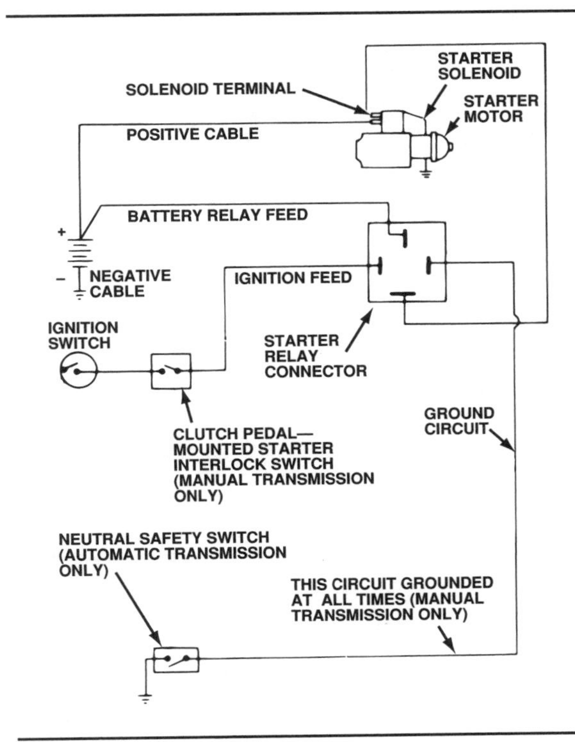

Figure 7-73. A typical Chrysler starting system with a five-terminal Bosch relay.

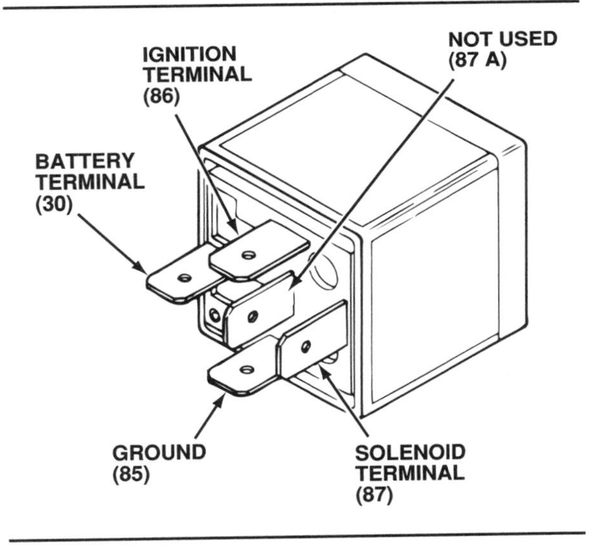

Figure 7-74. Only four of the five relay terminals are used when the Bosch relay is installed in the Chrysler starting system.

During the 1980s, Chrysler starting systems used a standard five-terminal Bosch relay, figure 7-73, but only four relay terminals are used in the circuit, figure 7-74. The relay is located at the front of the driver's side strut tower in a power distribution center or cluster.

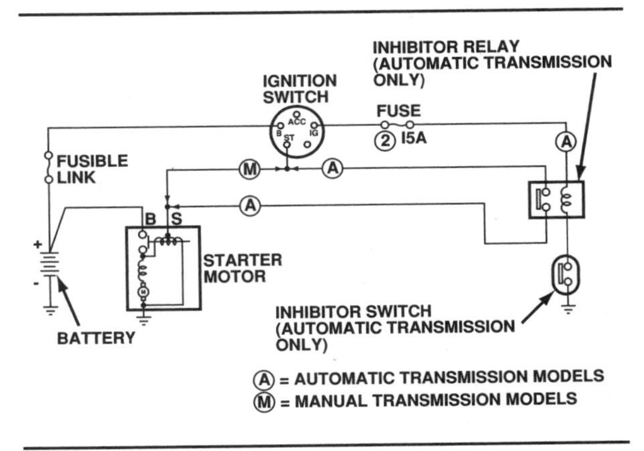

Figure 7-75. A typical Nissan starting system used on gasoline engines.

Chrysler automobiles with manual transmissions have a clutch interlock switch. Current from the starter relay can flow to ground only when the clutch pedal is fully depressed. Cars with automatic transmission have an electrical neutral start switch mounted on the transmission housing. When the transmission is out of gear, the switch provides a ground connection for the starter control circuit.

Toyota and Nissan

Toyota and Nissan use a variety of solenoid-actuated direct drive and reduction gear-starter designs manufactured primarily by Hitachi and Nippondenso, figure 7-75. The Park/Neutral position (PNP) switch (called an inhibitor switch by the Japanese) incorporates a relay in its circuit.

SUMMARY—EFFECTS OF BATTERY, CHARGING, AND STARTING SYSTEMS ON ENGINE PERFORMANCE

The symptoms caused by defects in these systems are often very similar. To make the relationships between these systems more clear, we have organized the possible defects according to the engine performance symptoms they may cause:

- No cranking or slow cranking
- Normal cranking—no starting or hard starting
- Engine quits under electrical load.

Before using these paragraphs to identify defects, make sure all other systems have been checked out first.

Solenoid-Actuated Starter: A starter that uses a solenoid both to control current flow in the starter circuit and to engage the starter motor with the engine flywheel.

No Cranking or Slow Cranking

If the engine does not turn over, or if it cranks too slowly, the starting system may be at fault. However, this could also be caused by a discharged battery, which in turn can be caused by a battery defect or a weak charging system.

Normal Cranking—No Starting or Hard Starting

When the engine is being cranked at normal speed but the engine will not start or is hard to start, the engine may not be getting enough ignition system voltage. A weak battery that does not have enough energy for both the starting and the ignition systems may cause starting problems.

The battery's condition is caused either by a battery defect or by a problem in the charging system. If the battery and the charging system are all right, the problem may be excessive current draw through the starting circuit, not leaving enough energy for the ignition system, even with a fully charged battery.

Quits Under Electrical Load

If the engine quits at idle or under acceleration when many accessories are being used, it is because the ignition voltage has dropped below the minimum requirement. This is caused by either a battery defect or a charging system problem that has weakened the battery.

Review Questions

Choose the single most correct answer.
Compare your answers to the correct answers on page 507.

1. Which of the following occurs within an automobile battery?
   a. The positive plate gains electrons and is positively charged
   b. The negative plate loses electrons and is negatively charged
   c. The positive plate loses electrons and the negative plate gains electrons
   d. The positive plate gains electrons and the negative plate loses electrons

2. The plates of a discharged battery are:
   a. Two similar metals in the presence of an electrolyte
   b. Two similar metals in the presence of water
   c. Two dissimilar metals in the presence of an electrolyte
   d. Two dissimilar metals in the presence of water

3. Which of the following does not occur during battery recharging?
   a. The lead sulfate on the plates gradually decomposes
   b. The sulfate is redeposited in the water
   c. The electrolyte is returned to full strength
   d. The negative plates change back to lead sulfate

4 At 80°F (27°C), the correct specific gravity of electrolyte in a fully charged battery is:
   a. 1.200 to 1.225
   b. 1.225 to 1.265
   c. 1.265 to 1.280
   d. 1.280 to 1.300

5. Which of the following materials is not used for battery separators?
   a. Lead
   b. Wood
   c. Paper
   d. Plastic

6. Maintenance-free batteries:
   a. Have individual cell caps
   b. Require water infrequently
   c. Have three pressure vents
   d. Use nonantimony lead alloys

7. Which of the following statements is not true of a replacement battery?
   a. It may have the same rating as the original battery
   b. It may have a higher rating than the original battery
   c. It may have a lower rating than the original battery
   d. It should be selected according to an application chart

8. The principle cause of battery water loss is:
   a. Spillage from the vent caps
   b. Leakage through the battery case
   c. Conversion of water to sulfuric acid
   d. Evaporation due to heat of the charging current

9. The electrolyte in a fully charged battery will generally not freeze until the temperature drops to:
   a. 32°F (0°C)
   b. 0°F (-18°C)
   c. -20°F (-29°C)
   d. -50°F (-46°C)

10. Recombinant batteries are:
    a. Rebuilt units
    b. Completely sealed
    c. Vented to release gassing
    d. Able to produce a higher cell voltage

11. Alternators induce voltage by rotating:
    a. A magnetic field inside a fixed conductor
    b. A conductor inside a magnetic field
    c. Both a or b
    d. Neither a nor b

12. Alternating current in an alternator is rectified by:
    a. Brushes
    b. Diodes
    c. Sliprings
    d. Transistors

13. An alternator consists of:
    a. A stator, a rotor, slip-rings, brushes, and diodes
    b. A stator, an armature, sliprings, brushes, and diodes
    c. A stator, a rotor, a commutator, brushes, and diodes
    d. A stator, a rotor, a field relay, brushes, and diodes

14. Automotive alternators that have three conductors generally use ___ diodes to rectify the output current.
    a. Two
    b. Three
    c. Four
    d. Six

15. Alternator output voltage is directly related to:
    a. Field strength
    b. Rotor speed
    c. Both field strength and rotor speed
    d. Neither field strength nor rotor speed

16. The shorting contacts of a double-contact regulator:
    a. Increase voltage creep
    b. Short the field circuit to the alternator
    c. React to battery temperature changes
    d. Reduce field current to zero

17. Which of the following cannot be used in a totally solid-state regulator?
    a. Zener diodes
    b. Thermistors
    c. Capacitors
    d. Circuit breakers

18. Warning lamps are installed so that they will not light when:
    a. The voltage on the battery side of the lamp is higher
    b. Field current is flowing from the battery to the alternator
    c. The voltage on both sides of the lamp is equal
    d. The voltage on the resistor side of the lamp is higher

19. The regulator is a charging system device that can control ____ circuit opening and closing.
    a. Ignition-to-battery
    b. Alternator-to-thermistor
    c. Battery-to-accessory
    d. Voltage source-to-battery

20. The starting system has ____ circuits to avoid excessive voltage drop.
    a. Two
    b. Three
    c. Four
    d. Six

21. Which of the following is not part of the starter control circuit?
    a. The ignition switch
    b. The starting safety switch
    c. The starter relay
    d. The starter motor

22. The starting safety switch is also called a:
    a. Remote-operated switch
    b. Manual-override switch
    c. Neutral-start switch
    d. Single-pole, double-throw switch

23. Starting safety switches on late-model vehicles with manual transmissions are usually:
    a. Electrical
    b. Mechanical
    c. Floor-mounted
    d. Column-mounted

Chapter

8 Ignition Primary Circuit and Components

This chapter explains the components of the low-voltage primary ignition circuit and its operation. Breaker points were used to open and close the low-voltage primary circuit until the mid-1970s, when solid-state electronic switching devices took their place. Whether breaker points or electronic switches are used, however, the principles of producing high voltage by electromagnetic induction remain the same.

NEED FOR HIGH VOLTAGE

Energy is supplied to the automotive electrical system by the battery. The battery can supply about 12 volts, but in breaker-point ignition systems, the voltage required to ignite the air-fuel mixture can range from 5,000 to more than 25,000 volts, depending on engine operating conditions.

This high voltage is required to cause an arc across the spark plug air gap. The required voltage level increases when the:

- Spark plug air gap increases
- Engine operating temperature increases (resistance increases with greater temperature)
- Air-fuel mixture contains less fuel (fewer volatile fuel particles)
- Air-fuel mixture is at a greater pressure (resistance increases with an increase in pressure).

Since part of the ignition system's job is to provide a high-voltage spark, battery voltage must be greatly increased to meet the needs of the ignition system. This can be done by using electromagnetic induction.

HIGH VOLTAGE THROUGH INDUCTION

We know that a current-carrying conductor or coil is surrounded by a magnetic field. As current in the coil increases or decreases, the magnetic field expands or contracts. If a second coiled conductor is placed within this magnetic field, figure 8-1, the expanding or contracting magnetic flux lines will cut the second coil, causing a voltage to be induced in the second coil. This transfer of energy between two unconnected conductors is called **mutual induction**.

Induction in the Ignition Coil

The ignition coil uses the principle of mutual induction to step up or transform low battery voltage to high ignition voltage. The ignition coil, figure 8-2, contains two windings of copper wire around a soft iron core. The primary winding is made of a hundred or so turns of heavy wire, and allows battery current to flow through

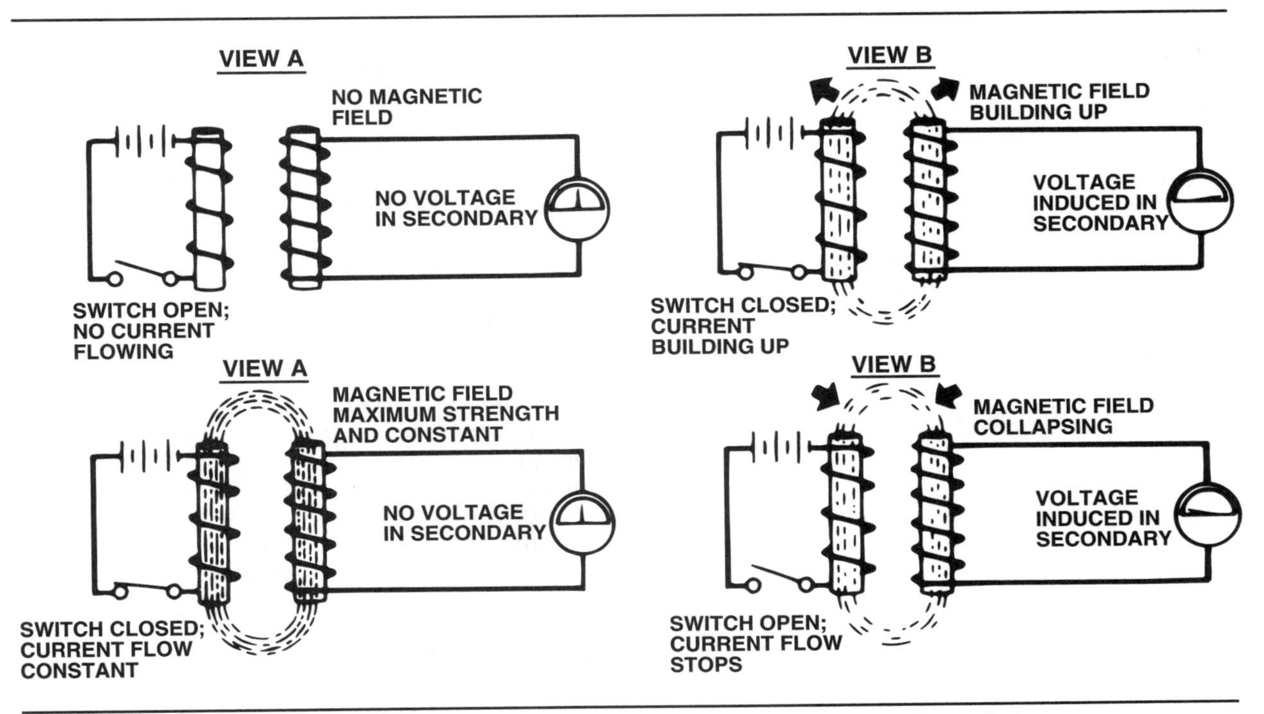

Figure 8-1. Mutual induction in the ignition coil supplies voltage to the spark plugs. (Prestolite)

it. The secondary winding is made of many thousand turns of fine wire. When current flow in the primary winding increases or decreases, a voltage is induced in the secondary winding, figure 8-2.

The ratio of the number of turns in the secondary winding to the number of turns in the primary winding can be between 100:1 and 200:1. This ratio is the voltage multiplier. That is, any voltage induced in the secondary winding will be 100 to 200 times the voltage present in the primary winding.

Several factors govern the coil's induction of voltage. Only two of these factors can be controlled easily in an ignition system. Induced voltage will increase with:

- More magnetic flux lines (a stronger magnetic field caused by greater current flow)
- More rapid movement of flux lines (faster collapse of the field caused by an abrupt end to current flow).

Normal voltage applied to the coil primary winding is about 9 to 10 volts (At high speeds, voltage may rise to 12 volts or more). This voltage causes from one to four amps of current to flow in the primary winding. When this current flows, a magnetic field builds up around the windings. Building up a complete magnetic field is called **magnetic saturation**, or coil saturation. When this current flow stops, the primary winding's magnetic field collapses. A greater voltage is self-induced in the primary winding by the collapse of its own magnetic field. This self-induction creates from 250 to 400 volts in the primary winding. If it develops 250 volts and the turns ratio multiplies this by 100, then 25,000 volts will be induced in the secondary winding. This is enough voltage to ignite the air-fuel mixture under almost all operating conditions for certain ignition systems.

This kind of ignition system, based on the induction of a high voltage in a coil, is called an **inductive-discharge ignition system.** An inductive-discharge system using a battery as the source of low-voltage current has been the standard automotive ignition system for about 80 years.

Mutual Induction: The transfer of energy between two unconnected conductors, caused by the expanding or contracting magnetic flux lines of the current-carrying conductor.

Magnetic Saturation: The condition when a magnetic field reaches full strength and maximum flux density.

Inductive-Discharge Ignition System: A method of igniting the air-fuel mixture in an engine cylinder. It is based on the induction of a high voltage in the secondary winding of a coil.

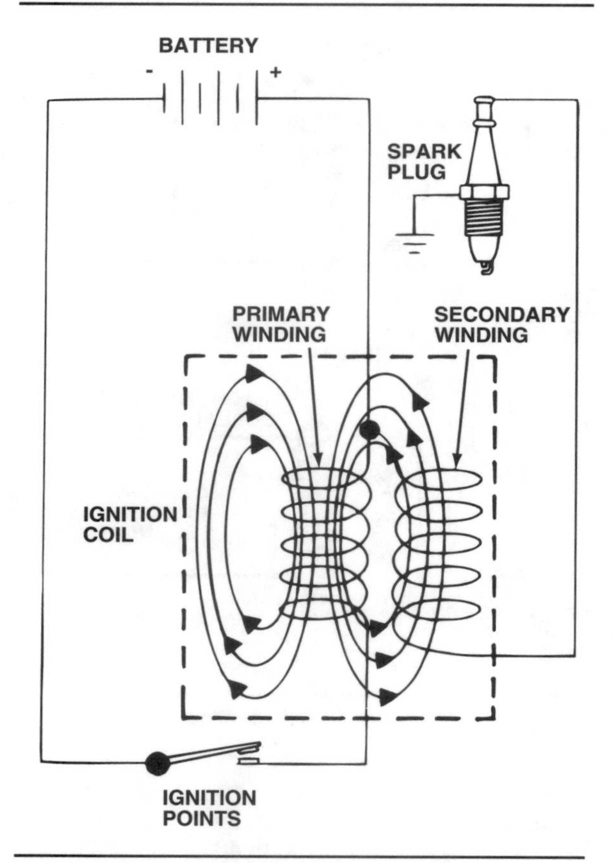

Figure 8-2. The ignition coil produces high-voltage current in the secondary winding when current flow is cut off in the primary winding.

We have seen that the ignition system can transform low battery voltage into high ignition voltage. Now we can look at the ignition system circuitry to see how the system works.

BASIC CIRCUITS AND CURRENT FLOW

The ignition system, figure 8-3, consists of two interconnected circuits:

- The primary (low-voltage) circuit
- The secondary (high-voltage) circuit.

When the ignition switch is turned on, battery current flows:

- Through the ignition switch and the ballast resistor, if one is included
- To and through the coil primary winding
- Through a switching device (breaker points or solid-state device)
- To ground and the grounded terminal of the battery.

Low-voltage current flow in the coil primary winding creates a magnetic field. When the switching device interrupts this current flow:

- A high-voltage surge is induced in the coil secondary winding.
- Current flows through an ignition cable from the coil to the distributor.
- Current passes through the distributor cap, rotor, across the rotor air gap, and through another ignition cable.
- Current flows to the spark plug, where it arcs to ground.

The inductive-discharge battery ignition system was invented by Charles F. Kettering in 1908. He used a set of contact points as an elecrical switch to open and close the circuit to the primary winding of the ignition coil. These contacts are opened by the rotation of a cam on the distributor shaft. They are called **breaker points** because they continually break the primary circuit.

Kettering's ignition system was quickly adopted as the standard for the automotive industry, and was used virtually unchanged for over 60 years. By the early 1970s, though, solid-state electronic components began to replace breaker points as switching devices for the primary circuit. The electronic or breakerless ignitions on late-model cars still use the inductive-discharge principles to produce a high-voltage spark. However, the primary circuits are controlled by electronic rather than mechanical switching devices.

PRIMARY CIRCUIT COMPONENTS

The primary circuit, figure 8-3, contains the:

- Battery
- Ignition switch
- Primary (ballast) resistor (in some systems)
- Starting bypass (in some systems)
- Switching device in the distributor
- Coil primary winding.

Battery

The battery supplies low-voltage current to the ignition primary circuit. This current flows when the ignition switch is in the START or RUN position.

Ignition Switch

The ignition switch controls low-voltage current through the primary circuit. This current can flow when the ignition switch is in the START or RUN position. Other switch positions route current to accessory circuits and lock the steering wheel in position.

Manufacturers use various types of ignition switch circuitry. The differences lie in how battery current is routed to the switch. Regardless of variations, full system voltage is always pre-

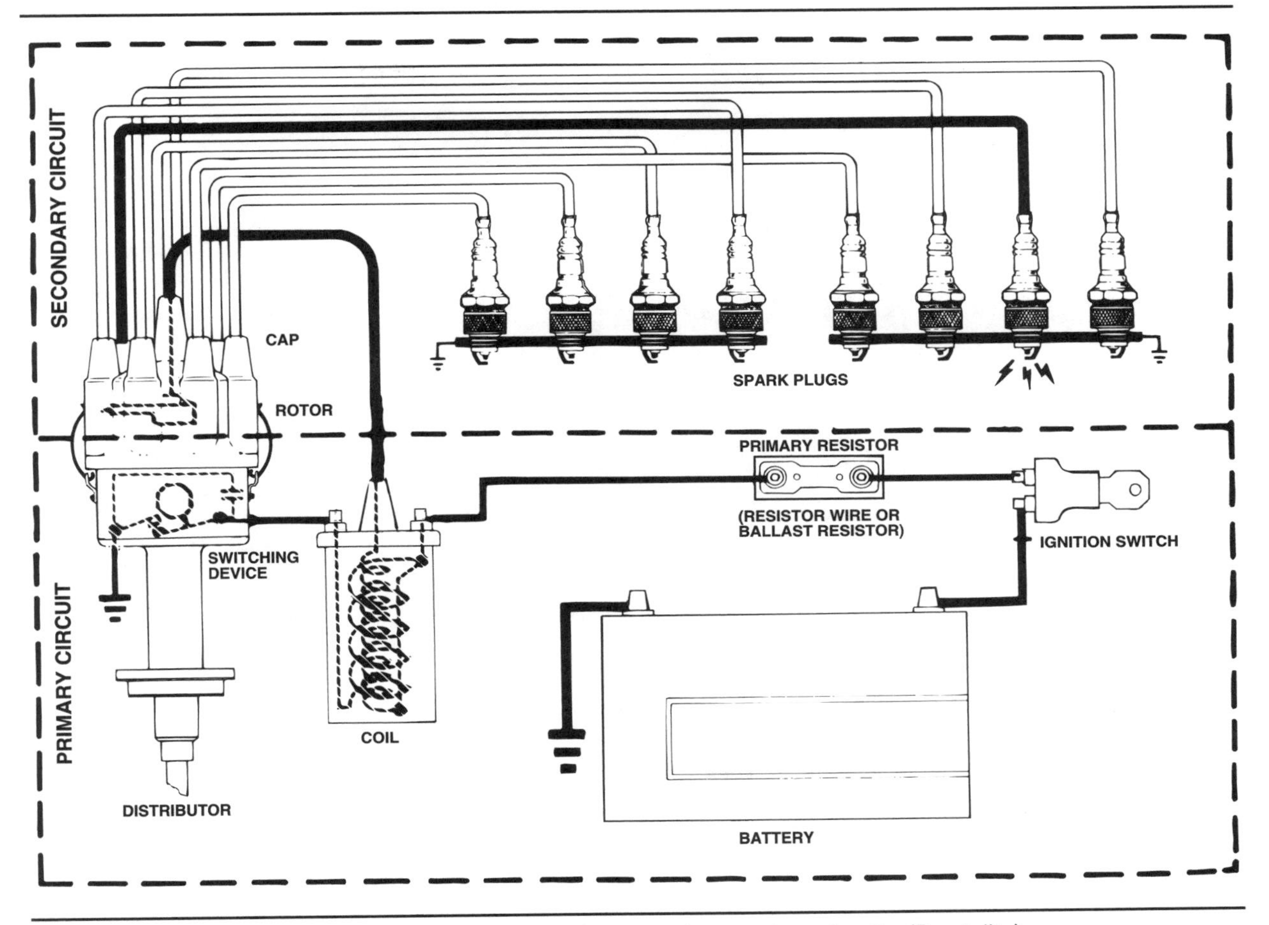

Figure 8-3. The ignition system is divided into the primary and secondary circuits. (Prestolite)

sent at the switch, as if it were connected directly to the battery.

Many General Motors (GM) automobiles draw ignition current from a terminal on the Delco-Remy starter motor solenoid, figure 8-4. Ford Motor Company systems draw ignition current from a terminal on the Motorcraft starter relay, figure 8-5. In Chrysler Corporation vehicles, ignition current comes through a wiring splice installed between the battery and the alternator, figure 8-6.

Ballast (Primary) Resistor

For an ignition coil to have uniform secondary voltage capabilities over a range of engine speeds, complete saturation of its magnetic fields must be developed at these varying speeds. Magnetic saturation depends on the amount of voltage applied to the coil, the amount of current flowing in the windings, and the length of time the current flows.

An automotive ignition system does not operate with uniform current and voltage or current flow time. When the starter is cranking, the high current draw of the starter motor drops system voltage to about 10 or 11 volts. For this reason, the ignition coil must be able to produce enough secondary voltage to fire the engine with only 10 volts of primary voltage applied. Also, at high engine speeds, primary current flow time is reduced to only a few milliseconds (ms). Therefore, uniform coil saturation must develop under extremes of low voltage and short current flow time. To achieve this, most 12-volt ignition coils are designed to operate on 9 or 10 volts under most conditions.

As we have mentioned, the starter motor drops ignition system voltage to about 10 volts when the engine is cranking. As soon as the engine starts, however, system voltage rises to 12 volts or more. To maintain the ignition primary voltage at the desired level, a resistor is

Breaker Points: The metal contact points that act as an electrical switch in a distributor. They open and close the ignition primary circuit.

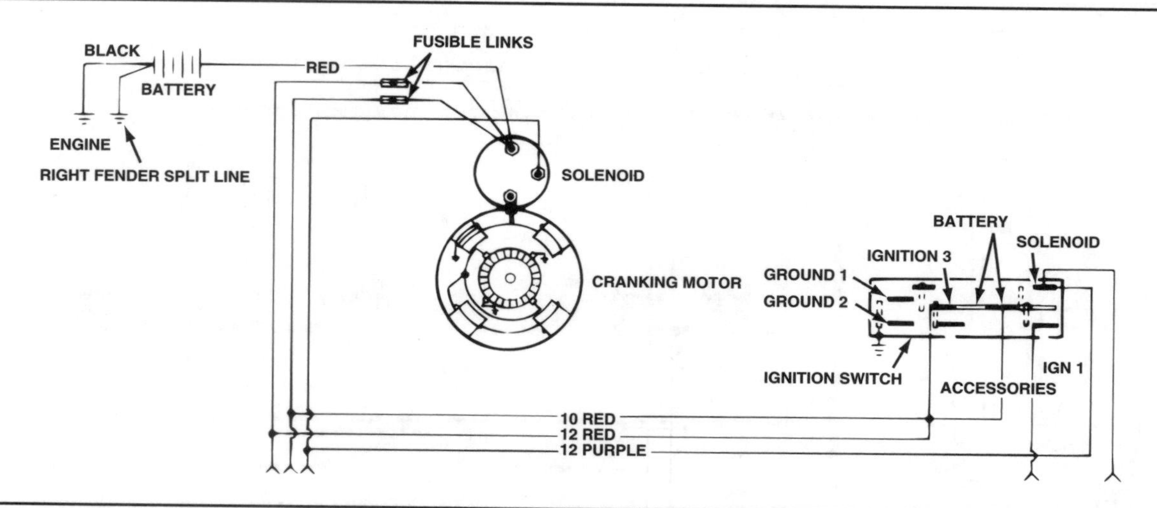

Figure 8-4. GM products draw current from a terminal on the starter solenoid. (Delco Remy)

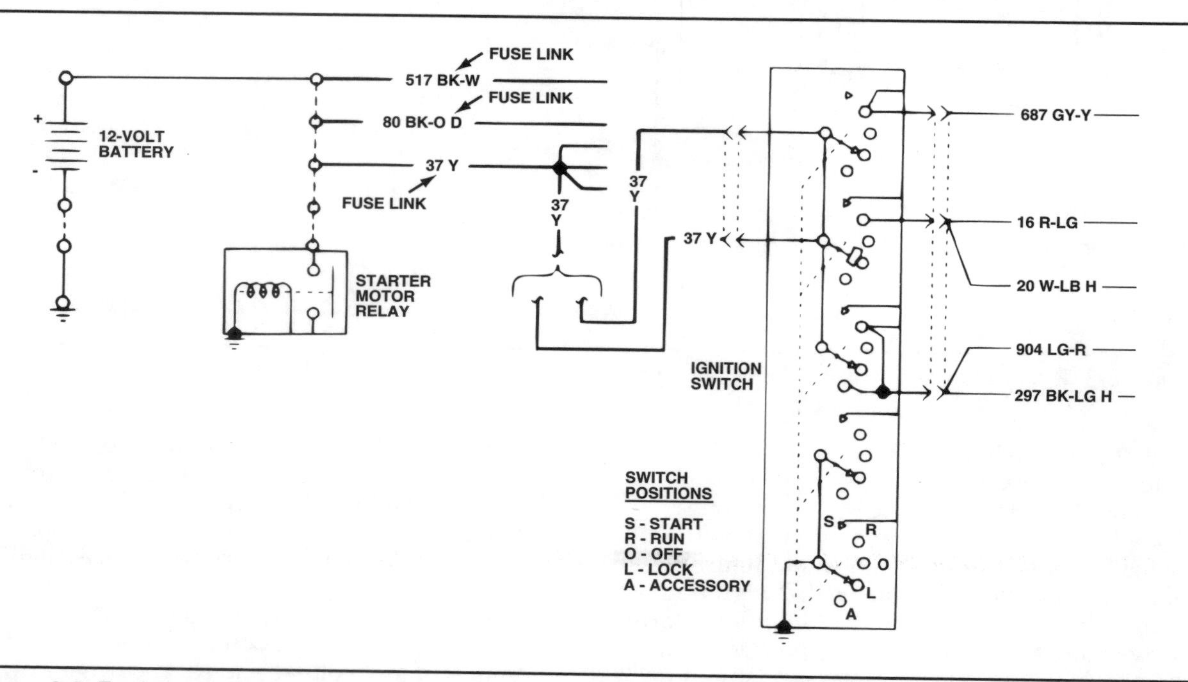

Figure 8-5. Ford products draw ignition current from a terminal on the starter relay. (Motorcraft)

installed in the primary circuit of all domestic automobile breaker-point ignitions. This is called the **ballast (primary) resistor**, figure 8-7.

The ballast resistor compensates for changes in voltage and current caused by engine speed and temperature changes. The resistor provides about half of the total primary circuit resistance (the coil is the other half), and is the only part of the primary circuit that is temperature-compensated.

At low speeds, current flows through the circuit for relatively long periods of time. As the current heats the resistor, its resistance increases, dropping the applied voltage at the coil. At higher speeds, the breaker points open more often and current flows for shorter periods of time. As the ballast resistor cools, its resistance drops. Higher voltage is applied to the coil, but the shorter current flow duration results in about the same magnetic saturation of the coil.

The ballast resistor simply evens out the voltage and current of the primary circuit. In doing so, it reduces peak voltage at the coil and thus reduces current that would burn the breaker points faster. This is its most noticeable effect.

During cranking, the ballast resistor is bypassed to provide full available battery volt-

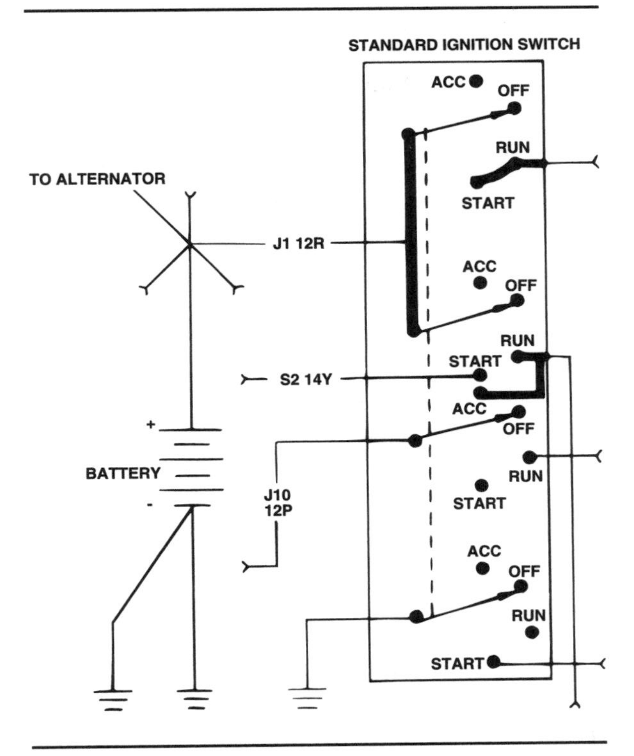

Figure 8-6. Chrysler products draw ignition current from a wiring splice between the battery and the alternator.

age to the primary circuit. This is done with a low-resistance starting bypass circuit in parallel with the ballast resistor, figure 8-7. When the engine starts, the bypass circuit opens and primary current flows through the resistor. Ignition primary voltage is reduced to the desired level. The ignition, breaker points, and coil are not damaged by this because:

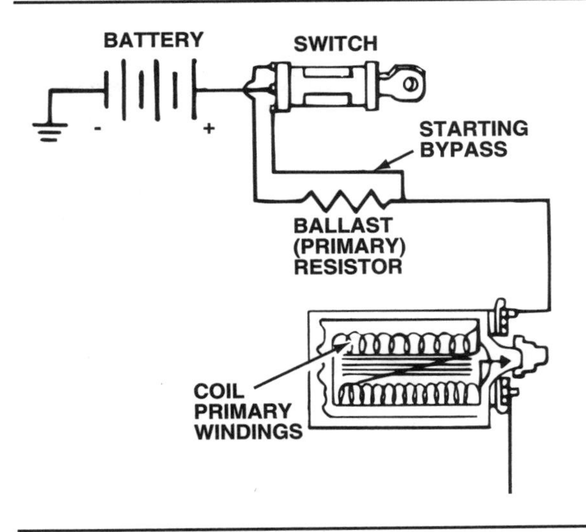

Figure 8-7. A ballast (primary) resistor can protect the primary circuit from excessive voltage.

- The ballast resistor is bypassed for a very short time.
- The battery voltage available during cranking is already reduced to a safe level.

Bypassing the ballast resistor at times other than cranking would cause rapid burning of the breaker points and could damage the coil primary winding.

Many electronic ignitions with fixed dwell use a ballast resistor in the primary circuit to limit current and voltage. Variable-dwell electronic ignitions do not require one, because the ignition module or computer regulates primary current and voltage to the coil. The following are descriptions of the installations in later-model domestic cars.

■ Charles Franklin Kettering (1876 –1958)

Charles F. Kettering was a leading inventor and automotive engineer. After graduating from Ohio State University, he became chief of the inventions department at the National Cash Register Company. While there, he designed a motor used in the first electrically operated cash register.

In 1909, he helped form Dayton Engineering Laboratories Company, later to be known as Delco. In 1917, he became president and general manager of General Motors Research Corporation.

Kettering invented both the automobile self-starter and the inductive-discharge battery ignition system. The accompanying illustration is an early sketch by Kettering of his design for an ignition system. He was involved in the invention and perfection of high-octane gasoline; improvements for engines, espe-

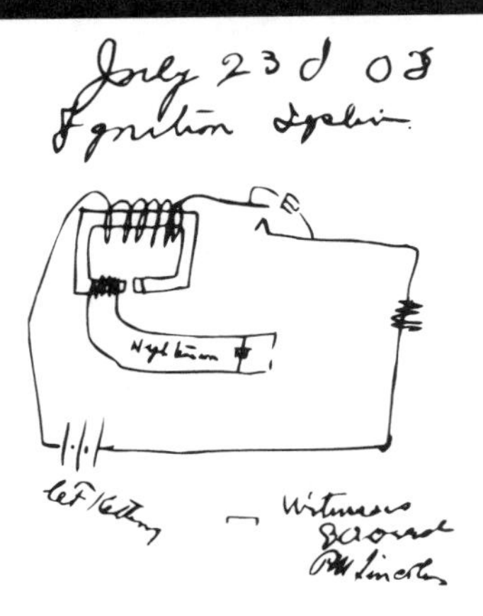

cially diesel engines; electric refrigeration; and much more. He also helped establish the Sloan-Kettering Institute for Cancer Research in New York City.

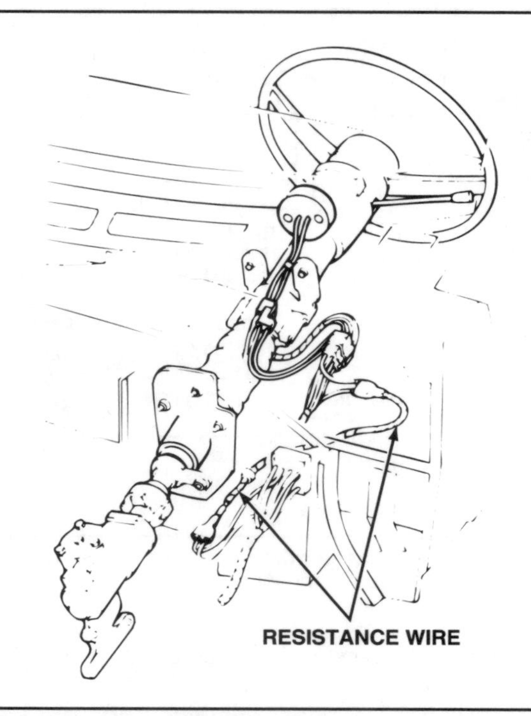

Figure 8-8. Ford products have a length of resistance wire installed near the ignition switch.

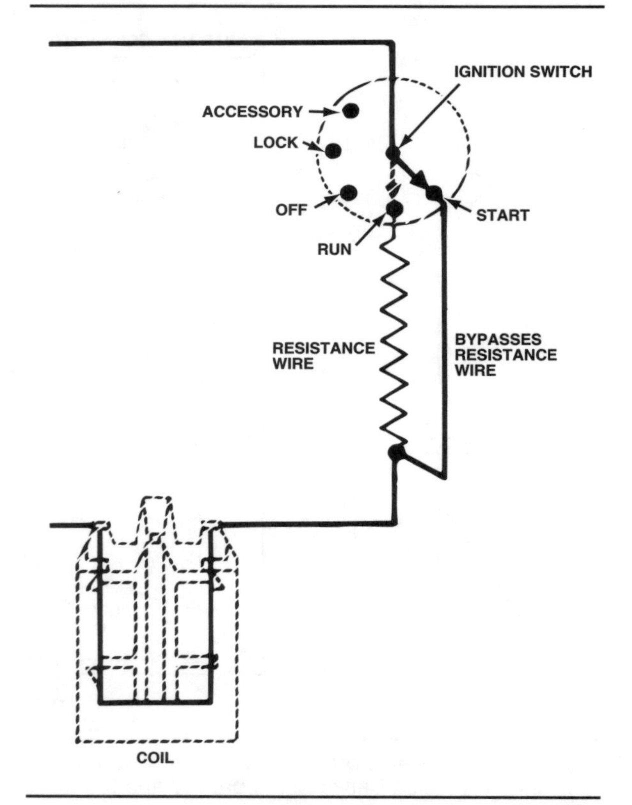

Figure 8-9. The ignition switch can control the starting bypass circuit.

- Ford products (except those with Dura Spark I or TFI electronic ignitions) using Ford ignitions have a length of resistance wire installed near the ignition switch, figure 8-8.
- Chrysler's 1977 318 CID V-8 and all 1977 and later Chrysler electronic lean burn (ELB) and electronic spark control (ESC) systems have only a 0.5-ohm ballast resistor. With the change from an analog to a digital computer in 1980, Chrysler eliminated a ballast resistor from its V6 and V8 ignitions.
- Chrysler front-wheel-drive (FWD) cars from 1978–79 (California) and 1980 (federal) used an analog computer and a 0.5-ohm ballast resistor. The resistor was eliminated on 1980 California and 1981 federal FWD cars with the introduction of a digital spark control computer.
- GM vehicles using the Prestolite BID or Delco-Remy HEI breakerless ignitions have no ballast resistor.

Starting Bypass

When the ballast resistor is bypassed during cranking, battery current flows to the primary circuit through a parallel circuit branch called the **starting bypass**. Current through this parallel branch can be controlled either by the ignition switch or by the starter relay or solenoid.

When the starting bypass is controlled directly by the ignition switch, figure 8-9, the ballast resistor is connected between the RUN position contacts and the coil primary winding. When the ignition switch is in the START position, full battery voltage is applied to the coil primary winding. Older larger Ford and Chrysler vehicles use this starting bypass method.

When the starting bypass is controlled by a starter relay or solenoid, figure 8-10, the ballast resistor is again connected between the RUN ignition-switch contacts and the coil. When the ignition switch is turned to START, current through the starter relay or solenoid closes a set of contact points. Battery current will flow through the relay or solenoid to the coil primary winding through a parallel circuit.

Small Ford automobiles and 1977 and later Chrysler products control the starting bypass through the starter relay. The starting bypass on GM cars is controlled through the starter solenoid.

Switching Devices

The magnetic field of the coil primary winding must collapse totally in order to induce a high voltage in the secondary winding. For the field to collapse, current through the primary wind-

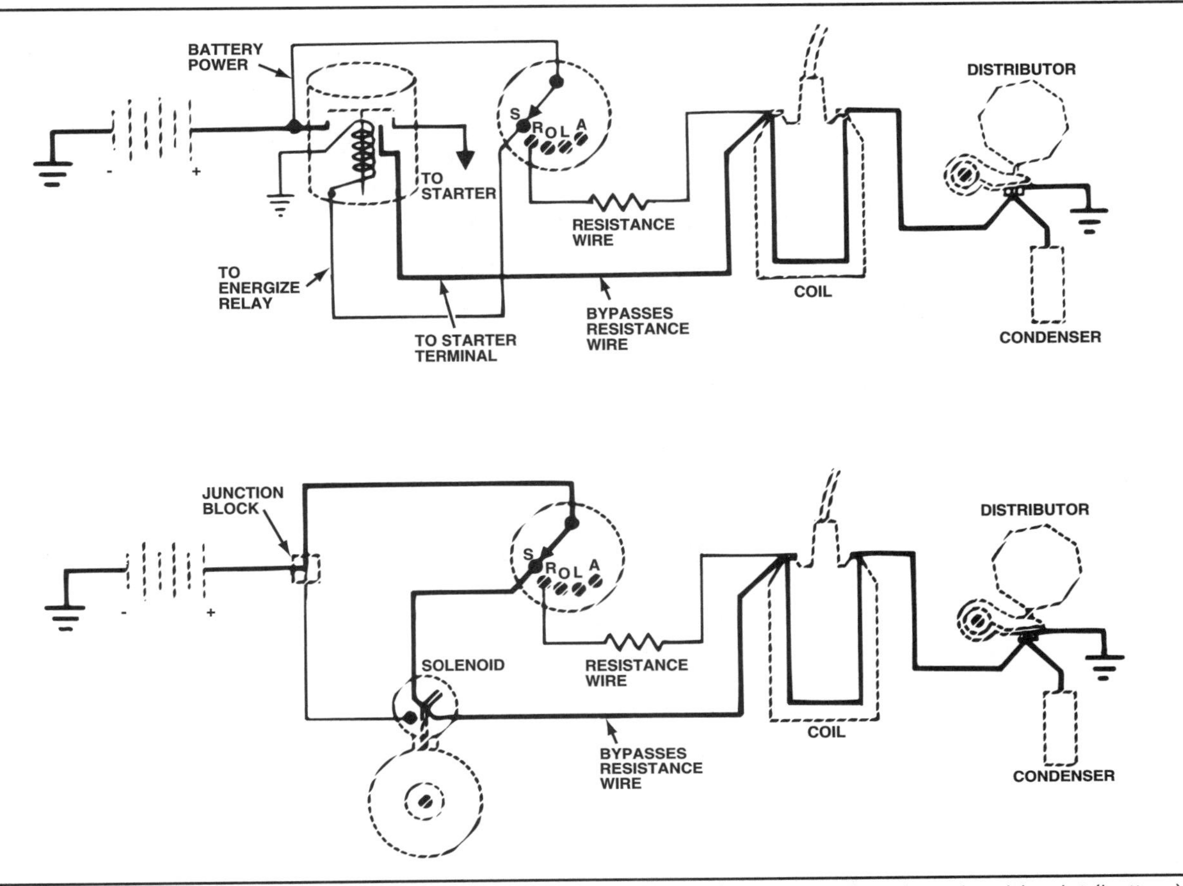

Figure 8-10. The starting bypass circuit can be controlled by the starter relay (top) or the solenoid point (bottom).

ing must stop very *rapidly*. Current must then start and stop again, to induce the next high-voltage discharge. The primary circuit needs a switching device that rapidly breaks and completes the circuit to start and stop the current. For more than 60 years, ignition systems used breaker points as a mechanical switch. Since the mid-1970s, solid-state electronic devices have replaced breaker points.

Breaker points

The ignition breaker-point assembly, figure 8-11, includes the:

- Fixed contact
- Movable contact
- Movable arm
- Rubbing block
- Pivot
- Spring
- Breaker plate.

Both breaker points are made from tungsten, an extremely hard metal with a high melting point. The fixed contact is grounded through the distributor housing. The movable contact is insulated from the distributor housing and is connected to the negative terminal of the coil primary winding. Because current flows from the movable to the fixed contact, the movable contact can be labeled (+) and the fixed contact labeled (-) in a negative-ground system.

The movable contact is mounted on a movable arm. The arm also holds a rubbing block, a small piece of plastic, or other synthetic nonconductive material that rides on the surface of the distributor cam. As the cam rotates, the lobes push the arm to open the breaker points. The spring closes the points when the cam lobes move away from the rubbing block.

The pivot and spring control the movement of the arm. The entire assembly is mounted on

Ballast (Primary) Resistor: A resistor in the primary circuit that stabilizes ignition system voltage and current flow.

Starting Bypass: A parallel circuit branch that bypasses the ballast resistor during engine cranking.

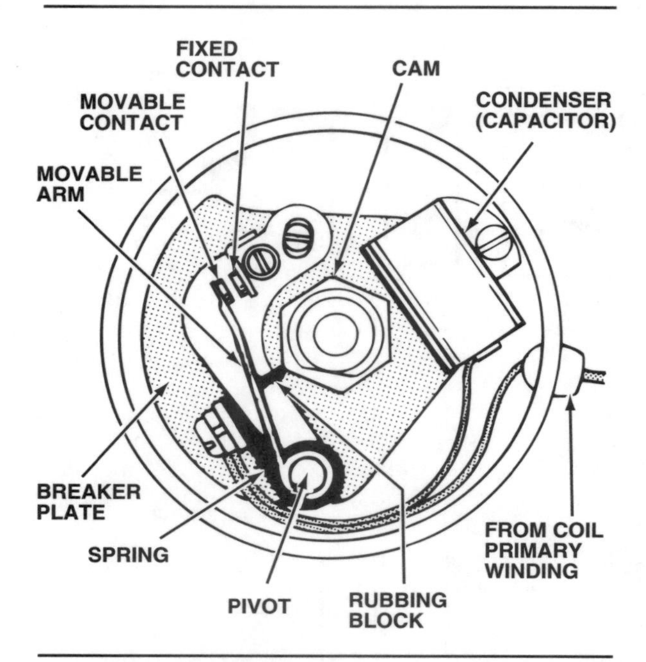

Figure 8-11. Ignition breaker-point (contact) assembly.

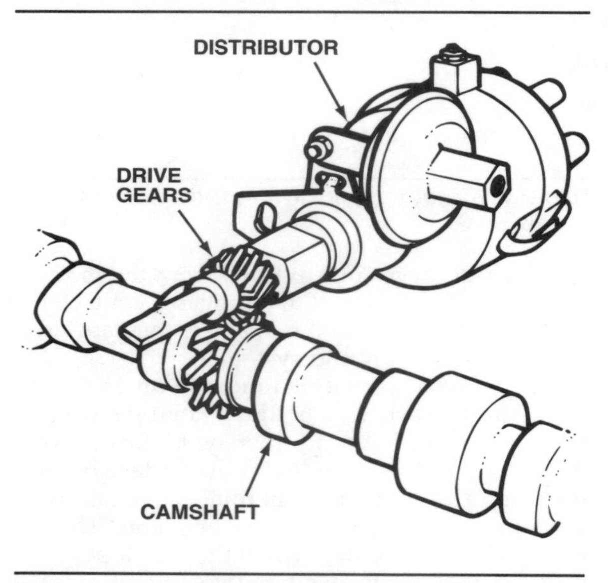

Figure 8-12. The gear on the distributor shaft is driven by another gear on the engine camshaft.

the breaker plate, which is attached to the distributor housing with screws.

The distributor cam is mounted on the centrifugal advance shaft, an assembly driven through the centrifugal advance weights by the distributor shaft. The distributor shaft is driven by the engine camshaft through gears, figure 8-12. Every opening of the breaker points induces a pulse of high-voltage current in the secondary circuit, and a spark plug fires. The breaker points must open as many times during one rotation of the cam as the engine has spark plugs. The cam, therefore, has as many lobes as the engine has spark plugs. The cam and points shown in figure 8-11 would be installed on a 6-cylinder engine.

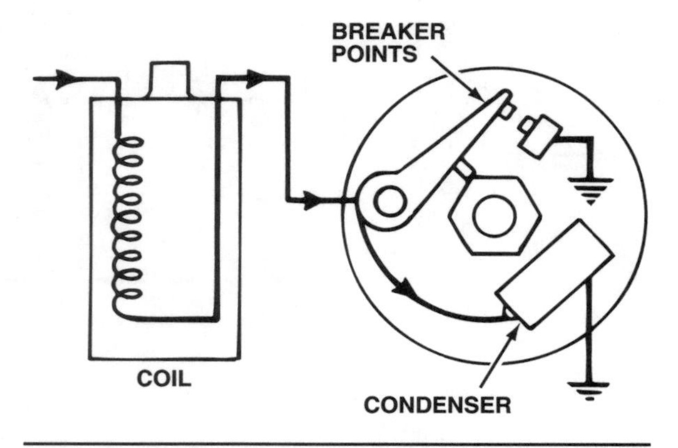

Figure 8-13. The condenser or capacitor prevents arcing as the breaker points open.

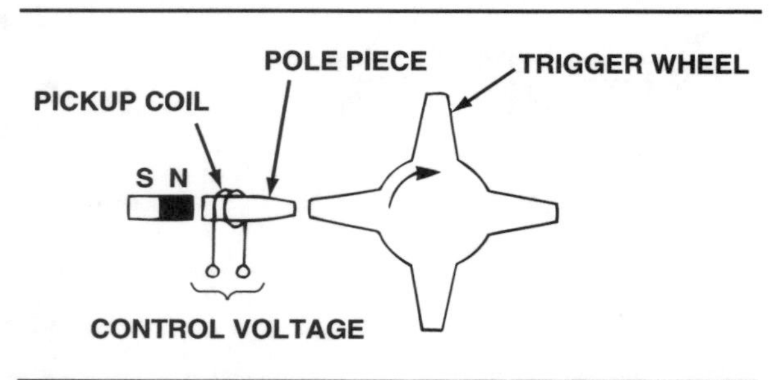

Figure 8-14. A simple magnetic pulse generator. (Bosch)

When the breaker points open, a voltage is self-induced in the coil primary winding, which would cause a damaging arc across the breaker points. To avoid this, a capacitor, or condenser, is wired in parallel with the breaker points, figure 8-13. The condenser prevents most arcing across the points, as we will see later.

Solid-state switching devices

Breaker-point ignitions make it difficult for an engine to meet today's exhaust emission control standards. Such standards not only require maximum system performance, they also require *consistent* performance. Because breaker points wear during normal operation, ignition system settings, and performance change.

Since solid-state switching devices do not wear, ignition system performance remains consistent and emission control can be effectively maintained. Virtually all automakers now use solid-state ignition systems. Although we will study the more common ignitions in detail in a later chapter, we will look now at the most widely used solid-state switching devices.

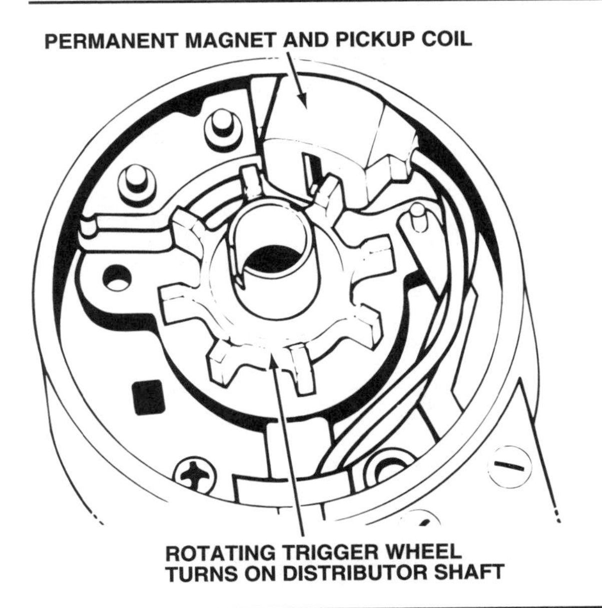

Figure 8-15. A magnetic pulse generator installed in the distributor housing. (Ford)

MANUFACTURER	STATIONARY PICKUP COIL	ROTATING TRIGGER WHEEL
BOSCH	PICKUP COIL & POLE PIECE	TRIGGER WHEEL
CHRYSLER	PICKUP COIL	RELUCTOR
FORD	STATOR	ARMATURE
GM	MAGNETIC PICKUP & POLE PIECE	TIMER CORE
NISSAN	STATOR	RELUCTOR
TOYOTA	PICKUP COIL	SIGNAL ROTOR

Figure 8-16. Manufacturers have different names for the trigger wheels and pickup coils or sensors in breakerless distributors, but all devices serve the same purpose.

A solid-state ignition control module (ICM) is responsible for switching the primary current on and off. The ICM must be signalled *when* to turn off the current. Of the four devices commonly used to do this, magneticpulse generators and Hall-effect switches are the most common. Chrysler and some Japanese automakers, such as Isuzu, use opticalsignal generators. A metal-detection switching device was used by Prestolite from 1975–77.

The **magnetic pulse generator** is installed in the distributor housing where the breaker points used to be. The pulse generator, figure 8-14, consists of a trigger wheel, a permanent magnet, a pole piece affected by the permanent magnet, and a pickup coil wound around the pole piece. The only moving part is the trigger wheel, which rotates as would a distributor cam.

The trigger wheel is made of steel with a low reluctance that cannot be permanently magnetized. Therefore, it provides a low-resistance path for magnetic flux lines.

As the trigger wheel rotates, its teeth come near the pole piece. Flux lines from the pole piece concentrate in the low-reluctance trigger wheel, increasing the magnetic field strength and inducing a voltage in the pickup coil. The pickup coil is connected to the ICM, or the powertrain control module (PCM), which senses this voltage and switches off the primary current. Each time a trigger wheel tooth comes near the pole piece, the ICM is signalled to switch off the primary current. Solid-state circuitry in the module determines when the primary current will be turned on again.

The simple pulse generator shown in figure 8-14 would be installed in a 4-cylinder engine. Figure 8-15 shows the typical construction of a pulse generator for an 8-cylinder engine.

Magnetic Pulse Generator: A signal-generating switch that creates a voltage pulse as magnetic flux changes around a pickup coil.

■ Dual-Point, Dual-Coil, Dual-Plug Ignition

Some early Nash automobiles used a "Twin-Ignition" system. This system had two ignition coils, two sets of spark plugs and cables, a distributor with 16-plug terminals and two coil-wire terminals, a rotor with offset tips, two sets of breaker points, and two condensers.

The in-line engines were designed with the spark plugs on both sides of the cylinders. The overhead valves were located directly over each cylinder in a vertical position.

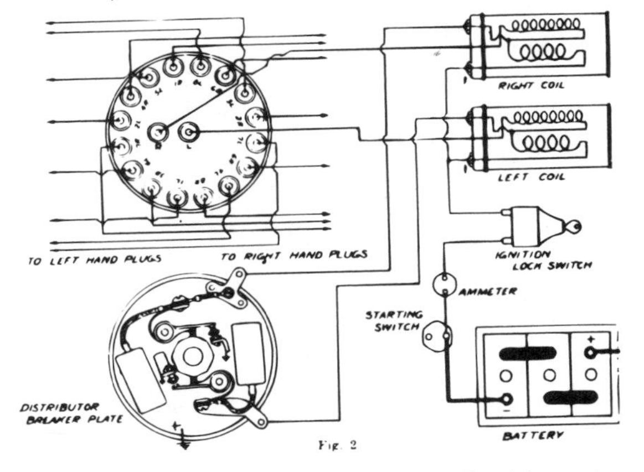

The breaker points were synchronized by using a dual-bulb test lamp. With the ignition switch on, the distributor cam was turned to just break a stationary set of points, thereby lighting one bulb. A movable set of points was then adjusted to break contact, and light the second bulb at the same instant. This adjustment assured that both sparks would occur in a cylinder at the same time.

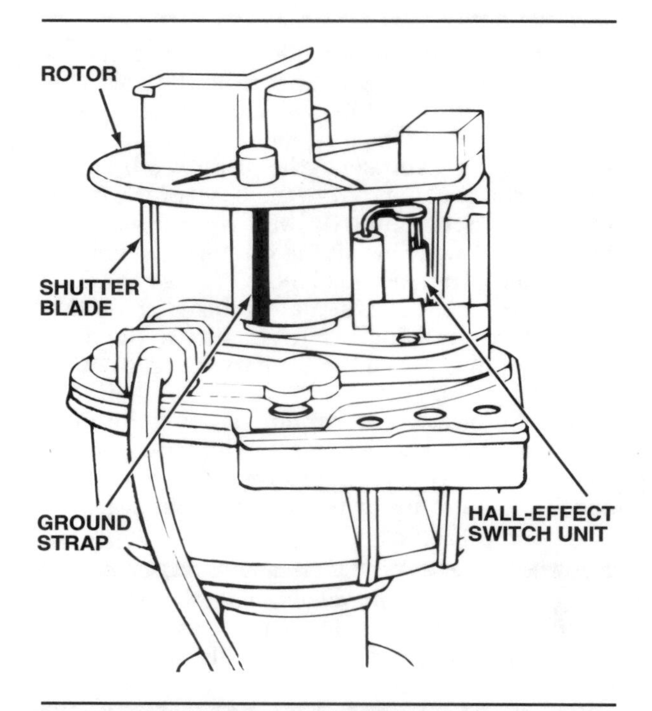

Figure 8-17. Shutter blades rotating through the Hall-effect switch air gap bypass the magnetic field around the pickup and drops the voltage output to zero. (Chrysler)

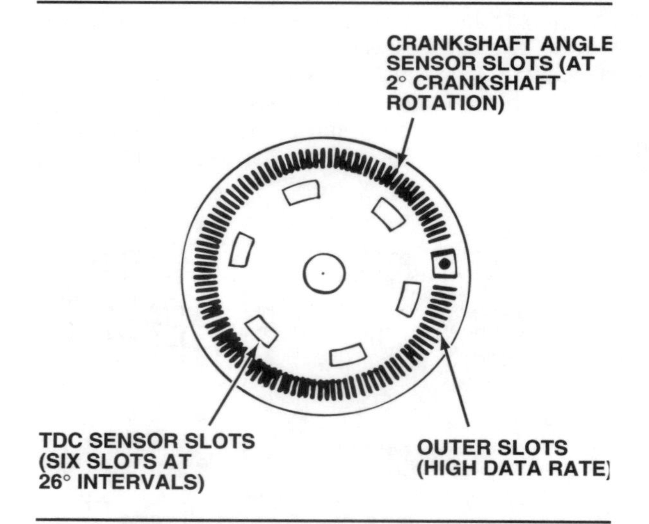

Figure 8-18. The optical signal generator works by interrupting a beam of light passing from the LED's to photodiodes. (Chrysler)

We have used the terms "trigger wheel" and "pickup coil" in describing the magnetic pulse generator. Various manufacturers have different names for these components, figure 8-16, but all serve the same purpose.

A **Hall-effect switch** also uses a stationary sensor and rotating trigger wheel (shutter), figure 8-17. Unlike the magnetic pulse generator, it requires a small input voltage to generate an output or signal voltage. As used in some Chrysler distributors, the rotor has a shutter blade for each cylinder (Ford and GM use a separate ring of metal blades). The pickup plate in the distributor housing contains a gate that the shutter blades pass as the distributor shaft rotates. An integrated circuit (IC) mounted on the plate faces the switch. As a shutter blade enters the air gap between the IC and the Hall-effect switch, it bypasses the magnetic field around the pickup, causing the Hall-effect output voltage to change. This changes the bias to the ICM, just as a magnetic pulse generator signal does.

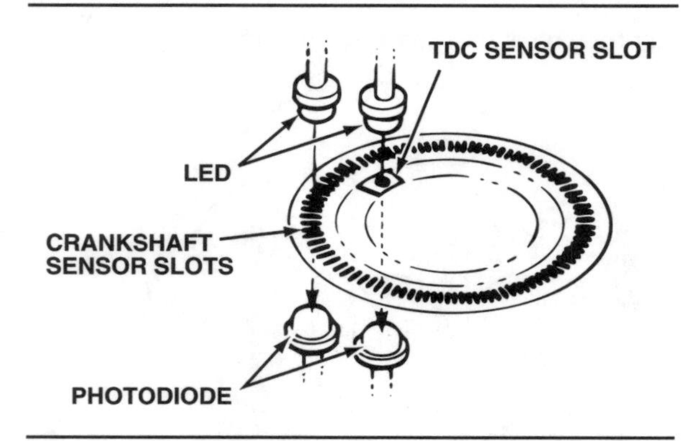

Figure 8-19. Each row of slots in the optical distributor disc acts as a separate sensor, creating signals used to control fuel injection, ignition timing, and idle speed. (Chrysler)

The optical signal generator uses the principle of light-beam interruption to generate voltage signals. As used by Chrysler with its dual-cam 3.0-liter engine, the optical signal distributor contains a pair of light-emitting diodes (LED's) and photodiodes installed opposite each other, figure 8-18. A disc containing two sets of chemically etched slots is installed between the LED's and photodiodes. Driven by the forward bank camshaft, the disc acts as a timing member and revolves at half engine speed. As each slot interrupts the light beam, an alternating voltage is created in each photodiode. A hybrid IC converts the alternating voltage into on/off pulses sent to the PCM.

The high-data-rate slots, or outer set, are spaced at intervals of 2 degrees of crankshaft rotation, figure 8-19. This row of slots is used for timing engine speeds up to 1,200 rpm. Certain slots in this set are missing, indicating the crankshaft position of the number 1 cylinder to the PCM. The low-data-rate slots, or inner set, consists of six slots correlated to the crankshaft top-dead-center (TDC) angle of each cylinder. The engine computer uses this signal for triggering

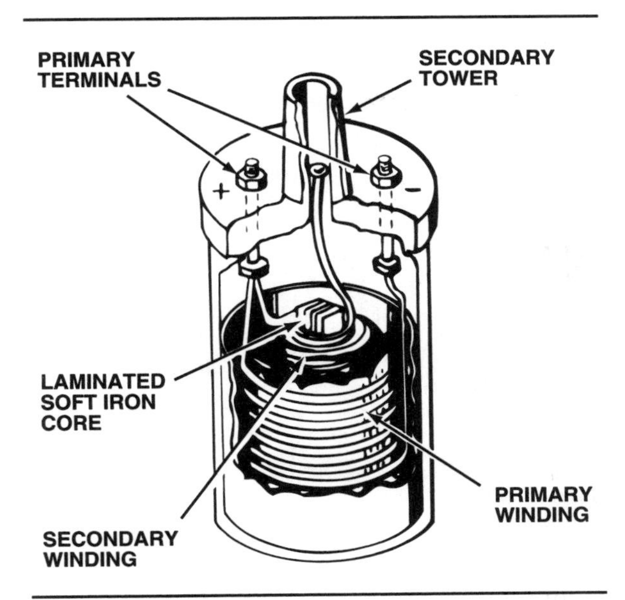

Figure 8-20. An ignition coil cutaway showing the primary and secondary windings.

the fuel injection system and for ignition timing at speeds above 1,200 rpm. In this way, the optical signal generator acts both as the crankshaft position sensor and TDC sensor, as well as a rpm switching device.

Coil Primary Winding

The coil primary winding, figure 8-20, is made of about 100 to 150 turns of a relatively heavy copper wire. The coil turns are insulated from each other by a thin coat of enamel. The two ends of the winding are connected to two terminals on the top of the coil. With a negative-ground electrical system, the coil's positive terminal is connected to the battery's positive terminal; the coil's negative terminal is connected to the ignition breaker points, and through the points, to ground. With a positive-ground system, the coil primary connections are reversed. We will look at the coil in greater detail in the next chapter.

Condensers

The condenser acts as an "electric shock absorber" to dampen any excessive voltage levels in a circuit.

Condenser construction and installation

Most automotive condensers are formed from two thin foil strips and separated by several layers of insulating paper. These layers, each more than eight feet long, are tightly rolled into a cylinder. The foil strips are offset, so that the top edge of one strip protrudes past the paper

Figure 8-21. A condenser installed inside the distributor housing.

on one end of the cylinder and the bottom edge of the other strip protrudes from the other end of the cylinder. These edges provide an electrical contact with foil strips. The edges are flattened, and the cylinder is installed in a metal canister. The bottom of the canister contacts one foil edge and grounds it. The other foil edge is connected to an insulated lead at the top of the canister.

To ensure ample insulation between the foil strips, the canister is placed in a vacuum. Wax or oil is drawn into the canister, and the entire unit sealed. Condensers cannot be adjusted or repaired, but must be replaced if defective.

An assembled condenser is usually installed inside the distributor housing on the breaker plate, figure 8-21. The bracket is held to the breaker plate by a screw. This is the ground connection. The insulated lead from the top of the condenser is attached to the spring of the movable point arm, giving it an electrical connection to the movable breaker point.

The condenser can also be attached to the outside of the distributor housing. The wiring connections look different, but provide the same electrical paths.

Some Delco-Remy distributors use a combined points and condenser unit called a Uni-Set, figure 8-22. This is a condenser, as described above, already attached to the point assembly. The assembly is installed and replaced as one unit. A similar product, the Prestolite Capaci-Point, attaches a ceramic capacitor to the point assembly. Both of these units eliminate the condenser wire lead, a principal cause of radio interference.

Hall-Effect Switch: A semiconductor that produces a voltage in the presence of a magnetic field. This voltage can be used to control a transistor for use as a switch.

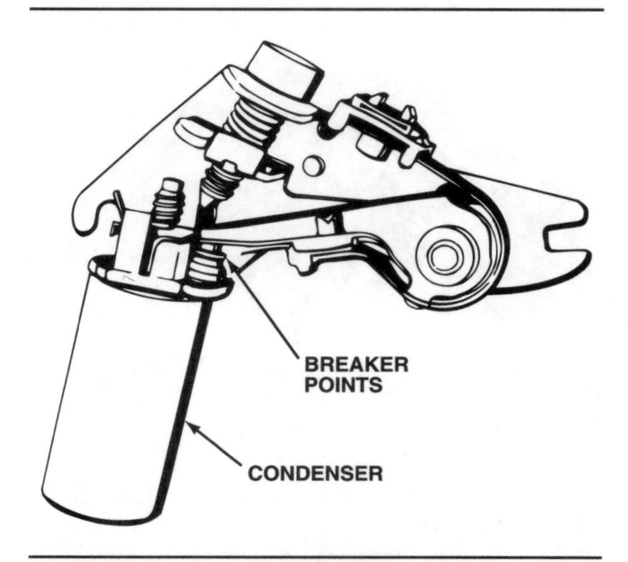

Figure 8-22. Delco-Remy's Uni-Set point and condenser assembly.

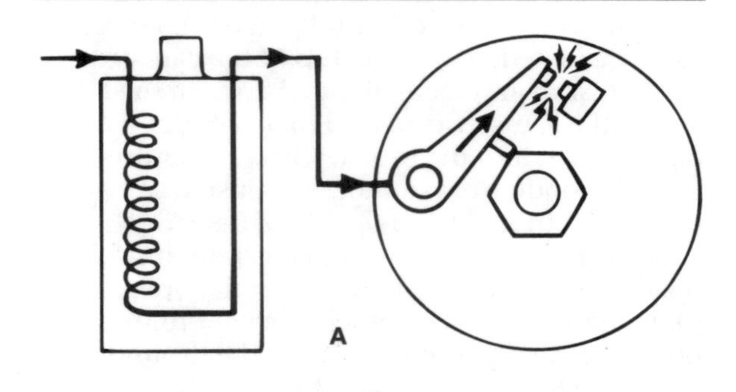

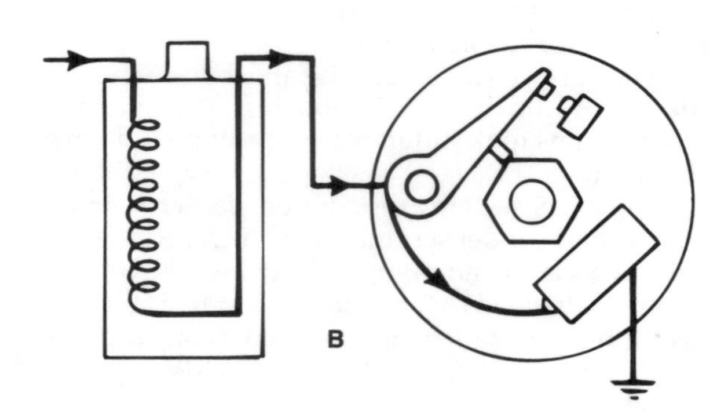

Figure 8-23. The primary current arcs at the breaker points as they open (A). The primary current flows to the condenser when the breaker points open, eliminating the arc (B).

Condenser purpose

Self-induction in the coil primary winding can increase the primary voltage to as much as 400 volts. When the breaker points open, figure 8-23 (top position), the high voltage would cause

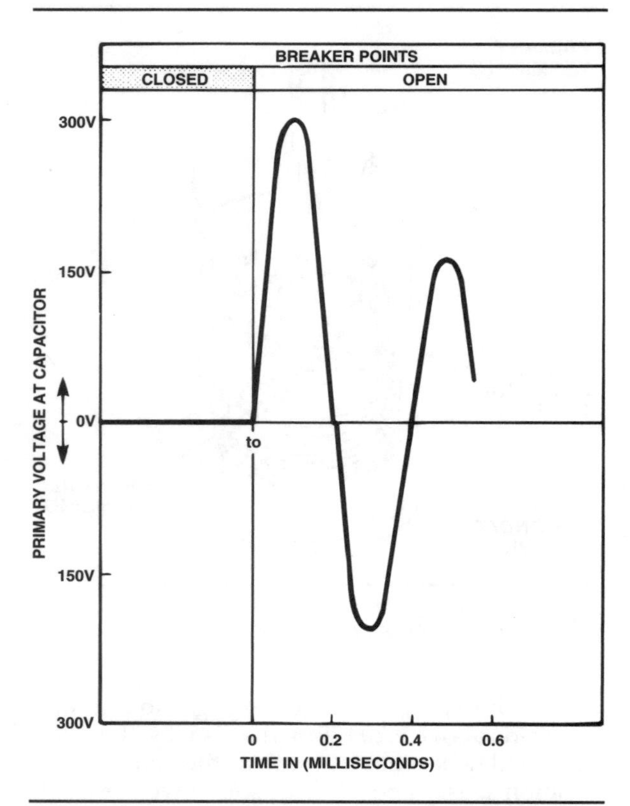

Figure 8-24. The energy stored in the condenser dissipates by oscillating through the primary circuit. (Bosch)

current to arc across the air gap. The unwanted spark would:

- Consume energy at the expense of the secondary circuit energy
- Burn and pit the contacts, causing rapid point failure
- Leave an oxidized coating on the points, increasing primary circuit resistance.

To prevent these problems, the ignition condenser is installed in parallel with the breaker points, figure 8-23 (bottom position). When the points open, the condenser is charged by the inductive current from the primary winding. It requires a small amount of time, about 0.1 ms, to charge the condenser to the peak voltage of the primary winding. By the time the condenser is fully charged, the contact points have opened far enough so that the current cannot arc across the air gap.

The energy in the condenser is then discharged, oscillating between the condenser and the coil primary winding, and dissipated as heat. These voltage oscillations in the primary circuit are shown in figure 8-24.

Since the condenser allows the primary circuit to be broken quickly and completely, the

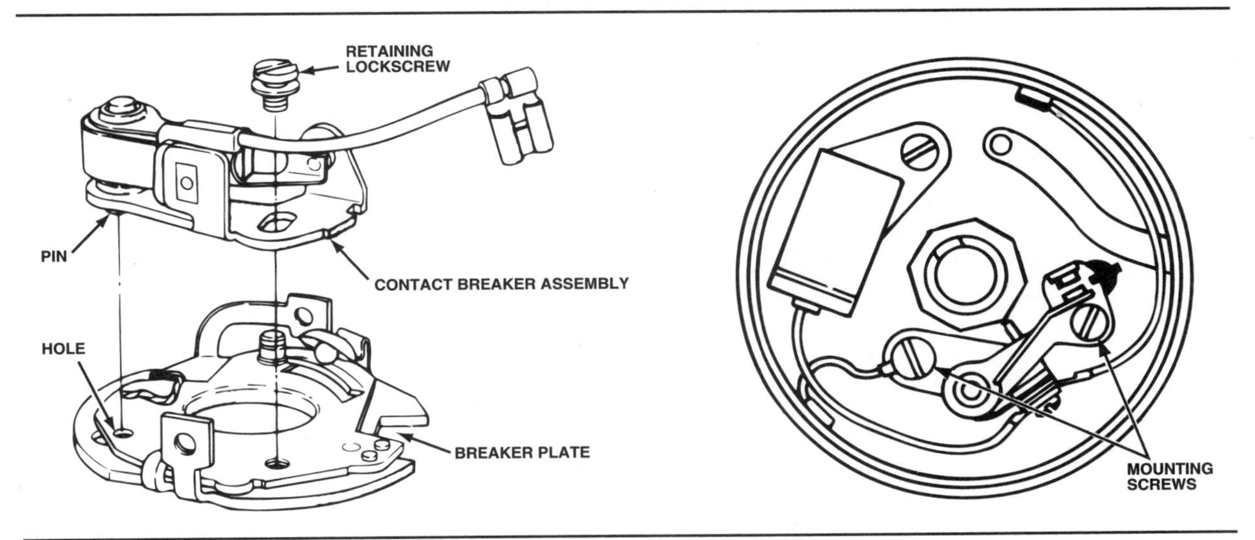

Figure 8-25. Point assemblies are usually mounted with either one screw and a locating pin (left) or with two screws (right). (Ford)

coil's magnetic field collapses rapidly. The field collapses about 20 times faster than if there was no condenser in the circuit. The faster the collapse, the greater the induced voltage in the secondary winding. The field has fully collapsed by the time the condenser begins its first discharge into the primary circuit.

At low engine speeds, the breaker points open slowly. At engine speeds requiring fewer than 3,000 sparks per minute, the primary winding's induced voltage is great enough to cause a slight arc across the slowly opening point air gap, despite the condenser. Automobiles driven mostly at low engine speeds will show more rapid point failure than those used for high-speed travel.

Condenser ratings

Condensers are rated in microfarads (μF). Typical automotive condensers have capacities from 0.18 to 0.32 μF.

It is important to follow manufacturers' recommendations when installing ignition condensers. If a condenser with too little capacity is used, primary current will charge the condenser and still be able to arc across the point gap. This causes pitting at the points, with metal transfer from the grounded (-) point to the movable (+) point. If the condenser has too great a capacity, pitting and metal transfer can occur in the opposite direction.

BREAKER-POINT DISTRIBUTORS

Great demands are placed on the ignition breaker points by the rest of the ignition system. Points must be correctly installed and accurately adjusted.

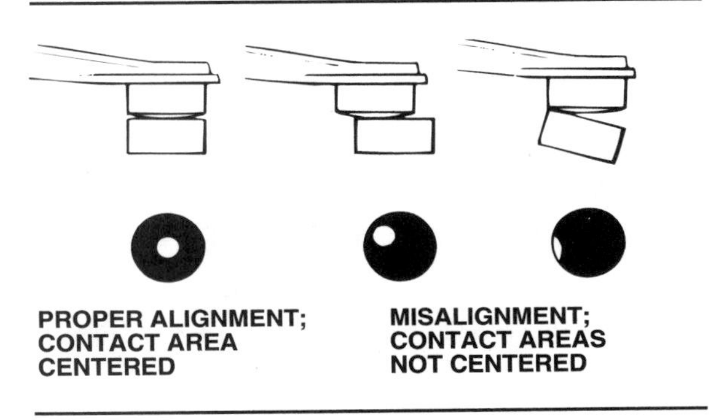

Figure 8-26. Breaker points must be properly aligned for proper ignition system performance.

Breaker-Point Installation

Ignition breaker points are normally supplied as a complete unit. This assembly is fastened to the distributor breaker plate, figure 8-25. The fasteners are either two screws or one screw and a small pin, as illustrated.

One of the mounting holes of the point assembly is elongated, so that the position of the points on the breaker plate can be adjusted. This adjustment has a great effect on breaker-point operation, as we will soon see.

The wiring connections between the points, the condenser, and the primary circuit can be made at one of two places. When the condenser is mounted inside the distributor, the connection is usually made at the bracket where the movable arm's spring is braced. In some distributors, a nut and bolt hold the two wiring terminals to the bracket. Other distributors have push-on terminals.

When the condenser is mounted outside the housing, the primary and condenser leads may

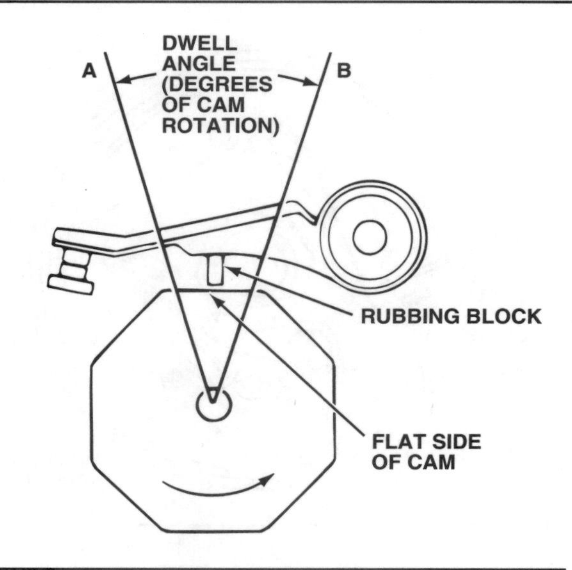

Figure 8-27. Dwell angle is the period during which the breaker points are closed. (Ford)

be attached to a single slide terminal. A lead from the movable breaker point attaches to this terminal.

The stationary breaker point and the condenser canister are grounded through the distributor housing and engine.

Correct ignition point alignment and spring tension are essential for proper ignition operation and long service life. Points are correctly aligned, figure 8-26, when the mating surfaces are in the center of both contacts, the faces are parallel, and the diameters are concentric. This ensures maximum contact area and precise switching action by the points.

Correct spring tension also ensures precise point action. Too much spring tension causes rapid cam and rubbing block wear. In some cases, it can even cause distributor shaft bushing wear or broken points. If spring tension is too light, the points will bounce as they open and close at high speed. This generally results in a loss of engine power. Spring tension is normally between 15 and 25 ounces (425 and 710 grams).

Point Dwell Angle

Point **dwell angle**, or cam angle, is a measurement of how far the distributor cam rotates while the points are closed. In figure 8-27, the points closed when line A was at the rubbing block. The points will open when line B reaches the rubbing block. Between lines A and B, the points stay closed. The number of degrees that the cam rotates between the point closing (line A) and the point opening (line B) is called the dwell angle.

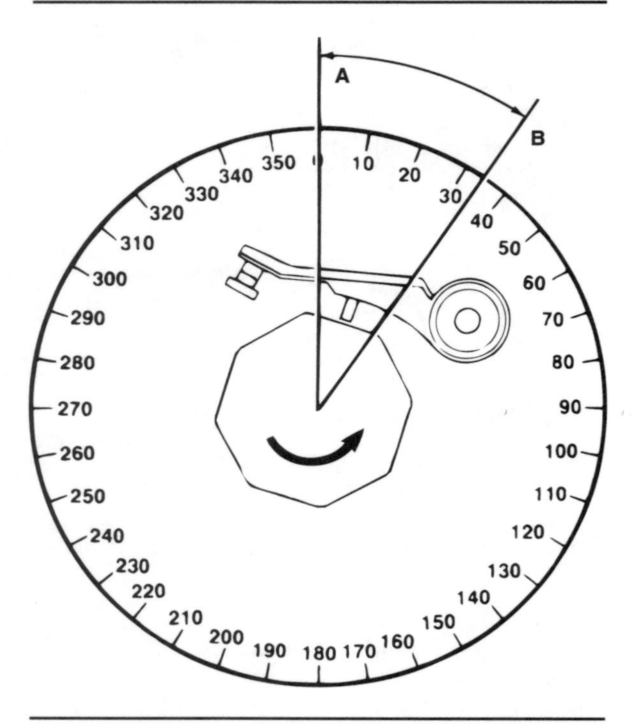

Figure 8-28. Dwell angle is measured in degrees of cam rotation. In this case, dwell is about 33 degrees.

Distributor rotation, like crankshaft rotation, is measured in degrees. If we superimpose a degree scale onto the point assembly, figure 8-28, we see that this particular dwell measures about 33 degrees.

The cam illustrated has eight lobes, so it is used with a V8 engine. Four-cylinder or six-cylinder engines require four-lobe or six-lobe cams respectively. This means that the points open four or six times, respectively, during each cam revolution. The more times the points must open, the less time they can remain closed.

The ignition dwell angle is directly related to an engine's firing intervals. There are 90 degrees of crankshaft rotation between the firing intervals in a V8 engine. Because the distributor rotates at half the crankshaft speed, the distributor cam rotates half as far as the crankshaft between firing intervals, or 45 degrees. The theoretical dwell angle of a V8 engine would be 45 degrees, but the points must be open during part of that time; and dwell equals the number of degrees during which the points are *fully* closed. About 12 to 17 degrees are required for the points to open and close on an engine with eight cylinders. Therefore, a typical dwell angle for a V8 engine is about 28 to 33 degrees.

An even-firing 6-cylinder engine fires every 120 degrees of crankshaft rotation (60 degrees of distributor rotation). A 4-cylinder engine fires every

180 degrees of crankshaft rotation (90 degrees of distributor rotation). Dwell on a 6-cylinder engine could be around 45 degrees; on a 4-cylinder engine, about 75 degrees. However, large dwell angles mean that the primary current flows for a long time at low engine speeds. This is not necessary for full coil saturation, and could lead to coil overheating. For this reason, 4- and 6-cylinder distributors are designed for less dwell than the maximum amount they could have. Dwell on a typical 4-cylinder engine is usually about 50 degrees; on a 6-cylinder engine, about 38 degrees.

Dwell and point gap

Point gap is the maximum distance between the breaker points when they are open. Figure 8-29 shows the relationship between point gap and dwell. A small point gap means a large dwell angle. A large point gap means a small dwell angle; large point gap is usually measured in thousandths of an inch or hundredths of a millimeter. Typical point gaps range from 0.015 to 0.025 in. (0.40 to 0.60 mm).

Point gap is adjusted by shifting the position of the breaker point assembly. This changes the position of the points relative to the cam, altering the point gap and the dwell. Point assemblies have elongated mounting holes so that the point gap can be adjusted.

Effect of dwell on the coil

As long as the points are closed, current flows through the coil primary winding. This creates a magnetic field within the coil and makes induction possible.

However, the primary winding magnetic field does not appear instantly. When current first flows through the primary winding, self-induction causes a countervoltage within the winding. This countervoltage opposes primary current flow. The magnetic field of the primary winding does not immediately reach full strength, because the primary current does not immediately reach full strength, figure 8-30. It generally takes from 10 to 15 ms for the primary current to reach full strength.

The breaker points must remain closed long enough for the primary winding's magnetic field to reach nearly full strength. If the field is

Dwell Angle: The measurement in degrees of how far the distributor cam rotates while the breaker points are closed. Also called cam angle, or dwell.

■ Dual Breaker Points

In the early days of the automobile, short dwell time was an obstacle to increasing the rpm limit and power of high-performance street engines. Racing engines used magnetos, which, by design, increase spark strength with rpm. Street engines, because of cost and starting requirements, were stuck with point-coil ignitions. At high engine speeds, above 4,500 or 5,000 rpm, a typical 33-degree dwell angle on a V8 engine did not allow enough coil saturation time. The result was insufficient voltage, making high-speed misfires a familiar problem. Reducing the point gap to increase dwell time drastically cut point life because the narrow gap caused point arcing and burning at low speeds.

By the 1930s, manufacturers built the first production dual-point distributors that eliminated this problem. The two-point sets had staggered openings and closings. The dwell on each set of points might be only 27 to 31 degrees; however, the combined dwell could be 36 to 40 degrees, which provided adequate coil saturation time and voltage for strong ignition at high speeds.

The point sets were connected in parallel, so the primary circuit was complete, or closed, whenever *either* or *both* point sets were closed. However, *both* sets of points had to be open to break the primary circuit and fire the spark plug. The first set of points to open did not break the circuit because the other set kept voltage flowing through the primary circuit. The second set to open was the "opening" set, since

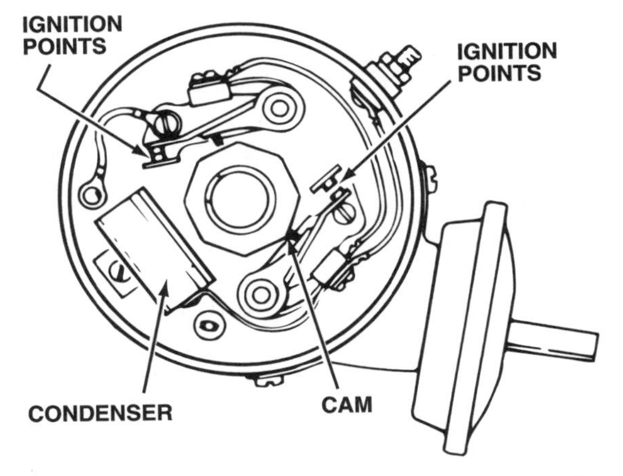

both sets were then open simultaneously for a short period. Because the first set to open was also the first to close, it was the "closing" set, and vital coil saturation time could begin. Dual-point distributors survived into the 1970s, but were not needed with the universal adoption solid-state switching devices and electronic ignition.

NORMAL DWELL

POINTS OPEN AND CLOSE AS SPECIFIED

SMALL DWELL

POINTS CLOSE LATE AND OPEN EARLY

LARGE DWELL

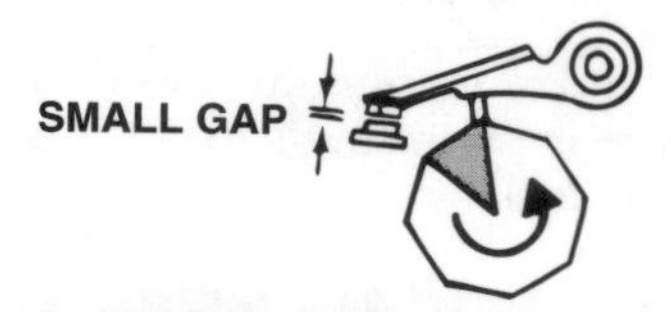

POINTS CLOSE EARLY AND OPEN LATE

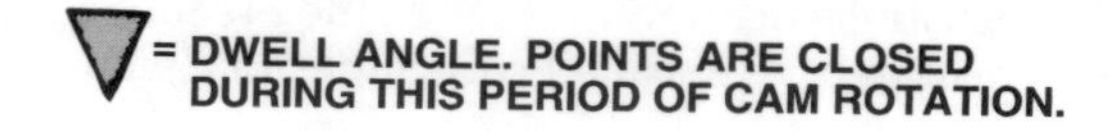

Figure 8-29. Point gap and dwell angle are inversely related; when one increases, the other decreases.

not at full strength, secondary voltage will be reduced. The point dwell must be great enough to allow the coil to reach full saturation.

However, if the point dwell is too great, the point gap will be very small. Primary current will be able to arc across the air gap, and the magnetic field will not collapse quickly and completely. Point dwell and gap must be adjusted exactly to the manufacturer's specifications if the ignition system is to perform most efficiently.

Dwell and engine speed

As engine speed increases, the distributor cam rotates faster. Dwell angle is unchanged, but the *time* it takes the cam to rotate through this angle is decreased. That is, the amount of time that the points are closed is reduced as engine speed increases.

We have seen that it takes a specific amount of time—about 10 ms—for primary current flow to reach full strength, figure 8-30. At a 625-rpm idle speed, the 33-degree dwell angle of a V8 engine lasts about 16.5 ms. Primary current flow can reach its full strength before the points open to interrupt it.

As engine speed increases to 1,000 rpm, the 33-degree dwell period decreases to 9.9 ms.

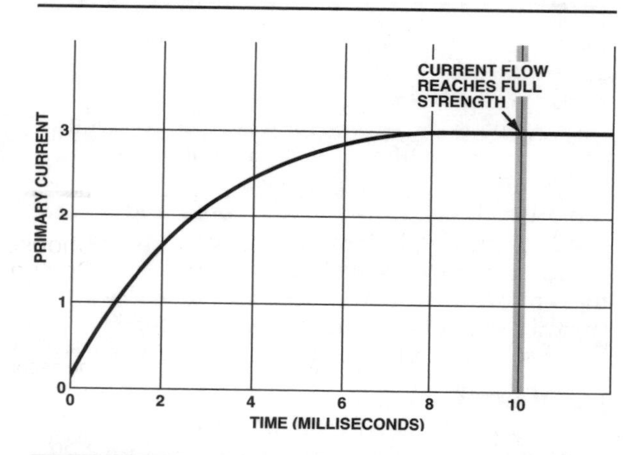

Figure 8-30. Primary winding current flow does not immediately reach full strength. (Bosch)

Primary current is interrupted just before it reaches its maximum strength. Available secondary voltage will be decreased slightly.

At 2,000 rpm, the dwell period is reduced to 4.5 ms. Primary current is interrupted well before it reaches its maximum strength. Available secondary voltage will be reduced considerably. If the engine and ignition system are not in excellent condition, the required voltage level may be greater than the available voltage level. If this occurs, the engine will misfire.

SUMMARY

Through electromagnetic induction, the ignition system transforms the low voltage of the battery into the high voltage required to fire the spark plugs. Induction occurs in the ignition coil where current flows through the primary winding to build up a magnetic field. When the field collapses rapidly, high voltage is induced in the coil secondary winding. All domestic original equipment ignitions are the battery-powered, inductive-discharge type.

The ignition system is divided into two circuits: primary and secondary. The primary circuit contains the battery, ignition switch, ballast (primary) resistor, starting bypass, coil primary winding, and a switching device in the distributor.

For over 60 years, mechanical breaker points were used as the primary circuit switching device. Solid-state electronic components replaced breaker points as the switching device in the mid-1970s. The two most common solid-state switching devices are the magnetic pulse generator and the Hall-effect switch.

The ignition condenser is a capacitor that absorbs primary voltage when the points open.

This prevents arcing across the points and premature burning. Typical ignition condensers are rated at 0.18 to 0.32 uF.

The breaker points are a mechanical switch that opens and closes the ignition primary circuit. The period during which the points are closed is called the dwell angle. The dwell angle varies inversely with the gap between the points when they are open. As the gap decreases, the dwell increases.

Review Questions

Choose the single most correct answer.
Compare your answers to the correct answers on page 507.

1. The voltage required to ignite the air-fuel mixture in a breaker-point ignition system can range from ___ volts.
   a. 5 to 25
   b. 50 to 250
   c. 500 to 2,500
   d. 5,000 to 25,000

2. Which of the following does *not* require higher voltage levels to cause an arc across the spark plug gap?
   a. Increased spark plug gap
   b. Increased engine operating temperature
   c. Increased fuel in the air-fuel mixture
   d. Increased pressure of the air-fuel mixture

3. The coil transforms low voltage from the primary circuit to high voltage for the secondary circuit through:
   a. Magnetic induction
   b. Capacitive discharge
   c. Series resistance
   d. Parallel capacitance

4. Voltage induced in the secondary winding of the ignition coil is how many times greater than the self-induced primary voltage?
   a. 1 to 2
   b. 10 to 20
   c. 100 to 200
   d. 1,000 to 2,000

5. The two circuits of the ignition system are the:
   a. Start and Run circuits
   b. Point and coil circuits
   c. Primary and secondary circuits
   d. Insulated and ground circuits

6. Which of the following components is part of both the primary and the secondary circuits?
   a. Ignition switch
   b. Distributor rotor
   c. Condenser
   d. Coil

7. Which of the following is not contained in the primary circuit of an ignition system?
   a. Battery
   b. Spark plugs
   c. Ignition switch
   d. Coil primary winding

8. When the cranking system is operating, the ballast resistor:
   a. Reduces coil primary voltage to about seven volts
   b. Heats up, increasing resistance and reducing voltage
   c. Is bypassed to provide full available voltage
   d. Cools, and increases primary current flow

9. Which of the following is true of the coil primary windings?
   a. They consist of 100 to 150 turns of very fine wire
   b. The turns are insulated by a coat of enamel
   c. The negative terminal is connected directly to the battery
   d. The positive terminal is connected to the breaker points and ground

10. In order to collapse the magnetic field of the coil, the primary circuit requires a:
    a. Ballast resistor
    b. Switching device
    c. Condenser
    d. Starting bypass circuit

11. Breaker points are usually made of:
    a. Silicon
    b. Tungsten
    c. Aluminum
    d. Copper

12. In the solid-state ignitions used as original equipment on later-model domestic vehicles, the breaker points and distributor cam have been replaced by:
    a. RFI filter capacitors
    b. Auxiliary ballast resistors
    c. Magnetic pickup triggering devices
    d. Integrated coil and distributor cap assemblies

13. Condensers are rated in:
    a. Ohms
    b. Milliohms
    c. Farads
    d. Microfarads

14. Dwell angle is the period during which the ignition points are:
    a. Fully open
    b. Fully closed
    c. Starting to open
    d. Starting to close

15. The point gap is ___ related to the dwell angle:
    a. Directly
    b. Inversely
    c. Proportionally
    d. Reciprocally

16. Almost all distributors rotate at ___ the speed of the crankshaft.
    a. one-half
    b. The same speed as
    c. Twice
    d. one-quarter

Chapter

Ignition Secondary Circuit and Components

We have explained the general operation of the ignition primary circuit and have studied some of the system components. In this chapter, we look at the secondary circuit and its components.

The secondary circuit must conduct surges of high voltage. To do this, it has large conductors and terminals and heavy-duty insulation. The secondary circuit, figure 9-1, consists of the:

- Coil secondary winding
- Distributor cap and rotor
- Ignition cables
- Spark plugs.

The secondary circuit has the same components and function whether the primary circuit uses breaker points or a solid-state switching device.

IGNITION COILS

As we have seen, the ignition coil steps up voltage in the same way as a transformer. When the magnetic field of the coil primary winding collapses, it induces a high voltage in the secondary winding.

Coil Secondary Winding and Primary-to-Secondary Connections

Two windings of copper wire compose the ignition coil. The primary winding of heavy wires consists of 100 to 150 turns; the secondary winding is 15,000 to 30,000 turns of a fine wire. The ratio of secondary turns to primary turns is usually between 100 and 200. To increase the strength of the coil's magnetic field, the windings are wrapped around a laminated core of soft iron, figure 9-2.

The coil must be protected from the underhood environment to maintain its efficiency. Three coil designs are used:

- Oil-filled coil
- Laminated E-core coil
- Distributorless ignition system (DIS) coil packs.

Oil-filled coil

In the oil-filled coil (used with both breaker-point and breakerless ignitions), the primary winding is wrapped around the secondary winding, which is wrapped around the iron core. The coil windings are insulated by layers of paper and the entire case is filled with oil for greater insulation. The top of the coil is molded from an insulating material such as Bakelite. Metal inserts for the winding terminals are installed in the cap. Primary terminals are generally marked with a + and -, figure 9-2. Leads are attached with nuts and washers on some coils; others use push-on lead connectors. The entire unit is sealed to keep out dirt and moisture.

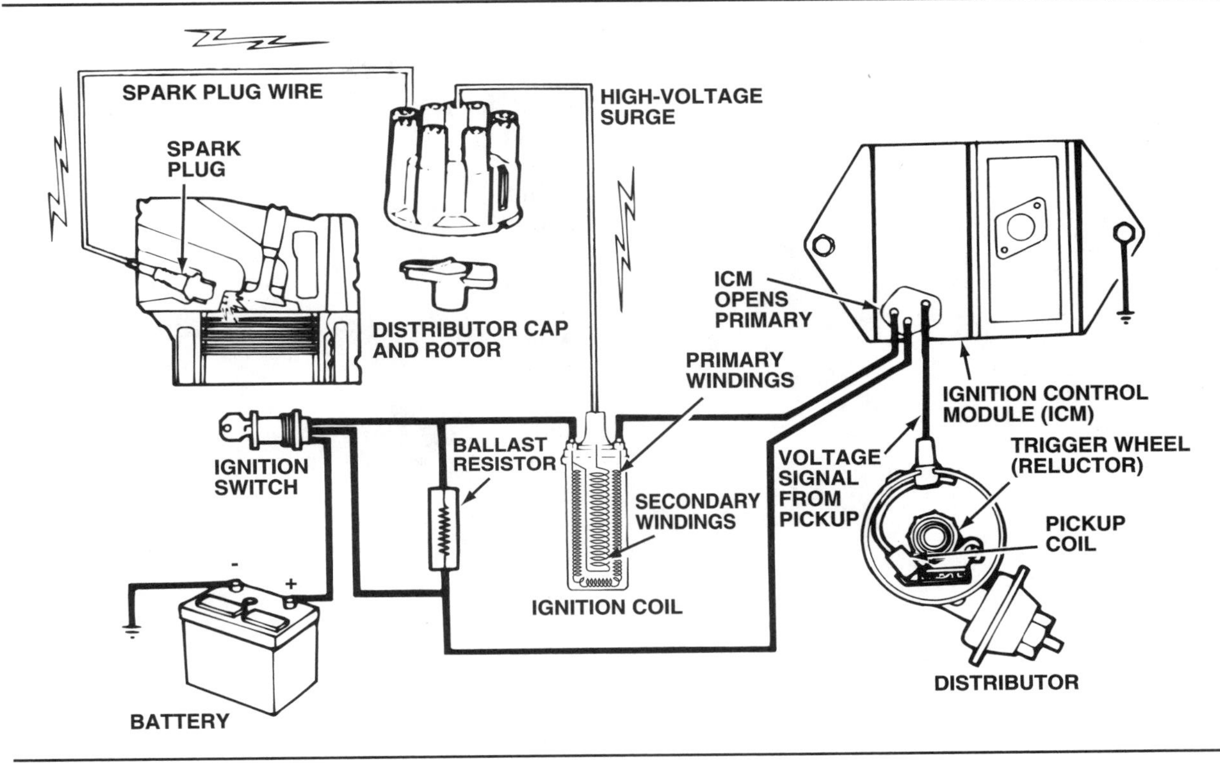

Figure 9-1. Operation of the ignition secondary circuit. (Chrysler)

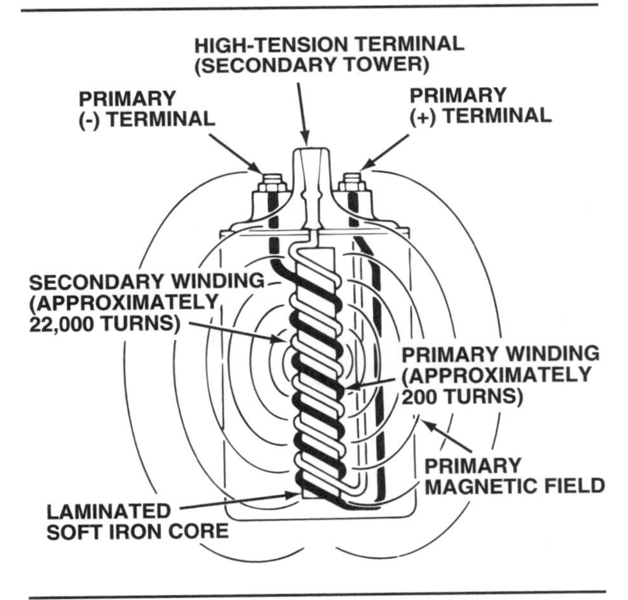

Figure 9-2. The laminated iron core within the coil strengthens the coil's magnetic field.

Laminated E-core coil

Unlike the oil-filled coil, the E-core coil uses an iron core laminated around the windings and potted in plastic, much like a small transformer, figure 9-3. The coil is named because of the "E" shape of the laminations making up its core.

Figure 9-3. The E-core coil is used without a ballast resistor.

Since the laminations provide a closed magnetic path, the E-core coil has a higher energy transfer. The secondary connection looks much like a spark plug terminal. Primary leads are housed in a single snap-on connector that attaches to the coil's blade-type terminals. The E-core coil has very low primary resistance and is used without a ballast resistor in Ford Thick-Film Ignition (TFI) and some GM High Energy Ignition (HEI) systems.

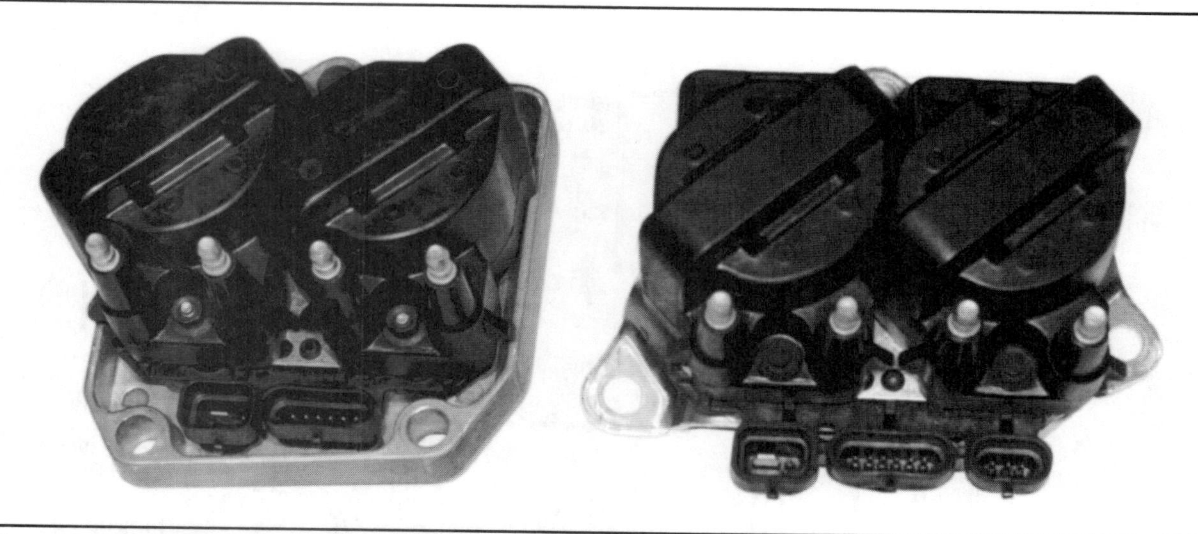
Figure 9-4. Typical DIS coil packs used on 4-cylinder engines.

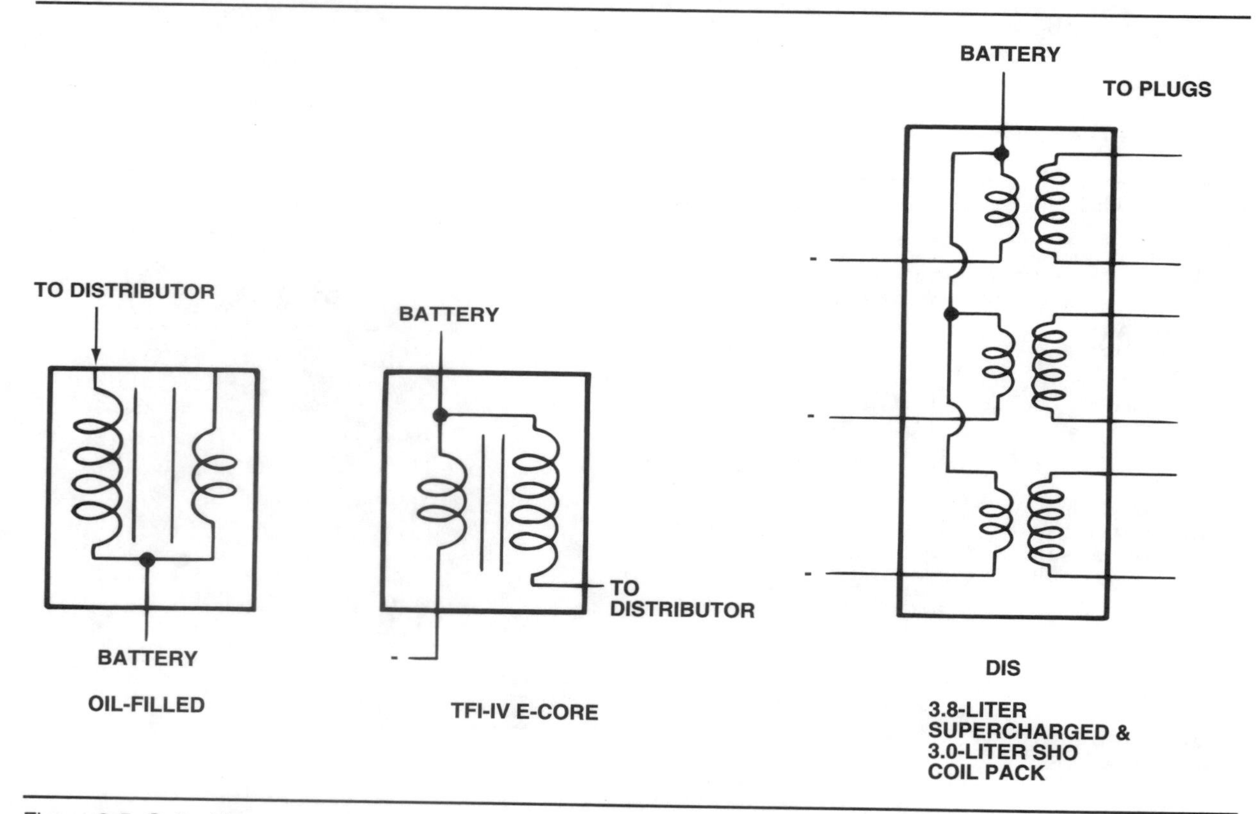

Figure 9-5. Coil wiring comparison. (Ford)

DIS coil packs

Distributorless ignitions use two or more coils in a single housing called a coil pack, figure 9-4. Figure 9-5 compares the internal windings of a typical coil pack with those of the oil-filled and E-core coils. Because the E-core coil has a primary and secondary winding on the same core, it uses a common terminal. Both ends of the E-core coil's primary winding connect to the primary ignition circuit; the open end of its secondary winding connects to the center tower of the coil, where the distributor high tension lead connects.

Coil packs are significantly different, using a closed magnetic core with one primary winding

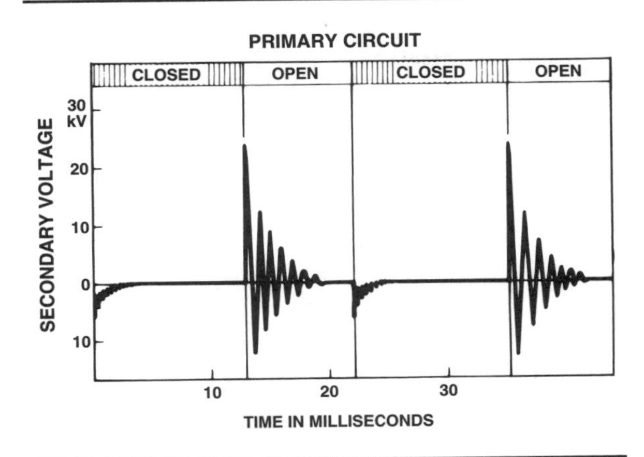

Figure 9-6. A secondary circuit no-load voltage trace. (Bosch)

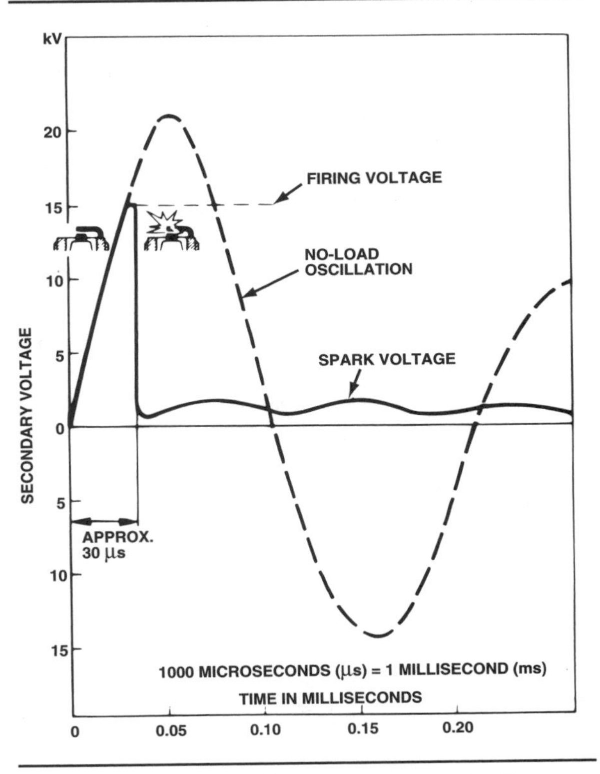

Figure 9-7. The dashed line shows no-load voltage; the solid line shows the voltage trace of firing voltage and spark voltage. (Bosch)

for each two high-voltage outputs. The secondary circuit of the coil pack is wired in series. Each coil in the coil pack directly provides secondary voltage for two of the spark plugs, which are wired in series with the coil secondary winding. Coil pack current is limited by transistors called output drivers in the ignition module attached to the bottom of the pack. The output drivers open and close the ground path of the coil primary circuit. Timing and sequencing of the output drivers are controlled by other module internal circuits.

Coil Voltage

A coil must supply the correct amount of voltage for any system. Since this amount of voltage varies, depending on engine and operating conditions, the coil's available voltage is generally more than the system's required voltage. If it is less, the engine may not run.

Available voltage

The ignition coil can supply much more secondary voltage than the average engine requires. The peak voltage that a coil can produce is called its available voltage.

Three important coil design factors determine available voltage level:

- Secondary-to-primary turns ratio
- Primary voltage level
- Primary circuit resistance.

The turns ratio is a multiplier that creates high secondary voltage output. The primary voltage level applied to a coil is determined by the ignition circuit's design and condition. Installing a ballast resistor of the wrong value will affect this voltage level, as will loose and corroded connections. Generally, a primary circuit voltage loss of one volt can decrease available voltage by 10,000 volts.

If there were no spark plug in the secondary circuit (that is, if the circuit were open), the coil secondary voltage would have no place to discharge quickly. The voltage would oscillate in the secondary circuit, dissipating as heat. The voltage would be completely gone in just a few milliseconds. Figure 9-6 shows the trace of this no-load, open-circuit voltage. This is called secondary voltage **no-load oscillation.** The first peak of the voltage trace represents the maximum available voltage from that particular coil. Available voltage is usually between 20,000 and 50,000 volts.

Required voltage

When there is a spark plug in the secondary circuit, the coil voltage creates an arc across the

Available Voltage: The peak voltage that a coil can produce.

No-Load Oscillation: The rapid, back-and-forth, peak-to-peak oscillation of voltage in the ignition secondary circuit when the circuit is open.

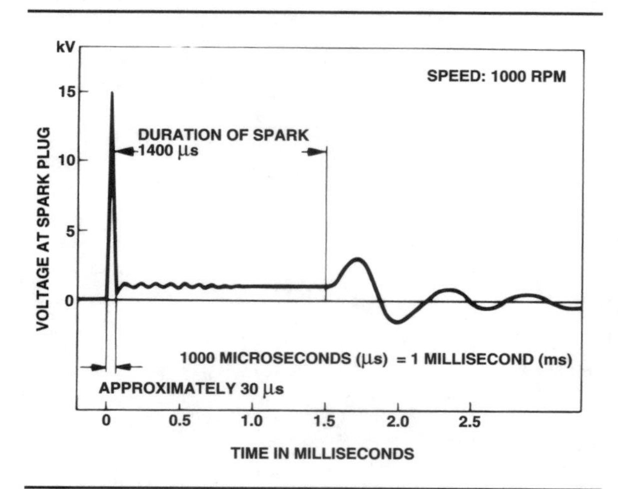

Figure 9-8. The voltage trace of an entire secondary ignition pulse. (Bosch)

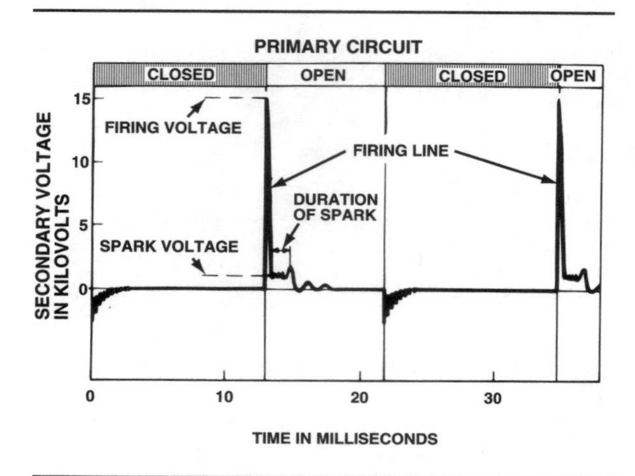

Figure 9-9. As the primary circuit opens and closes, the ignition cycle repeats. (Bosch)

plug air gap. Figure 9-7 compares a typical no-load oscillation to a typical secondary firing voltage oscillation. At about 15,000 volts, the spark plug air gap ionizes and becomes conductive. This is the ionization voltage level, also called the **firing voltage,** or **required voltage.**

As soon as a spark has formed, the energy demands of the spark cause the secondary voltage to drop to the much lower spark voltage level. This is the inductive portion of the spark. **Spark voltage** is usually about one-quarter of the firing voltage level.

Figure 9-8 shows the entire trace of the spark. When the secondary voltage falls below the inductive air-gap voltage level, the spark can no longer be maintained. The spark gap becomes nonconductive. The remaining secondary voltage oscillates in the secondary circuit, dissipating as heat. This is called secondary **voltage decay**. At this time, the primary circuit closes and the cycle repeats, figure 9-9. The traces shown in figures 9-8 and 9-9 are similar to the secondary circuit traces you see on an oscilloscope screen.

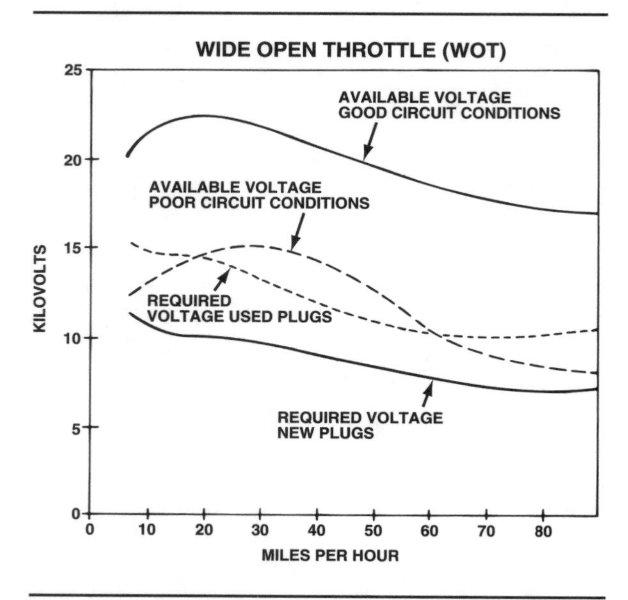

Figure 9-10. Available and required voltage levels under different system conditions.

Some conditions that cause required voltage levels to increase are:

- Eroded electrodes in the distributor cap, rotor, or spark plug
- Damaged ignition cables
- Reversed coil polarity
- High compression pressures
- A lean air-fuel mixture that is more difficult to ionize.

Voltage Reserve

The physical condition of the automotive engine and ignition system can affect both available and required voltage levels, as we have seen. Figure 9-10 shows available and required voltage levels in a particular ignition system under various operating conditions. **Voltage reserve** is the amount of coil voltage available in excess of the voltage required to fire the spark plug.

Under certain poor circuit conditions, there may be no voltage reserve. At these times, some spark plugs will not fire, and the engine will run poorly or not at all. Ignition systems must be properly maintained to ensure that there is always some voltage reserve. A well-tuned ignition system should have a voltage reserve of about 60 percent of available voltage under most operating conditions.

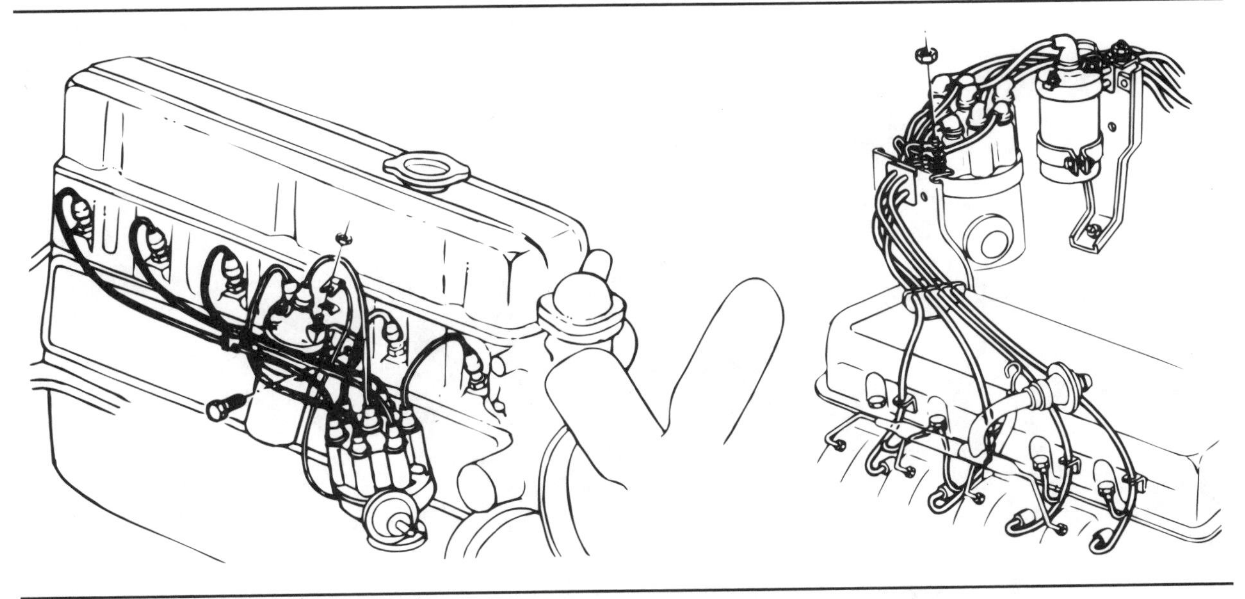

Figure 9-11. Ignition coils are commonly mounted on the engine (left) or on a fender panel (right) as shown on this Chevrolet.

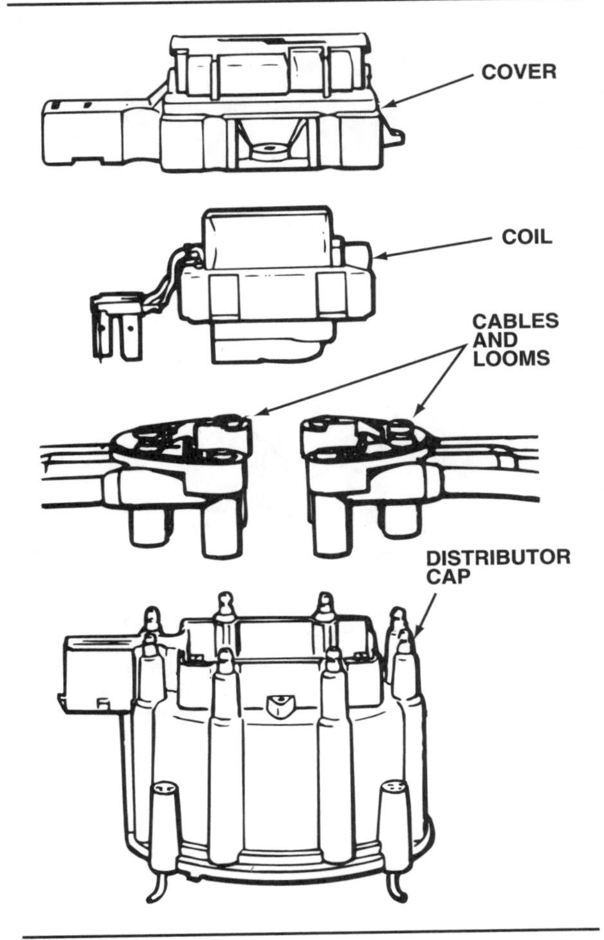

Figure 9-12. Many Delco-Remy HEI solid-state ignition systems used on V6 and V8 engines have a coil mounted in the distributor cap.

Coil Installations

Ignition coils are usually mounted with a bracket on a fender panel in the engine compartment or on the engine, figure 9-11. Some ignition coils have an unusual design and location. The Delco-Remy High-Energy Ignition (HEI) solid-state ignition system used on V6 and V8 engines has a coil mounted in the distributor, figure 9-12. The coil output terminal is connected directly to the center electrode of the distributor cap. The connections to the primary winding are made through a multiple-plug connector.

Distributorless ignitions use an assembly containing two or more separate ignition coils and an ignition control module (ICM), figure 9-13.

Firing Voltage (Required Voltage): The voltage level that must be reached to ionize and create a spark in the air gap between the spark plug electrodes.

Spark Voltage: The inductive portion of a spark that maintains the spark in the air gap between a spark plug's electrodes, usually about one-quarter of the firing voltage level.

Voltage Decay: The rapid oscillation and dissipation of secondary voltage after the spark in a spark plug air gap has stopped.

Voltage Reserve: The amount of coil voltage available in excess of the voltage required to fire the spark plugs.

Figure 9-13. The Buick C^{3}I distributorless ignition uses three separate ignition coils, each of which serves two cylinders 360 degrees apart in the firing order.

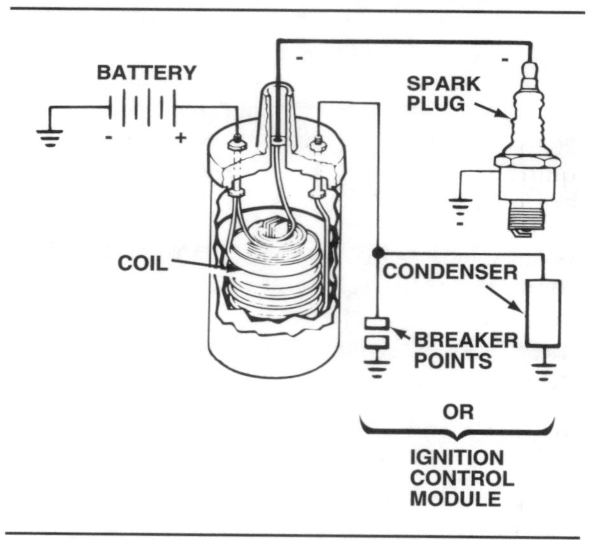

Figure 9-14. When coil connections are made properly, the spark plug center electrode is electrically negative.

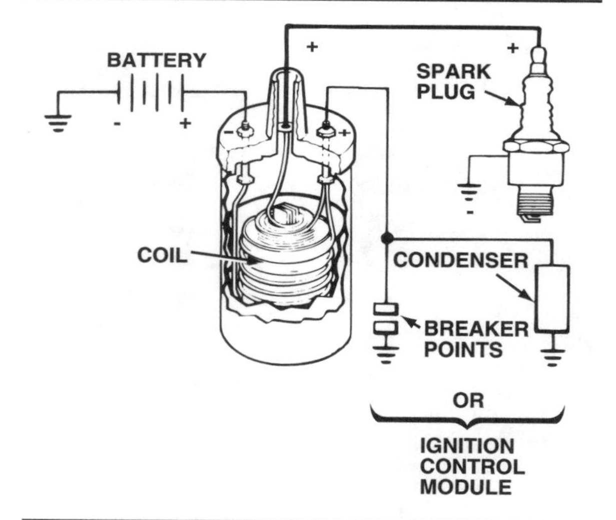

Figure 9-15. When coil connections are reversed, spark plug polarity is reversed.

Control circuits in the module discharge each coil separately in sequence, with each coil serving two cylinders 360 degrees apart in the firing order.

In any system, the connections to the primary winding must be made correctly. If spark plug polarity is reversed, greater voltage is required to fire the plug. Plug polarity is established by the ignition coil connections.

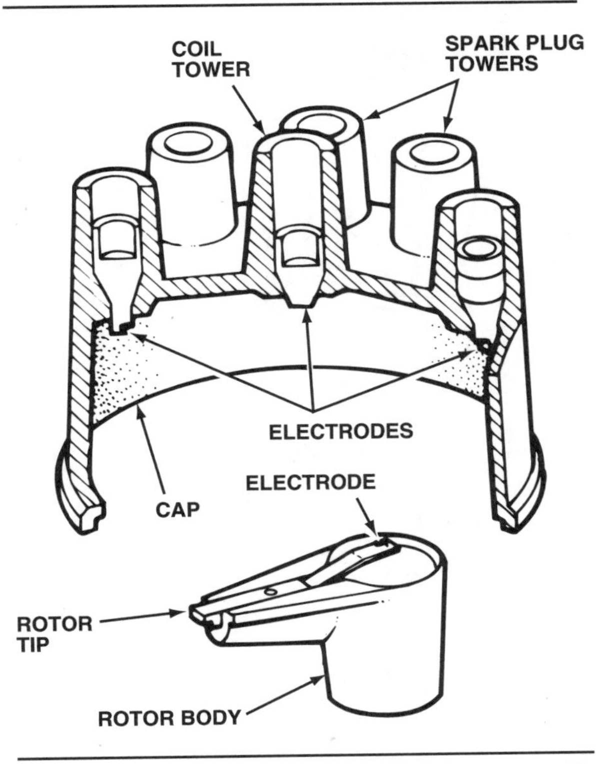

Figure 9-16. A distributor rotor and a cutaway view of the distributor cap.

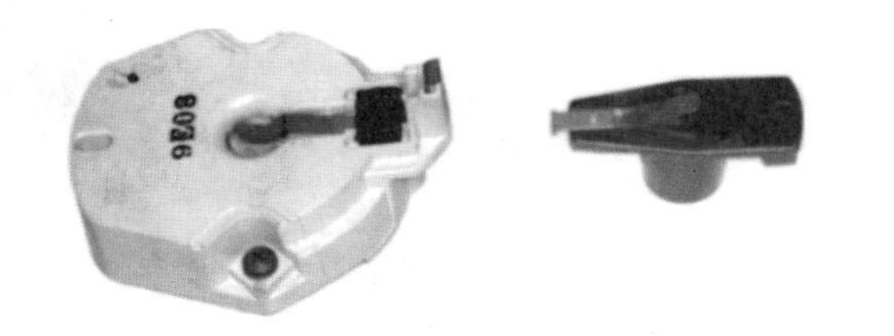

Figure 9-17. Typical distributor rotors.

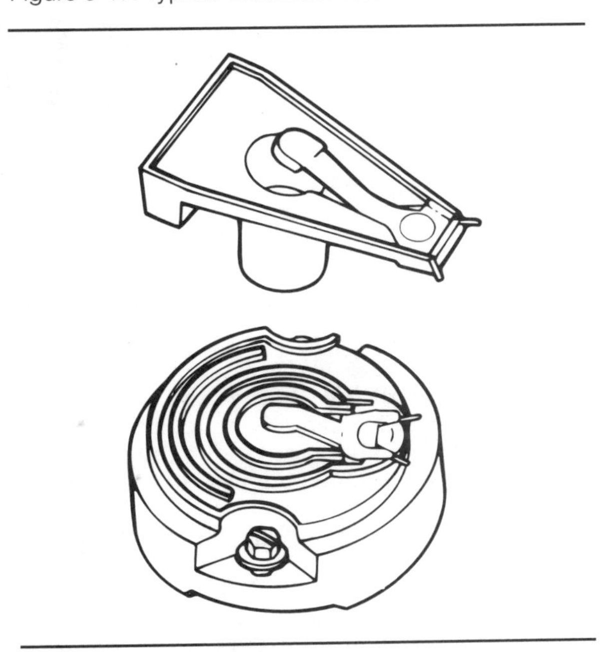

Figure 9-18. Typical Ford multiwire or "cat-whisker" rotors used in some 1983–86 systems.

For a non-DIS coil, one end of the coil secondary winding is connected to the primary winding, figure 9-14, so that the secondary circuit is grounded through the ignition primary circuit. When the coil terminals are properly connected to the battery, the grounded end of the secondary circuit is electrically positive. The other end of the secondary circuit, which is the center electrode of the spark plug, is electrically negative. The plug's grounded-side electrode is positive, and plug polarity is correct. Whether the secondary winding is grounded to the primary + or - terminal depends on whether the windings are wound clockwise or counterclockwise.

If the ignition coil is correctly connected, figure 9-14, voltage will be delivered to the spark plug so that the center electrode is negatively charged and the grounded electrode is positively charged. Electrons will move from the hotter center electrode to the cooler-side electrode at a relatively lower voltage. This is called **negative polarity.** If the coil is incorrectly connected, figure 9-15, electrons will be forced to move from the cooler-side electrode to the hotter center electrode. This is called **positive polarity,** or reverse polarity. When plug polarity is reversed, 20 to 40 percent more secondary voltage is required to fire the spark plug.

Coil terminals are usually marked BAT or +, and DIST or -. To establish the correct plug polarity with a negative-ground electrical system, the + terminal must be connected to the positive

Negative Polarity:The condition when current flows through the coil's primary windings in the correct direction. Coil voltage is delivered to the spark plugs so that the center electrode of the plug is negatively charged and the grounded electrode is positively charged. Also called ground polarity.

Positive Polarity: An incorrect polarity of the ignition coil caused by reversing the primary coil terminal connectors. Coil voltage is delivered to the spark plug so the center electrode of the plug is positively charged and the grounded electrode is negatively charged. Also called reverse polarity.

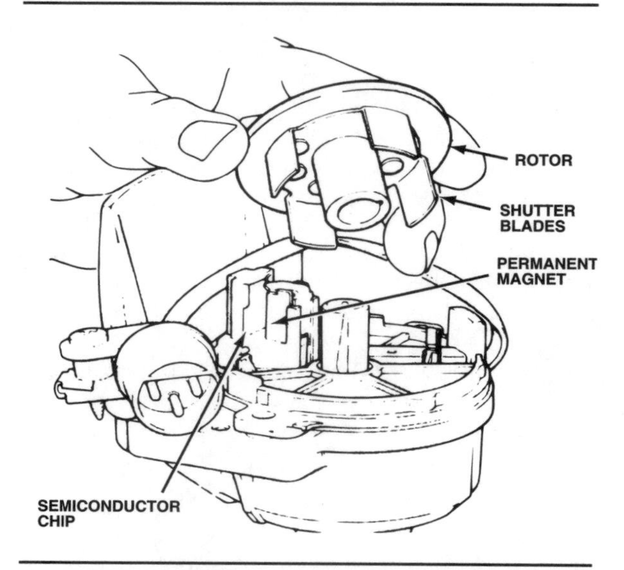

Figure 9-19. A Hall-effect triggering device attached to the rotor. (Chrysler)

terminal of the battery (through the ignition switch, starter relay, and other circuitry). The coil terminal must be connected to the distributor breaker points and condenser, or to the ICM.

DISTRIBUTOR CAP AND ROTOR

The distributor cap and rotor, figure 9-16, receive high-voltage current from the coil secondary winding. Current enters the distributor cap through the central terminal, called the coil tower. The rotor carries the current from the coil tower to the spark plug electrodes in the rim of the cap. The rotor is mounted on the distributor shaft and rotates with it, so that the rotor electrode moves from one spark plug electrode to another in the cap to follow the designated firing order.

Distributor Rotor

A rotor is made of silicone plastic, **Bakelite,** or a similar synthetic material that is a very good insulator. A metal electrode on top of the rotor conducts current from the carbon terminal of the coil tower.

The rotor is keyed to the distributor shaft to maintain its correct relationship with the shaft and the spark plug electrodes in the cap. The key may be a flat section or a slot in the top of the shaft. Delco-Remy V6 and V8 rotors, shown at the left in figure 9-17, and Ford TFI distributor rotors are keyed in place by two locators and secured by two screws. Most other rotors, shown at the right in figure 9-17, are pressed onto the shaft by hand. The rotor in Chrysler's optically

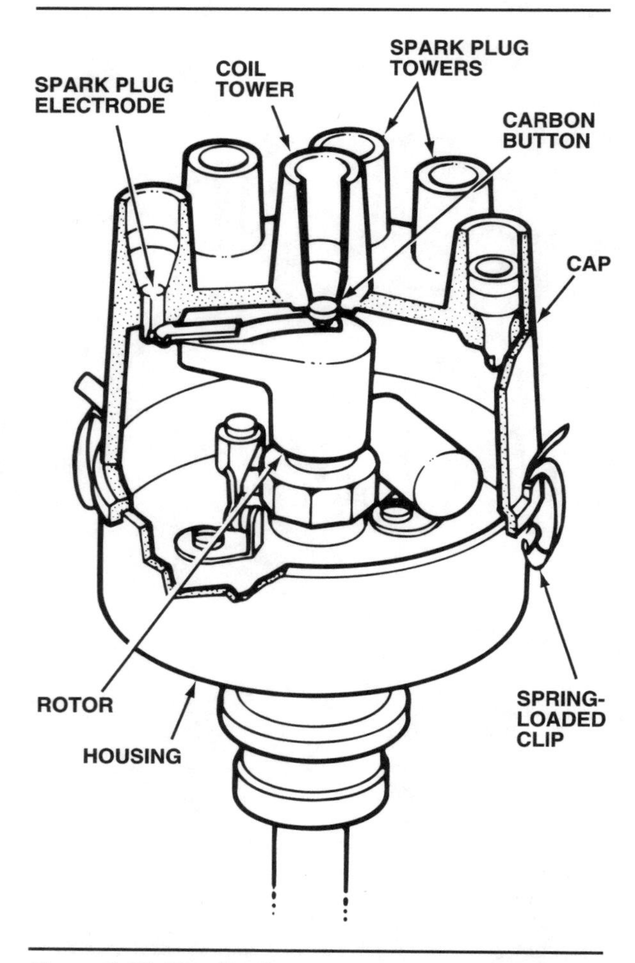

Figure 9-20. The distributor cap and rotor assembled with the distributor housing.

triggered distributor is retained by a horizontal capscrew.

Ford's basic rotor design uses a blade-type rotor tip. This was changed to a multiwire or "cat-whisker" rotor tip, figure 9-18, in some 1983–85 systems. This was an attempt to further reduce **radio frequency interference (RFI)** from the secondary circuit without using silicone grease on the rotor tip. However, arcing from the multiwire tip formed ozone and nitrogen oxides from the air inside the distributor. Over a period of time, these combined to form nitric acid that reacted with the distributor cap to create a short-circuit path for secondary voltage.

Ford released replacement caps of a different material in 1985 to counteract **crossfiring** problems and discontinued the cat-whisker rotors in 1986. Replacement rotors for 1983–85 models are the blade-type design.

Except for the multiwire-tip rotors, Ford and Chrysler breakerless distributor rotors are coated at the factory with a silicone grease. As the

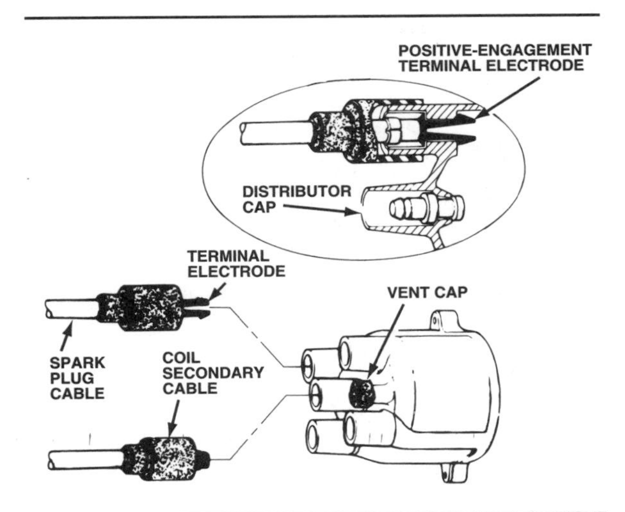

Figure 9-21. Chrysler 4-cylinder distributors have used positive locking terminal electrodes as part of the ignition cable since 1980.

silicone ages, it may look like contamination, but it is not. Do not remove or reapply any coating on a used Ford rotor. Chrysler recommends removing any excess on the tip of the rotor. When a new Ford rotor is installed, apply a 1/8-in. (3-mm) coating of silicone grease (Dow Corning 111, GE G-627, or equivalent) on all sides of the electrode, including the tip. The Ford multiwire-tip rotor does not require the silicone grease used on the blade-type rotors.

Rotors used with Hall-effect switches often have the shutter blades attached, figure 9-19, serving a dual purpose. In addition to distributing the secondary current, the rotor blades bypass the Hall-effect magnetic field and create the signal for the primary circuit to fire.

Rotor air gap

An air gap of a few thousandths of an inch, or a few hundredths of a millimeter, exists between the tip of the rotor electrode and the spark plug electrode of the cap. If they actually touched, both would wear very quickly. Because the gap cannot be measured when the distributor is assembled, it is usually described in terms of the voltage required to create an arc across the electrodes. Only about 3,000 volts are required to create an arc across most breaker-point distributor air gaps, but some Delco-Remy distributors require as much as 9,000 volts. The voltage required to jump the air gap in electronic distributors generally is higher than that required for breaker-point ignitions. As the rotor completes the secondary circuit and the plug fires, the rotor air gap adds resistance to the circuit. This raises the plug firing voltage, suppresses secondary current, and increases RFI.

Distributor Cap

The distributor cap is also made of silicone plastic, Bakelite, or a similar material that resists chemical attack and protects other distributor parts. Metal electrodes in the spark plug towers and a carbon insert in the coil tower provide electrical connections with the ignition cables and the rotor electrode, figure 9-20. The cap is keyed to the distributor housing and is held on by two or four spring-loaded clips or by screws.

Bakelite: A synthetic plastic material that is a good insulator. Distributor caps are often made of Bakelite.

Radio Frequency Interference (RFI): A form of electromagnetic interference created in the ignition secondary circuit that disrupts radio and television transmission.

Crossfiring: Ignition voltage jumping from the distributor rotor to the wrong spark plug electrode inside the distributor cap. Also, ignition voltage jumping from one spark plug cable to another due to worn insulation.

The Lincoln-Zephyr V12 Ignition System

The ignition system for the Lincoln-Zephyr V12 engines used two coils contained in a single housing mounted on top of a distributor with two distributor caps, two sets of breaker points, a rotor with two contacts, and two condensers. The distributor was mounted on the front of the engine and connected directly to the camshaft.

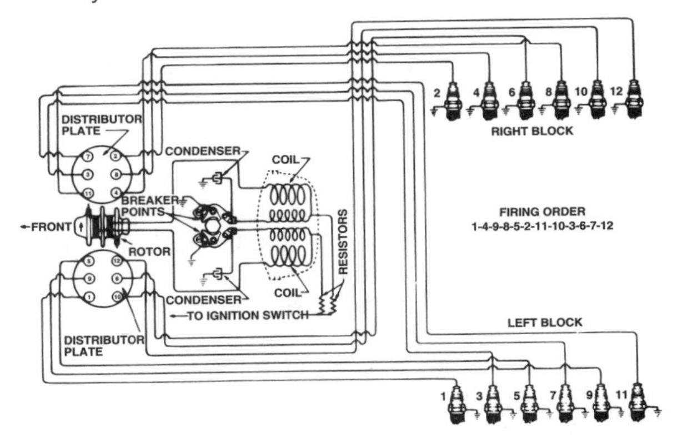

Looking at the distributor from the driver's seat position, the right-hand coil and a fixed set of breaker points fired the right bank of cylinders, numbers 2-4-6-8-10-12. The left-hand coil and an adjustable set of breaker points fired the left bank of cylinders, numbers 1-3-5-7-9-11.

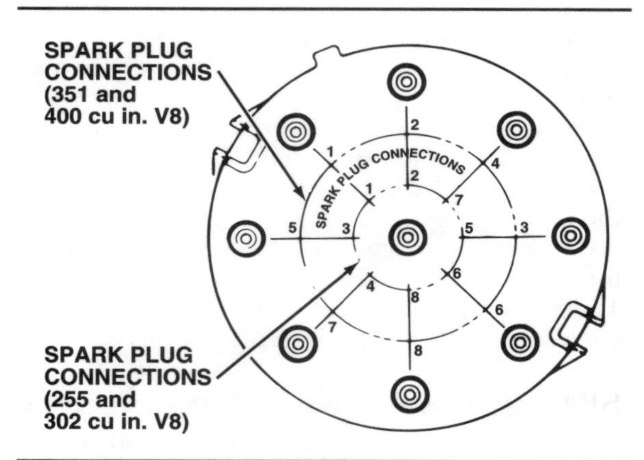

Figure 9-22. Spark plug cable installation order for V8 engines with EEC systems.

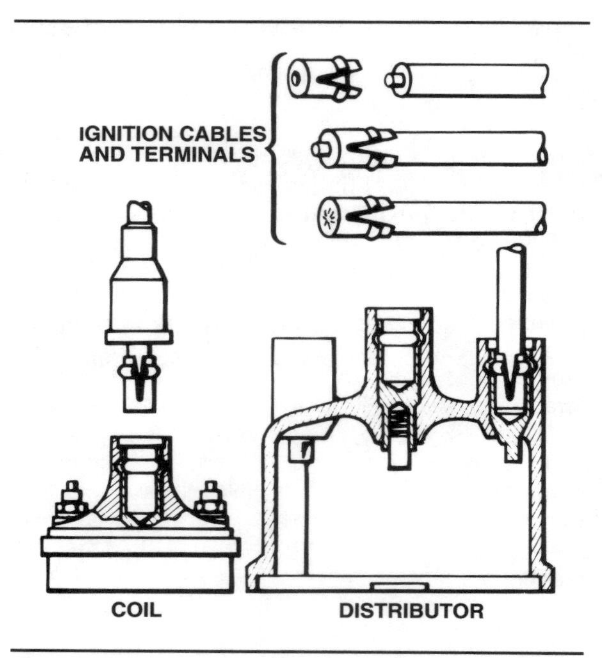

Figure 9-23. Ignition cables and terminals.

Delco-Remy HEI caps and all Ford Dura-Spark and TFI caps have male connectors rather than female spark plug towers. On later Dura-Spark distributors, the adapter ring is held to the body by two screws inside the ring.

HEI distributor caps with an integral coil, figure 9-12, are secured by four spring-loaded clips. When removing this cap, be sure that all four clips are disengaged and clear of the housing. Then lift the cap straight up to avoid bending the carbon button in the cap and the spring that connects it to the coil. If the button and spring are distorted, arcing can occur that will burn the cap and rotor.

Positive-engagement spark plug cables are used with some Chrysler and Ford 4-cylinder

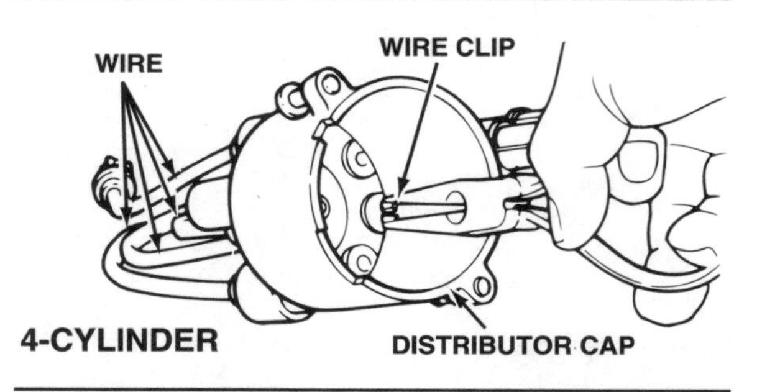

Figure 9-24. Positive locking terminal electrodes are removed by compressing the wire clips with pliers and removing the wire from the cap.

ignition systems. There are no electrodes in distributor caps used with these cables. A terminal electrode attached to the distributor-cap end of the cable locks inside the cap to form the distributor contact terminal, figure 9-21. The secondary terminal of the cable is pressed into the cap.

Ford Motorcraft Dura-Spark III distributors used with some of Ford's Electronic Engine Control (EEC) systems have caps and rotors with the terminals on two levels to prevent secondary voltage arcing. Spark plug cables are not connected to the caps in firing order sequence, but the caps are numbered with the engine cylinder numbers, figure 9-22. The caps have two sets of numbers, one set for 5.0-liter standard engines and the other for 5.7-liter and 302-cu in. high-performance engines. Cylinder numbers must be checked carefully when changing spark plug cables.

Distributor caps used on some later-model Ford and Chrysler vehicles have a vent to prevent moisture buildup and to reduce ozone accumulation inside the cap, figure 9-21.

IGNITION CABLES

Secondary ignition cables carry high-voltage current from the coil to the distributor (coil wire), and from the distributor to the spark plugs (spark plug cables). They use heavy insulation to prevent the high-voltage current from jumping to ground before it reaches the spark plugs. Ford, GM, and some other electronic ignitions use 8-mm cables; all others use 7-mm cables.

Conductor Types

Spark plug cables originally used a solid steel or copper wire conductor. Cables manufactured with these conductors were found to cause radio and television interference. While this type of cable is still made for special applications, such as racing, most spark plug cables have been made of a high-resistance, nonmetallic conductor for the past 30 years. Several nonmetallic conductors may be used, such as carbon, and linen or fiberglass strands impregnated with graphite. The nonmetallic conductor acts as a resistor in the secondary circuit, and reduces RFI and spark plug wear due to high current. Such cables are often called **television-radio-suppression (TVRS) cables,** or just suppression cables.

When replacing spark plug cables on vehicles with computer-controlled systems, be sure that the resistance of the new cables is within the automaker's specifications to avoid possible electromagnetic interference with the operation of the computer.

Terminals and boots

Secondary ignition cable terminals, figure 9-23, are designed to make a strong contact with the coil and distributor electrodes. They are, however, subject to corrosion and arcing if not firmly seated and protected from the elements.

Positive-engagement spark plug cable terminals, figure 9-21, lock in place inside the distributor cap and cannot come loose accidentally. They can only be removed with the cap off the distributor. The terminal electrode is then compressed with pliers and the wire is pushed out of the cap, figure 9-24.

The ignition cables must have special connectors, often called spark plug boots, figure 9-25. The boots provide a tight and well-insulated contact between the cable and the spark plug.

SPARK PLUGS

Spark plugs allow the high-voltage secondary current to arc across a small air gap. The three basic parts of a spark plug, figure 9-26, are:

Television-Radio-Suppression (TVRS) Cables: High-resistance, carbon-conductor ignition cables that suppress RFI.

■ Making Tracks

You often read instructions to inspect ignition parts for carbon tracks. Although you may have heard about or seen carbon tracks, have you ever thought about what they are and what causes them?

Carbon tracks are deposits or defects on distributor rotors, caps, spark plugs, and cables that create

a short-circuit path to ground for secondary high voltage. They also cause cross firing, in which the high voltage jumps from the distributor rotor to the wrong terminal in the cap.

The problems caused by carbon tracks are all pretty similar, but the causes for these defects are rather complex. A distributor cap or rotor may develop a hairline crack because of rough handling, a manufacturing defect, or some other problem. Under certain conditions, moisture can collect in the crack and create a lower-resistance path for high voltage. High-voltage arcing to ground in a distributor ionizes air molecules and can form conductive deposits along its path. If any dirt or grease is in the short-circuit path, the combination of high voltage and its accompanying current causes carbon deposits to form around the crack. Thus, a carbon track develops.

Carbon tracks can form even without a crack in a cap or rotor. High voltage ionizes air and oil molecules in the distributor and causes deposits to form. The deposits have high resistance, but if they are the least bit conductive, secondary voltage can arc to them. Over a period of time, the deposits build up and can create a short circuit.

Outside a distributor, similar carbon tracks can form on spark plug insulators and ignition cables due to grease deposits and weak points in damaged cable insulation.

Carbon tracks inside a distributor cap often can be tricky to diagnose. Sometimes an engine will run smoothly at idle but misfire at high speed. As the distributor advance mechanisms operate, the rotor moves farther away from the cap terminals as the coil discharges. The high-voltage current must cross an increasing air gap. If a nearby carbon track provides lower resistance, the voltage will jump to ground and the engine will misfire.

A typical carbon track has about the same, or a little less, resistance as a TVRS ignition cable. That's quite conductive enough to cause a misfire or a no-start problem. The accompanying photo shows a classic set of carbon tracks inside a distributor cap.

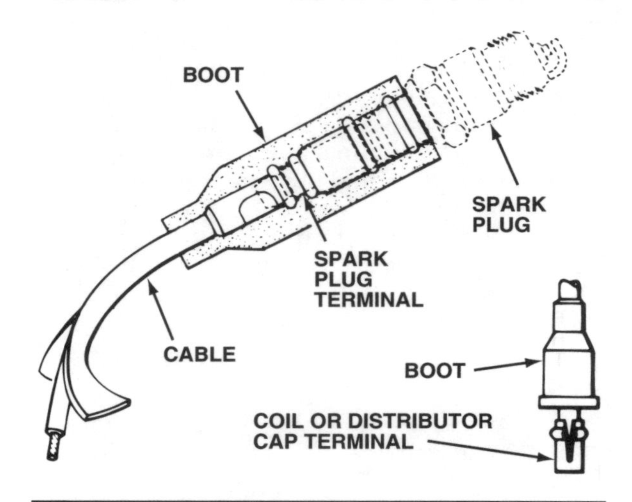

Figure 9-25. Ignition cables, terminals, and boots work together to carry the high-voltage secondary current.

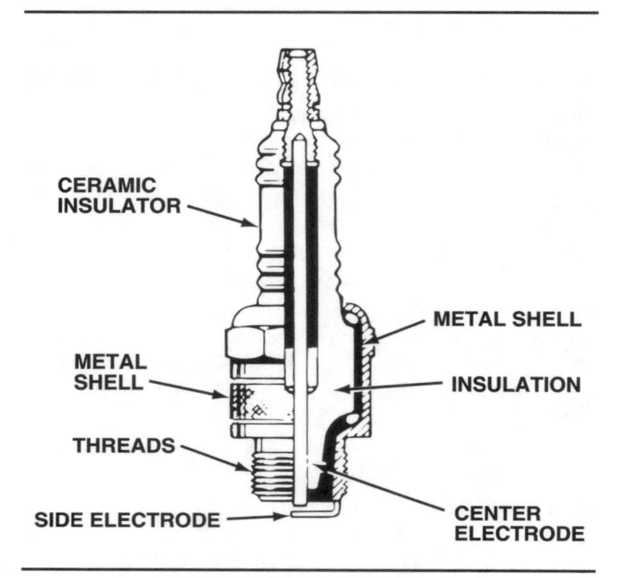

Figure 9-26. A cutaway view of the spark plug.

- A ceramic core, or insulator, which insulates the center electrode and acts as a heat conductor
- Two electrodes, one insulated in the core and the other grounded on the shell
- A metal shell that holds the insulator and electrodes in a gas-tight assembly and that has threads to hold the plug in the engine.

The metal shell grounds the side electrode against the engine. The other electrode is encased in the ceramic insulator. A spark plug boot and cable are attached to the top of the plug. High-voltage current flows through the center of the plug and arcs from the tip of the

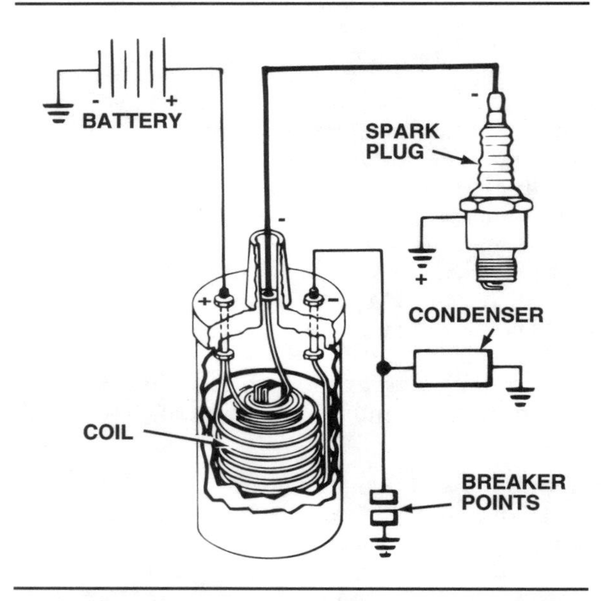

Figure 9-27. The spark plug should have a negative charge at the center electrode and a positive charge at the side electrode.

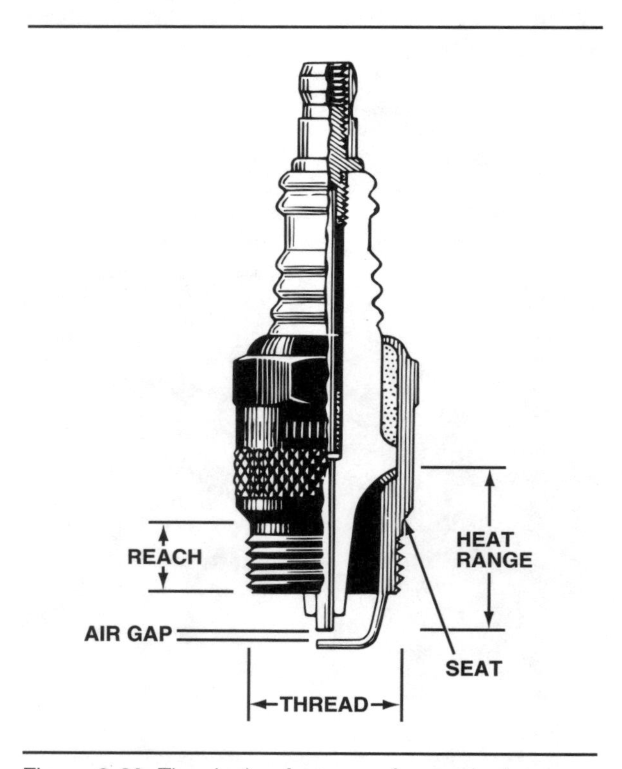

Figure 9-28. The design features of a spark plug.

insulated electrode to the side electrode and ground. This spark ignites the air-fuel mixture in the combustion chamber to produce power.

The burning gases in the engine can corrode and wear the spark plug electrodes. Electrodes are made of metals that resist this attack. Most

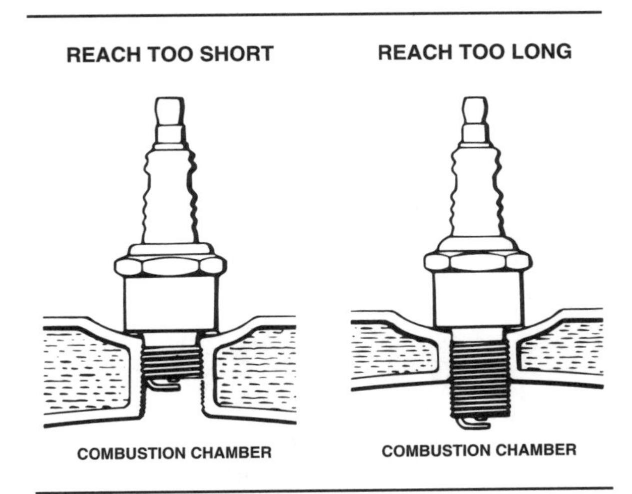

Figure 9-29. Spark plug reach.

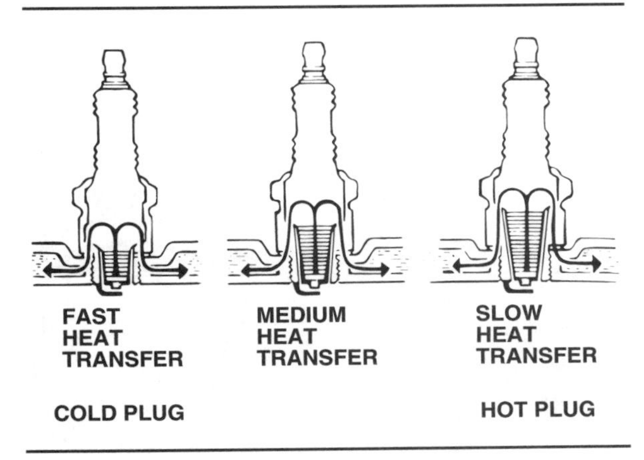

Figure 9-30. Spark plug heat range.

electrodes are made of high-nickel alloy steel. To increase the service interval, automobile manufacturers are installing plugs with platinum and silver-alloyed electrodes.

Spark Plug Firing Action

The arc of current across a spark plug air gap provides two types of discharge:

- Capacitive
- Inductive.

When a high-voltage surge is first delivered to the spark plug center electrode, the air-fuel mixture in the air gap cannot conduct an arc. The spark plug acts as a capacitor, with the center electrode storing a negative charge and the grounded-side electrode storing a positive charge. The air gap between the electrodes acts as a dielectric insulator. This is the opposite of the normal negative-ground polarity, and results from the polarity of the coil secondary winding, as shown in figure 9-27.

Secondary voltage increases, and the charges in the spark plug strengthen until the difference in potential between the electrodes is great enough to **ionize** the spark plug air gap. That is, the air-fuel mixture in the gap is changed from a nonconductor to a conductor by the positive and negative charges of the two electrodes. The dielectric resistance of the air gap breaks down, and current flows between the electrodes. The voltage level at this instant is called ionization voltage. The current that flows across the spark plug air gap at the instant of ionization is the capacitive portion of the spark. It flows from negative to positive and uses the energy stored in the plug itself when the plug was acting as a capacitor, before ionization. This is the portion of the spark that starts the combustion process within the engine.

The ionization voltage level is usually less than the total voltage produced in the coil secondary winding. The remainder of the secondary voltage (that voltage not needed to force ionization) dissipates as current across the spark plug air gap. This is the inductive portion of the spark discharge, which causes the visible flash or arc at the plug. It contributes nothing to the combustion of the air-fuel mixture, but is the cause of electrical interference and severe electrode erosion. High-resistance cables and spark plugs suppress this inductive portion of the spark discharge.

SPARK PLUG CONSTRUCTION

Spark Plug Design Features

Spark plugs are made in a variety of sizes and types to fit different engines. The most important differences, figure 9-28, among plugs are:

- Reach
- Heat range
- Thread and seat
- Air gap.

Reach

The **reach** of a spark plug is the length of the shell from the seat to the bottom of the shell, including both threaded and unthreaded por-

Ionize: To break up molecules into two or more oppositely charged ions. The air gap between the spark plug electrodes is ionized when the air-fuel mixture is changed from a nonconductor to a conductor.

Reach: The length of the spark plug shell from the seat to the bottom of the shell.

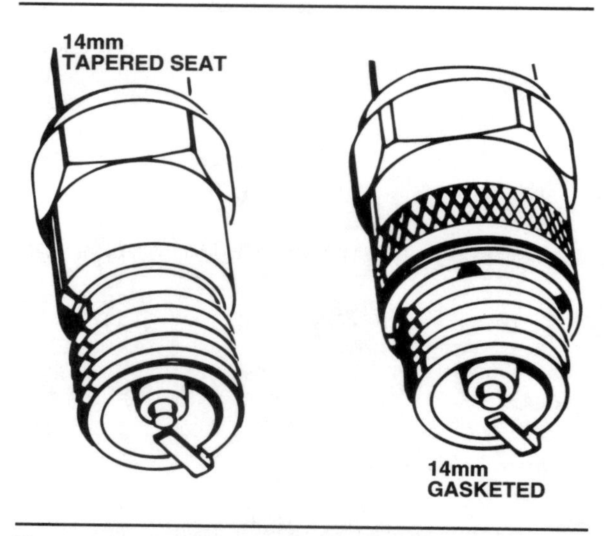

Figure 9-31. Spark plug thread and seat types.

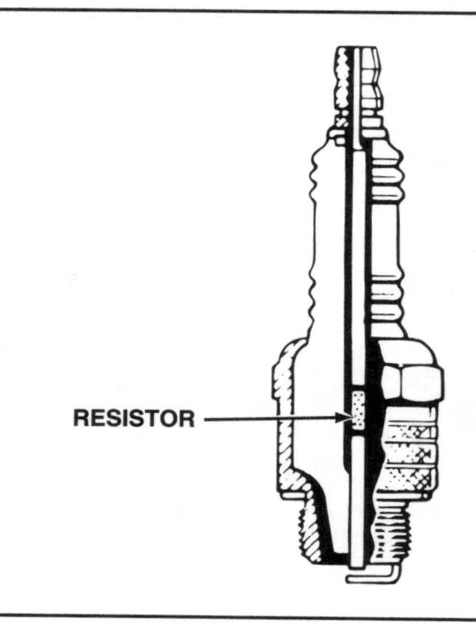

Figure 9-32. A resistor-type spark plug.

tions. If an incorrect plug is installed and the reach is too short, the electrode will be in a pocket and the spark will not ignite the air-fuel mixture very well, figure 9-29.

If the spark plug reach is too long, the exposed plug threads could get hot enough to ignite the air-fuel mixture at the wrong time. It may be difficult to remove the plug due to carbon deposits on the plug threads. Engine damage can also result from interference between moving parts and the exposed plug threads.

Heat range

The **heat range** of a spark plug determines its ability to dissipate heat from the firing end. The length of the lower insulator and conductivity of the center electrode are design features that primarily control the plug's rate of heat transfer, figure 9-30. A "cold" spark plug has a short insulator tip that provides a short path for heat to travel, and permits the heat to dissipate rapidly to maintain a lower firing tip temperature. A "hot" spark plug has a long insulator tip that creates a longer path for heat to travel. This slower heat transfer maintains a higher firing tip temperature.

Engine manufacturers choose a spark plug with the appropriate heat range required for the normal or expected service for which the engine was designed. Proper heat range is an extremely important factor because the firing end of the spark plug must run hot enough to burn away fouling deposits at idle, but must also remain cool enough at highway speeds to avoid preignition. It also is an important factor in the amount of emissions an engine will produce.

Current spark plug designations use an alpha-numeric system that identifies, among other factors, the heat range of a particular plug. Spark plug manufacturers gradually are redesigning and redesignating their plugs. For example, a typical AC Delco spark plug carries the alpha-numeric designation R45LTS6; the new all-numeric code for a similar AC spark plug of the same length and gap is 41-600. This will make it more difficult for those drivers who attempt to correct driveability problems by installing a hotter or colder spark plug than what the automaker specifies. Eventually, it no longer will be possible for drivers to affect emissions by their choice of spark plugs.

Thread and seat

Most automotive spark plugs are made with one of two thread diameters: 14 or 18 mm, figure 9-31. All 18-mm plugs have tapered seats that match similar tapered seats in the cylinder head. No gaskets are used. The 14-mm plugs are made either with a flat seat that requires a gasket or with a tapered seat that does not. The gasket-type, 14-mm plugs are still quite common, but the 14-mm tapered-seat plugs are now used in most late-model engines. A third thread size is 10 mm; 10-mm spark plugs are generally used on motorcycles, but some car engines also use them, specifically Jaguar's V12.

The steel shell of a spark plug is hex-shaped so that a wrench will fit it. The 14-mm, tapered-seat plugs have shells with a ⅝-in. hex; 14-mm gasketed and 18-mm tapered-seat plugs have shells with a $^{13}/_{16}$-in. hex.

Air gap

The correct spark plug air gap is important to engine performance and plug life. A gap that is too narrow will cause a rough idle and a change in the exhaust emissions. A gap that is too wide

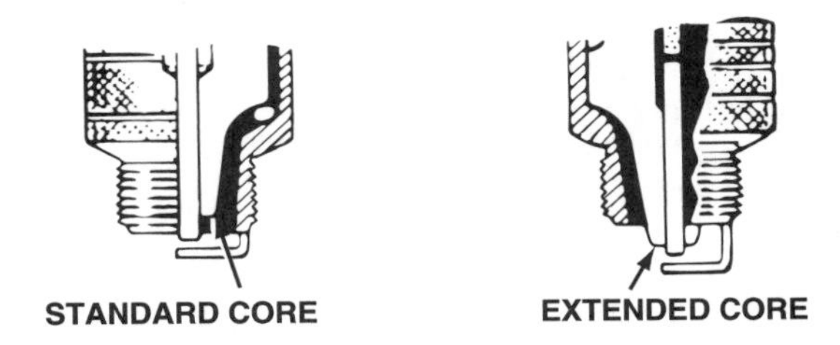

Figure 9-33. A comparison of a standard and an extended-core spark plug.

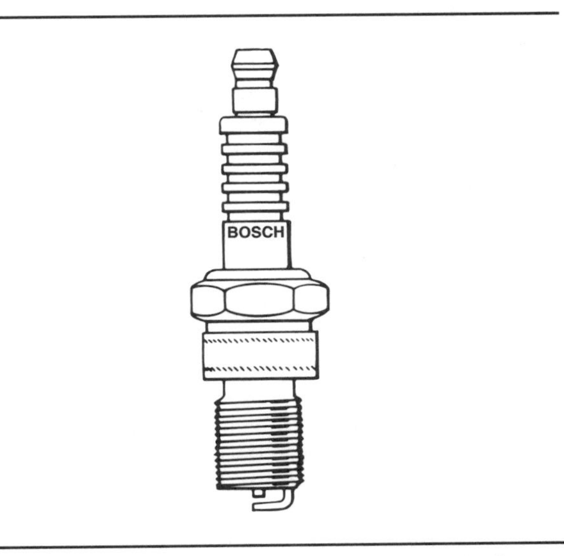

Figure 9-34. A long-reach, short-thread spark plug.

will require higher voltage to jump it; if the required voltage is greater than the available ignition voltage, misfiring will result.

Special-Purpose Spark Plugs

Specifications for all spark plugs include the design characteristics just described. In addition, many plugs have other special features to fit particular requirements.

Resistor-type spark plugs

This type of plug contains a resistor in the center electrode, figure 9-32. The resistor generally has a value of 7,500 to 15,000 ohms, and is used to reduce RFI. Resistor-type spark plugs can be used in place of nonresistor plugs of the same size, heat range, and gap without affecting engine performance.

Extended-tip spark plugs

Sometimes called an **extended-core spark plug,** extended-tip spark plugs use a center electrode and insulator that extend farther into the combustion chamber, figure 9-33. The extended tip operates hotter under slow-speed driving conditions to burn off combustion deposits, and cooler at high speed to prevent spark plug overheating. This greater efficiency over a wider temperature range has led to this plug's increased use in the smaller and less powerful engines manufactured during the 1980s.

Heat Range: The measure of a spark plug's ability to dissipate heat from its firing end.

Resistor-Type Spark Plug: A plug that has a resistor in the center electrode to reduce the inductive portion of the spark discharge, and RFI.

Extended-Core Spark Plug: The insulator core and the electrodes in this type of spark plug extend farther into the combustion chamber than they do on other types. Also called extended tip.

■ Spark Plug Design

Many people have tried to redesign the spark plug. Not all of the "new" designs have worked out. For example, a plug manufactured before World War I had an insulated handle at the top. By pulling this handle up, an auxiliary gap was opened, presumably to create a hotter spark and stop oil fouling. A window in the side of the plug showed whether the gap was open or closed.

Another "revolutionary" type of plug had a screw connector that allowed the inner core assembly to be removed and cleaned quickly.

Still another design had threads and electrodes at each end of the plug. The plug could be removed, the terminal cap installed on the other end, and then reinstalled upside down. All of the photos were provided by Champion Spark Plug Company.

Wide-gap spark plugs
The electronic ignition systems on some late-model engines require spark plug gaps in the 0.045- to 0.080-in. (1.0- to 2.0-mm) range. Plugs for such systems are made with a wider gap than other plugs. This wide gap is indicated in the plug part number. Do not try to open the gap of a narrow-gap plug to create the wide gap required by such ignitions.

Copper-core spark plugs
Many plug manufacturers are making plugs with a copper segment inside the center electrode. The copper provides faster heat transfer from the electrode to the insulator and then to the cylinder head and engine coolant. Copper-core plugs are also extended-tip plugs. The combined effects are a more stable heat range over a greater range of engine temperatures and greater resistance to fouling and misfire.

Platinum-tip spark plugs
Platinum-tip plugs are used in many later-model engines to increase firing efficiency. The platinum center electrode increases electrical conductivity, which helps prevent misfiring with lean mixtures and high temperatures. Since platinum is very resistant to corrosion and wear from combustion chamber gases and heat, the service maintenance interval is doubled.

Long-reach, short-thread spark plugs
Some later-model GM engines, Ford 4-cylinder engines, and Ford 5.0-liter V8 engines use 14-mm, tapered-seat plugs with a ¾-in. reach, but which only have threads for a little over half of their length, figure 9-34. The plug part number includes a suffix that indicates the special thread design, although a fully threaded plug can be substituted if necessary.

Advanced combustion igniters
This extended-tip, copper-core, platinum-tipped spark plug was introduced by GM in 1991. It combines all the attributes of the individual plug designs described earlier and uses a nickel-plated shell for corrosion protection. This combination delivers a plug life in excess of 100,000 miles (160 000 km). No longer called a spark plug, the GM advanced combustion igniter (ACI) has a smooth ceramic insulator with no cooling ribs. The insulator is coated with a baked-on boot release compound that prevents the spark plug wire boot from sticking and causing wire damage during removal.

SUMMARY

The ignition secondary circuit generates high voltage and distributes it to the engine's spark plugs. This circuit contains the coil secondary winding, the distributor cap and rotor, the ignition cables, and the spark plugs.

The ignition coil produces the high voltage necessary to ionize the spark plug gap through electromagnetic induction. Low-voltage current flow in the primary winding induces high voltage in the secondary winding. A coil must be installed with the same primary polarity as the battery to maintain proper secondary polarity at the spark plugs.

Available voltage is the amount of voltage the coil can produce. Required voltage is the voltage necessary to ionize and fire the spark plugs under any given operating condition. Voltage reserve is the difference between available voltage and required voltage. A well-tuned ignition system should have a 60-percent voltage reserve.

Spark plugs allow the high voltage to arc across an air gap and ignite the air-fuel mixture in the combustion chamber. Important design features of a spark plug are its reach, heat range, thread and seat size, and the air gap. Other special features are the use of resistors, extended tips, wide gaps, and copper cores. For efficient spark plug firing, ignition polarity must be established so that the center electrode of the plug is negative and the grounded electrode is positive.

Review Questions

Choose the single most correct answer.
Compare your answers to the correct answers on page 507.

1. Which of the following is used as insulation to protect the windings of coils?
   a. Steel
   b. Wood
   c. Iron
   d. Plastic

2. Which of the following statements is true about ignition coils?
   a. Easily repaired
   b. Adjustments made by setscrews
   c. Requires periodic adjustment
   d. Can be replaced

3. Many of Delco-Remy's solid-state ignition systems have a coil mounted:
   a. On the engine
   b. On a fender panel
   c. On the distributor cap
   d. On the radiator

4. When the coil terminals are properly connected to the battery, the grounded end of the secondary circuit is electrically:
   a. Positive before the air-fuel mix ture ionizes
   b. Negative
   c. Neutral
   d. Ionized

5. A loss of one volt in the primary circuit can decrease available secondary voltage by ___ volts.
   a. 10
   b. 100
   c. 1,000
   d. 10,000

6. The accompanying illustration shows:

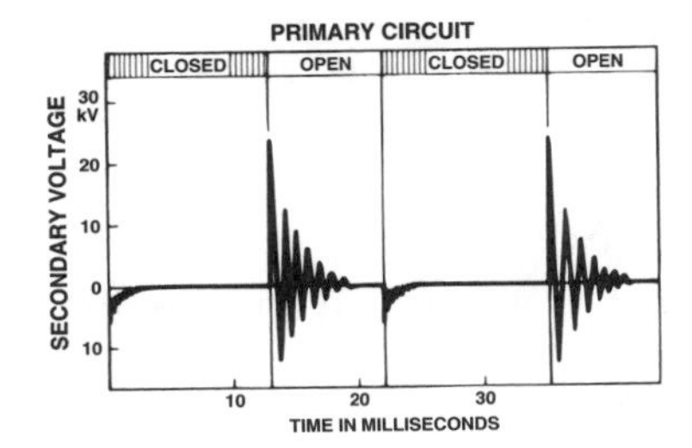

   a. Secondary circuit firing pulse trace
   b. Secondary circuit ignition pulse trace
   c. Secondary circuit no-load voltage trace
   d. Secondary circuit spark voltage trace

7. Firing voltage is usually about ___ as high as spark voltage.
   a. One-quarter
   b. One-half
   c. Four times
   d. Two times

8. The voltage delivered by the coil is:
   a. Its full voltage capacity under all operating conditions
   b. Approximately half of its full voltage capacity at all times
   c. Only the voltage necessary to fire the plugs under any given operating condition
   d. Its full voltage capacity only while starting

9. The voltage reserve is the:
   a. Voltage required from the coil to fire a plug
   b. Maximum secondary voltage capacity of the coil
   c. Primary circuit voltage at the battery side of the ballast resistor
   d. Difference between the required voltage and the available voltage of the secondary circuit

10. A well-tuned ignition system should have a voltage reserve of about ___ of available voltage, under most operating conditions.
    a. 30 percent
    b. 60 percent
    c. 100 percent
    d. 150 percent

11. Bakelite is a synthetic material used in distributors because of its good:
    a. Permeability
    b. Conductance
    c. Insulation
    d. Capacitance

12. Which of the following is a basic part of a spark plug?
    a. Plastic core
    b. Paper insulator
    c. Fiberglass shell
    d. Two electrodes

13. Which of the following is not an important design feature among types of spark plugs?
    a. Reach
    b. Heat range
    c. Polarity
    d. Air gap

14. In the illustration below, the dimension arrows indicate the:

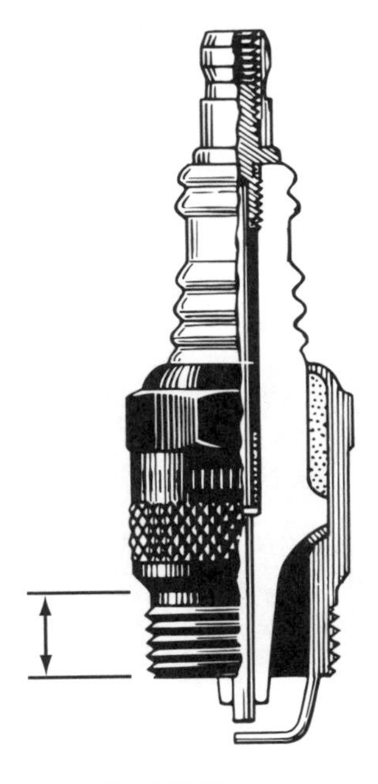

    a. Heat range
    b. Resistor portion of the electrode
    c. Extended core length
    d. Reach

15. All spark plugs have:
    a. A resistor
    b. An extended core
    c. A ceramic insulator
    d. A series gap

Chapter

10

Ignition Timing and Spark Advance Control

The "Engine Operating Principles" chapter dealt with ignition intervals and how the ignition must be synchronized with crankshaft rotation and cylinder firing order. We also have seen that the initial ignition timing is set for the best engine operation at a specific engine speed, usually at or near slow-idle speed. When engine speed changes, ignition timing also must change.

In this chapter, we will learn how ignition timing changes with engine speed and load and how the ignition system changes timing. This chapter will cover specific centrifugal and vacuum advance units, and the emission control systems that have been used to modify the vacuum advance operation. With the latest generation of electronic ignition systems, the computer has taken the place of mechanical and vacuum advance devices. The end of the chapter discusses early electronic spark timing controls.

REVIEWING BASIC TIMING AND BURN TIME

Engine speed and load changes require the ignition timing to advance or to retard. As we have seen, the burn time of an air-fuel mixture is about three milliseconds. Maximum combustion pressure should occur with the piston, connecting rod, and crankshaft in position to produce the most power. At low engine speeds, relatively little spark advance is required to achieve this. However, as engine speed increases, the process of combustion must be started earlier to provide enough burn time.

We have said that burn time is *about* three milliseconds. This means that it varies somewhat with engine load. When the fuel system provides a lean air-fuel mixture under light load, it takes longer to ignite and burn the mixture. Conversely, richer mixtures ignited under a heavy load burn a little faster.

SPARK ADVANCE

There are two basic factors that govern ignition timing: engine speed and load. All changes in timing are related to these two factors:

- Timing must increase, or advance, as engine speed increases; and it must decrease, or retard, as engine speed decreases.
- Timing must decrease, or retard, as load increases; and it must increase, or advance, as load decreases.

Optimum ignition timing under any given combination of these basic factors will result in maximum cylinder pressure. In turn, this delivers maximum power with a minimum of exhaust

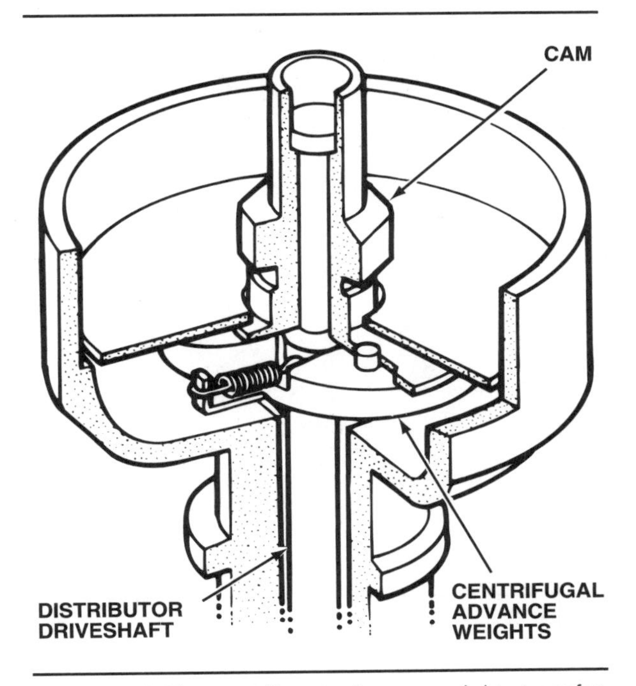

Figure 10-1. The centrifugal advance weights transfer the rotation of the distributor driveshaft to the cam, or trigger wheel, and the rotor.

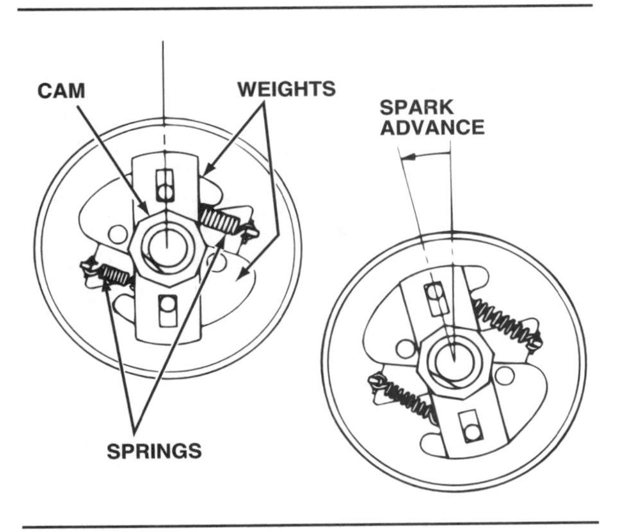

Figure 10-2. When the centrifugal advance weights move, the position of the cam, or trigger wheel, and the rotor changes.

emissions and the best possible fuel economy.

When ignition takes place too early, or is too far advanced, the increased combustion pressure slows down the piston, and may cause engine knock. When ignition takes place too late, the piston is too far down on its power stroke to benefit from the combustion pressure, resulting in a power loss.

Before the introduction of computer-controlled timing, most automotive distributors had two spark advance mechanisms to react to engine operating changes and alter ignition timing:

- The centrifugal advance changed ignition timing to match engine speed by altering the position of the distributor cam or trigger wheel on the distributor shaft.
- The vacuum advance changed ignition timing to match engine load by altering the position of the breaker points or the magnetic pickup coil (electronic sensor).

These changes in position altered the time, relative to crankshaft position, at which the primary circuit was opened.

Centrifugal Advance—Speed

The **centrifugal advance**, or **mechanical advance**, mechanism consists of two weights connected to the distributor driveshaft by two springs, figure 10-1. The distributor cam, or electronic trigger wheel, and the distributor rotor are mounted on another shaft. The second shaft fits over the driveshaft like a sleeve. Driveshaft motion is transmitted to the second shaft through the centrifugal advance weights. When the weights move, the relative position of the driveshaft and the second shaft changes.

As engine speed increases, distributor shaft rotation speed increases. The advance weights move outward because of **centrifugal force**. The outward movement of the weights shifts the second shaft and the cam or trigger wheel, figure 10-2. The primary circuit opens earlier in the compression stroke, and the spark occurs earlier.

Each advance weight is connected to the distributor driveshaft by a control spring. These springs are selected to allow the correct amount of weight movement and ignition advance for a particular engine.

At low engine speeds, spring tension holds the weights in, so that initial timing is maintained. As engine speed increases, centrifugal force overcomes spring tension and the weights move outward. The advance is not a large, rapid change, but rather a slow, gradual shift. Figure

Centrifugal (Mechanical) Advance: A method of advancing the ignition spark using weights in the distributor that react to centrifugal force.

Centrifugal Force: A force exerted by a rotating object that moves it away from the center of rotation.

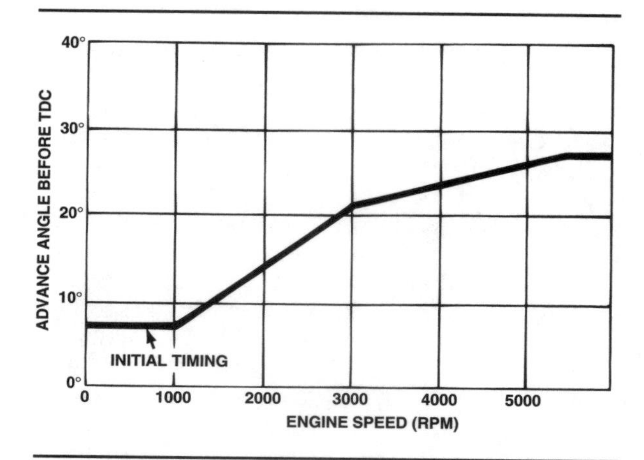

Figure 10-3. A typical advance curve showing a distributor's centrifugal advance.

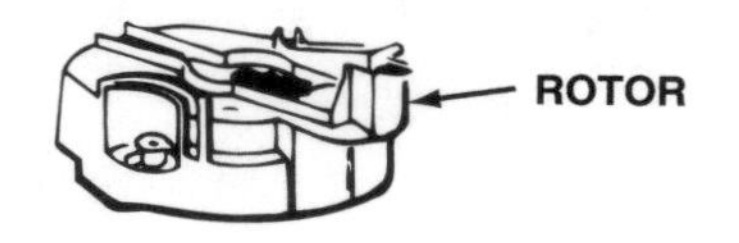

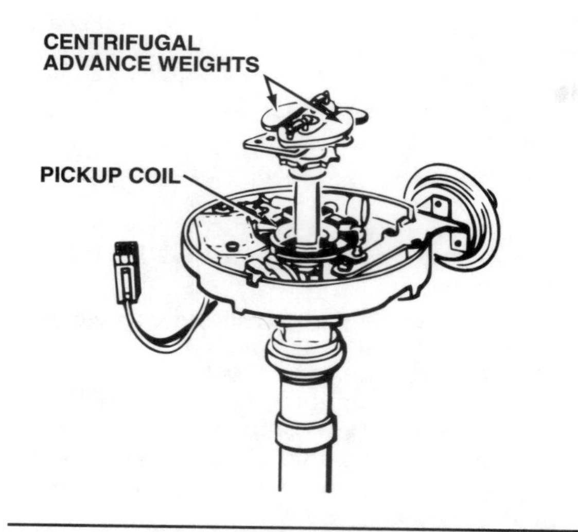

Figure 10-4. Delco-Remy V6 and V8 distributors have the centrifugal advance mechanism above the cam and rotor, or above the pickup coil and trigger wheel.

10-3 shows a typical centrifugal advance curve. The advance curve can be changed by changing the tension of the control springs. Remember that centrifugal advance responds to engine speed.

In most distributors, the centrifugal advance mechanism is mounted below the cam and breaker points, figure 10-1, or below the trigger wheel and pickup coil. Delco-Remy V6 and V8 distributors, as well as some Japanese distributors, have the advance mechanism above the cam or trigger wheel, just below the rotor, figure 10-4.

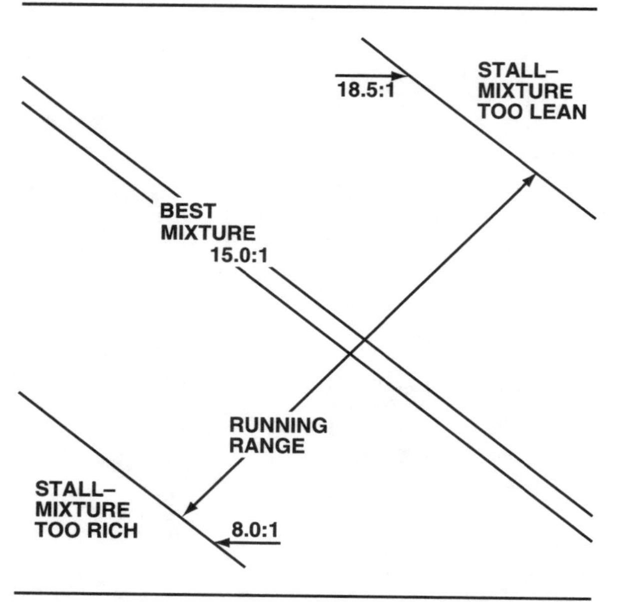

Figure 10-5. Air-fuel ratio limits for a four-stroke gasoline engine. (Chevrolet)

Vacuum Advance—Load

The **vacuum advance** mechanism allows efficient engine performance within a range of air-fuel ratios. These ratios are important, since there are limits to how rich or lean they can be and still remain fully combustible. The air-fuel ratio with which an engine can operate efficiently ranges from 8:1 to 18.5:1 by weight, figure 10-5.

These ratios are generally stated as eight parts of air combined with one part of gasoline (8:1), which is the richest mixture that most engines can tolerate and still fire regularly. A ratio of 18.5 parts of air mixed with 1 part of gasoline (18.5:1) is the leanest mixture that most engines can tolerate without misfiring.

An average air-fuel ratio is about 15:1. This mixture takes about three milliseconds to burn. A lean mixture (one with more air and less fuel) requires more time to burn. The ignition timing must be advanced to provide maximum combustion pressure at the correct piston position. A rich mixture (one with more fuel and less air) burns more quickly and emits more exhaust pollutants. As the air-fuel mixture richens, ignition timing should retard for complete combustion and emission control.

Engine Vacuum

The reciprocating engine can be considered an air pump. As a piston moves downward, air pressure in the cylinder decreases. Air from the atmosphere rushes in to fill the void.

The fuel delivery system uses this air movement to carry fuel to the cylinders. On older

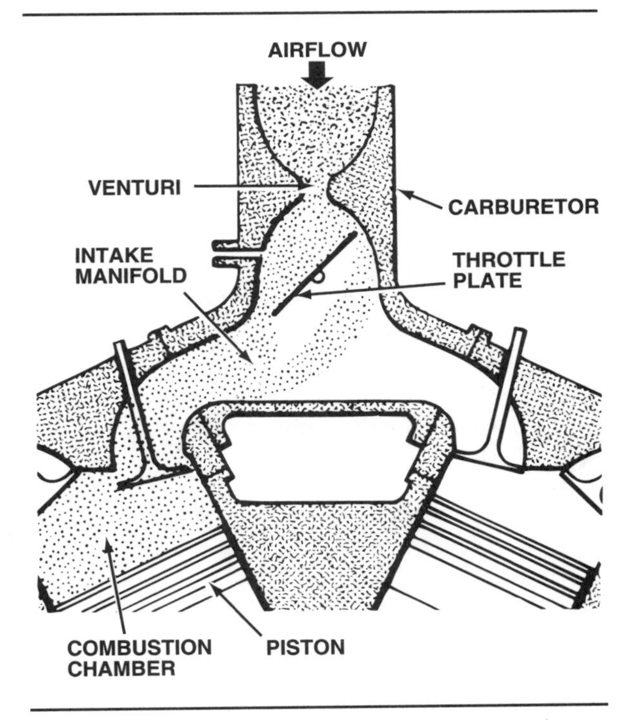

Figure 10-6. Air flows through the carburetor and intake manifold to reach the combustion chamber.

cars, the air must travel through a carburetor, figure 10-6, to reach the cylinders. Under several operating conditions, air movement caused by the downstroke of a piston is not great enough to draw fuel into the cylinder. The carburetor forces the air to flow through a restriction called a **venturi**. This increases the speed of the airflow and creates a low-pressure (vacuum) area. Fuel is drawn into the airflow by the vacuum, and the resulting air-fuel mixture enters the cylinder combustion chambers. The air-fuel ratio changes as the **vacuum** in the carburetor changes.

The ignition timing must be changed as the air-fuel ratio changes to allow enough time for the air-fuel mixture to burn. The vacuum advance mechanism connects to a small hole or port in the carburetor, figure 10-7, just above the throttle plate. This is called **ported vacuum**. When vacuum exists at the port, timing advances. When no vacuum exists at the port, the timing remains at a basic setting or is affected only by the centrifugal advance. The vacuum advance mechanism will be explained in more detail later. The following paragraphs explain the relationship between carburetor vacuum and the need for advanced timing.

The driver of a carburetor-equipped automobile controls engine load and carburetor vacuum through changing the position of the throttle plate, which varies the size of the restriction through the carburetor's bore. When the engine is at idle, figure 10-8, the throttle plate is almost closed. Very little air flows through the carburetor to mix with the fuel. With this rich air-fuel mixture, no spark advance is necessary. High vacuum exists in the intake manifold, but there is no vacuum at the port because it is above the closed throttle plate. The high vacuum present below the carburetor throttle is called **manifold vacuum.** There is no vacuum-controlled spark advance with a closed throttle.

When the throttle plate is partially open, figure 10-9, more air can flow through the carburetor. The air-fuel mixture becomes lean and requires

Vacuum Advance: The use of engine vacuum to advance ignition spark timing by moving the distributor breaker plate.

Venturi: A restriction in an airflow, such as in a carburetor, that increases the airflow speed, and creates a reduction in pressure.

Vacuum: A pressure less than atmospheric pressure.

Ported Vacuum: Vacuum immediately above the throttle plate in a carburetor.

Manifold Vacuum: Low pressure in the intake manifold below the carburetor throttle.

■ Air Pressure High and Low

You can think of an internal combustion engine as a big air pump. As the pistons move up and down in the cylinders, they pump in air and pump out the burned exhaust. They do this by creating a difference in air pressure. The air outside an engine has weight and exerts pressure, as does the air inside an engine.

As a piston moves down on an intake stroke with the intake valve open, it creates a larger area inside the cylinder for the air to fill. This lowers the air pressure inside the engine. Because the pressure inside the engine is lower than the pressure outside, air will flow in through the carburetor to try to fill the low-pressure area and equalize the pressure.

We call the low pressure in the engine "vacuum". You can think of vacuum as pulling air into the engine, but it is really the higher pressure on the outside that forces air into the low-pressure area inside. The difference in pressure between two areas is called a "pressure differential". The pressure differential principle has many applications in an automobile engine.

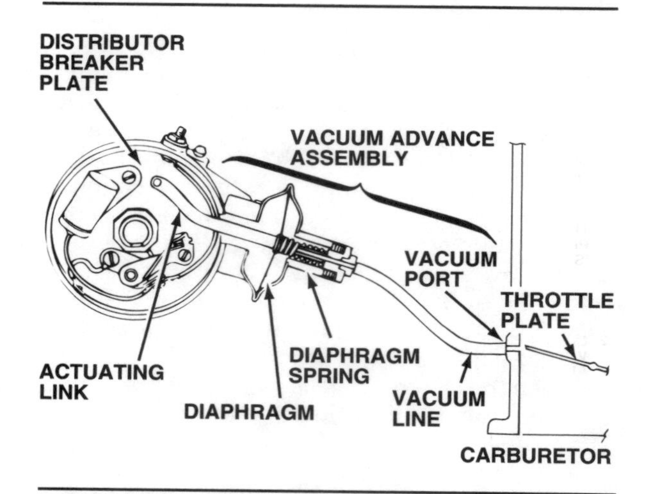

Figure 10-7. The vacuum advance assembly is connected to a port in the carburetor.

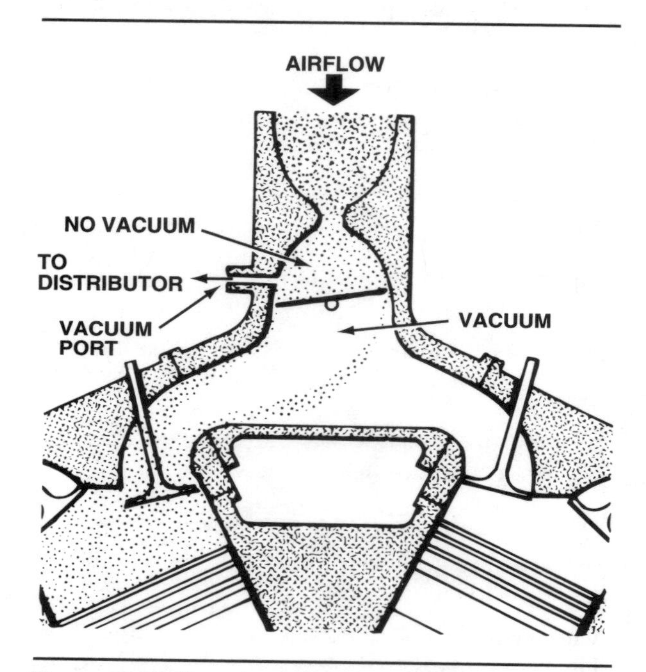

Figure 10-8. When the throttle is closed during idle or deceleration, there is no vacuum at the port.

an advanced spark. Since the port is now exposed to vacuum, ignition timing advances.

At medium cruising speeds, the engine operates with a partialy open throttle and an air-fuel ratio of approximately 15:1. Airflow velocity is high, and the vacuum signal at the carburetor port is strong enough to provide vacuum advance for this relatively lean air-fuel mixture.

A lean mixture is not the only factor that affects ignition timing requirements. During partial-throttle operation, the cylinders are only partially filled with the air-fuel mixture. Because

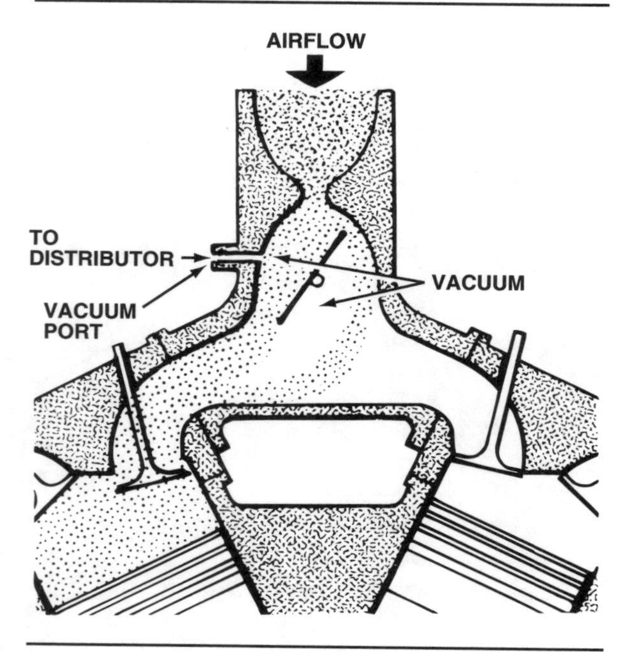

Figure 10-9. When the throttle is partially open, the port is exposed to manifold vacuum.

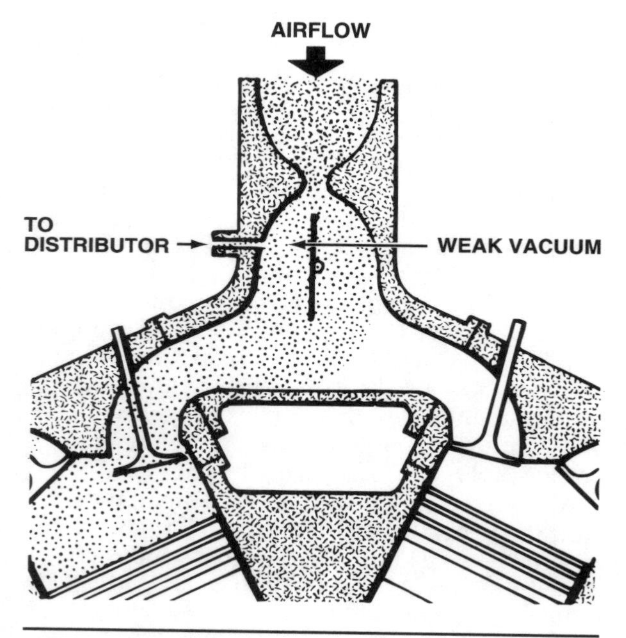

Figure 10-10. When the throttle is fully opened, the vacuum at the port is too weak to cause any vacuum advance.

there is less to compress, the compression pressure is less. A less highly compressed mixture takes a longer time to burn.

At wide-open throttle, figure 10-10, the power circuit in the carburetor provides a richer mixture than at partial-throttle cruising. This factor

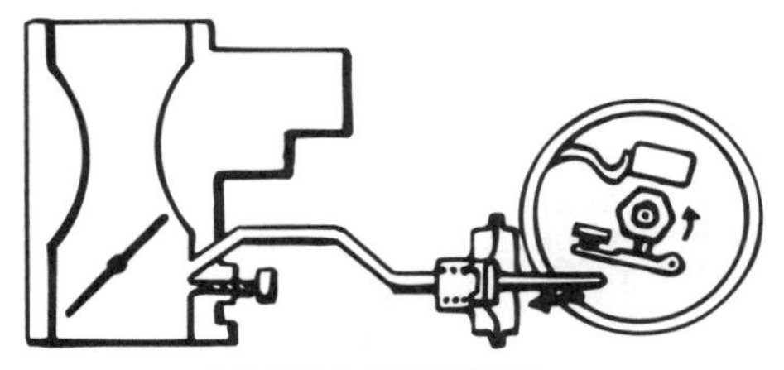

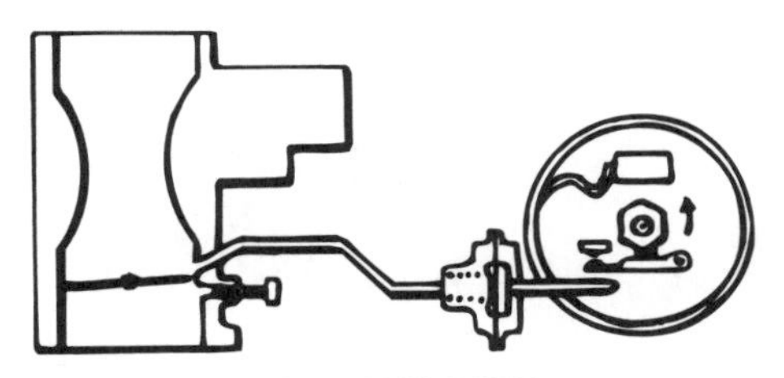

Figure 10-11. Vacuum at the carburetor port causes the position of the breaker plate to shift.

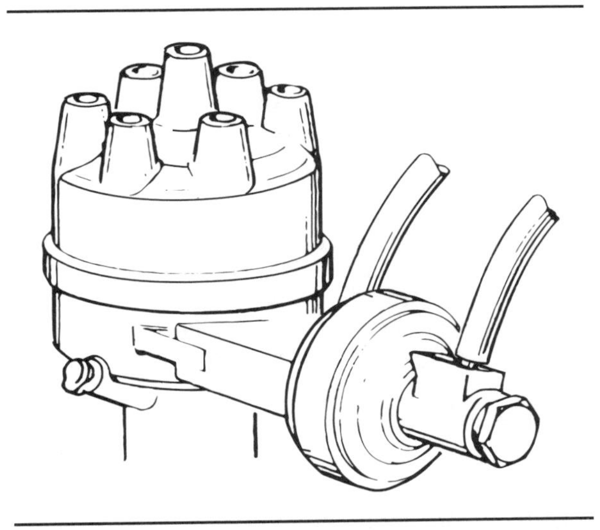

Figure 10-12. Dual-diaphragm vacuum advance units are easily recognized.

causes a faster burn that requires the ignition timing to retard or decrease slightly for proper efficiency. Manifold vacuum and ported vacuum drop at wide-open throttle and become approximately equal. Vacuum may drop to as little as 5 in-Hg (127 mm-Hg), and typically 7 in-Hg (178 mm-Hg) are required to operate a vacuum advance mechanism.

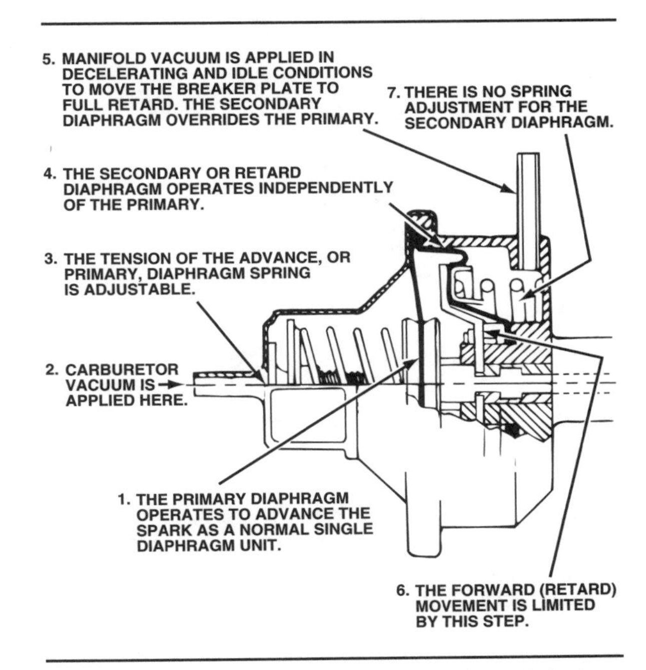

Figure 10-13. Operation of an advance/retard dual-diaphragm vacuum advance assembly. (Ford)

Vacuum advance also decreases when the throttle is opened quickly from idle or a partial-throttle position. This occurs because airflow velocity lags behind throttle opening. This reduced vacuum will not draw enough fuel through the carburetor's main metering circuit, so the accelerator pump supplies extra fuel. This momentarily rich mixture does not need as much spark advance for complete combustion. Because the vacuum at the carburetor's vacuum port is low when the throttle is first opened quickly, the vacuum advance decreases or retards to meet the needs of the momentarily rich mixture. Remember that vacuum advance responds to engine load.

Vacuum advance mechanism

The vacuum advance mechanism at the distributor, figure 10-7, consists of the:

- Movable breaker plate on which the points or electronic pickup coil are mounted
- Vacuum assembly, a housing with a flexible diaphragm and a spring
- Actuating link that connects the vacuum diaphragm to the breaker plate
- Tubing to connect the vacuum unit to the vacuum source.

When the **diaphragm** is pulled toward a vacuum, it pulls the actuating link, figure 10-11. This rotates the breaker plate to change the position of the point rubbing block in relation to the cam. Ignition timing increases or advances. The pick-up generator in electronic systems is shifted

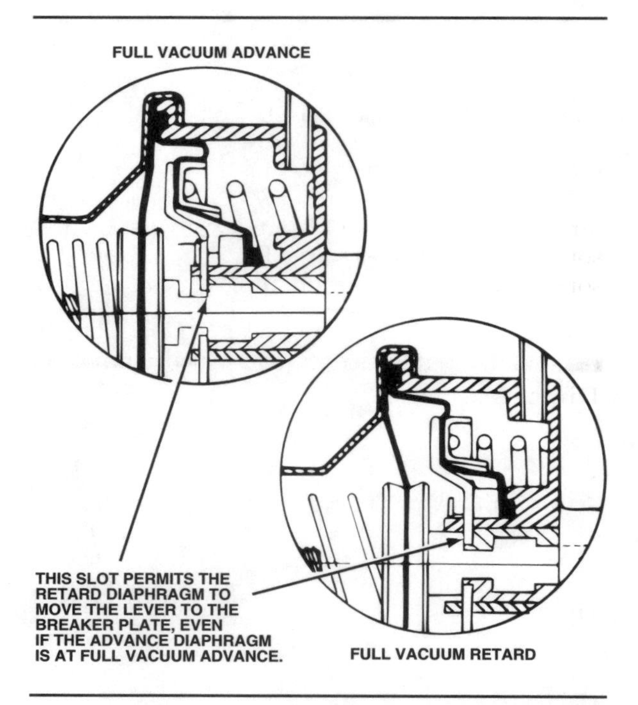

Figure 10-14. In an advance/retard dual-diaphragm unit, the retard diaphragm can override the action of the advance diaphragm. (Ford)

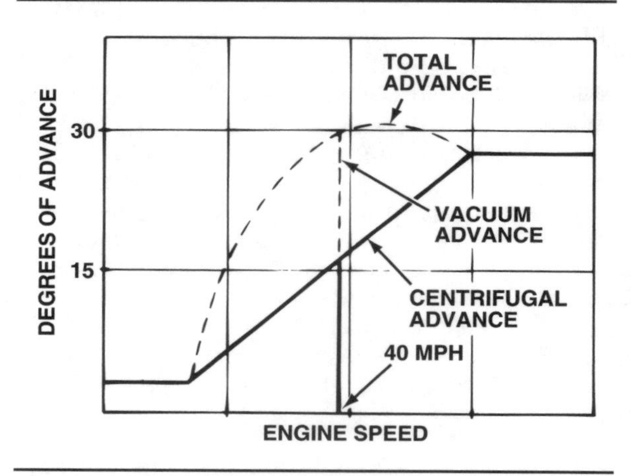

Figure 10-15. This typical advance curve shows that total advance equals the sum of vacuum plus centrifugal advances.

relative to the trigger wheel to advance timing.

Some distributors have a dual-diaphragm vacuum unit, figure 10-12. Dual-diaphragm distributor units can either use one diaphragm for advance and the other for retard, figure 10-13, or use one for low advance and the other for high advance. The advance/retard unit uses ported vacuum to advance the timing and manifold vacuum to allow the timing to retard further in the absence of ported vacuum. These vacuum

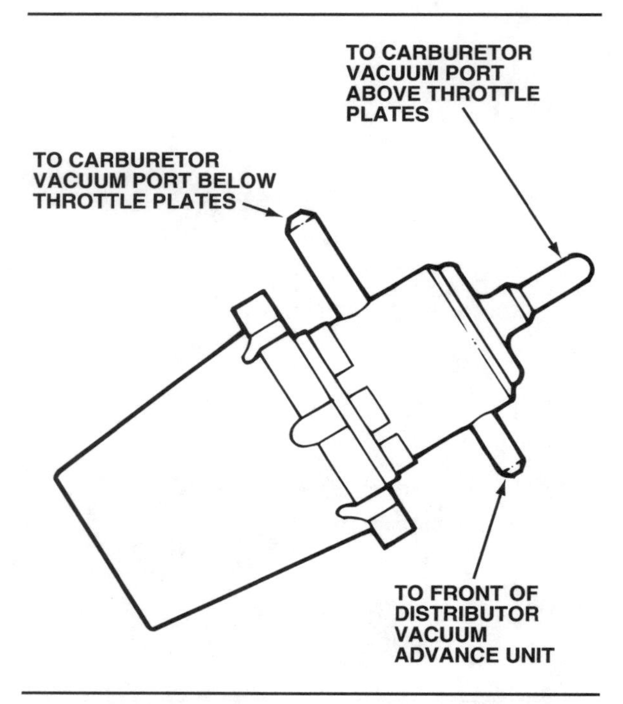

Figure 10-16. A deceleration vacuum advance valve.

advance units retard timing during idle and closed-throttle deceleration.

For exhaust emission control, the retard diaphragm can override the advance diaphragm. This causes timing to retard even when the advance diaphragm is in an advanced position, figure 10-14.

The dual diaphragm unit using two levels of advance uses manifold vacuum through a thermal vacuum switch to advance the timing on cold engines for better driveability. Ported vacuum, the second level of advance, is typically used to advance the timing during driving under load.

Total Ignition Advance

The two types of advance mechanisms we have described work on different parts of the distributor to advance timing. Therefore, their effects are additive: the **total ignition advance** is the sum of the centrifugal advance and the vacuum advance, *plus* the initial ignition timing. This formula determines actual ignition timing under all conditions. Figure 10-15 shows the advance curves of a typical ignition system.

EARLY SPARK-TIMING EMISSION CONTROLS

Spark-timing control systems were first introduced in the late 1960s to reduce exhaust emissions. Certain types of pollutants are produced under specific engine operating conditions.

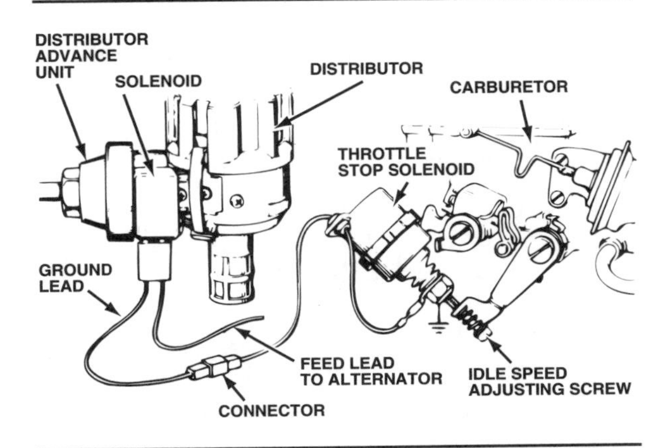

Figure 10-17. A distributor vacuum retard solenoid.

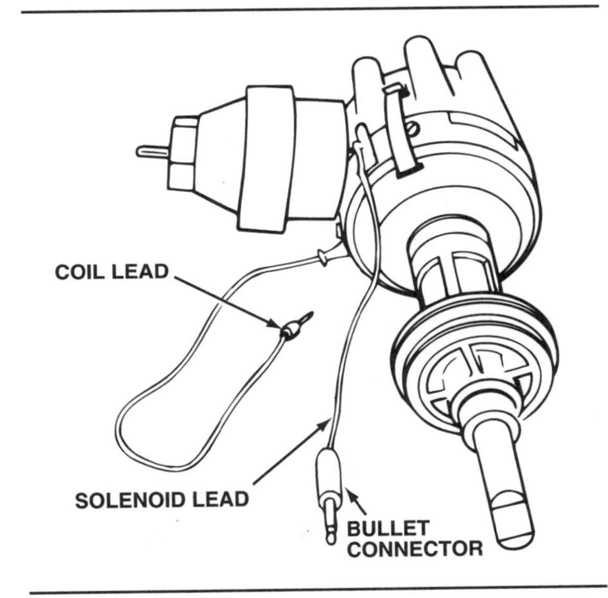

Figure 10-18. The installation of a distributor advance solenoid.

Spark-timing control systems help reduce **hydrocarbons (HC)** and **oxides of nitrogen (NO_x)** emissions. Retarded timing reduces combustion temperature to help reduce NO_x formation. At the same time, higher temperatures are created toward the end of combustion. This results in higher exhaust temperatures, which reduce the amount of HC in the exhaust.

Early Distributor Controls

Early emission control equipment advanced or retarded ignition timing under particular engine operating conditions, usually during starting, deceleration, and idle.

The deceleration vacuum advance valve, figure 10-16, was used during the mid-to-late 1960s on Chrysler, Ford, and Pontiac products with manual transmissions. In a manual-transmission car, the air-fuel mixture becomes extremely rich when decelerating or shifting gears.

This valve momentarily switches the vacuum for the vacuum advance from a low-vacuum source at the carburetor to a high-vacuum source during deceleration, then back to the

Diaphragm: A thin flexible wall separating two cavities, such as the diaphragm in a vacuum advance unit.

Total Ignition Advance: The sum of centrifugal advance, vacuum advance, plus initial timing; expressed in crankshaft degrees.

Hydrocarbon (HC): A major pollutant containing hydrogen and carbon produced by interrnal combustion engines. Gasoline is a hydrocarbon compound.

Oxides of Nitrogen (NO_x): Chemical compounds of nitrogen given off by an internal combustion engine. NO_x combines with hydrocarbons to produce ozone, a primary component of smog.

Does a Vacuum Pull Air In?

As you study automotive service, you see many applications for vacuum devices. The engine is a large air pump and the source of vacuum used to operate a variety of accessories. Vacuum diaphragms control ignition advance, exhaust gas recirculation (EGR) operation, carburetor chokes and power valves, air-conditioning doors and vents, heater water valves, concealed headlamp mechanisms, power brake boosters, cruise control servos, and many other devices.

It is very easy—and practical for day-to-day use—to think of vacuum as an independent force; it is not. Vacuum is simply air pressure that is lower than atmospheric pressure. Vacuum does not actually pull air in; air pressure pushes. We talk about connecting a diaphragm to a "vacuum source", but what actually moves the diaphragm? Low pressure (vacuum) on one side allows higher atmospheric pressure on the other to move the diaphragm.

Similarly, we often talk about finding and fixing a "vacuum leak". That concept works just fine for on-the-job troubleshooting, but ask yourself, "Can vacuum actually *leak?*" Not really. What you are actually fixing is an air leak. An unwanted opening allows air to leak *into* the low-pressure area that we call a vacuum.

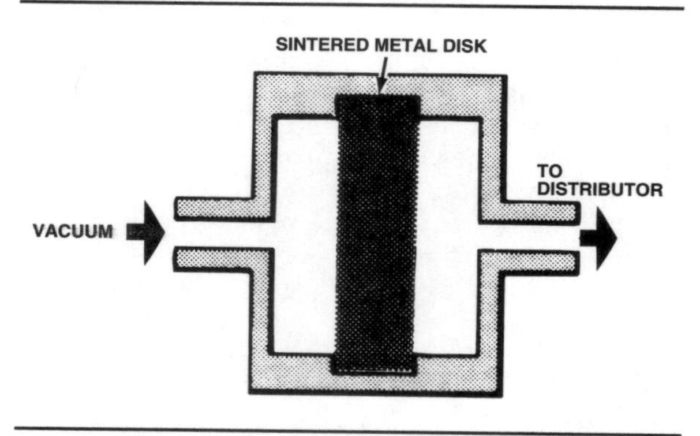

Figure 10-19. A cross-sectional view of a vacuum delay valve containing a sintered metal disc to slow the application of vacuum.

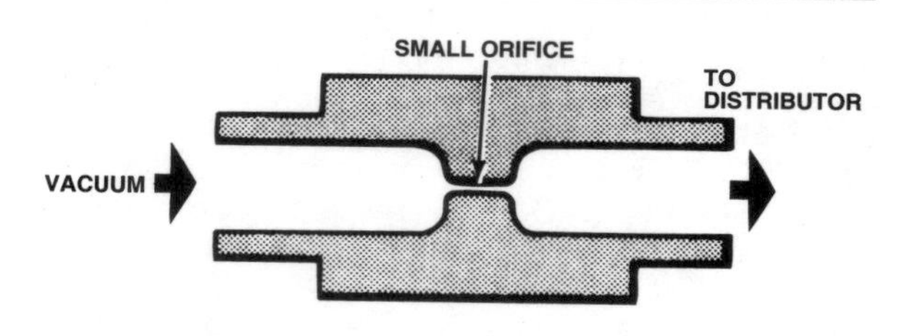

Figure 10-20. A cross-sectional view of a vacuum delay valve using a small orifice to delay the application of vacuum.

low-vacuum source. This prevents overly retarded timing during deceleration or gear shifting, which could cause some engines to emit a lot of **carbon monoxide (CO)**.

While the deceleration vacuum advance valve was effective against CO emissions, it did not limit HC and NO_x emissions. As emission limits for these pollutants became tighter in the early 1970s, use of the device ceased and manufacturers developed other devices that worked against all three major pollutants.

PRINCIPAL SPARK-TIMING CONTROLS OF THE 1970s

These systems were designed primarily to delay vacuum advance at low and intermediate speeds, and to allow advance during high-speed cruising. The most common types of control systems used in the early to mid-1970s are:

- Distributor solenoids
- Vacuum delay valves
- Speed- and transmission-controlled timing.

Distributor Solenoids

A distributor vacuum retard solenoid was used on some 1970–71 Chrysler products with V8 engines and automatic transmissions. This electric solenoid is attached to the distributor and controls the action of the distributor vacuum advance unit, figure 10-17.

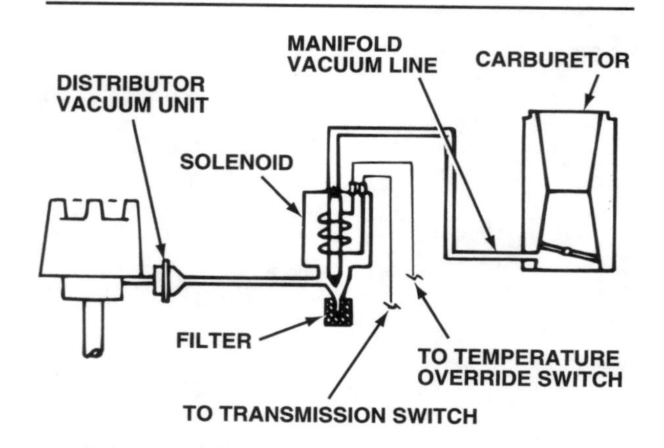

Figure 10-21. A simplified transmission-controlled spark system.

The solenoid is energized by contacts mounted on a carburetor throttle stop solenoid. When the throttle closes, the idle adjusting screw contacts the carburetor solenoid to complete the ground circuit. The contacts in the carburetor solenoid carry current to the distributor solenoid windings. Since the distributor solenoid plunger is connected to the vacuum diaphragm, solenoid movement shifts the breaker plate in the retard direction.

When engine speed increases, the idle adjusting screw breaks contact with the carburetor solenoid. Current to the distributor solenoid is stopped, and normal vacuum advance is allowed.

Some 1972–73 Chrysler V8 distributors had a spark timing advance solenoid that promoted better starting by providing a 7.5-degree spark advance. The solenoid is mounted in the distributor vacuum unit, figure 10-18, and is activated by power from the starter relay at the same terminal that sends power to the starter solenoid. The solenoid activates only while the engine is cranking.

The starting advance solenoid is not an emission control device by itself, but allows lower basic timing settings, which helps control NO_x emissions while providing advanced timing for quicker starting.

Vacuum Delay Valves

The vacuum delay valve "filters" the carburetor vacuum, slowing its application to the distributor vacuum advance unit. Generally, vacuum must be present in the system for 15 to 30 seconds before it is allowed to affect the advance mechanism.

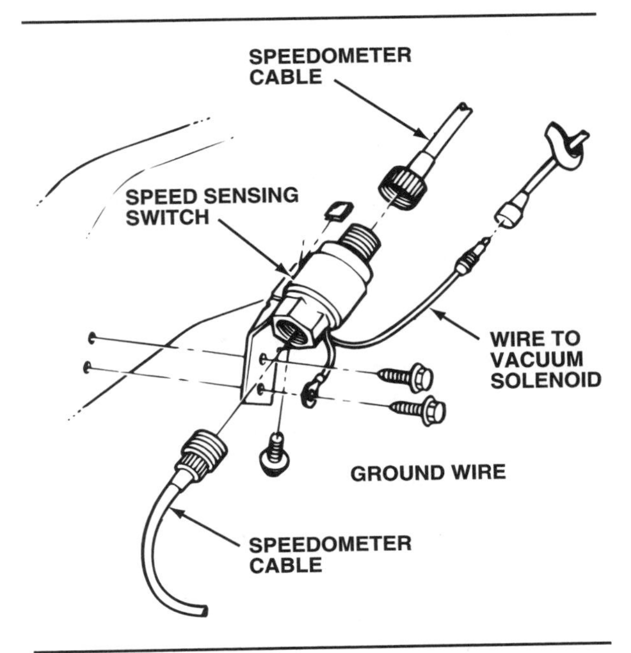

Figure 10-22. Most speed-controlled spark systems use a speed-sensing switch. (Cadillac)

One method of vacuum delay was used in Ford's spark delay valve (SDV) system, figure 10-19. In this design, vacuum must work its way through a **sintered**, or sponge-like, metal disk to reach the distributor. Many GM engines also used this type of spark delay valve.

Another method of vacuum delay was used in Chrysler's orifice spark advance control (OSAC) system, figure 10-20. A small **orifice** is placed in the vacuum line to delay vacuum buildup.

Manufacturers have often combined the use of vacuum delay valves with other emission control systems. All valves operate on one of the two principles just described.

Speed- and Transmission-Controlled Timing

These systems prevented any distributor vacuum advance when the vehicle is in a low gear or is traveling slowly. A solenoid controls the application of vacuum to the advance mechanism, figure 10-21. Current through the solenoid is controlled by a switch that reacts to various vehicle operating conditions.

A control switch used with a manual transmission reacts to shift lever position. A control switch used with an automatic transmission will usually react to hydraulic fluid pressure. Both systems prevent any vacuum advance when the car is in a low or intermediate gear.

A speed-sensing switch may be connected to the vehicle speedometer cable, figure 10-22. The switch signals an electronic control module when vehicle speed is below a predetermined level. The module triggers a solenoid that controls engine vacuum at the distributor.

Both vacuum-delay systems and speed- and transmission-controlled systems usually have an engine temperature bypass. This allows normal vacuum advance at high and low engine temperatures. Before March 1973, some systems had

Carbon Monoxide (CO): An odorless, colorless, tasteless poisonous gas. A pollutant produced by an internal combustion engine.

Sintered: A porous material welded together without using heat, such as the metal disk used in some vacuum delay valves.

Orifice: A small opening in a tube, pipe, or valve.

■ Starting the Locomobile

If you owned a 1927 or 1928 Locomobile, you had to know how to adjust the spark advance lever when you started your car. The spark lever allowed you to advance or retard spark timing for all conditions. Here are the owner's manual instructions for positioning the lever when starting your Locomobile:

1. Be sure the gear shift lever is in the neutral position, and that the hand brake is applied.
2. Retard the spark lever, which is at right center of steering wheel, about half way down from the full advance position.
3. The throttle control lever is to the left of center of the steering wheel. Move the lever up an inch or more from the closed position.
4. Turn the ignition switch arm to the "on" position.
5. Pull choke button as far as it will go, then place the right foot on the starting switch button and press down. On pushing down on the starting switch button, the starter gear will engage itself in the teeth of the flywheel, and the motor will start to turn over. As soon as the motor starts, release the starting button, and push the choke button part-way in. As the motor warms up a little, push the choke button all the way in. Do this as soon as possible to avoid excessive choking, as any excess gasoline thrown into the cylinders will pass by the rings into the oil in the crankcase, thus diluting and partly destroying its lubricating value.
6. Advance the spark by pushing the lever up as far as it will go and bring the throttle lever toward the closed position to slow down the motor.

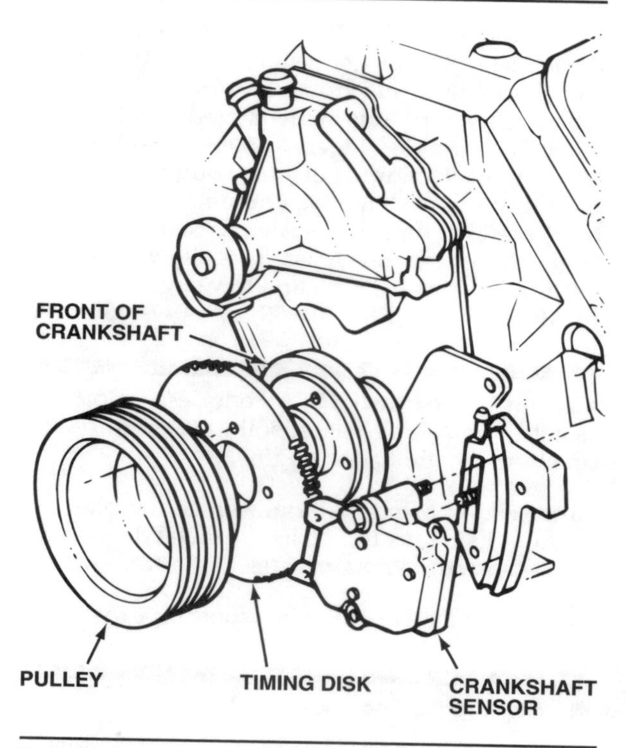

Figure 10-23. Crankshaft position signals can be taken directly from the crankshaft.

an **ambient temperature** override switch. Most of these switches were discontinued at the direction of the U.S. Environmental Protection Agency (U.S. EPA). Later, temperature-override systems sensed engine coolant temperature or underhood temperature.

DECLINE AND FALL OF EARLY SPARK-TIMING CONTROLS

Many spark-timing control systems were discontinued when oxidation catalytic converters were introduced in 1975. These converters reduce HC and CO emissions, and exhaust gas recirculation (EGR) systems provide more effective control of NO_x.

Emission standards continued to tighten during the mid-1970s. During this period, fuel mileage requirements also became important. Meeting the challenge of more stringent emission standards and better fuel economy required more accurate ignition system performance than mechanical devices could deliver. Centrifugal and vacuum advance mechanisms simply could not respond quickly enough to changes in engine operating conditions to provide the necessary accuracy, so major manufacturers turned to computer-controlled ignition systems.

EARLY ELECTRONICALLY CONTROLLED TIMING

In a computer-controlled ignition system, an electronic control module receives signals from various system sensors. These signals may include information on coolant temperature, atmospheric pressure and temperature, throttle position and rate of change of position, and crankshaft position. The central processing unit (CPU) in the control module is programmed to interpret this information and calculate the proper ignition timing for each individual spark.

Early systems worked with the manufacturers' existing solid-state ignition systems. Some changes were made to these ignition systems because they no longer had to control spark timing, but many components remained the same.

Early computer-controlled ignition systems used by domestic manufacturers can be divided into two types. In one type, rotation of the distributor shaft sends a crankshaft position signal to the control module. The other type receives crankshaft position information from a sensor mounted near the crankshaft, figure 10-23. The sensor reacts to a trigger attached to the crankshaft.

Signals taken directly from the crankshaft are more accurate than those taken from the distributor shaft. The gears or chain driving the camshaft and the gears driving the distributor shaft are manufactured within tolerances. Although the actual measurements are small, these tolerances can combine to cause a significant difference between crankshaft position and optimal ignition timing.

Early systems used by Chrysler, GM, and Ford were called:

- Chrysler Electronic Lean-Burn (ELB)
- GM Microprocessed Sensing and Automatic Regulation (MISAR)
- Ford Electronic Engine Control (EEC).

These early electronic timing control systems were partial-function engine control systems. They controlled ignition timing only. They were not fully integrated systems that controlled fuel metering with feedback signals from various engine sensors. Nevertheless, these early systems were ancestors of the more recent engine control systems that we will study in a later chapter.

The electronic timing-regulation function of all three systems is similar, but the electronics are fundamentally different. The ELB system uses an analog computer, while MISAR and EEC use digital microprocessors.

The practical difference between analog and digital electronics in this kind of application is that a digital computer can instantly alter timing from 1 to 65 degrees. An analog computer must calculate through all the points on a theoretical curve to make such an adjustment. Since an electronic spark advance adjustment takes only a few milliseconds, this fact is not really significant to the driver or technician. However, a digital system is more flexible and economical to build than an analog system.

SUMMARY

Ignition distributors have used two devices to advance or retard spark timing in response to changing conditions of engine operation. The two primary factors that determine spark advance are engine speed and load. The centrifugal advance alters the position of the distributor cam or trigger wheel on the distributor shaft and changes engine timing as engine speed changes. The vacuum advance alters the position of the breaker points or magnetic pickup coil with respect to the cam or trigger wheel. It changes timing as engine load changes. Many different spark timing emission control systems have been used. Regardless of their design and operation, these systems all regulated distributor vacuum advance. Vacuum advance is generally allowed only at cold startup, during high gear operation, or if the engine overheats.

The development of EGR systems and oxidation catalytic converters eliminated the use of transmission-controlled and speed-controlled spark systems. Electronic spark timing systems appeared because more accurate control was required. They regulate ignition timing for the best combination of emission control, fuel economy, and driveability. Early electronic spark timing systems were the forerunners of late-model, fully integrated electronic engine control systems that we will study in a later chapter.

Ambient Temperature: The temperature of the air surrounding a particular device or location.

■ Vacuum Measurement

When you work with vacuum devices on automobiles, you will encounter several different units of measurement used to gauge vacuum. The auto industry customarily has measured air pressure in pounds per square inch (psi) or kilopascals (kPa), and vacuum in inches of mercury (in-Hg) or millimeters of mercury (mm-Hg). This is confusing because we are using two different kinds of units to measure essentially the same thing—air pressure. The reason we use both is based on scientific tradition.

Air pressure and vacuum are measured in a laboratory with a device called a manometer. This is a U-shaped glass tube with each end connected to different pressure sources. One end can be opened to the atmosphere and the other can be connected to a pump or another source of low pressure. The tube is filled with liquid that moves up and down the two columns formed by the two legs of the "U". Using two columns joined in a "U" eliminates the effects of gravity and the weight of the liquid. Displacement caused by changes in pressure is read in marks graduated on the columns.

Laboratory manometers ordinarily are filled with mercury because it is stable, and flows freely when exposed to pressure differentials. When low pressure on one side drops enough to move the mercury column 1 inch (25.4 mm), vacuum (low pressure) equals 1 in-Hg (25.4 mm-Hg). When atmospheric pressure is removed completely from one side of the manometer, the mercury column is displaced 29.9 inches at 32°F (0°C). Thus, 29.9 in-Hg equals one atmosphere of negative pressure, or 14.696 psi. When the manometer column is graduated in millimeters instead of inches, one atmosphere of displacement equals 760 mm-Hg.

Another unit you will encounter is the "bar". A bar equals 0.99 of atmospheric pressure (ATM). This is barometric pressure, and that is where we get the term "bar". A standard bar equals one kilogram of force applied to one square centimeter. This equals 750 mm-Hg, or 14.5 psi. These values are close to the customary atmospheric pressure values of 760 mm-Hg and 14.7 psi. All standard pressures are calculated at 32°F (0°C) because pressure drops as temperature rises, and vice versa.

In automobile service, we work with psi and kPa to measure positive pressure. The most common units of vacuum measurement are "in-Hg", "mm-Hg", and "bar". Here are some handy conversion factors you can use to switch from one unit to another:

1 ATM = 14.7 psi = 29.92 in-Hg
1 in-Hg = 0.4912 psi = 3.3864 kPa
1 bar = 100 kPa = 14.5 psi
1 kPa = 0.1450 psi = 0.2953 in-Hg

All these equivalent measurements are at a standard temperature of 32°F (0°C), but the conversions are close enough for car service work.

Review Questions

Choose the single most correct answer.
Compare your answers to the correct answers on page 507.

1. Distributors commonly have two automatic spark advance mechanisms:
   a. Initial advance and mechanical advance
   b. Initial advance and dynamic advance
   c. Dynamic advance and static advance
   d. Centrifugal advance and vacuum advance

2. Centrifugal advance responds directly to engine:
   a. Load
   b. Horsepower
   c. Speed
   d. Torque

3. The purpose of the venturi in the carburetor is to:
   a. Mix the air and fuel mixture
   b. Speed the airflow and create a low-pressure area
   c. Carry fuel to mix with air
   d. Raise the air pressure sufficiently

4. Vacuum advance responds directly to engine:
   a. Load
   b. Horsepower
   c. Speed
   d. Torque

5. The richest air-fuel ratio with which an engine can operate efficiently is:
   a. 4:1
   b. 6:1
   c. 8:1
   d. 16:1

6. At medium cruising speeds, the engine operates with a partially open throttle and an air-fuel ratio of about:
   a. 8:1
   b. 12:1
   c. 15:1
   d. 18.5:1

7. Which of the following is not a component of the vacuum advance mechanism?
   a. Movable breaker plate
   b. Throttle valve
   c. Diaphragm
   d. Actuating link

8. Automotive emission control equipment is designed to reduce exhaust emissions of which pollutants?
   a. Hydrogens
   b. Nitrogens
   c. Carbons
   d. Hydrocarbons

9. Almost all spark-timing emission control systems work with the vacuum advance units to:
   a. Increase vacuum advance at low and intermediate vehicle speeds
   b. Cut off vacuum advance during high-speed acceleration
   c. Retard the timing during high-speed deceleration
   d. Cut off vacuum advance at low speeds

10. The dual-diaphragm vacuum unit can provide:
    a. Two levels of vacuum advance
    b. Faster vacuum retard than a single-diaphragm unit
    c. Both vacuum advance and vacuum retard
    d. Either a or c

Chapter

11

Solid-State Ignition Systems

During the 1960s, solid-state ignition (SSI) systems were used only on a few high-performance engines. SSI's were not standard on domestic automobiles until the early 1970s; foreign automakers followed a few years after with their versions. Within less than a decade, SSI's completely replaced breaker-point ignitions.

Breaker-point and electronic ignition systems do the same thing, except that a breaker-point system switches the primary circuit mechanically while an electronic system does so by means of a transistor. The major driving force behind the rapid transition to electronic systems was their ability to help meet stringent emission control standards.

Although SSI's are more expensive to produce than breaker-point systems, their advantages greatly outweigh the drawback of increased cost:

- Greater available voltage, especially at high engine speeds
- Reliable system performance at all engine speeds
- Potential for more responsive and variable advance curves
- Decreased maintenance.

In addition, engine operation and exhaust emissions are more accurately controlled by solid-state circuitry.

BASIC SOLID-STATE SYSTEMS

SSI's differ from breaker-point systems in the devices used to control the primary circuit current. Breaker-point systems use mechanical breaker points to open and close the primary circuit. The points are operated by a cam on the distributor shaft. Most SSI's use an electronic switch in the form of a sealed **ignition control module** (ICM) to control primary circuit current. The ICM contains one or more transistors, integrated circuits, or other solid-state control components. A triggering device in the distributor functions with the ICM. This triggering device is usually a magnetic pulse generator or a Hall-effect switch. SSI's also can be triggered by a set of breaker points, explained later in this chapter.

SSI's can be classified in terms of their primary circuit operation. There are two general types:

- Inductive discharge
- Capacitive discharge.

The difference between these types lies in how they use primary current to produce a high-voltage secondary current.

Inductive Discharge

The inductive discharge system uses battery voltage to create current through the coil primary winding. When a signal is sent by the trigger-

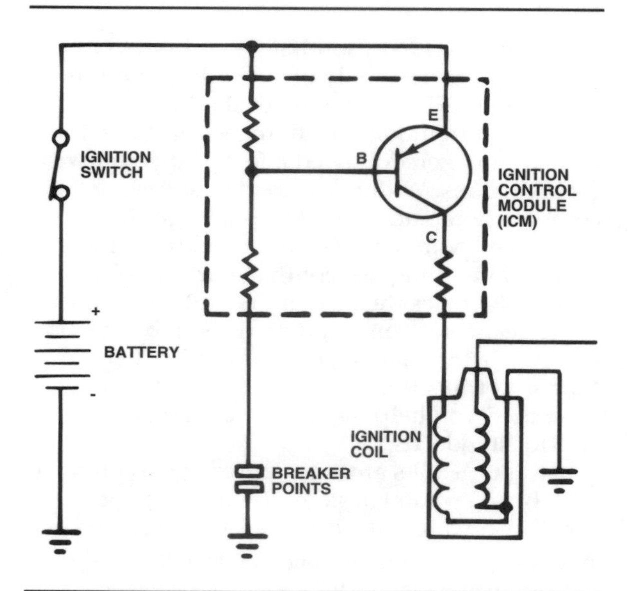

Figure 11-1. A typical breaker point-triggered electronic ignition system.

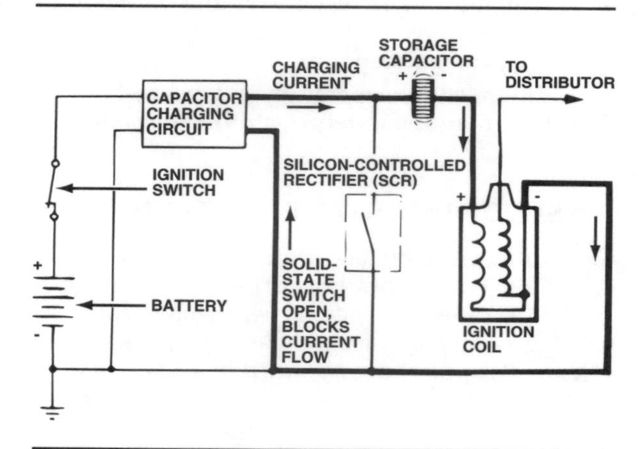

Figure 11-2. When the silicon-controlled rectifier (SCR) blocks current, the capacitor is charged. (Bosch)

ing device, primary current is interrupted. This sudden *decrease* in primary current collapses the coil's magnetic field, inducing a high-voltage surge in the secondary circuit.

In an electronic ignition, primary current passes through the ICM, not through the distributor. Most modules contain one or more large power transistors that switch the primary current. A power or a switching transistor can transmit as much as 10 amps of current—far more than a set of breaker points. The power transistor is controlled by a driver transistor that receives voltage signals from the distributor signal generator.

Original equipment SSI's use inductive discharge to provide at least 30,000 volts of available voltage, and sustain a spark for about 1.8 milliseconds (A millisecond is one-thousandth of a second, or 0.001 second).

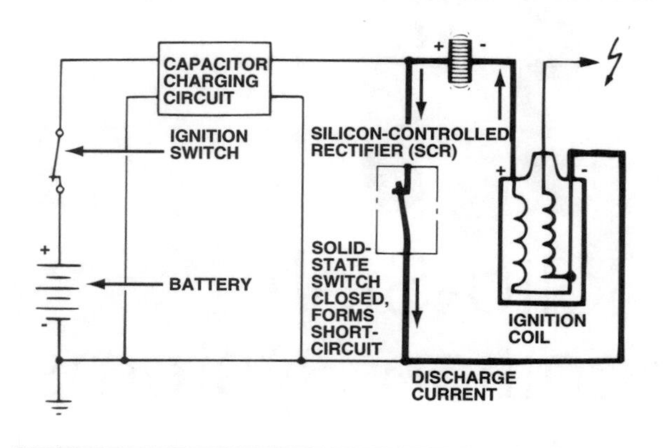

Figure 11-3. When a signal voltage makes the SCR conduct current, the capacitor discharges through the coil primary winding. (Bosch)

Breaker Points

Inductive discharge is used in both electronic and breaker-point systems; only the device used to open the circuit differs. The earliest transistorized ignitions used breaker points as a mechanical switch to control voltage applied to a power transistor. Battery voltage causes current to flow through the ICM to the breaker points, figure 11-1. This current flow biases the transistor base so that current can also flow through the coil primary winding. When the points open, current stops. The power transistor will no longer conduct current to the coil primary winding, and an ignition spark is produced.

While full primary current flowed through the transistor, the points carried less than one amp of current. This minimized the pitting and burning problems encountered with breaker points. However, they were still subject to mechanical wear and bounced at high speeds. Electronic ignitions replaced the points with a solid-state signal generator containing no moving parts.

Capacitive Discharge

The capacitive discharge system uses battery voltage to charge a large capacitor in the ICM. While current flows through the ICM, the capacitor charging time corresponds to dwell in the inductive discharge system. Current flows to the storage capacitor instead of to the coil during this period. The module-charging circuit uses a transformer to increase the voltage in the capacitor to as high as 400 volts. The ICM also contains a **thyristor**, or silicon-controlled rectifier (SCR),

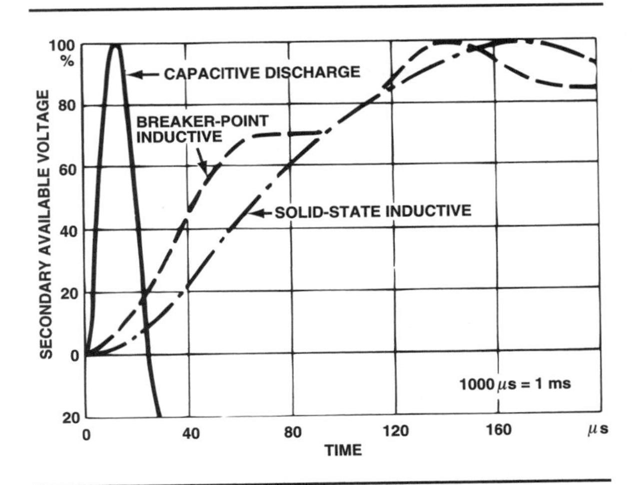

Figure 11-4. As shown by this Bosch graph, a distinguishing factor among different ignition systems is when they develop maximum secondary voltage after the opening of the primary circuit.

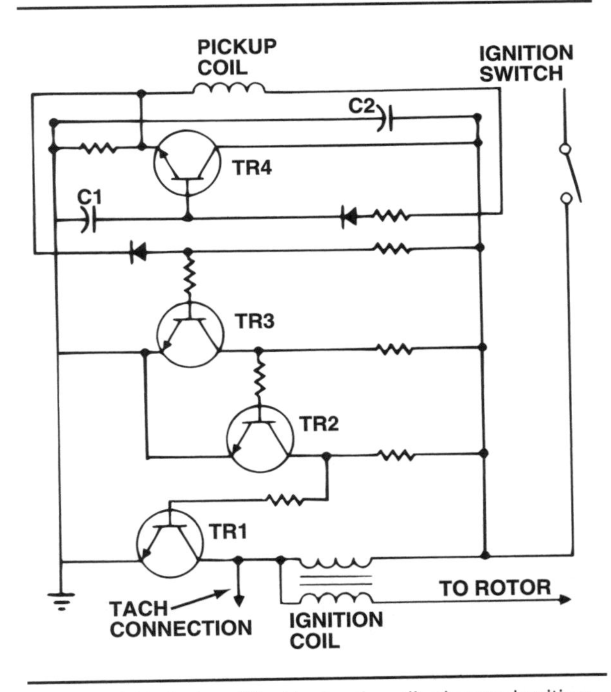

Figure 11-5. A simplified inductive discharge ignition control module. (Chevrolet)

which functions as an open switch to prevent the capacitor from discharging as it charges, figure 11-2.

When the triggering device signals the ICM, the SCR closes the capacitor discharge circuit, figure 11-3, which allows the capacitor to discharge through the coil primary winding. This sudden *increase* in primary current expands the coil's magnetic field, and induces a high-voltage surge in the secondary circuit.

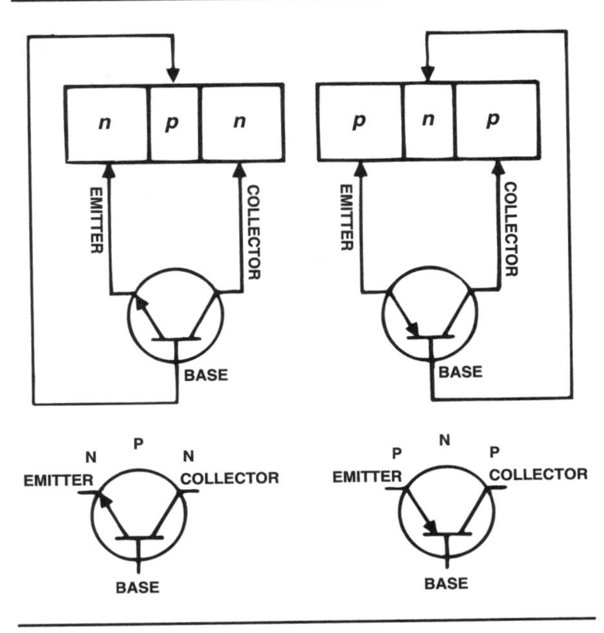

Figure 11-6. Transistor construction and symbols.

Capacitive discharge systems are available in the automotive aftermarket, but no major domestic manufacturer installs them as original equipment. A few imported cars, such as Audi and Mercedes-Benz, have used these ignition systems in the past. They provide a greater available voltage than inductive discharge systems, but can sustain a spark for only about 200 microseconds (μs). The spark is much more intense than that produced by an inductive discharge system, and will fire plugs that are in very poor condition. Under certain engine operating conditions, however, a longer spark time is required, or the air-fuel mixture will not burn completely.

Figure 11-4 shows the time and voltage characteristics of a breaker-point ignition system, a capacitive discharge system, and a solid-state inductive system.

Ignition Control Module: A self-contained sealed unit that houses the solid-state circuits that control ignition-related electrical or mechanical functions.

Thyristor: A silicon-controlled rectifier (SCR) that normally blocks all current flow. A slight voltage applied to one layer of its semiconductor structure will allow current flow in one direction while blocking current flow in the other direction.

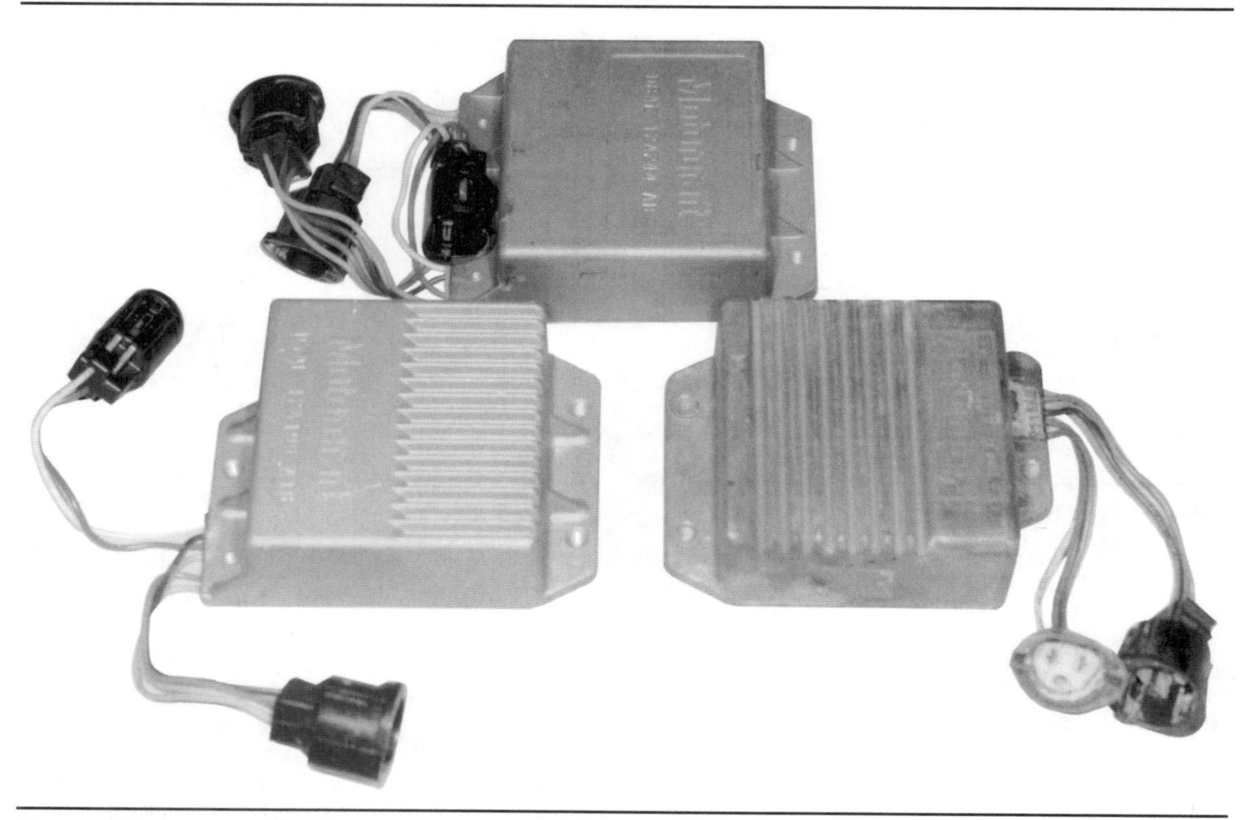

Figure 11-7. Typical external ignition control modules from Ford.

Figure 11-8. The ignition control module may be installed inside the distributor, as in this high-energy ignition (HEI) system.

ICM'S AND THE PRIMARY CIRCUIT

Although various manufacturers use different circuitry within their ICM's, the basic function of all modules is the same. The following paragraphs explain how these modules work. Because the electronic components within the module are delicate and complex, the modules are sealed during manufacture. If any individual component within the module fails, the entire unit is replaced rather than repaired. Our explanation of module circuitry and operation will be brief, because you will never need to service module components.

Inductive Discharge System

Figure 11-5 is a simple inductive discharge ICM circuit. The various triggering devices that can be used in this system all produce the same type of signal: a pulsating voltage. In the illustration, the signal comes from the pickup coil of a magnetic pulse generator. This pulsating voltage is applied to the ICM's transistor and signals the module to turn off the primary current.

Primary circuit control

Primary circuit control is made possible in an electronic ignition by the ability of transistors to control a high current flow in response to a very

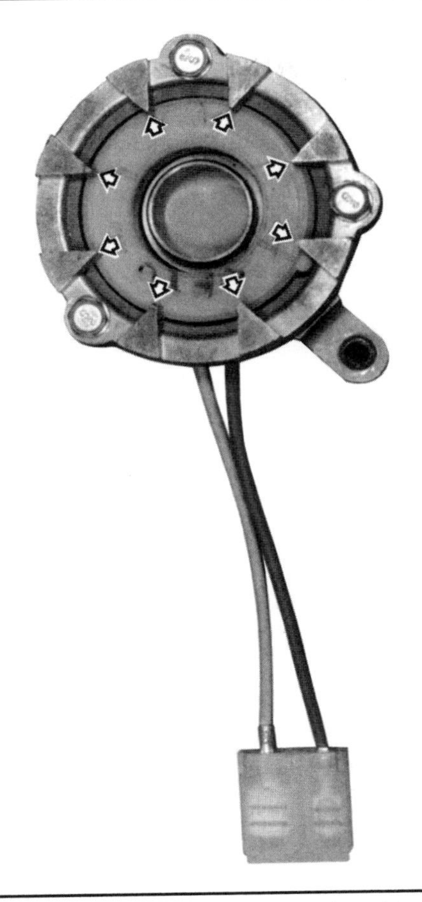

Figure 11-9. This Delco-Remy magnetic pulse generator has a pole piece for each cylinder (arrows), but operates in the same manner as single pole-piece units.

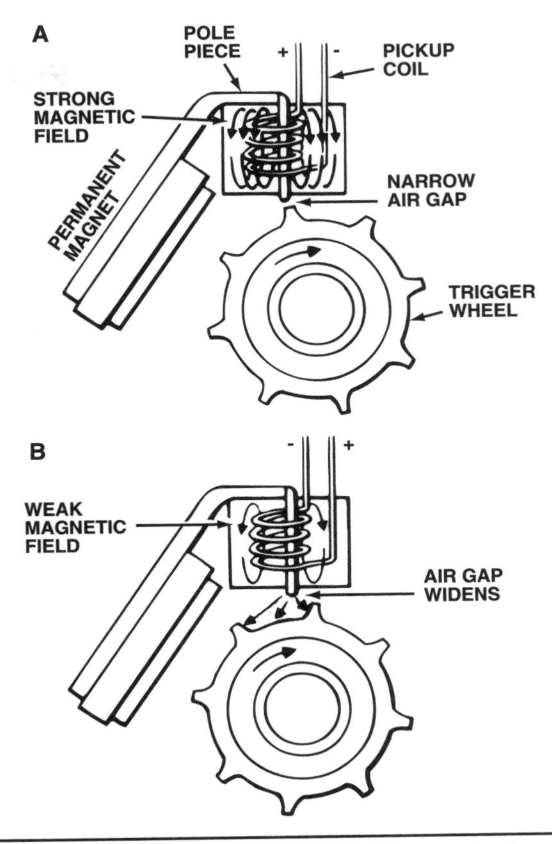

Figure 11-10. The magnetic field expands and collapses as the trigger wheel's teeth move past. This induces a varying strength voltage in the pickup coil.

small current. The transistor is like a solid-state relay. It has a base of one type of semiconductor material and an emitter and collector of the other type of material, figure 11-6. A certain amount of current must flow through the base-to-emitter or base-to-collector circuit before any current will flow through the emitter-to-collector circuit. From this, a small amount of current controls the flow of a large amount of current.

Ignition transistor operation

In figure 11-5, the ICM can be in the dwell mode (when primary current flows) or in the firing mode (when primary current stops).

During the dwell mode, TR1 and TR3 conduct current; TR2 does not. A signal from the pickup coil turns off TR3, charges C1, and turns on TR4. This results in the firing mode, as reduced primary current flow causes a high-voltage secondary surge.

TR4 stays on until C1 is discharged. When C1 is discharged, TR3 turns on, and the ICM returns to the dwell mode. At higher engine speeds, C1 will be charged less and less. This results in reduced firing times and longer dwell periods. C2 is a capacitor in the distributor for radio noise suppression.

Actual ICM's take various forms, but all work using these principles. Some modules are large units, figure 11-7. Others are relatively small integrated circuit units mounted in the distributor, figure 11-8.

TRIGGERING DEVICES AND IGNITION TIMING

We have seen that all triggering devices produce a pulsating voltage that signals the generation of an ignition spark. Four triggering devices are commonly used in a SSI:

- Magnetic pulse generator
- Hall-effect switch
- Metal detection
- Optical (light detection) signal generator.

Magnetic Pulse Generator

Magnetic pulse generators are the most common type of original-equipment triggering device.

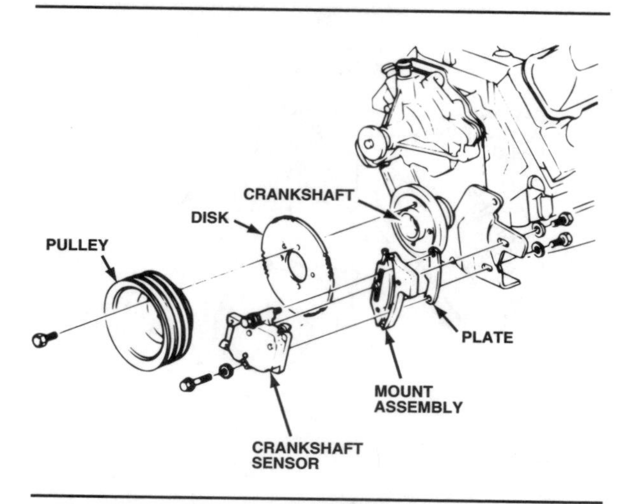

Figure 11-11. This magnetic pulse generator reacts to the movement of a disk mounted on the crankshaft. (Oldsmobile)

They were used in the original Chrysler, General Motors (GM), and Ford electronic ignitions, and are still used today by domestic and foreign automakers.

Many manufacturers use distributor shaft rotation to time voltage pulses. Distributor-mounted magnetic pulse generator designs may have a single pole piece, or as many pieces as the trigger wheel has teeth, figure 11-9. Regardless of the design, the pickup coil works in the same way. The pickup coil is wound around the permanent magnet, or between the magnet and the pole pieces.

As the nonmagnetic trigger wheel turns with the shaft, figure 11-10, teeth on the wheel approach, align with, and move away from the pole piece of the permanent magnet. The low reluctance of the trigger wheel teeth causes the magnet's field to expand as a tooth approaches, figure 11-10A. The field collapses as the tooth moves away, figure 11-10B. This motion of the magnetic field induces a signal voltage in the pickup coil wound around the pole piece.

Magnetic field strength is strongest when the trigger wheel tooth is aligned with the pole piece. Pickup coil voltage drops to zero at this point, changing the bias voltage on the ignition module driver transistor to turn it off. The driver transistor turns off the power transistors to interrupt the primary current. The coil then discharges secondary voltage to fire a spark plug.

The point where the trigger wheel tooth aligns with the pole piece is the ignition timing point for one cylinder, and corresponds to the point where breaker points open.

Since the late 1970s, GM, Ford, and some other automakers have used magnetic pulse

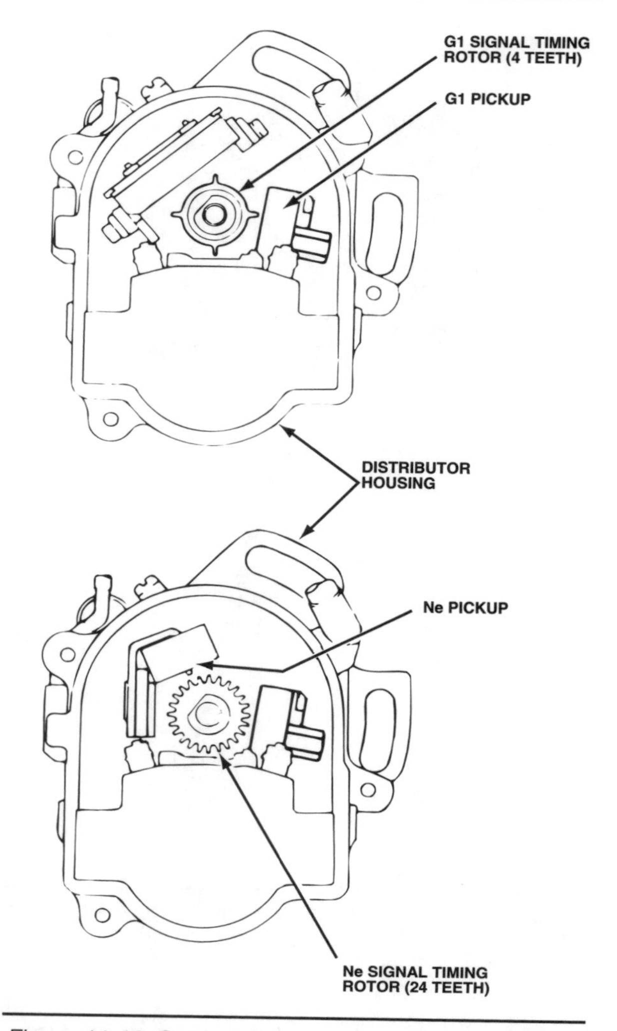

Figure 11-12. Some automakers, such as Toyota, use two pickup coils inside their distributors.

generators that rely on engine crankshaft rotation, instead of distributor shaft rotation, to produce a signal voltage. In GM's 1977 micro-processed sensing and automatic regulation (MISAR) system, the pickup coil, pole piece, and permanent magnet were mounted near a disk mounted on the end of the crankshaft, figure 11-11. Teeth on the disk acted like the teeth on a distributor-mounted trigger wheel to induce a single voltage in the pickup coil. In the 1978 MISAR system, the pickup coil and trigger wheel were installed in the distributor.

Some late-model Asian automakers use two or three magnetic pulse generators to provide rpm and crank position signals to the ICM or powertrain control module (PCM). Counting the number of teeth on each of the reluctor wheels helps to identify the application's signaling configuration, figures 11-12 and 11-13.

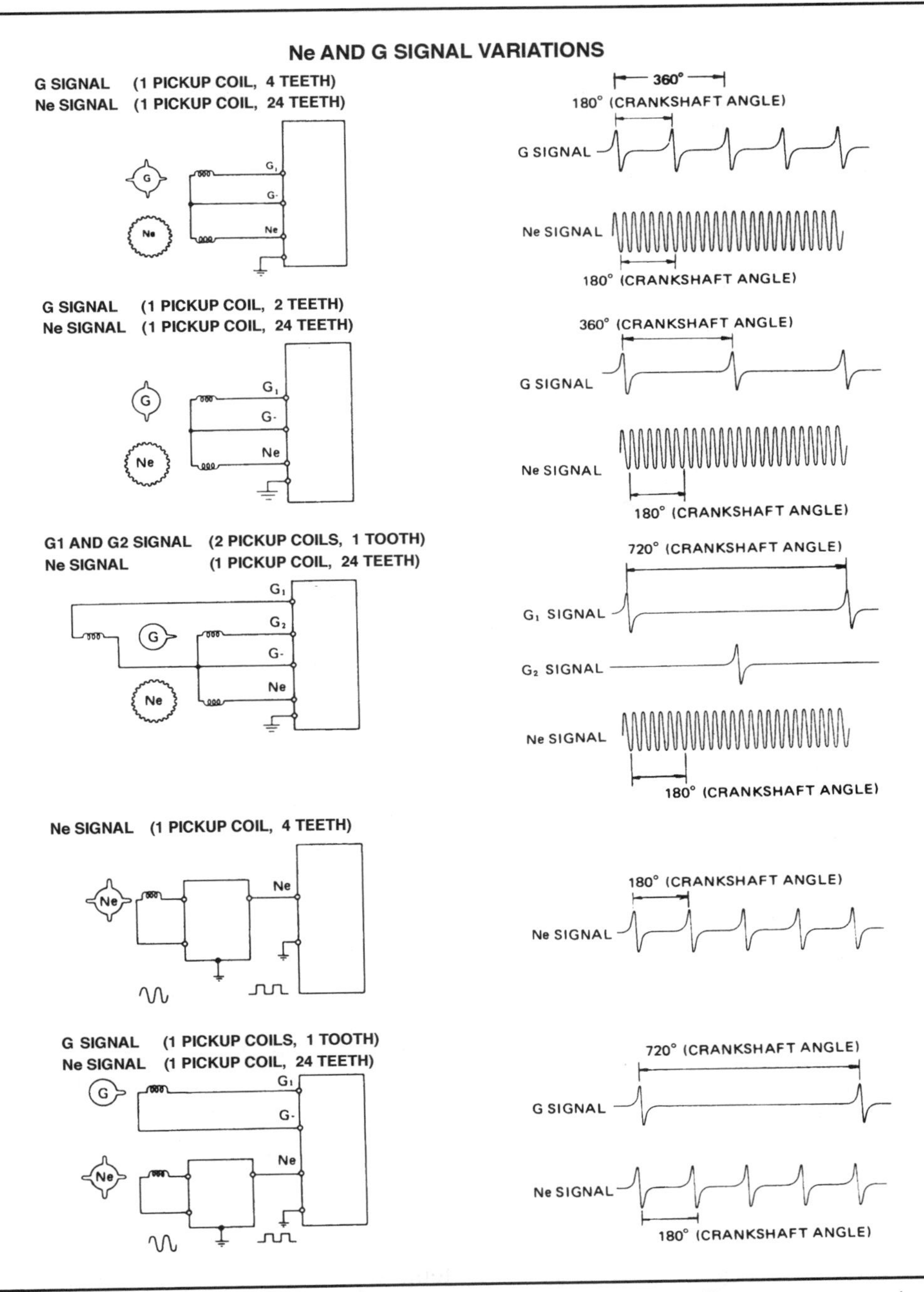

Figure 11-13. By using various reluctor wheel and pickup coil configurations, Toyota can generate many rpm and crankshaft signal combinations.

Hall-Effect Switch

Engineers developed the Hall-effect switch after the magnetic pulse generator. The Hall-effect triggering assembly uses a small chip of semiconductor material, a permanent magnet, and a ring of low-reluctance shutters, figure 11-14.

The Hall-effect switch does not generate a signal voltage in the same way as a magnetic pulse generator; it requires an input voltage to generate an output voltage.

The Hall effect is the generation of a small voltage in a semiconductor by passing current through it in one direction while applying a magnetic field at a right angle to its surface. When current flows from I_1 to I_2 through the semiconductor, and a magnetic field intersects the chip from M_1 to M_2, voltage develops across

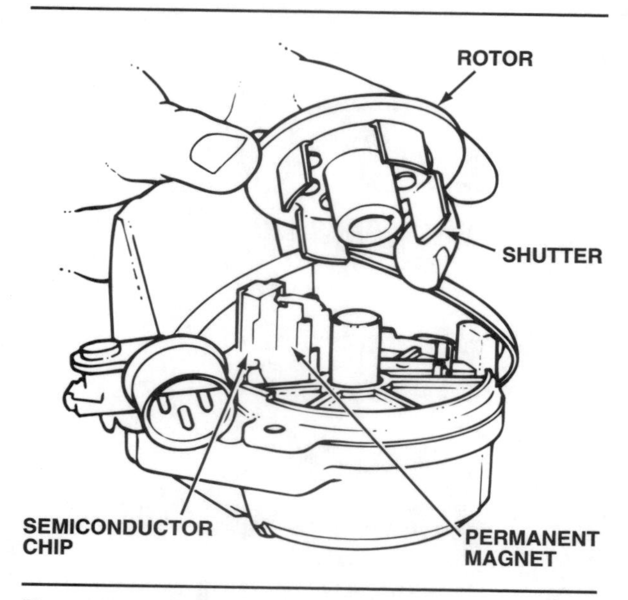

Figure 11-14. A Hall-effect triggering device.

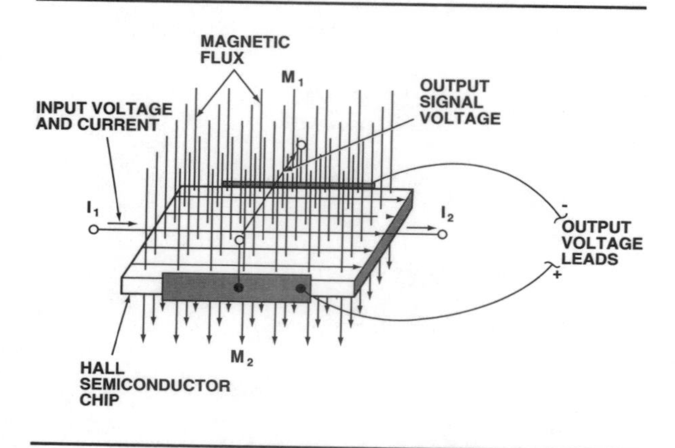

Figure 11-15. The Hall-effect output voltage varies with the strength of the magnetic field, while the input current remains constant.

Vane	Magnetic Field (B)	Hall Voltage	Generator Output Signal Voltage to Module	Ignition Module Transistors
Not in the air gap	Permeates the Hall layer	Maximum	Minimum	Switched off
Enters the air gap	Is deflected away from the Hall layer	Drops	Increases abruptly	Switched on
In the air gap	Very weak at the Hall layer	Minimum	Maximum	Switched on (energy storage)
Leaves the air gap	Permeates the Hall layer	Increases	Drops abruptly	Switched off (ignition point)

Figure 11-16. The relationship of Hall-effect signal voltage and ignition operation. (Bosch)

the semiconductor as shown in figure 11-15. If the input current remains constant, and the magnetic field is varied, the signal voltage will vary proportionally to the field strength.

When the shutter blade of a Hall-effect switch enters the gap between the magnet and the Hall semiconductor element, it creates a magnetic shunt that varies the Hall field strength. This changes the Hall signal voltage, which changes the bias on an ignition-driver transistor similarly to the signal from a magnetic pulse generator.

A Hall-effect switch is a complex electronic circuit. Figure 11-16 shows the relationships of the Hall-effect switch and ignition operation. An important point to understand is that ignition occurs when the Hall shutter *leaves* the gap between the Hall semiconductor element and the magnet.

The Hall-effect voltage is not affected by changing engine speed. Magnetic pulse generators (and metal-detection units) depend on induction to create the signal voltage. The strength of an induced voltage varies if the magnetic lines move more quickly or slowly. The Hall effect does not involve induction, and the speed of the magnetic lines has no effect on the signal voltage. This constant-strength signal voltage offers more reliable ignition system performance throughout a wide range of engine speeds. Moreover, a Hall-effect switch provides a uniform digital voltage pulse regardless of rotation speed. This makes a Hall-effect switch ideal as a digital engine sensor for fuel injection timing and other functions besides ignition control.

Metal Detection

This triggering device is a variation of the magnetic pulse generator. Instead of a permanent magnet affecting the pole piece, an electromagnet supplies the magnetic field, figure 11-17. The ICM applies a small amount of battery voltage to the electromagnet coil. The metal teeth on the trigger wheel affect the electromagnetic field and the voltage within the electromagnetic coil. These voltage changes are sensed by the ICM. This device produces a more reliable signal voltage at lower engine speeds than a magnetic pulse generator.

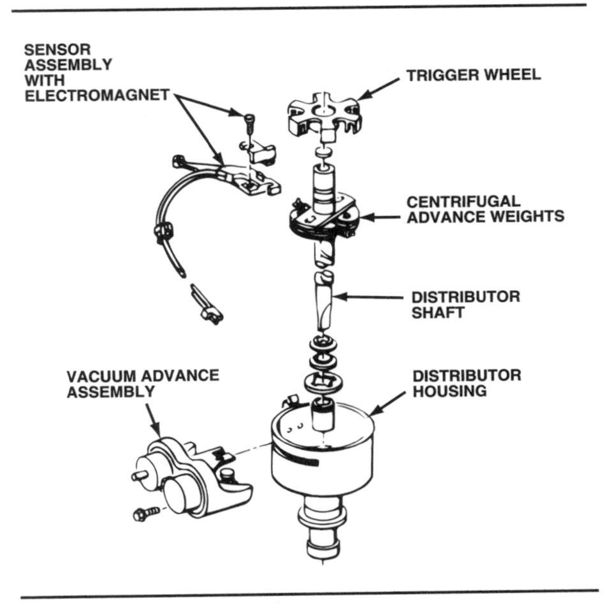

Figure 11-17. A metal-detection triggering device.

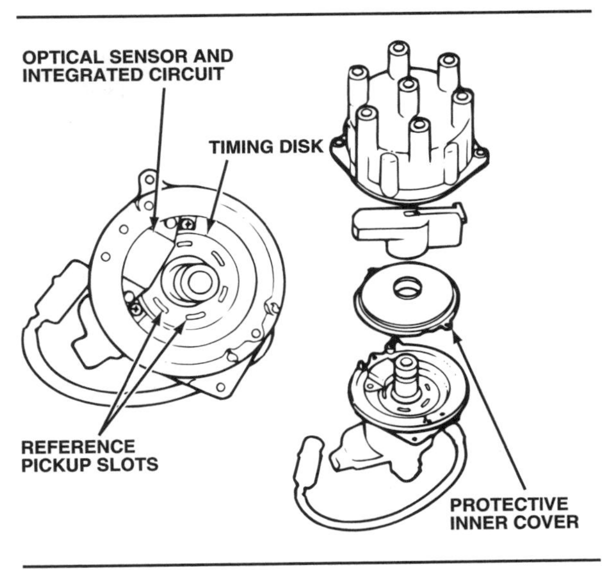

Figure 11-18. Components of the Chrysler optical distributor.

Optical (Light Detection) Signal Generator

The optical signal generator uses a light-emitting diode (LED) and a light-sensitive phototransistor (photocell) to produce signal voltage pulses. When the LED light beam strikes the photocell, voltage is generated. A slotted timing disk rotating on the distributor shaft, figures 11-18 and 11-19, interrupts the light beam, sending an on/off voltage signal to the ICM. Until Chrysler introduced an optical signal generator on some 1987 models, no domestic automaker had used one as original equipment. The Isuzu I-TEK and other Japanese ignition systems, however, rely on optical signal generators, which operate in essentially the same way as the Chrysler version.

The slotted timing disk used in the Chrysler optical distributor serves a dual purpose. Two sets of slots are etched along the inner and outer circumference of the disk. Signals produced by the slots also serve as the crankshaft angle sensor (outer slots) and the top dead center (TDC) sensor (inner slots).

Because the sensors are self-contained within the distributor, a protective inner cover is used to separate them from the high-tension distribution part of the distributor housing. The cover prevents actuation errors caused by electrical noise.

The optical signal generator provides a more reliable signal voltage at much lower engine speeds than either a magnetic pulse or metal-detection unit. However, periodic LED and photocell cleaning may be required. Several aftermarket, or add-on, ignition systems have been produced using this switching device.

ELECTRONIC IGNITION DWELL, TIMING, AND ADVANCE

As you learned in an earlier chapter, dwell is the period of time when the breaker points are closed and current is flowing through the coil's primary winding. In SSI's, a timing or current-sensing circuit in the ICM controls primary winding current flow.

The initial timing adjustment for most basic electronic ignitions is similar to that for breaker-point ignition systems. Timing is set by rotating the distributor housing with the engine idling at normal operating temperature.

The first generation of electronic ignitions used the same centrifugal and vacuum advance mechanisms to advance timing as breaker-point ignitions. As automakers began to equip their engines with computer-controlled systems in the late 1970s, however, timing advance became a function of the computer. Since a computer can receive, process, and send information rapidly, it

Mind Those Magnets

When you are servicing a distributor from an electronic ignition system, be sure that no metal particles or iron filings get inside. The magnetic pickup coils and pole pieces in the breakerless distributors used by many automakers will attract metal debris, which can really foul up ignition performance. Use a clean, soft-bristled brush or low-pressure compressed air to clean the inside of a distributor and keep scrap metal off the pickup coil.

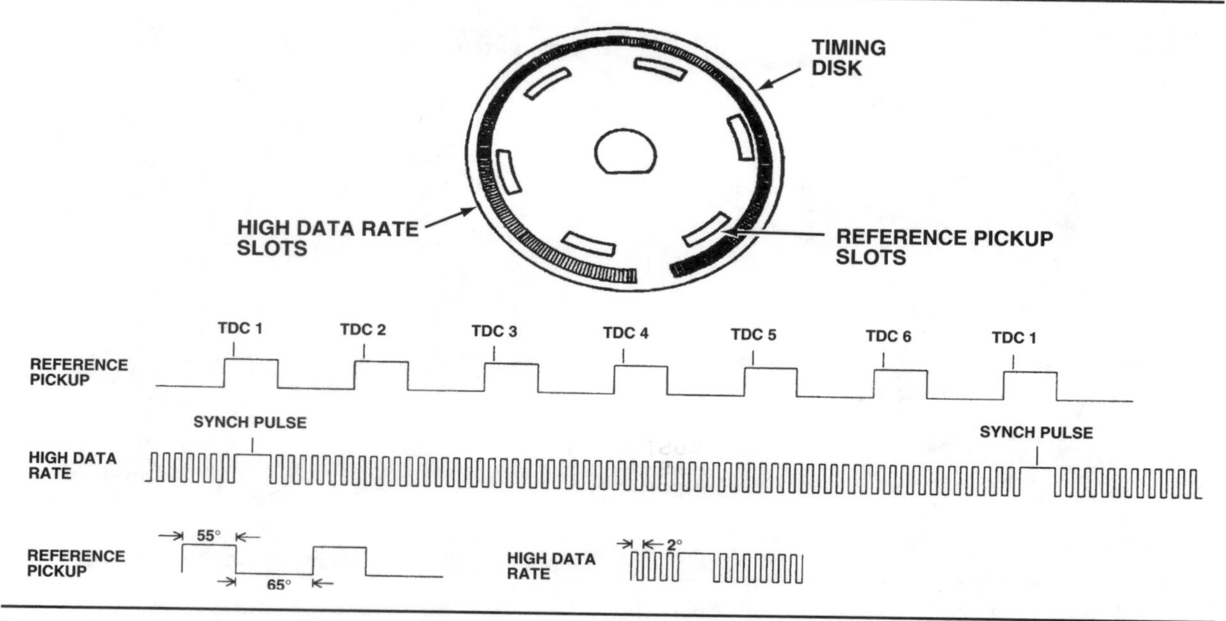

Figure 11-19. Chrysler's slotted timing disk interrupts the light beam in the optical sensor to produce on/off signals used to control fuel injection, idle speed, and ignition timing.

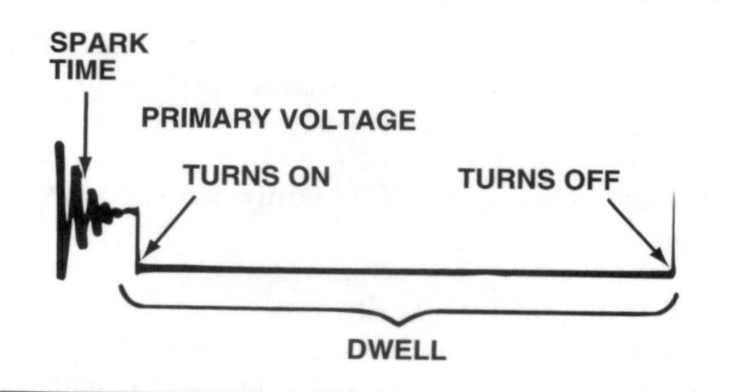

Figure 11-20. Dwell measured in degrees of distributor rotation remains relatively constant at all engine speeds in a fixed dwell system.

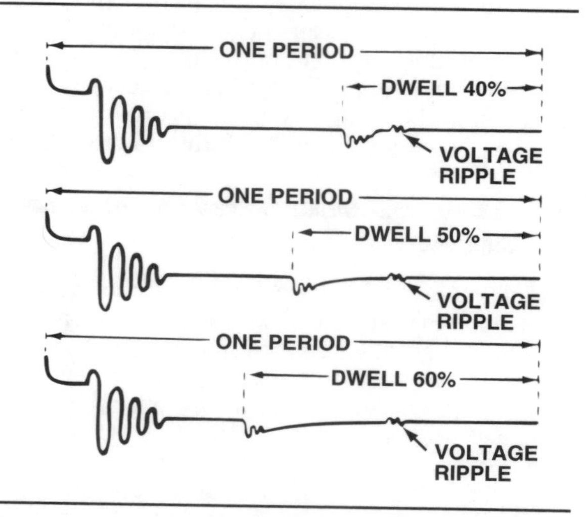

Figure 11-21. The dwell period increases with engine speed in the variable dwell systems.

can change ignition timing with far more efficiency and accuracy than any mechanical device.

The distributor in an electronic ignition used with a computer engine control system contains a triggering device for basic timing. Because the computer actually controls ignition timing, the distributor's primary function is to distribute secondary voltage.

Fixed Vs. Variable Dwell

Although dwell is not adjustable in electronic distributors, electronic ignitions may have one of two kinds of dwell control:

- Fixed dwell
- Variable dwell.

Fixed dwell

A ballast resistor is placed in the primary circuit of a **fixed dwell** electronic ignition to limit current and voltage. Dwell is the length of time the switching transistor sends current to the primary coil windings. It begins when the ICM allows current to flow through the primary circuit, which forms a magnetic field in the ignition coil's primary winding. The ballast resistor functions just as it does in a breaker-point system to control primary voltage and current. Dwell, measured in distributor degrees, remains constant at all engine speeds. Figure 11-20 shows the primary circuit oscilloscope pattern of a typical fixed dwell electronic ignition.

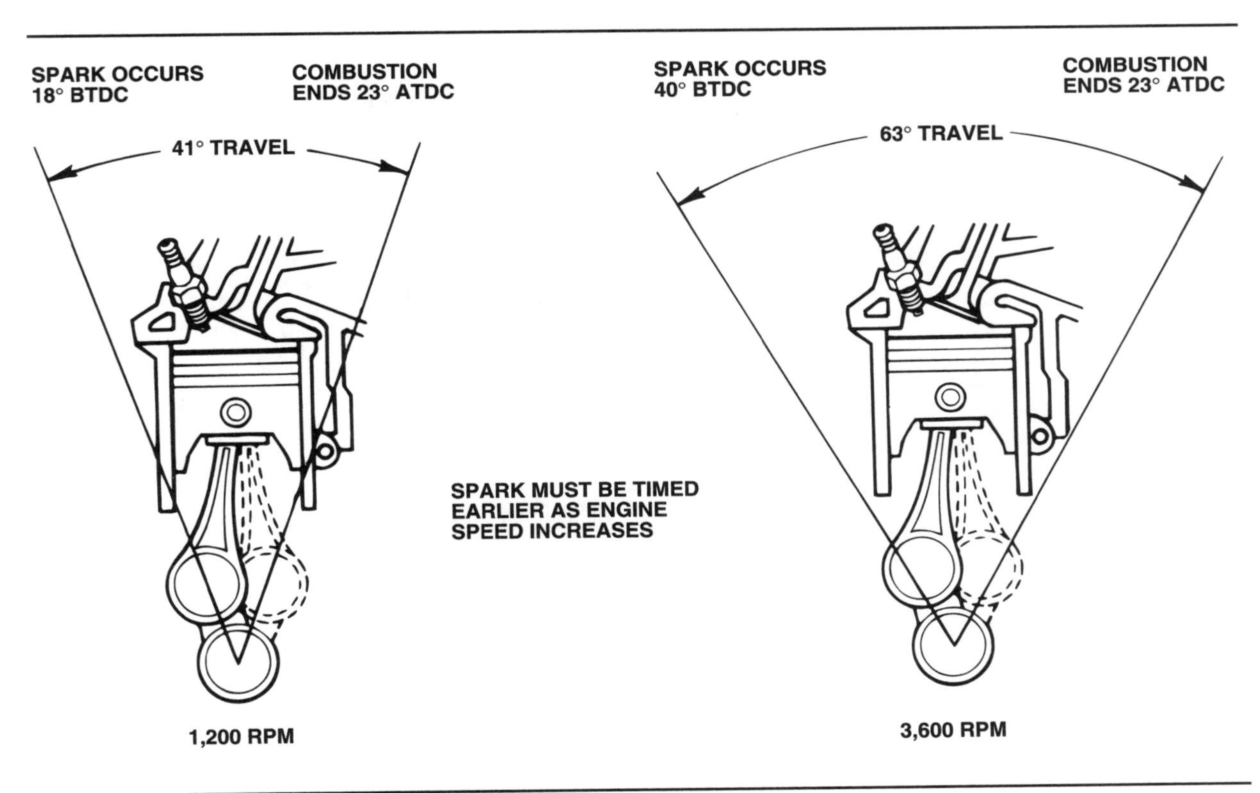

Figure 11-22. As rpm increases, the spark must be advanced to deliver the maximum spark for the best torque.

Chrysler's original electronic ignition is a good example of a fixed dwell system. The original Ford SSI and Dura-Spark II are other examples of fixed dwell systems with ballast resistors, as are some Bosch and Japanese electronic ignitions.

Variable dwell

A ballast resistor is not used in a **variable dwell** system. The coil and ICM receive full battery voltage. A module circuit senses primary current to the coil, reducing the current when the magnetic field is saturated. Unlike a breaker-point or fixed dwell electronic ignition, dwell measured in distributor degrees changes with speed in a variable dwell ignition, but dwell *time* remains relatively constant. Figure 11-21 shows the primary circuit oscilloscope pattern of a typical variable dwell electronic ignition.

In general, ignition coils used with variable dwell systems have a higher available voltage capability, lower primary resistance, and a higher turns ratio than the coils used with fixed dwell systems.

All variations of the GM Delco-Remy High Energy Ignition (HEI) system are variable dwell systems, as well as Ford Dura-Spark I and thick-film integrated ignition systems. Other examples are most Chrysler Hall-effect ignitions, some Bosch, Marelli, and several Japanese electronic ignitions.

Basic Timing and Advance Control

We learned in a previous chapter that ignition takes place at a point just before or just after TDC is reached during a piston's compression stroke. This time is measured in degrees of crankshaft rotation, and is established in distributor-type ignitions by the mechanical coupling between the crankshaft and distributor. Basic or initial ignition timing is usually set at idle speed.

As engine speed increases, however, ignition must take place earlier. This is necessary to ensure that maximum compression pressure from combustion develops as the piston starts downward on its power stroke, figure 11-22. This change in ignition timing is called spark advance, and is controlled in basic electronic ignitions (those not integrated with electronic

Fixed Dwell: A type of ignition system where the dwell period begins when the switching transistor turns on, and remains relatively constant at all speeds.

Variable Dwell: A type of ignition system where the dwell period varies in distributor degrees at different engine speeds, but remains relatively constant in duration or actual time.

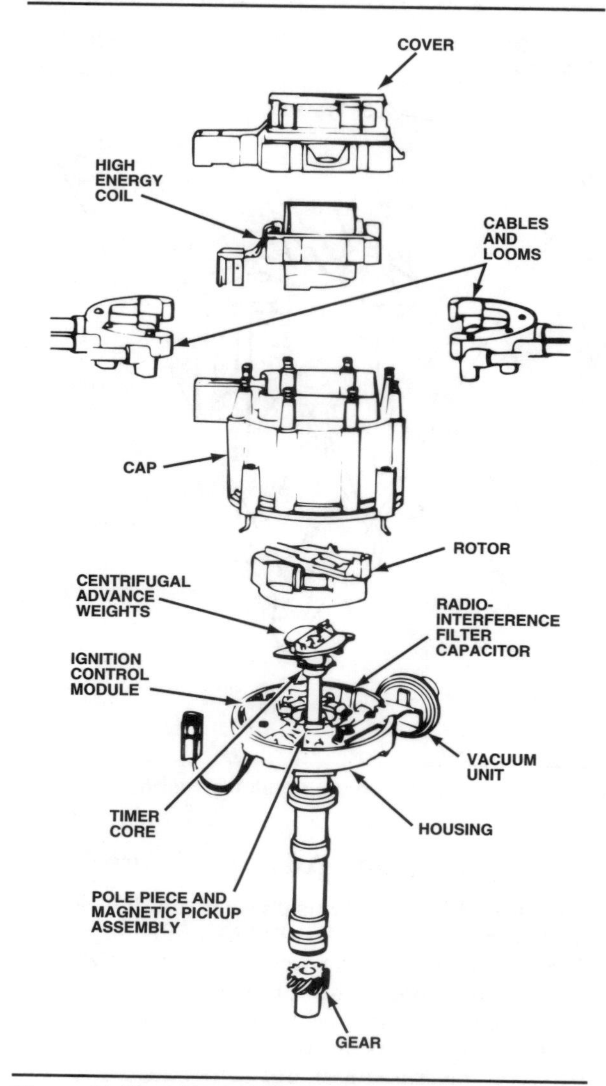

Figure 11-23. A basic HEI distributor.

engine controls) by the same mechanical devices used in breaker-point ignitions:

- Centrifugal advance weights
- Vacuum advance diaphragm.

The centrifugal advance mechanism responds to changes in *engine speed* and moves the position of the trigger wheel relative to the distributor shaft. The vacuum advance mechanism responds to changes in engine *load* and moves the position of the pickup coil. These changes in position alter the time, relative to crankshaft position, when the primary circuit opens.

When electronic ignitions are integrated with electronic control systems, the computer monitors engine speed and load changes, engine temperature, manifold pressure (vacuum), airflow, exhaust oxygen content, and other factors. The computer then changes ignition timing to produce the most efficient combustion.

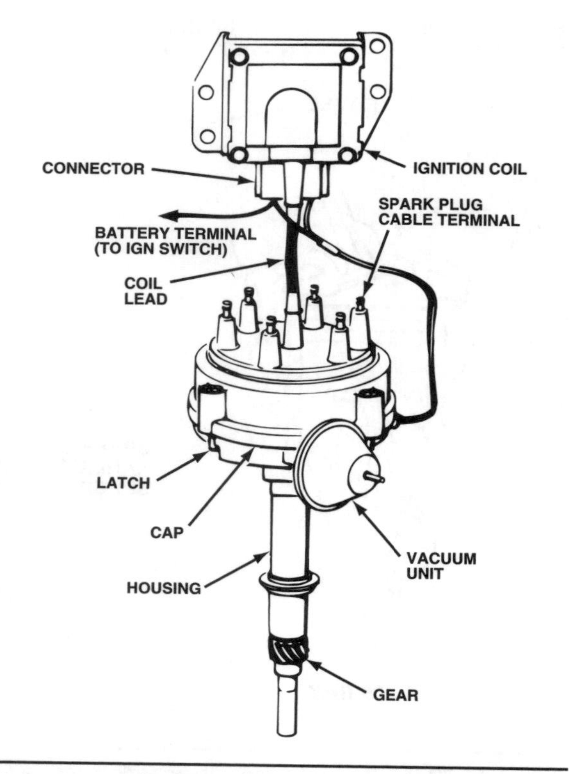

Figure 11-24. The HEI distributor used with some in-line and V6 GM engines.

ELECTRONIC DISTRIBUTOR-TYPE IGNITION SYSTEMS

This section contains brief descriptions of the major distributor-type electronic ignitions used by domestic and foreign automakers. All are inductive-discharge systems, but the triggering devices and module circuits vary somewhat from manufacturer to manufacturer.

Because these systems have been under constant development, they have been modified from year to year and model to model. The descriptions given in this section summarize the basic changes resulting from this ongoing development. Whenever an electronic ignition requires service, you should always refer to the automaker's shop manual or an appropriate repair manual to determine the exact specifications, and whether the system has any unique features.

Delco-Remy High Energy Ignition (HEI)

The Delco-Remy HEI system, figure 11-23, was introduced on some 1974 GM V8 engines, and became standard equipment on all GM engines in 1975. The HEI system was developed from an earlier Delco-Remy Unitized ignition used on a limited number of 1972–74 engines. The HEI and Unitized ignitions have all of the ignition components built into the distributor.

The HEI system was the first domestic original-equipment manufacturer (OEM) electronic ignition to use a variable dwell primary circuit and no ballast resistor. The ICM lengthens the dwell period as engine rpm increases to maintain uniform primary current and coil saturation throughout all engine speed ranges. The HEI ICM is installed on the breaker plate inside the distributor and has four terminals: two connected to the primary circuit and two attached to the pickup coil of a magnetic pulse ignition pickup. The back of the HEI ICM is coated with a silicone dielectric compound before installation for improved cooling.

The HEI pickup looks different from those of other manufacturers, but operates in the same basic manner. The rotating trigger wheel of the HEI distributor is called the "timer core", while the coil is attached to a fixed ring-shaped magnet called the "pole piece". The pole piece and trigger wheel have as many equally spaced teeth as the engine has cylinders, except on uneven-firing V6 engines, which have three teeth on the trigger wheel and six unevenly spaced teeth on the pole pieces.

The most common HEI system has an integral (built-in) coil mounted in the distributor cap, figure 11-23. Some 4- and 6-cylinder engines have HEI systems with a separate coil, figure 11-24, that provides additional distributor clearance on the engine. Both designs operate in the same way and have similar wiring connections. All HEI systems use 8-mm spark plug cables with silicone insulation to minimize cross firing caused by the high secondary voltage capability of the HEI coil. Some engines use wide-gap spark plugs to take advantage of the system's high-voltage capability.

In addition to distributing the spark to the appropriate cylinder, the HEI rotor serves as a fuse to protect the module. If an open circuit occurs in the ignition secondary and voltage rises above a certain level that causes a spark, the center of the rotor will burn through, allowing the spark to travel to ground rather than arc, and destroy the ICM. Early black rotors have a dielectric strength of approximately 70,000 volts; later white ones are designed to ground the secondary circuit at approximately 100,000 volts.

The newer design rotor may be used in place of the earlier one, but the cap and rotor must be replaced as a matched set. The air gap between the rotor tip and cap electrodes was 0.090 inch (2.29 mm) on the earlier parts, but the later design has an air gap of 0.125 inch (3.18 mm) to

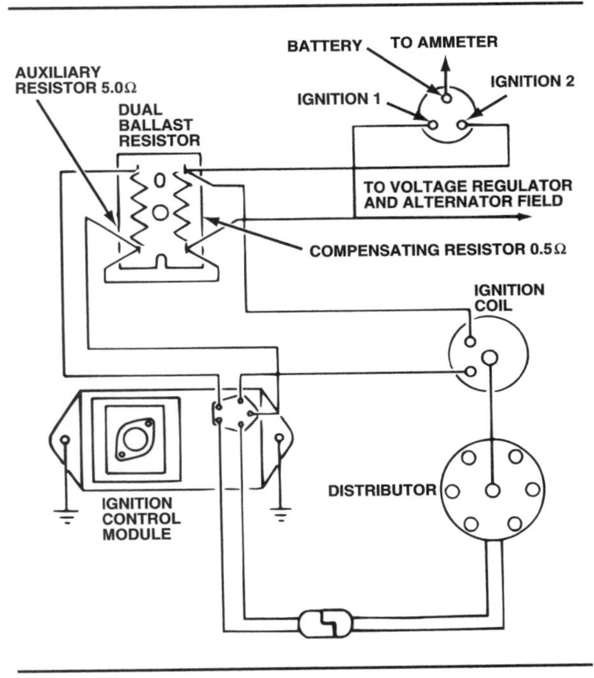

Figure 11-25. A primary circuit diagram of Chrysler's electronic ignition system.

better suppress radio frequency interference (RFI) that could interfere with the PCM. Caps and rotors from the early and late designs should never be mixed.

In addition to this basic HEI system, GM has used six other HEI versions with electronic engine controls and for certain spark-timing requirements. HEI systems include six variations:

- Electronic spark selection (ESS)
- Electronic spark control (ESC)
- Electronic module retard (EMR)
- Electronic spark timing (EST)
- EST and ESC
- EST and a Hall-effect switch.

The HEI system with ESS uses a five-terminal HEI module and was introduced by Cadillac on the Seville in 1978. The system has an electronic decoder that receives inputs from the:

- Exhaust gas recirculation (EGR) solenoid that signals coolant temperature
- Ignition switch that signals cranking
- The fuel economy switch that signals engine vacuum level.

The decoder uses these signals to determine the appropriate timing alterations. Timing retards by a fixed amount during cranking and cold-engine operation on California models. Timing advances by a fixed number of degrees during cruise.

The HEI system with ESC is a detonation-control system used on 1980 and later

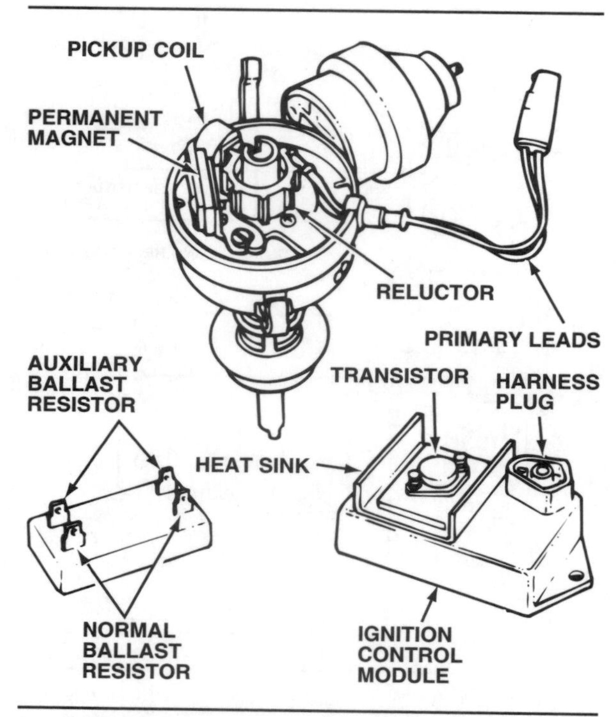

Figure 11-26. The basic Chrysler electronic ignition system.

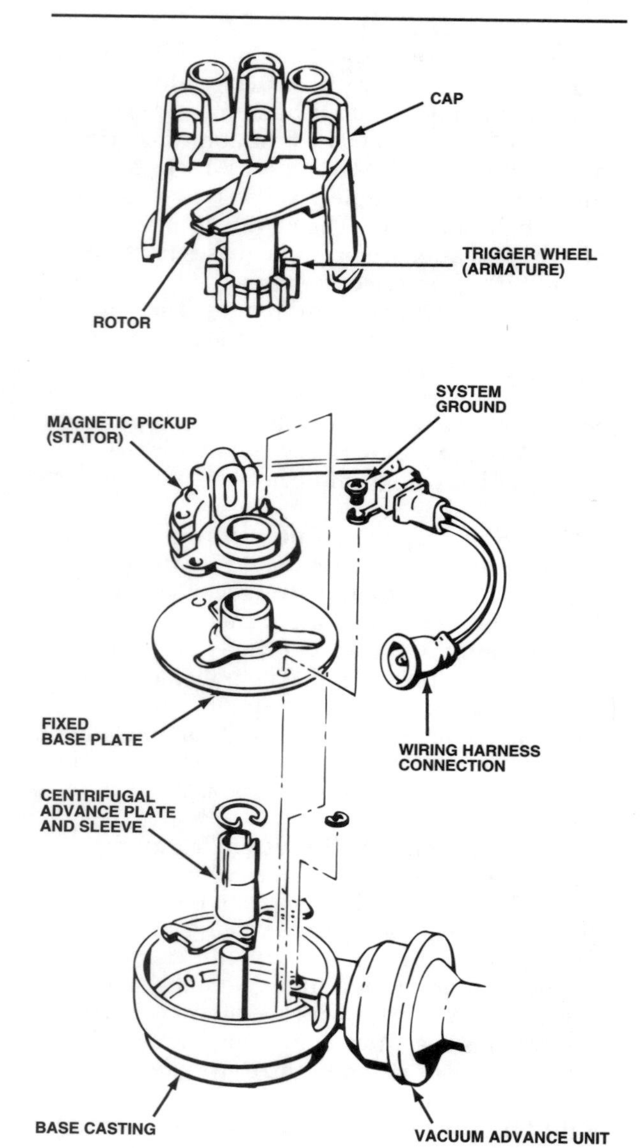

Figure 11-27. Ford's solid-state ignition (SSI) distributor.

turbocharged and high-compression engines. The system consists of a knock sensor, controller unit, and a special five-terminal HEI ICM.

When detonation occurs, the knock sensor sends a signal to the controller, which then instructs the ICM to retard the ignition timing by a small amount. If the knock sensor continues to detect detonation, the controller instructs the ICM to further retard timing by another small increment. This sensor-controller-module cycle goes on continuously, and the controller's instructions to the module are updated many times a second until the detonation is eliminated. As soon as that happens, the process reverses itself, advancing timing in small steps as long as detonation does not reoccur. Once the detonation-producing conditions have been eliminated, the timing will return to normal within 20 seconds.

The HEI system with EMR is a simple 10-degree timing retard system used for cold starts and was also introduced in 1980. The retard circuitry is contained within the special five-terminal HEI module, and is activated by a simple vacuum switch on most models. On cars with the computer-controlled catalytic converter (C-4) system, the module is controlled by the C-4 computer.

The HEI system with EST was introduced on 1981 engines with computer command control (CCC), except those with minimum-function CCC systems (Chevette, Pontiac T1000, and Acadian). The seven-terminal module converts the pickup coil signal into a crankshaft position signal used by the PCM to advance or retard ignition timing for optimum spark timing. HEI-EST distributors have no centrifugal or vacuum advance units.

The HEI system with EST and ESC was also introduced in 1981 and combines the electronic spark control of EST with the detonation sensor of ESC. It is used primarily with turbocharged engines.

The HEI system with EST and a Hall-effect switch combines the basic magnetic pulse generator of the HEI distributor with a Hall-effect

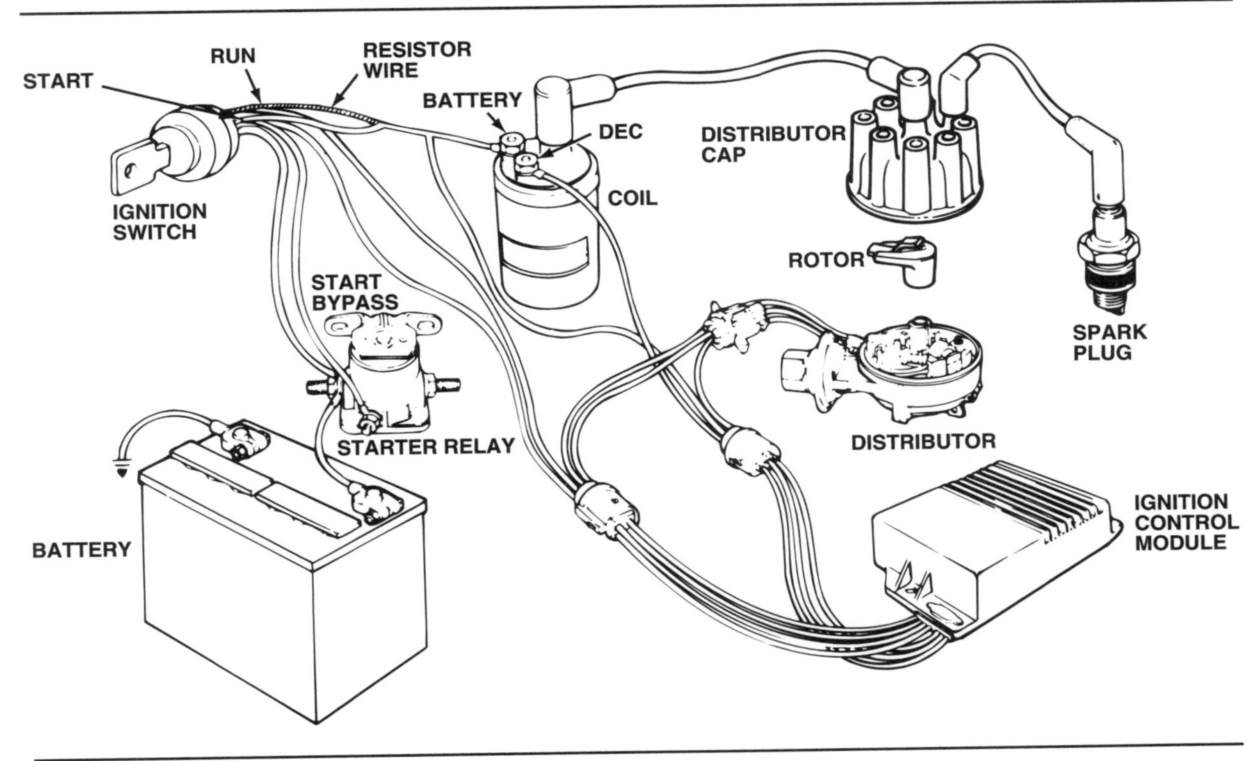

Figure 11-28. Components of Ford's solid-state ignition (SSI).

switch; it is used with CCC engine control systems. The pickup coil sends timing signals to the HEI ICM during cranking. Once the engine starts, the Hall-effect switch overrides the pickup coil and sends crankshaft position signals to the computer for electronic control of timing.

Chrysler

Electronic ignition

In 1971, Chrysler became the first domestic automaker to introduce a basic electronic ignition on some models, figure 11-25. The system became standard on all Chrysler cars in 1973.

The Chrysler electronic ignition system, figure 11-26, is a fixed dwell design using a magnetic pulse distributor, a remote-mounted ICM, and a unit-type ballast resistor. The ICM is mounted on the firewall or inner fender panel, and has an exposed switching transistor that controls primary current. Do not touch the transistor when the ignition is on, because enough voltage is present to give you a shock. The distributor housing, cap, rotor, and advance mechanisms are all similar to breaker-point components, as are the ignition coil and 7-mm spark plug cables.

All 1972–78 cars and some 1979–80 models have five-pin modules with a matching five-terminal wiring harness connector. Early five-pin module ignition systems used a dual ballast resistor that contained an exposed 0.5-ohm temperature-compensating resistor to control primary circuit current and voltage, and a 5.0-ohm temperature-compensating resistor to protect ICM circuitry from high voltage or current surges. This part was superseded in the mid-1970s by a revised dual resistor using a sealed 1.2-ohm nontemperature-compensating resistor in place of the earlier 0.5-ohm part. The later design is the approved replacement part for all dual-resistor systems.

Some 1979–80 and all 1981 and later vehicles with basic electronic ignition have a four-pin module and matching wiring harness connector. The four-pin module contains integral protection circuitry, eliminating terminal 3, which receives input voltage from the 5.0-ohm resistor. A single 1.2-ohm unit-type ballast resistor is used in four-pin module systems to regulate primary circuit current and voltage.

All early Chrysler electronic ignition systems have a single magnetic pickup in the distributor. However, the distributors used with some ELB and ESC systems use a dual-pickup distributor containing a start pickup and a run pickup. The run pickup is positioned to advance the ignition trigger signal compared to the start pickup. Under normal engine operation, the ICM uses the signal from the run pickup. When the ignition switch is in the START position, however, the retarded trigger signal from the start pickup is used to ensure faster starts.

Distributors with early ELB systems had only a centrifugal advance mechanism. Those for later ESC systems have neither centrifugal nor vacuum advance mechanisms; all spark advance is controlled by the computer.

The air gap between the reluctor and pickup coil, or coils, is adjustable, but has no effect on the dwell period, which is determined by the ICM. The air gap must be set to a specific clearance with a nonmagnetic feeler gauge when a new pickup unit is installed. Air gap specifications vary according to model year.

Hall-effect electronic ignition system

Chrysler introduced a different electronic ignition in 1978 on its first 4-cylinder, front-wheel-drive (FWD) cars. A Hall-effect switch is used instead of a magnetic pulse generator. The original fixed dwell ignition on 1978 4-cylinder engines was used with an analog computer, and had a 0.5-ohm ballast resistor to control primary current and voltage. The 1978 ELB and 1979 ESC system distributors had both centrifugal and vacuum advance mechanisms, and were similar in operation to the 6-cylinder and V8 versions described above.

A changeover to a digital spark-control computer in 1981 resulted in the electronic spark advance (ESA) system. This changed the system and its operation in several ways. Since the computer took over spark control timing, the distributor had no advance mechanisms. The system uses no ballast resistor, and dwell is variable; that is, it increases as engine speed increases. In 1984, the ESA system was incorporated into the electronic fuel injection spark control system used on fuel injected engines. A logic module and a power module replaced the spark control computer, but the ignition portion of this system works essentially the same as in the ESA system.

Ford

Motorcraft solid-state ignition (SSI)

The Ford Motorcraft SSI system was used in three forms from 1973 through 1976. It consists of a magnetic pulse distributor, figure 11-27, an ICM, and a special oil-filled coil that can be identified by its blue case or tower. The coil primary terminals are labeled BAT (battery) and DEC (distributor electronic control), and a standard primary circuit ballast resistor wire is used, figure 11-28.

The 1973–74 system ICM's are identified by a black grommet, and have seven wires that terminate in three- and four-wire connectors. The 1975 versions have a green grommet, and the three-wire connector shape differs from earlier systems. In mid-year, the blue wire for system protection was eliminated, and all later systems (and replacement modules) have a two-wire connector in place of the earlier three-wire part. The 1976 versions have a blue grommet and the black and purple wires are reversed in the four-wire connector.

Dura-Spark I and II

Ford introduced its second-generation electronic ignitions in 1977. They are direct descendants of the original SSI system. The Dura-Spark systems have higher secondary voltage capabilities, and the electrical values of some components differ from the earlier Ford systems.

The Dura-Spark I system is used on some 1977–79 California engines, and can be identified by a red grommet in the ICM. The system does not use a primary resistor, and has a new coil with a gray tower and unique terminals that prevent its use in other systems. The Dura-Spark I ICM adjusts ignition dwell relative to current through the coil at the time of a spark. The module also has a stall-shutdown feature for circuit protection. If the engine stalls, the module opens the primary circuit even though the ignition switch is in the ON position. The switch must be turned off, and then back to START to close the primary circuit.

Dura-Spark II systems more closely resemble the 1976 SSI system. They retain a primary resistor wire, but the resistance values are changed to provide higher secondary voltage. The coil is the same as that used with the 1973–76 systems.

Dura-Spark II systems have been produced in several variations. The most basic version, produced from 1977 on, has a blue grommet, just like the 1976 SSI system. Some 1979 and later 2.3-liter engines with automatic transmissions have a Dura-Spark II ICM with a white grommet. This module has "start-retard" circuitry that retards the timing up to 18 degrees at cranking speed. Certain 1979–80 engines use a "dual-mode" ICM with a yellow grommet. This module has three additional wires attached to a third connector. The connector plugs into one of two sensors that affect ignition timing in different ways, depending on the application. Some models have a vacuum switch for improved fuel economy, while others use a barometric pressure switch for altitude compensation.

From 1981 on, the dual-mode ICM is replaced by the universal ignition module (UIM), which also has three connectors and a yellow grommet. This newer module is smaller and more compact than previous Dura-Spark units, and contains a factory-programmable retard feature used for fuel economy calibrations, altitude compensa-

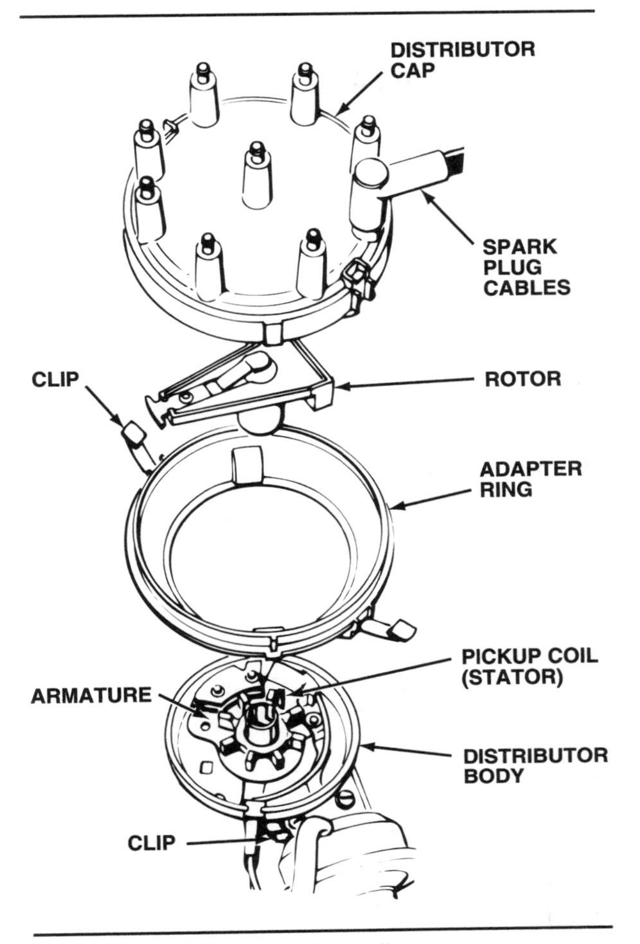

Figure 11-29. A Dura-Spark distributor.

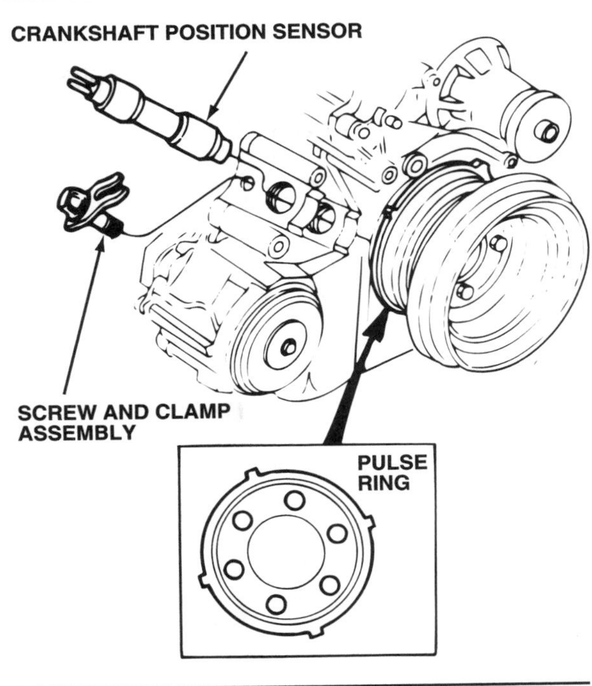

Figure 11-30. Ford's EEC-III crankshaft position sensor replaced the magnetic pulse generator in the Dura-Spark III distributor.

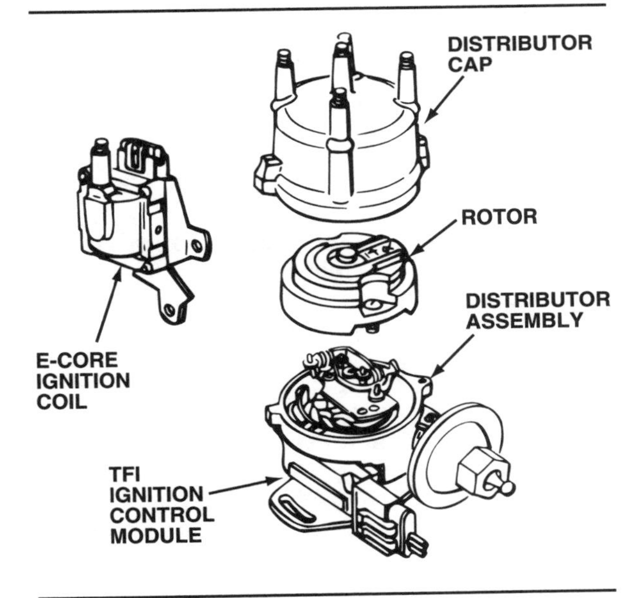

Figure 11-31. Ford's Thick-Film Integrated (TFI) ignition system components.

tion, or spark knock control, depending on the application.

All Dura-Spark I and II distributor caps, figure 11-29, use male terminals for the 8-mm spark plug cables. The caps on 6- and 8-cylinder distributors are much larger than previous models, and have adapter rings to mate with the distributor housing. Large rotors are also used with the bigger caps.

Dura-Spark III

This system was introduced in late 1979 as part of Ford's second-generation electronic engine control (EEC-II) system. The Dura-Spark III ICM can be identified by its brown grommet. Although it appears similar to other Dura-Spark modules, many of its control circuits were eliminated, and their functions incorporated in the EEC-II's PCM.

The Dura-Spark III distributor essentially is nothing more than a device to route the spark from the coil to the proper plug. It has no centrifugal or vacuum advance mechanisms. The magnetic pulse generator was removed from the distributor and relocated to the crankshaft as a type of magnetic pulse generator, called a crankshaft position (CKP) sensor, figure 11-30.

Thick-film integrated (TFI) ignition

The TFI-I ignition system, figure 11-31, was introduced in 1982 on the 1.6-liter Escort engine. It differs in many respects from the earlier Ford

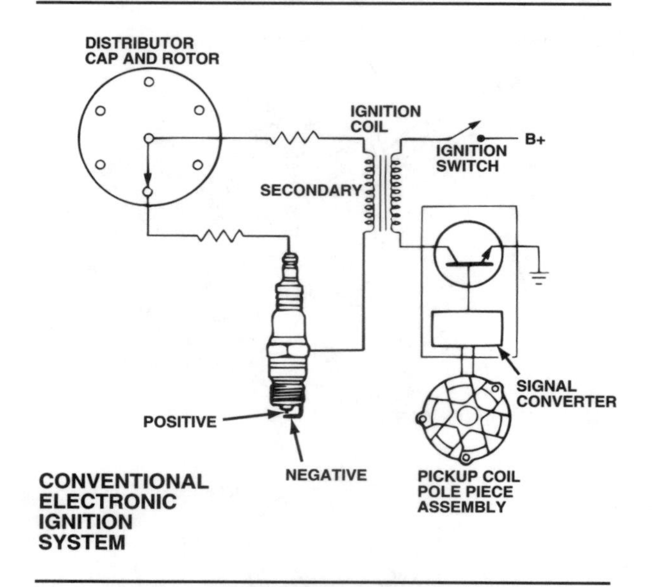

Figure 11-32. When a spark plug is fired with forward polarity, current flows from the center's positive to the side's negative electrode. (General Motors)

systems already discussed. Instead of a remote-mounted ICM, the TFI-I system uses an integrated circuit ICM attached to the outside of the distributor housing. The module connects directly to the distributor stator.

Inside the distributor is the familiar magnetic pulse generator, but TFI-I is a variable dwell system that operates without a ballast resistor. The conventional oil-filled coil is replaced with a special low-resistance E-core part. The distributor cap, rotor, and spark plug cables, however, are similar to those used with Dura-Spark systems.

There are two different versions of the TFI-I ICM. Early production parts were made of blue plastic and called "non-push-start" modules. They contain protection circuitry that will shut off voltage to the coil if an ignition trigger signal is not detected for 10 to 15 seconds. When shutdown occurs, the ignition switch must be turned off and back on again before the engine will start.

All later ignition ICM's are made of gray plastic and contain revised circuitry that still turns off the voltage to the coil after a preset period, but switches the power back on as soon as a trigger signal is detected.

A revised TFI ignition system appeared in 1983 as part of Ford's fourth generation of electronic engine controls (EEC-IV). Called TFI-IV, this design uses a gray ICM that appears similar to the TFI-I module, but has a six-wire connector instead of the three-wire connector. The additional three wires connect the ICM to the PCM.

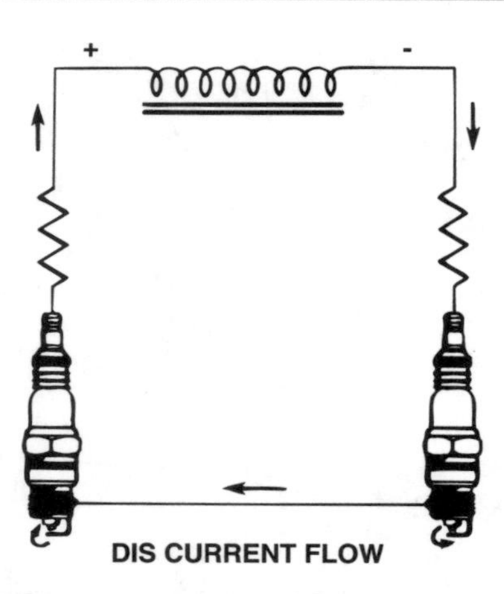

Figure 11-33. In a DIS, one spark plug in each pair always fires with forward polarity. The other plug fires with reverse polarity, with current flowing from the side to the center electrode. (General Motors)

The TFI-IV module is used with a Universal distributor that contains a Hall-effect switch. The TFI ignitions have replaced the Dura-Spark ignitions on virtually all new Ford vehicles.

The TFI module attaches to the distributor with several screws, and depends on good contact with the distributor housing for its cooling. Whenever an ICM is replaced, silicone dielectric compound must be applied to its back to improve heat conductivity to prevent premature module burnout. The ICM should never be used as a handle to turn the distributor when setting the initial timing. This kind of careless handling can cause the ICM to warp and fail soon after.

DISTRIBUTORLESS IGNITION SYSTEMS

This section contains brief descriptions of the distributorless electronic ignitions used by domestic automakers. Whenever a distributorless ignition requires service, you should always refer to the automaker's shop manual, or an appropriate repair manual to determine the exact specifications, and whether the system has any unique features you should know about.

Principles of Distributorless Ignitions

The term distributorless ignition system (DIS) refers to any ignition system without a distributor. A DIS fires the spark plugs using a multiple coil pack containing two or three separate ignition coils (according to the number of engine cylinders), and an ICM. Control circuits in the module discharge each coil separately in sequence, with each coil serving two cylinders

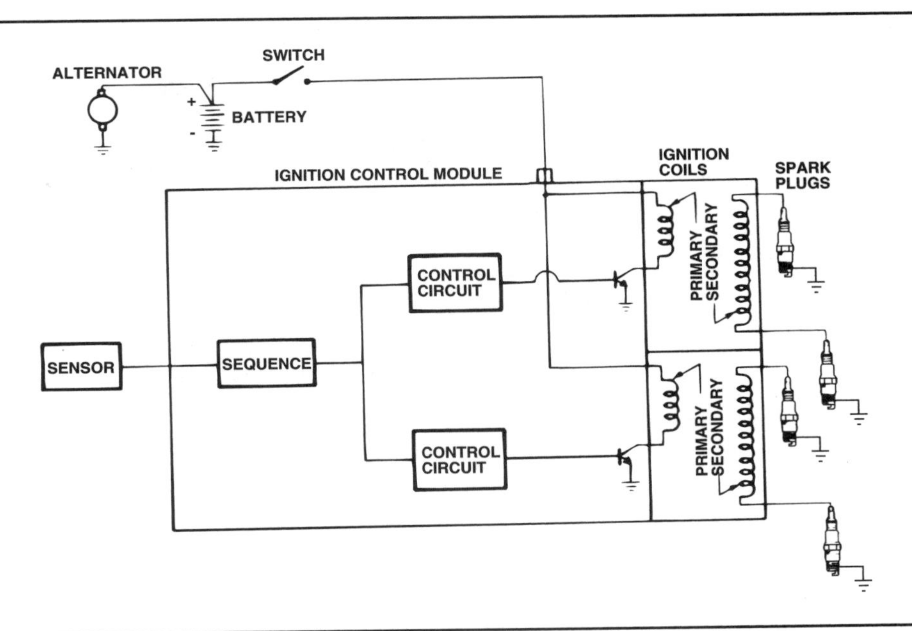

Figure 11-34. A functional schematic of a GM DIS ICM, showing the current-limiting control circuitry.

360 degrees apart in the firing order. Each coil fires two plugs simultaneously in what is called a **waste spark** method. One spark goes to a cylinder near TDC on the compression stroke, while the other fires the plug in a cylinder near TDC of the exhaust stroke. The plug in the cylinder on the exhaust stroke requires very little voltage (about 4 kilovolts) to fire, and has no effect on engine operation.

As we learned in an earlier chapter, the ignition coil secondary windings in a distributor ignition generally are wound to give the spark plug center electrode positive polarity and the side electrode negative polarity. When the spark plug fires, electrons flow from the coil secondary windings to the center electrode, across the plug gap to the side electrode where they return to the coil secondary windings through the engine block. A spark plug fired in this manner is said to have forward polarity, figure 11-32. If the electrons flow to the side electrode and across the plug gap to the center electrode, the plug is said to have reverse polarity.

In a distributorless ignition, each pair of spark plugs is connected to one coil. In this system, one plug is always fired with forward polarity, and the other is always fired with reverse polarity, figure 11-33. Because firing a spark plug with reverse polarity takes about 30 percent more energy, a misfire could result if the DIS coils did not have a different saturation time and primary current flow than a conventional coil. This provides more than 40 kilovolts of available energy—as much as 20 percent more than a conventional coil.

The ICM determines and maintains the coil firing order. When it orders the coil to fire, one spark plug fires forward and the other fires backward. The voltage drop across each plug is determined by firing polarity and cylinder pressure. The ICM also controls primary current flow and limits dwell time. The low resistance of the primary coil winding, combined with an applied voltage of 14 volts, results in a theoretical current flow greater than 14 amps, helping to decrease the coil's saturation time. Such a high current flow, however, will damage the system components unless it can be limited to a range of 8.5 to 10 amps. Limiting of the circuit current to the safe range is done by a control circuit inside the ICM, figure 11-34.

Some ICM's use a type of **closed-loop dwell control**, figure 11-35. In this system, the module continuously monitors coil buildup for maximum current. If maximum current was reached

Waste Spark: A type of ignition system without a distributor where one coil in a coil pack fires two spark plugs at the same time. On cylinder compression, the spark ignites the air-fuel mixture, while the spark in the cylinder on its exhaust stroke is wasted.

Closed-Loop Dwell Control: A type of distributorless ignition system in which the ignition control module varies dwell time in response to previous coil current buildup.

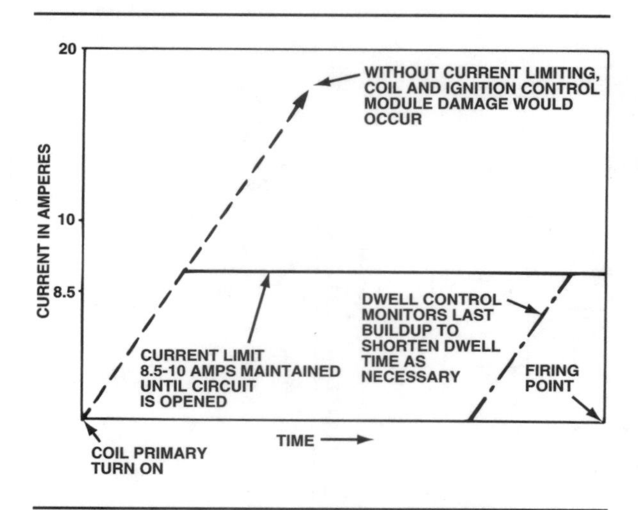

Figure 11-35. Closed-loop dwell allows full saturation of the ignition coil by increasing or decreasing dwell time. (General Motors)

during the previous buildup, the ICM shortens dwell time to lower the wattage used by the system. If minimum current was not reached during the previous buildup, the ICM lengthens dwell time to permit full saturation of the coil. When current limiting takes place before coil discharge, the module decreases dwell time for the next cycle.

Delco-Remy

Computer-controlled coil ignition (C³I)

This distributorless ignition was introduced in 1984, and is used on Buick-built V6 engines. There is no distributor with signal generator, rotor, and cap in the C³I system. The ignition trigger signal is provided by Hall-effect crankshaft and camshaft position sensors. The crankshaft sensor provides a signal that indicates basic timing, crankshaft position, and engine speed. The camshaft position sensor, figure 11-36, provides a firing order signal. On 3800 and 3.8-liter nonturbo engines, the camshaft position sensor is mounted in the timing cover. On 3.8-liter turbocharged engines, the camshaft position sensor replaces the normal distributor, and is mounted in the distributor location to drive the oil pump. On 3300 and 3.0-liter engines, the camshaft position sensor is combined with the crankshaft sensor at the front of the engine. Regardless of their appearance or location, all camshaft position sensors have the same purpose, and use the same electrical signals.

The crankshaft and camshaft position sensors use Hall-effect switches with revolving interrupter rings to synchronize and fire the coils at the required time. The ICM sends a reference voltage through a semiconductor wafer in the Hall switch. A permanent magnet mounted in line with the semiconductor wafer induces a voltage across the semiconductor, figure 11-37. As a metal blade on the interrupter ring passes between the permanent magnet and semiconductor wafer, the magnetic field is broken and Hall voltage drops, figure 11-38. The 3800 and 3.8-liter nonturbo engines reverse the process. On these engines, the permanent magnet is mounted on the camshaft sprocket, and the Hall-effect switch is part of the timing cover sensor, figure 11-39. As the camshaft sprocket revolves, it turns the Hall-effect switch on and off.

The camshaft position sensor serves only to establish the initial ignition firing sequence during engine cranking. The ICM synchronizes the initial camshaft position sensor signal with one of the crankshaft position sensor signals during cranking, and remembers the crankshaft position sensor sequence as long as the ignition remains on.

The operation of the C³I system is very similar to that of the HEI with EST system described earlier. During starting, the ICM controls both ignition timing and spark distribution. When the engine reaches a speed between 200 and 400 rpm, the PCM overrides the ICM, assuming control of timing based on signals from the crankshaft position sensor and other engine sensors. In case of PCM failure, the ICM reassumes timing control, and operates the ignition with a fixed advance of 10 degrees before top dead center (BTDC).

Since introducing the C³I system, GM has used three different variations:

- Type 1—all three coils are molded into a single housing with a smooth exterior surface, figure 11-40. Three spark plug cable terminals are provided on each side of the housing. If one coil malfunctions, the entire coil pack must be replaced.
- Type 1 Fast Start—the coil pack can be interchanged with a Type 1, but the ICM circuitry differs and connector plugs are not compatible.
- Type 2—similar to Type 1, but the coils can be replaced individually, figure 11-40.

The Type 1 Fast Start system measures crankshaft position sensor signals more precisely, resulting in a faster startup. A dual crankshaft position sensor is located beside the harmonic balancer/crankshaft pulley on the front of the engine. The harmonic balancer has two sets of interrupter rings. The outside ring consists of 18 evenly spaced interrupter blades that deliver 18 pulses every crankshaft revolution. These pulses are called the 18X signal. The inside ring consists

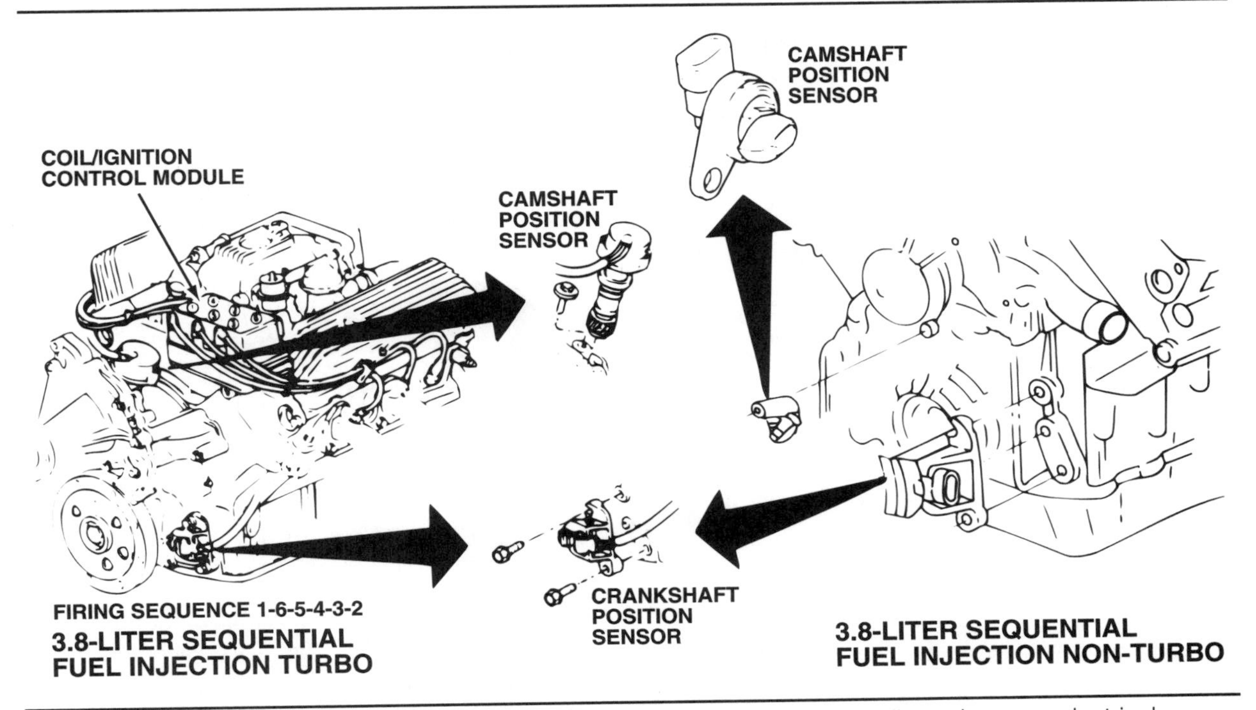

Figure 11-36. GM's C³I Hall-effect sensors vary in appearance and location, but all use the same electrical circuitry.

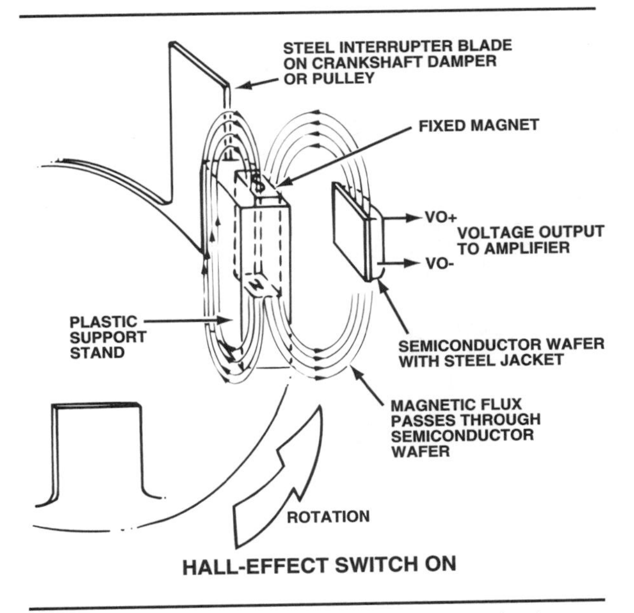

Figure 11-37. A Hall-effect switch with an unbroken field, generating a reference voltage. (General Motors)

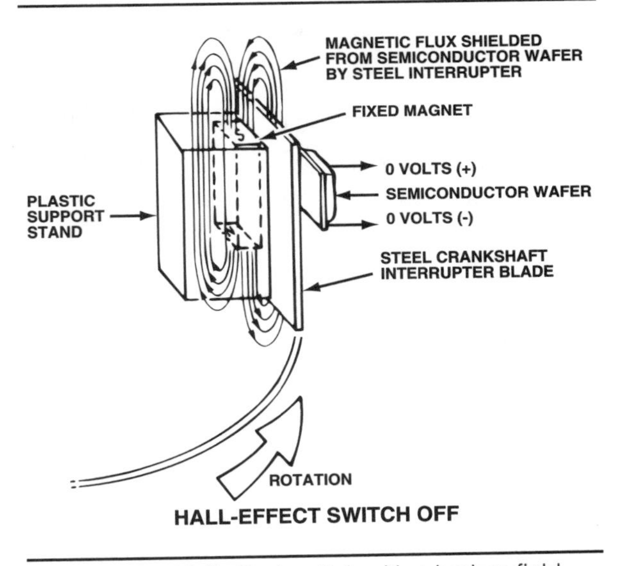

Figure 11-38. Hall-effect switch with a broken field and a voltage drop-off. (General Motors)

of three blades with gaps of 10, 20, and 30 degrees spaced at 100, 90, and 110 degrees apart, respectively. The inside ring pulses are called the 3X signal, figure 11-41.

Variations in the 3X signal allow the ICM to synchronize the correct coil without need of the camshaft position signal or the synchronization signal. Since the ICM can determine the correct coil within 120 degrees of crankshaft rotation, it starts firing on the first coil identified. The 18X pulse acts as a "clock pulse" to measure the length of each 3X pulse. The 18X pulse changes once during the 3X 10-degree gap, twice during the 20-degree gap, and three times during the 3X 30-degree gap. Once the ICM determines

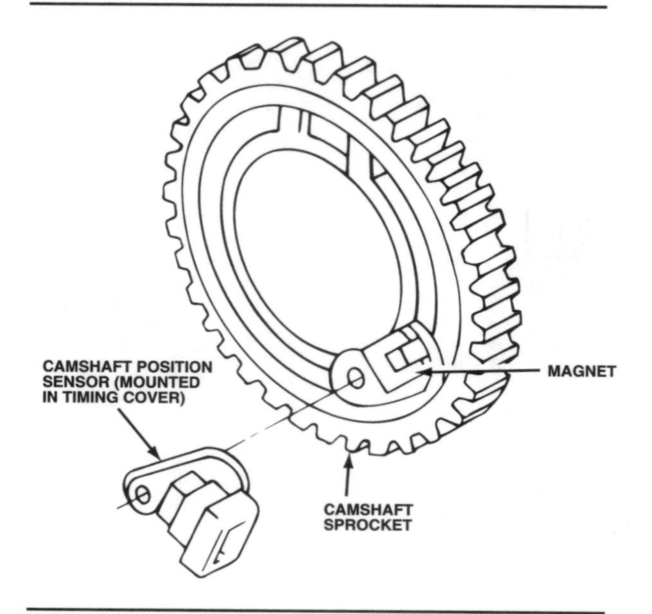

Figure 11-39. A variation of the Hall-effect switch camshaft position sensor where the magnet mounted on the camshaft sprocket rotates past the Hall-effect switch, turning it on and off. (General Motors)

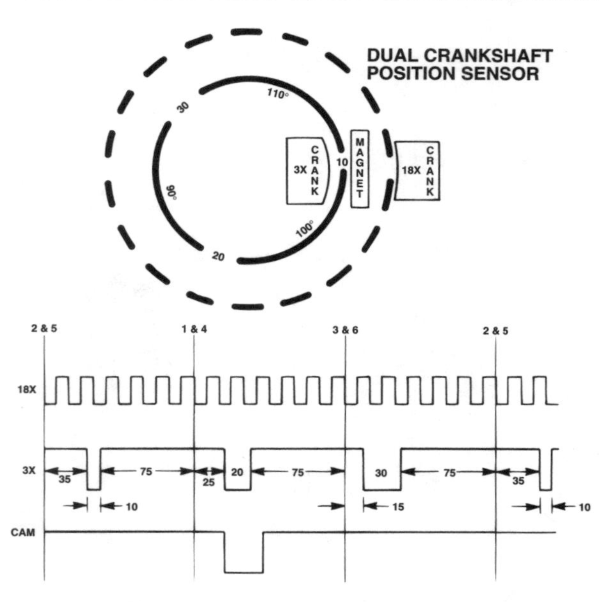

Figure 11-41. On the GM Fast-Start system, two Hall-effect switches and two sets of interrupter rings on the harmonic balancer detect crankshaft position with greater accuracy.

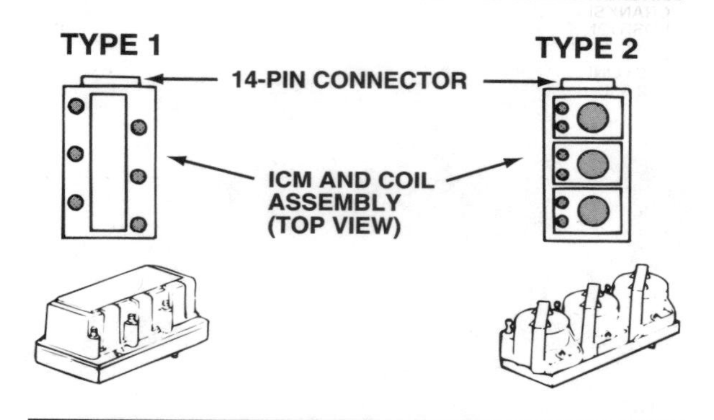

Figure 11-40. Identifying Delco's Type 1 and 2 coil and ignition control module assemblies.

which 3X pulse it is reading, it can energize the correct coil.

Direct ignition system (DIS)

The basic operation of the DIS used on many Chevrolet and Pontiac 4-cylinder and V6 engines is quite similar to the C³I system, except for the method used to sense crankshaft position. Instead of Hall-effect switches located on the front of the engine, the Chevrolet and Pontiac DIS uses a magnetic pulse generator sensor installed in the side of the engine block. When used with the 2.5-liter engine, the sensor is installed on the back of the ICM. With other 4-cylinder and V6 engines, the sensor is installed in the block below the ICM, and is connected externally, figure 11-42.

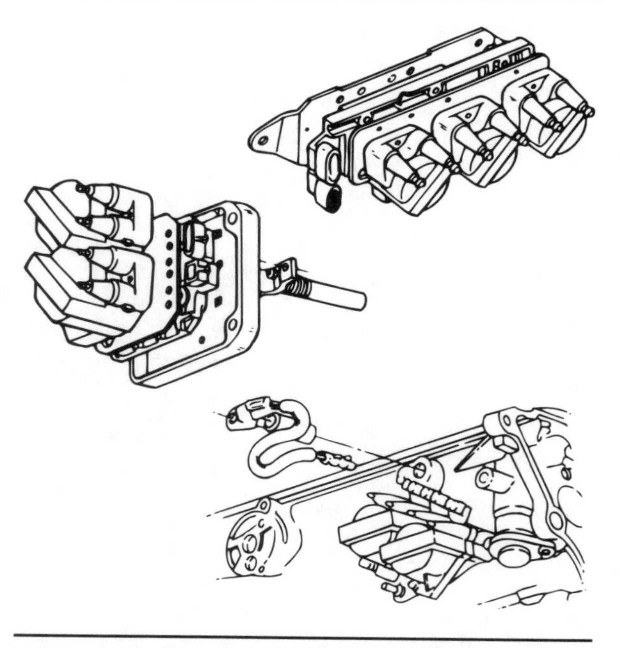

Figure 11-42. Location of GM DIS block-mounted crankshaft position sensors.

A notched wheel or reluctor is cast into the crankshaft. The crankshaft reluctor on both 4-cylinder and V6 engines is machined with seven notches or slots, and serves as the field interrupter. The sensor head consists of a perma-

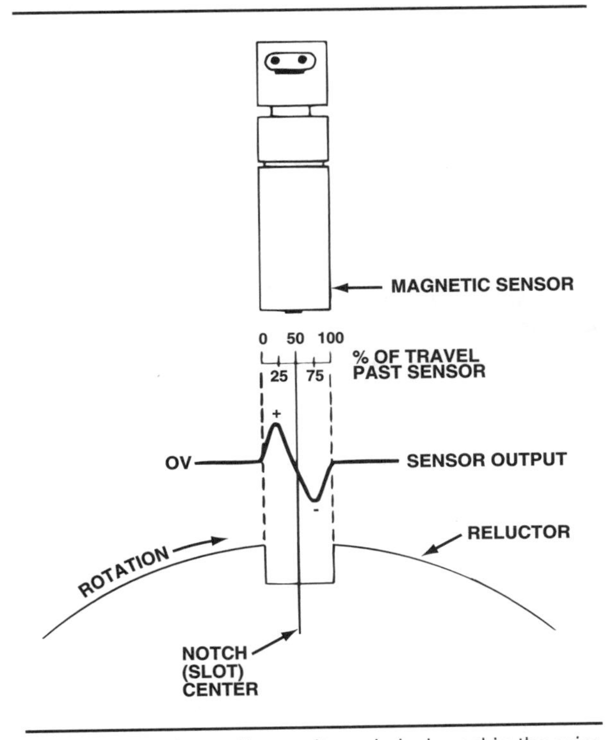

Figure 11-43. A small ac voltage is induced in the wire winding of this GM magnetic sensor when its magnetic field is interrupted by the reluctor. This voltage signal is used by the ignition control module to determine ignition firing.

nent magnet with a wire winding, and is positioned a specified distance from the reluctor. As the crankshaft rotates, the reluctor notches interrupt the sensor's magnetic field, causing a small alternating current (ac) voltage to be induced in the sensor's wire winding, figure 11-43. Because the sensor is installed in a fixed position in the engine block, and the reluctor is an integral part of the crankshaft, there is no timing adjustment possible, or required, with this system.

Six reluctor notches are evenly spaced around the reluctor surface at 60-degree intervals; the seventh notch is spaced 10 degrees from one of the six notches. The signal from the seventh notch is used by the ICM to synchronize coil firing sequence to the crankshaft position. While reluctor configuration is the same for 4-cylinder and V6 engines, the ICM calculates coil firing order and determination of crankshaft position differently.

In the V6 engine system, figure 11-44, the synchronization notch tells the ICM to ignore the first notch, and establish base timing for cylinders 2 and 5 with the second notch. The module ignores the third notch, and relies on the fourth notch to set base timing of cylinders 3 and 6. The fifth notch also is ignored in favor of the sixth notch for cylinders 1 and 4. After the seventh, or synchronization, notch passes, the entire sequence is repeated. As a result, the firing order of the first crankshaft revolution is 1-2-3, and 4-5-6 for the second revolution.

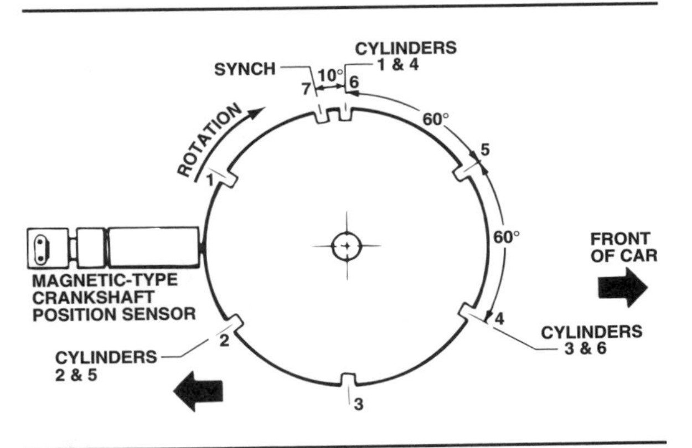

Figure 11-44. GM's V6 ignition control module recognizes notches 2, 4, and 6 as the signal to fire cylinders 2 and 5, 3 and 6, and 1 and 4, in that order.

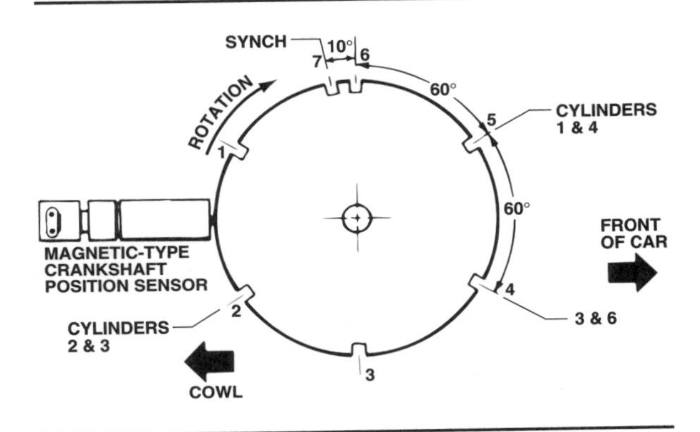

Figure 11-45. GM's 4-cylinder ignition control module recognizes notches 2 and 5 as the signal to fire cylinders 2 and 3 and 1 and 4, in that order.

With 4-cylinder engines, figure 11-45, the ICM starts the firing sequence on the seventh notch. If engine speed is below a predetermined value, the module fires each coil at a specified interval based only on engine speed. The synchronization notch tells the ICM to ignore the first notch, and to use the second notch to establish 10-degrees BTDC timing for cylinders 2 and 3. The ICM ignores the third and fourth notches, but uses the fifth notch to establish an equivalent timing setting for cylinders 1 and 4. In this way, the number 2 and 3 coil is fired first during startup.

The reference pulse in both systems is pulled low by the notch ahead of the one used to fire the cylinder, returning to its high state when the cylinder firing notch passes. The change in reference voltage is sent to the PCM for use in EST and fuel injection.

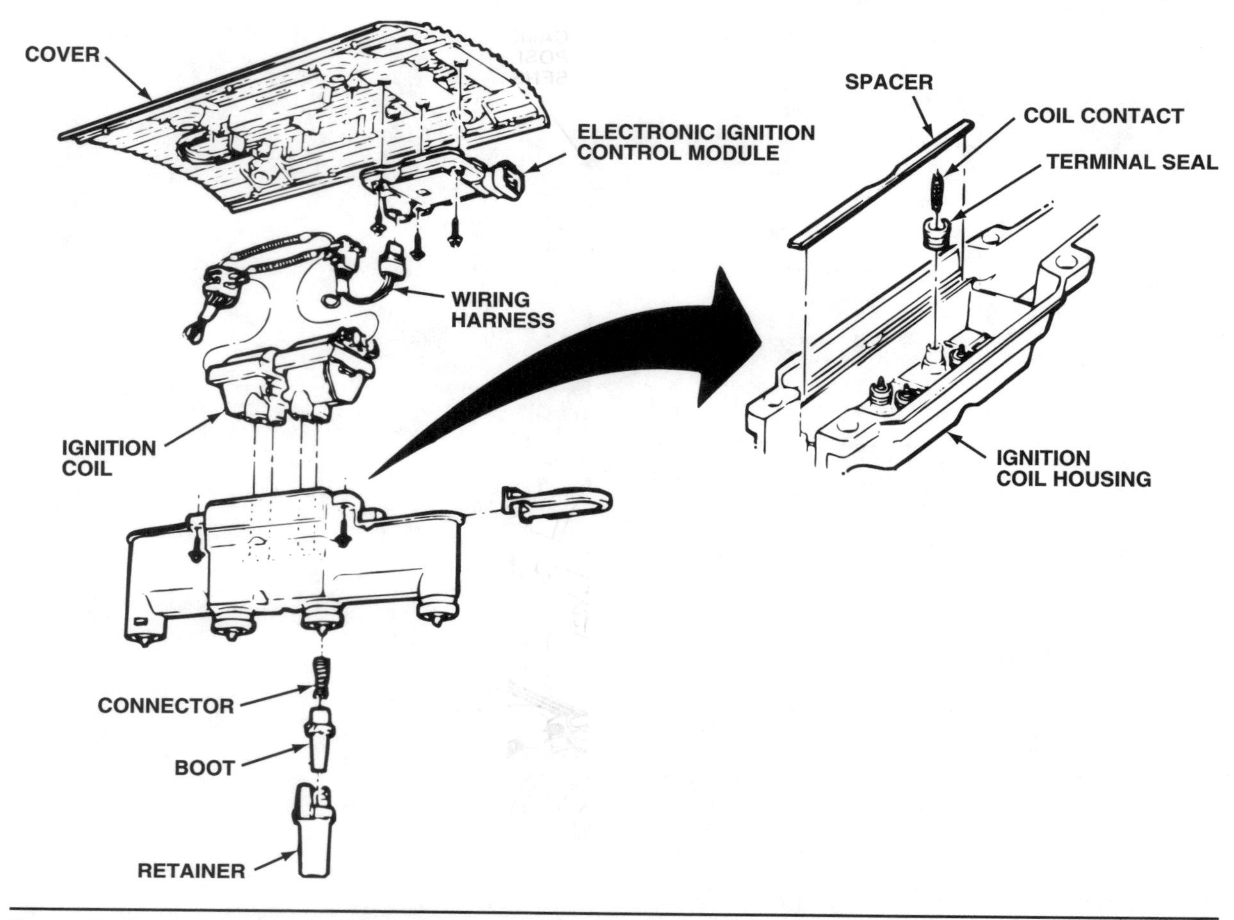

Figure 11-46. An exploded view of GM's IDI system components.

Integrated direct ignition (IDI)

This system is used only on the Oldsmobile-built Quad 4 engine, and differs from other 4-cylinder DIS's primarily in the configuration of the system's components, figure 11-46. The coil pack and ICM are contained in a unit that connects directly to the spark plugs. This eliminates the use of spark plug cables, but the entire housing must be removed when changing spark plugs. The coils can be replaced individually.

Chrysler

Direct ignition system (DIS)

Chrysler uses a DIS, figure 11-47, on its 3.3- and 3.5-liter engines. Crankshaft timing is determined by a magnetic sensor installed in the transaxle bellhousing. The single-board engine controller (SBEC), or PCM, sends an eight-volt reference signal to the sensor. The transaxle drive plate contains three groups of four slots. Each group of slots is positioned 20 degrees apart, and provides a signal for two spark plugs. Transsaxle drive plate rotation makes and breaks the sensor's magnetic field, causing sensor output voltage to the PCM to vary between near zero and five volts. The PCM uses this voltage signal to determine engine speed and calculate both timing advance and the required fuel delivery.

The camshaft sensor, also a magnetic type, is installed in the timing chain case cover and functions in the same way as the crankshaft sensor. The camshaft timing gear contains five areas with notched slots. Two areas have a single slot, two have two slots, and one has three slots. These are arranged around the timing gear to produce one long and four short solid unnotched surfaces. This arrangement of notched and solid area produces a predictable sequence of voltage signals sent to the PCM. The PCM uses the signals to determine crankshaft position, then calculates which coil and injector pair to energize within one crankshaft revolution during startup.

Ford (Motorcraft)

All Ford distributorless ignitions operate on the same basic principles as those used by GM and

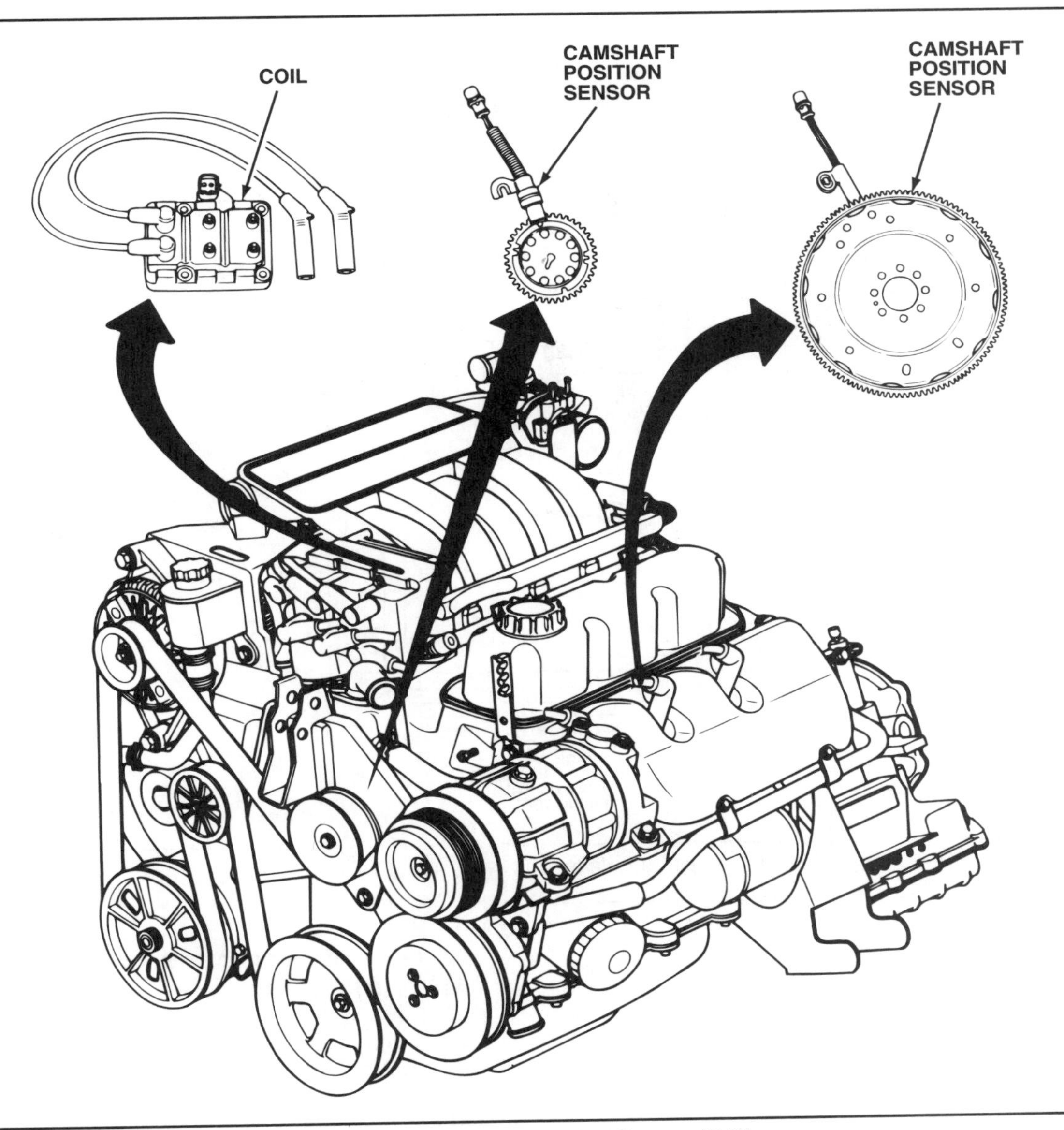

Figure 11-47. Major components of the Chrysler Direct Ignition System (DIS).

Chrysler. Each coil fires two spark plugs, with one spark igniting the mixture at the top of the compression stroke and the other spark wasted at the top of the exhaust stroke. One plug of each pair has positive polarity; the other has negative polarity.

4-2 distributorless ignition system (DIS)
Sometimes called the Dual Plug DIS, this unusual distributorless ignition was introduced on 1989 2.3-liter, 4-cylinder truck engines, figure 11-48. Each cylinder uses two spark plugs, with one plug installed on each side of the combustion chamber. Plugs on the right side of the engine form the primary system, and are responsible for engine operation at all times. Plugs on the left side of the engine form the secondary system, and are switched on and off by the EEC-IV computer, according to engine speed and load requirements.

Only the primary plugs fire when the engine is cranking. Once the engine is running, the EEC-IV PCM commands the DIS ICM through the dual plug inhibit (DPI) circuit to switch from single to dual-plug operation. The EEC-IV PCM also is responsible for ignition timing and dwell.

The 4-2 DIS uses two four-tower DIS coil packs, figure 11-49, with a single remote DIS module, figure 11-50. A bracket-mounted dual

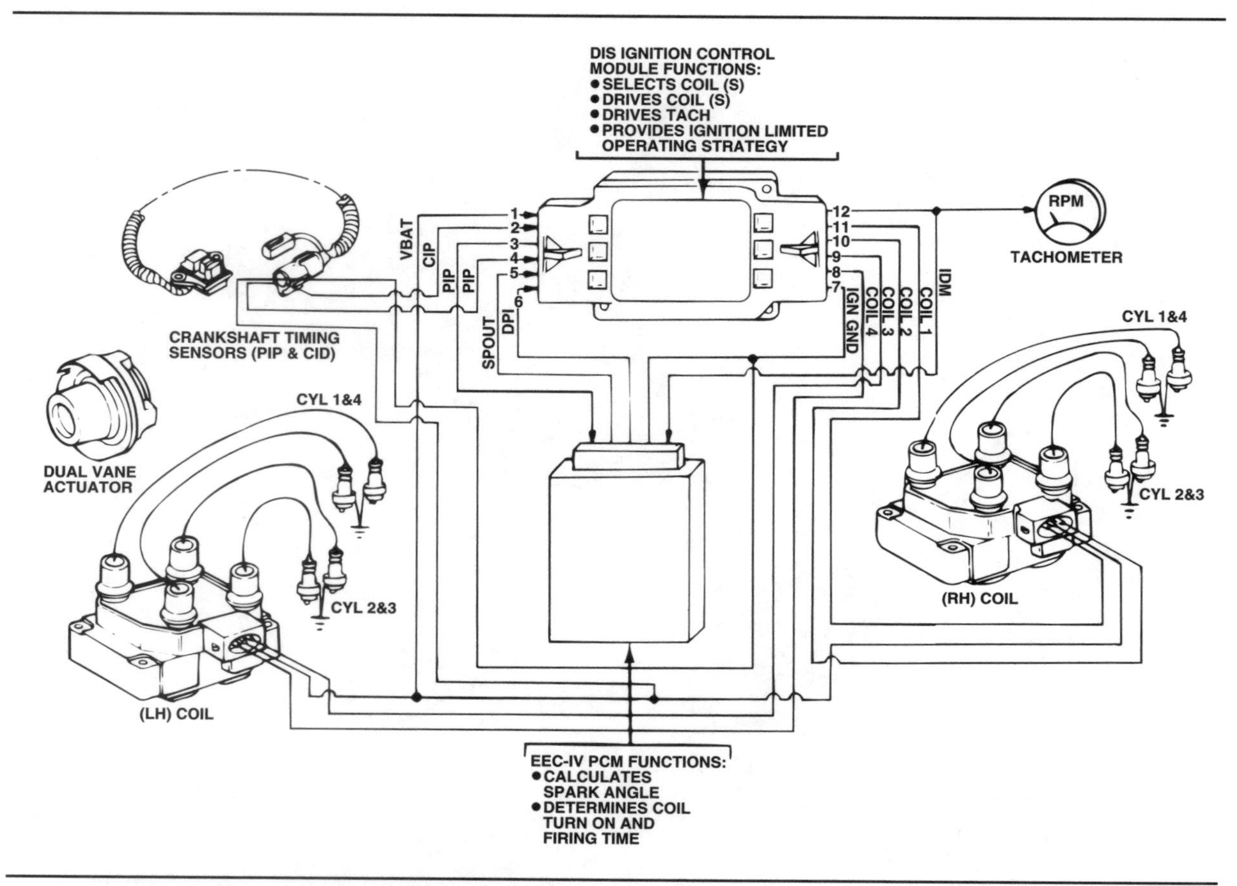

Figure 11-48. A diagram of the Ford 4-2 Distributorless Ignition System (DIS).

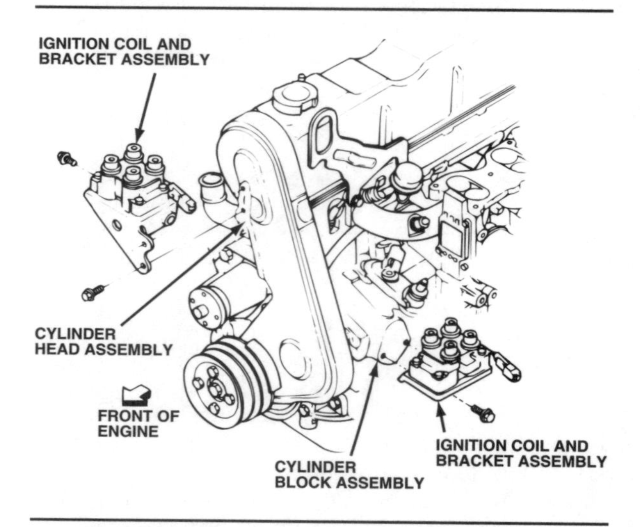

Figure 11-49. Ford 4-2 DIS coil pack locations.

Hall-effect crankshaft position sensor near the crankshaft damper completes the system, figure 11-51. The right coil fires the primary plugs during normal operating conditions; the left coil fires the secondary plugs as directed by the EEC-IV PCM and the DIS ICM.

The crankshaft position sensor works on the same principles as other Hall-effect switches you have studied, and is very similar in operation to the dual sensor used by GM in its C^3I ignition. A pair of rotating vane cups on the crankshaft damper produces a profile ignition pickup (PIP) signal for base timing data, and a cylinder identification (CID), or camshaft position sensor, signal used by the DIS ICM determines which coil to fire.

The EEC-IV PCM sends a spark output (SPOUT) signal to the DIS ICM. The leading edge triggers the coil, and the trailing edge controls dwell time. This feature is called computer-controlled dwell (CCD). A buffered tach signal called ignition diagnostic monitor (IDM) supplies ignition system diagnostic information used for self-test.

A CID sensor or circuit failure will not result in a no-start condition, for the DIS ICM randomly selects and fires one of the two coils under such circumstances. The result of the ICM's guess may be an engine that is hard to start. However, turning the key off and then back on to crank the engine again allows the ICM to

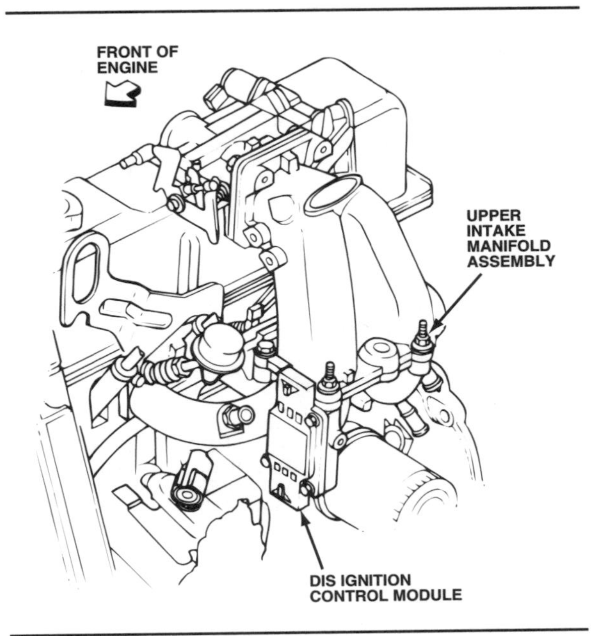

Figure 11-50. Ford 4-2 DIS ignition control module location.

make another guess. After a few tries, the module will make the right choice and select the correct firing sequence.

If an ignition failure results in the loss of the SPOUT signal, a failure effects management (FEM) program in the EEC-IV PCM memory prevents total driveability loss. The EEC-IV PCM opens the SPOUT line, allowing the DIS ICM to fire the coils directly from the PIP output. This results in a fixed spark angle of 10-degrees BTDC with fixed dwell.

V6 distributorless ignition system (DIS)
This DIS also was introduced in 1989 on the 3.0-liter SHO V6 engine and the 3.8-liter supercharged V6 engine. The system functions in essentially the same way as the 4-2 DIS discussed above, but its components differ, figure 11-52:

- A single six-tower coil pack contains three coils with an individual tach wire for each coil; the coil pack is serviced as an assembly.
- A single set of spark plugs is used, with one plug per cylinder.
- The 3.0-liter CID sensor is installed on the end of one camshaft.
- The 3.8-liter CID sensor is installed in the engine block where the distributor would normally be located.
- A single Hall-effect switch is used as the crankshaft sensor.

When the engine is cranked, the DIS ICM looks for a change in the CID signal, from high to low

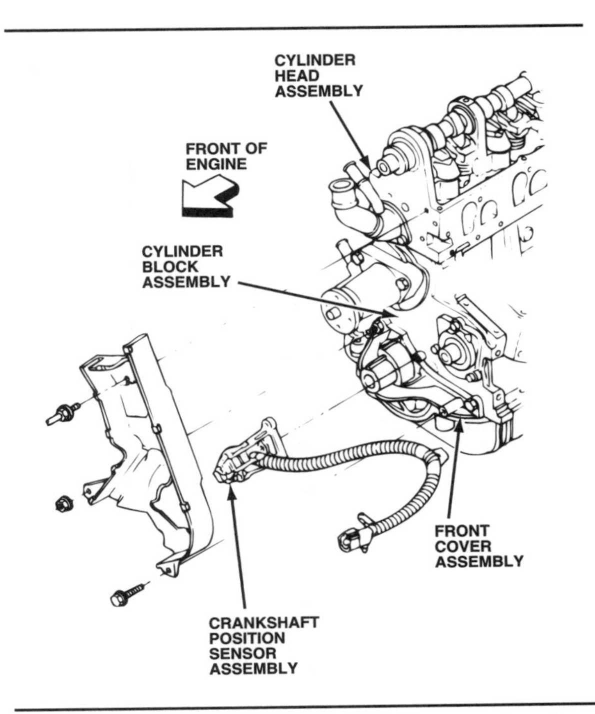

Figure 11-51. Ford 4-2 DIS crankshaft sensor location.

or low to high. As soon as the ICM sees the leading or trailing edge of the CID signal, it prepares to fire coil 2 of the coil pack. Once the change in the CID signal occurs, the ICM looks for the trailing edge of the SPOUT signal to turn on coil 2. When the ICM sees the next leading edge of the SPOUT signal, it turns the primary current off to coil 2, which fires coil 2, figure 11-53. The DIS module always fires the coils in one coil pack in a given order. At engine start-up, the coil firing sequence is always 2, 3, 1. Because the coils continue to fire in the same order, they are synchronized with compression and remain synchronized as long as the engine is running, even if the DIS module loses the CID signal. The SPOUT signal tells the DIS ICM when to fire the next coil as long as the engine is running. If the SPOUT circuit opens, the PIP signal is used by the ICM to fire the coils.

Electronic direct ignition system (EDIS)
A second-generation DIS, EDIS functions faster and with greater accuracy than the V6 DIS just described. Because the system has been used on 4-cylinder, V6, and V8 engines since its introduction in 1990, the number of coils and coil packs differs according to engine application. Other EDIS components also differ in appearance and location, but the system functions essentially the same regardless of engine application. Figure 11-54 shows the EDIS components used on the 1.9-liter SEFI engine.

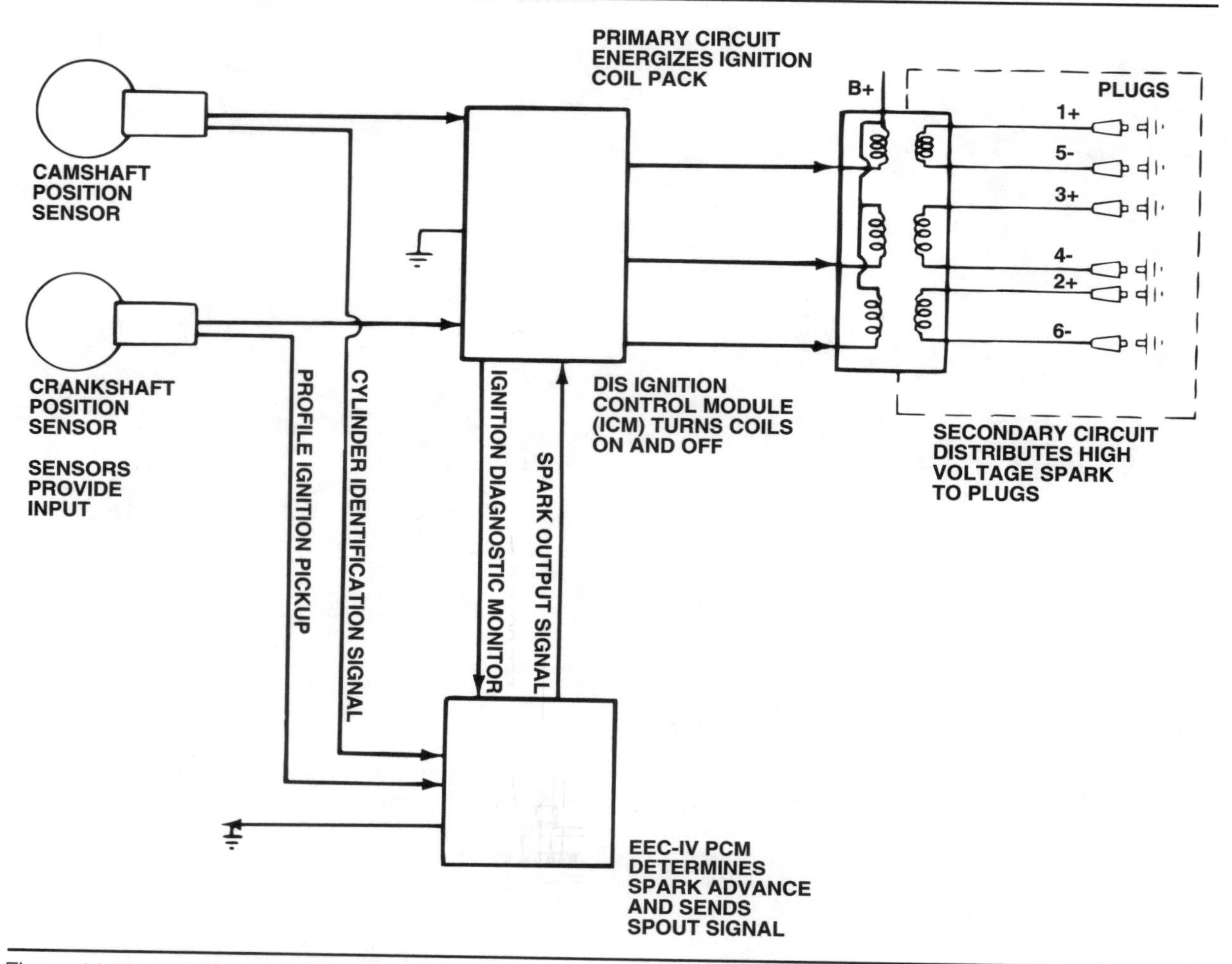

Figure 11-52. A diagram of the Ford V6 DIS.

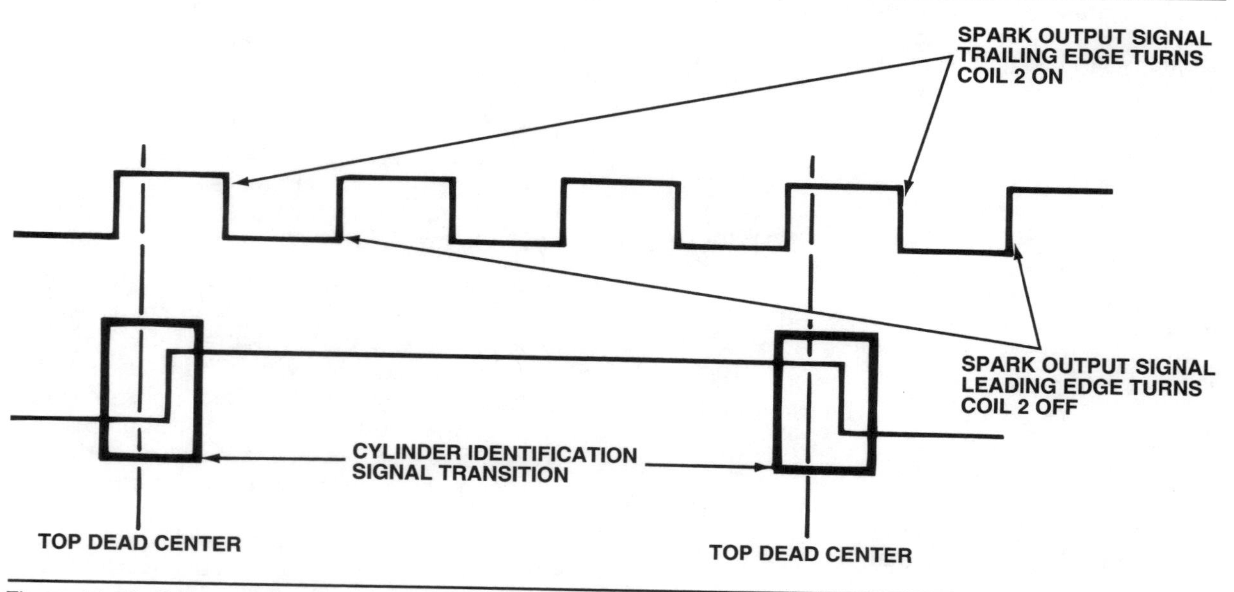

Figure 11-53. Coil selection with the engine cranking. (Ford)

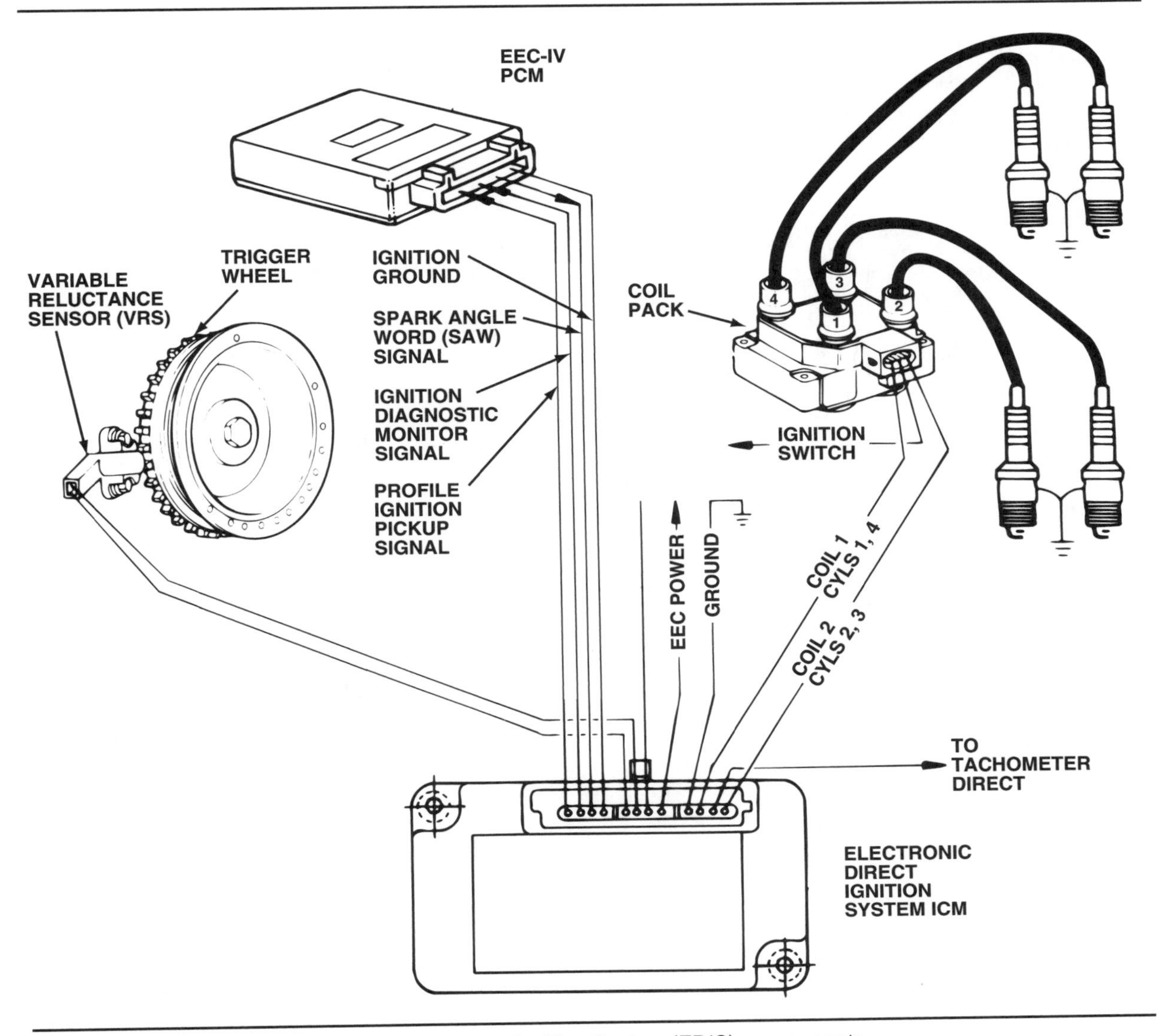

Figure 11-54. The Ford 1.9-liter Electronic Direct Ignition System (EDIS) components.

There are other differences between EDIS and DIS:

- EDIS does not use a CID (camshaft position) sensor or a crankshaft position sensor; a variable reluctance sensor and trigger wheel perform these functions.
- EDIS crankshaft position signals are more sophisticated and complex than those used on DIS's.
- EDIS uses a spark angle word (SAW) signal in place of the DIS SPOUT signal. The SAW signal also is more complex than a SPOUT signal.
- The EDIS ICM is smarter, faster, and has increased diagnostic ability compared with its predecessor, the DIS ICM.

The variable reluctance sensor (VRS) is a magnetic transducer containing a pole piece wrapped with fine wire, figure 11-55. If the transducer is

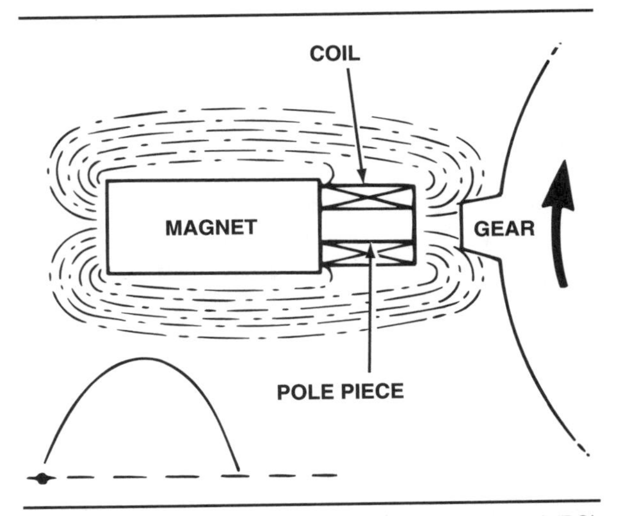

Figure 11-55. A Ford variable reluctance sensor (VRS).

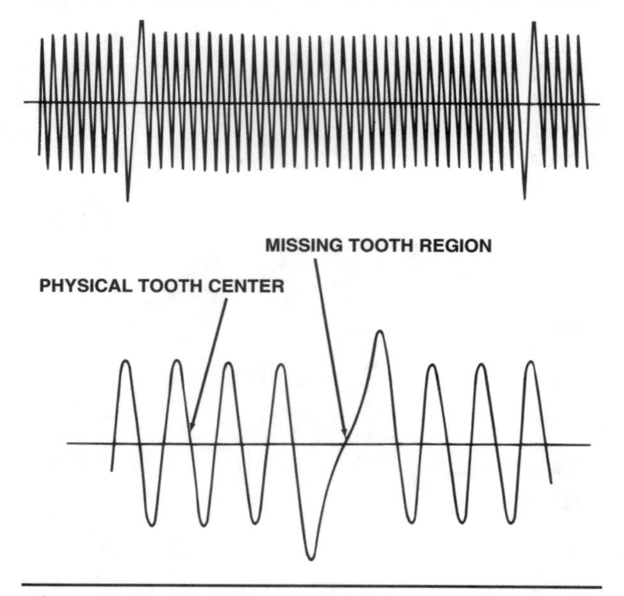

Figure 11-56. The VRS analog signal wave. (Ford)

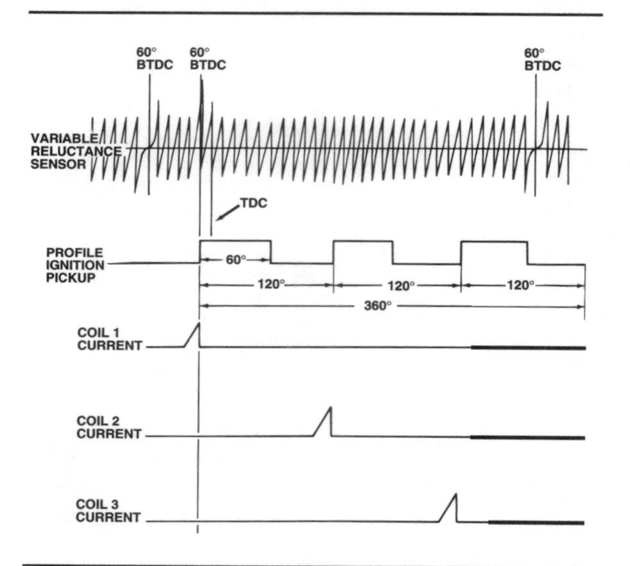

Figure 11-57. Comparison of analog VRS sensor signals with digital PIP signals created by the EDIS ignition control module. (Ford)

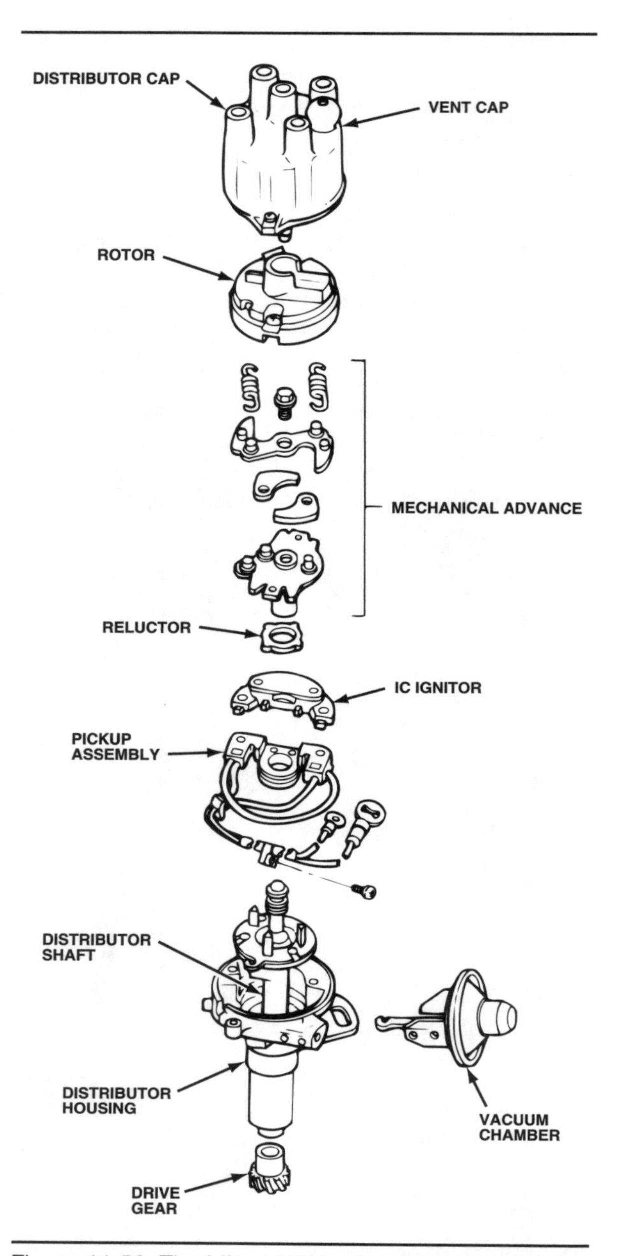

Figure 11-58. The Mitsubishi ignition system used on 2.6-liter engines is typical of the inductive discharge ignitions used on Japanese vehicles.

exposed to a change in flux lines, a differential voltage will be induced across the terminals of the wire windings. Thus, when a ferromagnetic toothed timing wheel on the crankshaft rotates in the presence of the VRS, the passing teeth cause the VRS reluctance to change, resulting in a varying analog voltage signal. The timing wheel has 35 teeth and a blank spot spaced at 10-degree increments. The blank spot for the 36th tooth serves as a fixed reference point for number 1 piston travel identification.

During cranking, the EEC-IV PCM refers to a predetermined fuel control strategy stored in its memory to supply the necessary fuel for starting the engine. At the same time, the EDIS ICM looks for any significant change in the VRS signal. When it recognizes the missing tooth, figure 11-56, the EDIS ICM has a reference point and is synchronized to fire the proper coil. While the engine continues to crank, the EDIS ICM fires the coil at 10-degrees BTDC (base timing). When the coil for piston number 1 is fired, a PIP signal is sent from the EDIS ICM to the EEC-IV PCM. The

PIP signal provides the EEC-IV PCM with crankshaft position and engine speed data. Because the PIP signal is generated by the EDIS ICM, it takes the form of a digital square wave with a 50-percent duty cycle, figure 11-57. When the EEC-IV computer recognizes the PIP signal, it enables fuel and spark functions. It also processes the PIP signal with other information to determine the SAW signal. This signal is sent to the EDIS module, which uses it to determine if the ignition timing should be advanced or retarded.

Imported Car Electronic Ignitions

Automakers of imported vehicles began equipping their vehicles with electronic ignitions shortly after domestic automakers. All major imported cars now use an inductive-discharge electronic ignition. In most of these systems, timing signals are sent to the ignition ICM from a magnetic pulse generator in the distributor, although Bosch makes a Hall-effect distributor used by Volkswagen.

The electronic ignition used on the Mitsubishi-built 2.6-liter engine in some Chrysler vehicles is typical of most Japanese designs, figure 11-58. The distributor contains a magnetic pickup with an integral IC ignitor (ICM), although some Toyota ignitions use an external ignitor mounted on the coil. The variable-dwell ignition contains no ballast resistor. Full battery voltage is supplied to both the coil and ignitor whenever the ignition switch is in the START or RUN position.

SUMMARY

Solid-state electronic ignition systems (SSI's) came into widespread use on both domestic and import vehicles in the early 1970s. Today, they are standard on all imported and domestically built cars and light trucks. SSI's provide greater available voltage and reliable performance for longer periods under varying operating conditions, with decreased maintenance.

SSI's use inductive discharge or capacitive discharge systems. The inductive discharge system is the most common. The primary circuits of SSI's can be triggered by breaker points, a magnetic pulse generator, a Hall-effect switch, a metal detector, or an optical triggering device. Original-equipment systems generally use a magnetic pulse generator, or a Hall-effect switch, in the distributor to trigger the primary circuit.

The primary circuit is switched on and off by transistors in the ICM. The solid-state ICM also controls the length of the ignition dwell period. Electronic ignitions with fixed dwell use a ballast resistor; those with variable dwell have no ballast resistor.

Basic electronic ignition systems use the same centrifugal and vacuum advance devices as breaker point ignitions. Those electronic ignitions used with electronic engine control systems have no advance mechanisms. Spark advance is controlled by the system's PCM.

Electronic ignitions have undergone major change and improvement since their introduction in the mid-1970s. Distributor-type ignitions are being replaced by distributorless ignitions, with increasing responsibility for their operation assigned to the electronic engine control system.

Individual-Coil DIS

The DIS's we have detailed in this chapter commonly employ half as many coils as they have spark plugs to fire. These systems fire each pair of spark plugs simultaneously: one with forward polarity and the other with reverse polarity. Depending on piston position, one spark is a waste spark. These systems eliminate moving parts, and reduce the associated friction and wear and tear. They are also economical to produce and quite adequate for their intended use. However, there is a better, though more expensive, way.

Some Ford, Acura, BMW, Nissan, and Saab models have DIS's that use a single ignition coil for each spark plug. In addition to the advantages of the waste-spark DIS, individual-coil DIS fire all plugs individually with forward polarity for a hotter spark, which provides better performance at high rpm or under load. These systems further reduce the length of the secondary cable or eliminate it altogether, decreasing the chance of secondary arcing or leakage, improving the long-term reliability of the system, and eliminating the chance of crossfire. Furthermore, shortening or eliminating the spark plug cable greatly reduces the amount of radio-frequency interference emitted by the ignition system.

Review Questions

Choose the single most correct answer.
Compare your answers to the correct answers on page 507.

1. Which of the following is true of solid-state ignition systems?
   a. Provide less available voltage than breaker-point systems
   b. Provide greater available voltage than breaker-point systems
   c. Give less reliable system performance at all engine speeds
   d. Require more maintenance than breaker-point systems

2. Inductive discharge systems use a triggering device to provide a sudden ___ in primary current.
   a. Increase
   b. Decrease
   c. High-voltage surge
   d. Low-voltage pulse

3. Which of the following is true of inductive discharge systems?
   a. Used as original equipment by most automakers
   b. A common aftermarket installation
   c. Provide about 50,000 volts of available voltage
   d. Sustains a spark for about 200 microseconds

4. Capacitive discharge systems provide ___ secondary voltage when compared to inductive discharge systems.
   a. Exactly the same
   b. About the same
   c. Greater
   d. Less

5. Transistors have:
   a. A base of one type of material and an emitter and collector of another
   b. An emitter of one type of material and a base and collector of another
   c. A collector of one type of material and a base and emitter of another
   d. A collector, base, and emitter all of the same material

6. The earliest solid-state ignition systems used which of the following triggering devices?
   a. Metal detectors
   b. Breaker points
   c. Magnetic pulse generators
   d. Light detectors

7. The most common type of original-equipment triggering devices are:
   a. Breaker points
   b. Light detectors
   c. Metal detectors
   d. Magnetic pulse generators

8. The dwell period of a solid-state system can be measured with:
   a. A voltmeter
   b. An ammeter
   c. An ohmmeter
   d. An oscilloscope

9. In the Delco High-Energy Ignition (HEI) system, the dwell period is controlled by the:
   a. Pole piece
   b. Timer core
   c. RFI filter capacitor
   d. Electronic ignition control module

10. The Delco HEI system uses:
    a. A single ballast resistor
    b. A dual ballast resistor
    c. No ballast resistor
    d. Calibrated resistance wire

11. The Chrysler electronic ignition system uses a:
    a. Dual ballast resistor
    b. Single ballast resistor
    c. No ballast resistor
    d. Calibrated resistance wire

12. The rotating component attached to the distributor shaft in the Chrysler electronic ignition is called the:
    a. Armature
    b. Reluctor
    c. Timer core
    d. Trigger wheel

13. The rotating component attached to the distributor shaft in the Ford electronic ignition is called the:
    a. Armature
    b. Reluctor
    c. Timer core
    d. Trigger wheel

14. Ford's Dura-Spark I system uses:
    a. A single ballast resistor
    b. A dual ballast resistor
    c. No ballast resistor
    d. A calibrated resistance wire

15. A Hall-effect switch requires ___ to generate an output voltage.
    a. A magnetic pulse generator
    b. A zener diode
    c. A 45-degree magnetic field
    d. An input voltage

16. Thick-film integrated (TFI) ignition control modules are:
    a. Mounted inside the distributor
    b. Mounted on the side of the distributor
    c. Mounted on the vehicle fenderwell
    d. Used as a handle to rotate the distributor

17. Dual-mode ignition control modules are used with:
    a. Dura-Spark I
    b. TFI-I
    c. Dura-Spark II
    d. TFI-IV

18. When a distributorless ignition coil fires its two spark plugs, one spark ignites the mixture at the ___ of the compression stroke and the other spark is wasted at the ___ of the exhaust stroke.
    a. Top, top
    b. Top, bottom
    c. Bottom, bottom
    d. Bottom, top

19. The initial firing sequence in a distributorless ignition is established by the ___ sensor.
    a. Vehicle speed
    b. Crankshaft
    c. Camshaft
    d. Either b or c, depending on the system

20. When a DIS ignition control module fires the spark plug in a cylinder on its exhaust stroke, there is ___ effect on engine operation.
    a. No
    b. A minor
    c. A stabilizing
    d. A detrimental

21. The Ford Dual Plug DIS is used on:
    a. 3.0-liter SHO V6 engines
    b. 3.8-liter supercharged V6 engines
    c. 2.3-liter truck engines
    d. 4-cylinder, V6, and V8 engines

PART THREE

The Fuel System

Chapter

12

Fuel and Air Intake System Operation

Creating a correct air-fuel mixture requires a properly functioning fuel and air delivery system. This system uses the following components:

- A fuel storage tank
- Fuel delivery lines
- Evaporative emission controls
- The fuel pump
- The fuel filters
- Air cleaners and filters
- Thermostatic air cleaner controls.

This chapter examines basic fuel and air intake system operation and its effects on engine performance, economy, and emissions.

TANKS AND FILLERS

The automobile fuel tank, figure 12-1, is made of two corrosion-resistant steel halves that are ribbed for additional strength and welded together. Exposed sections of the tank may be made of heavier steel for protection against road damage and corrosion.

Some cars, sports utility vehicles, and light trucks have an auxiliary fuel tank. Automakers make some of the auxiliary tanks to these vehicles of polyethylene plastic and most recently, of composites.

Tank design and capacity are a compromise between available space, filler location, fuel expansion room, and fuel movement. Some later-model tanks deliberately limit tank capacity by extending the filler tube neck into the tank low enough to prevent complete filling, figure 12-1. A vertical **baffle** in this same tank limits fuel sloshing as the car moves.

Regardless of size and shape, fuel tanks must have:

- An inlet or filler tube through which fuel can enter the tank
- A filler cap
- An outlet to the fuel line leading to the fuel pump
- A vent system.

Tank Location and Mounting

Most domestic cars use a horizontally suspended fuel tank, usually mounted below the rear of the floor pan, figure 12-2, between the frame rails and just ahead of or behind the rear axle. Station wagons may use a vertically positioned tank located on one side of the car between the outer and inner rear fender panels. To prevent squeaks, some cars have felt insulator strips cemented on the top or sides of the tank wherever it contacts the underbody.

Fuel inlet location depends on the tank design and filler tube placement. It is usually located behind a filler cap or a hinged door in the center

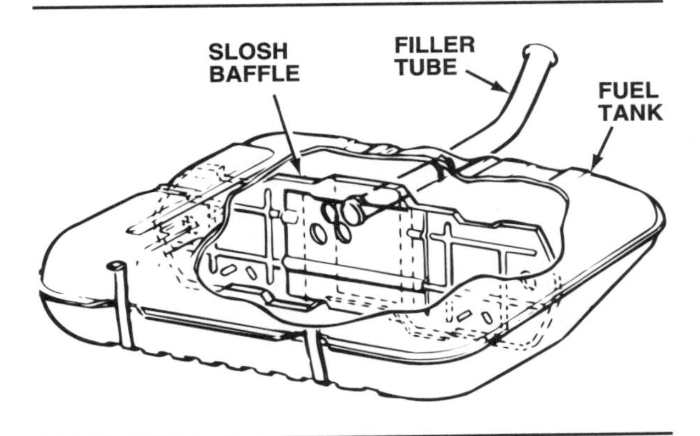

Figure 12-1. The filter tube is positioned in a way that the tank cannot be filled completely. The air space at the top of the tank allows room for fuel expansion. (Oldsmobile)

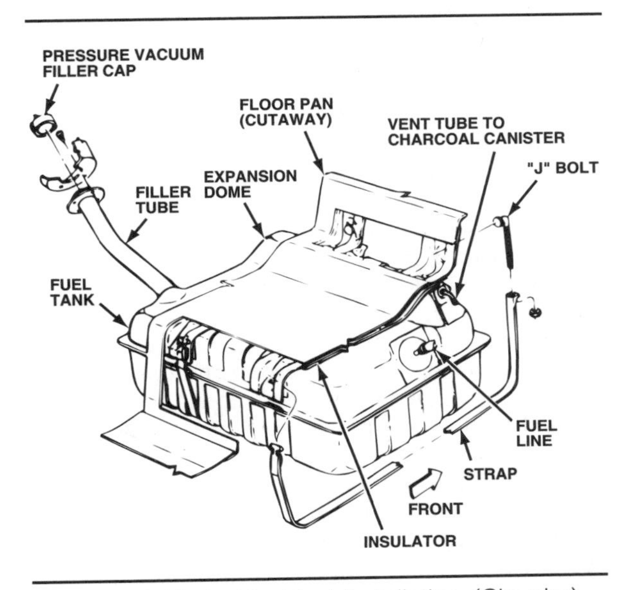

Figure 12-2. Typical fuel tank installation. (Chrysler)

of the rear panel or in the outer side of either rear fender panel. Some older cars have their fuel inlet in other positions. For example, the Type 1 Volkswagen had the fuel inlet under the front hood or in the front body panel. Today, all cars use catalytic converters and have decals reading "Unleaded Fuel Only" near their filler caps or on their instrumentation panels.

Generally, a pair of metal retaining straps holds a fuel tank in place. Underbody brackets or support panels hold the strap ends using bolts. The free ends are drawn underneath the tank to hold it in place, then bolted to other support brackets or to a frame member on the opposite side of the tank. The retaining straps fastened between the inner wheel well and quarter panel holds a station wagon's tank.

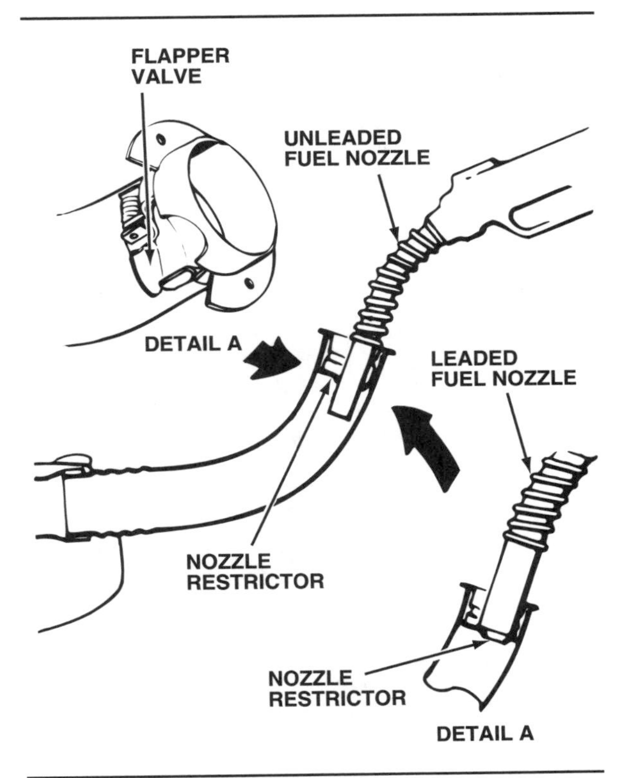

Figure 12-3. Vehicles have restrictors in their filler tubes so only the smaller unleaded fuel pump nozzles can fit. Restrictors may be spring-loaded flapper valves or simple ring-shaped pieces inside the tube.

Filler Tubes

Fuel enters the tank through a large tube extending from the tank to an opening on the outside of the vehicle. There are two types of filler tubes: a rigid, one-piece tube soldered to the tank, figure 12-2, and a three-piece unit. The three-piece unit has a lower neck soldered to the tank and an upper neck fastened to the inside of the body sheet metal panel. The two metal necks are connected by a length of hose that is clamped at both ends.

Federal regulations require that all vehicles with catalytic converters—and that is all late-model cars and light trucks—burn unleaded fuel, because lead coats their catalytic material and makes the converters useless. To prevent using leaded fuel, which is unavailable to most motorists, unleaded fuel vehicles have smaller openings at their filler tubes. The smaller-diameter unleaded fuel nozzles at gas station pumps

Baffle: A plate or obstruction that restricts the flow of air or liquids. The baffle in a fuel tank keeps the fuel from sloshing as the car moves.

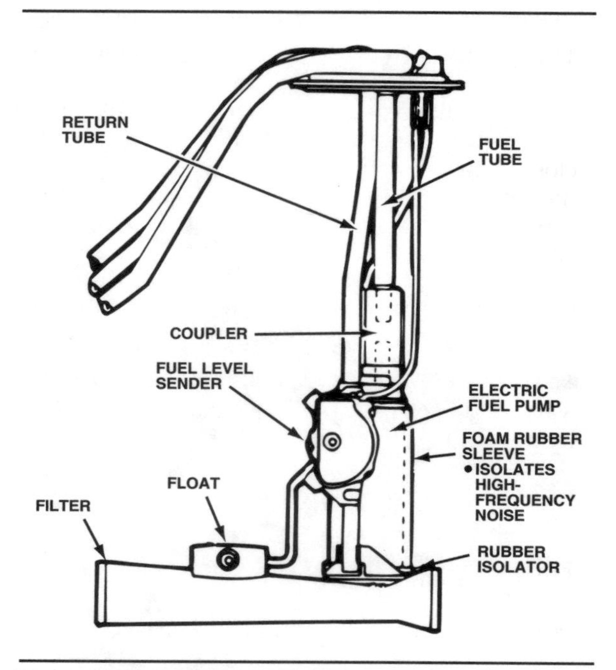

Figure 12-4. This GM fuel pickup tube is part of the fuel sender and pump assembly.

can fit into filler openings, but the larger nozzles for leaded fuel cannot, figure 12-3. In addition, a spring-loaded flapper valve in the filler tube prevents pouring leaded gasoline into a tank that is labeled for unleaded fuel. A deflector in the filler tube behind the flapper valve prevents fuel splash-back during filling. In 1996, the use of leaded fuel in almost all vehicles became illegal.

Effective September, 1993, federal regulations required automakers to install a device to prevent fuel from being siphoned through the filler neck. Federal authorities recognized methanol as a poison, and since the amount of methanol used in gasoline varies constantly in the United States, it was decided that siphoning fuel by mouth is a definite health hazard. To prevent siphoning, automakers have welded a filler neck check ball tube to 1993 and later model fuel tanks. The check ball inside the tube prevents the fuel in the tank from being siphoned. To drain such fuel tanks, a technician must disconnect the check ball tube at the tank and attach a siphon directly to the tank.

Fuel Pickup Tube

The fuel pickup tube is usually a part of the fuel sender assembly or the electric fuel pump assembly, figure 12-4. Since dirt and sediment eventually gather on the bottom of a fuel tank, the fuel pickup tube is fitted with a filter sock or strainer to prevent contamination from entering the fuel lines. The woven plastic strainer also

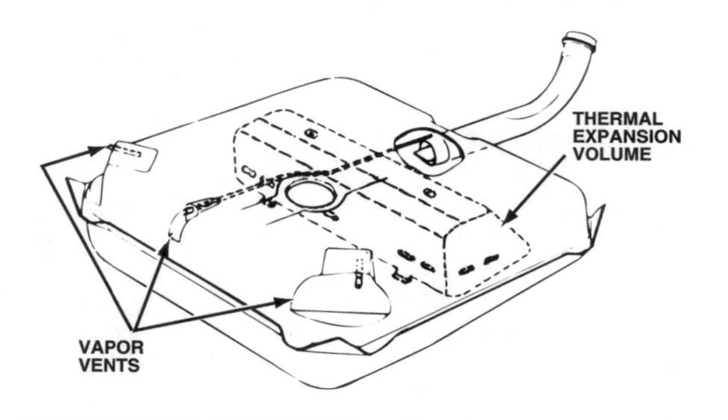

Figure 12-5. This fuel tank has an internal expansion tank to allow for changes in fuel volume due to temperature changes.

acts as a water separator by preventing water from being drawn up with the fuel.

Tank Venting Requirements

Fuel tanks must be vented or a **vacuum lock** will prevent fuel delivery. When fuel is used from the tank, its level drops and the space above the fuel increases. As the air in the tank expands to fill this greater space, its pressure drops. If the air pressure inside the tank drops below atmospheric pressure, a vacuum starts to develop and prevents the flow of fuel. Under extreme circumstances, atmospheric pressure could collapse the tank. Venting the tank allows more air to enter as the fuel level drops, preventing a vacuum from developing.

Before 1970, tanks were vented directly to the atmosphere with either a length of tubing (vent line) or through a vent in the filler cap. Both systems, however, added to air pollution by passing fuel vapors into the air. To reduce evaporative hydrocarbon (HC) emissions, evaporative emission control (EVAP) systems have been installed since 1970 on California cars and since 1971 on federal vehicles. In an EVAP system, the fuel tank vents directly to a vapor storage canister, using an unvented filler cap. Many filler caps contain pressure and vacuum relief valves that open to relieve pressure or vacuum above specified levels. Systems that use completely sealed caps have separate pressure and vacuum relief valves for venting.

Because fuel tanks are no longer vented directly to the atmosphere, the tank must allow for fuel expansion, contraction, and overflow that can result from changes in temperature or overfilling. One way to allow for this is to use a separate expansion tank, figure 12-5. Another method used was to provide a dome in the top of the tank, figure 12-2. As we mentioned earlier, some tanks are limited in capacity by the angle of the fuel filler tube, figure 12-1. The design used on many General Motors (GM) vehicles usually includes a vertical slosh baffle, and reserves up to 12 percent of the tank's total capacity for fuel expansion.

Rollover Leakage Protection

All 1976 and later cars require one or more devices to prevent fuel leaks in case of vehicle rollover. Automakers have met this requirement by using one of the following:

- Check valve
- Float valve.

Variations of the basic one-way **check valve** may be installed in any number of places between the fuel tank and the carburetor or throttle body injection unit. The valve can be installed in the fuel return line, vapor vent line, fuel tank filler cap, or carburetor fuel inlet filter, figure 12-6.

This type of rollover leakage protection is used primarily by Chrysler and GM.

Ford vehicles use a spring-operated **float valve** in the vapor separator, figure 12-7. The float valve closes whenever the car is at a 90-degree angle or more. Ford also redesigned its mechanical fuel pump on 1976 and later models to reduce fuel spills during an accident.

Vehicles with electric fuel pumps also use these same rollover protection devices, but they

Vacuum Lock: A stoppage of fuel flow caused by insufficient air intake to the fuel tank.

Check Valve: A valve that permits flow in only one direction.

Float Valve: A valve that is controlled by a hollow ball floating in a liquid, such as in the fuel bowl of a carburetor.

■ Safety Cells

Most automotive fuel tanks are simple steel tanks that hold liquid fuel. Formed to fit the chassis design and containing some baffles to reduce fuel sloshing, they are straightforward devices. A simple steel tank, however, has some serious disadvantages for vehicles used in hazardous operations. A common fuel tank can be punctured by an impact or leak fuel if overturned. Even with extensive baffling, off-road driving can cause fuel to slosh enough to upset vehicle balance or starve the fuel pickup line. All of these drawbacks to simple fuel tanks led to the development of fuel cells.

A fuel cell is a tank with a rigid shell of steel, aluminum, or some composite material. Inside the shell, a flexible rubber bladder forms a safety liner. The bladder is filled with low-density foam material that absorbs the liquid fuel. The fuel stays as a liquid within the foam and can be withdrawn easily by the fuel pump and pickup lines. The foam eliminates sloshing by distributing the fuel evenly throughout the tank regardless of the amount of fuel or vehicle motion. The combination of the rubber bladder and the foam prevents—or reduces—leakage in case of impact or tank rupture. Fuel cells also have check valves in the filler and vent lines to prevent leakage in case of rollover.

Fuel cells are mandatory safety equipment in most race cars and are used in many police cars, fire and rescue trucks, ambulances, and off-road equipment. Special cells with self-sealing, antiballistic ("bulletproof") liners are specified by the U.S. Secret Service as standard equipment in presidential limousines.

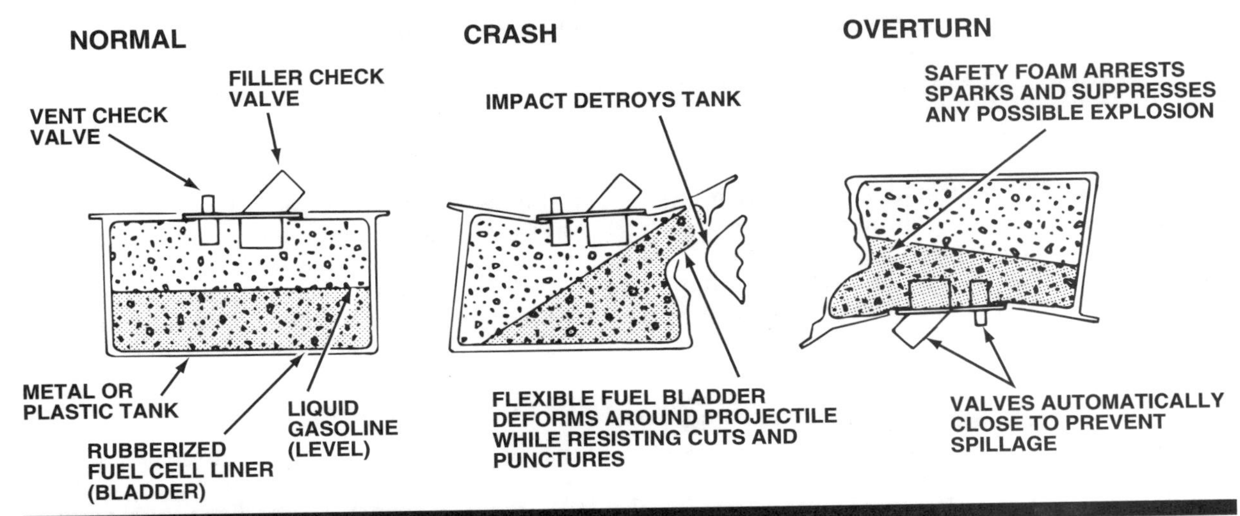

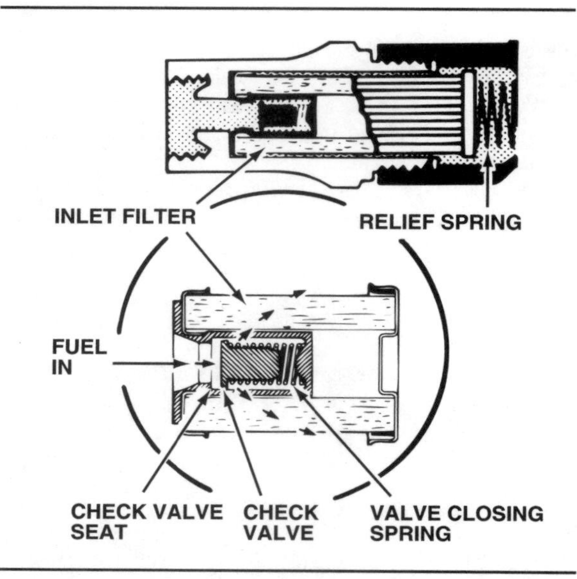

Figure 12-6. The rollover protection check valve is built into the fuel filter used on GM cars.

have additional features to ensure that the fuel pump shuts off when an accident occurs. Some pumps depend upon an oil pressure or an engine speed (rpm) signal to continue operating; these pumps turn off whenever the engine dies.

Later-model Ford vehicles with electronic engine controls and fuel injection have another form of rollover leakage protection. An inertia switch, figure 12-8, is installed in the rear of the vehicle between the electric fuel pump and its power supply. If the car is involved in any sudden impact, the inertia switch contacts open and shut off power to the fuel pump. The switch must be reset manually by pushing a button before power can be restored to the pump.

FUEL LINES

Fuel and vapor lines made of steel or nylon tubing and rubber hoses connect the parts of the fuel system. Fuel lines supply fuel to the carburetor, throttle body, or fuel rail. They also return excess fuel and vapors to the tank.

Fuel lines must remain as cool as possible. If any part of the line is located near too much heat, the gasoline passing through it vaporizes more rapidly than the fuel pump's suction capacity, and **vapor lock** occurs. When this happens, the fuel pump supplies only vapor that passes into the carburetor and out through the bowl vent without supplying gasoline to the engine. Depending on their function, fuel and vapor lines may be either rigid or flexible.

The fuel delivery pressure for most carbureted engines is about 5 to 8 psi (34 to 55 kPa). However, the delivery pressure in a low-pressure fuel injection system is 10 to 15 psi (69 to 103 kPa), and high-pressure systems often operate with 50 psi (345 kPa) or more. In addition, fuel injection systems retain residual pressure in the lines when the engine is off. Higher pressure systems such as these require special fuel lines.

Rigid Lines

All fuel lines fastened to the body, frame, or engine are made of seamless steel tubing. Steel springs may be wound around the tubing at certain points to protect against damage.

Only steel tubing should be used when replacing rigid fuel line. *Copper and aluminum tubing must never be substituted for steel tubing.* These materials do not withstand normal vehicle vibration and could combine with gasoline to cause a chemical reaction.

In some cars, rigid fuel lines are secured along the frame from the tank to a point close to the fuel pump. The gap between frame and pump is then bridged by a short length of flexible hose that absorbs engine vibrations. Other cars run a rigid line directly from tank to pump. To absorb vibrations, the line crosses 30 to 36 in. (750 to 900 mm) of open space between the pump and its first point of attachment to the frame.

Flexible Lines

Most carbureted fuel systems use synthetic rubber hose sections where flexibility is needed. Short hose sections often connect steel fuel lines to other system components. The fuel delivery hose inside diameter (ID) is generally larger (5⁄16 to 3⁄8 in. or 8 to 10 mm) than the fuel return hose ID (1⁄4 in. or 6 mm).

Replacement fuel hoses should be made of fuel-resistant material. Ordinary rubber hoses such as those used for vacuum lines deteriorate when exposed to gasoline. Similarly, vapor vent lines must be made of materials that resist fuel vapors. Replacement vent hoses are usually marked with the designation EVAP to indicate their intended use.

A metal or plastic restrictor often is used in vent lines to control the vapor flow rate. These may be installed either in the end of the vent pipe, or in the vapor vent hose itself. When used in the hose instead of the vent pipe, the restrictor must be removed from the old hose and installed in the new one whenever the hose is replaced.

Fuel Line Mounting

Fuel supply lines from the tank to a carburetor are routed to follow the frame along the underbody of the vehicle. Vapor and return lines may

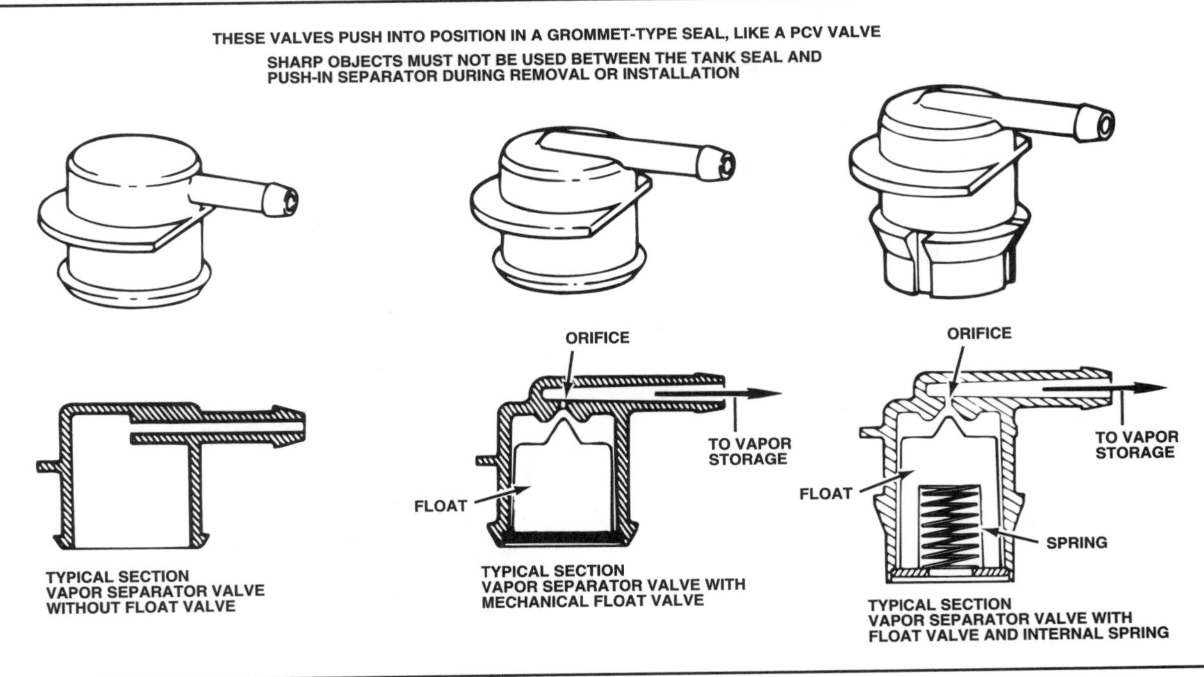

Figure 12-7. Orifice-type vapor separators used on Ford vehicles contain a mechanical check valve that also provides rollover leak protection.

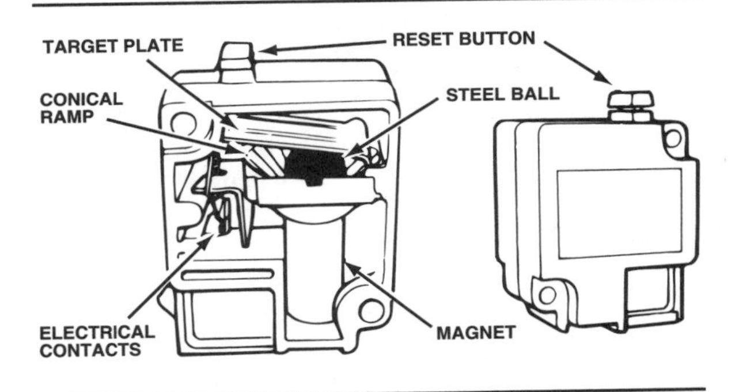

Figure 12-8. Ford uses an inertia switch to turn off the electric fuel pump in an accident.

be routed with the fuel supply lines, but usually are on the frame rail opposite the supply line. All rigid lines are fastened to the frame rail or underbody with screws and clamps, or clips.

Carbureted Fittings and Clamps

Brass fittings used in fuel lines are either the flared type or the compression type, figure 12-9; although, flared fittings are more common. The inverted, or Society of Automotive Engineers (SAE) 45-degree, flares slip snugly over the connectors to prevent leakage when the nuts are tightened. When replacement tubing is installed, a double flare should be used to ensure a good seal and to prevent the flare from cracking. Compression fittings use a separate sleeve, a

Vapor Lock: A condition in which bubbles in a car's fuel system stops or restricts fuel flow. High underhood temperatures sometimes causes fuel to boil within fuel lines.

Tools for Making Fuel Lines

With few exceptions, replacement fuel lines cannot be bought preformed. Tubing is stocked in large rolls and must be shaped and formed however the technician wants it. Ordinary hand tools cannot be used to properly make a replacement fuel line. Frequently, cutting a tube using a hacksaw leaves a distorted, jagged edge. To ensure a smooth cut, use these special tubing tools:

- Cutter
- Reamer
- Bender
- Flaring device.

The tube cutter uses sharpened metal disks to make a smooth, distortion-free cut. After cutting, a tapered reamer is necessary to remove any burrs that might prevent a good seal. The tube bender shapes the tubing without kinking or bending it. Flaring tools are available to make either single or double flares. It is essential to use them to properly shape the connecting ends of any new fuel line.

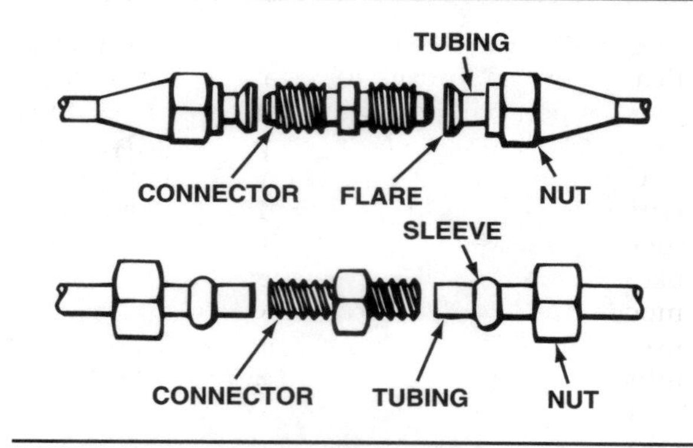

Figure 12-9. Fuel line fittings are either the flare type (top) or the compression type (bottom).

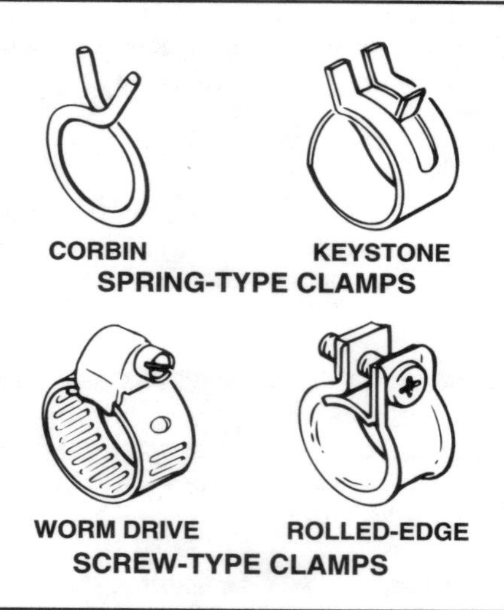

Figure 12-10. Various clamps are used on fuel system hoses.

tapered sleeve, or a half-sleeve nut to make a good connection.

Various types of clamps are used to secure fuel hoses on carbureted fuel systems, figure 12-10. Spring-type clamps are commonly used for original equipment installation, but only screw-type (aircraft) clamps should be reused when hoses are changed. Keystone, Corbin, and other spring-type clamps will not hold securely when reused and should be replaced with new ones if they are removed. Screw-type clamps are made in two styles: worm-drive clamps and those in which the screw and nut stand off from the clamp body, figure 12-10.

Fuel Injection Lines and Clamps

Hoses used for fuel injection systems are made of materials with high resistance to oxidation and deterioration. They also are reinforced to withstand higher pressures than carburetor system fuel hoses. Replacement hoses for injection systems should always be equivalent to original equipment manufacturer (OEM) hoses.

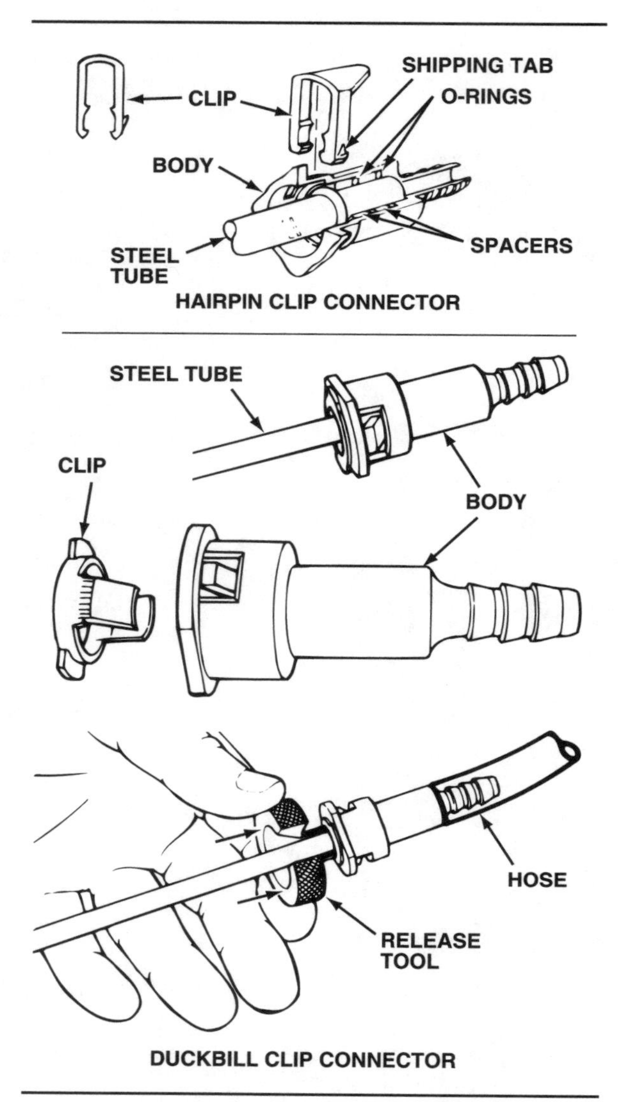

Figure 12-11. Late-model Ford products use these push-connect fuel line fittings. (Ford)

Spring-type clamps must *never* be used on fuel injected engines because they cannot withstand the fuel pressures involved. Screw-type clamps are essential on injected engines and should have rolled edges to prevent hose damage. Worm-drive clamps are satisfactory for use on carbureted engines, but should not be used on fuel injection systems. The screw teeth can cut and weaken the hose if overtightened.

Fuel Injection Fittings and Nylon Lines

Because of their higher operating pressures, fuel injection systems often use special kinds of fittings to ensure leakproof connections. Some high-pressure fittings on GM cars with port injection systems use O-ring seals instead of the traditional flare connections. Whenever you disconnect such a fitting, inspect the O-ring for damage and replace it if necessary. *Always* tighten O-ring fittings to the specified torque value to prevent damage.

Other automakers also use O-ring seals on fuel line connections. In all cases, the O-rings are made of special materials that can withstand contact with gasoline and alcohol-blend fuels. Some manufacturers specify that the O-rings should be replaced every time the fuel system connection is opened. Whenever you replace one of these O-rings, you *must* use a new part specifically designed for fuel system service. The O-rings used in air conditioning systems are *not* satisfactory.

Ford uses nylon fuel tubing with several unique push-connect fittings. Special barbed connectors are required to join sections of nylon tubing together. The ends of nylon tubing can be softened in hot water before sliding them onto the connectors. However, do not soak the tubing in boiling water for a long time. Tubing should not be heated over 212°F (100°C), or it will not return to its original shape and will not grip the connector tightly.

Connectors that join nylon to steel tubing also have barbed ends for the nylon lines, but they use O-rings to seal the steel line, figure 12-11. Ford uses two kinds of retainer clips on these connectors. Fittings for 5⁄16- and 3⁄8-in. lines have hairpin clips. Fittings for 1⁄4-in. lines have duck-bill clips.

To remove a duckbill clip, use the special tool shown in figure 12-11 or pliers with thin jaws to release the clip; then, pull the connector apart gently. To remove a hairpin clip, push the shipping tab down to clear the connector body and then spread the clip by hand. Pull the triangular tab to separate the clip from the connector and gently pull the connector apart. When you reassemble the connector, note that the prongs of the clip are tapered. The tapered sides must face the steel tubing or the connection will be forced apart by fuel pressure. Ford recommends that push connector clips be replaced with new ones whenever a connector is taken apart.

Ford also uses spring-lock connectors to join male and female ends of steel tubing. The coupling is held together by a garter spring inside a circular cage. The flared end of the female fitting slips behind the spring to lock the coupling together. To open these connectors, a special tool is required that fits around the connector and slides inside the cage to release the spring, figure 12-12. On some vehicles, an indicator ring is left on the fuel line at assembly. To aid reassembly, slide the ring into the cage after you separate the coupling. Reassemble the coupling by hand with a slight twisting motion. The indicator ring, if used, pops free when the connector is properly seated.

GM originally introduced nylon fuel lines with quick-connect fittings at the fuel tank and fuel filter on some 1988 models. Since then, their use has been increased to more vehicles each model year. Like the GM threaded couplings used with steel lines, nylon line couplings use internal O-ring seals. Unlocking the metal connectors requires a special quick-connector separator tool, figure 12-13. Plastic connectors can be released without the tool. Where access to metal connectors is restricted, a special tool is available.

Fuel System Development

The first automobiles relied on gravity to supply fuel to the engine. These gravity-feed systems mounted the fuel tank higher than the engine, allowing gravity to draw fuel from the tank to the engine. Because these tanks were front-mounted, they had limited capacity and were dangerous.

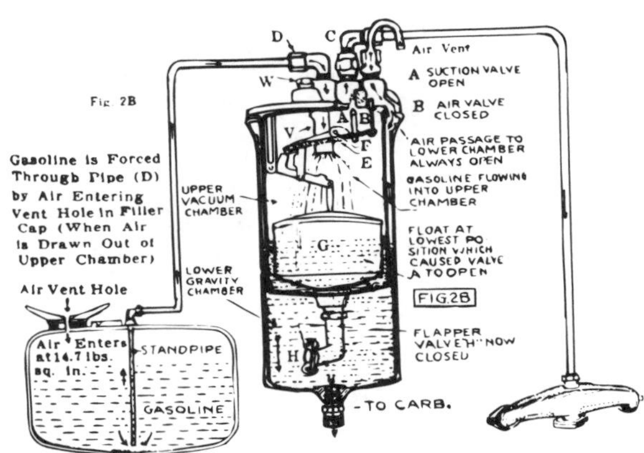

Moving the fuel tank to the rear of the car solved the problems of safety and storage capacity but required the use of a vacuum tank. This was a small fuel tank, still positioned above the engine in the cowl, but connected to the rear tank as well. Suction created by engine vacuum provided fuel for the vacuum tank from the larger rear-mounted tank.

If the car was not driven for a long time, the gasoline in the vacuum tank would eventually evaporate. In this case, it was necessary to prime the engine in order to start it and create vacuum which would move fuel through the system. With the appearance of the mechanical fuel pump after World War I, the vacuum tank was retired.

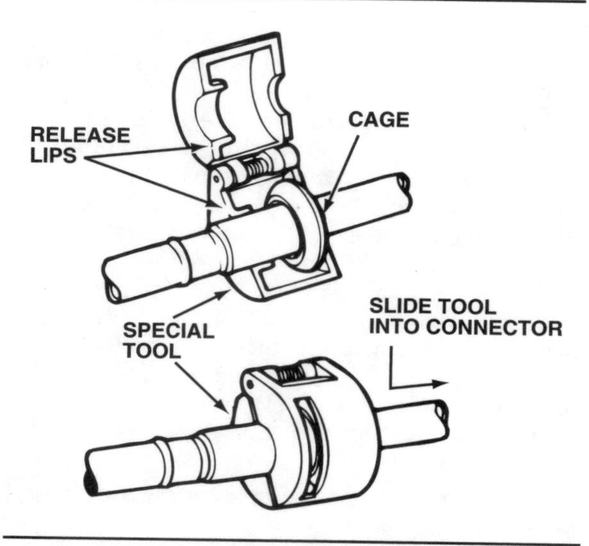

Figure 12-12. Ford spring-lock connectors require this special tool for disassembly.

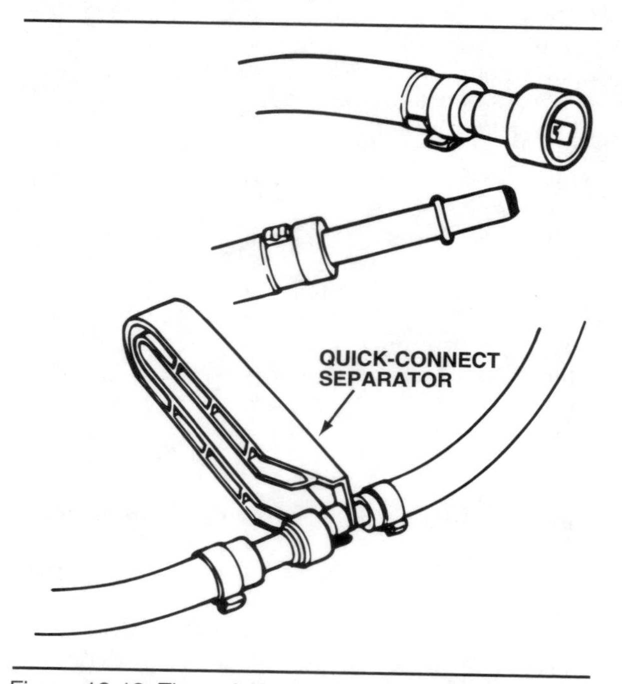

Figure 12-13. The quick-connect separator should be used with all metal GM quick-connect fittings. Plastic fittings can be released by hand without the tool.

EVAPORATIVE EMISSION CONTROL SYSTEMS

Evaporative emission controls (EVAP) have been an anti-pollution tool since the early 1970s. The purpose of the EVAP system is to trap gasoline vapors that would otherwise escape into the atmosphere and route them into the intake airflow so they can be burned in the engine.

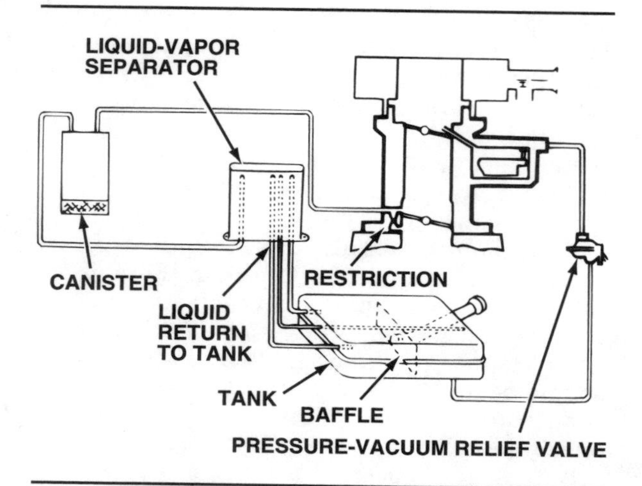

Figure 12-14. GM EVAP system with liquid-vapor separator and constant-purge canister.

Common Components

The fuel tank filler caps used on cars with EVAP systems are a special design. Some early GM EVAP systems used an unvented cap with a pressure-vacuum relief valve in the line between the fuel tank and carburetor, figure 12-14, but most EVAP fuel tank filler caps have pressure-vacuum relief built into them, figure 12-15. When pressure or vacuum exceeds a calibrated value, the valve opens. Once the pressure or vacuum has been relieved, the valve closes. If a sealed cap is used on an EVAP system that requires a pressure-vacuum relief design, a vacuum lock may develop in the fuel system, or the fuel tank may be damaged by fuel expansion or contraction.

Various methods protect fuel tanks against fuel expansion and overflow caused by heat. Temperature expansion tanks were used on many early 1970s EVAP systems to prevent filling the tanks completely. The expansion tank attaches to the inside of the fuel tank and contains small holes that open it to the fuel area. When the fuel tank appears to be completely full, and the fuel gauge reads full the expansion tank remains virtually empty. This provides enough space for fuel expansion and vapor collection if the vehicle is parked in the hot sun after filling the tank.

The dome design of the upper fuel tank section used in some late-model cars, or the overfill limiting valve contained within the vapor-liquid separator, eliminates the need for the overfill limiter tank used in earlier systems.

Some Ford cars use a **combination valve** which does three things:

- Isolates the fuel tank from engine pressures and lets vapor escape from the vapor separator tank to the vapor storage canister

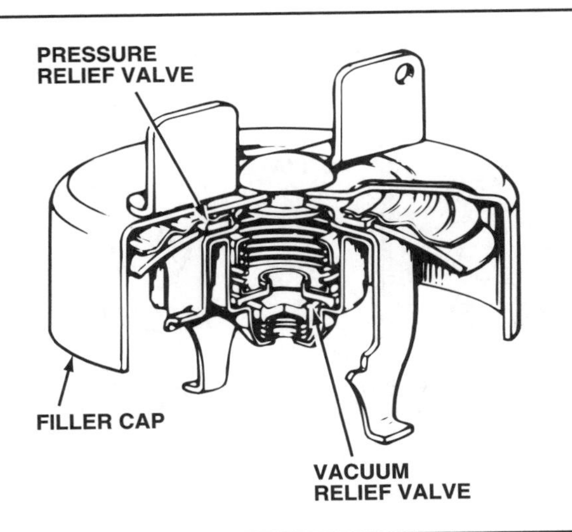

Figure 12-15. Fuel tank caps for EVAP systems have vacuum and pressure relief valves.

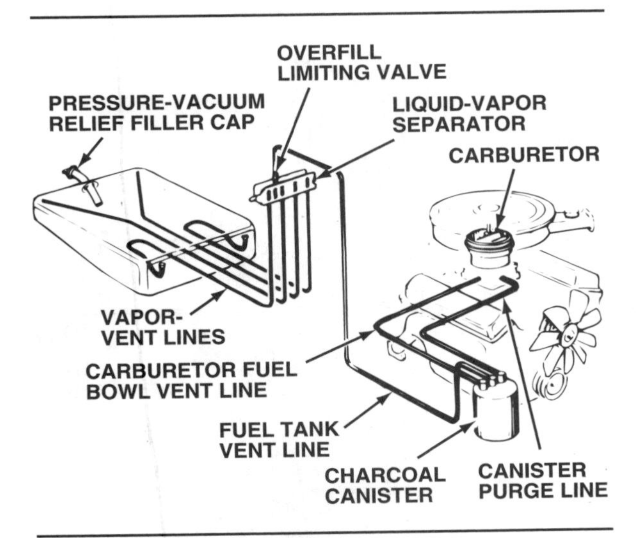

Figure 12-16. This EVAP system has a liquid-vapor separator mounted separately from the tank.

- Vents excess fuel tank pressure to the atmosphere in case of a block in the vapor delivery line
- Allows fresh air to be drawn into the fuel tank to replace the gasoline as it is used.

All EVAP systems use some form of **liquid-vapor separator** to prevent liquid fuel from reaching the engine crankcase or vapor storage canister. Some liquid-vapor separators are built into the tank and use a single vapor vent line from the tank to the vapor canister. When the separator is not built in, figure 12-16, it usually is mounted on the outside of the tank or on the frame near it. In this case, vent lines run from the tank to the separator and are arranged to vent the tank regardless if the car is level or not. Liquid fuel entering the separator returns to the tank through the shortest line.

Carburetor Venting

Carburetors must be vented to keep atmospheric pressure in the fuel bowl and provide the pressure differential needed for fuel metering. Carburetor vents may be internal or external.

Internal vents

Carburetors are usually vented internally through the vent tubes that connect the fuel bowl to the air intake. The main purpose of the tubes is to keep atmospheric pressure pushing down on the fuel bowl. This causes the fuel to flow from the bowl, through the circuits and jets, to the lower-pressure area created by the carburetor venturi. The vent tubes also compensate for any air pressure drop caused by a dirty air cleaner filter and help prevent a too-rich air-fuel mixture. The balance tubes also let vapors from the fuel bowl collect in the air cleaner when the engine is off to help control evaporative emissions.

External vents

Many carburetors have external vents for the fuel bowl. On old cars, without EVAP systems, these vents opened directly to the atmosphere. They released vapors from the fuel bowl to prevent the vapor pressure buildup that could cause **percolation**.

The carburetor bowl vent on cars with EVAP systems is connected to the vapor storage canister by a rubber hose. Often, carburetor linkage opens the external vents, figure 12-17, so they are closed when at open throttle and open at idle or when the engine is off.

Evaporative Emission Control (EVAP) System: A way of reducing HC emissions by collecting fuel vapors from the fuel tank and carburetor fuel bowl vents and directing them through an engine's intake system.

Combination Valve: A valve on the fuel tanks of some Ford cars that allows fuel vapors to escape to the vapor storage canister, relieves fuel tank pressure, and lets fresh air into the tank as fuel is withdrawn. Similar to a liquid-vapor separator valve.

Liquid-Vapor Separator Valve: A valve in some EVAP fuel systems that separates liquid fuel from fuel vapors.

Percolation: The bubbling and expansion of a liquid, similar to boiling.

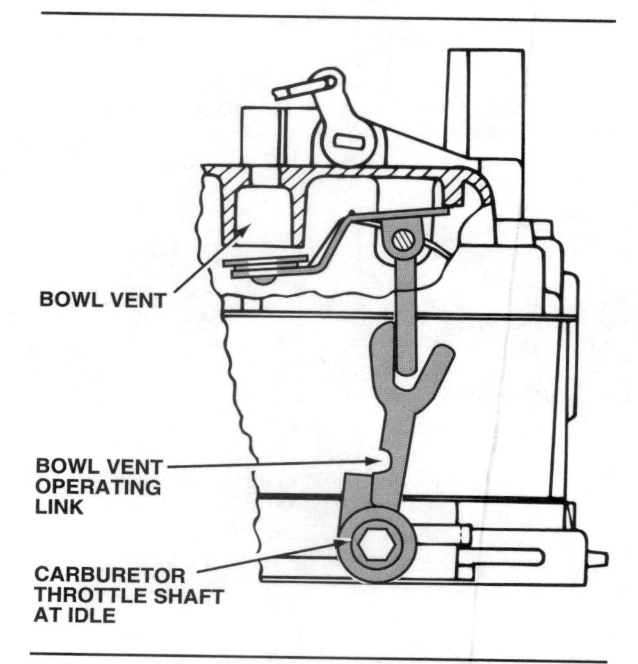

Figure 12-17. External carburetor vent operated by a link from the carburetor throttle shaft. (Chrysler)

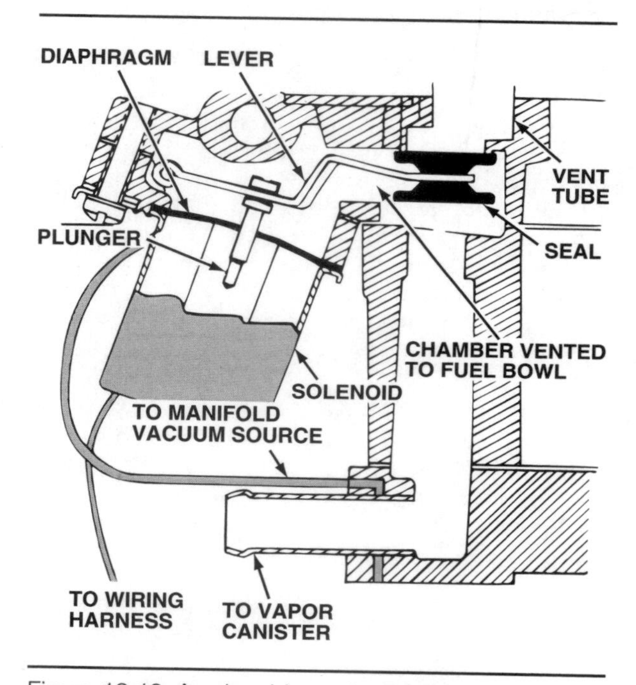

Figure 12-18. A solenoid may be used to switch bowl venting between an internal vent and the canister. (Chrysler)

Carburetors used solenoid-operated vent valves. The solenoid shown in figure 12-18 was used on a Carter Thermo-Quad to switch the vent passages between the external vent and the internal vent tube.

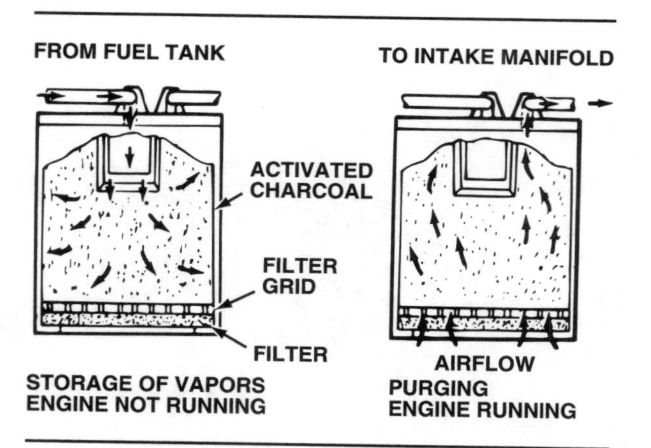

Figure 12-19. Typical vapor canister operation. (Chrysler)

Vapor Storage

As explained, an EVAP system traps gasoline vapors from the fuel tank and carburetor and feeds them into the engine intake system or stores them until the engine is started. Almost all late-model EVAP systems store the vapors in a charcoal-granule-filled canister. A few early EVAP systems stored the vapors in the engine crankcase.

Engine crankcase storage

Many early 1970s vehicles used the crankcase as a vapor storage area. When the engine was started, the **positive crankcase ventilation (PCV)** drew the stored vapors from the crankcase into the engine where they burned.

Vapor canister storage

Vapor storage canisters have been used on most domestic vehicles since 1972. The canister is located under the hood, figure 12-16, and is filled with activated charcoal granules that can hold up to ⅓ of their own weight in fuel vapors. A vent line connects the canister to the fuel tank. Carburetors with external bowl vents also were vented to the canister. Some Ford and Chrysler vehicles with large or dual fuel tanks may have dual canisters; GM engines may have an auxiliary canister connected to the primary canister purge air inlet to store vapor overflow.

Activated charcoal is an effective vapor trap because of its great surface area. Each gram of activated charcoal has a surface area of 1,100 square meters, or more than a quarter acre. Typical canisters hold either 300 or 625 grams of charcoal with a surface area equivalent to 80 or 165 football fields. **Adsorption** attaches the fuel vapor molecules to the carbon surface. This attaching force is not strong, so the system can purge the molecules quite simply by sending a fresh airflow through the charcoal.

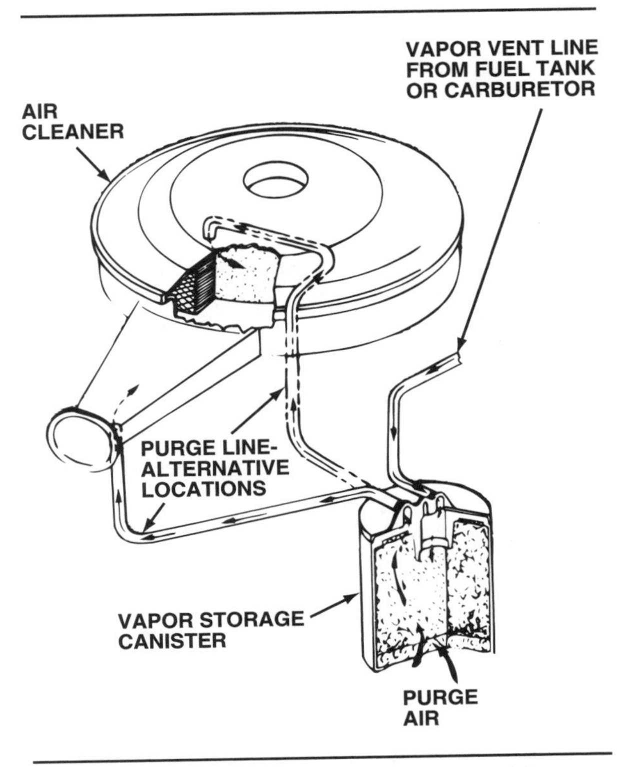

Figure 12-20. Purging the vapor storage canister can be done either through the air cleaner or the carburetor. (Ford)

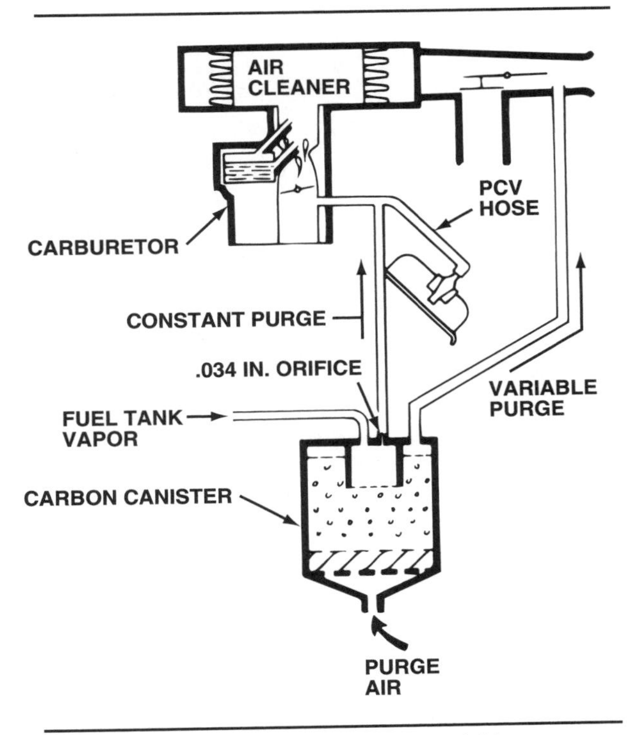

Figure 12-21. In this EVAP system, a variable-purge hose runs from the canister to the air cleaner, and a constant-purge hose runs to the intake manifold. (Ford)

There are two methods to provide fresh air to the canister for purging. In one design, the bottom of the canister is open to the atmosphere and air enters through a filter, figure 12-19. This design supplies purge air whenever the engine is running. In another design, canisters are closed to the atmosphere and obtain air from the air injection system. A solenoid controls closed-canister airflow to purge the vapors during specific engine operating conditions.

A small vapor separator in the supply line between the fuel pump and the carburetor reduces the amount of fuel vapors reaching the carburetor on many vehicles (particularly Ford products). A vapor return line connects this separator to the fuel tank. Vapors collected in the separator are routed back to the tank to recondense, or they may travel through the regular vent line to the canister. Continuously venting these vapors back to the fuel tank instead of allowing free travel to the carburetor prevents engine surging from over-rich fuel.

Vapor Purging

During engine operation, stored vapors are drawn from the canister into the engine through a hose connected to either the carburetor base or the air cleaner, figure 12-20. This "purging" process mixes HC vapors from the canister with the existing air-fuel charge. To compensate for the mixture enrichment, carburetors used with an EVAP system were calibrated to take vapor purging into account. If the purge rate was not properly controlled to maintain the correct air-fuel ratio under varying engine operating conditions, engine hesitation and surging resulted.

There are several ways to purge the canister. The purging flow rate and method are determined by two things that the process must accomplish:

- Reactivate the charcoal
- Minimize the effect on the air-fuel ratio and driveability.

Positive Crankcase Ventilation (PCV): A way of controlling engine emissions by directing crankcase vapors (blowby) back through an engine's intake system.

Adsorption: A chemical action when liquids or vapors gather on the surface of a material. In a vapor storage canister, chemical properties force fuel vapors to attach themselves (adsorb) to the surface of charcoal granules.

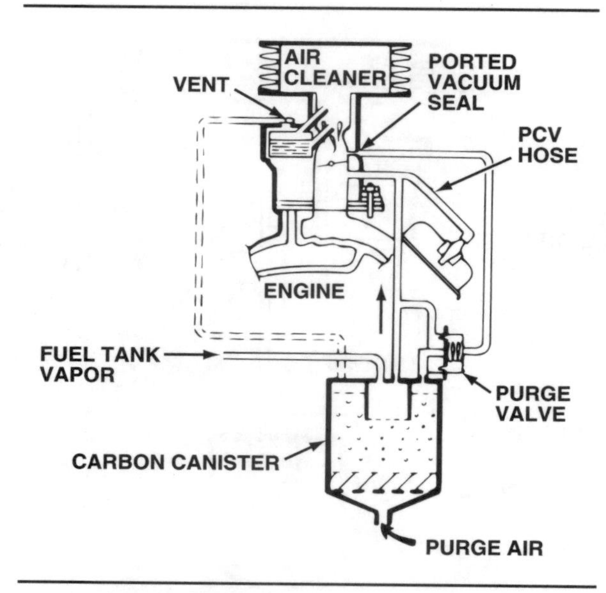

Figure 12-22. The two-stage purge arrangement in this EVAP system uses a vacuum-operated valve to open a second purge line from the canister to the manifold. (Ford)

Constant purge

In a constant purge system, the purge rate remains fixed, regardless of engine air consumption. Intake manifold vacuum draws vapor from the canister by a "tee" in the PCV line at the carburetor. Even though manifold vacuum fluctuates, an **orifice** in the purge line provides a constant flow rate.

Variable purge

In a variable purge system, the amount of purge air drawn through the canister is proportional to the amount of fresh air drawn into the engine. In other words, the more air the engine takes in, the more purge air is drawn through the canister. A simple variable purge system is shown in figure 12-20, which illustrates the system can draw the purge air through the canister by using either:

- A **pressure drop** across the air filter
- The velocity of the air moving through the air cleaner.

In both cases, airflow through the air cleaner varies the air flowing through the canister. The simple variable purge often is combined with a constant purge, figure 12-21.

Two-stage purge

If the air cleaner purge flow is not enough, the system may use a vacuum-operated **purge valve**, figure 12-22, in addition to the constant airflow to the manifold. **Ported vacuum** from the carburetor controls the purge valve line which opens a second passage from the canister to the intake manifold to provide additional purging.

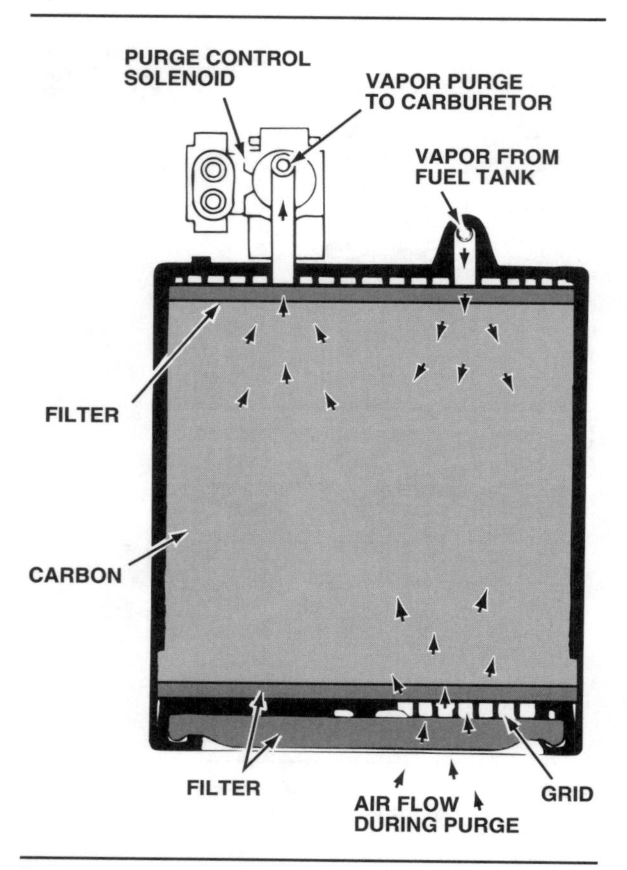

Figure 12-23. In a computer-controlled purging system, the microprocessor controls purge vacuum with a solenoid. (AC-Delco).

At idle and low engine speeds, spring tension inside the purge valve holds it closed. As the throttle valve moves beyond the carburetor vacuum port, vacuum is applied to the purge valve diaphragm, causing the valve to lift off its seat.

A carburetor purge port may also be used with the constant purge system. This port is located above the high side of the carburetor throttle plate so that there is no purge flow at idle, but the flow increases as the throttle opens.

Computer-controlled purge

Canister purging on engines with electronic fuel management control systems may be controlled by the engine control computer. Control of this function is particularly important because the additional fuel vapors sent through the purge line can upset the air-fuel ratio provided by a feedback carburetor or fuel injection system. Since air-fuel ratio adjustments are made many times per second, it is critical that vapor purging is controlled just as precisely.

This is done by a microprocessor-controlled vacuum solenoid mounted on top of the canister, figure 12-23, and one or more purge valves.

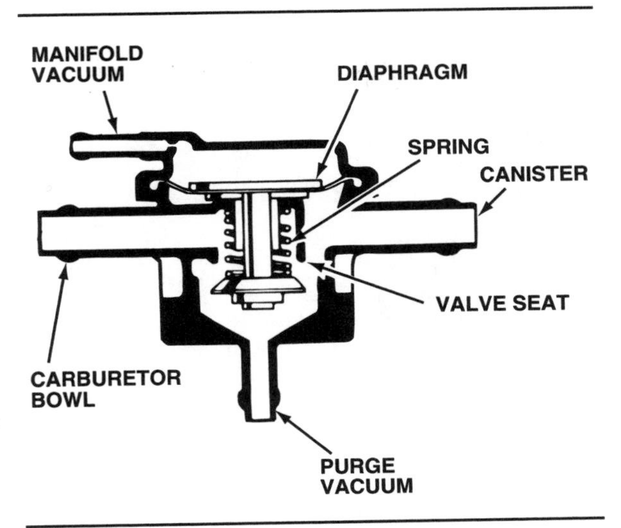

Figure 12-24. A Type 1 GM canister purge valve.

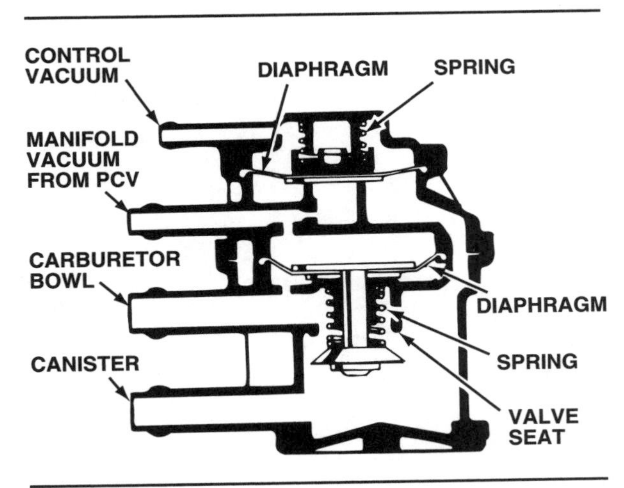

Figure 12-25. A Type 2 GM canister purge valve.

Under normal conditions, most engine control systems only permit purging during closed-loop operation at cruising speeds. During other engine operation conditions, such as open-loop mode, idle, deceleration, or wide-open throttle, the computer prevents canister purging.

GM uses various designs, but regardless of their configuration, all use one of two purge valves:

- The Type 1 valve, figure 12-24, uses spring tension to hold the valve open and permit fuel bowl venting with the engine off. When the engine is running, manifold vacuum closes the valve and permits canister purge.
- The Type 2 valve, figure 12-25, has two vacuum diaphragms. It does the same job as a Type 1 valve and also closes off fuel bowl venting when manifold vacuum, provided by the PCV system, activates the lower diaphragm with the engine running. When engine speed is increased above idle, control vacuum activates the upper diaphragm to permit canister purging through the PCV system.

Ford EVAP systems may use either the exhaust gas recirculation (EGR) valve or a separate purge solenoid to control purge valve vacuum. Ford purge valves may be mounted on the canister, remotely mounted, or installed in a vacuum line, figure 12-26. Regardless of their location, all Ford purge valves operate essentially the same as the GM valves just described.

Orifice: A small opening in a tube, pipe, or valve.

Pressure Drop: A reduction of pressure between two points.

Purge Valve: A vacuum-operated or electronically controlled-solenoid valve used to draw fuel vapors from a vapor storage canister.

Ported Vacuum: Vacuum immediately above the throttle valve in a carburetor.

■ Why Vapor Lock?

When gasoline vapors form in the fuel system, vapor lock occurs. This is the partial or complete stoppage of fuel flow to the carburetor. Partial vapor lock will lean the air-fuel mixture and reduce both the top speed and the power of an engine. Complete vapor lock will cause the engine to stall and make restarting impossible until the fuel system has cooled.

Four factors usually cause vapor lock:

- High gasoline temperature and pressure in the fuel system
- Vapor-forming characteristics of a particular gasoline
- The fuel system's inability to minimize vapors
- Poor engine operating conditions, such as overheating.

Vapor may form anywhere in the fuel system, but the critical temperature point is the fuel pump.

Engineers have improved fuel pumps and fuel systems to make today's cars unlikely to have vapor lock. Oil companies have succeeded in reducing the vapor-locking tendencies of gasoline by adjusting its volatility according to weather requirements. But vapor lock may still occur in older cars during long periods of idle (such as in heavy rush-hour traffic) or when the car's fuel system is not properly maintained. Periodically inspect the fuel system and correct all air leaks and defects to prevent vapor lock.

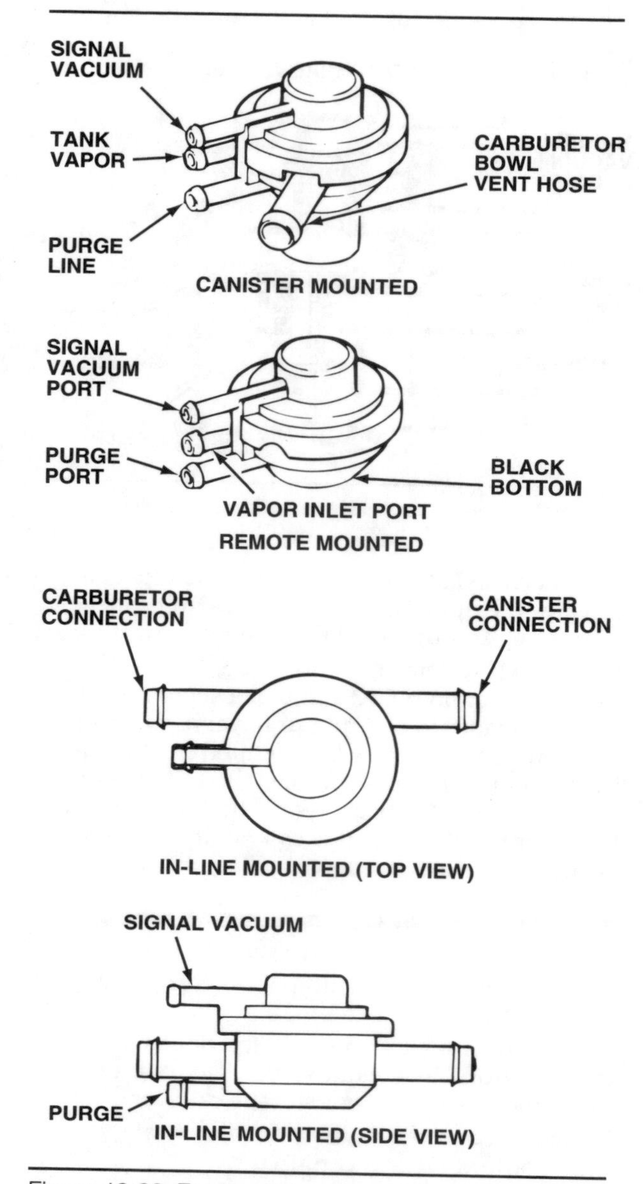

Figure 12-26. Ford canister purge valves are all similar, regardless of their mounting or positioning.

All automakers use variations of the basic EVAP system described in this section. The system configuration, components, and locations vary according to the specific fuel system and engine, but all function according to the general principles discussed in this chapter.

PUMP OPERATION OVERVIEW

The fuel pump and fuel lines deliver gasoline from the tank to the carburetor or injection system. The fuel pump moves the fuel with a mechanical action that creates a low-pressure or suction area at the pump inlet. This causes the higher atmospheric pressure in the fuel tank to force fuel to the pump. The pump spring also exerts a force on the fuel within the pump and delivers it under pressure to the carburetor or injection system. All fuel pumps, except electric turbine pumps, develop this mechanical action through a reciprocating "push-pull" motion.

Output pressure and volume are two measurements of a fuel pump's performance. When an output pressure or volume is specified, it represents the unrestricted output from the pump at a constant pumping speed. Mechanical fuel pumps used with carbureted engines generally develop 5 to 8 psi (34 to 55 kPa) output pressure. Although pump operating pressures used with electronic fuel injection systems can reach as high as 55 psi (379 kPa), depending on system design, most operate in the 41 to 47 psi (282 to 324 kPa) range. Such high pressures are provided by high-volume pumps and regulators that maintain high pressures. The following paragraphs describe various kinds of fuel pumps in detail.

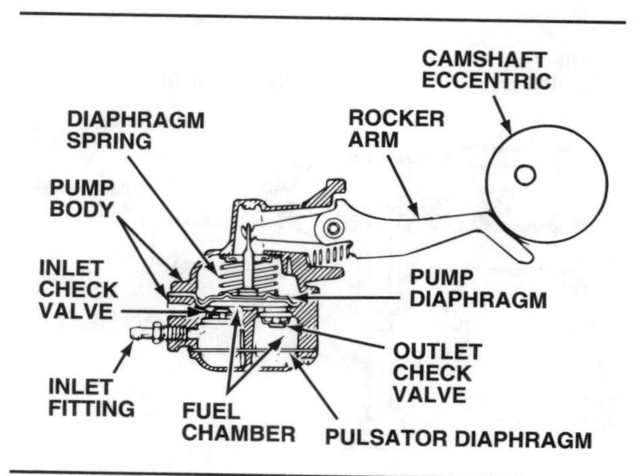

Figure 12-27. Typical diaphragm-type mechanical fuel pump.

Pump Types

While all pumps deliver fuel through mechanical action, they generally are divided into two groups:

- Mechanical—driven by the car engine
- Electrical—driven by an electric motor or vibrating **armature**.

Mechanical fuel pumps

Earlier model carbureted vehicles commonly use a single-action, diaphragm-type mechanical pump, figure 12-27. The rocker arm is driven by an eccentric lobe on the camshaft. On some overhead-cam 4-cylinder engines, the eccentric lobe may be on an accessory shaft. The pump makes one stroke with each revolution of the camshaft. The eccentric lobe—often called sim-

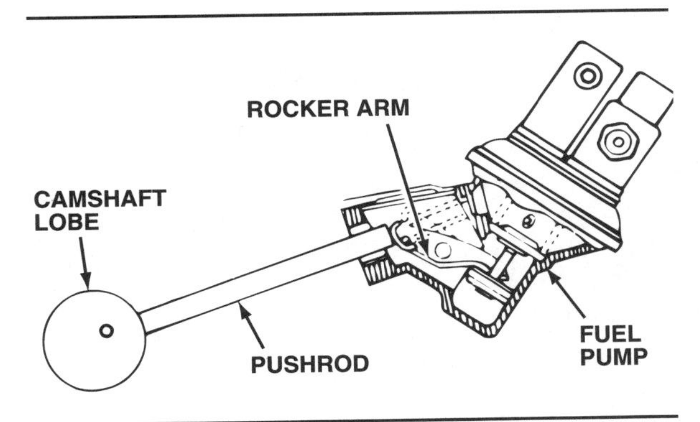

Figure 12-28. Some Chevrolet and Ford engines use a pushrod between the camshaft and pump rocker arm.

ply "the eccentric"—may be part of the camshaft or a pressed steel lobe that is bolted to the front of the camshaft, along with the drive gear.

In some applications, the rocker arm is driven directly by the eccentric, figure 12-27. Other engines have a pushrod between the eccentric and the pump rocker arm, figure 12-28. The most common example of this arrangement is the small-block Chevrolet V8 and some Ford 4-cylinder engines.

Mechanical pump operation

The fuel intake stroke begins when the rotating camshaft eccentric pushes down on one end of the pump rocker arm. This raises the other end that pulls the diaphragm up, figure 12-27, and tightens the diaphragm spring. Pulling the diaphragm up creates a vacuum, or low-pressure area, in the fuel chamber. The constant high pressure in the fuel lines forces open the inlet check valve and fuel enters the fuel chamber.

As the camshaft eccentric continues to turn, it allows the outside end of the rocker arm to "rock" back up. Along with the push given by the diaphragm spring, this allows the diaphragm to relax back down. This is the start of the fuel output stroke. As the diaphragm relaxes, it causes a pressure buildup in the fuel chamber.

This pressure closes the inlet check valve and opens the outlet check valve. The fuel flows out of the fuel chamber and into the fuel line on the way to the carburetor. The outlet check valve keeps a constant pressure in the outlet line and prevents fuel from flowing back into the pump.

We measure fuel pump output by the pressure and volume of the fuel it delivers. Delivery pressure is controlled by the diaphragm spring. Delivery rate is controlled by the float needle in the carburetor fuel bowl. The fuel pump delivery rate is proportional to the fuel required by the carburetor. When the carburetor inlet needle valve is open, fuel flows from the pump through the lines into the carburetor. When the carburetor fuel bowl is full, the needle valve closes and no fuel flows through the lines.

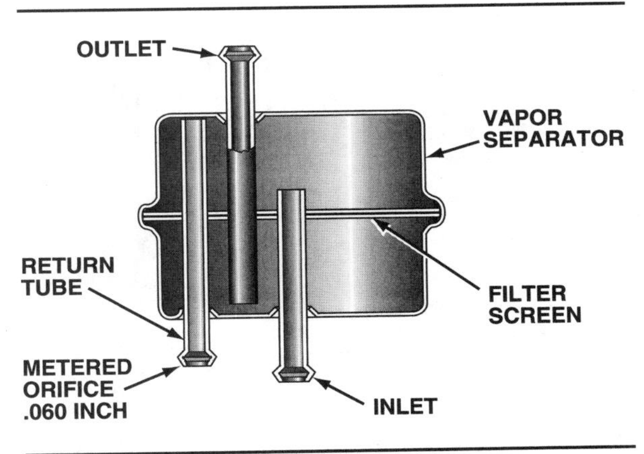

Figure 12-29. A vapor separator is installed between the pump and carburetor to relieve pressure in the fuel line. (Chevrolet)

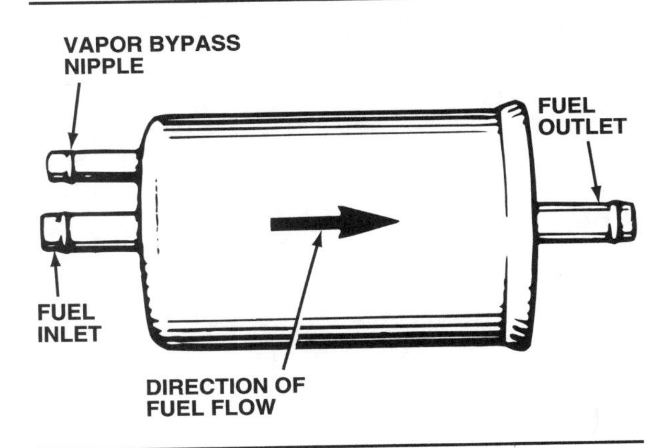

Figure 12-30. A vapor bypass filter combines the fuel filter and vapor relief functions in one unit.

With the needle valve closed and pressure in the fuel line increasing, the fuel pump diaphragm stays up, even though the rocker arm continues to move up and down in a "freewheeling" motion. No fuel is pumped until the fuel level in the carburetor bowl drops enough for the inlet needle valve to open again.

The fuel level in the carburetor bowl varies for different operating conditions, so the inlet needle valve's position varies between fully open and fully closed. The needle valve opening and the rate of fuel flowing into the carburetor is always controlled by and proportional to the fuel flow rate out of the carburetor.

Armature: The movable part in a relay. The revolving part in a generator or motor.

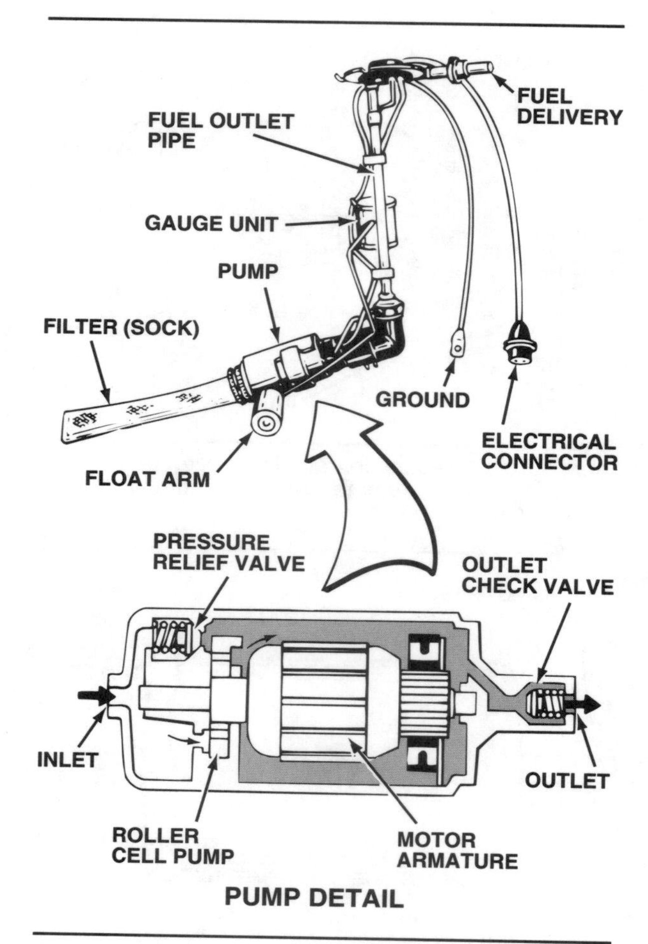

Figure 12-31. An impeller-type electric fuel pump.

Some fuel pumps have a slotted link that operates the diaphragm. The slot fits over the rocker arm and allows a partial stroke of the pump. As the diaphragm responds to outlet line pressure, the slotted link permits partial diaphragm movement and reduced fuel volume.

When an engine with a mechanical fuel pump is shut off, the pump diaphragm spring maintains pressure in the fuel line to the carburetor. If engine compartment heat expands the gasoline in the fuel line, the fuel pushes the carburetor inlet needle valve open and passes through. The result is too much fuel in the carburetor, known as **flooding** the carburetor. Also, since fuel expands when it is hot, fuel may turn from a liquid into a vapor and cause vapor lock in the pump and lines. Four methods—described in the following paragraphs—can maintain fuel pressure and prevent flooding and vapor lock.

Some older pumps use an air chamber on the outlet side of the pump to separate and recirculate vaporized and heated fuel to the fuel tank through a vapor return line.

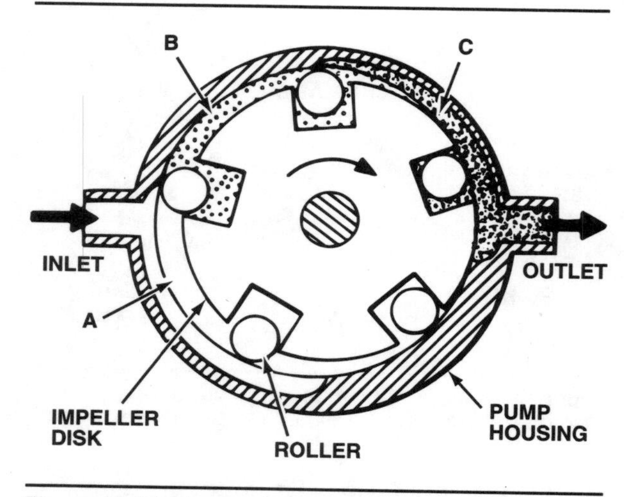

Figure 12-32. The pumping action of an impeller, or rotary vane, pump.

Pumps without an air chamber may use a vapor separator, figure 12-29, in the fuel line between the pump and carburetor. Fuel from the pump fills the vapor separator. The outlet tube in the separator picks up fuel from the bottom of the unit and passes it into the fuel line to the carburetor. Vapor that has gathered rises to the top of the separator where it is forced through a tube with a metering orifice, and then into a return line leading into the fuel tank.

The vapor bypass filter, figure 12-30, has been used mostly on cars with factory-installed air conditioning. It combines the fuel filter and vapor separator into a single unit. Like a vapor separator, a bypass filter uses a restricted nipple and fuel tank return line to relieve vapor pressure buildup. Both the vapor separator and the bypass filter are directional and must be properly installed to pass fuel to the carburetor.

Many pump designs have a bleed-down system in which tiny holes are drilled through each check valve. This permits pressure buildup in the fuel outlet line to bleed back to the fuel inlet line.

Mechanical pump applications

Mechanical fuel pumps used in domestic cars were manufactured by Carter, AC, or Airtex, all long-time industry suppliers. As a general rule, Ford used Carter and AC pumps, Chrysler used Carter and Airtex, and GM used AC pumps.

A diaphragm-type fuel pump is a simple device. Most of them operate the same way, and the main differences are usually in exterior appearance. The exterior design depends on which engine the pump uses and how much room the engine compartment has.

Pumps are so similar in some cases that a production run of the same engine block may use pumps from two different manufacturers.

However, replacement pumps must be identical in every respect. Installing a pump that just *looks* like the one removed can result in a broken camshaft or accessory shaft as soon as the engine starts.

Mechanical fuel pumps are quite dependable. If they break down, it is usually because of one of these factors:

- A leaking diaphragm
- A worn inlet or outlet check valve
- A worn or broken pushrod
- A worn linkage, which reduces the pump stroke.

Occasionally, the camshaft or accessory shaft eccentric may wear enough to reduce the pump stroke, or a bolt-on eccentric may come loose from the camshaft. In these cases, the camshaft or accessory shaft or the bolt-on eccentric must be replaced. It is possible to install an electric fuel pump to bypass a defective mechanical pump.

Electric fuel pumps

There are four basic kinds of electric fuel pumps:

- **Impeller** (turbine)
- Plunger
- Diaphragm
- Bellows.

The impeller, or turbine, pump, figure 12-31, is driven by a small electric motor. The other three are driven by an electromagnet and vibrating armature and are no longer used as original equipment on standard automobiles.

The impeller-type pump is sometimes called a turbine, roller cell, roller vane, or rotary vane pump. It draws fuel into the pump, then pushes it out through the fuel line to the carburetor or injection system. Since this type of pump uses no valves, the fuel is moved in a steady flow rather than the **pulsating** motion of other pumps.

Figure 12-32 shows the pumping action of a roller vane pump.The pump consists of a central impeller disk, several rollers that ride in notches in the impeler, and a pump housing that is offset from the impeller centerline. The impeller is mounted on the end of the motor armature and spins whenever the motor is running. The rollers are free to travel in and out within the notches in the im-peller to maintain sealing contact with the pump housing. Unpressurized fuel enters the pump and fills the spaces between the rollers, figure 12-32A. As the impeller rotates, a portion of the fuel is trapped between the impeller, the housing, and two rollers, figure 12-32B. Further rotation toward the offset side of the housing compresses the fuel and forces it out of the pump under pressure, figure 12-32C.

Electric pump location

The electric fuel pump is a pusher unit. It pushes the fuel through the supply line. Because it does not rely on the engine camshaft for power, an electric pump can be mounted in the fuel line anywhere on the vehicle—even inside the fuel tank.

Pusher pumps are most efficient when they are mounted as near as possible to the fuel tank and at or below its level, so it can use gravity to transfer fuel from the tank to the pump. This pump

Flooding: A condition caused by heat expanding the fuel in a fuel line. The fuel pushes the carburetor inlet needle valve open and fills up the fuel bowl even when more fuel is not needed. Also, the presence of too much fuel in the intake manifold.

Impeller: A rotor or rotor blade used to force a gas or liquid in a certain direction under pressure.

Pulsating: Expanding and contracting rhythmically.

■ No One Misses the Good Old Fuel Pump

Today's fuel pump may seem to be a simple device, but pump manufacturers have worked hard to make it so. Pump designs, capacities, pressures, and performance requirements make the modern pump a rather sophisticated device. This is especially true when you consider that a fuel pump is expected to transport large amounts of gasoline for thousands of miles or kilometers without failure.

Back in the thirties, fuel pump breakdown and replacement was a common occurrence every few thousand miles or several thousand kilometers. The fuel pump of the 1936 Ford V8 operated from a pushrod. As the pushrod wore, the pump stroke lessened. Most mechanics and a lot of owners kept the fuel pump operating with a wad of chewing gum or tinfoil stuffed into the pushrod cup to compensate for wear. Rather crude, but it worked.

The vacuum booster fuel pump was the first big change in pump design. Since no one could keep their windshield wipers running at a constant speed, pump designers provided additional vacuum with a dual pump design. But super highways, higher horsepower, and emission controls brought new approaches to pump design. Windshield wipers went electric and the vacuum booster fuel pump disappeared. Intake electric pumps have replaced the traditional mechanical pump design. Automakers now build modern fuel pumps to supply at least 30,000 trouble-free miles (48,000 km).

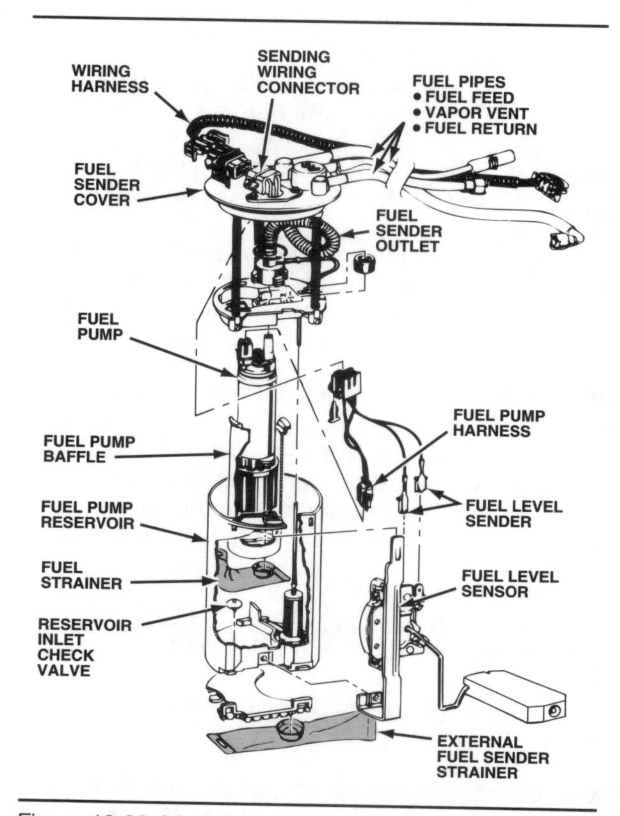

Figure 12-33. Modular fuel assembly components. (GM)

mounting also eliminates the problem of vapor lock under all but the most severe conditions. With the pump mounted at the tank, the entire fuel supply line to the carburetor is pressurized. Regardless of how hot the fuel line gets, it is unlikely that vapor bubbles will form to interfere with fuel flow. Having the pump close to or inside the tank also allows the pump to remain cooler because it is away from engine heat, so it is less likely to overheat during hot weather.

In-tank electric fuel pumps generally are combined with the fuel gauge sending unit, figure 12-31, to form a single assembly. Some GM and many Ford vehicles use two electric fuel pumps:

- A low-pressure, in-tank impeller pump
- A high-pressure, chassis-mounted impeller pump.

The low-pressure, in-tank pump provides fuel to the high-pressure pump to prevent vapor lock on the suction side of the fuel system. In a two-pump system, the in-tank pump sometimes is called a booster pump. The high-pressure pump provides the injection system operating pressure.

Modular fuel sender assembly

GM introduced the modular fuel sender assembly, figure 12-33, on some 1992 models in an effort to standardize fuel sender design. Previous in-tank fuel pump, fuel gauge, and fuel pickup unit design was influenced by the size and shape of the fuel tank to be used. The modular fuel sender assembly allows the use of a single design in a variety of different fuel tanks. This is possible because of its modular design, which allows the components to be assembled in three possible orientations, and because it is spring-loaded, allowing vertical self-adjustment to variations in fuel tank heights.

The modular fuel sender consists of a replaceable fuel level sensor bracket mounted to the side of a reservoir housing containing both a roller vane pump and a jet pump. The reservoir housing is capped by a support assembly that rests on three spring-loaded hollow support or guide pipes and is attached to the cover containing the three fuel pipes and the electrical connector. Fuel is transferred from the pump to the fuel pipe through a convoluted (flexible) fuel pipe. The convoluted fuel pipe allows alignment of the cover and housing in three different positions while eliminating the need for rubber hoses, nylon pipes, and clamps. The reservoir maintains a constant fuel level available to the roller vane pump and reduces fuel pump operation noise.

The roller vane pump, figure 12-34, has a two-stage pumping action with a low-pressure turbine section and a high-pressure roller vane section. The pump motor and end cap complete the pump assembly. The turbine impeller has a staggered blade design to minimize pump noise and separate vapor from the liquid fuel. The roller vane section creates the high pressure required for fuel injection. The end cap assembly contains a pressure relief valve and a radio frequency interference (RFI) module. The check valve generally used with roller vane pump designs has been relocated to the upper fuel pipe connector assembly, figure 12-35.

After filtering through the fuel strainers, fuel is drawn into the lower housing inlet port by the jet pump, which sends it to the first stage of the roller vane pump, where vapor is separated from the liquid fuel. The liquid fuel then is sent to the second stage of the roller vane pump, where it is pressurized and delivered to the convoluted fuel pipe for transfer through a check valve into the fuel feed pipe. A small portion of the flow, however, is returned to the jet pump for recirculation. Excess fuel in the fuel system is returned to the reservoir through one of the three hollow support pipes, figure 12-35. When the hot fuel returns to the reservoir, it mixes rapidly with cooler reservoir fuel due to the turbulence created by jet pump operation. This minimizes the possibility of vapor lock.

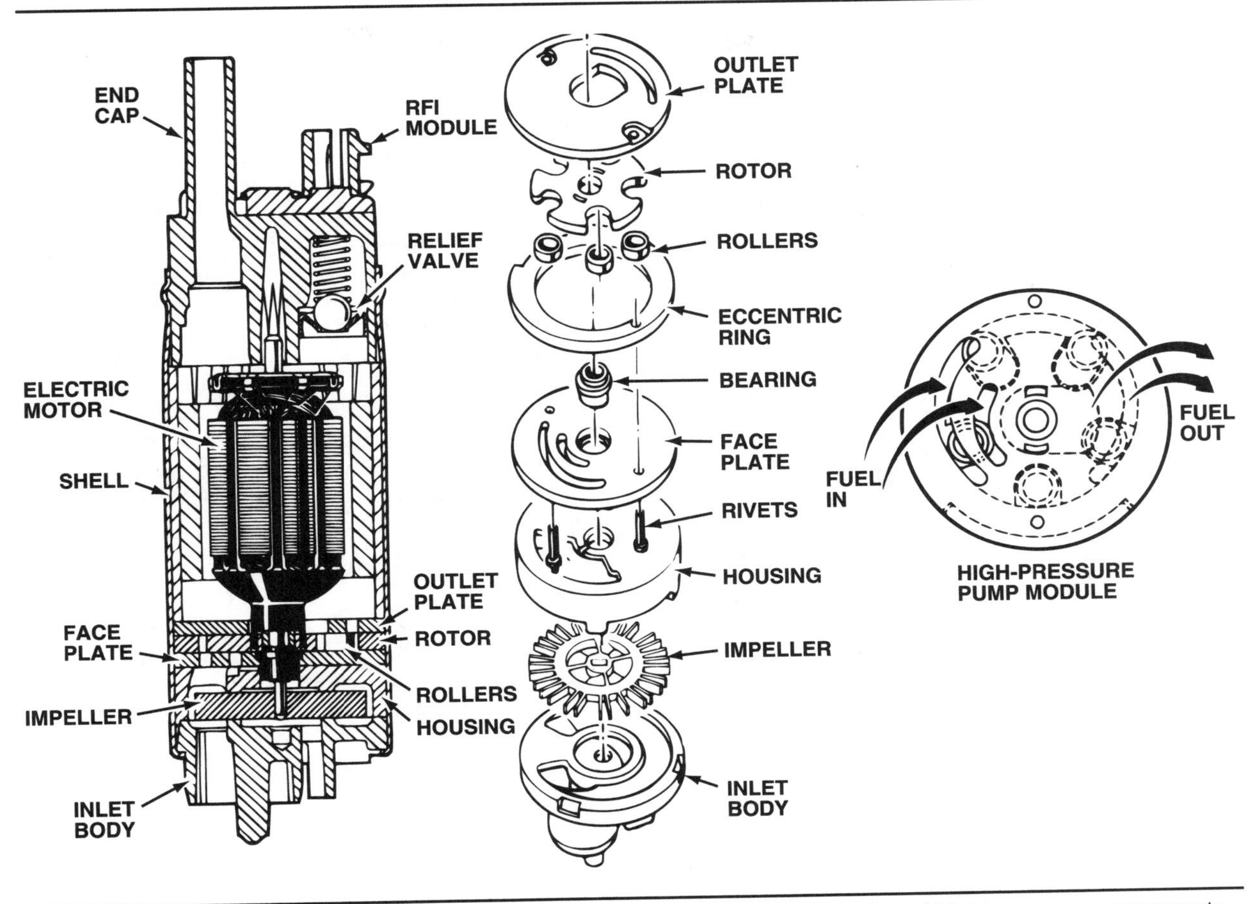

Figure 12-34. The modular fuel sender assembly roller vane pump showing low- and high-pressure components. (GM)

Electric pump control circuits

A pressure switch in the engine oil system controls most original equipment electric fuel pumps of the 1970s. This switch opens the electric circuit to the pump motor when the engine is off and controls the operation of the pump when the engine is started and while it is running.

The pressure switch has two sets of contact points. One set is normally closed and allows current to flow from the battery through the starter solenoid or relay to the fuel pump. The other set is normally open. When closed, it allows current to flow from the battery through the ignition switch to the fuel pump.

Turning the ignition key to START energizes the pump by providing current through the normally closed contact points. Once the engine is running, the pump receives current through the normally open contacts, which have been closed by engine oil pressure.

Engine oil pressure opens the normally closed contacts and closes the normally open contacts to keep the pump energized. When the ignition switch is turned off, the pump circuit de-energizes. If oil pressure drops below the specified level for any reason, electrical contact breaks at the pressure switch, stopping the fuel pump immediately.

Another type of pump, an electrical fuel pump, has a relay, figure 12-36, and uses the oil pressure switch circuit as a backup in case the relay malfunctions. These pumps are used on late-model fuel management systems.

On Chrysler cars, the logic module must receive an engine speed (rpm) signal during cranking before it can energize a relay inside the power module to activate the fuel pump, ignition coil, and injectors. If the rpm signal to the logic module is interrupted, the module signals the power module to activate the automatic shutdown relay (ASD) and turn off the pump, coil, and injectors.

Ford and GM systems energize the pump with the ignition switch to initially pressurize the fuel lines, but then de-activate the pump if a rpm signal is not received within one to two

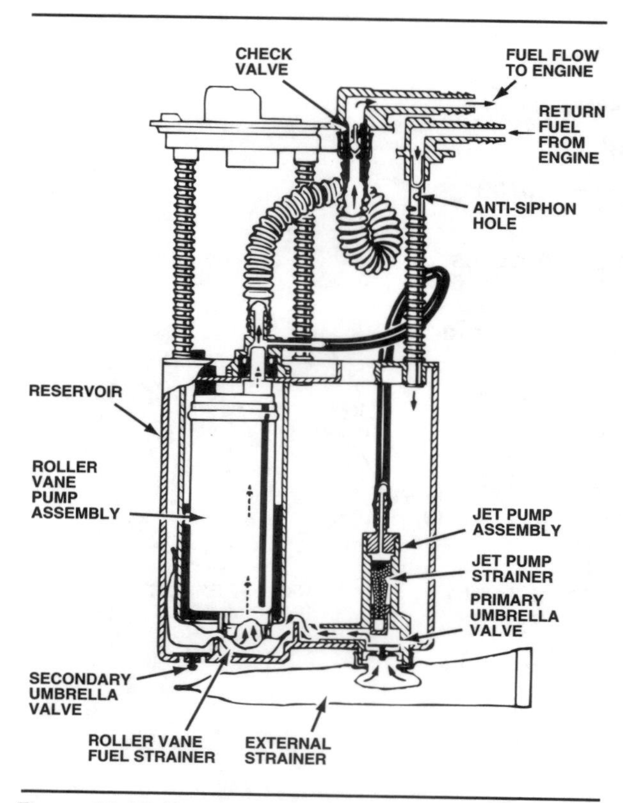

Figure 12-35. Fuel flow through the modular fuel sender assembly. (GM)

seconds. The pump is reactivated as soon as engine cranking is detected. The oil pressure sending unit serves as a backup to the fuel pump relay. In case of pump relay failure, the oil pressure switch will operate the fuel pump once oil pressure reaches about 4 psi (28 kPa).

Fords with fuel injection have an inertia switch between the fuel pump relay and fuel pump, figure 12-36. When the ignition switch is turned to the ON position, the electronic engine control (EEC) power relay energizes, providing current to the fuel pump relay and a timing circuit in the EEC module.

If the ignition key is not turned to the START position within about one second, the timing circuit opens the ground circuit to de-energize the fuel pump relay and shut down the pump. This circuit is designed to pre-pressurize the system. Once the key is turned to the START position, power to the pump is sent through the relay and inertia switch.

The inertia switch opens under a specified impact, such as a collision. When the switch opens, power to the pump shuts off. The switch must be reset manually by depressing the button on its top before current flow to the pump can be restored.

Fuel Filters

Despite the care generally taken in refining, storing, and delivering gasoline, some impurities get into the automotive fuel system. Fuel filters remove dirt, rust, water, and other contamination from the gasoline before it can reach the carburetor or injection system.

The useful life of all filters is limited, although Ford specifies that its filters used with some fuel injection systems should last the life of the vehicle. If fuel filters are not cleaned or replaced according to the manufacturer's recommendations, they become clogged and may restrict fuel flow.

In addition to using several different types of fuel filters, a single fuel system may contain two or more filters. Automakers can locate these filters in several places within the fuel system.

Fuel tank filters and strainers

A sleeve-type filter of woven Saran is usually fitted to the end of the fuel pickup tube inside the fuel tank. This filter "sock" prevents sediment, which has settled at the bottom of the tank, from entering the fuel line. It also protects against water contamination by plugging itself up. If enough water enters the fuel tank, it accumulates on the outside of the filter and forms a jelly-like mass. If this happens, the filter must be replaced. Otherwise, no maintenance is required for this filter.

In-line filters

The in-line filter, figure 12-37, is located in the line between the fuel pump and carburetor. This protects the carburetor from contamination, but does not protect the fuel pump. The in-line filter usually is a disposable plastic or metal container with a pleated paper element sealed inside.

Some fuel injection systems use in-line filter canisters. These are larger units than are generally used with carbureted engines, figure 12-38. They may be mounted on a bracket on the fender panel, a shock tower, or another convenient place in the engine compartment. They may also be installed on a frame rail under the vehicle near the electric fuel pump.

In-line filters must be installed so that gasoline flows through them in the direction shown by the arrow, figure 12-37. If an in-line filter is installed backwards, it restricts fuel delivery to the carburetor or injectors.

An in-line filter may have a built-in vapor bypass system. These filters have a third nipple, figure 12-30, that connects a fuel return line back to the fuel tank.

Some older domestic and imported cars have a sediment bowl between the fuel pump and carburetor. The bowl contains a pleated paper, ceramic, fiber, or metal filter element. The filter

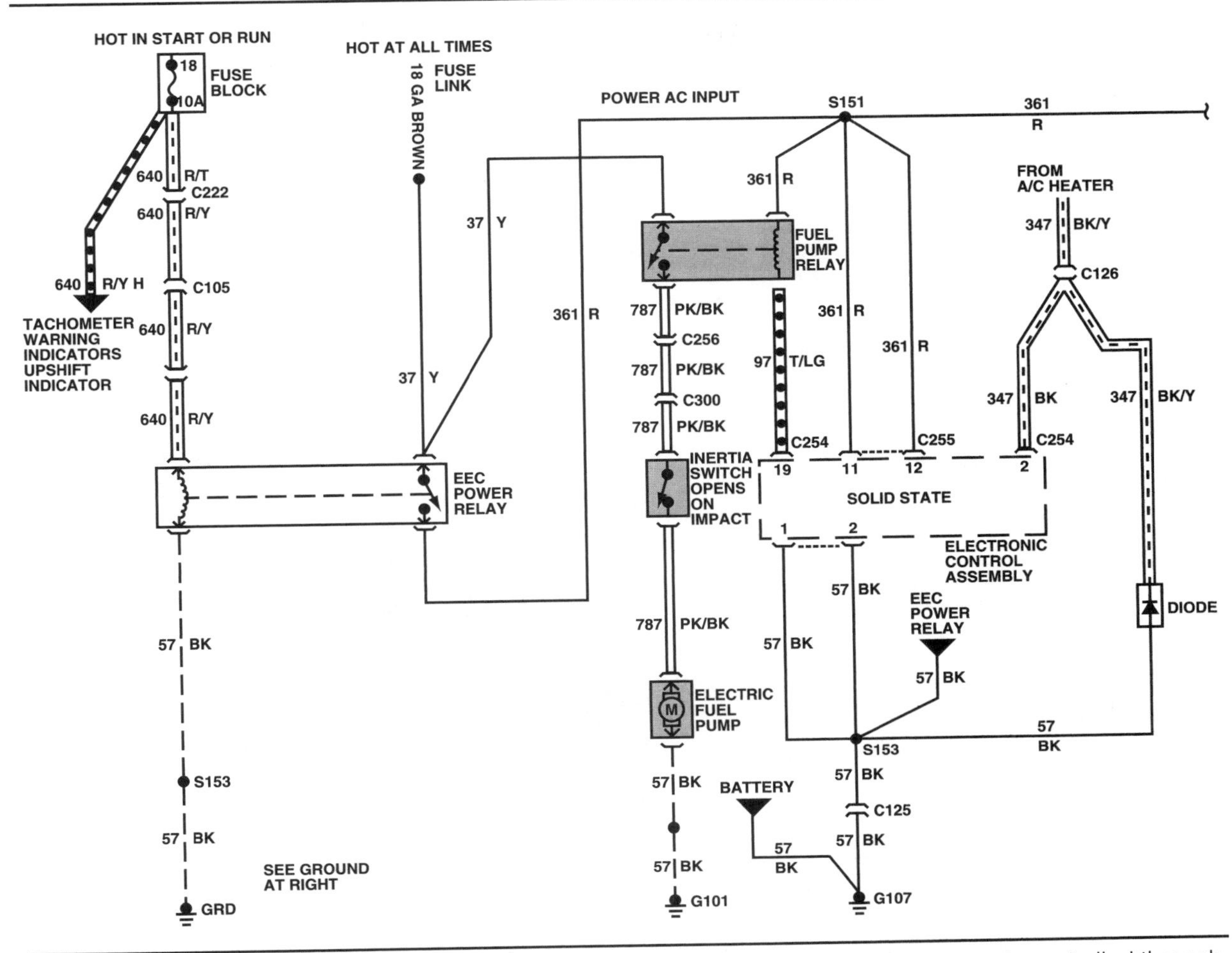

Figure 12-36. Later-model electric fuel pumps used with engine fuel management systems are controlled through a pump relay. This Ford system also uses an inertia switch.

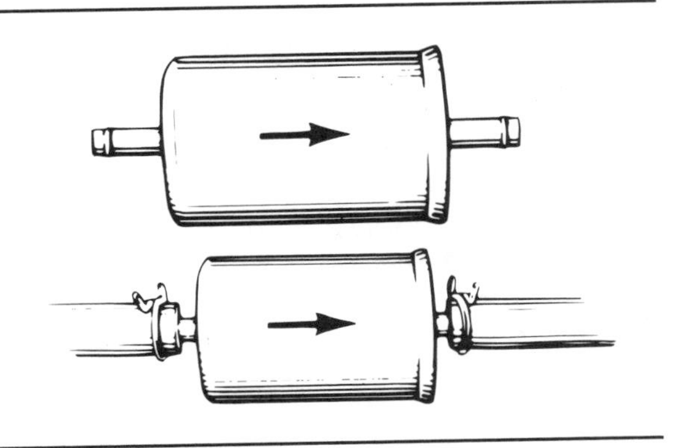

Figure 12-37. In-line fuel filters must be installed so that gasoline flows in the direction indicated by the arrow.

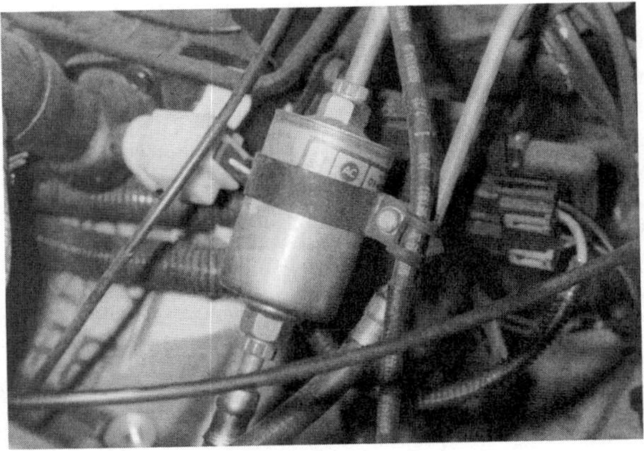

Figure 12-38. Fuel injection systems use large capacity in-line fuel filters.

element works much like an in-line filter. The bowl cover is held in place by a wire bail and clamp screw. It can be removed for filter cleaning or replacement. Ceramic and metal elements can be cleaned and reused, if necessary. Paper and fiber filter elements must be replaced when dirty.

Carburetor inlet filters

Ford and GM equip most of their carbureted engines with inlet filters, figure 12-39. The Ford filter is a one-piece throw-away metal unit containing a filter screen and magnetic washer to

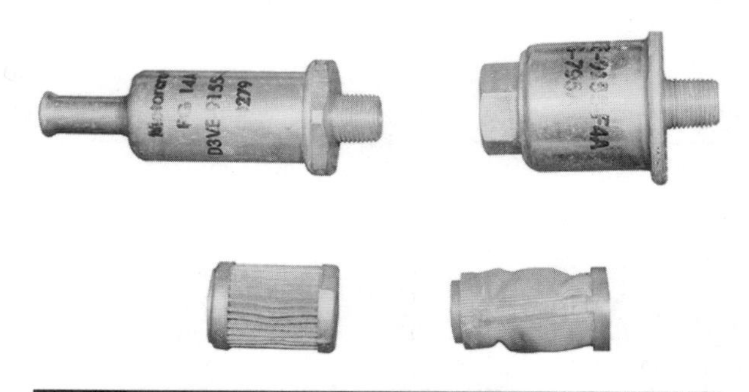

Figure 12-39. A variety of inlet filters used with Motorcraft and Holley carburetors.

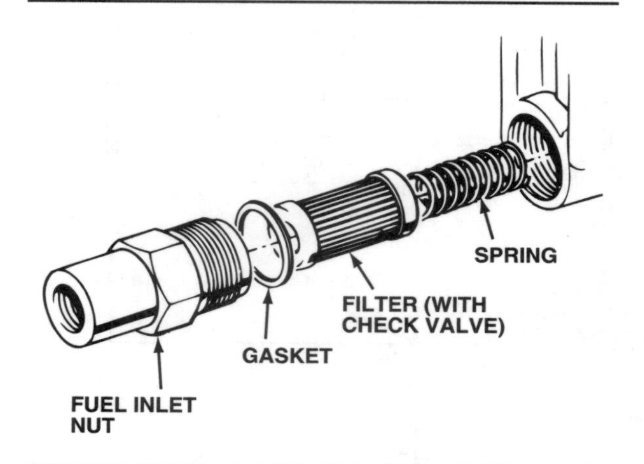

Figure 12-40. The check valve end of the filter must face the fuel line. (Buick)

trap dirt and metal particles. The filter generally screws into the carburetor fuel inlet at one end and clamps to the inlet hose at the other end. One Ford version used in the mid-1970s, however, was designed specifically as an in-line filter.

Some Motorcraft, Holley, and Rochester carburetors use a throw-away pleated paper element. Figure 12-39 (bottom) shows Motorcraft and Holley elements. Older cars had a bronze filter element, but reuse is not recommended. Filters on 1976 and later Rochester carburetors must contain the rollover check valve, described in an earlier chapter. Figure 12-40 shows the proper installation of the Rochester filter with check valve.

Pump outlet filters

Some cars have fuel filters in the outlet side of the fuel pump. Those pumps used on Chrysler 6-cylinder engines during the early 1970s contain a disposable filter element installed in the fuel outlet tower. Cadillacs through 1974 use a fuel pump outlet filter located on the bottom of the pump.

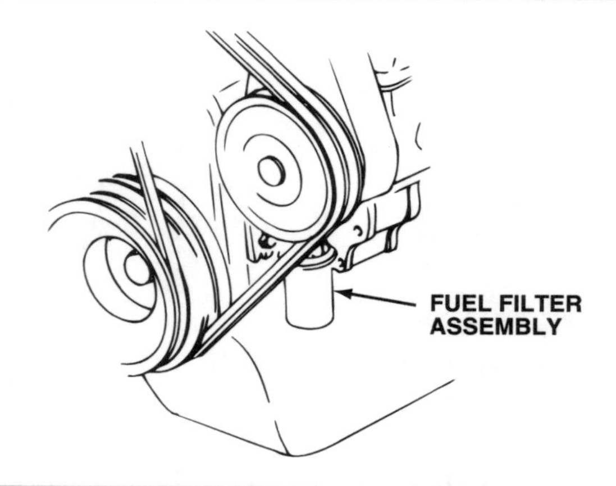

Figure 12-41. Disposable element filters may be mounted on the engine or near the fuel tank.

Disposable element filters

Screw-on, throw-away element filters, figure 12-41, look much like a replaceable oil filter. Ford has used this type on the fuel pump of some V8 engines. The first Cadillac Seville models used a disposable filter mounted to the frame near the left rear wheel. Other fuel injected Cadillacs have the filter mounted to a bracket on the lower left front of the engine.

Fuel injection and filters

Proper filtering of gasoline is essential to fuel injection operation, because particles smaller than one **micron** (0.000039 in.) can interfere with the close tolerances in injectors. Thus, fuel injection systems use various filters:

- The fuel tank filter removes particles larger than 50 microns (0.00197 in.) in size
- A large-capacity in-line filter, figure 12-38, removes particles greater than 10 to 20 microns (0.00039 to 0.00079 in.) in size.

In addition to these filters, some throttle body injection (TBI) units have a filter screen installed in the fuel inlet.

All injectors, throttle body or port, are fitted with one or more filter screens or strainers to remove any particles (generally 10 microns or 0.00039 in.) that might have passed through the other filters. These screens, which surround the fuel inlet, are external on throttle body injectors, figure 12-42, and internal on port injectors, figure 12-43.

Air Intake Filtration

The automotive engine burns about 9,000 gallons (34,069 liters) of air for every gallon of gasoline at an air-fuel ratio of 14.7:1. With many of today's engines operating on even leaner ratios, the quantity of air consumed per gallon of fuel is closer to 10,000 gallons (37,854 liters). This equals 200,000

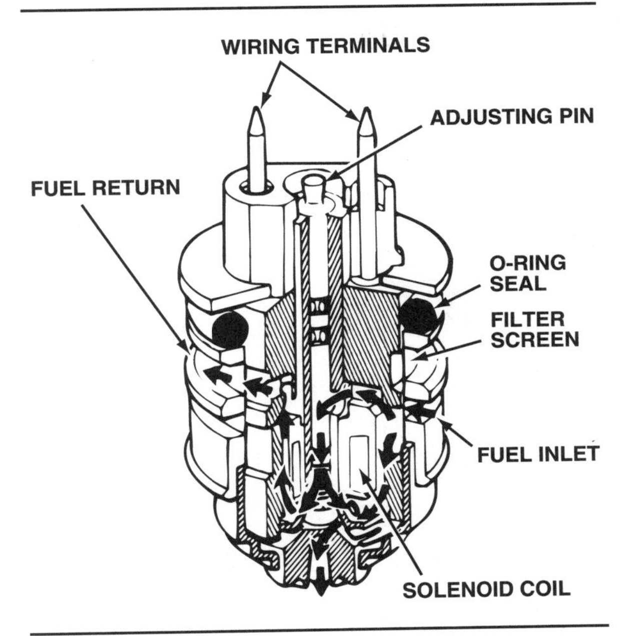

Figure 12-42. Injectors used in throttle body units have one or more external filter screens. (Chrysler)

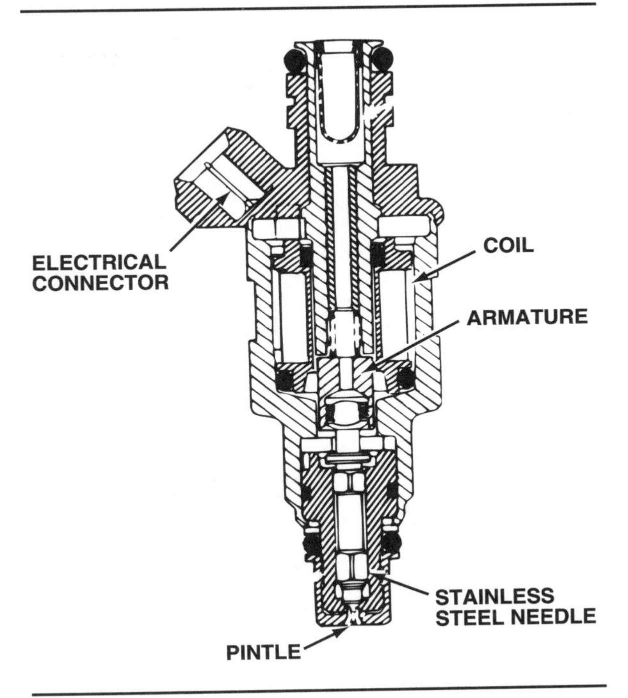

Figure 12-43. Port fuel injectors generally use an integral filter screen. (Ford)

gallons (757,082 liters) of air with every 20 gallons (76 liters) of fuel.

That is enough air to fill a large swimming pool, and—just like pool water—air is filled with particles of dust and dirt. Without proper filtering of the air before intake, these particles can affect the operation of the carburetor and upset the air-fuel ratio. Given enough time, they will seriously damage engine parts and shorten engine life.

While abrasive particles cause wear any place inside the engine where two surfaces move against each other, they first attack piston rings and cylinder walls. Contained in the **blowby** gases, they pass by the piston rings and into the crankcase. From the crankcase, the particles circulate throughout the engine in the oil. Large amounts of abrasive particles in the oil can damage other moving engine parts.

Although the basic airborne contaminants—dust, dirt and carbon particles—are found whenever a car is driven, they vary in quantity according to the environment. For example, engine air intake of abrasive carbon particles will be far greater in constant bumper-to-bumper traffic. Intake of dust and dirt particles will be greater in agricultural or construction areas.

The Air Cleaner

The filter that cleans the intake air is in a two-piece air cleaner housing made either of stamped steel or composite materials, figure 12-44. The

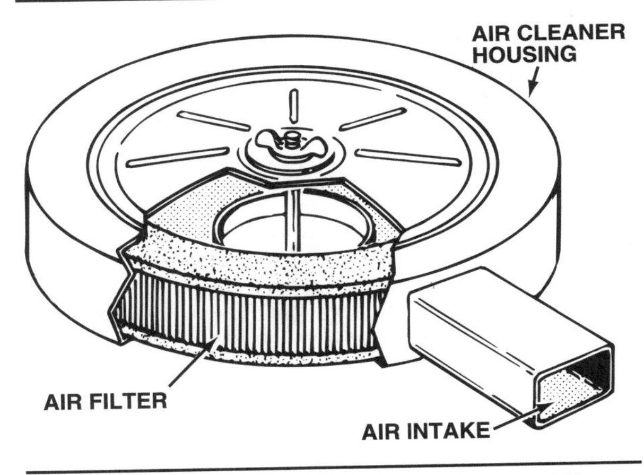

Figure 12-44. A simple air cleaner housing and filter assembly.

air cleaner housing is located on top of the carburetor or throttle body injection (TBI) unit or is positioned to one side of the engine, figure 12-45.

Older carbureted engine air cleaners had a snorkel or air intake tube which drew fresh air

Micron: A unit of length equal to one-millionth of a meter or one one-thousandth of a millimeter.

Blowby: Combustion gases that leak past the piston rings into the crankcase.

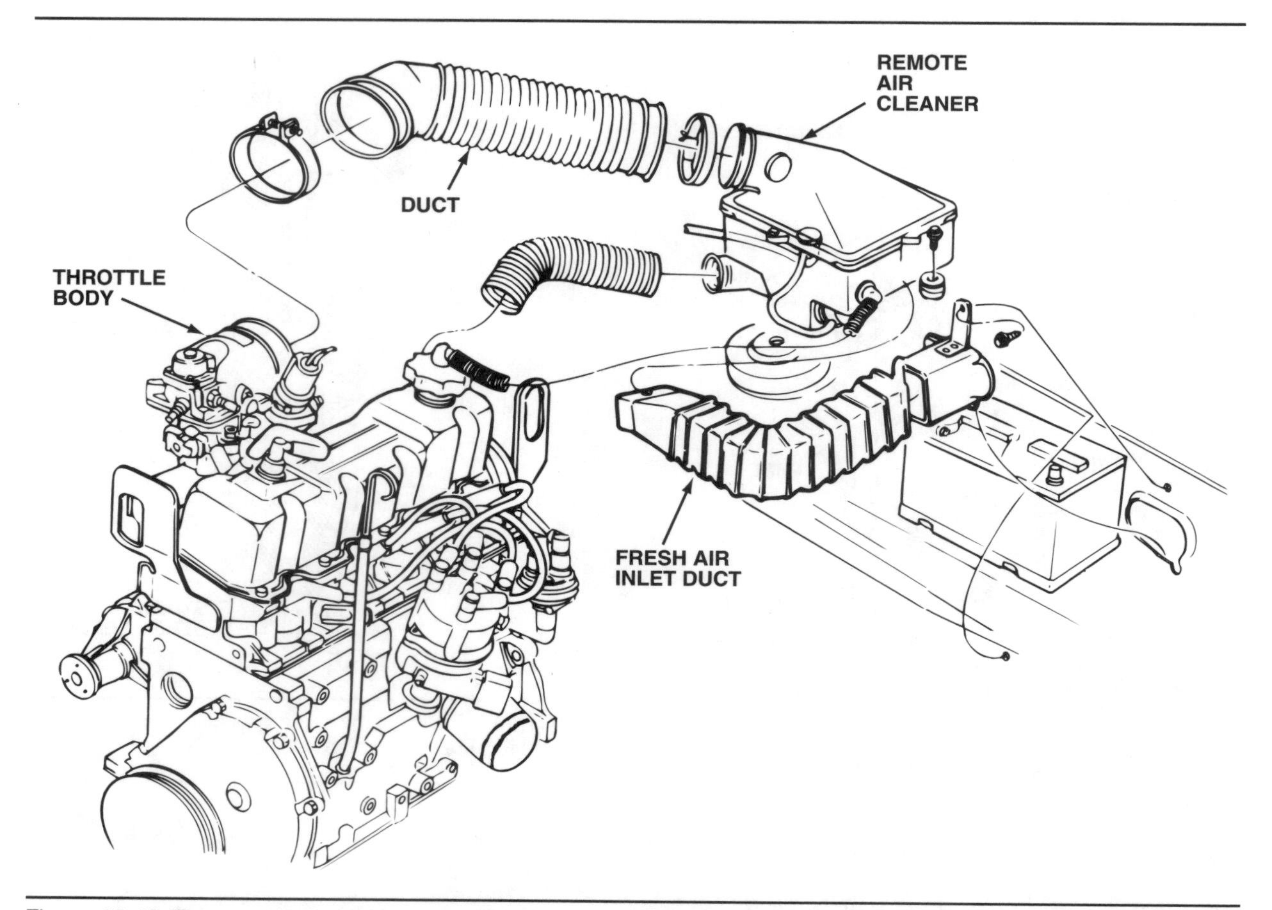

Figure 12-45. Remote air cleaners are positioned to one side of the engine and connected by ducting. (Ford)

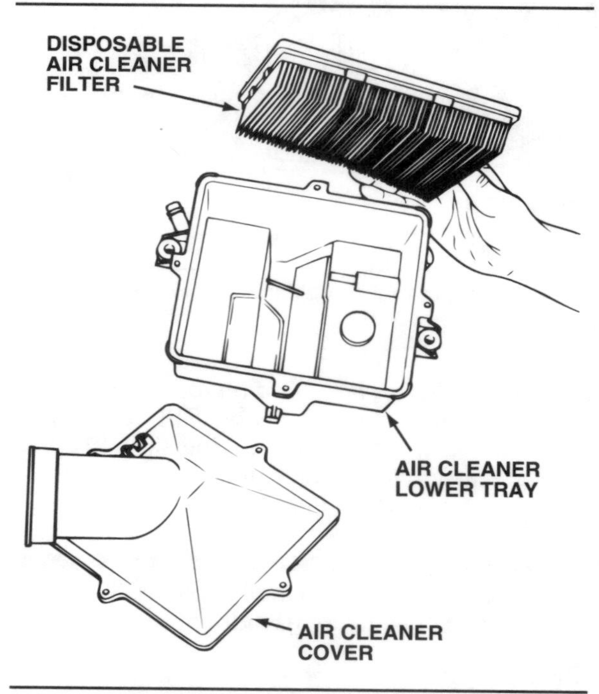

Figure 12-46. Paper air cleaner filter elements are disposable. (Ford)

into the housing from the engine compartment, figure 12-44. Snorkels are still used, but they are connected by ducting to a fresh air intake which draws air into the housing from outside the vehicle. Remote air cleaners connected to the carburetor or the throttle body by similar ducting also draw fresh air from outside the vehicle, figure 12-45. Additionly, the air cleaner housing has a removable top section.

Some air filter elements can be cleaned and reused, but most air filter elements are disposable, figure 12-46. Certain 4-cylinder GM engines, such as those used in the Chevette and Vega-class vehicles, have a disposable air cleaner housing containing a long-life filter. After 50,000 miles (80,000 km) the single-piece, welded air cleaner housing is removed from the air intake snorkel and carburetor airhorn and a new housing installed.

Filter replacement

Automakers recommend cleaning or replacing the air filter element at periodic intervals, usually listed in terms of distance driven or months of service. The distance and time intervals are based on so-called normal driving. More frequent air filter replacement is necessary more often when

Figure 12-47. Typical circular paper air filter element.

Figure 12-48. Some paper filters have an outer polyurethane wrapper.

the vehicle is driven under dusty, dirty, or other severe conditions.

It is best to replace a filter element before it becomes too dirty to be effective. A dirty air filter passes contaminants that cause engine wear and can change the air-fuel ratio and affect engine performance. The higher the engine speed, the greater the airflow required. Restricted or clogged filters greatly affect high-speed engine operation. If the element becomes so clogged that it does not let through enough air, it can act as a choke to increase fuel consumption. In severe cases, a clogged air filter can even keep the engine from running.

Air filter elements

Cars and light trucks use two types of air filter elements:

- Polyurethane filters
- Paper filters.

Polyurethane filters are available as aftermarket replacements for OEM filters. The paper air filter element, figure 12-47, on the other hand, is the most common type of filter used on late-model cars and light trucks. It is made of a chemically treated paper stock that contains tiny passages in the fibers. These passages form an indirect path for the airflow to follow. The airflow passes through several fiber surfaces, each of which traps microscopic particles of dust, dirt, and carbon.

Filter paper is pleated and formed into a circle, square, or rectangle (depending upon housing design). Circular filter elements have the top and bottom edges sealed with heat-resistant plastic, figure 12-47, to prevent unfiltered air from bypassing the filter. Square and rectangular filter elements generally seal only the top edges. A fine wire mesh screen may be used on the inside of the filter ring to reduce the possibility of the element catching fire from an engine backfire. A similar, coarser, wire mesh screen may be used on the outside of the filter ring for additional strength.

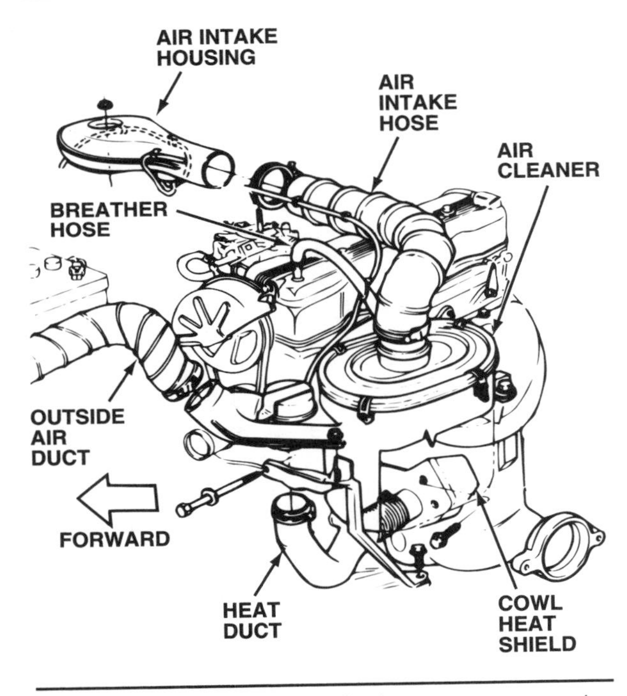

Figure 12-49. Some remote air cleaners are connected to the carburetor or TBI unit by an air intake housing. (Chrysler)

These filter elements generally are made of dry paper, although an oil-dipped paper stock is sometimes used. The light oil coating helps prevent contaminants from working their way through the paper. It also increases the dirt-holding capacity over the same area of dry paper stock. An outer wrapper of polyurethane, figure 12-48, is used sometimes to make the filter work

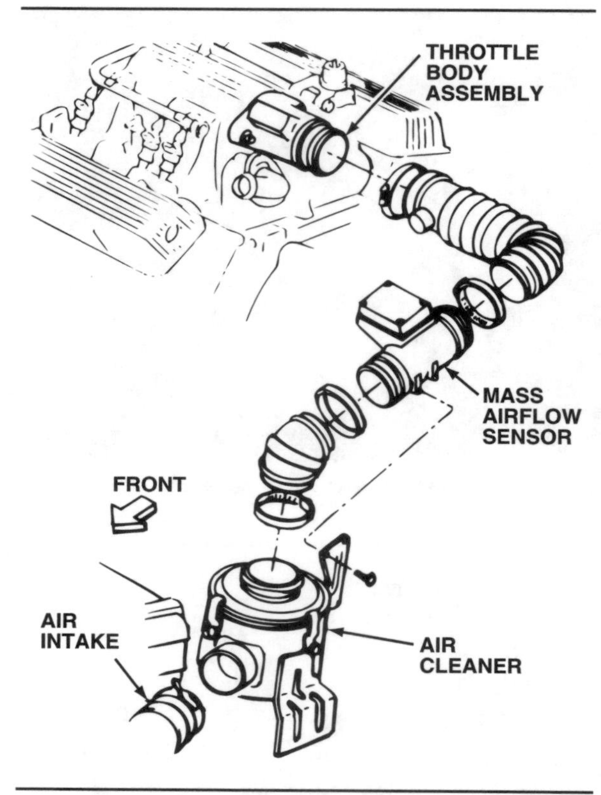

Figure 12-50. Port fuel injected (PFI) engines also use a remote air cleaner. If a mass airflow sensor is used, it is inserted in the ducting between the throttle body and the air cleaner. (AC-Delco)

better. Paper element filters are disposable and should be replaced at the recommended intervals. Do not attempt to clean a paper element filter by rapping it on a sharp object to dislodge the dirt, or blowing compressed air through the filter. This tends to clog the paper pores and further reduce the airflow capability of the filter.

Air Intake Ducts and Fresh Air Intakes

Most of the air entering the engine passes through the air cleaner snorkel or inlet tube. Some air cleaners use two snorkels to provide additional air intake at full throttle. High-performance engines of the 1960s and early 1970s commonly used these designs. They reappeared on a few high-performance carbureted engines in the 1980s.

The snorkel passes air to the filter and then to the carburetor from the engine compartment. The snorkel also increases the velocity of the air entering the air cleaner housing. Temperatures in the engine compartment often exceed 200°F (93°C) on a hot day, and hot air can lean the air-fuel mixture enough to cause detonation and possible engine damage.

Allowing the engine to breath cooler air from outside the engine compartment prevents such problems. Cooler air is provided by a cold air duct or induction (zip) tube. The tube runs from the snorkel to a fresh air intake at the front of the car. The fresh air intake normally is open at all times, but may have a screen to prevent insects and other foreign matter from being drawn into the air cleaner.

Air cleaner designs used on late-model high-performance engines may have two fresh air inlets, each of which is connected to the air cleaner housing by ducting. This is an updated version of the dual-snorkel air cleaner discussed earlier. One inlet provides airflow for general operation; the other opens to provide maximum airflow with the engine at wide-open throttle.

The fresh air intake may be in the cowl or in the rear area of the hood. Ducted hood air doors used on Corvettes in the mid-1970s open electrically at full throttle. Pedal linkage closes a switch when the accelerator is pushed to the floorboard. This switch operates a solenoid attached to the air door linkage. The air door provides more intake air at wide-open throttle, just as the second snorkel or intake duct does on some air cleaners.

Remotely mounted air filters and ducts
Air cleaner and duct design depend on a number of factors such as the size, shape, and location of other engine compartment components, as well as the vehicle body structure. Generally, the air cleaner housing is installed on top of a carburetor or TBI unit. However, it also can be located away from the engine and connected to the carburetor or TBI unit by an air intake housing, figure 12-49.

Port fuel injection systems generally use a horizontally mounted throttle body. Some systems also have a mass airflow (MAF) sensor between the throttle body and the air cleaner, figure 12-50. Because placing the air cleaner housing next to the throttle body would cause engine and vehicle design problems, it is more efficient to use this remote air cleaner placement.

Turbocharged engines present a similar problem. The air cleaner connects to the air inlet elbow at the turbocharger. However, the tremendous heat generated by the turbocharger makes it impractical to place the air cleaner housing too close to the turbocharger. For better protection, the MAF sensor is installed between the turbocharger and the air cleaner in some vehicles.

Turbocharger and fuel injection filters and ducts
Remote air cleaners are connected to the turbocharger air inlet elbow or fuel injection throttle body by composite ducting which is usually retained by clamps. The ducting used may be rigid or flexible, but all connections must be airtight.

Filters used in remote air cleaners vary widely in size and shape, but all are similar to the paper element filters described earlier and should be serviced in a similar manner.

AIR INTAKE TEMPERATURE CONTROL

Air temperature regulation requirements differ according to engine, carburetor, or fuel injection system used. Sensors are used to ensure proper intake air temperature. These sensors generally are installed in the air cleaner housing and are calibrated to open a vacuum bleed as low as 50°F (10°C) or as high as 120°F (49°C).

On carbureted systems, Ford and other automakers use a vacuum modulator or a retard delay valve to trap vacuum to the air cleaner vacuum motor and hold the air control door in the hot-air position at very low temperatures despite manifold vacuum. Both systems work essentially the same way to prevent low engine vacuum from overriding the temperature control. GM uses a temperature control valve on some engines for the same purpose. Figure 12-51 explains the operation of a vacuum modulator in a typical Ford system.

The retard delay valve, while similar to spark delay valves used in other emission control applications, has an umbrella-type check valve with a **sintered** steel restrictor to delay vacuum release. Most retard delay valve applications are color coded to indicate the release delay time.

Thermostatically Controlled Air Cleaners

Some form of **thermostatic** control has been used on automobile air cleaners since 1968 to control intake air temperature for improved driveability. These controls became even more important with the need to maintain the precise air-fuel ratios required for exhaust emission control.

The thermostatically controlled air cleaner has the usual sheet metal or composite housing described earlier. Another sheet metal duct, called a heat stove or shroud, is fastened around the exhaust manifold. The heat stove is connected to the air cleaner intake by a flexible hose or metal tube called a hot air tube or heat duct. Figures 12-49 and 12-52 show examples of various designs. Heat radiating from the exhaust manifold is retained by the heat stove and sent to the air cleaner inlet to provide heated air to the carburetor or the throttle body. However, fuel injection systems using a MAF sensor do not use temperature control.

An air control valve or damper permits the intake of:

- Heated air from the heat stove
- Cooler air from the snorkel or cold-air duct
- A combination of both.

While the air control valve generally is located in the air cleaner snorkel, it may be in the air intake housing or ducting of remote air cleaners. The air control valve maintains intake air at a specified temperature, usually 90° to 100°F (32° to 38°C). The air control valve is operated by a vacuum motor or diaphragm, although some older domestic and some imported cars use a thermostatic bulb.

Vacuum motors

Vacuum motor control of air intake temperature is used on most Chrysler products. Some Ford,

Sintered: Bonded together with pressure and heat, forming a porous material, such as the metal disk used in some vacuum delay valves.

Thermostatic: Referring to a device that automatically responds to temperature changes in order to activate a switch.

Bimetal Temperature Sensor: A sensor or switch that reacts to changes in temperature. It is made of two stripes of metal welded together that expand differently when heated or cooled, causing the strip to bend.

Air Cleaner Filter Maintenance

What causes engine wear? Contrary to popular belief, it is not how many miles the engine has been driven, nor how old the engine is. If the engine is kept properly lubricated, wear is mainly caused by the dust and dirt that enters it. A teaspoonful of gritty dirt will ruin the piston rings; a cupful will virtually destroy the entire engine. A properly maintained air cleaner filter is the primary line of defense against the gallons of dirty air sucked into the engine with each tiny sip of fuel.

To prevent dirt from entering the engine, a paper filter element should be discarded when dirty. Attempting to clean the element with low-pressure compressed air or rapping it on a hard object to dislodge the dirt and then reusing the filter is not a good idea. These cleaning methods get rid of the dirt which adheres to the outside of the element, but the dirt trapped inside the element cannot be removed and tends to plug up the pores of the paper.

You should avoid handling filters roughly or damaging them. Also, do not try to wash a paper filter element, since it cannot restore the air passages to normal. Air filter elements should be replaced. In most cases, they are inexpensive compared to the damage that dirt can do if it reaches the engine. At the same time, replace the crankcase ventilation filter, if the air cleaner uses one.

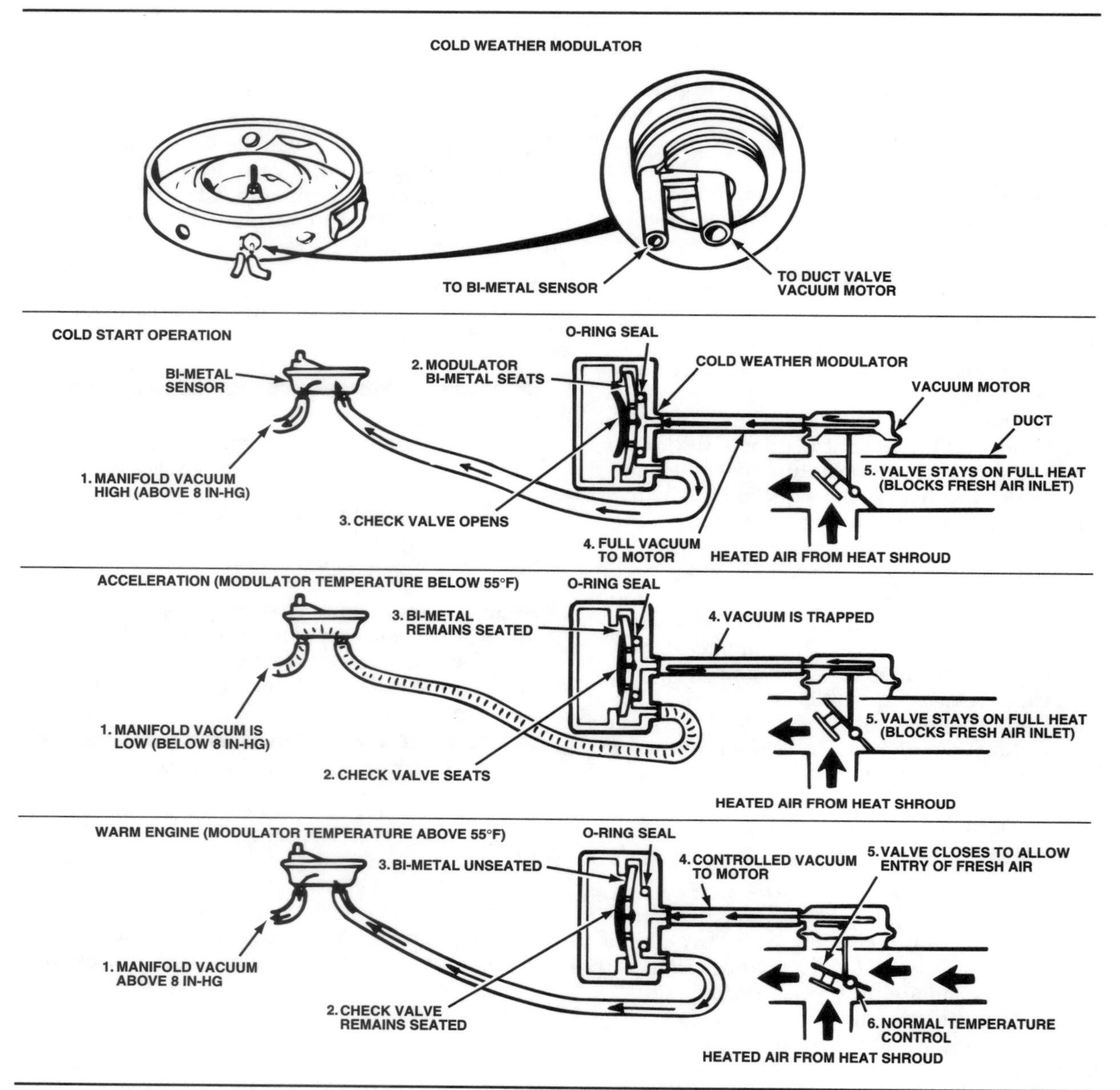

Figure 12-51. Operation of a thermostatically controlled vacuum modulator between the temperature sensor and the vacuum motor. (Ford)

and smaller GM engines had thermostatic-bulb air cleaners through the mid-1970s. By the 1980s, almost all engines used vacuum motors to control air cleaner damper operation.

In an air cleaner with a vacuum motor, a **bimetal temperature sensor** and a vacuum bleed in the air cleaner housing regulate vacuum supply to the vacuum motor. Vacuum is supplied from the intake manifold. When intake air temperature is below approximately 100°F (40°C), the temperature sensor holds the vacuum bleed closed and full manifold vacuum is applied to the vacuum motor. The motor holds the air control valve in the full hot-air position, figure 12-53A.

As intake air warms up, the sensor begins to open the vacuum bleed. This decreases the vacuum sent to the motor. A spring in the motor starts to move the air control valve from the hot-air to the cold-air position, figure 12-53B.

As air temperature continues to rise, the vacuum bleed continues to open, further reducing vacuum to the motor. At high air temperatures, the air cleaner motor receives no vacuum, placing the motor's plate in its full cold-air position,

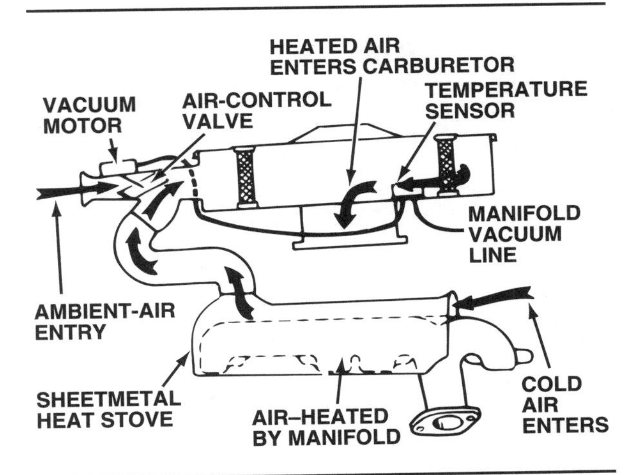

Figure 12-52. A thermostatically controlled air cleaner with a heat stove.

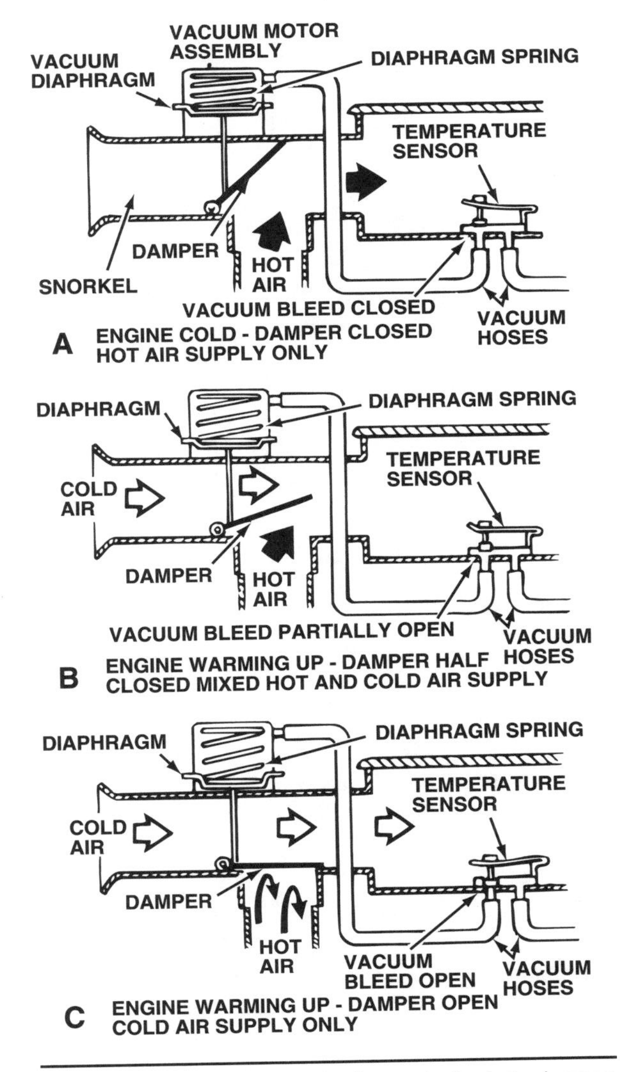

Figure 12-53. A thermostatically controlled air cleaner with a vacuum motor control.

figure 12-53C. On some engines, the air control valve also opens to the full cold-air position during heavy acceleration, regardless of air temperature. The opening of the valve provides maximum airflow through the air cleaner to the carburetor or throttle body when it is needed the most.

The operating requirements of other engines may be different, however. The vacuum modulator systems described previously trap vacuum in the air motor to hold the damper in the cold-air position. On a cold engine operating close to the 14.7:1 air-fuel ratio, a sudden charge of dense,

Foam and Oil Filtration

The disposable paper air filter element is now used as OEM equipment on all vehicles. Two decades ago, however, polyurethane or foam filter elements were used, and the oil bath filter was common on cars, trucks, and some off-road vehicles.

Foam filters consist of a polyurethane wrapper stretched over a metal support screen. Polyurethane contains thousands of pores and interconnecting strands that create a maze-like dirt trap while allowing air to flow through it. Properly maintained, a polyurethane element has a capacity and efficiency equal to that of the dry paper element. It can also be cleaned and reused if necessary. Today, they are popular aftermarket filters, although some automakers include an outer wrapper of polyurethane on their paper filters.

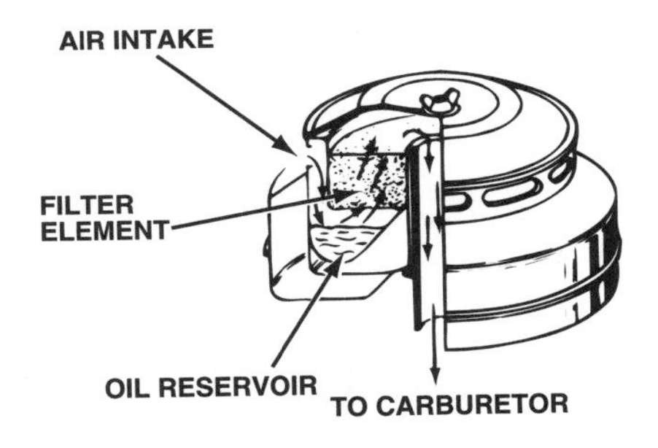

The oil bath cleaner rests in an oil reservoir in the air cleaner housing. Air entering the housing is deflected downward, where it strikes the oil in the reservoir and deposits heavier particles of dirt. Picking up an oil mist from the reservoir, the air flows back up and across the surface of the filter, where it leaves the mist with finer particles entrapped. The oil then drains back to the reservoir from the filter, carrying the entrapped dirt with it in a self-cleaning action. This type of filtration is most efficient at high airflow rates.

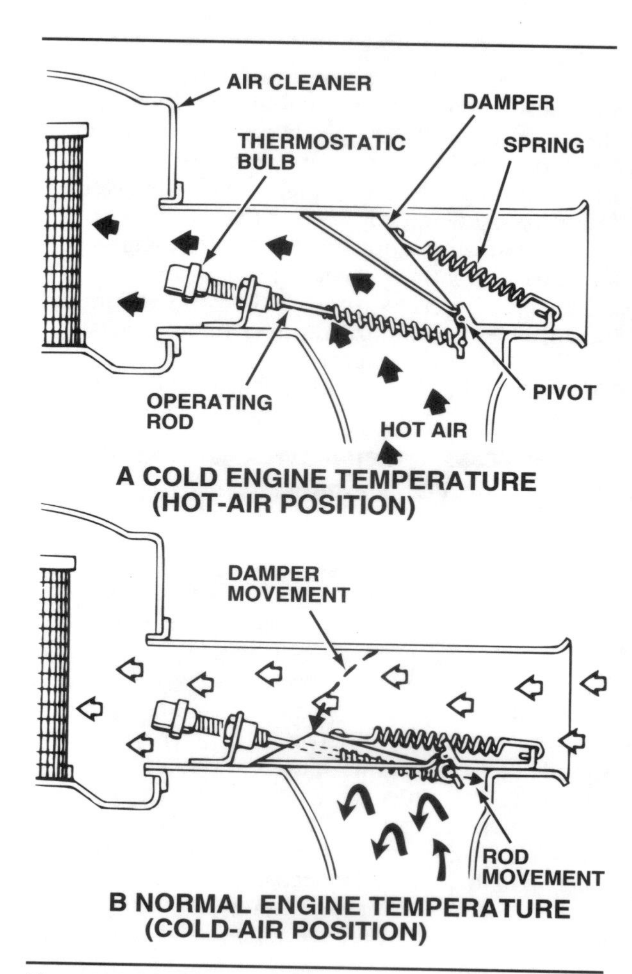

Figure 12-54. A thermostatically controlled air cleaner with thermostatic bulb control. As the bulb expands, it pushes the rod forward to move the damper downward into the hot-air position.

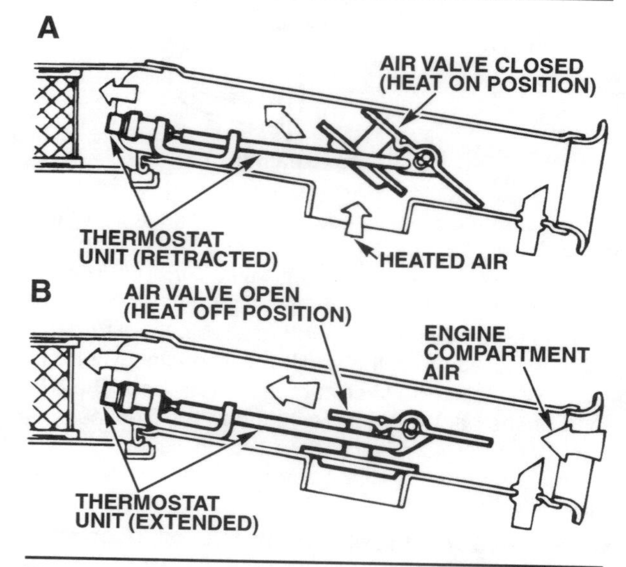

Figure 12-55. Operational cycle of a thermostatically controlled air cleaner.

Figure 12-56. A second vacuum motor is used on some air cleaners to operate a trap door or second air control valve in the snorkel.

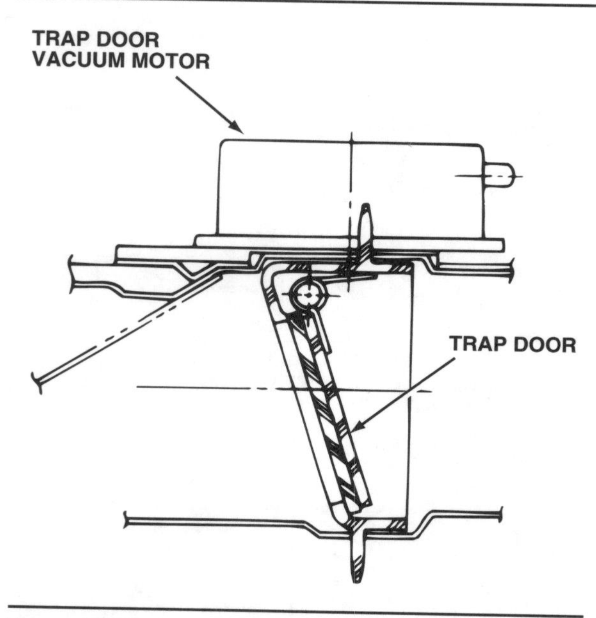

Figure 12-57. By sealing off the air cleaner duct from the atmosphere, the trap door prevents fuel vapors from escaping when the engine is off.

cold air can cause a lean condition and a hesitation on acceleration.

Thermostatic bulbs and coils

Thermostatic bulb operation of the air control was used on many Ford engines during the 1960s and 1970s. GM used a thermostatic coil on small 4-cylinder engines until the mid 1970s. Some import cars also used this method of regulating air intake temperature.

With this type of control, the thermostatic bulb or coil is inside the air cleaner snorkel and connected by linkage to a spring-loaded air con-

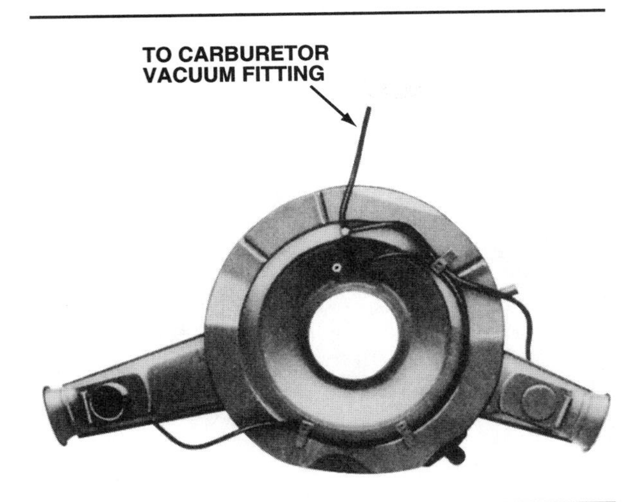

Figure 12-58. Chrysler's dual snorkel air cleaner.

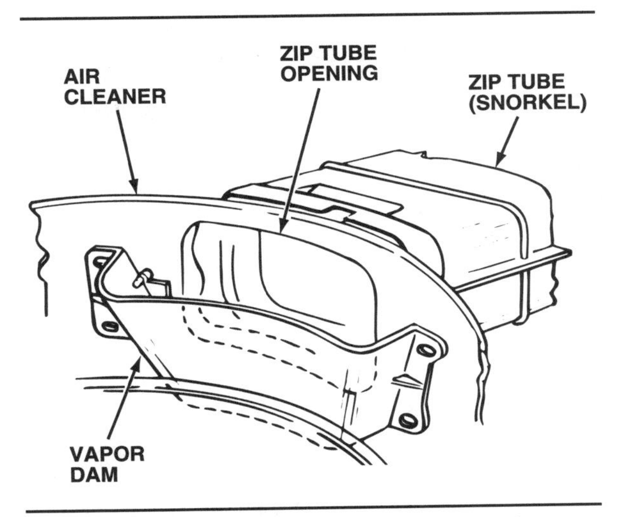

Figure 12-59. Some air cleaners use a vapor dam to trap carburetor fuel vapors when the engine is off. (Ford)

trol valve. The air control valve is normally held in its closed position by the spring, allowing heated intake air to enter the snorkel, figure 12-54A. As the temperature rises, the thermostatic bulb begins to expand. This expansion exceeds air valve spring tension and the valve gradually opens to its cold-air position, figure 12-54B.

Temperature calibration of the thermostatic bulb or coil differs according to manufacturer and application. On thermostatic bulb controlled air cleaners, the air control valve is held in the closed (heat on) position, figure 12-55A, when air entering the snorkel is less cold. When the air temperature warms, the air control valve opens partially, allowing a blend of heated manifold air and cooler engine compartment air to enter the air cleaner duct. When intake air temperature reaches its normal operating temperature, the air control valve is held in the fully open (heat off) position, figure 12-55B. Air now enters the air cleaner snorkel from the engine compartment or fresh air duct.

On 1966–67 Ford engines, the thermostatic bulb holds the air control valve in the closed position (heat on) when air temperature is less then 75°F (24°C). On 1968 and later models, the bulb holds the valve closed until intake air temperature reaches 95° to 100°F (35° to 38°C). At air temperatures between 85° to 105°F (29° to 41°C) on 1966–67 models, or 100° to 135°F (38° to 57°C) on later models, the bulb opens the valve to allow a blend of hot and cold air through the snorkel. Above these temperatures, the bulb opens the valve to the full cold-air position.

On GM air cleaners with thermostatic coil control, the valve is in the closed (heat on) position at temperatures under 50°F (10°C). Between 50° to 110°F (10° to 43°C), the valve is partly open to allow a blend of warm and cold air. Above 110°F (43°C), the valve is fully open (heat off).

Control variations

A number of design variations are used by automakers to tailor air cleaner operation for specific engines and operating conditions. Later-model Jeep 6-cylinder air cleaners have a second vacuum motor to control a second valve in the air cleaner snorkel called a trap door, figure 12-56. This trap door shuts off the air cleaner duct when the engine is off, figure 12-57, to prevent the escape of fuel vapors into the atmosphere. Retard delay valves may be used with either or both vacuum motors.

The dual snorkel air cleaner, figure 12-58, used on some Chrysler V8 engines, works in the same way as the single snorkel air cleaner. However, only one snorkel receives heated air from the heat stove. The vacuum motor in this snorkel is controlled by a temperature sensor as previously described. The vacuum motor in the second snorkel receives vacuum directly from the intake manifold without passing through a temperature sensor. The valve in the second snorkel is held closed under all operating conditions except heavy acceleration. When manifold vacuum drops under heavy acceleration, the springs in the vacuum motors of both snorkels open their valves to provide the maximum flow of cooler air. Ford has used a variation of this system on some later-model high-performance engines.

Some Ford air cleaners have an auxiliary air inlet valve controlled by a vacuum motor. Located on the rear of the air cleaner housing, it opens to provide more airflow under full-throttle.

Other Ford air cleaners with thermostatic bulb control have an auxiliary vacuum motor under the air cleaner snorkel. The motor is linked to the air control valve by a piston rod. Under full throttle, the vacuum motor takes over from the thermostatic bulb and opens the valve to air from both the heat stove and the cold air intake.

A ram air system is used on some Ford engines of the early 1970s. Under full-throttle, a vacuum motor opens an air valve in the functional hood scoop to let in extra outside air.

Many Ford engines after 1975 have a thermostatically controlled vacuum modulator between the temperature sensor and the vacuum motor. This keeps the spring in the vacuum motor from opening the air control valve under full throttle with a cold engine. The vacuum modulator, figure 12-51, uses a bimetal thermostatic disk and a check valve to trap vacuum in the vacuum motor. Above 55°F (13°C), the bimetal thermostat in the vacuum modulator opens the vacuum passage through the modulator so that the air cleaner vacuum motor works normally.

The GM temperature control valve (TCV) is located in the air cleaner and works similarly to Ford's vacuum modulator. At temperatures below 80°F (27°C), the valve traps vacuum in the vacuum motor to hold the air cleaner valve closed, even at full throttle. Above 95°F (35°C), the TCV is fully open to permit normal air cleaner operation.

Other Air Cleaner Uses

The air cleaner housing is a convenient place to locate a number of other emission control devices. Here are some that may be found on or in the air cleaner housing:

- The PCV system connects to the air cleaner housing to obtain a source of fresh air during most engine operating conditions. The crankcase ventilation filter usually is in the air cleaner, except on some later-model Ford 4-cylinder and V6 vehicles and Chrysler 6-cylinder and V8 engines, which have the filter in the oil filler cap or hose.
- GM and other automakers attach the manifold absolute pressure (MAP) sensor to the air cleaner housing.
- Chrysler's orifice spark advance control (OSAC) valve is attached to the air cleaner housing on many models.
- Ford has mounted an air injection thermal vacuum switch (TVS) in the air cleaner to control air injection operation.
- A vapor dam, figure 12-59, may be used inside some air cleaners to trap carburetor fuel vapors when the engine is off. Since the vapors are heavier than the air, they remain in the bottom of the air cleaner until the engine is started and they are purged.

SUMMARY

Automotive fuel tanks can be mounted either vertically or horizontally, depending on how much room there is under the vehicle. Filler tubes, besides allowing the tanks to be filled, are also used as fill limiters, leaving room for fuel to expand. Cars requiring unleaded gasoline do not accept a leaded gas nozzle in the tube. Tanks must be vented, but evaporative emission control (EVAP) requirements state that the fuel vapors must not be vented to the atmosphere.

Automakers have devised numerous ways to ensure that all vapors remain within the fuel system. They have provided rollover leak protection, pressure-vacuum relief valves, liquid-vapor separators, and positive crankcase ventilation.

The EVAP systems all use vapor storage canisters. Vapors stored in the canisters are purged into the engine. Each manufacturer has devised slightly different ways to do this, depending on the vehicle and engine requirements. Late-model vehicles with engine control systems have placed the purging function under the control of the computer to ensure that vapor flow does not interfere with air-fuel ratio control or driveability. The most important thing is that all of these EVAP systems prevent vapors containing unburned hydrocarbons from reaching the atmosphere, where they pollute the air.

Fuel pumps move the fuel from the tank to the carburetor or injection system. All pumps do this through a mechanical action that creates a low-pressure area into which the fuel flows. With check valves and high pressure, the fuel is then forced out of the pump and into the carburetor or injector throttle body or fuel rail.

There are two types of fuel pumps: mechanical and electrical. Mechanical pumps use the engine camshaft or auxiliary shaft eccentric for power. Although there are four types of electrical pumps (plunger, diaphragm, bellows, and impeller), only the impeller type has been used in recent fuel system designs. Electric pumps push the fuel rather than pull it, so they are frequently installed in the fuel tank.

Many types of filters are used in the fuel system. They remove contamination from the fuel before it reaches the carburetor or injectors. This is particularly important with fuel injection systems, because of the close tolerances within the injectors. Fuel filters must be replaced or cleaned as directed by the manufacturer. Filters are used in-line, at the fuel tank, at the carburetor or throttle body, and on fuel injectors.

Like fuel, the air used by an engine contains tiny particles of dirt and other contaminants that damage an engine if they are allowed to enter it. Air cleaners and their filters screen out this material, much like fuel filters clean the fuel before it gets to the carburetor or injectors. Air cleaners are also part of the emission control system since they help reduce emissions and increase performance and fuel economy.

Since 1960, most domestic cars and light trucks have used thermostatically controlled air cleaners that provide warm air to the carburetor or throttle body at low temperatures. Each automaker has a slightly different design, but all these devices work essentially the same.

Review Questions

Choose the single most correct answer.
Compare your answers to the correct answers on page 507.

1. All fuel tanks must:
   a. Be vertically mounted
   b. Be horizontally mounted
   c. Have a vent system
   d. Contain a vertical baffle

2. Fuel lines fastened to the frame, body, or engine are made of:
   a. Steel
   b. Aluminum
   c. Copper
   d. Rubber

3. Fuel line fittings:
   a. Are made of copper
   b. Crack easily
   c. Are used to secure rigid lines to the car frame
   d. Are either the flared or compression type

4. Clamps that can be reused when hoses are changed are:
   a. Keystone
   b. Corbin
   c. Screw type
   d. Flat spring

5. Carburetors are vented to:
   a. Allow fuel to return to the fuel tank
   b. Prevent overfill
   c. Maintain atmospheric pressure in the float bowl
   d. Encourage thermal expansion

6. Activated charcoal is used as a vapor trapping agent because:
   a. It has the area of a football field
   b. It has a huge surface area
   c. There are 625 grams of charcoal in a canister
   d. Charcoal is a light material

7. Variable purge:
   a. Takes place only when the engine is off
   b. Is proportional to the air drawn in by the engine
   c. Is controlled by intake manifold vacuum
   d. Requires a graduated orifice

8. Later-model Ford vehicles with electronic engine controls and fuel injection use ___ for rollover protection.
   a. Spring-operated float valves
   b. One-way check valves
   c. Inertia switches
   d. Combination valves

9. Canister purging on engines with electronic fuel management control systems is especially important because:
   a. The additional fuel vapors could upset the air-fuel ratio
   b. The additional fuel vapors could overload the vapor purge canister
   c. The additional fuel vapors could cause a false wide-open throttle signal to be sent to the computer
   d. None of the above

10. In a two-stage purge, what usually controls the purge valve line to provide additional purging?
    a. A signal from the computer
    b. Ported vacuum from the carburetor
    c. Pressure drop across the air filter
    d. Velocity of air through the air cleaner snorkel

11. The intake stroke in the fuel pump:
    a. Exerts pressure in the fuel tank line
    b. Creates a vacuum in the fuel chamber
    c. Opens the outlet check valve
    d. Closes the inlet check valve

12. The output stroke of the fuel pump:
    a. Increases pressure on the diaphragm spring
    b. Opens the inlet valve
    c. Increases pressure in the pump chamber
    d. Draws fuel into the fuel tank line

13. Fuel pump pressure is controlled by the:
    a. Carburetor inlet needle valve
    b. Strength of the diaphragm spring
    c. Carburetor float
    d. Diaphragm thickness and diameter

14. Safety control of an electric fuel pump can be provided by:
    a. An oil pressure switch
    b. The starter relay
    c. The ignition switch
    d. Fuel line pressure

15. The fuel pump in an electronic engine fuel management system will operate for one or two seconds and then shut down unless the computer:
    a. Receives an rpm signal
    b. Energizes the ignition switch
    c. Closes the inertia switch
    d. Switches to the back-up mode

16. An electric fuel pump has an advantage over a mechanical pump because it:
    a. Is lighter
    b. Requires no maintenance
    c. Overcomes vapor lock by rapid fuel delivery
    d. Does not cause wear to the crankshaft eccentric

17. Airborne contaminants can:
    a. Convert CO into carbon dioxide
    b. Act as a cleansing element in oil
    c. Be washed out of a paper filter
    d. Damage piston rings and cylinder walls

18. The primary source of air intake to the carburetor or throttle body is the:
    a. Air cleaner snorkel or intake duct
    b. Venturi
    c. Fuel bowl
    d. Heat stove

19. In the Chrysler dual snorkel air cleaner, the air control valve in the second snorkel is:
    a. Always closed
    b. Always open
    c. Partially open
    d. Open only under full throttle

20. Technician A says that heated intake air reduces carburetor icing in cold weather.
    Technician B says that a bimetal temperature sensor and vacuum bleed in the air cleaner regulate vacuum to the vacuum motor.
    Who is right?
    a. A only
    b. B only
    c. Both A and B
    d. Neither A nor B

Chapter

13 Basic Carburetion and Manifolding

So far in our study of a car's fuel system, we have examined enough hardware to be able to dump raw fuel into the combustion chambers. Adding a spark to that raw fuel will not produce the combustion needed to create the power to move the car.

We need something more: We need to change the fuel to a vapor, mix it with air, and feed it to the cylinders in precise air-fuel ratios. Until the mid-1980s, this was the job of the carburetor, figure 13-1.

Today, fuel injection has replaced carburetion on new-model cars. However, many cars with carbureted engines are still on the road, and the principles of carburetion are helpful in understanding engine operation.

There were tremendous variations in carburetor design, from the simple devices used on earlier cars to the wildly complex and expensive versions used on racing engines. Regardless of design, all carburetors used one basic principle: difference in air pressure.

In this chapter you will learn how differences in air pressure applied to a carburetor, how carburetors operate under all types of driving conditions, and how assist devices modified their operation. We will cover the similarities in carburetor designs, and discuss how proper carburetor and assist-device adjustments can improve driveability and lower exhaust emissions on vehicles equipped with carburetors.

If you understand how carburetors work and their similarities, you can make carburetor adjustments properly and diagnose carburetor problems more accurately.

PRESSURE DIFFERENTIAL

Since air is a substance, the air outside an engine has a specific weight, as does the air inside the engine. The weight of air exerts pressure on whatever it touches; the greater this weight, the greater the pressure. When the weight of air outside the engine is greater than the weight of air inside the engine, we say that there is a pressure difference, or differential, between the two.

Atmospheric Pressure

The weight of air is not always the same; it changes with temperature and pressure. For this reason, we must have a reference point when we talk about atmospheric pressure. At sea level, under what we call *standard conditions* of 32°F (0°C) with a barometric reading of 760 mm-Hg, one cubic foot of air (about 7.5 gallons or 28 liters) weighs about 1¼ ounces (36 grams). This seems light enough, but remember that the earth's

Figure 13-1. The modern carburetor is a complex device, but works on two simple principles: airflow and pressure differential.

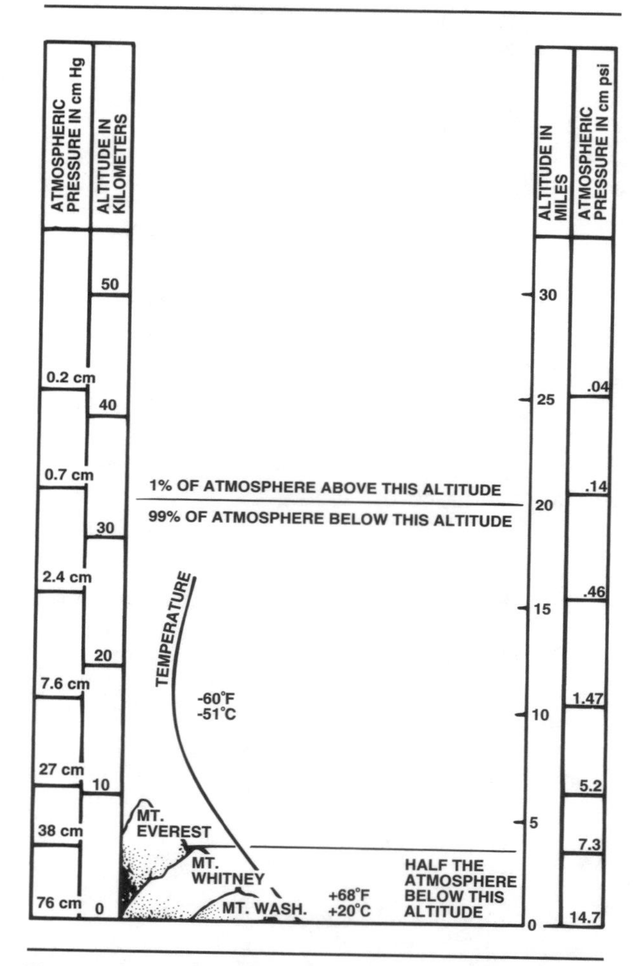

Figure 13-2. The blanket of air surrounding the earth extends many miles into the atmosphere. Atmospheric pressure decreases at higher altitudes.

atmosphere is quite thick, figure 13-2. Therefore, the column of air pressing down on an object at sea level is equal to about 14.7 psi (101 kPa).

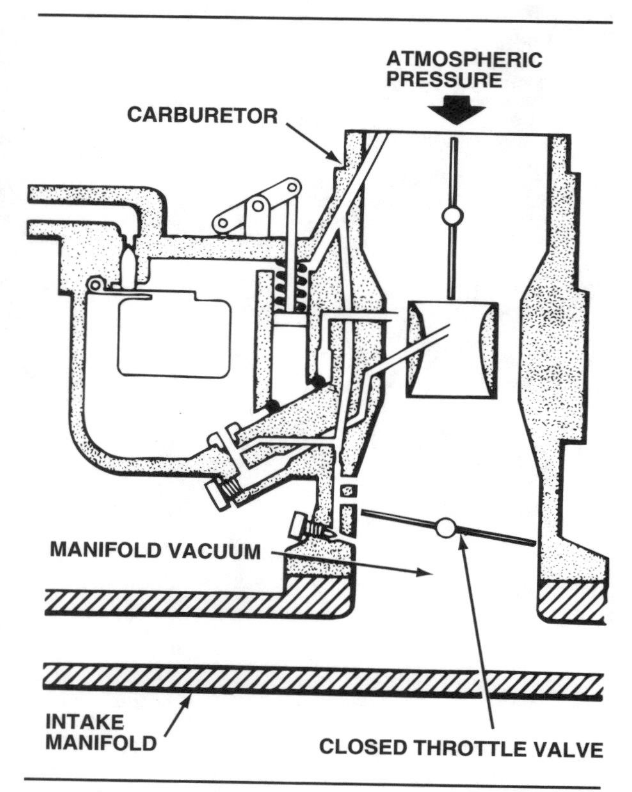

Figure 13-3. Manifold vacuum is the low pressure created in the intake manifold by the downward movement of the engine's pistons.

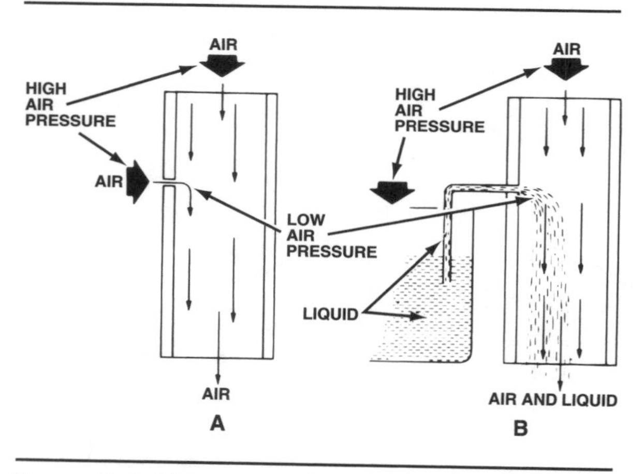

Figure 13-4. Airflow through a tube creates low pressure along the sides of the tube, allowing air to be drawn in through a bleed hole (A). This low pressure can also draw liquid into the tube (B).

Effect of temperature

Air expands and becomes lighter as its temperature rises. This reduces the pressure it exerts. As its temperature falls, air contracts; this makes it heavier and increases its pressure. Variations in air temperature account for changing weather conditions. Direct heat from the sun and reflect-

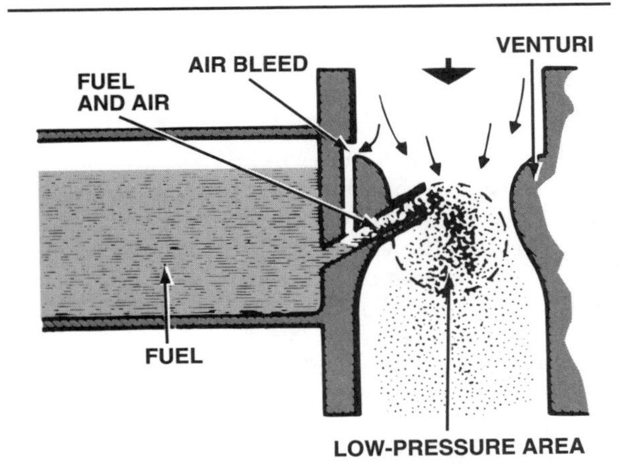

Figure 13-5. Air flowing through the venturi increases in speed. This lowers the pressure within the venturi to draw in more fuel.

ed heat from the earth's surface warm the air. As its temperature increases, air becomes lighter and rises. Cooler air sinks and takes its place, resulting in a constant motion. This motion creates wind and weather patterns.

Effect of altitude

As you climb above sea level, the amount of air pressing down on you becomes smaller. Since a smaller amount of air weighs less, it exerts less pressure. Air pressure gradually decreases with increased distance above sea level. At 30,000 feet or approximately 9 kilometers above sea level, air pressure is only about 5 psi (34 kPa). A few hundred miles or several hundred kilometers above the earth, the atmosphere ends and is replaced by a vacuum, or a complete lack of pressure.

Manifold Pressure—Vacuum

Each intake stroke of an engine piston in its cylinder produces a partial vacuum. As the piston moves down, it creates a larger space in which the air molecules can move. Since the molecules spread out to occupy this increased space, the distance between them increases. The greater the space between the air molecules, the greater the vacuum created.

As the piston moves farther down, it increases the vacuum and lowers the air pressure in the cylinder and intake manifold above it. This causes a **pressure differential** between the air inside and the air outside the engine. To offset this differential, outside air rushes into the engine. As it passes through the carburetor, it mixes with gasoline to form an air-fuel mixture. Vacuum draws this combustible vapor through the intake manifold and the open intake valve into the cylinder, figure 13-3. Here the engine compresses, burns, and exhausts the mixture.

AIRFLOW AND THE VENTURI PRINCIPLE

Opening the carburetor throttle valve causes air to move from the higher pressure area outside the engine, through the carburetor, and into the lower pressure area of the manifold. How wide the throttle valve opens determines how much air comes in and how fast it travels.

The pressure of air passing rapidly through a carburetor barrel is lower along the sides of the barrel than in the center. A small hole in the side of the barrel can draw more air into the air stream rushing through the barrel, figure 13-4A. When a hose connects the hole to a liquid-filled container or bowl, the liquid is forced through the hose and into the airstream, figure 13-4B.

The reason this happens is that the higher air pressure on the liquid forces it to the lower-pressure area inside the barrel. How much liquid passes through the hose depends on the airflow velocity—how fast the air is flowing through the inside of the barrel. The higher the airflow velocity, the lower the pressure at the inlet hole, and the more liquid flows.

If we want to make the carburetor work better, we must increase air velocity through the barrel. We can do this by placing a restriction called a **venturi** inside the barrel, figure 13-5. When air flows through the venturi restriction, it speeds up. The speed increase lowers the pressure inside the carburetor barrel and draws more liquid fuel into the airflow.

In addition to mixing liquid fuel with air, the carburetor must also vaporize the liquid as much as possible. A small opening, called an air bleed, in the fuel inlet passage helps break up the liquid fuel for better vaporization.

The carburetor must also change the air-fuel mixture automatically. It must deliver a rich mixture for starting, idle, and acceleration, and a lean mixture for part-throttle operation. Throttle valve position regulates engine speed and power by controlling the flow of the air-fuel mixture, figure 13-6.

Pressure Differential: A difference in pressure between two points.

Venturi: A restriction in an airflow, such as in a carburetor, that speeds the airflow and creates a vacuum.

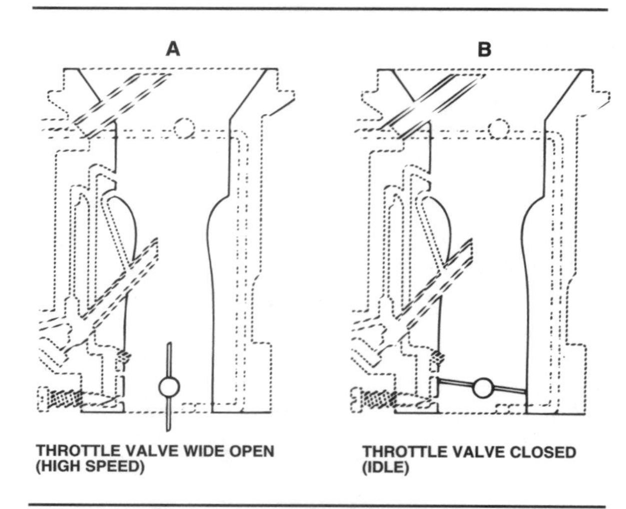

Figure 13-6. The carburetor throttle valve controls engine speed and power by regulating the amount of air and fuel entering the engine.

Carburetor Vacuum

There are four measurements of air pressure or vacuum that are important when discussing carburetors:

- Atmospheric pressure
- Manifold vacuum
- Venturi vacuum
- Ported vacuum.

Atmospheric pressure is the pressure of the air outside the carburetor. It is always present, and varies within a narrow range, depending upon altitude and atmospheric conditions.

Manifold vacuum is the low pressure beneath the carburetor throttle valve. The engine creates **manifold vacuum**, which is always present when the engine is running. Manifold vacuum decreases as the throttle valve opens.

Venturi vacuum is the low-pressure area created by airflow through the venturi restriction in the carburetor barrel. **Venturi vacuum** increases with the speed of the airflow through the venturi. It is present whenever the throttle valve is open and increases with throttle opening.

Ported vacuum is the low-pressure area just above the throttle valve. **Ported vacuum** is present whenever the throttle opens to expose the port in the lower portion of the carburetor barrel to manifold vacuum. Ported vacuum is absent at idle, high at small throttle openings, and decreases as the throttle opens farther. Vacuum taken from this point often operates distributor vacuum advance units and other vacuum-operated devices. Small ports, or holes, in the side of the carburetor are connected to hoses, which are connected to the vacuum devices.

CARBURETION OPERATING PRINCIPLES

As mentioned at the beginning of this chapter, all carburetors must perform three vital functions. They must break up the liquid gasoline into a fine mist, change the liquid into a vapor, and distribute the vapor evenly to the cylinders. These three principles of fuel atomization, vaporization, and distribution are important principles of carburetion.

Gasoline must be atomized, or broken up into a fine mist, to vaporize properly. Atomization takes place as the fuel travels from the carburetor discharge nozzles into the moving stream of air.

Vaporization starts as the atomized fuel passes the throttle and enters the intake manifold. Complete vaporization cannot occur unless the fuel is hot enough to boil. The following factors affect vaporization:

- Temperature—Vaporization increases as the fuel gets hotter.
- Volatility—The greater the volatility of the fuel, the lower the temperature at which it will vaporize, and the faster it will vaporize.
- Pressure—A decrease in pressure causes fuel to vaporize faster at a lower temperature.

Low volatility, cold intake air, or a cold manifold can cause poor vaporization. As you learned in an earlier chapter, thermostatic air cleaners and heated intake manifolds are ways to overcome the problem of a cold air-fuel mixture. Since manifold vacuum creates a low-pressure area, fuel vaporizes more efficiently in the intake manifold. A poorly designed manifold results in poor vaporization.

The throttle plate has a direct effect on distribution, since the angle of the throttle sends the mixture against one side of the intake manifold. This tends to feed some cylinders a rich mixture and other cylinders a lean mixture. Cylinders farther from the carburetor may get less of the mixture than those nearest the carburetor. Engineers must consider the fuel distribution or metering requirements of an engine when they design carburetors and intake manifolds.

Carburetion Air-Fuel Ratio Requirements

A carburetor must serve the varying air-fuel ratio needs of an engine for different operating conditions. During starting, an engine has low intake manifold vacuum and airflow velocity because the engine is turning slowly. Slow cranking speed and a cold engine combine to reduce fuel vaporization. With reduced vaporization, less gasoline reaches the combustion chamber, and the engine needs a richer air-fuel mixture for starting.

At idle, an engine also needs a rich air-fuel mixture. Manifold vacuum is high, but airflow velocity is low. The combined effects of these factors reduce vaporization. Also, some exhaust remains in the cylinders at idle, which dilutes the air-fuel mixture.

As a vehicle accelerates gradually at low speed, engine speed and airflow increase while vacuum rises in the carburetor. The engine gradually needs a leaner air-fuel mixture for smooth acceleration and economy with low emissions.

At steady cruising speeds with a light engine load, the engine needs a relatively lean air-fuel ratio of 15:1 or 16:1. At cruising speed, the engine operates at a relatively constant speed and load, with steady (relatively high) vacuum and airflow.

For extra power requirements, such as sudden acceleration, hill climbing, or full-throttle operation at any speed, the engine needs a richer air-fuel mixture. Air-fuel ratios of 12.5:1 to 13.5:1 allow an engine to develop maximum power for these conditions. Low vacuum accompanies full-power operation, and a carburetor must provide extra fuel with low airflow velocity for acceleration and low-speed, heavy-load conditions. Low vacuum and high airflow accompany high-speed, full-power operation, and a carburetor must respond to these needs as well. Air-fuel ratios for full power do not vary much except at low speed with low airflow velocity. Between the lean ratios of 15:1 or 16:1 and the rich ratios of 12.5:1 or 13.5:1, a modern carburetor must maintain the stoichiometric ratio of 14.7:1 for the best combination of power, economy, and emission control. On later vehicles, closed-loop feedback fuel systems maintain these requirements.

All of these variable requirements for carburetor operation may seem overly complex at first sight. As you study the basic carburetor systems in the following sections, however, you will see how the systems react to engine conditions and meet variable operating needs. Also, if you understand the engine requirements that a carburetor fulfills with its basic systems, you will understand the corresponding operations of fuel injection systems explained in Part Four of this *Classroom Manual.*

BASIC CARBURETOR SYSTEMS

To mix fuel and regulate engine speed, the carburetor has a series of fixed and variable passages, jets, ports, and pumps that make up the fuel metering systems of circuits. There are seven basic systems common to all carburetors:

- Float system
- Idle system
- Low-speed system
- High-speed (main metering) system
- Power system
- Accelerator pump system
- Choke system.

Float System

The fuel pump delivers gasoline from the fuel tank to the carburetor fuel bowl, where it is stored for use. Once in the fuel bowl, the gasoline must stay at a precise, nearly constant level. This level is critical, since it determines the fuel level in all the other passages and circuits within the carburetor. A fuel level that is too high in the bowl will produce an air-fuel mixture that is too rich; a fuel level that is too low will produce an overly lean mixture. For this reason, fuel level is one of the most critical adjustments required by the carburetor.

The main fuel discharge nozzle for the high-speed system is connected directly to the bottom

Manifold Vacuum: Low pressure in the intake manifold below the carburetor throttle.

Venturi Vacuum: Low pressure in a carburetor venturi caused by fast airflow through the venturi.

Ported Vacuum: The low-pressure area just above the throttle in a carburetor.

■ Funny-Looking Gaskets Are No Joke

It is easy to overlook gaskets. They are small and cheap, so they must be unimportant. . .NOT! Gaskets that do not fit right do not work—a bad gasket can rob your engine of a lot of power.

Here is a tip from the professionals: Be careful how you store gaskets. Store them flat, never standing on edge. Do not hang them on a nail—that pulls them out of shape. Do not store gaskets near heat, since that could warp them. All nonmetal gaskets have some water content, so do not let them dry up. If you do have a dry one, soak it in warm water for just a few minutes. Too much water will, again, cause it to warp.

The best thing to do is to store gaskets flat in a protective wrapper. Leave them in a drawer or cabinet where they will not be disturbed and will not have other parts set on top of them.

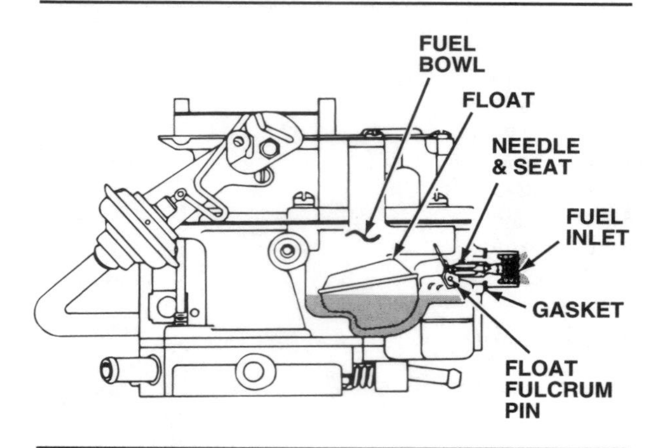

Figure 13-7. Fuel level in the fuel bowl is controlled by the float and needle valve acting against fuel pump pressure.

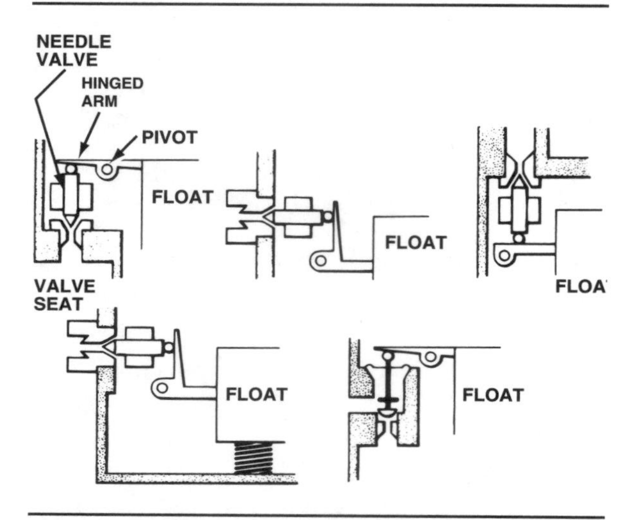

Figure 13-8. Float and needle valve designs vary with different carburetors.

of the fuel bowl. Because liquids seek their own level in any container, the fuel level in the bowl and in the nozzle is the same. If the level is too high, too much fuel will be drawn into the high-speed system. If the fuel level is too low, too little fuel will be drawn in.

The float and the inlet needle valve control fuel level, figure 13-7. As gasoline is drawn from the bowl, the float lowers in the remaining fuel. Fuel pump pressure then opens the needle valve and allows more fuel to enter the bowl. As the fuel level rises, so does the float, until it forces the inlet needle back against its seat. This closes the inlet valve and shuts off both fuel flow and pressure to the carburetor bowl. During many operating conditions, fuel flow into and out of the fuel bowl is about equal. The needle stays in

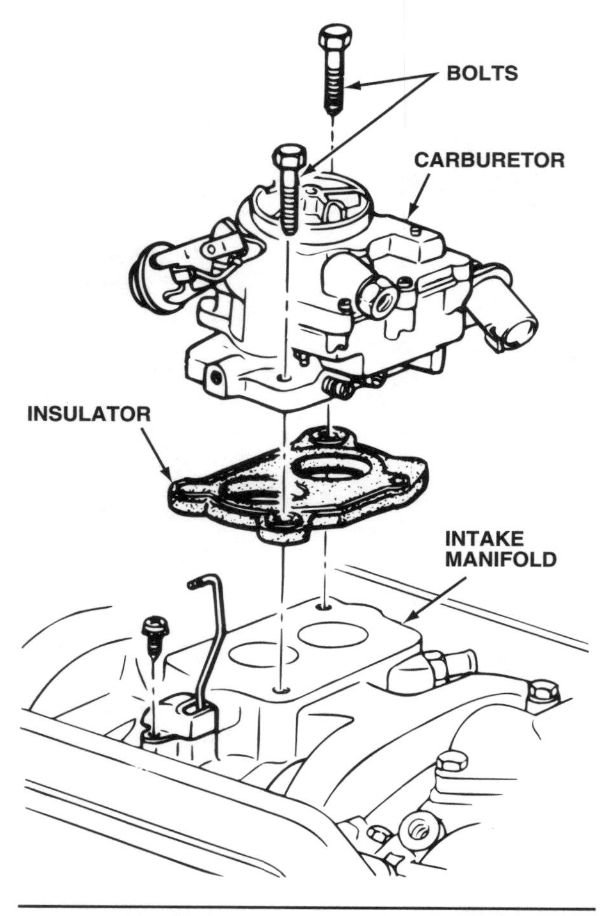

Figure 13-9. This insulator between the carburetor and the intake manifold reduces heat that causes fuel evaporation in the fuel bowl.

a partly open position to maintain the required flow rate.

The float and needle valve regulate fuel flow as well as fuel level. Since the needle valve is like a door between the carburetor fuel bowl and the fuel pump, it maintains an air space above the fuel in the bowl. This reduces pressure on the fuel to atmospheric pressure. A vent or balance tube venting the bowl to the carburetor airhorn maintains atmospheric pressure in the fuel bowl. Atmospheric pressure pushing down on the fuel in the bowl provides the pressure differential needed for precise fuel metering into the venturi vacuum area of the carburetor barrel. If the float and the needle valve do not maintain the correct fuel level in the bowl and too much fuel enters, the carburetor will flood.

Float and needle valve design and location in the fuel bowl vary with different carburetor designs, figure 13-8. Some floats have small springs to prevent them from bobbing up and down when the vehicle travels over rough

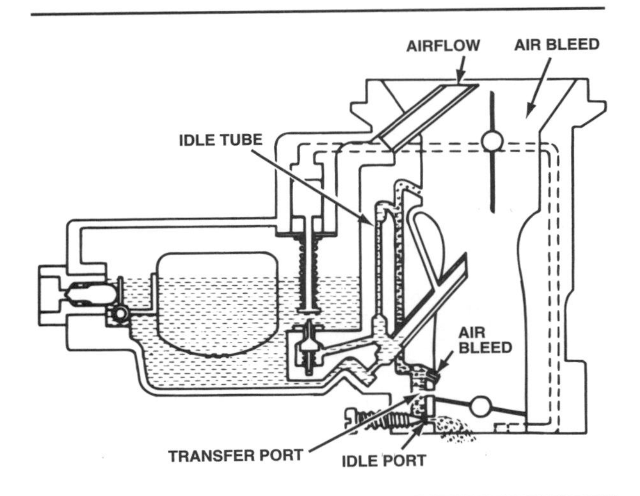

Figure 13-10. Air and fuel for the idle system are mixed inside the carburetor passages and delivered to the idle port below the throttle.

roads. Many fuel bowls have baffles, which keep the fuel from sloshing on rough roads and sharp turns. The needle valves and their seats in older carburetors were usually made of stainless steel. The steel often attracted metallic particles in the fuel. These particles would collect between the needle and seat, allowing the valve to leak. The needles and seats in most modern carburetors are made of brass, and the needles have tips made of Viton or other plastics that conform to any rough spots on the seat and still provide a good seal when the valve closes.

When the engine is shut off, engine heat causes the fuel in the bowl to evaporate. This was no problem in preemission control days, but with the installation of vapor canister systems, the amount of evaporation from a large fuel bowl can easily overload the canister. Therefore, emission carburetors use a somewhat smaller float bowl. Some carburetors, such as the Carter Thermo-Quad, use a molded plastic float bowl to reduce heat evaporation because plastic does not conduct heat as well as metal. Others use an insulator, figure 13-9, between the intake manifold and the carburetor to reduce heat.

Idle System

When an engine is idling, the throttle is open only slightly and airflow through the carburetor barrel and venturi decreases. Since there is little or no venturi effect, no fuel flows from the main discharge nozzle. The idle system, figure 13-10, supplies enough air and fuel to keep the engine running under these conditions.

Intake manifold vacuum is high at idle, so idle ports are located just below the closed throttle. The pressure differential between the fuel

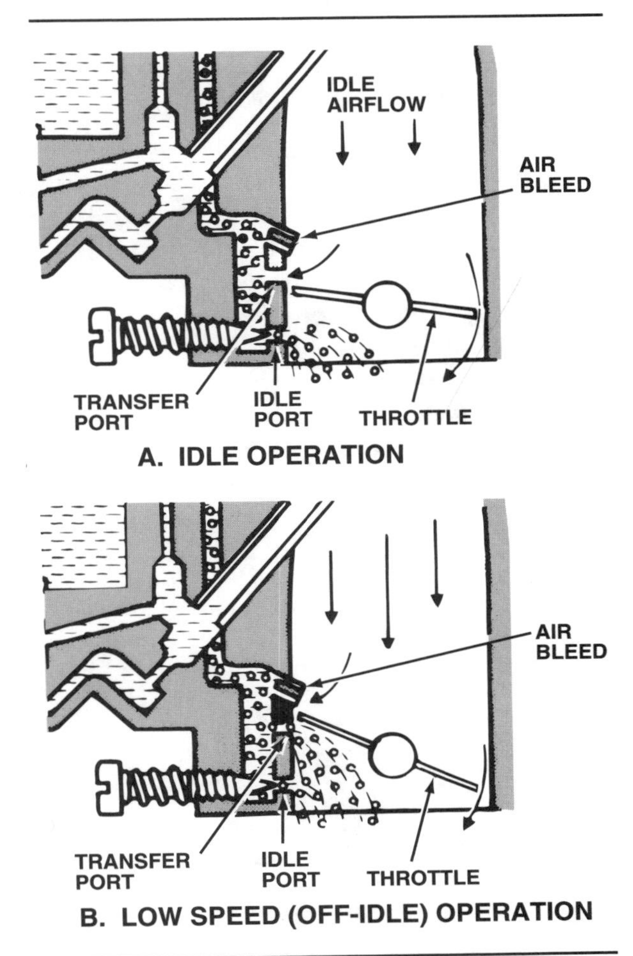

Figure 13-11. At idle, air flows in through the transfer port to mix with the idle air-fuel mixture. As the throttle opens, flow reverses through the transfer port. Fuel and air now flow out for low-speed operation.

bowl and the vacuum at the idle ports forces fuel through the ports. Gasoline flows from the bowl, through the main jet, to the idle tube. Because the fuel must mix well with air for proper distribution, air bleeds in the idle tube let

Check for a Cold Carburetor

Cold outside air can cause carburetors to freeze up, especially when low temperature combines with high humidity. Carburetor icing can, in turn, cause the choke valve to stick or bind in the carburetor. If you are servicing a car whose owner has complained about poor engine performance during cold weather, always be sure to check for carburetor icing. If it is your car, you can throw a blanket over the engine and carburetor when it is left to sit for a long time in cold weather.

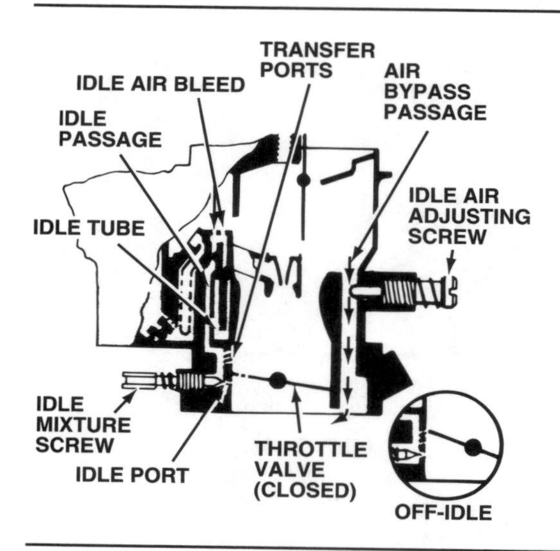

Figure 13-12. Air for the idle circuit in this carburetor passes through a bypass passage and is controlled by an idle air adjusting screw.

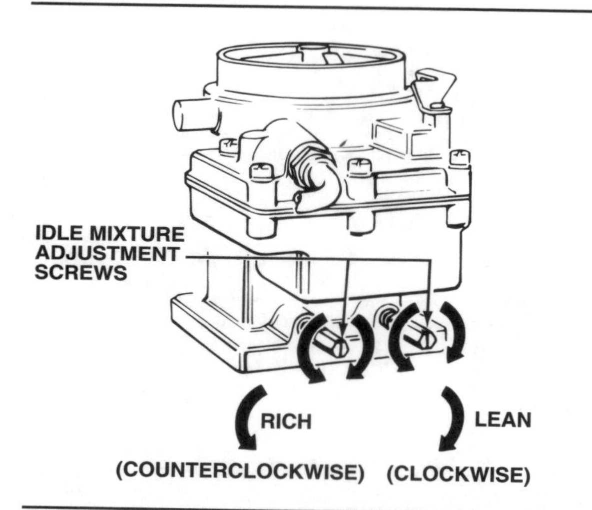

Figure 13-13. Idle mixture screws control the amount of gasoline in the air-fuel mixture at idle.

in air for the idle mixture. The air bleeds also prevent fuel **siphoning** at high speeds or when the engine is stopped.

Various methods can provide extra air for the idle air-fuel mixture. In many carburetors, the throttle valve remains slightly open to let in a small amount of air, figure 13-11. A few designs draw air for the idle circuit through a separate air passage in the carburetor body called the idle air bypass, figure 13-12.

Additional small openings called transfer ports, figure 13-11, are located just above the closed throttle valve in the carburetor barrel. At idle, the transfer ports suck air from the barrel

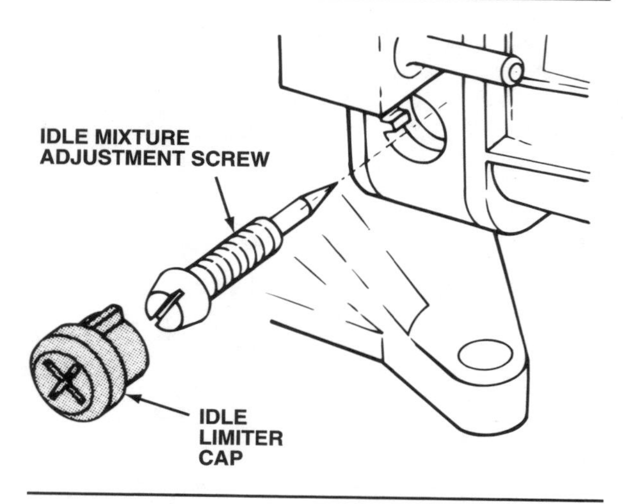

Figure 13-14. Idle limiter caps restrict the amount of adjustment allowed for the idle mixture.

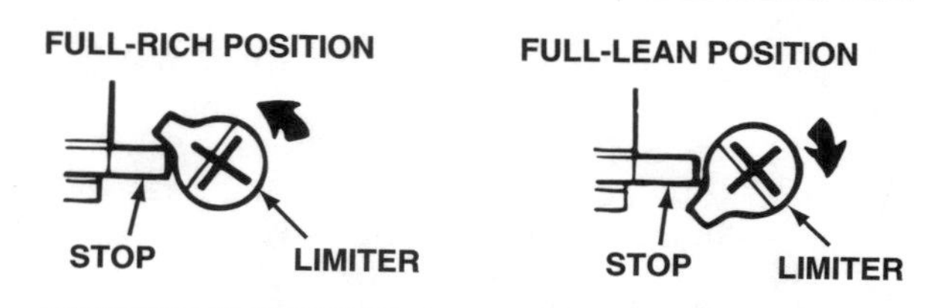

Figure 13-15. Limiter caps allow an adjustment of approximately one turn.

into the fuel flow in the idle system. A small amount of air and fuel is released just below the throttle valve. When the engine is under slight acceleration, the throttle valve opens a little and exposes the transfer port to manifold vacuum. This draws the fuel out into the barrel to mix with the air. We will discuss this more completely under the low-speed system.

Adjustable needle valves called idle mixture screws, figure 13-13, control the amount of gasoline used in the idle air-fuel mixture. Generally, each primary barrel has one adjustment screw. The screw tips stick out into the idle system passages and are turned inward (clockwise) to create a lean mixture, or outward (counterclockwise) to make the mixture richer.

Before emission control, technicians could adjust the idle mixture screws on carburetors from fully closed to fully open. Most carburetors built from the late 1960s to the late 1970s use plastic limiter caps on the idle mixture screws, figure 13-14. These caps restrict the amount of adjustment to about one turn in or out, figure 13-15. This prevents excessively rich idle mixtures that contain large amounts of hydrocarbons (HC) and carbon monoxide (CO). Always replace limiter caps when overhauling a carburetor.

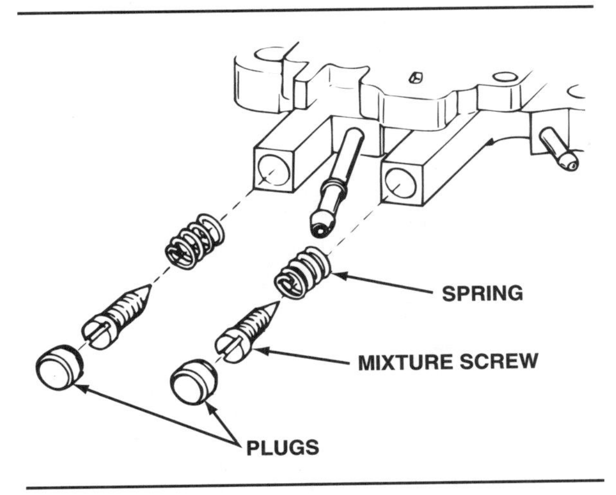

Figure 13-16. These Carter idle mixture screws on later-model carburetors are adjusted and then sealed with plugs or caps.

Regulations set by the U.S. Environmental Protection Agency (U.S. EPA) in 1979 required manufacturers to make carburetors tamperproof. To do this, the idle air-fuel ratio of most carburetors made after 1979 is calibrated at the factory, and the mixture screws are covered with caps or plugs, figure 13-16. Other carburetors use a tamper-resistant bracket riveted in place, figure 13-17. The concealment caps, plugs, or brackets prevent any changes in the air-fuel ratio that would affect idle emission control.

Technicians sometimes can adjust engine idle speed by changing the amount of air going to the idle system, usually by turning a screw that changes the position of the throttle valve in the carburetor, figure 13-18. The adjuster for carburetors that use idle air bypass passages is a large screw that varies the opening in the air passage to change the airflow, figure 13-19. Many feedback carburetors used with electronic engine control systems have an electric motor that controls idle speed and airflow as directed by the system microprocessor.

Low-Speed System

Once the throttle valve begins to open for low-speed operation, the engine needs more fuel than the idle port alone can provide. The airflow passing through the venturi is still not strong enough to develop fuel flow through the main discharge nozzle. To provide more fuel, the transfer port comes into operation as the low-speed system, figure 13-11.

The transfer port is located above the throttle at idle, and the air pressure there is about equal to atmospheric pressure. Air from the barrel flows *into* the transfer port to mix with the fuel going to the idle port. The throttle opening exposes the transfer port to intake vacuum, and the flow reverses. Extra fuel flows *out* of the transfer port to meet the engine's needs during the change from idle to low-speed operation. Fuel continues to flow from the idle port, but at

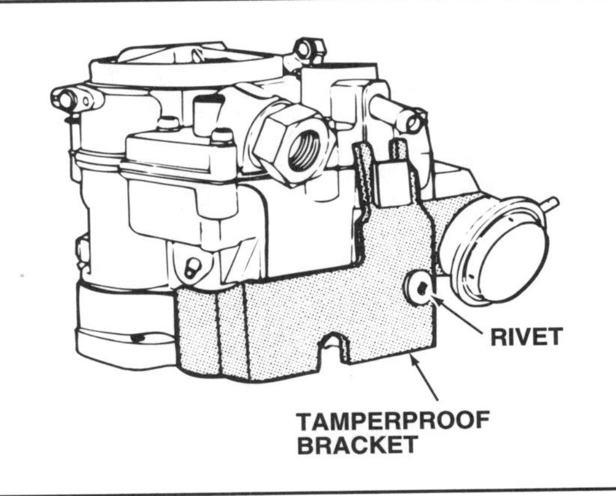

Figure 13-17. As illustrated on these AC-Delco carburetors, tamper-resistant brackets are used to prevent unauthorized adjustment.

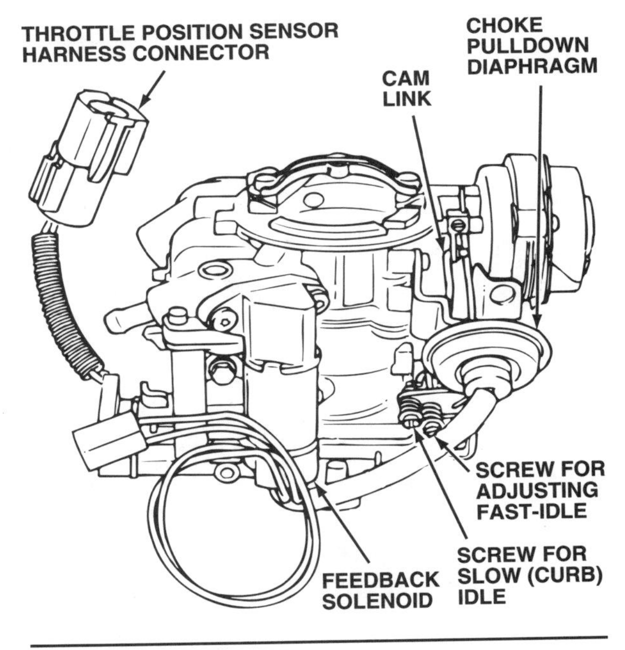

Figure 13-18. This Ford slow (curb) idle screw regulates idle airflow and engine speed by changing the throttle position.

Siphoning: The flowing of a liquid as a result of a pressure differential without the aid of a mechanical pump.

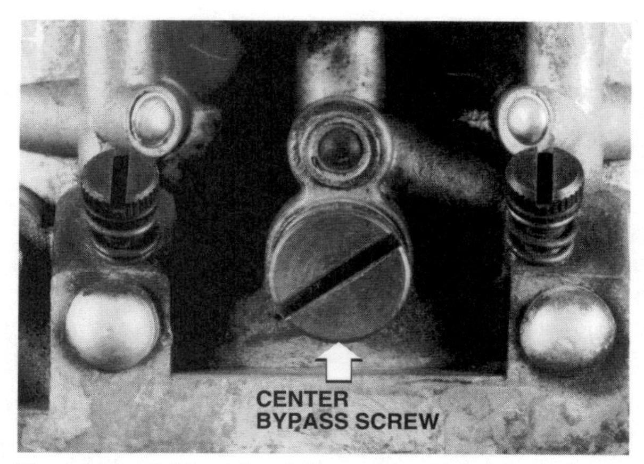

Figure 13-19. The idle air bypass speed adjustment screw controls the airflow through a bypass passage to regulate idle speed.

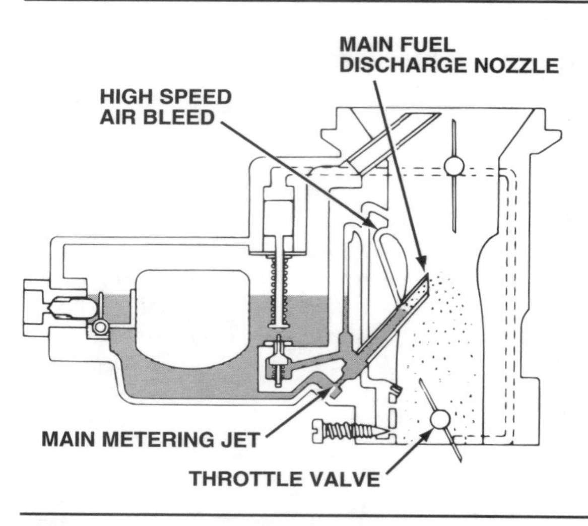

Figure 13-20. Fuel for the high-speed or main metering system flows through the main jet and out the fuel discharge nozzle in the venturi.

a reduced rate. This permits an almost constant air-fuel mixture during this transition period. Some engineers and technicians think of the idle and low-speed systems as two halves of a single system because both use the same carburetor passages. Whether you think of them as one system or two, the important points to understand are the operations that provide a smooth transition from idle to main-metering fuel flow.

High-Speed (Main Metering) System

When the throttle valve opens wider, airflow increases through the carburetor. At the same time, the partial vacuum (low-pressure) area of the intake manifold moves up in the carburetor barrel. This airflow and pressure change strengthens the venturi. This is the high-speed or main metering system, figure 13-20.

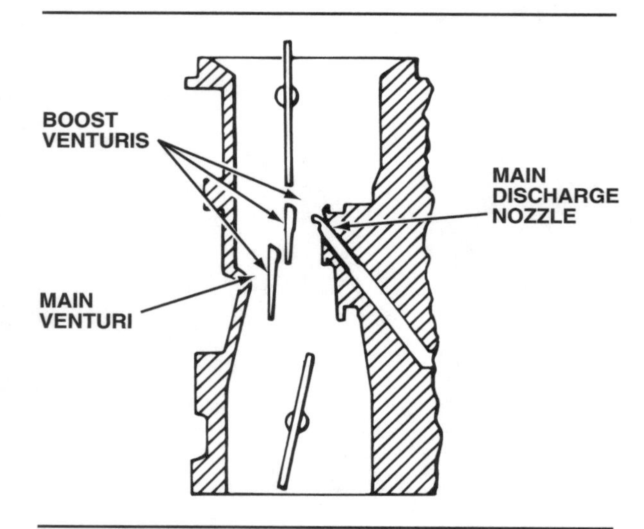

Figure 13-21. Most carburetors have multiple (boost) venturis for better air and fuel mixing.

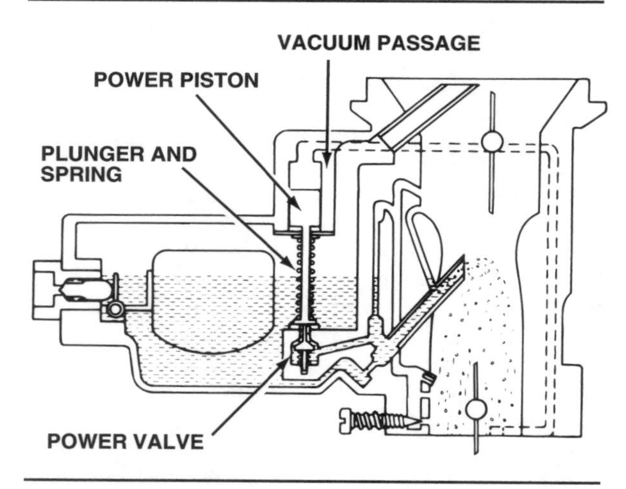

Figure 13-22. This power valve is operated by a vacuum-controlled piston and plunger. When vacuum decreases, the spring moves the plunger to open the power valve.

For better mixing of the fuel and air, most carburetors have multiple, or boost, venturis placed one inside another, figure 13-21. The main discharge nozzle is located in the smallest venturi to increase the partial vacuum effect on the nozzle. Fuel flows from the bowl through the main jet and main passage into the discharge nozzle. A high-speed air bleed, figure 13-20, mixes air into the fuel before the nozzle discharges it.

The primary or upper venturi produces vacuum, which causes the main discharge nozzle to spray fuel. The secondary venturi creates an airstream that holds the fuel away from the barrel walls where it would slow down and con-

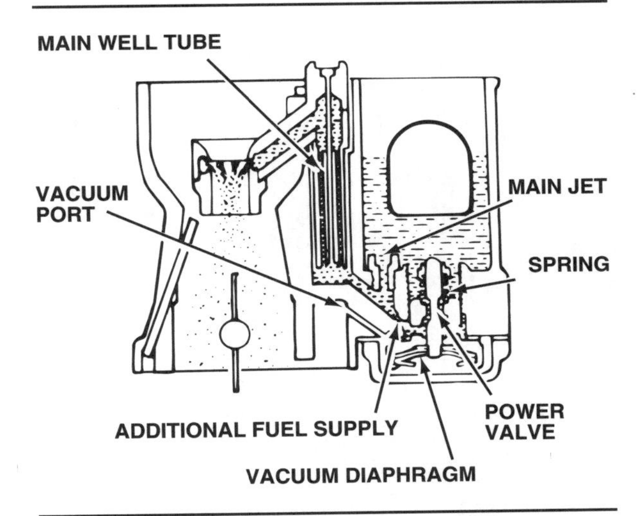

Figure 13-23. The vacuum diaphragm holds the power valve closed. When vacuum decreases, a spring opens the valve to allow more fuel into the main passage.

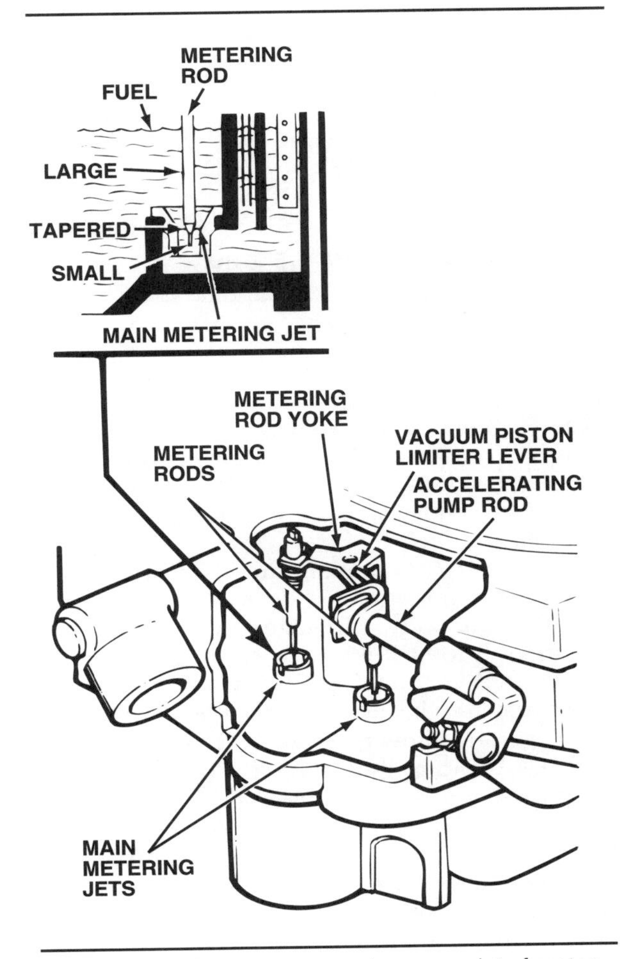

Figure 13-24. Some power systems consist of metering rods placed in the main jets. Mechanical or vacuum linkage moves the rods upward to allow more fuel to flow through the jets when required.

dense. The result is air turbulence, which causes better mixing and finer atomization of the fuel.

As the throttle continues to open wider, the fuel flow from the low-speed system tapers off, while the flow from the high-speed system increases. Now, the main discharge nozzle supplies the engine's fuel needs during high-speed, light-load cruising.

Power System

The main high-speed system delivers the leanest air-fuel mixture of all the carburetor systems. When engine load increases during high-speed operation, the air-fuel mixture is too lean to deliver the power required by the engine. In this case, the power system, or power valve, provides the extra fuel. The power system supplements main metering fuel delivery. Either vacuum or mechanical linkage can operate the power system or valve. The exact type differs according to carburetor design, but all provide a richer air-fuel mixture.

One type of power valve, figure 13-22, is located in the bottom of the fuel bowl with an opening to the main discharge tube. A spring holds a small poppet valve closed, while a vacuum piston holds a plunger above the valve. Since manifold vacuum decreases as the engine load increases, a large spring moves the plunger downward. This opens the valve and lets more fuel pass to the main discharge nozzle.

Another type of vacuum-operated power valve uses a diaphragm, figure 13-23. Manifold vacuum against the diaphragm holds the valve closed. As vacuum decreases under an increased load, a spring opens the valve. This sends more fuel through the power system and main discharge nozzle.

Metering rods also serve as a power system, figure 13-24, controlled by vacuum pistons and springs, or by mechanical linkage connected to the throttle. The ends of the rods installed in the main jet opening are tapered or stepped to increase the extra fuel gradually. The rods restrict the area of the main jets and reduce the amount of fuel that flows through them during light-load operation of the main metering system. Moving the rods out of the jets to increase the flow through the jets provides extra fuel for full-throttle power.

Manifold vacuum acts on pistons attached to vacuum-controlled metering rods, also called step-up rods, so the rods hold in the jets. When vacuum drops under heavy load, springs working against the pistons move the rods out of the jets. Mechanically operated metering rods are controlled directly by mechanical linkage connected to the throttle linkage control.

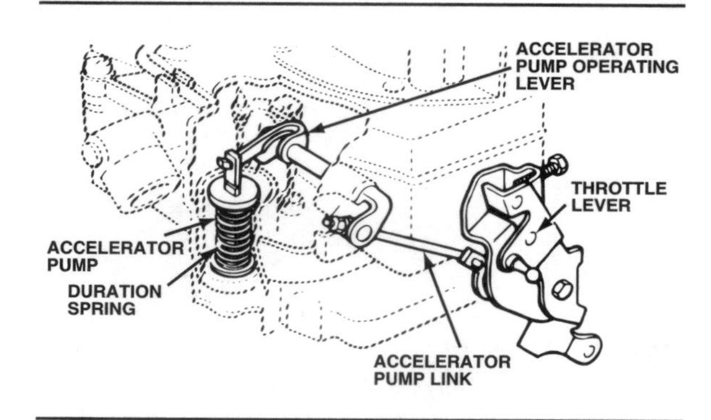

Figure 13-25. This Ford plunger-type accelerator pump is a design used by many manufacturers.

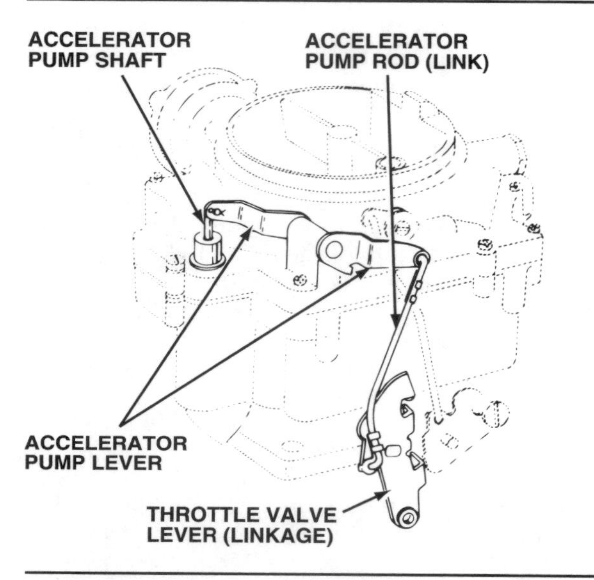

Figure 13-26. The accelerator pump linkage (lever and rod) is connected to the throttle linkage.

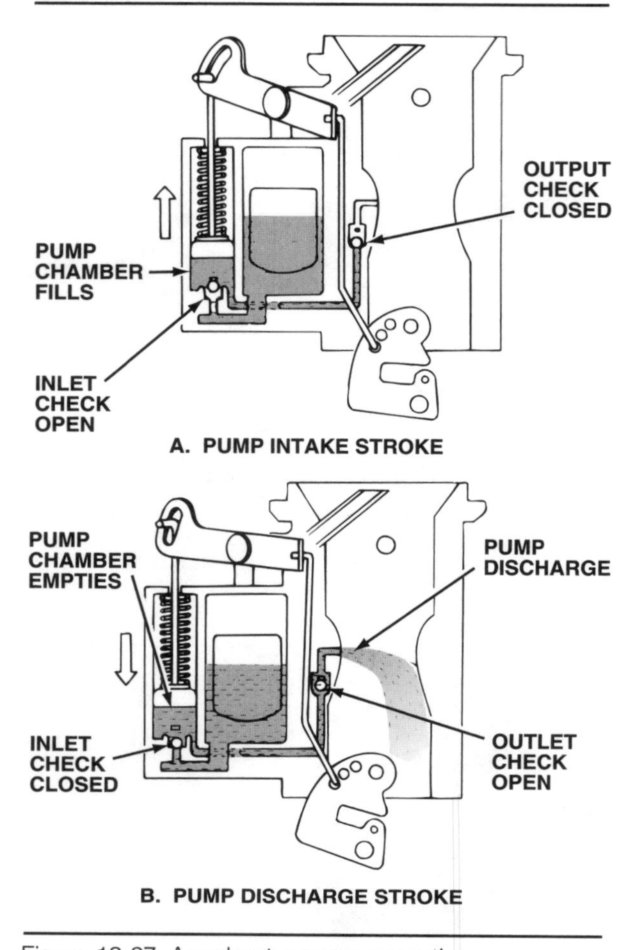

Figure 13-27. Accelerator pump operation.

Feedback carburetors used with electronic engine control systems generally do not have a separate power system. A mixture control (MC) solenoid operates the metering rods or air bleeds. When the engine needs a richer mixture for additional power, the powertrain control module (PCM) decreases the solenoid duty cycle, enriching the fuel mixture. See a later chapter for a more detailed explanation of feedback carburetors and duty cycle.

Accelerator Pump System

The accelerator pump system provides additional fuel for certain engine operating conditions. If the throttle opens suddenly from a closed or nearly closed position, airflow increases faster than fuel flow from the main discharge nozzle. This "dumping" of air into the intake manifold reduces manifold vacuum suddenly and causes a lean air-fuel mixture. This excessively lean mixture results in a brief hesitation or stumble. This is sometimes called a **flat spot**. To keep the mixture rich enough, the accelerator pump must provide extra fuel.

The accelerator pump, figure 13-25, is a plunger or diaphragm in a separate chamber in the carburetor body. A linkage connected to the carburetor throttle linkage, figure 13-26, operates the pump. When the throttle closes, the pump draws fuel into the chamber. An inlet check valve opens to allow fuel into the chamber, figure 13-27A, and an outlet check valve closes so that air will not be drawn through the pump nozzle. When the throttle opens quickly, the pump moves down or inward to deliver fuel to the nozzle in the barrel, figure 13-27B. The pump outlet check valve opens, and the inlet check valve closes. The inlet check ball is usually (but not always) larger than the outlet check ball.

The pump outlet check may be a steel ball or a plunger. The inlet check may be a steel ball, a

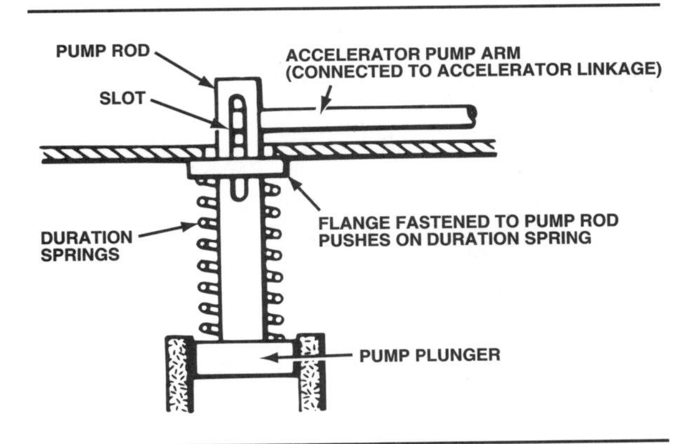

Figure 13-28. The duration spring provides uniform pump delivery regardless of the speed at which the throttle linkage moves.

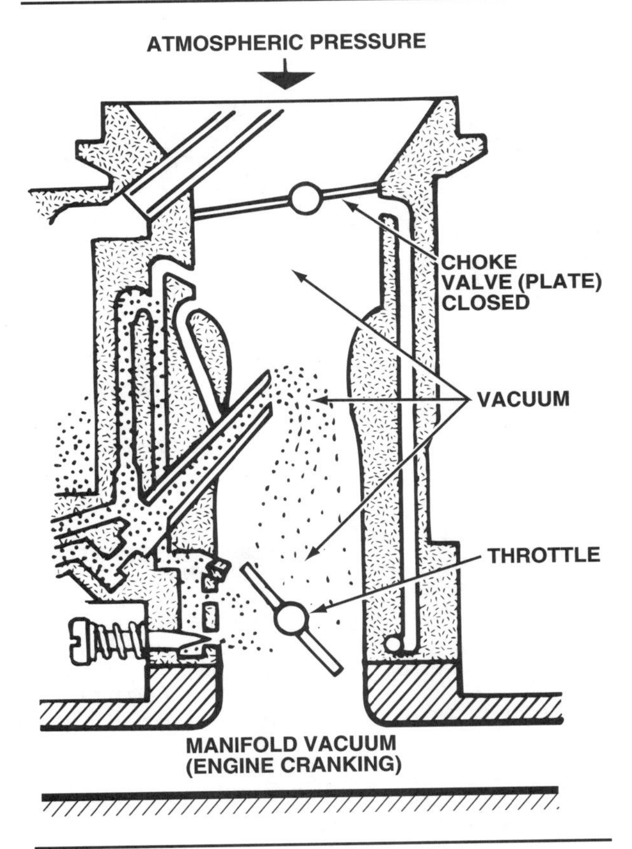

Figure 13-29. Vacuum is present throughout the carburetor barrel below the closed choke. This draws fuel from the idle, low-speed, and high-speed circuits for starting the engine.

rubber diaphragm, or part of the pump plunger. Not all pumps have inlet checks. Some rely on an inlet slot in the pump well, or chamber, that is closed by the plunger on the downward stroke.

A duration spring operates most pump plungers or diaphragms, figure 13-28. The throttle linkage holds the pump in the returned position. When the throttle opens, the linkage releases the pump and the spring moves the plunger for a steady and uniform fuel delivery. The accelerator pump operates during the first half of the throttle travel from the closed to the wide-open position.

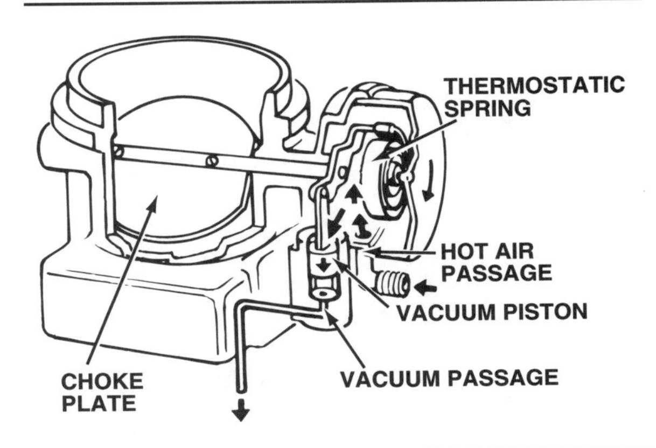

Figure 13-30. An older version of an integral or cap-type choke. The thermostatic bimetal spring is in a housing on the carburetor airhorn. A vacuum piston opens the choke when the engine starts.

During high-speed operation, the vacuum at the pump nozzle in the carburetor barrel may be strong enough to unseat the outlet check and siphon fuel from the pump. This is called pump pullover, or siphoning. A few carburetors include this extra fuel in the high-speed system adjustment. In most carburetors, air bleeds are placed in the pump discharge passages to prevent the siphoning. In other carburetors, an extra weight added to the outlet check resists siphoning. The pump plungers in some carburetors have anti-siphon check valves.

Choke System

The choke provides a very rich mixture for starting a cold engine. This extra-rich mixture is needed in the following situations:

- Engine cranking speed is slow
- Airflow speed is slow
- Cold manifold walls cause gasoline to condense from the air-fuel mixture, and less vaporized fuel reaches the combustion chambers.

To make the mixture richer, a choke plate or a butterfly valve is positioned above the venturi in the carburetor barrel. This choke plate can tilt at

Flat Spot: The brief hesitation or stumble of an engine caused when sudden opening of the throttle creates a momentary lean air-fuel condition.

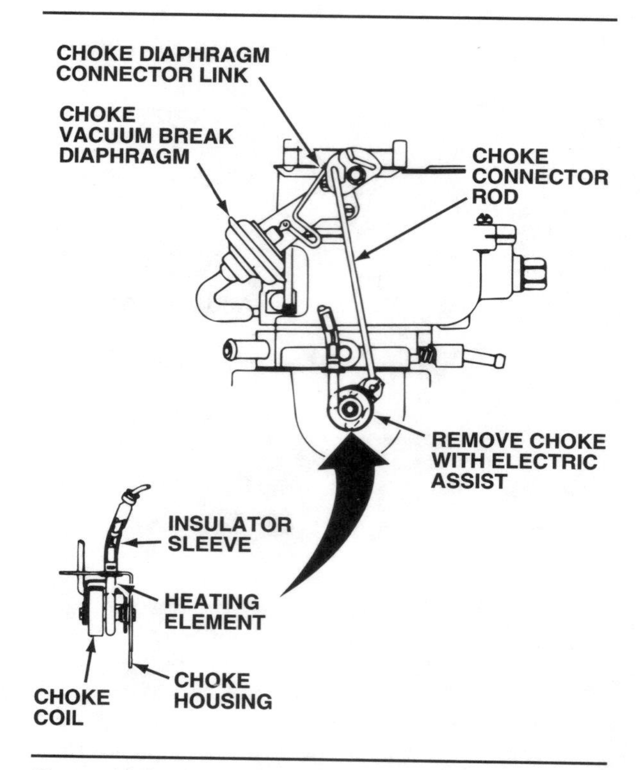

Figure 13-31. In a remote, or well-type choke, as seen on this Carter carburetor, the thermostatic bimetal spring is in a heated well on the intake manifold.

various angles to restrict the passage of air, figure 13-29. Cranking the engine with the choke plate closed creates a partial vacuum throughout the carburetor barrel below the plate. This airflow reduction and partial vacuum area work together to allow more fuel to be drawn into the mixture.

A cable running to the driver's compartment can control the choke plate manually, or a thermostatic coil spring can control it automatically. A bimetal thermostatic coil spring operates the chokes on most domestic carburetors built after 1950. Linkage connects the choke plate shaft to the spring. The bimetal spring normally is located in one of two places:

- In a round housing on the carburetor airhorn, called an integral choke, figure 13-30
- Off the carburetor, in a well on the intake manifold, figure 13-31, called a remote, or well-type, choke.

Regardless of type and location, the thermostatic coil spring forces the choke closed when the engine is cold. Running the engine heats up the spring, which then opens the choke. With an integral choke, hot air from a source near the exhaust manifold or hot coolant from the cooling system may flow to the choke housing to heat the spring. Normally, exhaust routed

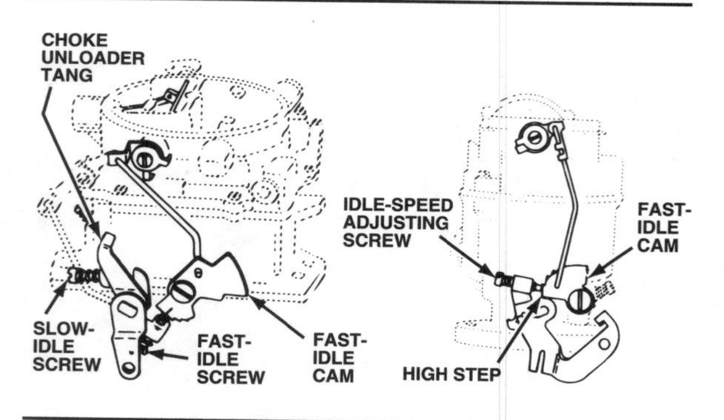

Figure 13-32. The fast-idle cam opens the throttle wider for faster engine speed when the choke is operating. It may work either on the slow-idle speed adjusting screw or on a separate fast-idle screw.

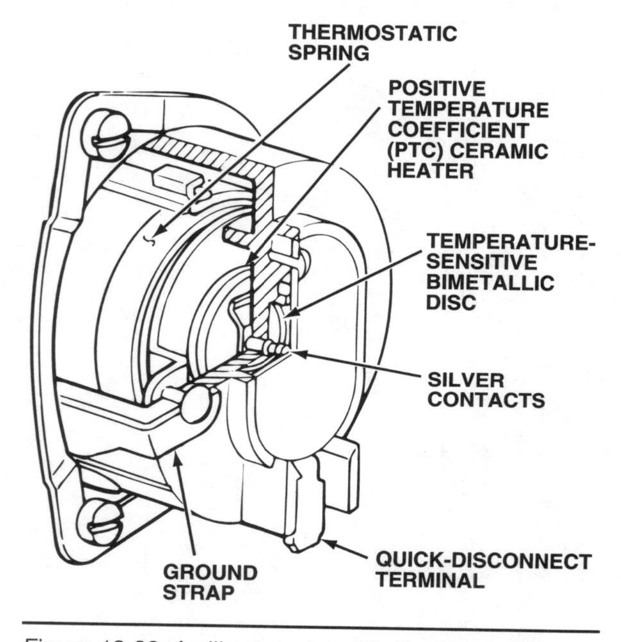

Figure 13-33. As illustrated on this Ford carburetor, an electric choke cap contains a ceramic heater and bimetallic disc to heat the thermostatic spring and release the choke as fast as possible.

through a crossover passage in the intake manifold heats a remote choke. Most late-model integral and remote chokes have electric heating elements to heat the spring faster and speed the choke opening.

When a cold engine is cranked, the choke must close completely. As soon as the engine starts, the choke must open slightly to provide enough airflow. This is done in two ways. First, the choke plate shaft is offset in the carburetor so that airflow will tend to open the plate. Second, manifold vacuum is applied to a vacuum piston or a vacuum-break diaphragm that pulls the choke open a few degrees.

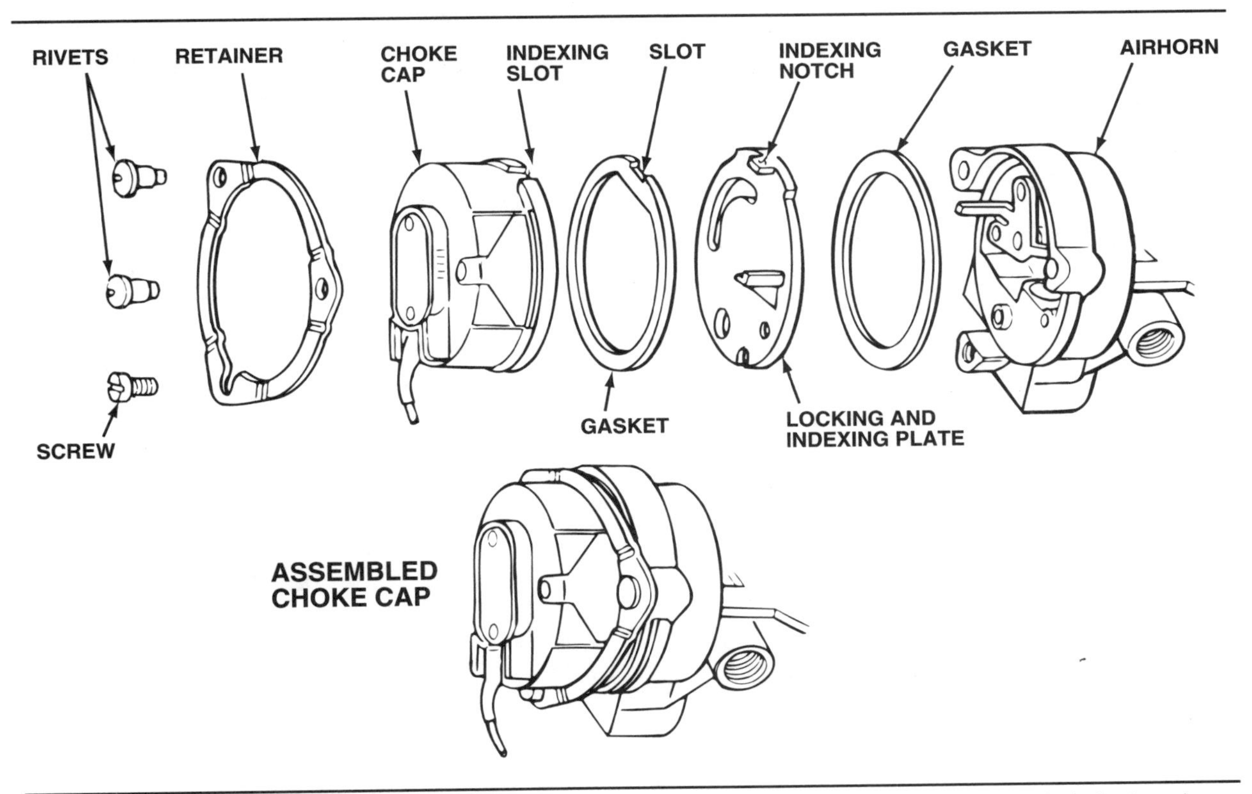

Figure 13-34. This Ford application shows that tamperproof carburetor design requires a sealed choke housing to prevent unauthorized adjustment.

Integral chokes once had a vacuum piston built into the choke housing, figure 13-30, but this design had a tendency to stick. The emission-control requirement for precise choke operation resulted in nearly universal use of the vacuum-break diaphragm on emission carburetors. Most chokes have the vacuum-break diaphragm mounted on the side of the carburetor, figure 13-31.

A cold engine must idle faster than a warm engine, or the lack of air and fuel flow will cause it to stall. A fast-idle cam and screw, figure 13-32, provide enough air and fuel to prevent engine stalling. The cam is linked to the choke plate, and the screw is located on the throttle valve shaft. Depressing the accelerator pedal to start a cold engine allows the choke to close. This moves the fast-idle cam to allow the screw to rest against a high step of the cam. The cam may contact the normal slow-idle adjusting screw or a separate fast-idle screw.

In both cases, the throttle is open slightly more than for a normal slow-idle, and idle speed increases from 400 to 800 rpm. As the engine gradually warms up, choke spring tension decreases and a weight pulls the fast-idle cam back, returning engine speed to idle rpm. On engines with computer control systems, a motor or a solenoid controlled by the system microprocessor often regulates fast-idle speed.

A mechanical link or choke unloader, figure 13-32, opens the choke about halfway when the throttle is fully open. If the engine is accidentally flooded during starting, this provides the extra airflow necessary to clear out the fuel.

Because the choke system provides a very rich mixture, it increases HC and CO emissions. To meet emission control standards, later-model engines must get off the choke as soon as possible. This is done in various ways, but a common method uses the electric choke cap, figure 13-33. This contains a small heating element connected to the alternator and a temperature-sensing switch. The switch lets the heating element warm the thermostatic coil spring and open the choke as quickly as possible. Depending upon their design, some carburetors may have two- or three-stage choke heaters to deliver specific amounts of heat according to ambient temperature.

On older chokes, technicians can adjust the spring tension to control the amount of choke closing and the rate at which it opens. Later-model choke housings, however, are sealed with breakaway screws, rivets, or brackets to prevent unauthorized choke adjustment that might change emissions, figure 13-34.

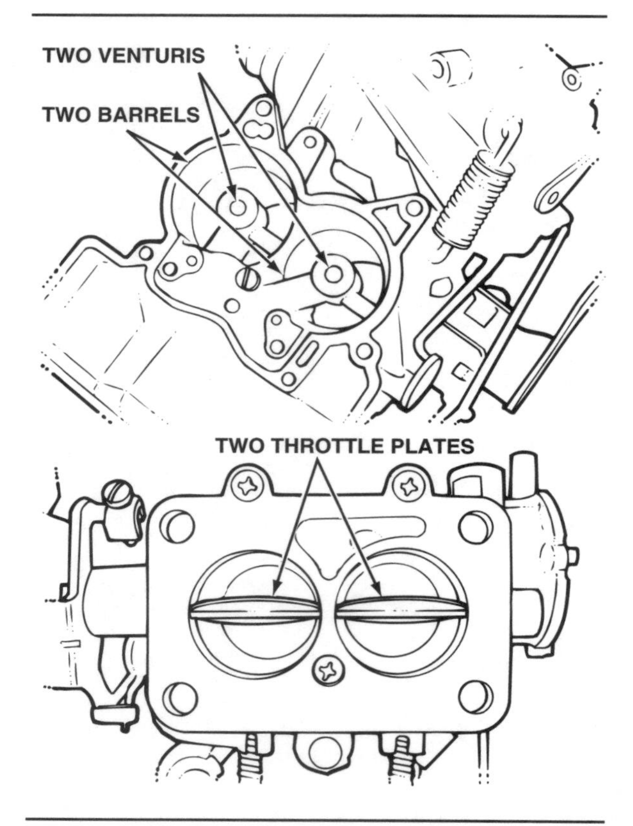

Figure 13-35. A two-barrel carburetor uses one airhorn but contains two venturis and throttle plates.

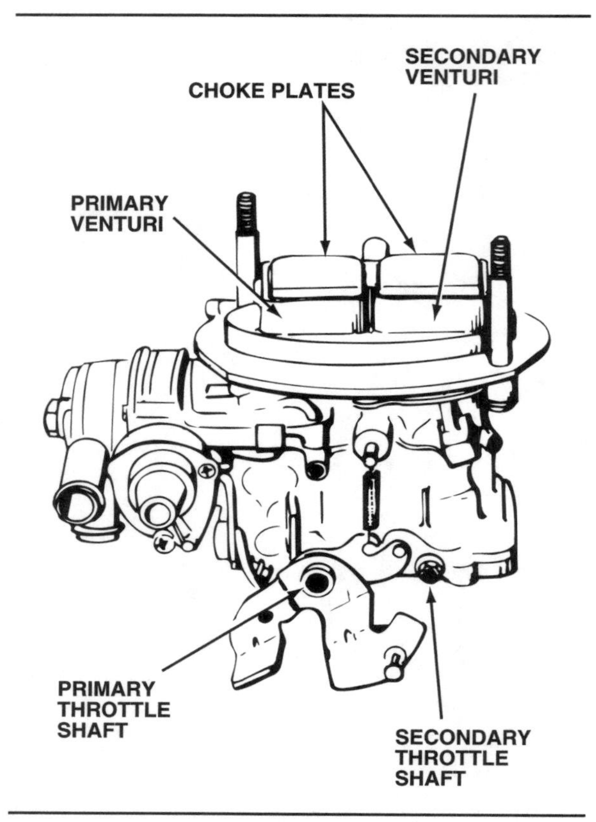

Figure 13-36. The two-stage, two-barrel carburetor has a primary and a secondary throttle, operating independently.

CARBURETOR TYPES

We have explained the operation of the basic carburetor systems in terms of a carburetor that uses a single barrel and throttle valve. But carburetors may have two or more barrels. Various carburetor types match fuel flow to engine requirements. Domestic engines all use downdraft carburetors; older imports often used sidedraft carburetor designs. Carburetors are usually classified by the number of barrels or venturis used. The differences are detailed below.

One-Barrel

The one-barrel carburetor has a single outlet through which all systems feed to the intake manifold. This type of carburetor may also be known as a single-venturi design. These carburetors have a flow capacity from 150 to 300 cubic feet per minute (cfm) and are generally used on 4-cylinder and smaller 6-cylinder engines.

Single-Stage, Two-Barrel

The single-stage, two-barrel carburetor design contains two barrels and two throttles that operate together, figure 13-35. Since the various fuel discharge passages in each barrel operate at the same time, we can consider it as two one-barrel carburetors sharing the same body. The two throttle plates are mounted on the same shaft and operate together. The two barrels share a common float, choke, power system, and accelerator pump. Many 6-cylinder and V8 engines use single-stage, two-barrel carburetors, and the carburetors generally have an airflow capacity from 200 to 550 cfm.

Two-Stage, Two-Barrel

The two-stage, two-barrel carburetor design resulted from emission control requirements, figure 13-36. It differs from the single-stage, two-barrel design in that its two throttles operate independently. The primary barrel is generally smaller than the secondary and handles engine requirements at low-to-moderate speeds and loads. The larger secondary opens when needed to handle heavier load requirements.

The primary stage usually includes the idle, accelerator pump, low-speed, main metering, and power systems. The secondary stage usually

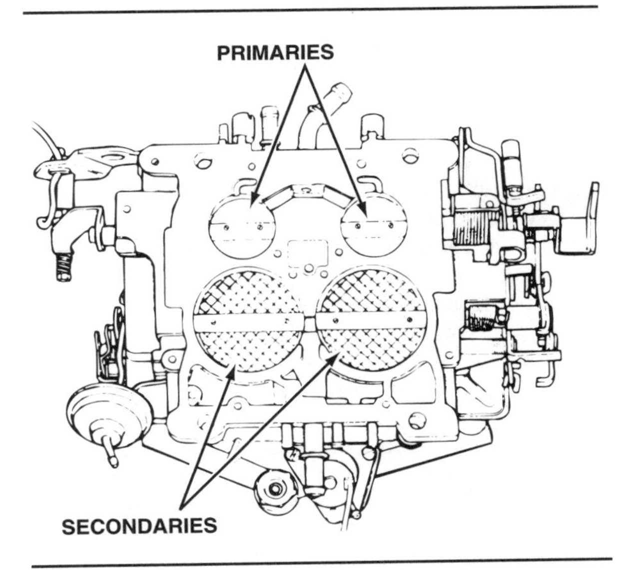

Figure 13-37. As shown on this Chrysler carburetor, most four barrels have primary and secondary systems that open progressively.

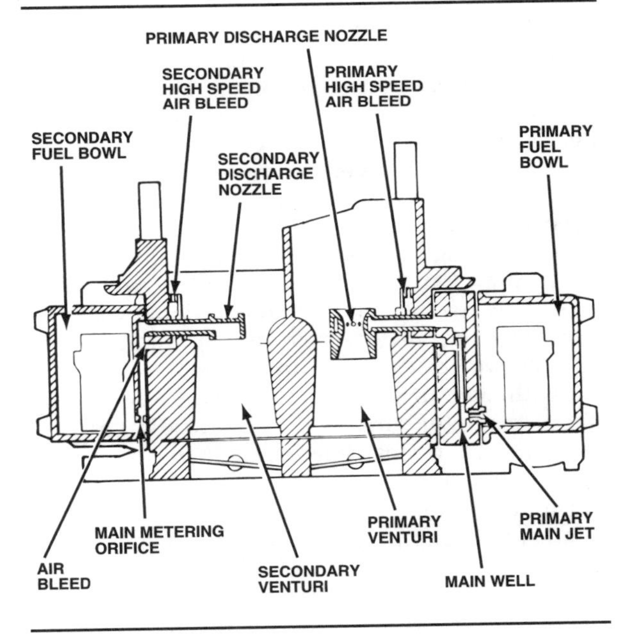

Figure 13-38. As shown on this Ford carburetor, venturi action controls airflow and fuel discharge through the secondary barrels of some four-barrel carburetors.

has a transfer, main metering, and power system. Both stages draw fuel from the same fuel bowl. Some designs use a common choke for both barrels; in others, only the primary stage is choked.

The two-stage, two-barrel carburetor has an airflow capacity from 150 to 300 cfm and is used primarily on 4-cylinder and smaller 6-cylinder engines.

Four-Barrel

Used primarily on V8 engines, the four-barrel, or quad, carburetor contains two primary and two secondary barrels in a single body, figure 13-37. The two primaries operate like a one-stage, two-barrel carburetor at low-to-moderate engine speeds and loads. The secondary barrels open at about one-half to three-quarters throttle to provide the increased fuel and airflow required for high-speed operation. The primary barrels contain the idle, low-speed, and high-speed systems, as well as an accelerator pump and a power system. The secondary barrels have their own high-speed and power systems, and may use their own accelerator system. Some four-barrel carburetors use separate fuel bowls and fuel supplies for the primary and secondary barrels; others work with a single fuel bowl and fuel supply for all four barrels.

Two methods can provide airflow through the secondary barrels: venturi action, figure 13-38, or air velocity valves, figure 13-39. Air velocity

■ Wick and Surface Carburetors

The float-type carburetor was standard on automobile engines for more than 80 years. Anyone could be excused for thinking that it was the only kind of carburetor ever used for automotive applications. A variety of techniques were used, however, to transmit an explosive air-fuel mixture for early internal combustion engines.

One of the first carburetors was a wick-type device in which fabric was suspended in the fuel tank with one end submerged in the fuel. Air passing through the saturated fabric vaporized the fuel and carried it into the engine. A control valve on the engine intake regulated the amount of air as a means of making the air-fuel mixture leaner or richer.

The surface carburetor also depended on vapors. Air could either pass through a fuel tank under pressure so that it bubbled to create vapors, or the fuel could be agitated for the same effect. A control valve on the intake similar to that used with a wick carburetor regulated the mixture ratio. Benzine, naphtha, and ether were commonly used for fuel with both wick and surface carburetors.

Karl Benz adapted the surface carburetor for automotive use in the late 1880s. His design bubbled air through benzine, using a float and needle valve to maintain the fuel level. Exhaust gas routed through a pipe on the bottom of the carburetor increased vaporization. An extractor prevented liquid fuel from entering the engine.

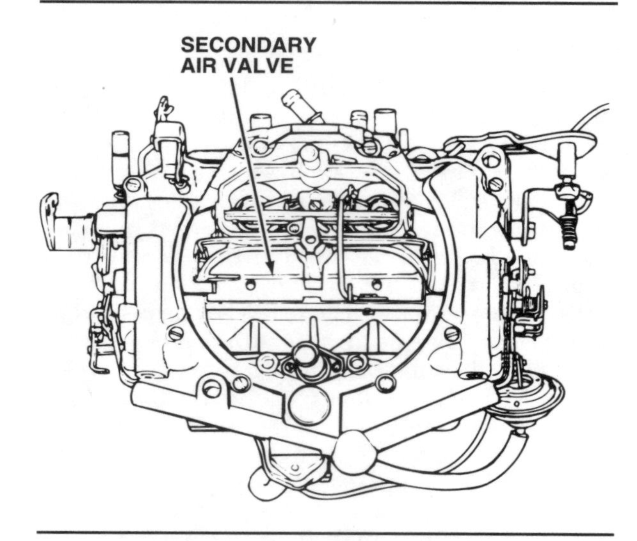

Figure 13-39. This Chrysler application illustrates that air valves are used to control secondary airflow in many Rochester and Carter carburetors.

valves look like large choke plates located in the secondary barrels. Low pressure created in the secondary barrels at open throttle opens the air velocity valves. Older Rochester and Carter four-barrel carburetors may have auxiliary air velocity valves inside the secondary barrels. Airflow through the barrels opens the velocity valves; counterweights hold them closed when the throttles close. Later-model Rochester Quadrajet carburetors use vacuum diaphragms to modulate air valve movement. A carburetor may also combine venturis and air valves to modulate the airflow through the barrels.

The primary barrels supply all eight cylinders during low-to-moderate speeds and loads. The secondary barrels provide additional fuel and airflow for high speeds and heavy loads. The four-barrel carburetor has a flow capacity of 400 to more than 900 cfm.

VARIABLE-VENTURI (CONSTANT DEPRESSION) CARBURETORS

As we have seen, standard carburetors meter fuel by using a venturi to create a partial vacuum in the barrel. Airflow through the venturi increases velocity, which decreases pressure. This pressure drop causes fuel to flow through the discharge nozzle into the barrel. Since the size of both the carburetor barrel and venturi are fixed, the volume and velocity of air passing through will be correct for some operating conditions, but not for others. To produce better performance under *all* operating conditions, auxiliary circuits such as the choke, power, and idle systems must supplement the main metering system.

A variable-venturi carburetor has a venturi whose size changes according to the demands of engine speed and load, and does not need these extra systems. Changing the size of the venturi relative to engine speed and load results in an even pressure drop across the venturi under all operating conditions. This enables a variable-venturi carburetor to control air-fuel mixtures more closely for better fuel economy and emission control. This also gives a variable-venturi carburetor its other name: a "constant depression" carburetor.

Certain imported cars used variable-venturi carburetors, such as those manufactured by SU, Solex, Hitachi, and Stromberg, for many years. In 1977, Ford Motor Company introduced the Motorcraft 2700 VV, the first variable-venturi carburetor used on domestic cars in 45 years. (Ford's first V8 engines in 1932 had variable-venturi, one-barrel carburetors.) This was followed by an electronically controlled feedback version of the same design, the 7200 VV.

The 2700 VV, figure 13-40, Ford's variable-venturi, two-barrel carburetor, has a fuel inlet system with a replaceable filter in the inlet housing. Throttle plates and an accelerator pump also are used. However, the variable venturis and the different fuel metering systems make this carburetor unique.

Two rectangular valve plates (actually a single casting) that slide back and forth across the tops of the two barrels form the variable venturis. A spring-loaded vacuum diaphragm controls movement, and a vacuum signal taken below the venturis (but above the throttle plates) in the carburetor barrels regulates the diaphragm. As the throttle opens, the vacuum increases, opening the venturis and allowing more air to enter.

The front edge of each venturi valve has a tapered metering rod. Each rod moves in and out of a fixed main jet on the other side of the barrel, figure 13-41. This arrangement meters fuel in proportion to the airflow through the venturis. Because of the variable venturis, this type of carburetor has fewer fuel metering systems than the fixed venturi design. However, the main metering system of the 2700 VV carburetor cannot handle all operating conditions without help from secondary systems. For example, at full throttle under heavy load, vacuum may not be strong enough to override the diaphragm spring. In this case, a limiter lever on the throttle shaft pushes the venturi valves fully open. Other auxiliary systems in this carburetor are the accelerator pump, idle trim, cranking enrichment, and cold enrichment systems.

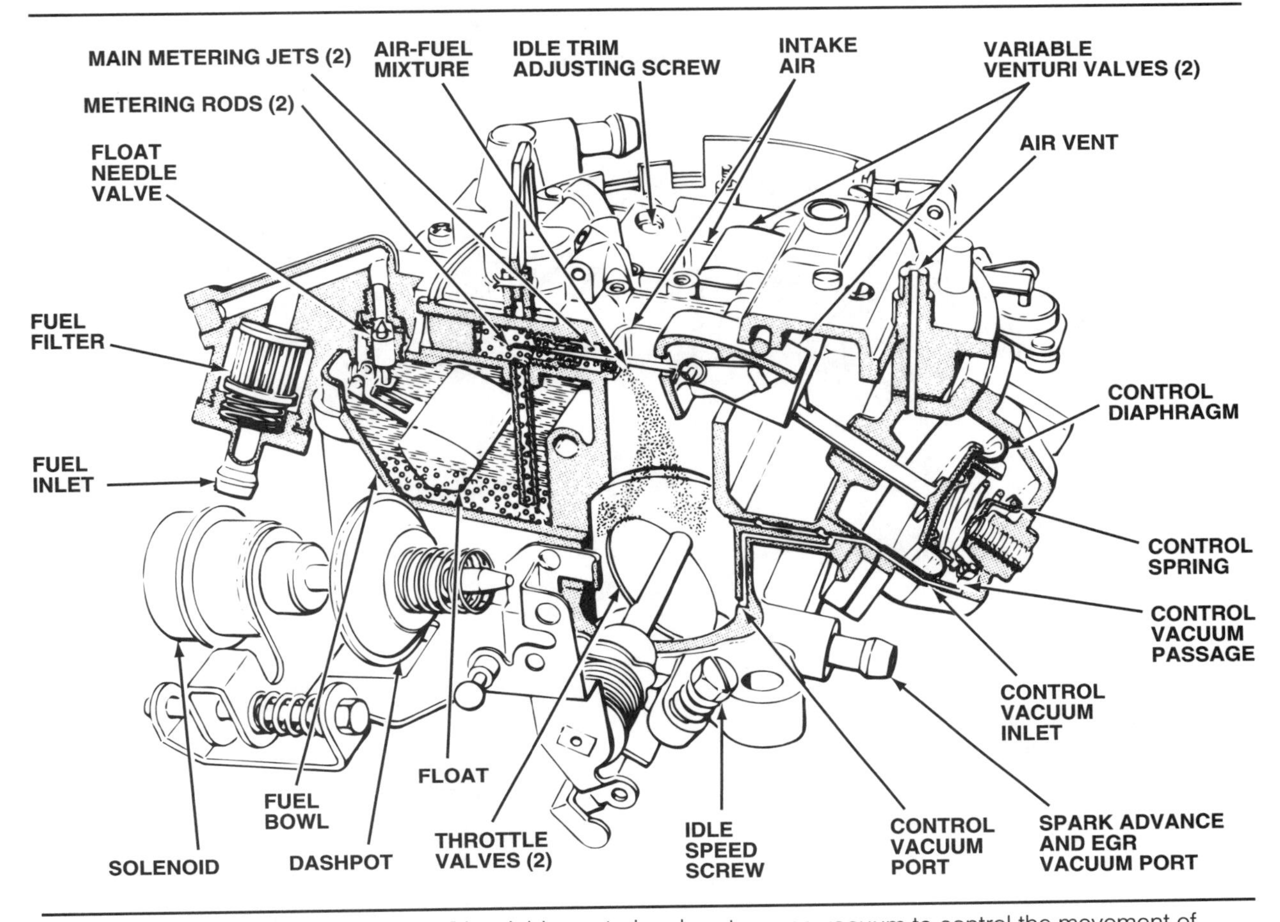

Figure 13-40. Ford's Motorcraft 2700 VV variable venturi carburetor uses vacuum to control the movement of the venturi valves.

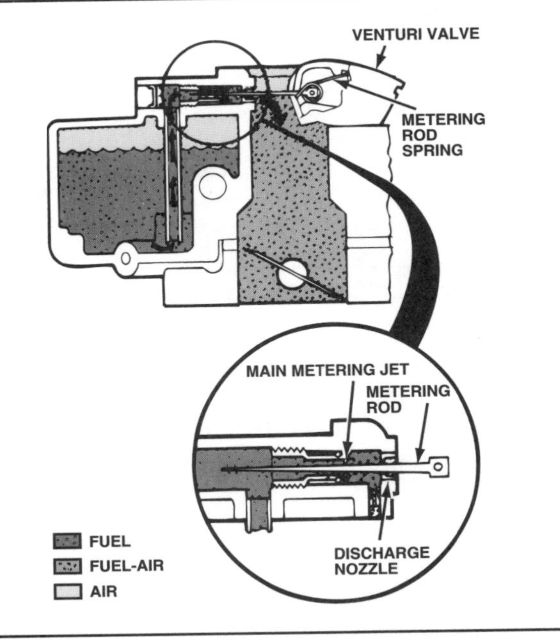

Figure 13-41. Ford's Motorcraft 2700 VV main metering system.

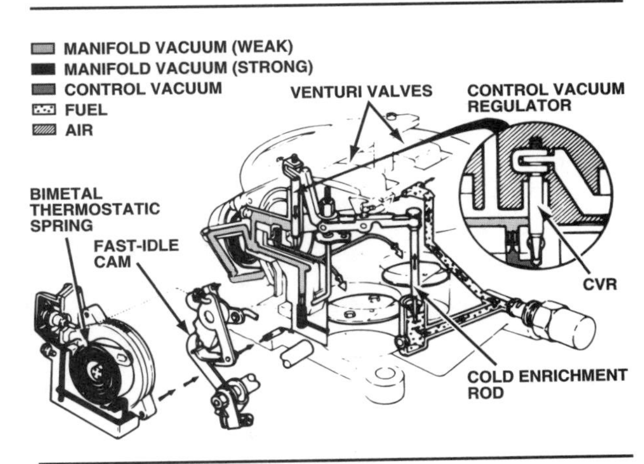

Figure 13-42. Ford's Motorcraft 2700 VV cold enrichment system operation with a cold engine.

The carburetor's design is innovative, but it requires extremely precise adjustments. A major redesign of some systems took place in 1980, with continuing refinements made through the mid-1980s, when Ford discontinued the 2700 VV and restricted use of the 7200 VV to police vehicles.

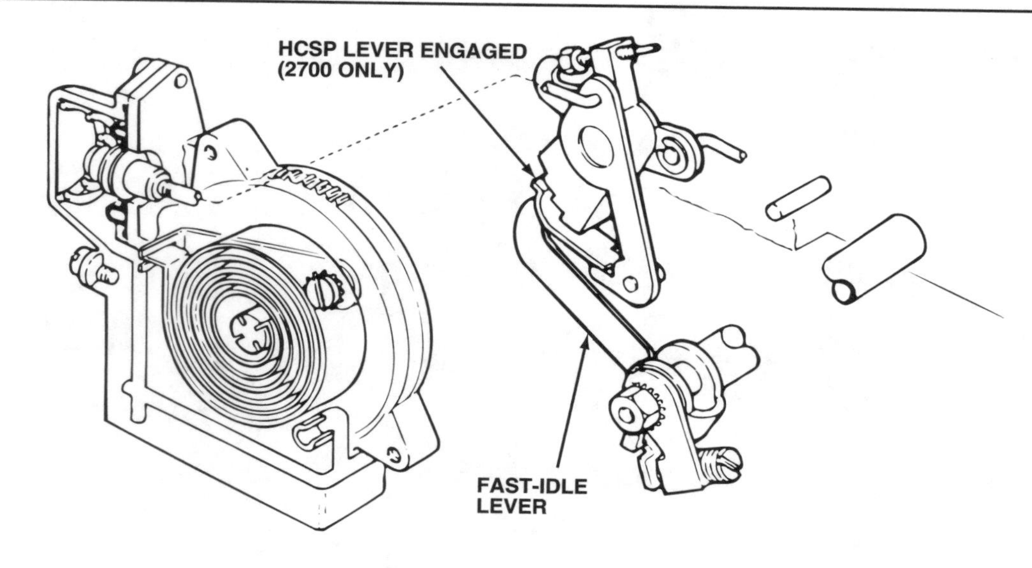

A. ENGINE OFF

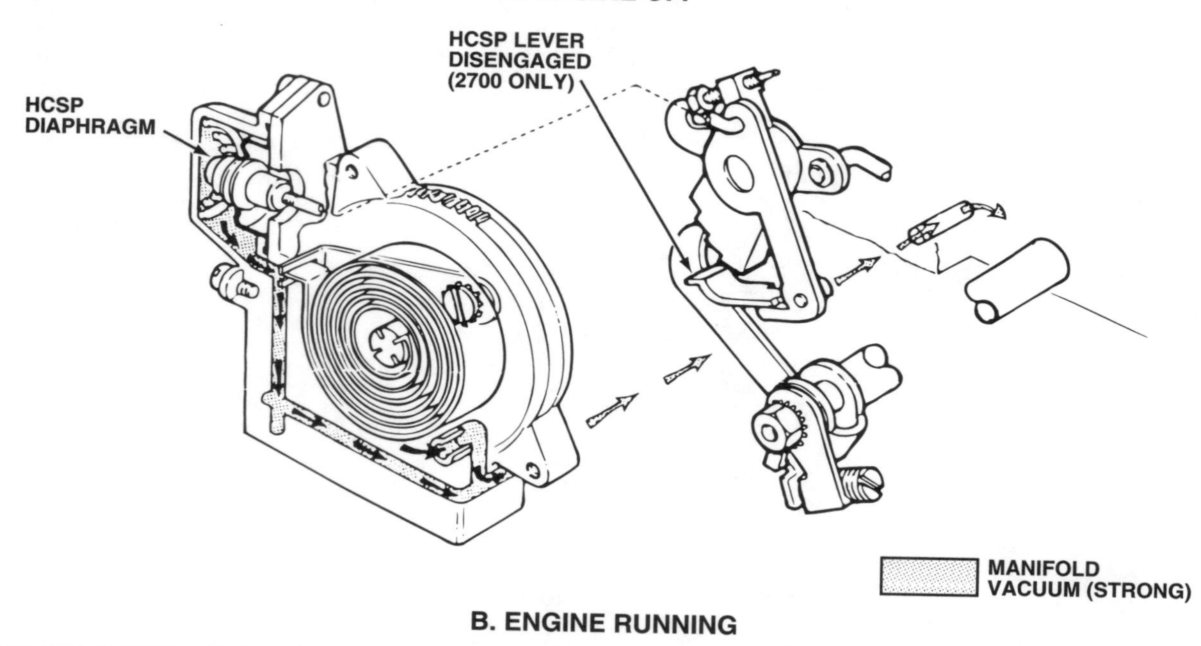

B. ENGINE RUNNING

Figure 13-43. Operation of the HCSP system on late 1970s Ford Motorcraft 2700 VV carburetors.

Cold Enrichment System

The 2700 VV carburetor does not have a traditional choke. Instead, a bimetal thermostatic choke control using an electric-assist heater manages the cold enrichment system, figure 13-42. This system contains a fast-idle cam, a cold enrichment auxiliary fuel passage and metering rod, and a control vacuum regulator rod.

A unique feature of the fast-idle cam on 1977–79 2700 VV models is a vacuum-operated high-cam-speed position (HCSP) lever. When a cold engine is started, a lever slides between the fast-idle cam and the fast-idle lever to provide more throttle opening, When the engine starts, vacuum applied to the HCSP diaphragm retracts the lever, figure 13-43B.

When the engine is started below 95°F (35°C), the choke spring pushes the control vacuum regulator rod to block the ported control vacuum and send manifold vacuum to the venturi valve diaphragm. This opens the venturis wider than normal. The HCSP lever touches the fast-idle

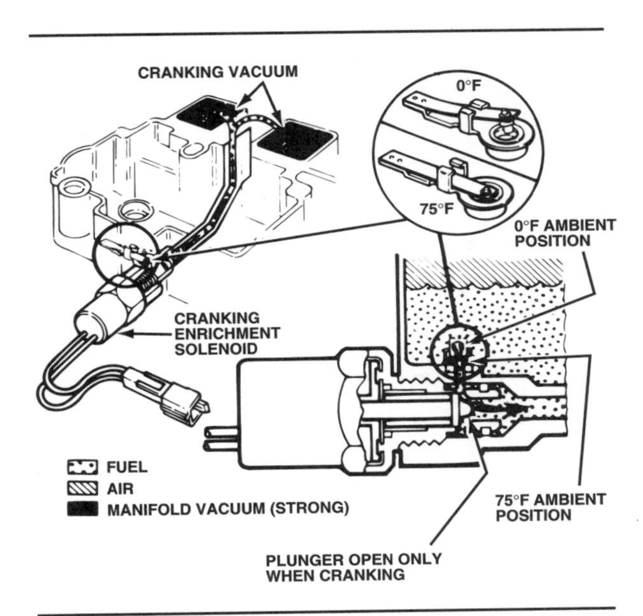

Figure 13-44. The cranking enrichment system used on 1970s Ford Motorcraft 2700 VV carburetors.

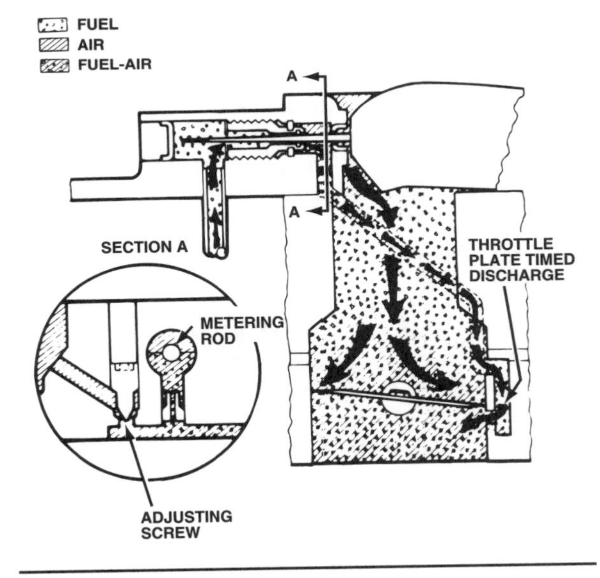

Figure 13-45. Ford's Motorcraft 2700 VV idle trim system.

cam for a wider throttle opening. Redesign of the cranking enrichment system eliminated this system on 1980 and later carburetors.

Cranking Enrichment System

The cranking enrichment system also provides extra fuel only for starting a cold engine. It uses an electric solenoid energized by the ignition switch to open an auxiliary fuel passage, figure 13-44. When the engine starts, the solenoid is deenergized and closes the cranking fuel passage. When this happens, the main metering and cold enrichment systems maintain the fuel flow.

Redesign of the system on 1980 and later models eliminated the solenoid. The cranking enrichment system was changed to provide fuel enrichment during cranking and cold engine running. This was done with a new linkage that provided an additional stroke for the cranking enrichment rod. The increase in rod travel supplies the additional enrichment fuel the solenoid formerly delivered.

Figure 13-46. This throttle linkage is a combination of solid rods and levers.

Figure 13-47. A cable from the accelerator to the carburetor is used for this throttle linkage.

Idle Trim System

On early models, the main jets control the idle fuel flow, but manifold vacuum draws an additional small amount of fuel through internal passages to discharge ports below the throttles. This is called the idle trim system, figure 13-45, and uses adjustable metering screws. The system was redesigned to eliminate the adjustable metering screws on 1978 carburetors.

CARBURETOR LINKAGE

The accelerator pedal, which moves rods, cables, levers, and spring to operate the carburetor throttle valve, controls vehicle speed. The throttle linkage on some vehicles is a combination of solid rods, levers, and links, figure 13-46. Other vehicles use a cable to link the accelerator to the carburetor throttle, figure 13-47.

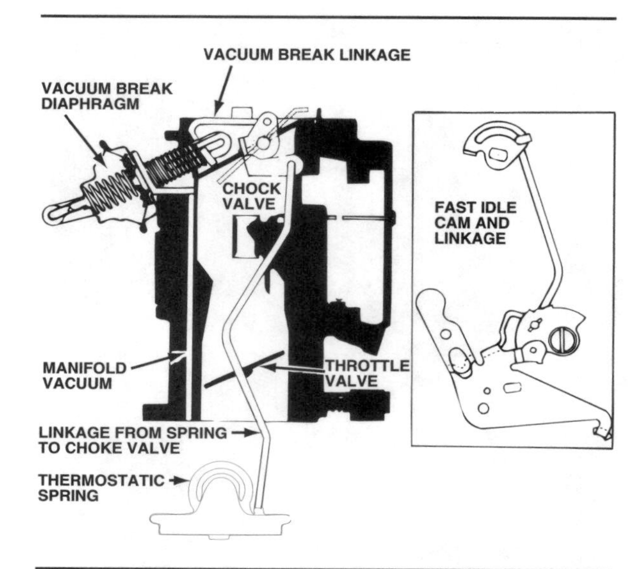

Figure 13-48. Common to many carburetors, this Chevrolet application illustrates that several pieces of linkage are used on the choke system.

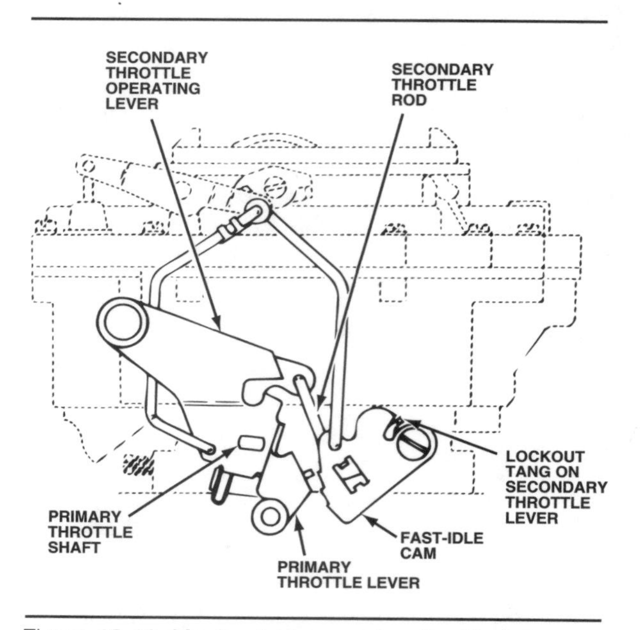

Figure 13-49. Mechanical secondary throttle linkage on a four-barrel carburetor.

Other kinds of linkage operate the accelerator pump, the automatic choke, the fast-idle cam, secondary throttle valves, and the automatic transmission downshift points on some cars.

Accelerator Pump Linkage

A small rod—or rods—and some levers connect the accelerator pump piston or plunger to the throttle. On some carburetors, the linkage holds the pump in the retracted position against a

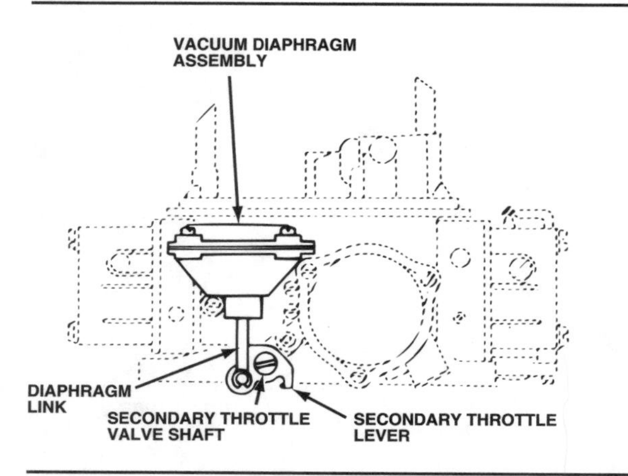

Figure 13-50. Vacuum-operated secondary throttle are controlled by a vacuum diaphragm.

compressed duration spring. When the throttle opens all the way, the linkage releases the spring and the duration spring moves the pump piston through its stroke.

On some carburetors, the accelerator pump linkage also opens a vent for the fuel bowl when the throttle closes. On other carburetors, the accelerator pump linkage operates a diaphragm to deliver fuel to the pump system. On yet other carburetors, mechanical linkage operates the accelerator pumps directly. In some cases, a duration spring and a return spring balance pump operation.

Accelerator pump and linkage designs vary from carburetor to carburetor, but all work on these same principles. On most carburetors, the accelerator pump ends its stroke at the half-throttle position. From this point to the full-throttle position, the high-speed and the power systems can supply enough fuel. The pump linkage on some carburetors can be installed in two or three positions on the carburetor throttle linkage to provide different pump strokes.

Automatic Choke Linkage

Automatic chokes vary in design, but all require connecting linkage between the thermostatic spring that closes the valve and the vacuum piston or diaphragm that opens it. Figure 13-48 shows a remote or well-type choke with a vacuum-break diaphragm. The rod linkage from the spring to the choke valve closes the choke. The vacuum-break diaphragm opens the choke through its linkage as soon as the engine starts.

With most integral piston-type chokes, the thermostatic spring acts directly on a lever on the end of the choke shaft inside the choke housing. The vacuum piston also acts directly on this lever to open the choke when the engine starts.

Some carburetors have an integral choke housing mounted away from the choke valve. This arrangement requires an external linkage rod from the thermostatic spring to the choke valve, as well as a separate vacuum-break diaphragm.

Fast-Idle Cam Linkage

The fast-idle cam linked to the choke valve, figure 13-48, provides a faster-than-normal idle and prevents stalling when the engine is cold. A fast-idle operating lever attaches to the choke shaft and a link connects the lever to the fast-idle cam. When the choke closes, the linkage turns the fast-idle cam so that a high step of the cam touches the idle-speed adjusting screw or a separate fast-idle screw. As the choke opens, the fast-idle cam continues to follow the choke movement and reduces idle speed step by step. When the choke is fully open, gravity keeps the fast-idle cam away from the idle-speed screw. However, when the engine is shut off, the idle-speed screw blocks the cam and keeps it from turning back to the fast-idle position as long as the throttle closes. This also keeps the choke thermostatic spring from closing the choke valve as the engine cools. To begin the choking operation with a cold engine, the accelerator must be depressed to release the fast-idle cam and linkage and allow the choke to close.

Secondary Throttle Linkage

Either mechanical linkage or vacuum may operate the secondary throttles of four-barrel and two-stage, two-barrel carburetors. Secondary throttle valves operated by mechanical linkage have an operating rod to connect the primary throttle shaft to the secondary throttle shaft, figure 13-49. Secondary throttles begin to open when the primary throttles are about half open. The secondaries continue to open along with the primaries, but at a faster rate. The primary and secondary throttles then reach the fully open position at the same time.

On some carburetors, the primary and secondary throttle shafts are mechanically linked, but the secondary throttles are not visible through the carburetor airhorn. This type of carburetor has secondary air velocity valves, called auxiliary throttle valves, figure 13-39. These valves have offset shafts and counter weights so that they remain closed until air velocity through the carburetor barrels is strong enough to open them. These auxiliary throttle valves operate only when the mechanical secondary throttles are open, but they do not rely on the mechanical movement of the secondary throttles.

A vacuum diaphragm mounted on the side of the carburetor, figure 13-50, controls vacuum-operated secondary throttle valves. The secondary throttle valves close at low cruising speeds, and the primary half of the carburetor meets the engine's air-fuel requirements, figure 13-51A. At higher speeds, when more fuel and air are needed,

■ Ford's First Variable Venturi Carburetor

Ford's first variable venturi carburetor was a one-barrel model made by Detroit Lubricator. It was used on the original flathead Model 18 V8 in 1932 and early 1933. The variable venturi was formed by two air vanes in the barrel that responded directly to airflow, rather than to vacuum linkage control.

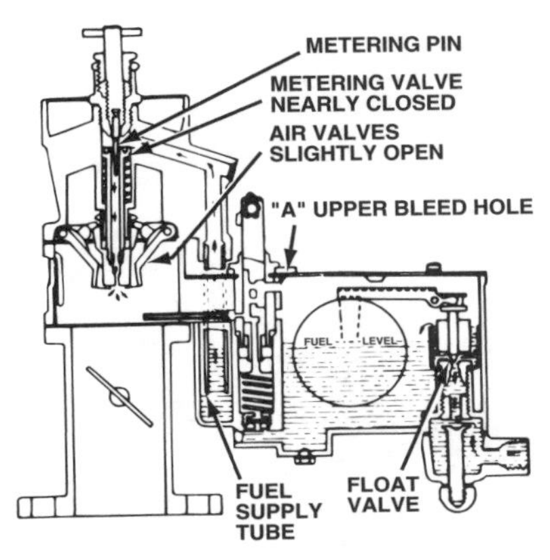

NORMAL RUNNING

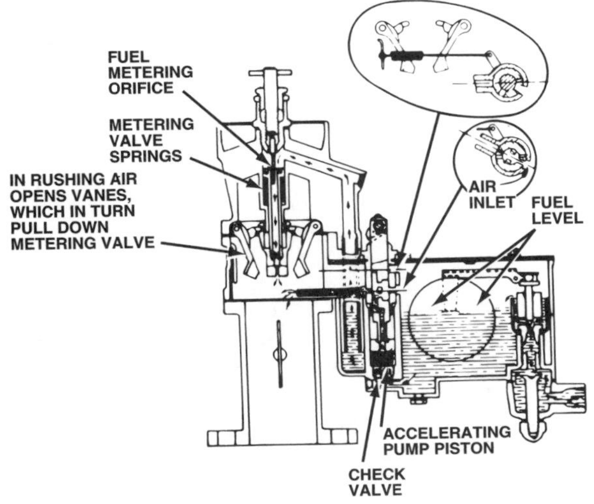

MAXIMUM POWER

The air vanes were linked to a movable main jet (metering valve) that slid up and down on a fixed metering rod (pin) in the center of the venturi. The metering rod was adjustable to obtain the proper idle mixture and main system fuel flow.

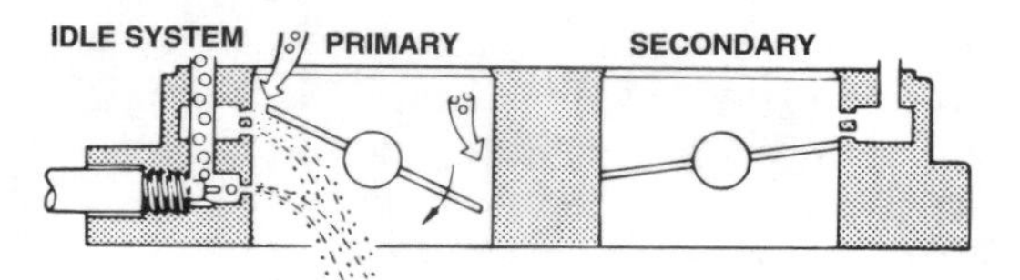

Figure 13-51. As shown on this Ford application, secondary throttle valves are closed during low cruising speeds but open progressively as more fuel and air are required at higher speeds.

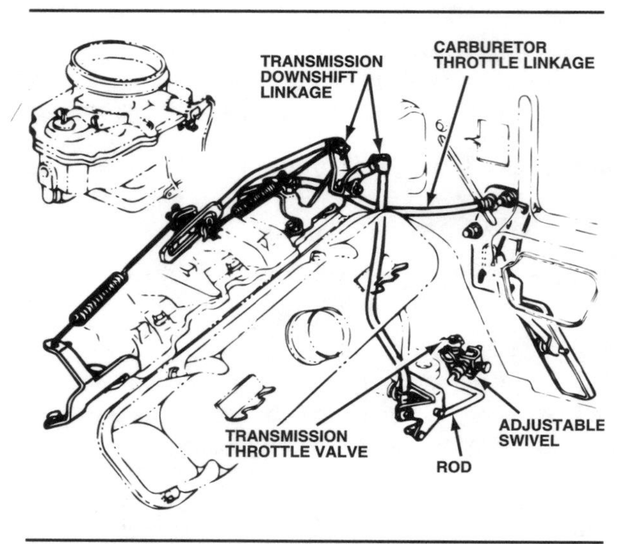

Figure 13-53. Chrysler products with automatic transmissions have mechanical linkage between the carburetor and transmission throttle valve to control shift points.

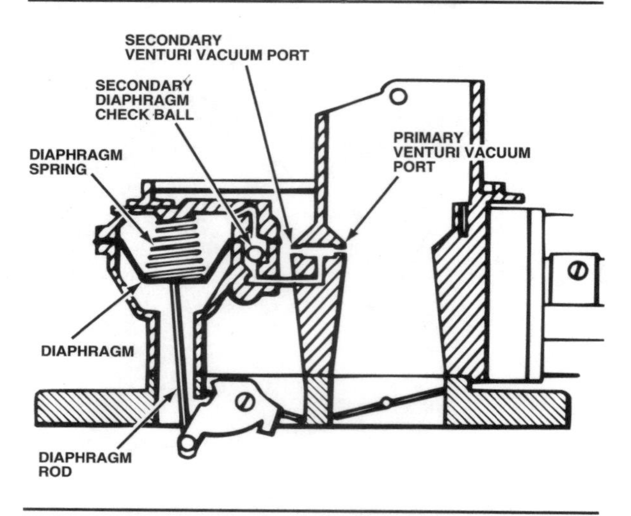

Figure 13-52. The secondary diaphragm responds to vacuum from ports within the carburetor barrels.

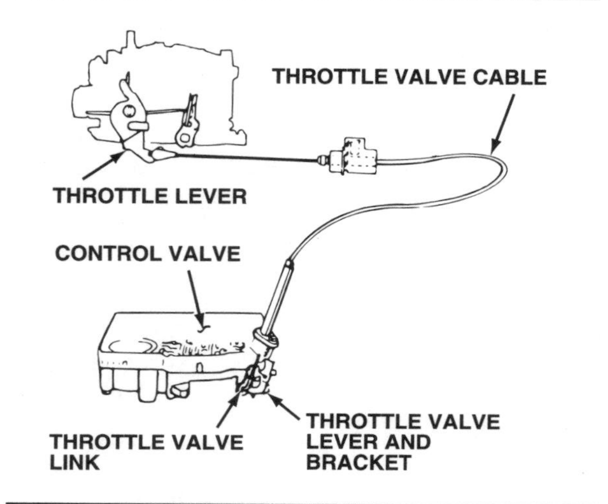

Figure 13-54. Some GM vehicles, as on this Buick, control automatic transmission shifting by a cable between the carburetor and the transmission throttle valve .

linkage connected to the vacuum diaphragm opens the secondary throttles, figure 13-51B.

The vacuum diaphragm responds to increasing vacuum within the primary venturi as engine airflow increases, figure 13-52. Linkage from the diaphragm opens the secondary throttles. Another vacuum port, or air bleed, in the secondary barrels changes diaphragm action.

The vacuum signal at the diaphragm determines the amount and rate at which the secondary throttles open. When the secondaries close, the air bleed in the secondary barrels weakens the vacuum signal at the diaphragm a set amount. As the secondaries open, the air bleed becomes a vacuum port as vacuum develops within the secondary barrels. Adding this vacuum signal to the vacuum from the port in the primary venturis opens the secondary throttles completely.

A ball check valve in the vacuum chamber passage that allows a gradual vacuum buildup prevents sudden secondary throttle opening. As engine speed decreases, the weaker vacuum signal allows the diaphragm spring to close the secondary throttles.

All secondary throttle linkage, whether mechanically or vacuum operated, includes a

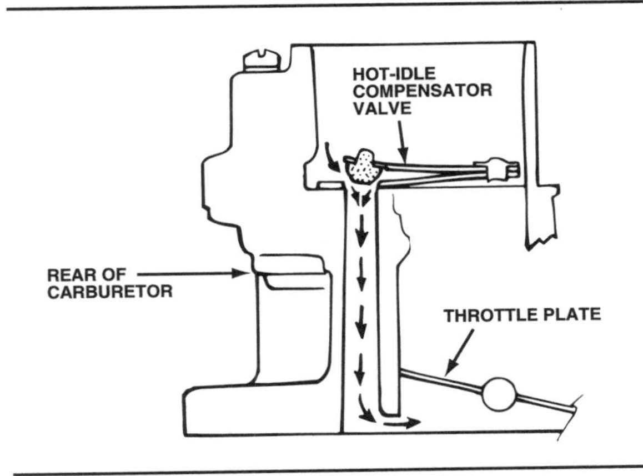

Figure 13-55. The hot-idle compensator is a thermostatic valve that opens at high temperature to admit more air to the idle circuit.

secondary throttle or air valve lockout device to keep the secondary throttle from opening when the choke closes.

Transmission Linkage

Some vehicles with an automatic transmission, or transaxle, may use an adjustable throttle rod, figure 13-53, or a cable linkage, figure 13-54, between the transmission and the carburetor or throttle body. This controls shift points and shift quality. Other automatic transmissions do this with vacuum control.

CARBURETOR CIRCUIT VARIATIONS AND ASSIST DEVICES

All automakers have used variations in the basic carburetor systems just discussed. One or more add-on devices also may improve economy, driveability, and emission control. The following section covers those most commonly used.

Hot-Idle Compensator Valves

High carburetor inlet air temperature causes gasoline to vaporize rapidly, which can cause an overly rich idle mixture. To prevent this, many carburetors use a hot-idle compensator valve, figure 13-55, a thermostatic valve consisting of a bimetal spring, bracket, and small poppet. The compensator valve usually is located either in the carburetor barrel or in a chamber on the rear of the carburetor bowl. A dust cover is placed over the chamber. A third location (used primarily in Autolite two-barrel carburetors on air-conditioned Ford vehicles of the late 1960s) is an external mounting in the positive crankcase ventilation (PCV) valve hose near the carburetor, figure 13-56.

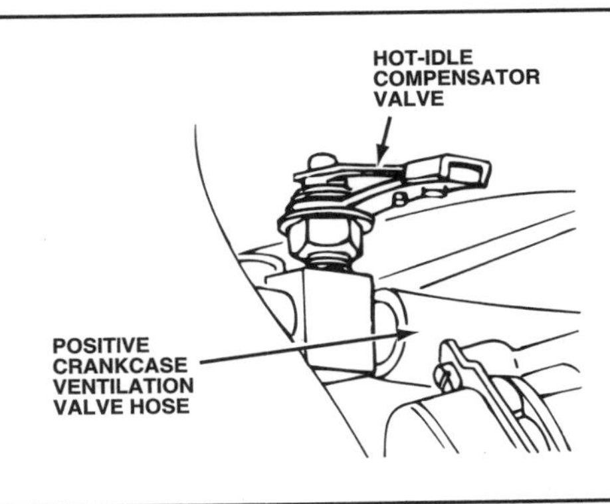

Figure 13-56. This hot-idle compensator is located in the PCV valve hose away from the carburetor.

Normally, spring tension and engine vacuum close the hot-idle compensator valve. As temperature rises, the bimetal strip bends. This uncovers an auxiliary air passage or air bleed through which air enters the carburetor below the throttle plates. As this extra air mixes with excess fuel to lean out the idle mixture, it prevents stalling and rough idling. Once the carburetor temperature returns to normal, the compensator valve closes to shut off the extra air. If the valve does not close fully, it causes a high idle speed.

Idle Enrichment Valves

Carburetors that have emission control devices run on leaner mixtures, but the idle mixture must be richer in some cases for good cold-engine operation. Some 1975 and later Chrysler carburetors have an idle enrichment system that works in a way opposite to a hot-idle compensator valve.

A small vacuum diaphragm mounted near the carburetor top, figure 13-57, controls idle circuit air. When control vacuum is applied, the diaphragm reduces idle system air. This increases fuel and reduces the air in the air-fuel mixture. A temperature switch in the radiator controls diaphragm vacuum. As the engine warms, this switch stops the vacuum signal, returning the air-fuel mixture to its normal lean level.

Fast-Idle Pulloff (Choke Pulloff)

The rich air-fuel mixtures resulting from long periods of choke and fast idle can damage catalytic converters. Some converter-equipped GM cars use a fast-idle pulloff to avoid converter overheating. In one system, manifold vacuum acts on a vacuum-break diaphragm at the rear of the carburetor. This diaphragm will drop the fast idle cam to a lower step 35 seconds after engine coolant temperature reaches 70°F (21°C). A vacuum solenoid operated by a coolant temperature

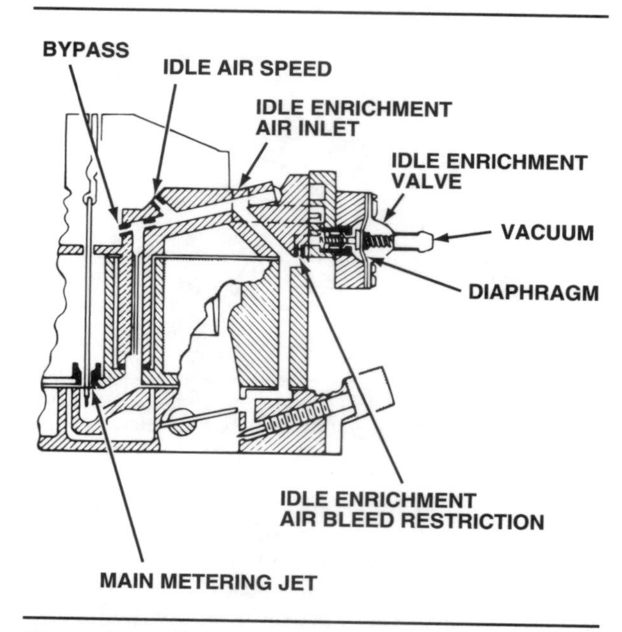

Figure 13-57. The Chrysler idle enrichment valve.

switch and timer controls delays vacuum to the diaphragm.

Another system uses the front vacuum diaphragm to pull the throttle down one step on the fast-idle cam as soon as engine coolant temperature reaches 150°F (66°C). A temperature vacuum switch controls vacuum to the diaphragm.

A third method uses an electric solenoid instead of a vacuum diaphragm to open the choke. This pulls the fast-idle screw off the cam whenever the engine is started with the coolant below a specified temperature. A temperature switch on the engine and a firewall-mounted relay provide current for the pulloff solenoid.

Regardless of the system used, fast-idle pulloff has no effect on engine warmup during ordinary operation because normal throttle movement will disengage the fast-idle cam. The fast-idle pulloff system only operates when the engine is warming up while parked.

Throttle Stop Solenoids

Engine **dieseling**, or after-run, results when combustion chamber temperatures remain hot enough to ignite an idle air-fuel mixture after the ignition is turned off. Several aspects of late-model engines can cause dieseling:

- Higher operating temperatures
- Faster idle speeds
- Retarded ignition timing at idle
- Lean air-fuel mixtures.

Closing the throttle more than it would close for the engine's normal slow-idle speed will prevent

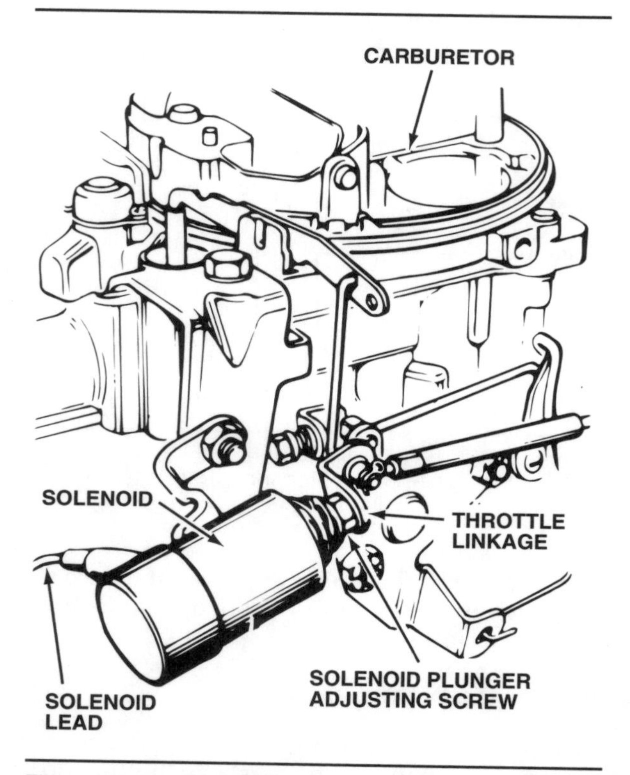

Figure 13-58. The throttle stop solenoid holds the throttle open for normal slow idle and allows the throttle to close farther when the engine is shut off.

dieseling. Shutting off airflow past the throttle valve closes the idle circuit.

A throttle stop solenoid, figure 13-58, provides the new stop position for the throttle during normal slow idle. Turning on the ignition energizes the solenoid, and its plunger moves out to contact the idle speed adjusting screw or a bracket on the throttle shaft. This holds the throttle open slightly for a normal slow idle until the ignition is shut off. The solenoid is then deenergized, its plunger retracts, and the throttle closes to block airflow.

Since engines have used a variety of throttle solenoids or positioners over the years, figure 13-59, you should check the manufacturer's adjustment procedures before attempting to service any throttle solenoid.

Air Conditioning Throttle Solenoids

Many late-model vehicles with air conditioning may use a solenoid that looks exactly like a throttle solenoid, figure 13-60. In many cases, it is the same solenoid and may even carry the same part number. However, you should not confuse it with the throttle stop solenoid just discussed since it performs an entirely different function.

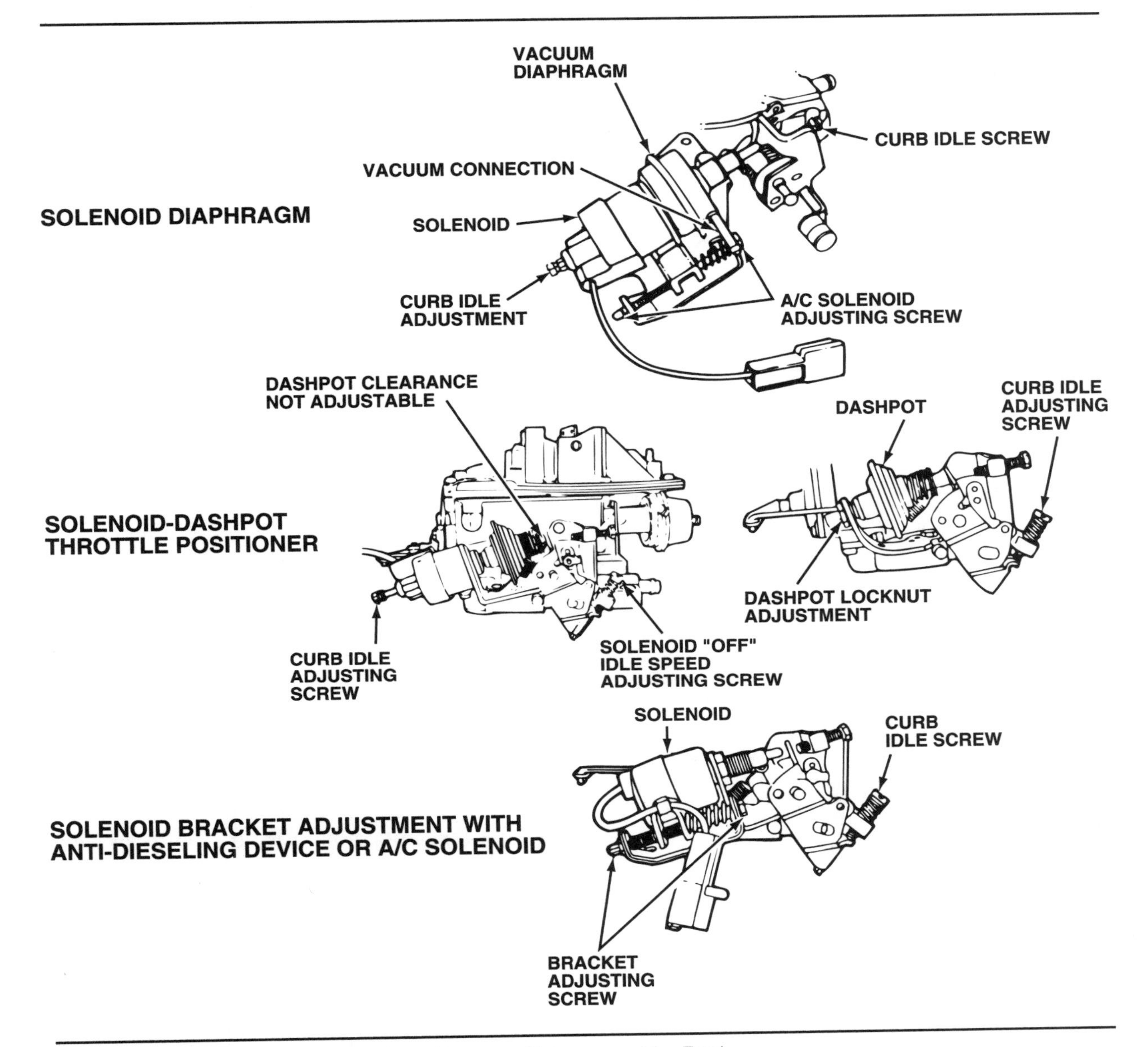

Figure 13-59. Some of the throttle solenoids or positioners used by Ford.

This solenoid is energized *only* through the air conditioner switch. Its plunger moves forward to contact a bracket on the throttle shaft *only* when the air conditioning is turned on. This maintains or slightly raises engine idle speed to prevent the engine from stalling due to the increased load. It also helps prevent overheating from the air conditioning condenser heat load by speeding up the radiator fan.

Choke Hot Air Modulators

Some GM engines of the mid-1970s used a choke hot air modulator check valve (CHAM-CV) in the air cleaner, figure 13-61. The valve closes at air cleaner temperatures below 68°F (20°C). Air to be heated by the heater coil passes through a tiny hole in the modulator. This restricts hot airflow over the bimetal thermostatic coil and results in a slower choke warmup. When air cleaner temperature rises above 68°F (20°C), the modulator opens to permit more airflow for a faster choke warmup.

Dieseling: A condition in which extreme heat in the combustion chamber continues to ignite fuel after the ignition has been turned off.

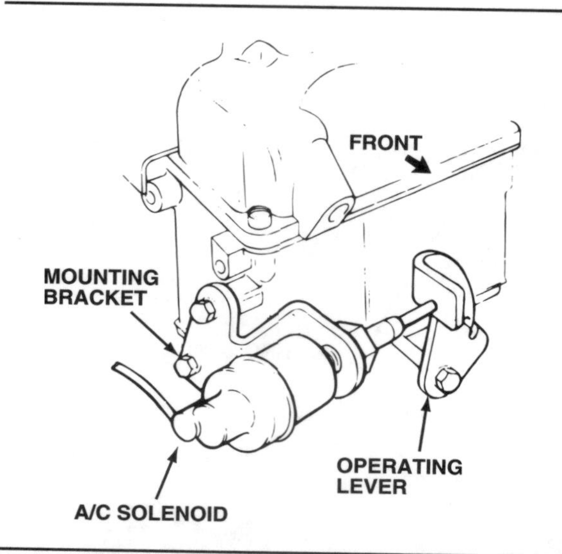

Figure 13-60. An air conditioning throttle solenoid opens the throttle slightly when the air conditioner is on. This maintains a uniform idle speed, even with the increased engine load.

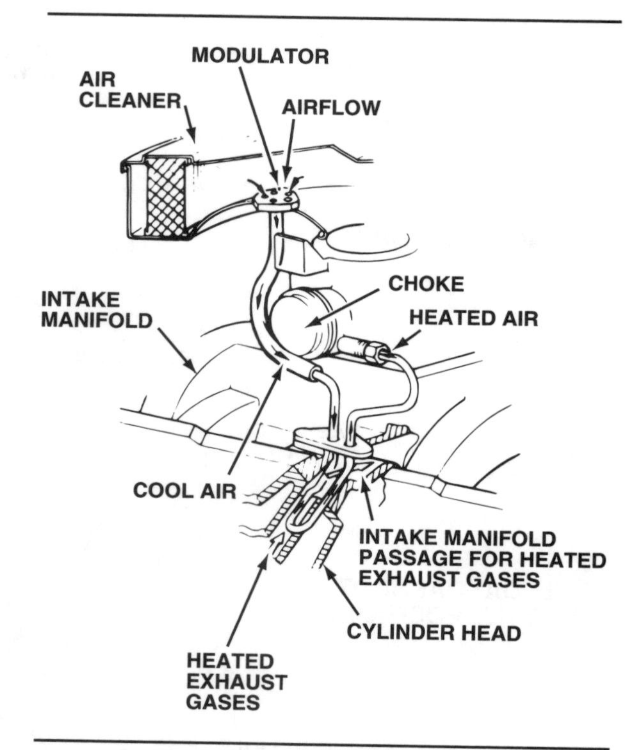

Figure 13-61. The General Motors choke hot-air modulator.

Delayed Choke Pulldowns

A delayed choke pulldown operated by a vacuum diaphragm is used on some Motorcraft carburetors beginning in 1975. This opens the choke to a wider setting from 6 to 18 seconds after the engine starts. As the pulldown diaphragm operates, the fast-idle screw is pulled from the top to the second step of the fast-idle cam to reduce cold-engine idle speed. Figure 13-62 shows the exact sequence of operation.

Temperature-Controlled Vacuum-Breaks

Many Rochester carburetors since 1975 use two vacuum-break diaphragms to provide better mixture control by choking the engine more when it is cold and less when warm. GM has used slightly different systems on V8 engines, figure 13-63; and 6-cylinder engines, figure 13-64.

On V8 engines, the primary (front) diaphragm opens the choke to keep the engine from stalling when first started. The secondary (rear) diaphragm opens the choke wider when air cleaner temperature exceeds 62°F (17°C). The rear diaphragm operates on manifold vacuum provided by a temperature vacuum valve on the air cleaner. The vacuum-break diaphragm has an inside restriction to delay diaphragm movement by several seconds.

On 6-cylinder engines, the primary diaphragm (choke coil side) opens the choke to keep the engine from stalling when first started. The auxiliary diaphragm (throttle lever side) opens the choke wider when engine coolant temperature exceeds 80°F (27°C). The auxiliary diaphragm operates on manifold vacuum provided by a two-nozzle temperature vacuum switch on the cylinder head.

Choke Delay Valves

Choke delay valves have various designs, but all do essentially the same thing as the vacuum-break diaphragm. They delay the opening of the choke for a period of time to improve driveability and cold engine warmup.

Ford uses the same valve for choke delay, figure 13-65, that it uses for spark delay in the ignition system. Color coding on the valve indicates delay interval, and you can use it for either purpose. When used as a choke delay valve, it is installed in a hose between the intake manifold and the choke vacuum piston or diaphragm. The black side must face the vacuum source (manifold) and the colored side faces the choke.

GM uses an internal bleed check valve in the rear vacuum-break diaphragm, figure 13-66. This delays choke opening beyond the amount allowed by the front vacuum-break for a few seconds until the engine can run at a leaner mixture. Vacuum acting on the diaphragm draws filtered air through the bleed hole to purge any fuel vapor or contamination.

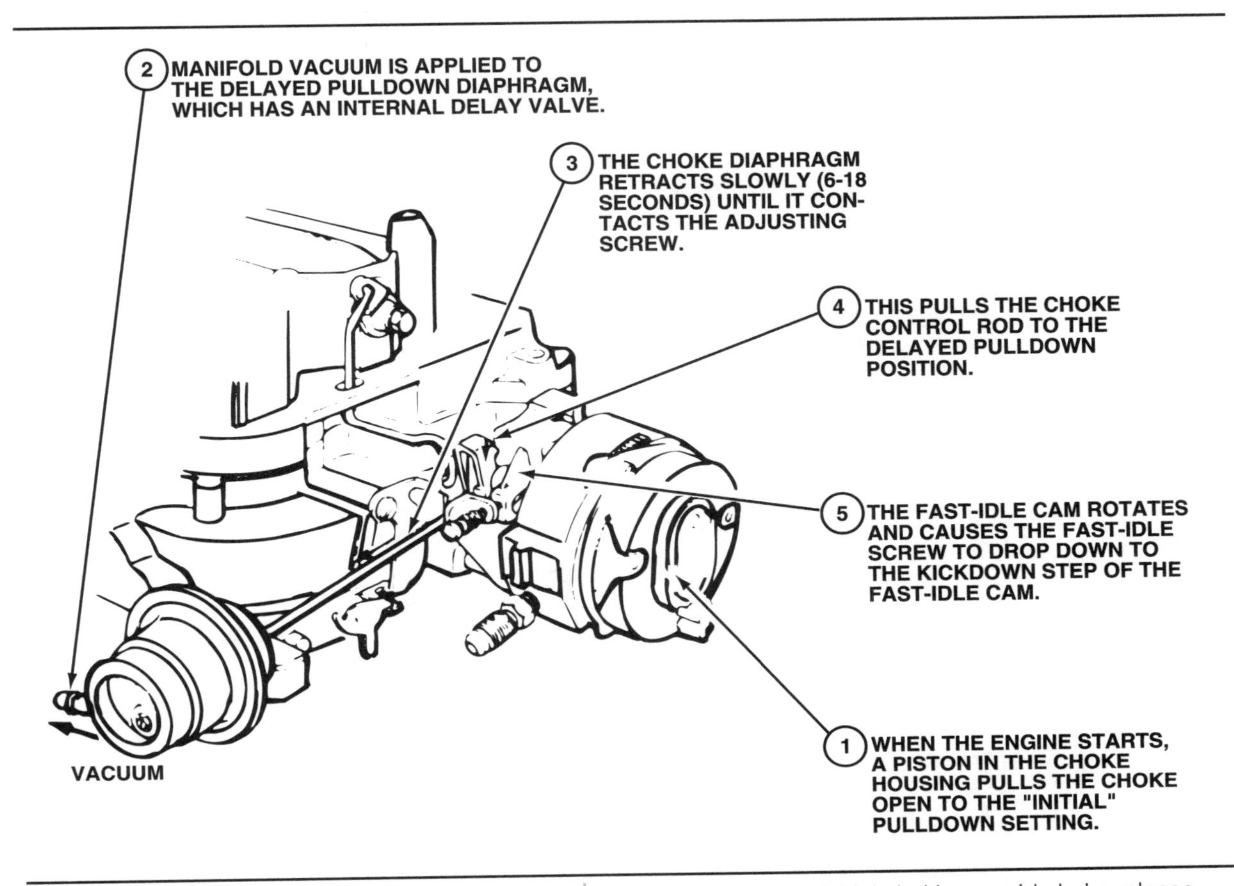

Figure 13-62. This Ford delayed choke pulldown diaphragm provides rich initial choking, rapid choke release, and fast-idle modulation.

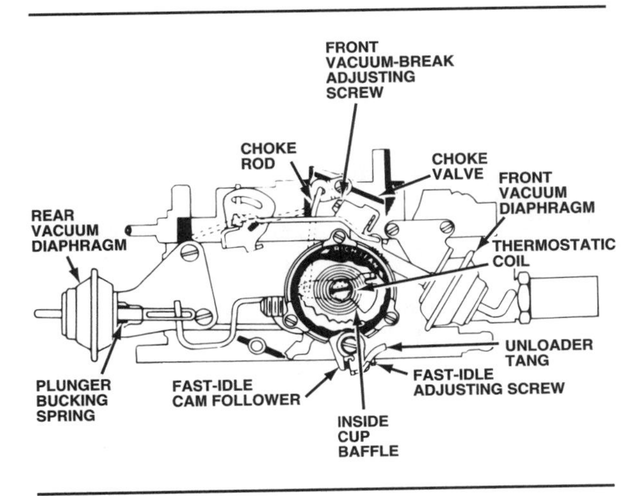

Figure 13-63. As shown on this Pontiac carburetor, many GM V-8 engines use this type of temperature-controlled vacuum break.

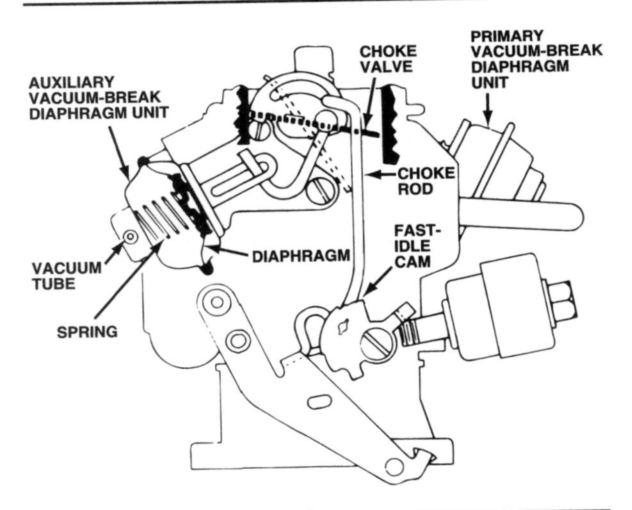

Figure 13-64. As shown on this Chevrolet carburetor, many GM 6-cylinder engines use this type of temperature-controlled vacuum break.

A slightly different system is used on some 2.5-liter (151-cu in.) 4-cylinder GM engines. The engine has a choke vacuum-break and a vacuum delay valve. The valve delays manifold vacuum against the choke vacuum-break unit for about 40 seconds when starting the engine. If the engine stalls after starting, a relief feature in the valve permits immediate release of vacuum to let the choke close quickly.

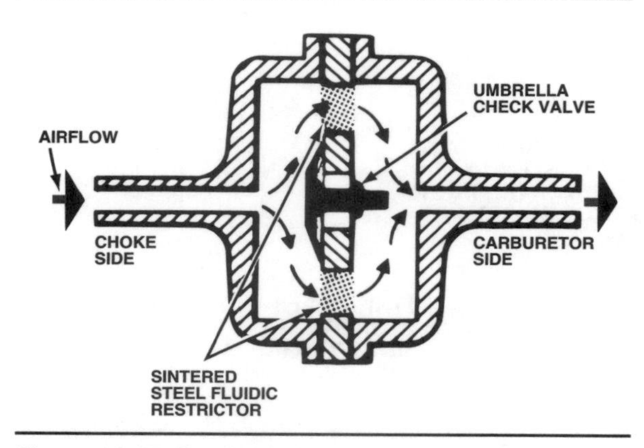

Figure 13-65. Ford's choke delay valve.

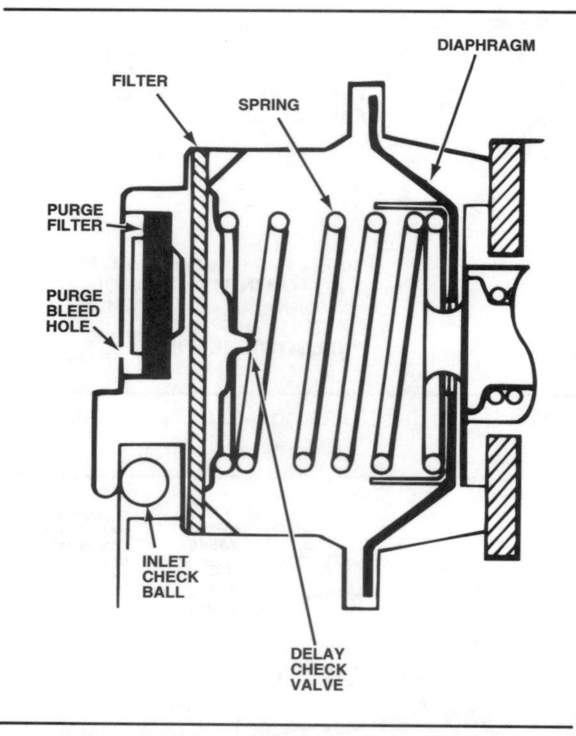

Figure 13-66. General Motors' choke delay valve.

Dashpots

Carburetors used dashpots—small chambers containing a spring-loaded diaphragm and plunger—for over 30 years. Before dashpots, carburetors used hydraulic and magnetic dashpots.

As the throttle closes, a link from the throttle contacts the dashpot plunger, figure 13-67. As force is applied to the plunger, air slowly bleeds out of the diaphragm chamber through a small hole.

Dashpots originally prevented an excessively rich mixture on deceleration, which can cause stalling. They now function as emission control devices on later-model engines by reducing HC emissions on deceleration.

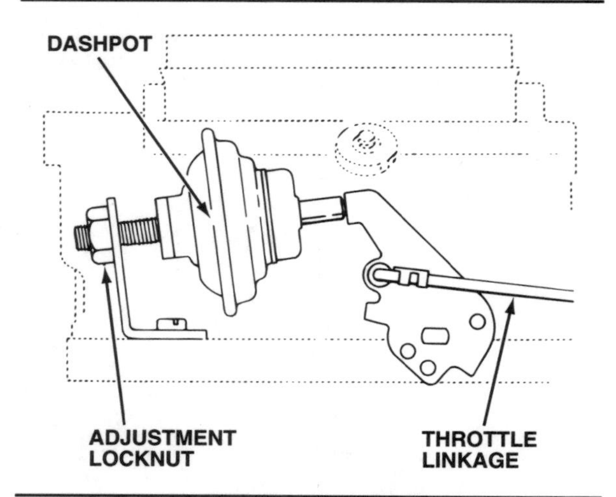

Figure 13-67. The dashpot slows throttle closing.

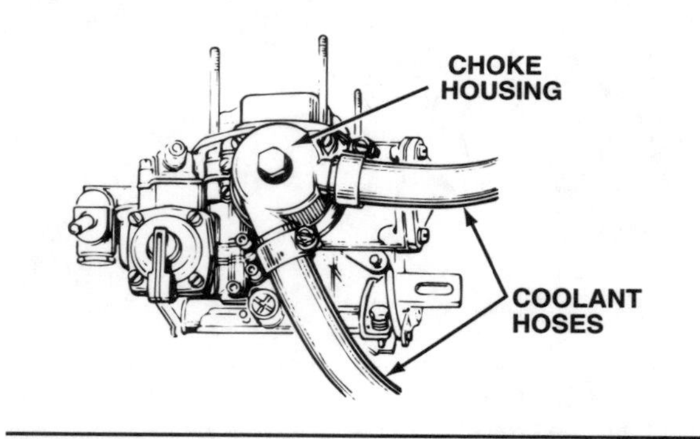

Figure 13-68. This type of hot-water-heated choke routes engine coolant through the choke housing.

Choke Heaters

The need to keep exhaust emissions low means that the system must choke the engine only for a brief time. There are several ways to release the choke promptly, all of which apply heat to the thermostatic coil spring to warm it up quickly.

Hot-water choke

One way to apply heat to the thermostatic coil spring is by routing part of the engine coolant through the choke housing, figure 13-68. Once the engine reaches normal operating temperature, coolant heat helps release the choke. Older Ford models used an external coolant bypass hose held against the choke housing cap by a spring clip, figure 13-69. This supplemented the exhaust manifold air passed through the choke housing.

Hot-water chokes reduce over-choking when an engine is restarted. Since water holds heat longer than air, the choke coil will remain warm longer when exposed to heat from hot coolant. This reduces the amount of choking on a restart.

Figure 13-69. Other chokes are heated by routing engine coolant through a hose clipped to the choke cap.

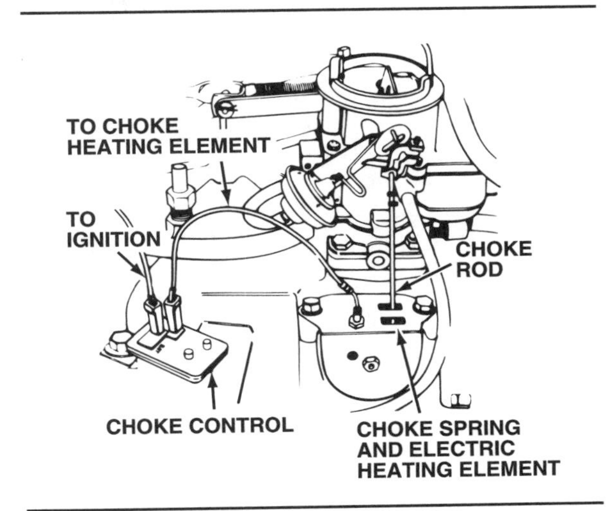

Figure 13-71. Chrysler's electric-assist choke.

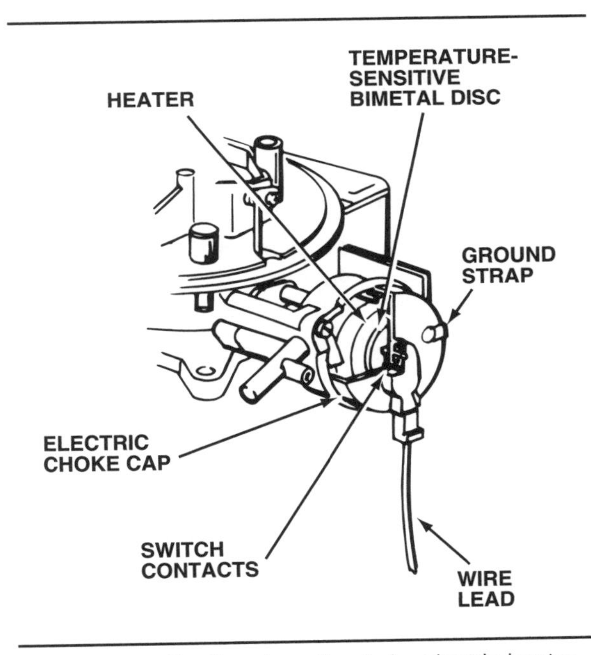

Figure 13-70. Ford's automatic choke electric heater.

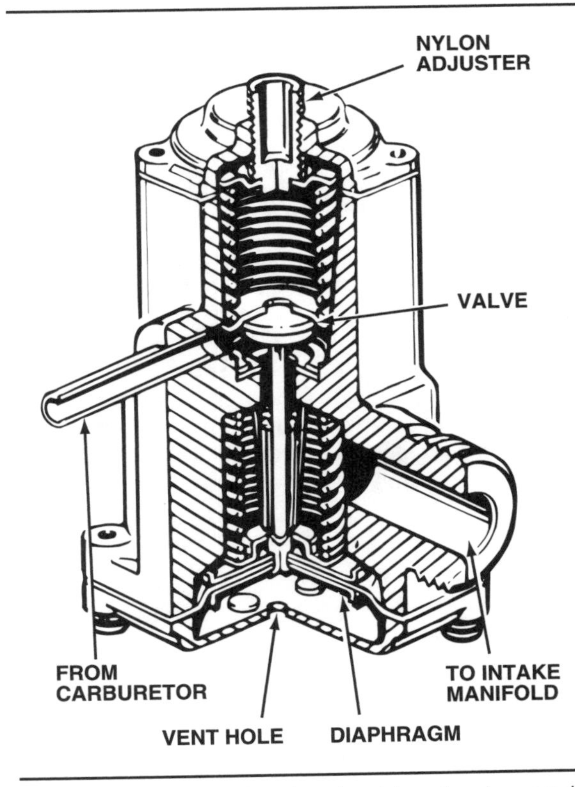

Figure 13-72. A fuel deceleration (decel) valve used on Ford carburetors.

Electric-assist choke

Later engines used electrically assisted heater elements. These speed up the choke opening when underhood temperature is above about 60°F (16°C) by heating the choke bimetal thermostatic coil spring.

Ford choke heater elements are located in the integral choke cap on the carburetor choke housing, figure 13-70. Electric current is supplied continuously from the alternator directly to the choke cover temperature-sensing switch. At underhood temperatures above 60° to 65°F (16° to 18°C), the switch closes to pass current to the choke heater. The circuit is grounded through a strap connected to the carburetor.

The Chrysler electric-assist choke heater is located in the intake manifold choke well; a control switch that receives power from the ignition switch, figure 13-71, regulates the heater. The control switch energizes the heater at temperatures above 63° to 68°F (17° to 20°C), and deenergizes it when the switch warms to 110° to

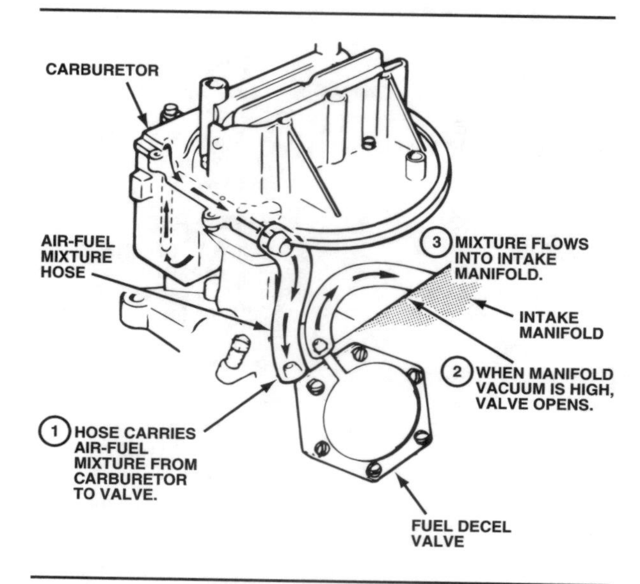

Figure 13-73. Ford decel valve operation.

130°F (43° to 54°C). A two-stage heater control is used on some engines to provide three levels of heat, depending upon ambient temperature. A resistor on the control switch provides the low heat level.

Some GM engines from 1975 on use an electric choke heater. This two-stage heater receives current from the engine oil pressure switch. Below specified air temperatures (usually 50° to 70°F or 10° to 21°C), a bimetal sensor in the cover turns off current to the larger heater stage. Both stages operate at higher temperatures. This choke heater receives current as long as the engine is running.

Fuel Deceleration (Decel) Valves

Ford used the decel valve, figure 13-72, primarily on some 4-cylinder and V6 engines during the 1970s to momentarily provide extra fuel and air during deceleration. Some GM and imported-car engines have also used decel valves. The valve prevents cylinders from misfiring during deceleration and sending an unburned charge of HC's through the exhaust. The extra air and fuel provided by the valve ensure complete combustion.

When the throttle closes, increased manifold vacuum opens the valve diaphragm, figure 13-73. This, in turn, opens a passage between the carburetor and the manifold, allowing an additional air-fuel charge to enter the intake system. The extra air-fuel charge slows deceleration speed and reduces HC emissions. When vacuum drops, spring action and an air bleed to the diaphragm reseat the valve.

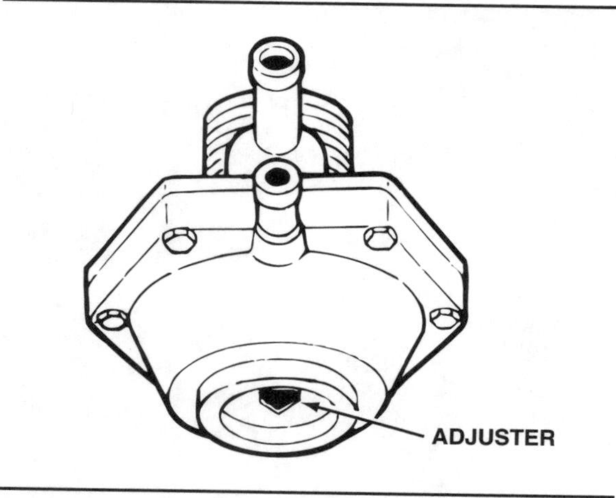

Figure 13-74. Ford's decel valve was redesigned in the 1970s, but its operation remained unchanged.

Some decel valves are adjustable. A nylon screw in the valve top, figure 13-72, controls when and for how long the valve opens. A round valve design, first used in 1974, figure 13-74, was not adjustable but was modified to permit adjustment on 1975 and later engines.

Deceleration Throttle Openers

A deceleration throttle opener is a solenoid that holds the throttle open during deceleration to prevent excess fuel from being pulled through the idle system by high manifold vacuum. It is used on catalytic converter-equipped cars to prevent converter overheating. GM cars with the combination emission control (CEC) system use a throttle positioning solenoid that is energized on deceleration. Some Chrysler cars with converters have a solenoid to keep the throttle from closing completely at engine speeds above 2,000 rpm.

ALTITUDE-COMPENSATING CARBURETORS

Earlier in this chapter you learned that atmospheric pressure is greatest at (or below) sea level. Suppose you drive a car from a low elevation into an area where the elevation is 7,000 feet (2130 meters). As the altitude increases, atmospheric pressure decreases, and less air enters the carburetor. This means that the air-fuel mixture passing into the engine becomes richer as altitude increases. The result is poor driveability and high CO emissions at the higher elevation. For the driver who is only passing through high-elevation areas, the poor driveability is mainly a temporary inconvenience. But for driving at that elevation for an extended time, the car will run better if the engine is tuned for high elevation by leaning the air-fuel mixture.

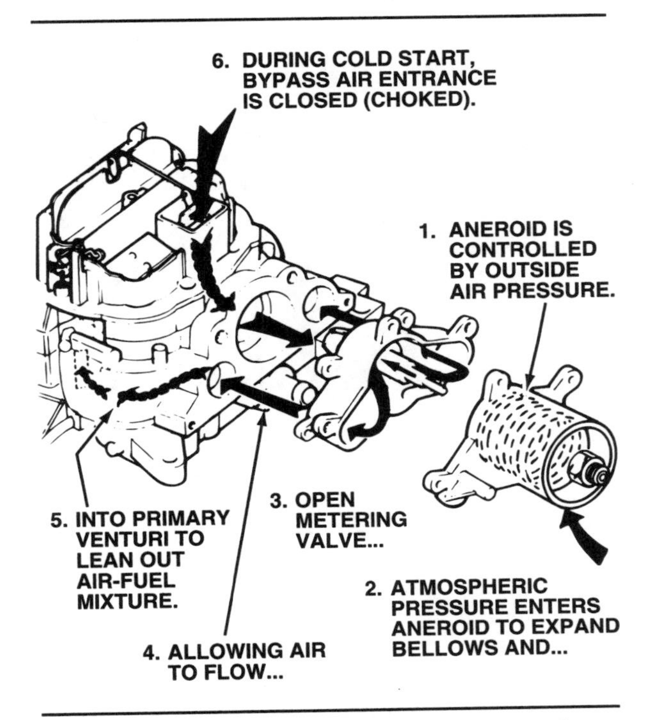

Figure 13-75. Motorcraft 2150 and 4350 automatic altitude compensator.

However, if the retuned engine is driven back to lower elevations without once again adjusting the mixture, performance will suffer. As the car descends from the higher elevation to sea level, the air-fuel mixture receives increasing amounts of air, leaning the mixture too much. If the engine is to operate properly at the lower elevation, it will have to be retuned to restore the proper air-fuel mixture. Unfortunately, it is not always possible or even desirable to tune the engine for such driving conditions. As a result, performance suffers, driveability is impaired, and emissions are excessive.

To maintain an appropriate air-fuel mixture while the car is driven in an altitude other than that for which it is adjusted, GM and Chrysler introduced the altitude compensating carburetor on some 1975 models. During the 1977 model year, the U.S. EPA designated 167 counties in 10 western states (not including California) as high-altitude emission control areas. These counties are entirely above 4,000 feet (1220 meters) in elevation. The U.S. EPA suspended these high-altitude emission control requirements in 1978, but reintroduced them later in a modified form.

Because the major problem at high altitude is CO emission due to richer mixtures, the center of the special emission controls is the carburetor. Feedback carburetors used with electronic engine control systems continuously adjust the air-fuel ratio and automatically compensate for changes in atmospheric pressure, as we will see when we study them later in this chapter. But nonfeedback carburetors used in high altitude areas require auxiliary systems or devices to provide more air or less fuel when operating at higher elevation than when operating at sea level. Most of the altitude compensating systems are automatic, responding to changes in atmospheric pressure. There are, however, systems that require manual adjustment or operation.

Automatic Compensation

The most widely used altitude compensating device is the **aneroid bellows,** an accordion-shaped bellows that responds to changes in

Aneroid Bellows: Accordion-shaped bellows that respond to changes in coolant pressure or atmospheric pressure by expanding or contracting.

■ What Is a Bar?

Seems like a simple question. A bar is a metal rod, a drinking establishment, a test that lawyers pass, or the Bureau of Automotive Repair (in California). A bar is also a unit of pressure measurement.

The term "bar" is short for barometric pressure, which means atmospheric pressure. One bar is one unit of atmospheric pressure. In the older kilogram-meter-second metric system, one standard bar is one kilogram of force applied to one square centimeter. This equals approximately 14.5 psi, which is close to the 14.7 psi of atmospheric pressure at sea level. Similar to 760 mm-Hg of atmospheric pressure, a standard bar is measured at 0°C at sea level. At 14.5 psi, one bar is close to the other units of atmospheric pressure you have learned: 750 mm-Hg or 100 kPa.

You will find fuel pressure and turbocharger pressure specifications given in bar units in many European vehicle service manuals. Using the conversion factor of 14.5 psi = 1 bar, fuel injection pressure of 2.5 bars equals 36.3 psi. Turbocharger boost pressure of 0.7 bar equals 10 psi above atmospheric pressure. (Boost pressure is the pressure increase above atmospheric pressure.)

The bar is just another way to measure the common factors of air and fuel pressure in engine operation. It is easy to switch from one measurement unit to another if you remember the conversion factors, and that all units generally represent the same quantities.

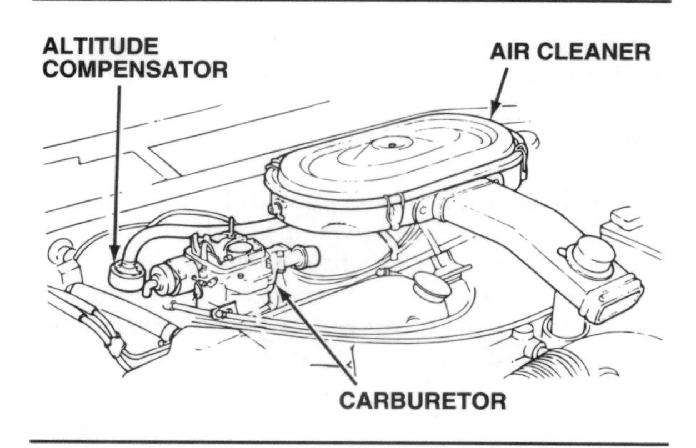

Figure 13-76. Some Ford and Chrysler engines use remote altitude compensators to bleed air from the clean side of the air cleaner into the idle fuel system.

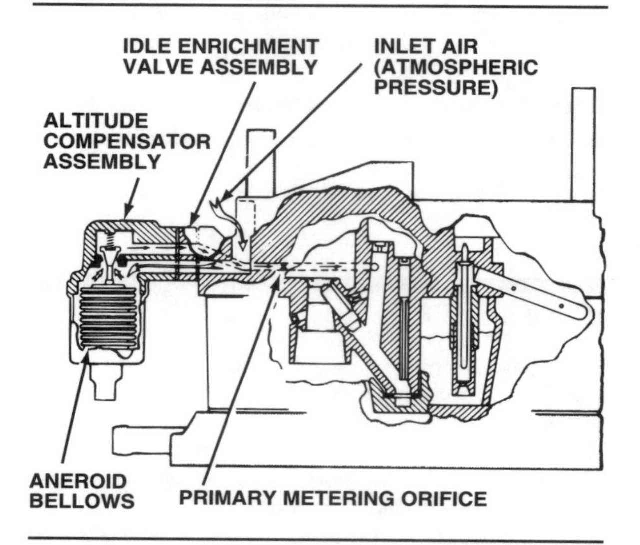

Figure 13-77. The Carter Thermo-Quad automatic altitude compensator found on Chrysler carburetors.

atmospheric pressure by expanding and contracting. As pressure decreases at high altitude, the bellows expands.

The air bypass passage on Motorcraft 2150 and 4350 altitude carburetors has its own air intake and choke valve, figure 13-75. The aneroid bellows on these carburetors have adjusting screws and locknuts that look like an adjuster for bellows tension. However, these screws were adjusted at the factory, and you should not change them while adjusting or overhauling the carburetor. You can remove the bellows and air valve for carburetor cleaning without upsetting the adjustment.

The altitude compensator used with Motorcraft 740 carburetors is a unit remotely mounted on the bulkhead, figure 13-76. It leans the mixture by drawing air from the clean side of the air cleaner filter and bleeding it into the primary

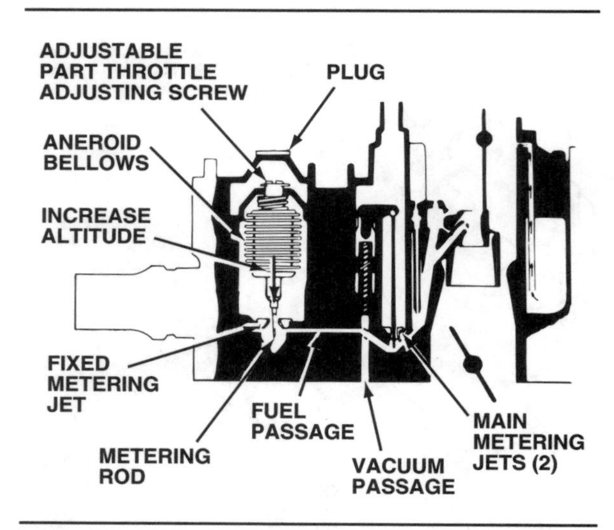

Figure 13-78. The Rochester Quadrajet automatic altitude compensator.

and idle fuel systems at altitudes above 3,000 feet (910 meters). Chrysler uses a similar system with many later-model 4-cylinder engines.

The altitude compensator on the Carter Thermo-Quad, figure 13-77, was introduced on 1975 models. It also is automatic and requires no service.

The aneroid bellows is also part of the Rochester M4MEA Quadrajet used on 1976 and later high-altitude engines. The aneroid bellows controls the position of the metering rods in the primary main jets, figure 13-78. At high altitude, the bellows expands and moves the rods into the jets to reduce fuel flow and keep the mixture from becoming excessively rich. This unit requires adjustment only when it is replaced.

Manual Compensation

High-altitude 6-cylinder engines built by AMC after the mid- and late 1970s use a Carter YF one-barrel carburetor with a manually adjusted auxiliary air bleed. The adjustment plug is located on the side of the airhorn near the fuel inlet.

Some Carter BBD two-barrel carburetors also have manually adjusted air bleeds. The adjustment screw for this air bleed is inside the airhorn above the venturi clusters.

Holley 5200 two-stage, two-barrel carburetors used on some 1977 Ford 2.3-liter engines have a driver-controlled fuel valve to compensate for altitude changes. A two-position lever (SEA LEVEL and ALTITUDE) mounted under the instrument panel controls the fuel valve, which is located in the carburetor base. A cable and swivel linkage opens and closes the fuel valve. Under normal use, the system requires no adjustment.

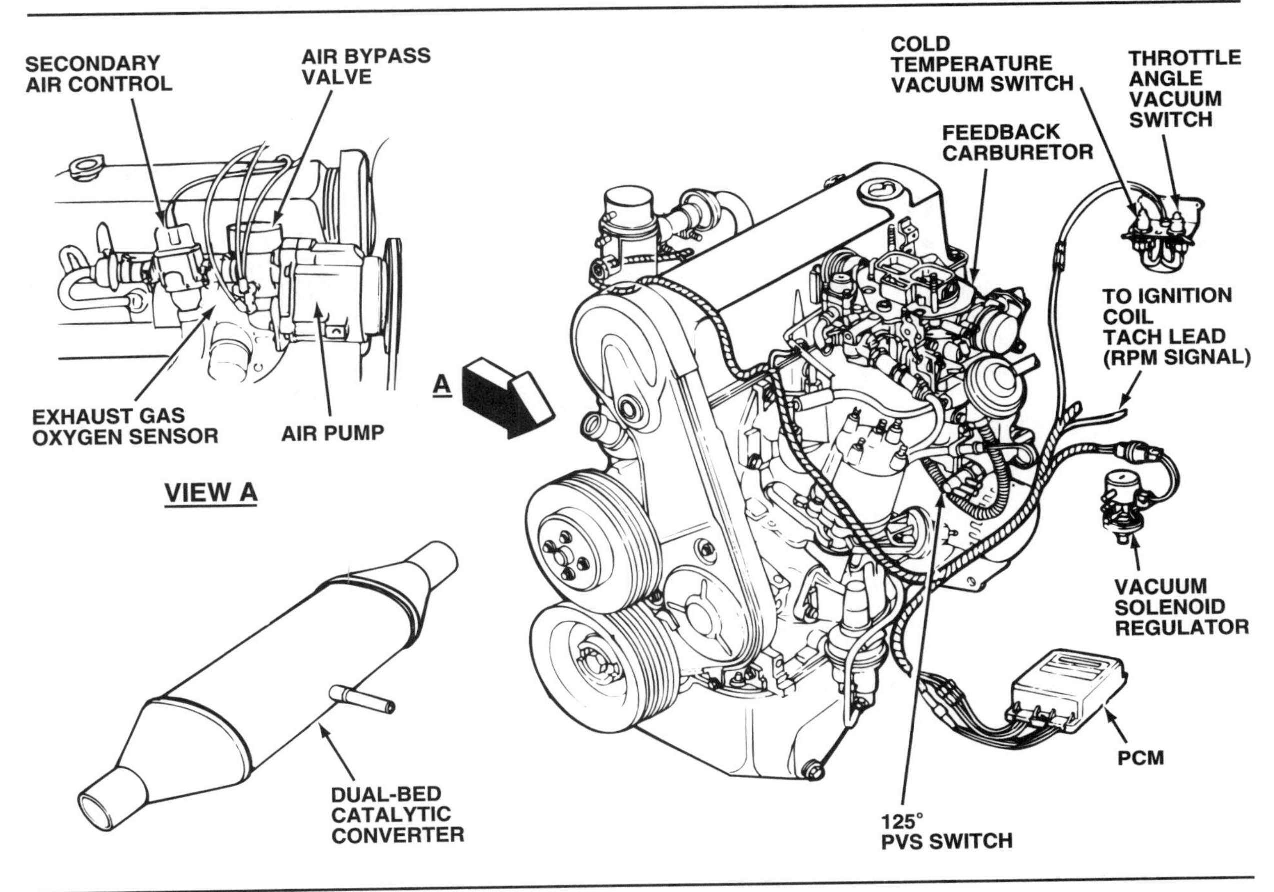

Figure 13-79. Ford's first electronic fuel management system appeared in 1978.

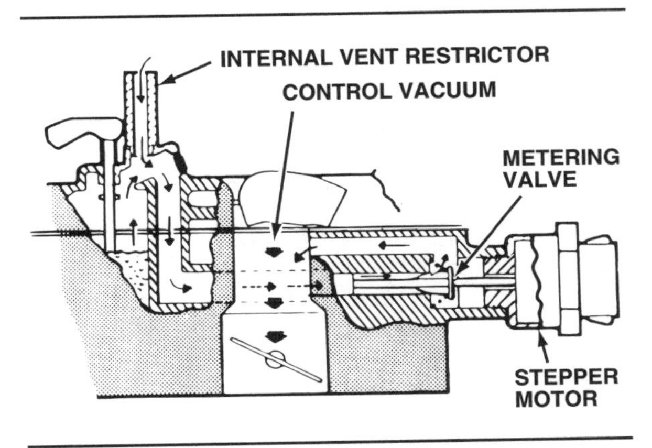

Figure 13-80. Early models of the Motorcraft 7200 VV carburetor use a stepper motor to vary air pressure on the fuel bowl.

FEEDBACK-CONTROLLED CARBURETORS

The first engine fuel management systems appeared in 1978 on California engines. Ford used the feedback carburetor electronic control system (FCECS), figure 13-79, on 2.3-liter Pinto and Bobcat engines. The system contained a modified Holley 5200 carburetor (redesignated the Model 6500) in which a metering valve and diaphragm controlled by the system's PCM replaced the power and fuel enrichment pistons. Other components were vacuum and temperature switches, a vacuum solenoid regulator, exhaust gas oxygen sensor (O2S), and a dual-bed catalytic converter.

The system was called "feedback" because the O2S installed in the exhaust manifold sent a voltage signal to the PCM indicating the amount of oxygen in the exhaust. The PCM responded to this signal by cycling a vacuum solenoid regulator on and off. The solenoid, in turn, regulated vacuum to the feedback metering valve diaphragm, which established the metering valve position, thus controlling the fuel flow into the carburetor's main well tube. This solenoid cycling action allowed the PCM to maintain the air-fuel ratio at approximately 14.7:1.

The GM electronic fuel control (EFC) system, which was introduced on 1978 California engines, contained similar parts and worked about the same to control the air-fuel ratio.

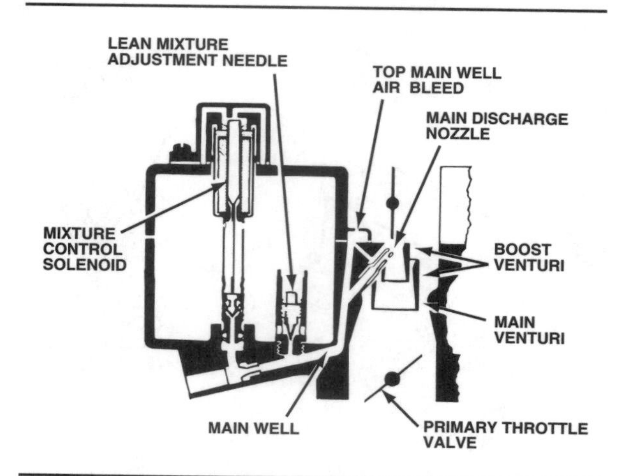

Figure 13-81. As shown on this Buick application, the mixture control solenoid in Rochester carburetors controls metering rod position.

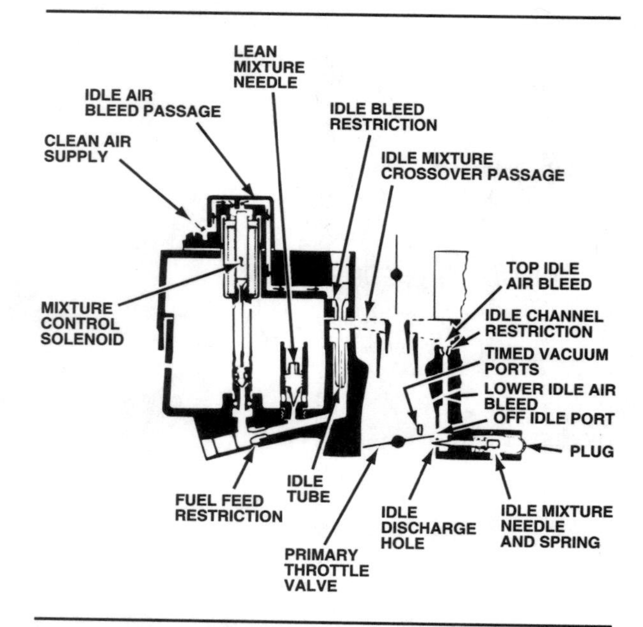

Figure 13-82. This carburetor found on a Buick engine shows that the same idle mixture solenoid may also control an idle air bleed passage.

Principles of Operation

Electronic control of fuel metering involves the ability of the PCM to turn a solenoid on and off more rapidly than any mechanical device can. This solenoid on-off sequence is called a cycle. The part of the time the solenoid is on is called the **duty cycle**. Some solenoids can operate at any number of cycles per second. A complete cycle requires a specific length of time, but the duty cycle is a variable percentage of the complete cycle.

For example, a solenoid may operate 10 times per second. Each operating cycle is thus one-tenth of a second. But of that one-tenth-second portion of the cycle, the solenoid may be on 10 percent of the time and off 90 percent, or on 90 percent of the time and off 10 percent, or any combination in between. This variable percentage of the complete cycle during which the solenoid is *on* is the solenoid's duty cycle.

The PCM uses the duty cycle to precisely meter air (air bleed control) or fuel (fuel flow control) into the mixture according to the input it receives from the O2S. Either a stepper motor or a solenoid can meter the air or fuel. The O2S tells the computer how much the mixture varies from the desired 14.7:1 ratio. The computer then signals the carburetor to correct its performance, if necessary.

This process repeats many times per second, allowing the computer to fine-tune carburetor operation and the air-fuel mixture, according to engine speed and load conditions. This makes it possible for the catalytic converter to work effectively to reduce exhaust emissions. Since the carburetor constantly responds to information about its past performance (feedback information from the O2S to the PCM), it is called a feedback carburetor.

The computer does not control the air-fuel mixture under all conditions at all times through the duty cycle. For example, when the engine is first started, the fuel management system operates in an **open-loop** mode. This means that the computer ignores any signals from the O2S and operates the stepper motor or solenoid with a programmed, fixed duty cycle. Various fixed duty cycles may be programmed into the computer for open-loop conditions, ranging from cold starts to hot wide-open throttle acceleration. When certain conditions are met, such as normal engine coolant temperature, the microprocessor switches system operation into the **closed-loop** mode. At this time, it evaluates the O2S signals and varies the duty cycle of the stepper motor or solenoid to manage the air-fuel ratio. You will learn more about the operation of an engine management system in Part Four.

Early Stepper Motors

A **stepper motor** is used with some feedback carburetors. This dc motor moves in a specified number of incremental steps (usually 100 to 120) according to the voltage applied. Each step is tiny (approximately 0.004 in. or 0.1 mm), with motor speed varying from 12 to 100 steps per second.

The stepper motor is connected to tapered metering pins positioned either in air bleed orifices or in the main metering jets. When power is

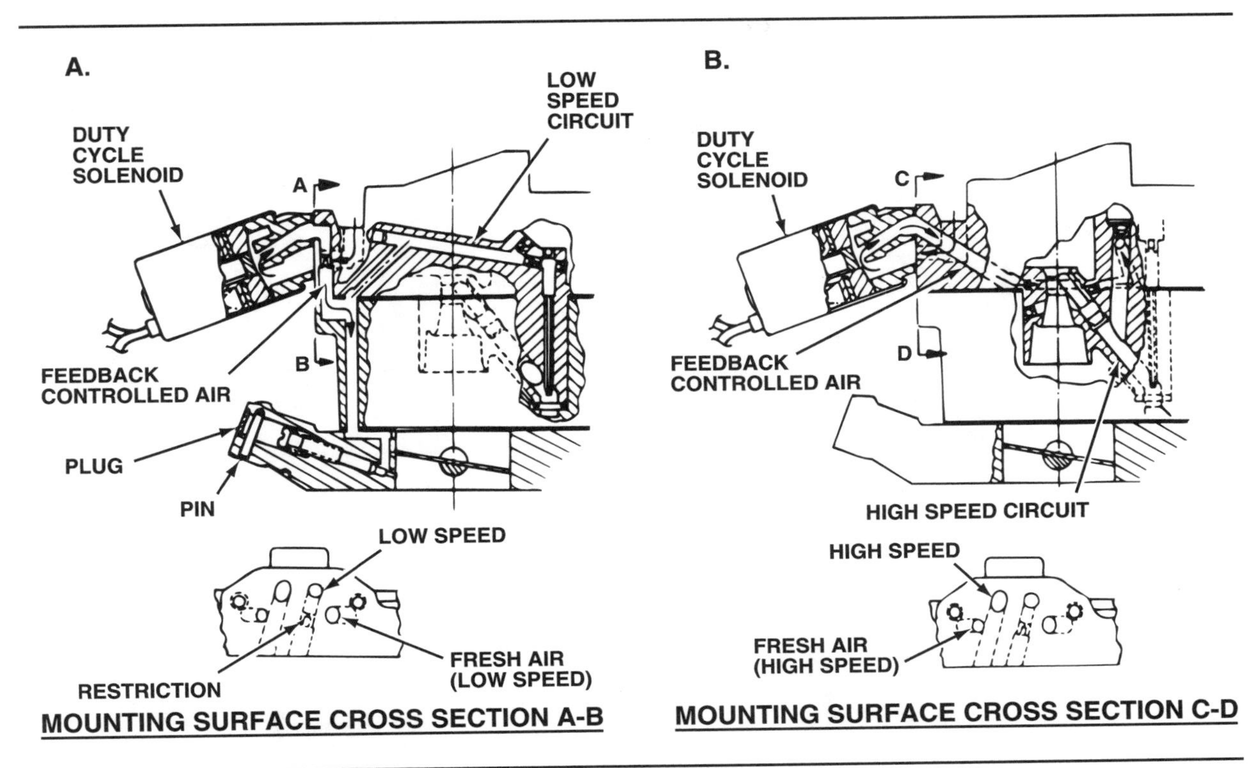

Figure 13-83. The Chrysler Thermo-Quad pulse solenoid opens and closes low-speed and high-speed air bleeds.

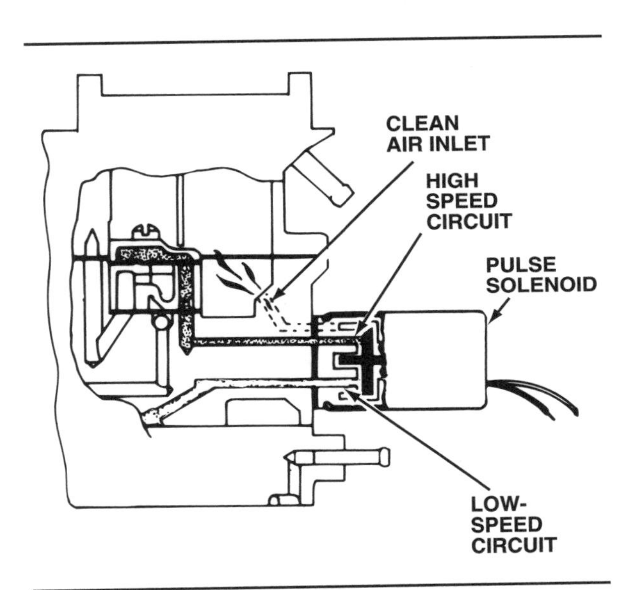

Figure 13-84. This Carter BBD pulse solenoid bleeds air to the low- and high-speed circuits.

applied to the stepper motor, it moves the metering pins inward (rich position) to an end stop. This gives the engine control computer a stable reference. The stepper motor then backs the pins out to a position calculated to deliver a 14.7:1 air-fuel ratio, and the computer takes over its operation. This is called initialization.

Early Motorcraft 7200 VV and some Carter carburetors used stepper motors. In the 7200 VV, the stepper motor moves a valve that lets control vacuum into the fuel bowl. Since this lowers the pressure above the fuel bowl, less fuel is pushed into the main metering system, figure 13-80. The result is a leaner air-fuel mixture. This design is sometimes called the "backsuction" system. The

Duty Cycle: The percentage of the complete cycle during which the solenoid is on, or the ratio of pulse width to complete cycle width.

Open Loop: Operational mode when the fuel delivery changes according to sensors other than the exhaust oxygen sensor (O2S).

Closed Loop: Operational mode when the fuel delivery changes according to changes in voltage from the O2S.

Stepper Motor: A direct-current motor that moves the armature in incremental movements or steps. When the field coils are energized, magnetic force can rotate the armature clockwise or counterclockwise.

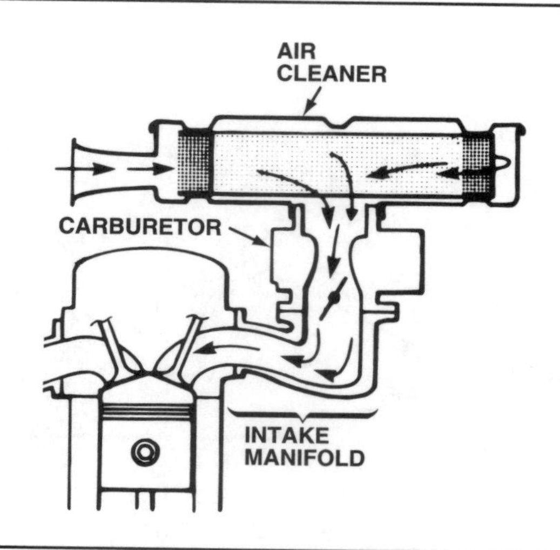

Figure 13-85. Intake manifold passages route the air-fuel mixture from the carburetor (or throttle body) to the intake valve ports.

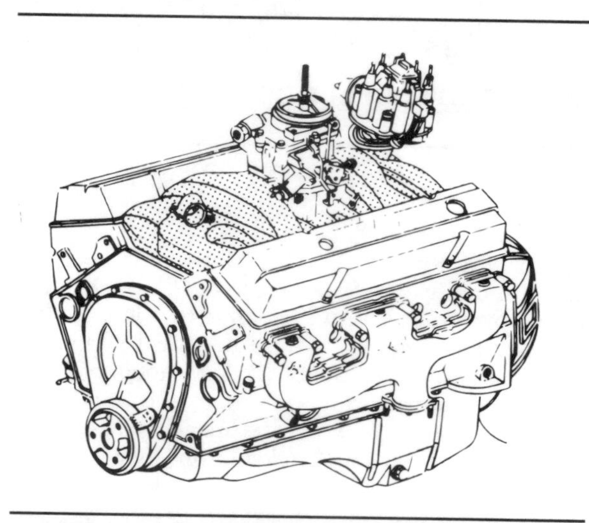

Figure 13-86. As shown on this Chevrolet engine, V-types normally have the intake manifold (shaded area) between the two banks of cylinders.

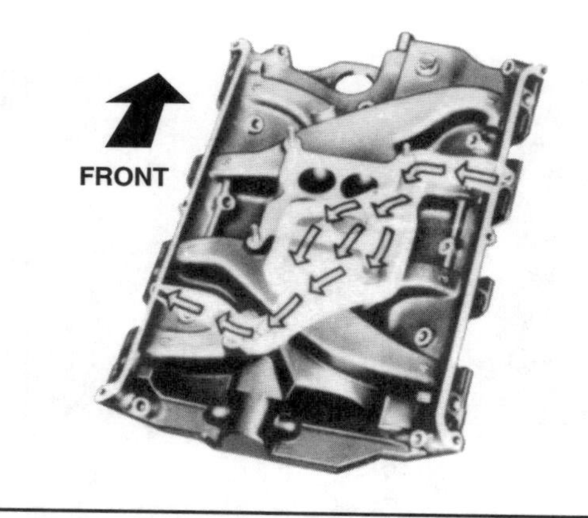

Figure 13-87. This Ford intake manifold illustrates that exhaust gases are routed from ports in the cylinder heads through separate passages in the manifold to form the manifold hot spot.

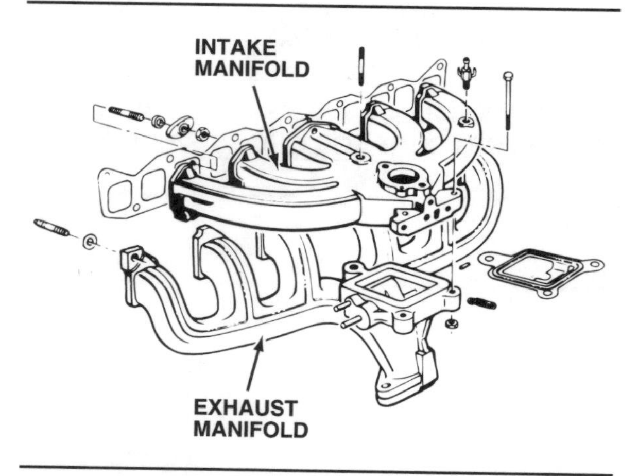

Figure 13-88. As shown on this Chrysler application, most in-line engines have both manifolds on one side of the engine.

Carter carburetors use a stepper motor to position the metering pins in the air bleeds. This controls the amount of air in the air-fuel ratio.

Mixture Control Solenoid

Mixture control (MC) solenoids control the air-fuel ratio in various ways, but all work on the variable duty cycle principle just discussed. The MC solenoid installed in Rochester carburetors regulates fuel flow directly by controlling a metering rod in the main jet, figure 13-81. The MC solenoid also controls a rod, figure 13-82, that opens and closes an idle air bleed.

Some Carter Thermo-Quad carburetors use the variable duty cycle of an MC solenoid to open and close the low-speed air bleed, figure 13-83A, and the high-speed air bleed, figure 13-83B. The Carter BBD carburetor uses a similar solenoid, figure 13-84, that bleeds air to the low-speed and high-speed circuits when energized, or shuts off air to both circuits when deenergized. Since solenoids used in this application are either on or off, they are called pulse solenoids.

Vacuum Control Solenoids

Other carburetors may use an external solenoid that controls vacuum to an internal diaphragm in the carburetor. The internal diaphragm may regulate fuel flow by changing the position of the metering rods in the carburetor jets. It can also open and close air bleeds used in the idle and main metering circuits.

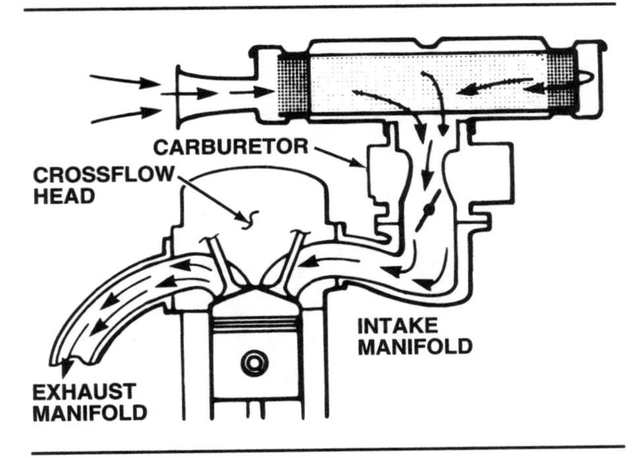

Figure 13-89. This Chrysler engine shows that a few in-lines have the two manifolds on opposite sides of the engine.

INTAKE MANIFOLD PRINCIPLES

The vaporized air-fuel mixture flowing through the carburetor or throttle body must be evenly distributed to each cylinder. The intake manifold does this with a series of carefully designed passages that connect the carburetor or throttle body with the engine's intake valve ports, figure 13-85. To do its job, the intake manifold must provide efficient vaporization and air-fuel delivery.

As you have learned, liquid gasoline is composed of various HC's that vaporize at different temperatures. If all the HC's in gasoline vaporized at the same rate, the intake manifold function would be simple. But they do not, so the intake manifold needs heat to keep the air-fuel mixture properly vaporized. Exhaust manifold heat, engine coolant heat, or both, depending on engine design, can heat the intake manifold. Some engines also use electric grid heaters.

V-type engines have the intake manifold in the valley between the cylinder banks, figure 13-86. An exhaust crossover passage inside the manifold carries hot exhaust gases near the base of the carburetor, figure 13-87. In a few manifolds, engine coolant is routed through the manifold near the carburetor or throttle body to heat the air-fuel mixture.

Most in-line engines have both manifolds on the same side of the engine, with the intake manifold on top of the exhaust manifold, figure 13-88. A chamber between the two manifolds fills with exhaust gases and creates a hot spot to improve fuel vaporization in the intake manifold passages. Some in-line engines have the two manifolds mounted on opposite sides of the cylinder head, figure 13-89. In this crossflow design, coolant passages and a heat jacket on the intake manifold supply the heat.

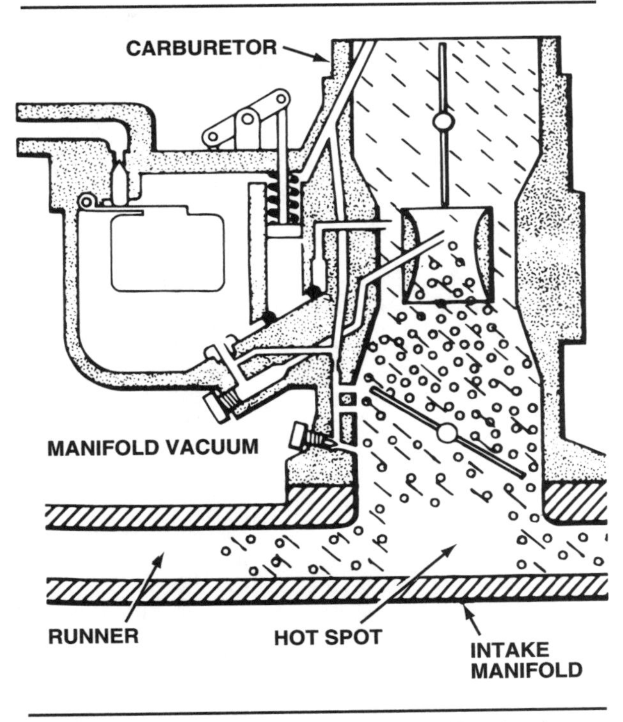

Figure 13-90. Intake manifold runners affect intake airflow velocity and temperature.

Even with the help of heat from the exhaust manifold engine coolant, the air-fuel mixture usually does not completely vaporize in the intake manifold. This results in unequal mixture distribution among the cylinders, with some cylinders receiving more fuel and developing more power than others. This problem is greater during engine warmup, when less-than-normal heat is available to vaporize the fuel.

Intake Manifold Design

Intake manifold design for a carburetor has a direct bearing on air-fuel mixture distribution and volumetric efficiency over the speed range of an engine. Both velocity and heating are also affected by the size of the manifold passages, or **runners**, through which the mixture must travel, figure 13-90. If the passages are large, the mixture travels slowly at low-engine rpm, letting fuel particles cling to the manifold wall and avoid vaporization. Small passages create a higher velocity but restrict the volume of the mixture that can pass through at high-engine rpm. The

Runners: The passages or branches of an intake manifold that connect the manifold plenum chamber to the engine inlet ports.

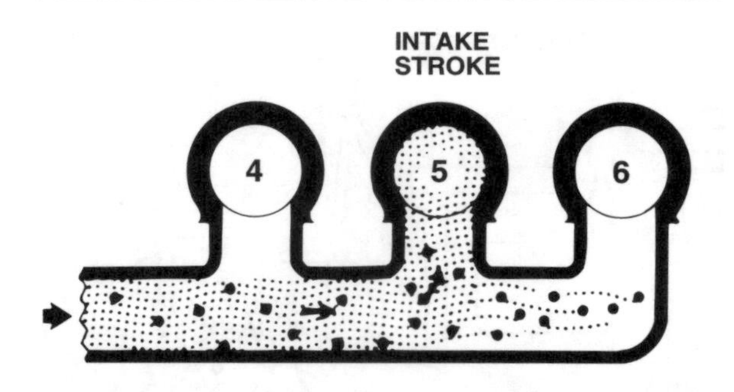

Figure 13-91. As shown on this Chevrolet manifold, an intake manifold with large passages and sharp angles will cause liquid fuel to separate out of the air-fuel mixture.

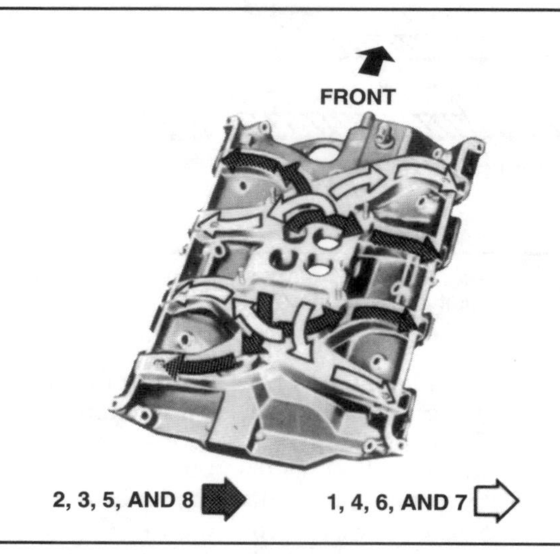

Figure 13-92. For a carbureted engine, this V8 has good air-fuel distribution because of the intake manifold design.

angles at which internal manifold passages turn are also critical, figure 13-91. When they are too sharp, fuel tends to separate out of the mixture, condensing and puddling in the runners.

The shape, interior surface, and size of the manifold runners determine manifold efficiency. Manifold design is usually a compromise—short, large-diameter runners produce the best power at high rpm, while long, smaller-diameter runners produce the best power at low to medium rpm. Passages must be just large enough in diameter to supply all cylinders with equal amounts of mixture at the necessary rpm. Too-large or too-small passages reduce efficiency. Manifold runners should be without sharp corners, bends, or turns to interfere with mixture flow. In carburetor manifolds the ports may be slightly rough to aid vaporization by reducing puddling and breaking up mixture flow.

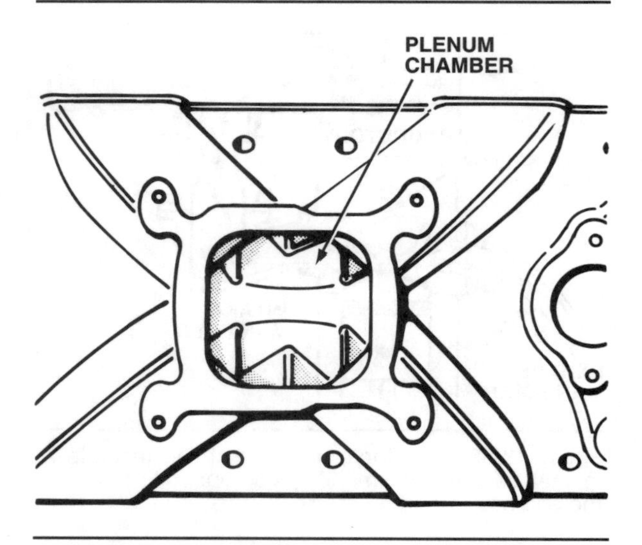

Figure 13-93. The floor of this single-plane manifold is shaped to equalize the mixture flow to all cylinders.

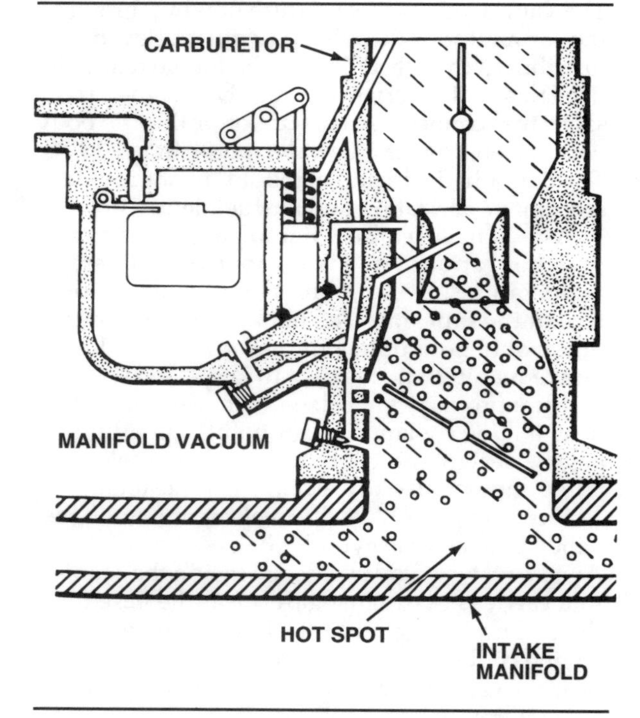

Figure 13-94. The angle of the carburetor throttle plate can affect the flow of the air-fuel mixture above the hot spot.

The air-fuel mixture should be distributed as evenly as possible among the cylinders. Figure 13-92 shows a carbureted V8 intake manifold design that promotes good distribution to all cylinders. If one or more cylinders receives a mixture that is too lean, the overall mixture must be richer for that cylinder to fire properly. In this case, the other cylinders receive a mixture

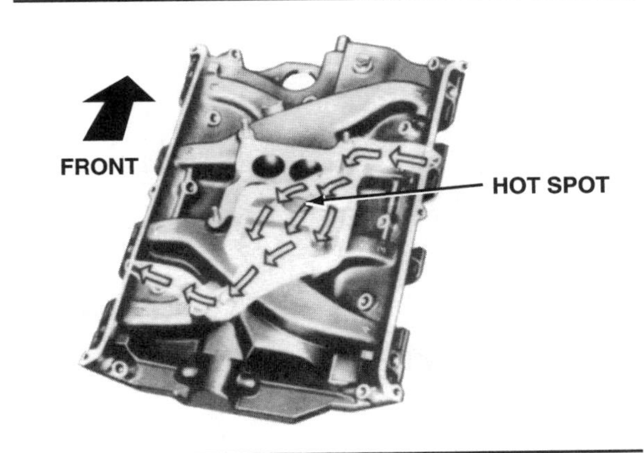

Figure 13-95. An exhaust passage near the base of the carburetor or throttle body creates a "hot spot" in the intake manifold.

that is too rich. Overly lean combustion produces oxides of nitrogen (NO_x), while overly rich combustion produces unburned HC's and CO. Neither condition is desirable, since they raise emissions and lower fuel economy.

Manifolds usually have an induction **plenum** chamber, a storage area for the air-fuel mixture that stabilizes the incoming air charge and actually allows it to rise slightly in pressure, figure 13-93. The mixture accumulates within the plenum until one of the cylinders draws it out. This provides a uniform mixture charge that can be distributed equally to each cylinder, unaffected by momentary variations in throttle position and intake turbulence. The plenum feeds the mixture into runners leading to the intake ports.

Unequal distribution and manifold design

The air-fuel mixture reaching the engine cylinders may vary between cylinders in amount and ratio for several reasons:

- The mixture flow is directed against one side of the manifold by the throttle valve in the carburetor, figure 13-94; to some extent, the carburetor choke plate can influence flow in a similar manner.
- Smaller liquid particles of the air-fuel mixture turn corners in the manifold more easily, while the larger particles tend to continue in one direction.
- Cylinders closer to the carburetor receive a richer mixture than those farther from the carburetor or throttle body.
- Charge-robbing can result between two cylinders that are located close together in the block and with intake strokes that occur at nearly the same time; the cylinder with the earlier intake stroke draws in a balanced mixture, but the later intake stroke draws a weaker, leaner mixture because less fuel is available.

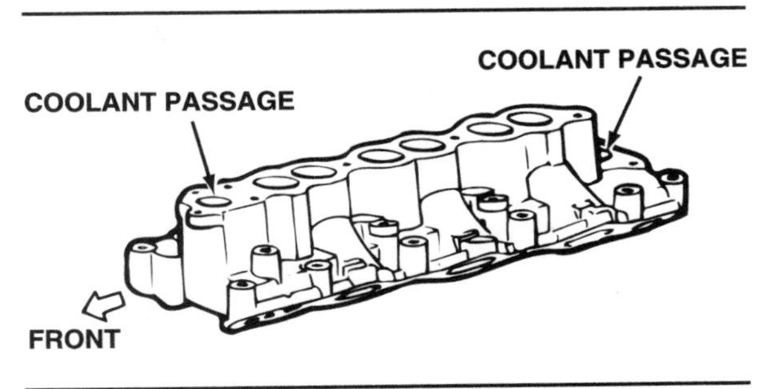

Figure 13-96. When the engine is warmed up, hot coolant in the coolant passages warms the intake manifold to improve fuel vaporization.

The tendency for cylinders closer to the carburetor to receive a richer mixture can be minimized by careful carburetor placement and good manifold design. Often, using a two- or four-barrel carburetor can improve distribution because each barrel or bore supplies fewer cylinders. Charge-robbing can be minimized by careful manifold design; however, the only real cure is port fuel injection.

Manifold and intake charge heat

As you have learned, liquid gasoline is composed of various HC's that vaporize at different temperatures. If all the HC's in gasoline vaporized at a fast rate, the job of the intake manifold would be simple. But they do not, so it is essential that the intake manifolds of engines with carburetors be heated to keep the air-fuel mixture properly vaporized.

Intake manifolds are heated with exhaust gas heat, engine coolant heat, exhaust manifold heat, or all three, depending on engine design. On V-type engines an exhaust crossover passage inside the manifold carries hot exhaust gases near the base of the carburetor, figure 13-95. In-line engines with both manifolds on the same side of the engine, position the intake manifold on top of the exhaust manifold. A passage between the two manifolds fills with exhaust gases to improve fuel vaporization in the intake manifold passages. On in-line engines with crossflow cylinder heads, a heat jacket on the intake manifold supplies the heat.

The area in the intake manifold heated by exhaust is called the "hot spot". The location, size, and surface area of the hot spot on the manifold floor affects vaporization. The hot spot is primarily a cold start feature. Once the manifold warms up, fuel vaporizes all through it.

Plenum: A chamber that stabilizes the air-fuel mixture and allows it to rise to slightly above atmospheric pressure.

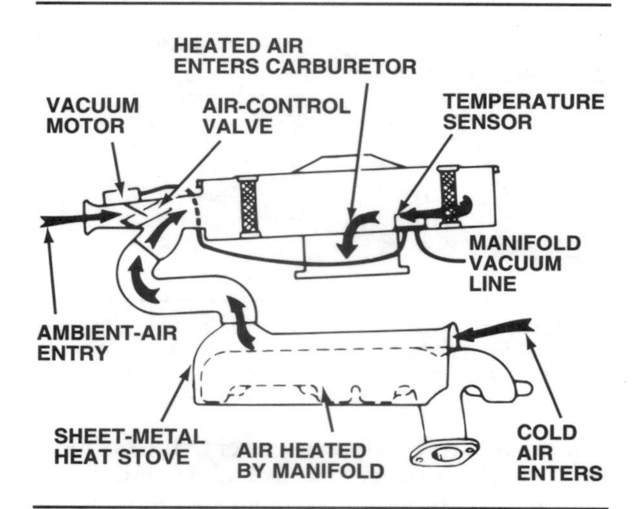

Figure 13-97. Heat from the exhaust manifold can be fed to the intake system as needed to warm the intake air.

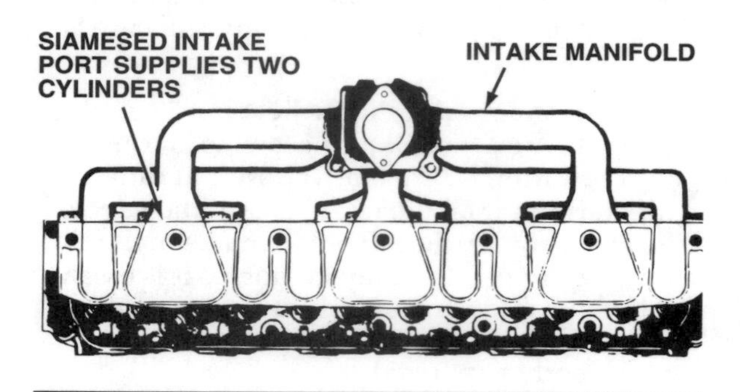

Figure 13-98. As shown on this Pontiac siamesed-ported manifold, this 6-cylinder engine requires only three intake manifold runners.

While exhaust gas heat is only used on manifolds for carburetors, many manifolds for all types of induction systems route engine coolant through the manifold, figure 13-96. Engine coolant provides little manifold heat during a cold startup. Its job is to provide manifold heat and improve vaporization at all times while the engine is warmed up and running.

Most carbureted engines also use an intake air temperature control system, figure 13-97. The heat radiating off the exhaust manifold is fed to the intake to warm the inlet air and help vaporize the fuel just after a cold startup. When the engine warms up, a thermostat closes the passage from the exhaust manifold to the intake. The thermostat constantly cycles to keep the intake air at around 125°F (52°C).

Even with the help of heat from the exhaust and engine coolant, a small amount of the air-fuel mixture usually does not completely vaporize in the intake manifolds of carbureted engines.

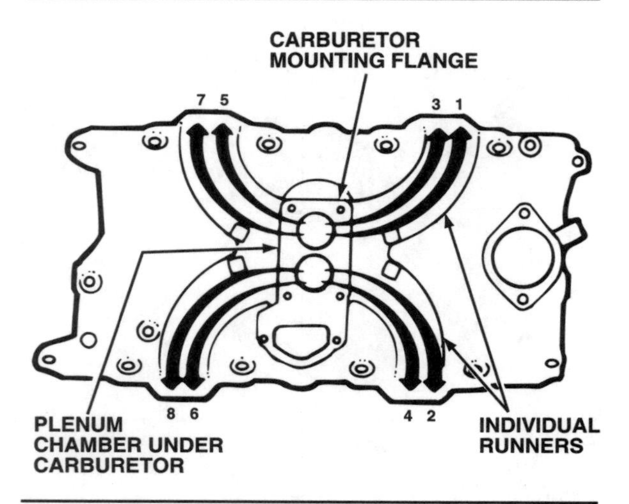

Figure 13-99. This Chrysler single-plane manifold feeds eight cylinders from a single plenum chamber.

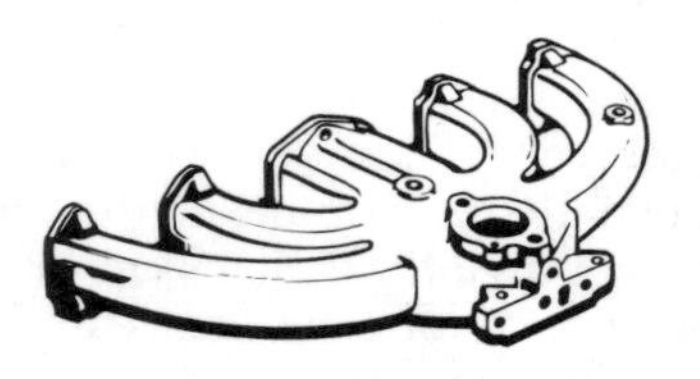

Figure 13-100. This Chrysler intake manifold has curved runners for better mixture distribution.

This results in slightly unequal mixture distribution among the cylinders, with some cylinders receiving more fuel and developing more power than others. This problem (although quite small) is greater during engine warmup, when less-than-normal heat is available to vaporize the fuel.

BASIC INTAKE MANIFOLD TYPES

Passenger car intake manifolds are made of cast iron or aluminum. The exact design and number of outlets to the engine depends on the engine type, number of cylinders, fuel delivery, and valve port arrangement. Cylinders may have individual intake and exhaust ports, or two cylinders can be supplied by a single port — called a **siamesed port**, figure 13-98. Siamesed intake ports are common on in-line engines but rare in V-type engines.

Manifold Planes

Production intake manifolds for cars are classified as either single-plane or dual-plane designs. A single-plane manifold, figure 13-99, uses short runners to connect all the engine's inlet ports to

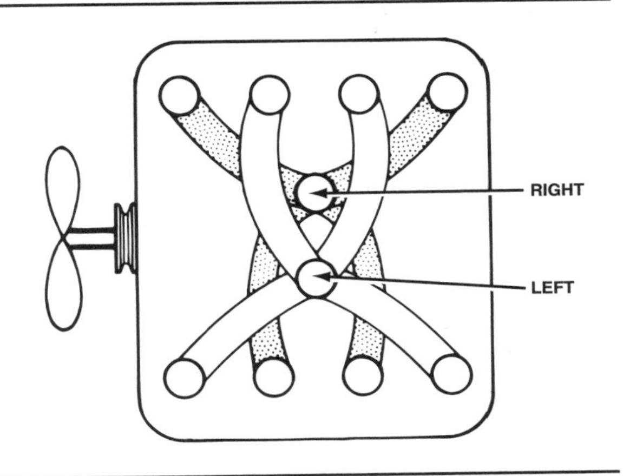

Figure 13-101. As illustrated by AC-Delco, the two chambers of a dual-plane manifold are of different heights.

a single chamber. This chamber, called the plenum chamber, is simply a storage area for the air-fuel mixture. The mixture accumulates within the chamber until one of the cylinders draws it out. The single-plane design permits more equal cylinder-to-cylinder mixture distribution at high rpm, but its large internal volume causes a drop in mixture velocity, reducing airflow at low-to-intermediate speeds. An engine with a single-plane manifold can produce more horsepower at high rpm than one with a dual-plane manifold because each cylinder can draw mixture from all throttle bores.

In-line engines use a single-plane manifold. The carburetor usually is in the middle of the manifold. Cylinders closer to the carburetor usually receive too much air in the mixture, while those farther away receive too much fuel. To address this problem, in-line engines may use siamesed ports, requiring fewer manifold runners. In addition, Chrysler uses in-line 6-cylinder engines that do not have siamesed intake ports, but intake manifolds having long, curved runners with no sharp turns, figure 13-100.

V-type engine intake manifolds may be either single- or dual-plane, but dual-plane designs are more common. A dual-plane intake manifold has two separate plenum chambers connected to the intake ports of the engine, figure 13-101. Each chamber feeds two central and two end cylinders. Mixture velocity is greater at low-to-intermediate engine speeds than with the single-plane manifold, but mixture distribution is usually even less than that produced by the single-plane manifold, especially at high speed. When a four-barrel carburetor is used with a dual-plane manifold, each side of the carburetor (one primary and one secondary barrel) feeds one plenum chamber. When a two-barrel carburetor is used, each barrel also feeds one chamber.

Manifolds for Spread-Bore Carburetors

Until emission control became an important factor in automotive design, all four throats of a four-barrel carburetor were the same size. Most later-model four-barrel carburetors are a spread-bore design, with smaller primaries and larger secondaries, figure 13-102. The way the flow goes through spread-bore designs makes it difficult to adapt these carburetors to manifolds not designed for them. It is generally true that dual-plane manifolds work best in the low-to-middle speed ranges, and single-plane designs work better at higher speeds. Engineers have discovered, however, that a single-plane manifold can be designed to incorporate the best features of both types. The design, length, and arrangement of the runners are all important in doing this.

A spread-bore carburetor and a dual-plane manifold distribute a fairly even mixture to all eight cylinders. In effect, each group of four cylinders is fed by similar but independent two-stage, two-barrel carburetors. When the same

Siamesed Port: On an intake manifold, a single port that supplies the air-fuel mixture to two cylinders.

■ The Torque Box

Designing manifolds for production engines is a series of compromises dictated by engine design, emission requirements, driveability, manufacturing costs, and various other factors. In general, the dual-plane manifold provides better throttle response, a broader torque curve, and higher maximum torque than a single-plane design. The single-plane manifold is responsible for high maximum horsepower (more efficiency) at high rpm, but is not good at idle, and can contribute to icing during cold weather.

An attempt to combine the best features of both designs resulted in what engineers call a "torque box". This design uses the long runners from the dual-plane manifold with the plenum chamber of the single-plane design. The result is good torque characteristics and higher maximum horsepower.

The best known example of a torque box used on production cars is the dual four-barrel manifold fitted to early Z28 Camaros.

The torque box became one of the most common racing manifold designs. Its use on production engines was limited by stricter emission requirements, the use of smaller and less powerful engines, and the emphasis on fuel economy.

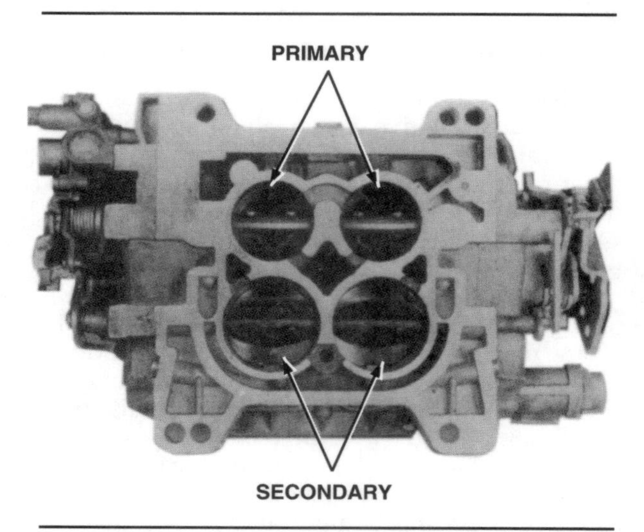

Figure 13-102. A spread-bore carburetor, showing the size difference between the primaries and the secondaries.

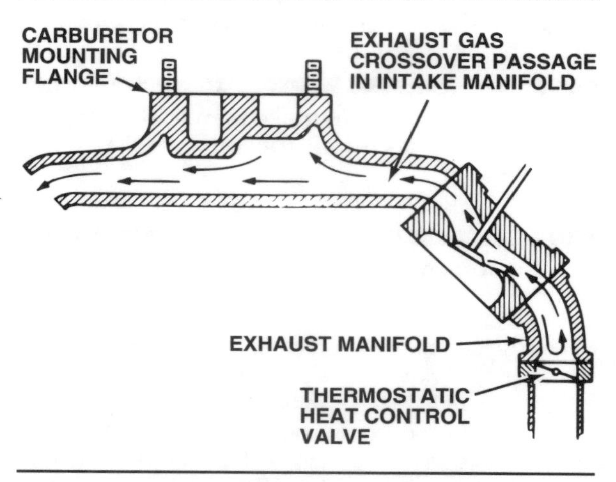

Figure 13-103. A thermostatic heat control valve or heat riser can be installed between the manifold and the exhaust pipe.

carburetor is fitted to a single-plane manifold, mixture flow tends to seek the shortest path. Each carburetor throat ends up feeding the two closest manifold runners. This means that four cylinders are fed by the larger secondaries, and the other four are fed by the smaller primaries. A lean mixture reaches the front four cylinders, and a rich mixture is fed to the rear four cylinders.

To use a spread-bore carburetor on a single-plane manifold and equalize the air-fuel flow, the plenum chamber floor is grooved and ridged. This speeds up the flow to the front cylinders closest to the primaries, and slows down the flow to the rear cylinders closest to the secondaries. This is the most efficient single-plane manifold design to use with a spread-bore carburetor.

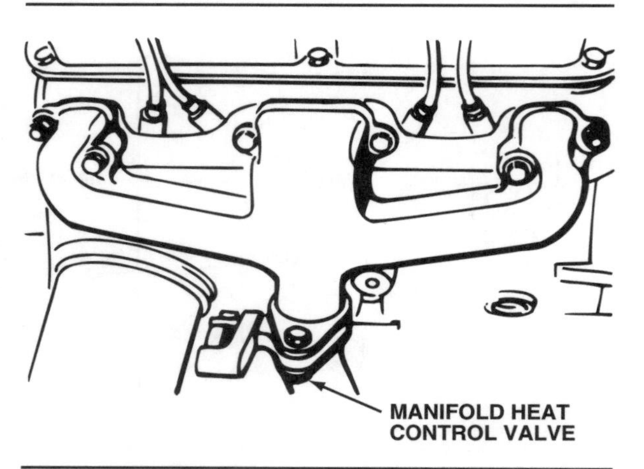

Figure 13-104. The heat control valve can also be within the manifold itself.

INTAKE MANIFOLD HEAT CONTROL

When the engine is cold, the incoming air-fuel mixture must be heated for complete vaporization. Hot exhaust gases are the most efficient heating source. A heat control valve, or heat riser, in the exhaust system routes a small part of the exhaust through passages in the intake manifold to provide heat. Heat control valves can be operated by a thermostatic spring or vacuum diaphragm. Each of these reacts to engine heat to control the valve operation, as you will see.

Thermostatic Heat Control Valve

Used by automakers for decades, the thermostatic heat control valve is located between the exhaust manifold and exhaust pipe, figure 13-103, or in the manifold itself, figure 13-104. A thermostatic spring holds the valve closed when the engine is cold, figure 13-105. This directs the hot exhaust gases around the intake manifold to preheat and help vaporize the air-fuel mixture, figure 13-106A. As the engine warms up, the thermostatic spring unwinds, and the counterweight and pressure of exhaust gas open the valve. Exhaust gases now pass directly out through the exhaust system, figure 13-106B.

The thermostatic heat control valve has a tendency to stick because the exhaust gases corrode it. If the valve sticks open, it can cause increased fuel consumption, poor performance during warmup, and excessive emissions because the choke remains on. If the valve sticks closed, it can cause poor acceleration, a lack of power, and poor high-speed performance.

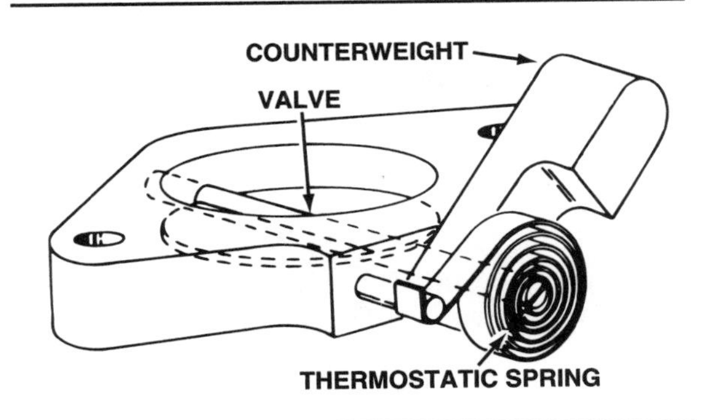

Figure 13-105. The manifold heat control valve can be operated by a thermostatic spring and a counterweight.

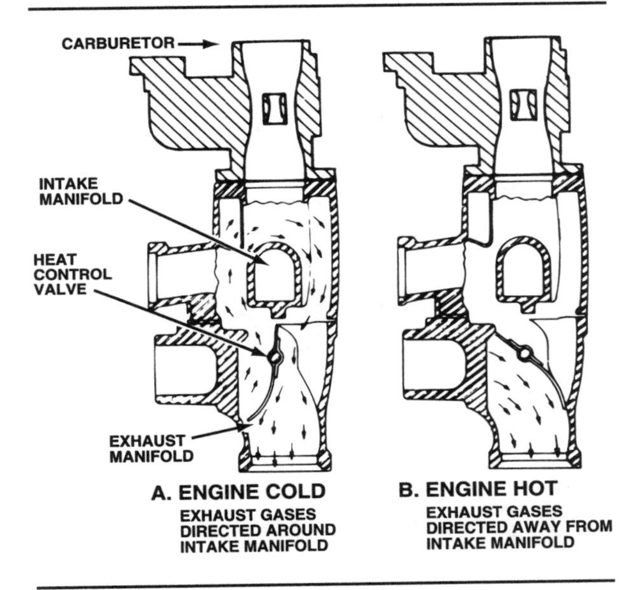

Figure 13-106. In this cross section of an in-line engine manifold, the heat control valve forces the exhaust gases to flow either around the intake manifold (A), or directly out the exhaust system (B).

Vacuum-Operated Heat Control Valve

General Motors and Ford introduced vacuum-operated manifold heat control valves, figure 13-107, on some 1975 engines. Chrysler followed in 1977 with vacuum-operated valves on its California V8 engines. GM calls its device a vacuum-servo early fuel evaporation (EFE) valve; Ford, a vacuum-operated heat control valve (HCV); and Chrysler, a power heat control valve. All work in the same way to provide more precise control of the manifold heat and to reduce emissions while improving driveability.

A rotating flapper valve is contained in a cast-iron body, figure 13-108. This valve body is installed between the manifold and the exhaust pipe. The valve shaft extends through the valve body and is linked to a diaphragm in a vacuum

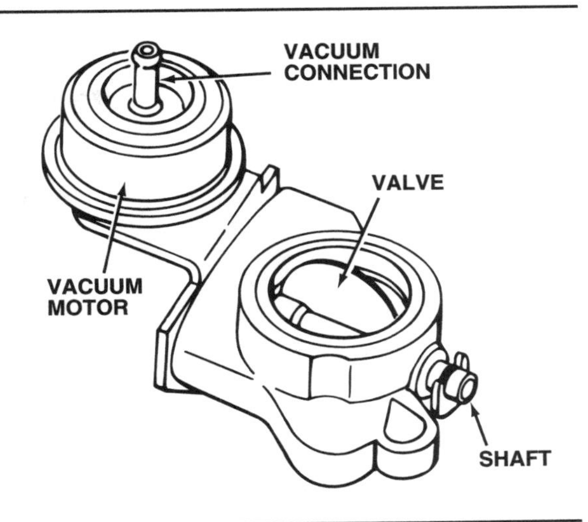

Figure 13-107. A vacuum-operated heat control valve.

motor. Intake manifold vacuum operates the vacuum diaphragm. The manifold vacuum source is controlled by a switch that reacts to either coolant or oil temperature. Ford, Chrysler, and some GM engines have a thermal vacuum valve (TVV) installed in the cooling system.

When an engine is cold, vacuum applied to the HCV diaphragm closes the valve. This directs exhaust gas through the intake manifold passage. As the engine reaches normal operating temperature, vacuum is shut off from the valve. In systems using a TVV, this valve closes when coolant temperature reaches a specified level, figure 13-109, to block vacuum from the valve. Where an oil temperature switch is used, the rising oil temperature opens a normally closed thermostatic switch. This opened switch deenergizes the vacuum solenoid, and a spring in the diaphragm opens the valve. Exhaust now flows through the exhaust system instead of bypassing to the intake manifold.

Coolant-Heated Intake Manifolds

Engine exhaust is not always used to heat intake manifolds. In some designs, hot engine coolant circulating through passages in the intake manifold preheats the air-fuel mixture, figure 13-110. A thermostat shuts off coolant flow to remove manifold heat when the engine is at normal operating temperature.

Electric Grid Heaters

A rubber insulator containing a ceramic heater grid is installed between the carburetor and the intake manifold of many late-model engines to improve cold engine driveability, figure 13-111. General Motors also calls this device an early fuel evaporation (EFE) system.

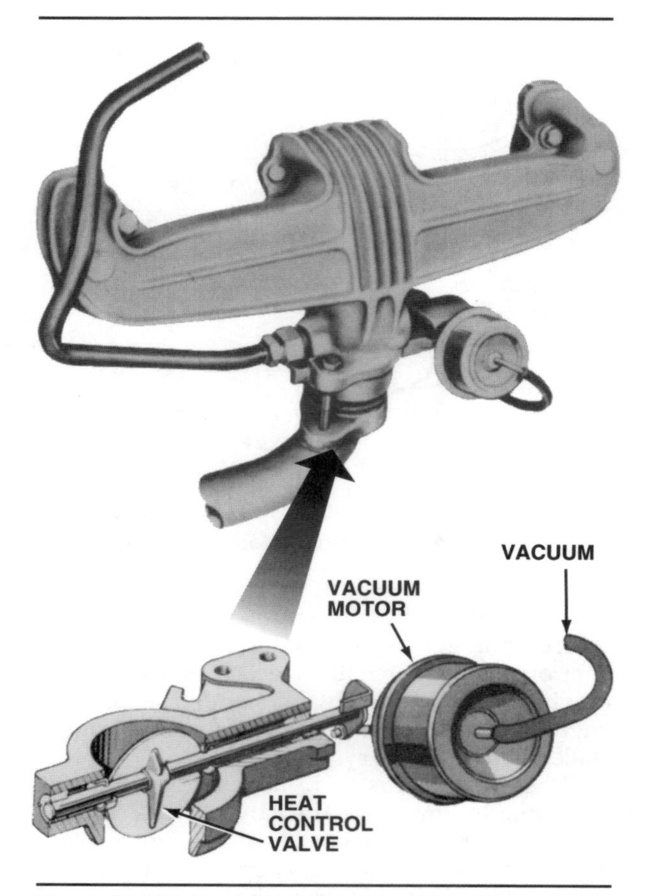

Figure 13-108. Here, Chrysler shows a vacuum-operated heat control valve installed in the exhaust manifold.

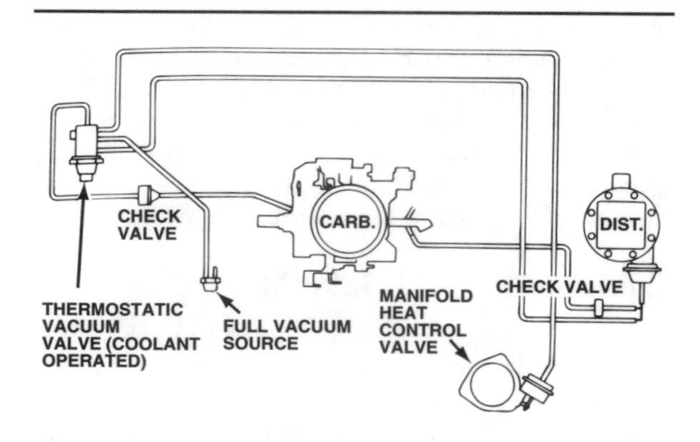

Figure 13-109. Buick's vacuum diagram of a heat control valve installation. This valve responds to engine coolant temperature.

The ceramic heater grid is underneath the primary throttle of the carburetor. When engine coolant is below a specified temperature, the PCM sends electrical current through a relay to the grid, or a TVV sends current to the grid on engines that are not computer-controlled. The heater grid temperature is self-regulating at a calibrated value and remains on until the engine coolant reaches a specified temperature, at which time current to the grid is shut off.

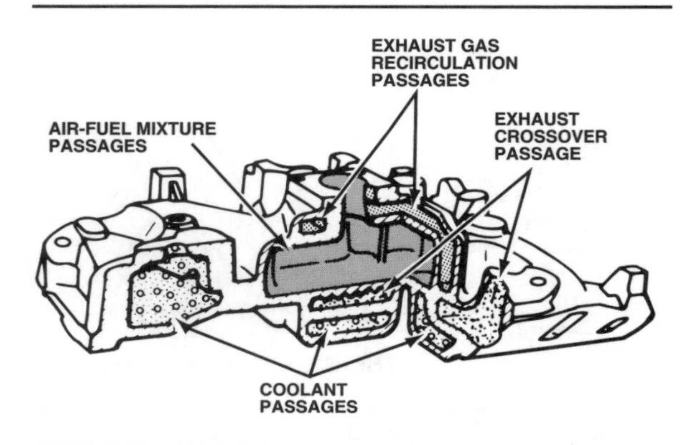

Figure 13-110. Here, Chrysler illustrates how hot engine coolant can also be used to preheat the air-fuel mixture.

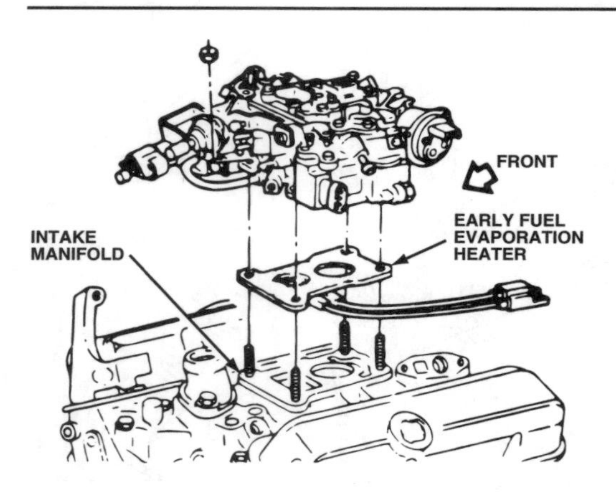

Figure 13-111. As shown on this Buick application, an electric heater grid can be installed between the carburetor and the intake manifold to improve cold engine driveability.

SUMMARY

Into the mid-1980s, the carburetor was the all-important device that converted air and gasoline into an air-fuel mixture that could burn in the cylinders. Carburetors must operate under all types of conditions and in all temperatures. Although there are dozens of carburetor designs available, all operate on the same basic principle of pressure differential. Pressure differential is the difference in air pressure between the relatively high pressure of the atmosphere outside the engine and the low pressure of the carburetor and intake manifold. The downward stroke of the piston creates a partial vacuum, which draws air in through the manifold and carburetor.

For a carburetor to operate efficiently, the gasoline must be properly atomized, vaporized, and distributed to the cylinders through the intake manifold.

In spite of different designs, most carburetors have the same seven basic systems of passages, ports, jets, and pumps. Venturis, or restrictions, help speed the airflow and draw more fuel into the airflow. For each barrel, or throat, on a carburetor, there is also a throttle valve.

Most carburetors made after 1979 use a tamperproof design in which the manufacturer sets the idle mixture and choke adjustment at the factory and seals the adjusters according to U.S. government regulations to prevent unauthorized adjustments that would affect emissions.

Various assist devices used with carburetors improve driveability, reduce emissions, help in cold-weather starting, and avoid overheating. Other assist devices provide better fuel mixture during the full range of driving conditions, including high-altitude driving.

Many late-model carburetors are part of an electronic fuel management system. An exhaust gas oxygen sensor (O2S) in the exhaust manifold measures the oxygen content in the exhaust gas and signals the engine control computer. To maintain the desired 14.7:1 air-fuel ratio, the computer signals a mixture control (MC) solenoid or stepper motor in the carburetor to make the mixture richer or leaner as required. Since carburetors respond to this feedback information from the O2S through the computer, they are called feedback carburetors.

Good intake manifold design is critical for smooth performance and low emissions. It must evenly distribute the air-fuel mixture to each cylinder. Runner length, smoothness of the inside surfaces, and arrangement of headers are important for the smooth flow of gases.

Intake manifold designs favor good low-to-middle speed operation, good high-speed operation, or a compromise of the two. Several methods are used to promote better vaporization in the manifold. Better vaporization can result from using a heat control valve that routes exhaust gases through the intake manifold while a cold engine warms up. Instead of exhaust gases, hot engine coolant may be routed through the manifold. Some later-model engines use an electric heater grid underneath the carburetor for this purpose.

Review Questions

Choose the single most correct answer.
Compare your answers to the correct answers on page 507.

1. Which of the following is *not* true of air pressure?
   a. It is measured in millimeters of mercury (mm-Hg)
   b. It is measured in pounds per square inch (psi)
   c. It is always constant regardless of temperature
   d. It results from the weight of air pressing on a surface

2. Air pressure:
   a. Increases at higher temperature
   b. Increases at lower altitudes
   c. Remains the same at all altitudes
   d. Remains the same at all temperatures

3. A pressure differential is created between outside air and engine air by:
   a. An increase in cylinder volume
   b. A decrease in cylinder volume
   c. Air rushing through the carburetor
   d. The intake manifold

4. The restriction in an airflow tube or barrel is called a(n):
   a. Throttle
   b. Air bleed
   c. Vaporizer
   d. Venturi

5. When air flows through a barrel, the air pressure along the sides of the barrel is:
   a. Higher than pressure at the center
   b. Higher than atmospheric pressure
   c. Lower than pressure at the center
   d. Lower than pressure at the barrel ends

6. Ported vacuum is the low-pressure area:
   a. Just above the throttle valve
   b. Beneath the throttle valve
   c. In the carburetor venturis
   d. In the combustion chamber

7. At idle:
   a. Venturi vacuum is high
   b. Manifold vacuum is high
   c. Ported vacuum is high
   d. All of the above

8. Which is *not* a fuel metering system?
   a. Float system
   b. Idle system
   c. Power system
   d. EEC system

9. The carburetor float controls fuel level:
   a. By closing the needle valve when the level is low
   b. By opening the needle valve when the level is high
   c. By closing the needle valve when the level is high
   d. All of the above

10. At engine idle:
    a. Venturi vacuum is high
    b. Fuel enters the carburetor barrel above the throttle
    c. Air is provided by an air bleed in the idle tube
    d. The throttle is one-third open

11. Which is *not* used to provide extra air at idle?
    a. The throttle valve
    b. The choke
    c. Transfer ports
    d. The idle air bypass

12. At low off-idle speeds, extra fuel is provided by:
    a. The transfer port
    b. The main nozzle
    c. The idle air bleed
    d. The idle air adjust screw

13. In a high-speed system, better fuel and air mixtures are obtained with:
    a. Transfer ports
    b. A single venturi
    c. Multiple venturis
    d. None of the above

14. Power circuits are operated by:
    a. Vacuum diaphragms
    b. Vacuum pistons
    c. Mechanical metering rods
    d. All of the above

15. Which is not part of the accelerator pump circuit?
    a. The metering rod
    b. The inlet check
    c. The outlet check
    d. The duration spring

16. Which carburetor circuit makes the fuel mixture richer when starting an engine?
    a. Power circuit
    b. Choke circuit
    c. High-speed circuit
    d. Accelerator pump circuit

17. The illustration below shows:

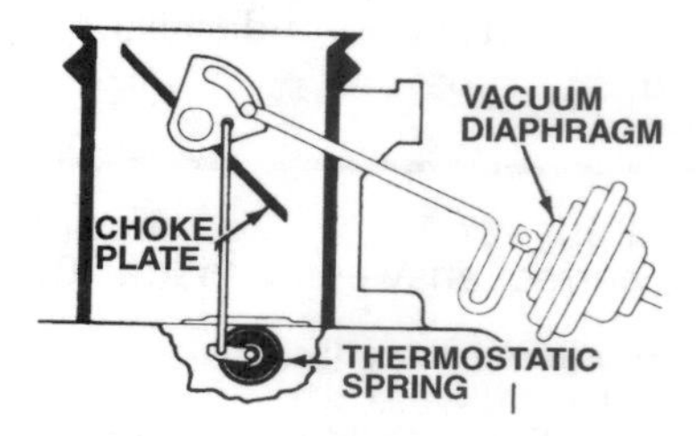

    a. A power valve diaphragm
    b. An integral choke
    c. A remote choke
    d. None of the above

18. The choke unloader:
    a. Releases the fast-idle cam after a specified time
    b. Directs warm air to the thermostatic spring
    c. Opens a vacuum bleed in the vacuum-break diaphragm
    d. Opens the choke valve when the throttle is open fully

19. The fast-idle cam:
    a. Opens the throttle
    b. Closes the throttle
    c. Opens the choke valve
    d. Closes the choke valve

20. Independently operated throttles are found in:
    a. One-barrel carburetors
    b. Single-stage, two-barrel carburetors
    c. Two-stage, two-barrel carburetors
    d. All of the above

21. Which is *not* true of four-barrel carburetors?
    a. The primaries feed four cylinders and the secondaries feed the other four
    b. Airflow through the secondary barrels is through venturi action or air velocity valves
    c. The primaries act like single-stage, two-barrel carburetors at low speeds
    d. The secondary barrels open between one-half and three-quarters throttle

22. Variable-venturi carburetors:
    a. Maintain constant throttle
    b. Maintain a uniform pressure drop across the venturi
    c. Have been used on domestic cars for 45 years
    d. Are formed by two parallel valve plates

23. Main fuel metering in a variable-venturi carburetor is done by:
    a. Metering rods
    b. A dashpot
    c. EGR port vacuum
    d. The idle trim system

24. Which is *not* part of the cold enrichment system of a 2700 VV carburetor?
    a. A fast-idle cam
    b. A solenoid
    c. A control vacuum regulator rod
    d. An auxiliary fuel passage and metering rod

25. The accelerator pump is controlled by:
    a. Vacuum-break linkage
    b. Thermostatic spring linkage
    c. Throttle linkage
    d. All of the above

26. Which of the following allows feedback carburetors used with electronic fuel management systems to compensate for changes in altitude?
    a. A remote compensator
    b. An aneroid bellows
    c. The powertrain control module
    d. An O2S

27. The early fuel management systems contained a microprocessor, a catalytic converter, vacuum and temperature switches, a solenoid vacuum regulator, an O2S, and:
    a. A feedback carburetor
    b. A nonfeedback carburetor
    c. An altitude compensating carburetor
    d. An emission carburetor

28. Feedback carburetors control fuel flow with some type of:
    a. Hot-idle compensator
    b. Mixture control solenoid
    c. Decel valve
    d. Aneroid bellows

29. The intake manifold is heated by:
    a. An intake manifold heater
    b. The blower motor
    c. Engine coolant heat
    d. The upper radiator hose

30. Unequal distribution of the air-fuel mixture can occur in a carbureted engine because of:
    a. Broken valve springs
    b. Misadjusted timing
    c. Exhaust gas blowby
    d. Charge robbing

31. The most efficient manifold design to use with a spread-bore carburetor is a:
    a. Single-plane
    b. Single-plane with siamesed intake ports
    c. Dual-plane
    d. Single-plane with grooved plenum chamber floor

32. Manifold heat control valves:
    a. Can be located between the carburetor and intake manifold
    b. Can be located between the exhaust manifold and exhaust pipe
    c. Route heat away from the intake manifold when the engine is cold
    d. Require no maintenance for the life of a car

33. GM's thermostatic vacuum switch reacts to:
    a. Oil temperature
    b. Coolant temperature
    c. Exhaust gases
    d. The vacuum solenoid

34. A ceramic heater grid is positioned under the carburetor:
    a. Primary throttle bore
    b. Secondary throttle bore
    c. Both of the primary and secondary bores
    d. Choke plate

35. A rotating flapper valve in the exhaust manifold or exhaust pipe is also called a(n):
    a. Ceramic insulator
    b. Heat riser
    c. Air-fuel preheater
    d. Coolant heater

36. Heat control valves tend to stick because of:
    a. High temperatures
    b. Rust and corrosion
    c. Poor fuel economy
    d. Exhaust backpressure

PART FOUR

Electronic Engine Management

Chapter

14 Principles of Electronic Control Systems

To meet stringent emission control requirements in the early 1970s, automotive engineers began to apply electronic control to basic automotive systems. The use of electronics was first applied to ignition timing and later to fuel metering. Electronic control introduced precision that electromechanical and vacuum-operated actuators alone could not achieve in matching fuel delivery and ignition timing with engine load and speed requirements. Electronic control resulted in a significant decrease in emission levels, major improvements in driveability, and increased system reliability.

Several years after the introduction of electronic ignitions, automakers began manufacturing engines using electronic fuel metering. Soon after that engineers combined ignition and fuel-metering control, forming early engine management systems. By the early 1980s, on-board computers controlled many automotive systems.

To understand electronic controls, you must know basic electricity, semiconductor theory, and computer principles.

BASIC ELECTRICAL CONCEPTS

Almost all automotive electronic and electrical circuits can be described completely with a few basic concepts: current, voltage, and resistance. A real familiarity with these fundamentals will allow you to understand any electrical problem.

Current

Current flow is one of the most fundamental concepts of electricity. Electric **current flow** is defined as the controlled, directed flow of electrons from atom to atom within a conductor.

Electron flow

Electrons constantly move in and out of all atoms in nature. This movement of drifting electrons is a current flow. The current flow is usually small, with only a few electrons moving in the same direction at any one time. However, if all the electrons are directed by being placed in an electric field, then a huge number of electrons are set in motion in a single direction, and the current flow can be large. This type of current flow can perform useful work if it is directed through a conductor.

We can compare current flow through a conductor to water flow through a pipe. To measure the water flow, we count how many gallons flow past some point in the pipe for a certain length of time. We measure current flow by counting electrons in the same manner. When 6.28 billion electrons pass a point in one second, we say that one **ampere** of current is flowing. The ampere is the unit that indicates the rate of electrical cur-

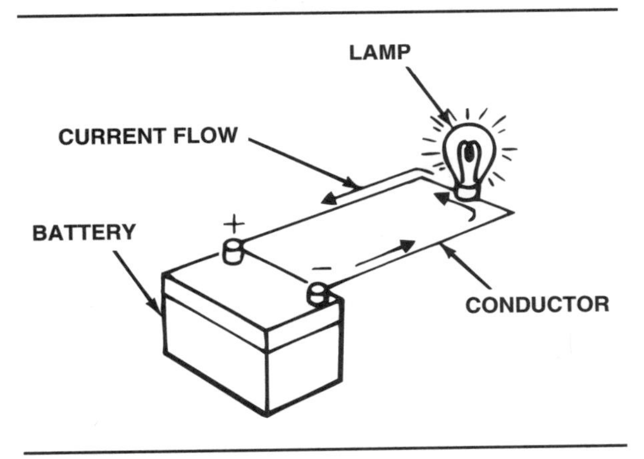

Figure 14-1. A simple electrical circuit.

rent flow. The ampere is commonly abbreviated as amp (A). Remember, current flows within a circuit.

Current flow theories
There are two theories that describe the *direction* of current flow in electrical circuits: the conventional flow theory and the electron flow theory. Either theory can be used, but it must be used *consistently*.

When scientists first began investigating electricity, they believed that current flowed from positive to negative. This belief persisted for many years, though it was later proved wrong, and became what is now called the **conventional flow theory** of current flow. The conventional flow theory is used to discuss automotive electrical systems.

The **electron flow theory** says that current flows from negative to positive. That is, from a source with more electrons to a point with fewer electrons. This is the theory used in electronics and the engineering fields today, and it is the theory we have used in describing basic electrical concepts. Auto electronics still uses conventional current flow.

Current Path

We discuss several circuit types, but each follows the same basic rule: *electric current can only flow through a closed, unbroken path*. This closed path is called a **complete circuit.** If the circuit is opened anywhere along the path, the circuit is "broken" and no current will flow, creating an **open circuit.** If the normal path is broken, but current can still find an incorrect path to another circuit or a "shorter" path to ground, this creates a **short circuit.**

One very simple electrical circuit contains a voltage source (battery), conductors (wires), and a lamp, figure 14-1. The battery provides a supply of electrons, the wires provide a controlled path for electrons to flow (the circuit), and the lamp lights when current flows through it.

Current flows from the negative battery terminal through the wires and lamp, back to the positive battery terminal, lighting the lamp. If a wire is taken off either battery terminal or either lamp terminal, the circuit is open and current will not flow. The lamp will not light.

Breaking a circuit by removing a wire from a terminal is the same as using a switch to break the circuit and interrupt current flow. The switch is simply a more convenient way to open or close a circuit.

Resistance

Every material opposes current flow to some extent. This electrical opposition is called **resistance,** and is measured in units called **ohms.** Some materials offer very little resistance, others offer a great deal of resistance, and some materials fall somewhere in between.

Resistance to current flow
Since every electrical device is built from materials that offer *some* degree of resistance, all electrical devices offer resistance. The amount of resis-

Current Flow: A controlled, directed flow of electrons from atom to atom within a conductor.

Ampere: The unit for measuring the rate of electrical current flow.

Conventional Flow Theory: The theory that says current flows from positive to negative.

Electron Flow Theory: The theory that says current flows from negative to positive.

Complete Circuit: A continuous, unbroken path in a circuit that allows current to flow from a source and return to that source.

Open Circuit: A discontinuous (broken) path in a circuit that does not allow current to flow to or from a source.

Short Circuit: A continuous path for current flow that changes the normal current path through the circuit. The short may be into another circuit or to ground.

Resistance: Opposition to electrical current flow.

Ohm: The standard unit for measuring electrical resistance.

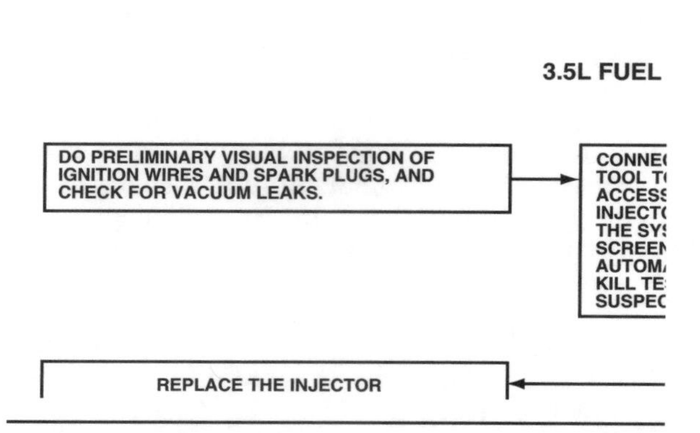

Figure 14-2. The infinite resistance caused by an open in a circuit stops current flow.

tance present in any part of an electrical circuit depends on the following five factors:

- The atomic structure of the material—Any material with few free electrons is a poor conductor, since resistance to current flow will be high. All conductors have some resistance, but the resistance of a good conductor is so small that a fraction of a volt will cause current flow.
- The length of the conductor—Electrons in motion are constantly colliding with the atoms of the conductor. The longer a piece of wire, the farther the flow must travel. The more collisions that occur, the greater the resistance of a conductor.
- The cross-sectional area of the conductor—The thinner a piece of wire, the higher its resistance.
- The temperature of the conductor—In most cases, the higher the conductor's temperature, the greater its resistance. That is why alternator regulators are tested at normal operating temperature for accurate readings.
- The condition of the conductor—If a wire is partially cut, it acts almost as if the entire wire were of a smaller diameter, offering a high resistance at the damaged point. Loose or corroded connections have the same effect. High resistance at connections is a major cause of electrical problems.

Resistance is present in electric circuits in places other than the conducting wires and devices of the electric circuit. An open circuit, such as in figure 14-2, creates an **infinite resistance** to current flow. Current simply stops flowing because the circuit is no longer closed. There is no longer any **continuity** in the circuit. Loose or corroded connections also cause electrical problems because they offer a greater resistance to current flow than normal.

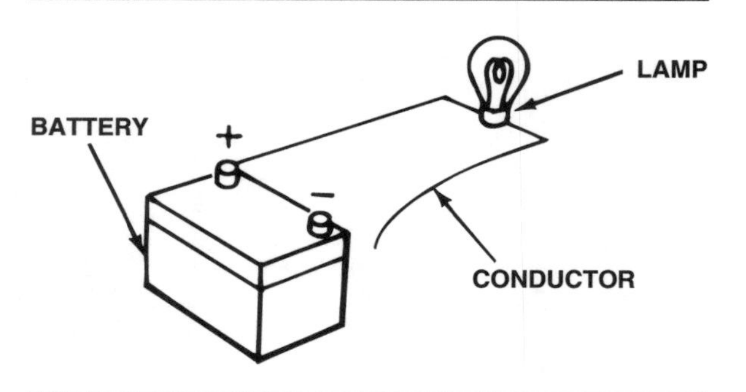

Figure 14-3. Opening this circuit prevents current from flowing, but does not remove the potential difference (voltage).

The resistance of any electrical part (load or conductor) can be measured in three ways:

- Direct measurement with an ohmmeter, which measures the ohms of resistance offered by the load
- Indirect measurement with a voltmeter, which measures the voltage drop through the load
- Indirect measurement with an ammeter, which measures the current through the load.

Conductors, Insulators, and Semiconductors

Materials that offer little resistance to electron flow are good **conductors.** Electric current can pass through these easily. Many metals, such as silver, copper, and aluminum, are good conductors.

Materials that do not conduct electricity well are called **insulators.** Rubber, glass, and certain plastics are good insulators. Insulators are used in electric circuits to ensure electron flow stays within the conductors and is properly directed within the circuit. Some materials are neither good conductors nor good insulators. These are called **semiconductors.** They are important in some parts of the electrical system, as we will see when we examine diodes, transistors, and other solid-state electronic devices.

Voltage

Current flow can only occur when some force aligns and pushes many drifting electrons in a single direction. The electrical (electromotive) force that does this is called **voltage,** and is measured in units called **volts.** Voltage is electrical force applied to a circuit.

Potential difference

Voltage exists even when there is no current flowing in a circuit. This is called an open-circuit condition. For example, when we disconnect a wire from the battery terminal in figure 14-3, voltage is still present at the battery terminals even though no current flows through the incomplete circuit. If you measure the voltage between the battery terminals, you still find 12 volts, but no voltage is applied across the circuit.

Voltage is a measure of the difference in potential, or **potential difference** in force that exists between two points. The *difference* in force exists because one point has a greater electrical charge than the other. A good example of this is the terminals of a battery because one is at a higher positive charge than the other. The other terminal is more negatively charged. The two terminals are distinguished from each other by calling one the positive terminal and the other the negative terminal. The *potential* in the force exists because there is energy stored in the battery that could *potentially* be used if the battery is connected to a complete circuit so that current can flow.

The strength of the voltage force depends on the strength of the charges at each point. If the difference between charges is great, voltage is high. If the difference between charges is small, voltage is low.

We can compare voltage to the hydraulic pressure that moves water through a garden hose. The hydraulic pressure is the driving force, but it is the water that flows. Likewise, voltage is the driving force, but it is the current that flows.

Voltage drop

Every device in a circuit (wires, lamps, motors, etc.) offers some resistance to the current flowing through the circuit. As current flows through the resistance of each device, some voltage is lost and dissipated as heat.

If you measure the voltage on both sides of a load, you can see how much voltage is used to move current through the load. This loss is called **voltage drop.** If you measure the voltage drop at every load in a circuit and add the measurements, they must equal the original voltage available to the circuit. No voltage actually disappears. It is changed into a different form (heat) by the resistance of the loads. Remember, all source voltage is dropped across any circuit.

Grounds

In a modern automotive electrical system, wiring provides only half of each electrical circuit. The other half of the circuit is provided by the automobile engine, frame, and body, which together provide a return path for current flow. This return side of the circuit is called the ground side, figure 14-4. Automotive electrical systems are called single-wire, or ground-return, systems.

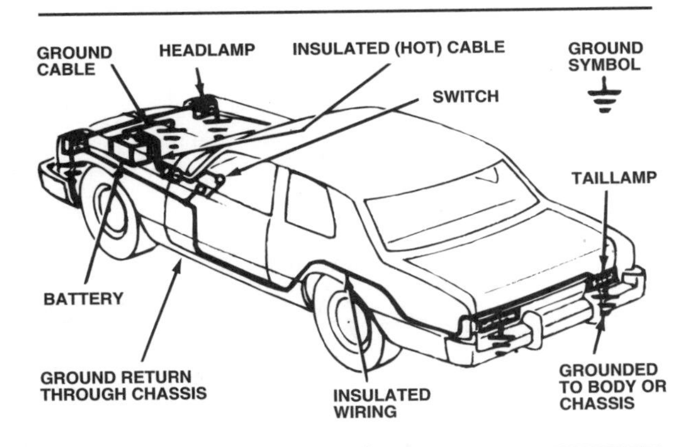

Figure 14-4. The vehicle chassis ground path makes up half of most automotive electrical circuits.

The cable from the negative battery terminal is bolted to the car engine or frame. This is called the ground cable. The cable from the positive battery terminal provides current for the vehicle's electrical loads. This is called the insulated, or hot, cable. The insulated side of every circuit in the vehicle is the wiring from the positive battery terminal to each device in the circuit. The ground side of every circuit is the vehicle chassis, ground straps, and return wiring.

Infinite Resistance: A resistance value that is too high to measure on a test instrument; usually the result of an open circuit. The true value is unknown.

Continuity: An uninterrupted connection. Used to describe an unbroken electrical circuit.

Conductors: Materials that allow for the easy flow of electrons.

Insulators: Materials that oppose electron flow because of their many bound electrons.

Semiconductors: Materials with four electrons in their valence shell that are neither good conductors nor good insulators.

Voltage: The electromotive force that causes current flow. The potential difference in electrical force between two points when one is negatively charged and the other is positively charged.

Volts: The standard unit for measuring electromotive force (voltage).

Potential Difference: The difference in electrical force between two points when one is positively charged and the other is negatively charged.

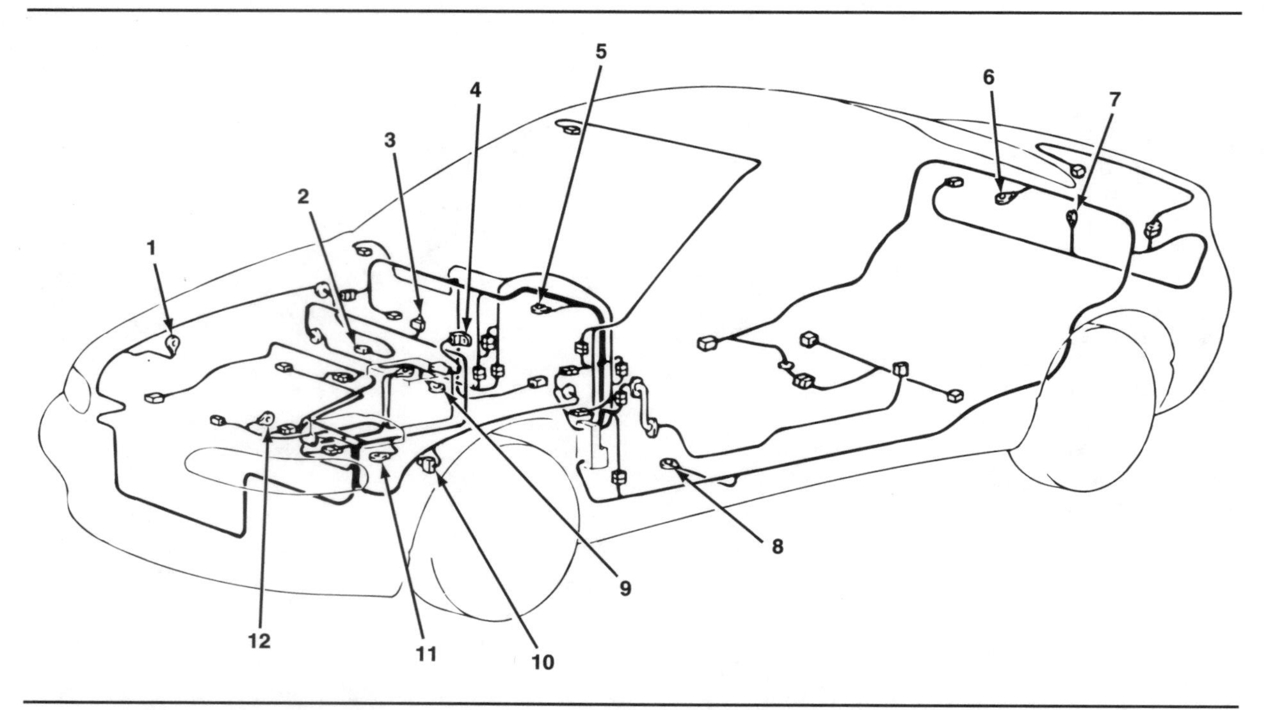

Figure 14-5. Late-model vehicles use multiple ground connections to minimize or suppress EMI.

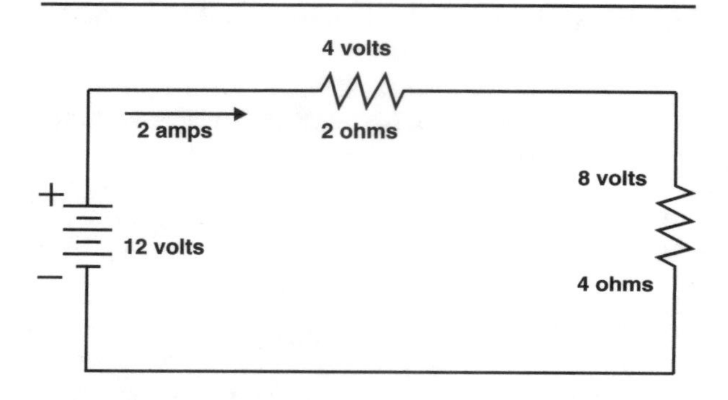

Figure 14-6. A series circuit. Although there are several resistors, there is only one current path.

Every electrical device is connected to the battery so that current passes from the positive battery terminal, through the device, through the ground strap, into the chassis, then back to the negative battery terminal. This connection design forms a complete circuit for all electrical devices in the system.

On many vehicles, additional ground straps or cables are connected between the engine and the vehicle body or frame. Extra cables may be connected to the engine or chassis and the transmission or transaxle. These additional ground cables ensure a low-resistance ground path between all components and the battery terminal. This is necessary for proper operation of engine circuits and those mounted elsewhere on the vehicle. Late-model vehicles, which rely heavily on computer components, often use several additional ground straps to minimize or eliminate **electromagnetic interference (EMI),** figure 14-5.

The resistance on the insulated side of most electric circuits in the vehicle depends on the load and the length of wiring. The resistance on the ground side of the circuits between the load and its **ground** connection should be close to zero ohms. This is because every ground connection on the vehicle is electrically the same as a direct connection to the ground terminal of the battery.

Ground connections must be clean, tight, and secure for a circuit to be complete. In older cars where plastics were rare, many parts had a direct connection to a metal ground. With the increased use of plastics in late-model cars, designers have had to add many additional ground wires to the system. Ground wires in many circuits are black for easy recognition.

Circuit Types

Circuit elements can be assembled in any number of ways, however, all of the circuit configurations can be separated into three different types of circuit structures: series, parallel, and series-parallel circuits. These circuit types are easy to recognize, and each has particular rules for determining voltage, current, and resistance values.

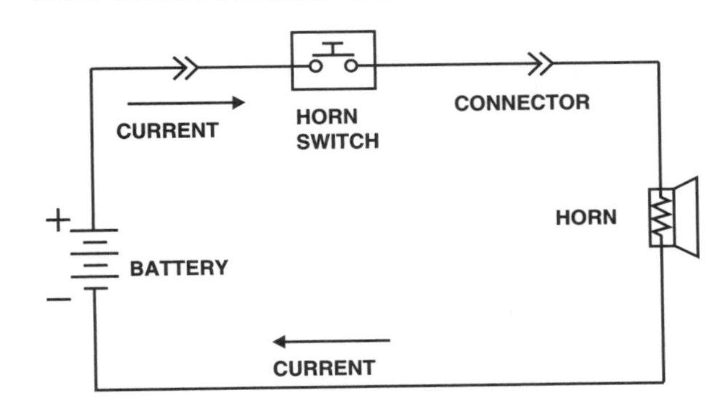

Figure 14-7. This horn circuit illustrates a series circuit with automotive components rather than resistors.

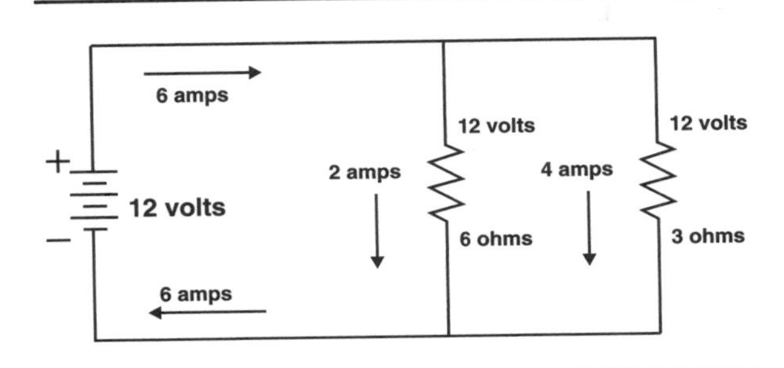

Figure 14-8. A parallel circuit. The voltage is the same across each resistor.

Series circuits

In a **series circuit,** figure 14-6, current has only one path to follow. Current flows from the battery, through the resistors, and back to the battery. The circuit is complete because current has a path to flow from and return to the battery. The circuit is continuous because it is unbroken. If any wire is disconnected from the battery or resistors, the circuit is open and no current flows.

This circuit is called a series circuit because all the devices are wired together, one after the other, in series. If electric devices are wired in series, they must *all* be switched on and operating, or the circuit will be open, and *none* of them will work. Figure 14-7 shows a simple vehicle series circuit. Current flows from the battery, through the horn switch, through the horn, then back to the battery. Some of the key electrical features of a series circuit are:

- Since there is only one path for current flow, the same current flows through all devices in the circuit.
- Voltage drops across each device according to its individual resistance, but the sum of all voltage drops in the circuit equals the original source voltage.
- The total resistance of the circuit is the sum of the individual resistances of each device.

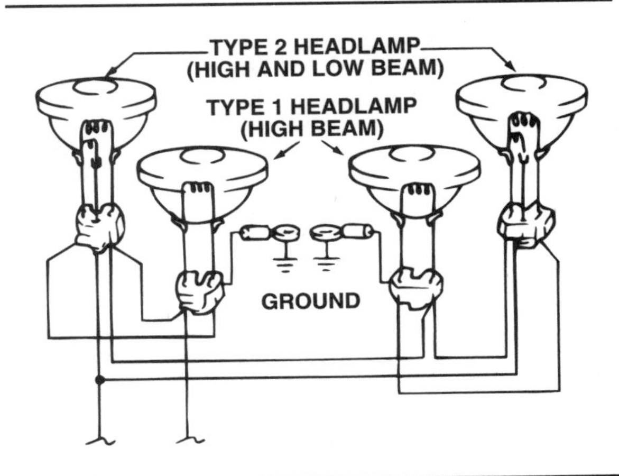

Figure 14-9. Automotive headlamps are wired in parallel. If one lamp burns out, the other still receives the same voltage.

Parallel circuits

If current can follow more than one path to complete its return to the voltage source, the circuit is called a **parallel circuit,** figure 14-8. The points where current paths split are called junction points or **nodes.** The separate paths that split and rejoin at junction points are called branches, or parallel paths. A good example of a parallel circuit in an automobile is the headlamp circuit, figure 14-9. The headlamps are wired in parallel to each other. Parallel circuits exhibit several characteristic features:

- The voltage across each branch of a parallel circuit is the same.
- The total current flow in a parallel circuit is the sum of the current flows in each of the branches. Individual current flows in each branch may be different, but they all add up

Voltage Drop: The loss of voltage caused by the resistance of a circuit load.

Electromagnetic Interference (EMI): Undesirable, high-frequency electromagnetic waves that interfere with electronic systems.

Ground: The wiring and connections that return current to the battery. The ground is common to all circuits in the electrical system.

Series Circuit: A circuit that has only one path through which current can flow.

Parallel Circuit: A circuit that has more than one path through which current can flow.

Nodes: The points where current paths split. Also called junction points.

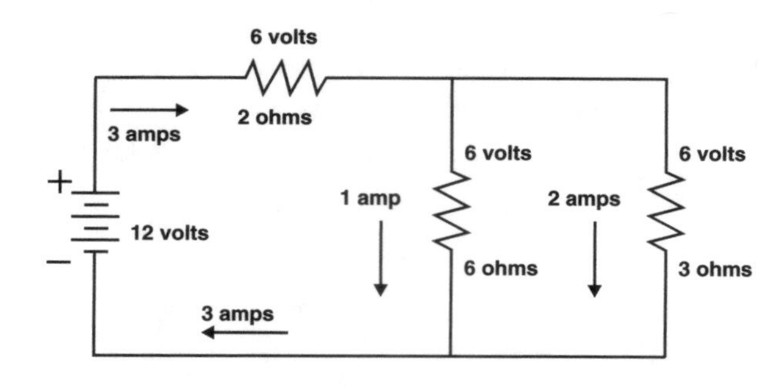

Figure 14-10. A series-parallel circuit. Here, the two parallel resistors are in series with the first resistor.

to the current delivered from, and returning to, the battery. This is the same situation as tributaries branching off from a river, then collecting again elsewhere and returning to the river at another point. The water is broken up for a time in use, but it does not disappear.

- The total resistance of a parallel circuit is less than the lowest individual resistance. This is because adding components in parallel provides more paths for current to flow, which reduces the total resistance of the circuit.

Series-parallel circuits

As the name suggests, a **series-parallel circuit,** figure 14-10, combines the two types of circuits already discussed and exhibits the operating characteristics of each. Some of the devices are wired in series and are complete branches wired in parallel. The full headlamp circuit of an automobile is a series-parallel circuit, figure 14-11. The headlamps are in parallel with each other, but the switches are in series with the battery and with each lamp. Both lamps are controlled by the switches, but one lamp will still light if the other is burned out.

Trace the circuit through for yourself a few times to make sure you understand the current flow and circuit operation. Most of the circuits in an automotive electrical system are series-parallel circuits, and many of them are no more difficult than this one. If you truly understand this circuit, then you understand automotive electrical circuits.

Ohm's Law

Ohm's Law expresses an important relationship between volts (electrical force), amperes (current flow), and ohms (electrical resistance). Ohm's law states:

> An electrical force of *one volt* will cause *one ampere* of current to flow through a circuit with *one ohm* of resistance.

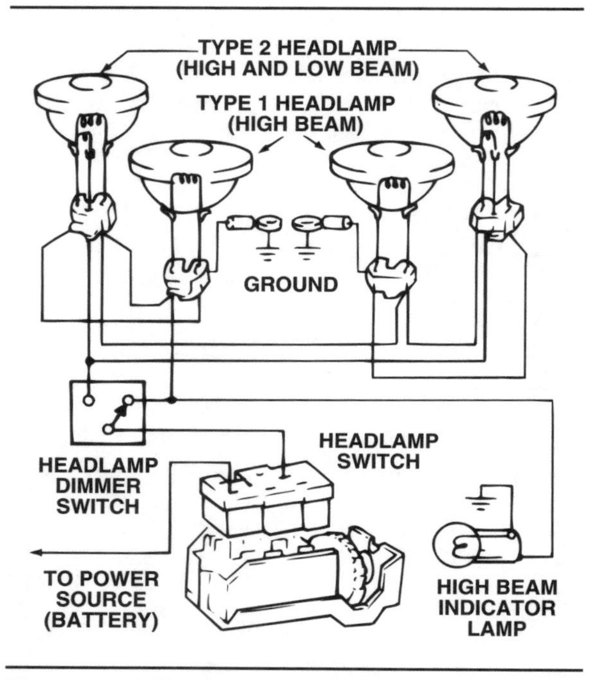

Figure 14-11. Complete automotive headlamp circuits are series-parallel circuits.

The formula for this is E = I x R, where: E = Voltage or Voltage Drop (Electromotive force), I= Current (Intensity), and R = Resistance.

The formula can be rewritten several ways to make it more convenient to use, but even more important than knowing how to use the formula is knowing how to use the *principle*. Anytime two of the three electrical values are known, the third can be found by applying Ohm's Law. The formula can be restated and used to find:

- Voltage Drop = the current flow multiplied by the resistance of the component
- Current Flow = the voltage drop divided by the resistance of the component
- Component Resistance = the voltage drop divided by the current flow.

Ohm's Law can also be used to gain a *feel* for the circuit and an understanding of what happens to the circuit when components fail or circuit conditions change. To illustrate, look again at Ohm's Law in its basic form: E = I x R. This is called a balanced equation. It states that one side of the equation (E) *is equal to* the other side of the equation (I times R) under all conditions, without listing any particular values. This means that whatever E is, I times R must work out to the same value. You can use that fact to understand any circuit by using the following method: hold one value steady, and change another to see what must happen to the third.

For example, what happens to a circuit if you increase the voltage? To examine the situation, hold one value steady. The components do not suddenly change physically, so hold resistance (R) steady and change the voltage (E). From the equation, we see that if E goes up, and R stays the same, then I (current) must also go up to keep the two sides equal.

Similarly, what happens to a circuit if the resistance drops (as in a short circuit)? There are two possible cases. First, if we hold I steady and lower R, then E must decrease for the two sides to remain equal. If we hold E steady and lower R, then I must increase. The second example is the more common case of a short circuit.

It is easy to check E with a voltmeter and see that it is still battery voltage; therefore E did not increase, but I (current) did increase. Here, we had two situations theoretically possible, but only one was practical. A little experience and common sense testing allows us to understand how a circuit reacts as operating conditions change.

PRINCIPLES OF SOLID-STATE ELECTRONICS

Up until the mid-1970s most radios, television sets, and stereo amplifiers used bulky, energy-consuming electron tubes. Whenever these electronic devices failed, the user tested every tube at a corner drug store's tube-testing machine. The advent of semiconductor technology during the 1950s replaced the cumbersome electron tube with diodes, rectifiers, and transistors. During the late 1960s, technologists decreased computer processing time and their physical size by using integrated circuits (IC's) on silicon chips. Engineers and consumers adopted the term "solid-state" to refer to the new semiconductor circuitry, to set it apart from the old electron tubes.

Understanding the basics about semiconductor-studded printed circuit boards commonly used today in automobiles is straightforward. To begin, we will discuss the differences between conductor, insulator, and semiconductor materials. The characteristics of doped N-type and P-type semiconductor material explains the behavior of current through devices such as a diode or transistor. In this section, we will be using electron flow theory to explain semiconductor operation. After describing the operation of two types of transistors, this section will conclude with a brief explanation of the IC chip and printed circuit board.

Semiconductor Theory

Electrons in an atom orbit around the nucleus in definite paths. These paths form shells, like concentric rings, around the nucleus. Only a specific number of electrons can orbit within each shell.

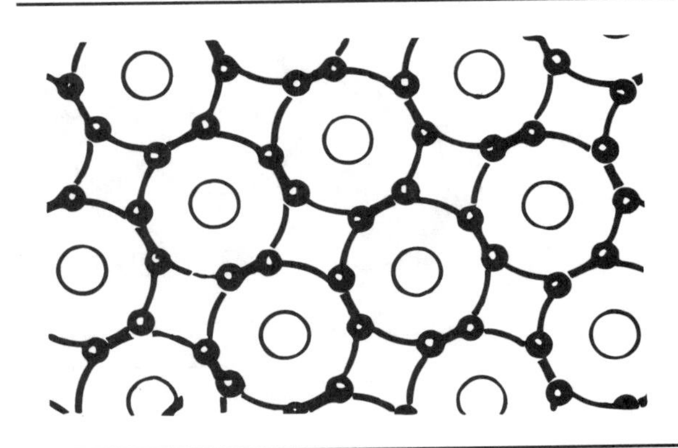

Figure 14-12. Crystalline silicon is an excellent insulator. (Delco-Remy)

If there are too many electrons for the first and closest shell to the nucleus, the others will orbit in additional shells until all electrons have an orbit within a shell. There can be as many as seven shells around a single nucleus.

The outermost electron shell, or ring, is called the **valence ring.** The number of electrons in this ring determines the valence of the atom and indicates its capacity to combine with other atoms.

If the valence ring of an atom contains three or fewer electrons, it has room for more. The electrons in its atoms are held very loosely, and it is easy for a drifting electron to join the valence ring and push another electron away. The material that consists of these types of atoms is considered a good conductor.

When the valence ring contains five or more electrons, it is fairly full. The electrons are held tightly, and it is hard for a drifting electron to push its way into the valence ring. The material that consists of these types of atoms is considered a good insulator.

Elements whose atoms have four electrons in their valence rings are neither good insulators nor good conductors. Their four valence electrons give them their special electrical properties deserving of the name "semiconductor". Germanium and silicon (Si) are two widely used semiconductor elements.

When semiconductor elements are in the form of a crystal, they bond together so that each atom has eight electrons in its valence ring. The

Series-Parallel Circuit: A circuit that has some parts in series with the voltage source and some parts in parallel with each other and the voltage source.

Valence Ring: The outermost electron shell of an atom.

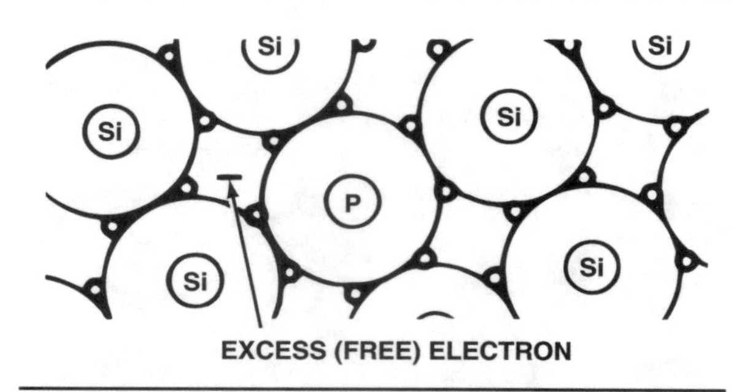

Figure 14-13. N-material has an extra, or free, electron. (Delco-Remy)

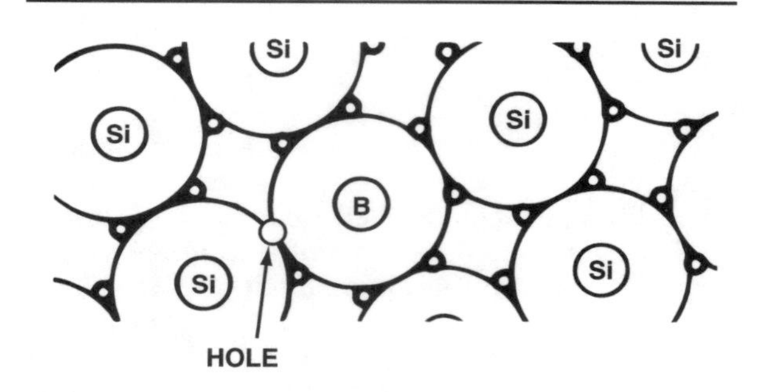

Figure 14-14. P-material has a hole in some of its valence rings. (Delco-Remy)

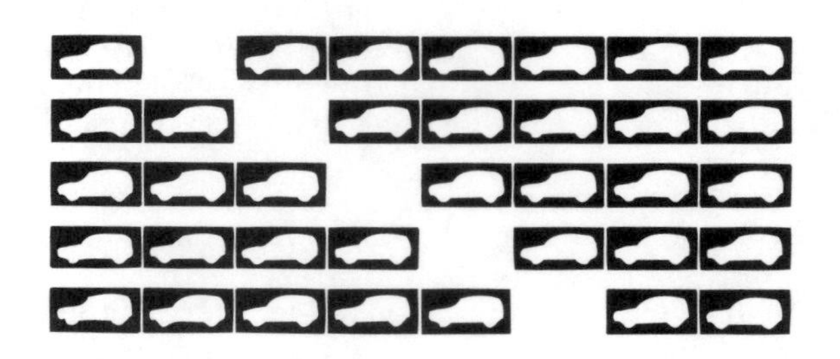

Figure 14-15. Space flow moves opposite to traffic flow just as hole flow moves opposite to electron flow.

atom has its own four electrons and shares four with surrounding atoms, figure 14-12. In this form it is an excellent insulator, because there are no free electrons to carry current flow.

Other elements can be added to silicon and germanium to change this crystalline structure. This is called **doping** the semiconductor. The ratio of doping elements to silicon or germanium is about 1 to 10,000,000. The doping elements are often called **impurities** because their addition to the silicon or germanium makes the semiconductor materials impure.

N-material

If silicon or germanium is doped with an element such as phosphorus (P), arsenic, or antimony, each having five electrons in its valence ring,

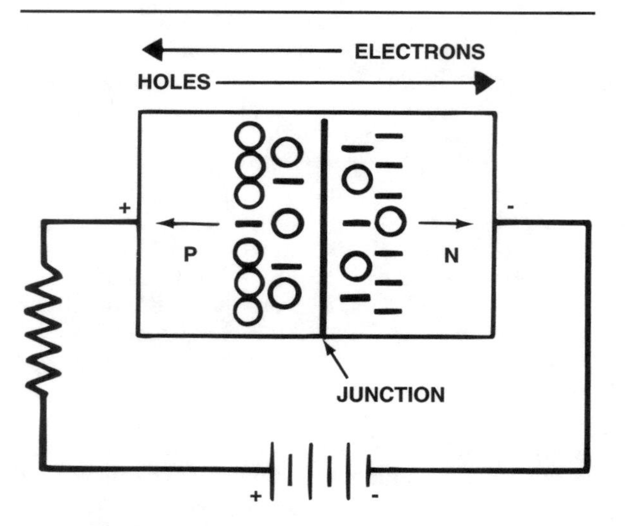

Figure 14-16. Current will flow through this forward-bias connection. (Delco-Remy)

there will not be enough space for the ninth electron in any of the shared valence rings. This extra electron is free, figure 14-13. This type of doped material is called negative, or **N-material,** because it already has excess electrons and will repel additional negative charges.

P-material

If silicon or germanium is doped with an element such as boron (B) or indium, each of which has only three electrons in its outer shell, some of the atoms will have only seven electrons in their valence rings. There will be a **hole** in these valence rings, figure 14-14. This type of doped material is called positive, or **P-material,** because it will attract a negative charge (an electron).

Hole flow

We have explained the electron flow theory of electricity as the movement of electrons from negative to positive. We can also visualize the movement of the holes that these electrons move into and out of. If you imagine a line of cars moving ahead one car length at a time, the car length of empty space can be seen to move from the front to the back of the line, figure 14-15. Just as the car length of empty space moves in the opposite direction from traffic flow, holes move in the direction opposite from electron flow, or from positive to negative. The hole can be thought of as a positive charge of electricity. Holes can move from atom to atom just as electrons move from atom to atom. The hole flow theory is an easy way to understand the operation of diodes and transistors, as we will see.

Semiconductors and voltage

Doping silicon or germanium causes it to behave in unusual but predictable ways when voltage is applied to it. The behavior depends on which charge of the voltage is connected to which type of doped material (P or N). To use this behavior in solid-state devices, P-material and N-material are placed side by side. The line along which they meet is called the **junction.** The application of voltage to the two doped semiconductor materials is called **biasing.** To begin our description of doped material behavior, let us look at a P-material and N-material junction, with positive voltage connected to the P-material and negative voltage connected to the N-material, figure 14-16. A simple device consisting of P-material and N-material joined at a junction is called a **diode.**

Diode operation

The negative battery voltage in figure 14-16 will repel the free electrons in the N-material, causing them to move toward the junction. The positive battery voltage will repel the holes in the P-material, moving them toward the junction also. With enough voltage, the electrons of the N-material will move across the junction into the holes of the P-material. This leaves behind positively charged holes in the N-material, which attract more electrons from the negative voltage source. At the same time, the free electrons that moved into the P-material continue to be attracted toward the positive battery voltage, leaving behind holes in the P-material at the junction. This area near the junction where the N-type material is depleted of electrons and the P-material is depleted of holes is called the **depletion region.** As long as battery voltage is maintained, this chain of events will be repeated, and current will flow through the doped materials. The application of voltage to maintain current flow as described here is called **forward bias.**

A diode cannot withstand unlimited forward bias voltage and current flow. If the forward current flow is too strong or flows for too long, the doped materials can be damaged or destroyed.

If the battery polarity is reversed, so that the negative battery voltage is connected to the P-material and the positive battery voltage is connected to the N-material, the doped materials' behavior changes, figure 14-17. The positive battery voltage attracts the free electrons in the N-material, causing them to move away from the junction. The negative battery voltage attracts the holes in the P-material, causing them to move away from the junction. At the junction, there will be no holes and no free electrons. No

Doping: The addition of a small amount of a second element to a semiconductor element.

Impurities: The doping elements added to pure silicon or germanium to form semiconductor materials.

N-material: A semiconductor material that has excess (free) electrons because of the type of impurity added. It has a negative charge that repels additional electrons.

Hole: The space in a valence ring where another electron could fit.

P-material: A semiconductor material that has holes for additional electrons because of the type of impurity added. It has a positive charge that attracts additional electrons.

Junction: The area where two types of semiconductor materials (P- and N-material) are joined.

Biasing: Applying voltage to a junction of semiconductor materials.

Diode: An electronic device made of P-material and N-material bonded at a junction. A diode allows current flow in one direction and blocks it in the other.

■ Semiconductor Doping

Adding impurities to a semiconductor material, or doping, must be a carefully controlled process. Manufacturing methods are constantly changing as better doping processes are developed. Early semiconductor technology called for painting layers of doping elements onto either side of a semiconductor wafer and baking it at temperatures above 2,000°F (1093°C). The ratio of impurity to the pure semiconductor was not easily controlled, and new uses for semiconductors demanded more precise production.

Some newer processes use a gas of the doping material. A process called diffusion passes this gas over the surface of a semiconductor wafer, which has been heated to 2,000°F. The doping elements are slowly absorbed into the wafer. The doping ratios can be carefully controlled during this slow absorption.

Ion implantation uses a 100-kV force to create an arc through the doping gas. The flow of electrons through the gas ionizes it, causing the gas particles to become positively and negatively charged. A semiconductor wafer is bombarded with the ionized gas, and doping elements are deeply implanted in the wafer. This method can be combined with the diffusion method to create a very deep but controlled layer of doped material.

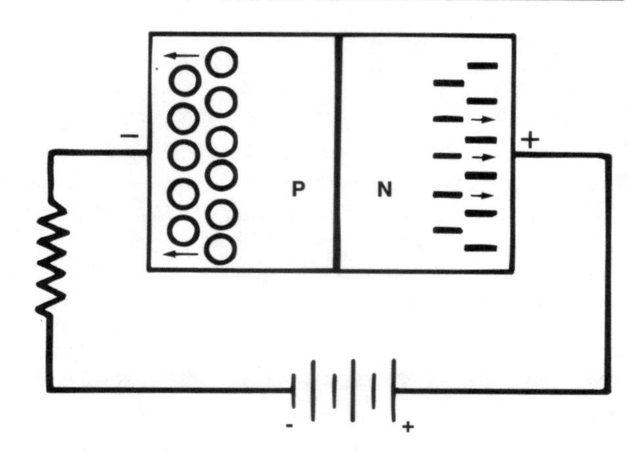

Figure 14-17. No current will flow through this reverse-bias connection. (Delco-Remy)

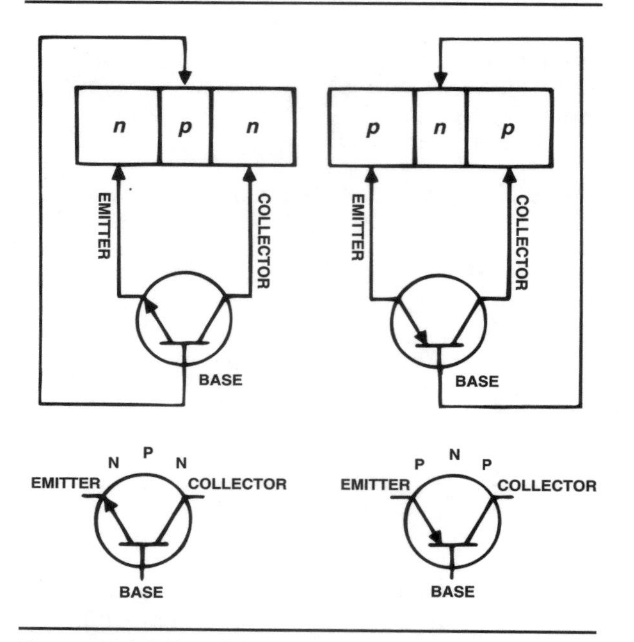

Figure 14-18. Transistor symbols and construction.

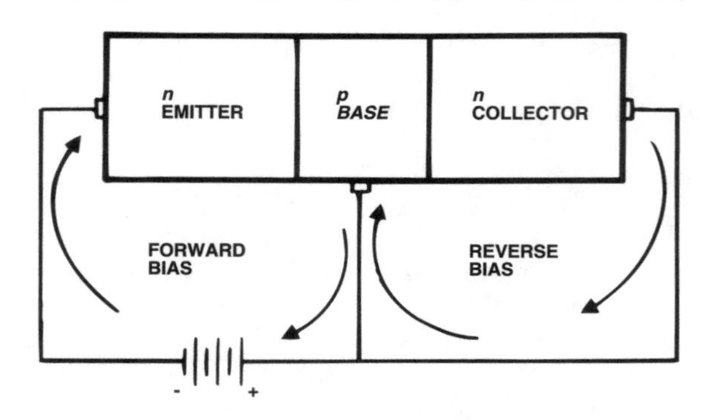

Figure 14-19. Forward and reverse biasing for an NPN transistor.

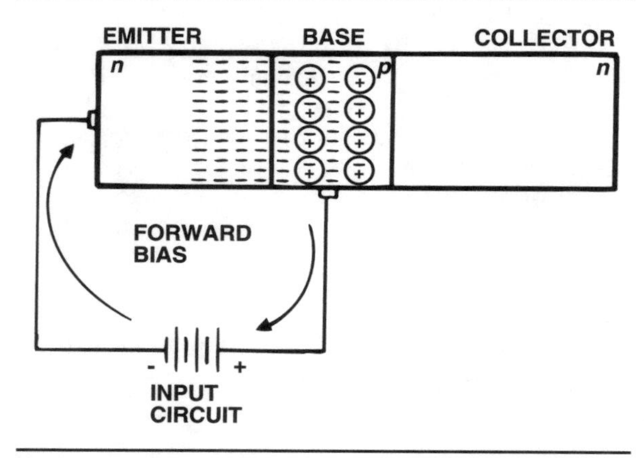

Figure 14-20. Free electrons cannot find enough holes to join within the base. Therefore, electrons accumulate in the base, and little forward-biased current flows.

current can flow under these conditions. Voltage applied in this way is called **reverse bias,** figure 14-17. The diode's ability to allow current to flow in only one direction allows engineers to use it as a one-way check valve.

If reverse bias voltage is very high, the diode will break down and current will flow. One of the most common ways to rate a diode is in terms of its **peak inverse voltage (piv).** This refers to the amount of voltage a diode can withstand before it breaks down and allows reverse current flow. A diode can withstand reverse bias voltage below its piv rating indefinitely, but reverse current flow caused by voltage higher than the piv rating can quickly damage the diode. This is the reason diodes in alternators can be damaged if a battery is hooked up backwards or if high voltage from a battery charger is applied to a battery without disconnecting it from a vehicle's electrical system.

Transistors

A **transistor** is a three-element semiconductor produced by adding another layer of doped material to a basic PN diode. The three materials are combined so that the two outer layers are the same (either P-material or N-material), and are doped with the same impurity. The inner layer is the opposite type of material and doped with a different impurity.

The three parts of a transistor are called the **base, emitter,** and **collector.** The emitter and collector are the outer layers and the base is the inner layer. If the emitter and collector are N-material, the base is P-material. If the emitter and collector are P-material, the base is N-material. Transistors are described as either PNP or

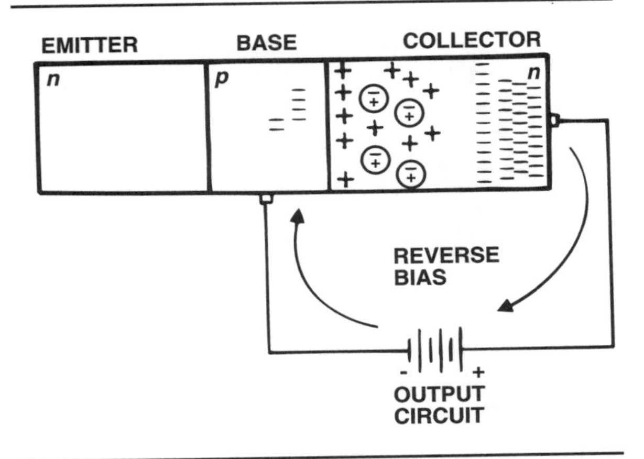

Figure 14-21. The base does not have enough free electrons to combine with holes in the collector, and little reverse current flows.

NPN construction. The circuit diagram symbols for the two kinds of transistors are shown in figure 14-18, along with illustrations of how the materials are joined in their construction. Note that the arrowhead is always shown on the emitter branch of the symbol and indicates the direction of hole flow. The emitter always produces the majority current carriers, either holes (P-material) or free electrons (N-material), and transfers them to the base. Because this type of transistor uses both holes and electrons to carry current, it is called a **bipolar** transistor.

Transistor operation

Transistors can be used to control current flow in solid-state electronic systems. They work like mechanical relays, which allow small amounts of current to trigger the flow of larger amounts of current, but with no moving parts. The most common automotive uses of transistors are as relays, although they also can be used to amplify voltage or current in electronic systems. In the following paragraphs, we will briefly look at how a transistor works as a solid-state relay. These are the basic principles on which many electronic systems operate.

To use a transistor as a simple solid-state relay, the emitter-base junction must be forward biased, and the collector-base junction must be reverse biased. Figure 14-19 shows this arrangement for an NPN transistor.

Input and output circuits

The input circuit for an NPN transistor is the emitter-base circuit, figure 14-20. Because the base is thinner and doped less than the emitter, it has fewer holes than the emitter has free electrons. Therefore, when forward bias is applied, the numerous free electrons from the emitter do not find enough holes to combine with in the base. In this condition, the free electrons accumulate in the base and eventually restrict further current flow.

The output circuit for this NPN transistor is shown in figure 14-21 with reverse bias applied. The base is thinner and doped less than the collector. Therefore, under reverse bias, it has few minority carriers (free electrons) to combine

Depletion Region: An area near the junction of a diode where P-material is depleted of holes and N-material is depleted of electrons.

Forward Bias: The application of voltage to produce current flow across the junction of a semiconductor.

Reverse Bias: The application of voltage so that normally no current will flow across the junction of a semiconductor.

Peak Inverse Voltage (PIV): The highest reverse bias voltage that can be applied to a junction of a diode before its atomic structure breaks down and allows current to flow.

Transistor: A three-element semiconductor device of NPN or PNP materials that transfers electrical signals across a resistance.

Base: The inner layer of semiconductor material in a transistor.

Emitter: The outside layer of semiconductor material in a transistor that conducts current to the base.

Collector: The outside layer of semiconductor material in a transistor that conducts current away from the base.

Bipolar: A transistor that uses both holes and electrons as current carriers.

■ Tran(sfer) + (Re)sistor

The word "transistor" was originally a business trademark used as a name for an electrical part that transferred electrical signals across a resistor. The point-contact transistor was invented by two scientists, John Bardeen and Walter H. Brattain, at Bell Telephone Laboratories in 1948. In 1951, the junction transistor was invented by their colleague, William Shockley. As a result of their work, these three men received the 1956 Nobel Prize for physics.

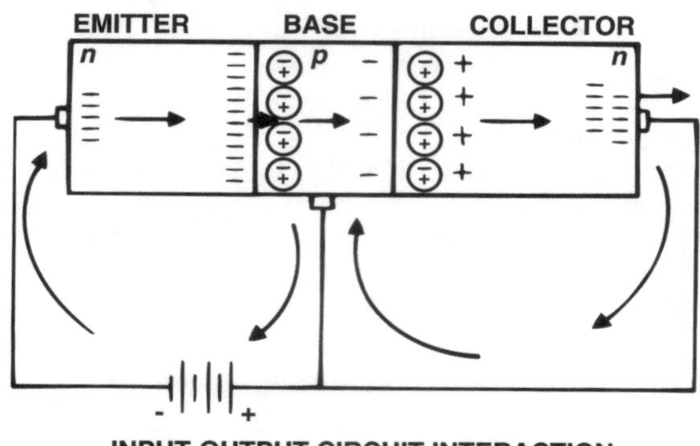

Figure 14-22. Electrons from the emitter pass through the base to the collector, and greater overall forward current flows. Note that greater reverse current flows in the output circuit because more holes in the collector are filled by free electrons.

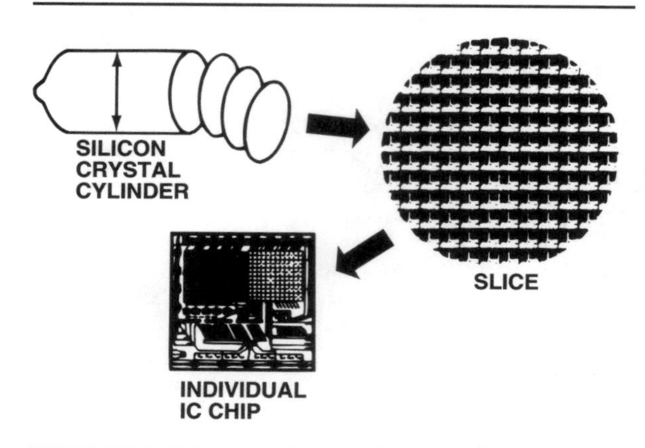

Figure 14-23. Each small square on the silicon slice or wafer becomes a complete IC chip. (Chrysler)

with many minority carriers (holes in the collector). When the collector-base output circuit is reverse biased, very little reverse current will flow. This is similar to the effect of forward biasing the emitter-base input circuit, in which very little forward current will flow.

When the input and output circuits are connected, with the forward and reverse biases maintained, the overall operation of the NPN transistor changes, figure 14-22. Now, the majority of free electrons from the emitter, which could not combine with the base, are attracted through the base to the holes in the collector. The free electrons from the emitter are moved by the negative forward bias toward the base, but most pass through the base to the holes in the collector, where they are attracted by a positive bias. Reverse current flows in the collector-base output circuit, but the overall current flow in the transistor is forward.

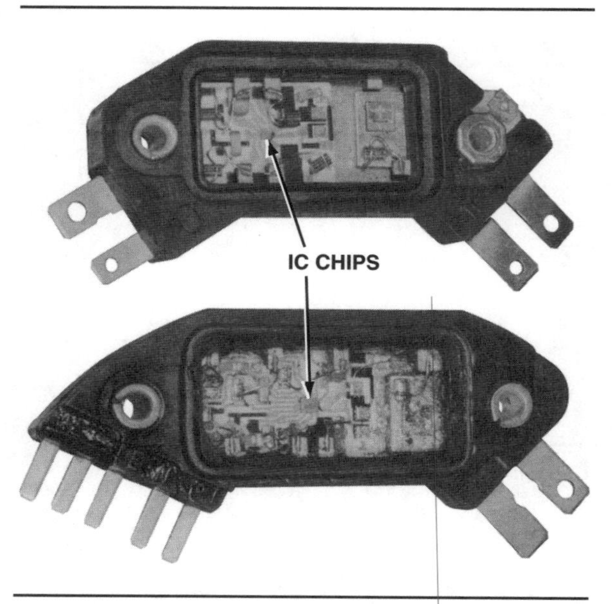

Figure 14-24. The Delco HEI and HEI-EST ignition modules use IC chips.

A slight change in the emitter-base bias causes a large change in emitter-to-collector current flow. This is similar to a small amount of current flow controlling a large current flow through a relay. As the emitter-base bias changes, either more or fewer free electrons are moved toward the base. This causes the collector current to increase or decrease. If the emitter-base circuit is opened or the bias is removed, no forward current will flow through the transistor because the collector-base junction acts like a PN diode with reverse bias applied.

Discrete Devices vs. Integrated Circuits

Each of the electronic devices we have seen so far is an individual component with its own leads for installing it in a circuit. Such solid-state components are often called **discrete devices.** Good examples are the rectifying diodes used on older alternators; each diode is an individual unit and is connected into the circuit by itself.

Although discrete devices are still in use, most electronic systems now use **integrated circuits (IC).** These are complex and small electronic circuits that contain thousands of transistors, diodes, and other devices on a silicon chip smaller than a fingernail. This has drastically reduced the size and cost of electronic components.

An IC is made from a slice or wafer of silicon crystal. Its circuit pattern is reproduced by a photographic and diffusion process to create hundreds of identical circuits, each containing transistors, diodes, conductors, and capacitors

Figure 14-25. A GM onboard computer module.

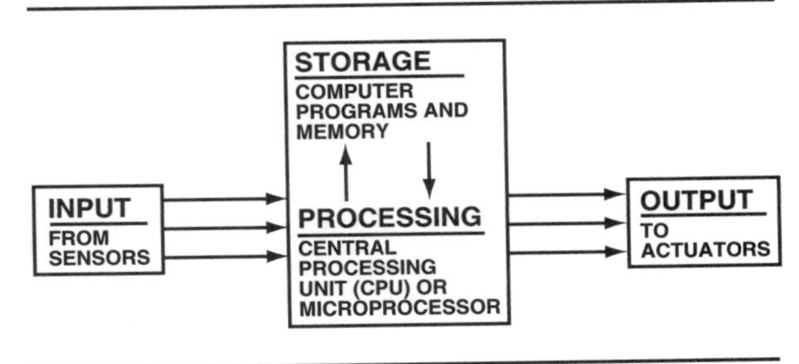

Figure 14-26. Any computer system performs four basic functions: input, processing, storage, and output.

within a tiny "chip" on the wafer. Once the IC's have been formed, each chip is cut from the wafer. Figure 14-23 summarizes the major steps in IC manufacture. The individual chip is then incased in its own packaging device and installed in a larger circuit. Figure 14-24 shows IC circuit use in ignition control modules.

Printed Circuits

Printed circuits are often used to connect individual solid-state components. An etched conductor is used on one or both sides of a circuit board to connect the various electronic components. The components are then soldered to the circuit board, as shown in figure 14-25.

COMPUTER CONTROL

Modern automotive control systems consist of a network of electronic sensors, actuators, and computer modules designed to regulate the powertrain and vehicle support systems. The **powertrain control module (PCM)** is the heart of this system. It coordinates engine and transmission operation, processes data, maintains communications, and makes the control decisions needed to keep the vehicle operating.

Automotive computers use voltage to send and receive information. As we have learned, voltage is electrical pressure and does not flow through circuits. It causes current, which does the real work in an electrical circuit. However, voltage can be used as a signal. A computer converts input information or data into voltage signal combinations that represent number combinations. The number combinations can represent a variety of information—temperature, speed, or even words and letters. A computer processes the input voltage signals it receives by computing what they represent, and then delivering the data in computed or processed form.

The Four Basic Computer Functions

The operation of every computer can be divided into four basic functions, figure 14-26:

- Input
- Processing
- Storage
- Output.

These basic functions are not unique to computers; they can be found in many noncomputer systems. However, we need to know how the computer handles these functions.

First, the computer receives a voltage signal (*input*) from an input device. The device can be as simple as a button or a switch on an instrument panel, or a sensor on an automotive engine. Typical types of automotive sensors are shown in figure 14-27. The keyboard on your personal computer or the programming keyboard of a video-cassette recorder are other examples of an input device.

Modern automobiles use various mechanical, electrical, and magnetic sensors to measure factors such as vehicle speed, engine rpm, air pressure, oxygen content of exhaust gas, airflow, and engine coolant temperature. Each sensor transmits its information in the form of voltage signals. The computer receives these voltage sig-

Discrete Devices: An independent, separately manufactured component with wire leads for connection into a circuit.

Integrated Circuits (IC): A very small, complex electronic circuit that contains thousands of transistors and other devices on a tiny silicon chip.

Powertrain Control Module (PCM): The main system computer module that controls the functions of the vehicle computer system.

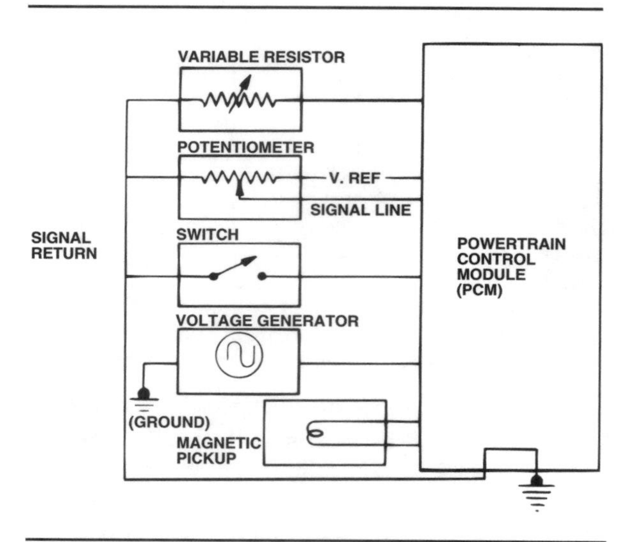

Figure 14-27. The PCM relies on sensors to monitor engine operating conditions. (Ford)

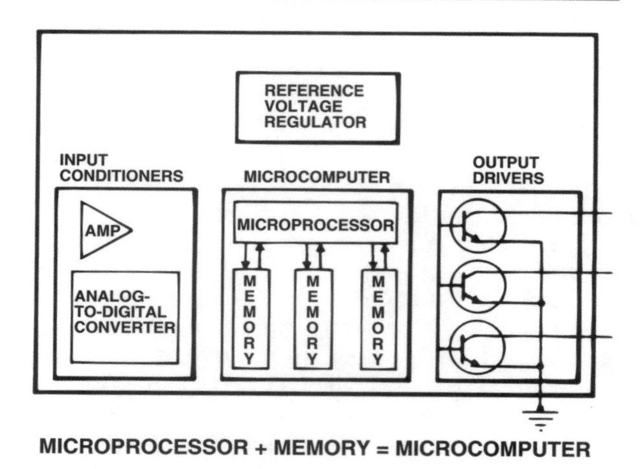

Figure 14-28. Input conditioners within the computer amplify weak voltage signals and convert analog signals to digital ones. (Ford)

nals, but before it can use them, the signals must undergo a process called **input conditioning.** This process includes amplifying voltage signals that are too small for the computer circuitry to handle. Input conditioners generally are located inside the computer, figure 14-28, but a few sensors have their own input-conditioning circuitry.

Second, input voltage signals received by a computer are *processed* through a series of electronic logic circuits maintained in its programmed instructions. These logic circuits change the input voltage signals, or data, into output voltage signals, or commands.

Third, the program instructions for a computer are *stored* in electronic memory. Some programs may require that certain input data be

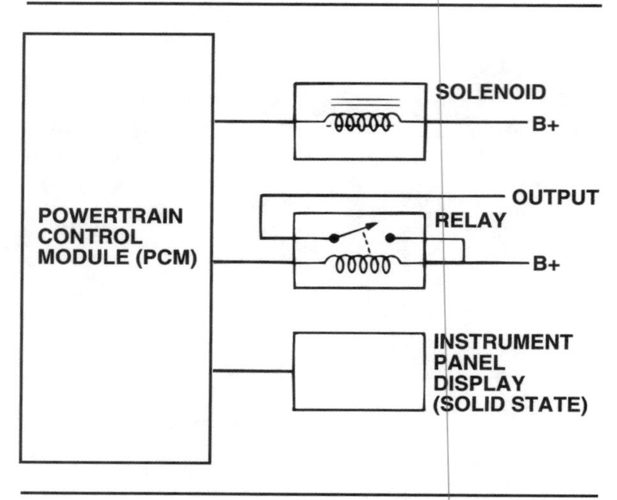

Figure 14-29. An actuator converts a PCM-generated voltage signal into mechanical action. (Ford)

stored for later reference or future processing. In others, output commands may be delayed or stored before they are transmitted to devices elsewhere in the system. Computers use a number of different memory devices, which we look at later in this chapter.

Fourth, after the computer has processed the input signals, it sends voltage signals or commands to other devices in the system, such as system actuators, figure 14-29. An **actuator** is an electrical or mechanical device that converts electrical energy into a mechanical action, such as adjusting engine idle speed, altering suspension height, or regulating fuel metering.

Computers also can communicate with, and control each other through their output and input functions. This means that the output signal from one computer system can be the input signal for another computer system. General Motors (GM) introduced one of the earliest body computer modules (BCM's) on some of its models in 1986. The BCM acts as a master control unit by managing a network containing all sensors, switches, and other vehicle computers, figure 14-30.

As an example, let us suppose the BCM sends an output signal to disengage the air conditioning compressor clutch. That same output signal can become an input signal to the PCM that controls engine operation. Based on the signal from the BCM, the PCM signals an actuator to reduce engine speed to account for the decreased load of the compressor. This in turn affects the fuel metering system.

The four basic functions described above are common to all computers, regardless of size or purpose. They also form an organizational pattern to troubleshoot a malfunctioning system.

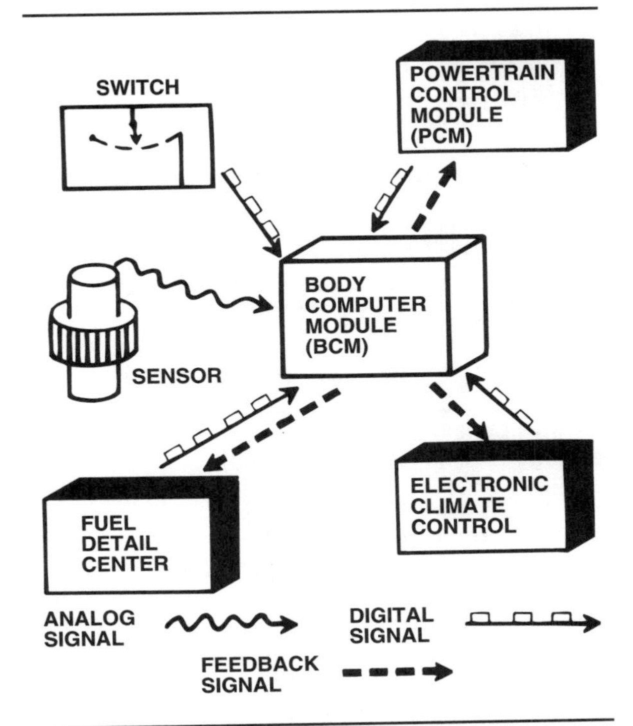

Figure 14-30. Cadillac's BCM accepts inputs from a variety of sources and manages the other on-board computer systems.

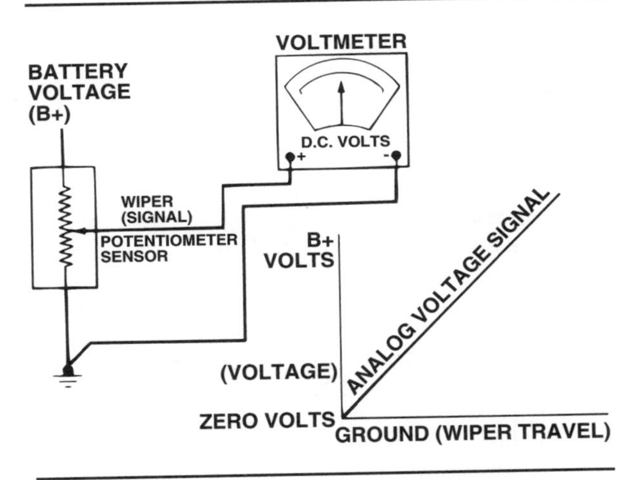

Figure 14-31. An analog signal is continuously variable. (Ford)

While most input and output devices can be adjusted or repaired, the processing and storage functions can only be replaced.

Analog and Digital Systems

A computer has to be told how to do its job. The instructions and data necessary to do this are called the program. Since a computer cannot read words, the information must be translated

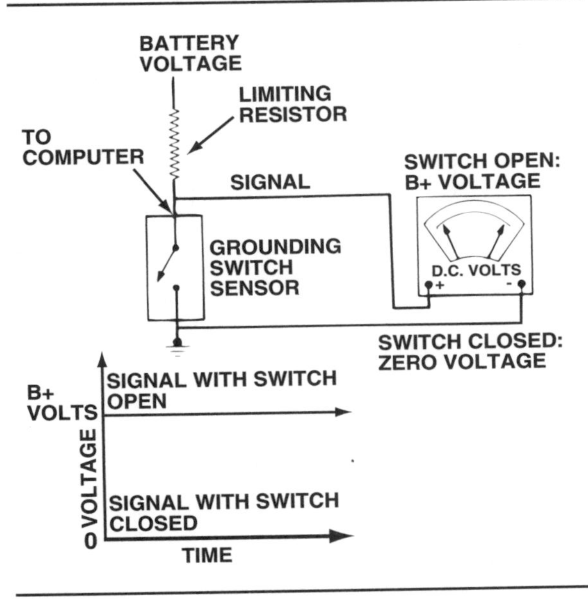

Figure 14-32. A digital signal is a simple on/off, or voltage/no voltage, signal. (Ford)

into a form the computer can understand—voltage signals. This can be done using an analog or a digital system.

An **analog** computer is one in which the voltage signal or processing function is continuously variable, relative to the function being measured or the adjustment required, figure 14-31. Most operating conditions affecting an automobile, such as engine speed, are analog variables. These operating conditions can be measured by sensors. For example, engine speed does not change abruptly from idle to wide-open throttle. It varies in clearly defined, finite, measurable steps—1,500 rpm, 1,501 rpm, 1,502 rpm, and so on. The same is true for temperature, fuel metering, airflow, vehicle speed, and other factors.

If a computer is to measure engine speed changes from 0 rpm through 6,500 rpm, it can be programmed to respond to an analog voltage that varies from 0 volts at 0 rpm to 4.9 volts at 6,500 rpm. Any analog signal between 0 and 4.9

Input Conditioning: The process of amplifying or converting a voltage signal into a form usable by the computer's central processing unit.

Actuator: An electrical or mechanical device that receives an output signal from a computer and performs an action in response to that signal.

Analog: A voltage signal or processing action that is continuously variable, relative to the operation being measured or controlled.

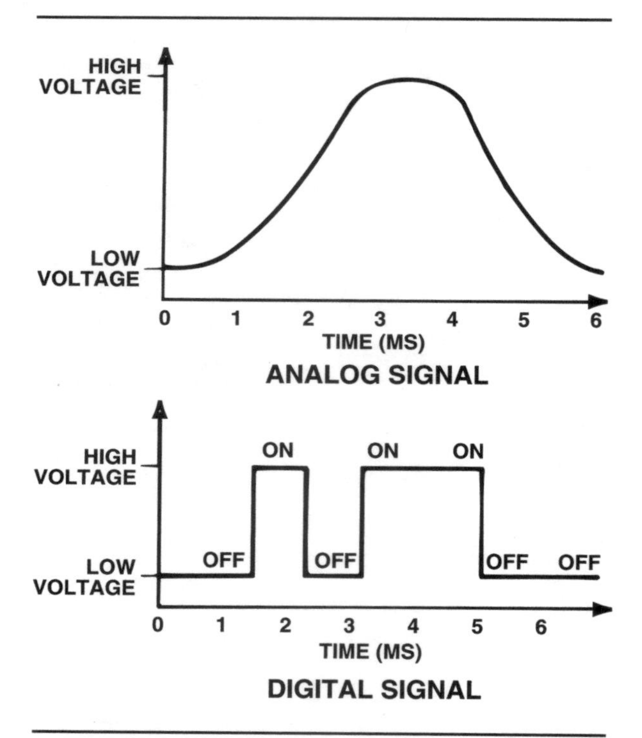

Figure 14-33. An analog signal may take the form of a sine wave; a digital signal is a "square" wave. (Ford)

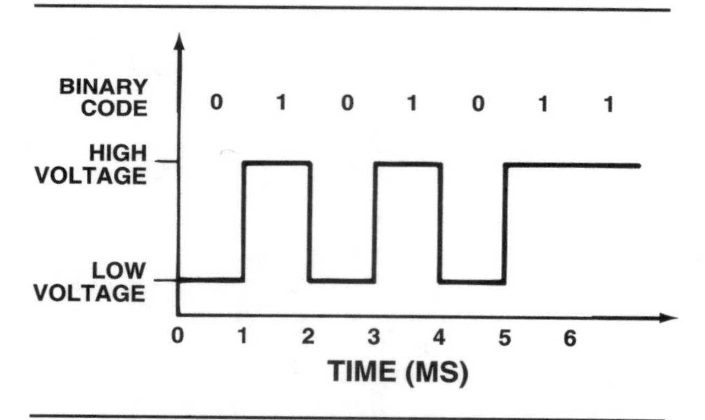

Figure 14-34. Digital computers receive information in a form called binary code. As shown here, a low-voltage signal is represented by 0; a high-voltage signal is 1. (Ford)

volts will represent a proportional engine speed between 0 and 6,500 rpm.

Analog computers have several shortcomings, however. They are affected by temperature changes, supply voltage fluctuations, and signal interference. They also operate slowly, and are costly to manufacture.

In a **digital** computer, the voltage signal or processing function is a simple high/low, yes/no, on/off signal. The digital signal voltage is limited to two voltage levels: high voltage and low voltage, figure 14-32. Since there is no stepped

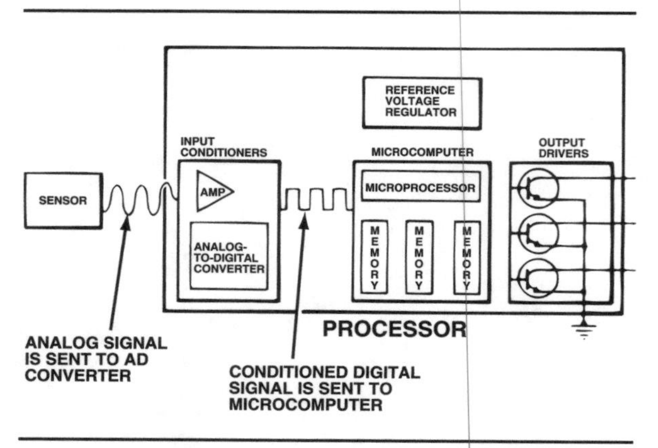

Figure 14-35. The processor must precondition an analog signal before using it. (Ford)

range of voltage or current in between, a digital binary signal is a "square wave", figure 14-33.

Using our engine speed example above, suppose the computer needs to know that engine speed is either above or below a specific level, say 1,800 rpm. Keep in mind that it does not need to know the exact engine speed, only whether the speed is above or below 1,800 rpm. In this case, the low-voltage portion of a digital signal can represent speeds below 1,800 rpm and the high-voltage portion, speeds above 1,800 rpm. As you can see, a digital signal functions like a simple switch, figure 14-32.

An engineer can reverse the switch functions to provide a high input signal below 1,800 rpm and a low (zero voltage) signal above 1,800 rpm. The result at the computer will be the same. The computer will get a simple digital input signal that represents a *change in operating conditions.*

The signal is called "digital" because the on and off signals are processed by the computer as the digits or numbers 0 and 1. The number system containing only these two digits is called the **binary** system. Any number or letter from any number system or language alphabet can be translated into a combination of binary 0's and 1's for the digital computer, figure 14-34.

A digital computer changes the analog input signals (voltage) to digital bits (BInary digiTS) of information through an **analog-to-digital (AD) converter** circuit, figure 14-35. The binary digital number, figure 14-36, is used by the computer in its calculations or logic networks. Output signals usually are digital signals that turn system actuators on and off. A digital signal can be changed to an analog output signal through a **digital-to-analog (DA) converter.** This is the opposite of the AD converter circuit that changes analog input signals. More often, however, a digital output signal is made to approximate an analog signal through a variable-duty cycle, discussed in a later chapter.

The digital computer can process thousands of digital signals per second because its circuits are able to switch voltage signals on and off in billionths of a second.

Analog-to-Digital Conversion

We mentioned earlier that most operating conditions that affect an automobile are analog variables. When our computer needs to know whether an operating condition is above or below a specified point, a digital sensor can be used to act as a simple off/on switch. Below the specified point, the switch is open. The computer receives no voltage signal until the condition reaches the specified point, at which time the switch closes. This is an example of a simple digital off/on circuit: off = 0, on = 1.

Using engine coolant temperature as an example, we specify that the computer needs to know the exact temperature within one degree. Suppose our sensor measures temperature from 0° to 300°F (-18° to 149°C) and sends an analog signal that varies from 0 to 6 volts. Each 1-volt change in the sensor signal is the equivalent of a 50°F (28°C) change in temperature. If 0 volts equals a temperature of 0°F (-18°F) and 6 volts equals 300°F (149°C):

1.0 volt = 50°F (28°C)
0.5 volt = 25°F (14°C)
0.1 volt = 5°F (3°C)
0.02 volt = 1°F (0.6°C)

In order for the computer to determine temperature within 1°F (0.6°C), it must react to sensor

Digital: A two-level voltage signal or processing function that is either on/off or high/low.

Binary: A mathematical system consisting of only two digits (0 and 1) that allows a digital computer to read and process input voltage signals.

Analog-to-Digital (AD) Converter: An electronic conversion process for changing analog voltage signals to digital voltage signals.

Digital-to-Analog (DA) Converter: An electronic conversion process for changing digital voltage signals to analog voltage signals.

Digital Logic

All digital computers handle data bits with three basic logic circuits called logic gates: the NOT, AND, and OR gates. This terminology is used to describe circuit switching functions only—it has nothing to do with their physical construction. Logic gates are called gates because the circuits act as routes or gates for output voltage signals according to different input signal combinations. The thousands of field-effect transistors (FET's) in a microprocessor are logic gates.

A digital computer does its job by switching output voltage on and off according to the input voltage signals. When input voltage enters a logic gate, its transistors can change from a cutoff state (no voltage) to full saturation (voltage). This is equal to an off (low) signal and an on (high) signal. By combining input and output signals in logical combinations, they can be made to equal binary numbers.

The most elementary logic gate is called a NOT gate and inverts the signal. When voltage to its single input terminal is high or on, the output voltage is low or off.

The AND gate has two inputs and one output. Its output is high only if both inputs are high. If one or both inputs are low, output is low.

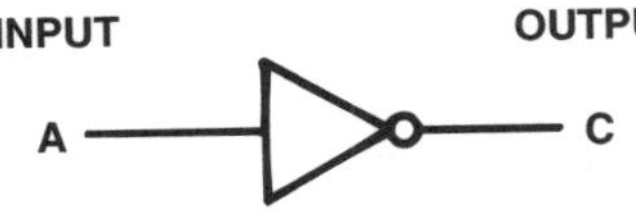

NOT GATE		AND GATE			OR GATE		
INPUT A	OUTPUT C	INPUTS A	INPUTS B	OUTPUT C	INPUTS A	INPUTS B	OUTPUT C
1	0	0	0	0	0	0	0
0	1	0	1	0	0	1	1
		1	0	0	1	0	1
		1	1	1	1	1	1

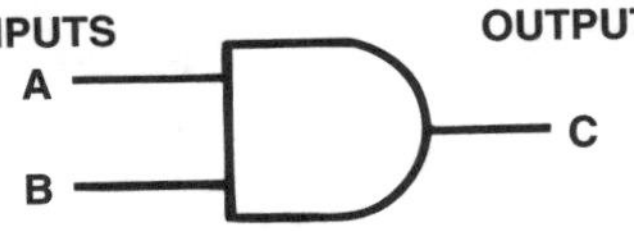

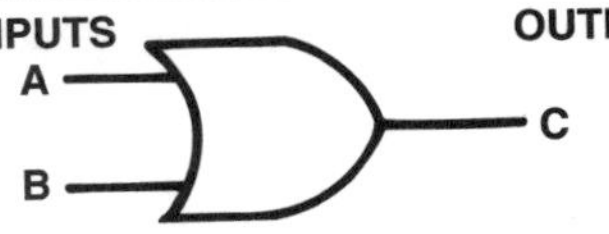

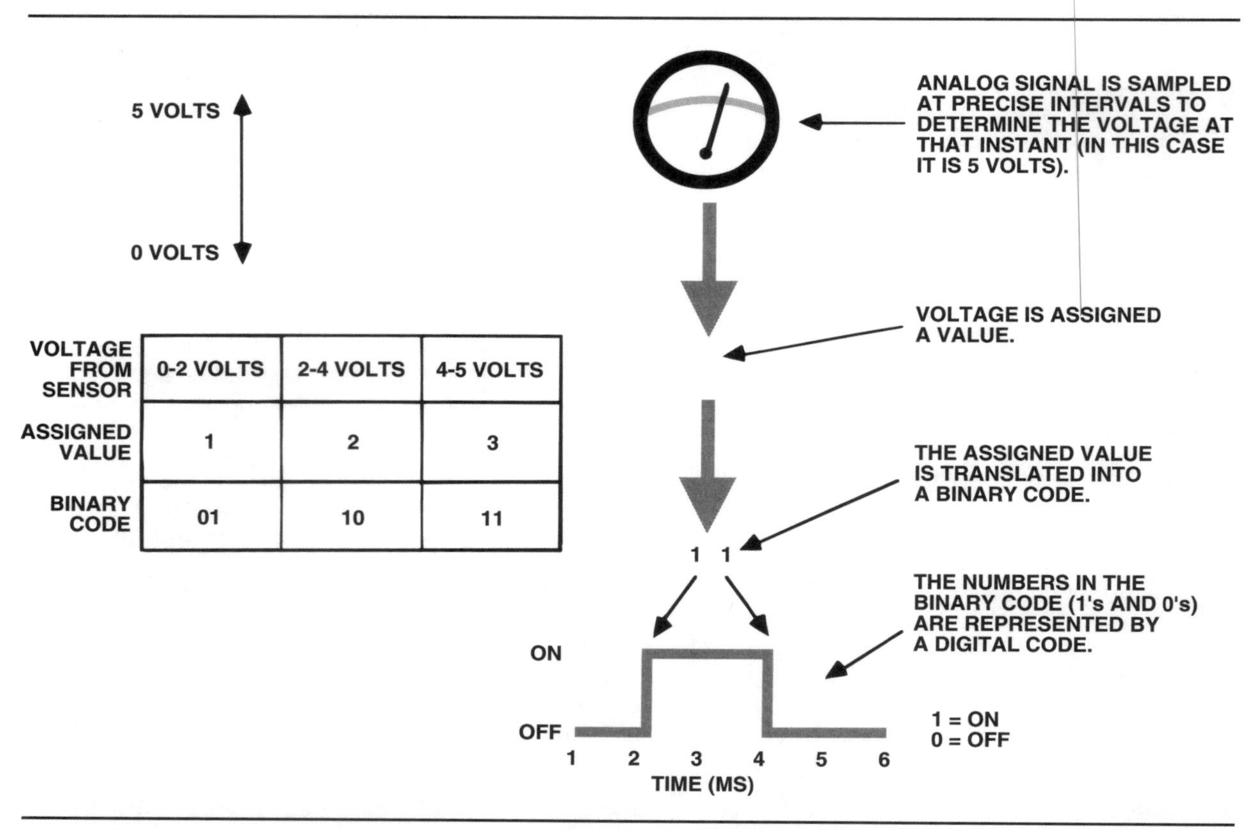

Figure 14-36. The process of analog-to-digital conversion performed by the converter section of the computer. (Ford)

voltage changes as small as 0.020 volt or 20 millivolts. For example, if the temperature is 125°F (51.7°C), the sensor signal will be 2.50 volts. If the temperature rises to 126°F (52.2°C), sensor voltage increases to 2.52 volts. In reality, temperature does not pass directly from one degree to another; it passes through many smaller increments, as does voltage as it changes from 2.50 to 2.52 volts. Our digital computer, however, processes only signals equaling 1°F (0.6°C) changes in temperature. To do so, the computer sends the signal through AD conversion circuits, where the analog sensor voltage is converted to a series of 0.020-volt changes for each degree. This is called "digitizing" an analog signal.

The AD conversion process brings us back to binary numbers. Transistors can be designed to switch on and off at different voltage levels or with differing combinations of voltage signals. In the computer we are discussing, transistor groups must switch from off to on at 20-millivolt increments. The input signal is created by varying the transistor combinations that are on or off. Since the computer can read the various voltage signal combinations as binary numbers, it performs its calculations. It does so almost instantly because the current travels through the miniature circuits at almost the speed of light.

Parts of a Computer

We have dealt with some of the basic functions and logical processes used in a computer. The software consists of the programs and logic functions stored in the computer's circuitry. The hardware is the mechanical and electronic parts of a computer. Figure 14-37 shows the physical location of the most important components of a computer.

Central processing unit (CPU)

Mentioned earlier, the microprocessor is the **central processing unit (CPU)** of a computer. Since it performs the essential mathematical operations and logic decisions that make up its processing function, the CPU can be considered the heart of a computer. Some computers use more than one microprocessor, called a coprocessor.

Computer memory

Other IC devices store the computer operating program, system sensor input data, and system actuator output data, information necessary for CPU operation, figure 14-38.

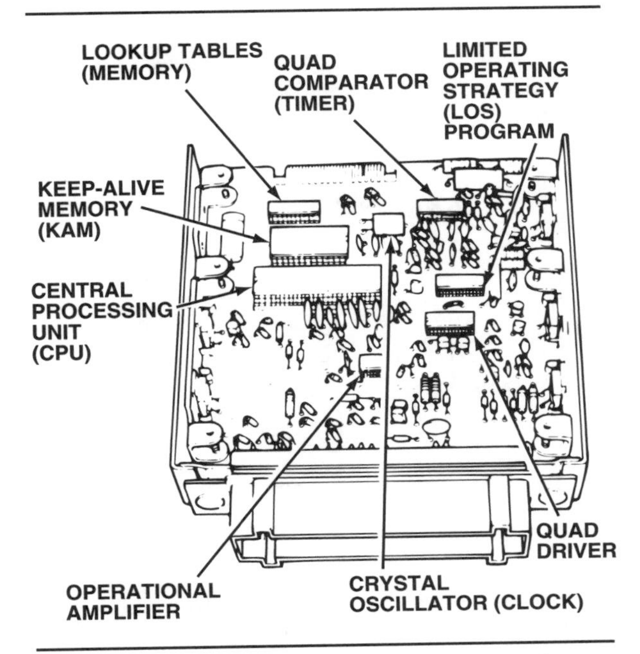

Figure 14-37. Some of the basic components of a PCM, housed in a metal box for protection. (Ford)

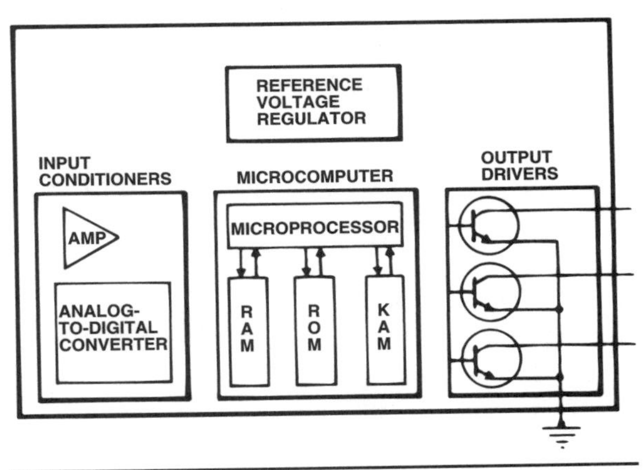

Figure 14-38. The microprocessor stores data into different types of memory —RAM, ROM, and KAM. (Ford)

Computers have two types of memory: permanent and temporary. Permanent memory is called **read-only memory (ROM)** because the computer can only read the contents; it cannot change the data stored in it. This data is retained even when power to the computer is shut off. Part of the ROM is built into the computer, and the rest is located in an IC chip called a **programmable read-only memory (PROM)** or calibration assembly.

Temporary memory is called **random-access memory (RAM)** because the microprocessor can write or store new data into it as directed by the computer program, as well as read the data already in it. Some computer people call it scratchpad memory because the CPU uses it as a notepad. Automotive computers use two types of RAM memory: volatile and nonvolatile. Volatile RAM memory is lost whenever the ignition is turned off. However, a type of volatile RAM called **keep-alive memory (KAM)** can be wired directly to battery power. This prevents its data from being erased when the ignition is turned off. Both RAM and KAM have the disadvantage of losing their memory when disconnected from their power source. One example of RAM and KAM is the loss of station settings in a programmable radio when the battery is disconnected. Since all the settings are stored in RAM, they have to be reset when the battery is reconnected. System trouble codes are commonly stored in RAM and can be erased by disconnecting the battery.

Nonvolatile RAM memory can retain its information even when the battery is disconnected. One use for this type of RAM is the storage of odometer information in an electronic speedometer. The memory chip retains the mileage accumulated by the vehicle. When speedometer replacement is necessary, the odometer chip is removed and installed in the new speedometer unit. KAM is used primarily in conjunction with adaptive strategies, which we study in the next chapter.

Computer programs

Every computer needs instructions to do its job. These instructions are called computer programs. The program for an engine control com-

Central Processing Unit (CPU): The processing and calculating portion of a computer.

Read-Only Memory (ROM): The permanent part of a computer's memory storage function. ROM can be read but not changed, and is retained when power is shut off to the computer.

Programmable Read-Only Memory (PROM): An integrated circuit chip installed in a computer that contains appropriate operating instructions and database information for a particular application.

Random-Access Memory (RAM): Temporary short-term or long-term computer memory that can be read and changed, but is lost whenever power is shut off to the computer.

Keep-Alive Memory (KAM): A form of long-term RAM used mostly with adaptive strategies. Requires a separate power supply circuit to maintain voltage when the ignition is off.

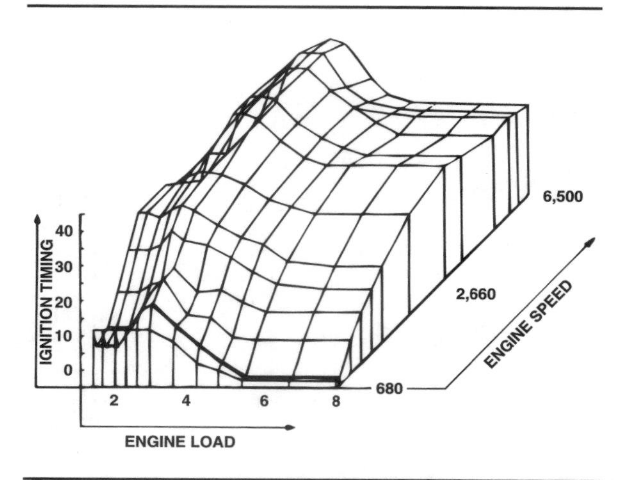

Figure 14-39. Automotive engineers program PROM's using the process of engine mapping. (Porsche)

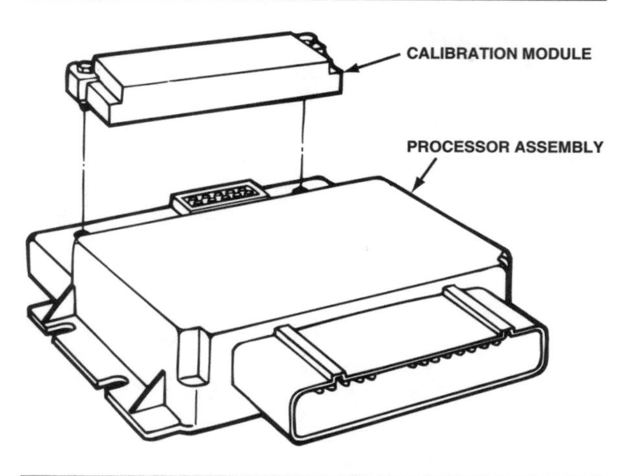

Figure 14-41. This Ford calibration module contains a system PROM.

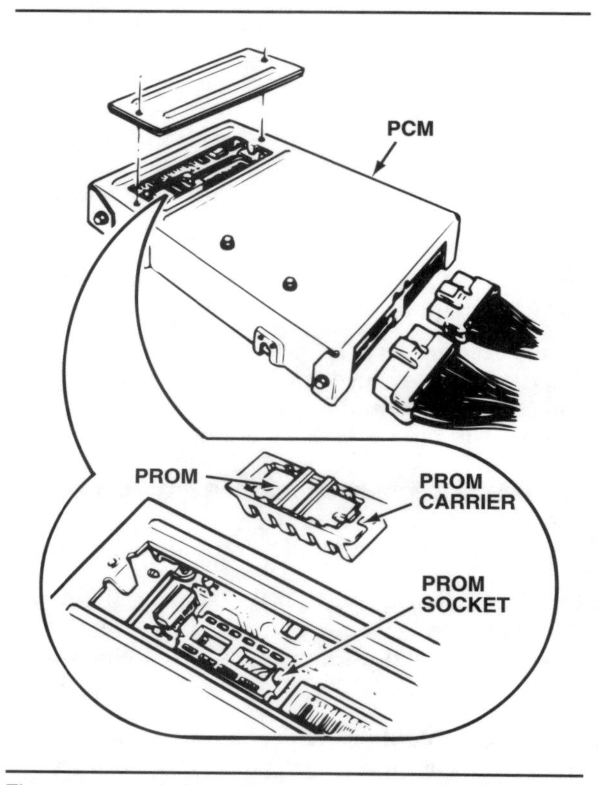

Figure 14-40. A GM PCM with a replaceable PROM.

puter consists of several elements. First are the mathematical instructions that tell the computer how to process, or "compute", the information it receives. Second is information that pertains to *fixed* vehicle values such as the number of cylinders, engine compression ratio, transmission type and gear ratios, firing order, and emission control devices. Finally, there is data that pertains to *variable* vehicle values such as engine speed, car speed, coolant temperature, intake airflow, fuel flow, ignition timing, and others.

Since the mathematical instructions and engine data are constant values, they are fixed and are easily placed into computer memory. To place the variable values into memory, it is necessary to simulate the vehicle and its system in operation. A large mainframe computer at the automaker's factory is used to calculate all of the possible variable conditions for any given system. This provides the control program for the individual on-board computers.

This is called system simulation and involves a process called **engine mapping.** By operating a vehicle on a dynamometer and manually adjusting the variable factors such as speed, load, and spark timing, it is possible to determine the optimum output settings for the best driveability, economy, and emission control.

Engine mapping creates a three-dimensional performance graph that applies to a given vehicle and powertrain combination, figure 14-39. Each combination is mapped in this manner to produce a PROM. This allows an automaker to use one basic computer for all models; a unique PROM individualizes the computer for a particular model. Also, if a driveability problem can be resolved by a change in the program, the carmaker can release a revised PROM to supersede the earlier part. Some PROM's are made in such a way that they can be erased by exposure to ultraviolet light and reprogrammed. These are called EPROM's, or erasable PROM's.

Most automakers use a single PROM that plugs into the computer, figure 14-40. Some Ford computers use a larger "calibration module" that contains the system PROM, figure 14-41. If the on-board computer needs to be replaced, the

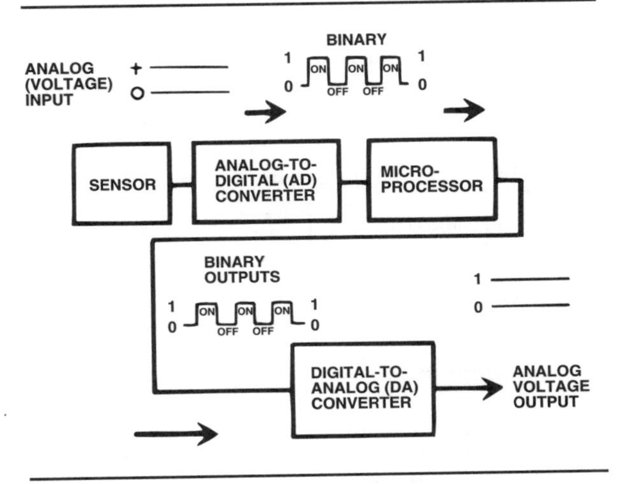

Figure 14-42. Input and output devices allow the microprocessor to communicate with the AD/DA converters. (Chrysler)

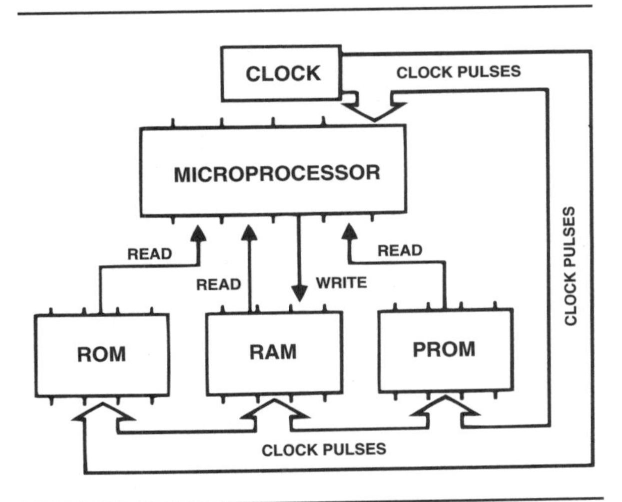

Figure 14-43. The computer's microprocessor and memories monitor the clock pulses. (GM)

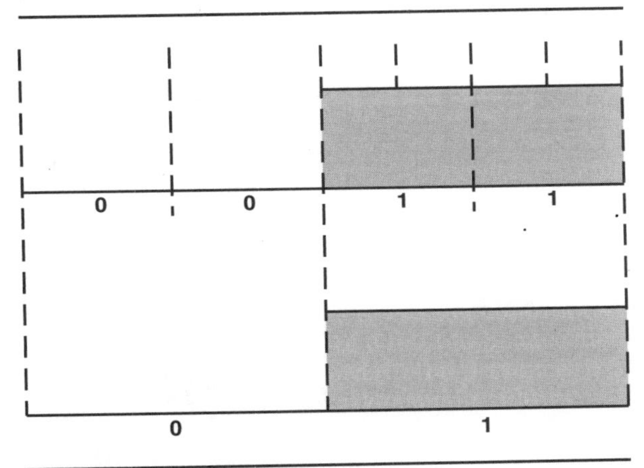

Figure 14-44. To separate voltage signals, the computer uses clock pulses. (GM)

PROM or calibration module must be removed from the defective unit and installed in the replacement computer.

Input and output circuits

A computer is not directly connected to every input or output device. The signals are received and sent by other IC devices, many of which provide the computer with parallel connections. This allows it to receive several input signals while it is sending several output signals.

Converter circuits

The computer must have circuits to convert input data into a form with which it can work. The analog signals we have discussed must be digitized, or changed to digital signals. This conversion is done by separate IC devices called AD converter circuits, figure 14-35.

The computer's output signals also must be converted into a form that the output device can recognize and act on. Since some of the output devices are analog, the digital signals must be changed to analog signals. This conversion is done by DA converter circuits, figure 14-42. Because the circuits that perform these functions "interface" the CPU with the input and output devices, they are sometimes called the input/output (I/O) interface.

Clock rates and timing

You have learned that the microprocessor receives sensor input voltage signals, processes them by using information from other memory units, and then sends voltage signals to the appropriate actuators. The microprocessor communicates by transmitting long strings of 0's and 1's in a language called binary code. But the microprocessor must have some way of knowing when one signal ends and another begins. That is the job of a crystal oscillator called a clock generator.

Clock pulses

The computer's crystal oscillator generates a steady stream of one-bit-long voltage pulses. Both the microprocessor and the memories monitor the clock pulses while they are communicating, figure 14-43. Because they know how long each voltage pulse should be, they can distinguish between a 01 and a 0011, figure 14-44. To

Engine Mapping: A process of simulating vehicle and engine operating conditions to establish values for the computer to use when exercising system control.

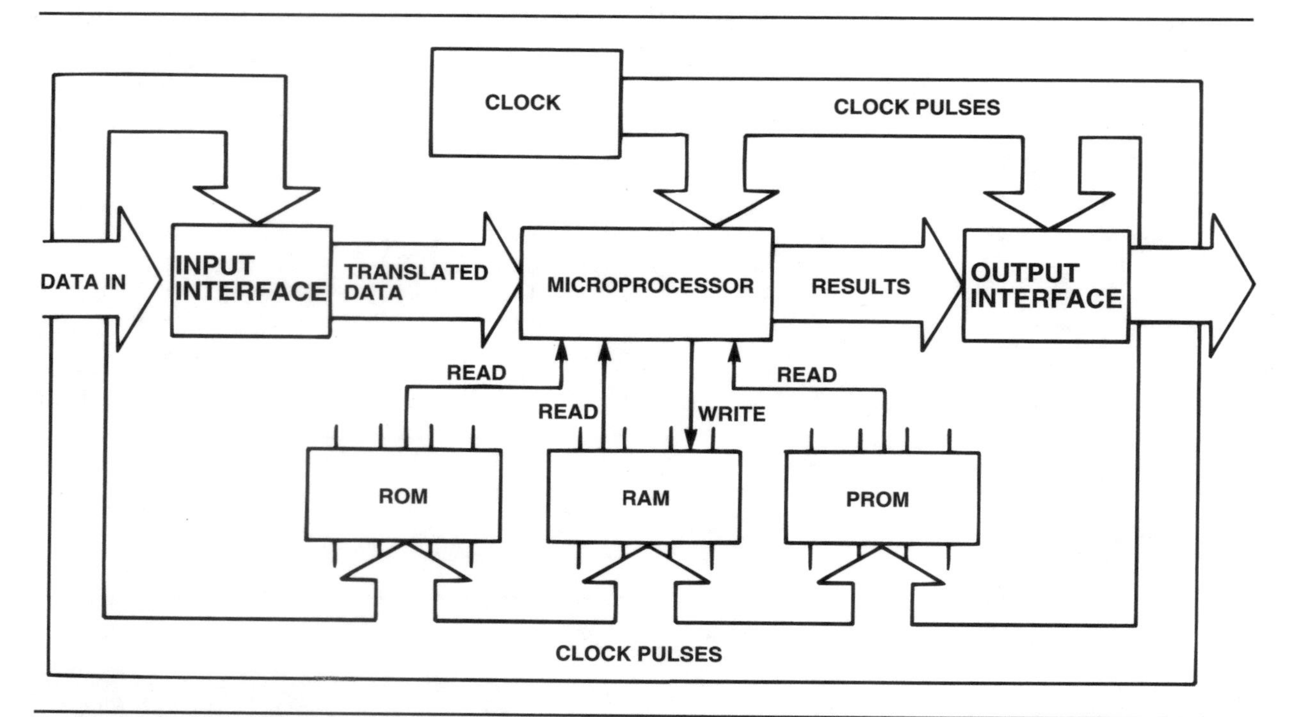

Figure 14-45. Input and output circuits monitor the computer's clock pulses. (GM)

Figure 14-46. Frequently, GM mounts PCM's in the passenger compartment.

complete the process, the input and output circuits also watch the clock pulses, figure 14-45.

Computer speeds

Not all computers operate at the same speed; some are faster than others. The speed at which a computer operates is specified by the cycle time, or clock speed, required to perform certain measurements. Cycle time or clock speed is measured in megahertz (4.7 MHz, 8.0 MHz, 15 MHz, 18 MHz, etc.), but a computer with an 18-MHz microprocessor is not necessarily faster than one running at a clock speed of 15 MHz. External factors such as a slower data transfer rate, an inefficient operating system, or a cumbersome program can cause a computer with a higher clock speed to operate slower than one that works at a lower clock speed. Current GM computers can handle more than 600,000 commands per second; Ford's application of EEC-IV used in the 1993 Mark VIII has 56 kilobytes of memory (versus 32 kilobytes in the previous version) and runs at an 18-MHz clock speed.

Baud rate

The computer transmits bits of a serial data stream at precise intervals. The computer's speed is called the baud rate, or bits per second. Just as mph helps in estimating the length of time required to travel a certain distance, the baud rate is useful in estimating how long a given computer will need to transmit a specified amount of data to another computer. Storage of a single character requires eight bits per byte, plus an additional two bits to indicate stop and start. This means that transmission of one character requires 10 bits. Dividing the baud rate by 10 tells us the maximum number of words per second that can be transmitted. For example, if the computer has a baud rate of 600, approximately 60 words can be received or sent per minute.

Automotive computers have evolved from a baud rate of 160 used in the early 1980s to a baud rate as high as 8,192. The speed of data transmission is an important factor both in system operation and in system troubleshooting.

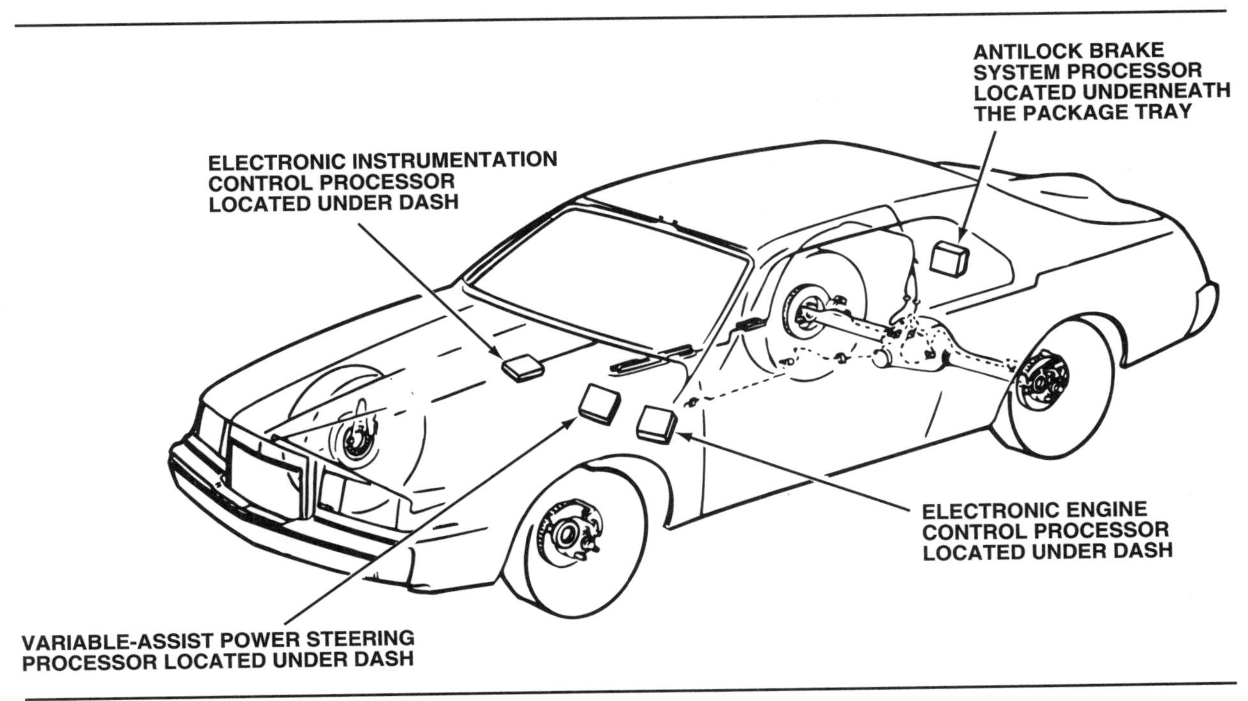

Figure 14-47. Ford locates most on-board computers in the passenger compartment.

The higher the baud rate, the more precisely the computer can control system operation and communicate with actuators or other computers.

Microprocessor bit size

The microprocessor bit size is another factor involved in how fast a computer can operate. Because early microprocessors could transmit only 4 bits at a time, they were called 4-bit processors. Since each byte contains 8 bits, these microprocessors required two passes to process or transmit 1 byte. They were followed by 8-bit processors that could transmit 1 byte in a single pass. Automotive computers generally use 4-bit, 8-bit, and 16-bit microprocessors, according to application.

Control Module Locations

The on-board automotive computer has many names. It may be called an electronic control unit, module, controller, or assembly, depending on the automaker and the computer application. The Society of Automotive Engineering (SAE) bulletin, J1930, standardizes the name as a "powertrain control module (PCM)". The computer hardware is all mounted on one or more circuit boards and installed in a metal case, figure 14-46, to help shield it from electromagnetic interference (EMI). The wiring harnesses that link the computer to sensors and actuators connect to multipin connectors or edge connectors on the circuit boards.

On-board computers range from single-function units that control a single operation to multifunction units that manage all of the separate (but linked) electronic systems in the vehicle. They vary in size from a small module to a notebook-sized box. Chrysler's early computers were attached to the air cleaner housing in the engine compartment. The computers used with Chrysler's modular control system (MCS) in the mid-1980s are two-piece units with the power module installed between the battery and the fenderwell in the engine compartment and the logic module behind a kick panel in the passenger compartment. Chrysler's engine controller that replaced the two-piece MCS computer is a one-piece assembly installed in the same location as the MCS power module. Most other engine computers are installed in the passenger compartment either under the instrument panel or in a side kick panel where they can be shielded from physical damage caused by temperature extremes, dirt and vibration, or interference by the high currents and voltages of various underhood systems, figure 14-47. Single-function units or modules are sometimes located underneath the driver's seat or in the luggage compartment.

FUEL CONTROL SYSTEM OPERATING MODES

A computer-controlled fuel metering system can be selective. Depending on the computer program, it may have different operating modes.

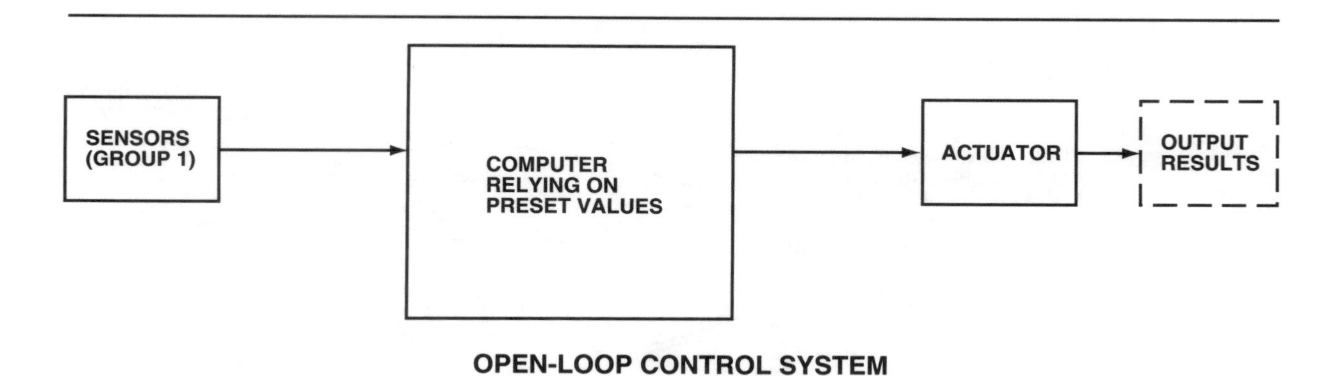

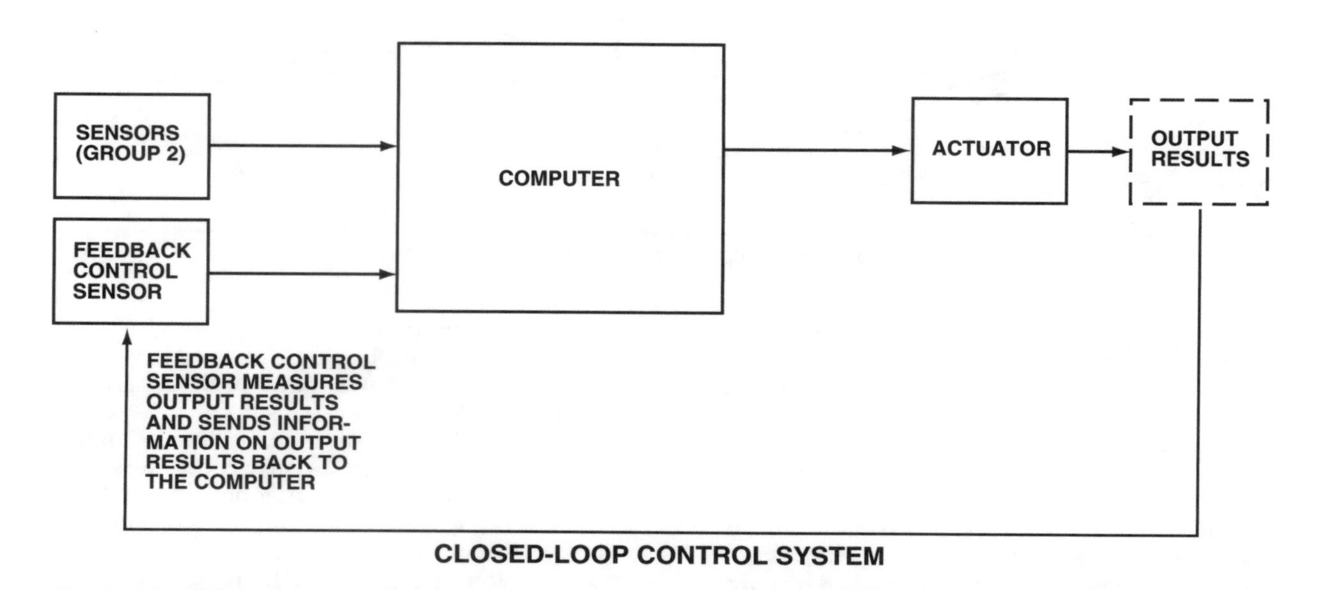

Figure 14-48. A computer-controlled engine system operates in closed or open loop. (Ford)

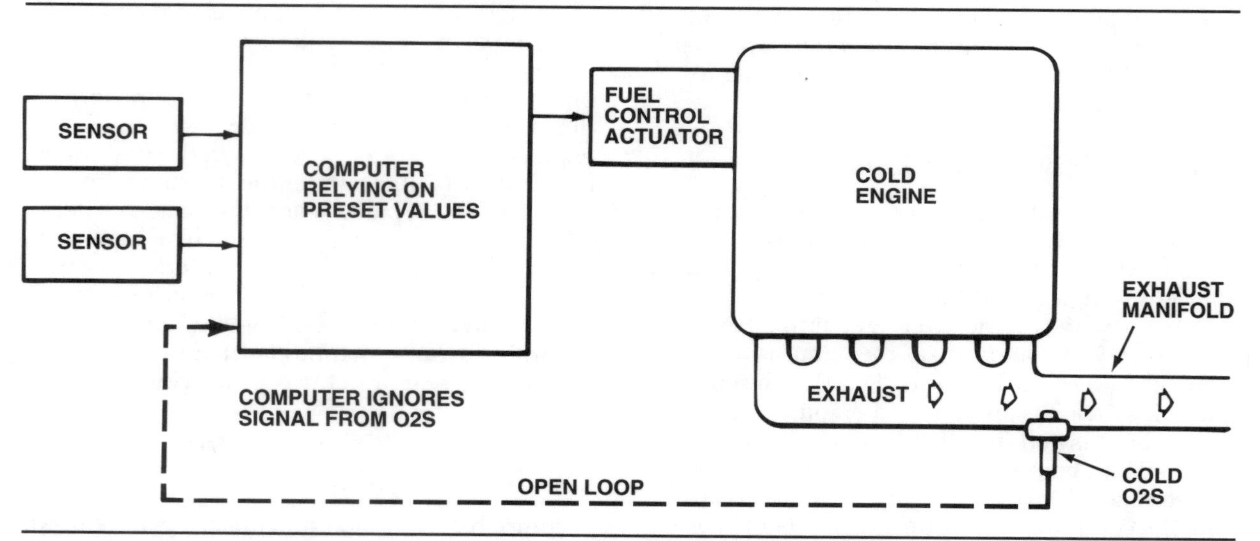

Figure 14-49. In open loop, the computer operates on a predetermined program, and ignores O2S signal input. (Ford)

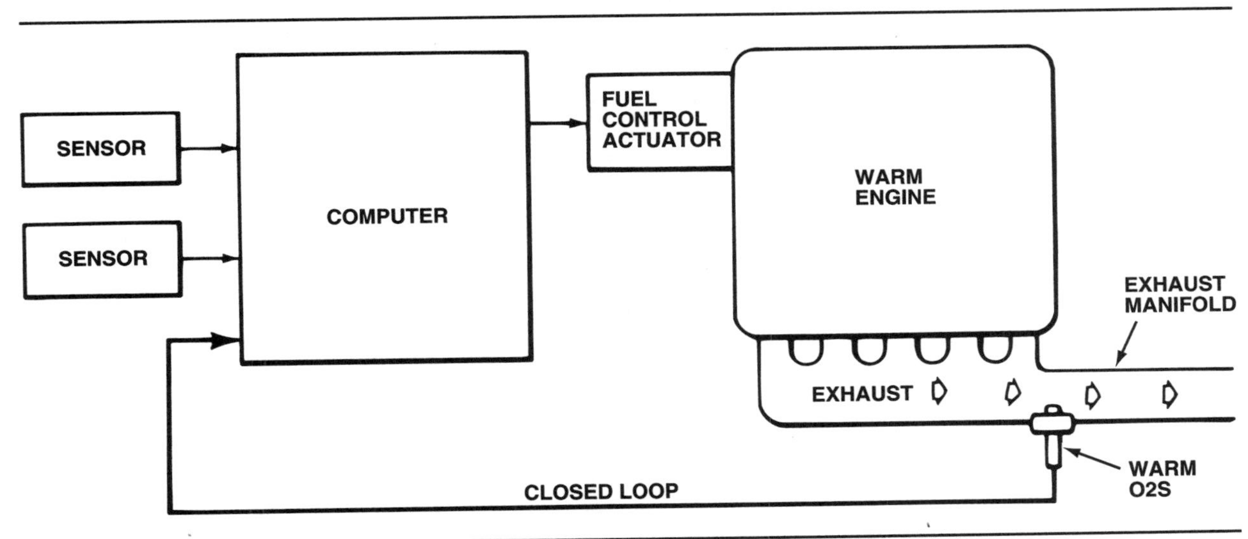

Figure 14-50. In closed loop, the computer accepts the O2S signal and readjusts the air-fuel mixture accordingly. (Ford)

The on-board computer does not have to respond to data from all of its sensors, nor does it have to respond to the data in the same way each time. Under specified conditions, it may ignore sensor input. Or, it may respond in different ways to the same input signal, based on inputs from other sensors. Most current control systems, figure 14-48, have two operating modes: open and closed loop. The most common application of these modes is in fuel-metering feedback control, although there are other open- and closed-loop functions. Air conditioning automatic temperature control is an example. Control logic programmed into the computer determines the choice of operating mode according to engine operating conditions.

Open-Loop Control

Open-loop control means that the on-board computer works according to established conditions in its program. It gives the orders and the output actuators carry them out. The computer ignores sensor feedback signals indicating an error or a deviation in the results of an actuator operation as long as the preestablished conditions exist.

For example, the computer is programmed to provide a specific amount of fuel and spark timing when the engine is first started. Since these factors are predetermined in the program (regardless of other factors), the computer will ignore signals (feedback) from the oxygen sensor (O2S) until enabling requirements such as coolant temperature reaches the predetermined level, figure 14-49.

Open-loop control is not restricted to computer systems. It is present in many other control systems. Suppose you are sitting in a comfortable living room chair reading when you begin to feel a chill in the room. You get up and turn on the switch controlling the furnace. The furnace warms the room to a point where you now feel too warm. You must get up again and shut off the furnace switch. The room cools back down until it gets chilly again and you must repeat the procedure. This is an example of an open-loop control system.

Closed-Loop Control

Closed-loop control means that once certain conditions (such as coolant temperature) have been met, the computer now reads and responds to feedback signals. When the engine is first started (open loop), the computer ignores input from the O2S. As soon as the preconditions specified in the computer program are met, the computer accepts the sensor input and adjusts the air-fuel ratio according to the O2S's signals. We say that the computer is responding to a "feedback" signal; that is, the sensor is telling the computer of an error factor in its operation that must be corrected, figure 14-50. Based on the feedback signals, the computer constantly corrects

Open-Loop Control: An operating mode in which the computer causes a system to function according to predetermined instructions.

Closed-Loop Control: An operating mode in which the computer responds to feedback signals from its sensors, and adjusts system operations accordingly.

its output signals in an effort to eliminate the error factor.

As with open-loop control, closed-loop operation is found in other forms of control systems. For example, suppose you had a temperature control thermostat instead of the on/off switch in the furnace control system discussed above. With the thermostat set at a certain temperature, the thermostat controller turns on the heat when room temperature drops below the setting, and shuts it off when the temperature rises above the setting. This is called a limit-cycle control system because it tries to maintain an average temperature. It does this by operating only when the temperature exceeds or falls under the preselected limits. The limits used depend on system design. For the simple temperature control system we have discussed, the limits fall within five percent. Thus, if the thermostat is set for a temperature of 70°F (21°C), the controller will turn the furnace on at approximately 66°F (19°C) and off at about 74°F (23°C). In this way, the system maintains an average room temperature of approximately 70°F (21°C).

Another type of closed-loop control is called proportional control. In this type of control system, the computer subtracts the feedback signal from the previous output signal to determine the error signal. The error signal then is used by the computer to change the next output signal. If the computer determines that its previous output was 10 percent greater than it should be, it will reduce the next output signal by 10 percent. Proportional closed-loop control is used with fuel metering, spark timing, and many other outputs in an engine control system.

Modified-Loop Control

Some computer control systems are designed to use a modified-loop operation under certain conditions. For example, if the O2S cools off enough during idle to make its feedback signal unreliable, the computer continues to control spark timing and idle speed according to its closed-loop program, but returns to preprogrammed values to control fuel metering. In this case, one part of the system operates in open-loop while the rest of the system continues to operate in closed-loop. When you are troubleshooting a control system, you must understand the types of operating modes used by the system on a specific vehicle and the conditions that control the modes.

SUMMARY

An electrical circuit includes an energy source, a means of conducting this energy, and an electrical load. A complete circuit is necessary for current to flow. Current is measured in amperes and flows *through* a circuit. Voltage is *applied to* a circuit. Resistance, measured in ohms, opposes current flow and is *contained within* the circuit.

There are two theories of electric current flow. The conventional current flow theory says that current flows from positive to negative. The electron current flow theory says that current flows from negative to positive. Either theory can be used to describe current flow in a circuit.

There are three kinds of electrical circuits: series, parallel, and series-parallel. In a series circuit, there is only one current path. In a parallel circuit, there are two or more current paths. A series-parallel circuit has several current paths in parallel with each other, but in series with other parts of the circuit.

Ohm's Law is the foundation of electrical circuit troubleshooting. It says, "When one volt forces one ampere of current through a circuit, one ohm of resistance is present". Ohm's Law can be expressed mathematically:

Voltage (E) = Current (I) x Resistance (R)
Current (I) = Voltage (E) ÷ Resistance (R)
Resistance (R) = Voltage (E) ÷ Current (I)

Electrical systems can use conductors, insulators, or semiconductors. Conductors have three or fewer free electrons in the outer shell of their atoms. Insulators have five or more bound electrons in their outer shells. Current does not flow easily in an insulator. Semiconductors have four electrons in their outer shells.

Current can be made to flow in a semiconductor under certain conditions. One condition includes the use of doped semiconductor material. To dope a semiconductor, small amounts of other elements must be added. Doping creates either N-material or P-material, depending on whether the doping element has five or three electrons in its valence ring, respectively. Doped semiconductor materials behave in unusual but predictable ways when exposed to voltage. A simple PN material junction, called a diode, acts as a one-way electrical check valve. It allows current flow in one direction but not the other. Applying voltage to a semiconductor device is called biasing.

Transistors are three-part semiconductor devices made of layers of positive-negative-positive (PNP) or negative-positive-negative (NPN) materials. The three parts of a transistor are the emitter, base, and collector. Transistors can be used to amplify current or voltage, but in automobiles transistors are generally used as solid-state relays.

Every computer works on four principle functions: input, processing, storage, and output. Computers can operate on analog or digital signals. An analog signal is infinitely variable. A digital signal is an on/off or high/low signal. Most variable measurements on an automobile produce analog signals that must be changed to digital signals for computer processing.

Digital computers use the binary system in which on/off or high/low voltage signals are represented by combinations of 0's and 1's.

Control systems use on-board computers. Computers monitor a vehicle's operating conditions using sensors. Computers activate actuators to adjust engine operation and other vehicle systems. A control system regulates the operation of a vehicle system in open- or closed-loop modes. In an open-loop mode, the system does not respond to the O2S output feedback signal. In a closed-loop mode, the computer responds to the O2S feedback signal and adjusts the output value accordingly.

Review Questions

Choose the single most correct answer.
Compare your answers to the correct answers on page 507.

1. A material with many free electrons is a good:
   a. Compound
   b. Conductor
   c. Insulator
   d. Semiconductor

2. A material with four electrons in the valence ring is a:
   a. Compound
   b. Insulator
   c. Semiconductor
   d. Conductor

3. The conventional theory of current flow says that current flows:
   a. Randomly
   b. Positive to negative
   c. Negative to positive
   d. None of the above

4. An ampere is a measure of:
   a. Charge
   b. Resistance
   c. Current flow
   d. Difference in potential

5. Voltage is:
   a. Applied to a circuit
   b. Flowing in a circuit
   c. Built into a circuit
   d. Flowing out of a circuit

6. The unit that represents resistance to current flow is:
   a. Ampere
   b. Volt
   c. Ohm
   d. Watt

7. Which of the following does not affect resistance?
   a. Diameter of the conductor
   b. Temperature of the conductor
   c. Atomic structure of the conductor
   d. Direction of current flow in the conductor

8. The resistance in a longer piece of wire is:
   a. Higher
   b. Lower
   c. Unchanged
   d. Higher, then lower

9. According to Ohm's Law, when one volt pushes one ampere of current through a conductor, the resistance is:
   a. Zero
   b. One ohm
   c. One watt
   d. One coulomb

10. What circuit does this figure illustrate?
    a. Series
    b. Parallel
    c. Series-parallel
    d. Broken

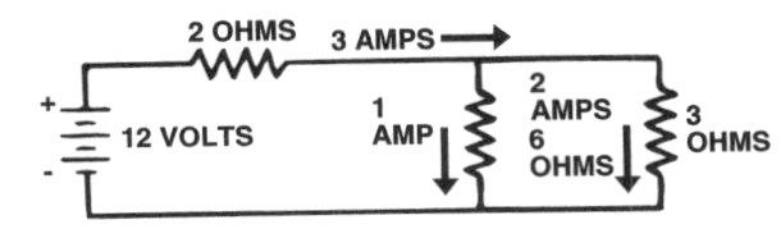

11. The amperage in a series circuit is:
    a. Always the same anywhere
    b. Always the same at certain points
    c. Sometimes the same, under some conditions
    d. Never the same anywhere

12. The sum of the voltage drops in a series circuit equals the:
    a. Amperage
    b. Resistance
    c. Source voltage
    d. Shunt-circuit voltage

13. The total resistance is equal to the sum of all the resistance in:
    a. Series circuits
    b. Parallel circuits
    c. Series-parallel circuits
    d. Series and parallel circuits

14. A diode is a simple device that joins:
    a. P-material and N-material
    b. P-material and P-material
    c. N-material and N-material
    d. P-material and a conductor

15. Which of the following is *not* true of forward bias in a diode?
    a. Free electrons in the N-material and holes in the P-material both move toward the junction
    b. N-material electrons move across the junction to fill the holes in the P-material
    c. Negatively charged holes left behind in the N-material attract electrons from the negative voltage source
    d. The free electrons that move into the P-material continue to move toward the positive voltage source

16. When reverse bias is applied to a simple diode, which of the following will result?
    a. The free electrons will move toward the junction
    b. The holes of the P-material move toward the junction
    c. No current will flow across the junction
    d. The voltage increases

17. Under normal use, a simple diode acts to:
    a. Allow current to flow in one direction only
    b. Allow current to flow in alternating directions
    c. Block the flow of current from any direction
    d. Allow current to flow from either direction at once

Continued

18. Which of the following combinations of materials can exist in the composition of a transistor?
    a. NPN
    b. PNP
    c. Both a and b
    d. Neither a nor b

19. Which of the following is *not* commonly used as a doping element?
    a. Arsenic
    b. Antimony
    c. Phosphorus
    d. Silicon

20. All of the following are characteristics of an integrated circuit (IC), except it:
    a. Is extremely small
    b. Contains thousands of individual components
    c. Is manufactured from silicon
    d. Is easily tested by the automotive technician

21. Technician A says that an engine computer receives input information from its actuators, processes data, stores data, and sends output information to its sensors.
    Technician B says that most late-model computers are based on analog microprocessors.
    Who is correct?
    a. A only
    b. B only
    c. Both A and B
    d. Neither A nor B

22. The operational program for a specific engine and vehicle is stored in the computer's:
    a. Logic module
    b. Programmable read-only memory (PROM)
    c. Random access memory (RAM)
    d. I/O interface

23. The result of engine mapping is:
    a. A clean engine compartment
    b. Easy-to-service component locations
    c. A three-dimensional performance graph
    d. Lower design profile of the vehicle hood

24. The computer can read but not change the information stored in:
    a. ROM
    b. RAM
    c. Adaptive memory
    d. Output circuits

25. Technician A says that analog input data must be digitized by an AD converter.
    Technician B says that output data must be changed to analog signals by a DA converter.
    Who is right?
    a. A only
    b. B only
    c. Both A and B
    d. Neither A nor B

Chapter

15

Electronic Engine Control: Sensors, Actuators, and Operation

Just like the human mind, an automotive computer would be useless without some mechanism to sense its surroundings and affect changes in its environment. Whereas humans have their five senses, engine control systems have electronic sensors to measure pressure, temperature, position, frequency, and voltage. For human muscle, automotive computers have devices called actuators which change positions of plungers and shafts.

After explaining the basic characteristics of sensors and actuators, this chapter describes how the automotive computer uses them to manage air-fuel ratio, ignition timing, and exhaust gas recirculation. In addition, this chapter describes various levels and types of system control, including On-Board Diagnostics Generation II (OBD-II).

SENSORS

Sensors change motion, temperature, light, and other kinds of energy into voltage signals. Automotive control sensors fall into several basic categories that include switches, timers, resistors, and voltage generators. Except for generators, all automotive sensors require voltage. This means that they are unable to create a voltage and can only modify an applied voltage. The computer supplies this applied voltage, also called **reference voltage.**

The computer sends reference voltage to a sensor and receives a different voltage back, figure 15-1. Changing conditions affecting the sensor determine the voltage level of the returning signal. The computer interprets the altered return voltage as a sign of specific changes in the engine operating condition and adjusts engine operation accordingly.

Most automotive electronic control systems use a 5- or 9-volt reference voltage. It must always be less than minimum battery voltage so that the computer can maintain it at a constant level at all times (even when battery power is low). This prevents faulty sensor signal inputs.

General Characteristics and Features

Automotive sensors must function in a severe environment. For this reason, they must be designed for long-term, dependable operation while providing reliable signals. Simpler sensor designs ensure reliability. Manufacturers vary sensor characteristics to satisfy requirements for many applications. All sensors, however, possess four basic operating features.

These characteristics affect the selection of a particular sensor for a given function and establish the specifications for troubleshooting and service. These four important characteristics include:

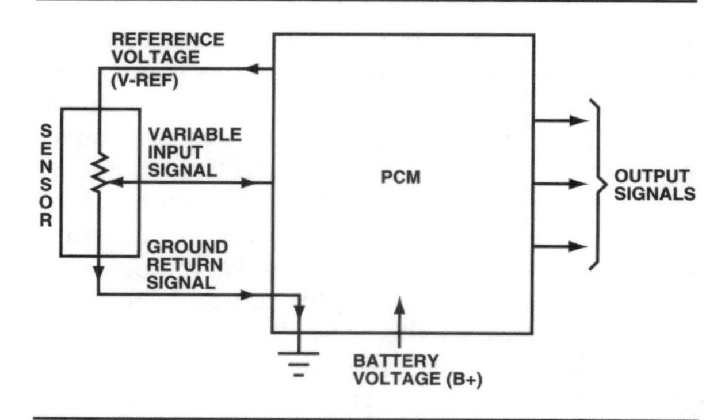

Figure 15-1. The computer reference voltage is sent out to a sensor and returns changed. This tells the computer how to readjust engine operation. (Ford)

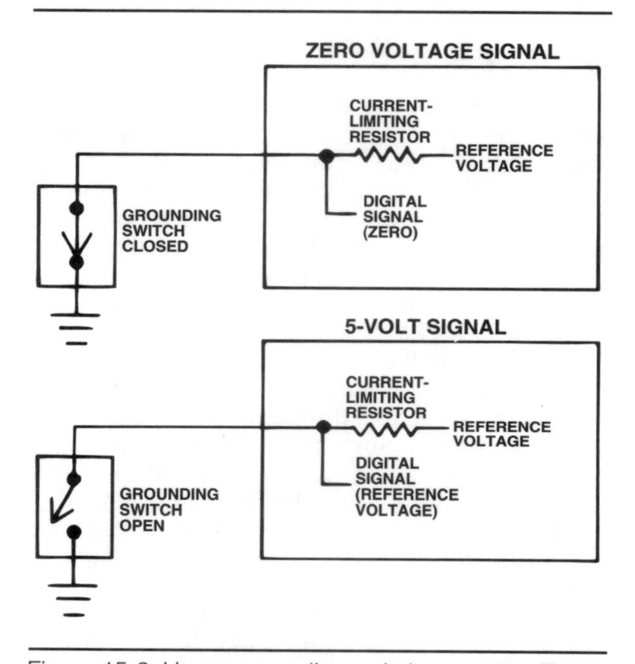

Figure 15-2. How a grounding switch operates. (Ford)

- Repeatability—A digital temperature sensor must open and close at the design points consistently. When it senses the specified point in the range of temperature values, it must switch from low to high or from high to low without deviation. From the previous chapter, you know that an analog sensor produces a varying signal proportional to the condition it measures. It must do so for each increment of change throughout its operating range. The signals produced at each increment must be repeatable in both directions, up and down.
- Accuracy—The sensor must work properly within the tolerances or limits designed into it. Our digital temperature switch may close at 195°F (91°C) within +/– 1°F (0.6°C), or it may close at 195°F (91°C) +/– 10°F (5.6°C). The tolerances depend on how the sensor is used, but once established, the sensor must work consistently within them. These design tolerances are used to establish the test specifications used to troubleshoot the sensor operation.
- Operating range—The sensor must operate within a certain range. The operating range of a digital sensor consists of only one or two switching points. An analog sensor, however, has a wider operating range and must produce accurate, proportional signals throughout that range. When the signal is not within its range limits, it is not proportional to the value being measured and is ignored by the computer.
- Linearity—This refers to sensor accuracy throughout its dynamic range. Within this range, an analog sensor must be as consistently proportional as possible to the measured value. An ideal linear sensor signal would appear as a straight diagonal line (positioned at a 45 degree angle) on a graph, but no sensor has perfect linearity. In actual practice, the sensor signal would deviate slightly above and below the 45 degree line on the graph, with the greatest accuracy appearing near the center of its dynamic range. The computer's program contains memory data to accommodate such minor deviations in a sensor's range, and its signal processing circuits make the necessary compensations to provide the computer with an accurate representation of the factor measured by the sensor.

Proper design and installation of sensors includes the use of electromagnetic interference (EMI) suppression and shielding, since sensors produce low-voltage, low-current signals that makes them prone to EMI degradation. The use of low-resistance connections and proper wiring location also are important factors in the transmission of sensor signals.

In the following paragraphs, you will learn about the various sensors used in an automobile. These include:

- Switches and timers
- Voltage generators
- Hall-effect switches
- Piezoresistive and capacitive-ceramic sensors
- Heated resistive sensors
- Exhaust oxygen sensors (O2S)
- Potentiometers
- Rheostats
- Thermistors.

Switching-Type Sensors

The simplest form of sensor is a switch. A signal from a switch returns either a high or low voltage signal to the automotive computer or pow-

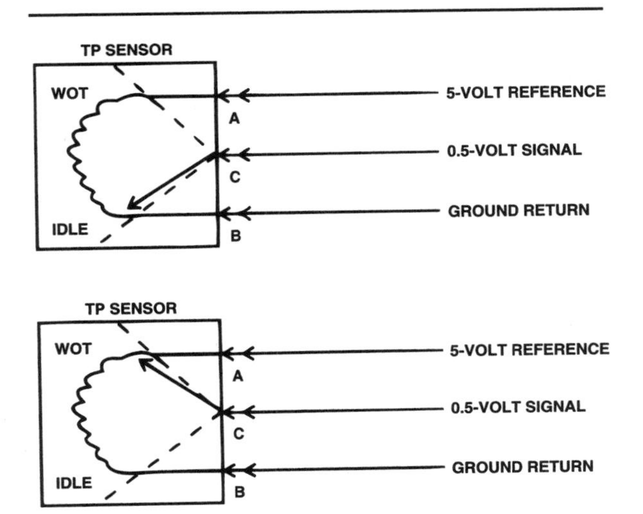

Figure 15-3. The signal from most throttle position potentiometers are low at idle and high at wide open throttle.

ertrain control module (PCM). One of two methods may be used to signal the PCM.

The first method returns full reference voltage to the computer when the switch is closed. When the switch is open, no return voltage signal is sent. Not all switches relay reference voltage back to the computer; some send a battery voltage signal directly to the computer when the condition they are monitoring is met.

The second method uses a grounding switch in series with a fixed, current-limiting resistor. Unlike the first method, when the switch is closed, no voltage signal is sent. When the switch is open, reference voltage returns to the computer, figure 15-2.

Commonly, switch sensors signal the PCM when the operator turns on or off a high-load accessory, such as an air conditioning compressor or rear window defogger. The computer then adjusts the idle speed to compensate for the added or reduced load. Some systems use engine coolant temperature switches that close when the engine reaches a certain temperature.

Combined with a switch, a timer can delay a signal for a specific and predetermined time. Timers prevent the PCM from compensating for momentary conditions that do not significantly affect engine operation. The timer may be built into the PCM or switch.

Resistive-Type Sensors

Resistors can send an analog voltage signal to the PCM in proportion to pressure, motion, heat, frequency, and voltage. Since they cannot generate a voltage, these types of sensors require a reference voltage from the PCM. Late-model vehicles may use three types of resistive sensors: the potentiometer, rheostat, and thermistor.

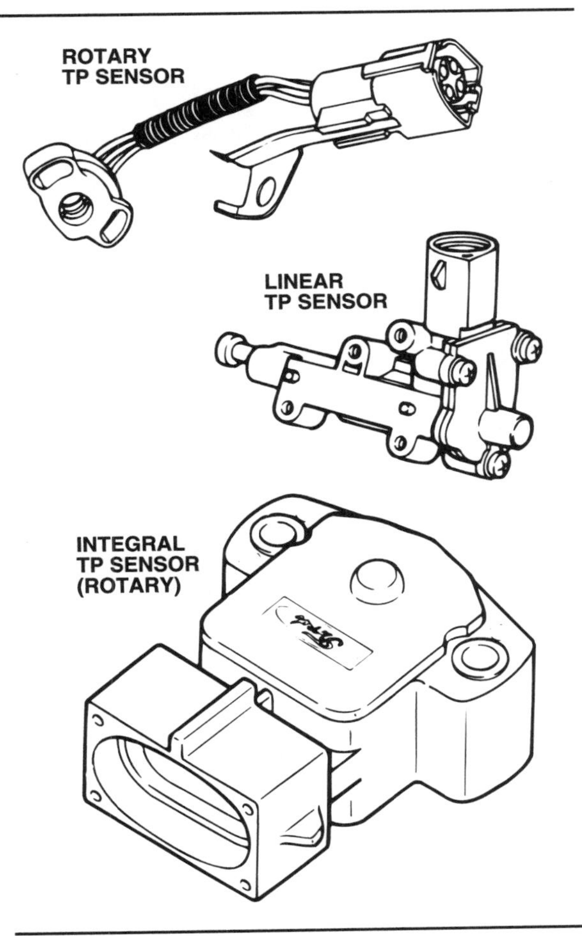

Figure 15-4. Typical throttle position sensors and their operation. (Ford)

Potentiometers

A potentiometer is a variable resistor that senses motion or position. Each of the potentiometer's three terminals has a specific function:

- Reference voltage
- Signal voltage
- Ground return.

The PCM applies a reference voltage, usually 5-volts, to the terminal at one end of a resistor. A terminal at the opposite end of the resistor con-

Reference Voltage: A constant voltage signal (below battery voltage) applied to a sensor by the computer. The sensor alters the voltage according to engine operating conditions and returns it as a variable input signal to the computer, which adjusts system operation accordingly.

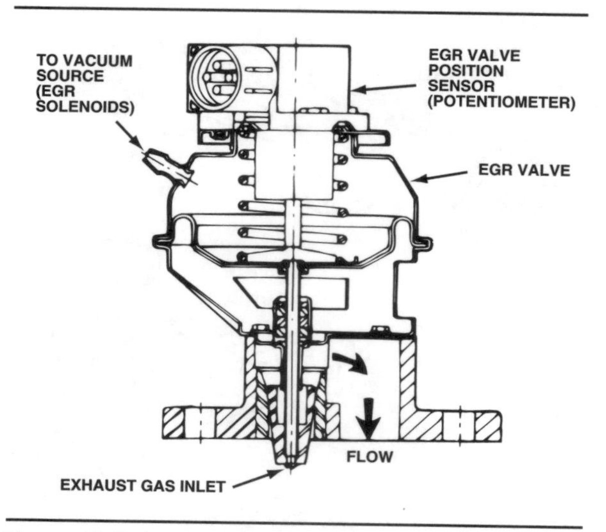

Figure 15-5. A potentiometer may be used to sense EGR flow. (Ford)

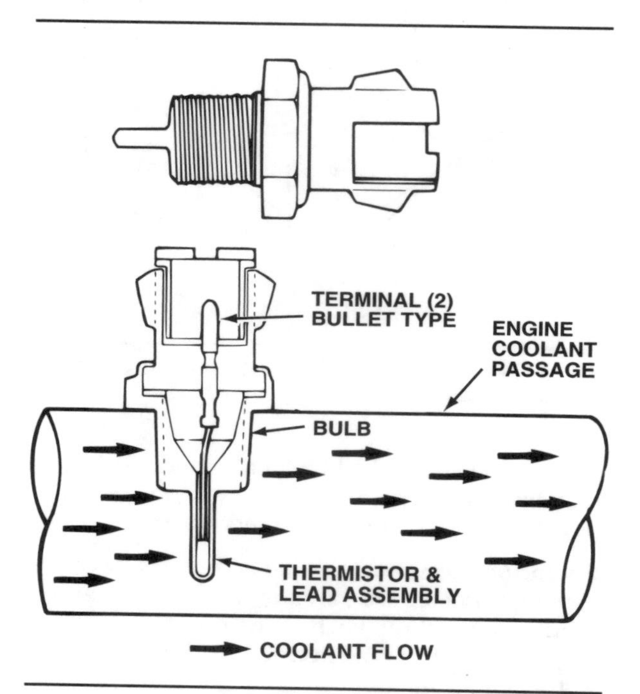

Figure 15-6. An ECT sensor is installed in a coolant passage. (Ford)

nects to ground and provides a return path. The third terminal, located between the other two, attaches to a movable wiper that sweeps back and forth across the resistor. This is the terminal that sends the variable signal voltage back to the computer. The wiper is mechanically attached to the device to be sensed: throttle linkage or an exhaust gas recirculation (EGR) valve stem, for example.

The signal voltage from the potentiometer is high or low, depending on whether the movable wiper is near the supply end or the ground end

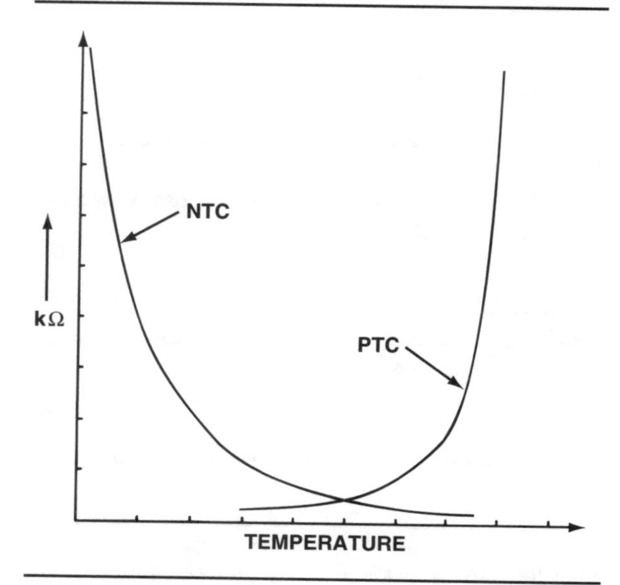

Figure 15-7. How temperature changes affect resistance for NTC and PTC thermistors.

of the resistor. Most potentiometer sensors are installed so that the linkage that they are sensing holds the wiper at the low-voltage (high-resistance) position. For example, when a throttle position (TP) or an EGR valve position (EVP) sensor is closed, the potentiometer signal voltage is at its lowest. As the linkage moves to its fully open position, signal voltage rises to its highest, figure 15-3.

Constant current through the potentiometer resistor maintains constant temperature, so resistance does not change due to temperature variation. This ensures a constant, uniform voltage drop across the entire resistor. The signal voltage varies only in relation to movement of the wiper.

TP sensors are among the most common potentiometer-type sensors, figure 15-4. The computer uses their input to determine the amount and rate of throttle opening. Vane airflow (VAF) sensors use a potentiometer to signal the amount of air entering the engine. EVP sensors, figure 15-5, use a potentiometer to indicate the valve's position. On some applications, the PCM uses this signal to verify proper EGR operation, control timing, fuel metering, and EGR valve operation.

Rheostats

Unlike the potentiometer, which regulates voltage, the rheostat regulates current. It is a variable-resistance sensor with two terminals. A fixed terminal on one end of the resistor receives reference voltage. The second terminal is attached to a moveable contact, providing a return path for the current. The position of the moveable contact determines the voltage drop across the resistor.

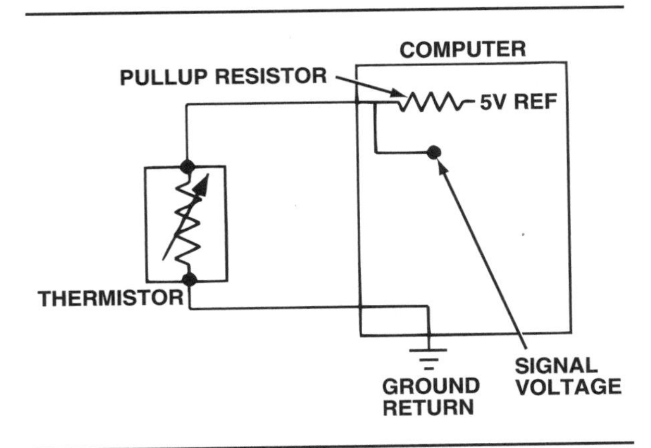

Figure 15-8. The computer applies a five-volt reference to a thermistor through a pull-up resistor.

TEMPERATURE DEGREES F(C)	SENSOR RESISTANCE OHMS	SENSOR VOLTAGE DROP
-40 (-40)	1000,000+	5.00
18 (-8)	3260	3.93
32 (0)	2940	3.56
50 (10)	2445	2.98
68 (20)	1956	2.41
86 (30)	1493	1.86
104 (40)	1115	1.40
SWITCH TO SHUNT		
122 (50)	786	3.69
140 (60)	566	3.27
158 (70)	426	2.87
176 (80)	308	2.44
194 (90)	226	2.05
212 (100)	170	1.70
230 (110)	128	1.39
248 (120)	98	1.15

Figure 15-9. Operating range chart for a dual-range thermistor with a shunt circuit.

Resistance increases and current decreases, or vice versa, according to the direction in which the contact is moved. Rheostats generally are used as a device to control current flow rather than as a computer system sensor. This device may be used as a dimmer control of an instrument panel lamp or in fuel tank sender circuits.

Thermistors

A **thermistor** is a solid-state variable resistor, usually with two connector terminals. The resistance of any resistor changes as temperature changes. The variable resistance curve of a thermistor throughout its operating range makes it an ideal analog temperature sensor. Automotive applications include engine coolant temperature (ECT) sensors, figure 15-6, and intake air temperature (IAT) sensors. Thermistors are divided into two groups, figure 15-7:

- **Negative temperature coefficient (NTC)** thermistors where resistance decreases as temperature increases
- **Positive temperature coefficient (PTC)** thermistors where resistance increases as temperature increases.

NTC thermistors are most commonly used as ECT and IAT sensors. Some NTC thermistors are also used to actuate low-fluid warning lamps on instrument panels. A few early Cadillac systems used a PTC thermistor to sense engine coolant temperature.

Sometimes, the PCM applies a reference voltage (usually 5 volts) through a **pullup resistor** to the thermistor. The computer then monitors the voltage drop across the pullup resistor and interprets any change in voltage as a signal of changing temperature, figure 15-8. High NTC thermistor resistance (cold) causes the reference voltage to drop, resulting in a higher signal voltage. Signal voltage is measured before the thermistor.

When NTC thermistor resistance is low (hot) it drops very little of the reference voltage. The pullup resistor drops most of the reference voltage, resulting in a low signal voltage.

Many late-model engines use a shunted or dual range circuit in the PCM to get more accurate temperature readings from the ECT thermistor. The sensor is the NTC type, so resistance when cold is high and progressively drops as the unit warms up. However, the voltage drop across the sensor occurs in two distinct stages.

Voltage drop is at or near five volts when resistance is highest (-40°F (-40°C) or below). Just before the midpoint of the temperature range (about 104°F (40°C)) voltage drop should be about 1.4 volts. As temperature crosses the midpoint, the PCM switches to the **shunt circuit**,

Thermistor (Thermal Resistor): A resistor that built to changes its resistance as the temperature changes.

Negative Temperature Coefficient (NTC) Resistor: A thermistor whose resistance decreases as the temperature increases.

Positive Temperature Coefficient (PTC) Resistor: A thermistor whose resistance increases as the temperature increases.

Pull-Up Resistor: A constant load internal to the PCM that feeds reference voltage to some sensors. The PCM measures the voltage drop across this load to determine signal voltage.

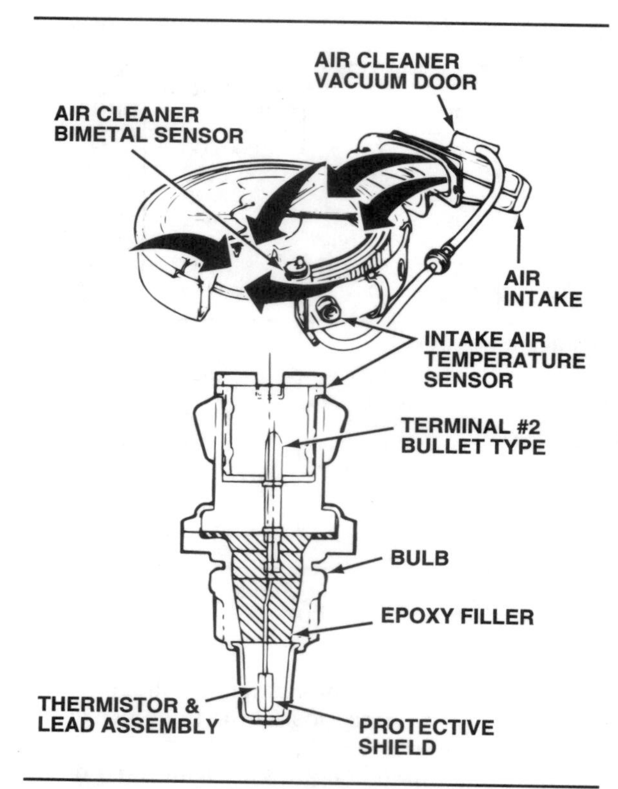

Figure 15-10. An IAT sensor may be installed in the air cleaner or intake manifold.

and once again the sensor drops most of the reference voltage. At a temperature of 122°F (50°C) voltage drop across the sensor is about 3.7 volts. Voltage readings gradually decrease as the sensor continues to warm up, figure 15-9.

Remember, this dual-ranging feature is an internal PCM function and not controlled by the sensor. This type of system has no effect on a scan tool reading. The scan tool displays all thermistor signals on the data stream as temperature readings. However, whether or not the system is dual ranging must be considered when using a voltmeter to take voltage drop readings across the sensor.

An IAT sensor is similar in construction to the ECT sensor, but provides a faster response time to air temperature changes. It may be located in the air cleaner, figure 15-10, to measure only air temperature, or in the intake manifold to measure the air-fuel mixture temperature.

Engine Speed, Crankshaft Position (CKP), and Camshaft Position (CMP) Sensors

These sensors, those that can provide an engine speed signal to the PCM, are the most important to the engine control system. Late-model vehicles typically use three different kinds: a type of generator called a magnetic pulse generator, the Hall-effect switch, and the optical signaling switch. These sensors' signals allow the PCM to manage ignition timing, fuel metering, EGR flow, automatic transmission converter lockup and shift patterns, and to detect misfires.

Magnetic pulse generators

Unlike the previously discussed sensors, magnetic pulse generators do not require a reference voltage to signal the PCM. They create an analog output signal by themselves by magnetic induction. The section on ignition triggering devices in the chapter entitled "Solid-State Electronic Ignition Systems" fully explains this process.

Automakers sometimes call magnetic pulse generators reluctance sensors because they use a rotating trigger wheel to vary the reluctance around a magnetic pole piece and pickup coil, figure 15-11. The output signal varies in voltage (amplitude) and frequency depending upon the speed with which the trigger wheel passes the inductive pickup. Often, these generators are used in electronic distributors to produce pulse signals used by the computer to control ignition timing. Besides engine speed, these sensors may signal crankshaft position, cylinder firing order, ABS wheel deceleration, vehicle speed, and engine misfire information to the PCM.

More complex systems use multiple speed sensors. For example, Chevrolet and GMC light-duty pickup trucks equipped with the Turbo Hydra-Matic 4L80-E transmission have both an output shaft speed sensor and an input speed sensor in addition to the main vehicle speed sensor. The PCM uses transmission input and output speeds to determine line pressure, transmission shift patterns, and torque converter clutch apply pressure. In addition, the PCM also calculates turbine speed, gear ratios, and torque converter clutch slippage for diagnostic purposes.

Optical generators

In an optical sensor, figure 15-12, a spinning blade passes through a light emitting diode (LED) beam. Each time the blade cuts through the beam, it reflects light back to a phototransistor. This creates a low-power signal which the sensor also amplifies and then sends to the PCM. For further information regarding optical ignition triggering applications, see the "Solid-State Electronic Ignition Systems" chapter.

Hall-effect switches

A Hall-effect switch rotates a trigger wheel with vanes through a stationary sensor, figure 15-13. When the PCM sends a regulated current through the switch's semiconductor Hall element and the

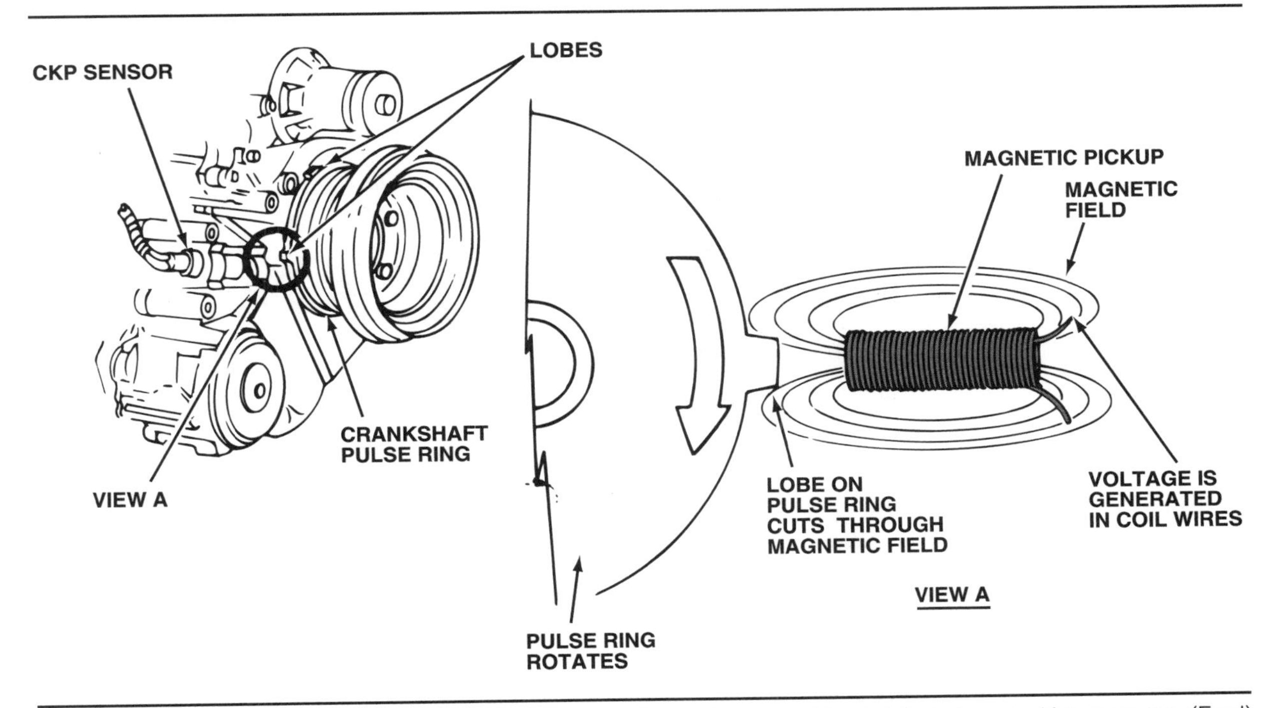

Figure 15-11. A CKP sensor uses a magnetic pickup similar to those used in distributorless ignition systems. (Ford)

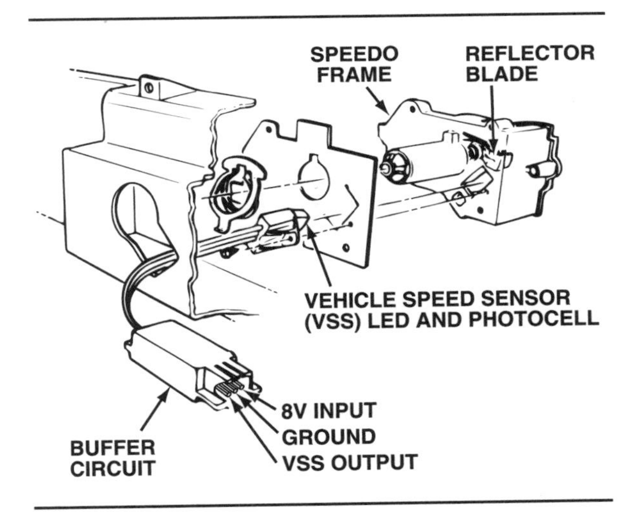

Figure 15-12. A light-emitting diode and a photocell in the speedometer act as a vehicle speed sensor (VSS). (GM)

permanent magnet applies a magnetic field to the element, the Hall-effect switch generates a low signal voltage, figure 15-14. As a blade or vane of an interrupter ring passes between the magnet and wafer, it interrupts the magnetic field and Hall voltage drops off, producing a high signal voltage figure 15-15. With constant rotation of the interrupter ring, the output voltage constantly changes in proportion to the strength of the magnetic field.

The Hall-effect switch uses a complex integrated circuit (IC) containing an output voltage generator. The generator converts the varying output voltage into a digitized square wave, which is either high or low. The PCM can then respond to the rising or falling edge of the signal.

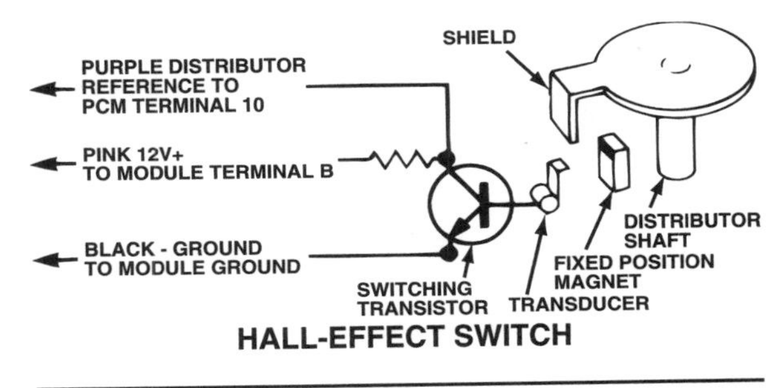

Figure 15-13. When the Hall-effect switch is on, a permanent magnet mounted beside the semiconductor wafer induces a voltage across the wafer. (GM)

Engine Load Sensors

These devices sense engine load so that the PCM can meter fuel properly. Additionally, the PCM uses their input to adjust ignition timing and EGR flow relative to load.

Shunt Circuit: An electrical connection or branch circuit in parallel with another branch circuit or connection.

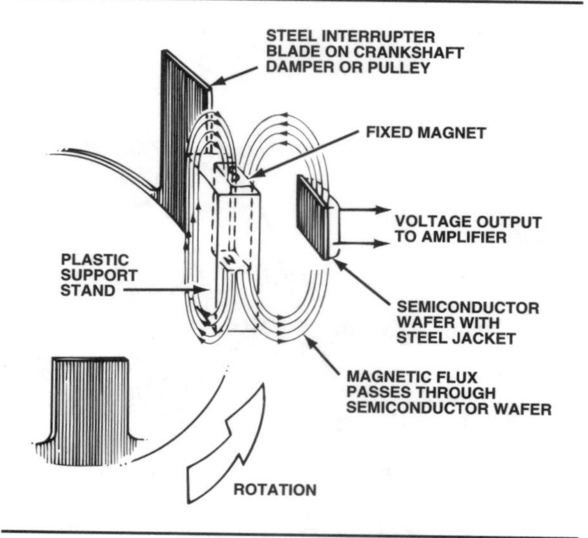

Figure 15-14. When the magnetic field in the Hall-effect switch is unbroken, the switch is off. (GM)

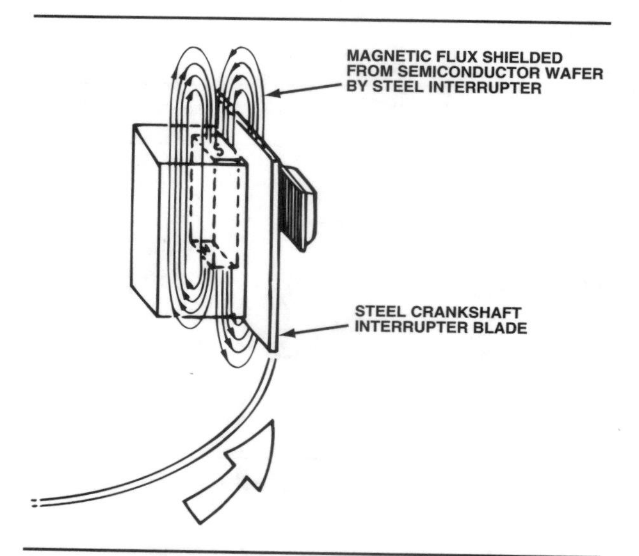

Figure 15-15. Interrupting the magnetic field in the Hall-effect switch turns the switch on. (GM)

Manifold Absolute Pressure (MAP) and Barometric (BARO) Sensors

MAP sensors measure manifold absolute pressure, which is equal to the difference between atmospheric pressure and intake manifold vacuum. As you can imagine, MAP sensors must be able to measure fluctuations in manifold pressure ranging from a high vacuum occurring during deceleration and at idle, to pressures exceeding atmospheric on turbocharged engines. Most late-model engines use a MAP sensor that either varies capacitance or resistance in proportion to pressure. Although of similar construction, BARO sensors differ in function. Their signals tell the

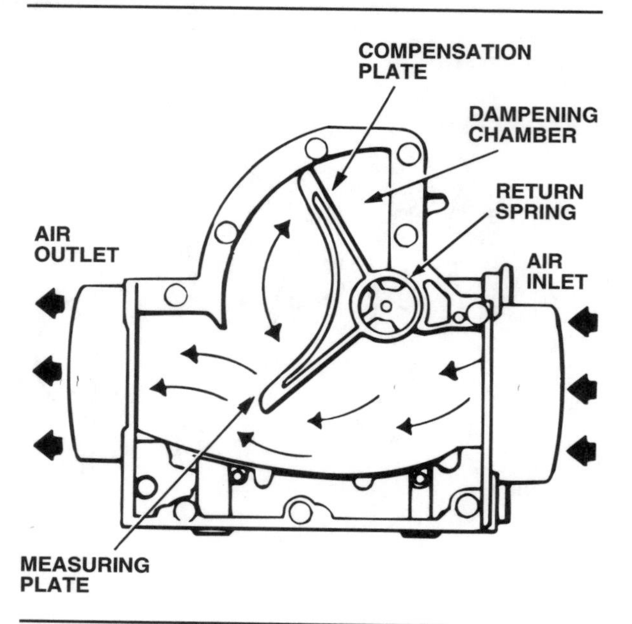

Figure 15-16. A vane airflow (VAF) sensor.

PCM how much to change fuel metering based on atmospheric pressure.

Capacitive-ceramic

The capacitance type uses manifold vacuum to vary the distance between two plates. This MAP sensor generates a digital signal whose frequency varies with intake manifold pressure. Ford commonly uses this type of load sensor in late-model vehicles.

Piezoresistive

The resistance-type of MAP sensor uses a **piezoresistive** crystal that receives a constant reference voltage and returns a variable signal in relation to its varying resistance. In addition to measuring pressure, they may also be used to sense engine misfire.

This sensor has its crystal mounted on a flexible silicon or ceramic diaphragm. A sealed chamber with a reference vacuum lies on one side. The opposite side is exposed to intake manifold pressure. This MAP sensor sends a voltage signal to the PCM in proportion to the intake manifold pressure. At idle (high vacuum), the signal voltage should be low, approximately 0.5 volts, and at atmospheric (low vacuum) with the key on, engine off, the signal voltage should be high, about 4.75 volts. Typically, these sensors fail due to vacuum leaks.

Vane airflow (VAF) sensors

The VAF sensor is located between the air cleaner and the throttle body. The sensor's major parts include the measuring plate, compensation plate, return spring, potentiometer, bypass passage, and

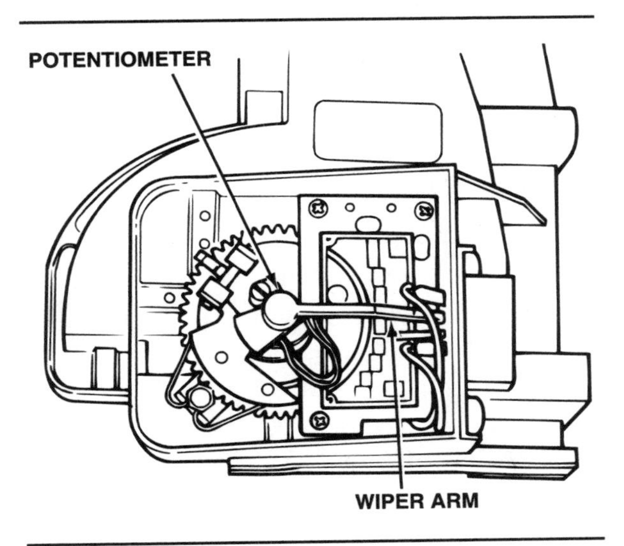

Figure 15-17. A potentiometer in the VAF sensor sends an analog signal to the PCM. (Ford)

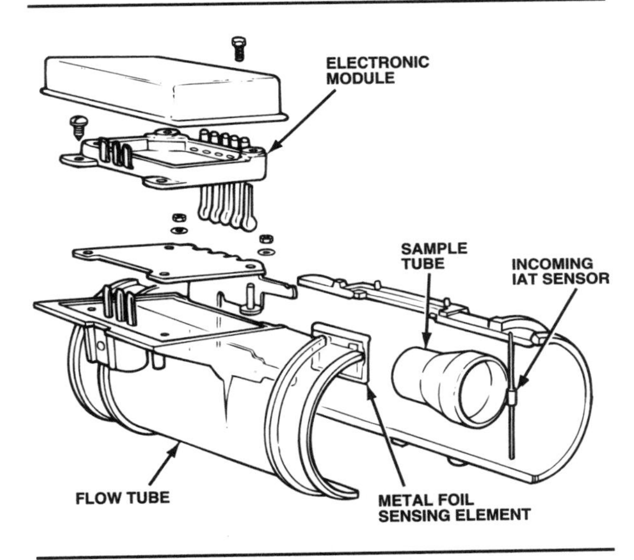

Figure 15-18. A heated thin-film MAF sensor consists of a screen or metal foil sensing element, an IAT sensor, and an electronic module.

idle carbon monoxide (CO) adjusting screw, figures 15-16 and 15-17. Also, it may have a fuel pump switch and an IAT sensor.

When the engine runs, air passes through the sensor, forcing the measuring plate open until it balances with the force of the return spring. The compensation plate and the chamber above it dampen vibrations caused by sudden engine speed changes. The wiper of the potentiometer moves with the measuring plate, and produces a variable voltage signal like that of a TP sensor.

Although most VAF signals increase in voltage as the measuring vane opens, some do not. The VAF sensors on Ford 1.8-liter and Toyota 2.8-liter 5M-GE engines, for instance, decrease their voltage as the measuring vane opens. Depending on the design, VAF sensors may depend on battery voltage or a PCM-regulated 5.0 volt supply as source voltage.

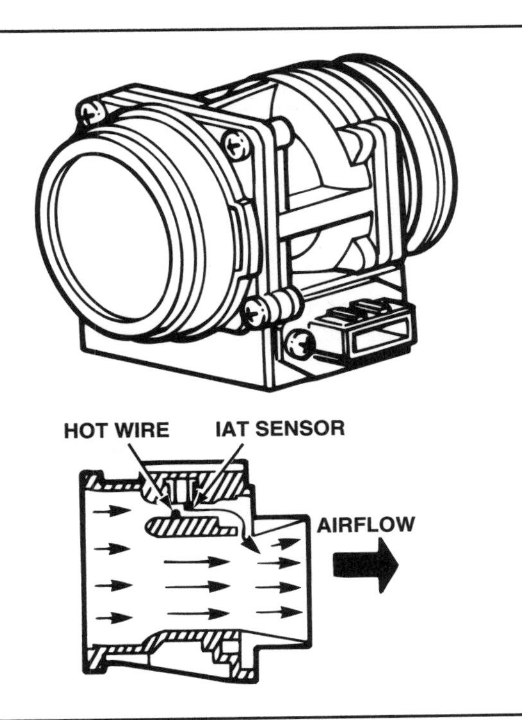

Figure 15-19. Operation of the hot-wire MAF sensor (Ford).

Mass airflow (MAF) sensors

Like the VAF sensor, the mass airflow sensor lies between the air intake and the throttle body to determine the molecular mass of the air entering the engine. Two types of MAF sensors are most commonly used: a heated platinum wire or a heated thin-film semiconductor. Current provided by the MAF module heats the wire or film that the incoming airflow cools. As the heated wire or film cools, the module increases current to keep the sensor temperature at a specified level. As the wire or film again heats up, current decreases. The module measures current changes and converts them into voltage signals sent to the engine control computer for use in determining injector pulse width. Figure 15-18 shows the components

Piezoresistive: A sensor whose resistance varies in relation to pressure or force applied to it. A piezoresistive sensor receives a constant reference voltage and returns a variable signal in relation to its varying resistance.

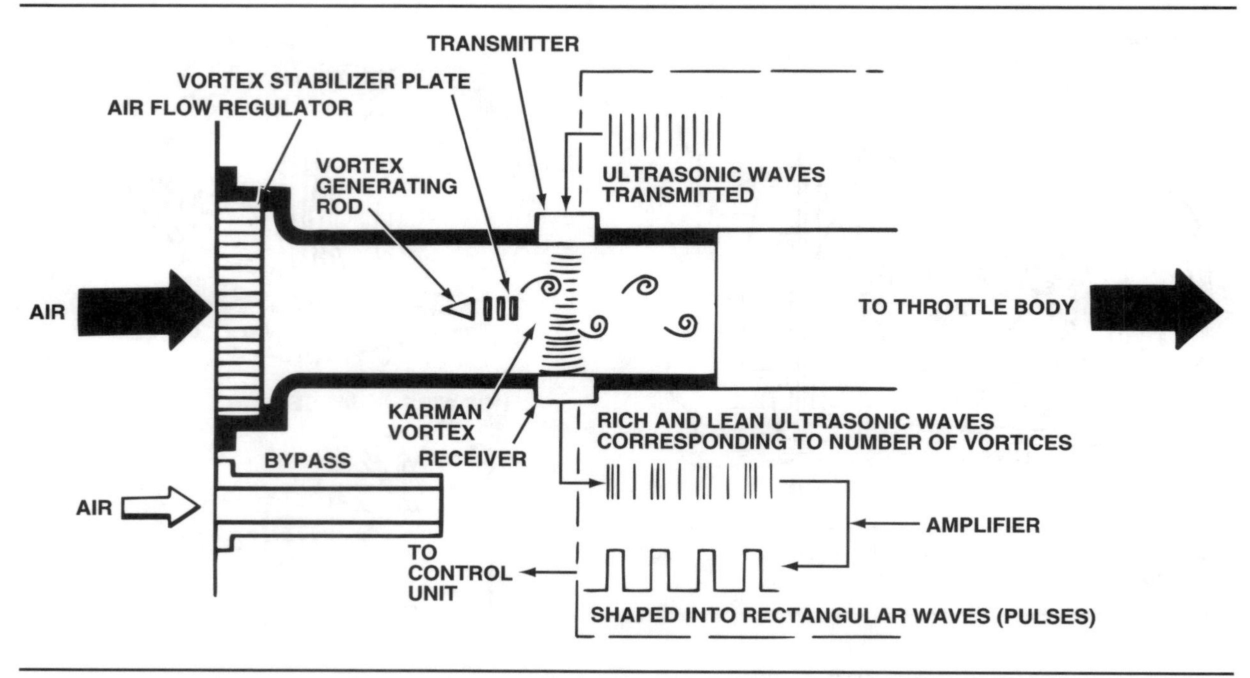

Figure 15-20. Operation of the Mitsubishi-Chrysler MAF sensor that measures mass airflow according to the Karman vortex theory.

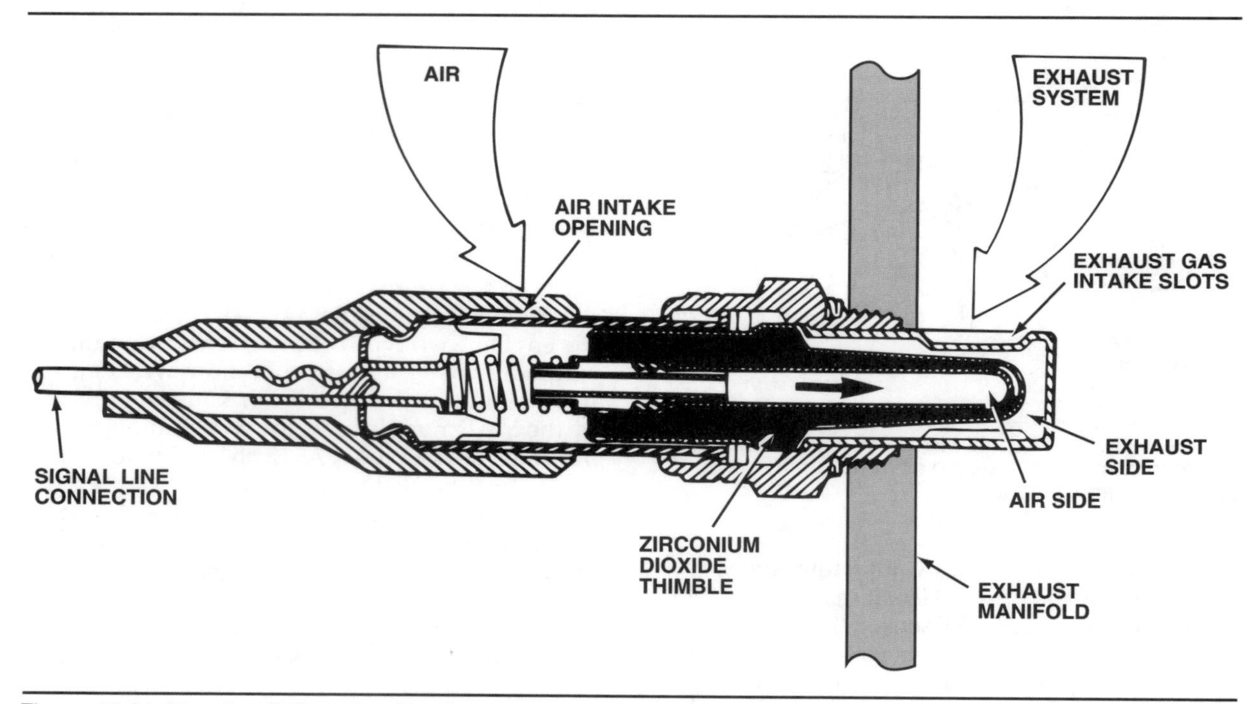

Figure 15-21. How the O2S works. (Ford)

of a typical heated thin-film MAF, and figure 15-19 shows operation of the heated wire design.

The Karman vortex type of MAF sensor is less common. Chrysler has used them on Mitsubishi-built engines with TBI injection for more than a decade. Although it also uses a 5-volt reference voltage and sends the same variable frequency signal as the heated-resistive MAF sensors, the principles behind its operation are quite different.

A triangular vortex generating rod is positioned in the middle of the intake airflow, figure 15-20, and interferes with the passage of the

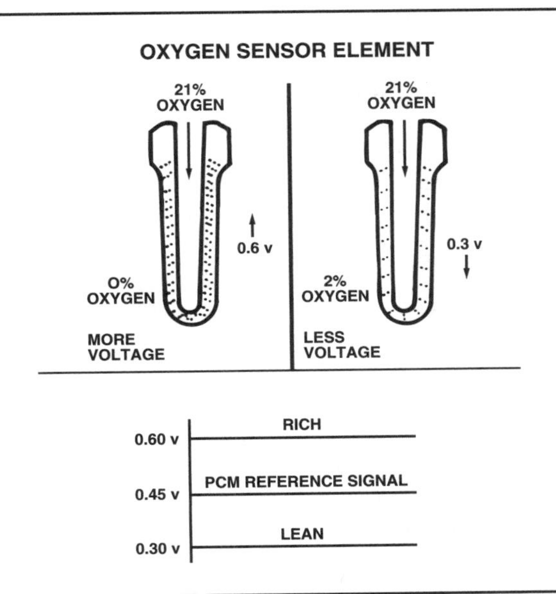

Figure 15-22. A difference in oxygen content between the atmosphere and exhaust gases enables an O2S to generate voltage.

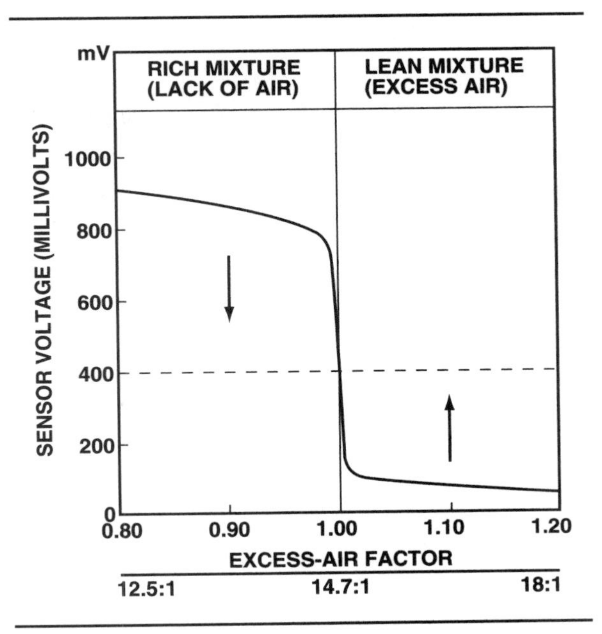

Figure 15-23. The O2S provides its fastest response at the stoichiometric air-fuel ratio of 14.7 to 1.

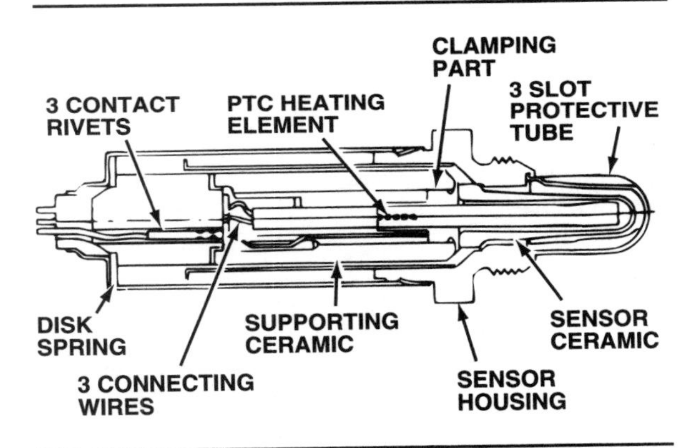

Figure 15-24. The components of a typical H02S sensor. (Ford)

incoming airflow column. This creates a disturbance called Karman vortices (named for the 19th century German physicist who is credited with discovery of the phenomenon).

An amplifier in the sensor uses the 5-volt reference voltage to drive a transmitter, causing ultrasound waves to pass across the sensor opening at right angles to the airflow to a receiver on the other side. The MAF sensor module uses the length of time required for sound wave reception by the receiver when no airflow passes through the sensor as a reference time signal. When air flows through the sensor, the rod creates vortices in the airflow. Sound waves transmitted through the vortices change speed, creating a variable signal that changes pitch. The signal frequency is directly proportional to the number of vortices, which are directly proportional to the amount of airflow. A low frequency equals low airflow; a high frequency equals high airflow. A modulator circuit changes the resulting ultrasound signal to a square wave signal, producing an input signal to the computer.

Exhaust Oxygen Sensors (O2S)

As explained in the last chapter, the O2S provides a feedback signal to the PCM regarding the amount of oxygen in the exhaust stream. It enables closed-loop operation. Usually it is installed in the exhaust manifold, although in some vehicles it may be located downstream from the manifold in the headpipe (but before the catalytic converter). The last section of this chapter discusses how the OBD-II system uses two O2S', the second located downstream of the catalytic converter.

Zirconia exhaust oxygen sensors

The most common type of O2S uses a sensor tip containing a thimble made of zirconium dioxide (ZrO_2). This compound is an electrically conductive material capable of generating a small voltage in the presence of oxygen.

Exhaust gases from the engine pass through the end of the sensor installed in the manifold where they contact the outer side of the thimble. Atmospheric air enters through the other end of the sensor and contacts the inner side of the thimble, figure 15-21. The inner and outer surfaces of

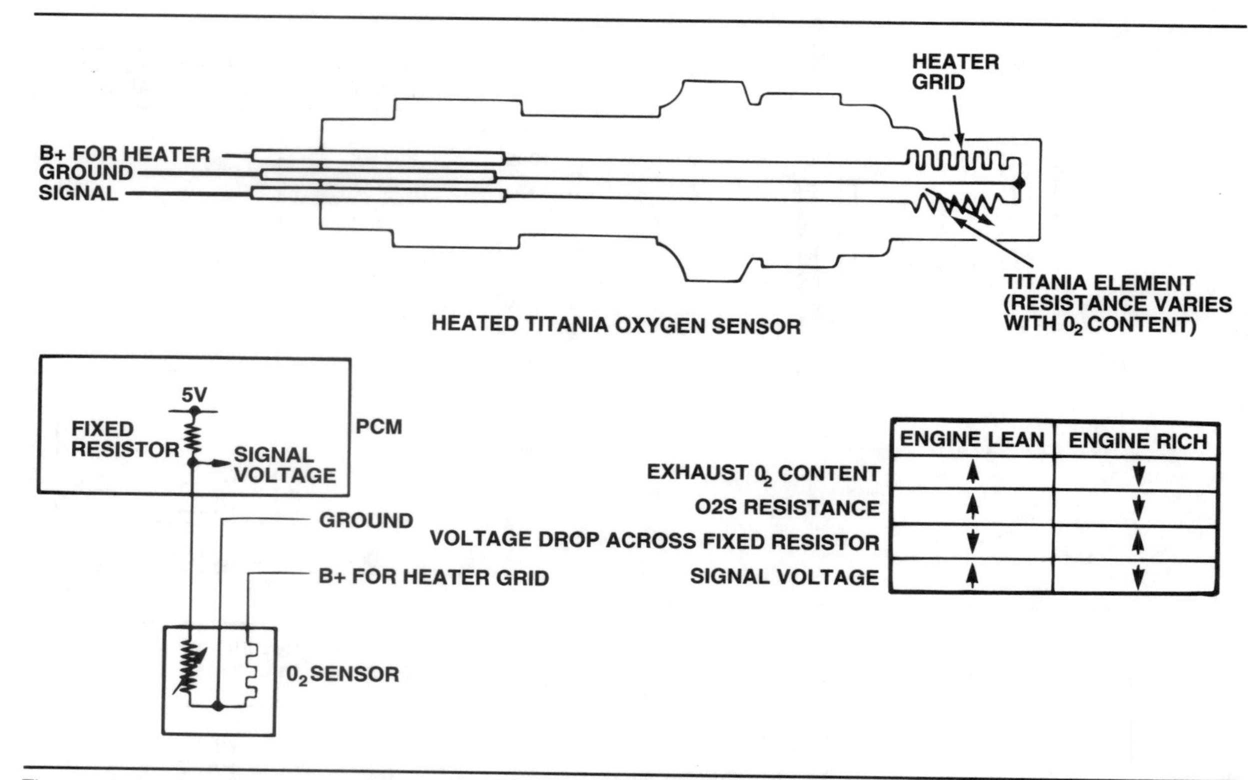

Figure 15-25. Operation of a titania O2S as used on 1987–90 Jeep 4.0-liter engines. (Chrysler)

the thimble are plated with platinum. The inner surface is a negative electrode; the outer surface is a positive electrode.

The atmosphere contains 21 percent oxygen. Rich exhaust gases contain virtually no oxygen. Exhaust from a lean mixture combustion or from a misfire contains more uncombined oxygen (still far less than the atmosphere).

Negatively charged oxygen ions are drawn to the thimble, where they collect on both the inner and outer surfaces, figure 15-22. Because the oxygen present in the atmosphere exceeds that in exhaust gases, the air side of the thimble draws more negative oxygen ions than the exhaust side. The difference between the two sides creates an electrical potential. When the concentration of oxygen on the exhaust gas side of the thimble is low, the difference between the electrodes generates a high voltage (0.60 to 1.00 volt). As the oxygen concentration on the exhaust side increases, the generated voltage drops to 0.00 to 0.40 volt.

This voltage signal is sent to the computer, where it passes through the input conditioner for amplification. The computer interprets the high-voltage signal as a rich air-fuel ratio, and a low-voltage signal as a lean air-fuel ratio. Based on the O2S signal, the computer makes the mixture either leaner or richer as required to maintain the air-fuel ratio within the 14.7:1 window. The O2S is therefore a key sensor of an electronically controlled fuel-metering system.

An O2S does not send a voltage signal until its tip reaches a temperature of about 572°F (300°C). O2S' provide their fastest response to mixture changes at about 1,472°F (800°C). These temperature requirements are the primary reason for open-loop fuel control on a cold engine.

Figure 15-23 shows the operating range of an O2S at 1,472°F (800°C). Sensor voltage changes fastest at an air-fuel ratio of 14.7:1 at this temperature.

Another important point about an O2S is that *it measures oxygen: it does not measure the air-fuel ratio.* A misfiring cylinder does not consume oxygen. The large amount of oxygen in the unburned exhaust mixture causes the sensor to deliver a false "lean mixture" signal. This is one reason why computer control of ignition timing and EGR is essential for effective fuel metering control.

Unlike resistive sensors, an O2S (as well as any other generator sensor) does not require a reference voltage. The PCM, however, uses an internal reference voltage as a comparison for the sensor signal. Because the O2S signal ranges from 0.1 to 0.4 volt (100 to 400 mV) with a lean mixture to 0.6 or 0.9 volt (600 to 900 mV) with a rich signal, the computer uses an internal reference of 0.45 volt (450 millivolts) as a reference, figure 15-22. The internal reference voltage also is the basis for fuel-metering signals during open-loop operation.

To make the system more responsive, automakers went first to the concept of installing a separate O2S in each manifold of a V-type engine. In addition to this arrangement, and after the introduction of OBD-II in 1996, the heated exhaust gas oxygen sensor (HO2S) became required, figure 15-24.

A HO2S operates the same as an O2S, but includes a built-in heater powered by the vehicle battery whenever the ignition is in the RUN position. A third wire to the sensor delivers battery current (1 amp or less) to the sensor electrode. This helps warm the sensor to operating temperature more quickly and permits the sensor to operate at a lower exhaust gas temperature (approximately 392°F or 200°C). The heating element also keeps the sensor from cooling off when exhaust temperature drops, such as during prolonged idling in cold weather. Turbocharged engines often use HO2S's installed downstream from the turbocharger, which absorbs much of the heat in the exhaust.

All zirconium dioxide O2S's work on the principles just discussed, but their specific designs differ. They may have one, two, or three wires that connect to the vehicle wiring harness. Early sensors use a single wire for signal output and their outer shell threads as a grounding path to the exhaust pipe or manifold.

Some sensors have a silicone boot to protect the sensor and to provide a vent for ambient air circulation. The positioning of a boot (when used) is important. Boots seated too far down on the sensor body can block the air vent, resulting in an inaccurate signal to the computer. Some late-model engines do not use a silicone boot because the silicone may emit fumes corrosive to electrical connections and terminals. O2S's can be divided into the following basic types:

- Standard unheated
- Water-resistant unheated
- Water-resistant heated
- Waterproof heated.

To make O2S's water-resistant and waterproof, it was necessary to design them to use a different air reference source. Older designs have their air reference source at the sensor, but the water-resistant and waterproof sensors receive their air reference input through the sensor lead wires and connector.

Titania exhaust oxygen sensors

Although Zirconia O2S's have proven themselves during more than a decade of use, some automakers, such as Jeep and Toyota, use O2S's containing titanium dioxide or titania-sensing elements. Both types of sensors provide a feed-

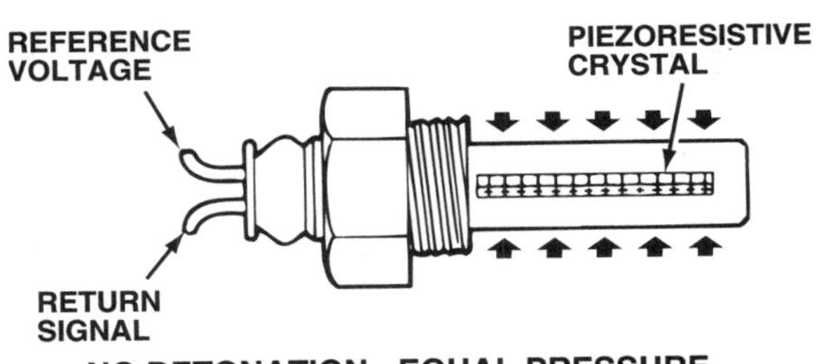

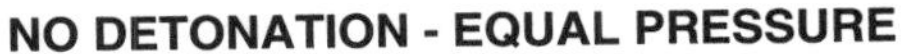

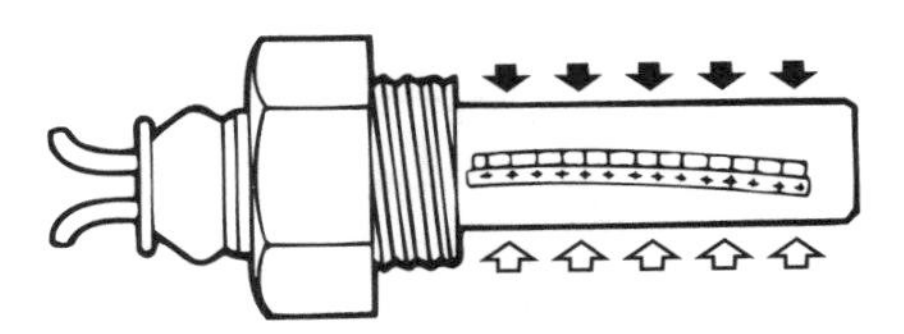

Figure 15-26. A piezoresistive crystal changes its resistance when pressure is applied. (Ford)

back signal indicating the relative rich or lean condition of the exhaust. Unlike the zirconia O2S, however, the titania sensor does not generate voltage. Instead, it detects exhaust oxygen content by acting as a variable resistor.

The computer provides a constant reference voltage to the sensor. As the oxygen content of the exhaust changes, the resistance of the titania sensing element varies. Since temperature may affect the sensing element's resistance, designers incorporate a heating element into the sensor that maintains a constant temperature of approximately 1,475°F (802°C).

Figure 15-25 shows the heated titania O2S as used on 1987–90 4.0-liter Jeep engines. The PCM provides the sensor with a 5-volt reference voltage, measuring the signal voltage between the sensor and a fixed resistor in series with the sensor. As sensor resistance changes, the signal voltage also changes. With a rich mixture (low exhaust oxygen content), the sensor signal voltage is below 2.5 volts. With a lean mixture (high exhaust oxygen content), sensor signal voltage exceeds 2.5 volts. Although signal voltage direction, relative to oxygen content of the exhaust gas, is opposite to that of a zirconia O2S, the PCM's program allows it to read the signal correctly and make the necessary adjustments in fuel metering.

Exhaust oxygen sensor contamination

O2S contamination causes signal degradation and failure. Investigate the following sources of contamination in cases of multiple or repeated O2S failures on the same vehicle.

- An overrich mixture can deposit black carbon

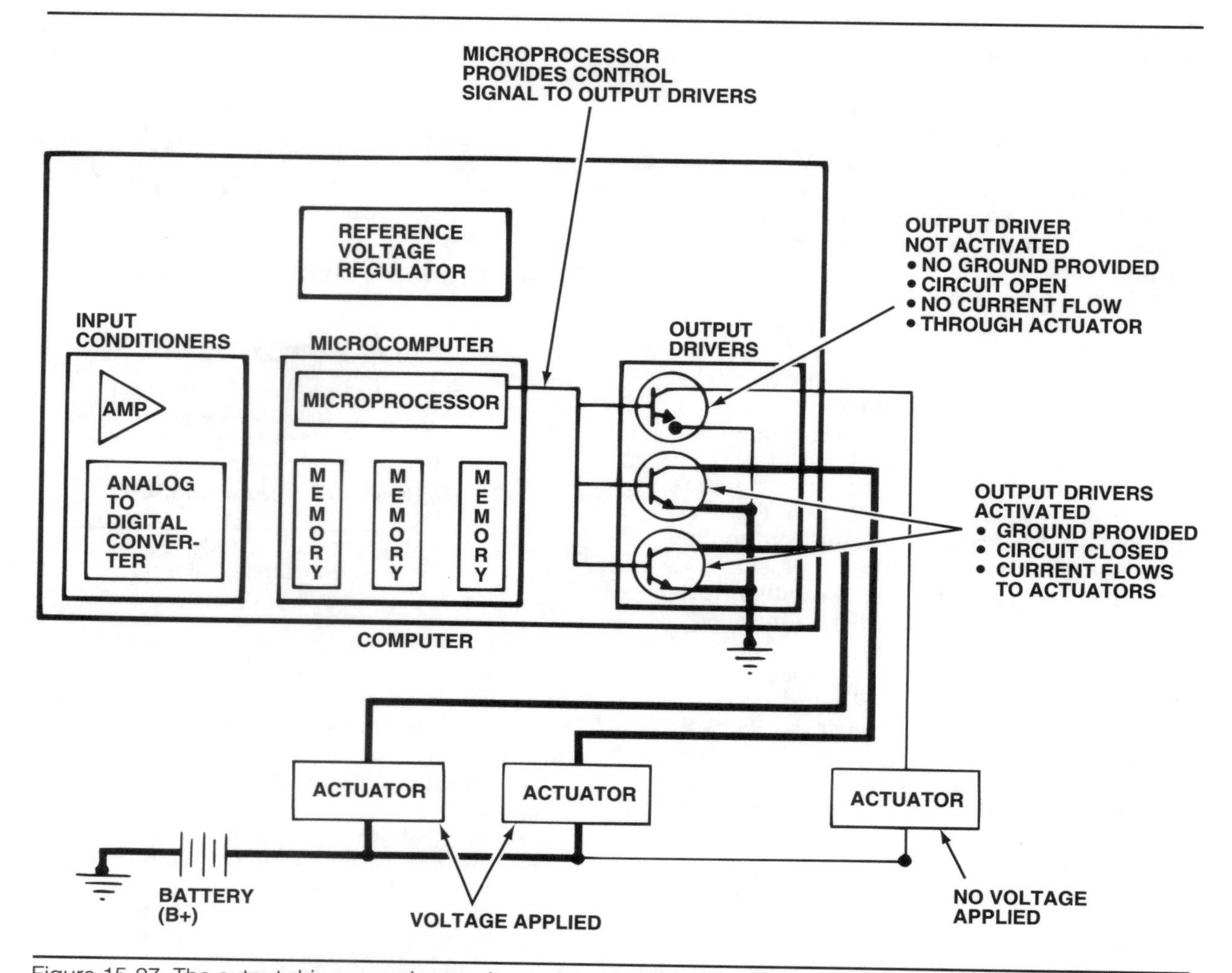

Figure 15-27. The output drivers receive a voltage signal from the microprocessor. This tells the drivers to open or close the ground circuit of the actuators they control. (Ford)

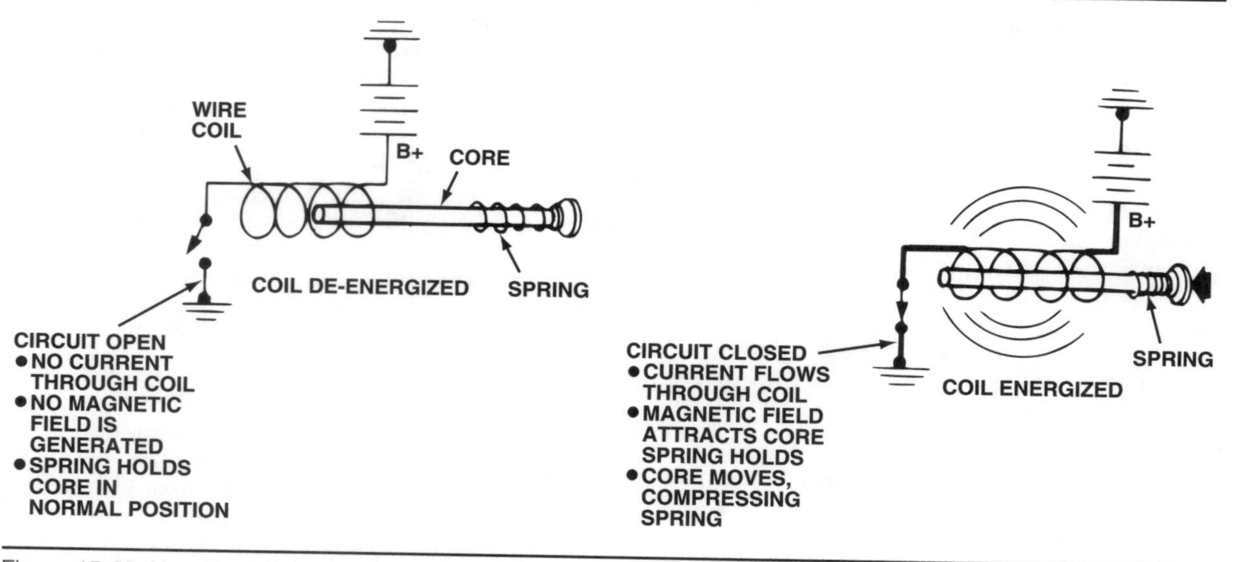

Figure 15-28. How a solenoid works. (Ford)

or soot on the sensor tip. Carbon deposits, however, do not cause sensor failure and can be burned off in the engine by running it at part-throttle for at least two minutes.

- Lead deposits are more serious, as they glaze the sensor element, rendering it useless. Lead deposits generally result from the use of leaded fuel, although methanol in gasoline also can cause lead contamination by dissolving the terne coat inside the fuel tank. Lead contamination is hard to detect by a visual inspection.
- Some RTV silicone gasket materials emit silica vapors that contaminate the O2S. Sand-like particles of silica in the vapors embed themselves in the sensor element and clog its surface. This type of contamination slows down the sensor's response time, affecting engine operation. Silica contamination produces a whitish appearance on the sensor element.
- Other deposits can eventually affect sensor operation. Oil produces a dark brown appearance on the element, while antifreeze results in a whitish appearance.

Detonation Sensors

Detonation or knock sensors (KS) typically use a piezoresistive crystal that changes resistance whenever pressure is applied to it, figure 15-26. One of the KS sensor's terminals receives reference voltage. The return signal voltage from the other terminal remains at its programmed value as long as there is no detonation and as long as pressure on the crystal is uniform. However, if detonation occurs, the unequal pressure on the crystal changes the sensor's resistance and the return voltage signal changes.

ACTUATORS

Actuators are output devices that change voltage and current into mechanical action. The PCM activates them after having received sensor inputs and completing any necessary calculations. Figure 15-27 illustrates this process.

Some of the computer's output signals, such as those that regulate ignition timing, control engine operation directly. However, whenever an output

The Rise and Fall of Gaskets in a Tube

In the 1970s, automakers replaced many traditional cork gasket installations with room-temperature-vulcanizing (RTV) silicone sealants. RTV sealants quickly gained an unchallenged reign as "gaskets in a tube" for installing such parts as water pumps, valve covers, oil pans, transmission and differential covers, and other components.

Cork gaskets had been the industry standard for decades in these applications. Cork, however, dries out with age and loses its shape. Parts departments often found themselves with an inventory of unusable over-age gaskets. Moreover, car dealers and garages had to stock a variety of gaskets for different car models. RTV sealant seemed a revolutionary breakthrough.

When properly applied, RTV is an excellent sealer, but mating surfaces must be perfectly clean of oil and grease because oil dissolves the RTV sealant. Additionally, RTV residue in bolt holes can cause a hydraulic effect that affects torque when a bolt is installed. These are minor problems, however, that were easily overcome by professional technicians. But RTV also has longer-term disadvantages that took time for the service industry to discover. For example, it:

- Has a short shelf life—about one year
- Does not cure properly when the shelf life is expired
- Spoils if the cap is left off the tube, because moisture in the air causes it to cure
- Is expensive to manufacture and stock for long periods.

Despite these disadvantages, automakers used and specified RTV sealants and created a virtual depression in the cork gasket industry. In the mid-1980s, though, GM discovered that RTV sealants can cause long-term problems that had been unforeseen a decade earlier.

Although RTV sealant cures sufficiently to provide a firm seal in 24 to 48 hours, it can require as long as one year after installation to cure *completely*. Final curing time depends on where it is used and how thickly it is applied. During the prolonged curing time, RTV sealant gives off acidic fumes that can corrode electrical connections and sensitive electronic parts, such as O2S's.

As solid-state electronic components increased in use during the 1980s, GM found that RTV sealant can contribute to failure of these sensitive devices. Now, GM and other automakers are providing cork-based or synthetic gaskets for applications where RTV sealant had been used. The gasket makers, meanwhile, have learned new methods of leak control. In one method, they bond cork to both sides of thin metal, producing gaskets that meet the needs of modern vehicles and eliminate the shrinkage and deterioration problems that plagued the cork gaskets of a generation ago. In a more recent method, the thin metal is encased in a rubberized material to produce a gasket that is reusable for the life of the vehicle (100,000 miles or 161,000 km).

Figure 15-29. A MC solenoid used in Rochester carburetors. (GM)

Figure 15-30. A pulse solenoid controls variable air bleeds in some Carter carburetors. (Carter)

must regulate a mechanical device, an actuator is required.

Actuator Types and Operation

There are three different types of automotive actuators: solenoids, relays, and stepper motors. All of them use magnetic induction to move either a plunger, pivot arm, or armature, respectively. In automotive electronic circuits actuators fulfill many functions.

Solenoids

A solenoid uses an electromagnetic coil to attract and retract a spring-loaded metallic plunger, figure 15-28. To move the plunger, the battery or PCM applies voltage to one terminal of the solenoid and the PCM opens and closes the ground circuit attached to the other terminal. When the switch is closed and current flows through the windings, the magnetic field of the coil attracts

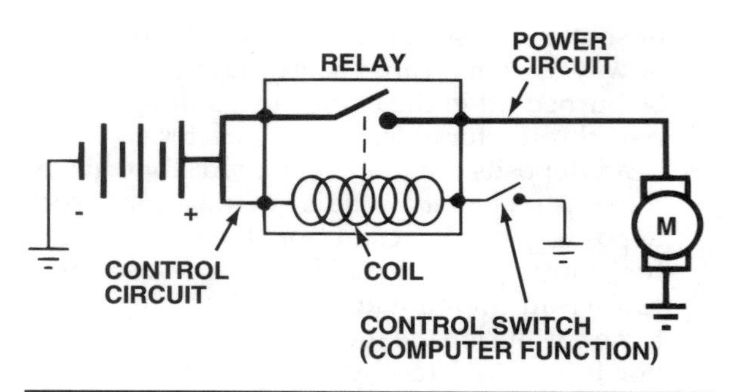

Figure 15-31. Applying current to the control circuit of a relay closes a switch allowing current to flow through the power circuit.

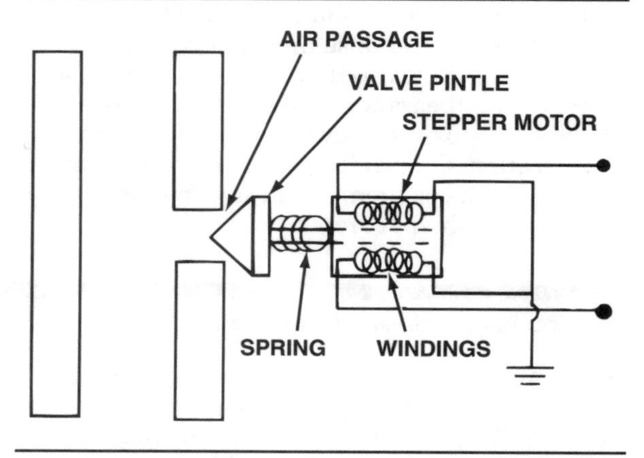

Figure 15-32. This IAC valve illustrates how a stepper motor can make incremental adjustments in the idle-air passage to regulate idle speed.

the movable plunger. This pulls it against spring pressure into the center of the coil toward the plate. Once current is shut off, the magnetic field collapses and spring pressure moves the plunger out of the coil.

In most applications, the PCM energizes the solenoid for varying periods of time. When energized, a solenoid may extend a plunger to richen the fuel mixture, as with a mixture control (MC) solenoid. These solenoids are crucial to many Rochester feedback carburetors, figure 15-29. It operates both a metering rod in the main jet and a rod that controls an idle air bleed passage. Refer to the chapter entitled "Basic Carburetion and Manifolding" for more information about MC solenoids.

Other systems use a solenoid that controls vacuum to a diaphragm installed in the carburetor. The vacuum diaphragm controls the operation of the metering rods and air bleeds. Some Carter carburetors use a pulse solenoid with a variable duty cycle to control the air bleeds, figure 15-30.

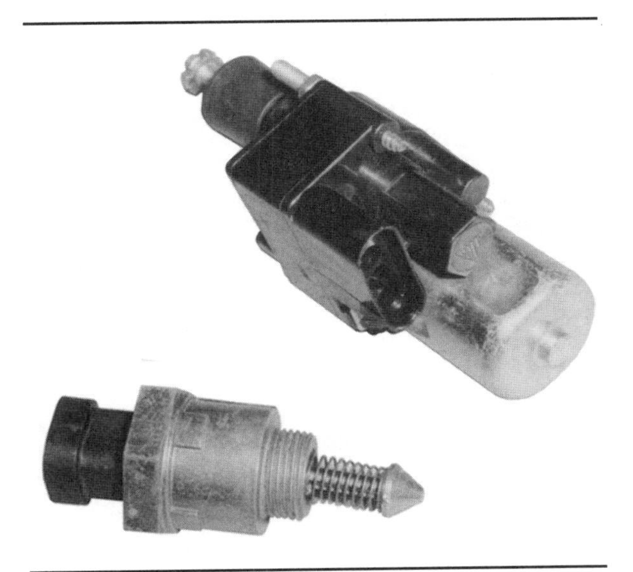

Figure 15-33. GM uses stepper motors as ISC motors for carburetors (top) and as IAC valves to control airflow in a fuel injection throttle body (bottom).

In addition to controlling fuel metering, late-model engine systems control several other functions:

- Torque converter lockup or engine shifting: Solenoid valves in the transmission or transaxle hydraulic circuits respond to computer signals based on vehicle speed and engine load sensors.
- Air injection switching: One or more solenoids operate a valve in the vacuum line to the air switching or air control valve.
- Vapor canister purge: A solenoid installed in the canister-to-carburetor or manifold valve line opens and closes as directed by the computer.

To understand how solenoids are used in fuel injectors and EGR valves, see the chapters entitled "Gasoline Fuel Injection Systems and Intake Manifolding" and "Exhaust Gas Recirculation", respectively.

Relays

A relay is an electromagnetic switch that uses a small amount of current in one circuit to open or close a circuit with greater current flow. To do this, it contains a control circuit and a power circuit, figure 15-31. The small flow of current through the relay coil moves an armature to open or close a set of contact points. This is called the control circuit. The PCM controls the operation of a relay through its output drivers, which open and close the relay control circuit. Typically, relays used in electronic engine control systems include: the fuel pump relay, lamp driver relays, and system power relays.

Stepper motors

A stepper motor uses direct current to move an armature in incremental steps. The magnetic force from its energized field coils can rotate its armature clockwise or counterclockwise, figure 15-32. Stepper motors have many discrete steps of motion that allows it to serve as an analog output operated by a digital signal.

A feature of almost all actuators controlled by an engine computer is that the computer switches the actuator on the ground side of the circuit. One side of the actuator coil receives system voltage and the ground side circuit connects to the computer.

Almost all circuit voltage is dropped across the actuator, and that reduces the power applied to the output driver transistors in the computer. This means that the coil windings of all actuators controlled by the computer have a minimum resistance specification. If coil resistance is too low, higher voltage and current results, applying more power to the computer. This increased power can overheat and destroy the transistors in the driver circuit to that device.

Actuator: An electrical or mechanical device that receives an output signal from a computer and performs an action in response to that signal.

Electronic Mufflers

Just when you thought electronics had penetrated every aspect of automotive technology that it possibly could, along come electronic mufflers.

Although it sounds like a joke, the concept is sound and actually pretty simple. Electronic mufflers use sensors and microphones to pick up the pressure waves of sound that the exhaust pipe emits. A computer analyzes the sound and produces a mirror image pattern of sound pulses, instantly sending them through a set of speakers mounted near the exhaust outlet. These computer-produced sounds are 180 degrees out of phase with the engine-produced sounds. The sounds from the speakers could be termed "anti-noise waves" and they help cancel out the total amount of noise. Since the system is computer controlled, it can be tailored to work over any rpm range or load condition.

With the better sound control that this system affords, the mufflers themselves can be less restrictive, resulting in greater power and fuel economy. With a little tuning of this system, muffler engineers may even be able to give a car a more pleasant exhaust note.

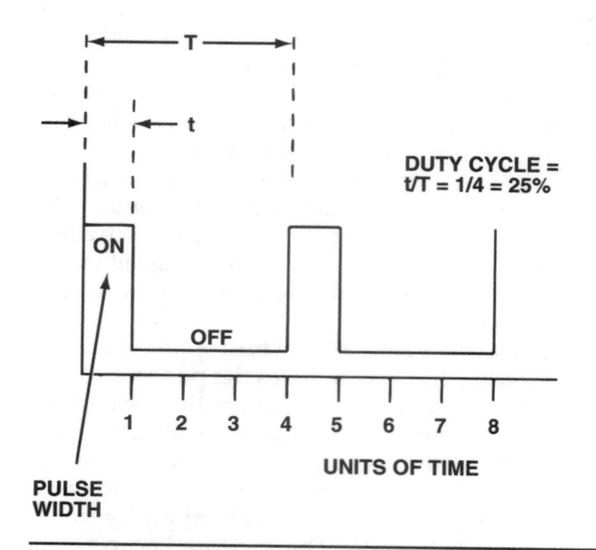

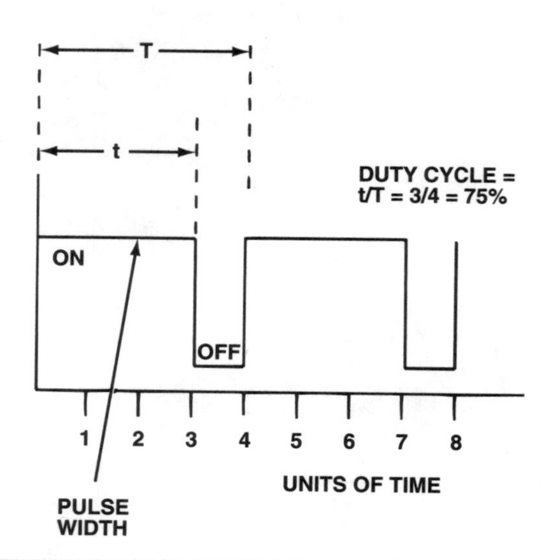

Figure 15-34. On a pulse width modulated solenoid, duty cycle is the percentage of on-time to total cycle time.

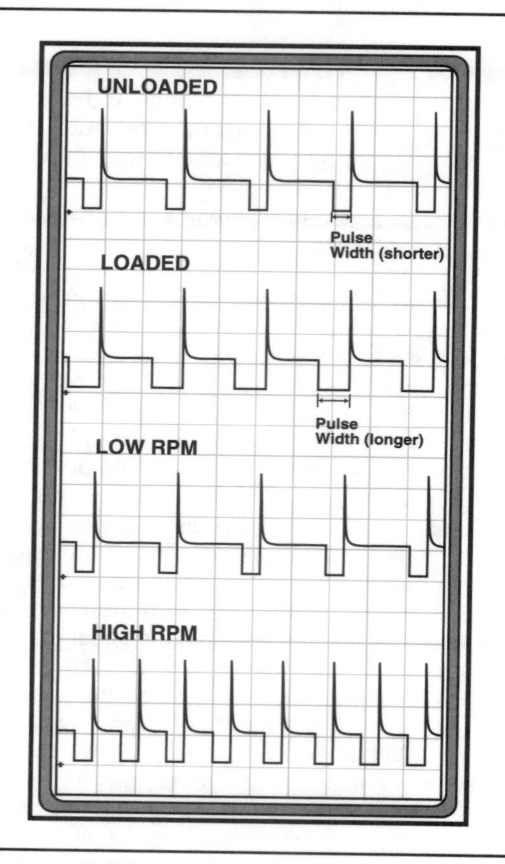

Figure 15-35. The top two waveforms illustrate varying pulse widths; the bottom two illustrate varying injector firing rates. Usually the PCM changes pulse width according to engine load and injector firing rates according to speed.

The most common uses for stepper motors are as idle speed controls. On carbureted engines, the motor often acts directly on the throttle linkage in the form of an idle speed control (ISC) motor, but in most fuel injection systems, it controls an idle air control (IAC) valve built into the throttle body. Figure 15-33 shows an example of an ISC motor and an IAC valve.

The IAC valve usually has two coils that move the armature in small steps by reversing the polarity of the coils. The valve can also be operated with four coils which share a common power source, but have separate grounds. During operation, each ground is opened and closed in sequence.

Pulse Width Modulation

A solenoid is a direct-current electrical device that is either fully on or fully off. Some solenoids in engine control systems, such as many canister purge solenoids and some EGR solenoids, are energized or de-energized for indefinite periods. Many other solenoids are pulsed on and off rapidly at a specific number of cycles per second, or **hertz (Hz).** The complete operating cycle of a pulsed solenoid is the sequence from off to on to back off again. Depending on its design and use, an actuator solenoid can operate at any number of cycles per second, or frequencies: 10 Hz, 20 Hz, 60 Hz, and so on.

Pulse width is the amount of time that a solenoid is energized (the "on" part of the cycle). Pulse width usually is measured in milliseconds (ms). An electronic fuel injector is a good example of a variable pulse width actuator. The computer changes injector pulse width in relation to the amount of fuel needed by the engine. A short pulse width delivers little fuel; a longer pulse width delivers more fuel. Fuel injectors, however, do not operate at a fixed frequency or specific number of cycles per second. Injector cycle varies with engine speed.

Many other solenoids do operate at a fixed frequency. Although the operating frequency may be fixed at 10 Hz or 60 Hz or some other frequency, the ratio of "on-time" to "off-time" can be varied. This is called **pulse width modulation** (PWM) or duty cycle modulation.

The **duty cycle** of a solenoid equals the pulse width in milliseconds divided by the complete off-on-off cycle time in milliseconds. Duty cycle is expressed in percentage, figure 15-34. A 25-percent duty cycle means that the solenoid is energized for 25 percent of the total off-on-off cycle time; a 75-percent duty cycle means that the solenoid is energized for 75 percent of the cycle. A 0- or 100-percent duty cycle indicates that the solenoid is fully off or fully on, respectively.

FUEL MANAGEMENT STRATEGY, IGNITION TIMING, AND EGR FLOW

By themselves, automotive sensors and actuators cannot run an engine control system. They rely on the programming information found within the PCM to carry out a fuel management strategy that takes into account ignition timing and EGR flow.

Fuel Management Approaches

Before discussing the three fuel-management strategies, it is important to review and understand a common fuel management principle shared by all engines operating in closed loop. Once these engines reach operating temperature, they mix the incoming air and fuel to form a ratio falling within a window centering around 14.7:1 during most driving conditions. At this proportion, the intake charge burns most completely, and the catalytic converter can reduce the most amount of hydrocarbons (HC's), CO, and oxides of nitrogen (NO_X). The PCM on these engines varies the pulse width or "on-time" of the fuel injectors and how fast the fuel injectors fire to control the amount of fuel entering the engine. The car's driver, by pressing down on the acceleration pedal, and the PCM, by opening and closing the IAC valve or ISC motor, control the amount of air flowing into the intake manifold. In order for the fuel management system to know how much fuel to spray into the cylinder, and still maintain the 14.7 to 1 ratio, the PCM must determine the density and rate of air flowing into the engine. Late-model electronic control systems make these determinations using either a MAP, VAF, or MAF sensor.

Three factors can affect the rate and amount of air molecules entering the engine during a constant driving condition: the engine's speed, the ambient temperature, and the manifold absolute pressure. As rpm increases, so does the number of times per minute the intake charge fills the cylinders. The PCM would then fire the injectors faster. Engine tuners measure the volume airflow rate in cubic feet per minute (CFM) or cubic meters per hour.

The two other variables, temperature and pressure, affect air density. If pressure remains constant and temperature increases, the distance between air molecules increases, causing them to loose density. The PCM would then command a shorter fuel injector pulse width, figure 15-35. If temperature remains constant and the atmospheric pressure were to drop due to a change in elevation, the air would also loose density. The PCM would then command a shorter fuel injector pulse width to maintain the 14.7:1 air-fuel ratio.

The speed-density approach

The first-developed fuel management system measures the density of the incoming air using a MAP sensor. Called a **speed-density system,** it relies on the following information to calculate air volume flow:

- The rpm signal from the engine speed sensor
- An estimate of EGR flow
- A PCM-held constant for the engine's displacement.

Hertz (Hz): A unit of frequency measurement equal to one cycle per second.

Pulse Width: The amount of time a solenoid is activated.

Pulse Width Modulation (PWM): A continuous on-off cycling of a solenoid at a specified number of times a second. The PCM can modulate the pulse width (on-time) to produce a variable duty cycle.

Duty Cycle: The amount of on-time of a solenoid relative to the total cycle time.

Speed-Density System: A fuel management method that *estimates* mass airflow rate so that the PCM can regulate fuel metering. This system relies primarily on the MAP sensor to determine density and on the engine speed sensor for volume airflow rate.

To calculate density, this approach looks at:

- The MAP sensor to measure intake manifold pressure
- The IAT sensor, or the ECT sensor, or both to measure temperature
- PCM-held constants dealing with the mass of air.

With these two factors, volume airflow and density, the PCM can then calculate the **mass airflow rate.** After reviewing the list above, know that the engine speed sensor and MAP sensor are the most important, thus the term speed-density.

At idle, a warmed-up engine's intake manifold has high vacuum, and its crankshaft turns relatively slowly. Pressure and speed is low; in other words, the distance between air molecules in the manifold is relatively large and the volume of air entering the engine is relatively small. During these conditions, the PCM commands a shorter fuel injector pulse width and a slower injector firing rate.

Now, think of how the MAP and engine speed sensors affect fuel management in a truck hauling gravel up a steep grade. Intake manifold vacvacuum is much lower than at idle; during heavy load, the throttle plate may be almost completely open to allow enough air-fuel mixture. Also, the rpm would be higher than at idle. During these conditions, the PCM commands a longer fuel injector pulse width and a faster injector firing rate.

How do you think the pressure and speed at cruise affects pulse width and injector firing rate? Compared to heavy load, the throttle does not open as wide, so intake manifold vacuum is higher, causing the PCM to decrease fuel injector pulse width. Also, since the engine's speed is greater, the PCM would fire the injectors faster.

The vane airflow approach

Like the speed-density approach, vane airflow (VAF) systems must calculate volume airflow. They do this by using:

- A VAF sensor
- The rpm signal from the engine speed sensor
- An estimate of EGR flow
- A PCM-held constant for the engine's displacement.

To calculate density, this approach considers:

- The voltage signal from the IAT sensor
- The voltage signal from the BARO sensor
- A PCM-held constant dealing with the mass of air.

Using these two factors, volume airflow and density, the PCM calculates mass airflow rate. Know that the VAF and engine speed sensors are the most important. At idle, a warmed-up engine produces a lower airflow because it turns its crank slowly and restricts the throttle opening. During these conditions, the PCM commands a shorter fuel injector pulse width, and a slower injector firing rate.

Now, think of how the VAF and engine speed sensors affect fuel management in a light-duty truck hauling gravel up a steep grade. Airflow is greater than at idle; during heavy load, the throttle plate may be almost completely open to allow enough air-fuel mixture. Also, the rpm is higher than at idle. During these conditions, the PCM commands a longer fuel injector pulse width and a faster injector firing rate.

How do you think the volume airflow rate at cruise affects pulse width and the injector firing rate? Compared to heavy load, the throttle does not open as wide, so airflow based upon the VAF sensor is lower, causing the PCM to decrease the fuel injector pulse width. Since the engine's speed is greater, the PCM would fire the injectors faster.

Just as the rate of volume airflow increases with load, so does it increase with engine speed. By comparing the VAF sensor signal with the engine speed sensor signal, the PCM can distinguish a loaded engine from a cruising engine. Using both inputs, the PCM varies injector pulse width and fuel injector firing rate accordingly.

The mass airflow approach

Compared to the speed-density and VAF airflow approaches, the mass airflow approach decreases the number of inputs and internally-held constants the PCM has to consider to determine the correct fuel injector pulse width. Using a MAF sensor, this fuel management method provides the PCM with a nearly direct measurement of the mass airflow rate. Compared to the speed-density and VAF approaches, MAF systems are quicker, more accurate, and more expensive. During the mid-1990s, approximately 20 percent of cars sold in the United States used MAF systems.

The hot-wire and thin-film sensors are most common, and work basically the same. As explained earlier in this chapter, the current-regulated, heated element is placed in the airflow stream. When air flowing past removes heat from the element, decreasing the resistance and increasing the amount of current, the MAF sensor decreases the current to maintain the element at a constant temperature. These current fluctuations vary according to changes in ambient temperature, intake manifold pressure, and air volume flow. The MAF sensor's signal to the PCM accounts for changes in all three simultaneously.

For temperature, most MAF sensors heat their hot wire or thin film to a certain number of degrees above the underhood temperature, based on a signal from an IAT sensor integral to the MAF sensor. As underhood temperature increases, the base current passing through the hot wire or film increases, causing the PCM to command a shorter fuel injector pulse width. When underhood temperatures decrease, the opposite occurs.

For pressure, it is important to remember that the MAF sensor lies between the throttle plate and the air intake opening. When the throttle is nearly closed and intake manifold pressure is low, the rate of volume airflow is small. When the throttle is wide open, the rate of volume airflow is larger. So, on a warm, idling engine the nearly closed throttle plate would block the higher atmospheric pressures from rushing past the hot wire or film, preventing the wire from cooling, causing the PCM to command a short fuel injector pulse width. At load, the wider throttle opening would allow more air under higher atmospheric pressure to flow past the hot wire or film, cooling the wire, causing the PCM to command a longer fuel injector pulse width.

Just as the rate of volume airflow increases with load, so does it increase with engine speed. By comparing the MAF sensor signal with the engine speed sensor signal, the PCM can determine whether a loaded or a high speed driving condition causes the higher rate of volume airflow. Using both inputs, the PCM varies injector pulse width and fuel injector firing rate accordingly.

Ignition Timing

Assuming a fixed air-fuel ratio and engine speed, variations in ignition timing dramatically affect fuel consumption and pollutants, figure 15-36. When timing is at top dead center (TDC) or slightly retarded, emissions are low and fuel consumption is high. As timing is advanced, fuel consumption drops off but emissions increase. Engine computer programs calculate the best timing for any combination of air-fuel ratio and engine speed that prevents detonation problems.

Exhaust Gas Recirculation

Exhaust gas recirculation is the most efficient way to reduce NO_x emissions without adversely affecting fuel economy, driveability, and levels of HC emissions. When exhaust gases are introduced into the intake charge, combustion temperatures decrease. Excessive EGR, flow how ever, leads to an increase in both HC emissions and fuel consumption. Again, the engine computer calculates the percentage of EGR that delivers the best compromise between NO_x control, HC emissions, and fuel economy without detonation problems. See the "Exhaust Gas Recirculation" chapter for more information on the EGR system.

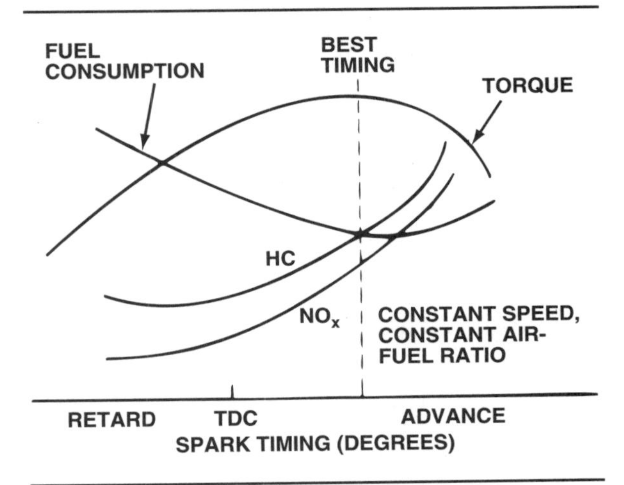

Figure 15-36. Changes in ignition timing produce these fuel consumption, emission, and torque curves.

Adaptive Learning Strategies

With special programming, engine control systems can learn from their own experience through **adaptive memory**. Adaptive memory allows the computer to adjust its memory for computing air-fuel ratio during closed-loop operation. Once the system is operating in closed loop, the computer compares its open-loop calculated air-fuel ratios against the average limit cycle values in closed loop. If there is a substantial difference, the computer stores a correction factor in its memory.

Adaptive learning strategies compensates for production variations and gradual wear in sys-

Mass Airflow Rate: The density and rate of air flowing into an engine, usually expressed in grams per second.

Vane Airflow System: A fuel management method that *estimates* mass airflow rate so that the PCM can regulate fuel metering. This method relies primarily on the engine speed and the VAF sensors to calculate volume airflow. For density, this system primarily uses the IAT sensor.

Mass Airflow System: A fuel management method that *directly calculates* mass airflow rate so that the PCM can regulate fuel metering. This system relies on one of three types of MAF sensors.

Adaptive Memory: A feature of computer memory that allows the microprocessor to adjust its memory for computing air-fuel ratio during closed-loop operation, based on changes in engine operation.

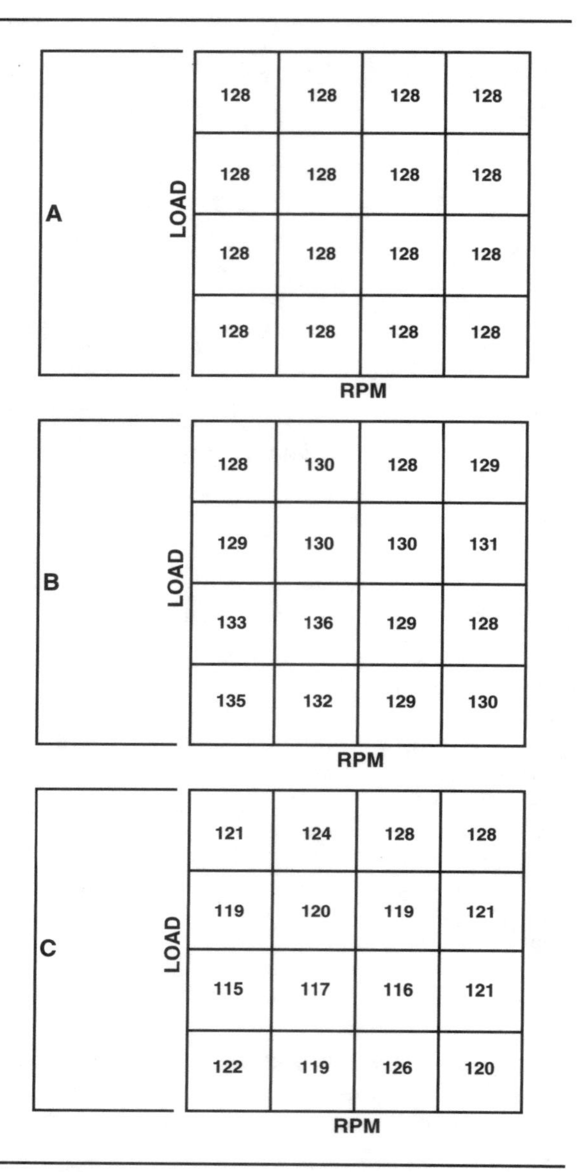

A (LOAD × RPM)

128	128	128	128
128	128	128	128
128	128	128	128
128	128	128	128

B (LOAD × RPM)

128	130	128	129
129	130	130	131
133	136	129	128
135	132	129	130

C (LOAD × RPM)

121	124	128	128
119	120	119	121
115	117	116	121
122	119	126	120

Figure 15-37. The block learn function divides engine operating range into 16 cells or blocks according to load and speed conditions. In A, all cells are running at the desired 14.7 to 1 ratio under all engine load and speed conditions. B shows typical readings during compensation for a slightly lean condition; C shows typical readings during compensation for a slightly rich condition. (GM)

tem components. When a controlled value sent to the computer is not within the original design parameters, the adaptive learning strategy modifies the computer program to accept the new value and restore proper operation of the system. Such modifications are stored in the PCM's random access memory (RAM) and remain in memory when the ignition is turned off, but not when the battery is disconnected. After restoration of battery power, vehicle driveability is

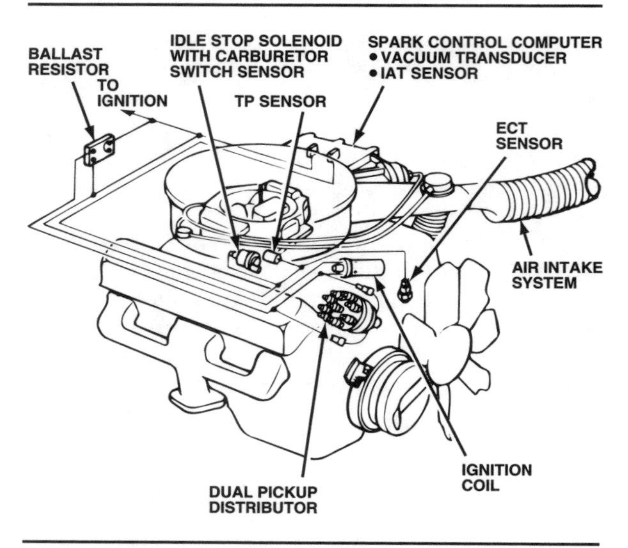

Figure 15-38. A Chrysler lean-burn system.

unsatisfactory. The computer must relearn a variety of parameters; this requires driving the vehicle several miles under varying conditions.

A malfunctioning sensor can affect vehicle driveability when replaced in a system with adaptive learning strategies. Since the computer had modified its original program to compensate for the unreliable sensor signals, the signals received from the new sensor are in conflict with the computer's modified program. This means that driveability remains unsatisfactory until the computer relearns the necessary parameters.

Short- and long-term fuel trim

The PCM used with GM late-model engines contains a pair of functions called integrator and block learn or **short- and long-term fuel trim (FT).** Both integrator and block learn are a form of adaptive memory designed to allow the computer to fine-tune air-fuel ratio of a fuel injected engine by changing injector on-time.

The term integrator, or short-term FT, applies to a method of temporarily changing the fuel delivery, and it functions only in closed-loop operation. The PCM program contains a base fuel calculation in memory. The short- and, long-term FT functions interface with this calculation, causing the PCM to add or subtract fuel from the base calculation according to the O2S feedback signal.

This information can be monitored as a number between 0 and 255, or as a percentage from 1 to 100, by connecting a scan tool to the serial data transmission line through the data link connector (DLC). Since the average short-term FT function is one-half the maximum, or 128, or 50 percent, the scan tool reads the base fuel calculation as the number 128 or 50 percent. The short-term FT function reads the O2S output voltage, adding or subtracting fuel as required to maintain the 14.7:1 ratio window. A short-term FT reading of 128 or 50 percent is neutral, which means that the O2S is telling the PCM that the cylinders have just combusted an intake charge with a 14.7:1 air-fuel ratio. If the scan tool reads a higher number or percentage, the PCM is adding fuel to the mixture. If the number is less than 128, or a lower percentage, fuel is being subtracted.

This corrective action is effective only on a short-term basis. The short-term FT allows the computer to make short-term minute-by-minute corrections in fuel metering. Such corrections may be necessary, for example, when driving a car from a low-lying area, across a high mountain pass and to a low-lying area in an hour or two.

Long-term FT represents the long-term effects of short-term FT corrections, although the corrections it makes are not as great as those made by the short-term FT. The function takes its name from division of the operating range of the engine for any given combination of rpm and load into 16 cells or "blocks", figure 15-37. Fuel delivery is based on the value stored in memory in the block corresponding to a given operating range. As with short-term FT, the number 128 or 50 percent, is the base fuel calculation and no correction is made to the value stored in that cell or block. Every time the engine is started, long-term FT starts at 128, or 50 percent, in every cell and corrects as required according to O2S. If the short-term FT increases or decreases, long-term FT makes a correction in the same direction. This causes a gradual reduction in the short-term FT correction until it returns to 128, or 50 percent.

Both short- and long-term FT have predetermined limits called "clamping" that vary according to engine design. If long-term FT reaches the limits of its control without correcting the condition, the short-term FT also goes to its limits. At this point, the engine starts to run poorly. Clamping is necessary, however, to prevent driveability problems that might result under certain conditions. For example, stop-and-go traffic on a crowded roadway could last long enough for long-term FT to make long-term corrections. Once the traffic clears and the vehicle returns to cruise speeds, a driveability problem would result in the absence of the clamping function.

LEVELS OF ELECTRONIC ENGINE CONTROL

Early electronic control systems manage fewer automotive systems than late-model control systems. Before troubleshooting, it is important to identify the control system's capabilities.

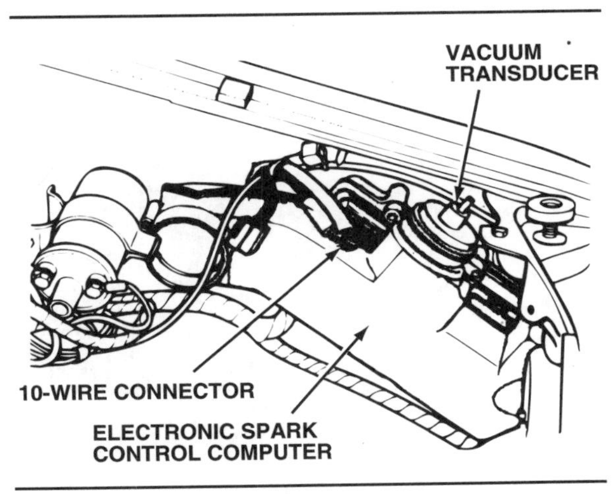

Figure 15-39. The Chrysler 4-cylinder electronic spark control computer.

Partial-Function Control Systems

Control systems of the late 1970s were partial-function systems. They regulated fuel metering or ignition timing, but not both. While older partial-function systems only had one or more of the following features, more recent full-function systems incorporate all of the following systems:

- Electronic spark timing control in place of traditional centrifugal and vacuum spark advance
- Electronic feedback control of fuel metering in the carburetor or fuel injection system
- A 3-way catalytic converter and an O2S for stoichiometric air-fuel ratio control
- Open- and closed-loop operating modes
- Electronic control of air injection switching, EGR, and vapor canister purging
- Electronic control of transmission shifting, torque converter lockup, and accessory operation.

Ignition timing control

Chrysler introduced its electronic lean burn (ELB) system in 1976, figure 15-38. It was an early example of a partial-function ignition timing control system. Although the name sounds like it was a fuel control system, ELB simply used a carburetor calibrated for lean air-fuel ratios. This system used no feedback, or variable, fuel control. The system

Short-Term Fuel Trim: A software function that allows the computer to make minute-by-minute changes in air-fuel ratio to maintain a 14.7:1 ratio.

Long-Term Fuel Trim: A software function that implements and stores the longer-term effects of short-term FT corrections.

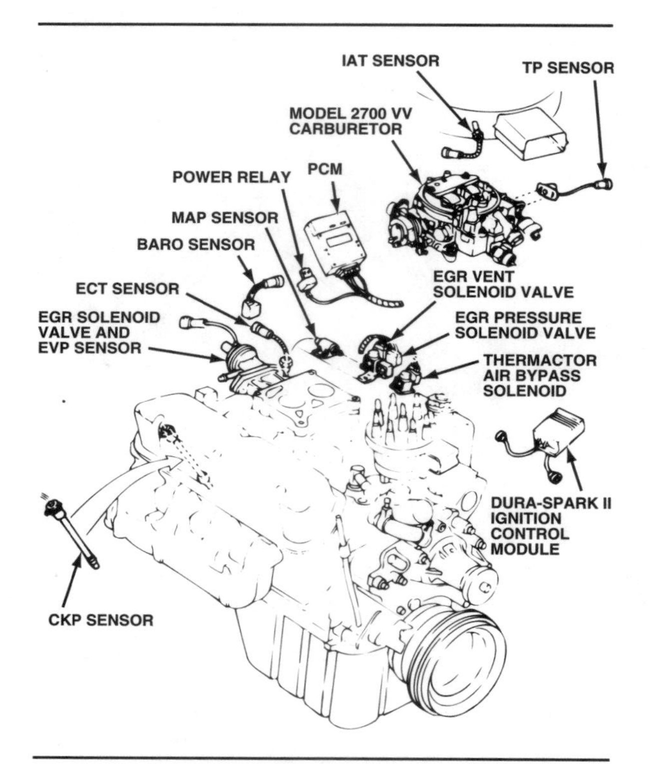

Figure 15-40. Ford's EEC-I engine control system.

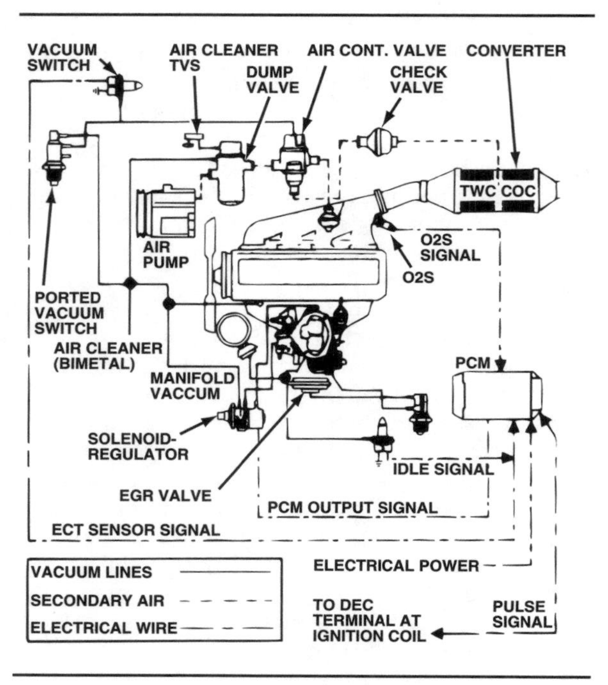

Figure 15-41. Ford's Feedback Electronic Engine Control System.

computer regulated ignition timing to allow the engine to operate smoothly on lean ratios. The first ELB systems used a distributor with centrifugal advance but no vacuum advance. In 1977, Chrysler eliminated centrifugal advance, and timing was under full electronic control. Chrysler's ELB system continued through 1978 on V8 and 4-cylinder engines and in 1979 was renamed electronic spark control (ESC), figure 15-39.

Ford introduced its first electronic engine control system (EEC-I), which included controlled spark timing, EGR, and air injection in 1978, figure 15-40. It did not include feedback fuel metering control. A crankshaft position (CKP) sensor provided ignition timing signals. On 1978 models, Ford installed the sensor in the rear of the engine block; in 1979, the automaker moved it to the front of the crankshaft behind the vibration damper.

The first partial-function timing control system GM used was the microprocessor sensing and automatic regulation (MISAR) system used on 1977 Oldsmobile Toronados. The system used a CKP sensor, a standard high energy ignition (HEI) distributor, and an electronic ignition control module to control spark advance. In 1978, the system was renamed electronic spark timing (EST).

Fuel metering control

In 1979, Chrysler modified its ELB, or electronic spark control system, to work with feedback carburetors and O2S's. Some 6-cylinder engines used this system, which approximated a full-function engine control system.

Ford used its earliest feedback control (FBC) system on some 1978 2.3-liter, 4-cylinder engines, figure 15-41. It included a 3-way catalytic converter (TWC) and an O2S. This system only controlled fuel metering through the Holley 6500 carburetor. It did not control ignition timing or other engine functions.

Ford changed the system computer from an analog to a digital processor on 1980 models and reidentified the system as the microprocessor control unit (MCU) system. The first versions controlled fuel metering in a way similar to the earlier FBC or TWC systems. Later versions in the mid-1980s controlled canister purging, idle speed, and detonation spark timing. Some MCU systems include simple self-diagnostic capabilities.

Ford's EEC-II system appeared on some 1979 V8's. This system controlled air injection switching and vapor canister purging, along with providing feedback fuel metering. Fully electronic spark timing control was not included. Ford's EEC-III, introduced in 1980, also was based on TWC's and O2S's. Ford installed this system on vehicles with feedback carburetors and throttle body fuel injectors (called central fuel injection (CFI)). EEC-III was the first Ford system to have a self-test program.

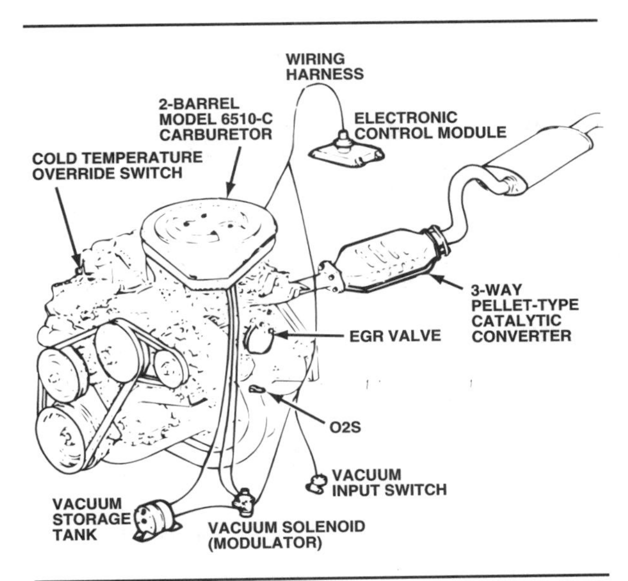

Figure 15-42. The GM electronic fuel control (EFC) system.

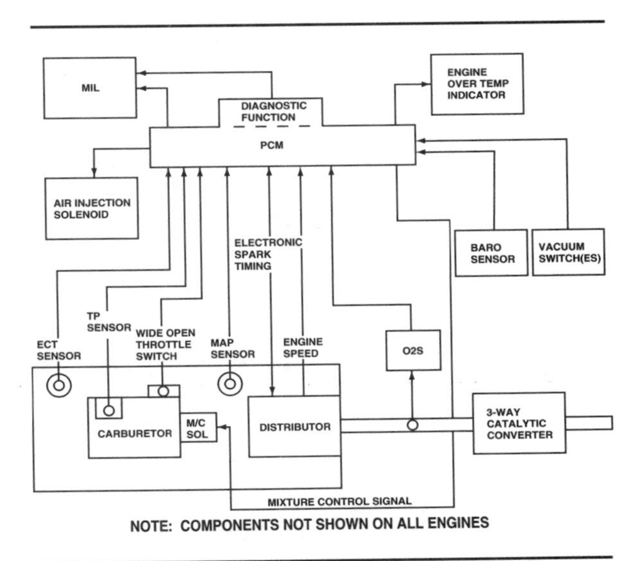

Figure 15-43. The GM C-4 system.

GM used electronic fuel control (EFC) on some 1978–79 2.5-liter engines, figure 15-42. The GM EFC system was quite similar to Ford's early FBC system. It used a TWC, an O2S, and a Holley 6510-C feedback carburetor. The early GM and Ford systems both controlled the carburetor through a vacuum solenoid that operated diaphragms to regulate metering rods and air bleeds.

In mid-1979, GM introduced the C-4 system which provided computer control and a catalytic converter. The 1980 X-cars and some Buick V6 engines used it, figure 15-43. Early C-4 systems were partial-function systems that controlled fuel metering and provided limited ignition timing control. The C-4 system developed into the GM full-function computer command control (CCC or C-3) system described later in this chapter.

Limitations of partial-function control

Partial function systems offer limited engine computer control. To illustrate, here are the specific features included in GM's 1982 Chevette and Pontiac T-1000 4-cylinder engines:

- It only controls fuel metering.
- It has only seven diagnostic trouble codes (DTC's).
- A coolant temperature switch closes when a predetermined engine coolant temperature is reached. At this time, the PCM switches system operation from open to closed loop, if the O2S is at operating temperature and a preprogrammed time interval has elapsed after the engine is started. Once the coolant switch closes, there is no effect on system operation if it opens due to a malfunction. GM replaced the temperature switch with an ECT sensor beginning with the 1983 model year.
- Since the PCM does not provide a fixed dwell during open-loop operation, the range of dwell control is limited. Open-loop time consists of four regions or areas. Dwell control varies according to which region the PCM is in at that time. For example, the system may be in open-loop, but if dwell was measured, it would give a reading as if the system was in closed-loop.
- DTC's set in memory are lost each time the ignition is turned off.

Full-Function Control Systems

Automakers gradually expanded beyond systems that managed only ignition or fuel to systems that control multiple engine functions. The early full-function control systems had one or more of the following characteristics; late-model systems have them all:

- The PCM controls timing electronically instead of relying on distributor vacuum and centrifugal advance mechanisms.
- The PCM controls air-fuel ratio as closely as possible to the stoichiometric value (14.7:1) through an O2S and a TWC.
- The PCM controls fuel metering by operating a MC solenoid or by pulsing fuel injectors according to data received from various sensors.
- Engine operation is divided into open- and closed-loop operational modes, with some systems also providing a modified closed-loop mode.

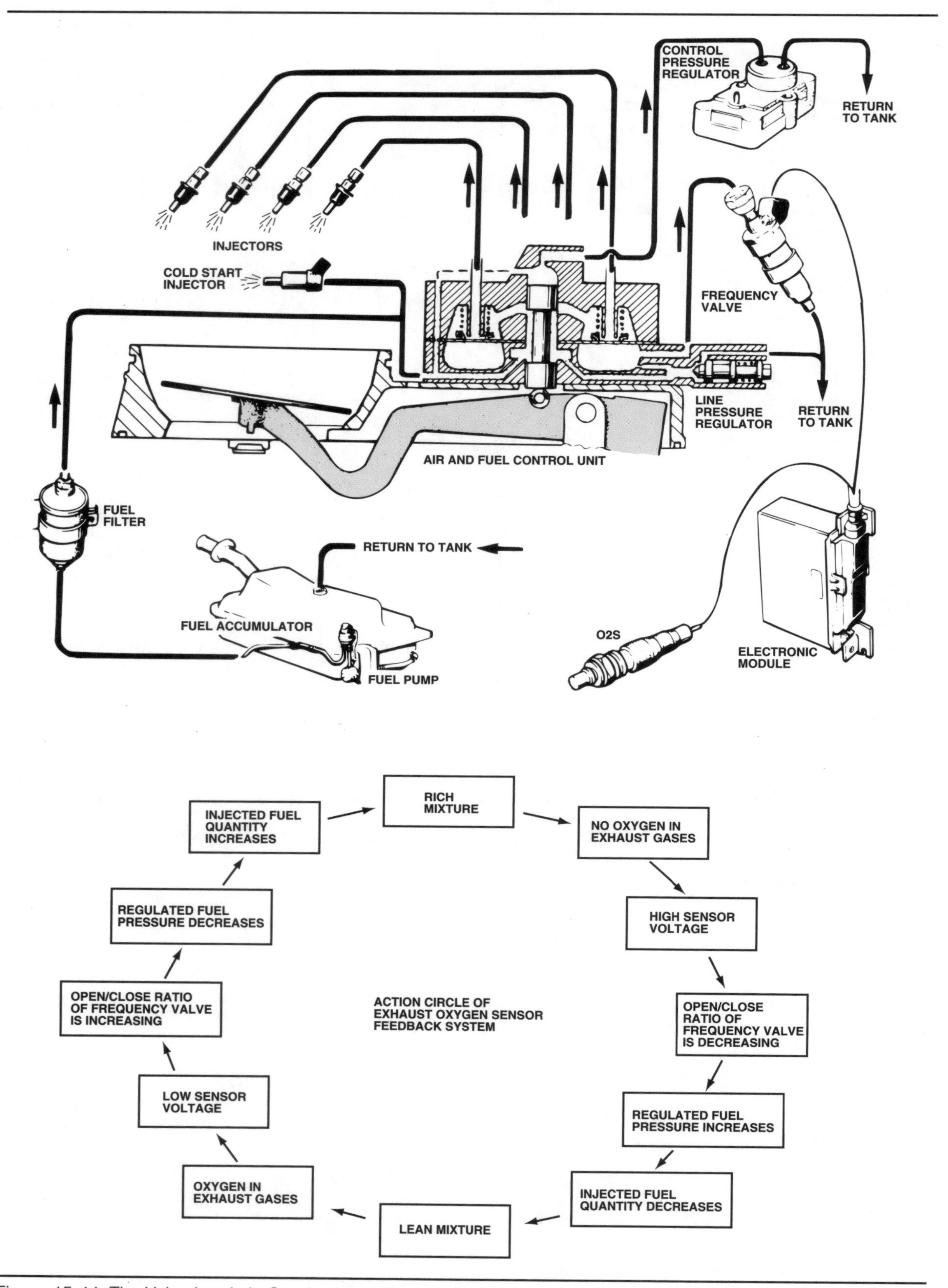

Figure 15-44. The Volvo Lambda-Sond system. (Volvo)

- When a serious system malfunction occurs, the PCM shifts into a "limp-in mode", or a "limited operational strategy", to allow the operator to drive the vehicle in for service.
- The PCM controls EGR, air injection switching, and vapor canister purging.
- The PCM controls automatic transmission shifting and torque converter lockup.
- The PCM can set a number of DTC's and has a self-diagnostic capability.
- The engine control system includes a malfunction indicator lamp (MIL) and monitors the O2S, the EGR valve, and the evaporative purge solenoid for proper operation.

ENGINE CONTROL SYSTEM DEVELOPMENT

The following section describes how manufacturer-specific engine control systems evolved from limited to full-function systems. While reading, keep in mind that advances in electronic engine control systems are environmentally driven. Automakers's efforts centered around decreasing emission levels while maintaining acceptable performance levels.

Bosch Systems

Volvo offered the first electronically controlled feedback fuel system on 1977 models sold in the United States. This application, called by Volvo the Lambda-Sond system, was produced by the Robert Bosch company, a pioneer European manufacturer of fuel injection systems and electronic components.

The Volvo system combined the use of a TWC and O2S with K-Jetronic continuous fuel injection, figure 15-44. The electronic control module processed sensor input, and then signaled a timing valve in the K-Jetronic system to vary injection control pressure and the amount of fuel delivered. Saab and other European automakers quickly followed with their own versions.

The Bosch Motronic or digital motor electronics (DME) system added ignition timing and electronic spark control to the fuel metering control of the Lambda-Sond system. In addition to the TWC and O2S, the DME system receives input signals from crankshaft speed and position sensors, a magnetic pulse generator in the distributor, and other sensors. This allows the computer to:

- Adjust injector pulse width for air-fuel ratio control
- Adjust ignition timing for combined speed and load conditions
- Shut off injection completely during closed-throttle deceleration.

Chrysler

As discussed earlier, Chrysler introduced its electronic lean-burn (ELB) spark timing control system in 1976 on some 6.6-liter V8 engines, figure 15-38. Two components form the basis for the ELB system: a special carburetor that provides air-fuel ratios as lean as 18 to 1, and a modified electronic ignition controlled by an analog spark control computer attached to the air cleaner housing. Two printed circuit boards inside the computer contain the spark control circuitry. The program schedule module receives the sensor inputs and interprets them for the ignition control module (ICM), which directs the spark timing output. The 1976 ELB distributor has dual ignition pickups and a centrifugal advance mechanism. The distributor secondary components are similar to those used with the basic Chrysler electronic ignition. A dual ballast resistor controls primary current and protects the spark control computer from voltage spikes.

In 1977, Chrysler made ELB systems available on all of its V8 engines, eliminating its centrifugal advance mechanism. The automaker used a second-generation ELB design on its 5.2-liter V8's. It dropped the start pickup in the distributor and redesigned the computer so all the circuitry fit on a single board.

In 1978, the second-generation system was adopted on all V8's. Omni and Horizon 4-cylinder engines received a new ELB version. The 4-cylinder system uses a Hall-effect distributor instead of a magnetic pickup. It has variable dwell to control primary current and does not use a ballast resistor. The 4-cylinder spark control computer is mounted on the left front fender, figure 15-39.

After the introduction of TWC's, Chrysler modified the ELB system in 1979 to work with revised carburetors that provided air-fuel ratios closer to 14.7:1. The automaker also integrated protective circuitry into the computer and replaced the dual ballast resistor with a single 1.2-ohm resistor. This third-generation system was renamed electronic spark control (ESC) and appeared on some 6-cylinder in-line engines with an O2S and a feedback carburetor.

The 1980 model year was a transitional one for Chrysler. All California engines received ESC with feedback fuel control. The 5.9-liter V8 and Canadian 5.2-liter 4-barrel V8's continued to use ESC without the feedback system. All other engines reverted to basic electronic ignition with mechanical and vacuum advance mechanisms. Chrysler also introduced KS sensors on some 1980 ESC systems. The most significant change, however, was their switch from an analog to a digital computer. You can identify these 1980 digital-equipped computer models by the dual ignition pickups in their distributors. Systems

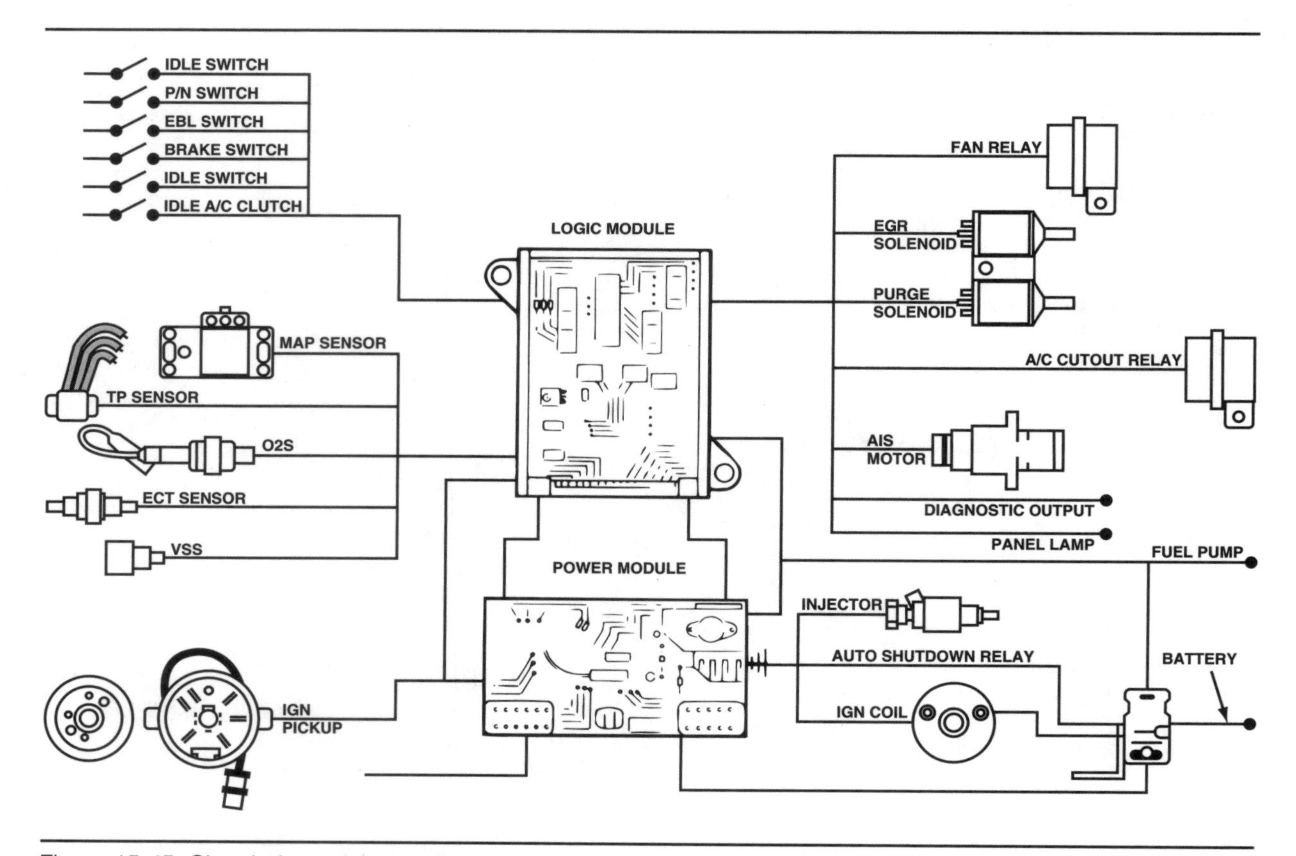

Figure 15-45. Chrysler's modular engine control system (single-point injection version shown).

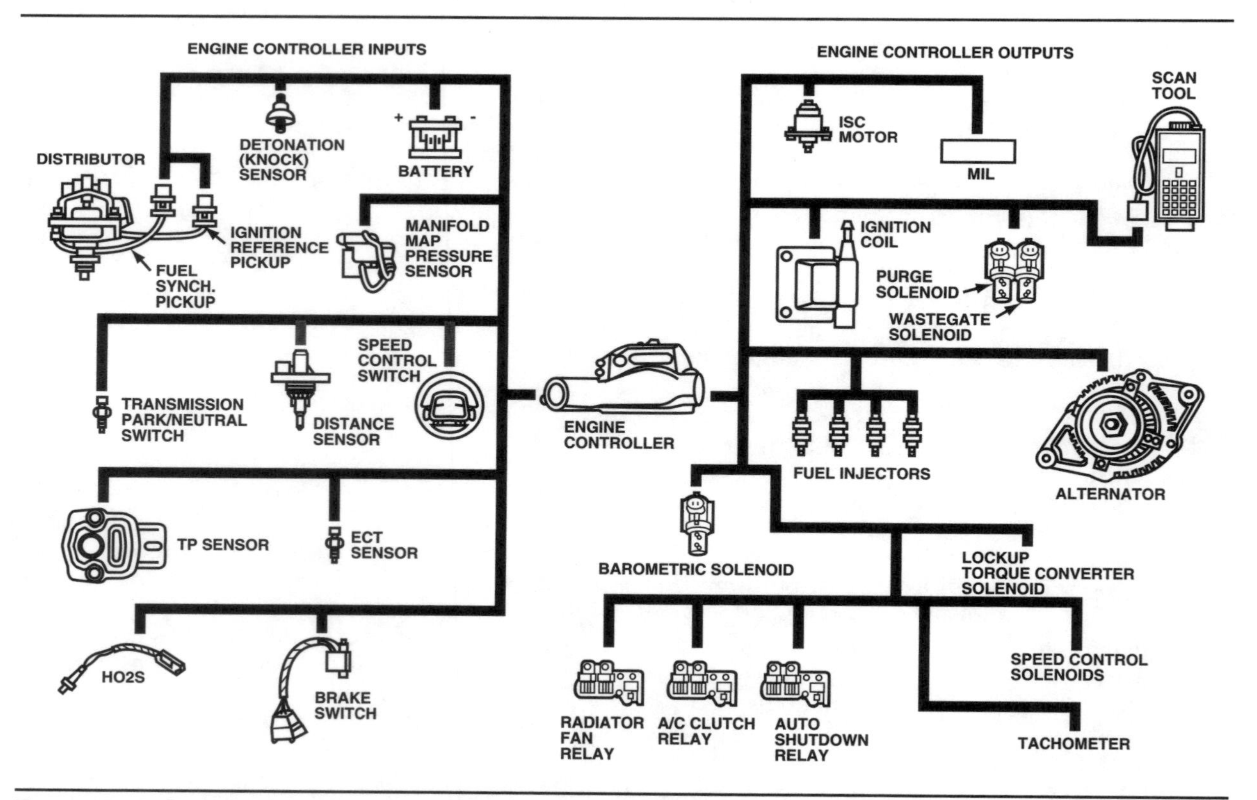

Figure 15-46. Chrysler's single-module engine control system (SMEC).

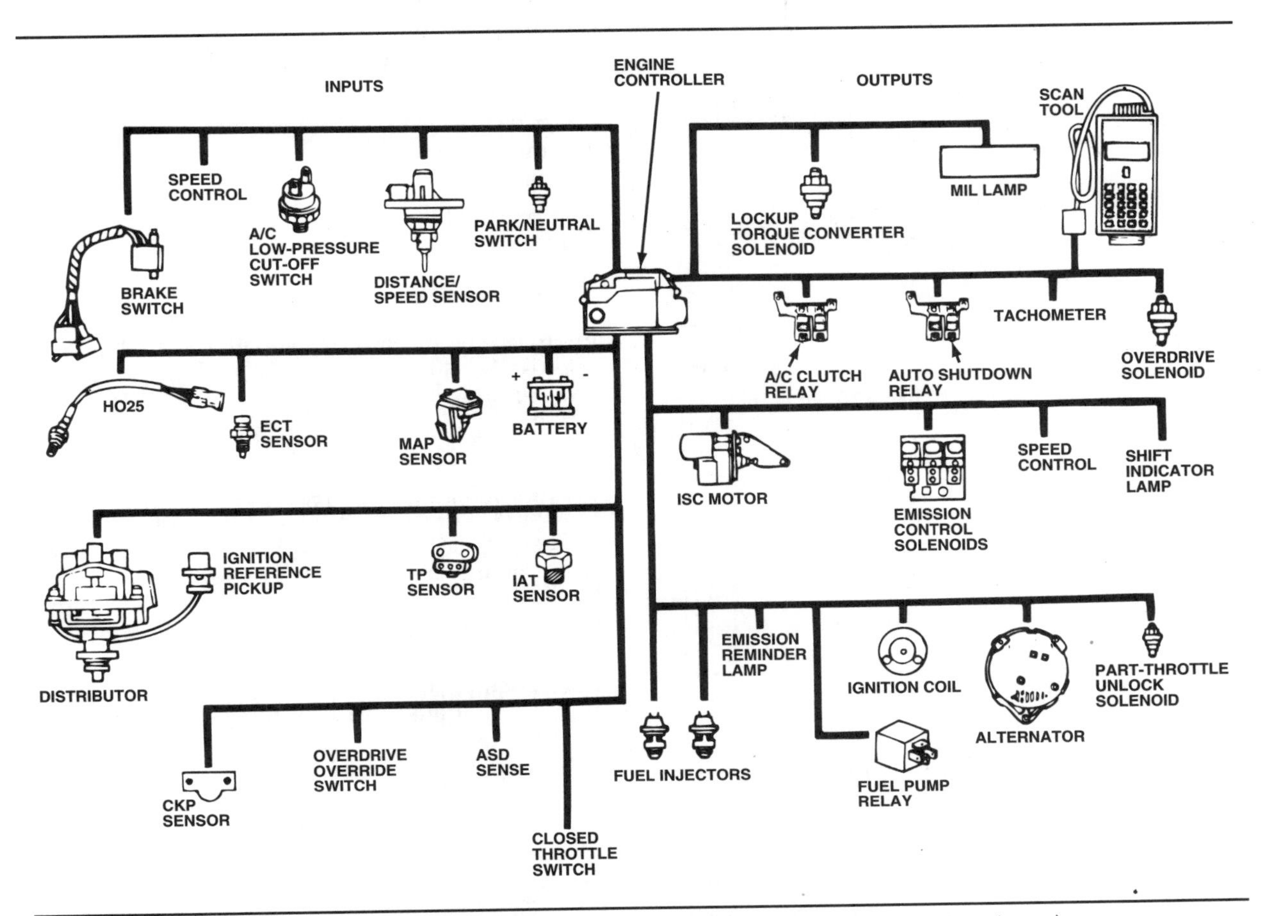

Figure 15-47. Chrysler's single-board engine control system (multi-point injection version shown).

with digital computers also eliminate the ballast resistor. Since 1981, all of Chrysler's domestic, carbureted, 4-cylinder engines and all 6-cylinder and V8 powerplants have feedback fuel control and digital ESC systems without ballast resistors.

Chrysler introduced the modular control system (MCS), figure 15-45, in late 1983 on throttle body fuel injected 4-cylinder engines. In 1984, turbocharged port-injected engines used MCS as well. The modular control system regulates vehicle functions using two separate modules whose functions are similar to the two circuit boards in the original ELB computer. The logic module handles all of the low-current tasks within the system, including receiving the inputs and making control decisions. The replaceable PROM mounted in the logic module housing includes a self-test program that aids in system diagnosis. To protect the logic module from underhood electrical interference, designers mount the module inside the car.

The power module, located in the left front fender, handles the high-current tasks. It looks similar to the spark control computer used in 4-cylinder ESC systems. The power module contains the regulated power supply for the entire control system, along with the switching controls for the ignition coil, fuel injectors, and auto-shutdown relay (ASD). The ASD supplies power to the coil, the fuel pump relay, and the power module when it detects a distributor cranking signal.

The modular control system (MCS) was replaced by the single-module engine control (SMEC) computer on some 1987 4-cylinder and V-6 engines, figure 15-46, which brought the two circuit boards used in MCS logic and power modules under one housing. Its advanced microprocessor is smaller, faster, and more powerful than the earlier MCS microprocessor, with electrically erasable memory (EEPROM) that can be programmed in the assembly plant. The SMEC processes instructions twice as fast as the older MCS system. This speed is important for use on the V6 engine, as the additional cylinders require the computer to process 50 percent more information per engine revolution than a 4-cylinder engine.

The use of complementary metal oxide semiconductor (CMOS) technology improves compo-

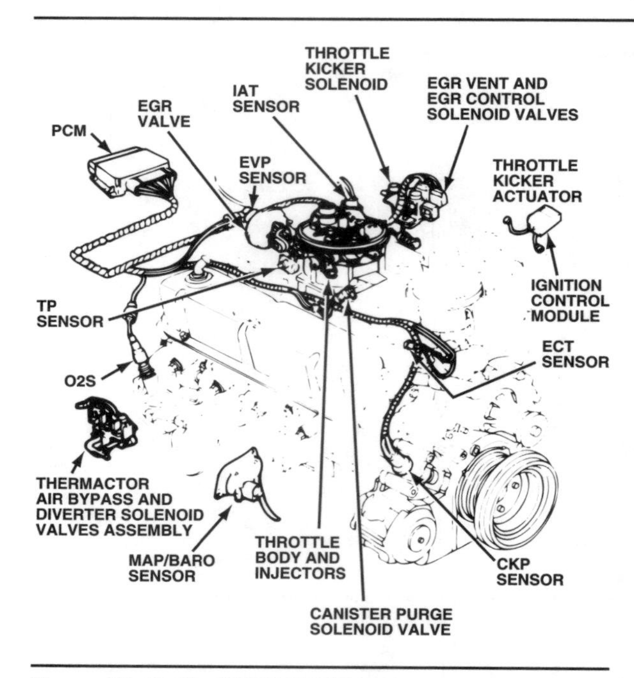

Figure 15-48. Ford EEC-III/CFI system components.

nent heat resistance. When combined with the reduction in overall size of the components, the manufacturer can install the SMEC in the engine compartment with far less heat-related problems than the separate logic and power modules. It also includes extensive diagnostic circuitry for use in determining and correcting problems.

Further refinement of Chrysler's engine computer came in 1989 with the introduction of the single-board engine controller (SBEC) and the SBEC II in 1992, figure 15-47. Both system's circuitry were simplified compared with previous engine controllers. Application-specific integrated circuits (ASIC's) replaced standard integrated circuits (IC's). The ASIC's are notably smaller in size than standard IC's, while providing an expanded number of functions. One ASIC takes the place of four separate input-output IC's previously used. Circuit simplification allowed the two circuit boards to be combined, forming a single board that requires less space. This resulted in a smaller, aerodynamically designed housing that provides greater airflow and a reduced number of external wiring connections.

As shown in figure 15-47, the SBEC engine controller accepts a larger number of input signals and produces more output signals than earlier controllers. In addition to the controller's increased number of functions, its enlarged memory capacity incorporates expanded diagnostic capabilities.

All of these Chrysler computer control systems include an emergency "limp-in" mode. In case of a system failure, the computer reverts to a fixed set of operating values. This allows the vehicle to be driven to a shop for repair. However, if the failure is in the start pickup or the coil triggering circuitry, the engine will not start.

Ford

Ford introduced its feedback electronic engine control system, figure 15-41, on some 1978 2.3-liter, 4-cylinder engines. The system contains a 3-way catalyst and conventional oxidation catalyst (TWC-COC) converter, an O2S, a vacuum control solenoid, and an analog computer. Its control was limited to fuel metering. In 1980, Ford replaced the analog computer with a digital computer, and renamed the system the microprocessor control unit (MCU) fuel feedback system. The major change in early applications is the addition of self-diagnostics, but later designs have expanded capabilities including control over idle speedup, canister purge, and detonation spark control. They might be considered complete engine control systems except that they lack continuous spark timing control.

Ford also introduced its first generation electronic engine control (EEC-I) system, figure 15-40, on the 1978 Lincoln Versailles. This system controls spark timing, EGR flow, and air injection. A digital microprocessor electronic control assembly (ECA) installed in the passenger compartment receives signals from various sensors. Then it determines the best spark timing, EGR flow rate, and air injection operation and sends signals to the appropriate control devices.

All 1978-79 California EEC-I systems use a variation of the blue-grommet Dura-Spark II ignition. The 1979 federal EEC-I system has a yellow-grommet dual-mode Dura-Spark II module. Although these modules appear similar to their non-EEC counterparts, they are controlled through the computer and cannot be tested with the same procedures.

The CKP sensor on 1978 EEC-I systems provides the ignition switching signal. It is located at the rear of the engine block, and uses four raised ridges on a magnetic pulse ring mounted to the end of the crankshaft. In 1979, Ford moved the pickup and pulse ring to the front of the engine immediately behind the vibration damper. The automaker also uses this design on the later EEC-II and EEC-III systems.

Late in 1979, Ford's second generation EEC-II system appeared on some 5.8-liter V8 engines. EEC-II added electronic controls for vapor canister purging and air injection switching. In addition, dual 3-way converters are used with a feedback carburetor for precise air-fuel mixture control.

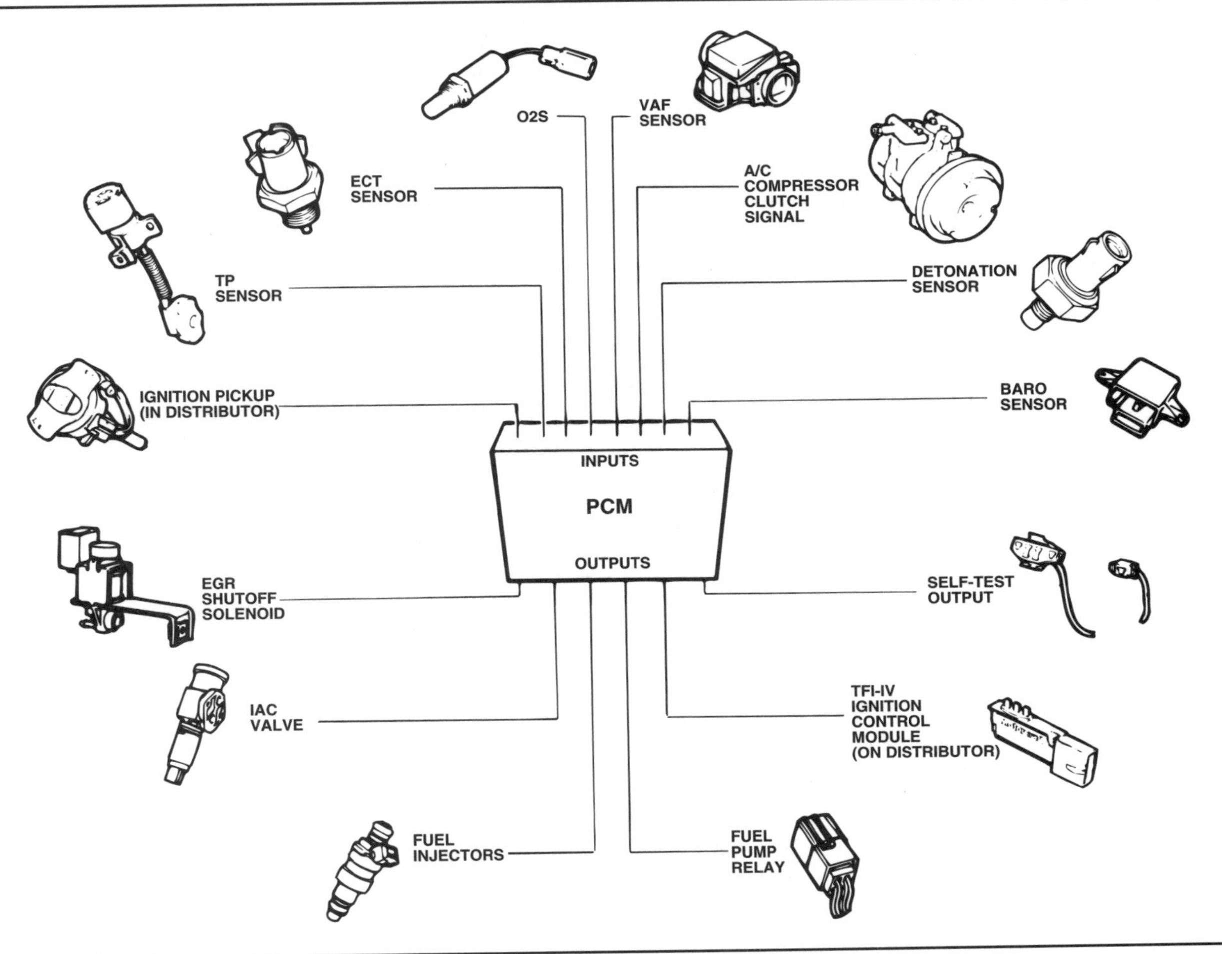

Figure 15-49. Typical components of a late-model Ford electronic engine control (EEC-IV) system.

Ford's third generation EEC-III system appeared in 1980 and is available in two versions through 1984. EEC-III/FBC incorporates a feedback carburetor similar to the EEC-II system; EEC-III/CFI, figure 15-48, has a throttle body type central fuel injection system. All EEC-III systems use the Dura-Spark III ignition module. EEC-III is the first Ford computer engine control system to have a self-test program.

In 1983, Ford introduced EEC-IV, figure 15-49. This system uses the thick-film integrated (TFI) ignition system. The two-microchip EEC-IV microprocessor is much more powerful than the four- or five-microchip PCM's used with earlier EEC systems. EEC-IV has both increased memory and the ability to handle almost one million computations per second. Unlike earlier EEC systems, however, the EEC-IV's calibration assembly is located inside the PCM and cannot be replaced separately. All EEC-IV systems have an improved self-test capability with trouble codes that are stored for readout at a later date.

Since 1986, Ford has made continuing and significant improvements in the processing speed, memory, and diagnostic capabilities of its EEC-IV microprocessor without changing the basic designation of the system. Compared to the EEC-IV microprocessor used during the early 1990s, the original EEC-IV microprocessor was a dinosaur. The most recent version introduced with the 1993 Mark VIII has 56 kilobytes of RAM (versus 32 kilobytes for the previous version) and processes data at a clock speed of 18 MHz (versus 15 MHz).

Unlike Chrysler and GM, Ford has not embraced the concept of multiplexing, but relies instead on smarter computers. Most Ford vehicles equipped with electronic automatic transmissions have the transmission control functions integrated in the ECA, since renamed a PCM.

All Ford EEC systems have a limited operating strategy (LOS) mode in case of a failure within the system. The exact nature of the LOS varies from one system to another, but generally,

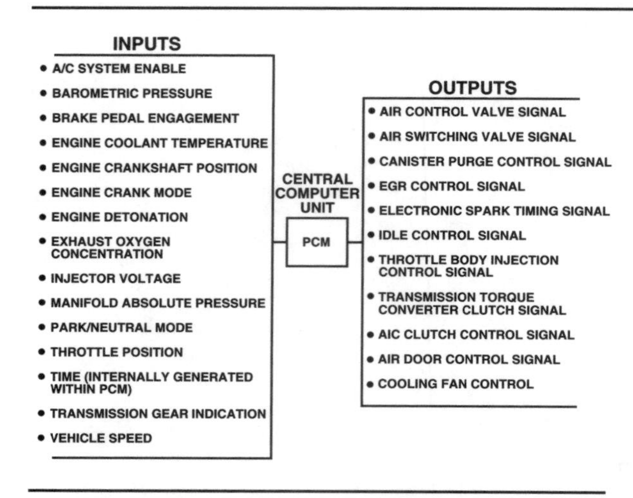

Figure 15-50. Typical CCC system sensors and actuators. (GM)

the timing is fixed and other PCM outputs are rendered inoperable.

General Motors (GM)

GM first introduced spark timing control on 1977-78 Oldsmobile Toronados. The 1977 system was called microprocessed sensing and automatic regulation (MISAR). MISAR is a basic spark timing system only. A rotating disk and stationary sensor on the front of the engine replace the pickup coil and trigger wheel of the distributor. Except for this change, the 1977 MISAR system uses a standard HEI distributor with a basic four-terminal ICM.

In 1978, GM modified the MISAR system, renaming it electronic spark timing (EST). Unlike previous systems, which used a crankshaft-mounted disk and stationary sensor, the MISAR system used a conventional pickup coil and trigger wheel in an HEI distributor. The ICM, however, is a special three-terminal design that is not interchangeable with any other HEI system.

In mid-1979, GM introduced the computer-controlled catalytic converter (C-4) system, figure 15-43. At first, the C-4 system was purely a fuel control system used with 3-way converters. But in 1980, Buick V6 engines with C-4 were also fitted with an electronic spark timing (EST) system. Although this is the same name applied to the MISAR system just discussed, the two systems are not the same. C-4 with EST has a single electronic control module (ECM) that regulates both fuel delivery and spark timing. It was the first complete computer engine control system General Motors used. The C-4 system was further upgraded in 1981 with EST in almost all applications, and additional control capabilities were added. The expanded system, figure 15-50, was renamed computer command control (CCC or C-3).

In 1986, GM began to update the CCC system through the introduction of a new high-speed PCM on certain vehicles. Systems using the new PCM often are called "P-4" systems. The high-speed PCM is smaller than previous models, but has more functional capabilities. It operates at twice the speed of previous PCM's, is capable of 600,000 commands per second, and contains fewer IC chips and internal connections. The high-speed PCM draws less current with the ignition off, provides more diagnostic functions, and operates reliably on battery voltage as low as 6.3 volts. Service procedures for the unit allow repair and reprogramming by replacing several different integrated circuits in the controller housing.

PCM's used on vehicles with electronic automatic transmissions have the transmission control circuitry integrated in the memory and calibration (MEM-CAL) unit. All C-4 and C-3 systems have self-diagnostic capabilities. The newer the system, the more comprehensive the diagnostic capabilities are. Several GM cars provide diagnostic readouts accessible through instrument panel displays, although the trend has clearly been to make such information available only through the Tech I scan tool in late-model diagnostic systems.

ON-BOARD DIAGNOSTICS GENERATION-II (OBD-II) SYSTEMS

As shown above, during the 1980s automakers began equipping their vehicles with full-function control systems capable of alerting the driver of a malfunction and of allowing the technician to retrieve codes that identify circuit faults. Ideally, these early diagnostic systems were meant to reduce emissions and speed up vehicle repair.

The automotive industry calls these systems On-Board Diagnostics (OBD). In 1985 the California Air Resources Board (CARB) developed the first regulation requiring automakers selling vehicles in that State to install OBD. Called OBD Generation I (OBD-I), it carried the following requirements:

1. An instrument panel warning lamp able to alert the driver of certain control system failures, now called a **malfunction indicator lamp (MIL)**
2. The system's ability to record and transmit **diagnostic trouble codes (DTC's)** for emission-related failures
3. Electronic system monitoring of the HO2S's, EGR valve, and evaporative purge solenoid.

OBD-I applies to all vehicles sold in California beginning with the 1988 model year. Although not U.S. EPA-required, during this time most automakers also equipped vehicles sold outside of California with OBD-I.

Representing only a beginning, these initial regulations failed to meet many expectations. By failing to monitor the catalytic converter, the evaporative system for leaks, and the presence of engine misfire, OBD-I did not do enough to lower automotive emissions. In addition, the OBD-I monitoring circuits that were installed lacked sufficient sensitivity. By the time an engine with a lit OBD-I MIL found its way into a repair shop, that vehicle would have had already emitted an excessive amount of pollutants.

Aside from OBD-I's lack of emission-reduction effectiveness, another problem existed. Automakers implemented OBD-I rules as they saw fit, resulting in a vast array of servicing tools and systems. Rather than simplifying the job of locating and repairing a failure, the aftermarket technician faced a tangled network of procedures often requiring the use of expensive special test equipment and dealer-proprietary information.

Soon it became apparent that more stringent measures were needed if the ultimate goal, reduced automotive emission levels, was to be achieved. This led the CARB to develop OBD Generation II (OBD-II). Ford calls its OBD-II compliant system EEC-V.

OBD-II Objectives

Generally, the CARB defines an OBD-II-equipped vehicle by their ability to:

1. Detect component degradation or a faulty emission-related system that prevents compliance with federal emission standards
2. Alert the driver of needed emission-related repair or maintenance
3. Use standardized DTC's, and accept a generic scan tool.

These requirements apply to all 1996 and later model light-duty vehicles; in 1997, these requirements apply to all light-duty trucks. Adopted by the U.S. EPA for use on a national level through the 1997 model year, new CARB-based federal standards take effect in 1998.

Based on these requirements, OBD-II is the most significant automotive emission-related development since the introduction of the catalytic converter during the mid-1970's. OBT-II systems are also unique because of their complexity. For example, OBD-II must be able to detect minute increases in emission levels that cause components to deteriorate. The MIL must light when catalytic converter degradation allows increases in hydrocarbons (HC's) greater than 0.4 gram/mile (g/m) above the 0.6 g/m federal test procedure (FTP) standards. (See the "Comprehensive Engine Tests" chapter in the *Shop Manual* for a discussion on g/m emission measurements and constant volume sampling.) Other similarly strict emission-level requirements stipulate when the MIL must light after the engine misfires or the HO2S degrades.

OBD-II Operation

An OBD-II vehicle tests its components and communicates these test results without your help. To use OBD-II, you must understand the types of OBD-II tests, test results, and communication methods.

The OBD-II's PCM-held information manager, the software program in charge of testing and indicating test results, is called the **Diagnostic Executive.** The two sections that follow describe this software program's functions.

OBD-II diagnostic tests

Depending on the OBD-II system, the Diagnostic Executive (usually referred to as the Executive) performs tests on up to seven emission systems

Malfunction Indicator Lamp (MIL): A type of OBD test result that lights to alert the driver of a faulty automotive circuit. Previously called a Check Engine lamp.

Diagnostic Trouble Code (DTC): A type of OBD test result that indicates a faulty circuit or system using an alphanumeric or numeric code.

Diagnostic Executive: The program within the PCM that coordinates the OBD-II self-monitoring system. It manages the comprehensive component and emission monitors, DTC and MIL operation, freeze-frame data, and scan tool interface.

Emission Monitor: One of eight OBD-II software programs that perform passive, active, and intrusive tests on emission-related components and systems once per trip.

Enable Criteria: Engine operating conditions that must be met before the Diagnostic Executive runs a test.

Trip: A key-on, engine running, key-off driving cycle where all of the enable criteria for a test have been met and where the Diagnostic Executive ran and completed that test.

Warm-Up Cycle: When used referring to an OBD-II system, a "warmed-up" engine has a specific meaning. For Ford, the engine temperature must reach a minimum of 158°F (70°C) and rise at least 36°F (20°C) over the course of a trip.

CATALYTIC CONVERTER EMISSION MONITOR

ENABLING CRITERIA

- ENGINE COOLANT TEMPERATURE GREATER THAN 170°F
- VEHICLE SPEED GREATER THAN 20 MPH FOR MORE THAN 2 MINUTES
- OPEN THROTTLE
- CLOSED LOOP OPERATION
- RPM BETWEEN 1,248 AND 1,952 (AUTO), OR BETWEEN 1,248 AND 2,400 (MANUAL)
- MAP VOLTAGE BETWEEN 1.50 AND 2.60

PENDING STATUS CONDITIONS

- MISFIRE DTC
- O2S MONITOR DTC
- UPSTREAM O2S HEATER DTC
- DOWNSTREAM O2S HEATER DTC
- FUEL SYSTEM RICH DTC
- FUEL SYSTEM LEAN DTC
- VEHICLE IS IN THE LIMP-MODE DUE TO MAP,TPS, OR TEMPERATURE DTC
- UPSTREAM O2S SENSOR RATIONALITY DTC
- DOWNSTREAM O2S SENSOR RATIONALITY DTC

CONFLICT STATUS CONDITIONS

- EGR MONITOR IS IN PROGRESS
- FUEL SYSTEM RICH INTRUSIVE TEST IS IN PROGRESS
- PURGE MONITOR IS IN PROGRESS
- TIME SINCE START IS LESS THAN 60 SECONDS
- ONE TRIP MISFIRE MATURING CODE
- ONE TRIP O2S MONITOR MATURING CODE
- ONE TRIP UPSTREAM O2S HEATER MATURING CODE
- ONE TRIP DOWNSTREAM O2S HEATER MATURING CODE
- ONE TRIP FUEL SYSTEM RICH MATURING CODE
- ONE TRIP FUEL SYSTEM LEAN MATURING CODE

SUSPENSION STATUS CONDITIONS

RESULTS OF THE MONITOR ARE NOT RECORDED UNTIL THE O2S MONITOR PASSES

Figure 15-51. This table specifies the conditions under which a Chrysler OBD-II catalytic converter emission monitor operates.

using dedicated **emission monitors.** In addition, an eighth emission monitor, called the comprehensive component monitor (CCM), tests components not included in the other seven monitors.

Each monitor can conduct tests during a key-on, engine running, key-off cycle when certain operating conditions called **enabling criteria** are met. A driving cycle that includes an emission monitor's test is called a **trip.** The criteria may include information such as elapsed time since start-up, engine speed, throttle position, and vehicle speed. Many tests require the vehicle to be **"warmed-up"** as part of the enabling criteria. Automakers may define warmed-up differently. For example, on most Ford vehicles this term means that the engine temperature has reached a minimum of 158°F (70°C) and rose at least 36°F (20°C) over the course of a key-on, engine running, key-off cycle.

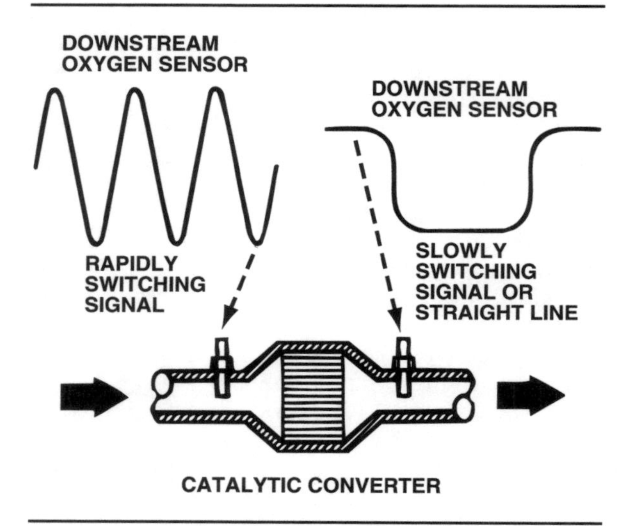

Figure 15-52. The OBD-II catalytic monitor compares the signals of the upstream and downstream O2S's to determine converter efficiency.

A variety of factors influence whether the Executive must delay a test, figure 15-51. They can be split into three groups:

- Pending tests—The Executive runs some secondary tests only after certain primary tests have passed. These delayed secondary tests are considered pending.
- Conflicting tests—Sometimes, different tests use the same circuits or components. In this case, the Executive would require each test to finish before allowing another to begin.
- Suspended tests—Each of the emission tests and monitors are prioritized. The Executive may suspend a test so that one of higher priority may run.

Generally, the Executive runs three types of tests. Passive tests simply monitor a system or component without affecting its operation. Active tests, conducted when a passive test fails, require that the monitor produce a test signal so that it can evaluate the response. Unlike the active tests, the third type, intrusive tests, do affect engine performance and emissions levels. The Executive performs these tests after both the passive and active tests have failed.

On completion of a test, the monitor reports a pass or fail to the Diagnostic Executive. For most monitors, the Executive does not set a DTC and light the MIL until the test fails during two consecutive trips. The following monitor descriptions include not only when that particular monitor lights the MIL and sets a DTC, but also a brief overview describing their operation and function.

The CCM checks the PCM's input and output signals for malfunctions affecting any component or circuit not already reviewed by another

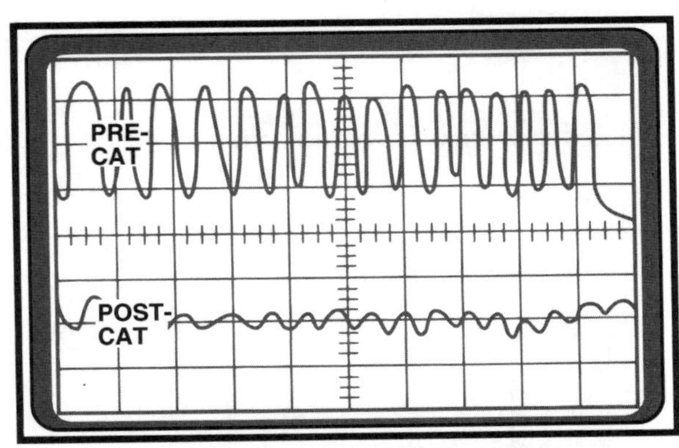

Figure 15-53. The waveform of an O2S downstream from a properly functioning converter shows little, if any, crosscounts.

monitor. Depending on the circuit or system, the CCM can conduct tests for open or short circuits and for out-of-range values. Two additional tests for "rationality" (inputs) and "functionality" (outputs) compare voltage signals from one sensor or actuator with those of another. For example, a rationality test on a speed-density system's TP sensor first examines TP sensor voltage and then compares it with MAP sensor voltage. Because a wider throttle opening decreases engine vacuum, and TP and MAP sensor voltages increase as the throttle opens and engine vacuum drops, the CCM can verify that the TP sensor works properly.

Depending on the system, the CCM may monitor the following components: MAF, IAT, ECT, TP, CMP, and CKP sensors; the fuel pump, IAC motor, and torque converter clutch (TCC). Typically, the Executive lights the MIL after the CCM detects the same emission-related fault during two consecutive trips.

In addition to the CCM, the Diagnostic Executive relies on dedicated monitors. These are designed to detect deteriorated systems that allow emission levels to exceed 1.5 times the FTP standard. These monitors evaluate the following systems and components:

- Catalytic converter
- HO2S
- Misfire detection
- System ability to correct the air-fuel ratio (fuel system monitor)
- EVAP system
- EGR system
- AIR system.

The catalytic converter monitor

OBD-II's catalytic converter monitor uses an up stream and downstream HO2S to *infer* catalytic efficiency, figure 15-52. When the engine combusts a lean air-fuel mixture, higher amounts

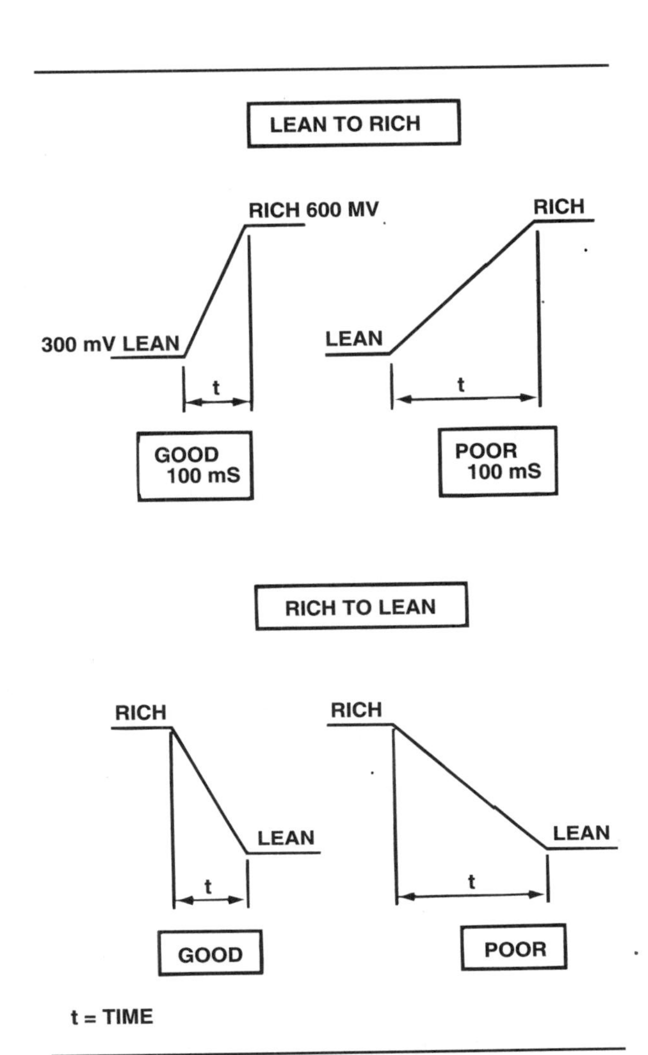

Figure 15-54. The OBD-II HO2S monitor tests for lean to rich and rich to lean transition response times. The Executive lights the MIL should the HO2S monitor fail these tests.

of oxygen flow through the exhaust into the converter. The catalyst materials absorb this oxygen for the oxidation process, thereby removing it from the exhaust stream. If a converter cannot absorb enough oxygen, oxidation does not occur. Engineers established a correlation between the amount of oxygen absorbed and converter efficiency.

The OBD-II system monitors how much oxygen the catalyst retains. A voltage waveform from the downstream HO2S of a good catalyst should have little or no crosscounts, figure 15-53. A voltage waveform from the downstream HO2S of a degraded catalyst shows many crosscounts. In other words, the closer the activity of the downstream HO2S matches that of the upstream HO2S, the greater the degree of converter degradation. In operation, the OBD-II monitor compares crosscounts between the two exhaust oxygen sensor's.

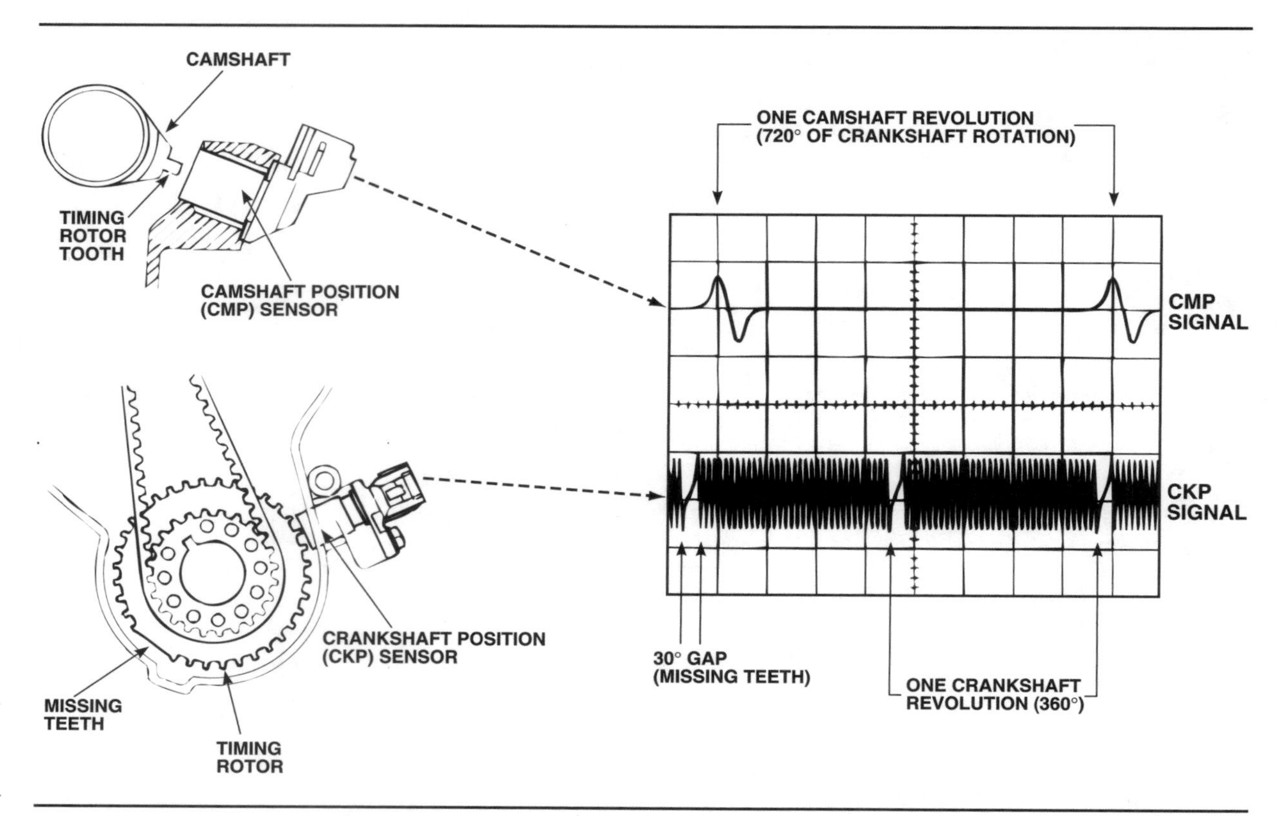

Figure 15-55. The OBD-II misfire monitor uses position sensors. In this application, each peak on the crankshaft position signal equals 10 degrees of rotation. A misfire momentarily slows the crankshaft and interrupts the waveform.

Depending on the system, the monitor may run statistical tests or wait until three consecutive failed trips, or both, before setting a DTC.

The exhaust oxygen sensor monitor

The second monitor, the HO2S monitor, can run several tests, depending on whether it is testing the up or downstream HO2S. This monitor checks all HO2S's for proper heater circuit and PCM reference signals. Pre-catalyst HO2S tests include a check for high and low threshold voltage and switching frequency. The switching frequency test counts the number of times that the signal voltage crosses the midpoint of 450 millivolts during a specific time and compares it to previously stored values. In addition, the monitor measures the time needed for a lean-to-rich and a rich-to-lean transition, figure 15-54. Usually, the lean-to-rich transition requires less time. Again, the monitor compares the average response times with a stored calibrated threshold value.

For the downstream HO2S, whose voltage signal fluctuates little, if at all, the HO2S monitor runs two tests. During rich running conditions, the monitor checks whether the HO2S signal waveform is fixed lower. Likewise, while running a lean mixture, the monitor checks for an HO2S signal fixed higher. For both HO2S's, the Diagnostic Executive only lights the MIL following two failed consecutive trips.

The misfire monitor

Another critical monitor detects poor cylinder combustion that causes engine misfire. The underlying problem may be a lack of compression, proper fuel metering or inadequate spark. Misfiring cylinders causes excess HC's to enter the converter. This accelerates converter degradation. In addition, misfire overloads the converter, causing an increase of HC emissions from the tailpipe.

Whenever a cylinder misfires, combustion pressure drops momentarily, slowing the piston. This retarded piston movement also slows the crankshaft. Using this principle, engineers are able to use CKP sensors to detect engine misfire. Misfire causes interruptions in the even spacing of the CKP sensor's waveform, alerting the monitor of a misfire. By comparing the CKP and CMP sensor signals, the monitor can also determine which cylinder misfired, figure 15-55.

Excessive driveline vibration caused by rough roads can mimic conditions that the monitor would recognize as a misfire. To prevent false

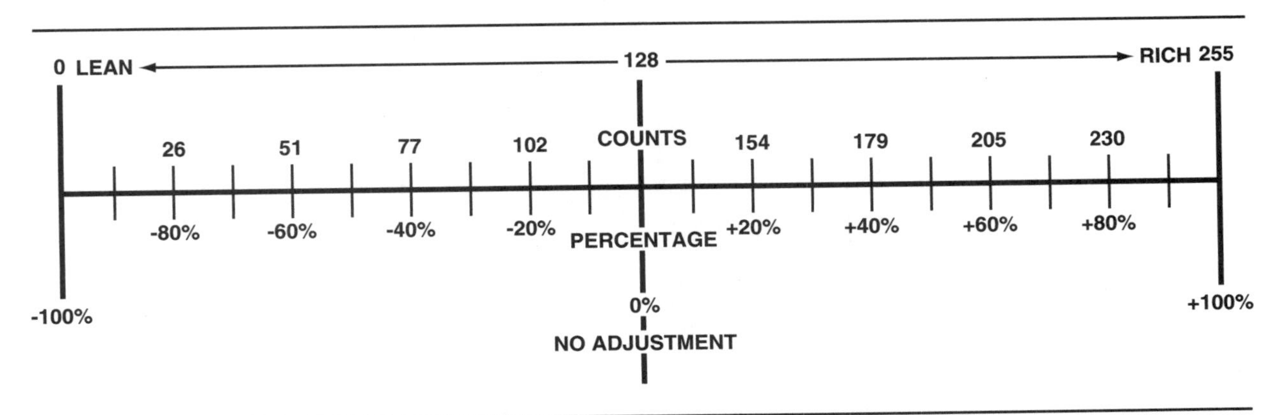

Figure 15-56. This chart shows GM fuel trim equivalencies between OBD-I and OBD-II systems.

DTC's and to aid the technician in isolating a problem, the monitor maintains a cylinder-specific misfire counter. Each cylinder's counter tallies the number of misfires that occurred during the past 200 and 1000 revolutions. Every time the misfire monitor reports a failure, the Diagnostic Executive reviews all of the cylinder's misfire counters. The Executive sets a DTC only if one or more of the counters has a significantly greater amount of misfire counts. In effect, by using cylinder counters, the Executive decreases the number of false DTC's by eliminating most of the "background noise" caused by driveline vibrations.

The misfire monitor recognizes two types of misfire: those that can damage the catalytic converter, and those that can cause exhaust emissions to exceed emission standards by 1.5 times. The Executive immediately sets a DTC and *causes the MIL to flash* when the monitor detects misfire in more than 15 percent of the cylinder firing opportunities during a 200-crank-revolution segment. In OBD-II language, this is a Type "A" misfire.

A Type "B" misfire occurs after two trips in which the monitor detects a misfire in two percent of the cylinder firing opportunities during a 1000-crank-revolution segment. The Executive would light the MIL continuously, and set one or more DTC's.

The fuel system monitor

The fourth monitor, the fuel system monitor, detects when the short- and long-term FT are no longer able to compensate for operating conditions that lead to an excessively rich or lean air-fuel mixture. The adaptive learning strategies discussed earlier in this chapter basically describe how this monitor functions. Unlike previous air-fuel mixture correction strategies that measured the correction in counts or dwell, the OBD-II fuel system monitor measures it in percentages, figure 15-56. The Executive lights the MIL after detecting a fault on two consecutive trips.

The evaporative systems monitor

The EVAP system monitor tests for purge volume and leaks. As you recall, the EVAP system collects and stores the HC vapors emitted from the fuel tank in a charcoal canister. Most applications purge the charcoal canister by venting the vapors into the intake manifold during cruise. To do this, the PCM typically opens a solenoid-operated purge valve installed in the purge line leading to the intake manifold.

A typical EVAP monitor first closes off the system to atmospheric pressure and opens the purge valve during criuse operation, figure 15-57. A fuel tank pressure sensor then monitors the rate with which vacuum increases in the system. The monitor uses this information to determine the purge volume flow rate. To test for leaks, the EVAP monitor closes the purge valve creating a completely closed system. The fuel tank pressure sensor then monitors the leak-down rate. If the rate exceeds PCM-stored values, a leak greater than or equal to the OBD-II standard of 0.040 in. (1.0 mm) exists. After two consecutive failed trips testing either purge volume or the presence of a leak, the Executive lights the MIL and sets a DTC.

The EGR system monitor

The OBD-II EGR emission monitor uses a variety of methods to test EGR flow, depending on the automaker and the application. All of the tests open and close the EGR valve, while measuring the amount of change in a test sensor's voltage signal. After comparing the test-sensor signal voltage with stored values from look-up tables that correlate with exhaust-gas flow, the monitor calculates EGR system efficiency. If the EGR efficiency level does not meet a predetermined standard after two consecutive trips, the

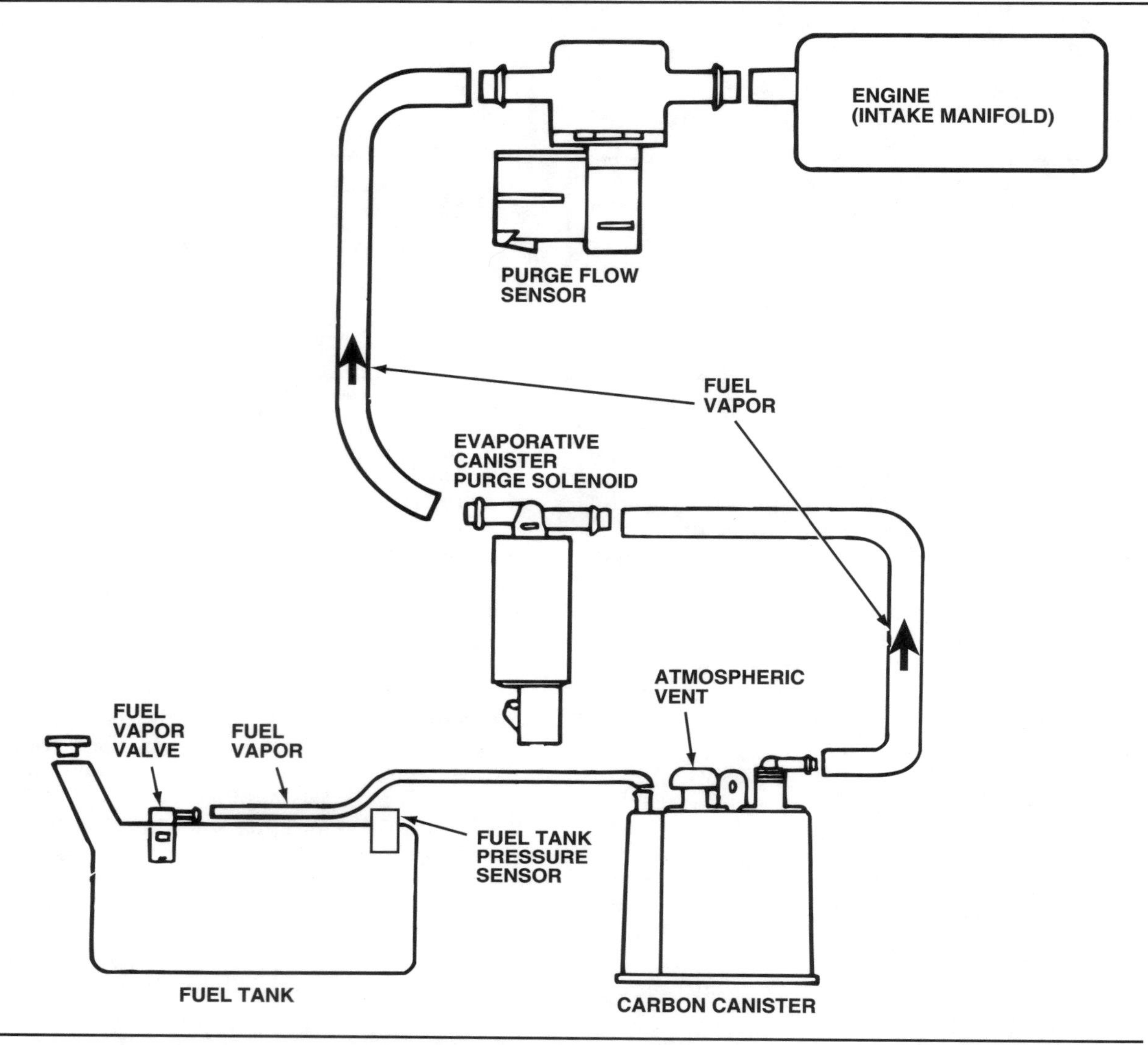

Figure 15-57. The Ford EVAP system uses fuel tank pressure and purge flow sensors to detect leaks and measure purge flow.

Executive lights the MIL and sets one or more DTC's. Later chapters of this *Classroom Manual* discuss EGR emission monitor test sensor strategy for each of the domestic automakers.

The air injection system monitor

Not all automakers presently include the seventh monitor, called the secondary air injection (AIR) system monitor. Usually it performs a variety of continuity and functionality tests on the bypass and diverter valves. If the air injection system uses an electronic pump, the monitor tests that component as well. Most importantly, the AIR monitor performs a functional test of the amount of air flowing into the exhaust stream from the AIR pump. Almost all automakers use the upstream HO2S to detect the excess oxygen from the AIR system. Of course, the Executive places the AIR monitor in a pending status until all of the HO2S monitor's tests have passed. Like most of the other emission monitors, the Executive lights the MIL and sets a DTC after failure of two consecutive trips.

OBD-II diagnostic test results

Having discussed all of the OBD-II emission monitors, this section now examines how the OBD-II Diagnostic Executive indicates test results to the technician. While reading, know that one of the goals of OBD-II is to standardize electronic diagnosis so that the same inexpensive scan tool can be used to test any compliant vehicle that comes into the shop. To aid the technician, the Society of

Figure 15-58. The standard OBD-II 16-pin data link connector.

Automotive Engineers (SAE) established OBD-II guidelines that provide:

- A universal data link connector (DLC) with dedicated pin assignments
- A standard location for the DLC, visible under the dash on the driver's side
- Vehicle identification automatically transmitted to the scan tool
- A standard list of DTC's used by all automakers
- The ability to record a "snapshot" of operating conditions when a fault occurs
- The ability to clear stored codes from vehicle memory with the scan tool
- A series of "flags" to alert the technician of a vehicle's readiness to take an inspection and maintenance (I/M) test
- A glossary of standard terms, acronyms, and definitions used for system components.

The data link connector

The DLC, a 16-pin connector, provides a compatible connection with any "generic" scan tool so that it can access the diagnostic data stream, figure 15-58. The female half of the connector is on the vehicle and the male end is on the scan tool cable. Pins are arranged in two rows of eight, numbered one to eight and nine to sixteen, figure 15-59. The connector is "D"-shaped and keyed so the two halves mate only one way.

Seven of the DLC's sixteen pin positions have mandatory assignments. The individual vehicle manufacturer may assign uses for the remaining nine pins. Most systems use only five of the seven dedicated pins. Terminals seven and fifteen are for systems that transmit data conforming to the European standards established by the International Standards Organization (ISO) regulation 1941-2. This is an alternate communications network to the **bus link,** terminals 2 and 10, defined by SAE J1850 recommended practices.

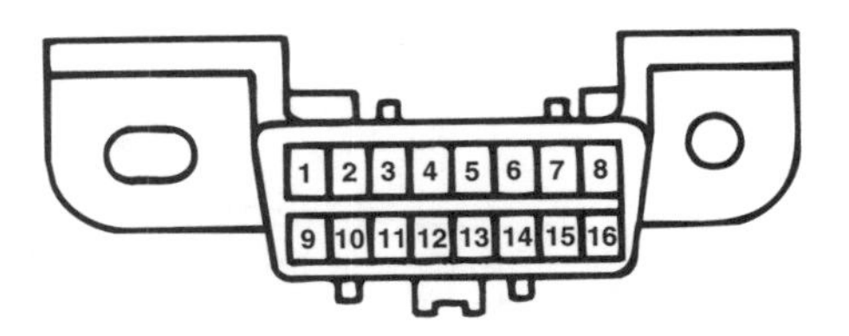

PIN NO.	ASSIGNMENTS
1.	MANUFACTURER'S DISCRETION
2.	BUS + LINE, SAE J1850
3.	MANUFACTURER'S DISCRETION
4.	CHASSIS GROUND
5.	SIGNAL GROUND
6.	MANUFACTURER'S DISCRETION
7.	K LINE, ISO 9141
8.	MANUFACTURER'S DISCRETION
9.	MANUFACTURER'S DISCRETION
10.	BUS – LINE, SAE J1850
11.	MANUFACTURER'S DISCRETION
12.	MANUFACTURER'S DISCRETION
13.	MANUFACTURER'S DISCRETION
14.	MANUFACTURER'S DISCRETION
15.	L LINE, ISO 9141
16.	VEHICLE BATTERY POSITIVE

Figure 15-59. OBD-II 16-pin DLC terminal assignments.

Diagnostic trouble codes and the malfunction indicator lamp

Once the Diagnostic Executive can communicate through the DLC with the scan tool, it can display DTC's. Under OBD-II guidelines, DTC's appear in a five-character, alpha-numeric format. The first character, a letter, defines the system where the code was set. The second character, a number, reveals whether it is an SAE or a manufacturer-defined code. The remaining three characters, all numbers, describe the nature of the malfunction, figure 15-60.

The first character of the DTC currently has four letters assigned for system recognition.

Data Link Connector (DLC): A special connector that allows a scan tool to communicate with the PCM. OBD II vehicles use a standardized 16 pin connector.

Bus Link: A common conductor or transmission path shared by several components in a computer system that transmits data and receives instructions.

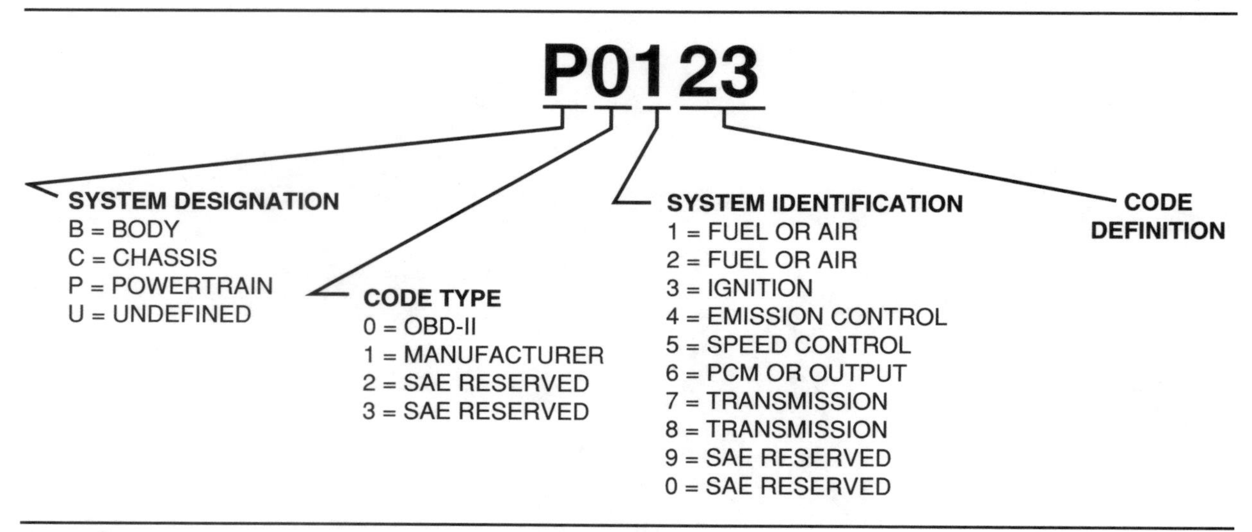

Figure 15-60. OBD-II five-character powertrain DTC structure.

These are "B" for body, "C" for chassis, "P" for powertrain, and "U" for undefined. Undefined codes are reserved for future assignment by SAE. Universal powertrain codes are defined and are used on current OBD-II systems.

The second character of the DTC is either a "0", "1", "2", or "3". A zero in this position indicates a malfunction that is defined and control-led by the SAE, the number one is a code defined by the vehicle manufacturer. The numerals two and three are designated for future use and both are reserved for SAE usage in powertrain codes. With body and chassis codes, the digit two is manufacturer controlled, and three is reserved for SAE.

The third character of a powertrain DTC indicates the system where the fault occurred. Both "1" and "2" designate fuel or air metering problems. A "3" in this position indicates an ignition malfunction or engine misfire. Number "4" is assigned to the auxiliary emission control system. Problems in the vehicle speed or idle control system set a number "5" in the DTC. Number "6" is used for computer or output circuit faults. Transmission control problems are indicated by either a "7" or "8" while "9" and "0" are reserved by SAE for future use.

The remaining two digits of the DTC tell you the exact condition that triggered the code. SAE assigns different sensors, actuators, and circuits to specific blocks of numbers. The lowest numeral of the block indicates a general malfunction in the monitored circuit. This is the "generic" DTC. Ascending numbers in the block provide more specific information, such as low or high circuit voltage, slow response or an out-of-range signal; these are know as enhanced OBD-II DTC's. The system should not allow duplicate codes. If the system has the capability, an enhanced code takes precedence over a generic code.

With regards to the emission monitors, OBD-II has divided DTC's into four categories:

- Type A—These DTC's indicate the presence of an emission-related failure that causes the Executive to light the MIL after only one trip. A Type "A" misfire would be an example.
- Type B—These DTC's indicate the presence of an emission-related fault that causes the Executive to light the MIL after a failure on two trips. This category includes many of the DTC's that set when faults occur during each of two consecutive trips. The misfire monitor can generate two kinds of Type B DTC's: one occurring after two consecutive trips, and the other after two non-consecutive trips. Each time the monitor detects a misfire, it stores engine load, speed, and coolant temperature information. If the misfire occurs during just one trip, the monitor waits until the next misfire, again stores the three items and compares them with one another. When the two sets of data indicate similar operating conditions, the Executive sets a DTC and lights the MIL.
- Type C—These DTC's indicate the presence of a nonemission-related failure that causes the Executive to light a service lamp after only one trip.
- Type D—These DTC's indicate the presence of a failure, not emission-related, that causes the Executive to light a service lamp after two consecutive trips.

When the Executive lights the MIL, it always stores a DTC. To clear a Type "A" or "B" DTC, the system in which the fault occurred must pass the test during 40 consecutive warm-up cycles

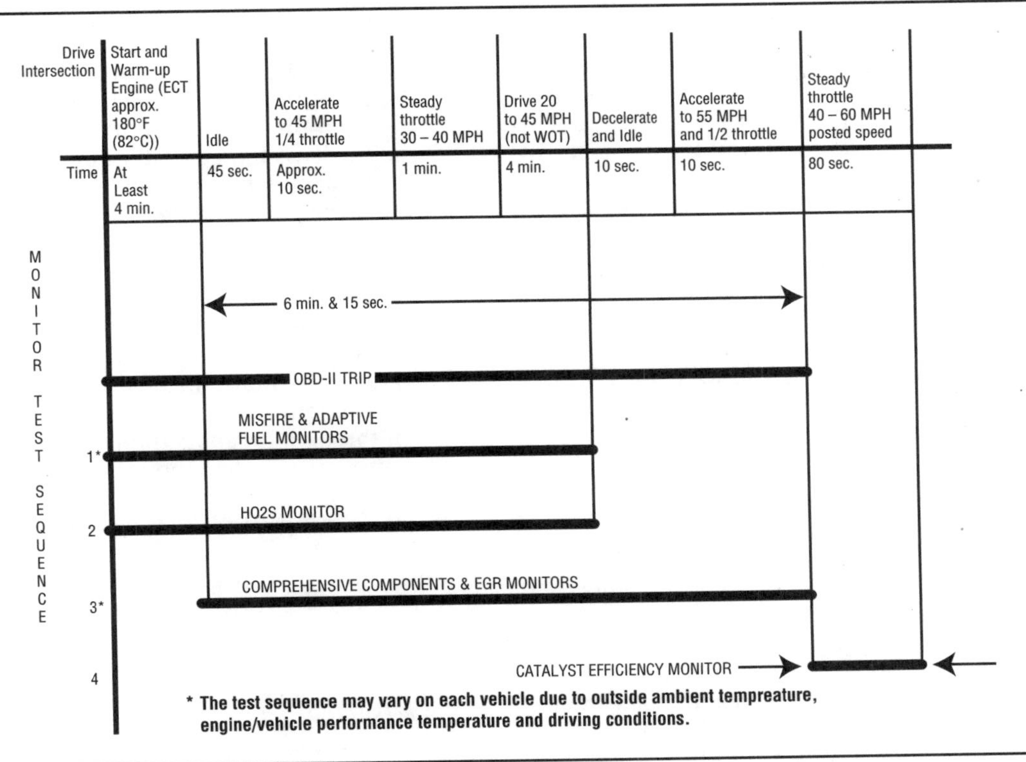

Figure 15-61. Driving an OBD-II vehicle through its I/M Readiness Drive Cycle Test allows the emission monitors to run all of their tests.

conducted after the MIL is turned off. For most monitors, the Executive does not turn off the MIL until the fault-finding emission monitor reports passes for three consecutive trips. MIL's set by a fuel trim or misfire-related DTC require special handling. The three consecutive trips must occur during engine speed, load, and temperature conditions similar to those that originally caused the fault.

The freeze-frame record

Another type of OBD-II test result indicator, the **freeze-frame record,** provides a wealth of diagnostic information to the technician. Whenever the MIL is lit by the Executive, it stores certain engine operating conditions into the freeze-frame record. The information usually includes:

- The reported DTC
- Air-fuel ratio
- Mass airflow rate
- Fuel trim (short- and long-term)
- Engine speed
- Engine load
- Engine coolant temperature
- Vehicle speed
- MAP/BARO sensor signal voltages
- Fuel injector base pulse width
- Loop status.

The Executive only allows operating conditions from the first occurrence of a fault to be recorded into the freeze-frame record. In the presence of multiple codes, the Executive usually allows only parameters for the first DTC to be recorded. However, if one of these multiple codes includes a fuel trim or misfire fault, the Executive allows the freeze-frame record to be overwritten. Erasure of a DTC causes the erasure of its accompanying freeze-frame record.

Freeze Frame Record: The part of the Diagnostic Executive that stores various vehicle operating data when setting an emission-related DTC and lighting the MIL.

Inspection and Maintenance (I/M) Readiness Drive Cycle: A drive cycle that runs all of the emission monitors and sets all of the I/M flags. Most states require that all I/M flags be set before running an I/M test.

Flags: Special OBD-II messages that indicate the completion of an emission monitor test. If the Executive has not set all of the flags, the technician may have to drive the vehicle through the I/M Readiness Drive Cycle.

Inspection and maintenance readiness test flags
I/M Readiness Drive Cycle Test flags provide the technician with the final type of test result indicator. The OBD-II emission monitors perform their diagnostic tests once during each trip. At the completion of each test, regardless of a pass or fail, the Executive sets a **flag.** Usually normal driving after a period of time causes all of the flags to be set; in some instances, however, the technician must drive the vehicle through a manufacturer-specific drive cycle to set all of the flags, figure 15-61. These drive cycles vary the time spent driving during idle, acceleration, steady throttle, and deceleration. Most require that the vehicle be started while cold.

The technician and OBD-II
After checking an OBD-II vehicles's MIL and I/M readiness inspection flags, and evaluating the DTC's and freeze-frame record, you can usually form a diagnosis. Some automakers then have you perform an OBD-II system check.

After completing emission-related repairs, it is important to verify their effectiveness. Use a scan tool to clear the set DTC's and perform an I/M Readiness Drive Cycle Test. Check whether all of the I/M flags have set, and then see if the Executive has reported additional DTC's. If none appear, you have made a successful repair.

SUMMARY

Electronic engine control systems use sensors to monitor vehicle operating conditions. Sensors can change motion, temperature, and light into voltage signals. Engineers characterize them according to four criteria: repeatability, accuracy, operating range, and linearity. These devices can be categorized into switches, resistors, and voltage generators.

Switches are the simplest form of a sensor. They provide the PCM with a digital high or low voltage signal. Resistive-type sensors include potentiometers, rheostats, and thermistors. All of these require a PCM-supplied reference voltage. Engine speed and load sensors provide the PCM with the most important control information. Exhaust oxygen sensors enable the PCM to maintain an air-fuel ratio near stoichiometric during closed-loop operation.

Automotive control systems use actuators to change certain vehicle operating conditions. These devices change voltage and current into mechanical action. There are three types: solenoids, relays, and stepper motors. Fuel injectors and IAC valves comprise the most important applications.

The PCM uses pulse width modulation (PWM) to operate many solenoids. By varying the on-time of a digital signal, PWM allows the PCM to approximate an analog response, achieving finer control.

All three fuel management strategies directly or indirectly calculate the mass airflow rate and use the engine's speed to determine the fuel injector's firing rate. The type of load sensor varies: the speed-density system uses a MAP sensor, the vane airflow system, a VAF sensor, and the mass airflow system, a MAF sensor.

Other factors, such as EGR flow and ignition timing, affect the fuel injector's base pulse width or MC solenoid duty cycle. Adaptive fuel learning strategies, which compensates for operating conditions that lead to an excessively rich or lean air-fuel mixture, also affects fuel input.

While early electronic systems control either ignition timing or fuel metering, later full-function systems control both. To further decrease emission-levels, and to facilitate repair, CARB developed OBD-I and OBD-II regulatory guidelines.

OBD-II detects system degradation and uses a standardized DLC, DTC's, and accepts a "generic" scan tool. This unique system, required of most light-duty vehicles in 1996, uses a Diagnostic Executive to manage the emission monitors that perform tests and to report test results. This sophisticated process takes place independent of the technician. Up to seven dedicated emission monitors test the following systems, components, and faults: the catalytic converter, HO2S, misfire detection, air-fuel ratio compensation, and the EVAP, EGR and AIR systems. The OBD-II system uses DTC's, the MIL, a freeze-frame record, and I/M Readiness Test flags to indicate diagnostic information to the technician.

Review Questions

Choose the single most correct answer.
Compare your answers to the correct answers on page 507.

1. A zirconia exhaust gas oxygen sensor (O2S) is an example of a:
   a. Resistor
   b. Potentiometer
   c. Generator
   d. Solenoid

2. An engine detonation sensor uses a:
   a. Piezoresistive crystal
   b. Voltage divider pickup
   c. Potentiometer
   d. Thermistor

3. When the on-board computer is in open-loop operation, it:
   a. Controls fuel metering to a predetermined value
   b. Ignores the temperature sensor signals
   c. Responds to the O2S signal
   d. Responds to the HO2S signal

4. The reference value sent to a sensor by the computer must be:
   a. Above battery voltage
   b. Exactly the same as battery voltage
   c. Less than minimum battery voltage
   d. Battery voltage plus or minus 5 volts

5. The simplest digital sensor is a:
   a. Solenoid
   b. Switch
   c. Timer
   d. Relay

6. The percentage of time a solenoid is energized relative to total cycle time is called the:
   a. Pulse width modulation (PWM)
   b. Frequency
   c. Duty cycle
   d. KAM

7. A stepper motor:
   a. Is continuously on
   b. Is one form of solenoid
   c. Is used to operate the EGR valve
   d. Can move clockwise and counterclockwise

8. Technician A says that TP sensors are potentiometers.
   Technician B says that TP sensors are analog devices.
   Who is right?
   a. A only
   b. B only
   c. Both A and B
   d. Neither A nor B

9. The first domestic engine control system was Chrysler's Electronic Lean-Burn. It controls:
   a. Ignition timing
   b. Fuel metering
   c. Both a and b
   d. Neither a nor b

10. NO_x emissions may be reduced by all except:
    a. Leaning the fuel mixture
    b. Lowering the engine's compression ratio
    c. Recirculating exhaust gases
    d. Retarding spark timing

11. An ECT sensor:
    a. Receives reference voltage from the computer
    b. Is a potentiometer
    c. Provides the computer with a digital signal
    d. Contains a piezoelectric crystal

12. Technician A says that an O2S with two wires is grounded to the exhaust manifold.
    Technician B says that the computer ignores an O2S in closed-loop operation.
    Who is right?
    a. A only
    b. B only
    c. Both A and B
    d. Neither A nor B

13. Technician A says that excessive EGR increases HC emissions and fuel consumption.
    Technician B says that NO_x emissions are low and fuel consumption is high when timing is advanced.
    Who is right?
    a. A only
    b. B only
    c. Both A and B
    d. Neither A nor B

14. A GM "minimum function" system:
    a. May use a coolant temperature switch
    b. Has long-term memory
    c. Provides fixed dwell
    d. Controls only ignition timing

15. Which of the following is not true of a fuel-injected engine?
    a. The computer controls the air-fuel ratio by switching injectors on or off
    b. The pulse width is increased to supply more fuel
    c. To lean the mixture, the computer decreases the fuel injector pulse width
    d. Engine speed determines the injector switching rate

16. The first GM ignition timing control was called:
    a. MISAR
    b. HEI
    c. EST
    d. TWC

17. Which of the following is not true of a late-model full-function control system?
    a. The computer controls timing electronically
    b. The computer changes the air-fuel ratio within the range of 14:1 to 17:1
    c. Fuel metering is controlled by pulsing fuel injectors
    d. The car can be driven in a limited operational strategy mode

18. Technician A says that MAF sensors may use either a hot-wire or heated film.
    Technician B says that automatic transmission operation is controlled by the engine control computer in late-model vehicles.
    Who is right?
    a. A only
    b. B only
    c. Both A and B
    d. Neither A nor B

19. A complete operating cycle of a pulsed solenoid includes:
    a. The sequence from on to off to back on again
    b. The sequence from off to on
    c. The sequence from off to on to back off again
    d. The sequence from on to off

20. Which of the following fuel management approaches determine mass airflow rate?
    a. Speed-density
    b. Vane airflow
    c. Mass airflow
    d. All of the above

21. Technician A says that an engine under load fires its fuel injectors faster compared to one not under load to enrich the air-fuel mixture.
    Technician B says that an engine at cruise fires its injectors using a longer base pulse width than an engine at idle.
    Who is right?
    a. A only
    b. B only
    c. Both A and B
    d. Neither A nor B

Continued

22. An OBD-II-equipped vehicle has just completed a trip. This means:
    a. The vehicle completed a warm-up cycle
    b. The vehicle completed a key-on, engine running, key-off cycle
    c. The vehicle completed a key-on, engine running, key-off cycle where an emission monitor reported a pass to the Diagnostic Executive
    d. The vehicle completed a key-on, engine running, key-off cycle where an emission monitor reported a pass or fail to the Diagnostic Executive

23. What can cause the MIL to flash on an OBD-II-equipped vehicle?
    a. A type "A" misfire
    b. A type "B" misfire
    c. Catalytic converter malfunction
    d. HO2S malfunction

24. Technician A says that OBD-II DTC's only cover engine and emission related faults. Technician B says that OBD-II DTC's may be automaker-designated, or follow a SAE standard. Who is right?
    a. A only
    b. B only
    c. Both A and B
    d. Neither A nor B

25. For most monitors, how long does it take for the Executive to light the MIL, and then to clear DTC's?
    a. Two consecutive trips; 40 consecutive warm-up cycles
    b. Four consecutive trips; 30 consecutive warm-up cycles
    c. Two consecutive trips; 30 consecutive warm-up cycles
    d. Four consecutive trips; 40 consecutive warm-up cycles

Chapter

16

Gasoline Fuel Injection Systems and Intake Manifolding

A carburetor is a mechanical device that is neither totally accurate nor particularly fast in responding to changing engine requirements. Adding electronic feedback mixture control improves a carburetor's fuel metering capabilities under some conditions, but most of the work is still done mechanically by the many jets, passages, and air bleeds. Adding feedback controls and other emission-related devices in recent years has resulted in very complex carburetors that are extremely expensive to repair or replace.

The intake manifold also is a mechanical device that, when teamed with a carburetor, results in less than ideal air-fuel control. Because of a carburetor's limitations, engineers locate the manifold centrally over the intake ports (V-engines) or next to the intake ports (in-line engines) while remaining within the space limitations under the hood. The manifold runners have to be kept as short as possible to minimize fuel delivery lag, and there cannot be any low points where fuel might puddle. These restrictions severely limit the amount of manifold tuning possible, and even the best designs still have problems with fuel condensing on cold manifold walls.

The solution to the problems posed by a carbureted fuel system is **electronic fuel injection (EFI)**. EFI provides precise mixture control over all speed ranges and under all operating conditions. Its fuel delivery components are simpler and often less expensive than a feedback carburetor. Some designs allow a wider range of manifold designs. Equally important is that EFI offers the potential of highly reliable electronic control. In this chapter, we discuss the:

- Advantages of fuel injection over carburetion
- Differences among various fuel injection systems
- Fundamentals of electronic fuel injection
- Subsystems and components of typical fuel injection systems now in use
- Intake manifold designs and fuel injection manifold systems.

FUEL INJECTION OPERATING REQUIREMENTS

The major difference between a carbureted and an injected fuel system is the method of fuel delivery. In a carbureted system, the carburetor mixes air and fuel. In a fuel injection system, one or more injectors meter the fuel into the intake air stream.

An electronic fuel injection system uses the same principle of pressure differential used in a carbureted system, but in a slightly different way. As you learned in an earlier chapter, in carbureted systems, the difference in pressure between the inside and the outside of the engine forces fuel out of the carburetor fuel bowl. In

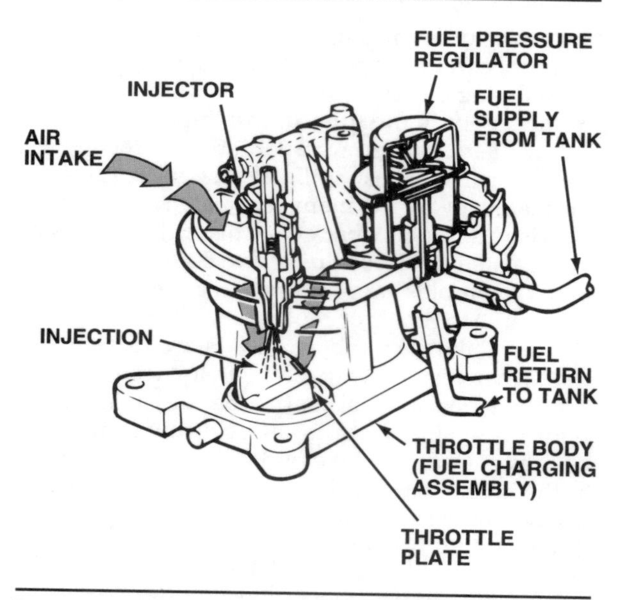

Figure 16-1. Air and fuel combine in a TBI unit at about the same point as in a carburetor. (Ford)

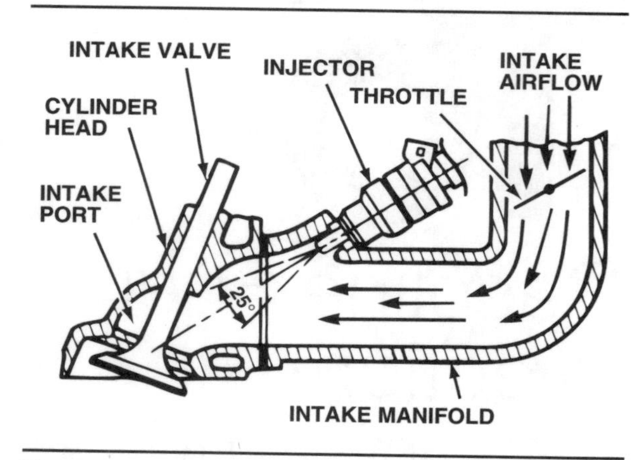

Figure 16-2. In a multipoint system, the air and fuel combine at the intake valve. (GM)

many EFI systems, the vane airflow (VAF) or manifold absolute pressure (MAP) sensor determines the pressure difference and informs the powertrain control module (PCM). The PCM evaluates the airflow sensor's input along with that of other sensors to decide how much fuel the engine requires, and then controls injector operation to provide the correct fuel quantity. The PCM makes it possible for the fuel injectors to do all the functions of the main systems used in a carburetor. However, throttle response in an injected system is more rapid than in a carbureted system because the fuel is under pressure at all times. The PCM calculates the changing pressure and opens the injector much quicker than a carburetor can react under similar conditions.

Carburetors must break up liquid gasoline into a fine mist, change the liquid into a vapor, and distribute the vapor evenly to the cylinders. Fuel injectors do the same. High temperatures instantly vaporize the fuel just before the intake valve after it is first atomized and delivered into the intake air stream by the fuel injector. In **throttle body injection (TBI)** systems, this occurs above the throttle at about the same point as in a carbureted system, figure 16-1. With **port fuel injection (PFI)** systems, the injectors are mounted in the intake manifold near the cylinder head and injection occurs as close as possible to the intake valve, figure 16-2. We look at both systems in greater detail later in the chapter. To review basic engine air-fuel requirements:

- When an engine starts, low airflow and manifold vacuum combined with a cold engine and poor fuel vaporization, require a rich air-fuel ratio.
- At idle, low airflow and high manifold vacuum combined with low vacuum and poor vaporization, still require a slightly rich air-fuel ratio.
- At low speed, the air-fuel ratio becomes progressively leaner as engine speed, airflow, and carburetor vacuum increase.
- At cruising speed, air-fuel ratios become the leanest for best economy with light load and high, constant vacuum, and airflow.
- For extra power when accelerating or pulling heavy loads, the engine needs a richer air-fuel ratio. Without added fuel during acceleration, the air-fuel mixture would become progressively leaner as the throttle plate opens wider.

A carburetor satisfies all of these requirements as its systems respond to changes in air pressure and airflow. A fuel injection system performs functions having the same effect. All of the sensors and other devices in an injection system respond to the same operating conditions as a carburetor, but an injection system responds faster and more precisely for a better combination of performance, economy, and emission control.

ADVANTAGES OF FUEL INJECTION

Fuel injection is not a new development. The first gasoline fuel injection systems were developed by the German Robert Bosch Company in 1912. Bosch limited their early work in fuel injection to aircraft applications, but in the 1930s, Bosch transferred that work to automotive applications. A decade later, two Americans named Hilborn and Enderle developed injection systems for use on racing engines. Chevrolet, Pontiac, and Chrysler offered mechanical fuel injection systems during the late 1950s and early 1960s. Some imported vehicles have had fuel injection since 1968. Among domestic manufacturers, Cadillac made electronic fuel injection standard on its 1976 Seville model and other automakers soon followed suit.

What really made fuel injection a practical alternative to the carburetor was the development of reliable solid-state components during the 1970s. Automakers were quick to apply these advances in electronics to fuel injection. The result was a more efficient and dependable system of fuel delivery.

Electronic fuel injection systems offer several major advantages over carburetors:

- Injectors can precisely match fuel delivery to engine requirements under all load and speed conditions. This reduces fuel consumption with no loss of engine performance.
- Since intake air and fuel are mixed at the engine port, keeping a uniform mixture temperature is not as difficult with PFI systems as with carburetors. There is no need for manifold heat valves.
- The manifold in a PFI system carries only air, so there is no problem of the air and fuel separating.
- Exhaust emissions are lowered by maintaining a precise air-fuel ratio according to engine requirements. The improved air-fuel flow in an injection system also helps to reduce emissions.
- Continuing improvements in electronic fuel injection design have allowed some automakers to eliminate other emission control systems such as heated intake air and air injection (AIR) systems.

AIR-FUEL MIXTURE CONTROL

Using an exhaust oxygen sensor (O2S), automakers have integrated almost all of the electronic fuel injection systems used on domestic vehicles in recent years into complete electronic engine control systems. The electronic controls of a fuel injected engine are similar to those of an engine with a feedback carburetor. The main difference is that the PCM, instead of regulating a vacuum solenoid, a stepper motor, or a mixture control (MC) solenoid, switches one or more solenoid-operated fuel injectors on and off. Engine rpm determines the rate of switching and engine load determines the injector's pulse width. The PCM varies pulse width to establish the air-fuel ratio, figure 16-3. The last chapter explained injector pulse width and duty cycle in greater detail.

This variable pulse width takes the place of all the different carburetor metering circuits. As the PCM receives data from its inputs, it can lengthen the pulse width to supply additional fuel for cold running (choke), heavy loads (power enrichment), fast acceleration (accelerator

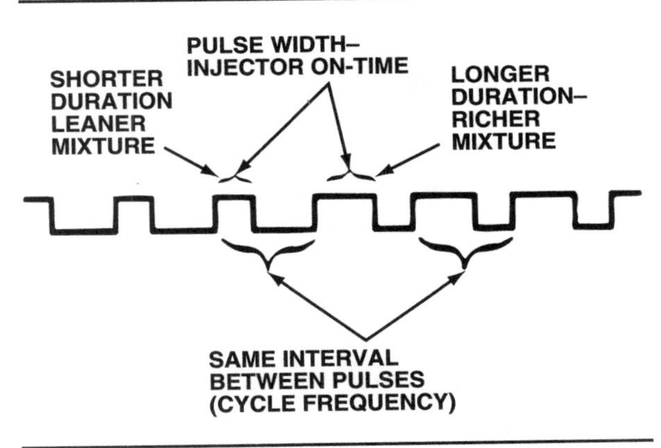

Figure 16-3. Fuel injector pulse width determines the air-fuel ratio.

pump), or several other situations. Similarly, it can shorten the pulse width to lean the mixture at idle (idle circuit), under cruise (main circuit), or during deceleration. On such systems, the O2S provides fine tuning of the mixture under most operating conditions.

By weight, fuel makes up only about one-fifteenth of the air-fuel mixture. The PCM is mainly concerned with regulating the fuel delivery, but to do this accurately, it must know the amount of air entering the engine. Because airflow and air pressure cannot act directly on the fuel as occurs in a carburetor, fuel injection systems use one of three basic kinds of intake air sensors:

- A MAP sensor
- A VAF sensor
- A mass airflow (MAF) sensor.

Manifold pressure, airflow velocity, and the molecular mass of the intake air are all directly related to the weight of intake air. (Remember that air-

Electronic Fuel Injection (EFI): A computer-controlled fuel injection system which gives precise mixture control and almost instant response to all operating conditions at all speed ranges.

Throttle Body Injection (TBI): A fuel injection system in which one or two injectors are installed in a carburetor-like throttle body mounted on a conventional intake manifold. Fuel is sprayed at a constant pressure above the throttle plate to mix with the incoming air charge.

Port Fuel Injection (PFI): A fuel injection system in which individual injectors are installed in the intake manifold at a point close to the intake valve. Air passing through the manifold mixes with the injector spray just as the intake valve opens.

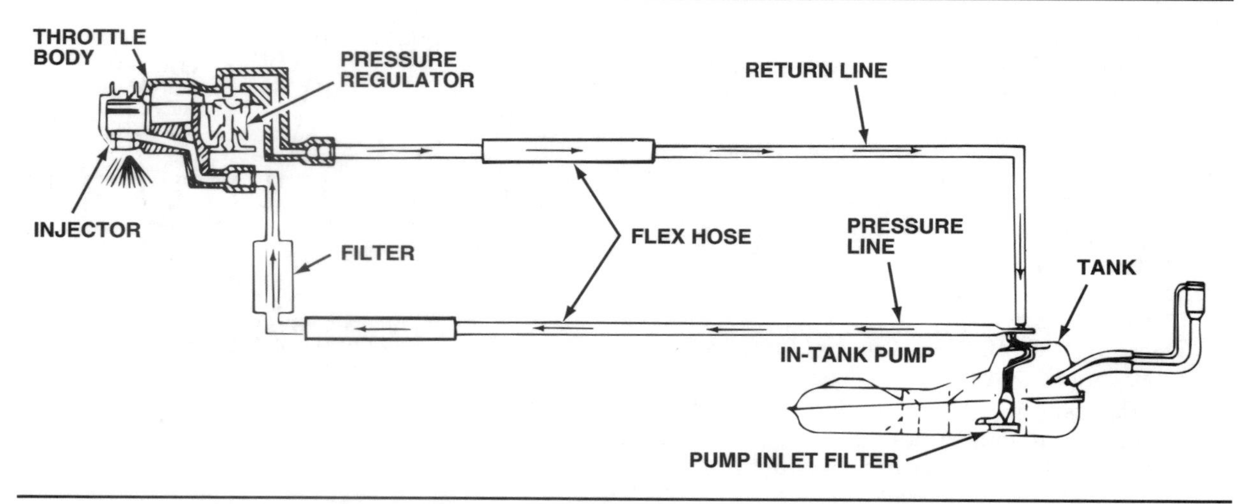

Figure 16-4. A typical throttle body injection system. (GM)

fuel ratios are expressions of air and fuel weights by volume.) The PCM has a set of values (similar to a graph) stored in memory. These values relate manifold pressure, airflow, or air mass to engine speed and the required fuel entering for any combination of conditions. Once the PCM knows the amount of air entering the engine, as well as other conditions, it adjusts injector pulse width to achieve the required air-fuel ratio.

TYPES OF FUEL INJECTION SYSTEMS

There are three general ways to categorize modern fuel injection systems:

- Mechanical or electronic injection
- Throttle body or multipoint (port or manifold) injection
- Continuous or intermittent injection.

However, the distinctions between the various types of fuel injection systems are not simple and clear-cut. All mechanical systems use some electronic components, and electronic systems may share some of the features of port and TBI in a single system. One example is Chrysler's continuous-flow TBI system introduced in 1981 and used exclusively on V8 engines in the Imperial. Although a technological dead end, it breaks the rules of operation we will discuss. It has two injectors mounted in the throttle body that respond to varying pressure from a control pump and deliver fuel continuously. A unique airflow sensor in the air cleaner inlet measures the volume of air passing into the system. The PCM controls the entire operation electronically.

Mechanical or Electronic Injection

A mechanical fuel injection system delivers gasoline by using its pressure to open the injector valve. Generally, mechanical injection systems are continuously operating. In other words, they inject fuel constantly while the engine is operating. The Bosch K-Jetronic is a typical example of a mechanical injection system.

An EFI system generally uses one or more solenoid-operated injectors to spray fuel in timed pulses, either into the intake manifold or near the intake port. EFI systems operate intermittently. All domestic manufacturers use some form of EFI.

Throttle Body or Port Fuel Injection

A common type of EFI system is TBI, figure 16-4. This design is something of a halfway measure between feedback carburetion and PFI. Automakers generally classify it as a single-point, pulse-time, modulated injection system. TBI injection, as its name implies, has one or two injectors in a carburetor-like TBI assembly that mounts on a traditional intake manifold. Fuel is kept at a constant pressure by a regulator built into the throttle body, figure 16-1.

The PCM controls injector pulsing in one of two ways: synchronized or nonsynchronized. If the system uses a synchronized mode, the injector pulses once for each distributor reference pulse. When dual injectors are used in a synchronized system, the injectors pulse alternately. In a nonsynchronized system, the injectors are pulsed once during a given period (which varies according to calibration) completely independent of distributor reference pulses.

A TBI system has certain advantages: it provides improved fuel metering over a carburetor, it is easier to service, and it is less expensive to manufacture. Its disadvantages are primarily related to

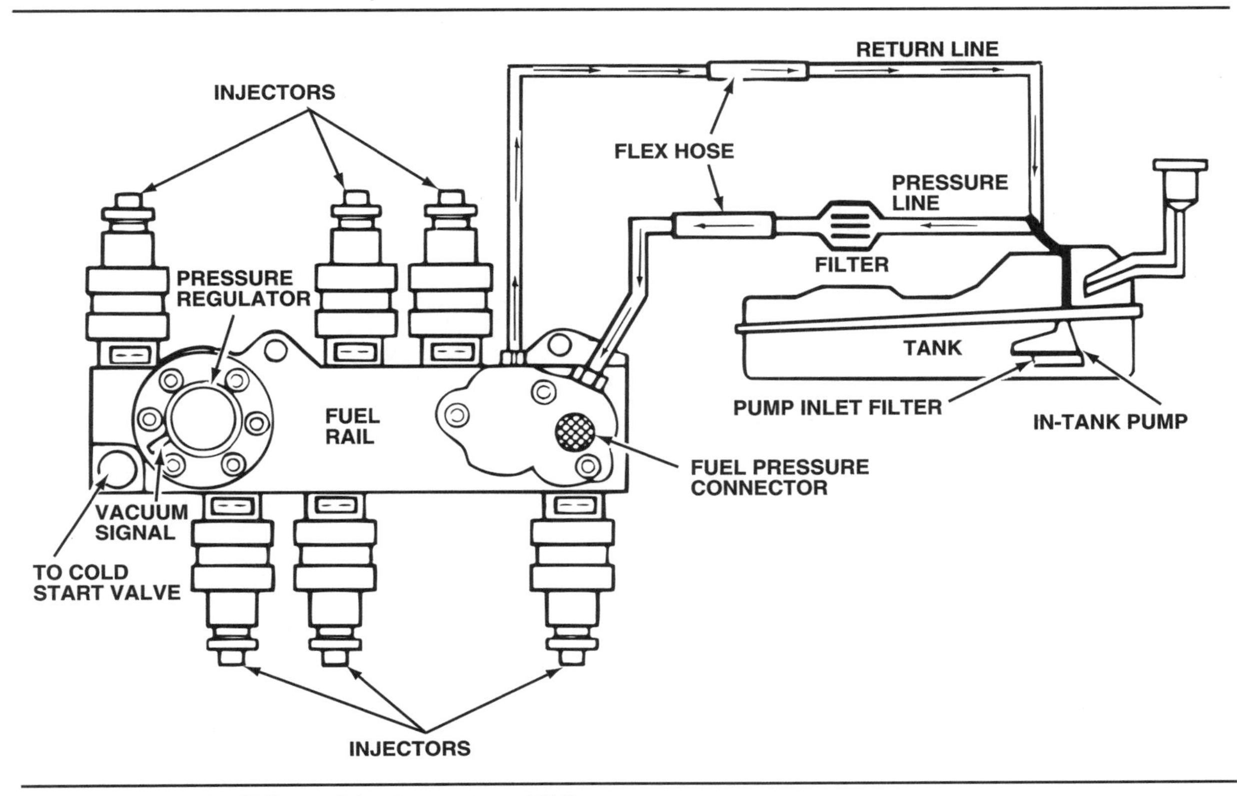

Figure 16-5. A typical V6 port injection system. (GM)

the manifold: fuel distribution is unequal and a cold manifold still causes fuel to condense and puddle. To compensate for this placement, some systems use two differently calibrated injectors. This results in a different amount of fuel being sprayed by each injector. Also, a TBI unit, like a carburetor, must be mounted above the combustion chamber level. This generally prevents the use of tuned intake manifold designs.

Multipoint or port fuel injection, figure 16-5, is older than TBI, and because of its many advantages, will probably continue to be the system of choice for all but economy vehicles. Port fuel systems have one injector for each engine cylinder. The injectors are mounted in the intake manifold near the cylinder head where they can inject fuel as close as possible to the intake valve, figure 16-2.

The advantages of this design also are related to characteristics of intake manifolds:

- Fuel distribution is equal to all cylinders because each cylinder has its own injector, figure 16-6.
- The fuel is injected almost directly into the combustion chamber, so there is no chance for it to condense on the walls of a cold intake manifold.
- Because the manifold does not have to carry fuel or properly position a carburetor or TBI unit, it can be shaped and sized to tune the intake airflow to achieve specific engine performance characteristics.

The primary disadvantage of PFI is the higher cost of individual injectors and other parts, as well as the PCM software to control their operation.

Continuous or Intermittent Injection

A **continuous injection system** constantly injects fuel whenever the engine is running. The most common variety used on production vehicles is the Bosch K-Jetronic mechanical continuous injection system. (Many automakers refer to K-Jetronic simply as CIS.)

The individual injectors, figure 16-7, operate on the opposing forces of fuel pressure and the spring-loaded valve in the injector tip. When inlet fuel pressure reaches about 45 psi (310 kPa), it overcomes spring pressure and forces the injector open. Each injector then delivers fuel continuously to the port near the intake valve. Fuel collects

Continuous Injection System (CIS): A fuel injection system in which fuel is injected constantly whenever the engine is running. An example of a CIS system is the Bosch K-Jetronic.

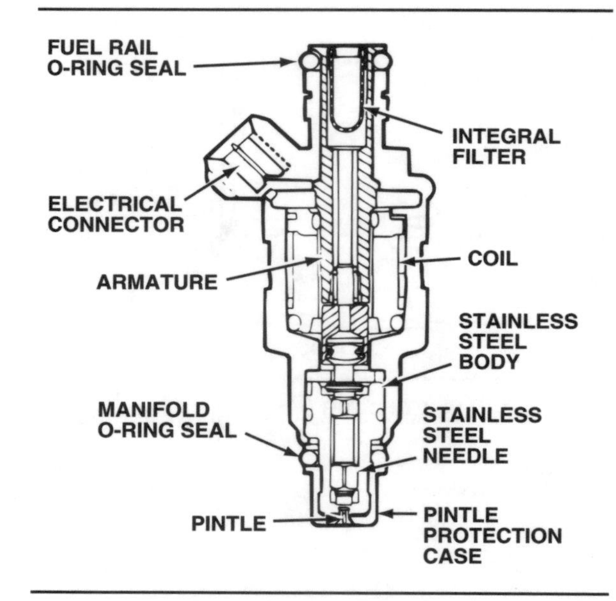

Figure 16-6. A multipoint (port) electronic fuel injector.

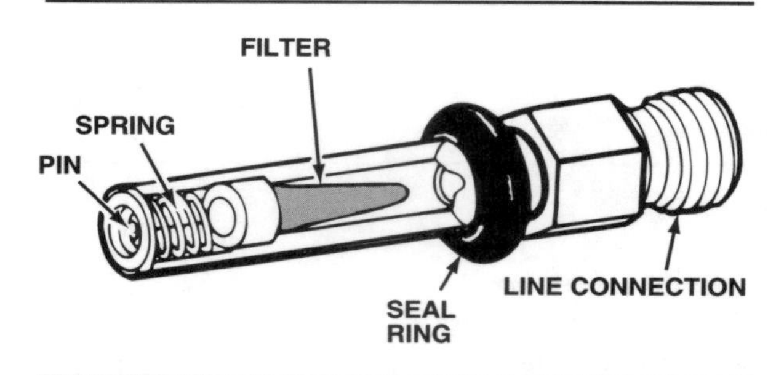

Figure 16-7. A Bosch K-Jetronic mechanical fuel injector. (Bosch)

or "waits" at the valve to mix with incoming air. When the intake valve opens, the air-fuel mixture enters the cylinder.

In the K-Jetronic system, fuel pressure controls injector opening and the amount of fuel that is injected. Fuel pressure varies continually during different engine operating conditions. Under heavy load, for example, fuel pressure may reach 70 psi (483 kPa). This forces the injector farther open to admit more fuel.

The main fuel distributor and regulator controls fuel pressure in response to airflow and preset regulator pressure. Injector spring force and fuel pressure cause the injector valve to vibrate. This constantly varies the amount of fuel in relation to engine requirements and helps to atomize the fuel as it is injected.

When the engine is shut off, fuel pressure drops below the injectors' tip spring pressure, causing the injectors to close. Residual fuel pressure is retained in the lines to ensure a ready fuel supply when the engine is restarted.

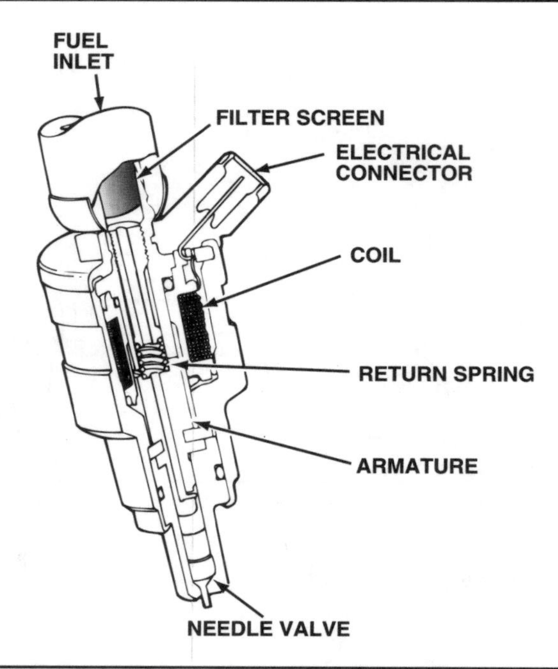

Figure 16-8. The solenoid-actuated needle valve intermittently allows fuel to pass through the injector. (Volvo)

Various Sequence Combinations

Electronic fuel injection systems use a solenoid-operated injector, figure 16-8, to spray atomized fuel in timed pulses (intermittently) into the manifold or near the intake valve. Injectors may be sequenced and fired in one of several ways, as we will see shortly, but the PCM determines and controls their duty cycle and pulse width.

Port systems contain an injector for each cylinder but they do not all fire the injectors in the same way. Domestic systems use one of five ways to trigger the injectors:

- Grouped single-fire
- Grouped double-fire
- Simultaneous double-fire
- Alternating synchronous double-fire
- Sequential.

The first four of these combinations are typical of a pulsed injection system, since the injectors are opened and closed at regular intervals. This differs from a continuous injection system, which provides an uninterrupted flow of atomized fuel when the injectors are open. The pulsed system controls fuel flow by varying the length of time during which the injectors are open. Sequential injection is similar, but automakers call it a timed injection system. The injector operates the same as in a pulsed system, but its firing is timed to inject fuel as the intake valve for each cylinder opens.

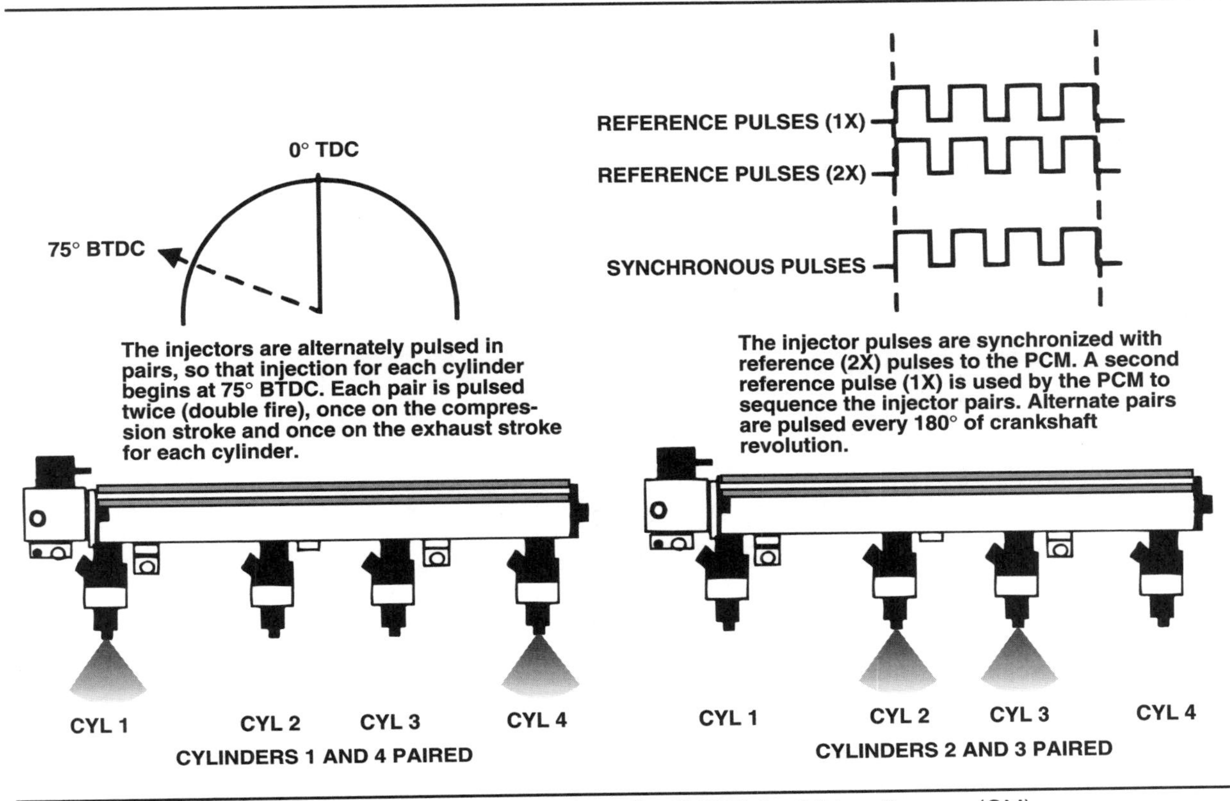

Figure 16-9. The Quad 4 alternating synchronous double-fire (ASDF) 2 x 2 firing diagram. (GM)

Grouped single-fire

Injectors in this system are split into two groups. The groups fire alternately with one group firing each engine revolution. Each combustion stroke uses only one injector pulse. Early Cadillac V8 systems use this design and while it works reasonably well, it is not as precise as newer designs. Since only two injectors can be fired relatively close to the point where the intake valve is about to open, the fuel charge for the remaining six cylinders must stand in the intake manifold for varying times. Since the fuel injectors release a new charge only once every two crankshaft revolutions, it is necessary to wait this long before the PCM can change the air-fuel mixture.

Grouped double-fire

This system again splits the injectors into two equal sized groups. The groups fire alternately, but each group fires once each revolution. Two injector pulses make up each intake charge, which means that the air-fuel mixture remains in the manifold for a shorter time and that mixture changes can be made sooner than with a single-fire system.

Simultaneous double-fire

This design fires all of the injectors at the same time once every engine revolution. Many PFI systems on 4-cylinder engines use this pattern of injector firing. It is easier for engineers to program and it can make relatively quick adjustments in the air-fuel ratio, but it still requires the intake charge to wait in the manifold for varying lengths of time.

Alternating synchronous double-fire

In this design, used in some GM Quad 4 engines, the four injectors are divided into two pairs: 1 and 4; 2 and 3. The PCM triggers alternate pairs every 180 degrees of crankshaft revolution, with each pair fired twice per combustion cycle, figure 16-9.

Sequential

Sequential firing of the injectors according to engine firing order is the most accurate and desirable method of regulating PFI. However, it is also the most complex and expensive to design and manufacture. In this system, the PCM controls the injectors individually. Each cylinder receives one charge every two revolutions just before the intake valve opens. This means that the mixture is never static in the intake manifold and that the PCM can make mixture adjustments almost instantaneously between the firing of one injector and the next. A camshaft sensor signal or a special distributor reference pulse informs the PCM when the no. 1 cylinder is on its compres-

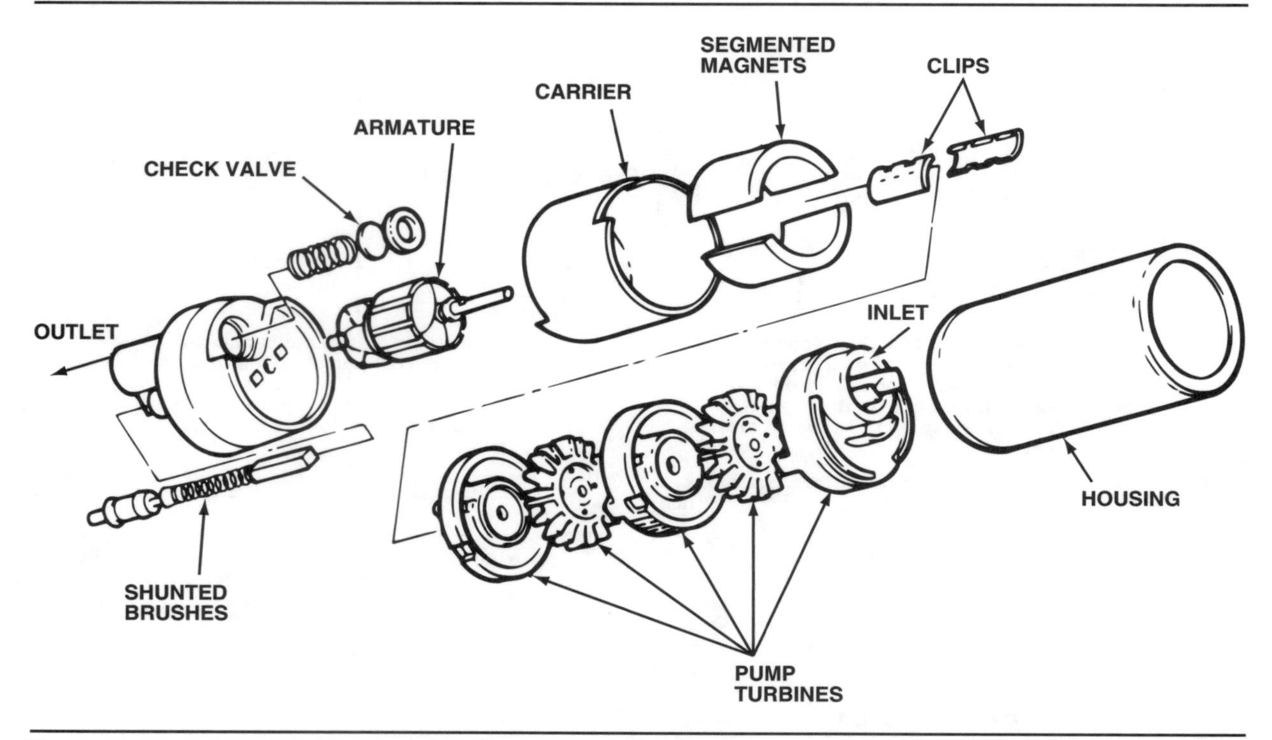

Figure 16-10. An in-tank turbine-type electric pump provides continuous low pressure (above 9 psi or 62 kPa). (GM)

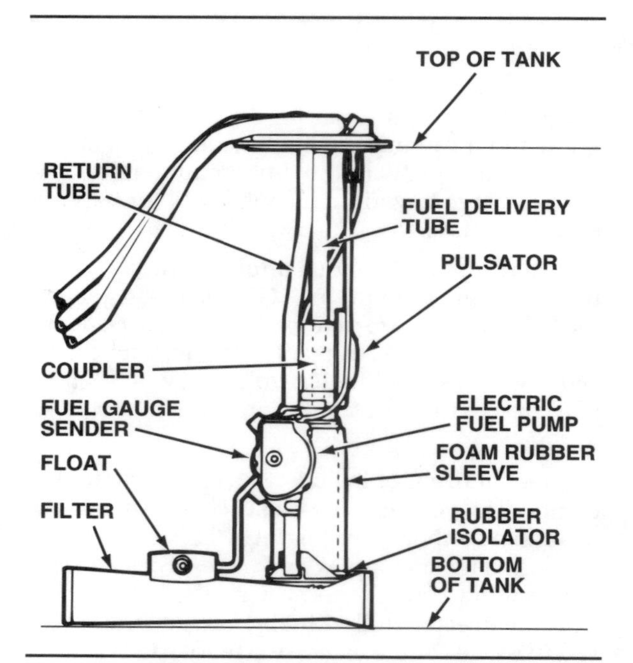

Figure 16-11. This in-tank roller vane fuel pump is combined with the fuel gauge sender unit. (GM)

sion stroke. If the sensor fails or the reference pulse is interrupted, some injection systems shut down, while others revert to pulsing the injectors simultaneously.

COMMON SUBSYSTEMS AND COMPONENTS

Regardless of the type, all fuel injection systems have three basic subsystems:

- Fuel delivery
- Air control
- Electronic control with auxiliary sensors or actuators.

Fuel Delivery System

The fuel delivery system consists of an electric fuel pump, a filter, a pressure regulator, one or more injectors, and the necessary connecting fuel distribution lines.

Fuel pump

An electric fuel pump provides constant and uniform fuel pressure at the injectors. Most systems use a positive-displacement vane, turbine, or roller pump, figure 16-10. Fuel injected vehicles generally use one of the following pump applications:

- High pressure, in-tank
- Low pressure, in-tank
- Low pressure, in-tank and high pressure, in-line.

In-tank fuel pumps may be separate units or may be combined with the fuel gauge sender assembly, figure 16-11. The pump supplies more fuel than

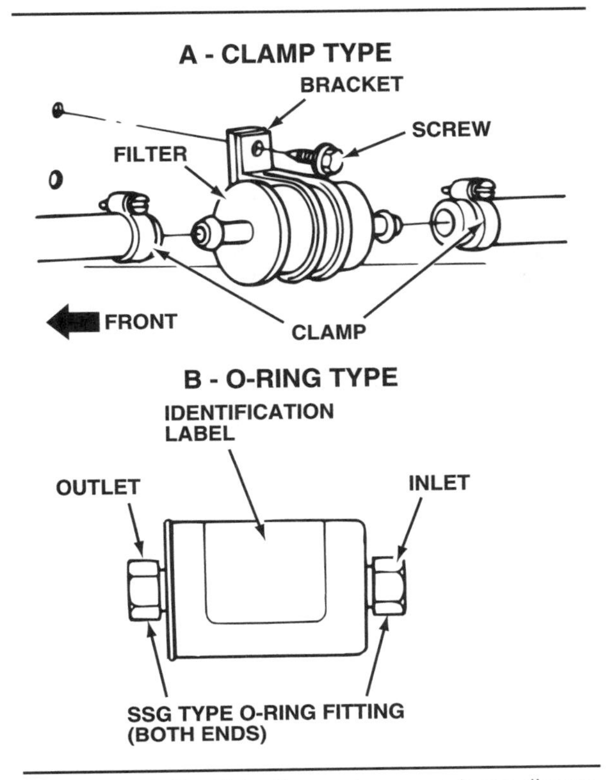

Figure 16-12. Fuel injection systems use large disposable filter canisters to ensure proper filtration. When threaded connections are used instead of clamps, O-ring seals are required. (GM)

the system requires through an in-line filter to the TBI unit or fuel rail, with the pressure regulator controlling volume and pressure. An internal relief valve protects the pump from excessive pressure if the filter or fuel lines become restricted.

Fuel filter

Clean fuel is extremely important in a fuel injection system because of the small orifice in the injector tip through which fuel must pass. Most injection systems use three forms of filters. The first is a fuel strainer or filter of woven plastic attached to the fuel pump inlet, figure 16-11. This prevents contamination from entering the fuel line and separates water from the fuel. The filter is self-cleaning and requires no maintenance. If a fuel restriction occurs at this point, the tank contains too much sediment or moisture and should be removed and cleaned.

Additional filtration is provided by a high-capacity in-line filter, figure 16-12, to remove contamination larger than 10 to 20 microns (0.000394 to 0.000787 in.). Automakers generally mount this filter on a frame rail under the vehicle or in the engine compartment.

The final line of defense against fuel contamination is in the fuel injector or TBI unit itself. Each injector contains its own inlet filter screen, figure

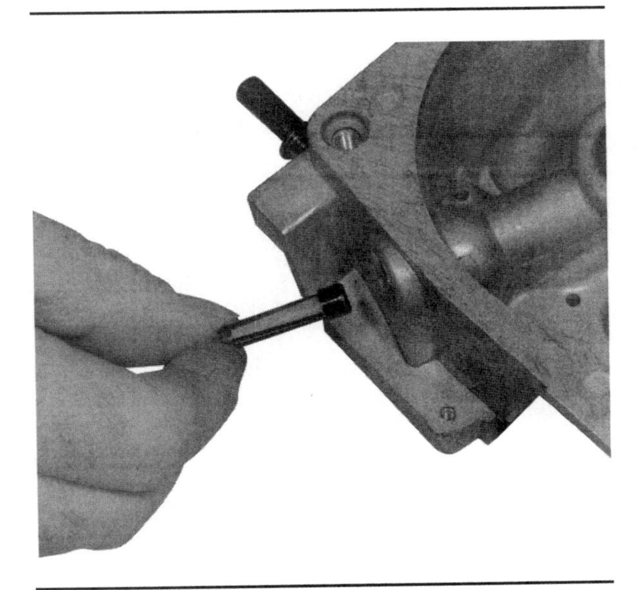

Figure 16-13. A filter screen may be installed in the fuel inlet of some throttle bodies.

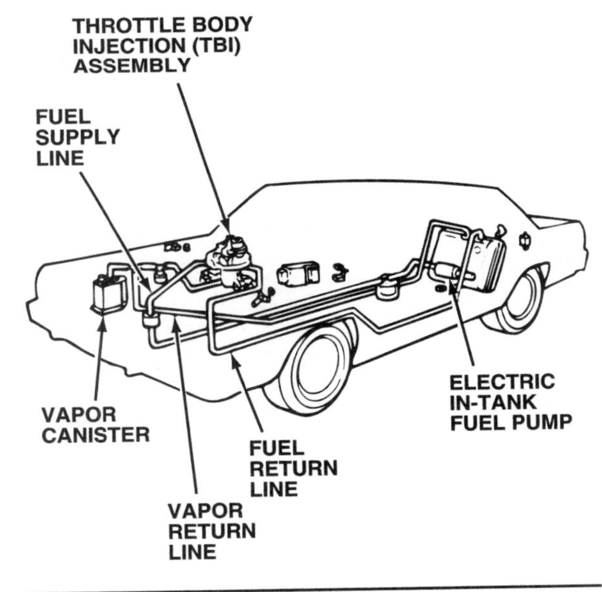

Figure 16-14. A typical fuel-injection system showing fuel line routing. (Ford)

16-6, to prevent contamination from reaching the tip. Some TBI units have a similar filter screen installed in the fuel inlet fitting, figure 16-13.

Fuel lines

A supply line carries fuel from the pump through the filter to the TBI unit or fuel rail. A return line carries excess fuel back to the tank. This allows the pump to continuously circulate the fuel from the tank to the injectors and back to the tank, figure 16-14. The system maintains constant fuel

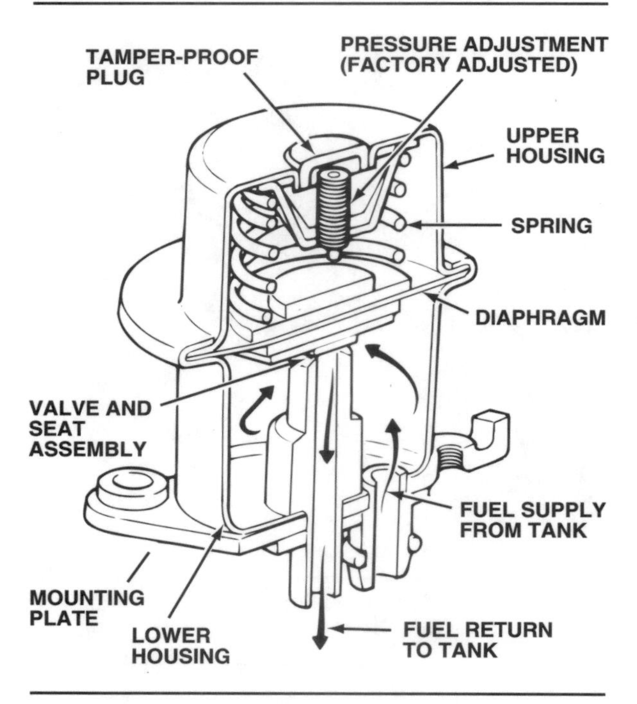

Figure 16-15. The pressure regulator commonly used with a TBI system functions by fuel pressure alone. (Ford)

pressure and volume while minimizing fuel heating and vapor lock. Most automakers make their injection system fuel lines from steel tubing, although Ford vehicles use nylon fuel line tubing. Some low-pressure return lines may be made out of flexible hose.

Pressure regulator

The pressure regulator and fuel pump work together to maintain the required constant pressure at the injector tips. The regulator consists of a spring-loaded diaphragm-operated valve in a metal housing. Two types of pressure regulators are used in injection systems.

Automakers install all fuel pressure regulators on current EFI systems on the return (downstream) side of the injectors. Downstream regulation minimizes fuel pressure pulsations caused by pressure drop across the injectors as the nozzles open. It also ensures positive fuel pressure at the injectors at all times and holds residual pressure in the lines when the engine is off.

Manufacturers install the TBI pressure regulator into or on the TBI unit. With this type, figure 16-15, fuel pressure must overcome spring pressure on the spring-loaded diaphragm to uncover the return line to the tank. This happens when system pressure exceeds operating requirements. The spring side of the diaphragm is exposed to inlet air pressure, or manifold vacuum, inside the TBI unit. Because it is close to the injector tip, the regulator senses essentially the same air pressure as the injector.

The pressure regulator used in a PFI system, figure 16-16, has an intake manifold connection on the regulator vacuum chamber. This allows spring pressure and manifold vacuum to manipulate the diaphragm, thus modulating fuel pressure.

In both systems, the regulator shuts off the return line when the fuel pump is not running. This maintains pressure at the injectors for easy restarting and reduced vapor lock.

Mechanical injection systems without electronic control operate on the differential between fuel pump pressure and the control pressure developed by the fuel distributor. The pressure regulator is located on the control pressure side of the fuel distributor and connected to the fuel return line, figure 16-17. Spring force and control pressure operate on one side of the regulator with pump pressure on the other side to modulate the return line opening. Bosch K-Jetronic, a type of CIS system, varies control pressure for different operation requirements with a primary regulator and a warmup regulator.

In later mechanical injection system designs, such as the CIS electronic system used on the Volkswagen Golf, a differential pressure regulator replaces the control pressure regulator. This device is a part of the fuel distributor, figure 16-18, and is operated by the PCM. A separate spring-loaded, diaphragm-operated pressure regulator governs system pressure. This eliminates the need for a pressure relief valve in the fuel distributor.

PFI systems generally operate with pressures at the injector of about 30 to 55 psi (207 to 379 kPa), while TBI systems work with injector pressures of about 10 to 20 psi (69 to 138 kPa). The difference in system pressures results from the difference in the systems operation. Remember that an injection system requires only enough pressure to move the fuel through the injector and help in atomizing it. Since injectors in a TBI system inject the fuel into the airflow at the manifold inlet (above the throttle), there is more time for atomization in the manifold before the air-fuel charge reaches the intake valve. This allows TBI injectors to work at lower pressures than injectors used in a port system.

Mechanical injection nozzles

The most common type of mechanical injection is the Bosch K-Jetronic system. Two types of injectors are used. Non-electronic systems use an injector with a spring-loaded tip to seal the nozzle against fuel pressure, figure 16-7. When pressure is greater than a specified level, it forces the nozzle open. The amount of fuel that enters the

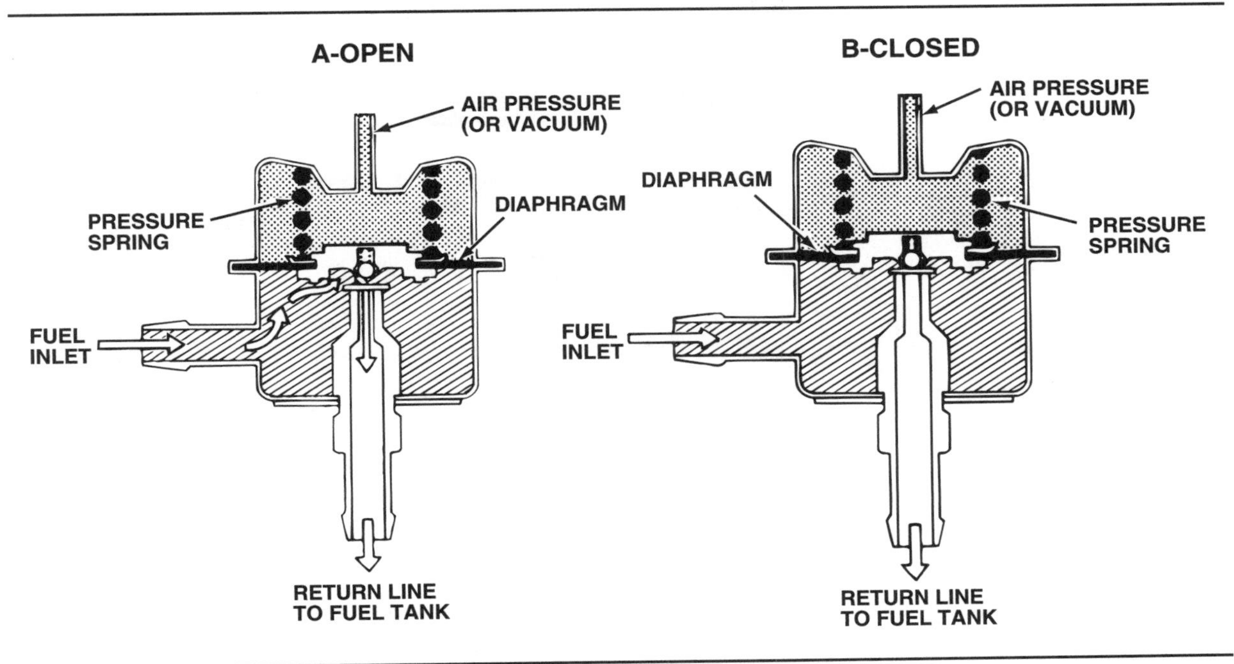

Figure 16-16. The pressure regulator commonly used with a multipoint system works by using a combination of fuel pressure and manifold vacuum or pressure. (GM)

manifold is controlled by the nozzle opening, which in turn depends on fuel pressure. The mixture control unit controls injection pressure by a combination of control pressure and pump pressure. K-Jetronic nozzles inject fuel constantly as long as the pump is operating. Bosch K-Jetronic nozzles are described in detail in the continuous injection system section, earlier in this chapter.

Electronically controlled K-Jetronic systems use an air-shrouded injector. These injectors work in the same way as those used with non-electronic systems, but design changes have made them more efficient. Air flows in through a cylinder head passage, passes between the injector and plastic shroud surrounding it, and then exits close to the injector tip. This improves atomization of the fuel as it leaves the injector, which reduces fuel condensation in the manifold. A second-generation air-shrouded injector, figure 16-19, uses a circlip to retain the seal ring. A separate plastic injector shroud and fluted air directional shield on the injector tip improve airflow around the injector tip. These changes further improve fuel atomization.

Electronic injection nozzles

EFI systems use a solenoid-operated injector. Figure 16-20 shows typical TBI injectors; figure 16-21 shows the two types of top-feed port fuel injectors currently in use. This electromagnetic device contains an armature and a spring-loaded needle valve or ball valve assembly. When the PCM energizes the solenoid, voltage is applied to

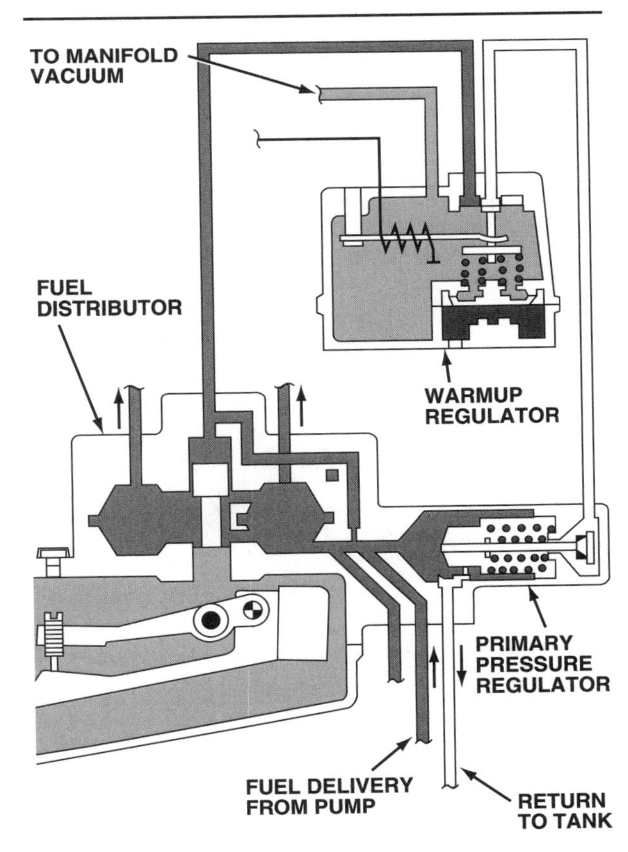

Figure 16-17. Injection pressure is controlled in a Bosch K-Jetronic system by a main pressure regulator and a separate warmup regulator. (Bosch)

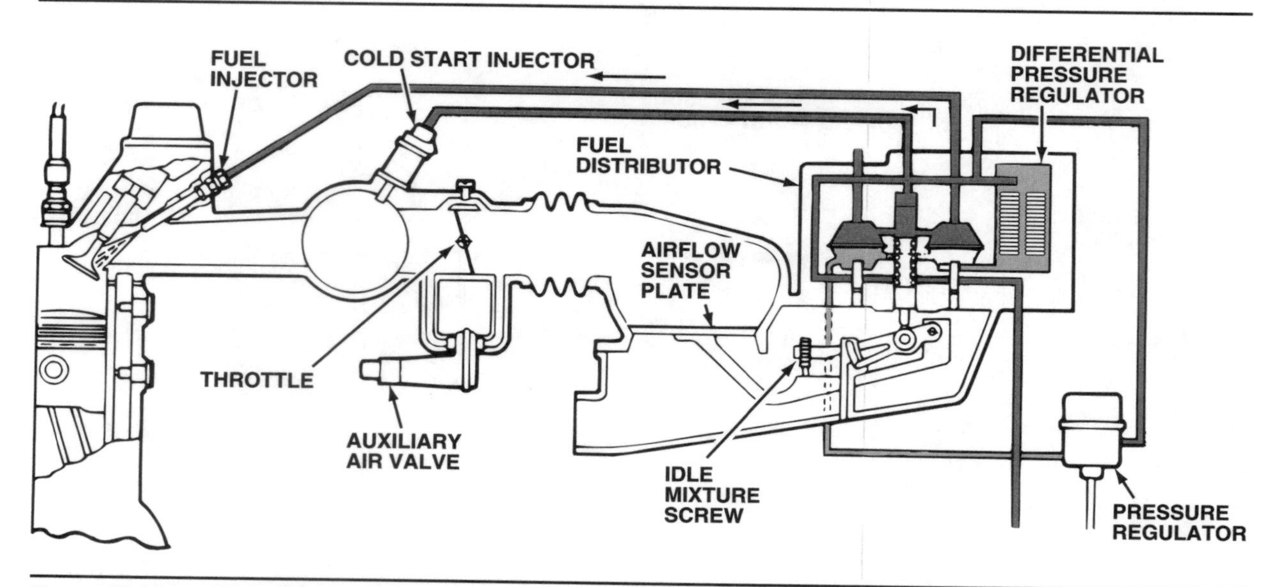

Figure 16-18. A differential pressure regulator is used instead of a control pressure regulator on some CIS late-model electronic systems. (Volkswagen)

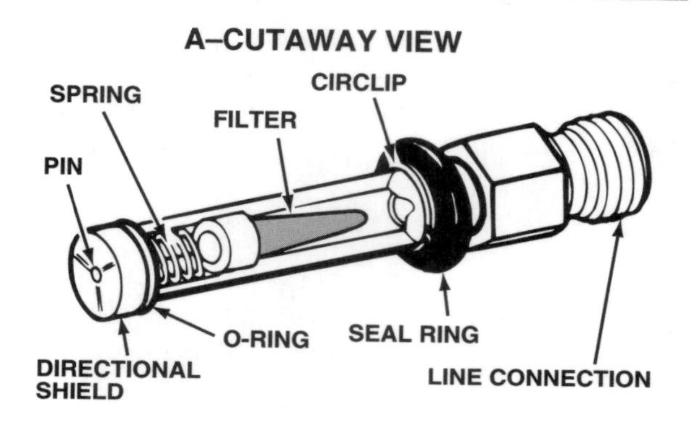

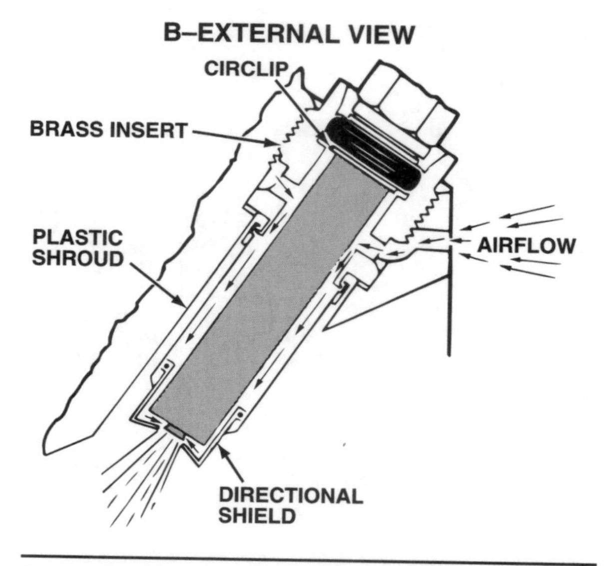

Figure 16-19. Air-shrouded injectors are similar to the K-Jetronic design, but flow air past the nozzle to improve atomization. (Volkswagen)

the solenoid coil until the current reaches a specified reference level (usually about five amperes). This permits a quick pull-in of the armature during turn-on. The armature is pulled off its seat against spring force, allowing fuel to flow through a 10-micron (0.000394-in.) screen to the spray nozzle, where it is sprayed in a conical pattern. When current reaches the reference level, it is regulated at a specified value (usually about one amp) until the injector is turned off. The low energy level during the holding state prevents overheating of the solenoid coil. The injector opens the same amount each time it is energized, so the amount of fuel injected depends on the length of time the injector remains open.

Most injectors are manufactured by Bosch (or made under Bosch license) and use a needle or pintle valve, figure 16-21A, for spray control. A diffuser positioned below the valve seat atomizes the fuel in a spray pattern of about 25 degrees. This design, however, has the disadvantage of allowing varnish deposits to gather on the pintle and seat, causing the injector to eventually malfunction unless it is cleaned regularly. Bosch reduced this deposit problem by extending the length of the protective cap or "chimney" on the injector tip.

Rochester Products took a different approach to the varnish buildup problem with the introduction of its Multec injector in 1987. The Multec injector uses a stainless steel ball and seat valve, figure 16-21B, instead of a needle valve. To assure a positive seal, the ball and seat are polished to a near-mirror finish. Spray pattern control is provided by a recessed director plate containing six

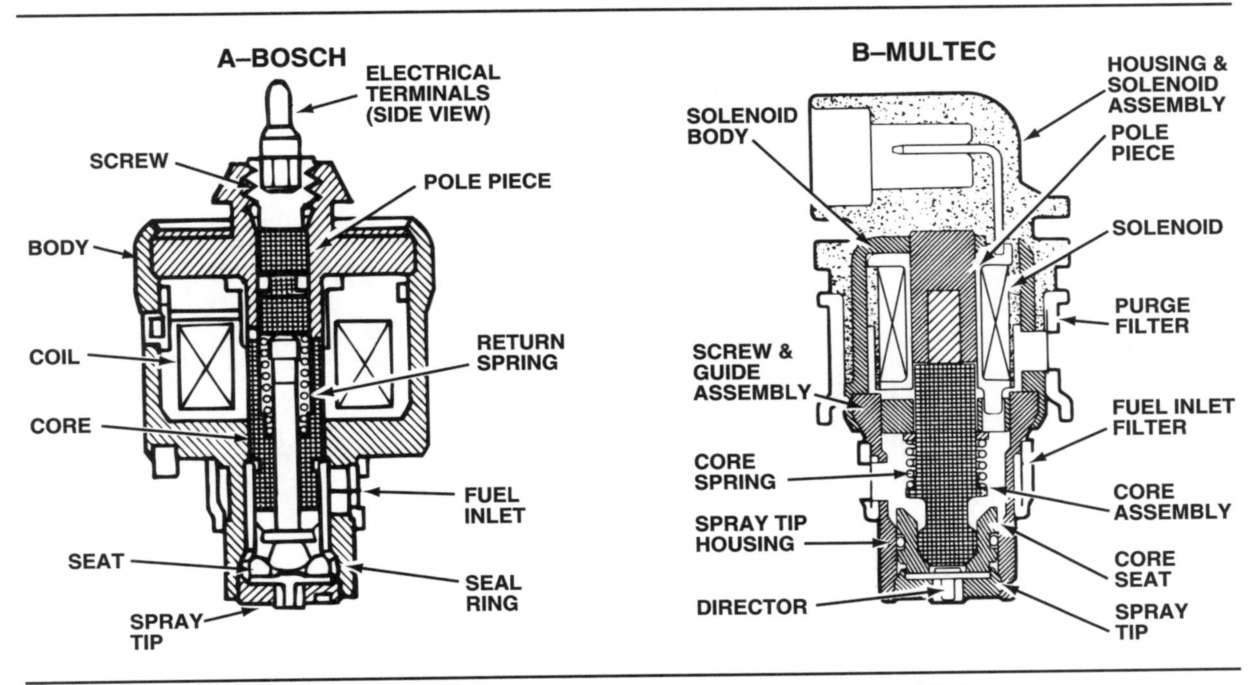

Figure 16-20. Bosch (left) and Rochester Multec (right) injectors used in a TBI unit. (GM)

machined holes. Since the Multec fuel metering area does not extend as deeply into the intake port as a pintle-type injector does, it is not as exposed to deposit-forming intake gases, figure 16-22.

By machining four holes in the director plate at an angle, Rochester created a dual spray Multec injector for use with 1991 and later 3.4-liter engines. By angling the director hole plates, the injector sprays fuel more directly at the intake valves, figure 16-23. This further atomizes and vaporizes the fuel before it enters the combustion chamber.

While PFI is a top-feed design in which fuel enters the top of the injector and passes through its entire length before being injected, recent changes in some manifold and fuel rail designs have led to the use of a Multec bottom-feed injector in GM PFI systems, figure 16-24, similar to that used in TBI systems.

A further refinement in Multec PFI injector design appeared on some 1993 GM engines. The new injector has a stamped spray tip with a larger bore to deliver an improved fuel spray pattern, figure 16-25. The stamped spray tip injectors are slightly shorter in length and have a plastic collar installed behind the lower O-ring for proper positioning in the manifold.

Ford introduced two basic designs of deposit resistant injectors on some 1990 engines. The design manufactured by Bosch uses a four-hole director/metering plate similar to that used by the Rochester Multec injectors. The design manufactured by Nippondenso uses an internal up stream orifice in the adjusting tube. It also has a redesigned pintle/seat containing a wider tip opening that tolerates deposit buildup without affecting injector performance.

■ Engine Modifications Cannot Do the Whole Job

Since the automobile was discovered to be the biggest source of air pollution, many changes in engine design have been made to "clean it up". These engine modifications have reduced exhaust emissions, but it is impossible to eliminate the major cause of HC emissions simply by changing the engine design. This is because the major cause of HC emissions is the effect of the "quench area" on combustion.

The quench area is the inner surface of the combustion chamber. When the ignition flame front passes through the combustion chamber, it burns the fuel charge as it goes until the quench area is reached. This is a thin layer between 0.002 and 0.010 in. (0.05 and 0.25 mm) thick at the edge of the combustion chamber. When the flame front reaches the quench area, it is snuffed out because the quench area is so close to the cylinder head water jacket that the temperature there is too low for combustion to continue. Consequently, hydrocarbons within the quench area do not burn. They are ejected from the cylinder on every exhaust stroke along with the exhaust gases formed by combustion and enter the atmosphere as pollutants.

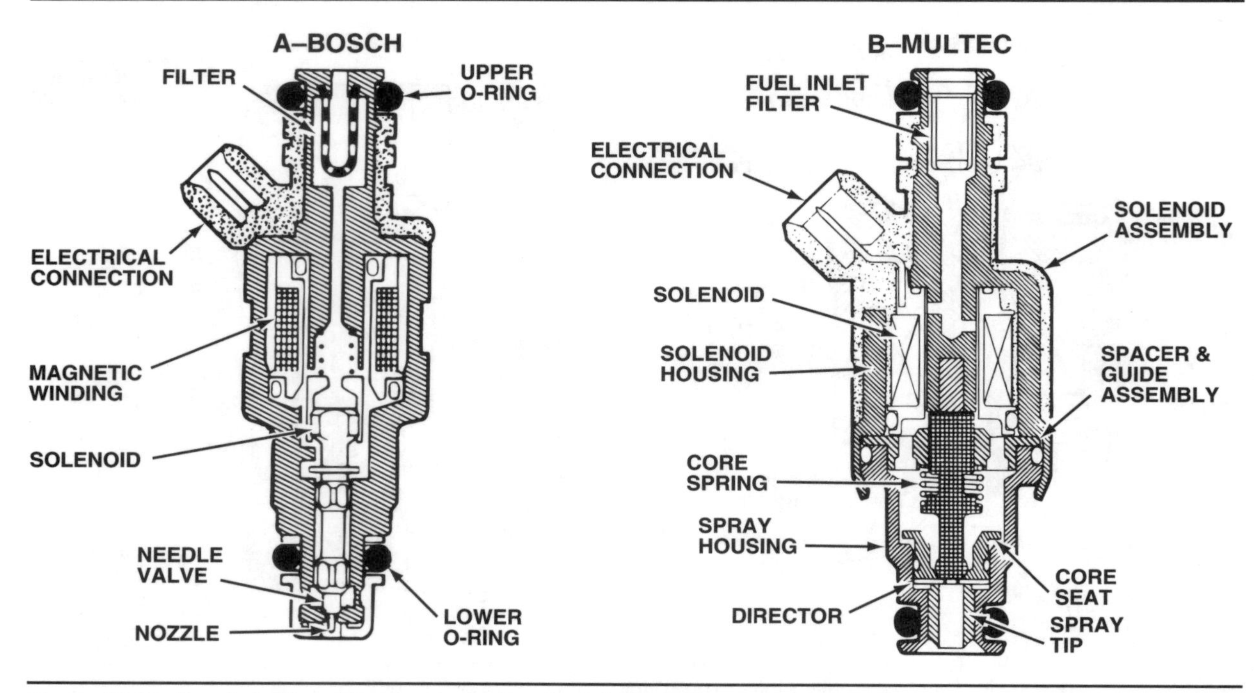

Figure 16-21. Bosch (left) and Rochester Multec (right) injectors used in multipoint systems. (GM)

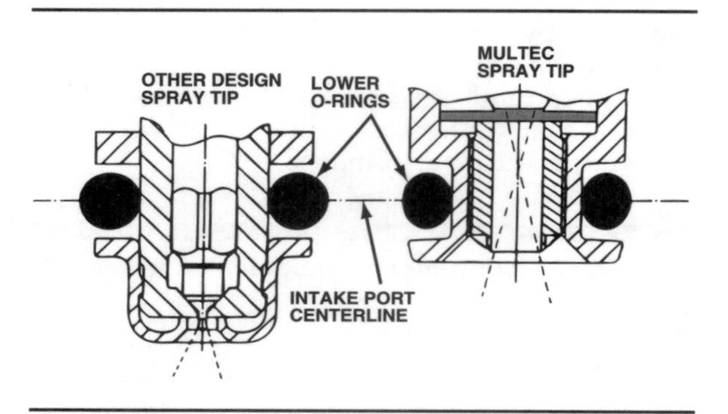

Figure 16-22. Since the Multec spray tip is not as exposed to hot intake gases as other injector designs, fewer deposits are formed to clog the injector. (GM)

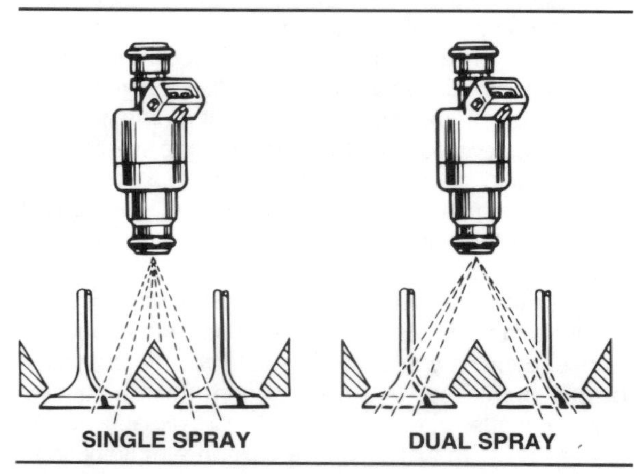

Figure 16-23. A comparison of single versus dual spray injectors. (GM)

Air Control System

As you learned in the "Basic Carburetion and Manifolding" chapter, the difference between fuel bowl air pressure and carburetor throttle body barrel air pressure controls fuel metering in a non-computerized carbureted system. Computerized fuel injection systems, on the other hand, control fuel metering by indirectly or directly measuring intake air volume, density and rate to determine the rate of mass airflow. These fuel injection systems accomplish this by using a VAF meter, a MAP sensor, or a MAF sensor. For further information on how these sensors function and their roles in a fuel management system, see the "Electronic Engine Control: Sensors, Actuators, and Operation"chapter.

Throttle

The throttle works exactly the same in both a carbureted and a fuel injected system. It is connected to the accelerator linkage and regulates the amount of air taken into the engine. With a TBI system, the throttle is mounted between the injectors and the manifold, figure 16-1. Since the fuel is introduced to the air above the throttle plate, the throttle regulates the amount of air-fuel mixture that enters the manifold. On port systems, the throttle is mounted horizontally in the air intake throttle body on the air inlet side of the injectors and controls only airflow, figure 16-26. A sensor determines the intake air volume. It can do this in one of three ways: by mea-

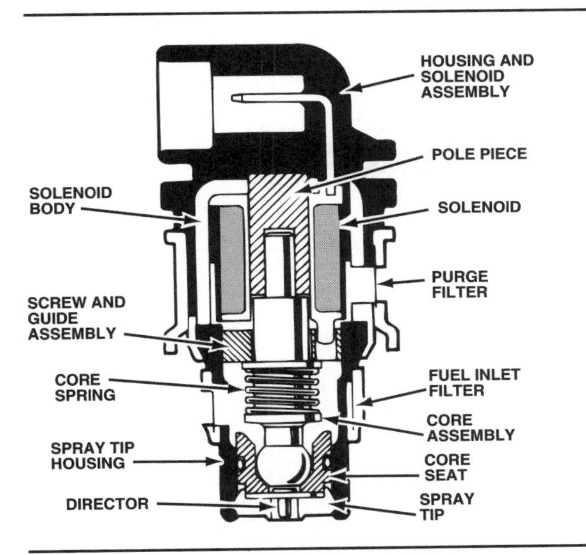

Figure 16-24. Cross-section of a bottom feed port (BFP) injector. (GM)

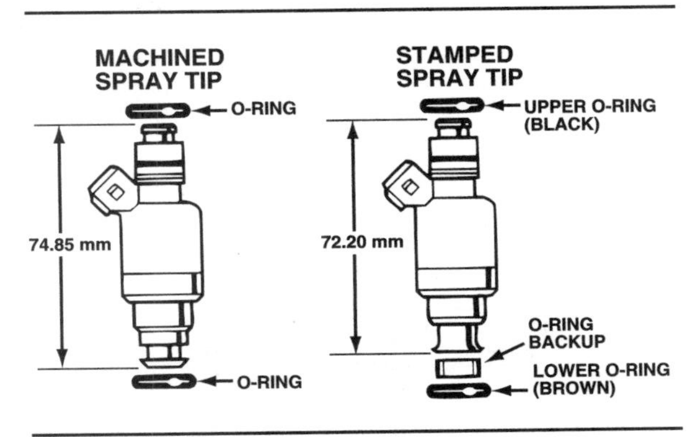

Figure 16-25. A comparison of machined spray tip and stamped spray tip injectors. (GM)

suring airflow speed, manifold pressure, or the molecular mass of the air drawn into the engine. All of these methods directly relates to the amount of fuel required from the injectors.

Electronic Control System with Auxiliary Sensors and Actuators

You may have noticed that to this point, we have not mentioned chokes, power valves, accelerator pumps, or idle speed and mixture controls in our discussion of fuel injection. Earlier in this chapter, however, you learned that a fuel injection system does the same job as a carburetor. The enrichment and mixture control functions of a carburetor are done by the devices just mentioned. Now you will learn how a fuel injection system does the same thing.

Computer control

An electronic fuel injection system uses engine sensors and auxiliary metering devices to perform the required enrichment and mixture control functions. Sensors provide input data to the PCM. After processing the data, the PCM sends output signals to actuators. In an EFI system, the actuators are solenoid-operated fuel injectors.

You should remember from the section on electronic fuel injectors that the injector opens the same amount each time it is energized, so the amount of fuel injected depends on the length of time the injector is energized. As you learned earlier, pulse width is measured in milliseconds. The PCM varies the injector pulse width according to the amount of fuel required under any operating condition, figure 16-27.

Cold starting

To start, a cold engine requires a richer air-fuel mixture. Port systems often use a separate cold-start injector to provide additional fuel during the crank mode. During engine cranking, the individual injectors send fuel into the cylinder ports. At the same time, the cold-start injector provides the additional fuel by injecting it into a central area of the manifold or into a separate passage in the manifold, figure 16-28. In either location, the extra fuel is distributed more or less equally to all cylinders. The colder the engine, the longer the cold-start injector remains on.

Cold-start injector duration can be regulated by a thermal-time switch in the thermostat housing. This switch grounds the cold-start injector circuit when the engine is cranked at a coolant temperature below a specified value. A heating element inside the switch also is activated during cranking and starts to heat a bimetallic strip in the ground circuit. Once the bimetallic strip reaches the specified temperature, it opens the ground circuit and shuts off power to the cold-start injector.

Idle control

PFI systems use an idle air control (IAC) valve, figure 16-26, or auxiliary air regulator, figure 16-29, to do the same job as the air bypass circuit in a carburetor. This IAC valve or regulator provides needed additional airflow by opening an intake air passage to let more air into the engine. The system is calibrated to maintain engine idle speed at a specified value regardless of engine temperature.

Most domestic PFI systems use a computer-controlled, solenoid-operated valve, figure 16-30, to regulate the airflow. The Bosch auxiliary air regulator used in several kinds of systems, figure 16-29, contains a heated bimetallic strip. When the engine is first started, air passes through the gate valve and current is applied to the bimetallic strip. As the engine warms up, the bimetallic strip starts

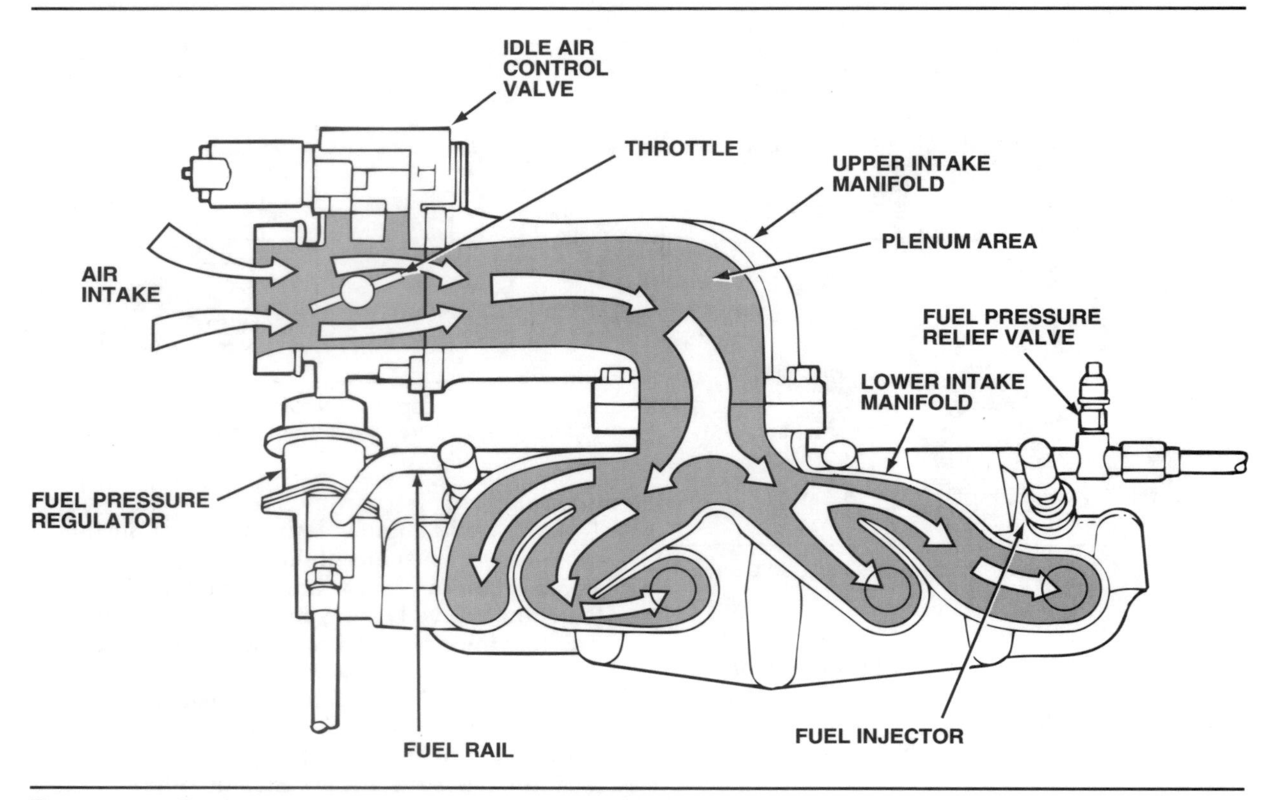

Figure 16-26. The horizontal throttle used in a multipoint system controls airflow only. (Ford)

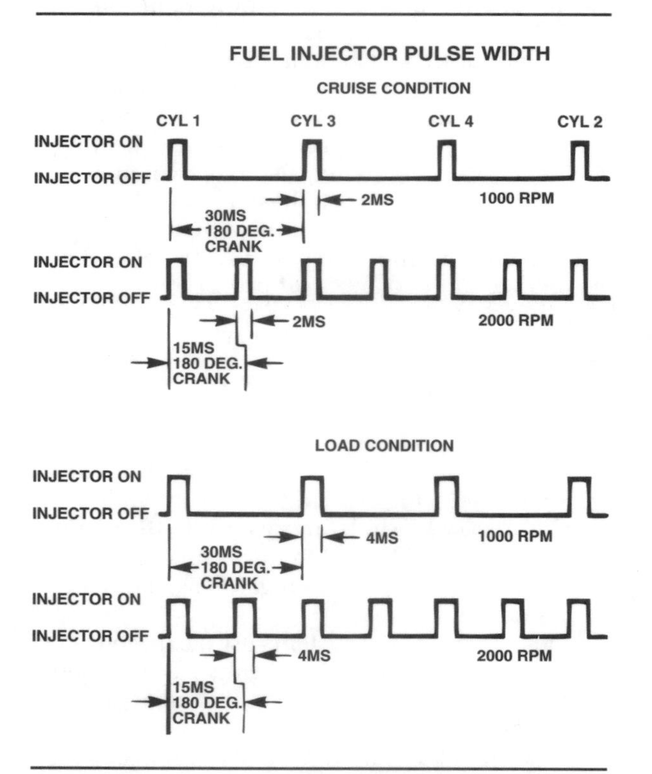

Figure 16-27. Fuel flow timing and pulse width. (Chrysler)

to deflect and gradually closes off the gate valve passage to shut off the additional airflow. By the time the engine has fully warmed up, the gate valve is completely closed and the idle bypass in the throttle chamber determines engine speed.

Idle airflow in a TBI system travels through a passage around the throttle and is controlled by a stepper motor. Idle fuel metering on a Bosch K-Jetronic system is controlled by a mechanical adjustment. Electronic fuel injection systems may control idle fuel metering either by a predetermined program in the PCM or from signals provided by a MAP sensor.

Acceleration and load enrichment

Airflow and manifold pressure change rapidly when an engine needs more fuel for additional power. With a mechanical injection system, these changes affect fuel pressure, causing the metering assembly to increase the volume of fuel flow. In an electronic injection system, signals from the VAF, MAP, or MAF sensors to the PCM result in an increase in injector pulse width.

Engine speed and crankshaft position

Engine speed is used by mechanical injection systems to operate the cold-start and cranking-enrichment subsystems. Electronic injection systems use engine speed and crankshaft position to time the injection pulses and determine the

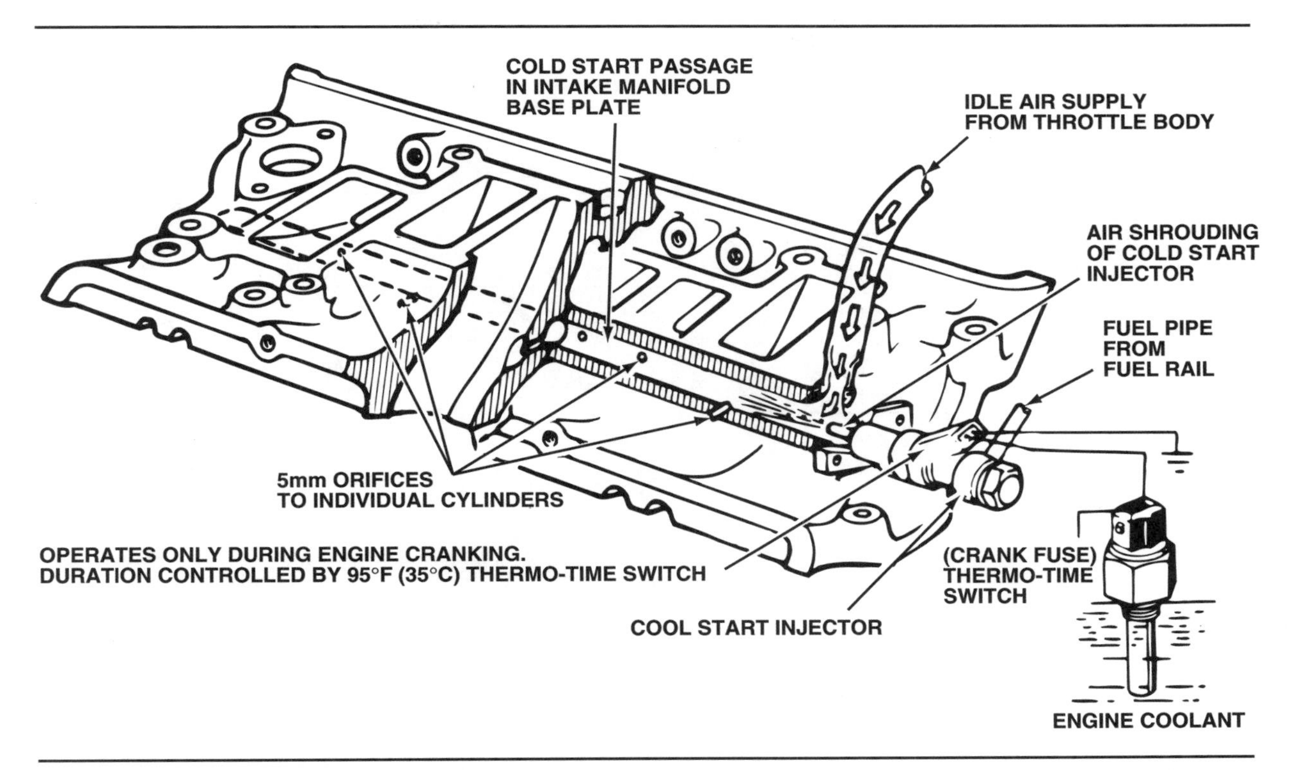

Figure 16-28. A cold-start injector provides extra fuel during cranking. In this design, it mixes with air and flows through small orifices into individual cylinders. (GM)

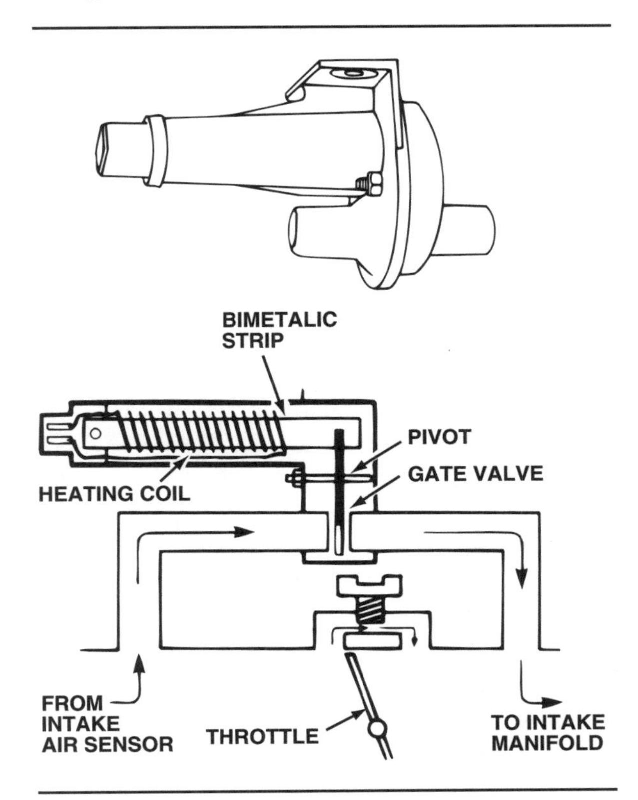

Figure 16-29. The auxiliary air regulator controls the air intake through a gate valve operated by a heated bimetallic strip. (Bosch)

pulse width. Speed and position signals can be provided either by a magnetic sensor in the engine or by Hall-effect switches or pulse generators in the distributor. Engine-mounted sensors generally are positioned in the bell housing. For further information, review the last chapter.

As mentioned previously, sequential port systems require a number 1 cylinder signal to time the injector firing order. This signal can be provid-

Poor Driving Means Poor Mileage

Automakers have succeeded in producing cars that use less fuel and are more efficient. Still, the driving patterns of individual drivers can make a big difference in how much fuel the car will use. Here are some helpful points to remember:

- Stopping and restarting a car engine can consume less fuel than idling for one minute.
- Driving at 55 mph (90 km/h) instead of 65 mph (105 km/h) reduces fuel consumption by about 12 percent.
- One failed spark plug in a V8 engine can cut mileage by 12 percent and increase HC emissions by as much as 300 percent.
- Turning off the air conditioner can improve gas mileage by up to 10 percent.

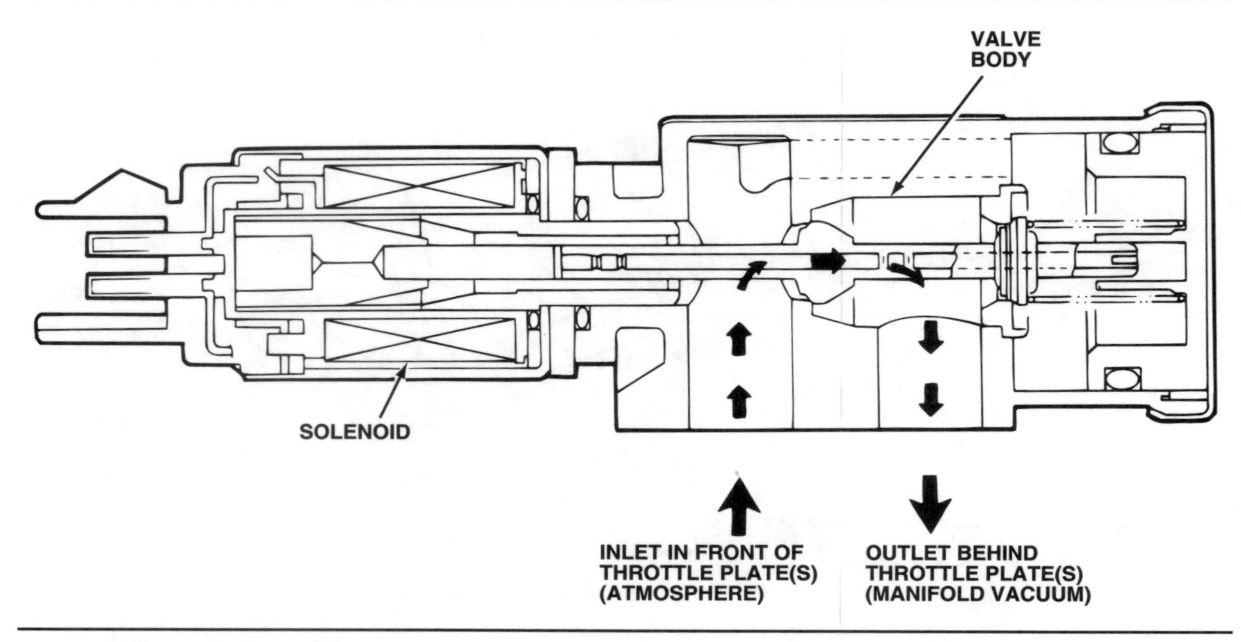

Figure 16-30. A solenoid-operated air bypass valve is used with most domestic multipoint systems to provide idle airflow around the throttle plate. (Ford)

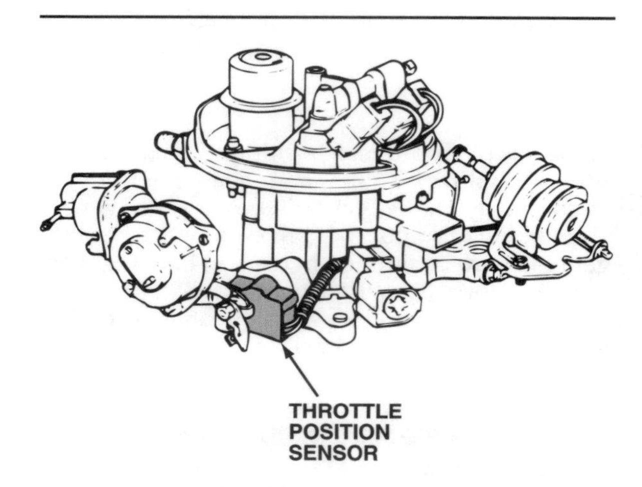

Figure 16-31. A typical throttle position sensor. (Ford)

ed by the same sensor that indicates crankshaft position or engine speed. Some engines use a separate sensor for the sequential timing signal, but it works on the same principles as other position or speed sensors. Without the number 1 cylinder signal, a sequential injection system will not run; it reverts to a programmed pulsed firing sequence that allows the car to be driven in for repair.

Throttle position

A sensor monitors throttle position for the PCM. Throttle position (TP) sensors may be a simple on/off switch, or a potentiometer that sends a variable resistance signal to the PCM, figure 16-31. By interpreting the signal, the PCM determines when the engine is at:

- Closed throttle (during idle and deceleration)
- Part throttle (normal operation)
- Wide-open throttle (acceleration and full-power operation).

Proper TP sensor adjustment is critical to ensure that fuel enrichment and shutoff occur at the appropriate times. Otherwise, driveability problems such as stalling, surging, and hesitation occur. Some TP sensors can be adjusted; others are nonadjustable and must be replaced if out of specifications.

Air and coolant temperature

The PCM must be kept aware of intake air temperature and engine coolant temperature. Intake air temperature is used as a density corrector for calculating fuel flow and proportioning cold enrichment fuel flow. Coolant temperature is used to determine the correct air-fuel ratio according to the engine's operating temperature. This information is provided to the PCM by engine coolant temperature (ECT) and intake air temperature (IAT) sensors. The PCM integrates their signals, changing the injector pulse width accordingly.

SPECIFIC FUEL INJECTION SYSTEMS

Fuel injection systems completely replaced the carburetor during the late 1980s and early 1990s.

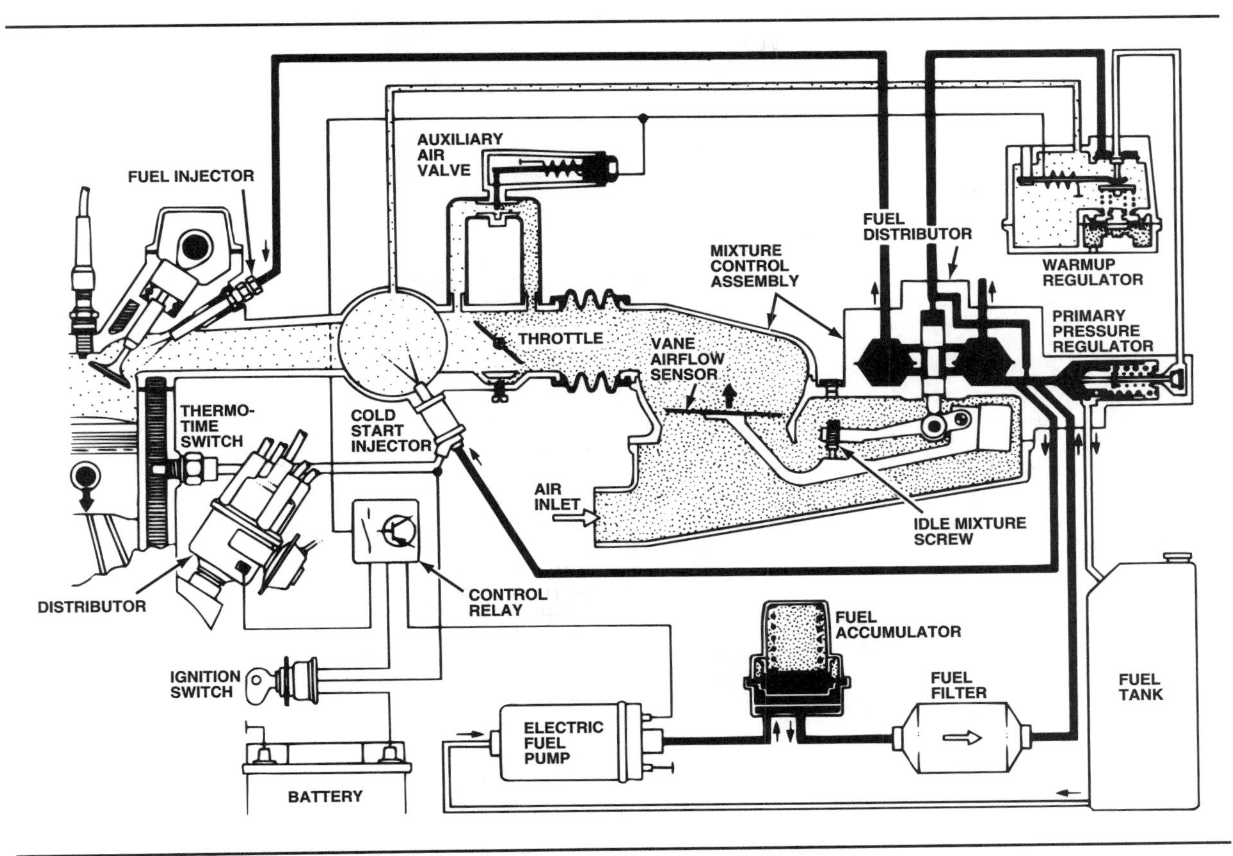

Figure 16-32. The Bosch K-Jetronic is a mechanical continuous injection system (CIS). (Bosch)

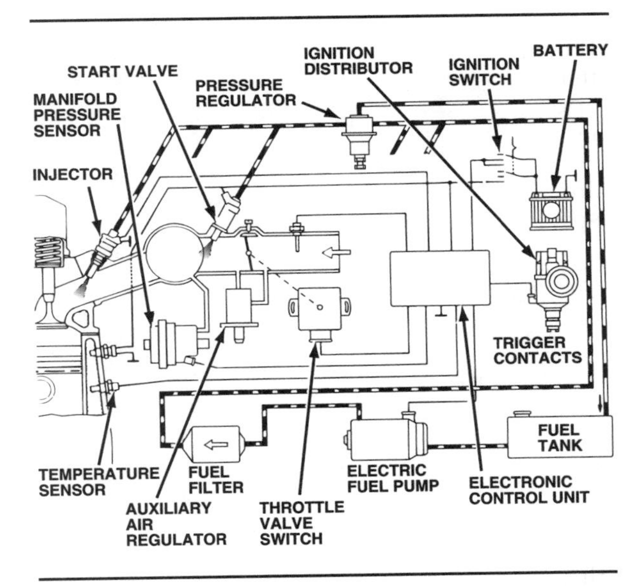

Figure 16-33. An electronic intermittent injection system, the Bosch D-Jetronic is controlled by manifold pressure. (Bosch)

The summary descriptions that follow provide the basic features of earlier fuel injection systems. These outlines can help you to recognize different components and understand how they perform similar tasks.

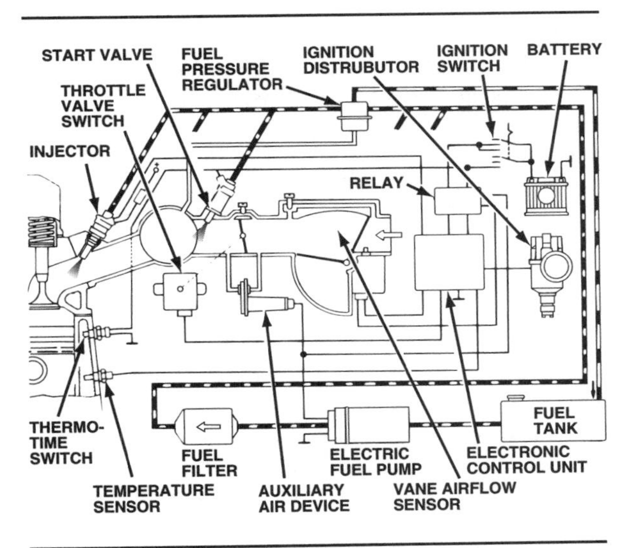

Figure 16-34. The Bosch L-Jetronic is controlled by engine speed and airflow measurement. (Bosch)

Bosch Jetronic Systems

A type of mechanical continuous injection system (CIS), the Bosch K-Jetronic, figure 16-32, was used on many European vehicles in the early 1970s. Air volume is measured by sensing airflow; fuel pres-

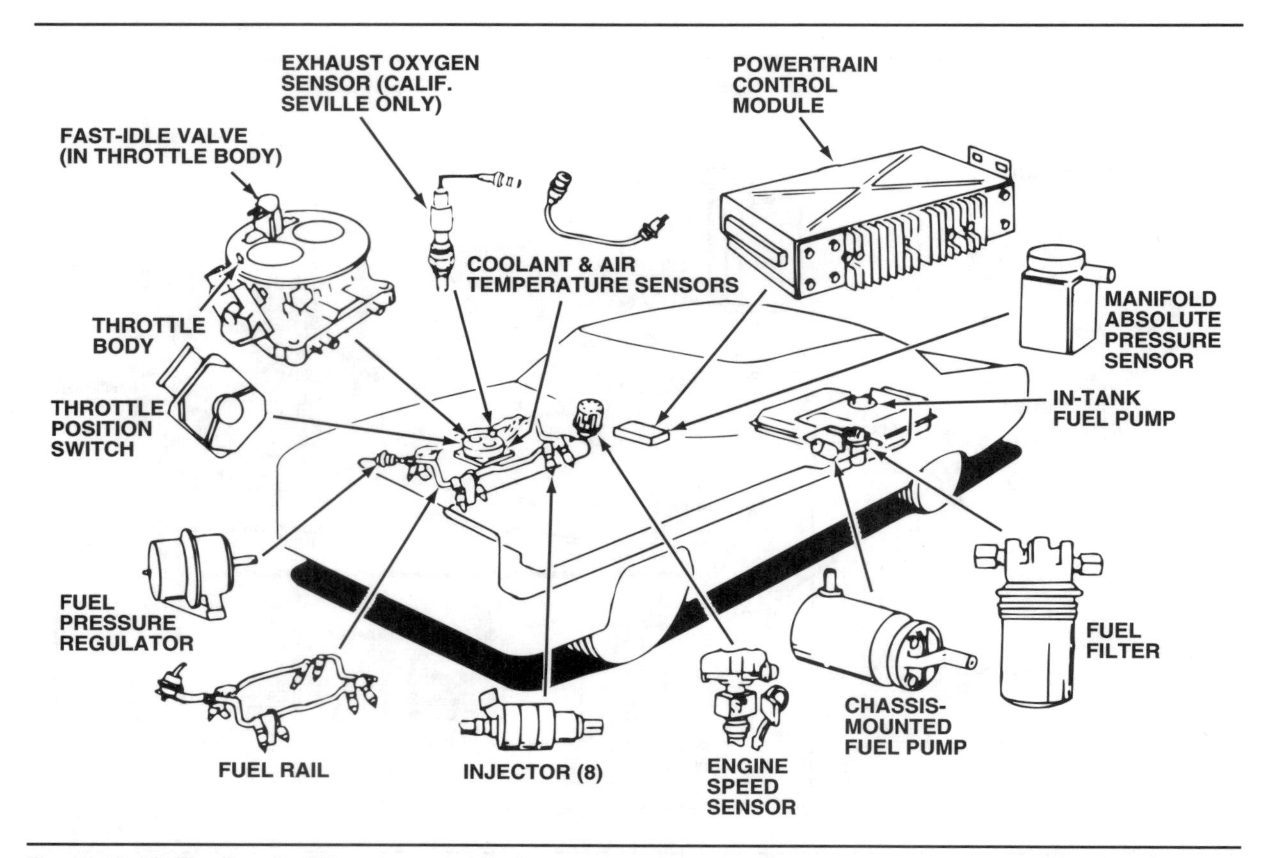

Figure 16-35. The Bendix EFI system was the first modern multi-point injection system used on domestic vehicles. (Cadillac)

sure is regulated relative to air volume and activates the individual injectors. A second-generation injection system, the KE-Jetronic system uses additional sensors to provide increased electronic control capabilities. An electrohydraulic pressure actuator within the fuel distributor controls fuel pressure (quantity) according to signals from the PCM.

The Bosch D-Jetronic system, figure 16-33, is an electronically controlled injection system used on 1968–73 Volkswagens and other European vehicles. Its solenoid-operated injectors operate in intermittent pulses; fuel is measured relative to engine speed and manifold air pressure.

The Bosch L-Jetronic system was introduced in 1974 and uses individual solenoid-operated injectors that operate intermittently. It is similar to the mechanical K-Jetronic system in that fuel metering is controlled relative to airflow, but the vane-type airflow sensor is connected to an electronic module which also regulates fuel metering relative to temperature and engine speed, figure 16-34. The L-Jetronic system sometimes is called airflow-controlled (AFC) injection. The system was revised in the 1980s to measure air volume more precisely by replacing the VAF meter with a hot-wire MAF sensor.

Figure 16-36. The Bendix EFI system on Cadillac Sevilles uses a carburetor-like throttle body for air intake and a fuel rail with individual injectors.

Cadillac EFI System

Used on 1976–79 Cadillac Sevilles, the Bendix EFI system was the first electronically controlled PFI system to be used on domestic vehicles, figure 16-35. Fuel is supplied to individual injectors in the intake manifold through a fuel rail, with air

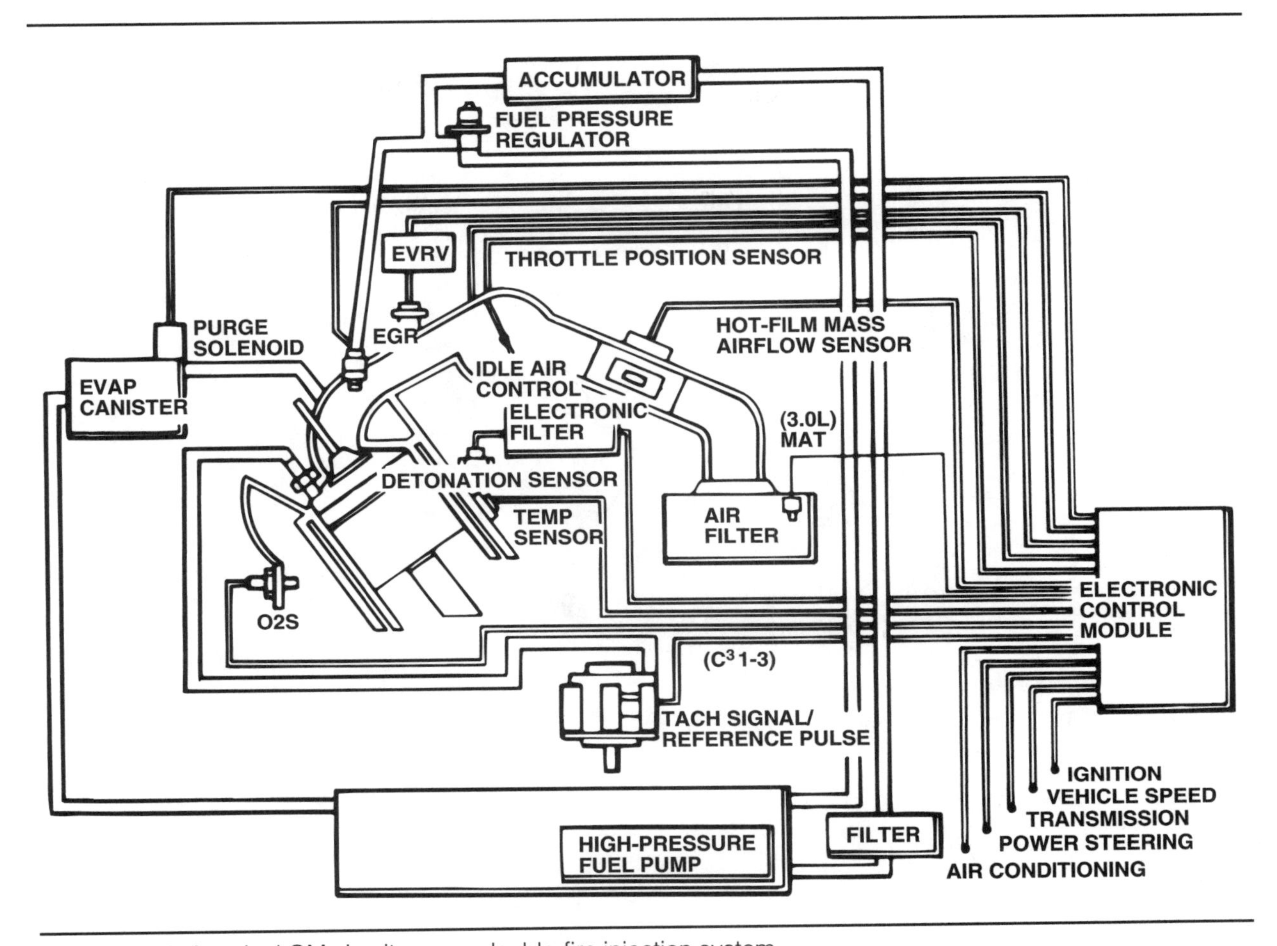

Figure 16-37. A typical GM simultaneous double-fire injection system.

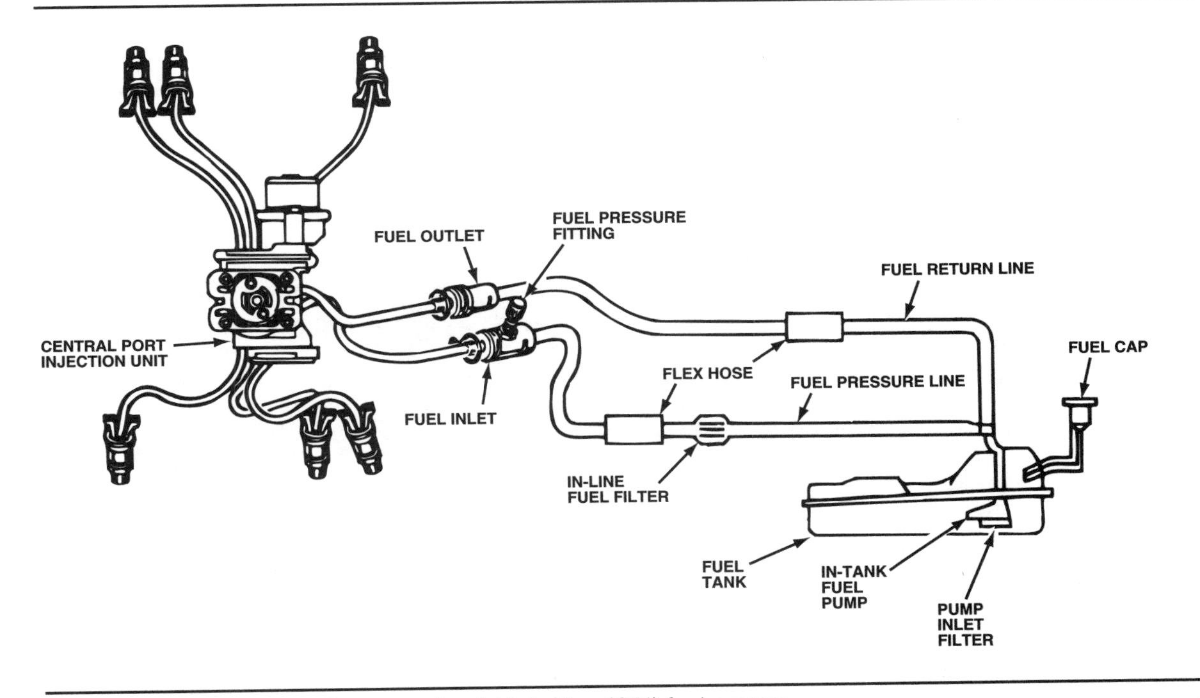

Figure 16-38. The GM 4.3-liter central point injection (CPI) fuel system.

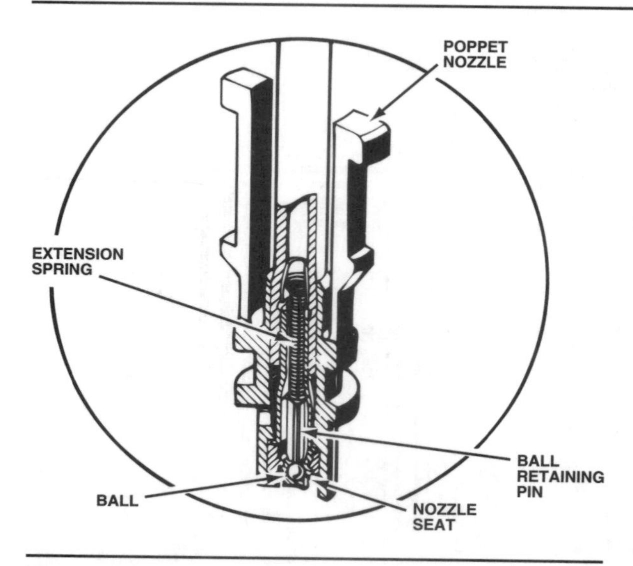

Figure 16-39. Cross-section of the CPI injection assembly. (GM)

intake controlled by a throttle body which looks much like a carburetor, figure 16-36. A PCM controls fuel delivery according to various sensor inputs. The injectors were pulsed intermittently in two banks. Feedback control of fuel metering with an O2S was used in later versions of this system.

Later-Model Port Fuel Injection Systems

The PFI systems used on domestic and many Japanese vehicles are based on Bosch designs and contain many components manufactured by or under license from Bosch. All have individual injectors installed in some form of fuel rail and mounted in the intake manifold.

General Motors

General Motors introduced PFI on some 1984 models and uses different types:

- Port fuel injection (PFI) or multiport fuel injection (MFI or MPFI)
- Tuned port injection (TPI)
- Sequential fuel injection (SFI)
- Central port fuel injection (CPI).

Port fuel injected and multi-port fuel injected systems are simultaneous double-fire systems in which all injectors fire once during each engine revolution, figure 16-37. In other words, two injections of fuel are mixed with incoming air for each combustion cycle. Some Quad 4 engines, however, use a variation called alternating synchronous double-fire (ASDF) injection. Two of the four injectors are triggered every 180 degrees of crankshaft revolution with each pair (1 and 4, 2 and 3) fired twice per combustion cycle.

TPI systems function in the same manner as PFI and MFI systems, but the intake air plenum runners are individually tuned to provide the best airflow for each cylinder. This system generally is used on high performance V8 engines in Corvettes, Camaros, and Firebirds.

Sequential fuel injection (SFI) systems are essentially the same design as PFI and MFI, but the injectors are triggered in firing order sequence and timed to the opening of the intake valves. SFI engines are used with a distributorless ignition system. This type of ignition system requires sequential fuel injection so that the "waste-spark" that fires in the cylinder on an exhaust stroke does not ignite any air-fuel mixture.

A cross between PFI and TPI, central port injection (CPI) was introduced by GM on 1992 4.3-liter V6 truck engines, figure 16-38. The CPI assembly consists of a single fuel injector, a pressure regulator, and six poppet nozzle assemblies with nozzle tubes, figure 16-39. When the injector is energized, its armature lifts off the six fuel tube seats and pressurized fuel flows through the nozzle tubes to each poppet nozzle. The increased pressure causes each poppet nozzle ball to also lift from its seat, allowing fuel to flow from the nozzle. This hybrid injection system combines the single injector of a TBI system with the equalized fuel distribution of a PFI system. It eliminates the individual fuel rail while allowing more efficient manifold tuning than otherwise possible with a TBI system.

Some GM PFI systems use a MAF sensor. Through the MAF sensor signals, the PCM is able to determine the air temperature, density, and humidity to calculate the "mass" of the incoming air. GM uses two types of MAF sensors: heated film and hot wire. Both types operate on the same principle: the resistance of a conductor varies with temperature.

Other GM PFI systems rely on a speed-density air measurement system instead of mass airflow. A speed-density system uses manifold absolute pressure and temperature along with an engine mapping program in the PCM to calculate mass airflow rate, as explained in the "Electronic Engine Control: Sensors, Actuators, and Operation" chapter. This type of system is sensitive to engine and EGR variations.

From 1987 through the early 1990s, the trend in fuel rail design for GM V6 and V8 port systems was toward dual or "split" fuel rails. Each bank of the engine has its own fuel distribution rail and injectors, figure 16-40. As a further refinement, some of these split-rail systems have "biased" injectors. The PCM controls each bank differently, and the injectors for right and left banks may have different flow rates for identical duty cycles and pulse widths. For example, with a pulse width of five milliseconds, the left bank injectors may flow 1.5 cm^3/s (0.092 in^3/s), and the right bank injectors

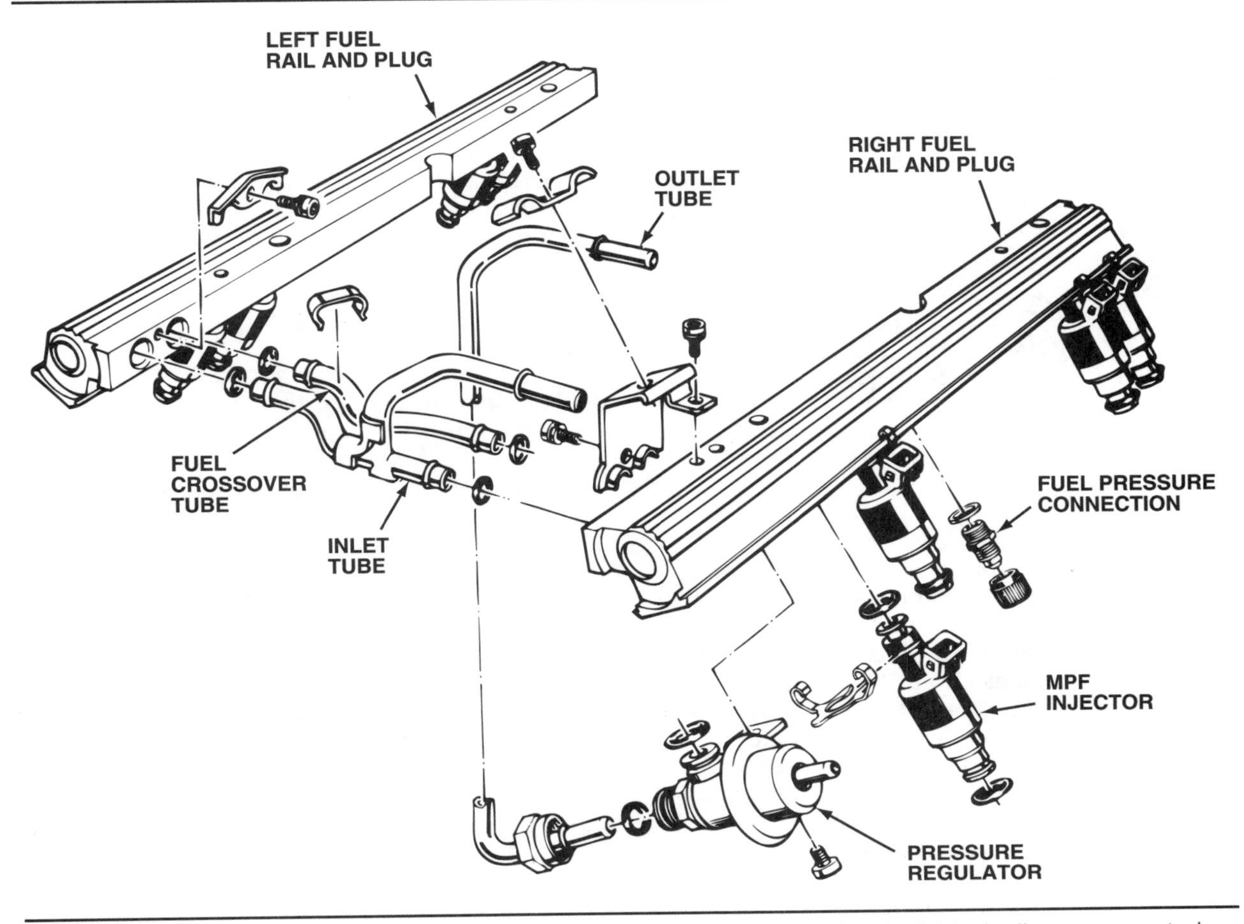

Figure 16-40. A typical GM dual fuel rail design used on the 5.7-liter engine. The individual rails are connected by fuel inlet, outlet, and crossover tubes.

may flow 1.6 cm^3/s (0.098 in^3/s). Biased injectors are color coded for identification. Many engines with split-rail injection also have two EGR sensors to allow the PCM to control fuel metering independently for each bank of the engine.

The split-rail design now is being challenged by lower intake manifold assemblies which contain a longitudinal fuel passage as an integral part of the casting, with intersecting injector bores in each runner, figure 16-41. This design completely eliminates the fuel rail. The injectors are held in place by a retaining plate, and the fuel feed and return lines connect directly to the manifold housing.

Ford

Ford refers to its PFI systems as Electronic Fuel Injection (EFI). Originally based on the Bosch L-Jetronic design, Ford EFI systems for 4-cylinder engines use simultaneous double-fire injector control. Ford introduced its first port EFI system on the 1.6-liter engine in 1983 Escort, Lynx, and EXP/LN7 models. The same basic system continued in use on some later 1.6- and 1.9-liter engines.

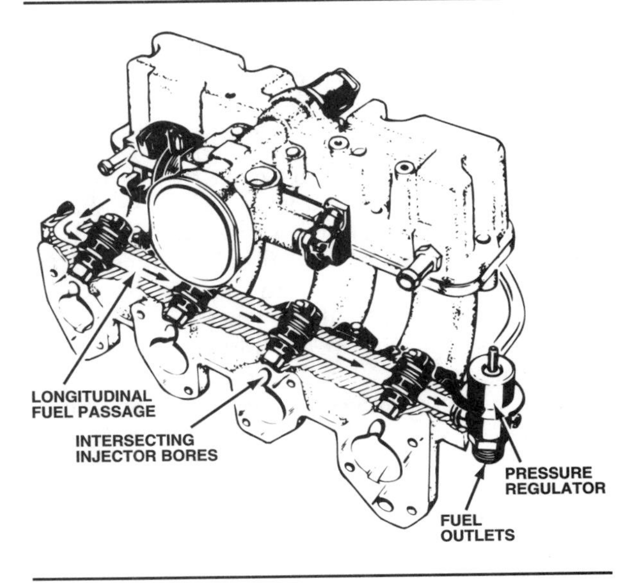

Figure 16-41. The fuel inlet and fuel passage are cast into the lower intake manifold in GM's integrated air fuel system (IAFS). The design requires the use of bottom feed port injectors.

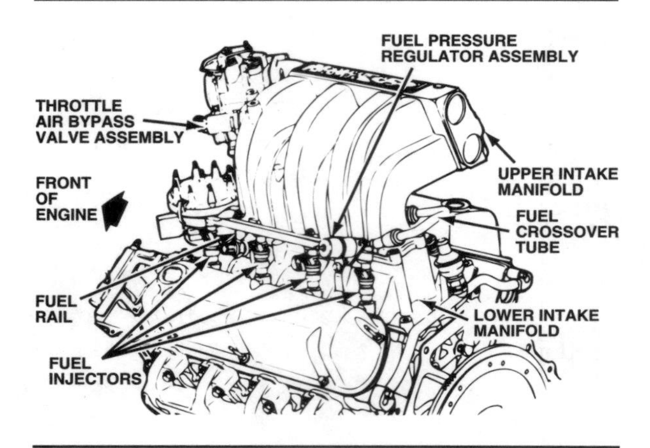

Figure 16-42. A typical Ford SEFI system.

Two new EFI systems were introduced in 1986. Taurus and Sable models with 3.0-liter V6 engines use a port simultaneous injection system similar to the one used on 4-cylinder engines. Starting in 1986, Ford introduced a new speed-density, sequential electronic fuel injection (SEFI) system, figure 16-42. In the SEFI system, the EEC-IV computer operates the injectors in firing order sequence. Each injector fires once every two crankshaft revolutions as the intake valve opens.

Ford EFI systems on 4-cylinder engines originally used a VAF meter to measure intake airflow. Similar to the Bosch L-Jetronic, the VAF meter is installed ahead of the throttle body in the air intake system. The vane moves a potentiometer sensor that sends a voltage signal to the PCM proportionate to intake airflow. When Ford introduced SEFI to its 4-cylinder engines, the VAF meter was replaced by a MAF sensor.

Chrysler

A grouped double-fire PFI system, figure 16-43, is used on some 1984 and later 2.2-liter 4-cylinder engines. This system is similar to a Bosch D-Jetronic system and uses Bosch-designed fuel injectors in combination with Chrysler-engineered electronics. The basic Chrysler 4-cylinder injection systems are speed-density designs.

Starting in 1989, Chrysler gradually has converted its turbocharged 4-cylinder engines, all V6 and most V8 engines to sequential PFI (SMFI). These are either speed-density or mass airflow designs. Figure 16-44 shows a typical SMFI system as used on 3.3- and 3.8-liter V6 engines.

Imports

The major Japanese automakers use PFI systems derived from the Bosch L-Jetronic or air flow controlled (AFC) system. Figure 16-45 shows common components used in both Nissan and Toyota EFI systems. Nissan introduced its system in 1975, with Toyota following in 1979. These early Japanese systems controlled only the injectors. Systems introduced in 1981 (Nissan) and 1983 (Toyota) are fully integrated into the engine management system, controlling idle speed, EGR, and ignition timing in relation to fuel injection.

Late-Model Throttle Body Injection Systems

TBI systems made their first appearance on domestic engines in the early 1980s. Once accounting for more than half of the EFI systems in use, they now are relegated primarily to economy cars and light-duty trucks where cost is important (TBI systems are less expensive to build than PFI systems). PFI and SFI systems are the dominant design, especially with the trend toward distributorless ignitions.

With the exception of the Chrysler V8 system introduced on 1981 Imperials, all TBI systems use similar components. Single or dual solenoid-operated injectors are positioned in a throttle body assembly that owes much of its inspiration to a carburetor base, figure 16-46. A MAP sensor measures intake air and the PCM controls injection volume proportionately.

General Motors

All GM TBI systems are manufactured by the Rochester Division of GM; its throttle body casting contain one or two injectors. There are several basic models in use:

- Models 100, 200, and 220 are two-barrel, two-injector assemblies. The Model 220 was introduced in 1985. Some Model 220 TBI units use injectors which are calibrated to flow fuel at a different rate in each throttle bore.
- Models 300, 500, and 700 are one-barrel, one-injector assemblies. With the exception of minor fuel and airflow differences, the Models 300 and 500 are the same. The Model 700 was introduced on 1987 4-cylinder engines as a replacement for both the 300 and 500.
- Model 400 is a unique crossfire assembly of two one-barrel, one-injector throttle bodies mounted on a single intake manifold, figure 16-47. This system was used on some 1982–84 Corvette, Camaro, and Firebird high-performance V8 engines. Like the Chrysler V8 TBI system, the GM crossfire fuel injection (CFI) system proved to be a technological dead end.

All GM throttle bodies use an idle speed control (ISC) motor or IAC valve, a TP sensor, and a fuel pressure regulator, figure 16-48. The

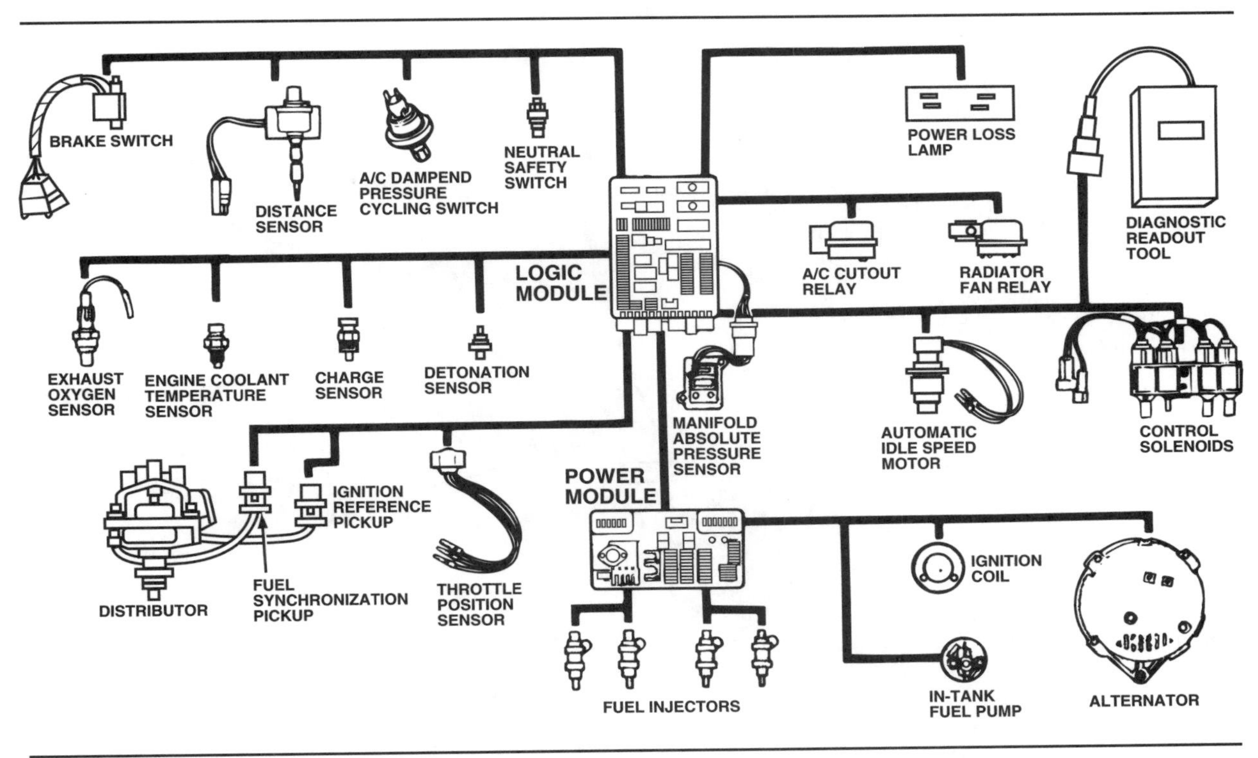

Figure 16-43. The Chrysler grouped double-fire injection system used on 2.2-liter engines.

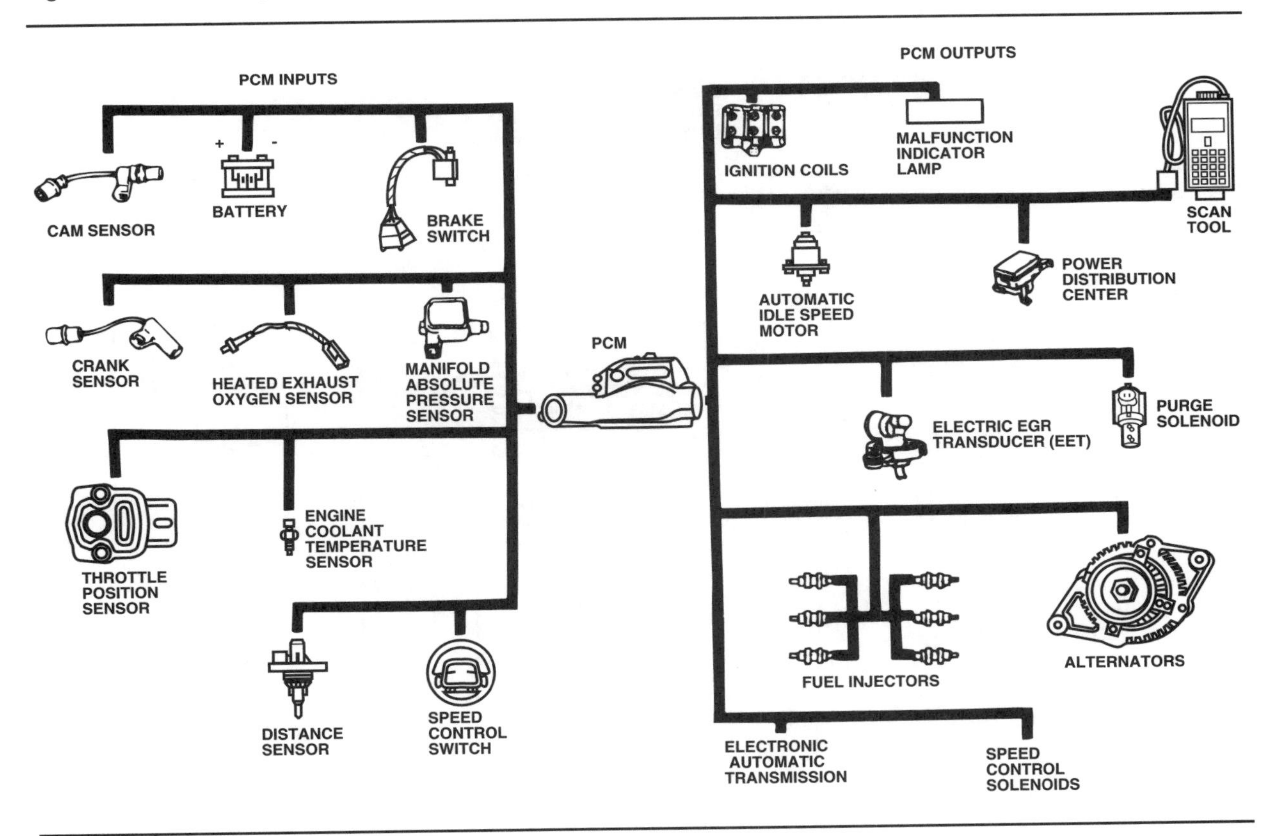

Figure 16-44. The Chrysler sequential multipoint fuel injection system used on 3.3- and 3.8-liter engines.

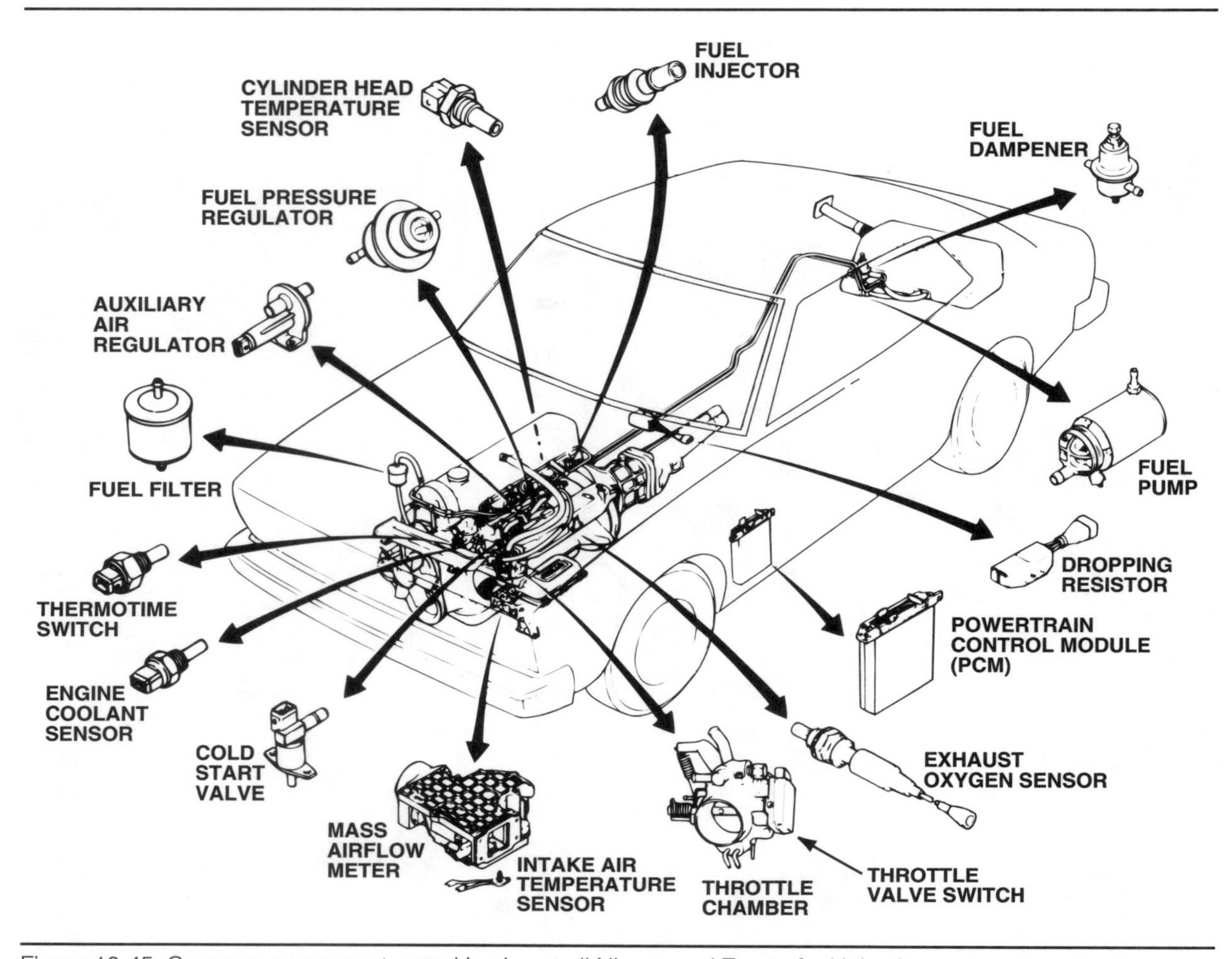

Figure 16-45. Common components used in almost all Nissan and Toyota fuel injection systems.

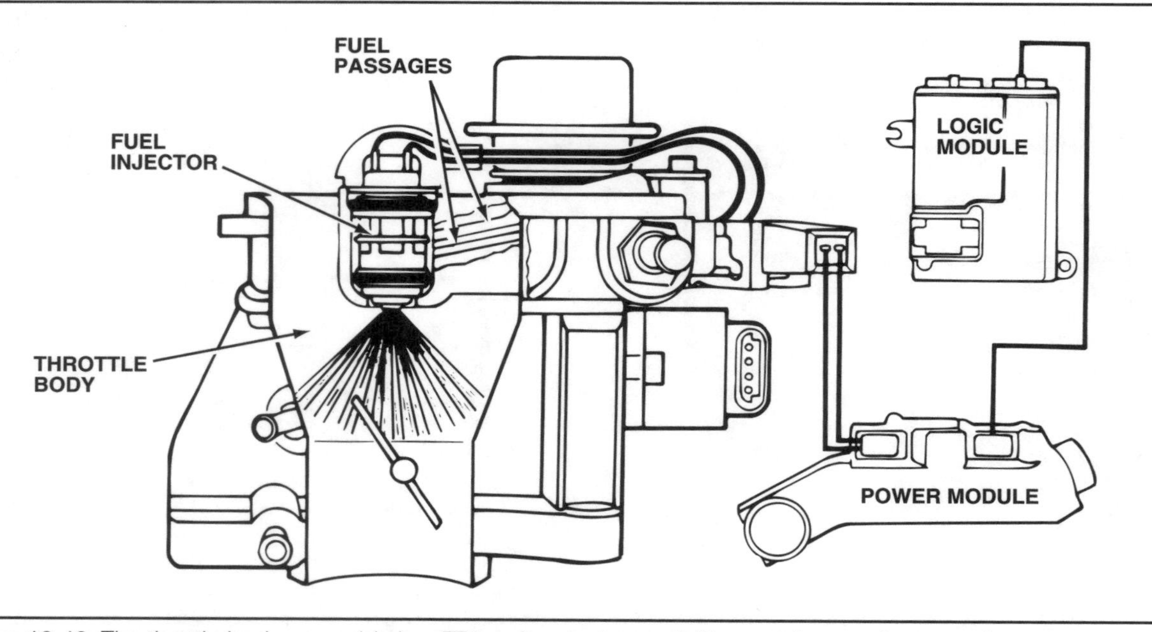

Figure 16-46. The throttle body assembly in a TBI system looks much like a carburetor, but contains only an injector, idle air control motor, and fuel pressure regulator. (Chrysler)

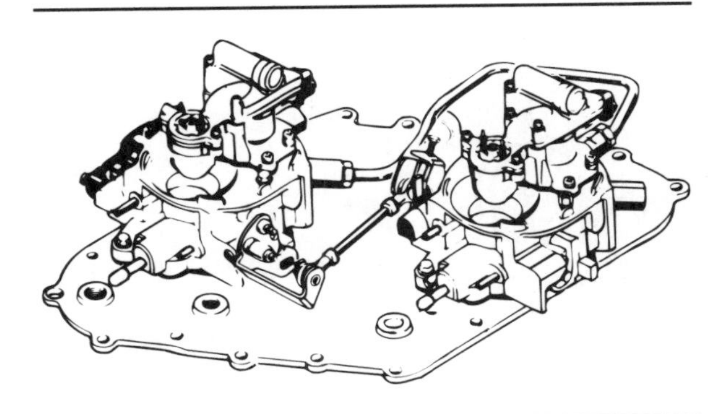

Figure 16-47. The Model 400 cross-fire injection (CFI) assembly consists of two TBI units mounted on a common manifold. (Chevrolet)

regulator senses intake manifold vacuum (or air pressure drop across the injector) at the same point as the injector nozzle tip. The regulator controls fuel pressure by opening and closing the fuel return line.

Ford

Through 1992, Ford called all of its throttle body injection systems central fuel injection (CFI) because the fuel is injected at a central point. Starting with 1993 models, it adopted the TBI designation used by other automakers. The high-pressure, dual injector fuel charging assembly (now called a throttle body), figure 16-49, has been used on V6 and V8 engines. A low-pressure, single injector unit was introduced in 1985 for use on 2.3- and 2.5-liter 4-cylinder engines. Like GM throttle bodies, the Ford TBI units also use TP sensors, ISC motors, and pressure regulators.

Chrysler

Chrysler's first TBI system was a continuous flow system used on its V8 engine in 1981–83 Imperials, figure 16-50. A special sensor in the air cleaner snorkel measures air intake volume, figure 16-51. Radial vanes swirl the air and a U-shaped pressure probe in the vortex of the swirling air determines the amount of air entering the engine.

Most 1981 and later nonturbocharged Chrysler-built 4-cylinder engines use a single injector TBI system. The original system used through 1985 had a Holley-designed throttle body, figure 16-52, with an injector supplied by Bendix. The TBI system was revised in 1986, with the newer version using a low-pressure injector designed by Bosch. The ball-type injector used through 1988 was replaced by a pintle-type injector in 1989 and later TBI units.

Imports

Most import vehicles equipped with TBI use systems that are quite similar in design and operation to those used on domestic vehicles. A

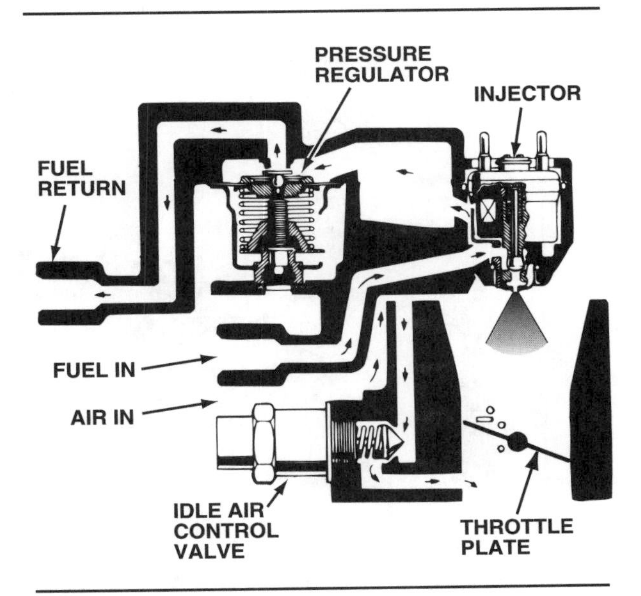

Figure 16-48. Throttle body injection fuel pressure regulation. (GM)

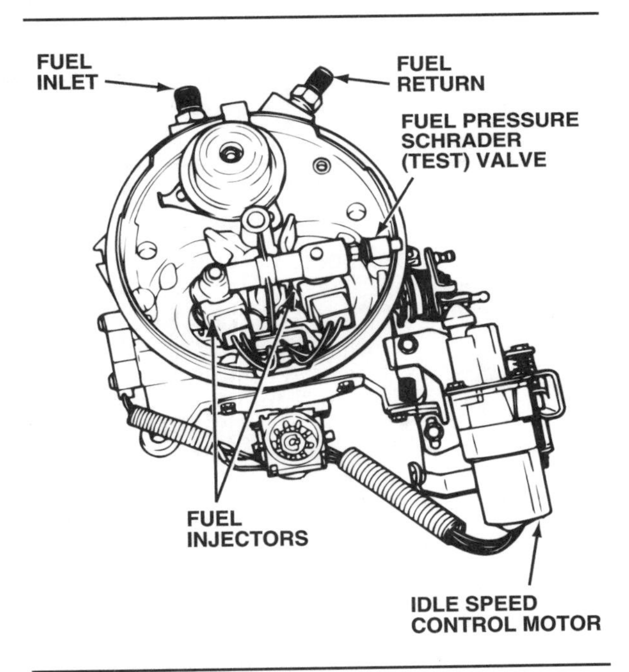

Figure 16-49. A Ford CFI fuel charging (throttle body) assembly. (Ford)

combination of Bendix and Renix components, the AMC/Renault system, figure 16-53, was a typical example.

Less typical is the TBI system used on turbocharged Japanese vehicles sold by Mitsubishi and Chrysler Corporation in the United States. This system measures intake airflow rate based on Karman's vortex theory, explained in the "Electronic Engine Control: Sensors, Actuators,

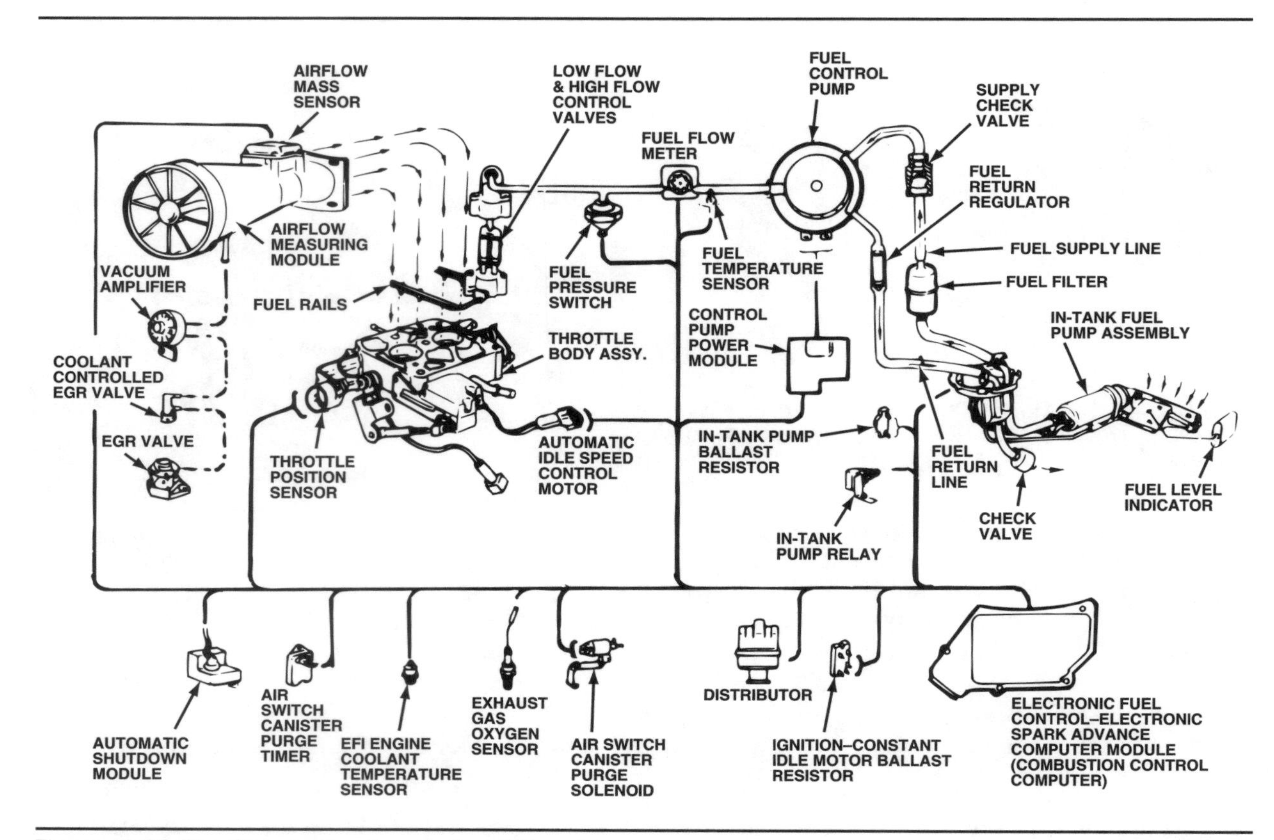

Figure 16-50. The Chrysler continuous-flow TBI system.

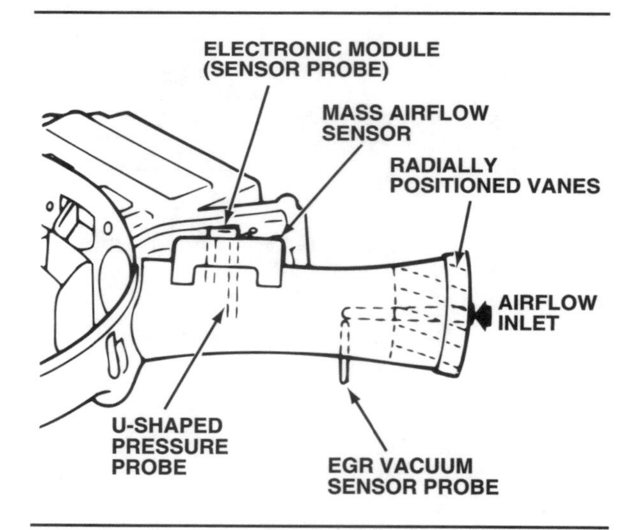

Figure 16-51. Chrysler's continuous-flow TBI airflow sensor.

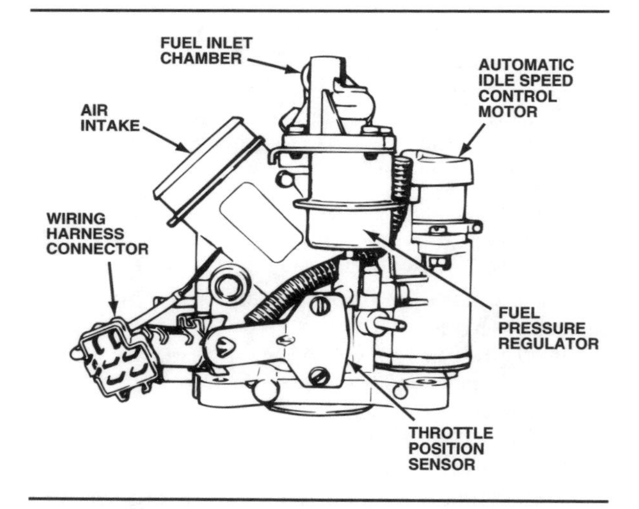

Figure 16-52. The Chrysler-Holly-Bendix high-pressure TBI unit. (Chrysler)

and Operation" chapter. Two injectors installed in the air intake throttle body are fired in an alternating sequence. A swirl nozzle design atomizes the fuel and "jiggles" the fuel spray, figure 16-54. This system is considerably more complex than the other TBI systems which we have seen.

FUEL INJECTION INTAKE MANIFOLD SYSTEMS

Fuel injected engines with TBI have manifolding requirements similar to those of carbureted engines. The TBI unit sits on the intake manifold, and its injector or injectors spray fuel into the manifold to be mixed with incoming air.

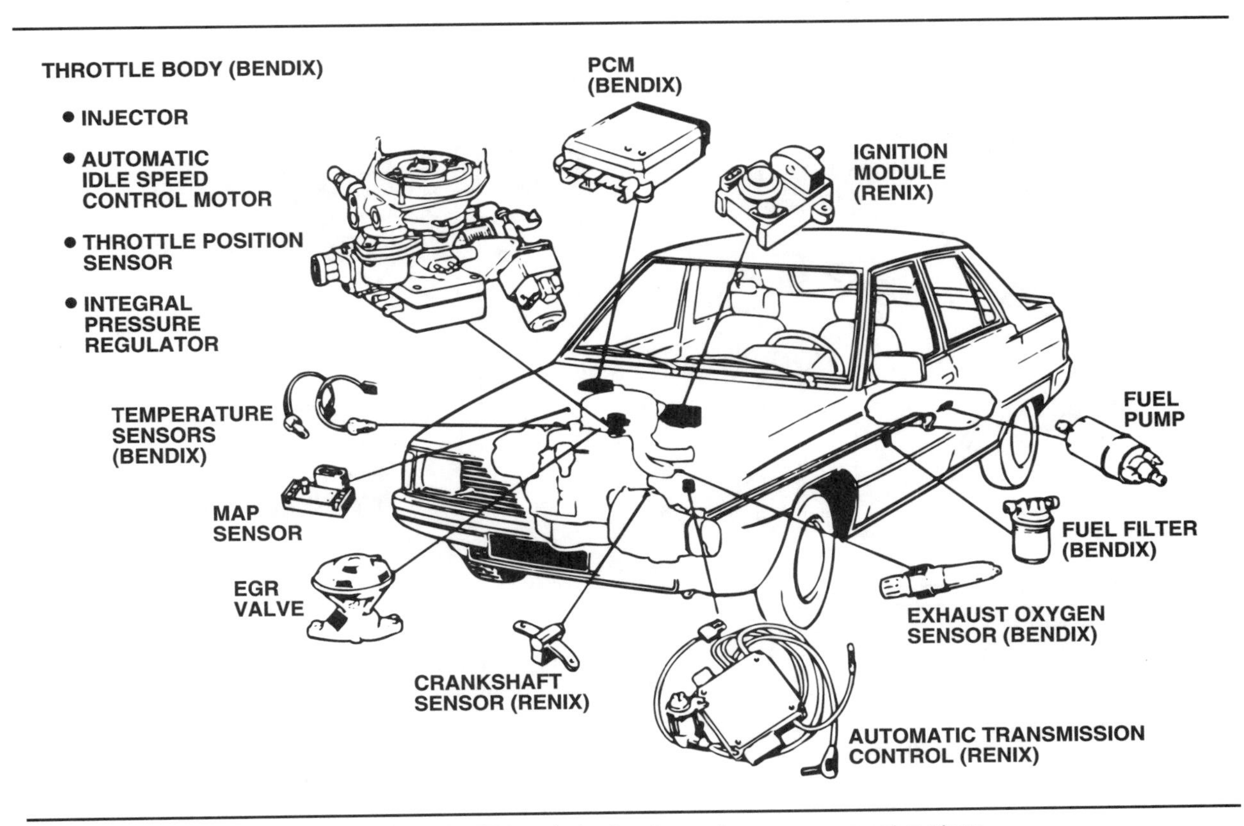

Figure 16-53. The major components of a TBI and electronic engine management system.

Theoretically, the injector sprays fuel into the area of maximum air velocity to ensure thorough atomization and ideal distribution. Some two-bore TBI units, however, use individually calibrated injectors to assist in proper distribution. When two TBI units are installed, as in the GM Cross Fire Injection system, the rear injector must be calibrated differently than the front injector for better distribution.

Fuel injected engines with PFI systems require special manifold designs. These may be one- or two-piece manifolds. The one-piece design is used with some V-type engines and has the plenum, intake passages, and runners within a single unit, figure 16-55. The two-piece design is used with in-line and some other V-type engines, figure 16-56, and consists of the upper intake manifold or plenum and a lower intake manifold with individual runners.

Since the PFI system does not deliver fuel through the manifold, each separate runner can be specifically tailored in size and length (ram-tuned). This helps induction flow to increase cylinder charging and improves fuel distribution among the cylinders. It also eliminates the need for manifold heating or inlet air heating, resulting in a denser air charge. Thus the name "ram tuning".

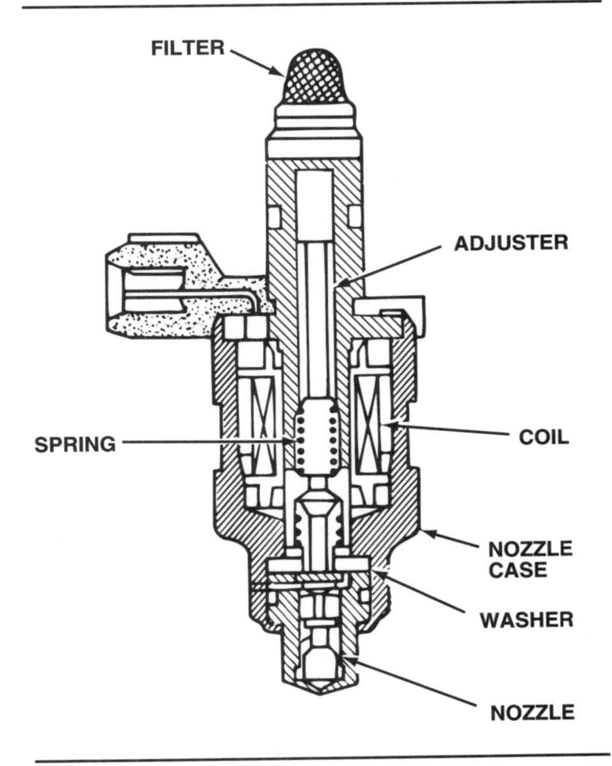

Figure 16-54. The Mitsubishi-Chrysler TBI injector uses a swirl nozzle for better fuel atomization, but operates like any other solenoid-actuated injector. (Chrysler)

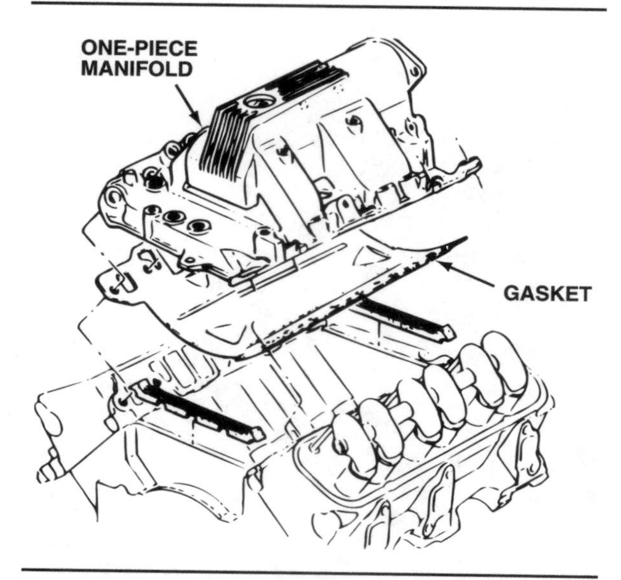

Figure 16-55. The one-piece manifold used on V-type engines with multi-point fuel injection incorporates the plenum, intake passages, and runners. (Buick)

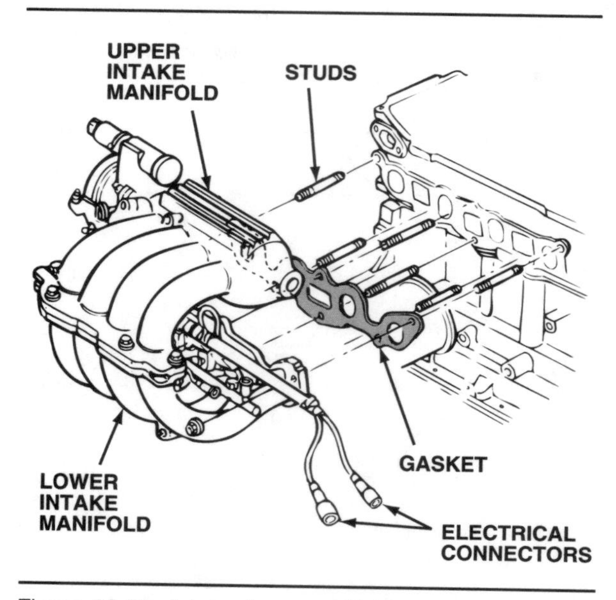

Figure 16-56. A two-piece manifold used with in-line and some V-type engines with multi-point fuel injection has an upper manifold (plenum) and lower manifold with individual runners. (Ford)

In a two-piece manifold design, airflow entering the throttle body passes into the upper intake manifold or intake plenum where it is distributed to the individual runners. These runners are designed to a specific length according to cylinder requirements. They route the airflow directly to the individual ports where it mixes with the fuel charge just before entering the combustion chamber. The difference in perfor-

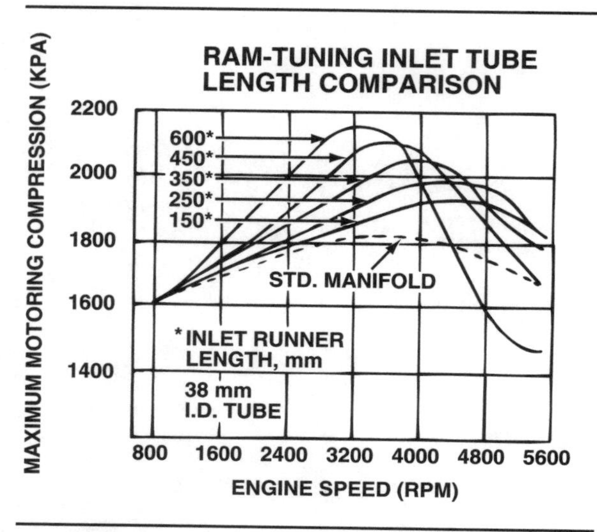

Figure 16-57. This graph shows the effect of ram-tuning the manifold runners. (Rochester)

mance between a standard manifold and the ram-tuned inlet runners on the Chevrolet 2.8-liter V6 is shown in figure 16-57. The 600-mm inlet runner, as an example, develops 20 percent more compression, and therefore power, than the standard manifold.

The one-piece manifold works in basically the same way, except that everything is built into the casting. The length of the intake passages and the shape and size of the main intake runner are designed to produce a denser air mass at each cylinder just before the intake stroke.

Some port fuel injected systems do not use an intake air temperature control system. The density of the air charge is increased by eliminating the heated inlet air and manifold heating. Heat is not as necessary for this design because there is no requirement to vaporize the fuel in this type of manifolding. However, many port-injected systems do use inlet air and manifold heat to speed engine warm-up.

Unlike carbureted or TBI systems, the PFI manifold must provide a place to mount the injectors and fuel rail assembly. This requires precision casting or drilling to properly locate each injector at its required position. Injectors generally are retained in the manifold by clamps and sealed with O-rings. The fuel rail is attached with capscrews.

Variable induction systems

Intake manifold design has always been a compromise. Traditionally, a manifold designer had to develop a manifold that produced best power at either high rpm, low rpm, or somewhere in between. The variable induction systems on some port fuel injection systems have changed this. These manifolds use multiple runners and

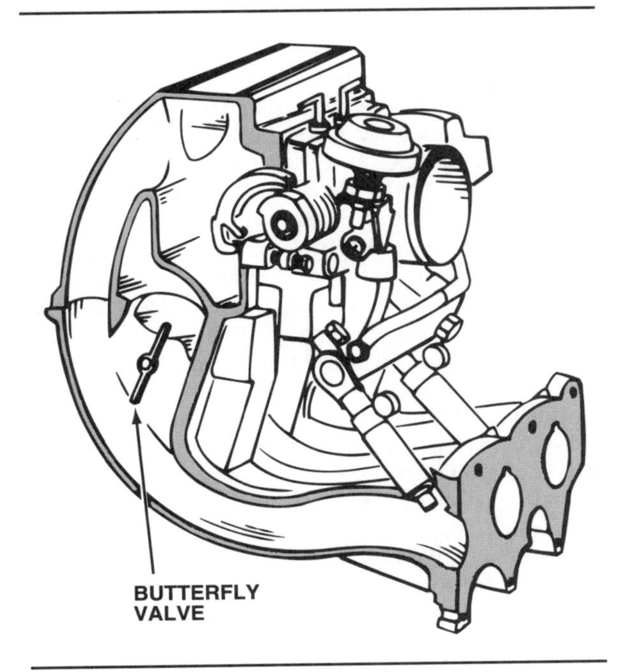

Figure 16-58. Mazda's VRIS system uses an inertia charge effect to improve torque output through interconnected, branched intake paths equipped with butterfly valves.

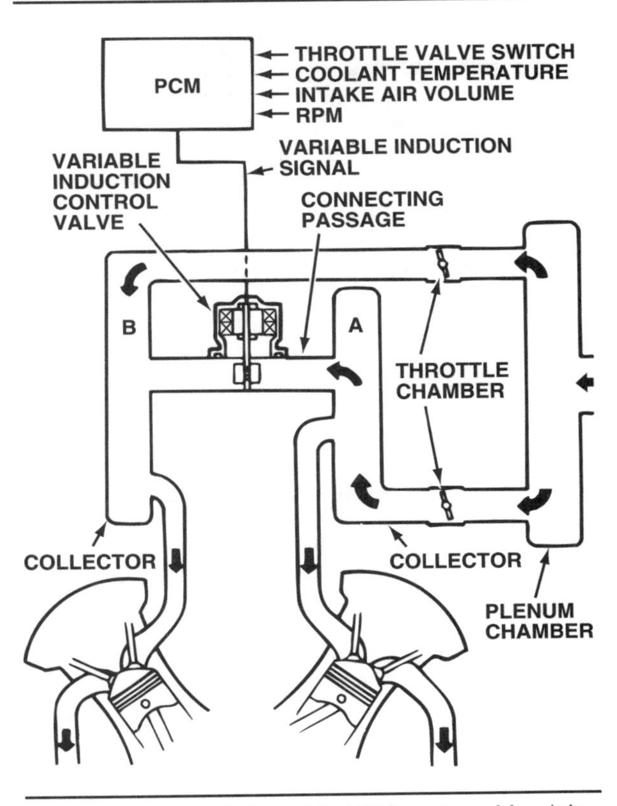

Figure 16-59. A variation of its VRIS system, Mazda's VICS system is used on 1.8-liter engines in some Ford Escorts and Tracers. (Ford)

additional throttle valves that, with the help of electronic controls, open at a specified rpm to change the manifold volume and length, providing power over very wide rpm bands. Some variable induction systems also work in conjunction with variable camshaft timing systems.

The tuned length intake runners used in the fuel injection manifolds just described are a low-cost alternative to a multivalve head for increasing an engine's volumetric efficiency. But the increasing number of 4-valve head designs appearing in production are being fitted with equally innovative manifolds designed to fine-tune air induction under the control of the PCM (also leading to greater volumetric efficiency). The 4-valve head designs generally use multiple camshafts and may provide variable valve timing, as required by engine speed and load. To take full advantage of this ability, numerous variable air induction systems have come off the drawing board and into production.

One Nissan design feeds each cylinder through two intake ports to two intake valves. One port in each pair has a butterfly valve which remains closed up to about 3,000 rpm in order to close off one of the intake paths. The velocity of the air flowing through the single intake path is, therefore, greater and promotes a high swirl effect. Above 3,000 rpm, the butterfly opens the other intake path and feeds the additional air needed at high engine rpm.

Mazda has been especially innovative in variable induction system design. One design is called a variable ram-effect induction system (VRIS). It uses a common plenum chamber with long curved runners (primary ports) for each cylinder. Each runner also has a second shorter branch (secondary ports) containing a butterfly valve, figure 16-58. The branches are interconnected by a passage and the valves remain closed below about 5,000 rpm. Above that speed, the PCM activates a vacuum actuator through a solenoid to open the secondary port control valve. This shortens the effective length of the passage and interconnects all of the branches. Closing the intake valves creates a positive pressure in the system, which has a domino effect moving from one cylinder to another to pack in more air. Since the interconnecting passage reduces air intake resistance, flow speed is reduced.

Another Mazda design used on Ford Escort and Tracer 1.8-liter engines is called a variable inertia charging system (VICS). This system also uses a common plenum chamber, but is designed to increase the length of the air intake path in the intake manifold at engine speeds above 5,000 rpm. The result is greater engine torque with a wider torque band at high engine speeds. The system consists of four normally-closed shutter valves mounted on a single rod inside a dual port intake plenum. The rod is externally connected to a vacuum-operated shutter valve actuator mounted on the passenger side of the plenum. A manifold port below the shutter valve actuator routes vacuum to a vacuum storage canister. The canister contains a check valve to prevent vacuum backflow and is connected to a three-way solenoid valve controlled by the PCM.

The system applies vacuum to the shutter valve actuator to hold the shutters closed at engine speeds below 5,000 rpm. When this engine speed is reached, the PCM signals the actuator to vent vacuum to the atmosphere. This opens the shutter valves, increasing the air intake path and altering the resonance-induced inertia charging effect. When engine speed drops below 5,000 rpm, the PCM reapplies vacuum to the actuator to close the shutter valves. The vacuum storage canister provides vacuum to hold the shutter valves closed during momentary periods of low vacuum, as during high engine load or wide open throttle conditions.

Nissan has modified its VG30 V6 engine used in several vehicles to incorporate variable valve timing and air induction controlled by the PCM, figure 16-59. Dual camshafts are used for each bank of cylinders, one for intake and one for exhaust valves. The intake cam has a helical gear and hydraulic coupling that changes intake valve timing according to a PCM-controlled solenoid.

The air induction system is just as ingenious. Air flows into a single plenum chamber, where it passes into a pair of collector boxes. Each collector box contains a throttle chamber which reduces air resistance at high rpm and load conditions. An interconnecting passage between the collector boxes contains a power valve controlled by the PCM. At low and mid-range rpm, this valve remains closed to lengthen the intake path. As rpm increases, the valve opens to connect the two collector boxes and increase the volume of airflow.

Toyota also builds several four-valve engines with dual-level, variable intake manifolds as part of the PCM-controlled fuel injection system. The Honda-Acura-Sterling V6 has a dual intake system with 12 manifold runners for six cylinders. A solenoid-controlled vacuum diaphragm opens the secondary runners for more airflow at high speed.

The Port-Throttle system used with Ford's Mark VIII 4.6-liter double overhead camshaft (DOHC) engine shows the design direction domestic automakers are likely to take. An integral part of the intake manifold is that the air intake manifold contains dual throttle plates, greatly increasing the amount of air entering the manifold. Throttle plate opening is staggered and operates much like a two-stage carburetor. The primary throttle plate opens about 25 degrees before the secondary plate starts to open, but both arrive at wide open throttle together, figure 16-60.

The intake manifold uses one long primary runner per cylinder. These primary runners provide constant airflow to the primary intake valves. Primary intake port shape and intake valve masking combine to induce swirl in the air-fuel mixture. A second short runner per cylinder contains a closed throttle plate controlled by the PCM. These short runners flow air to the secondary intake valve in each cylinder at speeds above approximately 3,200 rpm. Since these runners are shorter, the air passing through the secondary intake valves flows faster and without swirl. This helps control the combustion sequence by gradually changing the high swirl motion of the air-fuel mixture at lower rpm into a tumbling motion at higher rpm. The result is similar to using two completely independent low and high-speed induction systems.

By providing water crossover through a separate tube instead of using cast passages in the intake manifold, air entering the manifold remains cooler and denser. The cooler temperature combined with the higher air-fuel mixture velocity result in higher mass flow for better engine performance.

While such induction designs are increasing, the intent behind each is the same—to maintain engine torque across the power band—regardless of the different ways in which the engineers choose to implement it.

The important principles of split or dual-level intake manifolds are that the primary (low-speed) runners are longer and narrower than the secondary (high-speed) runners. Long, small cross section runners increase airflow velocity at low engine speed to deliver more air, faster at engine speeds below approximately 3,000 rpm. Short, large-cross-section secondary runners allow airflow to travel a shorter distance at engine speeds above 3,000 rpm. This delivers more air at higher speed than long, narrow runners can. Thus, dual-level manifolds avoid the traditional compromises of intake manifold design.

SUMMARY

Fuel injection systems must do the same tasks as carburetors. However, they do so in a slightly different way. Two types of injection systems are used: throttle body injection (TBI) and port fuel injection (PFI).

The TBI system uses a carburetor-like throttle body containing one or two injectors. This throttle body is mounted on the intake manifold in the same position as a carburetor. Fuel is injected above the throttle plate and mixed with incoming air.

In a port system, individual injectors are installed in the intake manifold at a point close to the intake valve where they inject the fuel to mix with the air as the valve opens. Intake air passes through an air intake throttle body containing a butterfly valve or throttle plate and travels through the intake manifold where it meets the incoming fuel charge at the intake valve. The difference in injector location permits more advance manifold designs in port systems to aid in fuel distribution.

Fuel injection has numerous advantages over carburetion. It can match fuel delivery to engine requirements under all load and speed conditions, maintain an even mixture temperature, provide more efficient fuel distribution, and reduce emissions while improving driveability.

Electronic fuel injection (EFI) is far more common than the earlier mechanical injection represented by the Bosch K-Jetronic. EFI integrates the injection system into a complete engine management system which includes control of EGR, ignition timing, canister purging, and various other functions.

The powertrain control module (PCM) uses various sensors to gather data on engine operation. Intake air volume is determined by measuring airflow speed, manifold pressure, or the mass of the air drawn into the engine. This data is added to information about throttle position, idle speed, engine coolant and air temperature, crankshaft position, and other operating conditions. After processing the data, the PCM signals the solenoid-operated injectors when to open and close. Injectors may be fired in various combinations, but all use the principle of pulse width modulation, in which the PCM varies the percentage of on-time in each full on/off cycle according to engine requirements. In this way,

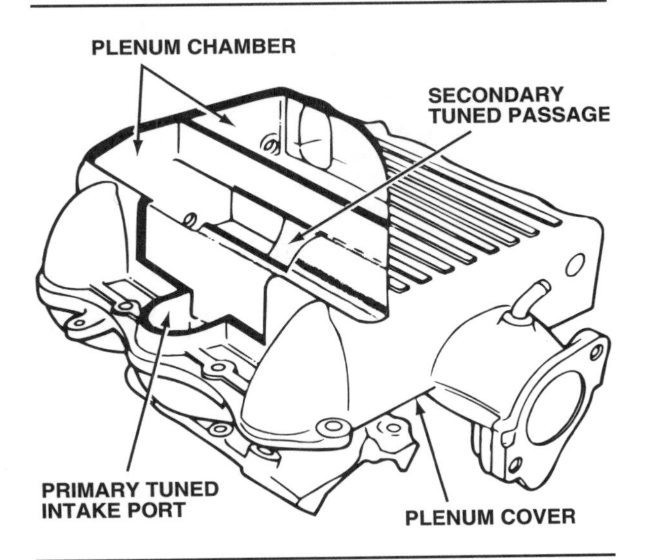

Figure 16-60. Secondary tuning in this Chrysler manifold is provided by a large-diameter passage inside the plenum chamber.

the amount of fuel injected can be varied instantly to accommodate changing conditions.

Port fuel injection (PFI) systems have special manifold requirements. Since this type of injection system does not deliver fuel through the manifold, each separate runner can be "tuned" to increase efficiency and ensure better distribution. PFI systems do not use intake air temperature control systems. They incorporate provisions for mounting the injectors and fuel rail assembly.

Intake manifold designs favor good low-to-middle speed operation, good high-speed operation, or a compromise of the two. A variety of variable air induction systems is beginning to appear on 4-cylinder engines, especially on those from Japan. While they differ in their approach, all are designed primarily to maintain torque across the engine's power band.

Review Questions

Choose the single most correct answer.
Compare your answers to the correct answers on page 507.

1. Fuel injection systems can lower emissions:
   a. By matching the air-fuel ratio to engine requirements
   b. Only at high speeds
   c. By using the intake manifold to vaporize fuel
   d. By matching engine speed to load conditions

2. All of the following are common kinds of gasoline fuel injection systems EXCEPT:
   a. Multipoint injection
   b. Throttle body injection
   c. Direct cylinder injection
   d. Continuous injection

3. Technician A says that Bosch K-Jetronic systems are mechanical injection systems that operate on differential fuel pressure.
   Technician B says that Bosch K-Jetronic systems are intermittent injection systems.
   Who is right?
   a. A only
   b. B only
   c. Both A and B
   d. Neither A nor B

4. Late-model Bosch L-Jetronic systems and similar systems measure airflow with a:
   a. MAP sensor
   b. Hot-wire MAF sensor
   c. VAF meter
   d. Intake air temperature sensor

5. Technician A says that TBI systems are electronic, intermittent injection systems.
   Technician B says that TBI systems require fuel pressure regulators.
   Who is right?
   a. A only
   b. B only
   c. Both A and B
   d. Neither A nor B

6. Technician A says that the throttle in a multipoint system is located between the injectors and the manifold.
   Technician B says that the throttle in an EFI system operates the same as a throttle on a carbureted engine.
   Who is right?
   a. A only
   b. B only
   c. Both A and B
   d. Neither A nor B

7. Mechanical fuel injectors are operated:
   a. Continuously by fuel pressure
   b. Intermittently by the computer
   c. Alternately by air pressure
   d. Sequentially by firing order

8. Modern fuel injection systems are based on work begun by:
   a. Rochester Products, Division of GM
   b. Ford Motor Company
   c. Hitachi
   d. Robert Bosch GmbH

9. A cold-start injector is a:
   a. Temperature-controlled detonation device
   b. Solenoid-operated auxiliary injector
   c. Mechanical intermittent injector
   d. Dual-fire electronic actuator

10. Multec injectors used in a multipoint injection system differ from those produced by Robert Bosch in that they use a:
    a. Pintle valve
    b. Swirl nozzle
    c. Pin and spring
    d. Ball-seat valve with director plate

11. All of the following are filter locations in a fuel injection system EXCEPT:
    a. In the pressure regulator
    b. At the fuel pump
    c. In the fuel line
    d. In the fuel injectors

12. Technician A says that an EFI system can use an on/off throttle position switch.
    Technician B says that a throttle position sensor is a variable resistor.
    Who is right?
    a. A only
    b. B only
    c. Both A and B
    d. Neither A nor B

13. A hot-wire or heated-film MAF sensor measures:
    a. Airflow velocity entering the engine
    b. Barometric absolute pressure
    c. Air temperature in a turbocharged system
    d. Rate of mass airflow

14. Variable-induction manifolds differ in design, but all have one common goal which is:
    a. To reduce production costs
    b. To reduce engine weight
    c. To maintain torque across the power band
    d. To help engine designs maintain a low hood profile

Chapter 17

Supercharging and Turbocharging

Engines with carburetors rely on atmospheric pressure to push an air-fuel mixture into the combustion chamber vacuum created by the downstroke of a piston. The mixture is then compressed before ignition to increase the force of the burning, expanding gases. The greater the mixture compression, the greater the power resulting from combustion. In this chapter, we study three ways to increase mixture compression: raising the mechanical compression ratio, adding a supercharger, or adding a turbocharger.

ENGINE COMPRESSION

One way mixture compression can be increased is by using a higher mechanical compression ratio. During the late 1960s, high-performance car engines used compression ratios as great as 12:1. However, compression ratios dropped to the range of 8:1 or 8.5:1 during the 1970s because of increasing emission control requirements, and the tendency of higher-compression engines to emit too much NO_x. Lower compression ratios were also necessary due to the greatly reduced lead content in gasoline. Using electronic engine management systems, however, makes it possible to raise compression ratios again to the 9:1 or 10:1 range.

Two major benefits of high compression ratios are:

- They increase volumetric efficiency because the piston displaces a larger percentage of the total cylinder volume on each intake stroke
- They increase thermal efficiency because they raise compression temperatures, resulting in more-complete combustion.

Although high compression ratios have advantages, they also have major disadvantages:

- Because the compression ratio remains unchanged throughout the engine's operating range, combustion temperature and pressure also are high, and cause increased emissions during deceleration, idle, and part-throttle operation.
- A high compression ratio increases NO_x emissions.
- High compression ratios require high-octane gasoline with effective antiknock additives.

Lead was the most effective antiknock additive, but is poisonous, produces harmful emissions, and destroys catalytic converters, so it is not used anymore.

THE BENEFITS OF AIR-FUEL MIXTURE COMPRESSION

The amount of force an air-fuel charge produces when it ignites is largely a function of the intake charge density. Density is the mass of a substance

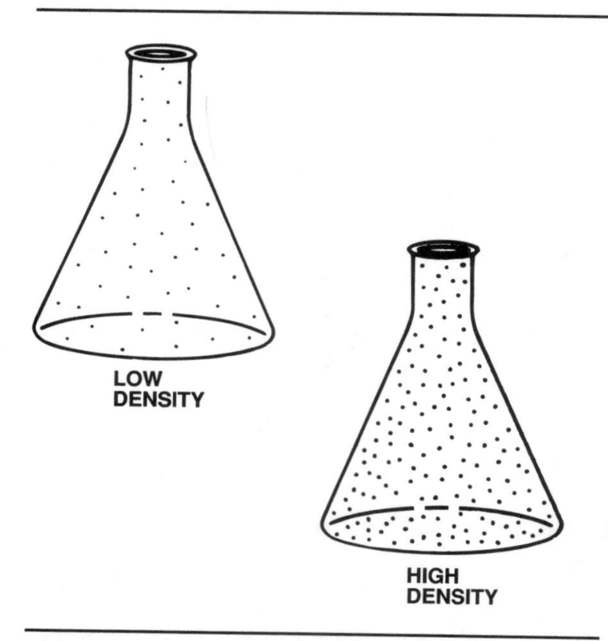

Figure 17-1. The more air and fuel that can be packed in a cylinder, the greater the density of the air-fuel charge. (Ford)

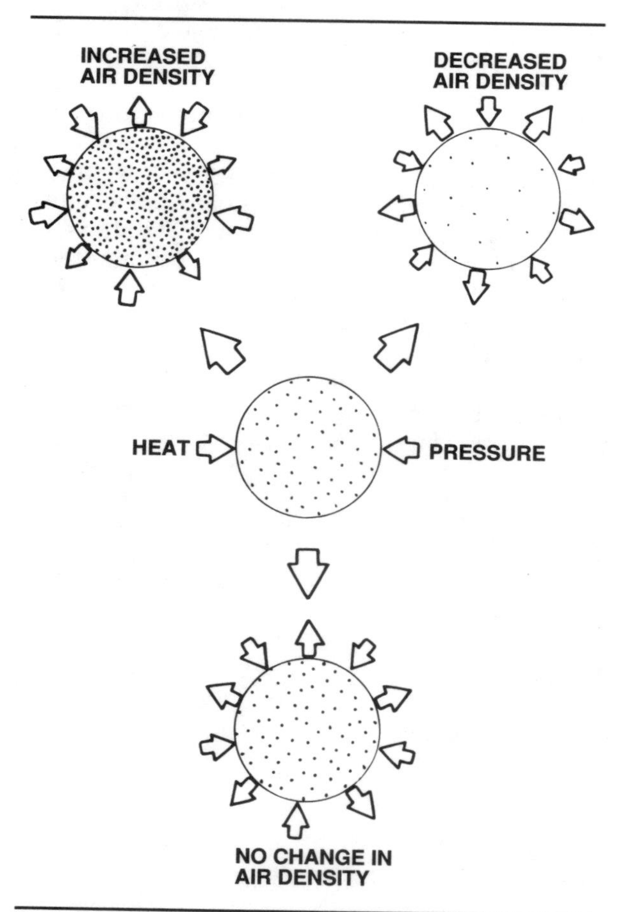

Figure 17-2. The effects of heat and pressure on density. (Ford)

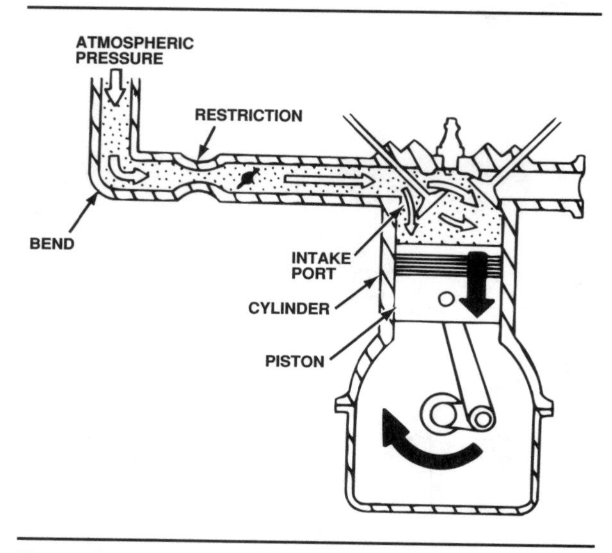

Figure 17-3. Air intake operation in a naturally aspirated engine. (Ford)

in a given amount of space, figure 17-1. The greater the density of the air-fuel charge forced into a cylinder, the greater the force it produces when ignited, and the greater the engine power.

Atmospheric pressure is actually a poor choice to push air through the intake system, as it has a predetermined limit to its efficiency. An engine that uses atmospheric pressure for intake is called a **naturally aspirated** engine. A better way to increase air density in the cylinder is to use a pump. When air is pumped into the engine, pressure and temperature assume a far more important role in determining density than in a naturally aspirated engine.

When air is pumped into the cylinder, the combustion chamber receives an increase of air pressure known as **boost.** This boost in air pressure generally is measured in pounds per square inch (psi) or bar. While boost pressure increases air density, friction heats air in motion and causes an increase in temperature. This increase in temperature works in the opposite direction by decreasing air density. Because of these and other variables, an increase in pressure does not always result in greater air density, figure 17-2.

SUPERCHARGING

Another way to achieve an increase in mixture compression is **supercharging**. This method uses a pump to pack a denser air-fuel charge into the engine's cylinders. Since the density of the air-fuel charge is greater, so is its weight—and power is directly related to the weight of an air-fuel charge consumed within a given time period. The result is similar to a high compression ratio, but the effect can be controlled during idle and deceleration to avoid high emissions.

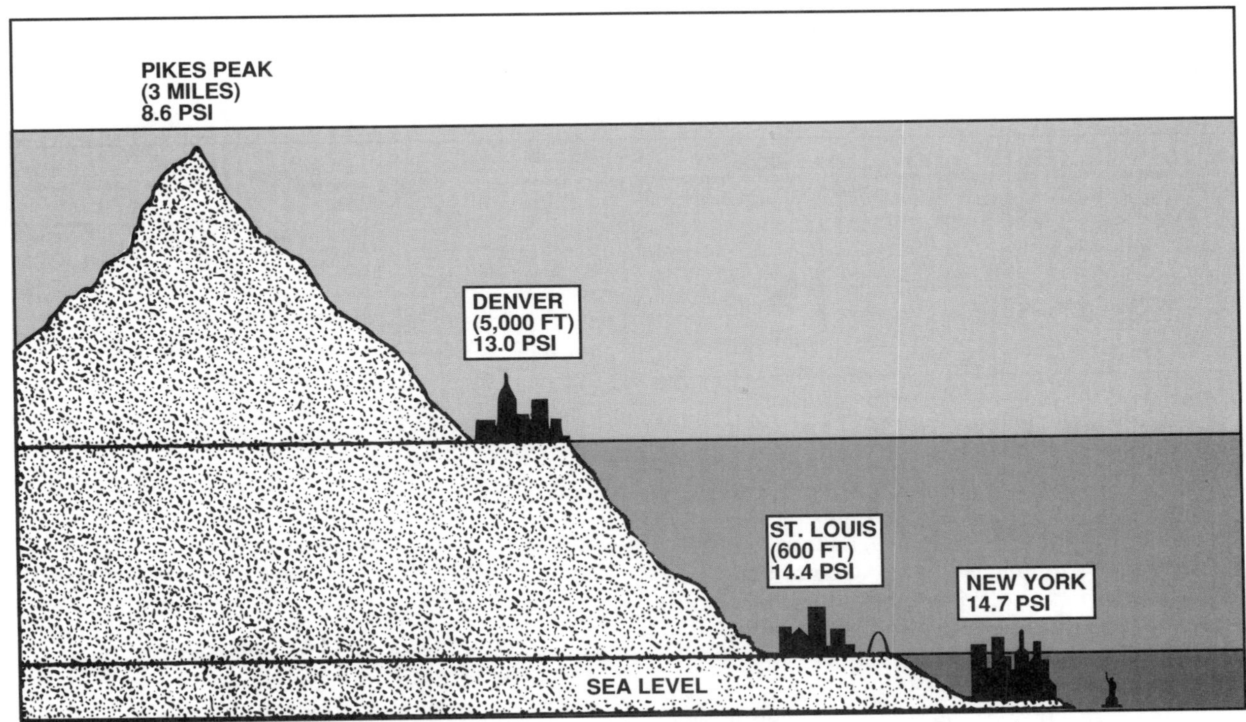

Figure 17-4. Air is a substance and has weight, which causes atmospheric pressure. Atmospheric pressure decreases as altitude increases. (Ford)

Supercharging Principles

As you recall from your study of how an automotive engine breathes (Part One of this manual), atmospheric pressure pushes air into the low-pressure area of the intake manifold on naturally aspirated engines. The low pressure or vacuum in the manifold results from the reciprocating motion of the pistons. When a piston moves downward during the intake stroke, it creates an empty space, or vacuum, in the cylinder. Although atmospheric pressure pushes air to fill as much of this empty space as possible, it has a difficult path to travel. The air must pass through the air filter, the carburetor or throttle body, the manifold, and the intake port before entering the cylinder. Bends and restrictions in this pathway limit the amount of air reaching the cylinder before the intake valve closes, figure 17-3. Consequently, the engine's volumetric efficiency is considerably less than 100 percent. Volumetric efficiency, as we have seen, is a percentage measurement of the volume of air drawn into a running engine compared to the maximum amount the engine could draw in (total displacement volume). While a stock engine averages approximately 70 to 80 percent volumetric efficiency, racing engines with the proper intake and exhaust tuning often can exceed 100 percent volumetric efficiency at their power curve peak.

The volumetric efficiency of any naturally aspirated engine is related to the density of the air drawn into it. Since atmospheric pressure and air density are greatest at (or below) sea level, both pressure and density normally decrease as altitude above sea level increases. For example, atmospheric pressure at sea level is about 14.7 psi (101 kPa); at higher elevations, atmospheric pressure may be only 8 or 9 psi (55 or 62 kPa), figure 17-4. Therefore, the volumetric efficiency of any engine will be greater at sea level than at an altitude above sea level.

Pumping air into the intake system under pressure forces it through the bends and restrictions at a greater speed than it would travel under normal atmospheric pressure, allowing more air to enter the intake port before it closes, figure 17-5. By increasing the engine's air intake in this manner, more fuel can mix with the air

Naturally Aspirated: An engine that uses atmospheric pressure and the normal vacuum created by the downward movement of the pistons to draw in its air-fuel charge. Naturally Aspirated engines are not supercharged.

Boost: A measure of the amount of air pressure above atmospheric that a supercharger or turbocharger can deliver. Boost remains constant regardless of altitude.

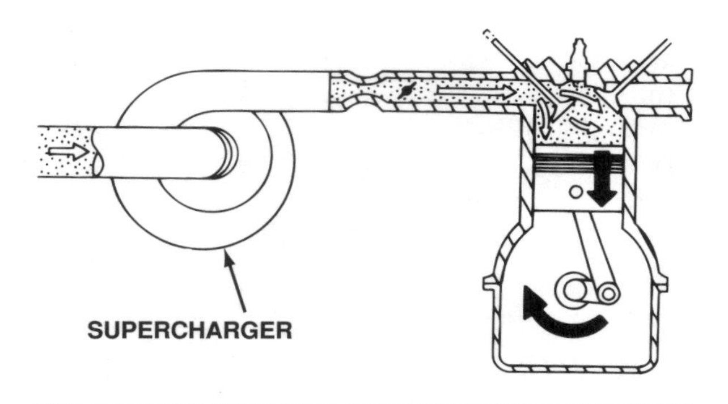

Figure 17-5. Air intake operation in a supercharged or turbocharged engine. (Ford)

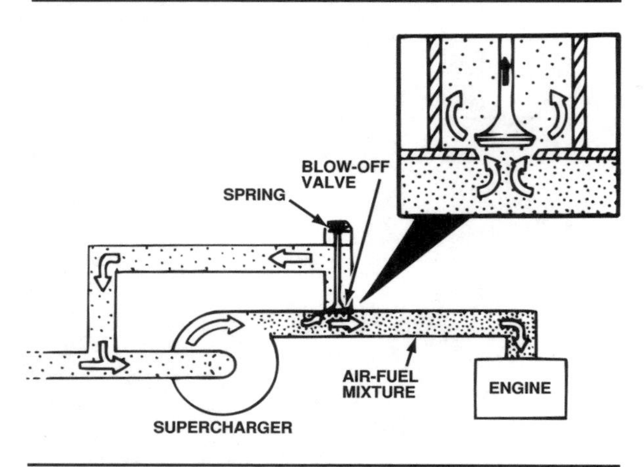

Figure 17-6. Blowoff valve operation. (Ford)

while still maintaining the same air-fuel ratio. The denser the air-fuel charge entering the engine during its intake stroke, the greater the potential energy released during combustion. In addition to increased power resulting from combustion, there are several other advantages of supercharging an engine:

- It increases the air-fuel charge density to provide high-compression pressure when power is required, but allows the engine to run on lower pressures when additional power is not required.
- The pumped air pushes the remaining exhaust from the combustion chamber during intake and exhaust valve overlap.
- The engine burns more of the air-fuel charge, lowering emissions.
- The forced airflow and removal of hot exhaust gases lowers the temperature of the cylinder head, pistons, and valves, and helps extend the life of the engine.

A supercharger pressurizes air to greater than atmospheric pressure. The pressurization above atmospheric pressure, or boost, can be measured in the same way as atmospheric pressure, which we studied in an earlier chapter. Atmospheric pressure drops as altitude increases, but boost pressure remains the same. If a supercharger develops 12 psi (0.8 bar) boost at sea level, it will develop the same amount at a 5,000-foot (1.5-km) altitude.

As we saw earlier, when air is both compressed and heated, its density may increase, decrease, or not change at all, figure 17-2. When a supercharger compresses the air it pumps into an engine, the friction caused by turbulence also heats the air. If the intake temperature becomes too great, the overheated air-fuel charge can cause premature detonation in the combustion chamber and result in engine damage. Superchargers generally use two devices to prevent overheating the air-fuel charge and premature detonation.

To prevent the air-fuel charge from overheating, it must be cooled down before reaching the combustion chamber. This is the function of an **intercooler,** or heat exchanger, installed between the supercharger and the engine. The intercooler may use air or engine coolant as a cooling medium. Air-to-liquid intercoolers are heavier and more complicated than the air-to-air type. How much the air or air-fuel charge cools as it passes through the intercooler depends on both the cooling medium's temperature and its flow rate. By cooling the air-fuel charge, an intercooler increases its density, allowing a greater quantity to enter the engine. Intercooling also lowers the thermal loading on engine components. However, premature detonation still can occur even when an intercooler is used.

To prevent premature detonation, a spring-loaded **blowoff valve** limits boost pressure to a predetermined amount, figure 17-6. The blowoff valve generally is installed between the supercharger and the engine to bleed off part of the air-fuel charge. If the maximum allowable boost pressure is reached, it overcomes the blowoff valve spring pressure and vents back to the supercharger. Superchargers that compress only air, however, can be vented to the atmosphere.

Superchargers

A supercharger is an air pump mechanically driven by the engine itself. Gears, shafts, chains, or belts from the crankshaft can be used to turn the pump, figure 17-7. This means the air pump or supercharger pumps air in direct relation to engine speed.

There are two general types of superchargers:

- *Positive displacement*—pumps the same volume of air on each cycle regardless of engine speed
- *Variable displacement*—the volume of air pumped varies with engine speed.

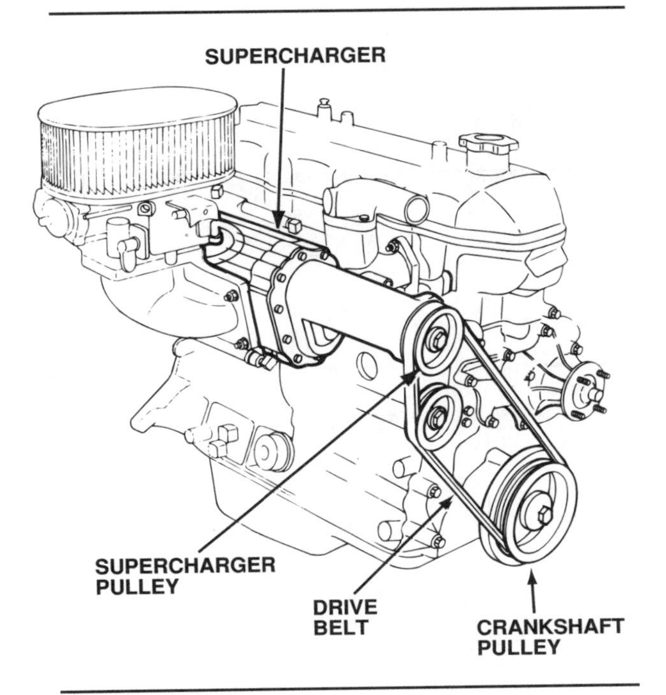

Figure 17-7. In a belt-driven operation, pulley size determines supercharger drive ratio. A small pulley will turn the supercharger faster than a large pulley; thus, supercharger operation can be controlled by pulley use. (Ford)

Positive displacement pumps include the reciprocating, lobe (Roots), and vane designs, figure 17-8. These operate by drawing in a large column of air, compressing it into a small area, and then forcing it through an outlet at high pressure. Unlike a turbocharger, the positive displacement supercharger provides substantial boost even when the engine is at idle.

The lobe-type, or Roots blower, is the most common positive displacement supercharger in use, figure 17-9. It was first developed in 1864 as a device to separate wheat from chaff, but was applied to automotive engines around the turn of the century. Roots blowers are used on many high-performance and racing engines, but also are found on two-stroke diesel engines, where they improve air intake and **exhaust scavenging** instead of acting as primary superchargers. Vane-type superchargers are not commonly used.

Variable displacement superchargers include the centrifugal, axial flow, and pressure-wave designs. The centrifugal pump is the most commonly used and most efficient design, figure 17-10. It does not heat the air as much as a lobe-type, and the resulting airflow is much smoother. As the impeller turns at high speed, air pulled into the center of the impeller is sped up, then thrown outward from the blades by centrifugal force. Air moved to the perimeter of the pump housing is forced through an outlet. Since centrifugal force increases as speed increases, a centrifugal supercharger will draw in more air and create a higher boost pressure as its speed increases. Its pumping output increases roughly as a square of the engine speed. When engine speed doubles, the centrifugal supercharger provides four times as much boost pressure.

Supercharging: Using a belt-driven air pump to deliver an air-fuel charge to the engine cylinders at a pressure greater than atmospheric pressure.

Intercooler: An air-to-air or air-to-liquid heat exchanger used to lower the temperature of the air-fuel mixture by removing heat from the intake air charge.

Blowoff Valve: A spring-loaded valve that opens when boost pressure overcomes the spring tension to vent excess pressure.

Exhaust Scavenging: The use of compressed air from an air pump entering the cylinders while both the exhaust and intake valves are opened at the same time (overlap) to force exhaust gases out through the exhaust valve, and provide more room for fresh air for the next combustion cycle.

A Challenge to the Roots Blower

The G-Lader blower, one of the most successful non-Roots blower designs to emerge from the drawing board and into actual production cars has appeared in recent years on some Volkswagen Corrado models. The G-Lader takes its name from the housing, which is shaped like the letter "G", and the German word for charger—lader. This spiral channel unit was derived from a French design patented in 1905, and is both lighter and quieter than a traditional Roots blower.

Concentric spiral ramps in both sides of a rotor mesh with similar ramps cast in the split casing. The rotor moves around an eccentric shaft instead of spinning on its axis, as in most other supercharger designs. As air is drawn into the casing, it is compressed by squeezing it through the spiral, and then forced through a cluster of ports in the center of the casing before passing into the engine. Boost pressure is slightly less than 12 psi (0.83 bar), yet the G-Lader uses approximately 17 horsepower (13 kW) under maximum throttle and rpm.

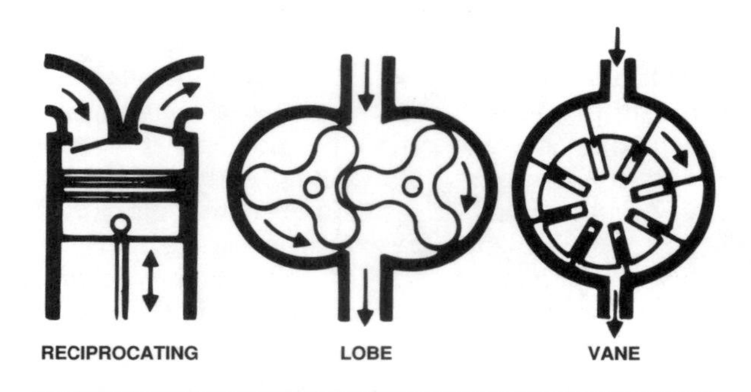

Figure 17-8. Three common positive-displacement pump designs.

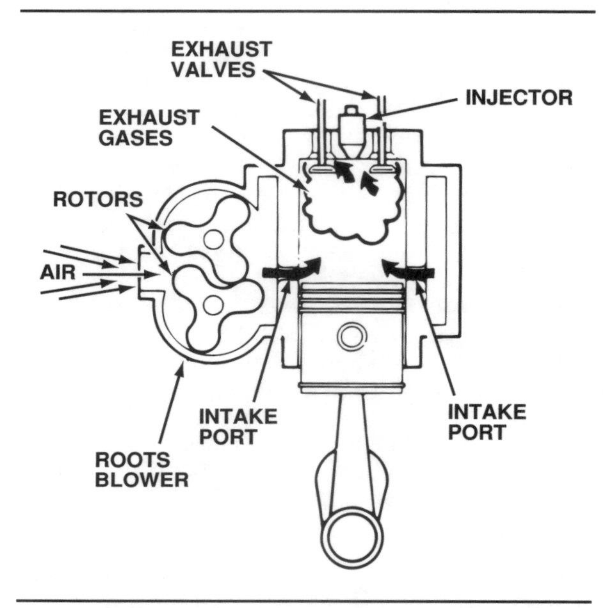

Figure 17-9. The Roots blower is the most common positive displacement pump used as a supercharger. (Ford)

If engine speed triples, the supercharger delivers nine times as much boost pressure.

Superchargers are installed to pump air directly into the intake manifold on diesel and fuel injected engines. With carbureted engines, two systems are possible: downstream or upstream. The downstream system is the most popular, with the supercharger mounted between the carburetor and the intake manifold, figure 17-11. This permits the supercharger to draw the air-fuel charge from the carburetor and then pump it into the intake manifold. This system helps atomize the fuel particles by mixing the air-fuel charge as it passes through the supercharger.

The upstream system locates the supercharger between the air filter and carburetor, figure 17-12. In this arrangement, the carburetor must be encased in a pressure box because the supercharger blows air through the carburetor into

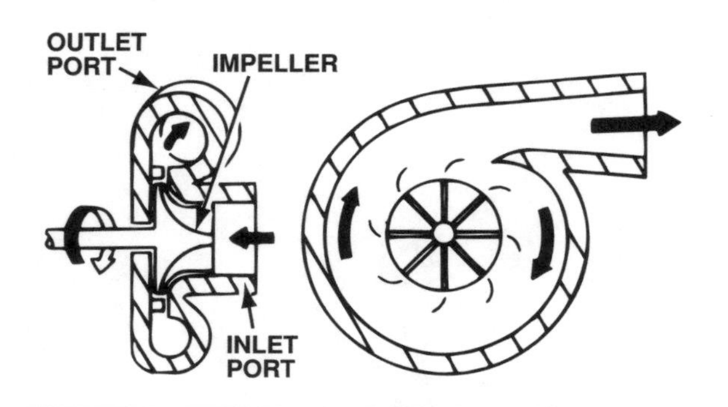

Figure 17-10. The centrifugal pump design accelerates and then slows the air to compress it. (Ford)

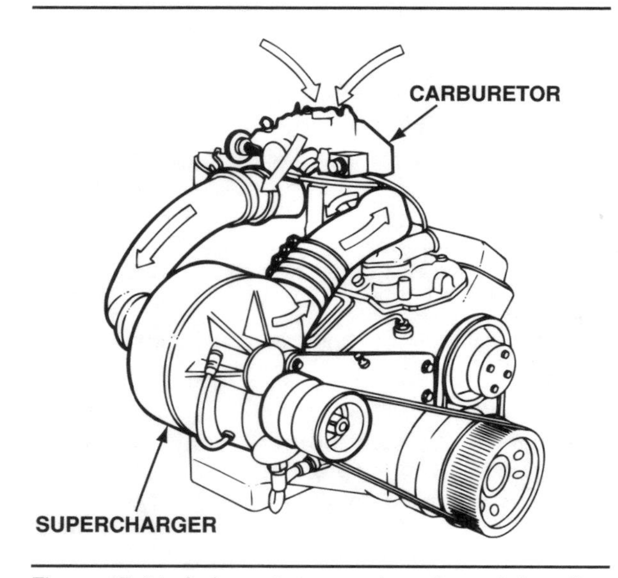

Figure 17-11. A downstream or draw-through installation. (Ford)

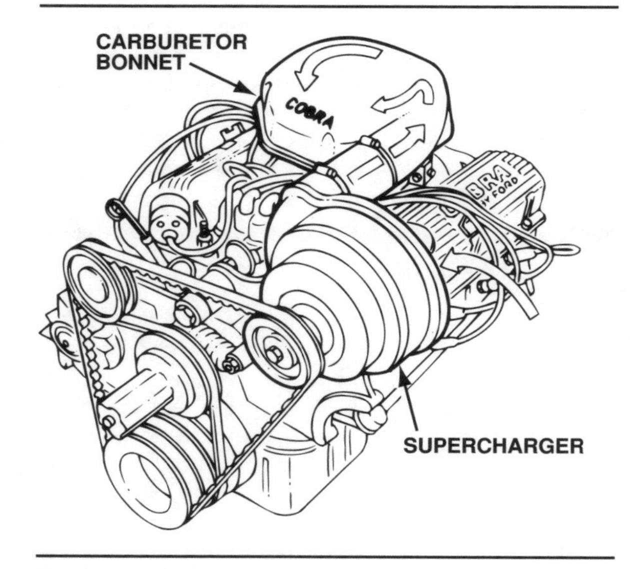

Figure 17-12. An upstream or blow-through installation. (Ford)

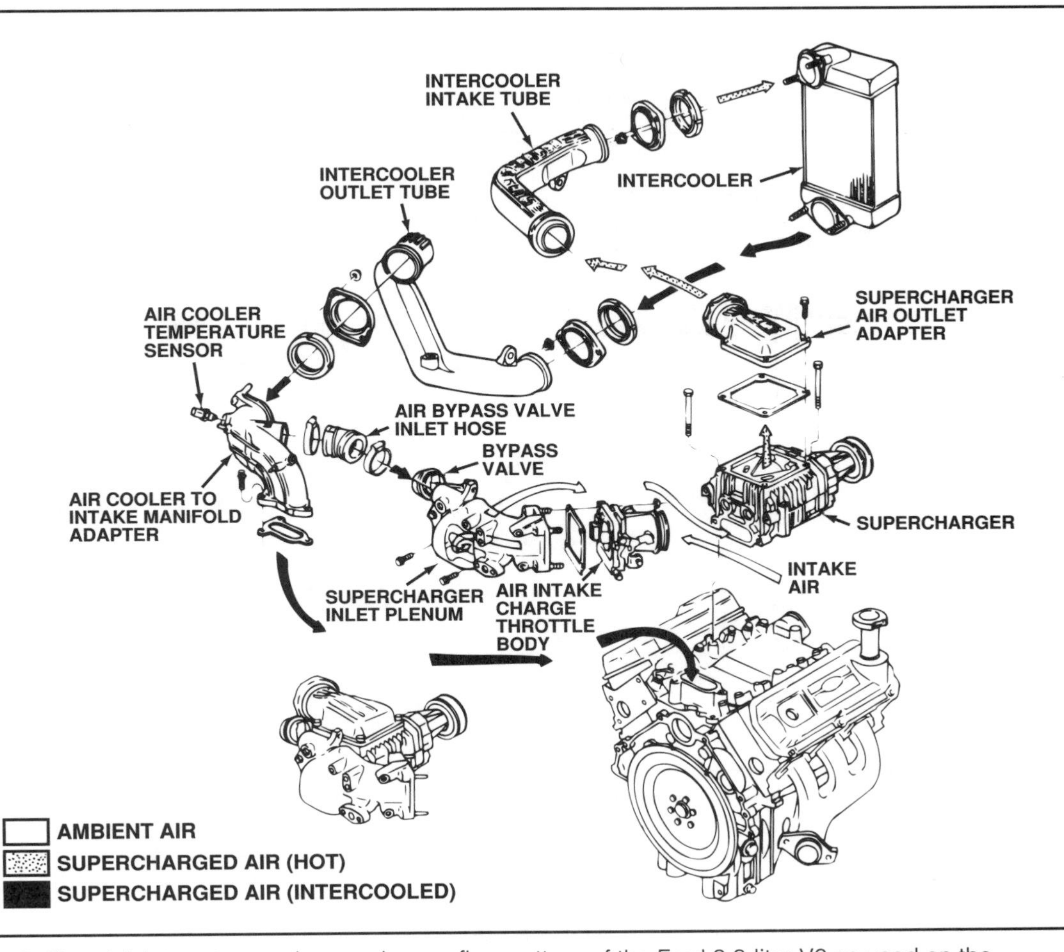

Figure 17-13. The air inlet system and supercharger flow pattern of the Ford 3.8-liter V6 as used on the Thunderbird Super Coupe.

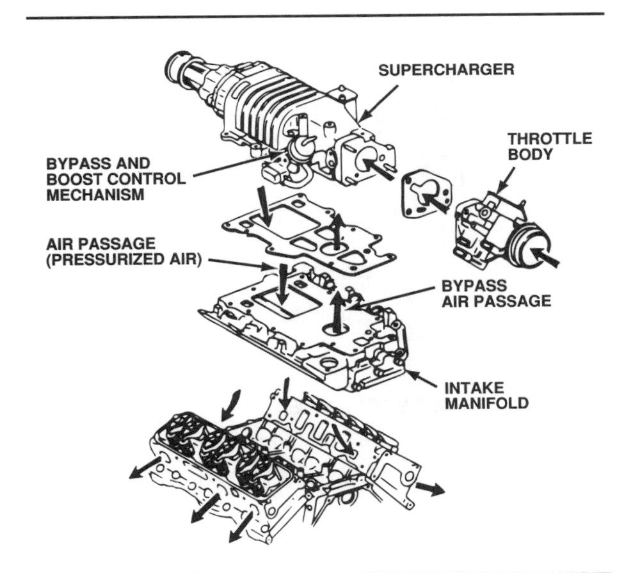

Figure 17-14. The GM application draws intake air through the throttle body at the rear of the supercharger and sends the compressed air directly into the intake manifold below it.

the intake manifold. Since a carburetor requires a pressure drop at the venturi to draw fuel from the bowl, the main jets will not deliver fuel into the intake air stream unless the air around the carburetor is pressurized.

The mechanical drive methods that operate a supercharger all have inherent limitations. They require up to 20 percent of the engine's power, even when the supercharger is not in use. They also need a high overdrive ratio to obtain a speed great enough to provide the desired boost. Since the drive belt or gear speed must be quite high, the mechanical components are subjected to heavy wear.

Supercharging and the 1990s

Although automakers experimented with a variety of superchargers on limited-production cars during the 1980s, none appeared on domestic vehicles until 1989, when Ford introduced a 3.8-liter V6 in the Thunderbird Super Coupe, with a Roots-type blower manufactured by Eaton

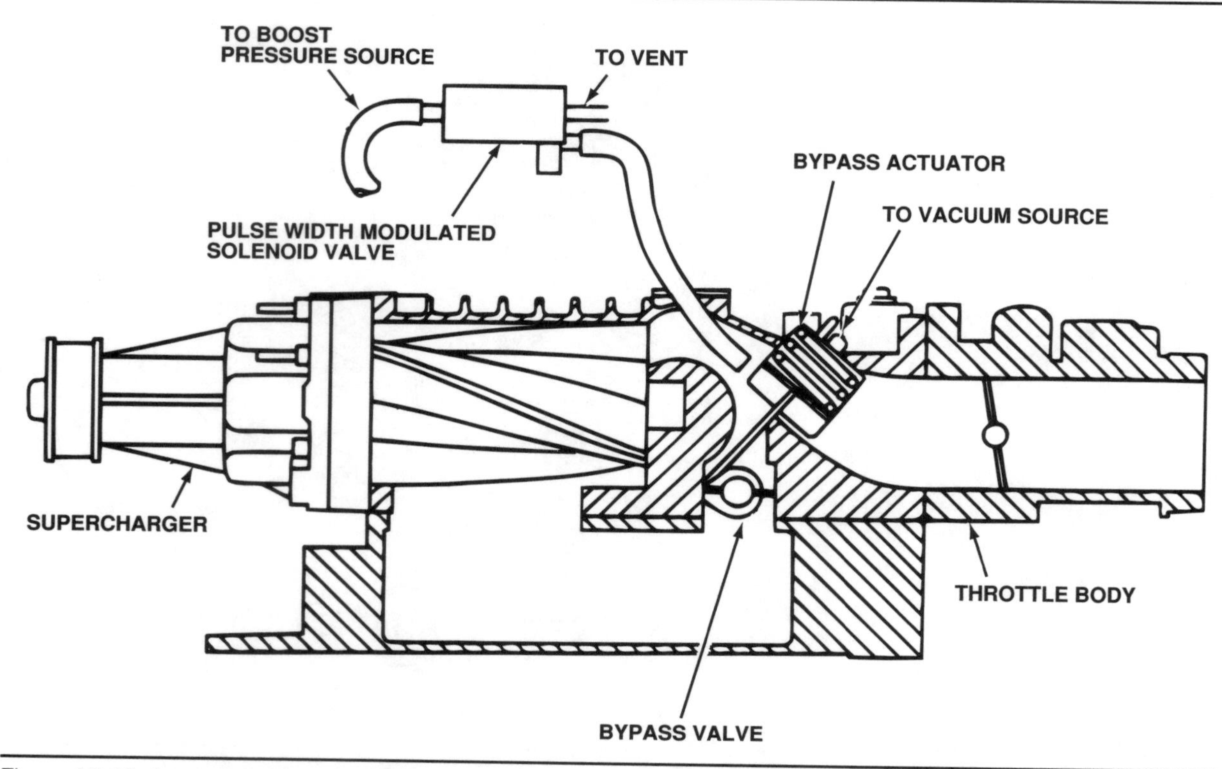

Figure 17-15. A cross section of a GM supercharger shows how the bypass valve operates to divert air to the blower inlet and limit boost pressure. Ford uses a similar bypass valve.

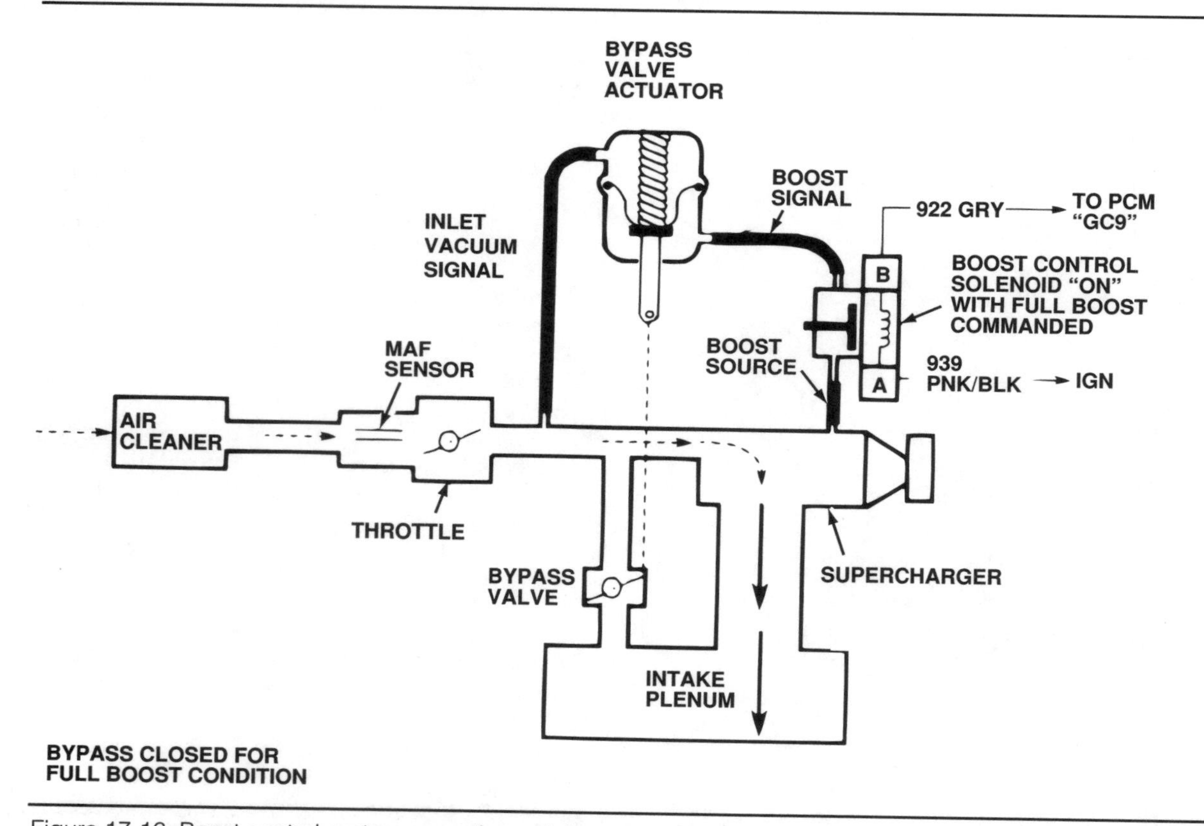

Figure 17-16. Boost control system operation with bypass valve closed for full boost. (Oldsmobile)

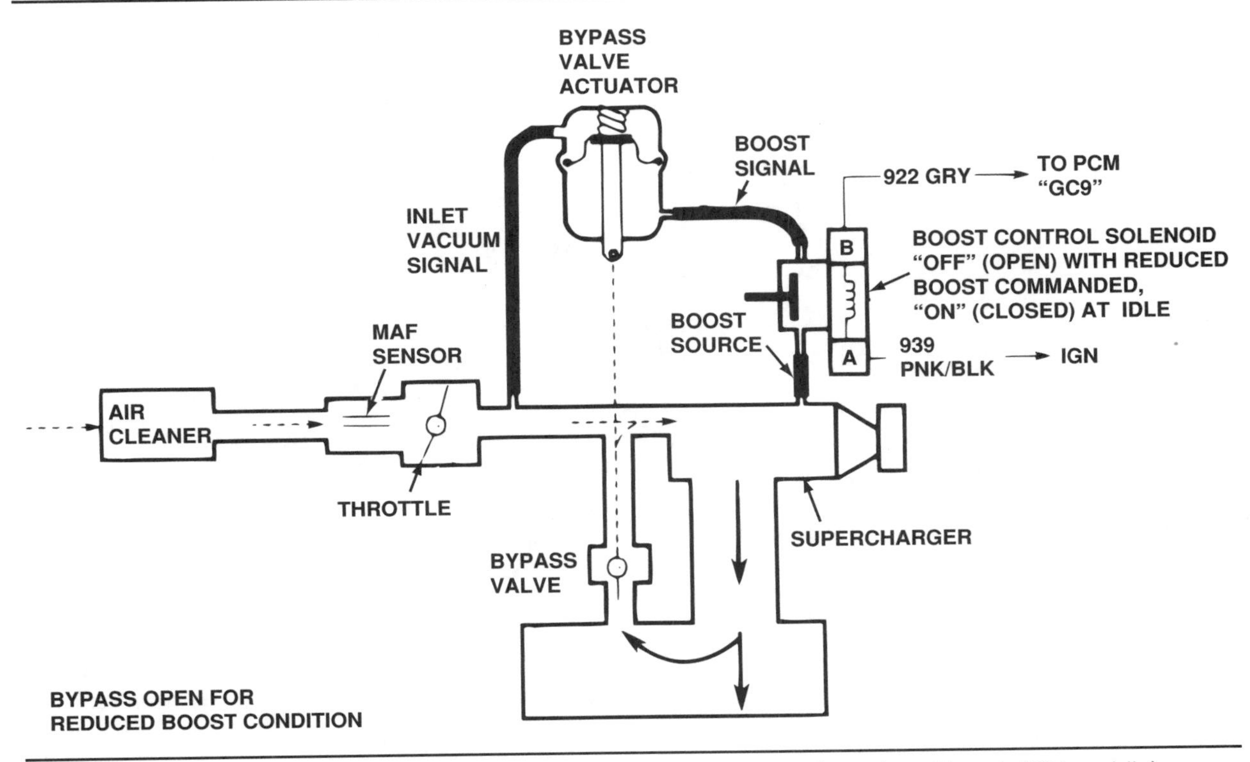

Figure 17-17. Boost control system operation with the bypass valve open for reduced boost. (Oldsmobile)

Corporation. This is Ford's first venture into supercharging since the late 1950s, when it also offered a belt-driven centrifugal supercharger.

As used by Ford, the intercooled Eaton blower develops 12 psi (0.83 bar) maximum boost at low manifold vacuum. When closed, a vacuum-operated bypass valve sends the full flow of air from the supercharger directly to the intake manifold. The butterfly valve starts to open at about 3 in-Hg (76.2 mm-Hg) vacuum to bleed boost pressure back to the blower inlet; at 7 in-Hg (177.8 mm-Hg) vacuum, the valve is fully open. Figure 17-13 shows this airflow pattern.

General Motors introduced a similar system without an intercooler on some 1992 3800-V6 engines. In this design, intake air enters at the manifold plenum from the bottom of the blower, figure 17-14. The system also uses a vacuum-operated bypass valve, figure 17-15, and a powertrain control module-controlled boost solenoid to regulate induction boost pressure during rapid deceleration, under high engine load, or in reverse gear. Boost system operation is shown in figures 17-16 and 17-17. Maximum boost pressure can range from 7 to 11 psi (0.48 to 0.76 bar), but is maintained at about 8 psi (0.55 bar) on applications with the 4T60E transaxle due to the transaxle's inability to deal with high torque.

TURBOCHARGERS

The major disadvantage of a supercharger is it relies on engine power to drive the unit. In some installations, a mechanical supercharger uses as much as 20 percent of the engine's power. However, by connecting a centrifugal supercharger to a turbine drive wheel and installing it in the exhaust path, the lost engine horsepower is regained to perform other work, and the combustion heat energy lost in the engine exhaust (as much as 40 to 50 percent) can be harnessed to do useful work. This is the concept of a **turbocharger.**

The turbocharger's main advantage over a mechanically driven supercharger is that it does not drain power from the engine. A naturally aspirated engine loses about half of the heat energy contained in the fuel through the exhaust system, figure 17-18. Another 25 percent is lost through radiator cooling. Only about 25 percent of the heat energy is actually converted into

Turbocharger: A supercharging device that uses exhaust gases to turn a turbine connected to a compressor that forces extra air-fuel mixture into the cylinders.

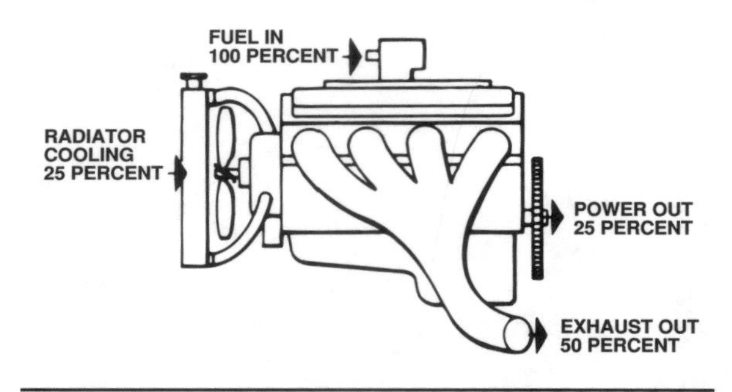

Figure 17-18. A turbine uses some of the heat energy that normally would be wasted.

mechanical energy. A mechanically driven pump uses some of this mechanical output, but a turbocharger gets its energy from the exhaust gases, converting more of the fuel's heat energy into mechanical energy.

A turbocharger turbine looks much like a typical centrifugal pump used for supercharging, figure 17-19. Hot exhaust gases flow from the combustion chamber to the turbine wheel. The gases heat and expand as they leave the engine. The hot gases expanding against the turbine wheel's blades force the turbine to spin, not the speed of force of the exhaust gases as is commonly believed.

Turbocharger Design and Operation

The modern turbocharger is both simple and compact, with few moving parts. Because its moving parts work at very high speeds and under extreme heat, a turbocharger must have very precise tolerances. A turbocharger consists of two chambers connected by a center housing. The two chambers contain a turbine wheel and a compressor wheel connected by a shaft that passes through the center housing, figure 17-20.

To take full advantage of the exhaust heat that provides the rotating force, a turbocharger must be positioned as close as possible to the exhaust manifold. This allows the hot exhaust to pass directly into the unit with a minimum of heat loss. As exhaust gas enters the turbocharger, it rotates the turbine blades, figure 17-21. The turbine and compressor wheels are on the same shaft, so they turn at the same speed. The compressor wheel's rotation draws air in through a central inlet, and centrifugal force pumps it through an outlet at the edge of the housing. A pair of bearings in the center housing supports the turbine and compressor wheel shaft, figure 17-22, and is lubricated by engine oil.

Both the turbine and compressor wheels must operate with extremely close clearances to minimize possible leakage around their blades. Any

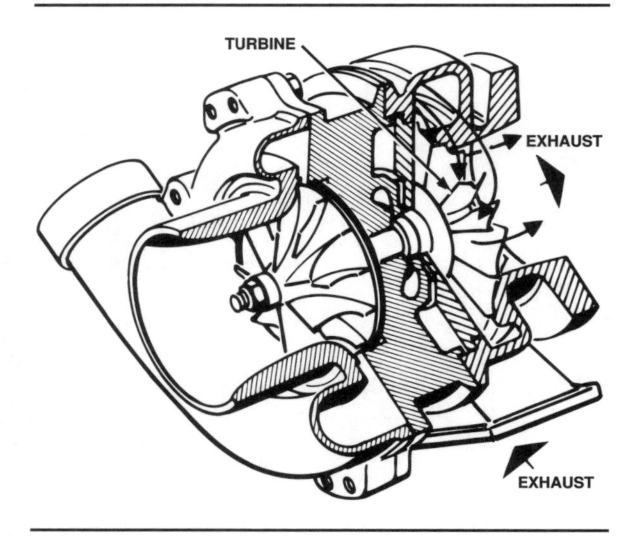

Figure 17-19. A turbine wheel is turned by the expansion of gases against its blades.

leakage around the turbine blades causes a dissipation of the heat energy required for compressor rotation. Leakage around the compressor blades prevents the turbocharger from developing its full boost pressure.

When the engine runs at low speed, both exhaust heat and pressure are low, and the turbine runs at a low speed (approximately 1,000 rpm). Because the compressor does not turn fast enough to develop boost pressure, air simply passes through it and the engine works like any naturally aspirated engine. As the engine runs faster or load increases, both exhaust heat and flow increase, causing the turbine and compressor wheels to rotate faster. Since there is very little rotating resistance on the turbocharger shaft, the turbine and compressor wheels accelerate as the exhaust heat energy increases. When an engine runs at full power, the typical turbocharger rotates at speeds between 100,000 and 150,000 rpm.

Engine deceleration from full power to idle requires only a second or two because of its internal friction, pumping resistance, and drive train load. The turbocharger, however, has no such load on its shaft, and is already turning many times faster than the engine at top speed. As a result, it can take as much as a minute or more after the engine returns to idle speed before the turbocharger also returns to idle. If the engine decelerates to idle and then is shut off immediately, engine lubrication stops flowing to the center housing bearings while the turbocharger is still spinning at thousands of rpm. The oil in the center housing is then subjected to extreme heat and can gradually "coke", or oxidize. If this happens, the turbocharger needs to be replaced.

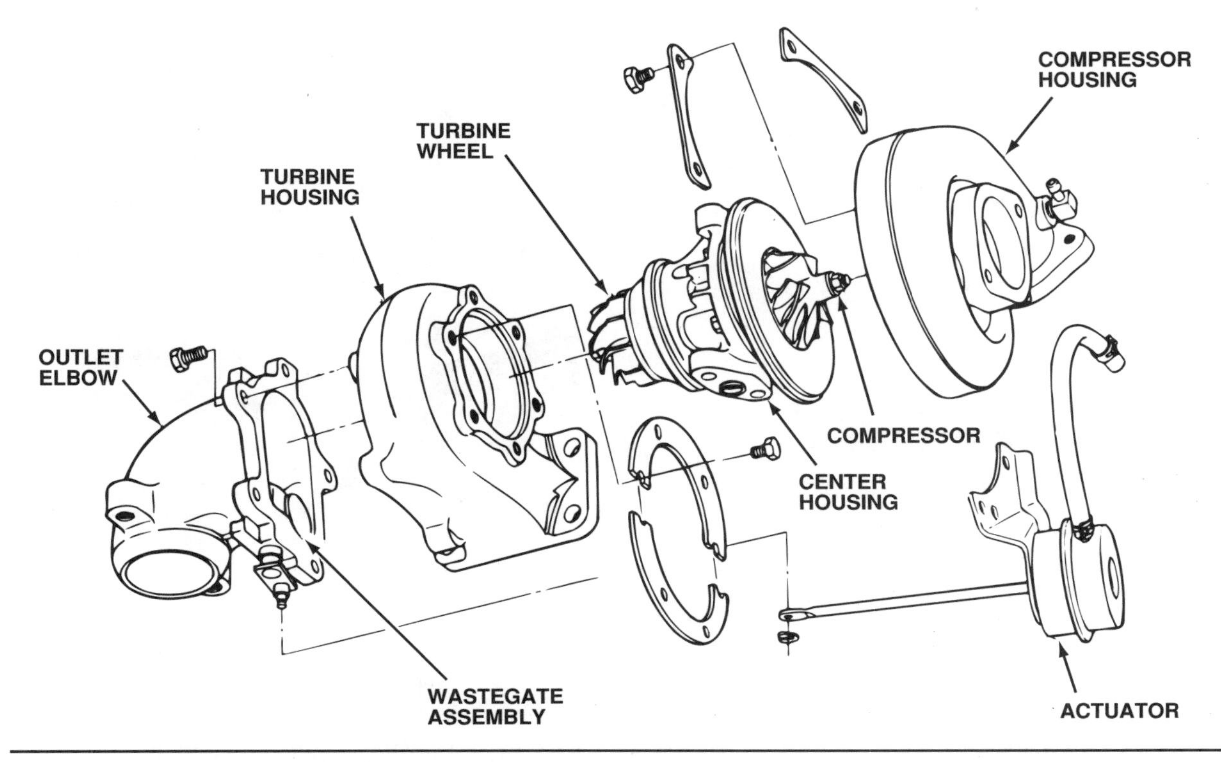

Figure 17-20. The components of a typical turbocharger. (Ford)

The high-rotating speeds and extremely close clearances of the turbine and compressor wheels in their housings require critical bearing clearances. The bearings must keep radial clearances of 0.003 to 0.006 in. (0.08 to 0.15 mm). Axial clearance (endplay) must be maintained at 0.001 to 0.003 in. (0.025 to 0.08 mm). Constant lubrication with clean oil is very important to bearings that must maintain these very critical clearances at extreme speeds. Such close clearances and high rotational speeds equal instant failure if a small dirt particle or other contamination enters the exhaust or intake housings.

The turbocharger is a simple but extremely precise device in which heat plays a critical role. If properly maintained, the turbocharger is also trouble-free. To prevent problems, three conditions must be met:

■ Switching the Emphasis of Turbos

Some drivers may think of turbochargers as exotic trapping for souped-up street cars and racers. That is natural when you consider that automotive engineers normally bring a turbo into play high up on the engine's performance curve to deliver maximum power at top rpm. The result is additional peak power that has won many races, and further convinced the general driving public that turbine-operated superchargers are not for them.

The reappearance of turbocharging as a factory option approached the subject from the opposite viewpoint. Engineers wanted 6-cylinder fuel economy with V8 performance, and the turbocharger was the way to get it. Since its appearance on 1978 domestic engines, the turbocharger has been used for low- and medium-speed passing, and acceleration from about 1,200 rpm up. As engine speed climbs, the wastegate begins to open, preventing an overload of the engine. This means that the turbocharger will be used for only about five percent of the time the engine is operating.

The performance-oriented 1980s brought turbocharging into favor with consumers. When used with 4-cylinder engines, turbochargers provide the power to make them perform like V6 and small V8 engines, while retaining good fuel economy. By the mid-1980s, seven percent of the 4-cylinder engines produced by domestic manufacturers were turbocharged. However, the upswing in turbocharger popularity has been challenged by the revival in popularity of V6 and small V8 engines, as well as the recent appearance of superchargers on some Ford and GM engines.

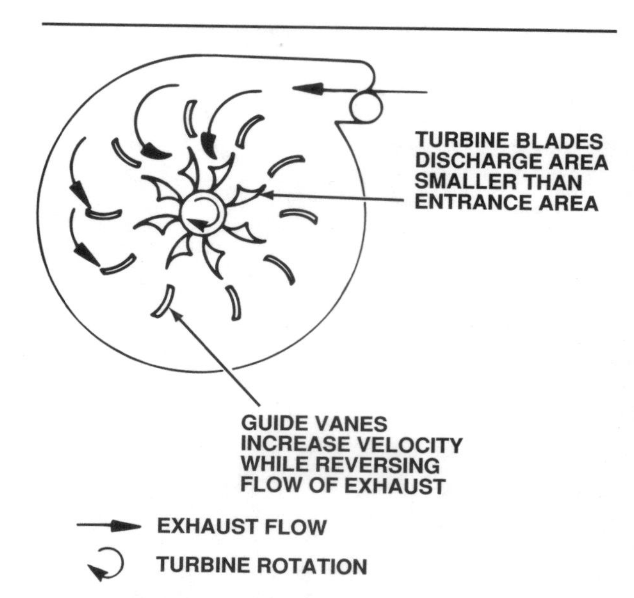

Figure 17-21. During turbine wheel rotation, the guide vanes increase velocity and reverse the flow of the gases. (Ford)

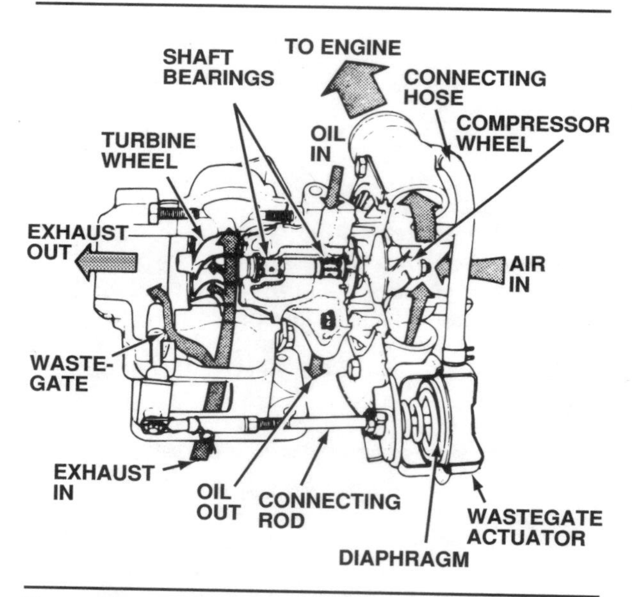

Figure 17-22. Basic operational cycle of a turbocharger. (Ford)

- The turbocharger bearings must be constantly lubricated with clean engine oil—turbocharged engines should have regular oil changes at half the time or mileage intervals specified for nonturbocharged engines.
- Dirt particles and other contamination must be kept out of the intake and exhaust housings.
- Whenever a basic engine bearing (crankshaft or camshaft) has been damaged, the turbocharger must be flushed with clean engine oil after the bearing has been replaced.

Figure 17-23. Water-cooled turbochargers induct engine coolant into passages in the center housing (arrow) to cool the center bearings.

- If the turbocharger is damaged, the engine oil must be drained and flushed, and the oil filter replaced as part of the repair procedure.

Late-model turbochargers all have liquid-cooled center bearings to prevent heat damage. In a liquid-cooled turbocharger, engine coolant is circulated through passages cast in the center housing to draw off the excess heat, figure 17-23. This allows the bearings to run cooler, and minimizes the probability of oil coking when the engine is shut down.

Turbocharger size and response time

As we have seen, there is a time lag between increase in engine speed and the turbocharger's ability to overcome inertia and spin up to speed as the exhaust gas flow increases. This delay between acceleration and turbo boost is called **turbo lag.** Like any material, moving exhaust gas has inertia. Inertia is also present in the turbine and compressor wheels, as well as in the intake airflow. Unlike a supercharger, the turbocharger cannot supply an adequate amount of boost at low speed.

Turbocharger response time is directly related to the size of the turbine and compressor wheels. Small wheels accelerate rapidly; large wheels accelerate slowly. While small wheels would seem to have an advantage over larger ones, they may not have enough airflow capacity for an engine. To minimize turbo lag, the engine's intake and exhaust breathing capacities must match the turbocharger's exhaust and intake airflow capabilities.

The turbocharger's location on the engine is another factor influencing response time. The most efficient arrangement is to position the

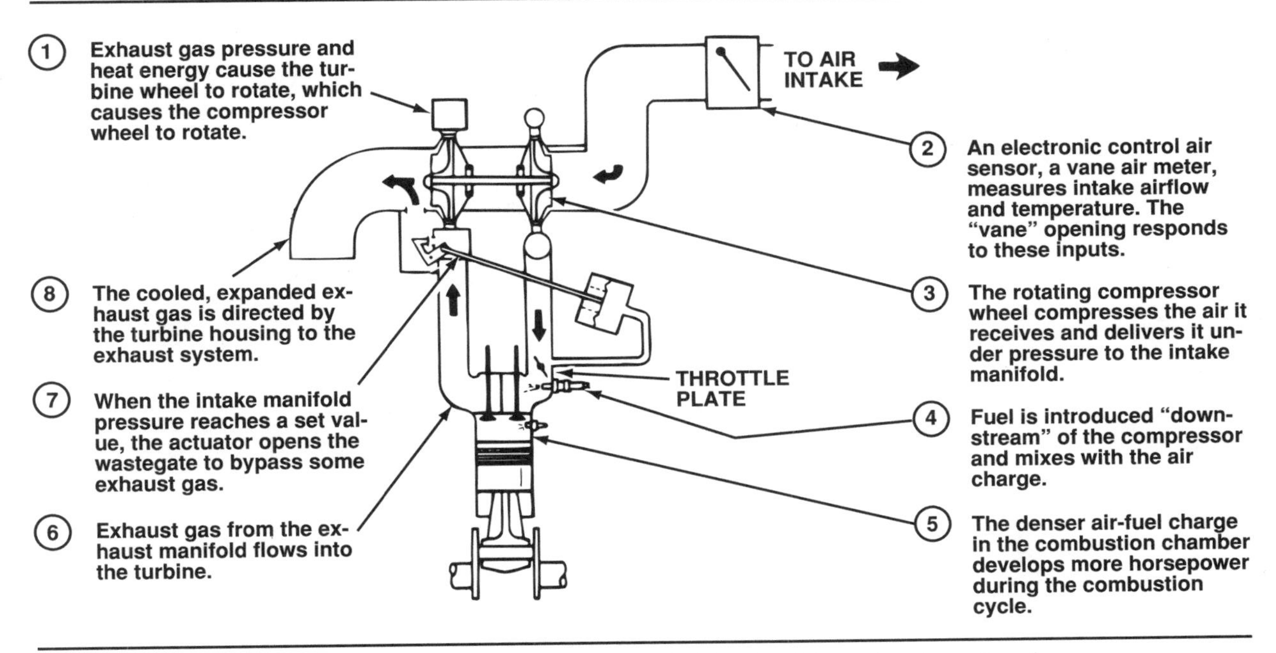

Figure 17-24. The operational cycle of a turbocharger used with a fuel-injected system. (Ford)

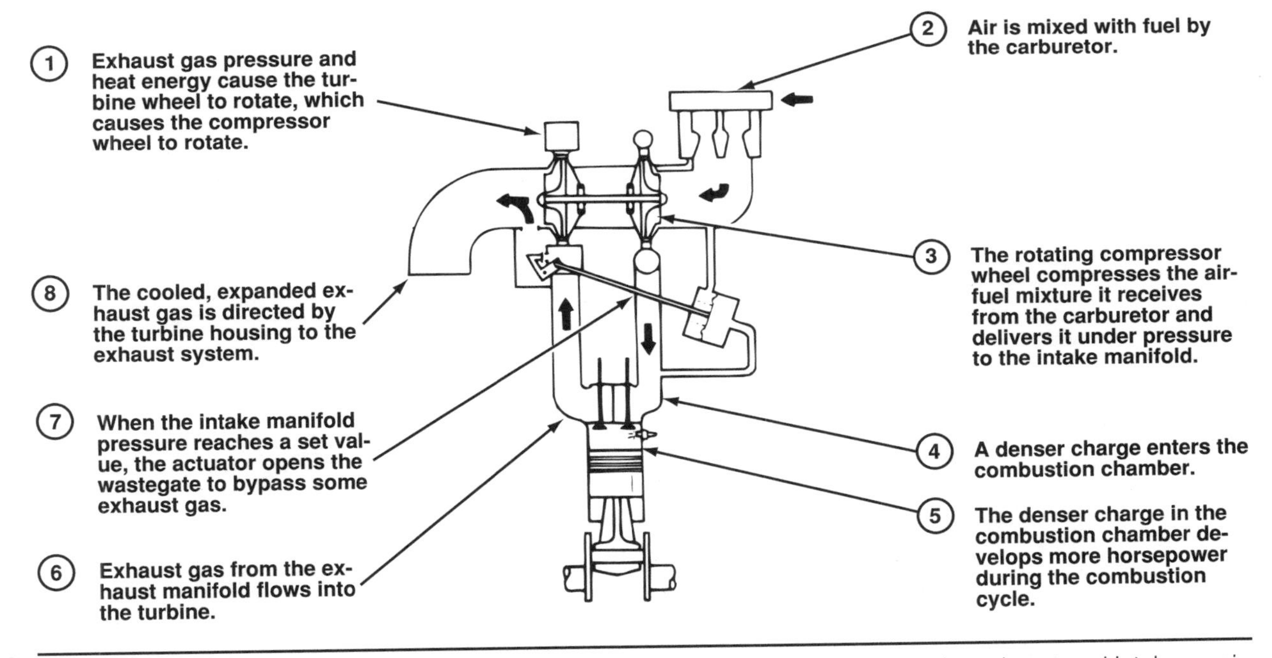

Figure 17-25. Turbo lag is reduced by locating the turbine and compressor close to the exhaust and intake manifolds, respectively. This is the operational cycle of a turbocharger used with a carbureted system. (Ford)

turbine outlet close to the intake manifold, as shown in the operational cycle for a fuel injected engine, figure 17-24, or a carbureted engine, figure 17-25.

Engineers continue to explore new ways to improve turbocharger response. One method is to reduce the weight of turbine and compressor wheels by using lightweight ceramic rotating parts. Another is to control the flow of inlet gas to the turbine through the use of a variable-nozzle turbine. This design uses a curved flap at the turbine inlet. When the flap is closed, exhaust gas entering the turbine is deflected to

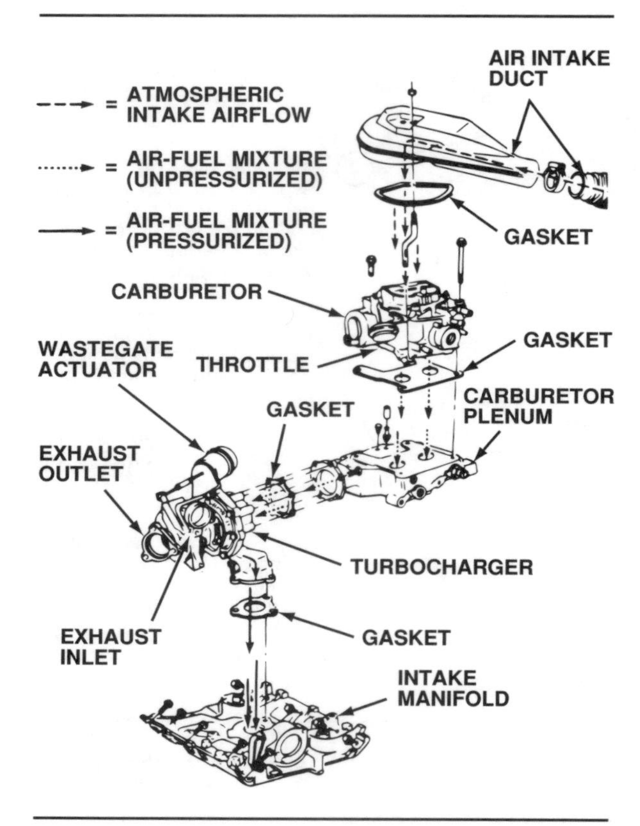

Figure 17-26. The turbocharger compresses the air-fuel mixture when the carburetor is installed at the compressor inlet. (Buick)

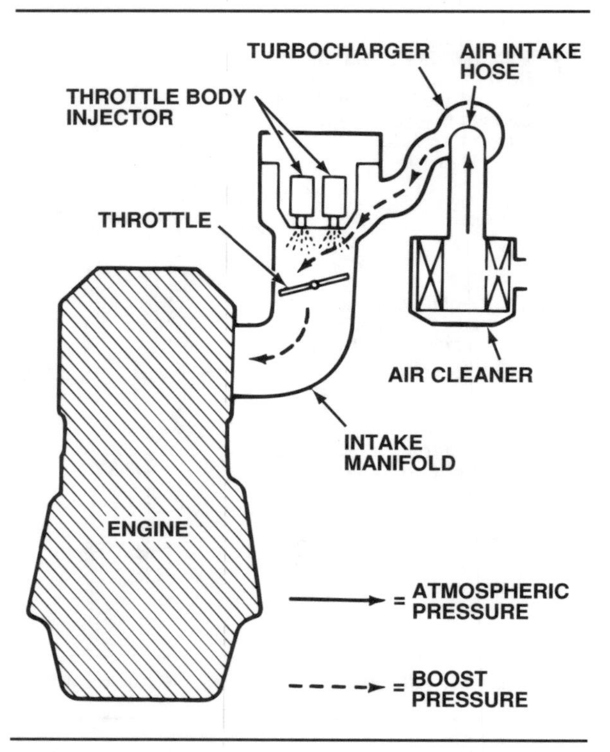

Figure 17-27. The throttle is placed downstream of the turbocharger and fuel injectors in this Mitsubishi TBI system.

strike the turbine at almost a 90-degree angle, producing a quicker response at low speed. As engine speed increases, the flap starts to open. This reduces exhaust backpressure and increases the amount of exhaust gas to keep the turbine spinning at high speed. The flap is controlled by a computer-operated vacuum solenoid. Chrysler used a variable-nozzle turbocharger on some 1990 models, but the cost and complexity of this design have not made it a popular choice.

Turbocharger Installation

Like superchargers, turbochargers can be installed either on the air intake side (upstream) of the carburetor or fuel injectors or on the exhaust side (downstream) of the carburetor. In an upstream installation, the turbocharger compresses and delivers a denser air charge to the carburetor or injectors. When installed downstream, the turbocharger delivers the compressed air-fuel mixture to the cylinders.

Downstream installations

Turbochargers are generally installed downstream with carbureted engines, as are superchargers, figure 17-11 (This is called a draw-through turbocharger). The carburetor is usually located on the intake side of the turbocharger, although it may be positioned on the outlet side. Positioning the carburetor at the compressor inlet allows the turbocharger to increase airflow and pressure to drop through the carburetor. This provides the required air-fuel mixture, which is compressed and sent to the cylinders through the manifold, figure 17-26.

Positioning the carburetor at the compressor inlet simplifies air-fuel control, and the carburetor does not have to be modified to withstand boost pressure. If the carburetor is located at the compressor outlet, it must be calibrated to meter fuel correctly under both atmospheric and above-atmospheric conditions, and pressurized to withstand boost pressure without leaking fuel. In addition, the fuel system must also move the fuel at a higher pressure to overcome the boost pressure present at the carburetor. These factors all make air-fuel ratio control difficult when the carburetor is located at the turbocharger outlet.

There are, however, minor drawbacks that must be overcome with carburetor placement at the compressor inlet:

- The carburetor is located away from the intake manifold, which may cause a slight hesitation when the throttle is opened rapidly.

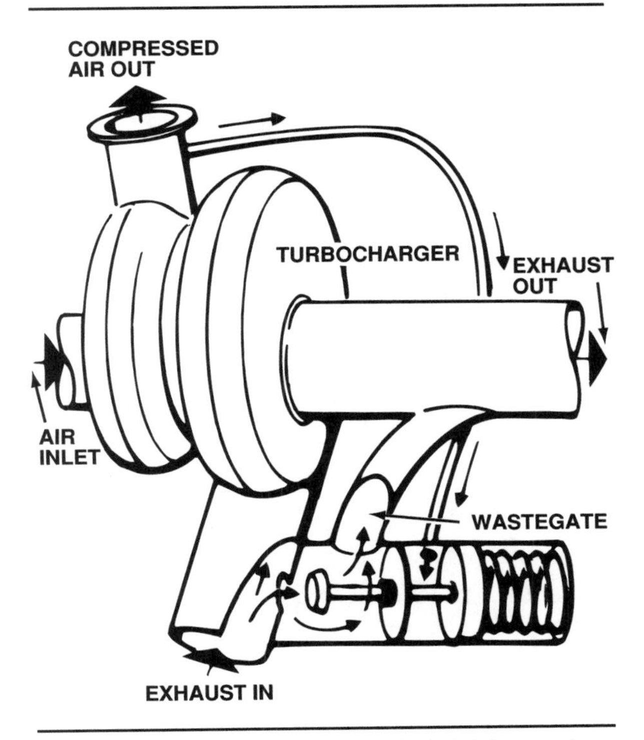

Figure 17-28. The wastegate reacts to intake manifold pressure, and controls the amount of exhaust gas reaching the turbine wheel.

- Throttle and choke linkages must be more complex in design and operation.
- Preheating the air-fuel mixture on a cold engine is more difficult.
- A special seal must be installed between the center housing and compressor to prevent the air-fuel mixture from entering the engine lubrication system.
- Fuel separation from the compressed mixture may occur before it reaches the manifold.

Upstream installations

Turbochargers are installed upstream with fuel injected engines (This is called a blow-through turbocharger). The unit is positioned at the manifold air intake, and compresses only intake air, figure 17-24. This reduces fuel delivery time and increases the amount of turbine energy available. Most turbocharged engines with fuel injection use a multipoint system, with an individual injector at each cylinder. Just before the compressed-air charge enters the cylinder, fuel injects into it, as shown in step 4, figure 17-24. The throttle plate installed between the turbocharger and injectors regulates airflow.

Mitsubishi uses a throttle body injection (TBI) system on some engines in which the turbocharger delivers compressed air to the injectors in the TBI unit. The throttle plate, however, is located between the injectors and the intake manifold,

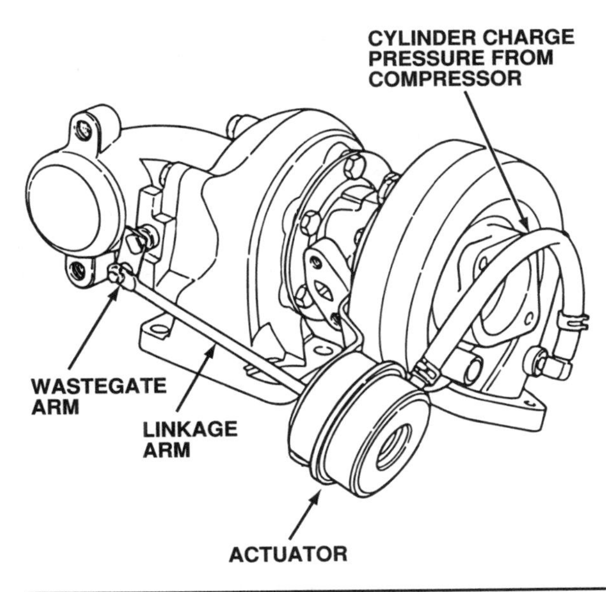

Figure 17-29. Wastegate operation is controlled by linkage connected to the actuator's vacuum diaphragm. (Ford)

figure 17-27. In this system, the throttle plate regulates the intake volume of compressed air-fuel mixture.

Twin turbo installations

Twin turbochargers with individual intercoolers and electronic controls are found on some expensive high-performance cars. Chrysler and Mitsubishi have used a complicated twin turbocharger system since 1991. In this type of installation, small and lightweight turbochargers are used, and each is driven independently by the exhaust from one cylinder bank of a V-type engine. Because the turbochargers are small, they can respond quicker at low engine speeds with a combined volume sufficient to deliver the required boost throughout the engine's operating range. Each turbocharger wastegate is computer-controlled through a pulse width modulated solenoid to bleed off boost pressure when necessary. Mazda introduced a twin turbocharger system on its 1993 RX-7 that provides boost throughout the engine's power range by using a small turbocharger for low-speed boost, and a larger one for use at higher engine speeds. Turbocharger engagement and disengagement, as well as wastegate operation, are computer controlled.

Turbo Lag: The time interval required for a turbocharger to overcome inertia and spin up to speed.

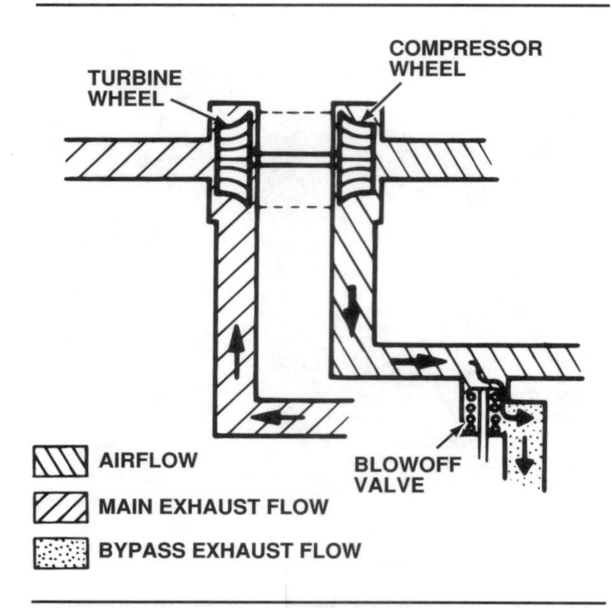

Figure 17-30. Blow-off valve location in an upstream installation. (Ford)

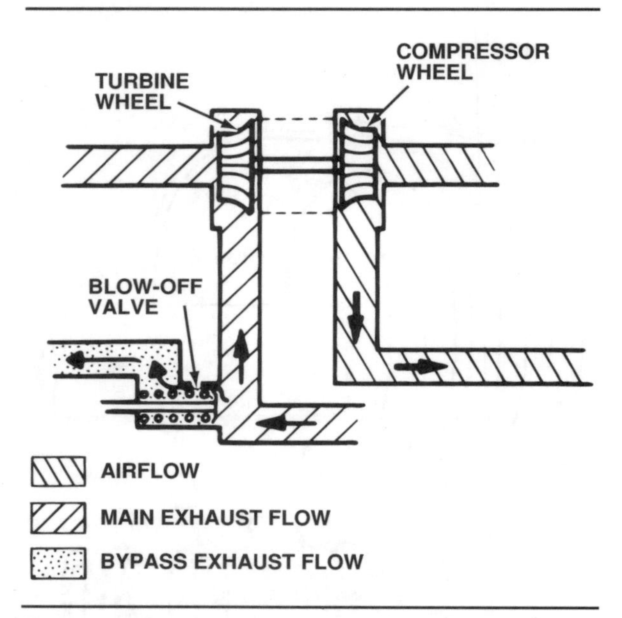

Figure 17-31. Blow-off valve location in a downstream installation. (Ford)

TURBOCHARGER CONTROLS

You cannot simply bolt a turbocharger onto an engine and automatically pick up free power. As we have seen, a turbocharger increases both compression pressure and loads on an engine's moving parts, as well as increasing underhood temperatures. Most late-model engines are strong enough to withstand the higher pressures without major modifications. Many turbo-charged engines, however, use oil coolers and heavy-duty radiators to deal with higher engine and underhood temperatures.

Three factors must be controlled because of the high pressures and temperatures created by a turbocharger:

- Boost pressure
- Air-fuel mixture temperature
- Detonation.

These factors are interrelated. Higher pressures (boost) raise intake mixture temperatures, which raises combustion temperatures. Higher combustion temperatures and pressures can combine to cause detonation.

Boost Pressure Control

Boost increases steadily as turbocharger rotation increases. Boost pressure must be limited to the maximum the engine can withstand without detonation or serious engine damage. This can be achieved either by limiting the amount of exhaust gas reaching the turbine, or by venting off some of the compressed air-fuel charge before it reaches the combustion chamber. This is done most efficiently by an exhaust **waste-gate,** or a blow-off valve.

The wastegate is a bypass valve at the exhaust inlet to the turbine. It allows all of the exhaust into the turbine, or it can route part of the exhaust past the turbine to the exhaust system, figure 17-28. The wastegate is operated by a vacuum-actuated diaphragm exposed to compressor outlet pressure. When pressure rises to a predetermined level, the diaphragm moves a linkage rod, figure 17-29, to open the wastegate. This diverts some or all of the exhaust to the turbine outlet, limiting the maximum turbine and compressor speed.

A pressure hose connects the actuator to the compressor outlet, figure 17-29. During deceleration, pressure rises at this point because the compressor is working against the closed throttle of an engine that requires little airflow. This causes the actuator to open the wastegate, eliminating overboost during a closed-throttle condition.

A turbocharger's location on the engine does not affect wastegate control of boost pressure. The turbocharger can be installed either upstream or downstream from the fuel injectors or carburetor. Location does not affect, nor is it affected by, intake air-fuel mixture.

Turbochargers installed on production engines during the 1970s were limited to 6 to 8 psi (0.41 to 0.55 bar) boost pressure. During the early 1980s, turbocharging was refined and integrated with electronic engine management systems. Many turbocharger systems now operate at boost pressures of 10 to 14 psi (0.69 to 0.97 bar). All turbochargers with wastegate control operate as described above, but some can be electronically controlled to modulate boost according to the octane content of the fuel being used.

Another form of boost pressure control is called a blow-off valve. As you learned when you studied superchargers, this is a large, spring-loaded relief valve that opens whenever pressure exceeds the desired maximum, figure 17-6. The blow-off valve generally is a poppet or flapper design, with a damper attached to prevent fluttering that could damage the valve. When a blow-off valve is used with a turbocharger, its location in the system differs according to whether the turbocharger is upstream, figure 17-30, or downstream, figure 17-31.

A blow-off valve can control exhaust emissions instead of boost pressure by operating it through intake manifold vacuum. During idle, closed-throttle deceleration, and choked operation, vacuum in the intake manifold causes a diaphragm to move and open the valve. The engine operates as a naturally aspirated unit, avoiding the excessive emissions that turbocharger compression would cause. Some blow-off valves are operated by vacuum routed to one side of the diaphragm and boost pressure to the other.

Intake Mixture Cooling

As turbocharging increases intake air pressure, the air temperature also increases. The higher temperature results in two unwanted effects:

- It reduces the air-fuel charge density, working against the boost pressure.
- It makes the air-fuel charge prone to premature detonation as it enters the cylinders.

Wastegate: A diaphragm-actuated bypass valve used to limit turbocharger boost pressure by limiting the speed of the exhaust turbine.

Water Injection: A method of lowering the air-fuel mixture temperature by injecting a fine spray of water that evaporates as it cools the intake charge.

■ The Next Generation of Turbochargers

Integrating the turbocharger into electronic engine management systems has brought it a well-deserved reputation for reliability and performance. To this point, all production turbocharger applications have been based on the centrifugal pump. A new generation of turbochargers has seen numerous design variations, the first of which was Nissan's N2-VN, or variable-nozzle (scroll) turbo.

A movable curved flap in the turbo housing changes the throat area to vary its output according to engine requirements. The flap can move 27 degrees in stepless increments. Flap position is determined by a vacuum-operated diaphragm controlled by a pressure-controlled modulator and the engine computer.

With the flap closed, the flow of exhaust gas that enters the turbine housing is sped up, increasing its pressure and rotating the turbine rapidly at low-speed, low-load conditions. At high-speed operation, the flap opens fully and the reduction in exhaust resistance aids in filling the cylinders.

The design provides a normal boost up to 15 psi (1.03 bar). However, if operating conditions will permit it without engine damage, the computer will allow boost up to 17 psi (1.17 bar) for short periods, such as overtaking another vehicle.

VARIABLE NOZZLE TURBOCHARGER

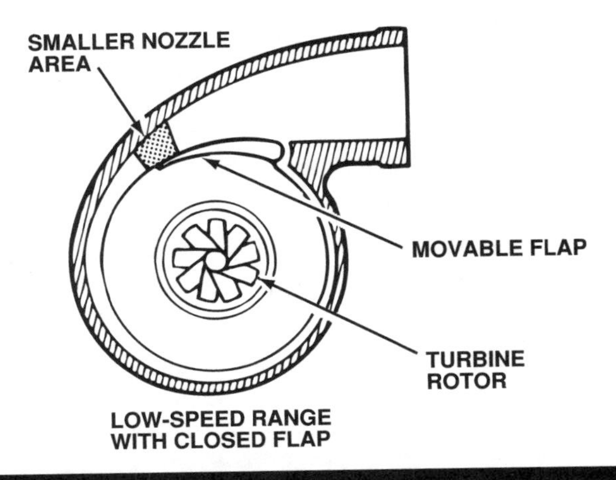

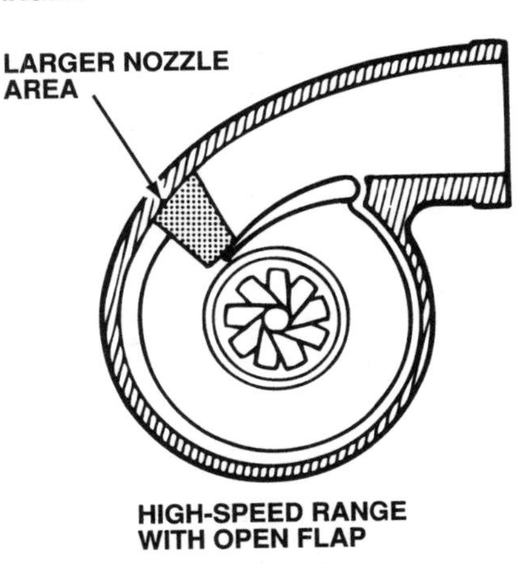

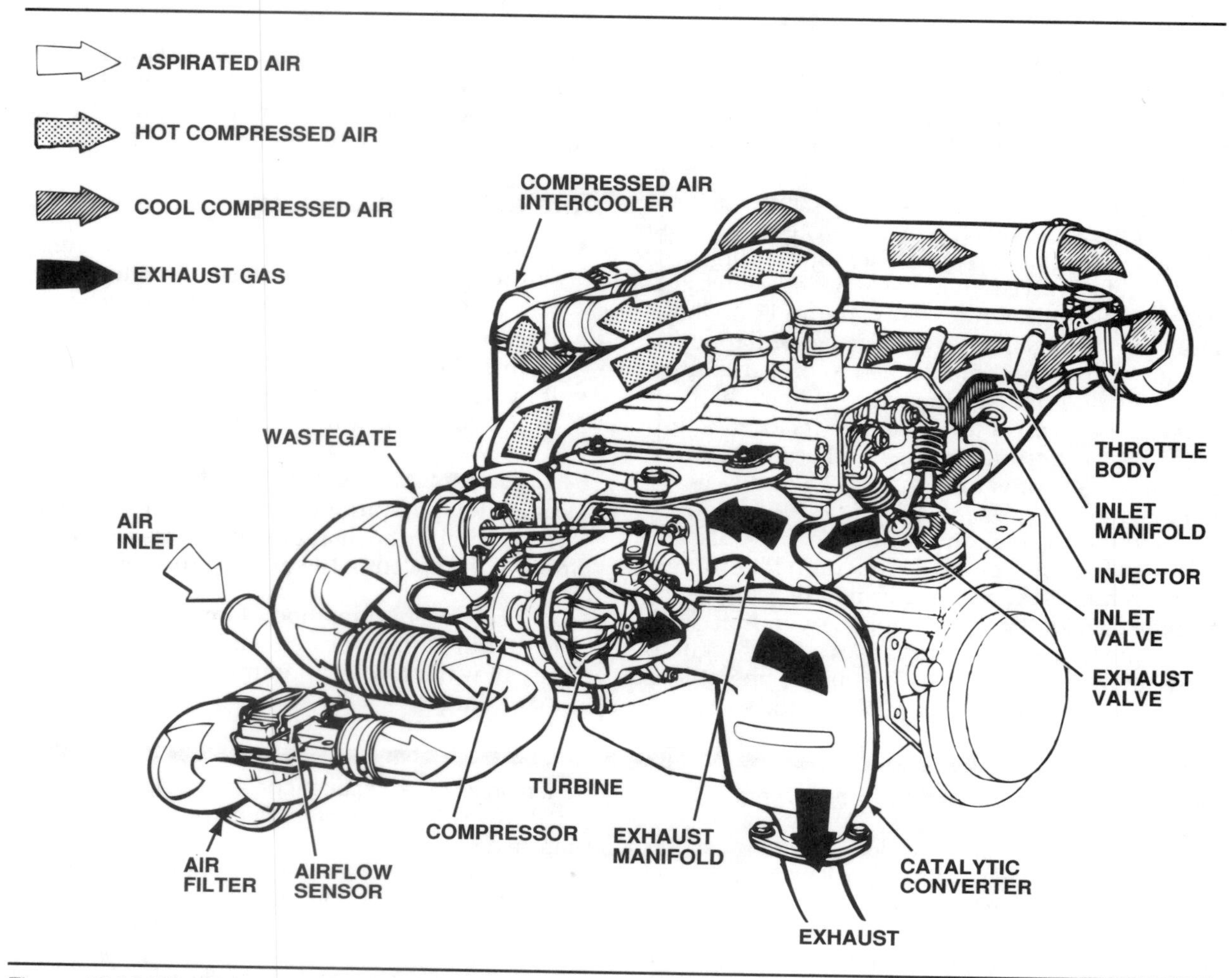

Figure 17-32. The Renault version of an air-to-air intercooler.

When detonation combines with the high pressures involved in a turbocharged engine, it can burn a piston, bend a connecting rod, or cause other damage to internal engine components. There are two ways to cool the compressed mixture before it reaches the combustion chamber: water injection and intercooling.

Water injection is rarely used on factory-installed turbochargers, but is common on aftermarket systems and generally used on carbureted engines when the turbocharger is downstream from the carburetor. At maximum boost, an electric- or vacuum-operated pump injects a fine spray of water into the carburetor inlet. The water vaporizes with the gasoline and has no affect on combustion, but cools the intake charge and leaves the engine as water vapor in the exhaust.

An intercooler is nothing more than a heat exchanger, figure 17-32, as you saw when you studied superchargers. The turbocharger sends compressed air through the intercooler, which transfers heat to the ambient airflow, much like a radiator. This is called an air-to-air intercooler. Air-to-liquid heat exchangers also have been used as intercoolers, but they are heavier and more complicated than the air-to-air type. Intercoolers work best on fuel injected systems or with carbureted systems in which the carburetor is downstream from the turbocharger. Since only air passes through the intercooler, fuel cannot separate from the mixture as it cools.

Spark Timing and Detonation

Excessive pressure and heating of air-fuel mixture are not the only causes of detonation in a turbocharged engine. Spark timing that is excessively advanced will ignite the dense air-fuel charge too soon. This causes the charge to burn unevenly, and maximum cylinder power develops too early. When a turbocharged engine is under maximum boost, ignition timing must be retarded from maximum advance.

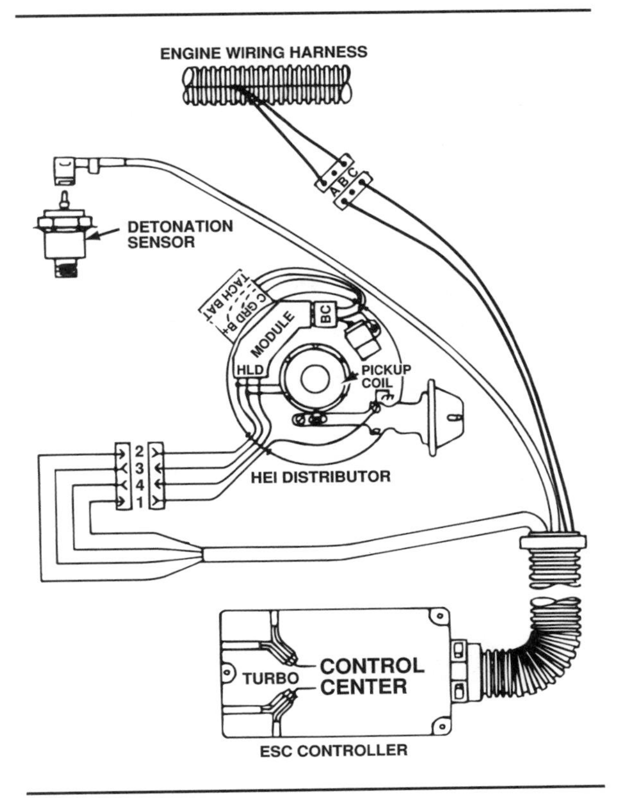

Figure 17-33. A detonation sensor signals the electronic module to retard timing when engine vibrations or knocking occurs. (Buick)

Turbocharged engines with breaker-point ignitions have a distributor vacuum advance unit, where manifold pressure acts on a diaphragm to retard ignition timing under high-boost conditions. This method of timing control was common on turbocharged engines of the 1960s and 1970s. The retard diaphragm is connected to the breaker plate to move it in the opposite direction. This works in a manner similar to a dual-diaphragm distributor, which retards timing with high manifold vacuum during idle and deceleration.

Retarding ignition through boost pressure was adequate for basic turbocharger systems used in the 1970s, but modern turbocharged engines with electronic controls use a detonation or knock sensor for precise control. The sensor is a piezoelectric crystal mounted in the engine block or intake manifold that generates a voltage or varies the return signal of a reference voltage from the computer when engine knock occurs, figure 17-33. The sensor detects detonation as physical vibration within the engine and instantly signals the computer to retard timing. Timing is retarded in 2- to 4-degree increments until either a maximum retard setting is reached or the detonation stops. The computer then gradually restores the proper spark advance required for optimum engine operation.

SUMMARY

Both supercharging and turbocharging are proven methods of mixture compression that increase combustion power without increasing exhaust emissions. A supercharger is an air pump mechanmechanically driven by the engine in direct relation to engine speed. A turbocharger is an air pump driven by exhaust gases. Except for their power source, superchargers and turbochargers perform essentially the same job in increasing the volumetric efficiency of an engine. Both increase intake air pressure above atmospheric pressure, resulting in a denser air-fuel charge.

Supercharging, however, can use up to 20 percent of the engine's power even when not in use. Turbochargers increase engine power only on demand, when additional power is required. One major disadvantage of a turbocharger is "turbo lag", the time interval between increasing engine speed and the turbocharger's ability to overcome inertia and spin up to speed.

Boost pressure, air-fuel mixture temperature, and detonation are all problems caused by pressurizing the intake air-fuel charge and must be controlled to prevent high exhaust emissions and possible engine damage. Boost pressure is controlled through use of a wastegate (bypass valve) or blow-off valve in the exhaust inlet to the turbine. The wastegate is vacuum-operated; the blowoff valve is a spring-loaded device. Mixture temperature can be controlled by injecting a fine water spray into the intake charge or by using an intercooler (heat exchanger) to remove heat from the air charge before it mixes with the fuel. Detonation can be controlled by retarding ignition timing mechanically through a vacuum diaphragm or electronically through a detonation sensor and the engine computer.

Review Questions

Choose the single most correct answer.
Compare your answers to the correct answers on page 507.

1. Supercharging delivers the air-fuel mixture to the cylinder at:
   a. Lower than atmospheric pressure
   b. Atmospheric pressure
   c. Higher than atmospheric pressure
   d. Three times atmospheric pressure

2. Which is *not* true of superchargers?
   a. They cannot operate at very low rpm
   b. They are mechanically driven
   c. There are two types
   d. Mechanical superchargers consume a lot of engine power

3. Positive displacement pumps:
   a. Contain impellers
   b. Pump the same volume of air each revolution regardless of engine speed
   c. Increase air pressure by decelerating
   d. Are highly efficient

4. A turbine wheel turns because of:
   a. Centrifugal force
   b. Exhaust gas speed
   c. Manifold vacuum
   d. Exhaust gas expansion

5. Of the heat energy contained in gasoline:
   a. 50 percent is converted to mechanical power
   b. 50 percent is lost to cooling
   c. 25 percent is converted to engine power
   d. 50 percent goes out the exhaust system

6. Despite the high compression pressures turbochargers achieve, they are okay for emission-controlled engines because:
   a. They have fixed compression ratios
   b. Turbocharger boost can be varied to meet engine needs
   c. They are placed between the air cleaner and carburetor
   d. They are placed between the carburetor and engine

7. Which is not a method of controlling a turbocharger system?
   a. Changing the amount of boost
   b. Cooling the compressed mixture
   c. Readjusting the carburetor idle
   d. Altering spark timing

8. The wastegate controls boost by controlling:
   a. Exhaust gas flow
   b. Compressed air
   c. Air-fuel mixture
   d. Exhaust emissions

9. Retarding spark time controls detonation by:
   a. Cooling the compressed mixture
   b. Cooling the exhaust manifold
   c. Reducing compression pressure
   d. Lowering peak temperature of combustion

10. Technician A says that turbochargers are constant-displacement pumps.
    Technician B says that Roots blowers are variable-displacement pumps.
    Who is right?
    a. A only
    b. B only
    c. Both A and B
    d. Neither A nor B

11. Boost control can be limited by a wastegate or by:
    a. An intercooler
    b. A blow-off valve
    c. A detonation sensor
    d. Manifold vacuum

12. Engine exhaust is used to drive a turbocharger:
    a. Turbine
    b. Wastegate
    c. Compressor
    d. Intercooler

13. The momentary hesitation between throttle opening and boost delivery by the turbocharger is called:
    a. Underboost lag
    b. Turbo lag
    c. Ignition lag
    d. Compression lag

14. Technician A says that a turbocharger installed upstream from the fuel source compresses only air.
    Technician B says that a turbocharger installed downstream from the fuel source compresses air and fuel.
    Who is right?
    a. A only
    b. B only
    c. Both A and B
    d. Neither A nor B

15. Technician A says that energy to drive a turbocharger comes from exhaust heat.
    Technician B says that energy to drive the turbocharger comes from exhaust pressure.
    Who is right?
    a. A only
    b. B only
    c. Both A and B
    d. Neither A nor B

16. Which is *not* a benefit of high compression ratios?
    a. Increased thermal efficiency
    b. Increased volumetric efficiency
    c. Decreased NO_x formation
    d. Increased compression pressure and temperature

17. A heat exchanger used to control intake air charge temperature is called a(n):
    a. Wastegate
    b. Blow-off valve
    c. Air charge temperature valve
    d. Intercooler

18. Technician A says that except for their power source, superchargers and turbochargers do essentially the same job.
    Technician B says that reducing turbocharger size and weight is one way of minimizing turbo lag.
    Who is right?
    a. A only
    b. B only
    c. Both A and B
    d. Neither A nor B

PART FIVE

Emission Control Systems

Chapter

18

Variable and Flexible Fuel Systems

All domestic carmakers have designed and produced variable fuel vehicles (VFV's) or flexible fuel (FF) vehicles. VFV's/FF's can operate on unleaded gasoline, or on a new fuel called **M85**, a blend of 85 percent **methanol** and 15 percent unleaded gasoline. In fact, they can operate equally well on any blend of methanol and gasoline between 0/100 percent and 85/15 percent.

GM uses the term VFV, while Ford calls its version a FF system. To avoid confusion throughout this chapter, we will use the GM VFV terminology when referring to those vehicles designed to run on an M85 methanol-blend fuel.

The use of M85 or other methanol blends in the vehicle's fuel does not change basic operation of the fuel injection system, but it does require a considerable redesign of system components for compatibility. In this chapter, we will learn about some of the changes required, as well as necessary precautions to be observed, when servicing a VFV fuel system.

VARIABLE OR FLEXIBLE FUEL VEHICLES

Vehicles capable of running on fuels other than gasoline will play an increasing role in meeting the emission requirements established by the federal government, as well as regulations promulgated by individual state governments. California, for example, has legislated that a specified percentage of vehicles sold within the state must, by the year 2000, meet emission standards so stringent that most of the vehicles sold in 1995 could not meet them. California's eventual goal is to eliminate vehicle emissions altogether.

During the late 1980s, the most feasible approach seemed to be the conversion of the internal combustion fuel system to one that will burn a combination of gasoline and methanol, figure 18-1. Methanol was a leading choice as a gasoline substitute for many reasons:

- Performance
- Cost
- Availability
- Ease of transition
- Reduced emissions.

Although there are many political factors involved concerning the final approach taken by the domestic automotive industry, both the fuel and cars to use it gradually are becoming available.

GM introduced an M85 methanol-blend system on the 3.1-liter V6 MFI engine used in some 1991 and later Chevrolet Lumina models, figure 18-1. Ford brought forth an M85 system on some 1992 and later Taurus, Sable, and Econoline

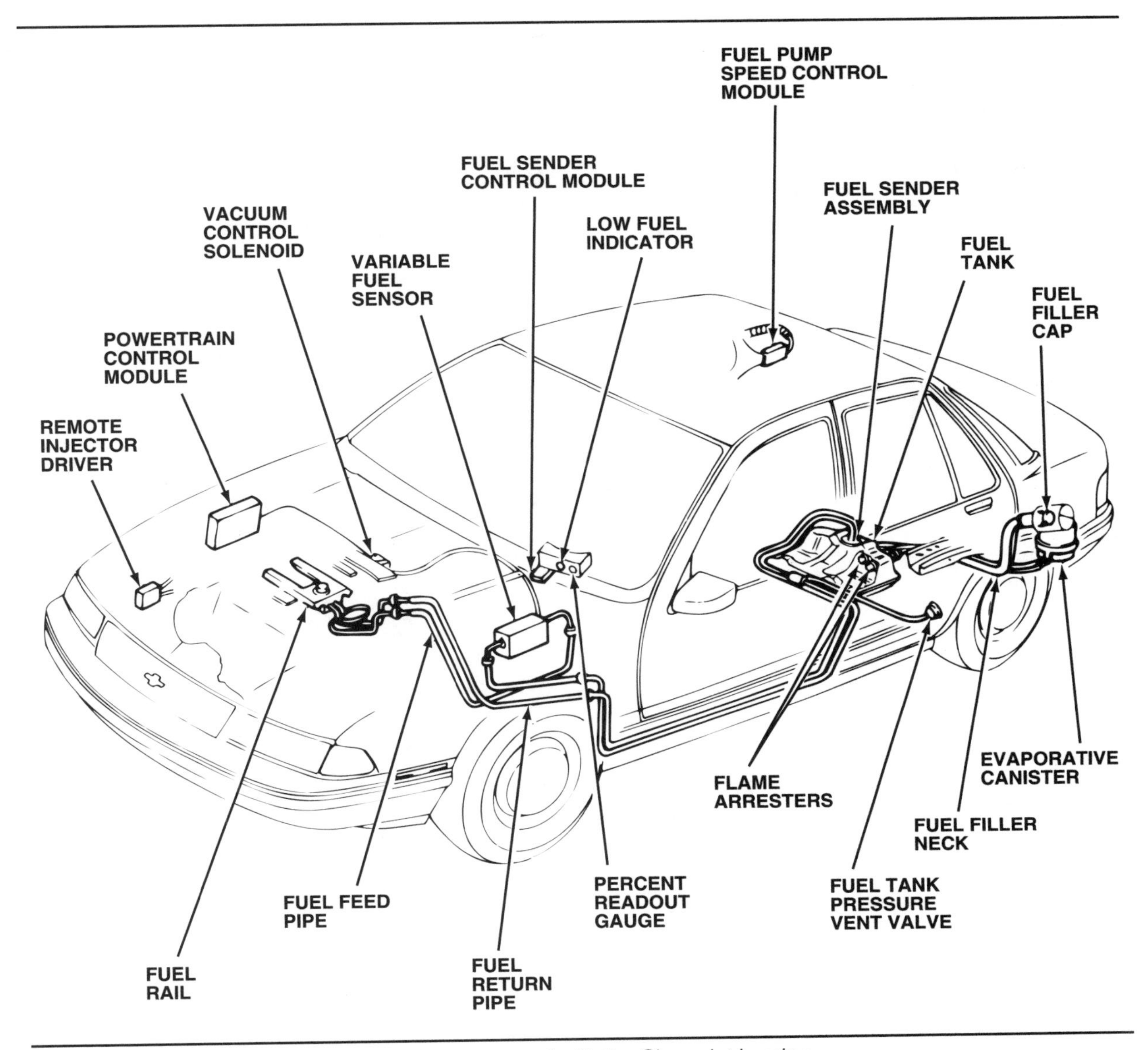

Figure 18-1. GM's methanol variable fuel system used with the Chevrolet Lumina.

models. Chrysler also has its own variable fuel version, but we will look only at the GM Lumina and Ford Econoline systems as representative examples.

Servicing the variable fuel system, however, will require additional knowledge on the part of the technician, since its components differ considerably from those used with gasoline fuel systems.

M85 AND METHANOL-BLENDED FUELS

One major difference between methanol and gasoline is the nature of methanol and its properties. Methanol can be manufactured from wood, coal, or natural gas and contains about 60 percent of the energy present in regular unleaded gasoline, figure 18-2. However, methanol is highly poisonous and cannot be made nonpoisonous. It is so dangerous that it can cause head-

M85: A clean-burning alternative fuel with an octane rating of 100 made by blending 85 percent methanol with 15 percent unleaded gasoline. When burned, it produces fewer hydrocarbons than pure gasoline.

Methanol: A clear, tasteless, and highly toxic form of alcohol made from natural gas, coal, or wood. Contains about 60 percent as much energy as gasoline.

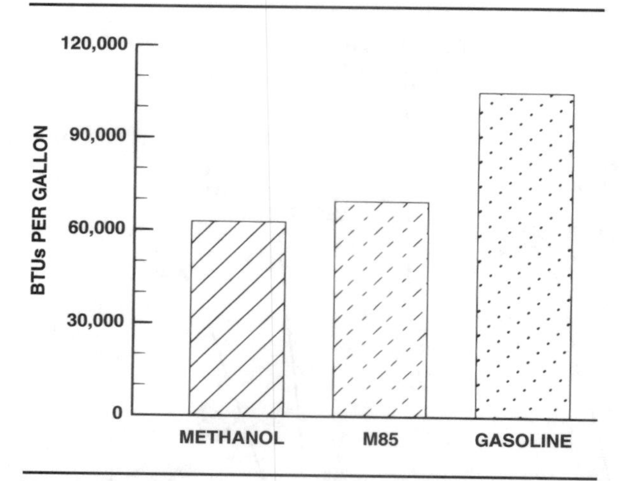

Figure 18-2. This GM illustration compares the energy content of methanol, M85, and gasoline.

aches, blindness, or even death. Unlike gasoline, methanol can be absorbed through your skin, and swallowing only a mouthful can result in death. For this reason, VFV's use either a check ball or screen in the fuel filler neck to prevent the siphoning of M85 by mouth.

Methanol is also colorless and clear (like water) with a faint alcohol odor, and breathing its vapors over a period of time can cause blindness or death. For these reasons, the following precautions should be observed whenever you are working around a VFV fuel system:

- Avoid methanol leaks, spills, and splashing.
- If any of the fuel gets on your skin, wash the affected parts immediately.
- If the fuel spills on your clothes, change them at once and then immediately wash any areas of skin the fuel might have touched. Let the clothes air dry, then machine wash to remove methanol contamination.
- If the fuel spills on a painted surface, flush the surface with water and air dry. If wiped off with a cloth, the paint surface can be permanently damaged.
- Always wear methanol-resistant gloves, such as nitrile gloves, and wrap-around safety glasses or goggles when working on a VFV fuel system.
- Do not use methanol fuel for cleaning parts. This would expose your skin to the substance; it is absorbed rapidly with a slight cooling effect, like the evaporation of rubbing alcohol.
- Like gasoline, methanol is highly flammable and burns violently. If methanol splashes on you and something ignites it, you could be burned severely. Do not smoke when working on a methanol fuel system, and do not work near any source of sparks or an open flame.
- While the exhaust of an engine burning gasoline contains carbon monoxide, an odorless poison, the exhaust of an engine burning methanol contains formaldehyde, which gives off a sharp odor. Exposure to the exhaust fumes from a VFV, however, will cause a severe burning of the eyes, nose, and throat. As with carbon monoxide, exposure to formaldehyde can cause death.
- If you remove the fuel filler cap on a VFV vehicle too quickly, fuel can spray out on you. This generally happens when the tank is nearly full, and is most likely to occur during hot weather. To avoid the possibility of a fuel spray, turn the fuel filler cap slowly, and wait for any hissing noise to stop before removing the cap completely.
- In case of a methanol fire, use a dry chemical extinguisher graded B, C, BC, or ABC, or an ARF- (alcohol-resistant foam) type extinguisher. Do not use water on a methanol-blend fire, as it causes gasoline in the blend to rise to the surface. Since gasoline burns hotter than methanol, spraying water on the fuel will make the situation worse, not better.

Symptoms of Methanol Exposure

Individuals react differently to methanol exposure. While some may experience symptoms immediately, others may not notice anything wrong for 10 to 15 hours after exposure, causing them to not associate the symptoms with methanol. Methanol exposure symptoms generally occur in three stages:

1. Headache, nausea, giddiness, stomach pains, chills, or muscle weakness
2. A 10-to 15- hour period free of any symptoms
3. Failing eyesight, intense nausea, dizziness, difficulty breathing, and intense headaches.

METHANOL AND AUTOMOTIVE ENGINES

Methanol is a highly corrosive agent when it comes in contact with materials such as aluminum, rubber, zinc, and low-grade steels. When used as a fuel, methanol causes corrosion problems with exhaust valves and seats, piston compression rings, and engine oil. Exhaust valves, valve seats, and compression rings can be made of stainless steel or chrome plated to reduce corrosion problems, but engine oil circulates throughout the engine. When standard engine oil is used, methanol tends to degrade it quickly. For this reason, M85-fueled engines require the use of a specific oil formulation, which should be changed at regular 3,000-mile intervals.

If an approved VFV oil is not available, any SAE 10W-30 SG engine oil can be used in an emergency, but only for a short duration. Prolonged use of standard engine oil in an engine running on M85 or other methanol blends can result in engine damage. If a standard engine oil has to be used, it should be changed as soon as possible. The oil filter also should be changed at the same time to prevent contaminating the VFV oil.

The ignition coil used with M85-fueled Luminas differs from those used with gasoline-fueled engines. The M85 coil must provide more current, a faster rise time, and a shorter duration spark. Using a standard coil on a VFV engine will result in cold-start problems because methanol vapor pressure at temperatures below about 50°F (10°C) is not great enough to produce an easily flammable air-fuel mixture. Ford M85 applications use a spark plug with a wider side electrode and a nontapered center electrode for increased heat transfer.

COMMON SUBSYSTEMS AND COMPONENTS

As we have seen, all fuel injection systems have three basic subsystems. The VFV air intake and electronic control systems are essentially the same as those used with a gasoline-powered vehicle. The fuel delivery and metering system, figure 18-3, also operates basically the same as in a gasoline system, with simultaneous injection used on the GM 3.1-liter V6 and sequential injection on Ford's 4.9-liter in-line 6-cylinder.

There are, however, numerous changes in the components required to deal with the methanol

■ The Future of Propane as an Alternative Fuel

As a result of the two energy crises during the 1970s, intense effort was devoted to developing a replacement fuel for gasoline. Ford Motor Company led the way in translating this effort into vehicles that would use such fuels. One result was the propane-fueled 2.3-liter in-line 4-cylinder engine that briefly graced 1982 Ford Granada/Mercury Cougar and 1983 Ford LTD/Mercury Marquis cars.

Propane, or liquified natural gas (LNG), is a liquid form of the same clean-burning natural gas used in the home. Liquified by chilling to -258°F (-161°C), it is stored in thermos-type containers. Extensive testing of propane in automobiles showed that exhaust pollutants were practically eliminated. Since many drivers of motor homes were accustomed to using propane gas, it was thought that cars could operate on the same fuel.

The Ford system included a unique air cleaner, a propane carburetor, a fuel lock, and a converter/regulator assembly—all in the engine compartment. Twin propane tanks mounted beneath the trunk floor provided the necessary fuel. The only external sign of the vehicle's power source was a small "PROPANE" emblem on each front fender.

Propane is a flammable substance like gasoline, but is a vapor at normal temperatures and barometric pressures. The Ford system had two relief valves to vent excessive pressure resulting from high ambient temperatures. If a leak developed or the system vented through the relief valves, the propane immediately vaporized and expanded to about 270 times its liquid volume. Since it is heavier than air, propane settles in low spots and gradually dissipates. This created the possibility of a fire hazard.

For this reason, there were many prohibitions for propane-fueled vehicles:

- Do not vent fuel unnecessarily.
- Do not drain the fuel tanks.
- Do not use a drying oven when refinishing the paint.
- Do not weld near the fuel system tanks or components.
- Do not service the vehicle near electrical equipment, such as motors or switches, that may discharge sparks.
- Do not store or service the vehicle over a confined area, such as a lube pit, where vapors might accumulate.

Technicians were not thrilled with such prohibitions, since the numerous safety precautions involved with propane vehicles interfered with their normal shop operation. Furthermore, these vehicles did not prove popular with the driving public, and the majority of these Ford cars ended up as fleet sales to companies interested in fuel conservation.

The lack of consumer response to Ford's effort was attributed to the difficulty in refueling the propane tanks and a general unavailability of the fuel in many areas. The cars required a greater-than-normal amount of care on the part of drivers and died a quick death in the marketplace.

Gasoline prices started to stabilize at about the same time as the propane-fueled Fords were made available to the public. Most drivers were not willing to cope with the particular problems presented by this alternative fuel source. Thus, the propane Fords passed into the pages of automotive history, much as the Chrysler turbine-powered vehicles had a decade earlier.

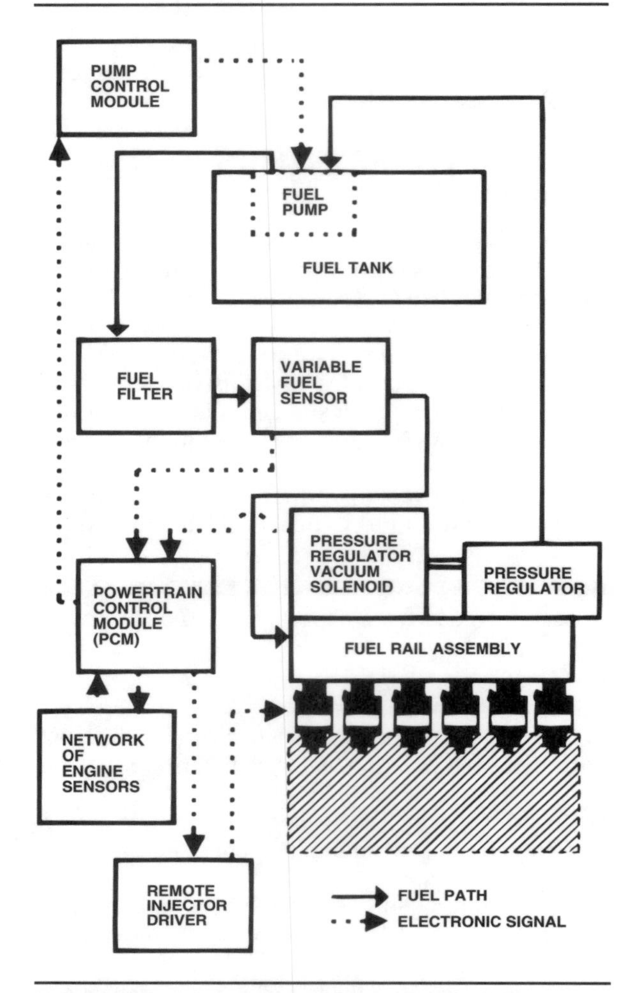

Figure 18-3. A functional diagram of a GM VFV fuel system.

content in the fuel. Methanol is very corrosive to aluminum, rubber, zinc, and low-grade steel. It is less corrosive to stainless steel, nylon, teflon, and chromium. All components that come in contact with methanol-blended fuel are made of methanol-resistant materials, or treated with methanol-tolerant coatings and paints.

GENERAL MOTORS VFV FUEL SYSTEM

The Chevrolet Lumina VFV fuel system, figure 18-1, consists of an in-tank electric fuel pump and fuel pump speed controller, flame arresters, an in-line filter, a variable fuel sensor, a fuel rail with pressure regulator and six injectors, and the necessary connecting fuel distribution hoses and lines. We will examine the fuel system, starting from the fuel rail and working back to the fuel tank.

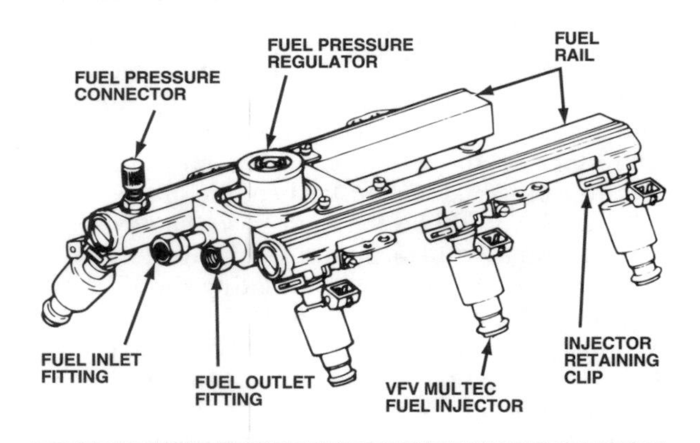

Figure 18-4. The GM R620 fuel rail has a replaceable pressure regulator assembly, but is otherwise the same as fuel rails used with gasoline fuel systems.

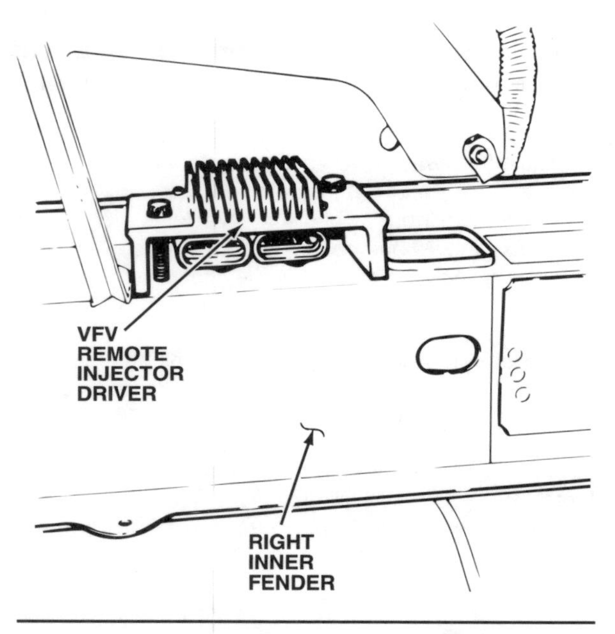

Figure 18-5. A remote injector driver is necessary to protect the PCM on this GM application by dissipating heat generated by the high current draw of the injectors.

Fuel Rails, Injectors, and Hoses

With two exceptions, the basic fuel rail design, figure 18-4, is the same as that used with gasoline fuel systems:

- The aluminum fuel rail is yellow anodized to prevent methanol corrosion, as well as identification as a VFV fuel rail.
- The replaceable fuel pressure regulator diaphragm is made from a methanol-resistant material.

The fuel rail is equipped with low-impedance Multec top-feed fuel injectors. Injector design is

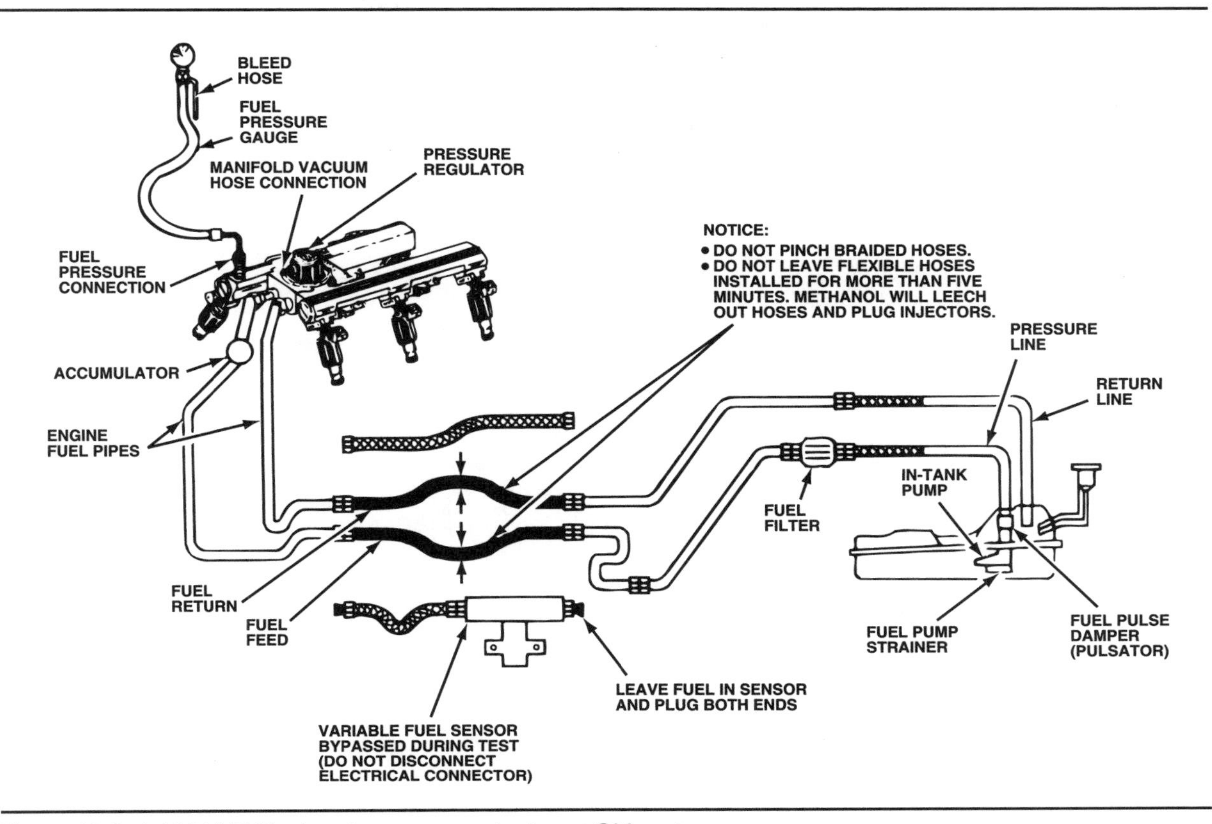

Figure 18-6. A 1991 VFV fuel system pressure test on a GM system.

the same as that used with gasoline systems, but the coil bobbin, coil wire insulation, and the encapsulated wire/bobbin assembly are manufactured with alcohol-resistant materials. However, since methanol contains only about one-half the energy per unit as gasoline, the VFV injectors deliver a higher fuel output (3.84 g/s) than gasoline injectors (1.95 g/s) when wide open. They also have a faster reaction time and a longer-duration wide-open pulse signal than gasoline injectors. Injectors used with 1992 and later models have an operating resistance of 1 ohm, reduced from the 2-ohm injectors used with 1991 models. All VFV injectors draw more current than gasoline injectors. Because this current draw creates more heat than the powertrain control module (PCM) can dissipate, a **remote injector driver** controlled by the PCM is located on the right front fender, figure 18-5. The driver energizes the injectors and dissipates the heat through its own heat sinks.

The fuel feed and return hoses in the engine compartment are braided hoses made of teflon wrapped with a stainless-steel mesh and encased in nylon sheaths at points where the hoses bend. This construction protects the hoses from engine heat as well as methanol corrosion. If standard flexible hoses are used as replacements for the braided hoses, methanol will deteriorate them within a few minutes, resulting in plugged injectors. Fuel hoses on 1991 models use O-rings in the fittings; quick-connect fittings are used on 1992 and later models.

The 1991 fuel system pressure of 41 to 47 psi (284 to 325 kPa) was raised on 1992 and later models to a range of 48 to 55 psi (333 to 376 kPa). The higher system pressure reduces vapor in the fuel lines and minimizes hot-starting problems. Refer to figure 18-6 (1991) or figure 18-7 (1992 and later) when performing a fuel system pressure test.

Variable Fuel Sensor

Because the methanol/gasoline ratio can vary from one tankful of fuel to another, the PCM must be able to determine the percentage of

Remote Injector Driver: The integrated circuit chip that controls fuel injector operation. Normally located in the PCM housing, the chip is installed in a separate housing to protect the PCM from excessive heat generated by the high current draw of the low-impedance injectors.

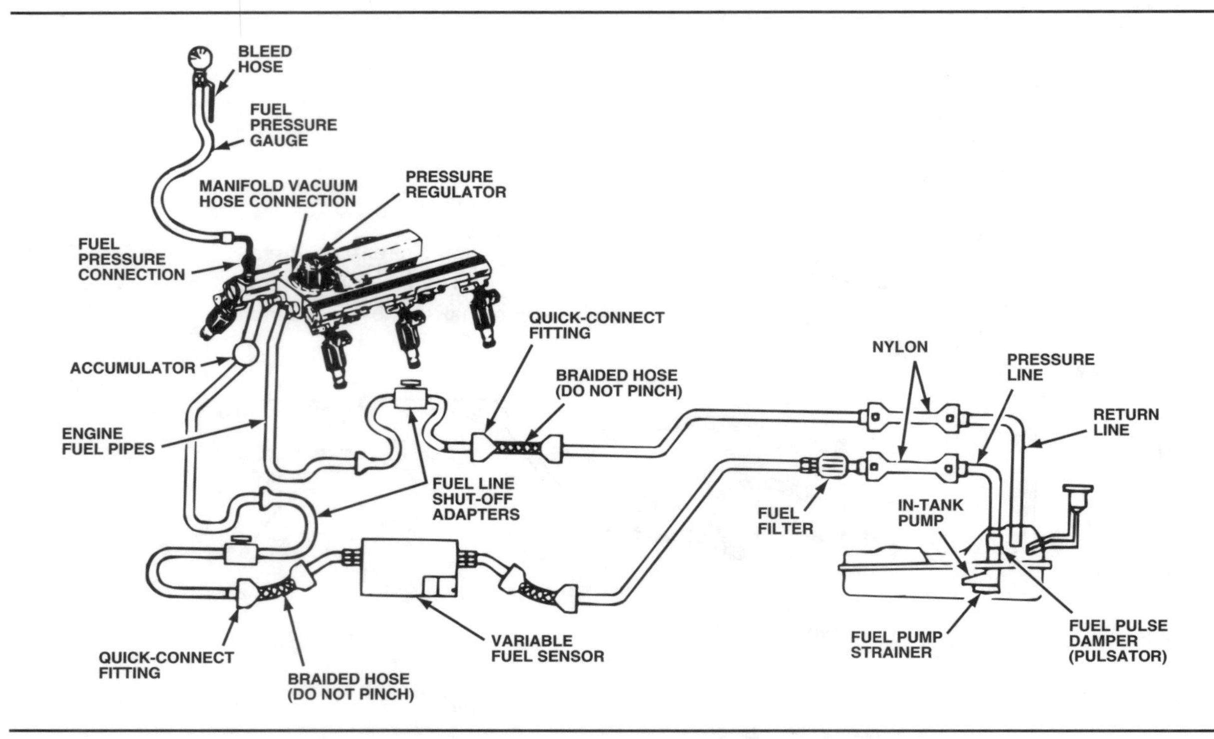

Figure 18-7. A 1992 and later VFV fuel system pressure test on a GM system.

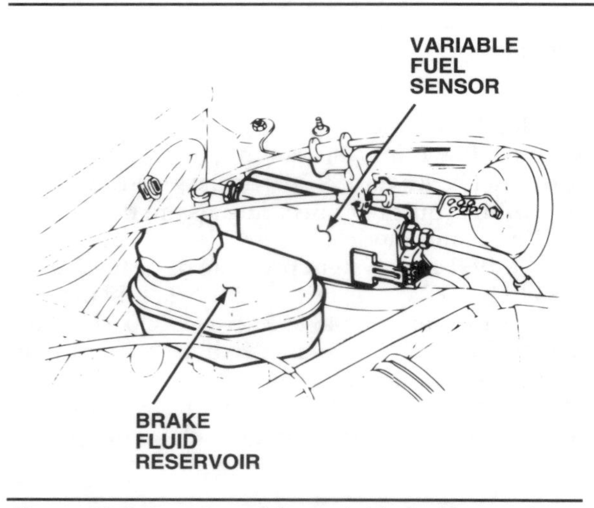

Figure 18-8. Location of the variable fuel sensor in GM's Lumina engine compartment.

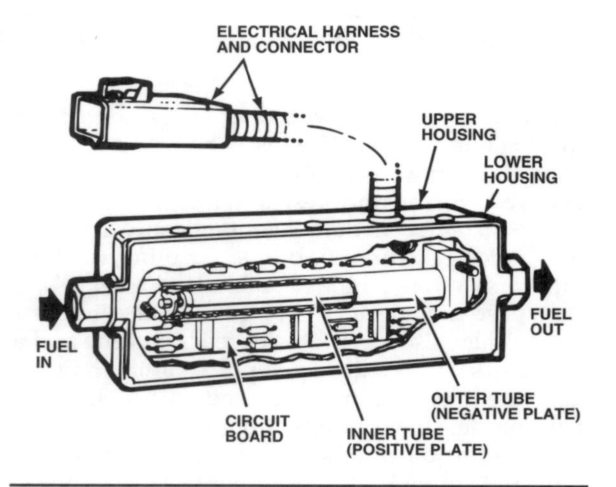

Figure 18-9. A cutaway of GM's 1991 Lumina variable fuel sensor.

methanol in the fuel at all times to control ignition timing and fuel delivery. The variable fuel sensor (VFS) consists of a variable coaxial capacitor sensor, a temperature sensor, and an electronic module contained in a housing and installed in the fuel feed line near the brake booster, figure 18-8. The VFS is powered by voltage through the ignition switch and grounded through the chassis.

When the ignition is turned on, the PCM sends a 5-volt reference voltage to the VFS module. As fuel passes through the sensor housing, its dielectric properties are monitored by the variable capacitor. The voltage buildup in the capacitor is proportional to the percentage of methanol contained in the fuel.

Since the capacitance reading varies according to fuel temperature, the temperature sensor allows the PCM to compensate for any capacitor

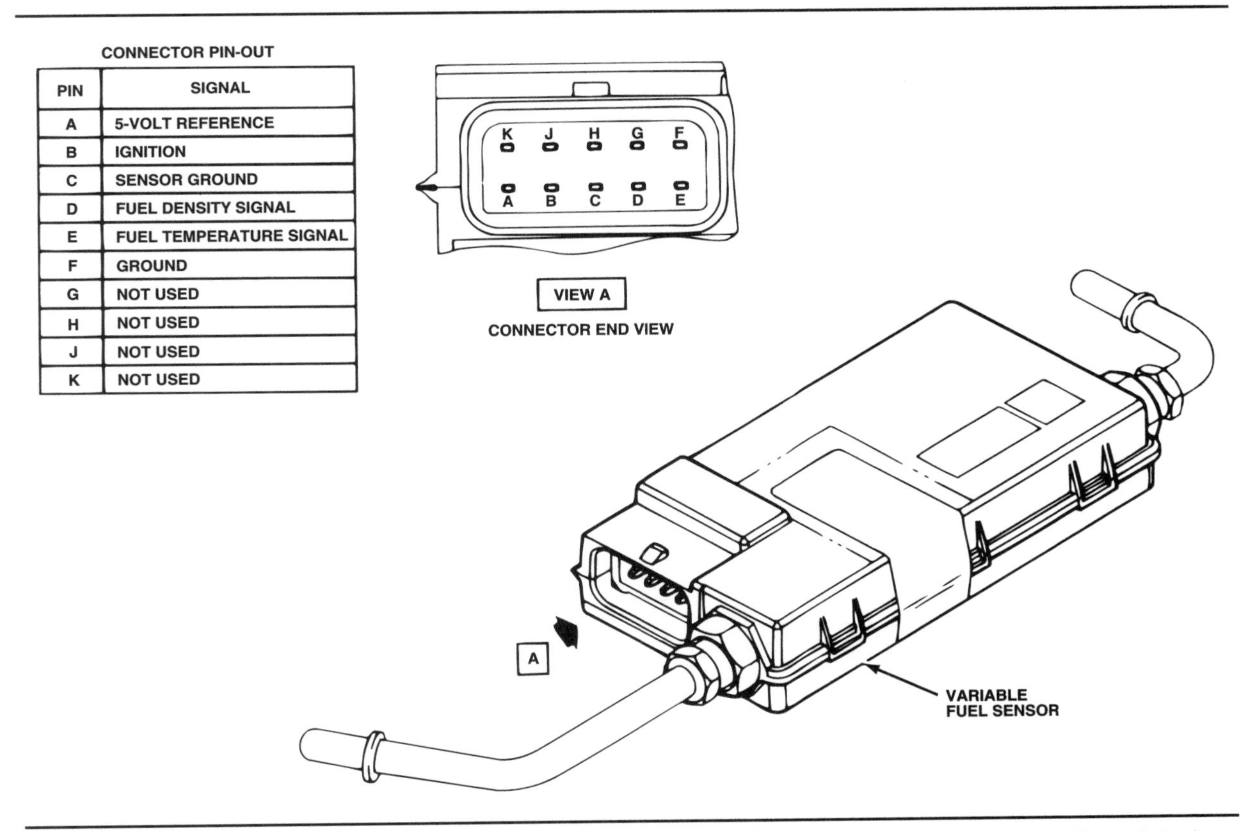

CONNECTOR PIN-OUT

PIN	SIGNAL
A	5-VOLT REFERENCE
B	IGNITION
C	SENSOR GROUND
D	FUEL DENSITY SIGNAL
E	FUEL TEMPERATURE SIGNAL
F	GROUND
G	NOT USED
H	NOT USED
J	NOT USED
K	NOT USED

Figure 18-10. Connector pin-out identification of GM's 1992 and later Lumina variable fuel sensor. The original two-piece pop-riveted case was abandoned in favor of a one-piece foldover case with three locking connectors.

variance. The sensor signals to the VFS module are returned to the PCM as fuel density and fuel temperature signals. The 1991 sensor shown in figure 18-9 uses a separate wiring harness connector and has fuel inlet and outlet nuts to connect into the fuel line. To remove the sensor, the threaded hose fittings are disconnected at the sensor. The housing was redesigned for 1992 and later models, figure 18-10, with an integral wiring harness connector and a length of fuel tubing on either end containing quick-connect fittings. Although the fuel tubes are connected to the sensor housing by inlet and outlet nuts, the threaded hose fittings should not be disconnected to remove the sensor assembly. If sensor removal is required, disconnect it from the system at the quick-connect fittings. The entire unit is manufactured of methanol-tolerant materials.

If the variable coaxial capacitor in the VFS fails, it will cause a code to be set in PCM memory. With the 1991 sensor, a code 56 is stored; the 1992 and later sensors will cause the PCM to set a code 56 (voltage low), code 57 (voltage high), or code 58 (degraded fuel sensor). A temperature sensor failure will set a code 64 (temperature high) or code 65 (temperature low). The Tech 1 scan tool can be used to determine the percentage of alcohol present in the fuel.

The VFS is serviced by replacement only, and should be handled with care. If dropped or otherwise subjected to a similar shock, it must be replaced.

In-Line Fuel Filter

The in-line fuel filter used with the variable fuel system works like any other fuel filter, but is made from methanol-tolerant stainless steel. Filters designed for use with gasoline fuel systems should not be used as a replacement.

Flame Arresters

Two flame arrester devices made of corrugated stainless steel are installed in the filler neck and vent hoses where they connect to the fuel tank, figure 18-11. The flame arresters act as filters to divide and quench any sparks or flames that originate outside the vehicle before they reach the fuel tank. They also serve as antisiphon devices to prevent accidental ingestion of fuel if someone should attempt to siphon the fuel tank.

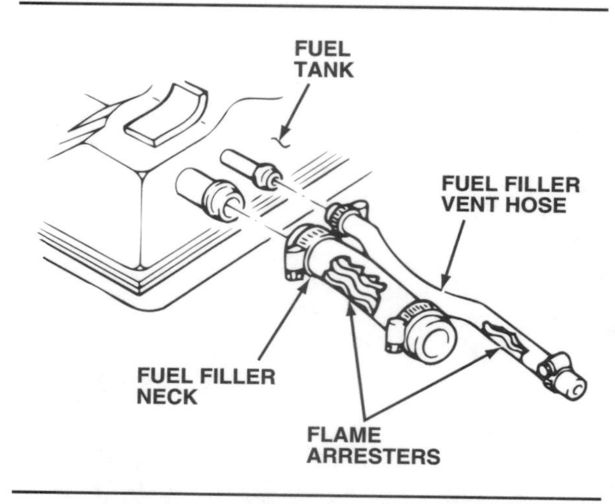

Figure 18-11. As shown on this GM application, flame arresters are installed in the Lumina filler neck and vent hoses.

Fuel Tank and Filler Cap

The size and shape of the fuel tank is identical to other Lumina tanks, but the VFV tank is made from stainless steel to resist the corrosive effects of methanol. The fuel tank filler tube contains a check ball to prevent siphoning fuel. The 1991 filler cap uses a pressure relief valve; 1992 and later filler caps contain a vacuum relief valve. A fuel tank pressure vent valve is used on 1992 and later VFVs to vent excessive tank pressure to the atmosphere.

Fuel Pump

Like most other in-tank fuel pumps used on fuel injected vehicles, the fuel pump is part of the fuel sending unit, figure 18-12. The sending unit and pump assembly also contain a pressure control and rollover valve.

Because the energy density of methanol is about 40 percent less than the energy density of gasoline, the high-pressure roller vane pump must run almost twice as fast as a comparable pump used with a gasoline fuel system if it is to deliver sufficient fuel for engine operation. For this reason, pump capacity has been nearly doubled. Except for the use of a check ball in the pump outlet line to prevent fuel backflow, figure 18-13, the pump design is the same as that installed in the modular fuel assembly and used with gasoline fuel systems.

The fuel pump has a two-stage pumping action. The low-pressure turbine section uses an impeller with a staggered blade design to minimize pump noise and to separate vapor from the liquid fuel. It is designed to prevent vapor lock resulting from the higher vapor pressures common with methanol blends. The roller vane section creates the high pressure required for fuel injection. The end cap assembly contains an outlet check valve, a pressure relief valve, and a radio frequency interference (RFI) module. The brushes are a methanol-tolerant design, with coated springs, insulator disc, plated shunt wire, and soldered connections. All metal parts and wire insulation are treated to resist electro-galvanic corrosion, and all elastomers and other seals are designed to minimize swelling caused by methanol.

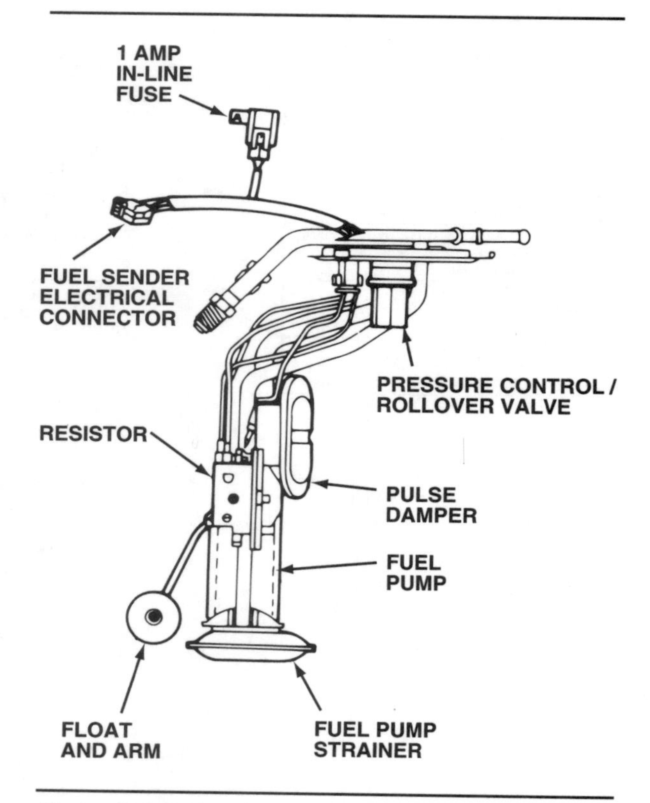

Figure 18-12. The GM variable fuel sender assembly.

If the fuel pump is removed from the sender assembly for service or replacement, always install a new fuel pump strainer before reinstalling the unit in the fuel tank.

Fuel Pump Speed Controller

Fuel pump operational speed on 1991–92 systems is controlled by the PCM and a fuel pump speed controller (FPSC). To prolong pump life, the controller reduces its speed whenever a high fuel flow is not necessary, as when the vehicle is operating on 100 percent gasoline. When the ignition switch is turned on, the PCM energizes the controller for two seconds, allowing the fuel pump to pressurize the system. If ignition reference

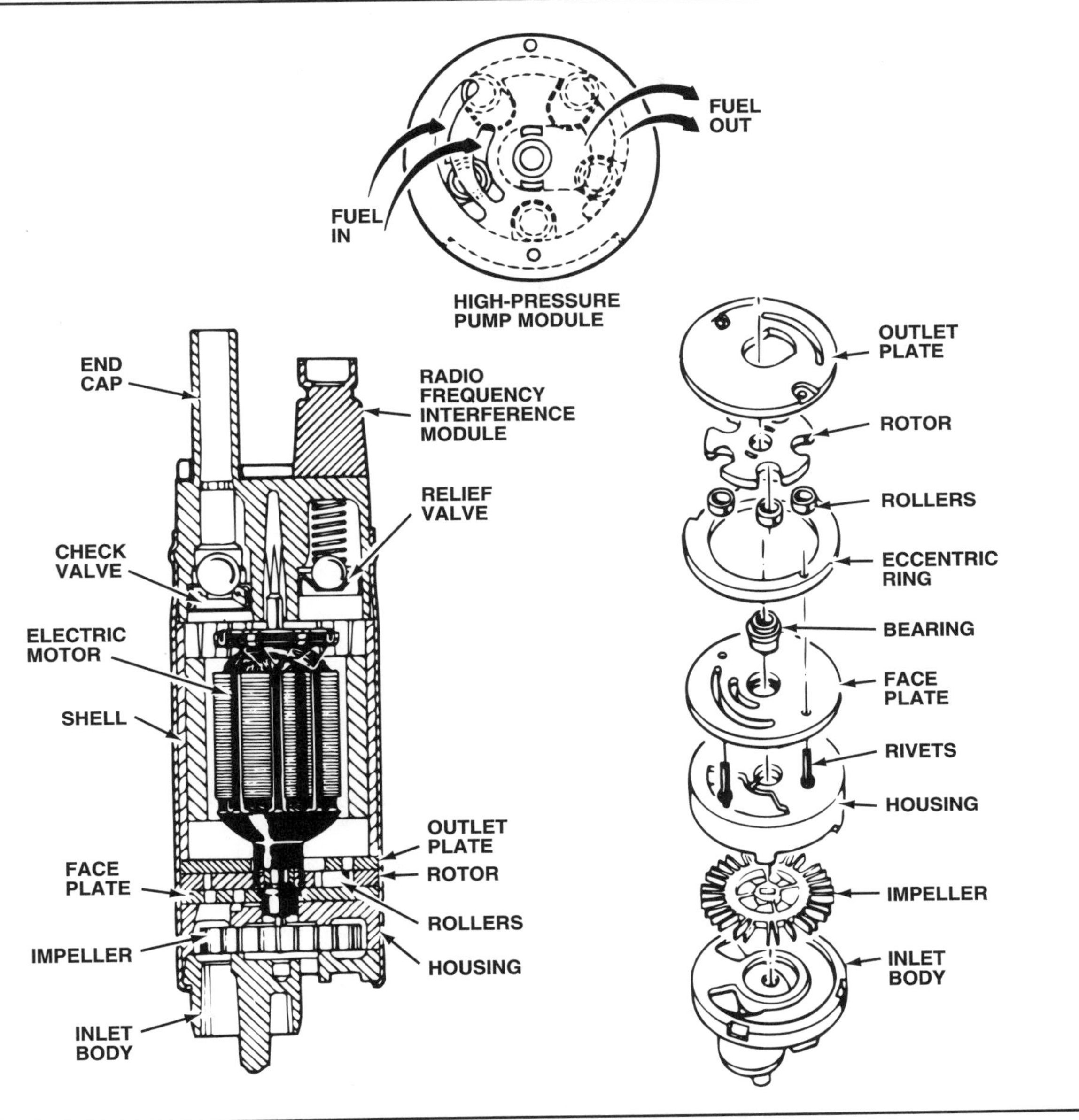

Figure 18-13. A cross section and components of the roller vane fuel pump used with GM's Lumina M85 fuel system.

pulses are not received by the PCM within the two seconds, it deenergizes the controller to shut the pump off. If the FPSC malfunctions, the PCM stores a code 54 (low voltage) in memory. Use of the FPSC was found to be unnecessary and was eliminated on 1993 and later models, allowing the fuel pump to operate at a constant speed.

Fuel Sender Unit and Control Module

Specifically designed for use with variable fuels, the fuel sender unit, figure 18-12, uses a unique electrical circuit. A pulse width modulated (PWM) signal generated by the fuel sender control module is sent to a straight strip resistor on the sender assembly. The resistor modifies the PWM signal according to the position of the float and returns it to the control module, where it is processed and sent to the instrument panel gauge. A low fuel indicator lamp on the instrument panel is controlled by the PCM according to signals from the control module.

As with other components of a variable fuel system, the sender unit is designed to be methanol-tolerant, with stainless steel used for all metal parts, fluorosilicone (orange) and Viton™ (black) seals, and an alcohol-resistant

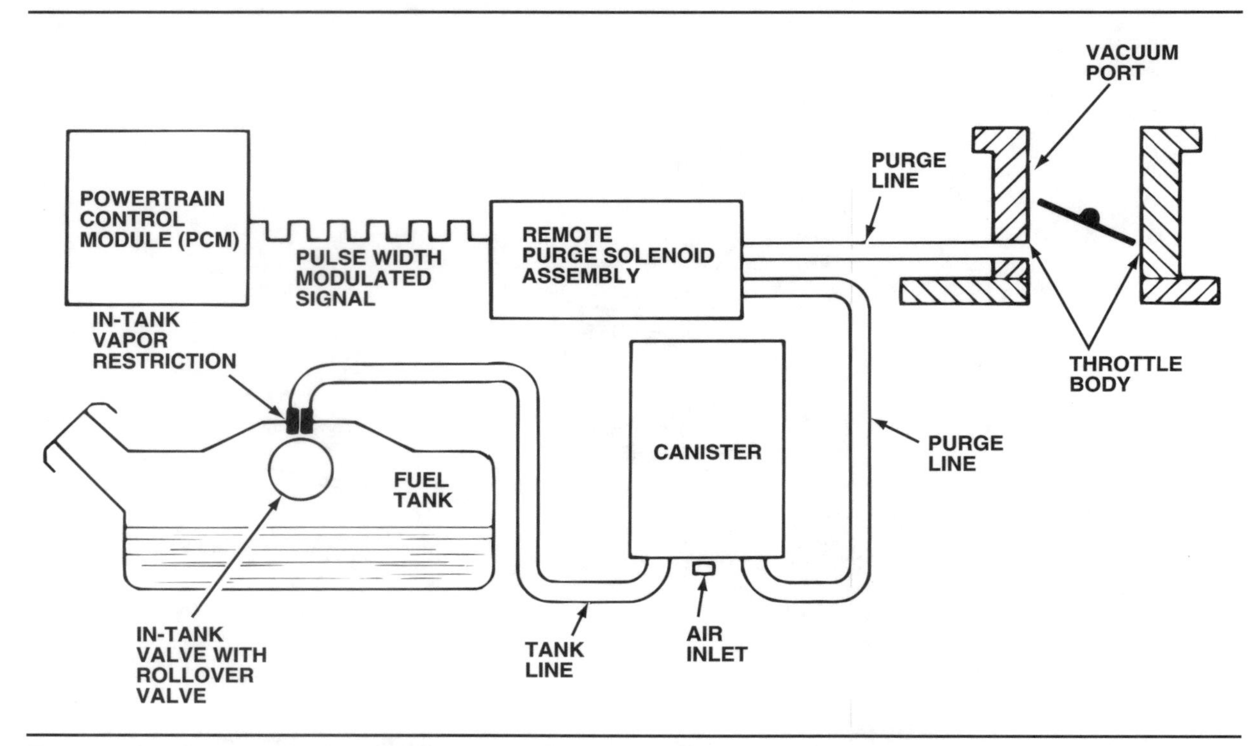

Figure 18-14. GM's Lumina M85 EVAP system with PCM pulse width modulation.

plastic float. All electrical connections have been modified to reduce positive-to-negative terminal current migration and potential shorting. A 1-amp fuse installed in the external sender harness provides circuit protection for the PWM circuit between the control module and strip resistor. If the fuse blows, the fuel gauge will go off-scale to the full side. The fuse is not serviced separately; if it blows, the entire sender assembly must be replaced.

When the sender unit is removed for sender or pump service, check the fuel pulse dampener or pulsator unit, figure 18-12, for degraded dampener seals. The 1991–92 dampeners use orange (fluorosilicone) seals; 1993 and later dampeners have black (Viton™) seals. If the dampener seals have deteriorated, they may have contaminated the in-line fuel filter. Install a new dampener and change the fuel filter.

The fuel pump strainer and in-line fuel filter are more likely to clog in a VFV fuel system than a gasoline fuel system. If the strainer requires replacement because of contamination, the in-line filter should also be replaced at the same time.

GM VFV EMISSION CONTROL SYSTEMS

The same emission control systems are used on both the gasoline and VFV versions of the 3.1-liter V6 MFI engine. However, the evaporative emission control system and the catalytic converter required modifications.

Evaporative Emission Control System

The EVAP used with methanol fuel, figure 18-14, is essentially the same as that used with a gasoline-powered engine. When the engine is off, fuel vapors from the tank are sent through the combination tank pressure control valve (TPCV) and roll-over valve to a charcoal canister for storage. When the engine is running, the vapors are purged from the canister by intake airflow and burned in the combustion chambers. Purging is controlled by the PCM.

The canister is made of methanol-tolerant materials and used in an inverted position, figure 18-15. The canister capacity is 2.3 quarts (2200 cc), or 0.7 quarts (700 cc) greater than the canister used with the 189 cu in. (3.1-liter) gasoline-powered Luminas. A unique remote high-purge solenoid valve is controlled by a PWM signal from the PCM. A limited amount of purging takes place during idle or closed-throttle operation. As engine speed increases, so does the amount of purging. Ambient air enters the bottom of the canister through the inlet air tube, where it mixes with the stored vapors. When the purge solenoid is modulated by the PCM, the vapor-air mixture

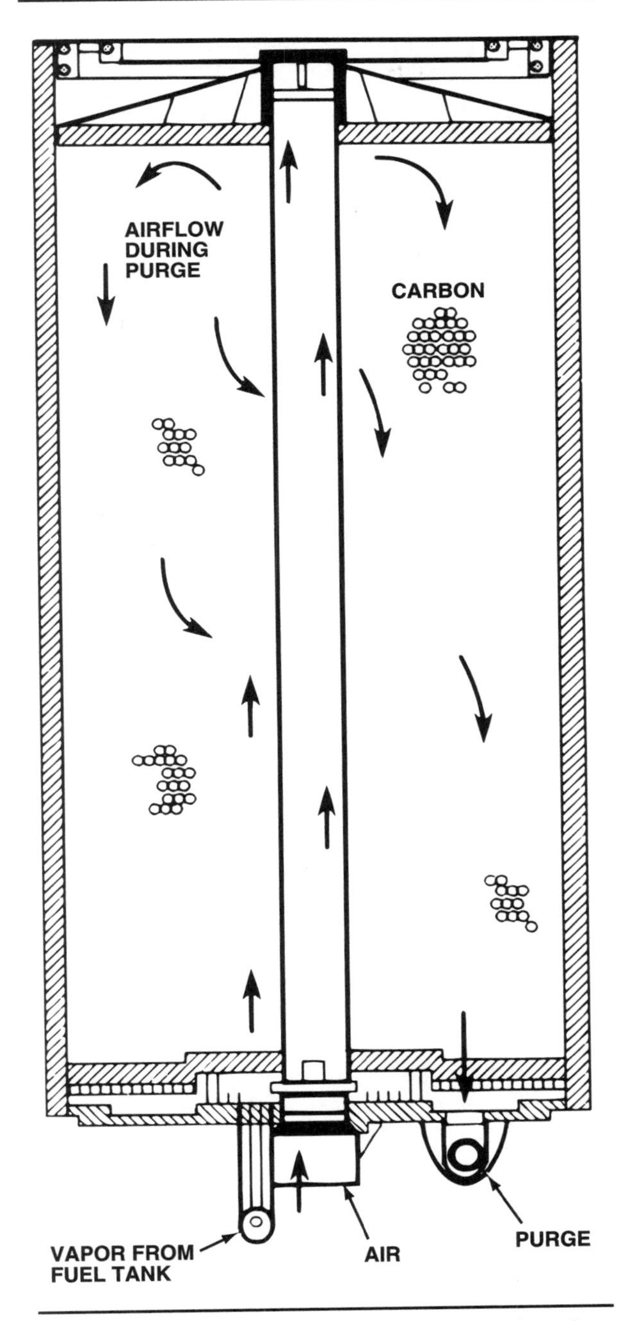

Figure 18-15. Operation of GM's inverted methanol vapor canister.

is drawn from the canister and passes through the air intake throttle body to the intake manifold for burning. If the purge solenoid malfunctions, the PCM will set a diagnostic trouble code.

The canister is color-coded to prevent its use in a gasoline EVAP system, or vice-versa. All GM canisters used with gasoline systems are black; the variable fuel canister is white, and is located underneath the rear of the vehicle, figure 18-16.

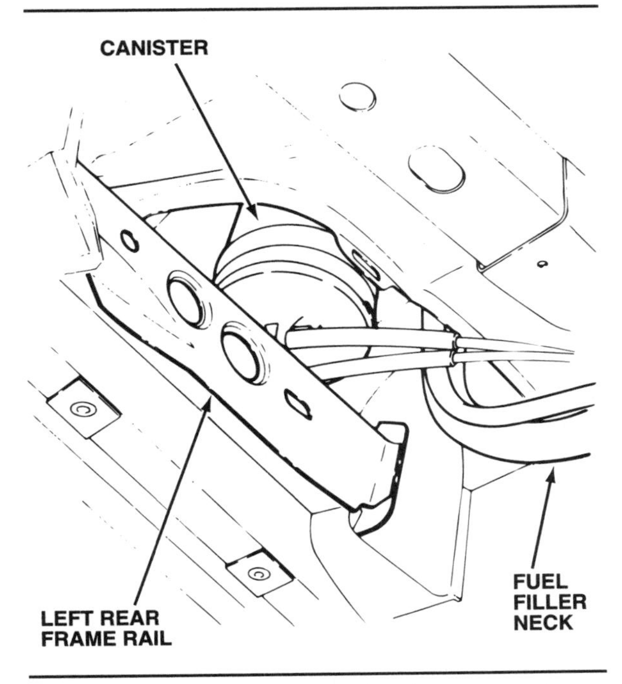

Figure 18-16. Location of GM's Lumina methanol vapor canister.

Catalytic Converter

The catalyst in a regular production converter cannot cope with the formation of formaldehyde emissions that result from methanol combustion. For this reason, the catalyst used in a VFV converter has been recalibrated to make it effective in controlling the combustion by-products of methanol-blended fuels.

FORD FLEXIBLE FUEL (FF) SYSTEM

The Ford FF system used on the Econoline consists of an in-tank electric fuel pump and sending unit, an in-line filter, a VFS sensor, a fuel rail with pressure regulator and six injectors, and the necessary connecting fuel distribution hoses and lines. We will examine the differences in the fuel system components from those of the GM application, starting in the engine compartment.

- The same fuel rail is used on M85 and gasoline engines. The pressure regulator diaphragm is made from a methanol-resistant material. The fuel rail is equipped with high-resistance (13 to 16 ohm) methanol-resistant injectors.
- The injector fuel nozzles are larger than those on gasoline injectors to provide the higher flow capacity required when using a methanol blend.
- A cold start injector (CSI) installed on top of the intake air throttle body near the idle air control (IAC) solenoid provides additional fuel for cold weather starting.

Figure 18-17. Ford uses a linear methanol monitor attached to the top of the Econoline steering column shroud to determine the percentage of methanol in the fuel.

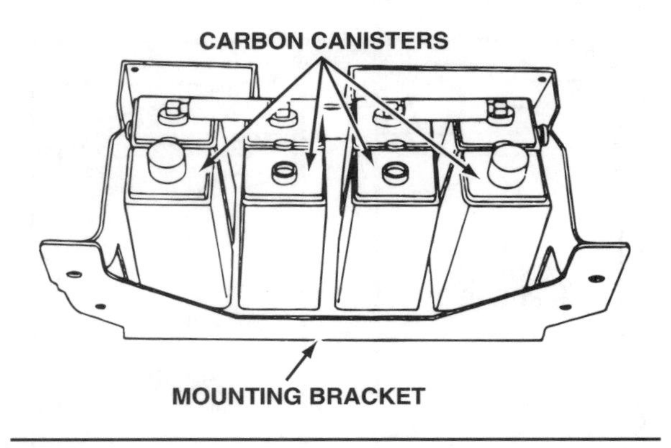

Figure 18-18. On Ford FF vehicles, the increased canister capacity provided by four canisters accommodates vapor pressures for extended idle periods.

- Stainless steel and nylon fuel lines are used to prevent corrosion damage. Standard rubber hoses or steel fuel lines should not be used as replacements, as methanol deteriorates these materials and will cause plugged injectors.
- The fuel system pressure of 45 to 60 psi (310 to 415 kPa) reduces vapor in the fuel lines and minimizes hot starting problems.
- The fuel sensor used on Ford M85 fuel systems performs the same function as that used by GM, and operates in a similar manner. Unlike the GM system, however, its reading is translated by a control module into a linear display positioned between the steering column shroud and instrument panel, figure 18-17.
- The in-line fuel filter works like other fuel filters, but is made of methanol-tolerant materials. Filters designed for use with gasoline fuel systems should not be used as a replacement.
- The size and shape of the fuel tank is the same as other Econoline tanks, but the steel M85 tank contains a methanol-resistant coating to resist the corrosive effects of methanol.
- The fuel tank filler tube uses a methanol-resistant coating and contains a screen to prevent siphoning fuel from the tank.
- The combination in-tank pump/sending unit assembly consists of nickel-plated metal parts. Composite parts of the assembly are made of materials that are compatible with methanol.

FORD FF EMISSION CONTROL SYSTEMS

The emission control systems used on gasoline and M85 versions of the 4.9-liter SFI engine are essentially the same. Like the GM VFV 3.1-liter V6, however, the EVAP and the catalytic converter are different.

Evaporative Emission Control System

The Ford M85 EVAP system functions in the same way as the EVAP system on a gasoline-powered vehicle, but there are two major differences in the components:

- Increased canister capacity
- A vapor management valve.

To provide increased vapor capacity, four carbon canisters are connected in pairs and tray-mounted, figure 18-18, to the outer side of the left frame rail. With the engine off, fuel vapors reach the canister assembly through vapor valves installed at each end of the tank. A vapor management valve (VMV) prevents vapor flow from the canister until the engine is started. Once the engine is running, vapor purge is controlled by Ford's Electronic Engine Control, Version IV (EEC-IV) module.

SUMMARY

Variable fuel vehicles (VFV's) or flexible fuel (FF) vehicles can operate on unleaded gasoline or M85, a blend of 85 percent methanol and 15 percent unleaded gasoline. Using M85 does not change basic operation of the fuel injection system, but it does require component redesign for compatibility.

Methanol is highly poisonous and can cause headaches, blindness, or even death. While some people exposed to methanol poisoning may experience symptoms immediately, others may not notice anything wrong for 10 to 15 hours.

Methanol is highly corrosive to aluminum, rubber, zinc, and low-grade steels, and tends to degrade ordinary engine oil. Methanol-system exhaust valves, valve seats, and compression rings are made of stainless steel or are chrome plated to reduce corrosion problems. Engines combusting methanol require the use of a specific oil formulation. All components that come in contact with methanol-blended fuel are made of methanol-resistant materials, or treated with methanol-tolerant coatings and paints.

During the early 1990s, GM and Ford had the most highly developed variable fuel systems for production cars. GM uses the term variable fuel vehicle (VFV), while Ford calls its version a flexible fuel (FF) system.

Review Questions

Choose the single most correct answer. Compare your answers to the correct answers on page 507.

1. The alternative fuel M85 consists of:
   a. 85 percent unleaded gasoline, 15 percent methanol
   b. 85 percent ethanol, 15 percent gasoline
   c. 85 percent methanol, 15 percent gasoline
   d. 85 percent unleaded gasoline, 15 percent ethanol

2. The energy content of methanol is approximately ___ that of gasoline.
   a. 60 percent
   b. 85 percent
   c. 25 percent
   d. 15 percent

3. Which of the following materials is least affected by methanol?
   a. Rubber
   b. Zinc
   c. Aluminum
   d. Teflon

4. Technician A says that symptoms of exposure to methanol fuel occur immediately after exposure. Technician B says that some symptoms may not occur for several hours after exposure. Who is right?
   a. A only
   b. B only
   c. Both A and B
   d. Neither A nor B

5. Technician A says that the Lumina VFV fuel rail is yellow anodized to make it impervious to methanol. Technician B says that the aluminum fuel rail is lined with teflon. Who is right?
   a. A only
   b. B only
   c. Both A and B
   d. Neither A nor B

6. The GM variable fuel sensor contains a variable capacitor and electronic circuitry to determine the percentage of:
   a. Unleaded gasoline
   b. Methanol
   c. Water contamination
   d. None of the above

7. The GM variable fuel sensor also senses fuel:
   a. Consumption
   b. Purity
   c. Temperature
   d. Flow

8. Technician A says that a remote injector driver is used because GM VFV injectors draw more current than gasoline injectors. Technician B says that VFV fuel system pressure on 1992 and later Luminas is 41 to 47 psi (284 to 325 kPa). Who is right?
   a. A only
   b. B only
   c. Both A and B
   d. Neither A nor B

9. Ford M85 fuel system pressure can be checked with a pressure tester that reads:
   a. 41 to 47 psi (284 to 325 kPa)
   b. 48 to 55 psi (333 to 376 kPa)
   c. 45 to 60 psi (310 to 415 kPa)
   d. 41 to 55 psi (284 to 376 kPa)

10. Technician A says that the operating resistance of Ford FF injectors is 2 ohms. Technician B says their resistance is 1 ohm. Who is right?
    a. A only
    b. B only
    c. Both A and B
    d. Neither A nor B

11. Which of the following devices prevents siphoning fuel from the tank of a VFV Lumina?
    a. An arrester in the filler hose
    b. A check ball in the tank filler tube
    c. A fuel tank pressure vent valve
    d. All of the above

12. Technician A says that regular production catalytic converters can be used with a VFV. Technician B says that VFV converters must be recalibrated to reduce formaldehyde emissions. Who is right?
    a. A only
    b. B only
    c. Both A and B
    d. Neither A nor B

13. Technician A says that the variable fuel sensor on a 1992 Lumina VFV is removed by disconnecting the threaded hose fittings. Technician B says that the variable fuel sensor on a 1991 Lumina VFV is removed by using the quick-connect fittings. Who is right?
    a. A only
    b. B only
    c. Both A and B
    d. Neither A nor B

14. Which of the following statements regarding the oil used in M85-fueled engines is false?
    a. It should be changed every 3,000 miles (4828 km)
    b. It is specially formulated for use in methanol-fueled engines
    c. The oil filter is manufactured of methanol-tolerant materials
    d. Any SAE 10W-30 engine oil can be used in an emergency

Chapter

19

Positive Crankcase Ventilation

Crankcase ventilation has been a problem since the beginning of the automobile. No piston ring, new or old, can provide a perfect seal between the piston and the cylinder wall. When an engine is running, the pressure of combustion forces the piston downward. This pressure also forces gases and unburned fuel from the combustion chamber past the piston rings and into the crankcase. These gases are called crankcase vapors.

These combustion by-products, particularly unburned hydrocarbons, form blowby, figure 19-1. The crankcase must be ventilated to remove these vapors and gases. However, the crankcase on modern engines cannot be ventilated directly to the atmosphere because the hydrocarbon vapors add to air pollution. Positive crankcase ventilation (PCV) systems were developed to ventilate the crankcase and recirculate the vapors to the engine's induction system.

DRAFT TUBE VENTILATION

Before the 1960s, most vehicles used a road draft tube, figure 19-2, to ventilate the engine crankcase. This is nothing more than a tube connected to the engine crankcase that allows vapors to pass into the air. Fresh air enters through a vented oil filler cap and passes into the crankcase to mix with the vapors. Airflow past the road draft tube creates a vacuum that draws the crankcase vapors out into the atmosphere. Road draft ventilation has three drawbacks:

- At vehicle speeds below 25 mph, not enough vacuum is created to remove the vapors from the crankcase.
- At high vehicle speeds, high crankcase pressure may combine with low pressure at the end of the tube to draw oil from the engine.
- It releases the unburned hydrocarbons directly into the atmosphere, causing air pollution.

OPEN PCV SYSTEMS

Open PCV systems use intake manifold vacuum to draw vapors from the crankcase, creating a positive airflow through the engine when it is running. The vapors mix with intake air and are sent to the combustion chambers for burning. These systems were first used on 1961 California cars and installed nationwide in 1963. Open crankcase ventilation systems can be divided into three types.

Type 1 System

In a Type 1 system, a hose and PCV valve connect the crankcase with the intake manifold, figure 19-3. Fresh air enters the crankcase

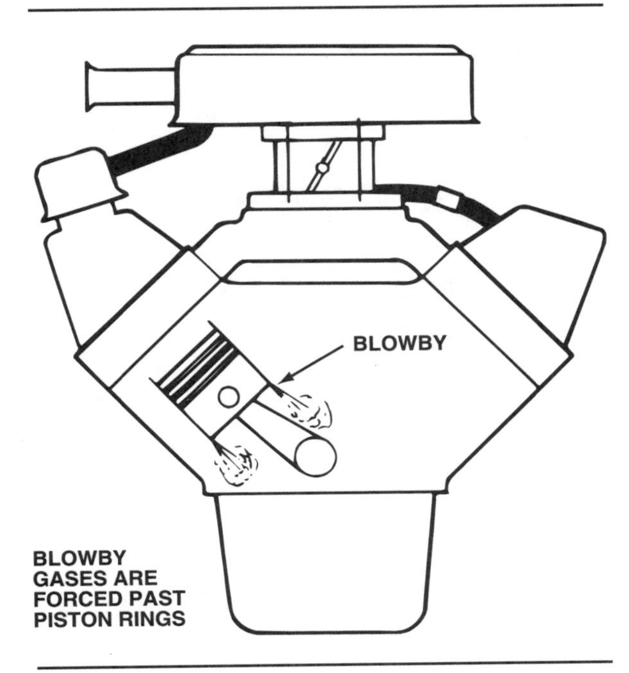

Figure 19-1. Piston rings do not provide a perfect seal. Combustion gases blow by the rings into the crankcase.

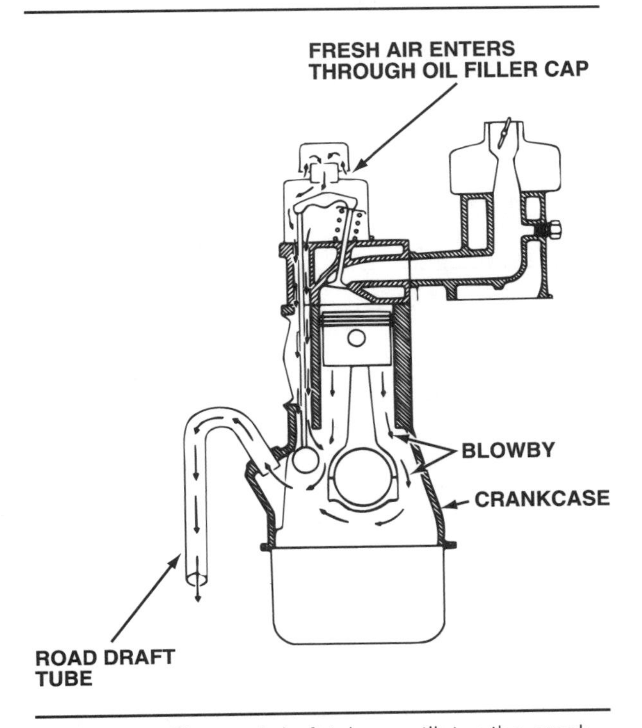

Figure 19-2. The road draft tube ventilates the crankcase to the atmosphere when the car is moving.

through a vented oil filler cap, mixes with the crankcase vapors, and is metered through the plunger-type PCV valve to the intake manifold, where it returns to the cylinders for burning.

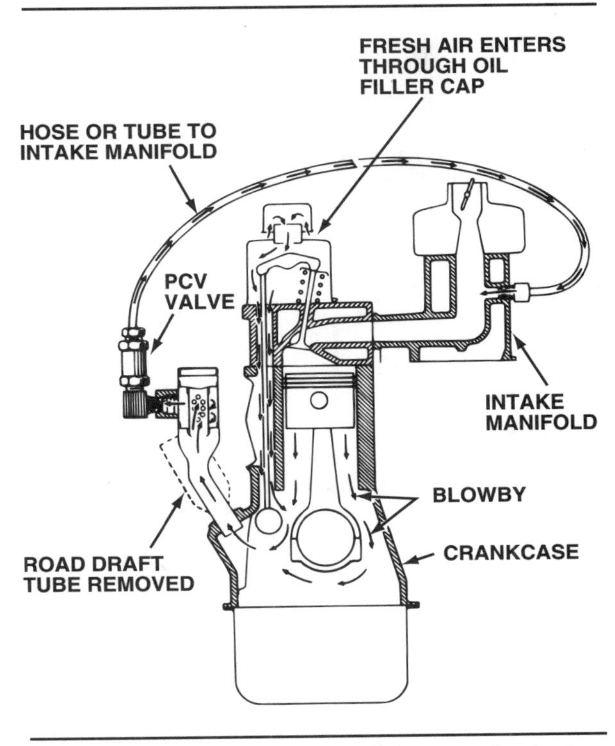

Figure 19-3. In a Type 1 open PCV system, fresh air enters through the oil filler cap. Crankcase vapors return to the intake manifold through a PCV valve and hose or tube.

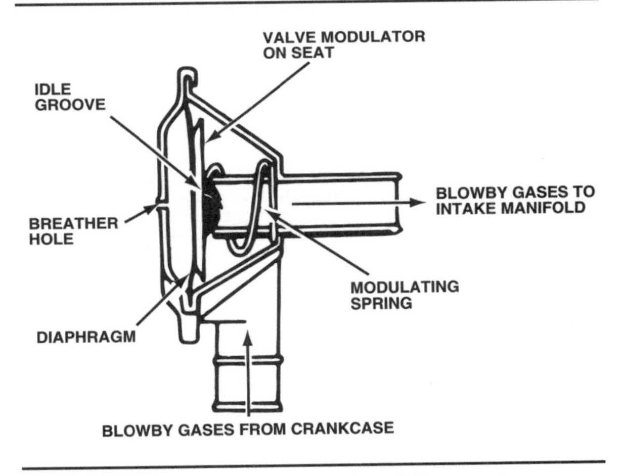

Figure 19-4. When the engine is at idle, crankcase vacuum closes the Type 2 PCV valve. Vapors flow through the idle groove at about three cubic feet (85 liters) per minute.

Type 2 System

The Type 2 system uses a diaphragm-type PCV valve that responds to crankcase vacuum, figures 19-4 and 19-5. The oil filler cap contains an orifice large enough to allow the right amount of air to enter, but small enough to maintain crankcase vacuum. The orifice must remain

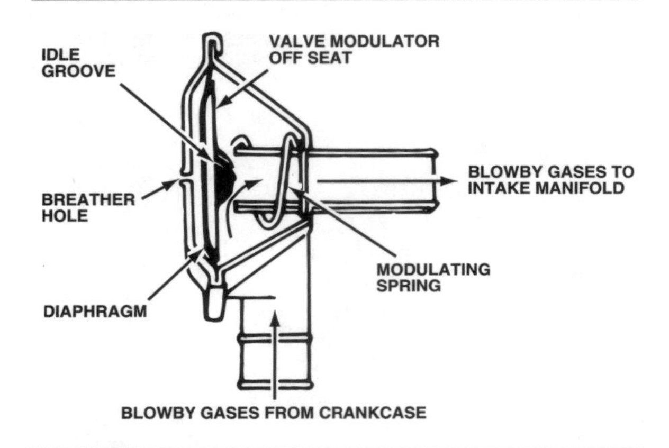

Figure 19-5. When the engine is at cruising speed, crankcase pressure opens the Type 2 PCV valve. The flow rate depends on the amount of blowby created by the engine.

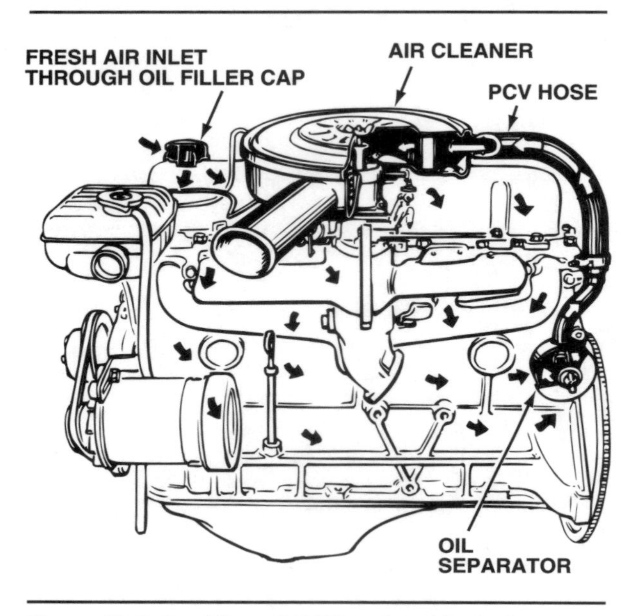

Figure 19-6. Ford Type 3 PCV systems had oil separators at the crankcase outlet.

open for the system to work correctly. The Type 2 system was mainly used on imported cars, particularly British models of the mid-1960s.

Type 3 System

The Type 3 system uses no PCV valve to meter blowby gases to the carburetor. This tends to enrich the air-fuel mixture and requires carburetor adjustment to accommodate the richer mixture. Figure 19-6 shows the operation of a typical Type 3 system as used on Ford 6-cylinder engines in the early 1960s. Oil separators at the crankcase outlets minimize the amount of oil drawn through the PCV hose to the air cleaner.

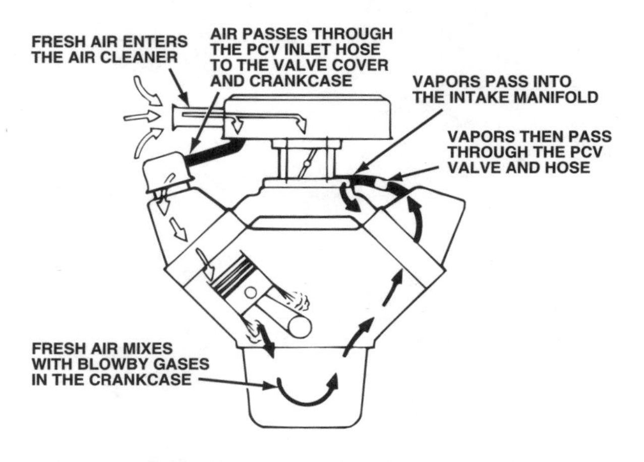

Figure 19-7. Closed PCV system operation under normal conditions.

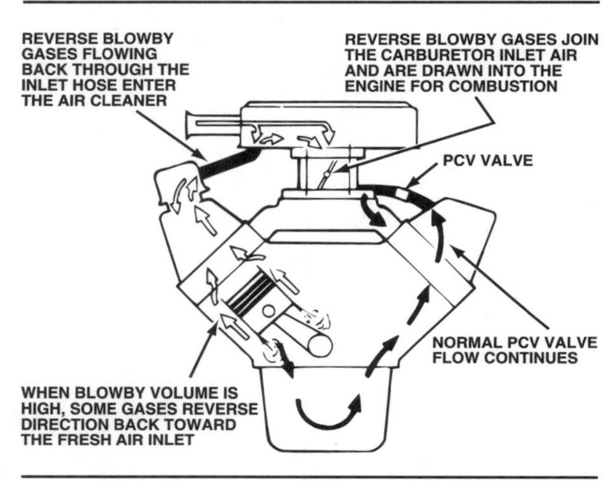

Figure 19-8. Closed PCV system operation under heavy load.

CLOSED PCV SYSTEMS

All new cars sold in the United States since 1968 have a Type 4 closed PCV system. The design of closed PCV systems is essentially the same, regardless of the manufacturer. All use a PCV valve, calibrated orifice or separator, an air inlet filter, and connecting hoses. Some systems use an oil/vapor or oil/water separator instead of a valve or orifice, particularly with turbocharged and fuel injected engines. The oil/vapor separator lets oil condense and drain back into the crankcase. The oil/water separator accumulates moisture and prevents it from freezing during cold engine starts.

Type 4 System

Unlike open PCV systems, the Type 4 system uses a sealed oil filler cap (not vented to the atmosphere). Type 4 systems were required on

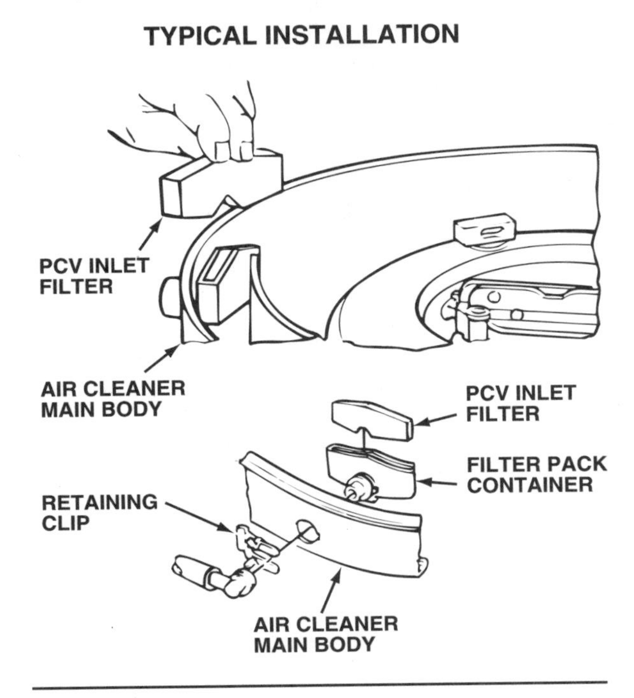

Figure 19-9. The PCV inlet filter on many vehicles is located in the air cleaner housing and is serviced as shown.

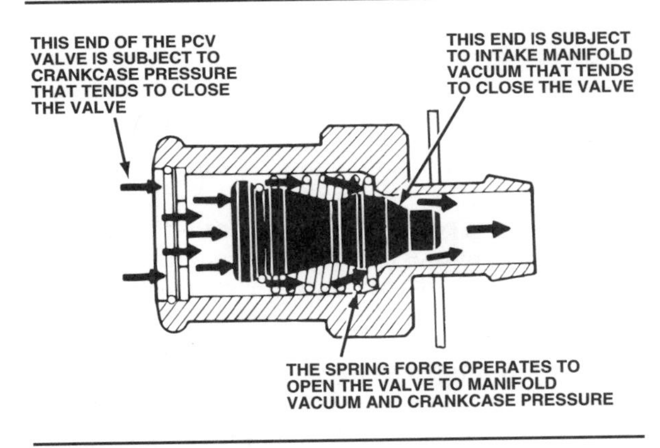

Figure 19-10. Spring force, crankcase pressure, and manifold vacuum work together in regulating PCV valve flow rate.

all new California cars in 1964, all nationwide cars in 1968, and are still used on all new cars sold in the United States. System operation is shown in figures 19-7 and 19-8.

In this system, crankcase ventilation air may come from either the clean side (inside) or the dirty side (outside) of the carburetor air filter. When air is drawn from the clean side, the air cleaner filter acts as a PCV filter, and a wire screen flame arrester is installed in the PCV air intake line to prevent a crankcase explosion if

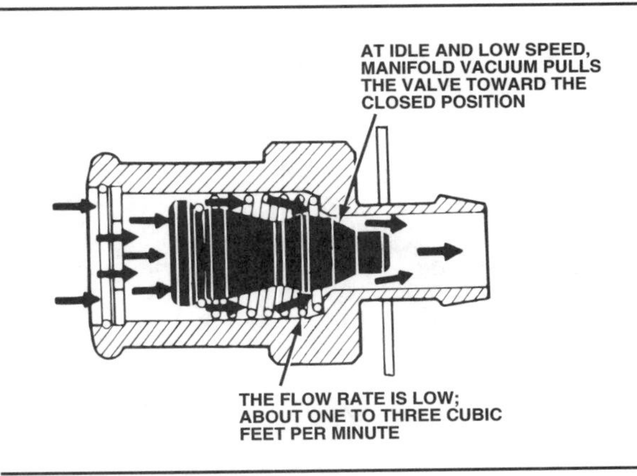

Figure 19-11. PCV valve airflow during cruising and light-load operation.

the engine backfires. When air is drawn from the dirty side of the air cleaner, a separate crankcase ventilation filter, or PCV inlet filter, is used. Generally, a polyurethane foam type is installed in the air cleaner, figure 19-9. The filter can, however, take the form of a wire gauze or mesh filter located either in the oil filler cap or the inlet air hose connection to the valve cover.

■ It Was Not Always as Simple

The early PCV systems caused a lot of grief and engine troubles for automakers. Many garages, even franchised dealers, ignored the PCV systems on 1963–64 cars. They required a lot of care and cleaning, and they clogged quickly when ignored. Contaminants remained in the crankcase, and sludge and moisture formed. This clogged oil lines and prevented adequate engine lubrication. The result was disaster for the engine, and major overhauls on engines still under warranty were often required.

The situation reached a crisis point for one major manufacturer, which stopped using PCV systems on its cars from the spring of 1964 until early in 1965. Auto engineers were frustrated by the problems PCV systems were creating. The systems had been designed to be simple and require only a minimum amount of service. But mechanics in the field completely ignored the emission control device, and engines began to fail.

These problems resulted in a crash project by the automakers. While engineers worked overtime developing a "better" PCV system, manufacturers started a program to educate dealers, servicemen, and car owners. The so-called "self-cleaning" PCV valve was developed and began appearing on mid-1965 models. This second-generation PCV system is practically the same one in use today.

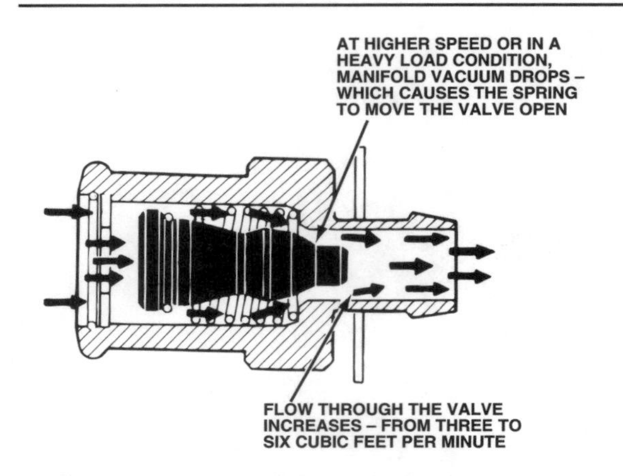

Figure 19-12. PCV valve airflow during acceleration and heavy-load operation.

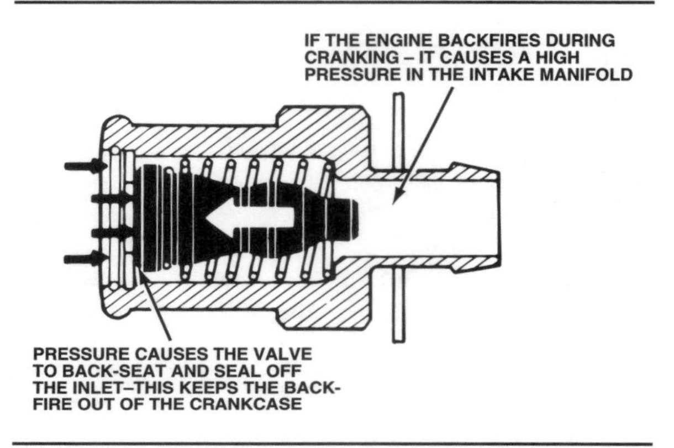

Figure 19-13. PCV valve operation in case of a backfire.

PCV Valves

The PCV valve in most systems is a one-way valve containing a spring-operated plunger that controls valve flow rate, figure 19-10. Flow rate is established for each engine, and a valve for a different engine should not be substituted. The flow rate is determined by the size of the plunger and the holes inside the valve. PCV valves usually are located in the valve cover or intake manifold.

The PCV valve regulates airflow through the crankcase under all driving conditions and speeds. When manifold vacuum is high (cruising and light-load operation), the PCV valve restricts the airflow to maintain a balanced air-fuel ratio, figure 19-11. It also prevents high intake manifold vacuum from pulling oil out of the crankcase and into the intake manifold. Under high speed or heavy loads, the valve opens and allows maximum airflow, figure 19-12. If the engine backfires, the valve will close instantly to prevent a crankcase explosion, figure 19-13.

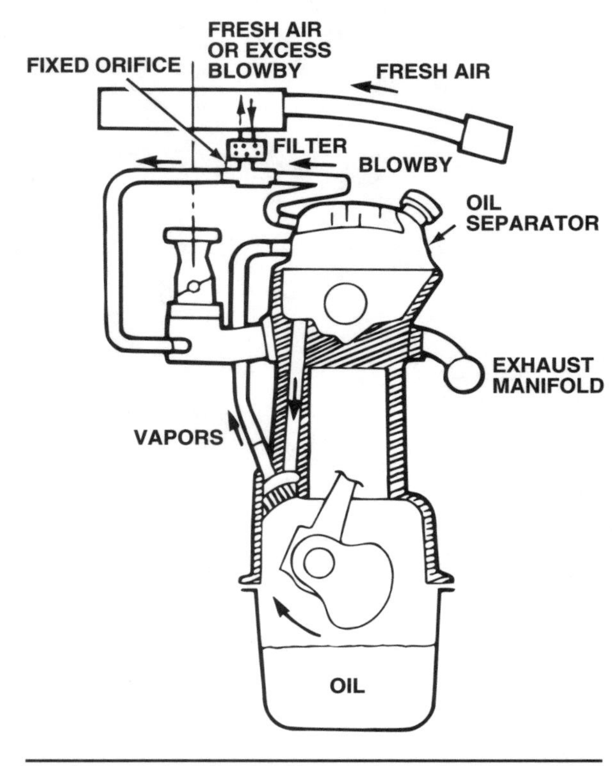

Figure 19-14. The orifice flow control system used on 1981–85 Ford 1.6- and 1.9-liter Escort engines does not use fresh-air scavenging of the crankcase.

ORIFICE-CONTROLLED SYSTEMS

The closed PCV system used on some 4-cylinder engines contains a calibrated orifice instead of a PCV valve. The orifice may be located in the valve cover, intake manifold, or in a hose connected between the valve cover, air cleaner, and intake manifold.

While most orifice flow control systems work the same as a PCV valve system, they may not use fresh-air scavenging of the crankcase. The carbureted 1981–85 Ford 1.6- and 1.9-liter Escort engines are good examples of this design, figure 19-14. Crankcase vapors are drawn into the intake manifold in calibrated amounts, depending on manifold pressure and orifice size. If vapor availability is low, as during idle, air is drawn in with the vapors. During off-idle operation, excess vapors are sent to the air cleaner.

A dual-orifice valve is used on carbureted 1986 and later Ford 1.9-liter Escort engines to increase PCV flow during off-idle engine operation. At idle, PCV flow is controlled by a 0.050-in. (1.3-mm) orifice. As the engine moves off-idle, spark port vacuum pulls a spring-loaded valve off its seat, allowing PCV flow to pass through a 0.090-in. (2.3-mm) orifice.

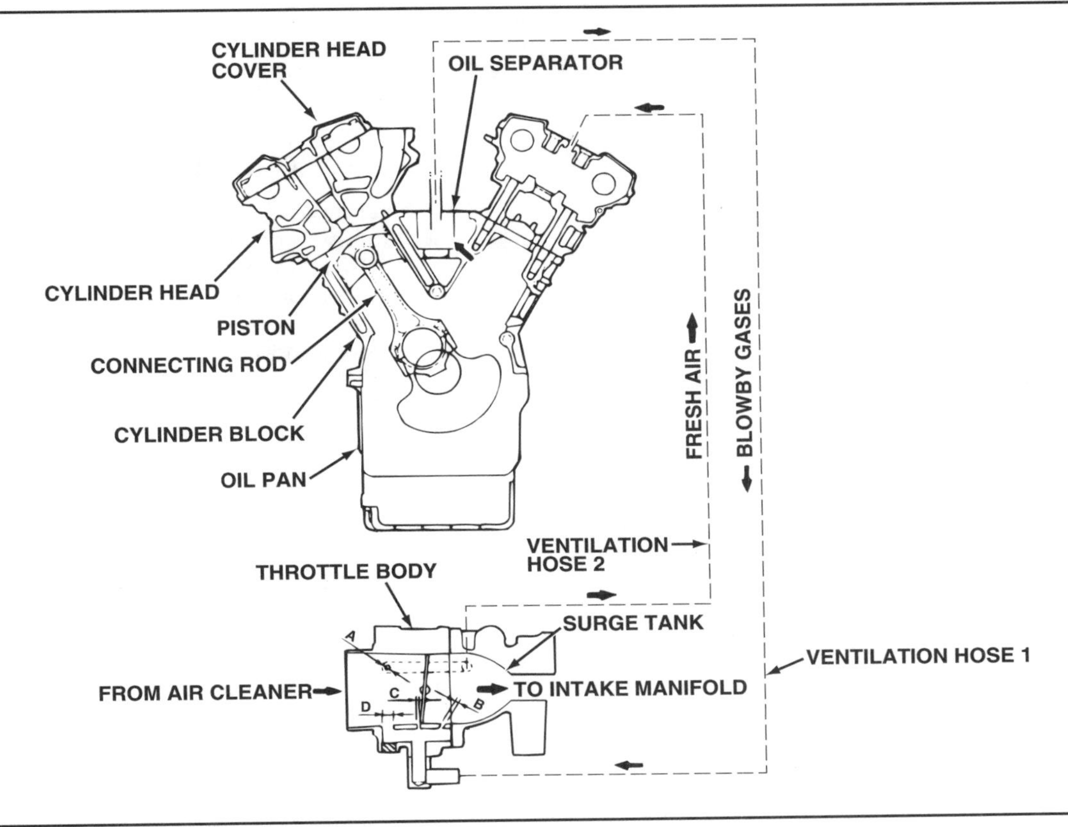

Figure 19-15. A typical PCV system with oil/vapor separator used with a fuel-injected engine. (Ford)

SEPARATOR-CONTROLLED SYSTEMS

Turbocharged and many fuel injected engines use an oil/vapor or oil/water separator instead of a PCV valve or calibrated orifice. In the most common applications, the air intake throttle body acts as the source for crankcase ventilation air and the metering device, figure 19-15. Fresh air from the air cleaner enters vent hose 2 at the throttle body and is sent to the crankcase through a cylinder head. The air and vapor mixture returns to the throttle body through the oil separator and vent hose 1.

Fresh air enters the throttle body through port A. When the throttle is closed, the air and vapor mixture flows to the intake manifold through port B. At the same time, the throttle plate diverts fresh air through ports C and D to enter the intake manifold with the vapor mixture through port B, figure 19-16A. As the throttle starts to open, the air and vapor mixture passes through ports B and C to the intake manifold. Fresh air is diverted by the throttle plate through port D and enters the intake manifold with the vapor mixture through ports B and C, figure 19-16B. At wide-open throttle, the vapor mixture enters the intake manifold through ports B, C,

■ PCV System Service

When a PCV system becomes restricted or clogged, the cause is usually an engine problem or the lack of proper maintenance. For example, scored cylinder walls or badly worn rings and pistons will allow too much blowby. Start-and-stop driving requires more frequent maintenance and causes PCV problems more quickly than highway driving, as will any condition allowing raw fuel to reach the crankcase. Using the wrong grade of oil, or not changing the crankcase oil at periodic intervals will also cause the ventilation system to clog.

When a PCV system begins to clog, the engine tends to stall, idle roughly, or overheat. As ventilation becomes more restricted, burned plugs or valves, bearing failure, or scuffed pistons can result. Also look for an oil-soaked distributor or points, or leaking out around valve covers or other gaskets. Do not overlook the PCV system while troubleshooting. A partly or completely clogged PCV valve, or one of the incorrect capacity, may well be the cause of poor engine performance.

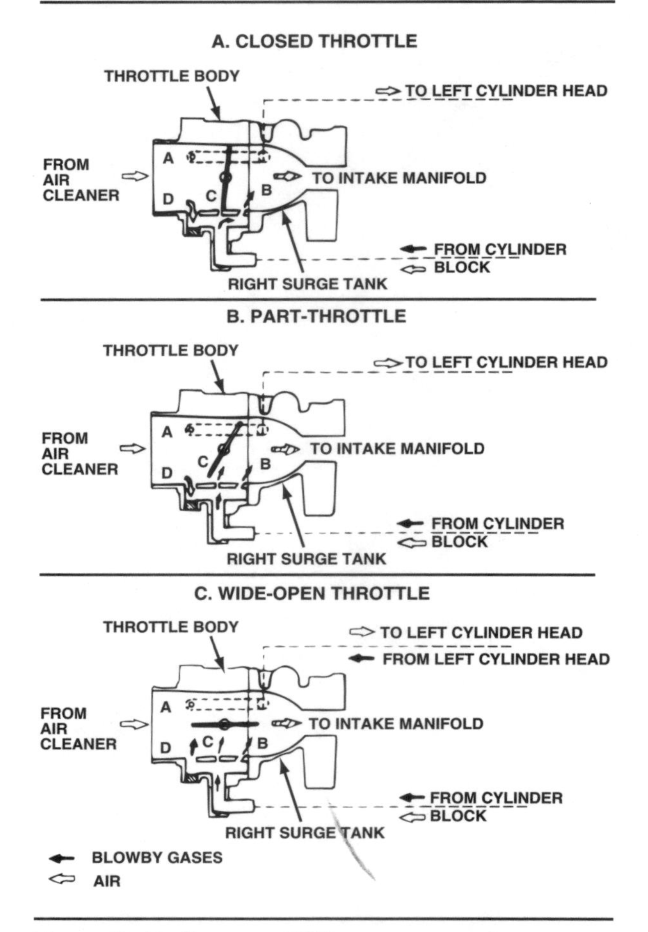

Figure 19-16. Common PCV system operation on turbocharged engines during closed, part-throttle, and wide-open throttle conditions. (Ford)

and D, figure 19-16C If the amount of vapor mixture is too great, it also flows through port A to the intake manifold.

PCV SYSTEM EFFICIENCY

When intake air flows freely, the PCV system will function properly, as long as the PCV valve or orifice is not clogged. Modern engine design includes the air and vapor flow as a calibrated part of the air-fuel mixture. For this reason, a flow problem in the PCV system will result in driveability problems.

A PCV system that is not properly vented or scavenged will cause oil dilution and sludge formation in the engine, causing premature engine wear. It will also cause oil deposits in the air cleaner, resulting in inefficient air cleaner operation.

SUMMARY

Pressure in the engine cylinders forces combustion gases past the pistons. These gases, called blowby, settle in the engine crankcase where they contaminate the lubricating oil and create harmful acids. Ventilation is necessary to remove the vapors from the crankcase. A positive crankcase ventilation (PCV) system recirculates crankcase vapors through the intake manifold to the cylinders, where they are burned. PCV systems use a metering valve or orifice to regulate airflow. A malfunctioning PCV system can cause driveability problems and create harmful acids and sludge that affect engine lubrication and result in premature wear.

Review Questions

Choose the single most correct answer.
Compare your answers to the correct answers on page 507.

1. In a PCV system, crankcase vapors are recycled to the:
   a. Exhaust system
   b. Road draft tube
   c. Oil-filler breather cap
   d. Intake manifold

2. In Type 3 PCV systems:
   a. The crankcase is connected to the intake manifold
   b. The PCV valve is a plunger type
   c. The PCV valve is a diaphragm type
   d. There is no PCV valve

3. The Type 3 PCV system:
   a. Is vented to the intake manifold
   b. Tends to make the air-fuel mixture leaner
   c. Has no effect on fuel mixture
   d. Tends to make the fuel mixture richer

4. A separate flame arrester is used in a closed PCV system when inlet air is drawn from the:
   a. Clean side of the carburetor air filter
   b. Dirty side of the carburetor air filter
   c. The intake manifold
   d. The oil filler cap on a valve cover

5. Which is not part of a Type 4 PCV system?
   a. A PCV valve
   b. A vented oil filler cap
   c. An air inlet filter
   d. A manifold vacuum hose

6. The PCV valve operates in which of the following ways?
   a. Restricts airflow when intake manifold vacuum is high
   b. Increases airflow when intake manifold vacuum is low
   c. Acts as a check valve in case of carburetor backfire
   d. All of the above

Chapter

20

Air Injection

Automakers have various names for their air injection systems, but all do about the same job. They provide additional air to the exhaust manifold where it mixes with the hot exhaust leaving the engine. This helps the oxidation, or burning reaction, necessary to lower hydrocarbon (HC) and carbon monoxide (CO) emissions.

In this chapter, you will learn:

- The reasons for, and principles of, **air injection**
- The components used in a typical pump air injection or pulse-air injection system
- The changes that created second-generation systems used with catalytic converters.

BASIC SYSTEM DESIGN AND OPERATION

Air injection was one of the first add-on devices used to help oxidize HC and CO exhaust emissions. An air injection system actually modifies the basic process of combustion in the engine, figure 20-1. Combustion, as you know, is an oxidation reaction, but usually an incomplete one. By providing additional air to the exhaust system as soon as the hot exhaust gases leave the cylinder, the injection system makes possible the continued oxidation of any HC and CO remaining in the exhaust. As a result, the HC and CO combine with oxygen (O_2) to form water vapor (H_2O) and carbon dioxide (CO_2). Air may be injected into the exhaust using an air-pump or pulse-air system.

The development of sophisticated electronic fuel management and engine control systems has made it possible for many of the new and smaller engines equipped with electronic fuel injection to meet emission standards without air injection. With the increasing precision of electronic controls, air injection will keep losing importance in future engine design.

Regardless of the various names used by automakers for their air-pump injection systems, all are relatively simple in design, function the same way, and use the same basic components, figure 20-1:

- A belt-driven pump with inlet air filter
- One or more air distribution manifolds and nozzles
- An antibackfire valve
- One or more exhaust check valves
- Connecting hoses for air distribution
- Air management valves and solenoids on late-model applications.

Air Injection Pump

Most air pumps used on domestic vehicles are manufactured by the Saginaw Division of General Motors (GM), figure 20-2. The pump is

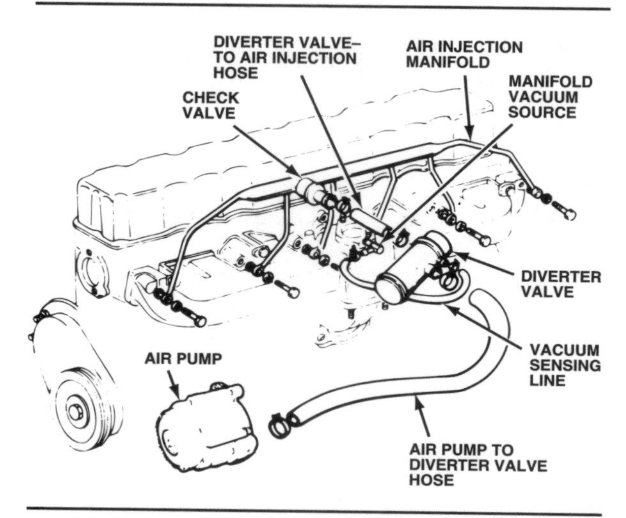

Figure 20-1. The basic components of an air pump injection system.

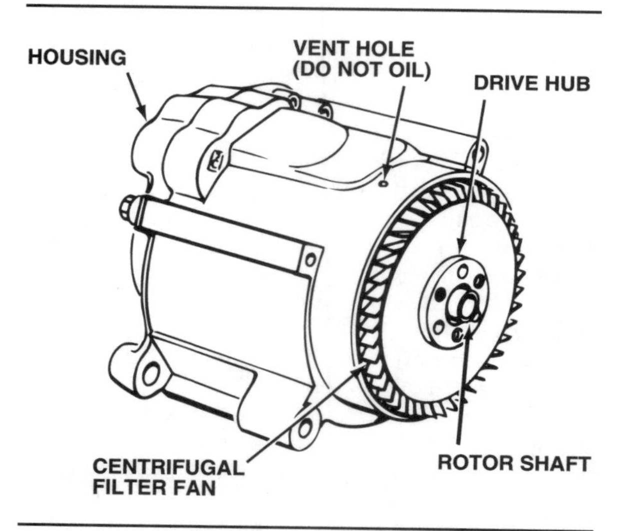

Figure 20-2. Later-model two-vane air pumps have an external centrifugal filter fan mounted on the front of the housing.

mounted at the front of the engine and driven by a belt from the crankshaft pulley. It pulls fresh air in through an external filter and pumps the air, under slight pressure, to each exhaust port through connecting hoses. Pumps used since 1968 contain two vanes and have an impeller-type centrifugal air filter fan mounted on the pump rotor shaft. This removes dirt particles from the air entering the pump by centrifugal force, figure 20-3. The heavier dust particles in the air are forced in the opposite direction to the inlet airflow. The lighter air is then drawn into the pump by the impeller-type fan. Most pumps use a pressure relief valve, figure 20-4, which

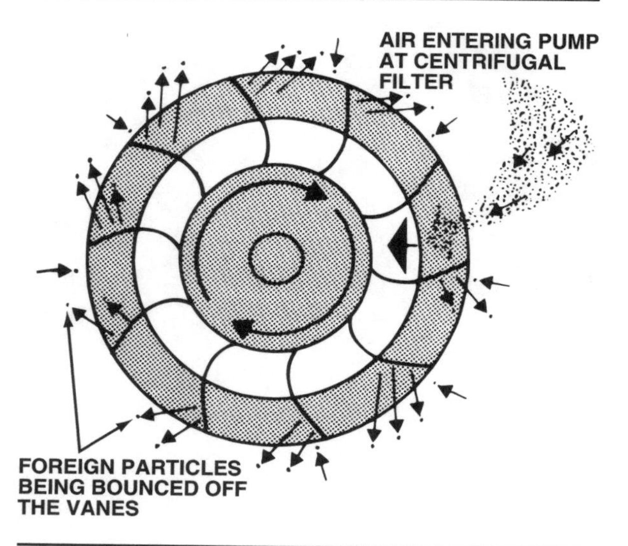

Figure 20-3. Centrifugal force created by the air pump's fan removes dust and dirt from the inlet air.

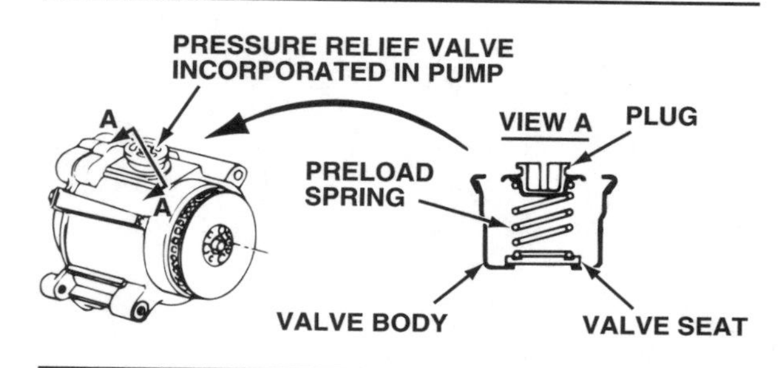

Figure 20-4. Some air pumps have built-in pressure relief valves. (AC-Delco)

opens at high engine speed. Pumps without a pressure relief valve use a diverter valve, explained later in this section.

Air Distribution Manifolds and Nozzles

Before the appearance of catalytic converters in 1975, the air injection system only sent air from the pump to nozzles installed in the cylinder head near each exhaust port. This provided equal air injection for each cylinder's exhaust, and made it available at a point in the system where exhaust gases were the hottest.

Air is delivered to the engine's exhaust system in one of two ways:

- An external air manifold or manifolds distributes the air through injection tubes with stainless-steel nozzles. The nozzles are threaded into the cylinder heads or exhaust manifolds close to each exhaust valve. This method is used primarily with smaller engines, figure 20-5.

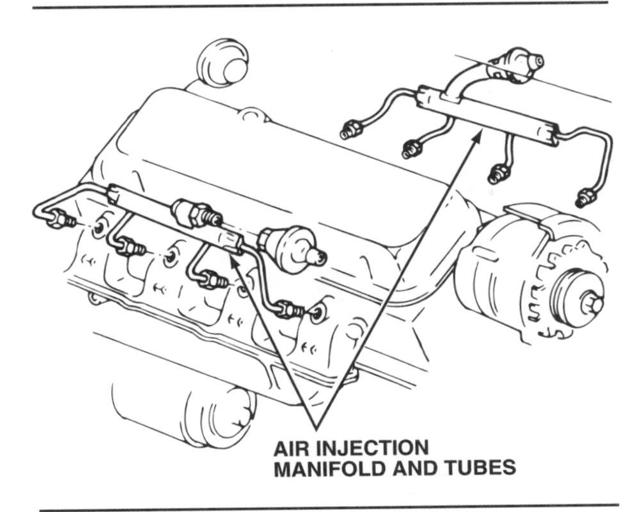

Figure 20-5. External air manifolds are used with many air injection systems. (Chevrolet)

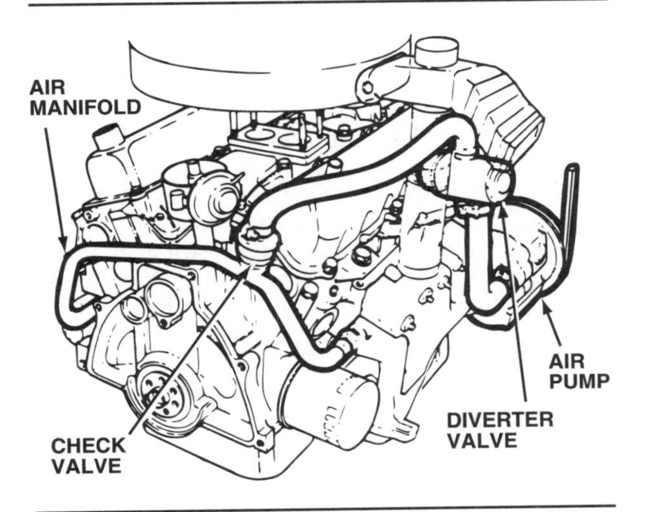

Figure 20-6. Some engines have air distribution passages built into the cylinder heads. (Ford)

- An internal air manifold distributes the air to the exhaust port near each exhaust valve through passages cast in the cylinder head or the exhaust manifold. This method is used mainly with larger engines, figure 20-6.

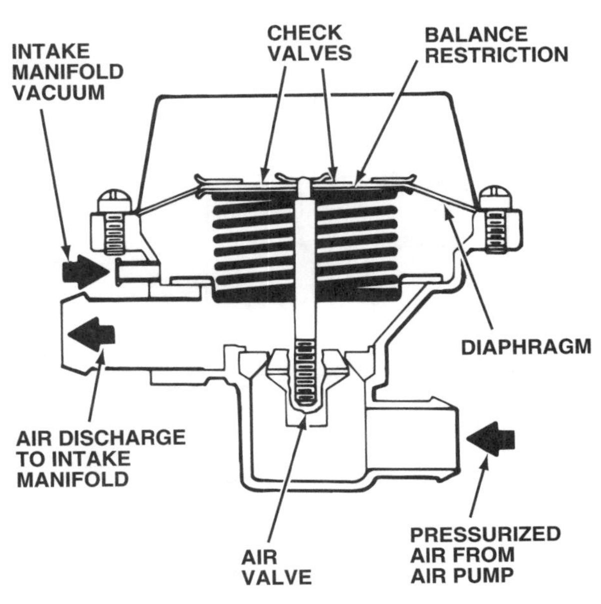

Figure 20-7. The gulp valve was used on early air injection systems.

The addition of catalytic converters as an emission control device changed the pattern of air distribution. Air injected upstream heats the exhaust gases, speeding converter warmup and efficiency. When the exhaust gas enters the converter, the additional air in the mixture helps the converter oxidize HC and CO more completely.

With the introduction of three-way converters capable of reducing oxides of nitrogen (NO_x), the air distribution pattern was changed again. Now

Air Injection: A method of reducing HC and CO emissions and providing quick warm-up of emission components. The air may be injected into the exhaust stream before the catalytic converter (upstream) or between the reduction and oxidation beds of a dual-bed converter (downstream), or both. When injected upstream, the air mixes with the hot exhaust, oxidizing the HC and CO to form H_2O and CO_2.

■ Do Not Oil the Air Pump!

No air injection system is completely quiet. Usually, pump noise increases in pitch as engine speed increases. If the drive belt is removed and the pump shaft turned by hand, it will squeak or chirp. Many who work on their own cars and even some technicians are not aware that air injection pumps are permanently lubricated, and require no periodic maintenance.

Suppose you pinpoint the air pump as the source of the noise. It would seem that a few squirts of oil would silence it. See those three small holes in the housing? While it is easy to mistake them for oiling points, these are actually vents. Do not oil them. More than a few pumps have failed because someone assumed that taking "good" care of the pump would make it last longer!

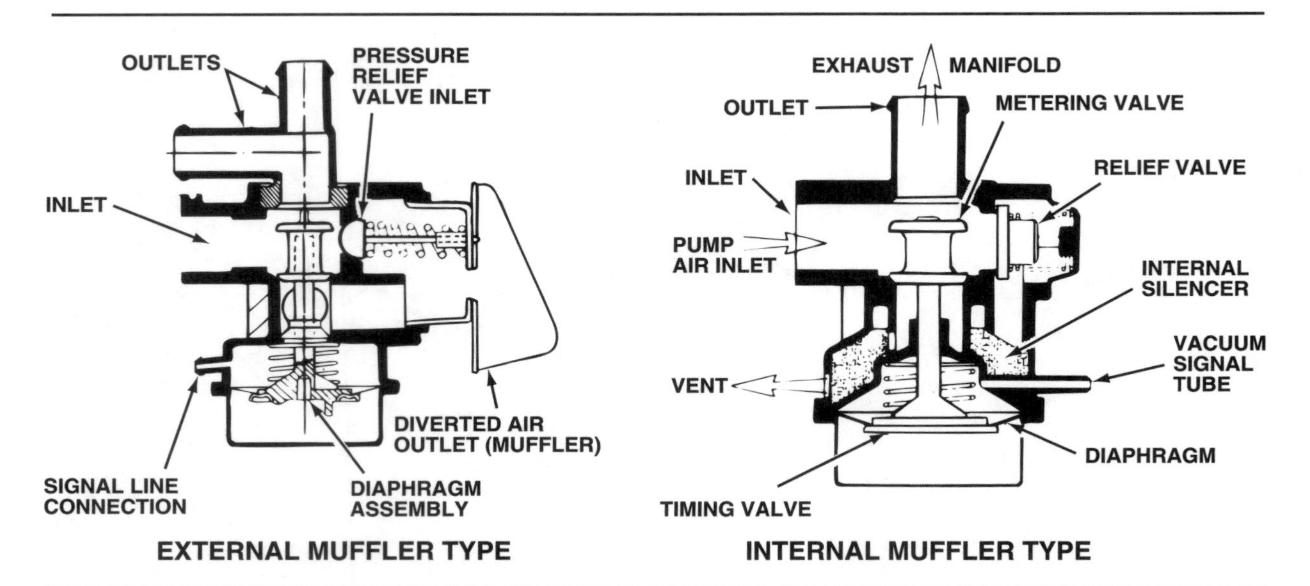

Figure 20-8. Two types of diverter valves. They differ mainly in placement of the air silencer. (AC-Delco)

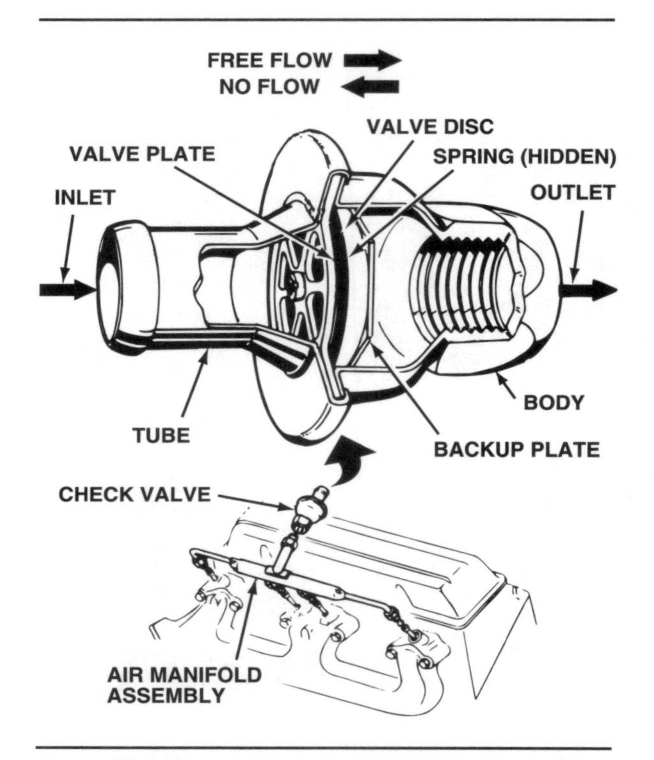

Figure 20-9. The check valve protects the system against the reverse flow of exhaust gases. (AC-Delco)

air could also be injected between the reduction and oxidation beds of the converter. The additional air in the exhaust allows the catalyst to reach operating temperature sooner, but once it becomes operational, it is most efficient with less air entering the converter. This brought about the use of air-switching devices, which will be explained later in this chapter.

Antibackfire Valves

During engine deceleration, high intake manifold vacuum enriches the air-fuel mixture. If air is allowed to flow into the exhaust manifold at this time, it will combine with excess unburned fuel in the exhaust. The result is engine **backfire**—a rapid combustion of the unburned gases that can destroy a muffler. To prevent engine backfire, the air pump flow must be shut off during deceleration. This is done by a diaphragm-operated antibackfire, or backfire suppression, valve. Two kinds of valves have been used: the gulp valve and the diverter valve. They differ in the direction in which they redirect the airflow.

Gulp valve

Early air injection systems used a **gulp valve,** figure 20-7. When intake manifold vacuum is applied to the valve diaphragm, it causes the air valve to move. This redirects the pump air to the intake manifold to lean out the enriched air-fuel mixture during deceleration.

The gulp valve is connected to the intake manifold by two hoses. The large hose is the air discharge hose. The small hose is the sensing hose that sends manifold vacuum to the gulp valve to operate the diaphragm. Balance restrictions or bleed holes inside the valve equalize pressure on both sides of the diaphragm after a few seconds. Even if manifold vacuum is high, the gulp valve only stays open for a few seconds until pressure equalizes.

Any sudden change in vacuum will operate the gulp valve. This is one of the undesirable

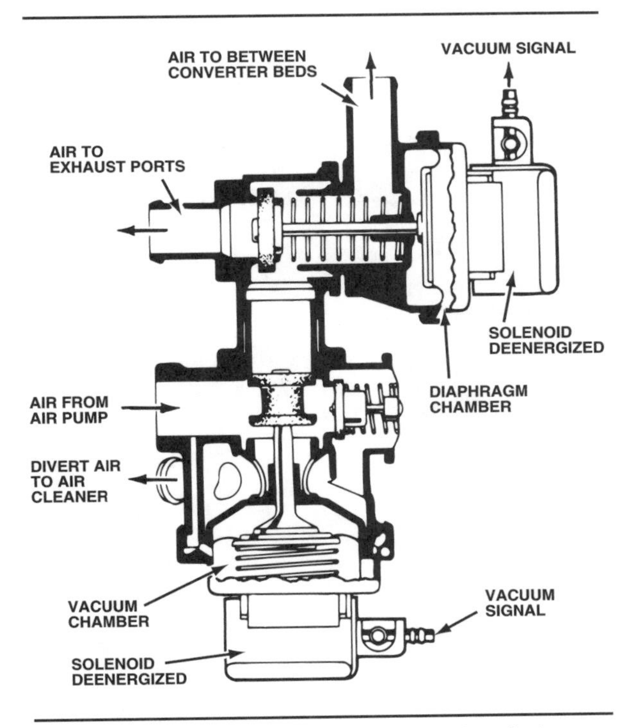

Figure 20-10. GM uses this electrically signaled diverter valve for switching and diverting tasks on air injection systems used with Computer Command Control (CCC) engines.

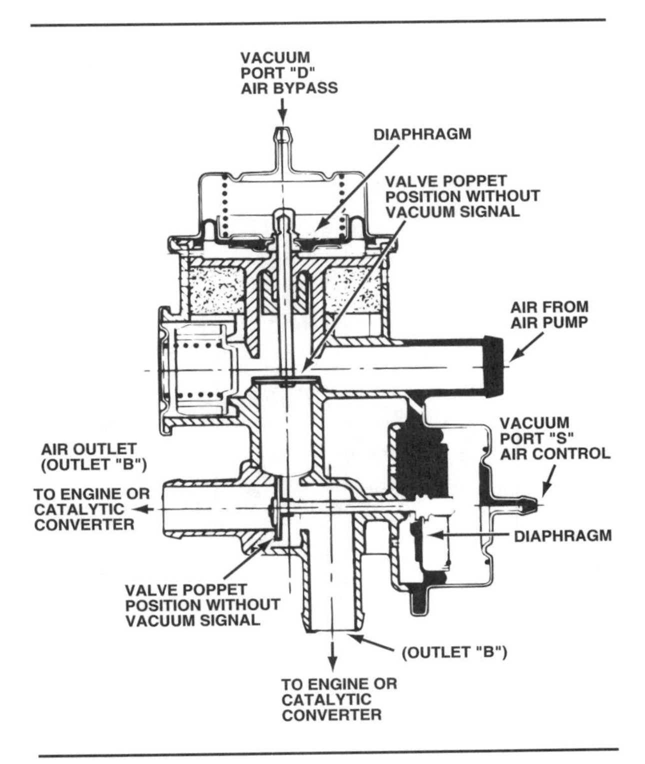

Figure 20-11. Ford's combination air bypass/air control valve handles air switching and diverting tasks.

features that led to its replacement. For example, the gulp valve will open when the engine starts. This can cause hard starting and a rough idle. Another problem with the gulp valve is that when the throttle is closed at high speeds while manifold vacuum is low, the valve may not open for a few seconds. During this time, an engine backfire can occur. The gulp valve has been replaced by the diverter valve on later-model vehicles.

Diverter valve

The **diverter valve,** figure 20-8, is also known as a dump or bypass valve. Like the gulp valve, the diverter valve uses a manifold-vacuum-operated diaphragm to redirect the airflow from the air pump. However, the pump air passes through the diverter valve continuously on its way to the air injection manifold. During deceleration, manifold vacuum operates the valve diaphragm to divert, or dump, the air directly to the atmosphere, not to the intake manifold.

Some diverter valves vent the air to the engine air cleaner for muffling. Others vent it through a muffler and filter built into the valve. Diverter valves are used with vacuum solenoids, vacuum differential valves, vacuum vent valves, and idle valves to fine-tune air injection. Some diverter valves contain the air injection system relief valve instead of the pump. Because diverter valves have no effect on the air-fuel mixture in the intake manifold, they are more trouble-free than gulp valves.

Exhaust Check Valves

All air injection systems use one or more one-way check valves, figure 20-9, to protect the air pump and other components from reverse exhaust flow. A check valve contains a spring-type metallic disc or reed that closes the air line under exhaust backpressure. Check valves are located between the air manifold and the gulp or diverter valve, figure 20-6. If exhaust pressure

Backfire: The accidental combustion of gases in an engine's intake or exhaust system.

Gulp Valve: A valve used in an air injection system to prevent backfire. During deceleration it redirects air from the air pump to the intake manifold, where the air leans out the rich air-fuel mixture.

Diverter Valve: A valve used in an air injection system to prevent backfire. During deceleration it "dumps" the air from the air pump into the atmosphere. Also called a dump valve.

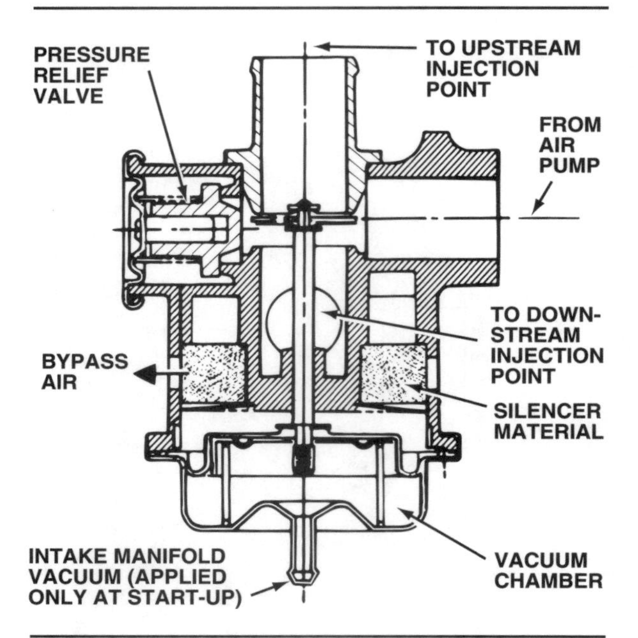

Figure 20-12. Cutaway of Chrysler's air switch/relief valve.

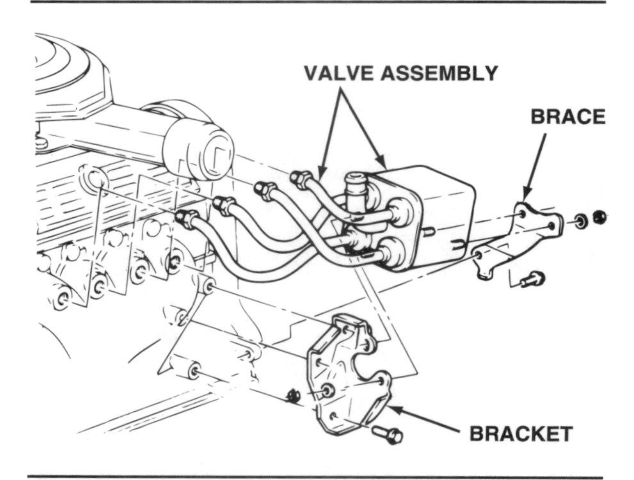

Figure 20-13. The GM pulse-air injection system is called PAIR, and was first used in 1975.

exceeds injection pressure, or if the air pump fails, the check valve spring closes the valve to prevent reverse exhaust flow.

Air Management Valves

Air management is a term used to describe the large variety of diverter, bypass, and air control or switching valves that domestic automakers have used to fine-tune air injection systems since 1975. Figure 20-10 shows one such valve used by GM. Ford uses a similar valve, figure 20-11, that combines the divert and switching functions in one valve. The divert, or air control, section of

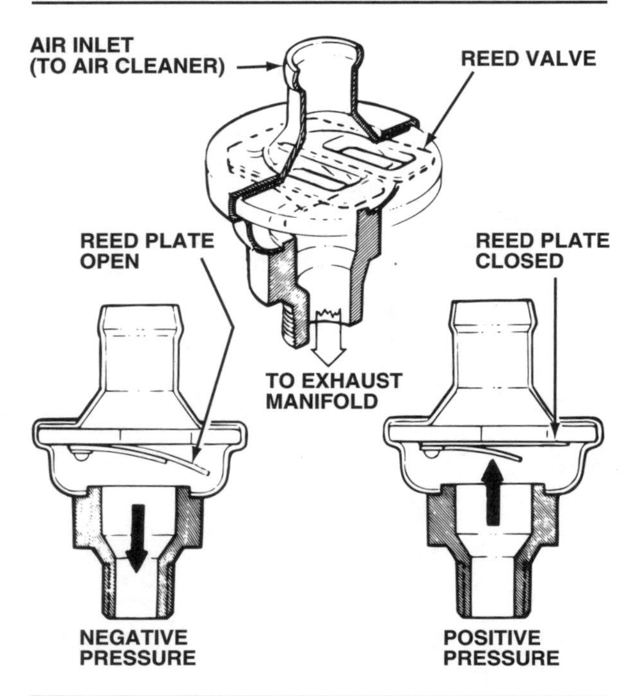

Figure 20-14. A pulse-air injection valve opens when negative pressure occurs in the exhaust manifold. (Ford)

the valve protects the converter during open-throttle operation and high temperatures. The air-switching section sends air to the exhaust ports during open-loop operation. When the engine control system moves into the closed-loop mode, the valve switches airflow to a point between the converter beds.

A combination switch/relief valve used by Chrysler, figure 20-12, is controlled by a coolant vacuum switch cold open (CVSCO) or by a vacuum solenoid. On a cold start, air injects as close as possible to the exhaust valves. When engine coolant temperature reaches the point where exhaust gas recirculation (EGR) begins, the CVSCO or vacuum solenoid shuts off the vacuum signal to the valve. This causes the valve to send most of the pump air downstream to the catalytic converter. The rest of the pump air continues to reach the exhaust ports by passing through slots in the upstream valve seat.

Regardless of manufacturer or system, most air-switching or control valves operate with:

- Manifold vacuum or ported vacuum working on a vacuum diaphragm in the valve
- Output pressure of the air pump working against vacuum or a spring in the vacuum chamber
- One or more solenoids in the vacuum line opening or closing the vacuum supply to the diaphragm.

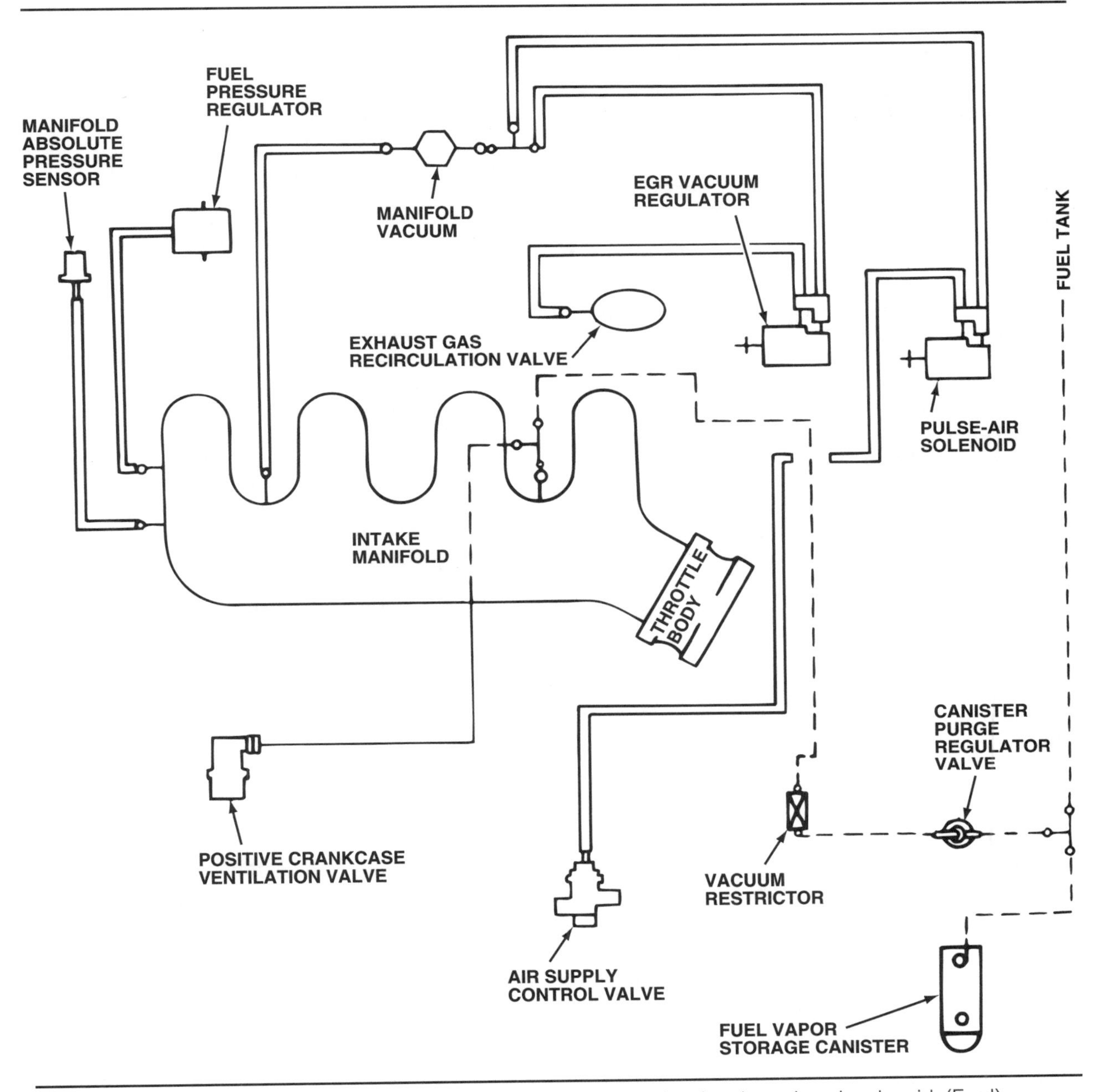

Figure 20-15. A 4-cylinder, computer-controlled engine schematic, showing the pulse-air solenoid. (Ford)

Air Injection and Catalytic Converters

Introduction of catalytic converters has modified the air distribution needs of engines. Air injection on later-model cars improves converter efficiency and accelerates catalyst warmup.

Air can be injected in several different places:

- Near each exhaust port (upstream)
- Into the exhaust manifold outlet (upstream)
- Into the exhaust pipe ahead of the converter (upstream)
- Directly into the converter (downstream).

When air flows into any of these points, it mixes with the exhaust gases and continues the oxidation process inside the converter. During initial engine operation, air injected upstream increases exhaust temperature, bringing the converter to operating temperature more quickly.

However, too much HC's in the exhaust caused by a hot idling engine overheat the converter, damaging the catalysts. Switching the injected air downstream from the exhaust ports helps to decrease the temperatures generated from burning the HC's upstream. The powertrain control module monitors the duration of air injection as well as engine temperature to prevent catalyst overheating.

Computer-controlled engines also use one or more exhaust oxygen sensors (O2S). These devices must reach at least 572°F (300°C) before they start to function. Air injected upstream during engine warmup helps the sensor reach operating temperature more rapidly.

With the introduction of NO_x reduction converters (also called dual-bed, TWC, or three-way converters) in 1977, the problem of air distribution became even more complex. The oxidation process adds O_2 to HC and CO, but the reduction process removes O_2 from NO_x compounds. Therefore, a reduction converter requires more air in the exhaust to bring it to operating temperature. When it reaches operating temperature, however, it works most efficiently with less air in the exhaust. This seeming paradox, along with the problems of catalyst damage and converter overheating, led to the development and use of air-switching systems.

Pulse-Air Injection

As part of the trend toward smaller engines during the early 1980s, automakers started replacing air pump systems with pulse-air injection. This type of air injection uses the negative pressure pulses generated in the exhaust to draw fresh air into the exhaust manifold. The pulse-air system eliminates the need for the belt-driven air pump, an air injection manifold or distribution tubes, and most of the injection system hoses. This results in both cost and weight savings.

Pulse air or aspirator valves are similar in design to exhaust check valves, figure 20-13. Each valve contains a spring-loaded diaphragm or reed valve and is connected by tubing to the exhaust port of each cylinder or to the exhaust manifold. Each time an exhaust valve closes, there is a period when manifold pressure drops below atmospheric pressure. During these low-pressure (slight vacuum) pulses, the pulse valve opens to admit fresh air to the exhaust, figure 20-14. When exhaust pressure rises above atmospheric pressure, the valve acts as a check valve and closes. As a result, air injection occurs without the power-consuming air pump.

Pulse-air injection works best at low engine speeds when the catalytic converter needs extra air. At high engine speeds, the vacuum pulses occur too rapidly for the valve to follow, and the internal spring simply keeps the valve closed. Pulse-air valves must be connected upstream in the exhaust system where negative pressure pulses are strongest. This means that pulse air cannot be switched downstream for use in the converter or between oxidation and reaction catalysts. The system is most effective on vehicles equipped with only an HC-CO oxidation catalyst. However, some models with three-way converters use pulse air to provide additional air in the exhaust of a cold engine. These systems have valves in the air inlet line to the pulse-air valves to shut off air supply when the engine is warm.

Late-model pulse-air systems are computer controlled. In the Ford system shown in figure 20-15, the computer allows pulse air at idle or below three mph, based on vehicle speed sensor input. The computer controls a normally closed on/off valve in the pulse-air silencer through a normally closed vacuum solenoid. This is similar to the system used by GM during open-loop mode in which the computer energizes a shutoff valve to open the fresh-air line at a time when the engine needs it most. Once the computer switches to closed-loop mode, it deenergizes the shutoff valve, closing off the fresh-air line.

SUMMARY

Air injection is one of the oldest methods used to control HC and CO exhaust emissions. The injected air mixes with hot exhaust gas as it leaves the combustion chambers to further oxidize HC and CO emissions. All air injection systems used on domestic cars operate in essentially the same way, regardless of automaker. Later-model air injection systems are no longer the major HC and CO control system; they now assist the converter by injecting air into the exhaust system to help warm up the catalyst and the O2S. When a reduction catalyst is used, air injection may be switched downstream once the engine is warm. Air-pump injection systems have given way to pulse-air systems on many late-model engines. Engine computers now control air injection systems to provide a quicker response to changing engine requirements, but many fuel injected engines no longer require air injection.

Review Questions

Choose the single most correct answer.
Compare your answers to the correct answers on page 507.

1. The main reason for an air injection system is to:
   a. Oxidize HC and CO exhaust emissions
   b. Reduce NO_x exhaust emissions
   c. Eliminate crankcase emissions
   d. Eliminate evaporative HC emissions

2. Oxidation of HC and CO emissions produces:
   a. HCO and CO_2
   b. H_2CO_3 and CO_2
   c. H_2O and CO_2
   d. HNO_3 and C_3PO

3. Two-vane pumps have:
   a. An impeller-type fan for filter
   b. An integral wire mesh filter
   c. A hose to the clean side of the air cleaner
   d. A separate air filter

4. Air injection nozzles are made of:
   a. Copper
   b. Stainless steel
   c. Aluminum
   d. Vanadium

5. The two types of air injection backfire suppressor valves are:
   a. The check valve and the gulp valve
   b. The gulp valve and the diverter valve
   c. The diverter valve and the relief valve
   d. The diverter valve and the check valve

6. The Chrysler air-switching valve is controlled by:
   a. A vacuum solenoid
   b. A reed valve
   c. A bypass timing orifice
   d. A coolant vacuum switch

7. Pulsed-air injection works best at:
   a. Low speeds
   b. High speeds
   c. Idle
   d. Deceleration

8. Technician A says that a gulp valve diverts air to the atmosphere.
   Technician B says that a diverter valve diverts air to the intake manifold.
   Who is right?
   a. A only
   b. B only
   c. Both A and B
   d. Neither A nor B

9. Technician A says that air injection helps an exhaust catalyst reach its operating temperature faster.
   Technician B says that a catalyst is most efficient when cooled by air injection.
   Who is right?
   a. A only
   b. B only
   c. Both A and B
   d. Neither A nor B

Chapter

21

Exhaust Gas Recirculation

Exhaust gas recirculation (EGR) is an emission control system that reduces the amount of oxides of nitrogen (NO_x) produced during combustion. In the presence of sunlight, NO_x can combine with hydrocarbons to form photochemical smog, the prime air pollutant. This chapter discusses:

- The principles of EGR systems
- How an EGR system reduces NO_x formation and prevents engine detonation
- Various domestic-manufacturer EGR systems.

NOx FORMATION

Under normal circumstances, nitrogen and oxygen do not combine unless temperatures exceed 2,500°F (1,370°C) in the combustion chamber. Since ignition timing controls peak combustion chamber temperature, the first attempts during the early 1970s to meet NO_x control requirements used spark-timing control systems. Slightly retarded spark produces less pressure and heat. This holds combustion chamber temperatures below the level where NO_x forms rapidly. In 1972, the U.S. EPA introduced new test procedures to determine NO_x levels. Auto manufacturers responded by developing more effective ways of controlling NO_x.

EGR and How It Works

The relatively small amounts of NO_x produced at temperatures below 2,500°F (1,370°C) can be easily controlled. To handle the amounts generated above 2,500°F (1,370°C), engineers have a choice from the following:

- Enrichment of the air-fuel mixture. More fuel lowers the peak combustion temperature, but it raises hydrocarbon (HC) and carbon monoxide (CO) emissions. The reduction in fuel economy also makes this solution unattractive.
- Lower the compression ratio. With the introduction of unleaded gasoline, automakers decreased compression ratios to avoid detonation. This decreased NO_x levels, but higher reduction levels were needed. When the compression ratio becomes too low, HC and CO emissions rise.
- Retard spark timing slightly. Automakers found they could meet NO_x emission standards prior to 1972 by retarding ignition timing. This method failed to cut emission levels enough to meet 1972 standards.
- Dilution of the incoming air-fuel mixture. To lower emission levels further, engineers developed a system that introduces small amounts of **inert exhaust gas** into the intake charge. This lowers combustion temperatures. Currently, this is one of the most efficient methods to meet NO_x emission level cutoff points without significantly affecting engine performance.

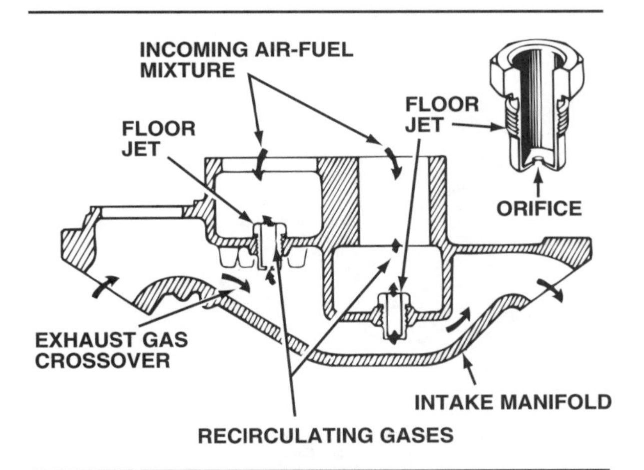

Figure 21-1. A Chrysler floor jet EGR system.

EGR routes small quantities (6 to 10 percent) of exhaust gas from the engine exhaust ports to the intake manifold. The inert gas dilutes the air-fuel mixture without affecting the air-fuel ratio. With less oxygen and fuel, combustion generates less power and heat.

EGR Operation During Different Driving Conditions

Since small amounts of exhaust gas can efficiently lower peak combustion temperatures, the orifice through which the gas passes is small.

Levels of NO_x emissions change according to engine speed, temperature, and load. While an engine runs at low speed, NO_x levels are low; EGR system operation is not desired. During high-speed driving at wide-open throttle, when NO_x emission levels are high, EGR systems usually stop diluting the intake charge to ensure superior engine performance. Maximum recirculation is required only during cruising and acceleration at speeds between 30 to 70 mph (48 to 113 kph), when NO_x formation is greatest. Engine temperature is also a determining factor in recirculation.

Typically, EGR systems prevent the introduction of exhaust gases while the engine is cold because NO_x formation is low, and the engine needs higher combustion temperatures to warm up quickly.

The power output of early engines equipped with EGR systems was less than that of the same engine without EGR, giving rise to the idea that EGR automatically meant a reduction in power output. During the early years of EGR, many drivers were convinced that they could increase engine performance by disconnecting the EGR system, a popular myth that still exists today. For many years now, engineers have designed

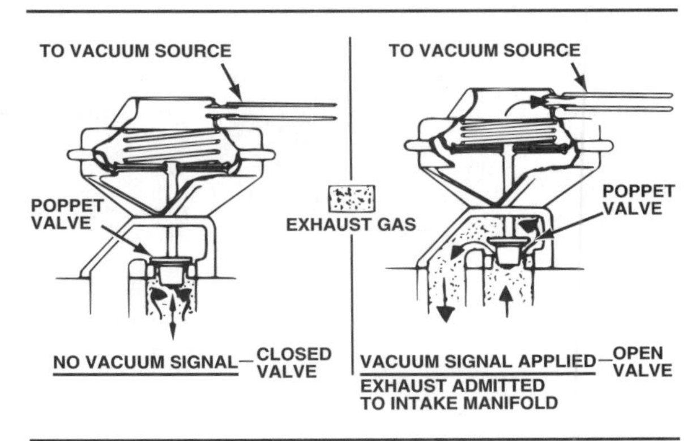

Figure 21-2. Typical spring-loaded EGR poppet valve using a single diaphragm. (AC-Delco)

engines that take into account the EGR dilution of the air-fuel charge into the fuel system calibration; maximum engine performance requires proper EGR functioning.

EGR and Engine Detonation

In addition to lowering NO_x levels, EGR also controls engine detonation. Detonation (engine ping) occurs when high pressure and heat explode the air-fuel charge after, and in addition to, the main spark. Detonation reduces engine power and efficiency. In some situations, detonation can severely damage the engine.

Engines running on tetraethyl-leaded gasoline detonated less frequently than those running on unleaded gasoline with catalytic converters.

Using EGR with electronically controlled later-model engines permits split-second changes in exhaust recirculation that minimizes the problem of detonation. In essence, this system incorporates the timing retard provided by the older spark-timing control systems, but more efficiently. You can prove this by simply disconnecting the EGR system on such an engine and listening to the pinging that results. Reconnect the system and the pinging disappears.

EGR SYSTEM DEVELOPMENT

EGR systems first appeared on 1972 Chrysler cars sold in California. The system used calibrated floor jets installed in the intake manifold underneath the carburetor to provide an open-

Exhaust Gas Recirculation (EGR): A way of reducing NO_x emissions by redirecting exhaust gases back into the intake mixture.

Inert Exhaust Gas: A gas that will not undergo chemical reaction.

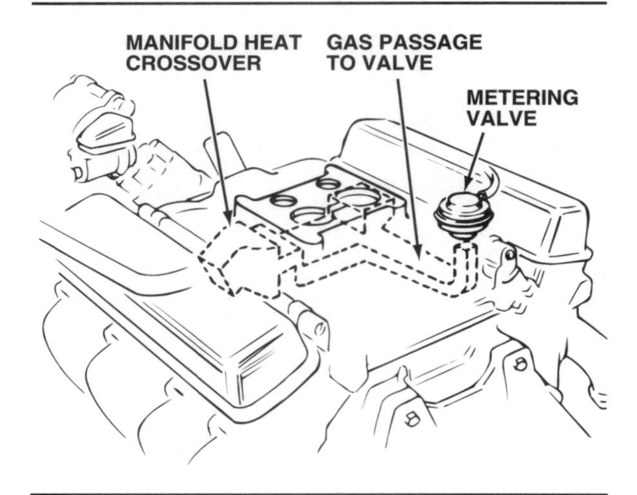

Figure 21-3. Exhaust-gas passages to the EGR valve.

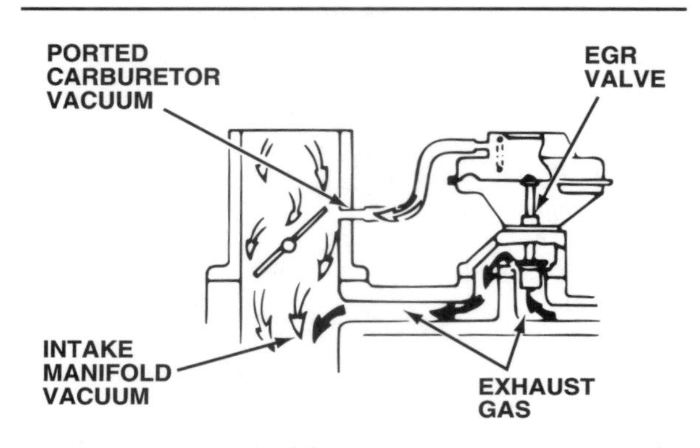

Figure 21-4. Ported vacuum operates this EGR valve.

ing between the exhaust crossover passage and the intake manifold, figure 21-1. A combination of exhaust pressure and manifold vacuum forced exhaust gases through calibrated orifices in the jets. This was the simplest of all EGR system designs, but proved to be the least efficient because the jets allowed exhaust gases to enter the intake manifold at all times. Since EGR is not necessary with a cold or idling engine, the result was rough engine operation and poor warmup under such conditions. After 1973, the floor jet EGR system became a technological dead end.

Buick also introduced an EGR system on its 1972 engines. The Buick system controlled EGR flow through a spring-loaded, vacuum-operated poppet valve, figure 21-2. The vacuum EGR valve was adopted by other automakers and soon became the basic component of all EGR systems. Although modulating and taper-stem valves also have been used, they operate essentially the same as the poppet valve.

In many EGR systems, the valve mounts on the intake manifold and connects to the intake

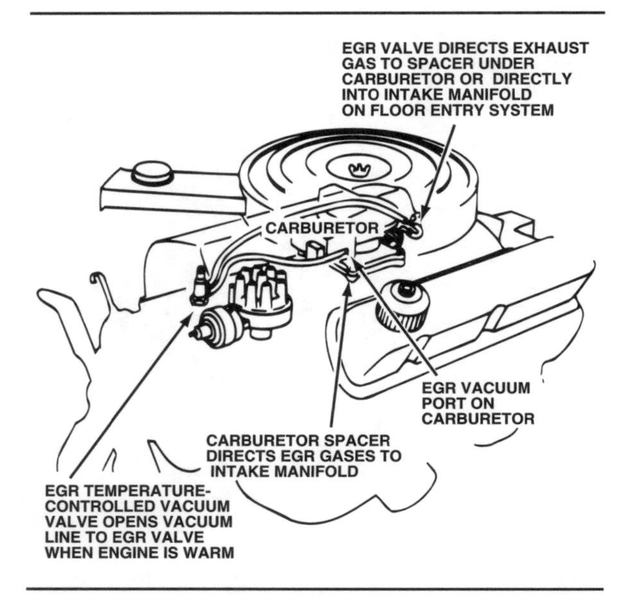

Figure 21-5. Basic Ford EGR system.

and exhaust systems through internal passages in the intake manifold, figure 21-3. In other systems, the valve may connect to the intake and exhaust systems through external steel tubing. The EGR valve remains in the closed position by the diaphragm spring. When vacuum is applied, the diaphragm overcomes the spring and opens the **pintle valve**. A vacuum line connects the diaphragm to a carburetor vacuum port located above the throttle. Using ported vacuum as a control signal keeps the valve closed at idle and during deceleration, when EGR is not necessary. Since ported vacuum is weak at wide-open throttle, the valve is designed to close as vacuum drops at wide-open throttle. During moderate acceleration and cruising, however, ported vacuum operates the valve to provide the necessary amount of EGR, figure 21-4.

Some Ford and Chrysler EGR systems used venturi vacuum as a control signal during the 1970s. Since venturi vacuum is very weak, a vacuum amplifier and storage reservoir was used to boost the control signal enough to operate the EGR valve. When the system was working perfectly, it produced an accurate and precise modulation of EGR flow proportional to engine load. However, the system was abandoned in the early 1980s because the amplifiers and storage reservoirs tended to leak in actual use.

The basic ported vacuum EGR valve works well enough, but it does not provide enough precision in controlling EGR flow to accommodate both engine performance and fuel economy. Since engine temperature has no effect on ported vacuum, the valve will operate whether the engine is hot or cold. Without some form of

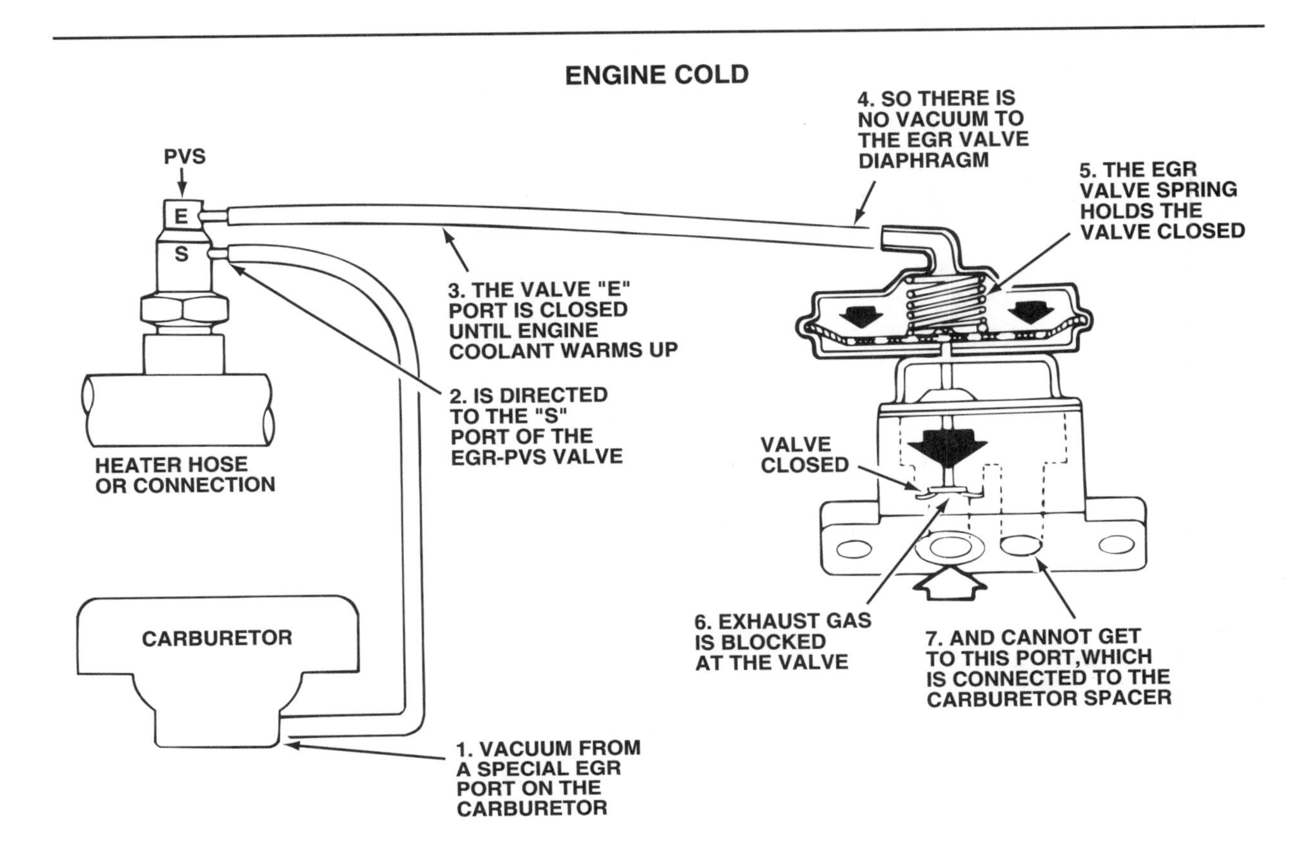

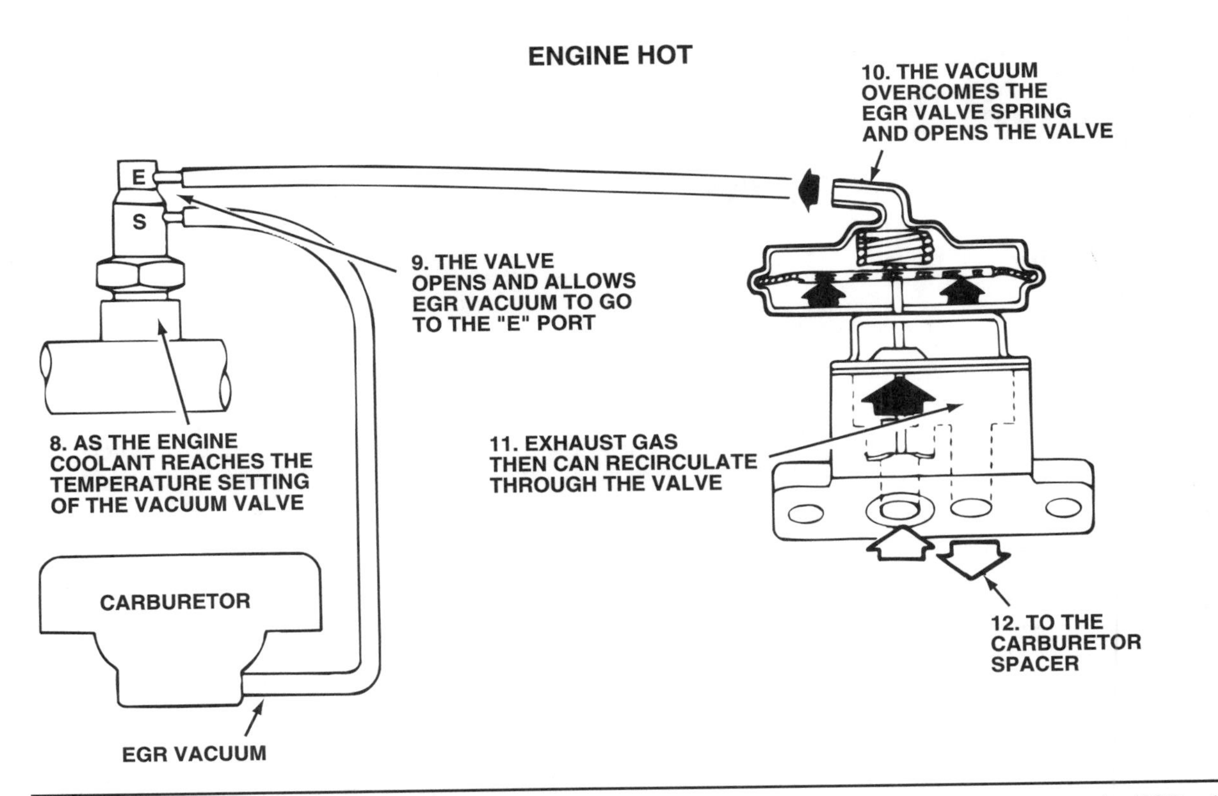

Figure 21-6. This Ford ported vacuum switch (PVS) valve controls whether a vacuum signal can open the EGR valve.

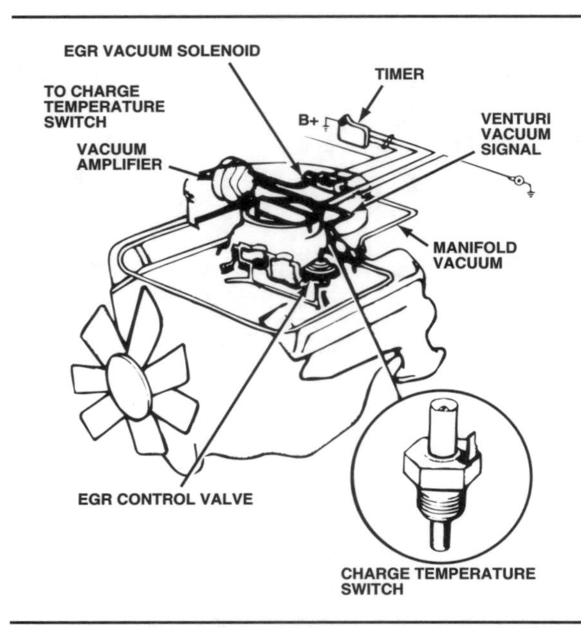

Figure 21-7. A Chrysler EGR system with charge temperature switch and timer.

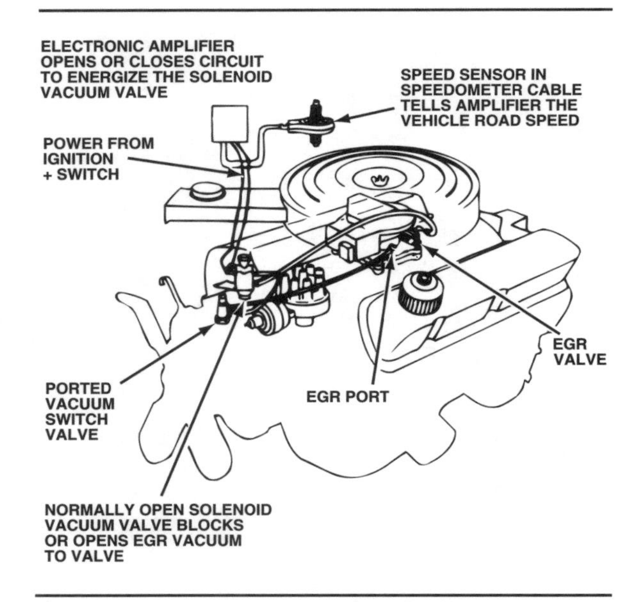

Figure 21-9. Ford high-speed EGR modulator subsystem.

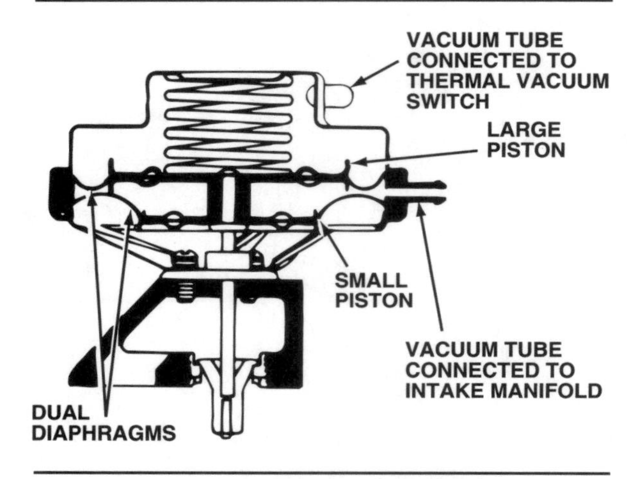

Figure 21-8. This dual-diaphragm EGR valve modulates the exhaust gas-flow.

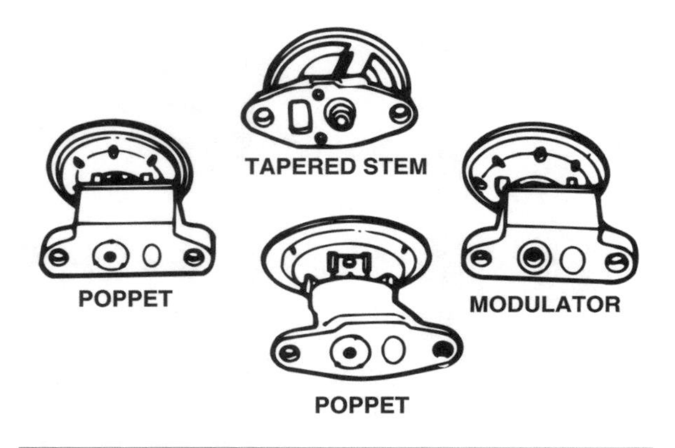

Figure 21-10. Types of Ford EGR valves operated by a ported-vacuum signal.

temperature control, EGR flow will occur during engine warmup, affecting driveability. To prevent the EGR valve from opening while the engine is cold, a thermostatic vacuum valve installed in the radiator or in an engine water jacket controls the carburetor's vacuum signal, figure 21-5. When a cold engine first starts, the temperature valve remains closed, blocking vacuum to the EGR valve. As the engine warms up, the temperature valve opens and allows vacuum to reach the EGR valve, figure 21-6.

Engineers have developed various methods for modulating, or controlling, EGR valve operation relative to engine operating conditions.

High- and low-ambient-temperature vacuum modulators may be used to bleed off vacuum to weaken the EGR signal at high and low temperatures. Chrysler has used a time delay switch and solenoids to prevent EGR flow when the engine first starts, figure 21-7. On this system, an intake charge temperature switch installed in the intake manifold turns the vacuum solenoid off when the intake charge (air-fuel mixture) temperature is cold. Other systems may use a vacuum-bias valve to bleed off part of the EGR vacuum signal under high-manifold conditions and eliminate **high-speed surge**. A dual-diaphragm EGR valve, figure 21-8, uses manifold vacuum to help the valve spring offset the carburetor vacuum under certain cruising conditions. Figure 21-9 shows an electronic speed-controlled system used during

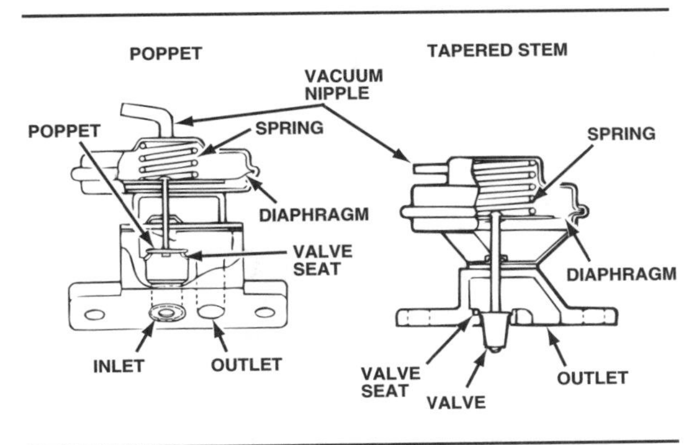

Figure 21-11. A cross-sectional view of Ford poppet and tapered stem EGR valves.

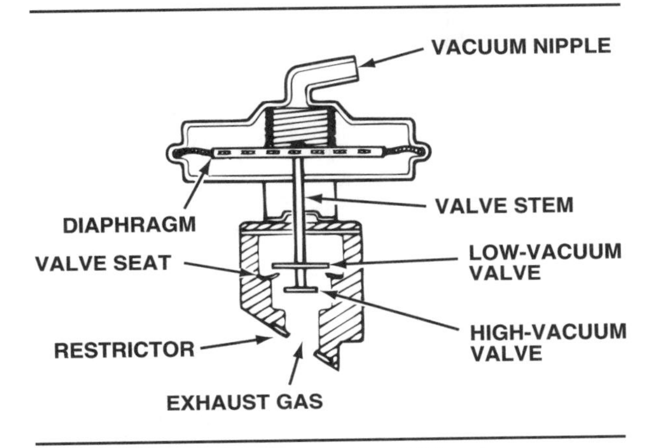

Figure 21-12. A Ford modulating EGR valve.

the 1970s to block EGR at high speeds. Most EGR systems now used with later-model engines are designed to modulate EGR flow through computer-controlled backpressure transducers and solenoids.

Ported EGR Valves

You have seen how a basic ported EGR valve, figure 21-2, operates. There are three types of ported EGR valves, figure 21-10, in use:

- Poppet
- Tapered stem
- Modulating.

The poppet valve contains a spring-loaded diaphragm, a valve and valve stem, and a flow restrictor. Ported carburetor vacuum opens the valve at a specific vacuum level and allows exhaust gas to enter the valve. Exhaust-gas flow to the combustion chambers is controlled by the flow restrictor in the valve body inlet port.

The tapered-stem valve, figure 21-11, operates the same way. It modulates the exhaust-gas flow as it gradually unseats to permit an increasing gas flow.

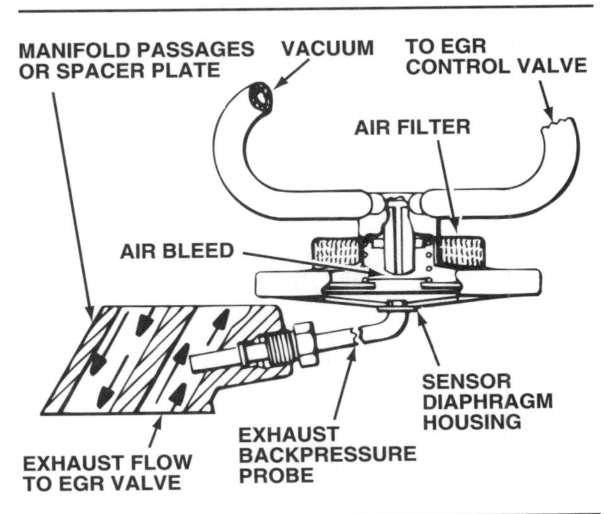

Figure 21-13. Typical EGR backpressure transducer.

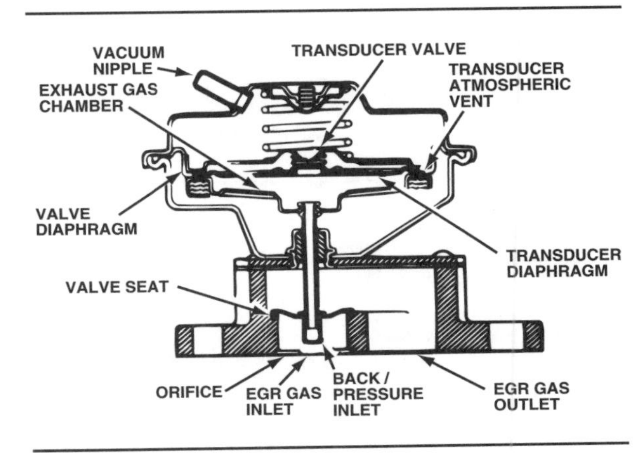

Figure 21-14. This Ford EGR valve has a built-in exhaust backpressure transducer.

The modulating valve, figure 21-12, has an extra disk on the stem below the main valve, and operates much like a poppet valve. However, at a specific vacuum level, the lower disk restricts exhaust-gas flow to the valve chamber. This modulating action improves the driveability of some engines.

Pintle Valve: A valve shaped like a hinge pin. In an EGR valve, the pintle is attached to a normally closed diaphragm. When ported vacuum is applied, the pintle rises from its seat and allows exhaust gas to be drawn into the engine intake system.

High-Speed Surge: A sudden increase in engine speed caused by high manifold vacuum pulling in too much air-fuel mixture.

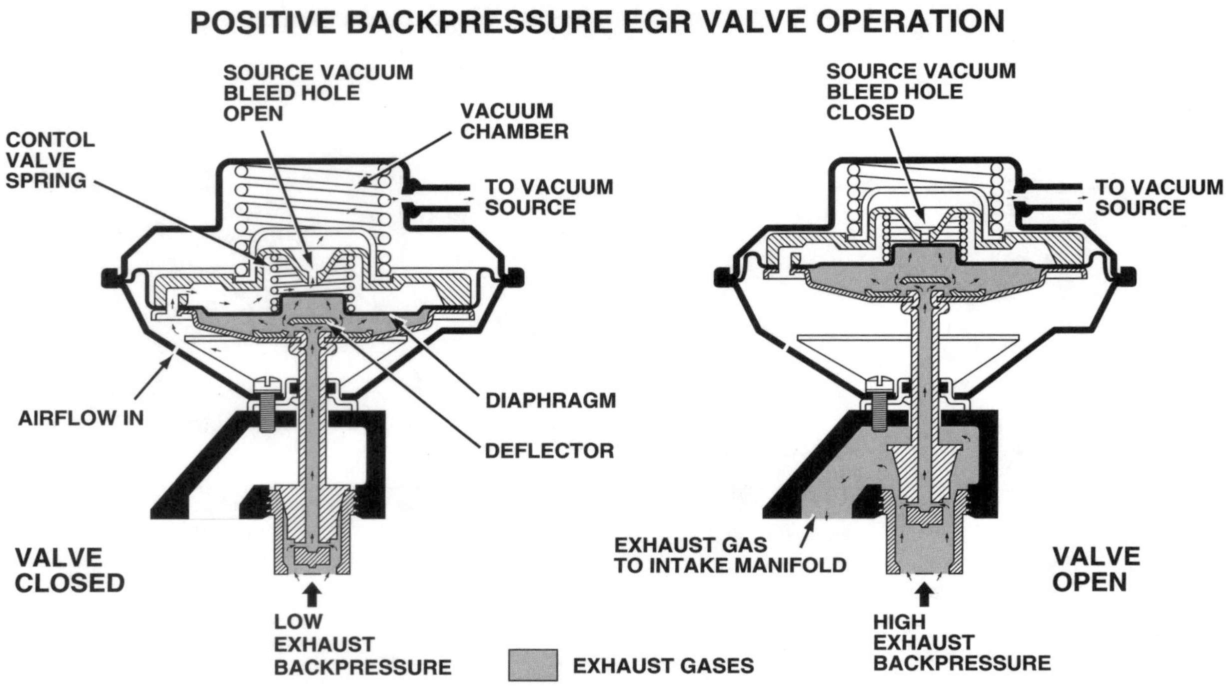

Figure 21-15. When a positive backpressure EGR valve senses high exhaust backpressure and receives a vacuum signal, the control valve spring collapses, causing the diaphragm to seal the EGR source vacuum bleed hole.

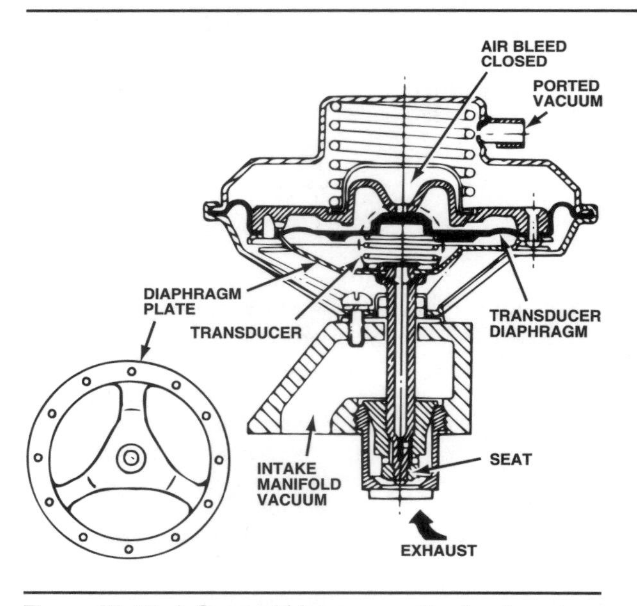

Figure 21-16. A General Motors negative backpressure EGR valve.

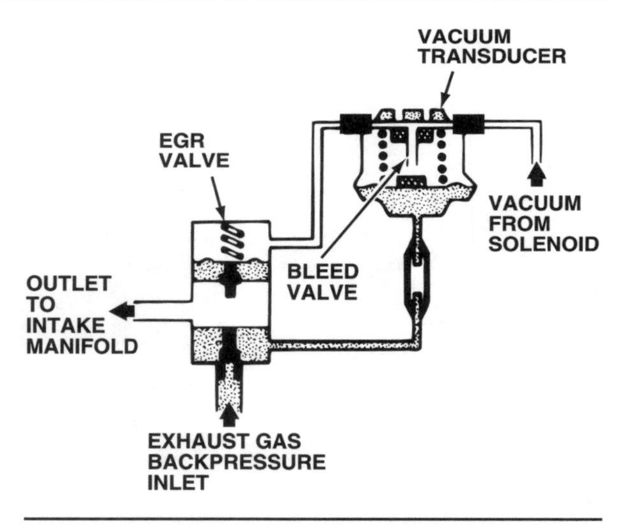

Figure 21-17. Chrysler uses a separate positive backpressure transducer and EGR valve on normally aspirated engines.

Backpressure Transducer EGR Valves

Chrysler was the first to add a separate backpressure transducer to the ported EGR system on some 1973 engines as a way of regulating the vacuum signal according to engine load. The transducer is installed between an exhaust-gas passageway and the EGR valve, figure 21-13. When high exhaust backpressure is sensed through the tube, the diaphragm closes an air bleed hole in the EGR vacuum line, resulting in maximum EGR during periods of heavy load, such as acceleration. As engine load reduces,

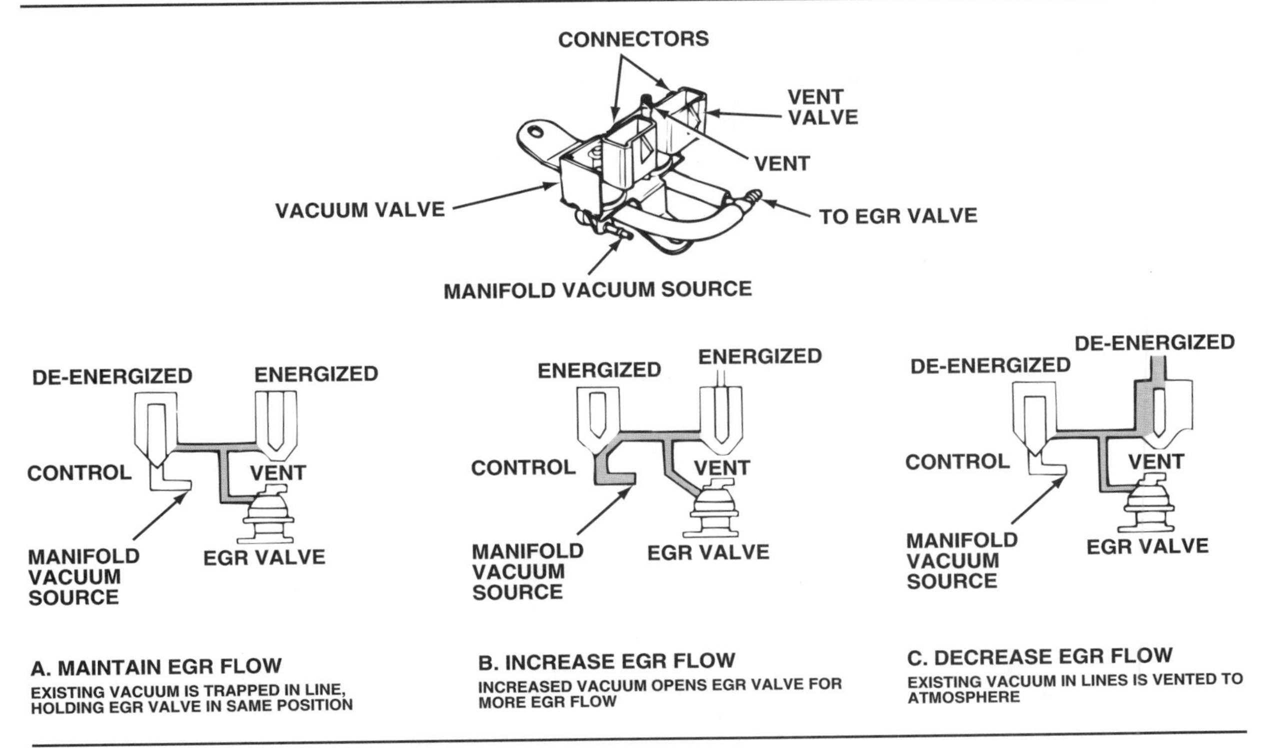

Figure 21-18. A Ford EGR solenoid vacuum valve assembly.

backpressure drops and a spring moves the transducer diaphragm to reopen the vacuum line bleed. Since this decreases the vacuum at the EGR valve, the amount of exhaust gas recirculated also reduces.

By 1977, the separate transducer had been replaced in many General Motors (GM) and Ford EGR systems by a transducer incorporated inside the EGR valve, figure 21-14. This design is called a positive backpressure EGR valve. The valve will not open just by applying vacuum to the diaphragm; there must be enough exhaust backpressure entering through the stem of the valve to close the normally open internal air bleed (control valve) before the valve will open, figure 21-15. Once the valve opens, backpressure drops and the spring reopens the air bleed, causing the valve to move back toward its closed position. Since this all happens rapidly, the valve pintle is modulated about 30 times per second relative to vacuum and exhaust pressure.

The negative backpressure EGR valve, figure 21-16, was introduced in 1979 for use with engines that have relatively little exhaust backpressure. In this design, the transducer air bleed is normally closed. As ported vacuum opens the EGR valve, a negative pressure signal from the vacuum in the intake manifold is buffered by the exhaust system pressure and travels up the inside of the EGR valve stem to the backside of the transducer diaphragm. When the pressure signal is low enough (high vacuum), it opens the air bleed and reduces the amount of EGR. Like the positive backpressure EGR valve, this modulating process goes on constantly. Since the negative backpressure valve has a slightly faster opening time, it can be used with engines prone to detonation under light acceleration.

Chrysler continued to use a separate positive backpressure transducer with a single-diaphragm EGR valve, figure 21-17, on all normally aspirated engines. The transducer is part of a dual (electric-vacuum) switch called an electronic EGR transducer (EET). The EET contains an electrically operated solenoid controlled by the engine computer and the vacuum-operated transducer controlled by exhaust system backpressure. The system operates in the same way as a positive backpressure EGR valve, but the engine computer determines when EGR flow will be allowed.

Solenoid Controls

EGR systems used with many computer-controlled engines have one or more solenoids in the EGR vacuum line to control valve operation, figure 21-18. The computer uses a solenoid to shut off vacuum to the EGR valve at certain temperatures or during idle and wide-open throttle operation. If two solenoids are used, one generally applies vacuum to the EGR valve

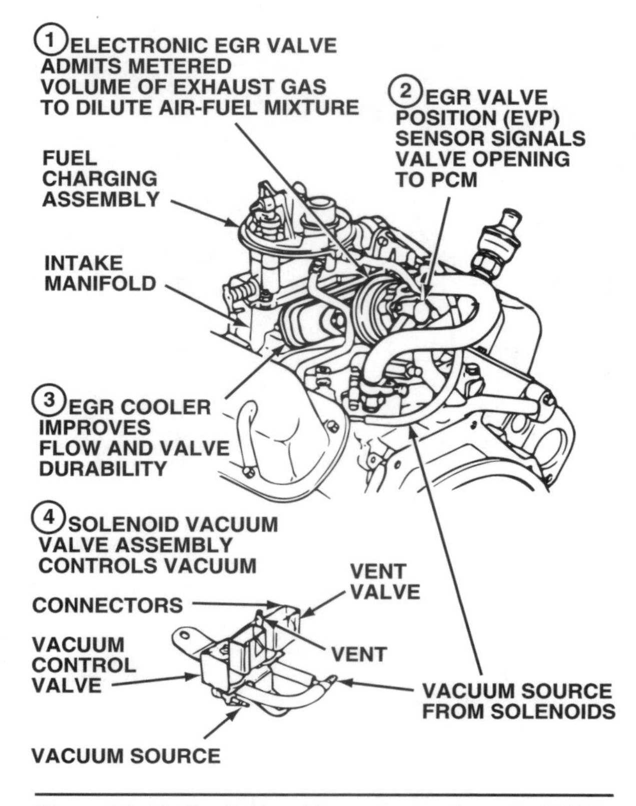

Figure 21-19. Ford closed-loop electronic control EGR system with solenoid vacuum valve control.

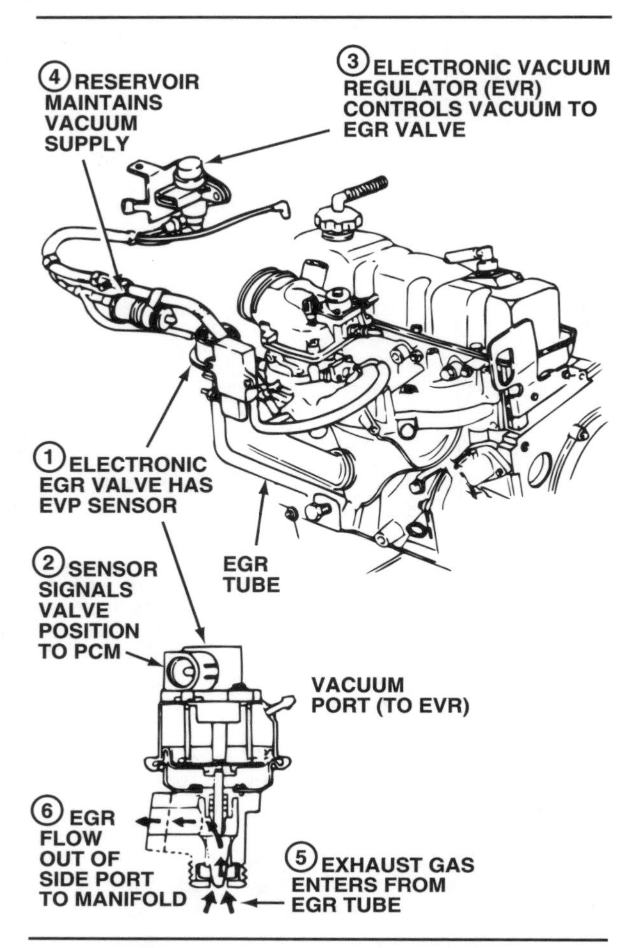

Figure 21-20. Ford closed-loop electronic control EGR system with an electronic vacuum regulator.

while the other vents vacuum when EGR flow is not desired. A solenoid also can be used to control an air bleed in the vacuum line as a way to reduce EGR flow under light engine loads.

COMPUTER-CONTROLLED EGR AND PULSE WIDTH MODULATION

In the early 1980s, automakers began integrating the EGR system into the electronic engine management system, using throttle position and manifold vacuum as the basic regulating inputs. Ford uses three slightly different types of electronic engine control (EEC) with its EGR systems. One type uses solenoid-vacuum valve control, figure 21-19. Another type uses an electronic vacuum regulator (EVR), figure 21-20. The sensor in both systems sends a position signal telling the powertrain control module (PCM) how far the valve is open. The third type is similar to that shown in figure 21-20, but uses a separate pressure feedback electronic (PFE) transducer to continually measure the amount of backpressure in the exhaust system.

In each case, the PCM compares the sensor or PFE transducer data with its program to determine whether the flow rate should be increased,

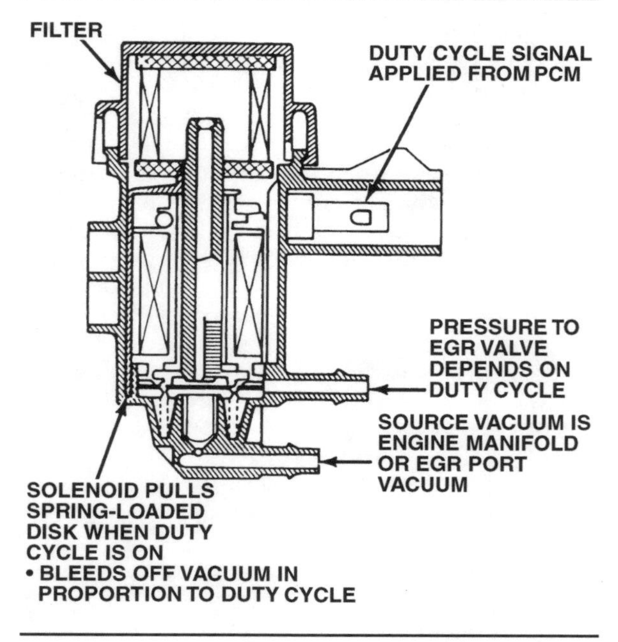

Figure 21-21. A Ford electronic vacuum regulator.

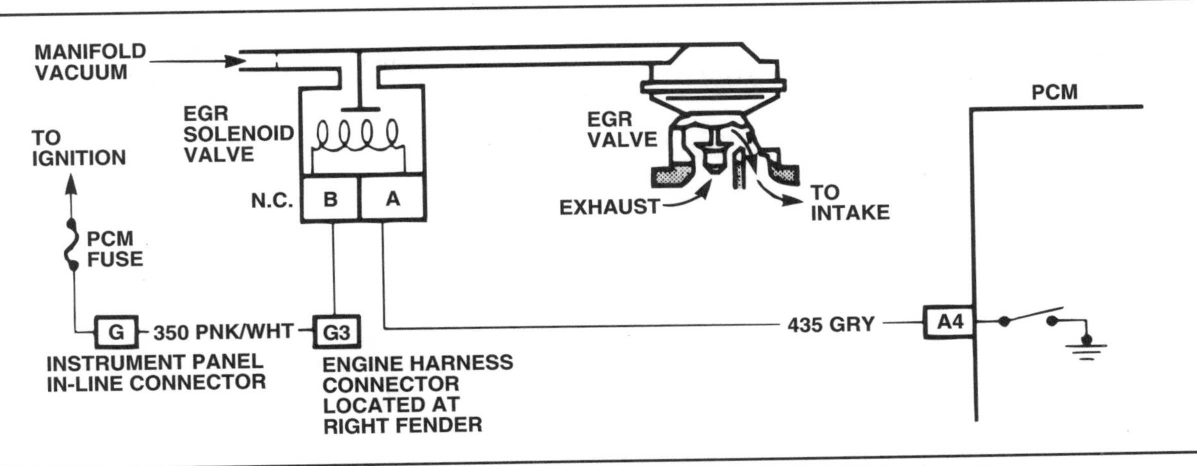

Figure 21-22. The PCM uses pulse width modulation to open and close this GM EGR solenoid.

maintained, or decreased. It then signals the solenoid-vacuum valve assembly, figure 21-19, or the EVR, figure 21-20, to maintain or change the vacuum on the EGR valve diaphragm as required. When using a solenoid assembly, figure 21-18, the two solenoids "dither", that is, they open and close rapidly to modulate the valve opening. The EVR, figure 21-21, is a spring-loaded solenoid that responds to the duty cycle established by the PCM by applying, trapping, or bleeding off vacuum to the EGR valve as required.

The typical PCM-controlled EGR system used by GM, figure 21-22, contains a pulse width modulated (PWM) solenoid and works much the same as the Ford system described above. In this design, the computer operates the solenoid continuously at a fixed frequency of 32 Hz. This is similar to the operation of a carburetor mixture control solenoid but at a faster rate. To regulate the amount of vacuum applied to the EGR valve, the computer varies the solenoid duty cycle (modulates the pulse width), the ratio of on-time to off-time, according to data from various sensors.

The EGR solenoid may be a normally open or normally closed version, depending on the system. A diagnostic switch senses exhaust-gas temperature as it enters the intake manifold through the EGR valve. If a vacuum circuit failure occurs, the diagnostic switch will set a code in the PCM memory and turn on the instrument panel malfunction indicator lamp (MIL).

Two basic variations of the PWM system were developed by Chevrolet and Buick, and used in many different GM models. The Chevrolet system uses a thermal sensor in the base of the EGR valve to delay vacuum application until engine temperature reaches a specified level. The Buick system includes a current-regulating module that maintains constant control voltage to the

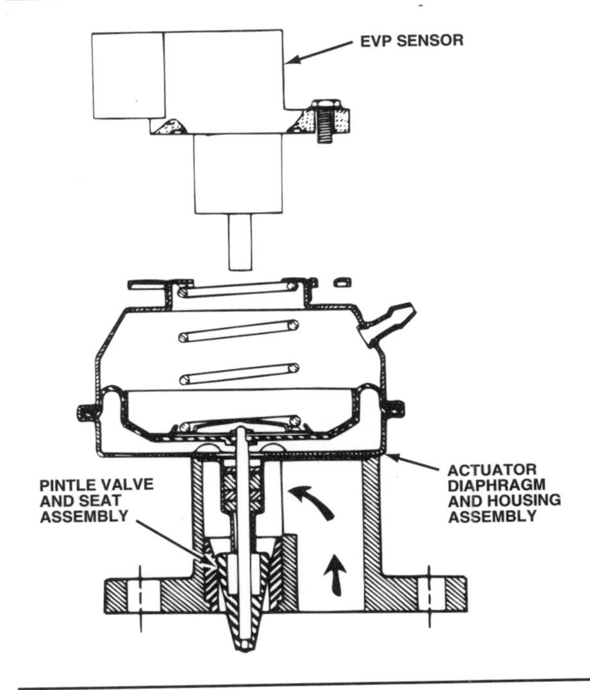

Figure 21-23. This top-mounted Ford EVP sensor uses a potentiometer to modify a reference voltage according to valve pintle position.

solenoid for more accurate EGR regulation. The Chevrolet system does not respond to vacuum changes as rapidly as does the Buick system, and is not fast enough to signal a loss of EGR on turbocharged engines.

Exhaust Valve Position (EVP) Sensors

Computer-controlled EGR systems increasingly use a feedback signal to indicate EGR valve position or exhaust gas-flow. On-Board Diagnostics Generation II (OBD-II) EGR system mon-

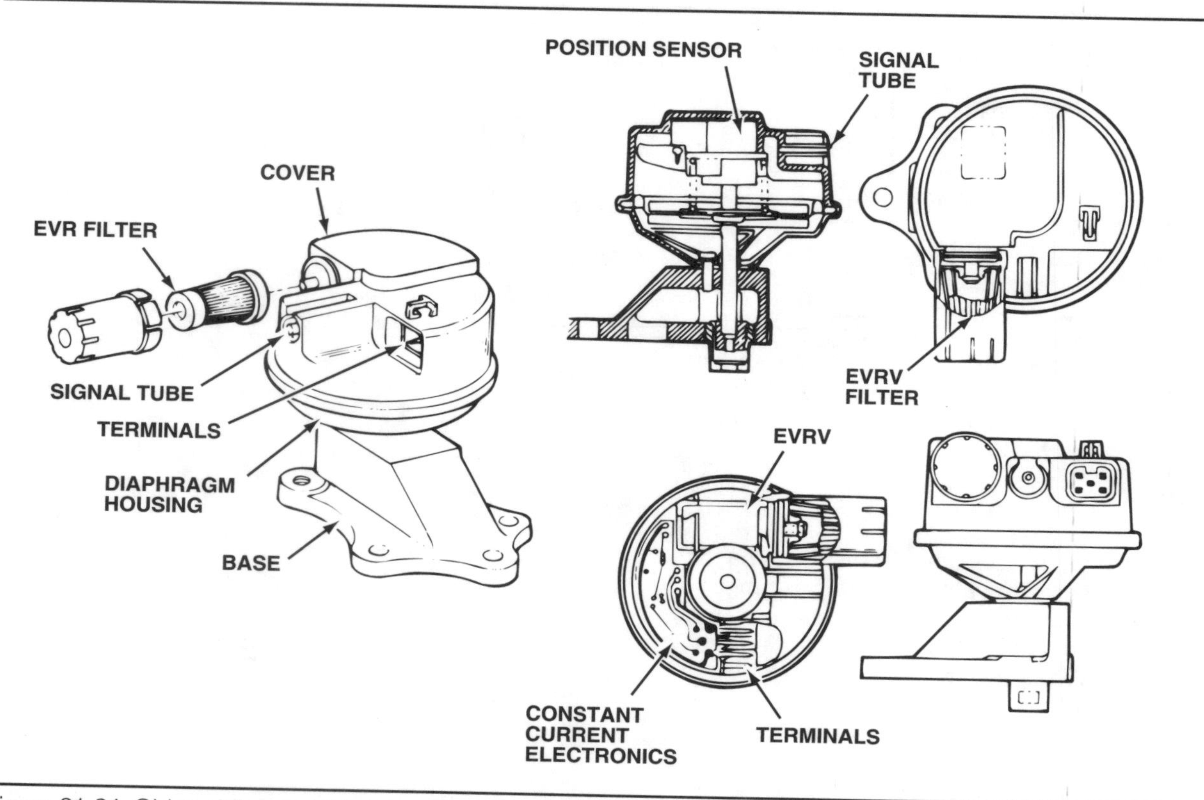

Figure 21-24. Oldsmobile used this integrated electronic EGR valve on some 1987–88 engines, but later abandoned it because of its complexity and cost.

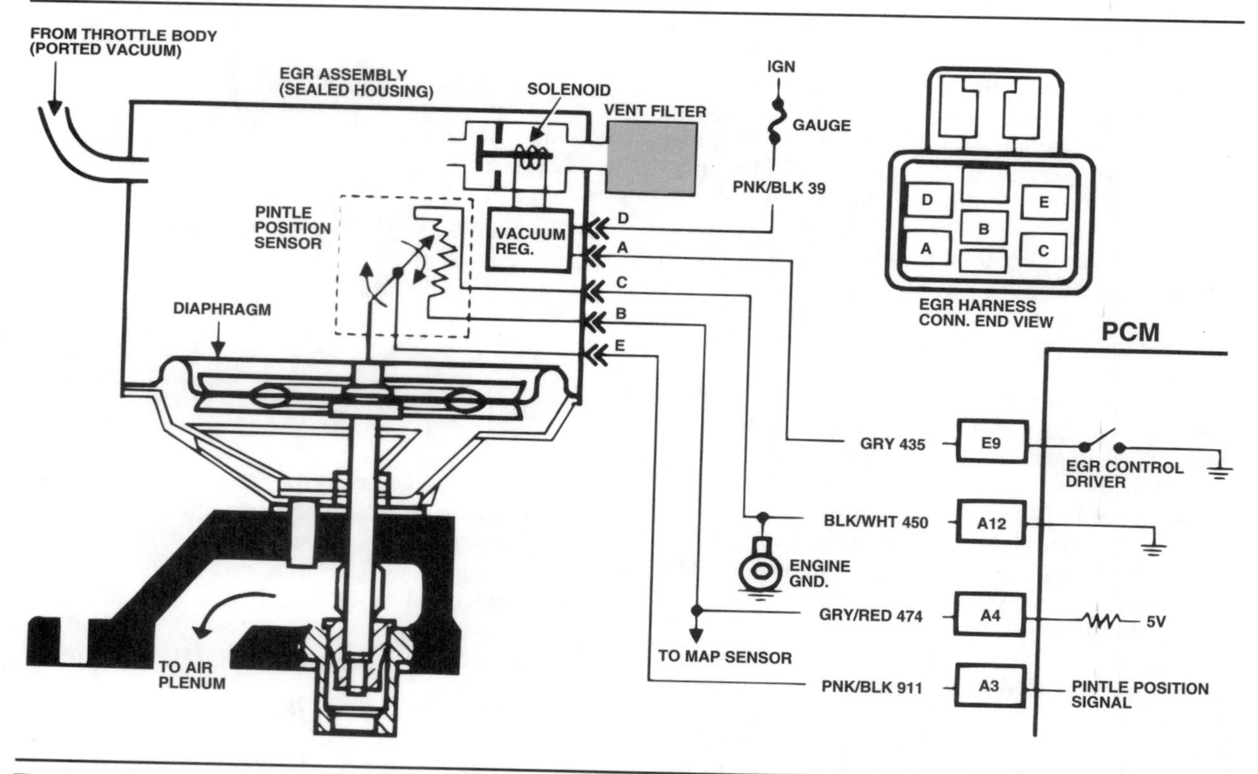

Figure 21-25. The integrated electronic EGR valve contains a built-in vacuum regulator, solenoid, and pintle position sensor. (Oldsmobile)

itors require a feedback signal. Ford introduced the concept of position sensors with its EEC systems in the late 1970s, figure 21-23. A potentiometer on the top of the EGR valve stem indicates valve position by sending a return voltage signal to the computer. Some later-model Ford EGR systems, however, use a feedback signal provided by a PFE transducer, which converts exhaust backpressure to a voltage signal.

The GM integrated electronic EGR valve uses a similar sensor, figure 21-24. The top of the valve contains a vacuum regulator and pintle-position sensor sealed inside a nonremovable plastic cover, combining all of the control devices in one assembly, figure 21-25. The pintle-position sensor provides a voltage output to the PCM, which increases as the duty cycle increases, allowing the PCM to monitor valve operation.

Digital EGR Valves

GM introduced a new EGR valve design on some 1990 engines. Unlike the previous vacuum-operated EGR valves you have studied, the digital EGR valve consists of two or three solenoids controlled by the PCM, figure 21-26. Each solenoid controls an orificed chamber in the base; each orifice is a different size. Since the PCM controls each solenoid ground individually, it can produce any of seven different flow rates. Because flow rate depends on orifice size instead of valve stem position, the digital EGR valve offers more precise control. Using a swivel pintle design improves valve sealing and makes proper alignment less critical during assembly. EGR leakage has less effect on idle quality because the floating seals are exposed to exhaust pressure rather than manifold vacuum.

On-Board Diagnostics Generation II (OBD-II) EGR Monitoring Strategies

In 1996 the U.S. EPA began requiring OBD-II systems in all passenger cars and most light-duty trucks. These systems include an EGR monitoring program that alerts the driver and the technician if the EGR system is not functioning efficiently. Simply stated, the OBD-II EGR system test begins by having the PCM open and close the EGR valve. The PCM then monitors for an expected change in signal voltage from a "test sensor". After comparing the test-sensor signal voltage with stored values from look-up tables that correlate with exhaust-gas flow, the PCM calculates EGR system efficiency. If the EGR efficiency level does not meet a predetermined standard, a diagnostic trouble code (DTC) sets. Domestic automakers use various test sensors to determine EGR system efficiency.

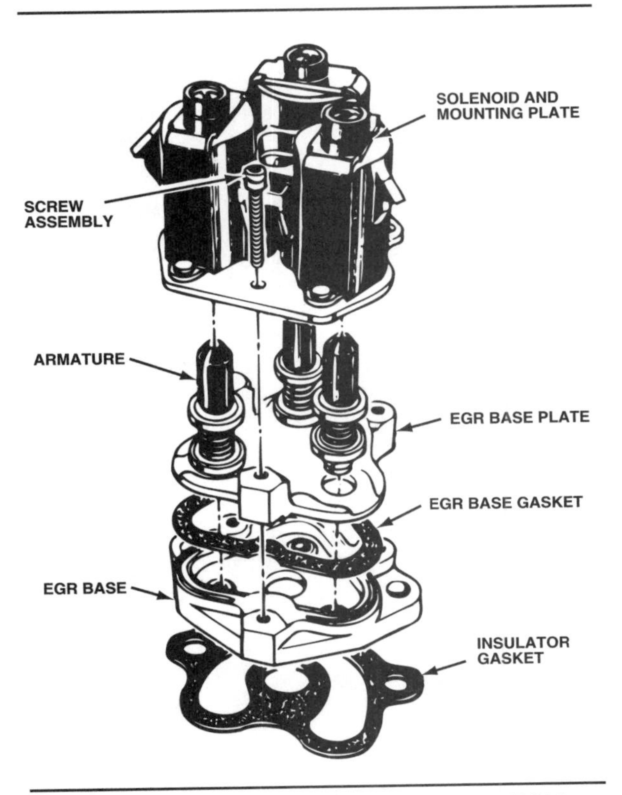

Figure 21-26. The digital EGR valve relies on PCM-modulated solenoids to control EGR flow.

Chrysler monitors the difference in exhaust oxygen sensor (O2S) voltage activity as the EGR valve switches from an open to a closed position. Normally, oxygen exhaust content rises when the EGR valve is closed, so O2S voltage should drop. The PCM sets a DTC if the oxygen content does not increase enough after the EGR valve closes.

Depending on the vehicle application, Ford uses one of at least two types of test sensors to evaluate exhaust-gas flow. The first uses a negative temperature coefficient thermistor mounted in the intake side of the EGR passageway. The PCM monitors changes in the temperature of the gas flowing through the EGR passageway while the EGR valve opens and closes. If the system functions properly, the thermistor voltage should decrease as the EGR valve opens. After comparing the change in sensor voltage with values stored in memory, the PCM calculates efficiency.

A second type of Ford EGR monitor test sensor is called a Delta Pressure Feedback EGR (DPFE) sensor. This sensor measures the pressure differential between two sides of a metered orifice positioned just below the EGR valve's exhaust side. When the EGR valve opens, pressure between the orifice and the EGR valve decreases as exhaust gas flows into the intake manifold. The DPFE sensor recognizes this pressure drop, compares it to the relatively higher pressure before the orifice, and transmits the value of the pressure difference to the PCM. When the EGR valve closes, the exhaust-gas pressure on both sides of the orifice should be equal.

In operation, Ford's second type of OBD-II EGR monitor first waits until the engine runs at a preset speed and under certain conditions. Then the monitor evaluates the pressure differential while the PCM commands the EGR valve to open. Like other systems, the monitor compares the measured value with the stored value to determine EGR system efficiency. If the pressure differential falls below a certain limit, a DTC sets.

GM uses the manifold absolute pressure (MAP) sensor as the test sensor on some applications. After meeting eight preconditions, EGR monitoring begins as the PCM commands the EGR valve open on deceleration. MAP sensor voltage should rise in response to lower manifold vacuum. If the voltage value does not meet those stored in the PCM's memory, a DTC sets.

SUMMARY

Combustion chamber temperatures that exceed 2,500°F (1,370°C) cause nitrogen and oxygen to combine readily, forming large quantities of oxides of nitrogen (NO_X), an air pollutant detrimental to human health. To reduce NO_X as much as possible, automakers use exhaust gas recirculation (EGR) systems to meter small quantities of exhaust gas into the incoming air-fuel mixture. This dilutes the fuel charge and results in lower combustion chamber temperatures.

Each automaker uses some variation of the basic system to achieve exhaust gas recirculation. Various temperature sensors, flow valves, and other control devices in the EGR system modulate the flow of exhaust gas to maintain driveability while reducing NO_X emissions.

With the advent of sophisticated engine control systems, EGR system control has become a function of the PCM. The computer determines when and how much EGR flow is required and activates a pulse width modulated (PWM) solenoid. The solenoid duty cycle established by the computer determines the flow rate and quantity.

Review Questions

Choose the single most correct answer.
Compare your answers to the correct answers on page 507.

1. Photochemical smog is a result of:
   a. Sunlight + NO_X + HC
   b. Sunlight + NO_X + CO_2
   c. Sunlight + CO + HC
   d. Sunlight + NO_X + CO

2. NO_X forms in an engine under:
   a. High pressure and low temperature
   b. Low pressure and low temperature
   c. High temperature and high pressure
   d. All of the above

3. Which of the following is not part of an exhaust gas recirculation system?
   a. Chrysler floor jets
   b. A Buick EGR valve
   c. Ported-vacuum systems
   d. Slow-idle solenoid

4. Which is not true of EGR valves?
   a. They operate on venturi-vacuum systems
   b. They operate on ported vacuum
   c. They may be mounted on the intake manifold
   d. They operate at wide-open throttle

5. Which is not one of the basic ways to reduce NO_X formation?
   a. Enrich the air-fuel mixture
   b. Lower the compression ratio
   c. Advance spark timing slightly
   d. Add a small amount of inert gas to the intake charge

6. In a closed-loop electronic control system, a sensor informs the PCM of how the system is functioning by sending a:
   a. Position signal
   b. Pressure signal
   c. Either a or b
   d. Neither a nor b

7. General Motors computer-controlled EGR systems use:
   a. A solenoid vacuum valve
   b. An electronic vacuum regulator
   c. An integrated electronic EGR valve
   d. A ported vacuum valve with a position sensor

8. Which is not a temperature control method used to control EGR flow?
   a. A coolant temperature sensor
   b. A temperature-sensitive vacuum switch
   c. A radiant temperature-sensitive valve
   d. A temperature delay valve or restrictor

9. Technician A says that the air bleed in a positive backpressure EGR valve is normally open.
   Technician B says the air bleed is normally closed.
   Who is right?
   a. A only
   b. B only
   c. Both A and B
   d. Neither A nor B

10. Technician A says that exhaust backpressure decreases as cruising load increases.
    Technician B says that exhaust backpressure increases as cruising load decreases.
    Who is right?
    a. A only
    b. B only
    c. Both A and B
    d. Neither A nor B

11. The Chevrolet PWM EGR system delays vacuum application until the engine temperature reaches a specified level by using a:
    a. Temperature switch in the engine block
    b. Thermal sensor in the EGR valve base
    c. Pintle position sensor in the EGR valve
    d. None of the above

12. A digital EGR valve does not:
    a. Contain two or three ECM-controlled solenoids
    b. Have solenoids that control differently sized orifices
    c. Produce only three different flow rates
    d. Allow exhaust gas recirculation when the PCM grounds each solenoid individually

13. A customer has just dropped off a 1996 Chrysler passenger car with a lighted MIL. One of the DTC's you retrieve indicates a malfunction in the EGR system. What test sensor does the OBD-II monitor most likely rely on to test EGR efficiency?
    a. MAP sensor
    b. O2S
    c. DPFE transducer
    d. EGR thermistor

Chapter

22

Catalytic Converters

To meet the exhaust emission limits first implemented in the late 1970s, automakers began installing **catalytic converters** to change harmful hydrocarbons (HC) and carbon monoxide (CO) into harmless by-products. The device, which looks from the outside like a small muffler or resonator, is installed in the exhaust system between the exhaust manifold and the muffler, and generally is positioned under the passenger compartment. Its location is important, since as much of the exhaust heat as possible must be retained for effective operation.

A simple device, the catalytic converter contains no moving parts; it simply forms a chamber in the exhaust system through which the exhaust gas passes. Inside the converter, the exhaust flows through a honeycomb monolith or pellet-type catalyst material, which turns the exhaust pollutants into water and carbon dioxide (CO_2). Chances are you will never see the inside of a converter unless you cut one in half, since there is nothing inside that can be repaired.

The link between a converter and a well-tuned engine is important. An engine that misfires or is improperly tuned can destroy a catalytic converter. If a spark plug misfires for a prolonged time, the temperature in the converter could be raised high enough to equal average cylinder combustion temperatures, shortening the life of the converter. The more you know about converters, the more you will understand the need for a well-tuned engine.

In this chapter you will learn:

- Catalytic converter operation
- Differences between oxidation and reduction converters
- Ways to make sure that the converter does its job efficiently.

REDUCING EMISSIONS

One way of lowering HC and CO is to increase the combustion temperature, which causes more complete burning. As combustion temperatures rise, however, so does the formation of the third major pollutant, oxides of nitrogen (NO_X). A second method of turning exhaust gas into nonpolluting materials is the use of the catalytic converter, figure 22-1. By passing the exhaust gas through a **catalyst** in the presence of oxygen, the HC and CO compounds unite with the oxygen, resulting in two harmless by-products of the catalytic reaction: water vapor (H_2O) and carbon dioxide (CO_2).

The Oxidation and Reduction Reactions

A catalyst is a substance that promotes a chemical reaction, but is not changed or affected by that reaction. An **oxidation**, or burning, reaction

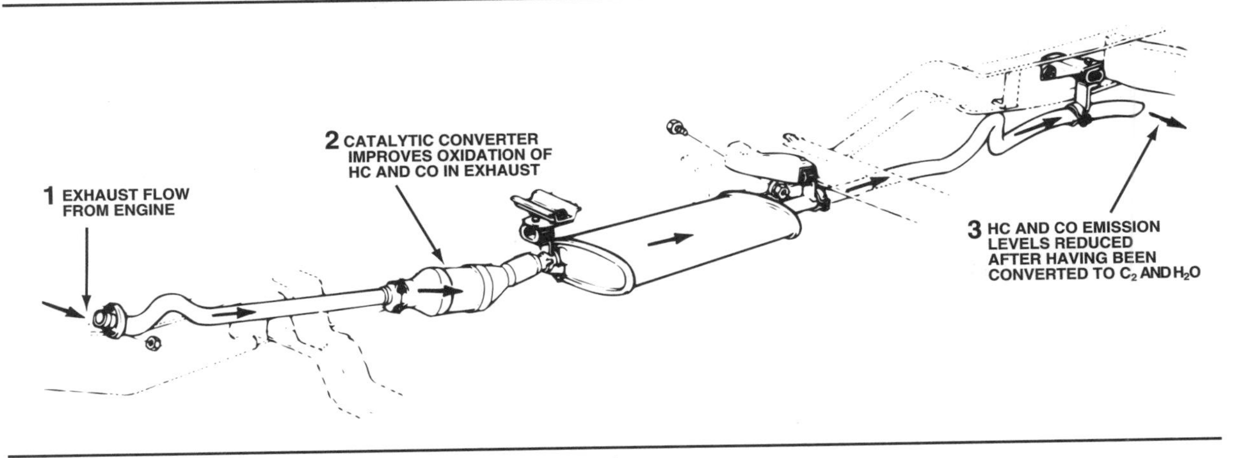

Figure 22-1. Typical catalytic converter installation. (Ford)

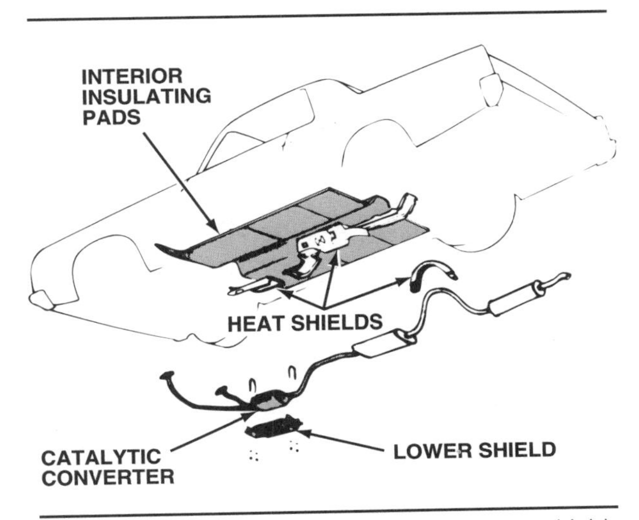

Figure 22-2. Monolithic converters require heat shielding. (Chrysler)

takes place when oxygen is added to an element or compound. Platinum and palladium are catalytic elements called **noble metals** that promote oxidation. If there is not enough oxygen in the exhaust, an air pump or aspirator valve provides extra air. The oxidation process mixes the HC and CO with oxygen to render them harmless by forming H_2O and CO_2.

Since the catalytic oxidation reaction has no effect on NO_X emissions, a separate reaction called **reduction** is required. The reduction reaction mixes NO_X with CO to form nitrogen gas (N_2) and CO_2. The most common reduction catalysts are platinum and another noble metal, rhodium.

A catalyst does not work when cold; it must be heated to its light-off temperature of 400° to 500°F (204° to 260°C) before it starts working at 50 percent effectiveness. When fully effective, the converter reaches a temperature range of 900° to 1,600°F (482° to 871°C). Because of this

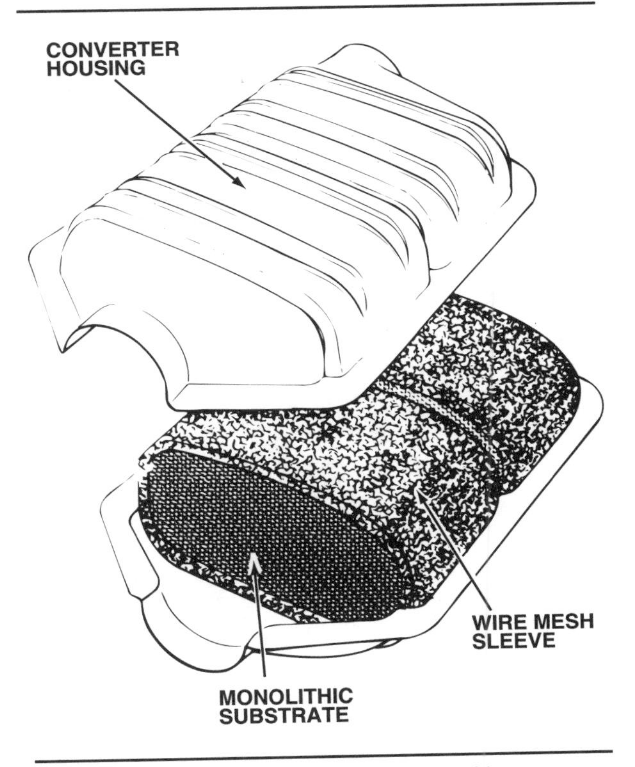

Figure 22-3. Typical catalytic converter with a monolithic substrate.

Catalytic Converter: A device installed in an exhaust system that converts up to three pollutants, hydrocarbons (HC), carbon monoxide (CO), and oxides of nitrogen (NO_X) into harmless by-products using a catalytic chemical reaction.

Catalyst: A substance that causes a chemical reaction, without being changed by the reaction.

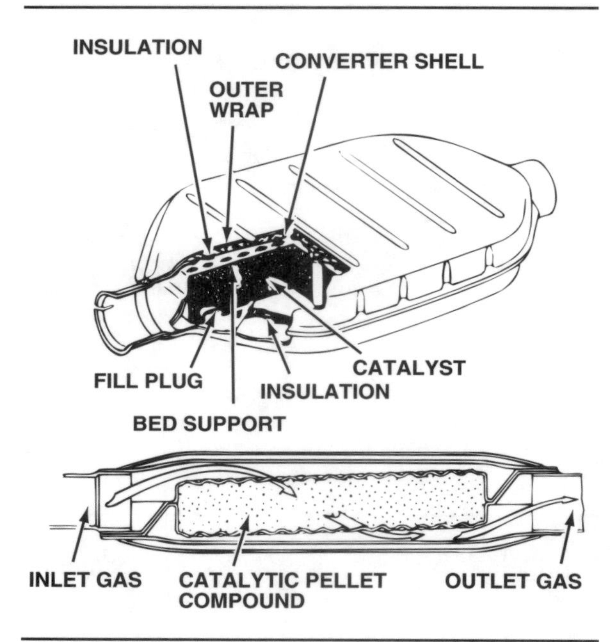

Figure 22-4. Typical catalytic converter with a pellet-type substrate.

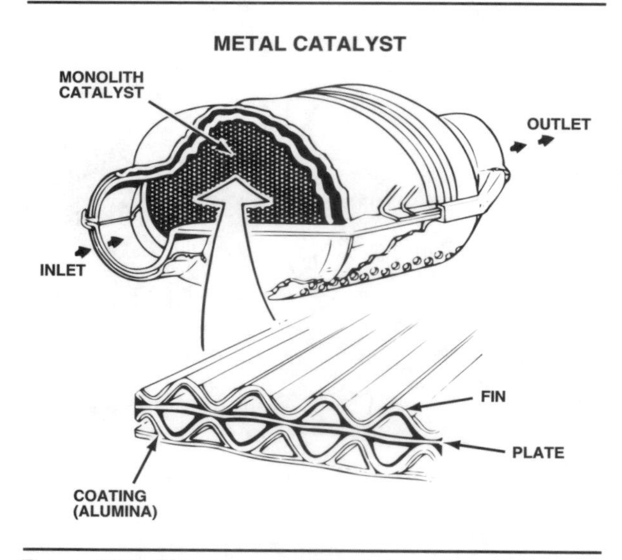

Figure 22-5. This cross-sectional view of a metal monolith substrate converter shows how the metal substrate is formed. (GM)

extreme heat (almost as hot as combustion-chamber temperatures), a converter remains hot long after the engine is shut off. Most vehicles use a series of heat shields, figure 22-2, to protect the passenger compartment, the automatic transmission, and other parts of the chassis from excessive heat. In spite of the intense heat, however, catalytic reactions do not generate the flame and radiant heat associated with a simple burning reaction.

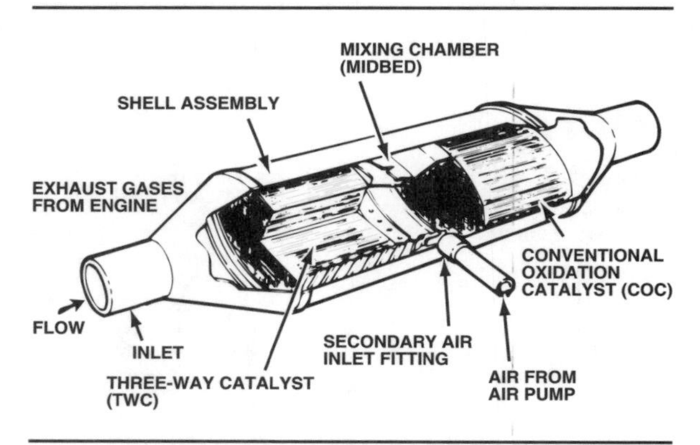

Figure 22-6. Three-way converters contain both catalysts in a single housing with an air inlet between the two catalysts. (Ford)

Converter Design

The platinum, palladium, and rhodium catalysts are not pure metal, but thin deposits on a ceramic or aluminum oxide **substrate**. Three kinds of substrates are used:

- Tiny pellets or beads
- A large porous ceramic block
- Formed metal.

Each type of substrate material creates several thousand square yards or meters of catalyst surface area for contact with the exhaust gases as they flow through the converter. In converters using a ceramic (monolith) substrate, figure 22-3, a diffuser inside the converter shell produces a uniform flow of exhaust gases over the entire substrate surface. If the converter uses a pellet substrate, figure 22-4, the gas flows over the top and down through the substrate layers. The metal substrate consists of alternating layers of corrugated fins and formed metal plates that resemble a honeycomb pattern, figure 22-5. The substrate is contained within a round or oval shell made by welding two stamped pieces of aluminum or stainless steel together. These metals are used because of their ability to withstand the high temperatures of oxidation.

Pellet-type converters create a fair amount of exhaust restriction, but are less expensive to manufacture, and the pellets in some models can be replaced if they become contaminated. The ceramic substrate in monolithic converters is much less restrictive, but breaks more easily when subjected to shock or severe jolts, and is more expensive to manufacture. Monolithic converters can only be serviced as a unit.

Converter Applications

A catalytic converter must be located as close as possible to the exhaust manifold to work effectively. The farther back the converter is posi-

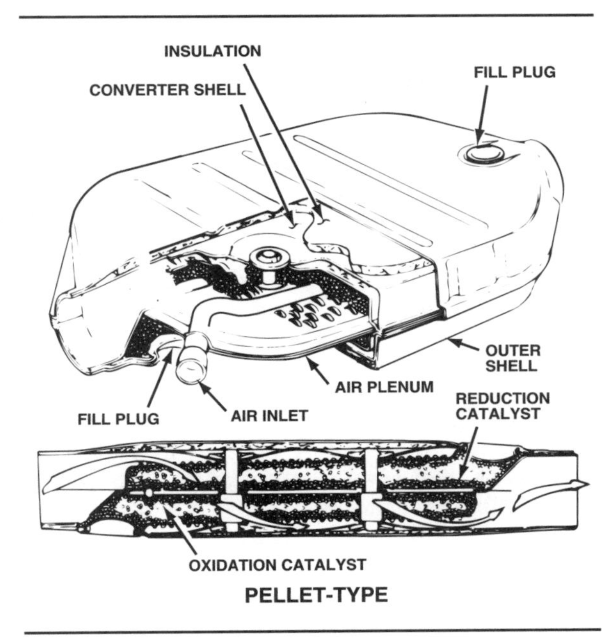

Figure 22-7. Three-way pellet-type catalytic converter construction. (GM)

tioned in the exhaust system, the greater the gases cool before reaching the converter.

Some vehicles use a small oxidation converter called a preconverter or mini-converter that connects directly to the exhaust manifold outlet. These converters have a small catalyst surface area close to the engine that heats up rapidly to start the oxidation process more quickly during cold engine warm-up. For this reason, they are often called light-off converters, or LOC. The oxidation reaction started in the LOC is completed by the larger main converter under the passenger compartment.

A vehicle requiring both oxidation and reduction catalysts may use a two-stage, three-way converter. The two catalysts can be housed in separate converters (dual-bed), or they can be located in separate chambers of the same housing. The reduction catalyst must be located ahead of the oxidation catalyst, since the reduction reaction requires lower levels of oxygen and higher levels of CO than that in the exhaust stream following an oxidation reaction. A three-way converter using a monolith substrate has the reduction catalyst located at the front of the converter, figure 22-6. A three-way converter with a pellet substrate has one placed above the other, with an air plenum separating the two catalysts, figure 22-7. The reduction catalyst is positioned on top of the plenum.

Some three-way converters contain a hybrid reduction and oxidation catalyst at the front of the housing, and a second oxidation catalyst installed behind it. The hybrid catalyst works

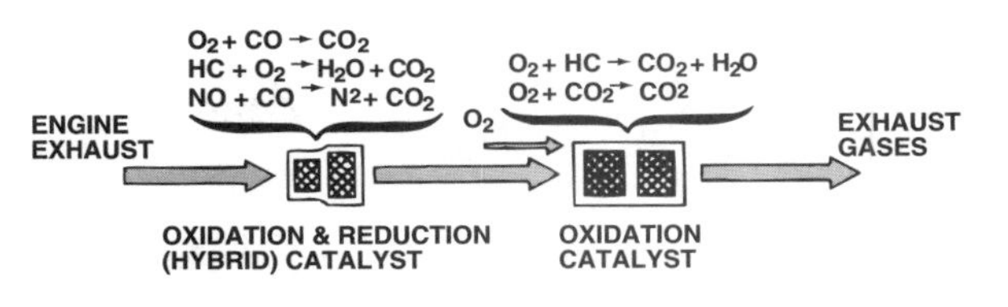

Figure 22-8. Catalytic oxidation and reduction reactions. (Chrysler)

best to reduce NO_x when the CO level in the exhaust is between 0.8 and 1.5 percent. The second catalyst completes the oxidation process. Figure 22-8 shows the two-stage reaction in this type of converter.

Oxidation and reduction converters work most efficiently when the air-fuel mixture is

Oxidation: The combining of an element with oxygen in a chemical process that often produces extreme heat as a by-product.

Noble Metals: Metals, such as platinum and palladium, that are inert.

Reduction: A chemical process in which oxygen is removed from a compound.

Substrate: The layer, or honeycomb, of aluminum oxide upon which the catalyst (platinum or palladium) in a catalytic converter is deposited.

■ Catalytic Converter Odors

Although catalytic converters control HC, CO, and NO_x emissions, they also produce other undesirable emissions in small quantities. For example, gasoline contains a little bit of sulfur that reacts with the water vapor inside a converter to produce hydrogen sulfide. This toxic by-product has the distinct odor of rotten eggs. The smell is usually most noticeable while the engine is warming up, or during deceleration.

When the odor is very strong at normal operating temperatures, it may mean that the engine is out-of-tune and is running too rich. In many such cases, the carburetor idle mixture screws are not set correctly, although the problem also may be caused by a high fuel level or some other carburetor problem. Since the odor does not necessarily mean an incorrect mixture adjustment, the carburetor should not be adjusted simply to get rid of the odor. Changing brands of gasoline may help control the odor in some cases, since the amount of sulfur present in gasoline varies from one brand to another.

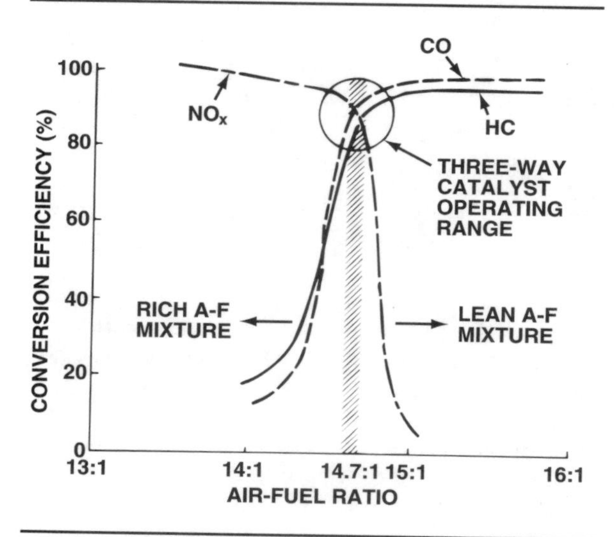

Figure 22-9. A three-way catalyst will only work properly in a narrow air-fuel ratio range. (GM)

maintained within a window of approximately 14.7:1, figure 22-9. The most complete combustion of air and fuel occurs exactly at 14.7:1, resulting in the least amount of harmful pollutants. When the air-fuel mixture is slightly leaner than 14.7:1, the converter receives excess oxygen needed for oxidation; when the air-fuel mixture is slightly richer than 14.7:1, the converter receives excess CO needed for reduction.

As we learned in Part Four of this manual, an exhaust oxygen sensor (O2S) during closed-loop operation monitors the level of oxygen in the exhaust stream, and then sends a corresponding voltage signal to the powertrain control module (PCM). With this signal, the PCM varies the pulse width of the fuel injector, or the duty cycle of the mixture control solenoid, to maintain the 14.7:1 air-fuel ratio.

Converter Operation

Converters have no moving parts and thus do not require periodic service. Prior to On-Board Diagnostics Generation II (OBD-II) systems, federal law required manufacturers to warranty catalyst effectiveness for 50,000 miles (80,000 km) or five years, whichever comes first. For OBD-II systems, the warranty is extended to 80,000 miles (128,700 km) or eight years. However, a catalyst will eventually wear out. When it does, the entire converter (monolith) or catalyst (pellet) must be replaced.

It is possible to damage a converter before the catalyst wears out. When leaded gasoline is used in the fuel system, the lead will plate the catalyst, preventing the exhaust gases from reaching the catalyst. A certain amount of lead in the exhaust will gradually burn off from the catalyst, allowing the converter to resume near-normal operation; but continued use of leaded fuel will eventually destroy the converter.

Although a converter operates at high temperature, it can be destroyed by excessive temperatures. This most often occurs when too much unburned fuel reaches the converter through engine misfiring or an excessively rich air-fuel mixture. When either of these circumstances occur, the converter can become glowing-red hot. Excessive temperatures also may be caused by long idling periods, since more heat develops when an engine runs at idle for long periods than when driving at normal highway speeds.

To avoid excessive catalyst temperatures and the possibility of fuel vapors reaching the converter, follow these rules:

- Use only unleaded gasoline in a vehicle equipped with a converter.
- Do not try to start the engine on compression by pushing the vehicle; use jumper cables instead.
- Do not crank an engine for more than 40 seconds when it is flooded or firing intermittently.
- Do not turn off the ignition switch when the car is in motion.
- Do not disconnect a spark plug for more than five seconds to test the ignition.
- Fix engine problems, such as dieseling or misfiring, that affect performance as soon as possible.

OBD-II Catalytic Converter Monitoring Strategy

Government regulations require OBD-II systems to monitor catalytic converter efficiency. Since it is too expensive to equip every vehicle with its own five-gas analyzer to measure how much HC, CO, or NO_x is being emitted, engineers developed an alternative method that *infers* catalytic efficiency.

When the engine combusts a lean air-fuel mixture, higher amounts of oxygen flow through the exhaust into the converter. The catalyst materials absorb this oxygen for the oxidation process, thereby removing it from the exhaust stream. If a converter cannot absorb enough oxygen, oxidation does not occur. Engineers established a correlation between the amount of oxygen absorbed and converter efficiency.

Using two O2S's installed before and after the converter, the OBD-II system monitors how much oxygen the catalyst retains. A voltage waveform from the post-catalyst O2S of a good catalyst shows a relatively small number of crosscounts. A voltage waveform from the post-catalyst O2S of a degraded catalyst shows a relatively large number of crosscounts. In other words, the closer the activity of the post-catalyst O2S matches that of the pre-catalyst O2S, the greater the degree of converter degradation. In operation, the OBD-II monitor compares crosscounts between the two O2S's, and then runs statistical tests before setting a diagnostic trouble code.

SUMMARY

Catalytic converters, first appearing in 1975, promote a chemical reaction to change HC and CO emissions into H_2O and CO_2. Two types of converters are used: oxidation and reduction. The oxidation type removes HC and CO emissions from the exhaust gases; reduction converters remove NO_X. Late-model engines with feedback fuel systems use a three-way converter that combines the oxidation and reduction reactions. Converters use a monolithic or pellet-type substrate coated with platinum, palladium, or rhodium to provide a surface for the reactions to occur. These converters can be damaged by too much heat, the use of leaded fuel, or too much unburned fuel.

Review Questions

Choose the single most correct answer.
Compare your answers to the correct answers on page 507.

1. Catalytic converters:
   a. Increase the HC content in exhaust emissions
   b. Neither add nor remove the oxygen from exhaust emissions
   c. Improve oxidation of HC and NO_x
   d. Improve oxidation of HC and CO
2. The catalyst material in an oxidation catalytic converter is:
   a. Platinum or palladium
   b. Aluminum oxide
   c. Stainless steel
   d. Ceramic lead oxide
3. A catalyst:
   a. Slows a chemical reaction
   b. Heats a chemical reaction
   c. Increases, but is not consumed by, a chemical action
   d. Combines with the chemicals in the reaction
4. A reduction reaction:
   a. Adds oxygen to a compound
   b. Removes oxygen from a compound
   c. Reduces HC in exhaust gases
   d. Removes H_2O from exhaust gases
5. The outer shell of the oxidation catalyst is made of:
   a. Aluminum oxide pellets
   b. Platinum or palladium
   c. A honeycomb monolith
   d. Aluminum or stainless steel
6. A catalyst is a metal deposited on:
   a. A stainless-steel shell
   b. A ceramic substrate
   c. A stainless-steel mesh
   d. Any of the above
7. Cars with catalytic converters are required to use:
   a. Premium unleaded fuel
   b. Unleaded fuel
   c. Low-lead fuel
   d. Any of the above
8. Which of the following will damage the converter?
   a. Push starting the engine
   b. Engine misfiring
   c. Long idling periods
   d. All of the above
9. Which of the following is *not* true for cars with catalytic converters?
   a. Engines should not be cranked for more than 60 seconds
   b. Spark plugs should always be disconnected to test ignition
   c. Engine dieseling, surging, and stalling should be fixed immediately
   d. Ignition should not be turned off while car is moving
10. The reduction converter must be located:
    a. Between the engine and the oxidation converter
    b. Between the oxidation converter and the muffler
    c. After the muffler
    d. Any of the above

ASE Technician Certification Sample Test

This sample test is similar in format to the series of eight tests given by the National Institute for Automotive Service Excellence (ASE). Each of these exams covers a specific area of automotive repair and service. The tests are given every fall and spring throughout the United States.

For a technician to earn certification in a particular field, he or she must successfully complete one of these tests, and have at least two years of "hands on" experience (or a combination of work experience and formal automobile technician training). Successfully finishing all eight tests earns the person certification as a Master Automobile Technician.

The questions in this sample test follow the format of the ASE exams. Learning to take this kind of test will help you if you plan to apply for certification later in your career. You can find the answers to the questions in this sample exam on page 507.

For test registration forms or additional information on the automobile technician certification program, write to:

National Institute for
AUTOMOTIVE SERVICE EXCELLENCE
13505 Dulles Technology Dr., Suite 2
Herndon, VA 22019-1502

1. Most OBD systems will recognize and store a hard DTC for all of the following **EXCEPT**:
   a. An absent signal
   b. An intermittent signal
   c. An improbable signal
   d. An out-of-range signal

2. Technician A says that to get accurate manifold vacuum readings you must correct for altitude. Subtract 1 inch for every 1000 feet above sea level.
   Technician B says manifold absolute pressure is not affected by altitude.
   Who is right?
   a. A only
   b. B only
   c. Both A and B
   d. Neither A nor B

3. Which oscilloscope pattern allows you to display the voltage traces for all cylinders one after the other from left to right across the screen in firing order?
   a. Raster
   b. Parade
   c. Stacked
   d. Superimposed

4. High levels of hydrocarbon (HC) emissions are often the result of a malfunction in the:
   a. Ignition system
   b. Fuel system
   c. PCV system
   d. Emission control system

5. Technician A says that an excessively high secondary voltage discharge can be caused by high primary circuit resistance.
   Technician B says a loss of secondary voltage will cause a greater loss of primary voltage.
   Who is right?
   a. A only
   b. B only
   c. Both A and B
   d. Neither A nor B

6. Technician A says that the battery must be able to provide at least 9 volts while cranking the engine.
   Technician B says a voltage drop of more than 0.2 volt across the battery ground cable will result in low cranking voltage.
   Who is right?
   a. A only
   b. B only
   c. Both A and B
   d. Neither A nor B

7. You can check the circuit resistance of a resistance-wire ballast resistor without disconnecting it by using:
   a. A voltmeter
   b. An ammeter
   c. An ohmmeter
   d. An oscilloscope

8. Technician A says that some electronic ignition distributors have an adjustable air gap between the pick-up and reluctor.
   Technician B says that most air gaps can be adjusted with a standard 0.008-in. feeler gauge and a screwdriver.
   Who is right?
   a. A only
   b. B only
   c. Both A and B
   d. Neither A nor B

9. An oscilloscope raster display of the secondary ignition system shows a dwell variation between cylinders of about 8 to 10 degrees. The most likely cause would be:

a. Worn sparkplugs
b. Worn distributor bushings
c. Defective ignition coil
d. Incorrect ignition timing

10. To check total, centrifugal, and vacuum advance on a conventional distributor, all of the following instruments are used **EXCEPT:**
a. Tachometer
b. Vacuum pump
c. Dwell meter
d. Timing light

11. A magnetic timing meter operates on the same principle as a:
a. Magnetic pulse generator in a distributor
b. Hall-effect sensor
c. Stroboscopic timing light
d. LED pickup coil

12. An electronic input device that provides a variable resistance signal across an operating range is called a:
a. Thermistor
b. Motion detector
c. Potentiometer
d. Piezoelectric generator

13. Technician A says that any solenoid that operates with a varying duty cycle can be checked with a dwell meter.
Technician B says the only way to check the variable duty cycle of a solenoid is with an oscilloscope.
Who is right?
a. A only
b. B only
c. Both A and B
d. Neither A nor B

14. Technician A says you can locate vacuum leaks on a running engine by spraying noncombustible solvent in the suspect area. When you find the leak, the idle will momentarily stabilize and manifold vacuum will increase.
Technician B says that if manifold vacuum is below 3 to 7 in-Hg (176-178 mm-Hg) with the throttle closed, all vacuum ports plugged, and the engine cranking, there is a vacuum leak.
Who is right?
a. A only
b. B only
c. Both A and B
d. Neither A nor B

15. An unmetered air leak downstream of the MAF sensor can cause:
a. A lean condition and poor engine performance
b. The engine management system to provide additional fuel to compensate for the excess air
c. The needle of a vacuum gauge to fluctuate between 10 to 15 in-Hg (254-381 mm-Hg) at idle
d. The electronic control system to remain in the closed-loop mode

16. Most turbocharger failures are caused by:
a. Excessive high-speed operation
b. Exhaust temperatures too high
c. Dirt and contamination
d. Operation with leaded gasoline

17. A turbocharger makes a light, steady whistling sound that rises in pitch as the engine accelerates. This is an indication of:
a. A loose exhaust pipe or outlet elbow
b. An intake air leak
c. Normal operation
d. Damaged bearings

18. Technician A says that if you disconnect the hose from the PCV valve on a running engine, the engine speed should drop slightly and you will hear a hissing noise from the valve.
Technician B says that with the engine at idle and the hose disconnected from the PCV valve, plug the valve with your finger and you should feel a strong vacuum and engine speed will drop if system is working properly.
Who is right?
a. A only
b. B only
c. Both A and B
d. Neither A nor B

19. When doing a cylinder balance test on a DIS system:
Technician A says the idle air control (IAC) valve may have to be disconnected.
Technician B says the secondary wires are shorted to ground for each cylinder.
Who is right?
a. A only
b. B only
c. Both A and B
d. Neither A nor B

20. The exhaust gas recirculation (EGR) system does all of the following **EXCEPT:**
a. Reduces oxides of nitrogen emissions
b. Slows down the combustion process
c. Dilutes the fuel mixture
d. Raises peak combustion temperatures

21. When testing an electronic ignition system with a ballast resistor, what would you expect for voltage at the coil negative connection with the ignition ON, and the engine not running?
a. 0 volts
b. Approximately 1 volt
c. 12 volts
d. It would depend on the position of the trigger wheel tooth in the distributor

22. An air injection system used to control emissions may use all of the following components **EXCEPT:**
a. A wastegate
b. A gulp valve
c. An aspirator
d. A check valve

23. Technician A says that a malfunction in the canister purge system can result in driveability problems and a loss of fuel economy.
Technician B says that incorrect canister purging can change the air-fuel ratio and result in increased exhaust emissions.
Who is right?
a. A only
b. B only
c. Both A and B
d. Neither A nor B

24. Technician A says that too much valve lash clearance can increase the effective camshaft duration and cause a rough idle.

Technician B says that too much valve lash clearance will prevent the valves from seating properly and cause poor performance that can lead to burnt valves.
Who is right?
a. A only
b. B only
c. Both A and B
d. Neither A nor B

25. When testing an electronic ignition system without a ballast resistor, what would you expect for voltage at the coil negative connection with the ignition ON, and the engine not running?
a. 0 volts
b. Approximately 1 volt
c. 12 volts
d. It would depend on the position of the trigger wheel tooth in the distributor

26. The valve lash may be adjusted on an overhead valve engine by all of the following methods **EXCEPT:**
a. An adjustment nut holding the rocker arm
b. An adjustment screw on the rocker arm
c. Replaceable adjustment shims
d. Selective length pushrods

27. Timing chain stretch can result in:
a. Excessive valve clearance
b. Erratic ignition timing
c. Advanced valve timing
d. Retarded valve timing

28. The first step in diagnosing an overheating complaint is to:
a. Verify the actual engine temperature
b. Check the temperature gauge or warning lamp circuit
c. Perform a visual inspection
d. Pressure test the cooling system

29. Technician A says that low results on a cranking voltage test can be caused by a weak battery, high circuit resistance, and internal engine problems.
Technician B says low current draw in the cranking system may be caused by the starter motor binding, an internal short circuit, or internal engine problems.
Who is right?
a. A only
b. B only
c. Both A and B
d. Neither A nor B

30. When checking the charging system with an oscilloscope, there is an uneven ripple in the waveform. The most probable cause would be:
a. An open diode
b. A shorted diode
c. High diode resistance
d. A shorted stator winding

31. A vacuum gauge is connected to an engine. The gauge reads 18 in-Hg (457 mm-Hg) at idle. When engine speed is increased the gauge momentarily drops to near zero, then slowly rises and stabilizes at 9 in-Hg (229 mm-Hg). The likely cause of these readings would be?
a. Weak piston rings
b. Incorrect fuel mixture
c. Restricted exhaust
d. Retarded ignition timing

32. When using an ohmmeter to check an ignition coil, you get a reading showing infinite resistance between a primary terminal and the coil case. This indicates:
a. The coil windings are shorted to the case
b. The case is shorted to the vehicle frame
c. The coil windings are in good condition
d. The ohmmeter is faulty, or set to the wrong scale.

33. A Hall-effect switch on a computer-controlled engine triggers spark plug firing through the use of:
a. A light emitting diode and a reluctor plate
b. A semiconductor and a permanent magnet
c. A pick-up coil and a trigger wheel
d. A magnetic pulse generator and an armature

34. The corrected specific gravity reading of a fully charged battery should be:
a. 1.250
b. 1.255
c. 1.265
d. 1.275

35. Technician A says that for a DIS ignition system to start the correct coil firing sequence it must receive a "synch" signal from the crank or camshaft. Technician B says that only one signal is necessary from the crankshaft.
Who is right?
a. A only
b. B only
c. Both A and B
d. Neither A nor B

36. To check base timing on a computer-controlled ignition system, it may be necessary to:
a. Disconnect the O2S
b. Disconnect the IAC motor
c. Connect or disconnect an electrical connector
d. Turn on the air conditioning to lower idle speed

37. Which should be true of a vehicle's stoichiometric air-fuel ratio?
a. HC is low; CO, O_2, and CO_2 are high
b. HC and CO are low; O_2 and CO_2 are high
c. HC, CO, and O_2 are low; CO_2 is high
d. HC, CO, O_2, and CO_2 are low

38. A car with an air pump emission control system backfires when decelerating. Which of these should the technician check?
a. The operation of the exhaust manifold check valve
b. The output pressure of the air pump
c. The operation of the diverter or gulp valve
d. The air manifolds for leakage

39. Which statement is true of a cylinder power balance test?
    a. An engine with electronic idle speed control should be tested at idle
    b. Exhaust emissions should noticeably increase when you cut out one cylinder
    c. Both a and b
    d. Neither a nor b

40. If the rpm gain measured during a propane enrichment test is less than specified:
    a. The air-fuel ratio is too lean
    b. The engine is overheated
    c. The air-fuel ratio is too rich
    d. The air injection system is not working correctly

41. During cranking, the voltage drop across the battery ground cable should not exceed:
    a. 0.5 volt
    b. 0.2 volt
    c. 1.5 volts
    d. 2.0 volts

42. On a fuel-injected engine, the fuel pressure regulator vacuum hose is unplugged.
    Technician A says that this will cause the fuel pressure to increase.
    Technician B says that this will cause the fuel pressure to decrease.
    Who is right?
    a. A only
    b. B only
    c. Both A and B
    d. Neither A nor B

43. The PCM has a faulty injector driver circuit.
    Technician A says to replace the PCM.
    Technician B says to check the injector resistance.
    Who is right?
    a. A only
    b. B only
    c. Both A and B
    d. Neither A nor B

44. Technician A says that air is injected into the catalyst to lower catalyst temperatures and limit NO_x formation.
    Technician B says that air injection to the oxidizing bed of a dual bed catalyst helps oxidize HC and CO, converting them to CO_2 and water.
    Who is right?
    a. A only
    b. B only
    c. Both A and B
    d. Neither A nor B

45. A vehicle which passes for both HC and CO, fails the I/M240 emission test for NO_x. What could be the cause?
    a. A plugged EGR port
    b. Carbon buildup on top of the piston heads
    c. Hot spots in the combustion chamber caused by a cool ing system blockage in the water jacket surrounding the cylinders
    d. All of the above

46. An engine idles well, but it is emitting higher than acceptable levels of HC. CO is within specs.
    Technician A says that over-advanced ignition timing could be causing the high HC.
    Technician B says a plugged air filter could be causing the high HC.
    Who is right?
    a. A only
    b. B only
    c. Both A and B
    d. Neither A nor B

47. A car fails the emission test for both high HC and CO. Oxygen readings at the tailpipe are above 5 percent. An examination of the vehicle shows that the catalyst is empty. With a new catalyst installed, it is discovered that the O2S voltage never goes above 200 mV.
    Technician A says that the O2S is bad and should be replaced.
    Technician B says that the O2S may be good, but a leak at the exhaust manifold is fooling the sensor into sending a lean signal.
    Who is right?
    a. A only
    b. B only
    c. Both A and B
    d. Neither A nor B

Glossary of Technical Terms

Actuator: An electrical or mechanical device that receives an output signal from a computer and performs an action in response to that signal.

Adaptive Memory: A feature of computer memory that allows the microprocessor to adjust its memory for computing closed-loop operation, based on changes in engine operation.

Adsorption: A chemical action when liquids or vapors gather on the surface of a material. In a vapor storage canister, chemical properties force fuel vapors to attach themselves (adsorb) to the surface of charcoal granules.

Air Injection: A method of reducing hydrocarbon (HC) and carbon monoxide (CO) emissions and providing quick warm-up of emission components. The air may be injected into the exhaust stream before the catalytic converter (upstream) or between the reduction and oxidation beds of a dual-bed converter (downstream) or both. When injected upstream, the air mixes with the hot exhaust, oxidizing the HC and CO to form H_2O and carbon dioxide (CO_2).

Air-Fuel Ratio: The ratio of air to gasoline in the air-fuel mixture that enters an engine.

Alternating Current (ac): A flow of electricity through a conductor, first in one direction, then in the opposite direction.

Ambient Temperature: The temperature of the air surrounding a particular device or location.

Amperage: The amount of current flow through a conductor.

Ampere: The unit for measuring the rate of electrical current flow.

Analog: A voltage signal or processing action that is continuously variable, relative to the operation being measured or controlled.

Analog-to-Digital (AD) Converter: An electronic conversion process for changing analog voltage signals to digital voltage signals.

Aneroid Bellows: Accordion-shaped bellows that responds to changes in coolant pressure or atmospheric pressure by expanding or contracting.

Antiknock Value: The characteristic of gasoline that helps prevent detonation.

Antioxidants: Chemicals or compounds added to motor oil to reduce oil oxidation, which leaves carbon and varnish in the engine.

API Service Classification: A system of letters signifying an oil's performance. The classification is assigned by the American Petroleum Institute.

Armature: The movable part in a relay. The revolving part in a generator or motor.

Atmospheric Pressure: The pressure that the earth's atmosphere exerts on objects. At sea level, this pressure is 14.7 psi (101 kPa) at 32°F (0°C).

Atomization: Breaking down into small particles or a fine mist.

Available Voltage: The peak voltage that a coil can produce.

Backfire: The accidental combustion of gases in an engine's intake or exhaust system.

Backpressure: The resistance, caused by turbulence and friction, that is created when a gas or liquid is forced through a restrictive passage.

Baffle: A plate or obstruction that restricts the flow of air or liquids. The baffle in a fuel tank keeps the fuel from sloshing as the car moves.

Bakelite: A synthetic plastic material that is a good insulator. Distributor caps are often made of Bakelite.

Ballast (Primary) Resistor: A resistor in the primary circuit that stabilizes ignition system voltage and current flow.

Base: The inner layer of semiconductor material in a transistor.

Biasing: Applying voltage to a junction of semiconductor materials.

Bifurcated: Separated into two parts. A bifurcated exhaust manifold has four primary runners that converge into two secondary runners; these converge into a single outlet into the exhaust system.

Bimetal Temperature Sensor: A sensor or switch that reacts to changes in temperature. It is made of two strips of metal welded together that expand differently when heated or cooled, causing the strip to bend.

Binary: A mathematical system consisting of only two digits (0 and 1) that allows a digital computer to read and process input voltage signals.

Bipolar: A transistor that uses both holes and electrons as current carriers.

Blowby: Combustion gases that leak past the piston rings into the crankcase; these include water vapor, acids, and unburned fuel.

Blowoff Valve: A spring-loaded valve that opens when boost pressure overcomes the spring tension to vent excess pressure.

Boost: A measure of the amount of air pressure above atmospheric that a supercharger or turbocharger can deliver. Boost remains constant regardless of altitude.

Bore: The diameter of an engine cylinder; to enlarge the diameter of a drilled hole.

Bottom Dead Center (BDC): The point when the piston is at the bottom of its stroke in the cylinder.

Breaker Points: The metal contact points that act as an electrical switch in a distributor. They open and close the ignition primary circuit.

Bus Link: A common conductor or transmission path shared by several components in a computer system that transmits data and receives instructions.

Camshaft Overlap: The period, measured in degrees of crankshaft rotation, during which both the intake and exhaust valves are open. It occurs at the end of the exhaust stroke and the beginning of the intake stroke.

Carbon Monoxide (CO): An odorless, colorless, tasteless poisonous gas. A major pollutant given off by an internal combustion engine.

Catalyst: A substance that causes a chemical reaction, without being changed by the reaction.

Catalytic Cracking: An oil refining process that uses a catalyst to break down (crack) the larger comcomponents of the crude oil. The gasoline produced by this method usually has a lower sulfur content than gasoline produced by thermal cracking.

Catalytic Converter: A device installed in an exhaust system that converts up to three pollutants, hydrocarbons (HC), carbon monoxide (CO), and oxides of nitrogen (NO_x) into harmless by-products using a catalytic chemical reaction.

Cavitation: An undesirable condition in the cooling system where the water pump blades are allowed to form air bubbles capable of forming small cavities in metal surfaces. Maintaining proper cooling system pressure prevents cavitation.

Cell: A case enclosing one element in an electrolyte. Each cell produces approximately 2.1 to 2.2 volts. Cells are connected in series.

Central Processing Unit (CPU): The processing and calculating portion of a computer.

Centrifugal Force: A force exerted by a rotating object that moves it away from the center of rotation.

Centrifugal (Mechanical) Advance: A method of advancing the ignition spark using weights in the distributor that react to centrifugal force.

Check Valve: A valve that permits flow in only one direction.

Circuit: A circle or unbroken path of conductors through which an electric current can flow.

Clearance Volume: The combustion chamber volume with the piston at TDC.

Closed Loop: Operational mode when the fuel delivery changes according to changes in voltage from the exhaust oxygen sensor (O2S).

Closed-Loop Control: An operating mode in which the computer responds to feedback signals from its sensors and adjusts system operations accordingly.

Collector: The outside layer of semiconductor material in a transistor that conducts current away from the base.

Combination Valve: A valve on the fuel tanks of some Ford cars that allows fuel vapors to escape to the vapor storage canister, relieves fuel tank pressure, and lets fresh air into the tank as fuel is withdrawn. Similar to a liquid-vapor separator valve.

Complete Circuit: A continuous, unbroken path in a circuit that allows current to flow from a source and return to that source.

Compression Ratio: A ratio of the total cylinder volume when the piston is at bottom dead center (BDC) to the volume of the combustion chamber when the piston is at top dead center (TDC).

Conductor: A material that allows easy electron flow because of its many free electrons.

Continuity: An uninterrupted connection. Used to describe an unbroken electric circuit.

Continuous Injection System (CIS): A fuel injection system in which fuel is injected constantly whenever the engine is running. An example of a CIS system is the Bosch K-Jetronic.

Conventional Flow Theory: The theory that says current flows from positive to negative.

Counterelectromotive Force (CEMF): An induced voltage that opposes the source voltage and any increase or decrease in source current flow.

Cranking Performance Rating: A battery rating based on the amperes of current that a battery can supply for 30 seconds at 0°F (-18°C) with no battery cell falling below 1.2 volts.

Cross Firing: Ignition voltage jumping from the distributor rotor to the wrong spark plug electrode inside the distributor cap. Also, ignition voltage jumping from one spark plug cable to another due to worn insulation.

Current Flow: A controlled, directed flow of electrons from atom to atom within a conductor.

Cycling: Battery electrochemical action and operation where one complete cycle is from fully charged to discharged and back to fully charged.

Data Link Connector (DLC): A special connector that allows a scan tool to communicate with the powertrain control module (PCM). On-Board Diagnostics Generation II (OBD-II) compliant vehicles use a standardized 16-pin connector.

Depletion Region: An area near the junction of a diode where P-material is depleted of holes and N-material is depleted of electrons.

Detented: Positions in a switch that allow the switch to stay in that position. In an ignition switch, the ON, OFF, LOCK, and ACCESSORY positions are detented.

Detergent Additives: A chemical compound added to motor oil that removes dirt or soot particles from surfaces, especially piston rings and grooves.

Detonation: An unwanted explosion of an air-fuel mixture caused by high heat and compression. Occurs *after* the spark plug fires. Also called knocking or pinging.

Diagnostic Trouble Code (DTC): A type of on-board diagnostics (OBD) test result that indicates a faulty circuit or system using a numeric or alphanumeric code.

Diagnostic Executive: The program within the powertrain control module (PCM) that coordinates the On-Board Diagnostics Generation II (OBD-II) self-monitoring system. It manages the comprehensive component and emission monitors, diagnostic trouble code (DTC) and malfunction indicator lamp (MIL) operation, freeze-frame data, and scan tool interface.

Diaphragm: A thin flexible wall separating two cavities, such as the diaphragm in a vacuum advance unit.

Dieseling: A condition in which extreme heat in the combustion chamber continues to ignite fuel after the ignition has been turned off.

Digital: A two-level voltage signal or processing function that is either on/off or high/low.

Digital-to-Analog (DA) Converter: An electronic conversion process for changing digital voltage signals to analog voltage signals.

Diodes: Electronic devices made of P-material and N-material bonded at a junction. A diode allows current flow in one direction and blocks it in the other.

Direct Current (dc): A flow of electricity in one direction through a conductor.

Discrete Devices: An independent, separately manufactured component with wire leads for connection into a circuit.

Dispersant: A chemical added to motor oil that keeps sludge and other undesirable particles picked up by the oil from gathering and forming deposits in the engine.

Displacement: A measurement of engine volume. It is calculated by multiplying the piston displacement of one cylinder by the number of cylinders. The total engine displacement is the volume displaced by all the pistons.

Diverter Valve: A valve used in an air injection system to prevent backfire. During deceleration it "dumps" the air from the air pump into the atmosphere. Also called a dump valve.

Doping: The addition of a small amount of a second element to a semiconductor element.

Duty Cycle: The percentage of the complete cycle during which the solenoid is on, or the ratio of pulse width to complete cycle width.

Dwell Angle: The measurement in degrees of how far the distributor cam rotates while the breaker points are closed. Also called cam angle or dwell.

Eccentric: Off center. A shaft lobe that has a center different from that of the shaft.

Electrolyte: The chemical solution in a battery that conducts electricity and reacts with the plate materials.

Electromagnetic Interference (EMI): Undesirable, high-frequency electromagnetic waves that interfere with electronic systems.

Electron Flow Theory: The theory that says current flows from negative to positive.

Electronic Fuel Injection (EFI): A computer-controlled fuel-injection system that precisely controls mixture for all operating conditions and at all speed ranges.

Element: A complete assembly of positive plates, negative plates, and separators making up one cell of a battery.

Emission Monitor: One of eight On-Board Diagnostics Generation II (OBD-II) software programs that perform passive, active, and intrusive tests on emission-related components and systems once per trip.

Emitter: The outside layer of semiconductor material in a transistor that conducts current to the base.

Enable Criteria: Engine operating conditions that must be met before the Diagnostic Executive runs a test.

Engine Mapping: A process of simulating vehicle and engine operating conditions to establish values for the computer to use when exercising system control.

Ethanol: Ethyl alcohol distilled from grain or sugar cane.

Evaporative Emission Control (EVAP) System: A way of reducing hydrocarbon (HC) emissions by collecting fuel vapors from the fuel tank and carburetor fuel bowl vents and directing them into an engine's intake system.

Exhaust Gas Recirculation (EGR): A way of reducing oxides of nitrogen (NO_X) emissions by redirecting exhaust gases back into the intake mixture.

Exhaust Scavenging: The use of compressed air from an air pump entering the cylinders while both the exhaust and intake valves are opened at the same time (overlap) to force exhaust gases out through the exhaust valve, and provide more room for fresh air for the next combustion cycle.

Extended-Core Spark Plug: The insulator core and the electrodes in this type of spark plug extend farther into the combustion chamber than they do on other types. Also called extended tip.

External Combustion Engine: An engine, such as a steam engine, that burns fuel outside of the engine.

Fast-Burn Combustion Chamber: A compact combustion chamber with a centrally located spark plug. The chamber is designed to shorten the combustion period by reducing the distance of flame front travel.

Field Circuit: The charging system circuit that delivers current to the alternator field.

Firing Order: The sequence by cylinder number in which combustion occurs in the cylinders of an engine.

Firing Voltage (Required Voltage): The voltage level that must be reached to ionize and create a spark in the air gap between the spark plug electrodes.

Fixed Dwell: A type of ignition system where the dwell period begins when the switching transistor turns on, and remains relatively constant at all speeds.

Flags: Special On-Board Diagnostics Generation II (OBD-II) messages that indicate the completion of an emission monitor test. If the Diagnostic Executive has not set all of the flags, the technician may have to drive the vehicle through the Inspection and Maintenance (I/M) Readiness Drive Cycle.

Flat Spot: The brief hesitation or stumble of an engine caused when sudden opening of the throttle creates a momentary lean air-fuel condition.

Float Valve: A valve that is controlled by a hollow ball floating in a liquid, such as in the fuel bowl of a carburetor.

Flooding: A condition caused by heat expanding the fuel in a fuel line. The fuel pushes the carburetor inlet needle valve open and fills up the fuel bowl even when more fuel is not needed. Also, the presence of too much fuel in the intake manifold.

Forward Bias: The application of a voltage to produce current flow across the junction of a semiconductor.

Four-Stroke Engine: The Otto-cycle engine. An engine in which a piston must complete four strokes to make up one operating cycle. The strokes are intake, compression, power, and exhaust.

Freeze Frame Record: The part of the Diagnostic Executive that stores various vehicle operating data when setting an emission-related diagnostic trouble code (DTC) and lighting the malfunction indicator lamp (MIL).

Full-Wave Rectification: A process by which all of an alternating current (ac) sine wave voltage is rectified and allowed to flow as direct current (dc).

Gasohol: A blend of ethanol and unleaded gasoline, usually at a 1:9 air-fuel ratio. It is also referred to as an oxygenated fuel.

Ground: The wiring and connections that return current to the battery. The ground is common to all circuits in the electrical system.

Ground Cable: The battery cable that provides a ground connection from the vehicle chassis to the battery.

Group Number: A battery identification number that indicates battery dimensions, terminal design, holddown location, and other physical features.

Gulp Valve: A valve used in an air injection system to prevent backfire. During deceleration it redirects air from the air pump to the intake manifold, where the air leans out the rich air-fuel mixture.

Half-Wave Rectification: A process by which only one-half of an alternating current (ac) sine wave voltage is rectified and allowed to flow as direct current (dc).

Hall-Effect Switch: A semiconductor that produces a voltage in the presence of a magnetic field. This voltage can be used to control a transistor for use as a switch.

Heat Range: The measure of a spark plug's ability to dissipate heat from its firing end.

Hertz (Hz): A unit of frequency measurement equal to one cycle per second.

High-Speed Surge: A sudden increase in engine speed caused by high manifold vacuum pulling in too much air-fuel mixture.

High-Swirl Combustion Chamber: A combustion chamber in which the intake valve is shrouded or masked to direct the incoming air-fuel charge and create turbulence that will circulate the mixture more evenly and rapidly.

Hole: The space in a valence ring where another electron could fit.

Hydrocarbon (HC): A major pollutant made up of hydrogen and carbon given off by internal combustion engines. Gasoline is a hydrocarbon compound.

Ignition Interval (Firing Interval): The number of degrees of crankshaft rotation between ignition sparks. Sometimes called firing interval.

Ignition Control Module: A self-contained sealed unit that houses the solid-state circuits that control ignition-related electrical or mechanical functions.

Impeller: A rotor or rotor blade used to force a gas or liquid in a certain direction under pressure.

Impurities: The doping elements added to pure silicon or germanium to form semiconductor materials.

Induction: The production of an electrical voltage in a conductor or coil, by moving the conductor or coil through a magnetic field, or by moving the magnetic field past the conductor or coil.

Inductive-Discharge Ignition System: A method of igniting the air-fuel mixture in an engine cylinder. It is based on the induction of a high voltage in the secondary winding of a coil.

Inert Exhaust Gas: A gas that will not undergo chemical reaction.

Infinite Resistance: A resistance value that is too high to measure on a test instrument; usually the result of an open circuit. The true value is unknown.

Injection Pump: A pump used on diesel engines to deliver fuel under high pressure at precisely timed intervals to the fuel injectors.

Input Conditioning: The process of amplifying or converting a voltage signal into a form usable by the computer's central processing unit.

Inspection and Maintenance (I/M) Readiness Drive Cycle: A drive cycle that runs all of the emission monitors and sets all of the I/M flags. Most states require that all I/M flags be set before running an I/M test.

Insulated (Hot) Cable: The battery cable that conducts battery current to the automotive electrical system.

Insulators: Materials that oppose electron flow because of their many bound electrons.

Integrated Circuits (IC): A very small, complex electronic circuit that contains thousands of transistors and other devices on a tiny silicon chip.

Intercooler: An air-to-air or air-to-liquid heat exchanger used to lower the temperature of the air-fuel mixture by removing heat from the intake air charge.

Internal Combustion Engine: An engine, such as a gasoline or diesel engine that burns fuel inside the engine.

Ionize: To break up molecules into two or more oppositely charged ions. The air gap between the spark plug electrodes is ionized when the air-fuel mixture is changed from a nonconductor to a conductor.

Isolated Field Circuit: A variation of the A-circuit. Field current is drawn from the alternator output *outside* of the alternator and sent to an insulated brush. The other brush is grounded through the voltage regulator.

Junction: The area where two types of semiconductor materials (P- and N-material) are joined.

Keep-Alive Memory (KAM): A form of long-term random-access memory (RAM) used mostly with adaptive strategies. Requires a separate power supply circuit to maintain voltage when the ignition is off.

Liquid-Vapor Separator Valve: A valve in some evaporative emission control (EVAP) fuel systems that separates liquid fuel from fuel vapor.

Long-Term Fuel Trim: A software function that implements and stores the longer-term effects of short-term fuel trim (FT) corrections.

M85: A clean-burning alternative fuel with an octane rating of 100 made by blending 85-percent methanol with 15-percent unleaded gasoline. When burned, it produces fewer hydrocarbons than pure gasoline.

Magnetic Pulse Generator: A signal-generating switch that creates a voltage pulse as magnetic flux changes around a pickup coil.

Magnetic Saturation: The condition when a magnetic field reaches full strength and maximum flux density.

Malfunction Indicator Lamp (MIL): A type of on-board diagnostics (OBD) test result that lights to alert the driver of a faulty automotive circuit. Previously called a Check Engine lamp.

Manifold Vacuum: Low pressure in the intake manifold below the carburetor throttle.

Mass Airflow Rate: The density and rate of air flowing into an engine, usually expressed in grams per second.

Mass Airflow System: A fuel management method that *directly calculates* mass airflow rate so that the powertrain control module (PCM) can regulate fuel metering. This system relies on one of three types of mass airflow (MAF) sensors.

Methanol: A clear, tasteless, and highly toxic form of alcohol made from natural gas, coal, or wood. Contains about 60 percent less energy than gasoline.

Micron: A unit of length equal to one-millionth of a meter or one one-thousandth of a millimeter.

Multigrade: An oil that meets viscosity requirements at more than one test temperature, designated by dual Society of Automotive Engineers (SAE) viscosity numbers.

Mutual Induction: The transfer of energy between two unconnected conductors, caused by the expanding or contracting magnetic flux lines of the current-carrying conductor.

N-material: A semiconductor material that has excess (free) electrons because of the type of impurity added. It has a negative charge that repels additional electrons.

Naturally Aspirated: An engine that uses atmospheric pressure and the normal vacuum created by the downward movement of the pistons to draw in its air-fuel charge. Naturally aspirated engines are not supercharged.

Negative Polarity: The condition when current flows through the coil's primary windings in the correct direction. Coil voltage is delivered to the spark plugs so that the center electrode of the plug is negatively charged and the grounded electrode is positively charged. Also called ground polarity.

Negative Temperature Coefficient (NTC) Resistor: A thermistor whose resistance decreases as the temperature increases.

No-Load Oscillation: The rapid, back-and-forth, peak-to-peak oscillation of voltage in the ignition secondary circuit when the circuit is open.

Noble Metals: Metals, such as platinum and palladium, that are inert.

Nodes: The points where current paths split. Also called junction points.

Octane Rating: The measurement of the antiknock value of a gasoline.

Ohm: The standard unit for measuring electrical resistance.

On-Board Diagnostics (OBD) Systems: A type of automotive diagnostic system mandated by the California Air Resources Board (CARB) or the U.S. EPA. Although the system's abilities vary depending on the specific version, generally, OBD seeks to have the vehicle serviced sooner, and to improve the technician's ability to repair emission-related problems.

Open Circuit: A discontinuous (broken) path in a circuit that does not allow current to flow to or from a source.

Open Loop: Operational mode when the fuel delivery changes according to sensors other than the exhaust oxygen sensor (O2S).

Open-Loop Control: An operating mode in which the computer causes a system to function according to predetermined instructions.

Orifice: A small opening in a tube, pipe, or valve.

Output Circuit: The charging system circuit that sends voltage and current to the battery and other electrical systems and devices.

Oxidation: The combining of an element with oxygen in a chemical process that often produces extreme heat as a by-product.

Oxides of Nitrogen (NO_x): Chemical compounds of nitrogen given off by an internal combustion engine. NO_x combines with hydrocarbons to produce ozone, a primary component of smog.

Ozone: A gas with a penetrating odor, and a primary component of smog. Ground-level ozone forms when hydrocarbons (HC's) and oxides of nitrogen (NO_x), in certain proportions, react in the presence of sunlight. Ozone irritates the eyes, damages the lungs, and aggravates respiratory problems.

P-material: A semiconductor material that has holes for additional electrons because of the type of impurity added. It has a positive charge that attracts additional electrons.

Parallel Circuit: A circuit that has more than one path through which current can flow.

Particulate Matter (PM_{10}): Microscopic particles of materials such as lead and carbon that are given off by an internal combustion engine as pollution.

Peak Inverse Voltage (PIV): The highest reverse bias voltage that can be applied to a junction of a diode before its atomic structure breaks down and allows current to flow.

Percolation: The bubbling and expansion of a liquid, similar to boiling.

Photochemical Smog: A combination of pollutants that form harmful chemical compounds when exposed to sunlight.

Piezoresistive: A sensor whose resistance varies in relation to pressure or force applied to it. A piezoresistive sensor receives a constant reference voltage and returns a variable signal in relation to its varying resistance.

Pintle Valve: A valve shaped like a hinge pin. In an exhaust gas recirculation (EGR) valve, the pintle is attached to a normally closed diaphragm. When ported vacuum is applied, the pintle rises from its seat and allows exhaust gas to be drawn into the engine intake system.

Plenum: A chamber that stabilizes the air-fuel mixture and allows it to rise to slightly above atmospheric pressure.

Poppet Valve: A valve that plugs and unplugs its opening by linear movement.

Port Fuel Injection (PFI): A fuel injection system in which individual injectors are installed in the intake manifold at a point close to the intake valve. Air passing through the manifold mixes with the injector spray just as the intake valve opens.

Ported Vacuum: The low-pressure area just above the throttle in a carburetor.

Positive Crankcase Ventilation (PCV): A way of controlling engine emissions by directing crankcase vapors (blowby) back through an engine's intake system.

Positive Polarity: An incorrect polarity of the ignition coil caused by reversing the primary coil terminal connectors. Coil voltage is delivered to the spark plug so the center electrode of the plug is positively charged and the grounded electrode is negatively charged. Also called reverse polarity.

Positive Temperature Coefficient (PTC) Resistor: A thermistor whose resistance increases as the temperature increases.

Potential Difference: The difference in electrical force between two points when one is positively charged and the other is negatively charged.

Pour-Point Depressants: Chemical compounds added to motor oil to help the oil flow at colder temperatures.

Powertrain Control Module (PCM): The main system computer module that controls the functions of the vehicle computer system.

Preignition: A premature ignition of the air-fuel mixture before the spark plug fires. It is caused by excessive heat or pressure in the combustion chamber.

Pressure Differential: A difference in pressure between two points.

Pressure Drop: A reduction of pressure between two points.

Primary Battery: A battery in which chemical processes destroy one of the metals necessary to create electrical energy. Primary batteries cannot be recharged.

Programmable Read-Only Memory (PROM): An integrated circuit chip installed in a computer that contains appropriate operating instructions and database information for a particular application.

Pull-Up Resistor: A constant load internal to the powertrain control module (PCM) that feeds reference voltage to some sensors. The PCM measures the voltage drop across this load to determine signal voltage.

Pulsating: Expanding and contracting rhythmically.

Pulse Width Modulation (PWM): A continuous on/off cycling of a solenoid at a specified number of times a second. The powertrain control module (PCM) can modulate the pulse width (on-time) to produce a variable duty cycle.

Pulse Width: The amount of time a solenoid is activated.

Purge Valve: A vacuum operated or electronically controlled solenoid valve used to draw fuel vapors from a vapor storage canister.

Radio Frequency Interference (RFI): A form of electromagnetic interference created in the ignition secondary circuit that disrupts radio and television transmission.

Random-Access Memory (RAM): Temporary short- or long-term computer memory that can be read and changed, but is lost whenever power is shut off to the computer.

Reach: The length of the spark plug shell from the seat to the bottom of the shell.

Read-Only Memory (ROM): The permanent part of a computer's memory storage function. ROM can be read but not changed, and is retained when power is shut off to the computer.

Reciprocating Engine: An engine in which the pistons move up and down or back and forth as a result of combustion of an air-fuel mixture at one end of the piston cylinder. Also called a piston engine.

Recombinant: A nongassing battery design in which the oxygen released by the electrolyte recombines with the negative plates.

Rectified: Electrical current changed from alternating to direct current.

Reduction: A chemical process in which oxygen is removed from a compound.

Reference Voltage: A constant voltage signal (below battery voltage) applied to a sensor by the computer. The sensor alters the voltage according to engine operating conditions and returns it as a variable input signal to the computer, which adjusts system operation accordingly.

Remote Injector Driver: The integrated circuit chip that controls fuel injector operation. Normally located in the powertrain control module (PCM) housing, the chip is installed in a separate housing to protect the PCM from excessive heat generated by the high current draw of the low-impedance injectors.

Reserve Capacity Rating: A battery rating based on the number of minutes a battery can supply 25 amps at 80°F (27°C), with no battery cell falling below 1.75 volts.

Resistance: Opposition to electrical current flow.

Resistor-Type Spark Plug: A plug that has a resistor in the center electrode to reduce the inductive portion of the spark discharge and RFI.

Reverse Bias: The application of a voltage so that normally no current will flow across the junction of a semiconductor.

Road Draft Tube: The earliest type of crankcase ventilation; it vents blowby gases to the atmosphere.

Runners: The passages or branches of an intake manifold that connect the manifold plenum chamber to the engine inlet ports.

SAE Viscosity Grade: A system of numbers signifying an oil's viscosity at a specific temperature. The viscosity grade is assigned by the Society of Automotive Engineers (SAE).

Secondary Battery: A battery in which chemical processes can be reversed. A secondary battery can be recharged so that it will continue to supply voltage.

Semiconductors: Materials with four electrons in their valence shell that are neither good conductors nor good insulators.

Series Circuit: A circuit that has only one path through which current can flow.

Series-Parallel Circuit: A circuit that has some parts in series with the voltage source and some parts in parallel with each other and with the voltage source.

Short Circuit: A continuous path for current flow that changes the normal current path through the circuit. The short may be into another circuit or to ground.

Short-Term Fuel Trim: A software function that allows the computer to make minute-by-minute changes in air-fuel ratio to maintain a 14.7:1 ratio.

Shunt Circuit: An electrical connection or branch circuit in parallel with another branch circuit or connection.

Siamesed Port: On an intake manifold, a single port that supplies the air-fuel mixture to two cylinders.

Sine Wave Voltage: The constant change, first to a positive peak and then to a negative peak, of an induced alternating voltage in a conductor.

Single Grade: An oil that has been tested at only one temperature, designated by one SAE viscosity number.

Single-Phase Current: Alternating current caused by a single-phase voltage.

Single-Phase Voltage: The sine wave voltage induced within one conductor by one revolution of an alternator rotor.

Sintered: A porous material welded together without using heat, such as the metal disk used in some vacuum delay valves.

Siphoning: The flowing of a liquid as a result of a pressure differential without the aid of a mechanical pump.

Sludge: A thick black deposit caused by the mixing of blowby gases and oil.

Solenoid-Actuated Starter: A starter that uses a solenoid both to control current flow in the starter circuit and to engage the starter motor with the engine flywheel.

Solid-State: A method of controlling electrical current flow, using parts primarily made of semiconductor materials.

Spark Voltage: The inductive portion of a spark that maintains the spark in the air gap between a spark plug's electrodes, usually about one-quarter of the firing voltage level.

Specific Gravity: The weight of a volume of liquid divided by the weight of the same volume of water at a given temperature and pressure. Water has a specific gravity of 1.000.

Speed-Density System: A fuel management method that *estimates* mass airflow rate so that the powertrain control module (PCM) can regulate fuel metering. This system relies primarily on the manifold absolute pressure (MAP) sensor to determine density and on the engine speed sensor for volume airflow rate.

Starting Safety Switch: A neutral start switch. It keeps the starting system from operating when a car's transmission is in gear.

Starting Bypass: A parallel circuit branch that bypasses the ballast resistor during engine cranking.

Stepper Motor: A direct-current motor that moves the armature in incremental movements or steps. When the field coils are energized, magnetic force can rotate the armature clockwise or counterclockwise.

Stoichiometric Ratio: An ideal air-fuel mixture for combustion, in which all oxygen and fuel will burn completely.

Stratified-Charge Engine: An engine that combusts the air-fuel mixture in two stages. The first occurs when a rich intake charge burns in the precombustion chamber; the second, when a leaner intake charge burns in the main combustion chamber.

Stroke: One complete top-to-bottom or bottom-to-top movement of an engine piston.

Substrate: The layer, or honeycomb, of aluminum oxide upon which the catalyst (platinum or palladium) in a catalytic converter is deposited.

Sulfation: The crystallization of lead sulfate on the plates of a constantly discharged battery.

Sulfur Oxides (SO_x): Sulfur given off by processing and burning gasoline and other fossil fuels. As it decomposes, sulfur dioxide combines with water to form sulfuric acid, or "acid rain".

Supercharging: Using a belt-driven air pump to deliver an air-fuel charge to the engine cylinders at a pressure greater than atmospheric pressure.

Synthetic Motor Oils: Lubricants formed by artificially combining molecules of petroleum and other materials.

Television-Radio-Suppression (TVRS) Cables: High-resistance, carbon-conductor ignition cables that suppress radio frequency interference.

Temperature Inversion: A weather pattern in which a layer, or "lid", of warm air keeps a layer of cooler air trapped beneath it.

Tetraethyl Lead (TEL): A gasoline additive that helps prevent detonation.

Thermal Cracking: A common oil refining process that uses heat to break down (crack) the larger components of the crude oil. The gasoline produced by this method usually has a higher sulfur content than gasoline produced by catalytic cracking.

Thermistor (Thermal Resistor): A resistor that changes its resistance as the temperature changes.

Thermostatic: Referring to a device that automatically responds to temperature changes in order to activate a switch.

Throttle Body Fuel Injection (TBI): A fuel-injection system in which one or two injectors are installed in a carburetor-like throttle body mounted on a conventional intake manifold. Fuel is sprayed at a constant pressure above the throttle plate to mix with the incoming air charge.

Thyristor: A silicon-controlled rec- rectifier (SCR) that normally blocks all current flow. A slight voltage applied to one layer of its semiconductor structure allows current flow in one direction while blocking current flow in the other direction.

Top Dead Center (TDC): The point when the piston is at the top of its stroke in the cylinder.

Total Ignition Advance: The sum of centrifugal advance, vacuum advance plus initial timing; expressed in crankshaft degrees.

Transistor: A three-element semiconductor device of NPN or PNP materials that transfers electrical signals across a resistance.

Trip: A key-on, engine running, key-off driving cycle where all of the enable criteria for a test have been met and where the Diagnostic Executive ran and completed that test.

Turbo Lag: The time interval required for a turbocharger to overcome inertia and spin up to speed.

Turbocharger: A supercharging device that uses exhaust gases to turn a turbine connected to a compressor that forces extra air-fuel mixture into the cylinders.

Vacuum: Low pressure within an engine created by a downward piston intake stroke with the intake valve open.

Vacuum Advance: The use of engine vacuum to advance ignition spark timing by moving the distributor breaker plate.

Vacuum Lock: A stoppage of fuel flow caused by insufficient air intake to the fuel tank.

Valence Ring: The outermost electron shell of an atom.

Vane Airflow System: A fuel management method that *estimates* mass airflow rate so that the powertrain control module (PCM) can regulate fuel metering. This method relies primarily on the engine speed and the vane airflow (VAF) sensors to calculate volume airflow. For density, this system primarily uses the intake air temperature (IAT) sensor.

Vapor Lock: A condition in which bubbles in a car's fuel system stops or restricts fuel flow. High underhood temperatures sometimes causes fuel to boil within fuel lines.

Vaporization: Changing from a liquid into a gas (vapor).

Vaporize: To change from a solid or liquid into a gaseous state.

Variable Dwell: A type of ignition system where the dwell period varies in distributor degrees at different engine speeds, but remains relatively constant in duration or actual time.

Varnish: An undesirable deposit, usually on the engine pistons, formed by oxidation of fuel and motor oil.

Venturi: A restriction in an airflow, such as in a carburetor, that speeds the airflow and creates a vacuum.

Venturi Vacuum: Low pressure in a carburetor venturi caused by fast airflow through the venturi.

Viscosity: The tendency of a liquid, such as oil, to resist flowing.

Viscosity Index (VI) Improvers: Chemical compounds added to motor oil to help the oil resist thinning at high temperatures.

Volatility: The ease with which a liquid changes from a liquid to a gas or vapor.

Volt: The unit for measuring the amount of electrical force.

Voltage: The electromotive force that causes current flow. The potential difference in electrical force between two points when one is negatively charged and the other is positively charged.

Voltage Decay: The rapid oscillation and dissipation of secondary voltage after the spark in a spark plug air gap has stopped.

Voltage Drop: The loss of voltage caused by the resistance of a circuit load.

Voltage Reserve: The amount of coil voltage available in excess of the voltage required to fire the spark plugs.

Volts: The standard unit for measuring electromotive force (voltage).

Volumetric Efficiency: The *actual* volume of air-fuel mixture an engine draws in compared to the *theoretical maximum* it could draw in, written as a percentage.

Warm-Up Cycle: When used referring to an On-Board Diagnostics Generation II (OBD-II) system, a "warmed-up" engine has a specific meaning. For Ford, the engine temperature must reach a minimum of 158°F (70°C) and rise at least 36°F (20°C) over the course of a trip.

Waste Spark: A type of ignition system without a distributor where one coil in a coil pack fires two spark plugs at the same time. On cylinder compression, the spark ignites the air-fuel mixture, while the spark in the cylinder on its exhaust stroke is wasted.

Wastegate: A diaphragm-actuated bypass valve used to limit turbocharger boost pressure by limiting the speed of the exhaust turbine.

Water Jackets: Passages in the head and block that allow coolant to circulate throughout the engine.

Water Injection: A method of lowering the air-fuel mixture temperature by injecting a fine spray of water that evaporates as it cools the intake charge.

Index

Answers to Review Questions and ASE Questions

Chapter 1: Engine Operating Principles
1-D 2-B 3-C 4-A 5-B 6-D 7-B 8-B 9-C 10-A 11-D 12-C 13-B 14-D 15-C 16-A 17-D 18-D 19-A 20-C

Chapter 2: Engine Air-Fuel Requirements
1-C 2-C 3-D 4-A 5-C 6-B 7-D 8-B 9-D 10-B 11-B 12-D 13-C 14-B 15-B 16-A 17-D 18-A

Chapter 3: Engine Lubrication
1-C 2-D 3-B 4-D 5-A 6-B 7-C 8-A 9-D 10-B 11-D 12-A 13-D 14-C 15-A 16-C 17-C 18-B

Chapter 4: Cooling and Exhaust Systems
1-A 2-A 3-B 4-C 5-D 6-A 7-C 8-D 9-C 10-D 11-D 12-B 13-C 14-C 15-A 16-C 17-C 18-A 19-B 20-D 21-C 22-D 23-B

Chapter 5: Introduction to Emission Controls
1-C 2-C 3-D 4-A 5-C 6-D 7-B 8-C 9-B 10-C 11-C 12-D 13-A 14-A

Chapter 6: What Is A Tune-Up?
1-C 2-A 3-C 4-D 5-D 6-D

Chapter 7: Battery, Cranking, and Charging Systems
1-C 2-B 3-D 4-C 5-A 6-D 7-C 8-D 9-D 10-B 11-A 12-B 13-A 14-D 15-C 16-D 17-D 18-C 19-D 20-A 21-D 22-C 23-C

Chapter 8: Ignition Primary Circuit and Components
1-D 2-C 3-A 4-C 5-C 6-D 7-B 8-C 9-B 10-B 11-B 12-C 13-D 14-B 15-B 16-A

Chapter 9: Ignition Secondary Circuit and Components
1-D 2-D 3-C 4-A 5-D 6-C 7-C 8-B 9-D 10-B 11-C 12-D 13-C 14-D 15-C

Chapter 10: Ignition Timing and Spark Advance Control
1-D 2-C 3-B 4-A 5-C 6-C 7-B 8-D 9-D 10-D

Chapter 11: Solid-State Ignition Systems
1-B 2-B 3-A 4-C 5-A 6-B 7-D 8-D 9-D 10-C 11-A 12-B 13-A 14-C 15-D 16-B 17-C 18-A 19-D 20-A 21-C

Chapter 12: Fuel and Air Intake System Operation
1-C 2-A 3-D 4-C 5-C 6-B 7-B 8-A 9-C 10-B 11-B 12-C 13-B 14-A 15-A 16-C 17-D 18-A 19-D 20-C

Chapter 13: Basic Carburetion and Manifolding
1-C 2-B 3-A 4-D 5-C 6-A 7-B 8-D 9-C 10-C 11-B 12-A 13-C 14-D 15-A 16-B 17-C 18-D 19-A 20-C 21-A 22-B 23-A 24-B 25-C 26-C 27-A 28-B 29-C 30-D 31-D 32-B 33-B 34-A 35-B 36-B

Chapter 14: Principles of Electronic Control Systems
1-B 2-C 3-B 4-C 5-A 6-C 7-D 8-A 9-B 10-C 11-A 12-C 13-A 14-A 15-C 16-C 17-A 18-C 19-D 20-D 21-D 22-B 23-C 24-A 25-C

Chapter 15: Electronic Engine Control: Sensors, Actuators, and Operation
1-C 2-A 3-A 4-C 5-B 6-C 7-D 8-C 9-A 10-A 11-A 12-A 13-A 14-A 15-A 16-A 17-B 18-C 19-C 20-D 21-D 22-D 23-A 24-B 25-A

Chapter 16: Gasoline Fuel Injection Systems and Intake Manifolding
1-A 2-C 3-A 4-B 5-C 6-B 7-A 8-D 9-B 10-D 11-A 12-C 13-D 14-C

Chapter 17: Supercharging and Turbocharging
1-C 2-A 3-B 4-D 5-D 6-B 7-C 8-A 9-D 10-D 11-B 12-A 13-B 14-C 15-A 16-C 17-D 18-C

Chapter 18: Variable and Flexible Fuel Systems
1-C 2-A 3-D 4-C 5-A 6-B 7-C 8-A 9-C 10-D 11-B 12-B 13-D 14-C

Chapter 19: Positive Crankcase Ventilation
1-D 2-D 3-D 4-A 5-B 6-D

Chapter 20: Air Injection
1-A 2-C 3-A 4-B 5-B 6-A 7-A 8-D 9-A

Chapter 21: Exhaust Gas Recirculation
1-A 2-C 3-D 4-D 5-C 6-C 7-C 8-D 9-A 10-D 11-B 12-C 13-B

Chapter 22: Catalytic Converters
1-D 2-A 3-C 4-B 5-D 6-D 7-B 8-D 9-B 10-A

Answers to ASE Technician Certification Sample Test
1-B 2-C 3-B 4-A 5-D 6-C 7-A 8-A 9-B 10-C 11-A 12-C 13-A 14-C 15-A 16-C 17-C 18-B 19-C 20-D 21-B 22-A 23-C 24-D 25-C 26-C 27-D 28-A 29-D 30-C 31-C 32-C 33-B 34-C 35-A 36-C 37-C 38-C 39-C 40-C 41-B 42-A 43-C 44-B 45-D 46-A 47-B

NOTES

NOTES

NOTES

NOTES

NOTES

NOTES

NOTES

NOTES

NOTES